FORMULAS/EQUATIONS

Distance Formula If $P_1 = (x_1, y_1)$ and $P_2 = (x_2, y_2)$, the distance from P_1 to P_2 is

$$d(P_1, P_2) = \sqrt{(x_2 - x_1)^2 + (y_2 - y_1)^2}$$

Equation of a Circle The equation of a circle of radius r with center at (h, k) is

$$(x - h)^2 + (y - k)^2 = r^2$$

Slope Formula The slope m of the line containing the points $P_1 = (x_1, y_1)$ and $P_2 = (x_2, y_2)$ is

$$m = \frac{y_2 - y_1}{x_2 - x_1} \qquad \text{if } x_1 \neq x_2$$

$$m \text{ is undefined} \qquad \text{if } x_1 = x_2$$

Point–Slope Equation of a Line The equation of a line with slope m containing the point (x_1, y_1) is

$$y - y_1 = m(x - x_1)$$

Slope–Intercept Equation of a Line The equation of a line with slope m and y-intercept b is

$$y = mx + b$$

Quadratic Formula The solutions of the equation $ax^2 + bx + c = 0$, $a \neq 0$, are

$$x = \frac{-b \pm \sqrt{b^2 - 4ac}}{2a}$$

If $b^2 - 4ac > 0$, there are two real unequal solutions.
If $b^2 - 4ac = 0$, there is a repeated real solution.
If $b^2 - 4ac < 0$, there are two complex solutions.

FUNCTIONS

Constant Function	$f(x) = b$
Linear Function	$f(x) = mx + b$, m is slope, b is y-intercept
Quadratic Function	$f(x) = ax^2 + bx + c$
Polynomial Function	$f(x) = a_n x^n + a_{n-1}x^{n-1} + \cdots + a_1 x + a_0$
Rational Function	$R(x) = \dfrac{p(x)}{q(x)} = \dfrac{a_n x^n + a_{n-1}x^{n-1} + \cdots + a_1 x + a_0}{b_m x^m + b_{m-1}x^{m-1} + \cdots + b_1 x + b_0}$
Exponential Function	$f(x) = a^x$, $a > 0, a \neq 1$
Logarithmic Function	$f(x) = \log_a x$, $a > 0, a \neq 1$

College Algebra and Trigonometry

The Precalculus Series by Michael Sullivan:

College Algebra with Review
This text contains an in-depth review of intermediate algebra topics for those students who either may not be adequately prepared to begin a traditional college algebra course or desire to become reacquainted with certain prerequisite topics before beginning college algebra. After completing this book, the student will have covered the same mathematics at the same level as the student who has completed a traditional college algebra text.

College Algebra, Second Edition
This text contains a traditional approach to college algebra, with three chapters of review material preceding the chapter on functions and their graphs. After completing this book, the student will be adequately prepared to handle subsequent courses in finite mathematics, business mathematics, and engineering calculus.

College Algebra and Trigonometry, Second Edition
This text, in addition to all the features of *College Algebra*, also develops the trigonometric functions using a right triangle approach and showing how it leads to the unit circle approach. Graphing techniques are emphasized, including a thorough discussion of polar coordinates, parametric equations, and conics using polar coordinates.

Precalculus, Second Edition
This text contains one review chapter before covering the traditional precalculus topics of functions and their graphs, polynomial and rational functions, and exponential and logarithmic functions. The trigonometric functions are introduced using a unit circle approach and showing how it leads to the right triangle approach. Graphing techniques are emphasized, including a thorough discussion of polar coordinates, parametric equations, and conics using polar coordinates.

Trigonometry, Second Edition
This text, designed for stand-alone courses in trigonometry, develops the trigonometric functions using a unit circle approach and showing how it leads to the right triangle approach. Graphing techniques are emphasized, including a thorough discussion of polar coordinates, parametric equations, and conics using polar coordinates.

College Algebra and Trigonometry

Second Edition

Michael Sullivan
Chicago State University

Dellen Publishing Company
San Francisco, California

Collier Macmillan Publishers
London

Divisions of Macmillan, Inc.

On the cover: The "Untitled" oil on paper monoprint on the cover is a diptych executed by Los Angeles artist Charles Arnoldi at the Garner Tullis Workshop in Santa Barbara, California, on November 18, 1988. Arnoldi's work may be seen at the Fuller Gross Gallery in San Francisco, the James Corcoran Gallery in Santa Monica, the Texas Gallery in Houston, and the Charles Cowles Gallery in New York City. His work is included in the permanent collections of the Albright-Knox Museum in Buffalo, New York, the Chicago Art Institute, the Museum of Modern Art in New York City, the Metropolitan Museum, the Los Angeles County Museum, and the San Francisco Museum of Modern Art.

Printed in the United States of America

Permissions: Dellen Publishing Company
400 Pacific Avenue
San Francisco, California 94133

Orders: Dellen Publishing Company
c/o Macmillan Publishing Company
Front and Brown Streets
Riverside, New Jersey 08075

Collier Macmillan Canada, Inc.

Library of Congress Cataloging-in-Publication Data

Sullivan, Michael, 1942–
College algebra and trigonometry / Michael Sullivan. —2nd ed.
p. cm. —(Precalculus series)
ISBN 0-02-418291-5
1. Algebra. 2. Trigonometry. I. Title. II. Series: Sullivan, Michael, 1942– Precalculus series.
QA154.2.S844 1990 89-48839
512′.13—dc20 CIP

Printing: 1 2 3 4 5 6 7 8 9 Year: 0 1 2 3 4

ISBN 0-02-418291-5

For an unnamed person

Contents

Preface

Intent/Purpose

As a professor at an urban public university for over 20 years, I am aware of the varied needs of students—students who range from being less well-prepared and insufficiently motivated to those who are well-prepared and highly motivated. As the author of an engineering calculus text as well as texts in finite mathematics and business mathematics, I understand what students must know in order to be successful in such courses. At the same time, I am also aware of, and very concerned about, the student who has decided this course in algebra and trigonometry is to be a terminal mathematics course. Based on these experiences, I have written a book to serve students with varied backgrounds and goals.

For the under-prepared student, a review of basic material is contained in Chapter 1. To help further, concepts are viewed from different perspectives whenever possible, and alternate methods of solutions are given. Every effort has been made to be clear, precise, and consistent. Whenever it seemed appropriate, special encouragement has been offered; and whenever necessary, warnings have been given. To provide motivation, I have included understandable, realistic applications that are consistent with the abilities of the student. The problem sets have been carefully graded to build the student's confidence.

At the same time, a conscious attempt has been made to preview material that will be seen by students who intend to take courses such as finite mathematics, business mathematics, or engineering calculus. These previews relate to important concepts, applications, examples, and exercises that will be encountered in later courses and can be viewed now as algebra or trigonometry problems.

To summarize, my purpose in writing this book is to serve the different needs of each type of student through suitable motivation and material.

New to This Edition

The elements of the previous edition that proved so successful remain in this edition. Nevertheless, many changes—some obvious, others subtle—have been made. Virtually every change is the result of thoughtful comments and suggestions from colleagues and students who used the previous edition. As a result of this input, for which I am sincerely grateful, this edition will be an improved teaching device for professors and a better learning tool for students.

The list below, which is by no means exhaustive, should give you an idea of what has been done.

Illustrations. Every piece of art has been redrawn to improve accuracy and clarity. Many new illustrations have been inserted, especially in the exercises, to assist the student in visualizing concepts and to reinforce the graphical orientation of the text. This edition now contains over 1200 illustrations.

New Material. Section 4.6, "Constructing Functions: Applications," has been added to Chapter 4, Functions and Their Graphs. This section contains abundant applications in geometry, business, economics, consumer affairs, engineering, and other fields of interest, and should provide additional motivation and rationale for studying and learning college algebra.

Section 7.6, "Trigonometric Functions: Unit Circle Approach," has been added to Chapter 7. This section illustrates how the right triangle approach leads to the unit circle approach. Emphasis is placed on the properties of the trigonometric functions.

For emphasis, the chapter on conics now appears as Chapter 11, Analytic Geometry. Now included are Section 11.5, "Rotation of Axes; General Form of a Conic," Section 11.6, "Polar Equations of Conics," and Section 11.7, "Plane Curves and Parametric Equations."

A discussion of infinite geometric series has been added to Chapter 13, Induction; Sequences.

A discussion of vector projections and its application to the work done by a constant force now appears in Section 14.5, "The Dot Product."

As part of the chapter review section at the end of each chapter, a new component called "Things to Know" has been added. This capsule summary of the important elements of the chapter should be of value to students as they prepare for examinations. The new "How To" section in the chapter review also can be used to identify the objectives of the chapter.

Revised/Reorganized Material. The change-of-scale method for graphing certain functions has been abandoned in favor of a compression/stretch approach in Section 4.3.

Section 5.3, "Rational Functions," now appears earlier in the text. While finding the zeros of a polynomial and graphing polynomials remain the thrust of Section 5.5, more emphasis has been placed on solving polynomial equations.

Section 5.4, "Synthetic Division," has been rewritten and is now consistent with the generally accepted addition method.

The definition of a logarithmic function is now based on properties of inverse functions (Chapter 6).

The chapter on trigonometric functions has been reorganized into two chapters. Chapter 7 introduces the trigonometric functions, placing emphasis on their properties. Then Chapter 8 deals with graphing the trigonometric functions and their inverses. The former Section 7.6 now becomes two sections: 8.1, "Graphs of the Six Trigonometric Functions," and 8.2, "Sinusoidal Graphs." This will allow for easier coverage of one section per class period.

Section 10.5, "Polar Equations and Graphs," has been completely rewritten. It now contains a clear and complete coverage of graphing in polar coordinates.

Section 12.1 now includes the material on solving systems of linear equations by substitution and by elimination combined into a single section.

Examples. New examples have been inserted where the level of content or the diversity of approach warranted. This edition now contains over 650 examples.

Exercises. Many new exercises have been added. The majority fall into the categories of applied word problems, problems involving geometry, and problems to challenge the better student. This edition now contains over 6000 exercises.

Calculator Usage. Earlier emphasis (see Section 1.2) has been given to the use of a calculator to provide concrete examples of certain properties of real numbers. Additional examples of calculator use in college algebra and trigonometry have been inserted where appropriate. Also, references have been added to the new guides to using graphics calculators, which explain the use of a graphics calculator in algebra and trigonometry (see Supplementary Material).

Contents and Organization

The topics contained in this book were selected and organized with the different types of students who take this course in mind. Thus, throughout the book, a deliberate attempt is made to preview topics and techniques that will be used in later courses.

The first part of the book, Chapters 1–3, provides a detailed presentation of material needed for later topics. Chapter 1 mostly consists of review material. Although complex numbers are introduced in Section 1.9, this section may be postponed at the discretion of the instructor. If the instructor chooses to cover complex numbers later, Section 2.4, "Quadratic Equations with a Negative Discriminant," also can be postponed and then covered along with Section 1.9. Whether complex numbers are covered early or late, all exercises dependent upon complex numbers are clearly identified in the directions for the exercise.

The coverage of geometry in Chapter 1 should also be noted. No review would be complete without a discussion of the Pythagorean Theorem and the formulas for perimeter and area of a rectangle, area and circumference of a circle, and volume of a sphere.

For some students, Chapters 2 and 3 will represent new material; for others, they will be review. The large majority, of course, fall somewhere in between. In Section 2.3, quadratic equations are treated solely from the real number point of view, while in Section 2.4, complex number solutions are discussed. This ensures flexibility of coverage for the instructor and clarity of understanding for the student.

The middle part of the book, Chapters 4–10, covers the essential topics of college algebra and trigonometry. In Chapter 4, special emphasis is placed on functions and the graphs of functions. Graphing is usually done in steps, all of which are illustrated. The graphing techniques introduced in Chapter 4 are utilized and reinforced in Chapters 5, 6, and 8 using the new functions introduced there.

Because of the need to evaluate polynomial functions, both synthetic division and the nested-form technique are utilized in Chapter 5 as alternatives to substitution. The nested-form technique will be especially appreciated by students of computer science. Additional methods for finding the zeros of a polynomial, such as the Rational Root Theorem and Descartes' Rule of Signs, are introduced to facilitate the graphing of polynomial and rational functions. Approximating results, such as upper and lower bounds on zeros and partition techniques, are discussed in a separate section to provide a clear sense of the difference between numerical and nonnumerical methods.

Chapter 6 treats the exponential and logarithmic functions in the detail necessary and with language consistent for subsequent use in calculus. Applications to compound interest, earthquake magnitude, and growth and decay are among those given.

With Chapter 7, we begin the study of trigonometry. The trigonometric functions are introduced using the right triangle approach, followed by the trigonometric functions of a general angle. Following a section on applications, the unit circle approach and the properties of the trigonometric functions are discussed. (For those who prefer to introduce the trigonometric functions using the unit circle approach, see the description of *Precalculus*, Second Edition, on page ii.)

Chapter 8 continues the study of trigonometric functions with a detailed presentation of their graphs. Here, we use the graphing concepts introduced in Chapter 4, as well as some additional techniques. A discussion of inverse trigonometric functions rounds out the chapter. Chapter 9 treats the analytic side of trigonometry: identities; sum and difference, double-angle, half-angle, sum-to-product, and product-to-sum formulas; and trigonometric equations.

In Chapter 10, additional applications of trigonometry to solving a general triangle are given, including a rather complete discussion of the area of a triangle. Two full sections have been devoted to polar coordinates and graphing polar equations to provide a thorough development of this subject. The final section treats DeMoivre's Theorem.

Chapter 11 contains topics from analytic geometry. Beginning with a detailed presentation of the conics (Sections 11.1–11.4), the parabola, ellipse, and hyperbola are defined using geometric (distance-based) means. After a discussion of rotation of axes and the general form of a conic, a unified definition of the conics using eccentricity is given. Lastly, plane curves and parametric equations are studied.

Chapter 12 discusses a variety of methods (four in all) for solving systems of linear equations. Sections on systems of nonlinear equations and systems of linear inequalities are also included.

The remaining chapters of the book cover topics that may be selected according to the specific needs of the students. For example, Chapter 13 presents mathematical induction, the Binomial Theorem, and arithmetic and geometric sequences. Chapter 14 includes an introduction to matrices and linear programming, topics of use to students intent on taking finite mathematics. Also included are sections on vectors and the dot product that would be of help to students who may go on to study calculus. Note that partial fraction decomposition is presented in Section 14.3, after systems of equations have been covered, rather than following rational functions in Chapter 5. In this way, it provides not only an application for solving systems of linear equations, but also allows for a full and detailed discussion. This should prove useful to students going on to calculus.

Finally, Chapter 15 discusses sets, counting, and probability.

Applications

As we mentioned earlier, every opportunity has been taken to present understandable, realistic applications consistent with the abilities of the student, drawing from such sources as tax rate tables, the *Guinness Book of World Records*, and newspaper articles. For added interest, some of the applied exercises have been adapted from textbooks the students may be using in other courses (for example, economics, chemistry, physics, etc.). See, for example, Sections 2.2 and 4.6.

Historical Notes

William Schulz of Northern Arizona University has provided historical context and information in anecdotes that appear as introductory material and at the ends of many sections. In some cases, these comments also include exercises and discussion of comparative techniques. See, for example, Problems 73–81 on page 336.

Examples, Exercises, and Illustrations

The text includes over 650 examples and 6000 exercises, of which approximately 900 are applied problems. The examples are worked out in appropriate detail, starting with simple, reasonable problems and working gradually up to more challenging ones.

Exercises are numerous, well-balanced, and graduated. They usually begin with problems designed to build confidence, continue with drill-type problems that mimic worked out examples, and conclude with problems that are more challenging. They include a number of exercises where the student will need a calculator for their solution; these are clearly marked with the symbol [C]. A few computer problems also have been included. Answers are given in the back of the book for all the odd-numbered exercises.

Illustrations are abundant, numbering over 1200. Full use is made of a second color to help clarify and highlight.

A Word about Format and Design

Each chapter-opening page contains a table of contents for that chapter and an overview—often historical—of the contents.

New terms appear in boldface type where they are defined. The most important definitions are shown in color.

Theorems are set with the word "Theorem" in the margin for easy identification; if it is a named theorem, the name also appears in the margin. When a theorem has a proof given, the word "Proof" appears in the margin to mark clearly the beginning of the proof, and the symbol ■ is used to indicate the end of the proof.

All important formulas and procedures are enclosed by a box and shown in color.

Examples are numbered within each section and identified clearly with the word "Example" in the margin. The solution to each example appears immediately following the example with the word "Solution" in the margin to identify it. The symbol ■ indicates the end of each solution.

Each section ends with an exercise set. Each chapter ends with a chapter review containing a list of "Things to Know" (a summary of the key elements of the chapter), a selection of fill-in-the-blank items (to test vocabulary and formulas), and a set of review exercises.

Supplementary Material

For the Instructor:

Instructor's Solution Manual to Accompany College Algebra and Trigonometry, by Katy Sullivan and Marsha Vihon, contains worked-out solutions to all the even-numbered problems. If used in conjunction with the *Student's Solution Manual*, the instructor will have worked-out solutions for every problem in the book.

Instructor's Resource Material contains approximately 200 transparency masters that duplicate important illustrations in the text.

Instructor's Test Battery, Dellen Test III, contains several printed samples of chapter and cumulative tests. Each chapter test contains 20–30 questions. Also included are the appropriate answer keys. From these, an instructor can construct appropriate quizzes, chapter tests, and cumulative examinations. Also included are the question models, documentation, and diskettes for both the Free Response and the Multiple Choice versions of Dellen Test III, an algorithm-based test generation system that utilizes an IBM PC™ (or compatible). Using it, it is possible to generate a virtually unlimited number of test items comparable to the question models, each different from the other, yet each testing the same concept at the same level. This system thus provides a virtually unlimited number of quizzes, chapter tests or final examinations. The system also generates an answer key and a student worksheet with an answer column that exactly matches the column on the answer key. Graphing grids are included for all problems requiring graphs. The Multiple Choice version of Dellen Test III utilizes distractors that are representative of commonly made student errors; this version is machine-gradable.

Precalculus Explorer is IBM PC compatible software that enables the user to enter a function and see the graph on the screen.

For the Student:

Student's Solution Manual to Accompany College Algebra and Trigonometry, by Katy Sullivan and Marsha Vihon, contains complete step-by-step solutions to all the odd-numbered problems.

A Guide to Using the Casio Graphics Calculators with Michael Sullivan's Precalculus Series, by Joan Girard, provides a collection of activities that demonstrate the use of the Casio series of graphics calculators (FX7000G, FX8000G, FX7500G) to solve selected examples and exercises from the text.

A Guide to Using the Hewlett-Packard Graphics Calculator with Michael Sullivan's Precalculus Series, by Joan Girard, contains the same activities as the guide described above using the HP-28S graphics calculator.

Acknowledgments

Textbooks are written by an author, but evolve from an idea into final form through the efforts of many people. Before initial writing began, a survey was conducted which drew nearly 250 instructor responses. The manuscript then underwent a thorough and lengthy review process, including class-testing at Chicago State University. I would like to thank my colleagues and students at Chicago State, who cooperated and contributed to this text while it was being class-tested.

The following contributors to the first edition and its revision, and I apologize for any omissions, have my deepest thanks and appreciation:

James Africh, Brother Rice High School
Steve Agronsky, Cal Poly State University
Joby Milo Anthony, University of Central Florida
James E. Arnold, University of Wisconsin, Milwaukee
Wilson P. Banks, Illinois State University
Dale R. Bedgood, East Texas State University
William H. Beyer, University of Akron
Richelle Blair, Lakeland Community College
Trudy Bratten, Grossmont College
Lois Calamia, Brookdale Community College
Roger Carlsen, Moraine Valley Community College
Duane E. Deal, Ball State University
Vivian Dennis, Eastfield College
Karen R. Dougan, University of Florida
Louise Dyson, Clark College
Don Edmondson, University of Texas, Austin
Christopher Ennis, University of Minnesota
Garret J. Etgen, University of Houston
W. A. Ferguson, University of Illinois, Urbana/Champaign
Merle Friel, Humboldt State University
Richard A. Fritz, Moraine Valley Community College
Wayne Gibson, Rancho Santiago College
Joan Goliday, Santa Fe Community College
Frederic Gooding, Goucher College
Ken Gurganus, University of North Carolina
James E. Hall, University of Wisconsin, Madison
Brother Herron, Brother Rice High School
Kim Hughes, California State College, San Bernardino
Ron Jamison, Brigham Young University
Sandra G. Johnson, St. Cloud State University
Arthur Kaufman, College of Staten Island
Thomas Kearns, North Kentucky University
Keith Kuchar, Manatee Community College
H. E. Lacey, Texas A & M University

Christopher Lattin, University of Florida
Stanley Lukawecki, Clemson University
Virginia McCarthy, Iowa State University
Laurence Maher, North Texas State University
James Maxwell, Oklahoma State University, Stillwater
Eldon Miller, University of Mississippi
James Miller, West Virginia University
Michael Miller, Iowa State University
Jane Murphy, Middlesex Community College
James Nymann, University of Texas, El Paso
E. James Peake, Iowa State University
Thomas Radin, San Joaquin Delta College
Ken A. Rager, Metropolitan State College
Jane Ringwald, Iowa State University
Stephen Rodi, Austin Community College
Howard L. Rolf, Baylor University
John Sanders, Chicago State University
John Spellman, Southwest Texas State University
Becky Stamper, Western Kentucky University
Neil Stephens, Hinsdale South High School
Tommy Thompson, Brookhaven College
Richard J. Tondra, Iowa State University
Marvel Townsend, University of Florida
Richard G. Vinson, University of Southern Alabama
Carlton Woods, Auburn University
George Zazi, Chicago State University

Recognition and thanks are due particularly to the following individuals for their invaluable assistance in the preparation of this edition: Don Dellen, for his patience, support, and commitment to excellence; Phyllis Niklas, for her many talents as supervisor of production; Beth Anderson and the entire Macmillan sales staff for their sincere interest in this revision; Lorraine Bierdz, for struggling with my penmanship while typing the revision; Katy Sullivan and Marsha Vihon, for checking my answers to the odd-numbered problems; and my wife Mary and our children, Katy, Mike, Danny, and Colleen, for giving up the dining room table and helping in the preparation of the revision in more ways than they would like to remember.

College Algebra and Trigonometry

1 Preliminaries

The word *algebra* is derived from the Arabic word *al-jabr*. This word is a part of the title of a ninth century work, "Hisâb al-jabr w'al-muqâbalah," written by Mohammed ibn Mûsâ al-Khowârizmî. The word *al-jabr* means "a restoration," a reference to the fact that if a number is added to one side of an equation, then it must also be added to the other side in order to "restore" the equality. The title of the work, freely translated, means "The science of reduction and cancellation." Of course, today, algebra has come to mean a great deal more.

1.1 Real Numbers

Algebra can be thought of as a generalization of arithmetic in which letters may be used to represent numbers.

Operations

We will use letters such as x, y, a, b, and c to represent numbers. The symbols used in algebra for the operations of addition, subtraction, multiplication, and division are $+$, $-$, $\cdot$, and $/$. The words used to describe the results of these operations are **sum**, **difference**, **product**, and **quotient**. Table 1 summarizes these ideas.

Table 1

OPERATION	SYMBOL	WORDS
Addition	$a + b$	Sum: a plus b
Subtraction	$a - b$	Difference: a less b
Multiplication	$a \cdot b$, $(a) \cdot b$, $a \cdot (b)$, $(a) \cdot (b)$, ab, $(a)b$, $a(b)$, $(a)(b)$	Product: a times b
Division	a/b or $\frac{a}{b}$	Quotient: a divided by b

In algebra, we generally avoid using the multiplication sign $\times$ and the division sign $\div$ so familiar in arithmetic. Notice also that when two expressions are placed next to each other without an operation symbol, as in ab, or in parenthesis, as in $(a)(b)$, it is understood that the expressions, called **factors**, are to be multiplied.

The four operations of addition, subtraction, multiplication, and division are called **binary operations**, since each of them is performed on a pair of numbers.

The symbol $=$, called an **equal sign** and read as "equals" or "is," is used to express the idea that the number or expression on the left of the equal sign is equivalent to the number or expression on the right.

Example 1 (a) The sum of 2 and 7 equals 9. In symbols, this statement is written as $2 + 7 = 9$.

(b) The product of 3 and 5 is 15. In symbols, this statement is written as $3 \cdot 5 = 15$. ■

Equality

We have used the equal sign to convey the idea that one expression is equivalent to another. Three important properties of equality are listed at the top of the next page. In this list, a and b represent numbers.

Reflexive Property

1. The **reflexive property** states that a number always equals itself; that is, $a = a$. Although this result seems obvious, it forms the basis for much of what we do in algebra.

Symmetric Property

2. The **symmetric property** states that if $a = b$, then $b = a$.

Principle of Substitution

3. The **principle of substitution** states that if $a = b$, then we may substitute b for a in any expression containing a.

Sets

When we want to treat a collection of similar but distinct objects as a whole, we use the idea of a **set**. For example, the set of *digits* consists of the collection of numbers 0, 1, 2, 3, 4, 5, 6, 7, 8, and 9. If we use the symbol D to denote the set of digits, then we can write

$$D = \{0, 1, 2, 3, 4, 5, 6, 7, 8, 9\}$$

In this notation, the braces { } are used to enclose the objects, or **elements**, in the set. This method of denoting a set is called the **roster method**. A second way to denote a set is to use **set-builder notation**, where the set D of digits is written as

$$D = \{ \quad x \quad | \quad x \text{ is a digit}\}$$

Read as "*D* is the set of all *x* such that *x* is a digit."

Example 2 (a) $E = \{x|x \text{ is an even digit}\} = \{0, 2, 4, 6, 8\}$

(b) $O = \{x|x \text{ is an odd digit}\} = \{1, 3, 5, 7, 9\}$ ■

In listing the elements of a set, we do not list an element more than once because the elements of a set are distinct. Also, the order in which the elements are listed is not relevant. Thus, for example, {2, 3} and {3, 2} both represent the same set.

If every element of a set A is also an element of a set B, then we say that A is a **subset** of B. If two sets A and B have the same elements, then we say that A **equals** B. For example, {1, 2, 3} is a subset of {1, 2, 3, 4, 5}; and {1, 2, 3} equals {2, 3, 1}.

Classification of Numbers

It is helpful to classify the various kinds of numbers we deal with in algebra. The **counting numbers**, or **natural numbers**, are the numbers 1, 2, 3, 4, ... (The three dots, called an **ellipsis**, indicate that the pattern continues indefinitely.) As their name implies, these numbers are often used to count things. For example, there are 26 letters in our alphabet; there are 100 cents in a dollar. The **whole numbers** are the numbers 0, 1, 2, 3, . . . , that is, the counting numbers together with 0.

Integers

The **integers** are the numbers . . . , −3, −2, −1, 0, 1, 2, 3, ...

These numbers prove useful in many situations. For example, if your checking account has $10 in it and you write a check for $15, you can represent the current balance as −$5.

Notice that the counting numbers are included among the whole numbers. Each time we expand a number system, such as from the whole numbers to the integers, we do so in order to be able to handle new, and usually more complicated, problems. Thus, the integers allow us to solve problems requiring both positive and negative counting numbers, such as profit/loss, height above/below sea level, temperature above/below 0°F, and so on.

But integers alone are not sufficient for *all* problems. For example, they do not answer the questions, "What part of a dollar is 38 cents?" or "What part of a pound is 5 ounces?" To answer such questions, we enlarge our number system to include *rational numbers*. For example, $\frac{38}{100}$ answers the question, "What part of a dollar is 38 cents?" and $\frac{5}{16}$ answers the question, "What part of a pound is 5 ounces?"

Rational Number

A **rational number** is a number that can be expressed as a quotient a/b of two integers. The integer a is called the **numerator**, and the integer b, which cannot be 0, is called the **denominator**.

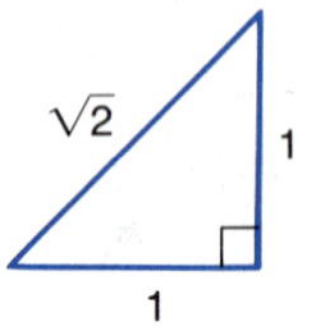

Figure 1

Examples of rational numbers are $\frac{3}{4}, \frac{5}{2}, \frac{0}{4}, -\frac{2}{3}$, and $\frac{100}{3}$. Since $a/1 = a$ for any integer a, it follows that the rational numbers contain the integers as a special case.

Although rational numbers occur frequently enough in applications, there is also a need for numbers that are not rational. For example, consider the isosceles right triangle whose legs are each of length 1; see Figure 1. The number that equals the length of the hypotenuse is the positive number whose square is 2. It can be shown that this number, symbolized by $\sqrt{2}$, is not a rational number.

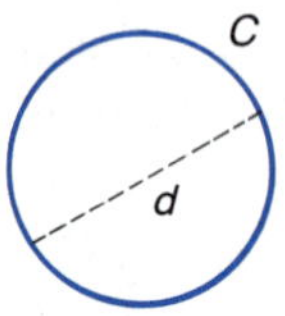

Figure 2
$\pi = \frac{C}{d}$

Also, the number that equals the ratio of the circumference C to the diameter d of any circle, denoted by the symbol π (the Greek letter pi), is not a rational number. See Figure 2.

Other symbols for numbers that are not rational numbers are $\sqrt{3}$, $\sqrt{5}$, $\sqrt{7}$, $\sqrt[3]{2}$, $\sqrt[3]{3}$, and so on.

Irrational Numbers; Real Numbers

Numbers that represent lengths and are not rational are called **irrational numbers**. Together, the rational numbers and irrational numbers form the **real numbers**.

See Figure 3.

Figure 3

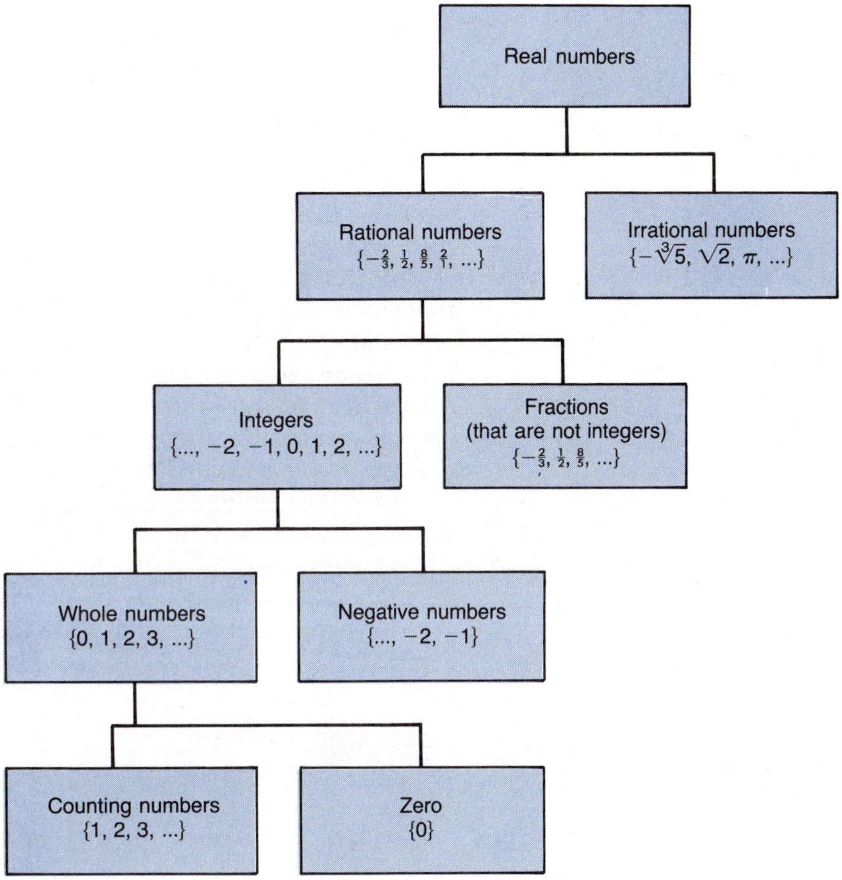

Decimals

Real numbers may be reperesented as **decimals**. For example, the rational numbers $\frac{3}{4}$, $\frac{5}{2}$, $-\frac{2}{3}$, and $\frac{7}{66}$ may be represented as decimals by merely carrying out the indicated division:

$$\frac{3}{4} = 0.75 \qquad \frac{5}{2} = 2.5 \qquad -\frac{2}{3} = -0.666\ldots \qquad \frac{7}{66} = 0.1060606\ldots$$

Notice that the decimal representations of $\frac{3}{4}$ and $\frac{5}{2}$ terminate, or end. The decimal representations of $-\frac{2}{3}$ and $\frac{7}{66}$ do not terminate, but they do exhibit a pattern of repetition. For $-\frac{2}{3}$, the 6 repeats indefinitely; for $\frac{7}{66}$, the block 06 repeats indefinitely. It can be shown that every rational number may be represented by a decimal that either terminates or is nonterminating with a repeating block of digits.

On the other hand, there are decimals that do not fit into either of these categories. For example, the decimal $0.12345678910111213\ldots$, in which we write down the positive integers successively one after the other, will neither repeat (think about it) nor terminate. In fact, every irrational number may be represented by a decimal that neither repeats nor terminates. For example, the irrational numbers $\sqrt{2}$ and π have decimal representations that begin as follows:

$$\sqrt{2} = 1.414213\ldots \qquad \pi = 3.14159\ldots$$

Thus, every decimal may be represented by a real number (either rational or irrational) and, conversely, every real number may be represented by a decimal. It is this feature of real numbers that gives them their practicality. In the physical world, many changing quantities such as the length of a heated rod, the velocity of a falling object, and so on, are assumed to pass through every possible magnitude from the initial one to the final one as they change. Real numbers in the form of decimals provide a convenient way to measure such quantities as they change.

In practice, irrational numbers are generally represented by approximations. For example, using the symbol $\approx$ (read as "approximately equal to"), we can write

$$\sqrt{2} \approx 1.4142 \qquad \pi \approx 3.1416$$

We shall discuss the idea of approximation in more detail in the next section.

Some Conventions

In algebra, we agree in advance to the following three conventions:

1. Given $a \cdot b + c$ or $c + a \cdot b$, we agree to multiply a by b first and then add the result to c. Thus, we have $2 \cdot 4 + 5 = 8 + 5 = 13$ and $3 + 4 \cdot 5 = 3 + 20 = 23$.
2. In arithmetic, when mixed numbers are used, addition is understood; for example, $2\frac{3}{4}$ means $2 + \frac{3}{4}$. In algebra, use of a mixed number may be confusing because the absence of an operation symbol between two terms is generally taken to mean multiplication. Thus, we prefer not to use mixed numbers in algebra. The expression $2\frac{3}{4}$ is written instead as 2.75 or as $\frac{11}{4}$.
3. Parentheses are used in algebra to group terms and indicate precedence of operations. Thus, $(2 + 3) \cdot 4$ is an instruction to add 2 and 3 and then multiply by 4, so that $(2 + 3) \cdot 4 = 5 \cdot 4 = 20$. When parentheses are nested, the inside parentheses are evaluated first, followed in order by the outer ones. For example,

$$\begin{aligned} 3 - \{2 - [4 + 5 \cdot (8 - 2)]\} &= 3 - \{2 - [4 + 5 \cdot 6]\} \\ &= 3 - \{2 - [4 + 30]\} \\ &= 3 - \{2 - 34\} \\ &= 3 - \{-32\} \\ &= 3 + 32 = 35 \end{aligned}$$

Inside first: $(8 - 2)$; Second: $[\ldots]$; Third: $\{\ldots\}$

Notice in the above expression that we used braces { } and brackets [ ] to assist us in seeing the relative order of operations. Calculators and computers, needing no such help, will merely nest parentheses. Thus, a computer "sees" the expression

$$3 - \{2 - [4 + 5 \cdot (8 - 2)]\}$$

as

$$3 - (2 - (4 + 5 \cdot (8 - 2)))$$

and will perform the operations from inside to outside.

Properties of Real Numbers

Certain basic properties of real numbers are given as equations involving addition and multiplication. In the list that follows, a, b, and c represent real numbers:

Commutative Properties

$$a + b = b + a \qquad a \cdot b = b \cdot a \tag{1}$$

In words, the order in which addition or multiplication takes place will not affect the final result.

Associative Properties

$$a + (b + c) = (a + b) + c = a + b + c \tag{2a}$$
$$a \cdot (b \cdot c) = (a \cdot b) \cdot c = a \cdot b \cdot c \tag{2b}$$

The way one adds or multiplies three real numbers will not affect the final result. Thus, expressions such as $2 + 3 + 4$ and $3 \cdot 4 \cdot 5$ present no ambiguity, even though addition and multiplication are performed on one pair of numbers at a time.

Distributive Property

$$a \cdot (b + c) = a \cdot b + a \cdot c \tag{3a}$$
$$(a + b) \cdot c = a \cdot c + b \cdot c \tag{3b}$$

The distributive property is perhaps the most important. With it, we justify such statements as

$$2(x + y) = 2x + 2y \qquad \text{and} \qquad 3x + 5x = (3 + 5)x = 8x$$

The two real numbers 0 and 1 have the following properties:

Identity Properties

$$0 + a = a + 0 = a \qquad a \cdot 1 = 1 \cdot a = a \tag{4}$$

We call 0 the **additive identity** and 1 the **multiplicative identity**.

For each real number a, there is a real number $-a$, called the **additive inverse** of a, having the following property:

Additive Inverse Property

$$a + (-a) = -a + a = 0 \tag{5a}$$

For each *nonzero* real number a, there is a real number $1/a$, called the **multiplicative inverse** of a, having the following property:

Multiplicative Inverse Property

$$a \cdot \frac{1}{a} = \frac{1}{a} \cdot a = 1 \qquad a \neq 0 \tag{5b}$$

The multiplicative inverse $1/a$ of a nonzero real number a is also referred to as the **reciprocal** of a.

The additive inverse of a, namely, $-a$, is often called the *negative* of a. But this practice can be dangerous, because use of the term suggests that the additive inverse is a negative number, which it may not be. For example, the additive inverse of -3, namely, $-(-3)$, equals 3, a positive number.

With these properties for adding and multiplying real numbers, we can now define the operations of subtraction and division as follows:

Difference

The **difference** $a - b$, also read "a less b" or "a minus b," is defined as

$$a - b = a + (-b) \tag{6}$$

Quotient

If b is a nonzero real number, the **quotient** a/b, also read as "a divided by b" or "the ratio of a to b," is defined as

$$\frac{a}{b} = a \cdot \frac{1}{b} \qquad b \neq 0 \tag{7}$$

For example,

$$8 - 5 = 8 + (-5) = 3 \qquad 4 - 9 = 4 + (-9) = -5$$

$$\tfrac{5}{8} = (5)\left(\tfrac{1}{8}\right) \qquad \tfrac{20}{-5} = (20)\left(\tfrac{1}{-5}\right) = -4$$

The further properties of real numbers given below also will be useful. Although these can be derived or proved using equations (1)–(5), we shall not attempt to prove them here. As before, a, b, and c denote real numbers in the properties listed.

For any number a, the product of a times 0 is always 0; that is,

$$a \cdot 0 = 0 \qquad (8)$$

For a nonzero number a,

Division Properties

$$\frac{0}{a} = 0 \qquad \frac{a}{a} = 1 \qquad a \neq 0 \qquad (9)$$

Note: Division by 0 is *not allowed*. One reason is to avoid the following difficulty: $\frac{2}{0} = x$ means to find x such that $0 \cdot x = 2$. But $0 \cdot x$ equals 0 for all x, so there is *no* number x such that $\frac{2}{0} = x$.

Rules of Signs

$$a(-b) = -(ab) \qquad (-a)b = -(ab) \qquad (-a)(-b) = ab$$

$$-(-a) = a \qquad \frac{a}{-b} = \frac{-a}{b} = -\frac{a}{b} \qquad \frac{-a}{-b} = \frac{a}{b} \qquad (10)$$

For example,

$$2(-3) = -(2 \cdot 3) = -6 \qquad (-3)(-5) = 3 \cdot 5 = 15$$

$$\frac{3}{-2} = \frac{-3}{2} = -\frac{3}{2} \qquad \frac{-4}{-9} = \frac{4}{9}$$

If c is a nonzero number, then

Cancellation Properties

$$ac = bc \quad \text{implies} \quad a = b \qquad c \neq 0$$

$$\frac{ac}{bc} = \frac{a}{b} \qquad b \neq 0,\ c \neq 0 \qquad (11)$$

Product Law

$$\text{If } ab = 0, \text{ then } a = 0 \text{ or } b = 0 \text{ or both.} \qquad (12)$$

The product law is used often to draw conclusions. For example, if twice a number x is 0 (that is, if $2x = 0$), then the number x must be 0.

Arithmetic of Quotients

$$\frac{a}{b} + \frac{c}{d} = \frac{ad + bc}{bd} \qquad b \neq 0, d \neq 0 \qquad (13)$$

$$\frac{a}{b} \cdot \frac{c}{d} = \frac{ac}{bd} \qquad b \neq 0, d \neq 0 \qquad (14)$$

$$\frac{\frac{a}{b}}{\frac{c}{d}} = \frac{a}{b} \cdot \frac{d}{c} = \frac{ad}{bc} \qquad b \neq 0, c \neq 0, d \neq 0 \qquad (15)$$

Example 3

$$\frac{2}{3} + \frac{5}{2} \underset{\uparrow}{=} \frac{2 \cdot 2 + 3 \cdot 5}{3 \cdot 2} = \frac{4 + 15}{6} = \frac{19}{6}$$

By equation (13) ■

Example 4

$$\frac{3}{5} - \frac{2}{3} \underset{\uparrow}{=} \frac{3}{5} + \left(-\frac{2}{3}\right) \underset{\uparrow}{=} \frac{3}{5} + \frac{-2}{3}$$

By equation (6) By equation (10)

$$\underset{\uparrow}{=} \frac{3 \cdot 3 + 5 \cdot (-2)}{5 \cdot 3} = \frac{9 + (-10)}{15} = \frac{-1}{15}$$

By equation (13) ■

Example 5

$$\frac{8}{3} \cdot \frac{15}{4} \underset{\uparrow}{=} \frac{8 \cdot 15}{3 \cdot 4} = \frac{2 \cdot \cancel{4} \cdot \cancel{3} \cdot 5}{\cancel{3} \cdot \cancel{4} \cdot 1} \underset{\uparrow}{=} \frac{2 \cdot 5}{1} = 10$$

By equation (14) By equation (11) ■

Note: We follow the common practice of using slash marks to indicate cancellation. Slanting the cancellation marks in different directions for different factors, as in Example 5, is a good practice to follow, since it will help in checking for errors.

Example 6

$$\frac{\frac{3}{5}}{\frac{7}{9}} \underset{\uparrow}{=} \frac{3}{5} \cdot \frac{9}{7} \underset{\uparrow}{=} \frac{3 \cdot 9}{5 \cdot 7} = \frac{27}{35}$$

By equation (14)

By equation (15) ■

Note: In writing quotients, we shall follow the usual convention and write the quotient in **lowest terms**—that is, we write it so that any common divisors of the numerator and the denominator have been removed using the cancellation properties, equation (11). Thus,

$$\frac{90}{24} = \frac{3 \cdot 5 \cdot \cancel{2} \cdot \cancel{3}}{2 \cdot 2 \cdot \cancel{2} \cdot \cancel{3}} = \frac{15}{4}$$

$$\frac{24x^2}{18x} = \frac{\cancel{2} \cdot 2 \cdot 2 \cdot \cancel{3} \cdot \cancel{x} \cdot x}{\cancel{2} \cdot \cancel{3} \cdot 3 \cdot \cancel{x}} = \frac{4x}{3} \qquad x \neq 0$$

The Real Number Line

It can be shown that there is a one-to-one correspondence between real numbers and points on a line. That is, every real number corresponds to a point on the line and, conversely, each point on the line has a unique real number associated with it. We establish this correspondence of real numbers with points on a line in the following manner.

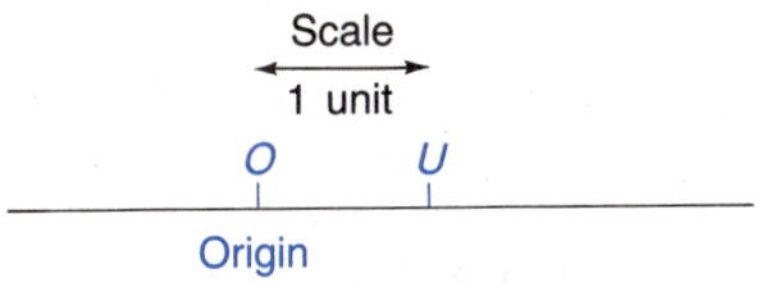

Figure 4
Horizontal line

We start with a line that is, for convenience, drawn horizontally. Pick a point on the line and label it O, for **origin**. Then pick another point some fixed distance to the right of O and label it U, for unit, as shown in Figure 4.

The fixed distance, which may be 1 inch, 1 centimeter, 1 light-year, or any unit distance, determines the **scale**. We associate the real number 0 with the origin O and the number 1 with the point U. Refer now to Figure 5. The point to the right of U that is twice as far from O as U is associated with the number 2. The point to the right of U that is three times as far from O as U is associated with the number 3. The point midway between O and U is assigned the number 0.5, or $\frac{1}{2}$. Corresponding points to the left of the origin O are assigned the numbers $-\frac{1}{2}$, -1, -2, -3, and so on, depending on how far they are from O. Notice in Figure 5 that we placed an arrowhead on the right end of the line to indicate the direction in which the assigned numbers increase. Figure 5 also shows the points associated with the irrational numbers $\sqrt{2}$ and π.

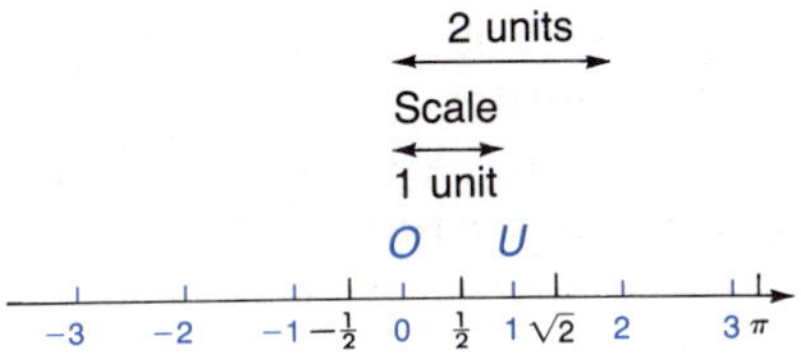

Figure 5
Real number line

Coordinate; Real Number Line

The real number associated with a point P is called the **coordinate** of P, and the line whose points have been assigned coordinates is called the **real number line**.

The real number line divides the real numbers into three classes, as shown in Figure 6:

Figure 6

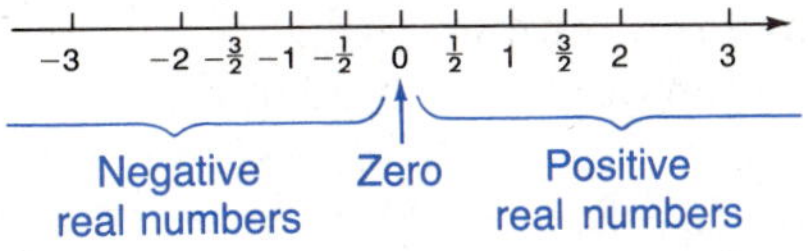

Negative Real Numbers

1. The **negative real numbers** are the coordinates of points to the left of the origin O.

Zero

2. The real number **zero** is the coordinate of the origin O.

Positive Real Numbers

3. The **positive real numbers** are the coordinates of points to the right of the origin O.

Negative and positive numbers have the following multiplication properties:

Properties of Positive and Negative Numbers

1. The product of two positive numbers is a positive number.
2. The product of two negative numbers is a positive number.
3. The product of a positive number and a negative number is a negative number.

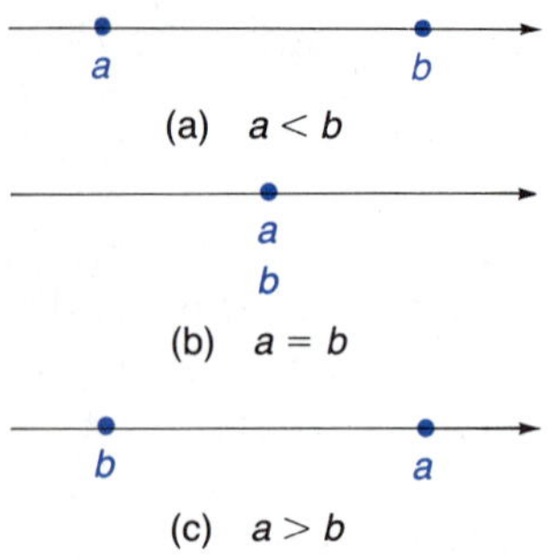

Figure 7

Inequalities

An important property of the real number line follows from the fact that given two numbers (points) a and b, either a is to the left of b, a equals b, or a is to the right of b. See Figure 7.

If a is to the left of b, we say "a is less than b" and write $a < b$. If a is to the right of b, we say "a is greater than b" and write $a > b$. Of course, if a equals b, we write $a = b$. If a is either less than or equal to b, we write $a \leq b$. Similarly, $a \geq b$ means a is either greater than or equal to b. Collectively, the symbols $<$, $>$, $\leq$, $\geq$ are called **inequality symbols**. A formal definition of *less than* and *greater than* follows.

Less Than; Greater Than

Let a and b be two real numbers. We say that a **is less than** b, written $a < b$, or, equivalently, that b **is greater than** a, written $b > a$, if the difference $b - a$ is positive.

Note that $a < b$ and $b > a$ mean the same thing. Thus, it does not matter whether we write $2 < 3$ or $3 > 2$.

Example 7

(a) $3 < 7$ (b) $-8 > -16$ (c) $-6 < 0$
(d) $-8 < -4$ (e) $4 > -1$ (f) $8 > 0$ ■

In Example 7(a), we conclude that $3 < 7$ either because 3 is to the left of 7 on the real number line or because the difference $7 - 3 = 4$, a positive real number.

Similarly, we conclude in Example 7(b) that $-8 > -16$ either because -8 lies to the right of -16 on the real number line or because the difference $-8 - (-16) = -8 + 16 = 8$, a positive real number.

Look again at Example 7. You may find it useful to observe that the inequality symbol always points in the direction of the smaller number.

Statements of the form $a < b$ or $b > a$ are called **strict inequalities**, while statements of the form $a \le b$ or $b \ge a$ are called **nonstrict inequalities**. An **inequality** is a statement in which two expressions are related by an inequality symbol. The expressions are referred to as the **sides** of the inequality.

Based on the discussion thus far, we conclude that:

$a > 0$ is equivalent to a is positive
$a < 0$ is equivalent to a is negative

Thus, we sometimes read $a > 0$ by saying that "a is positive." If $a \ge 0$, then either $a > 0$ or $a = 0$, and we may read this as "a is nonnegative."

Absolute Value

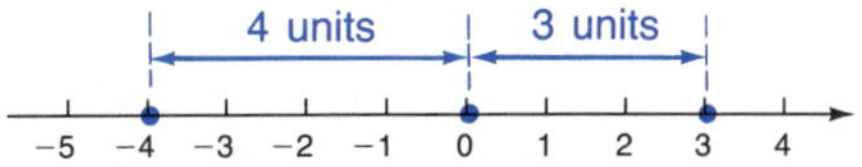

−4 is a distance of 4 units from the origin
3 is a distance of 3 units from the origin

Figure 8

The *absolute value* of a number x is the distance from the point whose coordinate is x to the origin. For example, the point whose coordinate is -4 is 4 units from the origin. The point whose coordinate is 3 is 3 units from the origin. See Figure 8. Thus, the absolute value of -4 is 4, and the absolute value of 3 is 3.

A more formal definition of absolute value is given below.

Absolute Value

The **absolute value** of a real number x, denoted by the symbol $|x|$, is defined by the rules

$$|x| = x \quad \text{if } x \ge 0 \qquad \text{and} \qquad |x| = -x \quad \text{if } x < 0$$

For example, since $-4 < 0$, then the second rule must be used to get $|-4| = -(-4) = 4$.

Example 8 (a) $|8| = 8$ (b) $|0| = 0$ (c) $|-15| = 15$ ■

Look again at Figure 8. The distance from the point whose coordinate is -4 to the point whose coordinate is 3 is 7 units. This distance is the difference $3 - (-4)$, obtained by subtracting the smaller coordinate from the larger. However, since $|3 - (-4)| = |7| = 7$ and $|-4 - 3| = |-7| = 7$, we can use absolute value to calculate the distance between two points without being concerned about which coordinate is smaller.

Distance between P and Q

If P and Q are two points on a real number line with coordinates x and y, respectively, the **distance between P and Q**, denoted by $d(P, Q)$, is

$$d(P, Q) = |y - x|$$

Since $|y - x| = |x - y|$, it follows that $d(P, Q) = d(Q, P)$.

Example 9 Let P, Q, and R be points on a real number line with coordinates -5, 7, and -3, respectively. Find the distance:

(a) Between P and Q (b) Between Q and R

Solution (a) $d(P, Q) = |7 - (-5)| = |12| = 12$ (See Figure 9.)

(b) $d(Q, R) = |-3 - 7| = |-10| = 10$

Figure 9

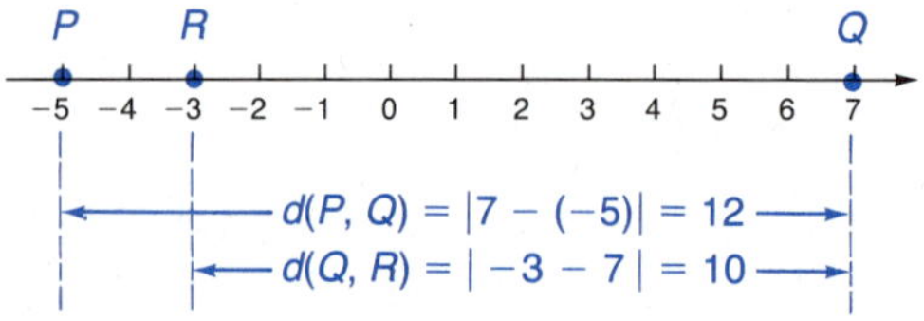

■

Constants and Variables

We said earlier that in algebra we use letters such as x, y, a, b, and c to represent numbers. If the letter used is to represent *any* number from a given set of numbers, it is called a **variable**. A **constant** is either a fixed number, such as 5, $\sqrt{3}$, etc., or a letter that represents a fixed (possibly unspecified) number.

Example 10 The formula for the area of a circle is

$$A = \pi r^2$$

In this formula, r is a variable representing the radius of the circle, A is a variable representing the area of the circle, and π is a constant (≈ 3.14). See Figure 10. ■

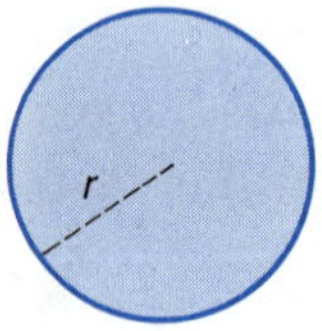

Figure 10
$A = \pi r^2$

In working with expressions or formulas involving variables, the variables may be allowed to take on values from only a certain set of numbers. For example, in the formula for the area A of a circle of radius r, $A = \pi r^2$, the variable r is necessarily restricted to the positive real numbers. In the expression $1/x$, the variable x cannot take on the value 0, since division by 0 is not allowed.

Domain of the Variable

The set of values that a variable may assume is called the **domain of the variable**.

Example 11

(a) The domain of the variable x in the expression

$$\frac{5}{x - 2}$$

is $\{x|x \neq 2\}$ since, if $x = 2$, the denominator becomes 0, which is not allowed.

(b) In the formula for the circumference C of a circle of radius r,

$$C = 2\pi r$$

the domain of the variable r, representing the radius of the circle, is the set of positive real numbers. The domain of the variable C, representing the circumference of the circle, is also the set of positive real numbers. ■

In describing the domain of a variable, we may use either set notation or words, whichever is more convenient.

Historical Comment

■ The real number system has a history that stretches back at least to the ancient Babylonians (1800 BC). It is remarkable how much the ancient Babylonian attitudes resemble our own. As we stated in the text, the fundamental difficulty with irrational numbers is that they cannot be written as quotients of integers or, equivalently, as repeating or terminating decimals. Because the Babylonians wrote their numbers in a system based on 60 in the same way we write ours based on 10, it did not upset the Babylonians that π or $\sqrt{2}$ were not quotients of two integers. They would carry as many places for π as the accuracy of the problem demanded, just as we now use

$$\pi \approx 3\tfrac{1}{7} \quad \text{or} \quad \pi \approx 3.1416 \quad \text{or} \quad \pi \approx 3.14159 \quad \text{or} \quad \pi \approx 3.141592653589$$

depending on how accurate we need to be.

Things were very different for the Greeks, whose number system allowed only rational numbers. When it was discovered that $\sqrt{2}$ was not a rational number, this was regarded as a fundamental flaw in the number concept. So serious was the matter that the Pythagorean Brotherhood (an early mathematical society) is said to have drowned one of its members for revealing this terrible secret. Greek mathematicians then turned away from the number concept, expressing facts about whole numbers in terms of line segments.

In astronomy, however, Babylonian methods, including the Babylonian number system, continued to be used. Simon Stevin (1548–1620), probably using the Babylonian system as a model, invented the decimal system, complete with rules of calculation, in 1585. [Others—for example, al-Kashi of Samarkand (d. 1424)—had made some progress in the same direction.] The decimal system so effectively conceals the difficulties that the need for more logical precision began to be felt only in the early 1800's. Around 1880, Georg Cantor (1845–1918) and Richard Dedekind (1831–1916) gave precise definitions of real numbers. Cantor's definition, though more abstract and precise, has its roots in the decimal (and hence, Babylonian) numerical system.

Sets and set theory were a spin-off of the research that went into clarifying the foundations of the real number system. Set theory has developed into a large discipline of its own, and many mathematicians regard it as the foundation upon which modern mathematics is built. Cantor's discovery that infinite sets also can be counted and that there are different sizes of infinite sets are among the most astounding results of modern mathematics.

Historical Problems

■ *The Babylonian number system was based on* 60. *Thus* 2,30 *means* $2 + \frac{30}{60} = 2.5$, *and* 4,25,14 *means*

$$4 + \frac{25}{60} + \frac{14}{60^2} = 4 + \frac{1514}{3600} = 4.42055555\ldots$$

1. What are the following numbers in Babylonian notation?
   (a) $1\frac{1}{3}$ (b) $2\frac{5}{6}$
2. What are the following Babylonian numbers when written as fractions and as decimals?
   (a) 2,20 (b) 4,52,30 (c) 3,8,29,44 ■

EXERCISE 1.1

In Problems 1–12, express each rational number as a decimal, either repeating or terminating.

1. $\frac{2}{3}$ **2.** $\frac{2}{5}$ **3.** $\frac{1}{8}$ **4.** $\frac{10}{3}$ **5.** $-\frac{8}{5}$ **6.** $-\frac{16}{3}$

7. $\frac{1}{9}$ **8.** $\frac{1}{10}$ **9.** $\frac{4}{25}$ **10.** $-\frac{5}{6}$ **11.** $-\frac{3}{7}$ **12.** $\frac{8}{11}$

In Problems 13–30, evaluate each expression.

13. $-6 + 4 \cdot 2$ **14.** $8 - 4 \cdot 3$ **15.** $4 + 5 - 10$

16. $8 - 3 - 2$ **17.** $4 + \frac{1}{3}$ **18.** $2 - \frac{1}{2}$

19. $6 - [3 \cdot 5 + 2 \cdot (3 - 2)]$ **20.** $2 \cdot [8 - 3(4 + 2)] - 3$

21. $2 \cdot (3 - 5) + 8 \cdot 2 - 1$ **22.** $1 - (4 \cdot 3 - 2 + 2)$

23. $10 - [6 - 2 \cdot 2 + (8 - 3)] \cdot 2$ **24.** $2 - 5 \cdot 4 - [6 \cdot (3 - 4)]$

25. $5 - 3 \cdot \frac{1}{2}$ **26.** $18 + 4 \cdot \frac{1}{3}$ **27.** $\frac{4}{5} + \frac{8}{3}$

28. $\frac{2}{5} - \frac{4}{3}$ **29.** $\dfrac{\frac{16}{5} - \frac{1}{2}}{\frac{1}{3}}$ **30.** $\dfrac{\frac{14}{3} + \frac{1}{2}}{\frac{3}{4}}$

In Problems 31–42, use the distributive property to remove the parentheses.

31. $3(x + 4)$ **32.** $5(2x - 1)$ **33.** $x(x - 3)$

34. $3x(x + 4)$ **35.** $(x + 2)(x + 4)$ **36.** $(x + 5)(x + 1)$

37. $(x - 2)(x + 1)$ **38.** $(x - 4)(x + 1)$ **39.** $(x - 8)(x - 2)$

40. $(x - 4)(x - 2)$ **41.** $(x + 2)(x - 2)$ **42.** $(x - 3)(x + 3)$

43. Is subtraction commutative? Support your conclusion.

44. Is subtraction associative? Support your conclusion.

45. Is division commutative? Support your conclusion.

46. Is division associative? Support your conclusion.

47. If $2 = x$, why does $x = 2$?

48. If $x = 5$, why does $x^2 + x = 30$?

49. On a real number line, label the points with coordinates of 0, 1, −1, $\frac{5}{2}$, −2.5, $\frac{3}{4}$, and 0.25.

50. Repeat Problem 49 for the coordinates 0, −2, 2, −1.5, $\frac{3}{2}$, $\frac{1}{3}$, and $\frac{2}{3}$.

In Problems 51–56, write each statement as an inequality.

51. x is positive **52.** z is negative

53. x is less than 2 **54.** y is greater than −5

55. The absolute value of x is less than or equal to 1.

56. The reciprocal of x is greater than or equal to 2.

In Problems 57–66, $a > 0$ and $b < 0$. Tell whether the given expression is positive, negative, or 0.

57. $|b|$ **58.** $a \cdot b$ **59.** b^2 **60.** a/b

61. $a - b$ **62.** $b - a$ **63.** $|a| + |b|$ **64.** $|ab|$

65. $b + |b|$ **66.** $a + |a|$

In Problems 67–76, find the value of each expression if $x = 2$ and $y = -3$.

67. $|x + y|$ **68.** $|x - y|$ **69.** $|x| + |y|$ **70.** $|x| - |y|$

71. $\dfrac{|x|}{x}$ **72.** $\dfrac{|y|}{y}$ **73.** $|4x - 5y|$ **74.** $|3x + 2y|$

75. $||4x| - |5y||$ **76.** $3|x| + 2|y|$

In Problems 77–86, determine the domain of the variable x in each expression.

77. $\dfrac{5}{x - 3}$ **78.** $\dfrac{-8}{x + 5}$ **79.** $\dfrac{x}{x + 4}$ **80.** $\dfrac{x - 2}{x - 6}$

81. $\dfrac{1}{x} + \dfrac{1}{x - 1}$ **82.** $\dfrac{3}{x(x - 4)}$ **83.** $\dfrac{x + 1}{x(x + 2)}$ **84.** $\dfrac{2}{x + 1} + \dfrac{3}{x + 2}$

85. $\dfrac{x + 1}{x^2 + 4}$ **86.** $\dfrac{x - 3}{x^2 + 6}$

In Problems 87–96, express each statement as an equation involving the indicated variables and constant.

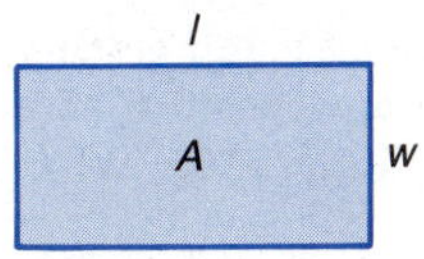

87. The area A of a rectangle is the product of its length l times its width w.

88. The perimeter P of a rectangle is twice the sum of its length l and its width w.

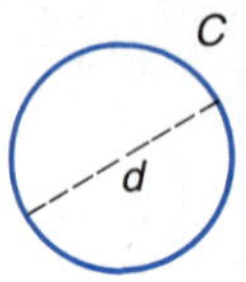

89. The circumference C of a circle is the product of π times its diameter d.

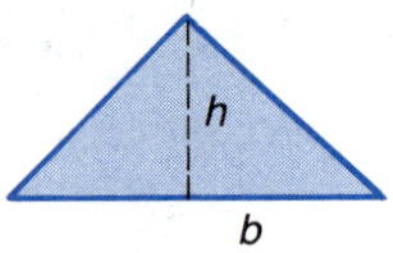

90. The area A of a triangle is one-half the product of its base b times its height h.

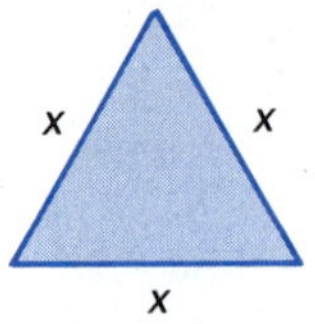

91. The area A of an equilateral triangle is $\sqrt{3}/4$ times the square of the length x of one side.

92. The perimeter P of an equilateral triangle is 3 times the length x of one side.

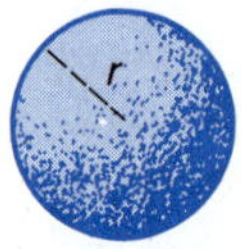

93. The volume V of a sphere is the product of $\frac{4}{3}$ times π times the cube of the radius r.

94. The surface area S of a sphere is the product of 4 times π times the square of the radius r.

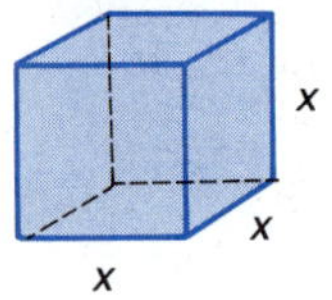

95. The volume V of a cube is the cube of the length x of a side.

96. The surface area S of a cube is 6 times the square of the length x of a side.

In Problems 97–100, use the formula $C = \frac{5}{9}(F - 32)$ for converting degrees Fahrenheit into degrees Celsius to find the Celsius measure of each Fahrenheit temperature.

97. $F = 32°$ **98.** $F = 212°$ **99.** $F = 68°$ **100.** $F = -4°$

101. The weekly production cost C of manufacturing x watches is given by the formula $C = 4000 + 2x$, where the variable C is in dollars.

(a) What is the cost of producing 1000 watches?

(b) What is the cost of producing 2000 watches?

102. The weekly production cost C of manufacturing x hand calculators is given by the formula

$$C = 3000 + 6x - \frac{x^2}{1000}$$

(a) What is the cost of producing 1000 hand calculators?

(b) What is the cost of producing 3000 hand calculators?

103. In the first quarter of its fiscal year, a company posted earnings of \$1.20 per share. During the second and third quarters, it posted losses of \$0.75 per share and \$0.30 per share, respectively. In the fourth quarter, it earned a modest \$0.20 per share. What were the annual earnings per share of this company?

104. At the beginning of the month, Mike had a balance of \$210 in his checking account. During the next month, he deposited \$80, wrote a check for \$120, made another deposit of \$25, wrote two checks for \$60 and \$32, and was assessed a monthly service charge of \$5. What was his balance at the end of the month?

In Problems 105–110, give an example to show that each statement is, in general, false.

105. $\dfrac{a}{b} + \dfrac{c}{d} = \dfrac{a+c}{b+d}$ **106.** $\dfrac{1}{a} + \dfrac{1}{b} = \dfrac{1}{a+b}$

107. $\dfrac{a+b}{a+c} = \dfrac{b}{c}$ **108.** $a \cdot (b \cdot b) = (a \cdot b) \cdot (a \cdot b)$

109. $a \cdot a = 2a$ **110.** $(a+b)^2 = a^2 + b^2$

111. Find the value of $1 + 2 + 3 + 4 + \cdots + 99$.
[*Hint:* Regroup the terms as $(1 + 99) + (2 + 98) + \cdots$ and proceed from there.]

112. Find the value of $1 + 3 + 5 + 7 + \cdots + 99$.

113. Find the value of $-2 - 4 - 6 - 8 - \cdots - 98$.

114. Find the value of $1 - 2 + 3 - 4 + 5 - 6 + \cdots + 97 - 98 + 99$.

115. A recent Gallup poll on "body image" found that two-thirds of college-educated women want a more muscular, "hard-bodied" look. If 300 college-educated women were involved in this poll, how many would have answered "yes" to the question, "Do you want a more muscular hard-bodied look."

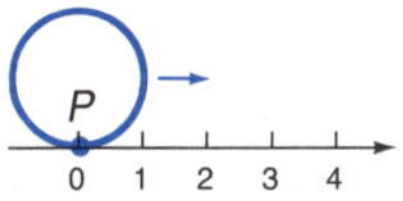

116. On the real number line in the figure, we have placed a circle of radius 1 so that the circle touches the line at 0. This point on the circle is labeled P. If the circle now rolls without slippage in the positive direction along the real number line, at what coordinates will the point P again come into contact with the line?

1.2 Approximations; Calculators*

In approximating decimals, we either *round off* or *truncate* to a given number of decimal places. This number of places establishes the location of the *final digit* in the decimal approximation.

> **Truncation**: Drop all the digits that follow the specified final digit in the decimal.
>
> **Rounding**: Identify the final digit in the decimal. If the next digit is 5 or more, add 1 to the final digit; if the next digit is 4 or less, leave the final digit as it is. Now truncate following the final digit.

Example 1 Approximate 20.98752 to two decimal places by:

(a) Truncating (b) Rounding

Solution For 20.98752, the final digit is the 8, since it is two decimal places from the decimal point.

* C Refer to Graphics Calculator Supplement: Getting Acquainted.

(a) To truncate, we remove all digits following the final digit 8. Thus, the truncation of 20.98752 to two decimal places is 20.98.

(b) The digit following the final digit 8 is the digit 7. Since 7 is 5 or more, we add 1 to the final digit 8. The rounded form of 20.98752 to two decimal places is 20.99. ■

Example 2

	NUMBER	ROUNDED TO TWO DECIMAL PLACES	ROUNDED TO FOUR DECIMAL PLACES	TRUNCATED TO TWO DECIMAL PLACES	TRUNCATED TO FOUR DECIMAL PLACES
(a)	3.14159	3.14	3.1416	3.14	3.1415
(b)	0.056128	0.06	0.0561	0.05	0.0561
(c)	893.46125	893.46	893.4613	893.46	893.4612 ■

Calculators

Calculators are finite machines. As a result, they are incapable of handling decimals that contain a large number of digits. For example, a Texas Instruments TI57 calculator is capable of displaying only eight digits. When a number requires more than eight digits, the TI57 approximates it by rounding. To see how your calculator handles decimals, divide 2 by 3. How many digits do you see? Is the last digit a 6 or a 7? If it is a 6, your calculator truncates; if it is a 7, your calculator rounds.

When the arithmetic involved in a given problem in this book is messy, we shall mark the problem with a [C], indicating that you should use a calculator. Calculators are not required for this course and should not be used except on problems labeled with a [C]. The answers provided in the back of the book for calculator problems have been found using a TI57. Due to differences among calculators in the way they do arithmetic, your answers may vary slightly from those given in the text.

If you are about to purchase a calculator, be sure to choose a **scientific**, as opposed to **arithmetic**, calculator. An arithmetic calculator can only add, subtract, multiply, and divide numbers. Scientific calculators contain keys labeled ln x, log x, sin x, cos x, tan x, y^x, inv, and others, called **function keys**. As you proceed through this text, you will discover how to use many of the function keys.

Another difference among calculators is the order in which various operations are performed. Some employ an **algebraic system**, whereas others use **reverse Polish notation (RPN)**. Either system is acceptable, and the choice of which system to get is a matter of individual preference. Of course, no matter what scientific calculator you purchase, be sure to study the instruction manual so that you use the calculator efficiently and correctly. In this book, our examples are worked using an algebraic system.

Calculator Operations

Arithmetic operations on your calculator are generally shown as keys labeled as follows:

[+] Addition [−] Subtraction [×] Multiplication [÷] Division

Your calculator also has the following keys:

[x^2] To multiply a number by itself.

[AC] or [C] To clear the memory and the display window of previous entries.

Now, let's look at a few examples.

C **Example 3** Approximate $(8.61)^2$. Round off the answer to two decimal places.

Solution

	Press [AC]	Enter [8.61]	Press [x^2]
Display:	0	8.61	74.1321

Thus, $(8.61)^2 \approx 74.13$. ■

C **Example 4** Evaluate: $8 + 9 \cdot 6$

Solution

	Enter [8]	Press [+]	Enter [9]	Press [×]
Display:	8		9	

	Enter [6]	Press [=]
Display:	6	62

■

C **Example 5** Evaluate: $\frac{12}{6} + 5$

Solution

	Enter [12]	Press [÷]	Enter [6]	Press [+]
Display:	12		6	

	Enter [5]	Press [=]
Display:	5	7

■

C **Example 6** Evaluate: $3 + 2 \cdot \frac{9}{3} - 6$

Solution

Enter	3	Press	+	Enter	2	Press	×
Display:	3				2		

Enter	9	Press	÷	Enter	3
Display:	9				3

Press	−	Enter	6	Press:	=
Display:			6		3

■

These examples should convince you that a calculator that uses the algebraic system does arithmetic according to the agreed-upon conventions regarding order of operations. However, there are times when you need to be careful.

The next example illustrates a use of the keys for parentheses,

(and)

C **Example 7** Evaluate: $\dfrac{22 + 8}{8 + 2}$

Solution We treat this expression as if parentheses enclose the numerator and the denominator.

Press	(	Enter	22	Press	+
Display:			22		

Enter	8	Press	)
Display:	8		

Press	÷	Press	(	Enter	8	Press	+
Display:					8		

Enter [2] Press [)] Press [=]

Display: [2] [3] ■

Be careful! If you work the problem in Example 7 without using parentheses, you obtain 22 + 8/8 + 2 = 22 + 1 + 2 = 25.

EXERCISE 1.2

In Problems 1–12, approximate each number (a) rounded and (b) truncated to three decimal places.

1. 18.9526 **2.** 25.86134 **3.** 28.65349 **4.** 99.05229

5. 0.06291 **6.** 0.05388 **7.** 9.9985 **8.** 1.0006

C **9.** $\frac{3}{7}$ C **10.** $\frac{5}{9}$ C **11.** $\frac{521}{15}$ C **12.** $\frac{81}{5}$

C *In Problems 13–32, use a calculator to approximate each expression. Round off your answer to two decimal places.*

13. $(8.51)^2$ **14.** $(9.62)^2$ **15.** $4.1 + (3.2)(8.3)$

16. $(8.1)(4.2) + 6.1$ **17.** $(8.6)^2 + (6.1)^2$ **18.** $(3.1)^2 + (9.6)^2$

19. $8.6 + \dfrac{10.2}{4.2}$ **20.** $9.1 - \dfrac{8.2}{10.2}$ **21.** $\pi + \dfrac{\sqrt{2}}{8}$

22. $\sqrt{5} - 8\pi$ **23.** $\dfrac{\pi + 8}{10.2 + 8.6}$ **24.** $\dfrac{21.3 - \pi}{6.1 + 8.8}$

25. $(22.6 + 8.5)/81.3 + 21.2$ **26.** $(16.5 - 7.2) \cdot 51.2 - 18.6$

27. $22.6 + 8.5/81.3 + 21.2$ **28.** $16.5 - 7.2/51.2 - 18.6$

29. $(22.6 + 8.5)/(81.3 + 21.2)$ **30.** $(16.5 - 7.2)/(51.2 - 18.6)$

31. $22.6 + 8.5/(81.3 + 21.2)$ **32.** $16.5 - 7.2/(51.2 - 18.6)$

33. In 1900, the level of carbon dioxide in the atmosphere was 290 parts per million parts of atmosphere.* What fraction of the atmosphere was carbon dioxide in 1900? Express this fraction as a decimal.

34. In 1988, the level of carbon dioxide in the atmosphere was 350 parts per million parts of atmosphere.* What fraction of the atmosphere was carbon dioxide in 1988? Express this fraction as a decimal.

35. Women have only four-fifths as many red cells in each drop of their blood as men do. If a typical man has 1000 red cells in a drop of blood, how many red cells does a typical woman have in a drop of blood?

C **36.** On your calculator, divide 8 by 0. What happens?

**Source:* Testimony of Syukuro Manabe, an atmospheric scientist at Princeton's Geophysical Fluid Dynamics Laboratory, before the U.S. Senate Energy and Natural Resources Committee, June 1988.

1.3 Integer Exponents

Integer exponents provide a shorthand device for representing repeated multiplications of a real number.

a^n

If a is a real number and n is a positive integer, then the symbol $\mathbf{a^n}$ represents the product of n factors of a. That is,

$$a^n = \underbrace{a \cdot a \cdot \cdots \cdot a}_{n \text{ factors}} \tag{1}$$

where it is understood that $a^1 = a$.

In particular, we have

$$a^1 = a$$
$$a^2 = a \cdot a$$
$$a^3 = a \cdot a \cdot a$$

and so on.

In the expression a^n, a is called the **base** and n is called the **exponent**, or **power**. We read a^n as "a raised to the power n" or as "a to the nth power." We usually read a^2 as "a squared" and a^3 as "a cubed."

Example 1 (a) $2^3 = 2 \cdot 2 \cdot 2 = 8$ (b) $5^2 = 5 \cdot 5 = 25$ (c) $10^1 = 10$

(d) $(-2)^4 = (-2)(-2)(-2)(-2) = 16$

(e) $-2^4 = -(2 \cdot 2 \cdot 2 \cdot 2) = -16$

In words, 2 cubed equals 8; 5 squared equals 25; 10 to the first power equals 10; etc. ■

Notice the difference between Examples 1(d) and 1(e): The exponent applies only to the number or parenthetical expression immediately preceding it.

We define a raised to a negative power as follows:

a^{-n}

If n is a positive integer and if a is a nonzero real number, then we define

$$a^{-n} = \frac{1}{a^n} \qquad a \neq 0 \tag{2}$$

Example 2 (a) $2^{-3} = \dfrac{1}{2^3} = \dfrac{1}{8}$ (b) $\left(\dfrac{1}{5}\right)^{-2} = \dfrac{1}{\left(\frac{1}{5}\right)^2} = \dfrac{1}{\frac{1}{25}} = 25$ ■

a^0 If a is a nonzero number, we define

$$a^0 = 1 \qquad a \neq 0 \tag{3}$$

Notice that we do not allow the base a to be 0 in a^{-n} or in a^0.

Laws of Exponents

There are several general rules that can be used when dealing with exponents. The first one we consider is used when multiplying two expressions that have the same base:

$$3^2 \cdot 3^4 = \underbrace{(3 \cdot 3)}_{\text{2 factors}}\ \underbrace{(3 \cdot 3 \cdot 3 \cdot 3)}_{\text{4 factors}} = \underbrace{3 \cdot 3 \cdot 3 \cdot 3 \cdot 3 \cdot 3}_{\text{6 factors}} = 3^6$$

Multiplying (pointing to $\cdot$); Same base (pointing to the 3s in $3^2 \cdot 3^4$); Sum of powers 2 and 4 (pointing to the exponent 6); Same base (pointing to the 3 in 3^6)

In general, if a is any real number and if m and n are positive integers, we have

$$a^m \cdot a^n = \underbrace{(a \cdot a \cdot \cdots \cdot a)}_{m \text{ factors}}\underbrace{(a \cdot a \cdot \cdots \cdot a)}_{n \text{ factors}} = \underbrace{a \cdot a \cdot \cdots \cdot a}_{m\,+\,n \text{ factors}} = a^{m+n}$$

Thus, to multiply two expressions having the same base, retain the base and add the exponents. That is,

$$a^m a^n = a^{m+n} \tag{4}$$

It can be shown that equation (4) is true whether the integers m and n are positive, negative, or 0. Of course, if m, n, or $m + n$ is 0 or negative, then a cannot be 0, as stated earlier.

Example 3 (a) $2 \cdot 2^4 = 2^5 = 32$ (b) $(-3)^2(-3)^{-3} = (-3)^{-1} = \dfrac{1}{-3}$

(c) $4^{-3} \cdot 4^5 = 4^2 = 16$ (d) $(-2)^{-1}(-2)^{-3} = (-2)^{-4} = \dfrac{1}{(-2)^4} = \dfrac{1}{16}$ ■

Another law of exponents applies when an expression containing a power is itself raised to a power:

$$(2^3)^4 = \underbrace{2^3 \cdot 2^3 \cdot 2^3 \cdot 2^3}_{4 \text{ factors}} = \underbrace{\underbrace{(2 \cdot 2 \cdot 2)}_{3 \text{ factors}}\underbrace{(2 \cdot 2 \cdot 2)}_{3 \text{ factors}}\underbrace{(2 \cdot 2 \cdot 2)}_{3 \text{ factors}}\underbrace{(2 \cdot 2 \cdot 2)}_{3 \text{ factors}}}_{3 \cdot 4 = 12 \text{ factors}} = 2^{12}$$

In general, if a is any real number and if m and n are positive integers, we have

$$(a^m)^n = \underbrace{a^m \cdot a^m \cdot \cdots \cdot a^m}_{n \text{ factors}} = \underbrace{\underbrace{(a \cdot a \cdot \cdots \cdot a)}_{m \text{ factors}}\underbrace{(a \cdot a \cdot \cdots \cdot a)}_{m \text{ factors}}\underbrace{(a \cdot a \cdot \cdots \cdot a)}_{m \text{ factors}}}_{m \cdot n \text{ factors}} = a^{mn}$$

Thus, if an expression containing a power is raised to a power, retain the base and multiply the powers. That is,

$$(a^m)^n = a^{mn} \qquad (5)$$

It can be shown that equation (5) is true whether the integers m and n are positive, negative, or 0. Again, if m or n is 0 or negative, then a must not be 0.

Example 4 (a) $[(-2)^3]^2 = (-2)^6 = 64$ (b) $(3^{-4})^0 = 3^{(-4)(0)} = 3^0 = 1$

(c) $(2^{-3})^2 = 2^{-6} = \dfrac{1}{2^6} = \dfrac{1}{64}$ (d) $(5^{-1})^{-2} = 5^{(-1)(-2)} = 5^2 = 25$ ■

The next law of exponents involves raising a product to a power:

$$(2 \cdot 5)^3 = (2 \cdot 5)(2 \cdot 5)(2 \cdot 5) = (2 \cdot 2 \cdot 2)(5 \cdot 5 \cdot 5) = 2^3 \cdot 5^3$$

In general, if a and b are real numbers and if n is a positive integer, we have

$$(a \cdot b)^n = \underbrace{(ab)(ab) \cdot \cdots \cdot (ab)}_{n \text{ factors}} = \underbrace{(a \cdot a \cdot \cdots \cdot a)}_{n \text{ factors}}\underbrace{(b \cdot b \cdot \cdots \cdot b)}_{n \text{ factors}} = a^n \cdot b^n$$

Thus, if a product is raised to a power, the result equals the product of each factor raised to that power. That is,

$$(a \cdot b)^n = a^n \cdot b^n \qquad (6)$$

It can be shown that equation (6) is true whether the integer n is positive, negative, or 0. If n is 0 or negative, neither a nor b can be 0.

Example 5

(a) $(2x)^3 = 2^3 \cdot x^3 = 8x^3$

(b) $(-2x)^0 = (-2)^0 \cdot x^0 = 1 \cdot 1 = 1 \qquad x \neq 0$

(c) $(ax)^{-2} = a^{-2}x^{-2} = \dfrac{1}{a^2} \cdot \dfrac{1}{x^2} = \dfrac{1}{a^2x^2} \qquad x \neq 0, a \neq 0$

(d) $(ax)^{-2} = \dfrac{1}{(ax)^2} = \dfrac{1}{a^2x^2} \qquad x \neq 0, a \neq 0$ ■

Equations (4) and (6) both involve products. Two analogous laws involve quotients. If a and b are real numbers and if m and n are integers, then

$$\frac{a^m}{a^n} = a^{m-n} = \frac{1}{a^{n-m}} \qquad a \neq 0$$
$$\left(\frac{a}{b}\right)^n = \frac{a^n}{b^n} \qquad b \neq 0 \tag{7}$$

Example 6

(a) $\dfrac{2^6}{2^4} = 2^{6-4} = 2^2 = 4$

(b) $\left(\dfrac{2}{3}\right)^4 = \dfrac{2^4}{3^4} = \dfrac{16}{81}$

(c) $\dfrac{3^{-2}}{3^{-5}} = 3^{-2-(-5)} = 3^3 = 27$

(d) $\left(\dfrac{5}{2}\right)^{-2} = \dfrac{1}{\left(\dfrac{5}{2}\right)^2} = \dfrac{1}{\dfrac{5^2}{2^2}} = \dfrac{1}{\dfrac{25}{4}} = \dfrac{4}{25}$ ■

You may have observed a shortcut for problems like Example 6(d), namely,

$$\left(\frac{a}{b}\right)^{-n} = \left(\frac{b}{a}\right)^n \qquad a \neq 0, b \neq 0 \tag{8}$$

For example,

$$\left(\frac{2}{3}\right)^{-2} = \left(\frac{3}{2}\right)^2 = \frac{9}{4} \qquad \left(\frac{3}{2}\right)^{-3} = \left(\frac{2}{3}\right)^3 = \frac{8}{27}$$

Example 7 Write each expression so that all exponents are positive.

(a) $\dfrac{x^5y^{-2}}{x^3y}$, $x \neq 0, y \neq 0$ (b) $\dfrac{xy}{x^{-1} - y^{-1}}$, $x \neq 0, y \neq 0$

Solution (a) $\dfrac{x^5y^{-2}}{x^3y} = \dfrac{x^5}{x^3} \cdot \dfrac{y^{-2}}{y} = x^{5-3} \cdot y^{-2-1} = x^2y^{-3} = x^2 \cdot \dfrac{1}{y^3} = \dfrac{x^2}{y^3}$

(b) $\dfrac{xy}{x^{-1} - y^{-1}} = \dfrac{xy}{\dfrac{1}{x} - \dfrac{1}{y}} = \dfrac{xy}{\dfrac{y - x}{xy}} = \dfrac{(xy)(xy)}{y - x} = \dfrac{x^2y^2}{y - x}$ ■

Calculator Use

Your calculator should have the key [y^x], which is used for computations involving exponents. The next example shows how this key is used.

[C] **Example 8** Evaluate: $(2.3)^5$

Solution

Enter	[2.3]	Press	[y^x]	Enter	[5]	Press	[=]
Display:	2.3				5		64.36343

■

Negative exponents may be handled in a variety of ways on your calculator.

[C] **Example 9** Evaluate: $(8.1)^{-3}$

Solution A For this solution, we use the [±] key, which is used to change the sign of a number.

Enter	[8.1]	Press	[y^x]	Enter	[3]	Press	[±]
Display:	8.1				3		−3

Press	[=]
Display:	0.00188

Solution B For this solution, we use the reciprocal key [1/x].

Enter	[8.1]	Press	[y^x]	Press	[3]	Press	[=]
Display:	8.1				3		531.441

Press [1/x]

Display: [0.00188]

Notice that solution B uses the fact that $a^{-n} = 1/a^n$. ■

Scientific Notation

Measurements of physical quantities can range from very small to very large. For example, the mass of a proton is approximately 0.00000000000000000000000000167 kilogram and the mass of the Earth is about 5,980,000,000,000,000,000,000,000 kilograms. These numbers obviously are tedious to write down and difficult to read, so we use exponents to rewrite each one. When a number has been written as the product of a number x, where $1 \leq x < 10$, times a power of 10, it is said to be written in **scientific notation**. In scientific notation:

$$\text{Mass of a proton} = 1.67 \times 10^{-27} \text{ kilogram}$$
$$\text{Mass of the Earth} = 5.98 \times 10^{24} \text{ kilograms}$$

> To change a positive number into scientific notation, do the following:
>
> 1. Count the number N of places the decimal point must be moved in order to arrive at a number x, where $1 \leq x < 10$.
> 2. If the original number is greater than or equal to 1, the scientific notation is $x \times 10^N$. If the original number is between 0 and 1, the scientific notation is $x \times 10^{-N}$.

Example 10 Write each number in scientific notation.

(a) 9582 (b) 1.245 (c) 0.285 (d) 0.000561

Solution (a) The decimal point in 9582 follows the 2. Thus, we count

9 5 8 2 .

3 2 1

stopping after three moves because 9.582 is a number between 1 and 10. Since 9582 is greater than 1, we write

$$9582 = 9.582 \times 10^3$$

(b) The decimal point in 1.245 is between the 1 and the 2. Since the number is already between 1 and 10, the scientific notation for it is $1.245 \times 10^0 = 1.245$.

(c) The decimal point in 0.285 is between the 0 and the 2. Thus, we count

$$0\;.\;2\;8\;5$$
$$1$$

stopping after one move because 2.85 it a number between 1 and 10. Since 0.285 is between 0 and 1, we write

$$0.285 = 2.85 \times 10^{-1}$$

(d) The decimal point in 0.000561 is moved as follows:

$$0\;.\;0\;0\;0\;5\;6\;1$$
$$1\;\;2\;\;3\;\;4$$

Thus,

$$0.000561 = 5.61 \times 10^{-4}$$ ■

Example 11 Write each number as a decimal.

(a) 2.1×10^4 (b) 3.26×10^{-5} (c) 1×10^{-2}

Solution (a) $2.1 \times 10^4 = 2\;.\;1\;0\;0\;0 \times 10^4 = 21{,}000$
$\qquad 1\;\;2\;\;3\;\;4$

(b) $3.26 \times 10^{-5} = 0\;0\;0\;0\;0\;3\;.\;2\;6 \times 10^{-5} = 0.0000326$
$\qquad 5\;\;4\;\;3\;\;2\;\;1$

(c) $1 \times 10^{-2} = 0\;0\;1\;. \times 10^{-2} = 0.01$
$\qquad 2\;\;1$ ■

On a calculator, a number such as 3.615×10^{12} is displayed either as `3.615 12` or as `3.615 E12`, depending on the calculator.

Summary

We close this section by summarizing the laws of exponents. In the list that follows, a and b are real numbers and m and n are integers. Also, we assume that no denominator is 0 and that all expressions are defined.

Laws of Exponents

$$a^{-n} = \frac{1}{a^n} \qquad a^0 = 1$$

$$a^m a^n = a^{m+n} \qquad (a^m)^n = a^{mn} \qquad (ab)^n = a^n b^n$$

$$\frac{a^m}{a^n} = a^{m-n} = \frac{1}{a^{n-m}} \qquad \left(\frac{a}{b}\right)^n = \frac{a^n}{b^n}$$

Historical Comment

■ Our method of writing exponents originated with René Descartes, although the concept goes back in various forms to the ancient Babylonians. Even after its introduction by Descartes in 1637, the method took a remarkable amount of time to become completely standardized, and expressions like $aaaaa + 3aaaa + 2aaa - 4aa + 2a + 1$ remained common until 1750. The concept of a rational exponent (see Section 1.8) was known by 1400, although inconvenient notation prevented the development of any extensive theory. John Wallis, in 1655, was the first to give a fairly complete explanation of negative and rational exponents, and Sir Isaac Newton's use of them made exponents standard in their current form. ■

EXERCISE 1.3

In Problems 1–20, simplify each expression.

1. 3^0 **2.** 3^2 **3.** 4^{-2} **4.** $(-3)^2$ **5.** $\left(\frac{2}{3}\right)^2$

6. $\left(\frac{-4}{5}\right)^3$ **7.** $3^0 \cdot 2^{-3}$ **8.** $(-2)^{-3} \cdot 2^0$ **9.** $2^{-3} + \left(\frac{1}{2}\right)^3$ **10.** $3^{-2} + \frac{1}{3}$

11. $3^{-6} \cdot 3^4$ **12.** $4^{-2} \cdot 4^3$ **13.** $\frac{8^2}{2^3}$ **14.** $\frac{4^3}{2^2}$ **15.** $\left(\frac{2}{3}\right)^{-2}$

16. $\left(\frac{3}{2}\right)^{-3}$ **17.** $\frac{2^3 \cdot 3^2}{2 \cdot 3^{-2}}$ **18.** $\frac{3^{-2} \cdot 5^3}{3 \cdot 5}$ **19.** $\left(\frac{9}{2}\right)^{-2}$ **20.** $\left(\frac{6}{5}\right)^{-3}$

In Problems 21–56, simplify each expression so that all exponents are positive. Whenever an exponent is negative or 0, we assume the base does not equal 0.

21. x^0y^2 **22.** $x^{-1}y$ **23.** $x^{-2}y$ **24.** x^4y^0 **25.** $(8x^3)^{-2}$

26. $(-8x^3)^{-2}$ **27.** $-4x^{-1}$ **28.** $(-4x)^{-1}$ **29.** $5x^0$ **30.** $(5x)^0$

31. $\frac{x^{-2}y^3}{xy^4}$ **32.** $\frac{x^{-2}y}{xy^2}$ **33.** $x^{-1}y^{-1}$ **34.** $\frac{x^{-2}y^{-3}}{x}$ **35.** $\frac{x^{-1}}{y^{-1}}$

36. $\left(\frac{2x}{3}\right)^{-1}$ **37.** $\left(\frac{4x}{5y}\right)^{-2}$ **38.** $(xy^2)^{-2}$ **39.** $x^{-1} + y^{-2}$ **40.** $x^{-1} + y^{-1}$

41. $\dfrac{x^{-1}y^{-2}z}{x^2yz^3}$ **42.** $\dfrac{3x^{-2}yz^2}{x^4y^{-3}z}$ **43.** $\dfrac{(-2)^3x^4(yz)^2}{3^2xy^3z^4}$ **44.** $\dfrac{4x^{-2}(yz)^{-1}}{(-5)^2x^4y^2z^{-2}}$

45. $\dfrac{x^{-2}}{x^{-2}+y^{-2}}$ **46.** $\dfrac{x^{-1}+y^{-1}}{x^{-1}-y^{-1}}$ **47.** $\dfrac{\left(\frac{x}{y}\right)^{-2}\cdot\left(\frac{y}{x}\right)^{4}}{x^2y^3}$ **48.** $\dfrac{\left(\frac{y}{x}\right)^{2}}{x^{-2}y}$

49. $\left(\dfrac{3x^{-1}}{4y^{-1}}\right)^{-2}$ **50.** $\left(\dfrac{5x^{-2}}{6y^{-2}}\right)^{-3}$ **51.** $\dfrac{(xy^{-1})^{-2}}{xy}$ **52.** $\dfrac{(3xy^{-1})^2}{(2x^{-1}y)^3}$

53. $\dfrac{\left(\frac{x^2}{y}\right)^{3}}{\left(\frac{x}{y^2}\right)^{2}}$ **54.** $\dfrac{\left(\frac{2x}{3y^2}\right)^{-2}}{\frac{1}{y^3}}$ **55.** $\left(\dfrac{x}{y^2}\right)^{-2}\cdot(y^2)^{-1}$ **56.** $\dfrac{(x^2)^{-3}y^3}{(x^3y)^{-2}}$

57. Find the value of the expression $2x^3 - 3x^2 + x - 5$ if $x = 2$. If $x = -1$.

58. Find the value of the expression $3x^3 + 2x^2 - x - 2$ if $x = 1$. If $x = 2$.

C *In Problems 59–66, evaluate each expression. Round off your answers to three decimal places.*

59. $(8.2)^5$ **60.** $(3.7)^4$ **61.** $(6.1)^{-3}$ **62.** $(2.2)^{-5}$

63. $(-2.8)^6$ **64.** $-(2.8)^6$ **65.** $(-8.11)^{-4}$ **66.** $-(8.11)^{-4}$

In Problems 67–74, write each number in scientific notation.

67. 454.2 **68.** 32.14 **69.** 0.013 **70.** 0.00421

71. 32,155 **72.** 21,210 **73.** 0.000423 **74.** 0.0514

In Problems 75–82, write each number as a decimal.

75. 2.15×10^4 **76.** 6.7×10^3 **77.** 1.214×10^{-3} **78.** 9.88×10^{-4}

79. 1.1×10^8 **80.** 4.112×10^2 **81.** 8.1×10^{-2} **82.** 6.453×10^{-1}

C **83.** One light-year is defined by astronomers to be the distance a beam of light will travel in 1 year (365 days). If the speed of light is 186,000 miles per second, how many miles are in a light-year? Express your answer in scientific notation.

C **84.** How long does it take a beam of light to reach the Earth from the Sun, when the Sun is 93,000,000 miles from Earth? Express your answer in seconds, using scientific notation.

1.4 Polynomials

We have described algebra as a generalization of arithmetic in which letters are used to represent real numbers. From now on, we shall use the letters at the end of the alphabet, such as x, y, and z, to represent variables and the letters at the beginning of the alphabet, such as a, b, and c, to represent constants. Thus, in the expressions $3x + 5$ and $ax + b$, it is understood that x is a variable and that a and b are constants, even though the constants a and b are unspecified. As you will find out, the context usually makes the intended meaning clear.

Now we introduce some basic vocabulary.

Monomial

A **monomial** in one variable is the product of a constant times a variable raised to a nonnegative integer power. Thus, a monomial is of the form

$$ax^k$$

where a is a constant, x is a variable, and $k \geq 0$ is an integer. The constant a is called the **coefficient** of the monomial. If $a \neq 0$, then k is called the **degree** of the monomial.

Examples of monomials are:

MONOMIAL	COEFFICIENT	DEGREE	
$6x^2$	6	2	
$-\sqrt{2}x^3$	$-\sqrt{2}$	3	
3	3	0	Since $3 = 3 \cdot 1 = 3x^0$
$-5x$	-5	1	Since $-5x = -5x^1$
x^4	1	4	Since $x^4 = 1 \cdot x^4$

Two monomials ax^k and bx^k with the same degree are called **like terms**. Such monomials when added or subtracted can be combined into a single monomial by using the distributive property. For example,

$$2x^2 + 5x^2 = (2 + 5)x^2 = 7x^2 \qquad \text{and} \qquad 8x^3 - 5x^3 = (8 - 5)x^3 = 3x^3$$

The sum or difference of two monomials having different degrees is called a **binomial**. The sum or difference of three monomials with three different degrees is called a **trinomial**. For example,

$x^2 - 2$ is a binomial

$x^3 - 3x + 5$ is a trinomial

$2x^2 + 5x^2 + 2 = 7x^2 + 2$ is a binomial

Polynomial

A **polynomial** in one variable is an algebraic expression of the form

$$a_nx^n + a_{n-1}x^{n-1} + \cdots + a_1x + a_0 \tag{1}$$

where $a_n, a_{n-1}, \ldots, a_1, a_0$ are constants* called the **coefficients** of the polynomial, $n \geq 0$ is an integer, and x is a variable. If $a_n \neq 0$, it is called the **leading coefficient** and n is called the **degree** of the polynomial.

*The notation a_n is read as "a sub n." The number n is called a **subscript** and should not be confused with an exponent. We use subscripts in order to distinguish one constant from another when a large or undetermined number of constants is required.

The monomials that make up a polynomial are called its **terms**. If all the coefficients are 0, the polynomial is called the **zero polynomial**, which has no degree.

Polynomials are usually written in **standard form**, beginning with the nonzero term of highest degree and continuing with terms in descending order according to degree. Examples of polynomials are:

POLYNOMIAL	COEFFICIENTS	DEGREE
$3x^2 - 5 = 3x^2 + 0 \cdot x + (-5)$	$3, 0, -5$	2
$8 - 2x + x^2 = 1 \cdot x^2 - 2x + 8$	$1, -2, 8$	2
$5x + \sqrt{2} = 5x^1 + \sqrt{2}$	$5, \sqrt{2}$	1
$3 = 3 \cdot 1 = 3 \cdot x^0$	3	0
0	0	No degree

Although we have been using x to represent the variable, letters such as y or z are also commonly used. Thus,

$3x^4 - x^2 + 2$ is a polynomial (in x) of degree 4.

$9y^3 - 2y^2 + y - 3$ is a polynomial (in y) of degree 3.

$z^5 + \pi$ is a polynomial (in z) of degree 5.

Algebraic expressions such as

$$\frac{1}{x} \quad \text{and} \quad \frac{x^2 + 1}{x + 5}$$

are not polynomials. The first is not a polynomial because $1/x = x^{-1}$ has an exponent that is not a nonnegative integer. Although the second expression is the quotient of two polynomials, the polynomial in the denominator has degree greater than 0, so the expression cannot be a polynomial.

Adding and Subtracting Polynomials

Polynomials are added and subtracted by combining like terms.

Example 1 Find the sum of the polynomials

$$8x^3 - 2x^2 + 6x - 2 \quad \text{and} \quad 3x^4 - 2x^3 + x^2 + x$$

Solution We shall find the sum in two ways.

Horizontal Addition: The idea here is to group the like terms and then combine them.

$$\begin{aligned} &(8x^3 - 2x^2 + 6x - 2) + (3x^4 - 2x^3 + x^2 + x) \\ &\qquad = 3x^4 + (8x^3 - 2x^3) + (-2x^2 + x^2) + (6x + x) - 2 \\ &\qquad = 3x^4 + 6x^3 - x^2 + 7x - 2 \end{aligned}$$

Vertical Addition: The idea here is to vertically line up the like terms in each polynomial and then add the coefficients.

$$\begin{array}{rrrrrr} & x^4 & x^3 & x^2 & x^1 & x^0 \\ & 8 & x^3 & -\,2x^2 & +\,6x & -\,2 \\ (+) & 3x^4 & -\,2x^3 & +\,x^2 & +\,x & \\ \hline & 3x^4 & +\,6x^3 & -\,x^2 & +\,7x & -\,2 \end{array}$$

■

The difference of two polynomials also can be found in either of the ways outlined above.

Example 2 Find the difference: $(3x^4 - 4x^3 + 6x^2 - 1) - (2x^4 - 8x^2 - 6x + 5)$

Solution *Horizontal Subtraction:*

$$(3x^4 - 4x^3 + 6x^2 - 1) - (2x^4 - 8x^2 - 6x + 5)$$

$$= 3x^4 - 4x^3 + 6x^2 - 1 \underbrace{- 2x^4 + 8x^2 + 6x - 5}$$

Be sure to change the sign of each term in the second polynomial.

$$= (3x^4 - 2x^4) + (-4x^3) + (6x^2 + 8x^2) + 6x + (-1 - 5)$$

↑ Combine like terms.

$$= x^4 - 4x^3 + 14x^2 + 6x - 6$$

Vertical Subtraction:

$$\begin{array}{rrrrrr} & x^4 & x^3 & x^2 & x^1 & x^0 \\ & 3x^4 & -\,4x^3 & +\,6x^2 & & -\,1 \\ (-) & 2x^4 & & -\,8x^2 & -\,6x & +\,5 \\ \hline & x^4 & -\,4x^3 & +\,14x^2 & +\,6x & -\,6 \end{array}$$

■

The choice of which of these methods to use for adding and subtracting polynomials is left to you. To save space, we shall most often use the horizontal format.

Multiplying Polynomials

The product of two monomials ax^n and bx^m is obtained using the laws of exponents and commutative and associative properties. Thus,

$$ax^n \cdot bx^m = abx^{n+m}$$

Products of polynomials are found by repeated use of the distributive property and the laws of exponents. Again, there is a choice of horizontal or vertical format.

Example 3 Find the product: $(2x + 5)(x^2 - x + 2)$

Solution *Horizontal Multiplication:*

$$
\begin{aligned}
(2x + 5)(x^2 - x + 2) &= 2x(x^2 - x + 2) + 5(x^2 - x + 2) && \text{Distributive property}\\
&= 2x \cdot x^2 - 2x \cdot x + 2x \cdot 2 + 5 \cdot x^2 - 5 \cdot x + 5 \cdot 2 && \text{Distributive property}\\
&= 2x^3 - 2x^2 + 4x + 5x^2 - 5x + 10 && \text{Law of exponents}\\
&= 2x^3 + 3x^2 - x + 10 && \text{Combine like terms}
\end{aligned}
$$

Vertical Multiplication: The idea here is very much like multiplying a two-digit number by a three-digit number.

$$
\begin{array}{rrrrl}
 & x^2 - & x + & 2 & \\
 & & 2x + & 5 & \\
\hline
2x^3 - & 2x^2 + & 4x & & \text{This line is } 2x(x^2 - x + 2).\\
(+) & 5x^2 - & 5x + & 10 & \text{This line is } 5(x^2 - x + 2).\\
\hline
2x^3 + & 3x^2 - & x + & 10 & \text{The sum of the above two lines.}
\end{array}
$$

■

Special Products

Certain products, which we call **special products**, occur frequently in algebra. In the list that follows, x, a, b, c, and d are real numbers:

Difference of Two Squares

$$(x - a)(x + a) = x^2 - a^2 \tag{2}$$

Squares of Binomials, or Perfect Squares

$$(x + a)^2 = x^2 + 2ax + a^2 \tag{3a}$$

$$(x - a)^2 = x^2 - 2ax + a^2 \tag{3b}$$

Miscellaneous Trinomials

$$(x + a)(x + b) = x^2 + (a + b)x + ab \tag{4a}$$

$$(ax + b)(cx + d) = acx^2 + (ad + bc)x + bd \tag{4b}$$

Cubes of Binomials, or Perfect Cubes	$(x+a)^3 = x^3 + 3ax^2 + 3a^2x + a^3$ $(x-a)^3 = x^3 - 3ax^2 + 3a^2x - a^3$	(5a) (5b)
Difference of Two Cubes	$(x-a)(x^2+ax+a^2) = x^3 - a^3$	(6)
Sum of Two Cubes	$(x+a)(x^2-ax+a^2) = x^3 + a^3$	(7)

We will verify equations (4b) and (5b); derivations of the remaining rules are left as exercises.

Equation (4b)

$$(ax+b)(cx+d) = ax(cx+d) + b(cx+d) = (acx^2+adx) + (bcx+bd)$$
$$= acx^2 + (ad+bc)x + bd \quad ■$$

Equation (5b)

$$(x-a)^3 = (x-a)(x-a)^2 \underset{\uparrow \text{(3b)}}{=} (x-a)(x^2-2ax+a^2)$$
$$= x(x^2-2ax+a^2) - a(x^2-2ax+a^2)$$
$$= x^3 - 2ax^2 + a^2x - ax^2 + 2a^2x - a^3$$
$$= x^3 - 3ax^2 + 3a^2x - a^3 \quad ■$$

The formulas in equations (2)–(7) are used often and should be committed to memory. But if you forget one or are unsure of its form, you should be able to derive it as needed.

Let's look at some examples.

Example 4

$$(x-2)^2 \underset{\uparrow \text{(3b)}}{=} x^2 - 4x + 4 \quad ■$$

Example 5

$$(x-3)(x+1) \underset{\uparrow \text{(4a) with } a=-3,\ b=1}{=} x^2 - 2x - 3 \quad ■$$

Example 6

$$(2x+3)(3x-5) \underset{\uparrow \text{(4b) with } a=2,\ b=3,\ c=3,\ d=-5}{=} 6x^2 - x - 15 \quad ■$$

Example 7

$$(x+2)^3 \underset{\uparrow \text{(5a)}}{=} x^3 + 3 \cdot 2x^2 + 3 \cdot 2^2x + 2^3 = x^3 + 6x^2 + 12x + 8 \quad ■$$

Example 8

$$(x - 2)(x^2 + 2x + 4) \underset{\substack{\uparrow \\ (6)}}{=} x^3 - 2^3 = x^3 - 8$$ ■

Polynomials in Two or More Variables

Our discussion thus far has involved only polynomials in a single variable. A **polynomial in two variables** x and y is the sum of one or more monomials of the form $ax^n y^m$, where a is a constant called the **coefficient**, x and y are variables, and n and m are nonnegative integers. The **degree** of the monomial $ax^n y^m$ is $n + m$. The **degree** of a polynomial in two variables x and y is the highest degree of all the monomials with nonzero coefficients that appear.

Polynomials in three variables x, y, and z and polynomials in more than three variables are defined in a similar way. Here are some examples:

$3x^2 + 2x^3y + 5$	$\pi x^3 - y^2$	$x^4 + 4x^3y - xy^3 + y^4$
Two variables, degree is 4	Two variables, degree is 3	Two variables, degree is 4
$x^2 + y^2 - z^2 + 4$	x^3y^2z	$5x^2 - 4y^2 + z^3y + 2w^2x$
Three variables, degree is 2	Three variables, degree is 6	Four variables, degree is 4

Adding and multiplying polynomials in two or more variables is handled in the same way as for polynomials in one variable.

Example 9

$$\begin{aligned}(x^3 - 3x^2y + 3xy^2 - y^3) - (x^3 - x^2y + y^3)\\ &= (x^3 - x^3) + (-3x^2y + x^2y) + 3xy^2 + (-y^3 - y^3)\\ &= -2x^2y + 3xy^2 - 2y^3\end{aligned}$$ ■

Example 10

$$\begin{aligned}(xy - 2)(2x^2 - xy + y^2) &= xy(2x^2 - xy + y^2) - 2(2x^2 - xy + y^2)\\ &= 2x^3y - x^2y^2 + xy^3 - 4x^2 + 2xy - 2y^2\end{aligned}$$ ■

EXERCISE 1.4

In Problems 1–10, tell whether the expression is a polynomial. If it is, give its degree.

1. $3x^2 - 5$ **2.** $1 - 4x$ **3.** 5 **4.** $-\pi$ **5.** $3x^2 - \dfrac{5}{x}$

6. $\dfrac{3}{x} + 2$ **7.** $2y^3 - \sqrt{2}$ **8.** $10z^2 + z$ **9.** $\dfrac{x^2 + 5}{x^3 - 1}$ **10.** $\dfrac{3x^3 + 2x - 1}{x^2 + x + 1}$

In Problems 11–68, perform the indicated operations. Express each answer as a polynomial.

11. $(x^2 + x + 5) + (3x - 2)$ **12.** $(x^3 - 3x^2 + 2) + (x^2 - x + 1)$

13. $(x^3 - 3x^2 + 5x + 10) - (2x^2 - x + 1)$ **14.** $(x^2 - 3x - 1) - (x^3 - 2x^2 + x + 5)$

15. $(6x^5 + x^3 + x) + (5x^4 - x^3 + 3x^2)$
16. $(10x^5 - 8x^2) + (3x^3 - 2x^2 + 6)$
17. $3(x^2 - 3x + 1) + 2(3x^2 + x - 4)$
18. $-2(x^2 + x + 1) + 6(-5x^2 - x + 2)$
19. $6(x^3 + x^2 - 3) - 4(2x^3 - 3x^2)$
20. $8(4x^3 - 3x^2 - 1) - 6(4x^3 + 8x - 2)$
21. $(x^2 - x + 2) + (2x^2 - 3x + 5) - (x^2 + 1)$
22. $(x^2 + 1) - (4x^2 + 5) + (x^2 + x - 2)$
23. $9(y^2 - 2y + 1) - 6(1 - y^2)$
24. $8(1 - y^3) + 2(1 + y + y^2 + y^3)$
25. $(x + a)^2 - x^2$
26. $(x - a)^2 - x^2$
27. $(x + a)^3 - x^3$
28. $(x - a)^3 - x^3$
29. $(x + 4)(x - 3)$
30. $(x + 6)(x - 2)$
31. $(x + 8)(2x + 1)$
32. $(2x - 1)(x + 2)$
33. $(-3x + 1)(x + 4)$
34. $(-2x - 1)(x + 1)$
35. $(1 - 4x)(2 - 3x)$
36. $(2 - x)(2 - 3x)$
37. $(x - 1)(x^2 + x + 1)$
38. $(x + 1)(x^2 - x + 1)$
39. $(2x + 3)(x^2 - 2)$
40. $(3x - 4)(x^2 + 1)$
41. $(3x - 5)(2x + 3)$
42. $(2x + 3)(-5x + 1)$
43. $(2x + 5)(x^2 + x + 1)$
44. $(3x - 4)(x^2 - x + 1)$
45. $(3x - 1)(2x^2 - 3x + 2)$
46. $(2x + 3)(3x^2 - 2x + 4)$
47. $(x^2 + x - 1)(x^2 - x + 1)$
48. $(x^2 + 2x + 1)(x^2 - 3x + 4)$
49. $(2x^2 - 3x + 4)(3x^2 - x + 4)$
50. $(3x^2 - 2x + 4)(2x^2 + 4x - 5)$
51. $(x^3 - 3x^2 + 5)(2x^2 - 1)$
52. $(3x^3 - x + 4)(3x^2 - 4x + 1)$
53. $(x + 2)^2(x - 1)$
54. $(x - 3)^2(x + 2)$
55. $(x - 1)^2(2x + 3)$
56. $(x + 5)^2(3x - 1)$
57. $(x - 1)(x - 2)(x - 3)$
58. $(x + 1)(x - 2)(x + 4)$
59. $(x + 1)^3 - (x - 1)^3$
60. $(x + 1)^3 - (x + 2)^3$
61. $(x - 1)^2(x + 1)^2$
62. $(x - 1)^3(x + 1)$
63. $(x^4 - 3x^3 + 2x^2 - x + 1)(x^2 - x + 1)$
64. $(x^4 + 4x^3 - 3x^2 + 2x - 5)(x^2 + 5x - 3)$
65. $(2x + 3)^3$
66. $(3x + 2)^3$
67. $(2x - 3a)^3$
68. $(3x - 2a)^3$

In Problems 69–84, perform the indicated operations. Express each answer as a polynomial.

69. $(x + y)(x - 2y)$
70. $(x - 2y)(x - y)$
71. $(x^2 + 2x + y^2) + (x + y - 3y^2)$
72. $(x^2 - 2xy + y^2) - (x^2 - xy)$
73. $(x - y)^2 - (x + y)^2$
74. $(x - y)^2 - (x^2 + y^2)$
75. $(x + 2y)^2 + (x - 3y)^2$
76. $(2x - y)^2 - (x - y)^2$
77. $[(x + y)^2 + z^2] + [x^2 + (y + z)^2]$
78. $[(x - y)^2 + z^2] + [x^2 + (y - z)^2]$
79. $(x - y)(x^2 + xy + y^2)$
80. $(x + y)(x^2 - xy + y^2)$
81. $(2x^2 + xy + y^2)(3x^2 - xy + 2y^2)$
82. $(3x^2 - xy + y^2)(2x^2 + xy - 4y^2)$
83. $(x + y + z)(x - y - z)$
84. $(x + y - z)(x - y + z)$
85. Explain why the degree of the product of two polynomials equals the sum of their degrees.
86. Explain why the degree of the sum of two polynomials equals the larger of their degrees.
87. Derive equations (2), (3a), (3b), and (6) (pages 36–37).

88. Derive equations (4a), (5a), and (7) (pages 36–37).

89. Develop a formula for $(x + a)^4$.

1.5 Factoring Polynomials

Consider the following product:

$$(2x + 3)(x - 4) = 2x^2 - 5x - 12$$

The two polynomials on the left side are called **factors** of the polynomial on the right side. Expressing a given polynomial as a product of other polynomials—that is, finding the factors of a polynomial—is called **factoring**.

We shall restrict our discussion here to factoring polynomials in one variable into products of polynomials in one variable, where all coefficients are integers. We call this **factoring over the integers**. There will be times, though, when we will want to **factor over the rational numbers** and even **factor over the real numbers**. Factoring over the rational numbers means to write a given polynomial whose coefficients are rational numbers as a product of polynomials whose coefficients are also rational numbers. Factoring over the real numbers means to write a given polynomial whose coefficients are real numbers as a product of polynomials whose coefficients are also real numbers. Unless specified otherwise, we will be factoring over the integers.

Any polynomial can be written as the product of 1 times itself or as -1 times its additive inverse. If a polynomial cannot be written as the product of two other polynomials (excluding 1 and -1), then the polynomial is said to be **prime**. When a polynomial has been written as a product consisting only of prime factors, then it is said to be **factored completely**. Examples of prime polynomials are

$$2, \quad 3, \quad 5, \quad x, \quad x + 1, \quad x - 1, \quad 3x + 4$$

The first factor to look for in a factoring problem is a common monomial factor present in each term of the polynomial. If one is present, use the distributive property to factor it out. For example:

POLYNOMIAL	COMMON MONOMIAL FACTOR	REMAINING FACTOR	FACTORED FORM
$2x + 4$	2	$x + 2$	$2x + 4 = 2(x + 2)$
$3x - 6$	3	$x - 2$	$3x - 6 = 3(x - 2)$
$2x^2 - 4x + 8$	2	$x^2 - 2x + 4$	$2x^2 - 4x + 8 = 2(x^2 - 2x + 4)$
$8x - 12$	4	$2x - 3$	$8x - 12 = 4(2x - 3)$
$x^2 + x$	x	$x + 1$	$x^2 + x = x(x + 1)$
$x^3 - 3x^2$	x^2	$x - 3$	$x^3 - 3x^2 = x^2(x - 3)$
$6x^2 + 9x$	$3x$	$2x + 3$	$6x^2 + 9x = 3x(2x + 3)$

Notice that, once all common monomial factors have been removed from a polynomial, the remaining factor is either a prime polynomial of degree 1 or a polynomial of degree 2 or higher. (Do you see why?) Thus, we concentrate on techniques for factoring polynomials of degree 2 or higher that contain no monomial factors.

Special Factors

The list of special products (2)–(7) given on pages 36–37 provides a list of factoring formulas when the equations are read from right to left. For example, equation (2) states that if the polynomial is the difference of two squares, $x^2 - a^2$, it can be factored into $(x - a)(x + a)$.

Example 1 Factor completely: $x^2 - 4$

Solution We notice that $x^2 - 4$ is the difference of two squares, x^2 and 2^2. Thus, using equation (2), we find

$$x^2 - 4 = (x - 2)(x + 2)$$ ■

Example 2 Factor completely: $x^3 - 1$

Solution Equation (6) states that the difference of two cubes, $x^3 - a^3$, can be factored as $(x - a)(x^2 + ax + a^2)$. Because $x^3 - 1$ is the difference of two cubes, x^3 and 1^3, we find

$$x^3 - 1 = (x - 1)(x^2 + x + 1)$$ ■

Example 3 Factor completely: $x^3 + 8$

Solution Equation (7) states that the sum of two cubes, $x^3 + a^3$, can be factored as $(x + a)(x^2 - ax + a^2)$. Because $x^3 + 8$ is the sum of two cubes, x^3 and 2^3, we have

$$x^3 + 8 = (x + 2)(x^2 - 2x + 4)$$ ■

Example 4 Factor completely: $x^4 - 16$

Solution Using equation (2) for the difference of two squares, $x^4 = (x^2)^2$ and $16 = 4^2$, we have

$$x^4 - 16 = (x^2 - 4)(x^2 + 4)$$

But $x^2 - 4$ is also the difference of two squares. Thus,

$$x^4 - 16 = (x^2 - 4)(x^2 + 4) = (x - 2)(x + 2)(x^2 + 4)$$ ■

Whenever the first term and third term of a trinomial are both positive and are perfect squares, such as x^2, $9x^2$, 1, 4, and so on, check to see whether either of the special products (3a) or (3b) applies.

Example 5 Factor completely: $x^2 + 4x + 4$

Solution The first term, x^2, and the third term, $4 = 2^2$, are perfect squares. Because the middle term is twice the product of x and 2, we use equation (3a) to obtain

$$x^2 + 4x + 4 = (x + 2)^2$$

Example 6 Factor completely: $9x^2 - 6x + 1$

Solution The first term, $9x^2 = (3x)^2$, and the third term, $1 = 1^2$, are perfect squares. Because the middle term is twice the product of $3x$ and 1, we use equation (3b) to obtain

$$9x^2 - 6x + 1 = (3x - 1)^2$$

Example 7 Factor completely: $25x^2 + 30x + 9$

Solution The first term, $25x^2 = (5x)^2$, and the third term, $9 = 3^2$, are perfect squares. Because the middle term is twice the product of $5x$ and 3, we use a form of equation (3a) to obtain

$$25x^2 + 30x + 9 = (5x + 3)^2$$

If a trinomial is not a perfect square, it may be possible to factor it using the technique discussed next.

Factoring Second-Degree Polynomials

Factoring a second-degree polynomial, $Ax^2 + Bx + C$, where A, B, and C are integers, is a matter of skill, experience, and often some trial and error. The idea behind factoring $Ax^2 + Bx + C$ is to see whether it can be made equal to the product of two, possibly equal, first-degree polynomials. Thus, we want to see whether there are integers a, b, c, and d so that

$$Ax^2 + Bx + C = (ax + b)(cx + d)$$

We start with second-degree polynomials that have a leading coefficient of 1. Such a polynomial, if it can be factored, must follow the form of the special product (4a),

$$x^2 + Bx + C = (x + a)(x + b) = x^2 + (a + b)x + ab \qquad B = a + b,\ C = ab$$

Note the pattern in this formula: The correct choice of the factors a and b of the constant term C must add up to the coefficient of the middle term B.

Example 8 Factor completely: $x^2 + 7x + 12$

Solution First, determine all possible integral factors of the constant term 12, and compute their sums:

Factors of 12	1, 12	−1, −12	2, 6	−2, −6	3, 4	−3, −4
Sum	13	−13	8	−8	7	−7

The factors of 12 that add up to 7, the coefficient of the middle term, are 3 and 4. Thus,

$$x^2 + 7x + 12 = (x + 3)(x + 4)$$ ■

Example 9 Factor completely: $x^2 - 6x + 8$

Solution First, determine all possible integral factors of the constant term 8, and compute each sum:

Factors of 8	1, 8	−1, −8	2, 4	−2, −4
Sum	9	−9	6	−6

Since -6 is the coefficient of the middle term, we have

$$x^2 - 6x + 8 = (x - 2)(x - 4)$$ ■

Example 10 Factor completely: $x^2 - x - 12$

Solution First, determine all possible integral factors of -12, and compute each sum:

Factors of −12	1, −12	−1, 12	2, −6	−2, 6	3, −4	−3, 4
Sum	−11	11	−4	4	−1	1

Since -1 is the coefficient of the middle term, we have

$$x^2 - x - 12 = (x + 3)(x - 4)$$ ■

Example 11 Factor completely: $x^2 + 4x - 12$

Solution The factors -2 and 6 of -12 have the sum 4. Thus,

$$x^2 + 4x - 12 = (x - 2)(x + 6)$$ ■

To avoid errors in factoring, always check your answer by multiplying it out to see if the result equals the original expression.

When none of the possibilities work, the polynomial is prime.

Example 12 Show that $x^2 + 9$ is prime.

Solution First, list the integral factors of 9, and compute their sums:

Factors of 9	1, 9	−1, −9	3, 3	−3, −3
Sum	10	−10	6	−6

Since the coefficient of the middle term in $x^2 + 9$ is 0 and none of the sums above equals 0, we conclude that $x^2 + 9$ is prime. ■

When the leading coefficient is not 1, a somewhat longer list of possibilities may be required. Observe the pattern in the following formula:

$$Ax^2 + Bx + C = (ax + b)(cx + d) = acx^2 + (ad + bc)x + bd$$

$$A = ac, B = ad + bc, \quad C = bd$$

Our task is to find factors a and c of the leading coefficient A and factors b and d of the constant term C so that the expression $ad + bc$ equals the coefficient B of the middle term.

Example 13 Factor completely: $2x^2 + 5x + 3$

Solution The positive integral factors of the leading coefficient $ac = 2$ are $a = 2$, $c = 1$. We begin the factorization by writing

$$2x^2 + 5x + 3 = (2x \qquad)(x \qquad)$$

The positive integral factors of the constant term $bd = 3$ are $b = 1$, $d = 3$, or $b = 3$, $d = 1$. This suggests the following possibilities:

$$2x^2 + 5x + 3 \begin{cases} (2x \quad 1)(x \quad 3) \\ (2x \quad 3)(x \quad 1) \end{cases}$$

Next, we select the signs to be placed inside the factors. Since the constant term is positive and the coefficient of the middle term is positive, the only signs that can possibly work are + signs. (Do you see why?) Thus,

$$2x^2 + 5x + 3 \begin{cases} (2x + 1)(x + 3) = 2x^2 + 7x + 3 \\ (2x + 3)(x + 1) = 2x^2 + 5x + 3 \end{cases}$$

Thus, we conclude that

$$2x^2 + 5x + 3 = (2x + 3)(x + 1)$$ ■

Example 14 Factor completely: $2x^2 - x - 6$

Solution The positive integral factors of 6 are 1, 6, and 2, 3. This suggests the following possibilities:

$$2x^2 - x - 6\begin{cases} (2x \quad 1)(x \quad 6) \\ (2x \quad 6)(x \quad 1) \\ (2x \quad 2)(x \quad 3) \\ (2x \quad 3)(x \quad 2) \end{cases}$$

Since the constant term is negative, the signs chosen for each of the possible factors must be opposite. This leads to the possibilities

$$(2x - 1)(x + 6) = 2x^2 + 11x - 6 \qquad (2x - 2)(x + 3) = 2x^2 + 4x - 6$$
$$(2x + 1)(x - 6) = 2x^2 - 11x - 6 \qquad (2x + 2)(x - 3) = 2x^2 - 4x - 6$$
$$(2x - 6)(x + 1) = 2x^2 - 4x - 6 \qquad (2x - 3)(x + 2) = 2x^2 + x - 6$$
$$(2x + 6)(x - 1) = 2x^2 + 4x - 6 \qquad (2x + 3)(x - 2) = 2x^2 - x - 6$$

Thus, $2x^2 - x - 6 = (2x + 3)(x - 2)$. ■

Note: $(2x \pm 3)(x \mp 2)$ is a notation that can be used to save space. It actually represents two products: $(2x + 3)(x - 2)$ and $(2x - 3)(x + 2)$.

Study the patterns illustrated in these examples carefully. Practice will give you the experience needed to use this factoring technique skillfully and efficiently.

Other Factoring Techniques

Sometimes a common factor occurs, not in every term of the polynomial, but in each of several groups of terms that together make up the polynomial. When this happens, the common factor can be factored out of each group by means of the distributive property. This technique is called **factoring by grouping**.

Example 15 Factor completely by grouping: $(x^2 + 2)x + (x^2 + 2) \cdot 3$

Solution Notice the common factor $x^2 + 2$. By applying the distributive property, we have

$$(x^2 + 2)x + (x^2 + 2) \cdot 3 = (x^2 + 2)(x + 3)$$

Since $x^2 + 2$ and $x + 3$ are prime, the factorization is complete. ■

Example 16 Factor completely by grouping: $x^3 - 4x^2 + 2x - 8$

Solution To see if factoring by grouping will work, group the first two terms and the last two terms. Then look for a common factor in each group. In this example, we can factor x^2 from $x^3 - 4x^2$ and 2 from $2x - 8$. The remaining factor in each case is the same, namely, $x - 4$. This means factoring by grouping will work, as follows:

$$\begin{aligned} x^3 - 4x^2 + 2x - 8 &= (x^3 - 4x^2) + (2x - 8) \\ &= x^2(x - 4) + 2(x - 4) \\ &= (x^2 + 2)(x - 4) \end{aligned}$$

Since $x^2 + 2$ and $x - 4$ are prime, the factorization is complete. ■

Example 17 Factor completely by grouping: $3x^3 + 4x^2 - 6x - 8$

Solution Here, $3x + 4$ is a common factor of $3x^3 + 4x^2$ and of $-6x - 8$. Hence,

$$\begin{aligned} 3x^3 + 4x^2 - 6x - 8 &= (3x^3 + 4x^2) - (6x + 8) \\ &= x^2(3x + 4) - 2(3x + 4) \\ &= (x^2 - 2)(3x + 4) \end{aligned}$$

Since $x^2 - 2$ and $3x + 4$ are prime (over the integers), the factorization is complete. ■

The final factoring technique we discuss, which involves adding and subtracting an appropriate term, is usually left as a final alternative.

Example 18 Factor completely the polynomial: $x^4 + 7x^2 + 16$

Solution If the middle term were $8x^2$ instead of $7x^2$, the polynomial would be the perfect square $x^4 + 8x^2 + 16 = (x^2 + 4)^2$. We shall make it so by adding x^2, but we must subtract x^2 at the same time. Let's see how this works.

$$\begin{aligned} x^4 + 7x^2 + 16 &= x^4 + 7x^2 + x^2 + 16 - x^2 \\ &= (x^4 + 8x^2 + 16) - x^2 \\ &= (x^2 + 4)^2 - x^2 \\ &\underset{\substack{\uparrow \\ \text{Difference of squares}}}{=} (x^2 + 4 - x)(x^2 + 4 + x) \\ &= (x^2 - x + 4)(x^2 + x + 4) \end{aligned}$$

The latter two polynomials are prime, as you can verify. ■

Summary

We close the section with the capsule summary given at the top of the next page.

TYPE OF POLYNOMIAL	METHOD	EXAMPLE
Any polynomial	Look for common monomial factors. (Always do this first!)	$6x^2 + 9x = 3x(2x + 3)$
Binomials of degree 2 or higher	Check for a special product:	
	Difference of two squares, $x^2 - a^2$	Examples 1, 4
	Difference of two cubes, $x^3 - a^3$	Example 2
	Sum of two cubes, $x^3 + a^3$	Example 3
Trinomials of degree 2	Check for a perfect square $(x \pm a)^2$.	Examples 5–7
	List possibilities.	Examples 8–14
Three or more terms	Grouping	Examples 15–17
	Addition and subtraction of an appropriate term	Example 18

EXERCISE 1.5

In Problems 1–10, factor each polynomial by removing the common monomial factor.

1. $3x + 6$ **2.** $7x - 14$ **3.** $ax^2 + a$ **4.** $ax - a$
5. $x^3 + x^2 + x$ **6.** $x^3 - x^2 + x$ **7.** $2x^2 + 2x + 2$ **8.** $3x^2 - 3x + 3$
9. $3x^2y - 6xy^2 + 12xy$ **10.** $60x^2y - 48xy^2 + 72x^3y$

In Problems 11–82, factor completely each polynomial. If the polynomial cannot be factored, say it is prime.

11. $x^2 - 1$ **12.** $x^2 - 4$ **13.** $4x^2 - 1$ **14.** $9x^2 - 1$
15. $x^2 + 7x + 10$ **16.** $x^2 + 3x - 4$ **17.** $x^2 - 10x + 21$ **18.** $x^2 - 4x - 21$
19. $x^2 - 7x - 8$ **20.** $x^2 - 6x + 5$ **21.** $x^2 + 2x + 1$ **22.** $x^2 - 4x + 4$
23. $x^2 + 4x + 4$ **24.** $x^2 - 2x - 1$ **25.** $x^2 - 2x - 15$ **26.** $x^2 - 6x - 14$
27. $3x^2 - 12x - 36$ **28.** $x^3 + 8x^2 - 20x$ **29.** $y^4 + 11y^3 + 30y^2$
30. $3y^3 - 18y^2 - 48y$ **31.** $16x^2 + 8x + 1$ **32.** $25x^2 + 10x + 1$
33. $4x^2 + 12x + 9$ **34.** $9x^2 - 12x + 4$ **35.** $ax^2 - 4a^2x - 45a^3$
36. $bx^2 + 14b^2x + 45b^3$ **37.** $x^3 - 27$ **38.** $x^3 + 27$
39. $8x^3 + 27$ **40.** $27 - 8x^3$ **41.** $2x^4 + 16x$
42. $3x^5 - 3x^2$ **43.** $3x^2 + 4x + 1$ **44.** $4x^2 + 3x - 1$
45. $x^4 - 81$ **46.** $x^4 - 1$ **47.** $x^6 - 2x^3 + 1$
48. $x^6 + 2x^3 + 1$ **49.** $x^7 - x^5$ **50.** $x^8 - x^5$
51. $2z^2 + 5z + 3$ **52.** $6z^2 - z - 1$ **53.** $16x^2 - 24x + 9$
54. $9x^2 + 24x + 16$ **55.** $16x^2 - 16x - 5$ **56.** $16x^2 - 11x - 5$
57. $4y^2 + 16y + 15$ **58.** $9y^2 + 9y - 4$ **59.** $18x^2 - 9x - 27$
60. $8x^2 - 6x - 2$ **61.** $8x^2 + 2x + 6$ **62.** $9x^2 - 3x + 3$
63. $x^2 - x + 4$ **64.** $x^2 + 6x + 9$ **65.** $4x^3 - 10x^2 - 6x$
66. $27x^3 - 9x^2 - 6x$ **67.** $x^4 + x^2 + 1$ **68.** $x^4 - 7x^2 + 9$

69. $9x^4 + 8x^2 - 1$ **70.** $8x^4 + 14x^2 - 4$ **71.** $x(x + 3) - 6(x + 3)$

72. $5(3x - 7) + x(3x - 7)$ **73.** $(x + 2)^2 - 5(x + 2)$ **74.** $(x - 1)^2 - 2(x - 1)$

75. $3(x^2 + 10x + 25) - 2(x + 5)$ **76.** $7(x^2 - 6x + 9) + 3(x - 3)$ **77.** $x^3 + 2x^2 - x - 2$

78. $x^3 - 3x^2 - x + 3$ **79.** $x^4 - x^3 + x - 1$ **80.** $x^4 + x^3 + x + 1$

81. $x^5 + x^3 + 8x^2 + 8$ **82.** $x^5 - x^3 + 8x^2 - 8$

83. Show that $x^2 + 4$ is prime. **84.** Show that $x^2 + x + 1$ is prime.

1.6 Rational Expressions

If we form the quotient of two polynomials, the result is called a **rational expression**. Some examples of rational expressions are

$$\text{(a)}\ \frac{x^3 + 1}{x} \qquad \text{(b)}\ \frac{3x^2 + x - 2}{x^2 + 5} \qquad \text{(c)}\ \frac{x}{x^2 - 1} \qquad \text{(d)}\ \frac{xy^2}{(x - y)^2}$$

Expressions (a), (b), and (c) are rational expressions in one variable, x, whereas (d) is a rational expression in two variables, x and y.

Rational expressions are described in the same manner as rational numbers. Thus, in expression (a), the polynomial $x^3 + 1$ is called the **numerator**, and x is called the **denominator**. When the numerator and denominator of a rational expression contain no common factors (except 1 and -1), we say the rational expression is **reduced to lowest terms**, or **simplified**.

A rational expression is reduced to lowest terms by completely factoring the numerator and the denominator and cancelling any common factors by using the cancellation property,

$$\frac{ac}{bc} = \frac{a}{b} \qquad b \neq 0,\ c \neq 0 \tag{1}$$

Example 1 Reduce each rational expression to lowest terms.

$$\text{(a)}\ \frac{x^2 + 4x + 4}{x^2 + 3x + 2} \qquad \text{(b)}\ \frac{x^3 - 8}{x^3 - 2x^2} \qquad \text{(c)}\ \frac{8 - 2x}{x^2 - x - 12}$$

Solution

(a) $$\frac{x^2 + 4x + 4}{x^2 + 3x + 2} = \frac{\cancel{(x + 2)}(x + 2)}{\cancel{(x + 2)}(x + 1)} = \frac{x + 2}{x + 1}, \quad x \neq -2, -1$$

(b) $$\frac{x^3 - 8}{x^3 - 2x^2} = \frac{\cancel{(x - 2)}(x^2 + 2x + 4)}{x^2\cancel{(x - 2)}} = \frac{x^2 + 2x + 4}{x^2}, \quad x \neq 0, 2$$

(c) $$\frac{8 - 2x}{x^2 - x - 12} = \frac{2(4 - x)}{(x - 4)(x + 3)} = \frac{2(-1)\cancel{(x - 4)}}{\cancel{(x - 4)}(x + 3)} = \frac{-2}{x + 3}, \quad x \neq -3, 4$$ ■

Multiplying and Dividing Rational Expressions

The rules for multiplying and dividing rational expressions are the same as the rules for multiplying and dividing rational numbers, namely,

$$\frac{a}{b} \cdot \frac{c}{d} = \frac{ac}{bd} \qquad b \neq 0, d \neq 0 \qquad (2)$$

$$\frac{\frac{a}{b}}{\frac{c}{d}} = \frac{a}{b} \cdot \frac{d}{c} = \frac{ad}{bc} \qquad b \neq 0, c \neq 0, d \neq 0 \qquad (3)$$

In using equations (2) and (3) with rational expressions, be sure first to factor each polynomial completely so that common factors can be cancelled. We shall follow the practice of leaving our answers in factored form.

Example 2 Perform the indicated operation and simplify the result. Leave your answer in factored form.

(a) $\dfrac{x^2 - 2x + 1}{x^3 + x} \cdot \dfrac{4x^2 + 4}{x^2 + x - 2}$ (b) $\dfrac{\frac{x + 3}{x^2 - 4}}{\frac{x^2 - x - 12}{x^3 - 8}}$

Solution (a)
$$\frac{x^2 - 2x + 1}{x^3 + x} \cdot \frac{4x^2 + 4}{x^2 + x - 2} = \frac{(x - 1)^2}{x(x^2 + 1)} \cdot \frac{4(x^2 + 1)}{(x + 2)(x - 1)}$$
$$= \frac{(x - 1)^{\not 2}(4)\cancel{(x^2 + 1)}}{x\cancel{(x^2 + 1)}(x + 2)\cancel{(x - 1)}} = \frac{4(x - 1)}{x(x + 2)},$$
$$x \neq -2, 0, 1$$

(b)
$$\frac{\frac{x + 3}{x^2 - 4}}{\frac{x^2 - x - 12}{x^3 - 8}} = \frac{x + 3}{x^2 - 4} \cdot \frac{x^3 - 8}{x^2 - x - 12}$$
$$= \frac{x + 3}{(x - 2)(x + 2)} \cdot \frac{(x - 2)(x^2 + 2x + 4)}{(x - 4)(x + 3)}$$
$$= \frac{\cancel{(x + 3)}\cancel{(x - 2)}(x^2 + 2x + 4)}{\cancel{(x - 2)}(x + 2)(x - 4)\cancel{(x + 3)}} = \frac{x^2 + 2x + 4}{(x + 2)(x - 4)},$$
$$x \neq -3, -2, 2, 4$$ ■

Adding and Subtracting Rational Expressions

The rules for adding and subtracting rational expressions are the same as the rules for adding and subtracting rational numbers. Thus, if the denominators of two rational expressions to be added (or subtracted) are equal, we add (or subtract) the numerators and keep the common denominator. That is, if a/b and c/b are two rational expressions, then

$$\frac{a}{b} + \frac{c}{b} = \frac{a+c}{b} \qquad \frac{a}{b} - \frac{c}{b} = \frac{a-c}{b} \qquad b \neq 0 \qquad (4)$$

Example 3 Perform the indicated operation and simplify the result. Leave your answer in factored form.

(a) $\dfrac{2x^2-4}{2x+5} + \dfrac{x+3}{2x+5}, \quad x \neq -\frac{5}{2}$ (b) $\dfrac{x}{x-3} - \dfrac{3x+2}{x-3}, \quad x \neq 3$

Solution (a)

$$\frac{2x^2-4}{2x+5} + \frac{x+3}{2x+5} = \frac{(2x^2-4)+(x+3)}{2x+5}$$
$$= \frac{2x^2+x-1}{2x+5} = \frac{(2x-1)(x+1)}{2x+5}$$

(b)

$$\frac{x}{x-3} - \frac{3x+2}{x-3} = \frac{x-(3x+2)}{x-3} = \frac{x-3x-2}{x-3}$$
$$= \frac{-2x-2}{x-3} = \frac{-2(x+1)}{x-3}$$

■

Example 4 Perform the indicated operation and simplify the result. Leave your answer in factored form.

$$\frac{2x}{x-3} + \frac{5}{3-x} \qquad x \neq 3$$

Solution Notice that the denominators of the two rational expressions are different. However, the denominator of the second expression is just the additive inverse of the denominator of the first. That is,

$$3 - x = -x + 3 = -1 \cdot (x-3) = -(x-3)$$

Thus,

$$\frac{2x}{x-3} + \frac{5}{3-x} \underset{\uparrow}{=} \frac{2x}{x-3} + \frac{5}{-(x-3)} \underset{\uparrow}{=} \frac{2x}{x-3} + \frac{-5}{x-3}$$

$3 - x = -(x-3)$ $\qquad$ $\dfrac{a}{-b} = \dfrac{-a}{b}$

$$= \frac{2x+(-5)}{x-3} = \frac{2x-5}{x-3}$$

■

If the denominators of two rational expressions to be added or subtracted are not equal, we can use the general formulas for adding and subtracting quotients:

$$\frac{a}{b} + \frac{c}{d} = \frac{a \cdot d}{b \cdot d} + \frac{b \cdot c}{b \cdot d} = \frac{ad + bc}{bd} \qquad b \neq 0, d \neq 0$$
$$\frac{a}{b} - \frac{c}{d} = \frac{a \cdot d}{b \cdot d} - \frac{b \cdot c}{b \cdot d} = \frac{ad - bc}{bd} \qquad b \neq 0, d \neq 0 \tag{5}$$

Example 5 Perform the indicated operation and simplify the result. Leave your answer in factored form.

(a) $\dfrac{x-3}{x+4} + \dfrac{x}{x-2}, \quad x \neq -4, 2$ (b) $\dfrac{x^2}{x^2-4} - \dfrac{1}{x}, \quad x \neq -2, 0, 2$

Solution (a) $\dfrac{x-3}{x+4} + \dfrac{x}{x-2} = \dfrac{(x-3)(x-2) + (x+4)(x)}{(x+4)(x-2)}$

$$= \frac{x^2 - 5x + 6 + x^2 + 4x}{(x+4)(x-2)} = \frac{2x^2 - x + 6}{(x+4)(x-2)}$$

(b) $\dfrac{x^2}{x^2-4} - \dfrac{1}{x} = \dfrac{x^2(x) - (x^2-4)(1)}{(x^2-4)(x)} = \dfrac{x^3 - x^2 + 4}{(x-2)(x+2)(x)}$ ■

Least Common Multiple (LCM)

If the denominators of two rational expressions to be added (or subtracted) have common factors, we usually do not use the general rules given by equation (5), since, in doing so, we make the problem more complicated than it needs to be. Instead, just as with fractions, we apply the **least common multiple (LCM) method** by using the polynomial of least degree that contains each denominator polynomial as a factor. Then we rewrite each rational expression using the LCM as the common denominator and use equation (4) to do the addition (or subtraction).

To find the least common multiple of two or more polynomials, first factor completely each polynomial. The LCM is the product of the different prime factors of each polynomial, each factor appearing the greatest number of times it occurs in each polynomial. The next two examples will give you the idea.

Example 6 Find the least common multiple of the following pair of polynomials:

$$x(x-1)^2(x+1) \quad \text{and} \quad 4(x-1)(x+1)^3$$

Solution The polynomials are already factored completely as

$$x(x - 1)^2(x + 1) \qquad \text{and} \qquad 4(x - 1)(x + 1)^3$$

Start by writing the factors of the left-hand polynomial. (Alternatively, you could start with the one on the right.)

$$x(x - 1)^2(x + 1)$$

Now look at the right-hand polynomial. Its first factor, 4, does not appear in our list, so we insert it:

$$4x(x - 1)^2(x + 1)$$

The next factor, $x - 1$, is already in our list, so no change is necessary. The final factor is $(x + 1)^3$. Since our list has $x + 1$ to the first power only, we replace $x + 1$ in the list by $(x + 1)^3$. The LCM is

$$4x(x - 1)^2(x + 1)^3$$

Notice that the LCM is, in fact, the polynomial of least degree that contains $x(x - 1)^2(x + 1)$ and $4(x - 1)(x + 1)^3$ as factors. ■

Example 7 Find the least common multiple of the following pair of polynomials:

$$2x^2 - 2x - 12 \qquad \text{and} \qquad x^3 - 3x^2$$

Solution First, we factor completely each polynomial:

$$2x^2 - 2x - 12 = 2(x^2 - x - 6) = 2(x - 3)(x + 2)$$
$$x^3 - 3x^2 = x^2(x - 3)$$

Now we write the factors of the first polynomial:

$$2(x - 3)(x + 2)$$

Looking at the factors that appear in the second polynomial, we see that the first factor, x^2, does not appear in our list, so we insert it:

$$2x^2(x - 3)(x + 2)$$

The remaining factor, $x - 3$, is already in our list. Thus, the LCM of $2x^2 - 2x - 12$ and $x^3 - 3x^2$ is $2x^2(x - 3)(x + 2)$. ■

The next example illustrates how the LCM is used for adding and subtracting rational expressions.

Example 8 Perform the indicated operation and simplify the result. Leave your answer in factored form.

$$\frac{x}{x^2 + 3x + 2} + \frac{2x - 3}{x^2 - 1} \qquad x \neq -2, -1, 1$$

Solution First we find the LCM of the denominators:

$$x^2 + 3x + 2 = (x + 2)(x + 1)$$
$$x^2 - 1 = (x - 1)(x + 1)$$

The LCM is $(x + 2)(x + 1)(x - 1)$. Next, we rewrite each rational expression using the LCM as the denominator:

$$\frac{x}{x^2 + 3x + 2} = \frac{x}{(x + 2)(x + 1)} = \frac{x(x - 1)}{(x + 2)(x + 1)(x - 1)}$$

↑ Multiply numerator and denominator by $x - 1$ to get the LCM in the denominator.

$$\frac{2x - 3}{x^2 - 1} = \frac{2x - 3}{(x - 1)(x + 1)} = \frac{(2x - 3)(x + 2)}{(x - 1)(x + 1)(x + 2)}$$

↑ Multiply numerator and denominator by $x + 2$ to get the LCM in the denominator.

Now we can add by using equation (4).

$$\begin{aligned}\frac{x}{x^2 + 3x + 2} + \frac{2x - 3}{x^2 - 1} &= \frac{x(x - 1)}{(x + 2)(x + 1)(x - 1)} + \frac{(2x - 3)(x + 2)}{(x + 2)(x + 1)(x - 1)} \\ &= \frac{(x^2 - x) + (2x^2 + x - 6)}{(x + 2)(x + 1)(x - 1)} \\ &= \frac{3x^2 - 6}{(x + 2)(x + 1)(x - 1)} = \frac{3(x^2 - 2)}{(x + 2)(x + 1)(x - 1)}\end{aligned}$$

■

If we had not used the LCM technique to add the quotients in Example 8, but decided instead to use the general rule of equation (5), we would have obtained a more complicated expression, as follows:

$$\begin{aligned}\frac{x}{x^2 + 3x + 2} + \frac{2x - 3}{x^2 - 1} &= \frac{x(x^2 - 1) + (x^2 + 3x + 2)(2x - 3)}{(x^2 + 3x + 2)(x^2 - 1)} \\ &= \frac{3x^3 + 3x^2 - 6x - 6}{(x^2 + 3x + 2)(x^2 - 1)} = \frac{3(x^3 + x^2 - 2x - 2)}{(x^2 + 3x + 2)(x^2 - 1)}\end{aligned}$$

Now we are faced with a more complicated problem of expressing this quotient in lowest terms. It is always best to first look for common factors in the denominators of expressions to be added or subtracted and use the LCM if any common factors are found.

Example 9 Perform the indicated operations and simplify the result. Leave your answer in factored form.

$$\frac{1}{x^2 + x} - \frac{x + 4}{x^3 + 1} + \frac{3}{x^2} \qquad x \neq -1, 0$$

Solution Again, we start by finding the LCM of the denominators:

$$x^2 + x = x(x + 1)$$
$$x^3 + 1 = (x + 1)(x^2 - x + 1)$$
$$x^2 = x^2$$

The LCM is $x^2(x + 1)(x^2 - x + 1)$. Thus,

$$\begin{aligned}
&\frac{1}{x^2 + x} - \frac{x + 4}{x^3 + 1} + \frac{3}{x^2} \\
&= \frac{1}{x(x + 1)} - \frac{x + 4}{(x + 1)(x^2 - x + 1)} + \frac{3}{x^2} \\
&= \frac{x(x^2 - x + 1)}{x^2(x + 1)(x^2 - x + 1)} - \frac{x^2(x + 4)}{x^2(x + 1)(x^2 - x + 1)} + \frac{3(x + 1)(x^2 - x + 1)}{x^2(x + 1)(x^2 - x + 1)} \\
&= \frac{(x^3 - x^2 + x) - (x^3 + 4x^2) + 3(x^3 + 1)}{x^2(x + 1)(x^2 - x + 1)} = \frac{3x^3 - 5x^2 + x + 3}{x^2(x + 1)(x^2 - x + 1)}
\end{aligned}$$ ■

Mixed Quotients

When sums and/or differences of rational expressions appear as the numerator and/or denominator of a quotient, the quotient is called a **mixed quotient**. For example,

$$\frac{1 + \frac{1}{x}}{1 - \frac{1}{x}} \quad \text{and} \quad \frac{\frac{x^2}{x^2 - 4} - 3}{\frac{x - 3}{x + 2} - 1}$$

are mixed quotients. To **simplify** a mixed quotient means to write it as a rational expression reduced to lowest terms. This can be accomplished in either of two ways:

METHOD 1: Treat the numerator and denominator of the mixed quotient separately, performing whatever operations are indicated and simplifying the results. Follow this by simplifying the resulting rational expression.

METHOD 2: Find the LCM of the denominators of all rational expressions that appear in the mixed quotient. Multiply the numerator and denominator of the mixed quotient by the LCM, and simplify the result.

We will use both methods in the next example. By carefully studying each method, you can discover in which situations one method may be easier to use than the other.

Example 10 Simplify: $\dfrac{\frac{1}{2} + \frac{3}{x}}{\frac{x + 3}{4}} \qquad x \neq -3, 0$

Solution METHOD 1: First, we perform the indicated operation in the numerator, and then we divide:

$$\frac{\frac{1}{2}+\frac{3}{x}}{\frac{x+3}{4}} \underset{\uparrow}{=} \frac{\frac{1\cdot x+2\cdot 3}{2\cdot x}}{\frac{x+3}{4}} = \frac{\frac{x+6}{2x}}{\frac{x+3}{4}} \underset{\uparrow}{=} \frac{x+6}{2x}\cdot\frac{4}{x+3}$$

Rule for adding quotients Rule for dividing quotients

$$\underset{\uparrow}{=} \frac{(x+6)\cdot 4}{2\cdot x\cdot(x+3)} = \frac{\not{2}\cdot 2\cdot(x+6)}{\not{2}\cdot x\cdot(x+3)} = \frac{2(x+6)}{x(x+3)}$$

Rule for multiplying quotients

METHOD 2: The rational expressions that appear in the mixed quotient are

$$\frac{1}{2},\quad \frac{3}{x},\quad \frac{x+3}{4}$$

The LCM of their denominators is $4x$. Thus, we multiply the numerator and denominator of the mixed quotient by $4x$ and then simplify:

$$\frac{\frac{1}{2}+\frac{3}{x}}{\frac{x+3}{4}} \underset{\uparrow}{=} \frac{4x\cdot\left(\frac{1}{2}+\frac{3}{x}\right)}{4x\cdot\left(\frac{x+3}{4}\right)} \underset{\uparrow}{=} \frac{4x\cdot\frac{1}{2}+4x\cdot\frac{3}{x}}{\frac{4x\cdot(x+3)}{4}}$$

Multiply by $4x$ Distributive property

$$= \frac{\not{2}\cdot 2x\cdot\frac{1}{\not{2}}+4\not{x}\cdot\frac{3}{\not{x}}}{\frac{\not{4}x\cdot(x+3)}{\not{4}}} \underset{\uparrow}{=} \frac{2x+12}{x(x+3)} \underset{\uparrow}{=} \frac{2(x+6)}{x(x+3)}$$

Simplify Factor ■

Example 11 Simplify: $\dfrac{\frac{x^2}{x^2-4}-3}{\frac{x-3}{x+2}-1} \qquad x \neq -2, 2$

Solution We shall use Method 1:

$$\frac{\frac{x^2}{x^2-4}-3}{\frac{x-3}{x+2}-1} = \frac{\frac{x^2}{x^2-4}-\frac{3(x^2-4)}{x^2-4}}{\frac{x-3}{x+2}-\frac{x+2}{x+2}} = \frac{\frac{x^2-3(x^2-4)}{x^2-4}}{\frac{(x-3)-(x+2)}{x+2}} = \frac{\frac{-2x^2+12}{x^2-4}}{\frac{-5}{x+2}} = \frac{\frac{-2(x^2-6)}{(x-2)(x+2)}}{\frac{-5}{x+2}}$$

$$= \frac{-2(x^2-6)}{(x-2)(x+2)}\cdot\frac{x+2}{-5}$$

$$= \frac{-2(x^2-6)\cancel{(x+2)}}{(x-2)\cancel{(x+2)}(-5)} = \frac{2(x^2-6)}{5(x-2)}$$

■

EXERCISE 1.6

In Problems 1–20, reduce each rational expression to lowest terms.

1. $\dfrac{5x + 10}{x^2 - 4}$
2. $\dfrac{4x + 8}{12x + 24}$
3. $\dfrac{x^2 - 2x}{3x - 6}$
4. $\dfrac{15x^2 + 24x}{3x^2}$
5. $\dfrac{24x}{12x^2 - 6x}$
6. $\dfrac{3x - 12}{x^2 - 16}$
7. $\dfrac{y^2 - 25}{2y - 10}$
8. $\dfrac{3y + 2}{3y^2 + 5y + 2}$
9. $\dfrac{x^2 + 4x - 5}{x - 1}$
10. $\dfrac{x - x^2}{x^2 + x - 2}$
11. $\dfrac{x^2 - 4}{x^2 + 5x + 6}$
12. $\dfrac{x^2 + x - 6}{9 - x^2}$
13. $\dfrac{x^2 + 5x - 14}{2 - x}$
14. $\dfrac{2x^2 + 5x - 3}{1 - 2x}$
15. $\dfrac{2x^3 - x^2 - 10x}{x^3 - 2x^2 - 8x}$
16. $\dfrac{4x^4 + 2x^3 - 6x^2}{4x^4 + 26x^3 + 30x^2}$
17. $\dfrac{(x - 4)^2 - 9}{(x + 3)^2 - 16}$
18. $\dfrac{(x + 2)^2 - 8x}{(x - 2)^2}$
19. $\dfrac{6x(x - 1) - 12}{x^3 - 8}$
20. $\dfrac{3(x - 2)^2 + 17(x - 2)}{2(x - 2)^2 + 7(x - 2)}$

In Problems 21–34, perform the indicated operation and simplify the result. Leave your answer in factored form.

21. $\dfrac{3x - 6}{5x} \cdot \dfrac{x^2 - x - 6}{x^2 - 4}$
22. $\dfrac{9x - 25}{2x - 2} \cdot \dfrac{1 - x^2}{6x - 10}$
23. $\dfrac{4x^2 - 1}{x^2 - 16} \cdot \dfrac{x^2 - 4x}{2x + 1}$
24. $\dfrac{12}{x^2 - x} \cdot \dfrac{x^2 - 1}{4x - 2}$
25. $\dfrac{4x - 8}{-3x} \cdot \dfrac{12}{12 - 6x}$
26. $\dfrac{6x - 27}{5x} \cdot \dfrac{2}{4x - 18}$
27. $\dfrac{x^2 - 3x - 10}{x^2 + 2x - 35} \cdot \dfrac{x^2 + 4x - 21}{x^2 + 9x + 14}$
28. $\dfrac{x^2 - x - 6}{x^2 - 4x - 5} \cdot \dfrac{x^2 - 25}{x^2 + 2x - 15}$
29. $\dfrac{\dfrac{2x^2 - x - 28}{3x^2 - x - 2}}{\dfrac{4x^2 + 16x + 7}{3x^2 + 11x + 6}}$
30. $\dfrac{\dfrac{9x^2 + 3x - 2}{12x^2 + 5x - 2}}{\dfrac{9x^2 - 6x + 1}{8x^2 - 10x - 3}}$
31. $\dfrac{\dfrac{8x^2 - 6x + 1}{4x^2 - 1}}{\dfrac{12x^2 + 5x - 2}{6x^2 - x - 2}}$
32. $\dfrac{\dfrac{3x^2 + 2x - 1}{5x^2 - 9x - 2}}{\dfrac{2x^2 - x - 3}{10x^2 - 13x - 3}}$
33. $\dfrac{5x^2 - x}{3x + 2} \cdot \dfrac{2x^2 - x}{2x^2 - x - 1} \cdot \dfrac{10x^2 + 3x - 1}{6x^2 + x - 2}$
34. $\dfrac{x^2 - 7x - 8}{x^2 + 2x - 15} \cdot \dfrac{x^2 + 8x + 12}{x^2 - 6x - 7} \cdot \dfrac{x^2 + 4x - 5}{x^2 + 11x + 18}$

In Problems 35–54, perform the indicated operations and simplify the result. Leave your answer in factored form.

35. $\frac{x+1}{x-3} + \frac{x-4}{x-3}$

36. $\frac{2x-5}{3x+2} + \frac{x+3}{3x+2}$

37. $\frac{3x+5}{2x-1} - \frac{2x-4}{2x-1}$

38. $\frac{5x-4}{3x+4} - \frac{x+1}{3x+4}$

39. $\frac{4}{x-2} + \frac{x}{2-x}$

40. $\frac{6}{x-1} - \frac{x}{1-x}$

41. $\frac{4}{x-1} - \frac{1}{x+2}$

42. $\frac{2}{x+5} - \frac{3}{x-5}$

43. $\frac{x}{x+1} + \frac{2x-3}{x-1}$

44. $\frac{3x}{x-4} + \frac{2x}{x+3}$

45. $\frac{x-2}{x+2} - \frac{x+2}{x-2}$

46. $\frac{2x-1}{x-1} - \frac{2x+1}{x+1}$

47. $\frac{x}{x^2-4} + \frac{1}{x}$

48. $\frac{x-1}{x^3} + \frac{x}{x^2+1}$

49. $\frac{x^3}{(x-1)^2} - \frac{x^2+1}{x}$

50. $\frac{3x^2}{4} - \frac{x^3}{x^2-1}$

51. $\frac{x}{x+1} + \frac{x-2}{x-1} - \frac{x+1}{x-2}$

52. $\frac{3x+1}{x} + \frac{x}{x-1} - \frac{2x}{x+1}$

53. $\frac{1}{x} + \frac{1}{x+1} - \frac{1}{x-1}$

54. $\frac{1}{x} - \frac{1}{x-1} - \frac{1}{x-2}$

In Problems 55–62, find the LCM of the given polynomials.

55. $x^2 - 4,\ \ x^2 - x - 2$

56. $x^2 - x - 12,\ \ x^2 - 8x + 16$

57. $x^3 - x,\ \ x^2 - x$

58. $3x^2 - 27,\ \ 2x^2 - x - 15$

59. $4x^3 - 4x^2 + x,\ \ 2x^3 - x^2,\ \ x^3$

60. $x - 3,\ \ x^2 + 3x,\ \ x^3 - 9x$

61. $x^3 - x,\ \ x^3 - 2x^2 + x,\ \ x^3 - 1$

62. $x^2 + 4x + 4,\ \ x^3 + 2x^2,\ \ (x+2)^3$

In Problems 63–74, perform the indicated operations and simplify the result. Leave your answer in factored form.

63. $\frac{x}{x^2-7x+6} - \frac{x}{x^2-2x-24}$

64. $\frac{x}{x-3} - \frac{x+1}{x^2+5x-24}$

65. $\frac{4}{x^2-4} - \frac{2}{x^2+x-6}$

66. $\frac{3}{x-1} - \frac{x-4}{x^2-2x+1}$

67. $\frac{3}{(x-1)^2(x+1)} + \frac{2}{(x-1)(x+1)^2}$

68. $\frac{2}{(x+2)^2(x-1)} - \frac{6}{(x+2)(x-1)^2}$

69. $\frac{x+4}{x^2-x-2} - \frac{2x+3}{x^2+2x-8}$

70. $\frac{2x-3}{x^2+8x+7} - \frac{x-2}{(x+1)^2}$

71. $\frac{1}{x} - \frac{2}{x^2+x} + \frac{3}{x^3-x^2}$

72. $\frac{x}{(x-1)^2} + \frac{2}{x} - \frac{x+1}{x^3-x^2}$

73. $\frac{1}{h}\left(\frac{1}{x+h} - \frac{1}{x}\right)$

74. $\frac{1}{h}\left[\frac{1}{(x+h)^2} - \frac{1}{x^2}\right]$

In Problems 75–86, perform the indicated operations and simplify the result. Leave your answer in factored form.

75. $\dfrac{x - \frac{1}{x}}{x + \frac{1}{x}}$

76. $\dfrac{1 - \frac{x}{x+1}}{2 - \frac{x-1}{x}}$

77. $\dfrac{3 - \frac{x^2}{x+1}}{1 + \frac{x}{x^2-1}}$

78. $$\frac{3x - \dfrac{3}{x^2}}{\dfrac{1}{(x-1)^2} - 1}$$

79. $$\frac{\dfrac{x+4}{x-2} - \dfrac{x-3}{x+1}}{x+1}$$

80. $$\frac{\dfrac{x-2}{x+1} - \dfrac{x}{x-2}}{x+3}$$

81. $$\frac{\dfrac{x-2}{x+2} + \dfrac{x-1}{x+1}}{\dfrac{x}{x+1} - \dfrac{2x-3}{x}}$$

82. $$\frac{\dfrac{2x+5}{x} - \dfrac{x}{x-3}}{\dfrac{x^2}{x-3} - \dfrac{(x+1)^2}{x+3}}$$

83. $$1 - \frac{1}{1 - \dfrac{1}{x}}$$

84. $$1 - \frac{1}{1 - \dfrac{1}{1-x}}$$

85. $$\frac{\dfrac{x+h-2}{x+h+2} - \dfrac{x-2}{x+2}}{h}$$

86. $$\frac{\dfrac{x+h+1}{x+h-1} - \dfrac{x+1}{x-1}}{h}$$

87. The expressions below are called **continued fractions**:

$$1 + \frac{1}{x}, \quad 1 + \frac{1}{1 + \dfrac{1}{x}}, \quad 1 + \frac{1}{1 + \dfrac{1}{1 + \dfrac{1}{x}}}, \quad 1 + \frac{1}{1 + \dfrac{1}{1 + \dfrac{1}{1 + \dfrac{1}{x}}}}, \quad \ldots$$

Each one simplifies to an expression of the form

$$\frac{ax + b}{bx + c}$$

Trace the successive values of a, b, and c as you "continue" the fraction. Can you discover the patterns these values follow?

1.7 Radicals

Suppose $n \geq 2$ is an integer and a is a real number. An ***n*th root of *a*** is a number which, when raised to the power n, equals a.

Example 1
(a) A 3rd root of 8 is 2, since $2^3 = 8$.
(b) A 2nd root of 16 is 4, since $4^2 = 16$.
(c) A 3rd root of -64 is -4, since $(-4)^3 = -64$.
(d) A 2nd root of 16 is -4, since $(-4)^2 = 16$.
(e) A 4th root of $\frac{1}{16}$ is $-\frac{1}{2}$, since $\left(-\frac{1}{2}\right)^4 = \frac{1}{16}$.
(f) An nth root of a is x if $x^n = a$. ■

Consider an nth root of a. If n is an odd integer, there is only one real number x for which $x^n = a$. Further, if a is negative, so is the nth root of a; if a is positive, so is the nth root of a. Look back at Examples 1(a) and 1(c).

If n is an even integer and a is positive, there are two real numbers x for which $x^n = a$: one is positive and the other is negative. Look at Examples 1(b) and 1(d).

If n is an even integer and a is negative, there is no real number x for which $x^n = a$. Thus, if n is even and a is negative, there is no nth root for a; that is, it does not exist in the real number system.

Based on this discussion, we state the following definition:

Principal nth Root of a

The **principal nth root of a number a**, symbolized by $\sqrt[n]{a}$, is defined as follows:

(a) If a is positive and n is even, then $\sqrt[n]{a}$ is the *positive* nth root of a.

(b) If a is negative and n is even, then $\sqrt[n]{a}$ does not exist in the real number system.

(c) If n is odd, then $\sqrt[n]{a}$ is the nth root of a.

(d) $\sqrt[n]{0} = 0$.

The symbol $\sqrt[n]{a}$ for the principal nth root of a is sometimes called a **radical**; the integer n is called the **index**, and a is called the **radicand**. If the index of a radical is 2, we call $\sqrt[2]{a}$ the **square root** of a and omit the index 2 by simply writing $\sqrt{a}$. If the index is 3, we call $\sqrt[3]{a}$ the **cube root** of a.

To summarize,

$$\sqrt[n]{a} = b \qquad \text{means} \qquad b^n = a \qquad \text{where } b \geq 0 \text{ if } n \text{ is even}$$

Thus, when it is defined, the principal nth root of a number is unique.

Example 2

$$\sqrt[3]{8} = 2 \qquad \sqrt{64} = 8 \qquad \sqrt[3]{-64} = -4 \qquad \sqrt[4]{\tfrac{1}{16}} = \tfrac{1}{2}$$ ■

These are examples of **perfect roots**. Thus, 8 and -64 are perfect cubes, since $8 = 2^3$ and $-64 = (-4)^3$; 64 is a perfect square, since $64 = 8^2$; and $\frac{1}{2}$ is a perfect 4th root of $\frac{1}{16}$, since $\frac{1}{16} = \left(\frac{1}{2}\right)^4$.

In general, if $n \geq 2$ is a positive integer and a is a real number, we have

$$\sqrt[n]{a^n} = a \qquad \text{if } n \text{ is odd} \tag{1a}$$

$$\sqrt[n]{a^n} = |a| \qquad \text{if } n \text{ is even} \tag{1b}$$

Notice the need for the absolute value in equation (1b). If n is even, then a^n is positive whether $a > 0$ or $a < 0$. But if n is even, the principal nth root must be nonnegative. Hence, the reason for using the absolute value—it gives a nonnegative result.

Example 3 (a) $\sqrt[3]{4^3} = 4$ (b) $\sqrt[5]{(-3)^5} = -3$ (c) $\sqrt[4]{2^4} = 2$
(d) $\sqrt[4]{(-3)^4} = |-3| = 3$ (e) $\sqrt{x^2} = |x|$ ■

Radicals provide a way of representing many irrational real numbers. For example, there is no rational number whose square is 2. Thus, using decimals, we can only approximate the positive number whose square is 2. Using radicals, we can say that $\sqrt{2}$ *is* the positive number whose square is 2.

Properties of Radicals

Let $n \geq 2$ and $m \geq 2$ denote positive integers, and let a and b represent real numbers. Assuming all radicals are defined, we have the following properties:

$$\sqrt[n]{ab} = \sqrt[n]{a}\sqrt[n]{b} \tag{2a}$$

$$\sqrt[n]{\frac{a}{b}} = \frac{\sqrt[n]{a}}{\sqrt[n]{b}} \tag{2b}$$

$$\sqrt[n]{a^m} = (\sqrt[n]{a})^m \tag{2c}$$

$$\sqrt[m]{\sqrt[n]{a}} = \sqrt[mn]{a} \tag{2d}$$

When used in reference to radicals, the direction to "simplify" will mean to remove from the radicals any perfect roots that occur as factors. Let's look at some examples of how the rules listed above are applied to simplify radicals.

Example 4

$$\sqrt{32} = \sqrt{16 \cdot 2} = \sqrt{16}\sqrt{2} = 4\sqrt{2}$$

(first equality: 16 is a perfect square.; second equality: (2a)) ■

Example 5

$$\sqrt[3]{8x^4} = \sqrt[3]{8} \cdot \sqrt[3]{x^4} = 2\sqrt[3]{x^3 \cdot x} = 2\sqrt[3]{x^3}\sqrt[3]{x} = 2x\sqrt[3]{x}$$

(first equality: (2a); second equality: Factor out perfect cube.; last equality: (1a)) ■

Example 6

$$\sqrt[3]{-16x^4y^7} = \sqrt[3]{-8 \cdot 2 \cdot x^3 \cdot x \cdot y^6 \cdot y} = \sqrt[3]{(-2)^3 \cdot x^3 \cdot (y^2)^3 \cdot 2xy}$$

(first equality: Factor perfect cubes inside radical; second equality: Combine perfect cubes.)

$$= \sqrt[3]{(-2xy^2)^3 \cdot 2xy} = \sqrt[3]{(-2xy^2)^3} \cdot \sqrt[3]{2xy}$$

$$= -2xy^2\sqrt[3]{2xy}$$ ■

Example 7

$$\sqrt{\sqrt[3]{x^7}} \underset{\uparrow \atop (2d)}{=} \sqrt[6]{x^7} = \sqrt[6]{x^6 \cdot x} = \sqrt[6]{x^6} \cdot \sqrt[6]{x} \underset{\uparrow \atop (1b)}{=} |x| \sqrt[6]{x}$$ ■

Example 8

$$\sqrt[3]{\frac{8x^5}{27y^2}} = \sqrt[3]{\frac{2^3x^3x^2}{3^3y^2}} = \sqrt[3]{\left(\frac{2x}{3}\right)^3 \cdot \frac{x^2}{y^2}} = \sqrt[3]{\left(\frac{2x}{3}\right)^3} \cdot \sqrt[3]{\frac{x^2}{y^2}} = \frac{2x}{3}\sqrt[3]{\frac{x^2}{y^2}}$$ ■

Example 9

$$\frac{\sqrt{x^5y}}{\sqrt{x^3y^3}} = \sqrt{\frac{x^5y}{x^3y^3}} = \sqrt{\frac{x^2}{y^2}} = \sqrt{\left(\frac{x}{y}\right)^2} = \left|\frac{x}{y}\right|$$ ■

Two or more radicals can be combined, provided they have the same index and the same radicand.

Example 10

$$\begin{aligned} 4\sqrt{27} - 8\sqrt{12} + \sqrt{3} &= 4\sqrt{9 \cdot 3} - 8\sqrt{4 \cdot 3} + \sqrt{3} \\ &= 4 \cdot \sqrt{9}\sqrt{3} - 8 \cdot \sqrt{4}\sqrt{3} + \sqrt{3} \\ &= 12\sqrt{3} - 16\sqrt{3} + \sqrt{3} \qquad \text{Factor out } \sqrt{3}. \\ &= (12 - 16 + 1)\sqrt{3} = -3\sqrt{3} \end{aligned}$$ ■

Example 11

$$\begin{aligned} \sqrt[3]{8x^4} + \sqrt[3]{-x} + 4\sqrt[3]{27x} &= \sqrt[3]{2^3x^3x} + \sqrt[3]{-1 \cdot x} + 4\sqrt[3]{3^3x} \\ &= \sqrt[3]{(2x)^3} \cdot \sqrt[3]{x} + \sqrt[3]{-1} \cdot \sqrt[3]{x} + 4\sqrt[3]{3^3} \cdot \sqrt[3]{x} \\ &= 2x\sqrt[3]{x} - 1 \cdot \sqrt[3]{x} + 12\sqrt[3]{x} \\ &= (2x + 11)\sqrt[3]{x} \end{aligned}$$ ■

The special products listed earlier are also helpful for finding products involving sums or differences of radicals.

Example 12 Perform the indicated operation and simplify the result. The domain of the variable x is $\{x | x \geq 0\}$.

(a) $(\sqrt{x} - 3)(\sqrt{x} + 3)$ (b) $(\sqrt{x} + \sqrt{2})^2$ (c) $(\sqrt{x} - \sqrt{3})(\sqrt{x} + \sqrt{2})$

Solution

(a) $(\sqrt{x} - 3)(\sqrt{x} + 3) = (\sqrt{x})^2 - 3^2 = x - 9$

(b) $(\sqrt{x} + \sqrt{2})^2 = (\sqrt{x})^2 + 2\sqrt{x}\sqrt{2} + (\sqrt{2})^2 = x + 2\sqrt{2x} + 2$

(c) $$\begin{aligned}(\sqrt{x} - \sqrt{3})(\sqrt{x} + \sqrt{2}) &= (\sqrt{x})^2 - \sqrt{3}\sqrt{x} + \sqrt{2}\sqrt{x} - \sqrt{6} \\ &= x + (\sqrt{2} - \sqrt{3})\sqrt{x} - \sqrt{6}\end{aligned}$$ ■

Rationalizing

When radicals occur in quotients, it has become common practice to rewrite the quotient so that the denominator contains no radicals. This process is referred to as **rationalizing the denominator**.

The idea is to find an appropriate expression so that, when it is multiplied by the radical in the denominator, the new denominator that results contains no radicals. For example:

IF RADICAL IS	MULTIPLY BY	TO GET PRODUCT FREE OF RADICALS
$\sqrt{3}$	$\sqrt{3}$	$\sqrt{9} = 3$
$\sqrt[3]{4}$	$\sqrt[3]{2}$	$\sqrt[3]{8} = 2$
$\sqrt{3} + 1$	$\sqrt{3} - 1$	$(\sqrt{3})^2 - 1^2 = 3 - 1 = 2$
$\sqrt{2} - 3$	$\sqrt{2} + 3$	$(\sqrt{2})^2 - 3^2 = 2 - 9 = -7$
$\sqrt{5} - \sqrt{3}$	$\sqrt{5} + \sqrt{3}$	$(\sqrt{5})^2 - (\sqrt{3})^2 = 5 - 3 = 2$
$\sqrt[3]{2} + 2$	$\sqrt[3]{2^2} - 2\sqrt[3]{2} + 4$	$\sqrt[3]{2^3} + 8 = 2 + 8 = 10$
$\sqrt[3]{5} - 1$	$\sqrt[3]{5^2} + \sqrt[3]{5} + 1$	$\sqrt[3]{5^3} - 1 = 5 - 1 = 4$

You are correct if you observed in this list that, after the second type of radical, the special products for differences of squares $[(a - b)(a + b) = a^2 - b^2]$, sums of cubes $[(a + b)(a^2 - ab + b^2) = a^3 + b^3]$, and differences of cubes $[(a - b)(a^2 + ab + b^2) = a^3 - b^3]$ are the bases for determining by what to multiply.

Example 13 Rationalize the denominator: $\dfrac{4}{\sqrt{2}}$

Solution

$$\frac{4}{\sqrt{2}} = \frac{4}{\sqrt{2}} \cdot \frac{\sqrt{2}}{\sqrt{2}} = \frac{4\sqrt{2}}{(\sqrt{2})^2} = \frac{4\sqrt{2}}{2} = 2\sqrt{2}$$

Multiply by $\dfrac{\sqrt{2}}{\sqrt{2}}$. ■

Example 14 Rationalize the denominator: $\dfrac{\sqrt{3}}{\sqrt[3]{2}}$

Solution

$$\frac{\sqrt{3}}{\sqrt[3]{2}} = \frac{\sqrt{3}}{\sqrt[3]{2}} \cdot \frac{\sqrt[3]{4}}{\sqrt[3]{4}} = \frac{\sqrt{3}\sqrt[3]{4}}{\sqrt[3]{8}} = \frac{\sqrt{3}\sqrt[3]{4}}{2}$$

Multiply by $\dfrac{\sqrt[3]{4}}{\sqrt[3]{4}}$. ■

Example 15 Rationalize the denominator: $\dfrac{5}{\sqrt{3} + 2}$

Solution

$$\frac{5}{\sqrt{3} + 2} = \frac{5}{\sqrt{3} + 2} \cdot \frac{\sqrt{3} - 2}{\sqrt{3} - 2} = \frac{5(\sqrt{3} - 2)}{(\sqrt{3})^2 - 2^2}$$

$$= \frac{5(\sqrt{3} - 2)}{3 - 4} = -5(\sqrt{3} - 2)$$ ■

Example 16 Rationalize the denominator: $\dfrac{\sqrt{x} - 2}{\sqrt{x} + 2}, \quad x \geq 0$

Solution

$$\frac{\sqrt{x}-2}{\sqrt{x}+2}=\frac{\sqrt{x}-2}{\sqrt{x}+2}\cdot\frac{\sqrt{x}-2}{\sqrt{x}-2}=\frac{(\sqrt{x}-2)^2}{(\sqrt{x})^2-2^2}$$
$$=\frac{(\sqrt{x})^2-4\sqrt{x}+4}{x-4}=\frac{x-4\sqrt{x}+4}{x-4}$$ ■

Example 17 Rationalize the denominator: $\dfrac{1}{\sqrt[3]{x}-1}$, $x \neq 1$

Solution

$$\frac{1}{\sqrt[3]{x}-1}=\frac{1}{\sqrt[3]{x}-1}\cdot\frac{\sqrt[3]{x^2}+\sqrt[3]{x}+1}{\sqrt[3]{x^2}+\sqrt[3]{x}+1}$$
$$=\frac{\sqrt[3]{x^2}+\sqrt[3]{x}+1}{\sqrt[3]{x^3}-1}=\frac{\sqrt[3]{x^2}+\sqrt[3]{x}+1}{x-1}$$ ■

In calculus, sometimes the numerator must be rationalized.

Example 18 Rationalize the numerator: $\dfrac{\sqrt{x}-2}{\sqrt{x}+1}$, $x \geq 0$

Solution We multiply by $\dfrac{\sqrt{x}+2}{\sqrt{x}+2}$:

$$\frac{\sqrt{x}-2}{\sqrt{x}+1}=\frac{\sqrt{x}-2}{\sqrt{x}+1}\cdot\frac{\sqrt{x}+2}{\sqrt{x}+2}=\frac{(\sqrt{x})^2-2^2}{(\sqrt{x}+1)(\sqrt{x}+2)}=\frac{x-4}{x+3\sqrt{x}+2}$$ ■

Historical Comment

■ The radical sign, $\sqrt{\ }$, was first used in print by Coss in 1525. It is thought to be the manuscript form of the letter *r* (for the Latin word *radix* = *root*), although this is not quite conclusively proved. It took a long time for $\sqrt{\ }$ to become the standard symbol for a square root and much longer to standardize $\sqrt[3]{\ }$, $\sqrt[4]{\ }$, $\sqrt[5]{\ }$, and so on. The indices of the root were placed in every conceivable position, with

$$\sqrt{8}^{\,3}, \qquad \sqrt{\ }\,③\,8, \qquad \text{and} \qquad \underset{3}{\sqrt{8}}$$

all being variants for $\sqrt[3]{8}$. The notation $\sqrt{\sqrt{16}}$ was popular for $\sqrt[4]{16}$. By the 1700's, the index had settled where we now put it.

The bar on top of the present radical symbol, as shown below,

$$\sqrt{a^2+2ab+b^2}$$

is the last survivor of the *vinculum*, a bar placed atop an expression to indicate what we would now indicate with parentheses. For example,

$$a\overline{b+c}=a(b+c)$$

Historical Problems

■ **1.** Christian Dibuadius, a Dane, in 1605 used the early symbols $\surd$ for square root and $\surd C$ for cube root. Change the following into modern notation:

(a) $\surd C8$ (b) $\surd C \surd C512$ (c) $\surd \surd C64$ (d) $\surd C \surd 64$

(e) $\surd \surd 81$ (f) $\surd \surd 256x^8$ (g) $\surd C\,\overline{-4 + \surd 8}$

2. Oresme (1360) used

$$\tfrac{1}{2}2^p \text{ for } 2^{1/2} \qquad \text{and} \qquad \boxed{1\tfrac{1}{2}}\,2^p \text{ for } 2^{1+1/2} = 2^{3/2}$$

What are the following in modern notation?

(a) $\frac{1}{3}8^p$ (b) $\frac{1}{4}256^p$ (c) $\boxed{2\tfrac{1}{3}}\,8^p$ (d) $\boxed{-1\tfrac{1}{4}}\,64^p$ ■

EXERCISE 1.7

In Problems 1–12, evaluate each perfect root.

1. $\sqrt{64}$ **2.** $\sqrt{81}$ **3.** $\sqrt[3]{27}$ **4.** $\sqrt[3]{125}$ **5.** $\sqrt[3]{-1}$

6. $\sqrt[3]{-8}$ **7.** $\sqrt{\frac{1}{4}}$ **8.** $\sqrt[3]{\frac{8}{27}}$ **9.** $\sqrt{9x^2}$ **10.** $\sqrt[3]{27x^6}$

11. $\sqrt[3]{8(1 + x)^3}$ **12.** $\sqrt{4(x + 4)^2}$

In Problems 13–40, simplify each expression. Assume all variables are positive when they appear.

13. $\sqrt{8}$ **14.** $\sqrt[4]{32}$ **15.** $\sqrt[3]{16x^4}$ **16.** $\sqrt{27x^3}$ **17.** $\sqrt[3]{\sqrt{x^6}}$

18. $\sqrt{\sqrt{x^6}}$ **19.** $\sqrt{\frac{32x^3}{9x}}$ **20.** $\sqrt[3]{\frac{x}{8x^4}}$ **21.** $\sqrt[4]{x^{12}y^8}$ **22.** $\sqrt[5]{x^{10}y^5}$

23. $\sqrt[4]{\frac{x^9y^7}{xy^3}}$ **24.** $\sqrt[3]{\frac{3xy^2}{81x^4y^2}}$ **25.** $\sqrt{36x}$ **26.** $\sqrt{9x^5}$ **27.** $\sqrt{3x^2}\sqrt{12x}$

28. $\sqrt{5x}\sqrt{20x^3}$ **29.** $\dfrac{\sqrt{3xy^3}\sqrt{2x^2y}}{\sqrt{6x^3y^4}}$ **30.** $\dfrac{\sqrt[3]{x^2y}\sqrt[3]{125x^3}}{\sqrt[3]{8x^3y^4}}$

31. $\sqrt{\frac{4}{9x^2y^4}}$ **32.** $\sqrt{\frac{9}{16x^4y^6}}$ **33.** $(\sqrt{5}\sqrt[3]{9})^2$

34. $(\sqrt[3]{3}\sqrt{10})^4$ **35.** $\sqrt{\frac{2x - 3}{2x^4 + 3x^3}}\sqrt{\frac{x}{4x^2 - 9}}$ **36.** $\sqrt[3]{\frac{x - 1}{x^2 + 2x + 1}}\sqrt[3]{\frac{(x - 1)^2}{x + 1}}$

37. $\sqrt{\sqrt[3]{\sqrt[4]{x}}}$ **38.** $\sqrt[4]{\sqrt[3]{x}}$ **39.** $\sqrt{\frac{x - y}{x + y}}\sqrt{\frac{x^2 + 2xy + y^2}{x^2 - y^2}}$

40. $\sqrt{\frac{x^2 + y^2}{xy}}\sqrt{\frac{xy(x)}{x^4 - y^4}}$

In Problems 41–50, simplify each expression.

41. $3\sqrt{2} + 2\sqrt{2} - \sqrt{2}$ **42.** $6\sqrt{5} - \sqrt{5} - 4\sqrt{5}$ **43.** $3\sqrt{2} - \sqrt{18} + 2\sqrt{8}$

44. $5\sqrt{3} + 2\sqrt{12} - 3\sqrt{27}$ **45.** $\sqrt[3]{16} + 5\sqrt[3]{2} - 2\sqrt[3]{54}$ **46.** $9\sqrt[3]{24} - \sqrt[3]{81}$

47. $\sqrt{8x^3} - 3\sqrt{50x} + \sqrt{2x^5}$, $x \geq 0$ **48.** $\sqrt{x^2y} - 3x\sqrt{9y} + 4\sqrt{25y}$, $x \geq 0, y \geq 0$

49. $\sqrt[3]{16x^4y} - 3x\sqrt[3]{2xy} + 5\sqrt[3]{-2xy^4}$ **50.** $8xy - \sqrt{25x^2y^2} + \sqrt[3]{8x^3y^3}$, $x \geq 0$, $y \geq 0$

In Problems 51–64, perform the indicated operation and simplify the result.

51. $(3\sqrt{6})(2\sqrt{2})$ **52.** $(5\sqrt{8})(-3\sqrt{3})$ **53.** $(\sqrt{3} + 3)(\sqrt{3} - 1)$

54. $(\sqrt{5} - 2)(\sqrt{5} + 3)$ **55.** $(3\sqrt{7} + 3)(2\sqrt{7} + 2)$ **56.** $(2\sqrt{6} + 3)^2$

57. $(\sqrt{x} - 1)^2$, $x \geq 0$ **58.** $(\sqrt{x} + \sqrt{5})^2$, $x \geq 0$ **59.** $(\sqrt[3]{x} - 1)^3$

60. $(\sqrt[3]{x} + \sqrt[3]{2})^3$ **61.** $(2\sqrt{x} - 3\sqrt{y})(2\sqrt{x} + 5\sqrt{y})$, $x \geq 0, y \geq 0$

62. $(4\sqrt{x} - \sqrt{y})(\sqrt{x} + 3\sqrt{y})$, $x \geq 0, y \geq 0$

63. $\sqrt{1 - x^2} - \dfrac{1}{\sqrt{1 - x^2}}$, $x \neq -1, 1$ **64.** $\sqrt{1 - x^2} + \dfrac{x^2}{\sqrt{1 - x^2}}$, $x \neq -1, 1$

In Problems 65–86, rationalize the denominator of each expression.

65. $\dfrac{1}{\sqrt{2}}$ **66.** $\dfrac{6}{\sqrt[3]{4}}$ **67.** $\dfrac{-\sqrt{3}}{\sqrt{5}}$ **68.** $\dfrac{-\sqrt[3]{3}}{\sqrt{8}}$

69. $\dfrac{1}{\sqrt{x}}$, $x > 0$ **70.** $\dfrac{1}{\sqrt[3]{x}}$, $x > 0$ **71.** $\dfrac{\sqrt{3}}{5 - \sqrt{2}}$ **72.** $\dfrac{\sqrt{2}}{\sqrt{7} + 2}$

73. $\dfrac{2 - \sqrt{5}}{2 + 3\sqrt{5}}$ **74.** $\dfrac{\sqrt{3} - 1}{2\sqrt{3} + 3}$ **75.** $\dfrac{\sqrt{3} - \sqrt{2}}{\sqrt{3} + \sqrt{2}}$ **76.** $\dfrac{\sqrt{5} + \sqrt{3}}{\sqrt{5} - \sqrt{3}}$

77. $\dfrac{1}{\sqrt{x} + 2}$, $x \geq 0$ **78.** $\dfrac{1}{\sqrt{x} - 3}$, $x \geq 0, x \neq 9$ **79.** $\dfrac{1}{2\sqrt{x} + 1}$, $x \geq 0$

80. $\dfrac{1}{3\sqrt{x} - 2}$, $x \geq 0, x \neq \frac{4}{9}$ **81.** $\dfrac{\sqrt{x + h} - \sqrt{x}}{\sqrt{x + h} + \sqrt{x}}$ **82.** $\dfrac{\sqrt{x + h} + \sqrt{x - h}}{\sqrt{x + h} - \sqrt{x - h}}$

83. $\dfrac{1}{\sqrt[3]{x} + 1}$, $x \neq -1$ **84.** $\dfrac{1}{\sqrt[3]{x} - 1}$, $x \neq 1$ **85.** $\sqrt[3]{\dfrac{-1}{9x^2y^5}}$, $x \neq 0, y \neq 0$

86. $\sqrt[3]{\dfrac{-8x}{3y^4}}$, $y \neq 0$

87. Rationalize the numerator: $\dfrac{\sqrt{x + h} - \sqrt{x}}{h}$

88. Rationalize the numerator: $\dfrac{\dfrac{1}{\sqrt{x + h}} - \dfrac{1}{\sqrt{x}}}{h}$

1.8 Rational Exponents

Our purpose in this section is to give a definition for "a raised to the power m/n," where a is a real number and m/n is a rational number. However, we want the definition we give to ensure that the laws of exponents stated for integer exponents in Section 1.3 remain true for rational exponents. For example, if the law of exponents $(a^r)^s = a^{rs}$ is to hold, then it must be true that

$$(a^{1/n})^n = a^{(1/n)n} = a^1 = a$$

That is, $a^{1/n}$ is a number that, when raised to the power n, is a. But this was the definition we gave of the principal nth root of a in Section 1.7. Thus,

$$a^{1/2} = \sqrt{a} \qquad a^{1/3} = \sqrt[3]{a} \qquad a^{1/4} = \sqrt[4]{a}$$

and we can state the following definition:

$a^{1/n}$

If a is a real number and $n \geq 2$ is an integer, then

$$a^{1/n} = \sqrt[n]{a} \tag{1}$$

provided $\sqrt[n]{a}$ exists.

Example 1 (a) $4^{1/2} = \sqrt{4} = 2$ (b) $(-27)^{1/3} = \sqrt[3]{-27} = -3$
(c) $8^{1/2} = \sqrt{8} = 2\sqrt{2}$ (d) $16^{1/3} = \sqrt[3]{16} = 2\sqrt[3]{2}$ ■

Note that if n is even and $a < 0$, then $\sqrt[n]{a}$ and $a^{1/n}$ do not exist.

We now seek a definition for $a^{m/n}$, where m and n are integers containing no common factors (except 1 and -1) and $n \geq 2$. Again, we want the definition to obey the laws of exponents stated earlier. For example,

$$a^{m/n} = a^{m(1/n)} = (a^m)^{1/n} \qquad \text{and} \qquad a^{m/n} = a^{(1/n)m} = (a^{1/n})^m$$

$a^{m/n}$

If a is a real number and m and n are integers containing no common factors with $n \geq 2$, then

$$a^{m/n} = \sqrt[n]{a^m} = (\sqrt[n]{a})^m \tag{2}$$

provided $\sqrt[n]{a}$ exists.

We have two comments about equation (2):

1. The exponent m/n must be in lowest terms and n must be positive.
2. In simplifying the rational expression $a^{m/n}$, either $\sqrt[n]{a^m}$ or $(\sqrt[n]{a})^m$ may be used, the choice depending on which one is easier to simplify. Generally, taking the root first, as in $(\sqrt[n]{a})^m$, is preferred.

Example 2 (a) $4^{3/2} = (\sqrt{4})^3 = 2^3 = 8$ (b) $(-8)^{4/3} = (\sqrt[3]{-8})^4 = (-2)^4 = 16$
(c) $(32)^{-2/5} = (\sqrt[5]{32})^{-2} = 2^{-2} = \frac{1}{4}$ (d) $4^{6/4} = 4^{3/2} = (\sqrt{4})^3 = 2^3 = 8$ ■

Based on the definition of $a^{m/n}$, no meaning is given to $a^{m/n}$ if a is a negative real number and n is an even integer.

The definitions in equations (1) and (2) were stated so that the laws of exponents would remain true for rational exponents. For convenience, we list again the laws of exponents.

If a and b are real numbers and r and s are rational numbers, then

Laws of Exponents

$$a^r a^s = a^{r+s} \qquad (a^r)^s = a^{rs} \qquad (ab)^r = a^r \cdot b^r$$

$$a^{-r} = \frac{1}{a^r} \qquad \left(\frac{a}{b}\right)^r = \frac{a^r}{b^r} \qquad \frac{a^r}{a^s} = a^{r-s} = \frac{1}{a^{s-r}}$$

where it is assumed that all expressions used are defined.

Rational exponents sometimes can be used to simplify radicals.

Example 3 Simplify each expression.

(a) $\sqrt[9]{x^3}$ (b) $\sqrt[3]{4}\sqrt{2}$ (c) $\sqrt{a\sqrt[3]{a}}$

Solution

(a) $\sqrt[9]{x^3} = x^{3/9} = x^{1/3} = \sqrt[3]{x}$

(b) $\sqrt[3]{4}\sqrt{2} = 4^{1/3} \cdot 2^{1/2} = (2^2)^{1/3} \cdot 2^{1/2} = 2^{2/3} \cdot 2^{1/2}$
$= 2^{7/6} = 2 \cdot 2^{1/6} = 2\sqrt[6]{2}$

(c) $\sqrt{a\sqrt[3]{a}} = (a \cdot a^{1/3})^{1/2} = (a^{4/3})^{1/2} = a^{2/3} = \sqrt[3]{a^2}$ ■

Example 4 Simplify each expression. Express your answer so that only positive exponents occur. Assume that the variables are positive.

(a) $\left(\dfrac{2x^{1/3}}{y^{2/3}}\right)^{-3}$ (b) $(x^{2/3}y^{-3/4})(x^{-2}y)^{1/2}$

(c) $\left(\dfrac{9x^2y^{1/3}}{x^{1/3}y}\right)^{1/2}$ (d) $\dfrac{(2x+5)^{1/3}(2x+5)^{-1/2}}{(2x+5)^{-3/4}}$

Solution

(a) $\left(\dfrac{2x^{1/3}}{y^{2/3}}\right)^{-3} = \left(\dfrac{y^{2/3}}{2x^{1/3}}\right)^{3} = \dfrac{(y^{2/3})^3}{(2x^{1/3})^3} = \dfrac{y^2}{2^3(x^{1/3})^3} = \dfrac{y^2}{8x}$

↑ Equation (8), page 27

(b) $(x^{2/3}y^{-3/4})(x^{-2}y)^{1/2} = (x^{2/3}y^{-3/4})[(x^{-2})^{1/2}y^{1/2}]$
$= x^{2/3}y^{-3/4}x^{-1}y^{1/2} = (x^{2/3}x^{-1})(y^{-3/4}y^{1/2})$
$= x^{-1/3}y^{-1/4} = \dfrac{1}{x^{1/3}y^{1/4}}$

(c) $\left(\dfrac{9x^2y^{1/3}}{x^{1/3}y}\right)^{1/2} = \left(\dfrac{9x^{2-(1/3)}}{y^{1-(1/3)}}\right)^{1/2} = \left(\dfrac{9x^{5/3}}{y^{2/3}}\right)^{1/2} = \dfrac{9^{1/2}(x^{5/3})^{1/2}}{(y^{2/3})^{1/2}} = \dfrac{3x^{5/6}}{y^{1/3}}$

(d) $$\frac{(2x+5)^{1/3}(2x+5)^{-1/2}}{(2x+5)^{-3/4}} = (2x+5)^{(1/3)-(1/2)-(-3/4)}$$
$$= (2x+5)^{(4-6+9)/12} = (2x+5)^{7/12}$$ ■

The next three examples illustrate some algebra you will need to know for certain calculus problems.

Example 5 Write the expression below as a single quotient in which only positive exponents appear.

$$(x^2+1)^{1/2} + x \cdot \frac{1}{2}(x^2+1)^{-1/2} \cdot 2x$$

Solution $$(x^2+1)^{1/2} + x \cdot \frac{1}{\not 2}(x^2+1)^{-1/2} \cdot \not 2x$$
$$= (x^2+1)^{1/2} + \frac{x^2}{(x^2+1)^{1/2}} = \frac{(x^2+1)^{1/2}(x^2+1)^{1/2} + x^2}{(x^2+1)^{1/2}}$$
$$= \frac{(x^2+1)+x^2}{(x^2+1)^{1/2}} = \frac{2x^2+1}{(x^2+1)^{1/2}}$$

We can also solve this problem by factoring out the common term with the smallest negative power, namely, $(x^2+1)^{-1/2}$:

$$(x^2+1)^{1/2} + x \cdot \frac{1}{\not 2}(x^2+1)^{-1/2} \cdot \not 2x = (x^2+1)^{-1/2}[(x^2+1)+x^2]$$
$$= (x^2+1)^{-1/2}[2x^2+1] = \frac{2x^2+1}{(x^2+1)^{1/2}}$$ ■

Example 6 Write the expression below as a single quotient in which only positive exponents appear.

$$\frac{(x^2+4)^{1/2} \cdot 2x - x^2 \cdot \frac{1}{2}(x^2+4)^{-1/2} \cdot 2x}{x^2+4}$$

Solution $$\frac{(x^2+4)^{1/2} \cdot 2x - x^2 \cdot \frac{1}{\not 2}(x^2+4)^{-1/2} \cdot \not 2x}{x^2+4}$$
$$= \frac{2x(x^2+4)^{1/2} - x^3(x^2+4)^{-1/2}}{x^2+4} = \frac{2x(x^2+4)^{1/2} - \dfrac{x^3}{(x^2+4)^{1/2}}}{x^2+4}$$
$$= \frac{\dfrac{2x(x^2+4)^{1/2}(x^2+4)^{1/2} - x^3}{(x^2+4)^{1/2}}}{x^2+4} = \frac{2x(x^2+4) - x^3}{(x^2+4)^{1/2}} \cdot \frac{1}{(x^2+4)}$$
$$= \frac{2x^3+8x-x^3}{(x^2+4)^{3/2}} = \frac{x^3+8x}{(x^2+4)^{3/2}} = \frac{x(x^2+8)}{(x^2+4)^{3/2}}$$ ■

Example 7 Factor: $4x^{1/3}(2x + 1) + 2x^{4/3}$

Solution We begin by looking for factors that are common to the two terms. Notice that 2 and $x^{1/3}$ are common factors. Thus,

$$\begin{aligned} 4x^{1/3}(2x + 1) + 2x^{4/3} &= 2x^{1/3}[2(2x + 1) + x] \\ &= 2x^{1/3}(5x + 2) \end{aligned}$$

EXERCISE 1.8

In Problems 1–20, simplify each expression.

1. $8^{2/3}$ **2.** $4^{3/2}$ **3.** $(-27)^{1/3}$ **4.** $16^{3/4}$ **5.** $16^{3/2}$

6. $64^{3/2}$ **7.** $9^{-3/2}$ **8.** $25^{-5/2}$ **9.** $\left(\frac{9}{8}\right)^{3/2}$ **10.** $\left(\frac{27}{8}\right)^{2/3}$

11. $\left(\frac{8}{9}\right)^{-3/2}$ **12.** $\left(\frac{8}{27}\right)^{-2/3}$ **13.** $4^{1.5}$ **14.** $16^{-1.5}$ **15.** $\left(\frac{1}{4}\right)^{-1.5}$

16. $\left(\frac{1}{9}\right)^{1.5}$ **17.** $(\sqrt{3})^6$ **18.** $(\sqrt[3]{4})^6$ **19.** $(\sqrt{2})^{-2}$ **20.** $(\sqrt[4]{5})^{-8}$

In Problems 21–30, simplify each expression. Express your answer so that only positive exponents occur. Assume that the variables are positive.

21. $x^{3/4}x^{1/3}x^{-1/2}$ **22.** $x^{2/3}x^{1/2}x^{-1/4}$ **23.** $(x^3y^6)^{1/3}$ **24.** $(x^4y^8)^{3/4}$

25. $(x^2y)^{1/3}(xy^2)^{2/3}$ **26.** $(xy)^{1/4}(x^2y^2)^{1/2}$ **27.** $(16x^2y^{-1/3})^{3/4}$ **28.** $(4x^{-1}y^{1/3})^{3/2}$

29. $\left(\frac{x^{2/5}y^{-1/5}}{x^{-1/3}y^{2/3}}\right)^{15}$ **30.** $\left(\frac{x^{1/2}}{y^2}\right)^4\left(\frac{y^{1/3}}{x^{-2/3}}\right)^3$

In Problems 31–44, write each expression as a single quotient in which only positive exponents and/or radicals appear.

31. $\frac{x}{(1 + x)^{1/2}} + 2(1 + x)^{1/2}$ **32.** $\frac{1 + x}{2x^{1/2}} + x^{1/2}$

33. $2x(x^2 + 1)^{1/2} + x^2 \cdot \frac{1}{2}(x^2 + 1)^{-1/2} \cdot 2x$ **34.** $(x + 1)^{1/3} + x \cdot \frac{1}{3}(x + 1)^{-2/3}$

35. $\sqrt{4x + 3} \cdot \frac{1}{2\sqrt{x - 5}} + \sqrt{x - 5} \cdot \frac{1}{5\sqrt{4x + 3}}$

36. $\frac{\sqrt[3]{8x + 1}}{3\sqrt[3]{(x - 2)^2}} + \frac{\sqrt[3]{x - 2}}{24\sqrt[3]{(8x + 1)^2}}$

37. $\frac{\sqrt{1 + x} - x \cdot \frac{1}{2\sqrt{1 + x}}}{1 + x}$

38. $\frac{\sqrt{x^2 + 1} - x \cdot \frac{2x}{2\sqrt{x^2 + 1}}}{x^2 + 1}$

39. $\dfrac{(x+4)^{1/2} - 2x(x+4)^{-1/2}}{x+4}$

40. $\dfrac{(9-x^2)^{1/2} + x^2(9-x^2)^{-1/2}}{9-x^2}$

41. $\dfrac{\dfrac{x^2}{(x^2-1)^{1/2}} - (x^2-1)^{1/2}}{x^2}$

42. $\dfrac{(x^2+4)^{1/2} - x^2(x^2+4)^{-1/2}}{x^2+4}$

43. $\dfrac{\dfrac{1+x^2}{2\sqrt{x}} - 2x\sqrt{x}}{(1+x^2)^2}$

44. $\dfrac{2x(1-x^2)^{1/3} + \frac{2}{3}x^3(1-x^2)^{-2/3}}{(1-x^2)^{2/3}}$

In Problems 45–52, factor each expression.

45. $(x+1)^{3/2} + x \cdot \frac{3}{2}(x+1)^{1/2}$

46. $(x^2+4)^{4/3} + x \cdot \frac{4}{3}(x^2+4)^{1/3} \cdot 2x$

47. $6x^{1/2}(x^2+x) - 8x^{3/2} - 8x^{1/2}$

48. $6x^{1/2}(2x+3) + x^{3/2} \cdot 8$

49. $3(x^2+4)^{4/3} + x \cdot 4(x^2+4)^{1/3} \cdot 2x$

50. $2x(3x+4)^{4/3} + x^2 \cdot 4(3x+4)^{1/3}$

51. $4(3x+5)^{1/3}(2x+3)^{3/2} + 3(3x+5)^{4/3}(2x+3)^{1/2}$

52. $6(6x+1)^{1/3}(4x-3)^{3/2} + 6(6x+1)^{4/3}(4x-3)^{1/2}$

53. If $u = \dfrac{1}{2}\left(x^3 - \dfrac{1}{x^3}\right)$, show that $\sqrt{1+u^2} = \dfrac{1}{2}\left(x^3 + \dfrac{1}{x^3}\right)$.

54. If $u = 2\sqrt{3x}\sqrt{3x^2+1}$, show that $1 + u^2 = (6x^2+1)^2$.

55. If $u = \dfrac{1}{2}\left(x^2 - \dfrac{1}{x^2}\right)$, show that $1 + u^2 = \dfrac{1}{4}\left(x^2 + \dfrac{1}{x^2}\right)^2$.

1.9 Complex Numbers

One of the properties of a real number is that its square is nonnegative. For example, there is no real number x for which

$$x^2 = -1$$

To remedy this situation, we introduce a number called the **imaginary unit**, which we denote by $\boldsymbol{i}$ and whose square is -1. Thus,

$$i^2 = -1$$

This should not surprise you. If our universe were to consist only of integers, there would be no number x for which $2x = 1$. This unfortunate circumstance was remedied by introducing numbers such as $\frac{1}{2}$, $\frac{2}{3}$, etc.—the *rational numbers*. If our universe were to consist only of rational numbers, there would be no number x whose square equals 2. That is, there would be no number x for which $x^2 = 2$. To remedy this, we introduced numbers such as $\sqrt{2}$, $\sqrt[3]{5}$, etc.—the *irrational numbers*. The *real numbers*, you will recall, consist of the rational numbers and the irrational numbers. Now, if our universe were to consist only of real numbers, then there would be no number x whose square is -1. To remedy this, we introduce a number i, whose square is -1.

In the progression outlined above, each time we encountered a situation that was unsuitable, we introduced a new number system to remedy this situation. And each new number system contained the earlier number system as a subset. The number system that results from introducing the number i is called the **complex number system**. See Figure 11.

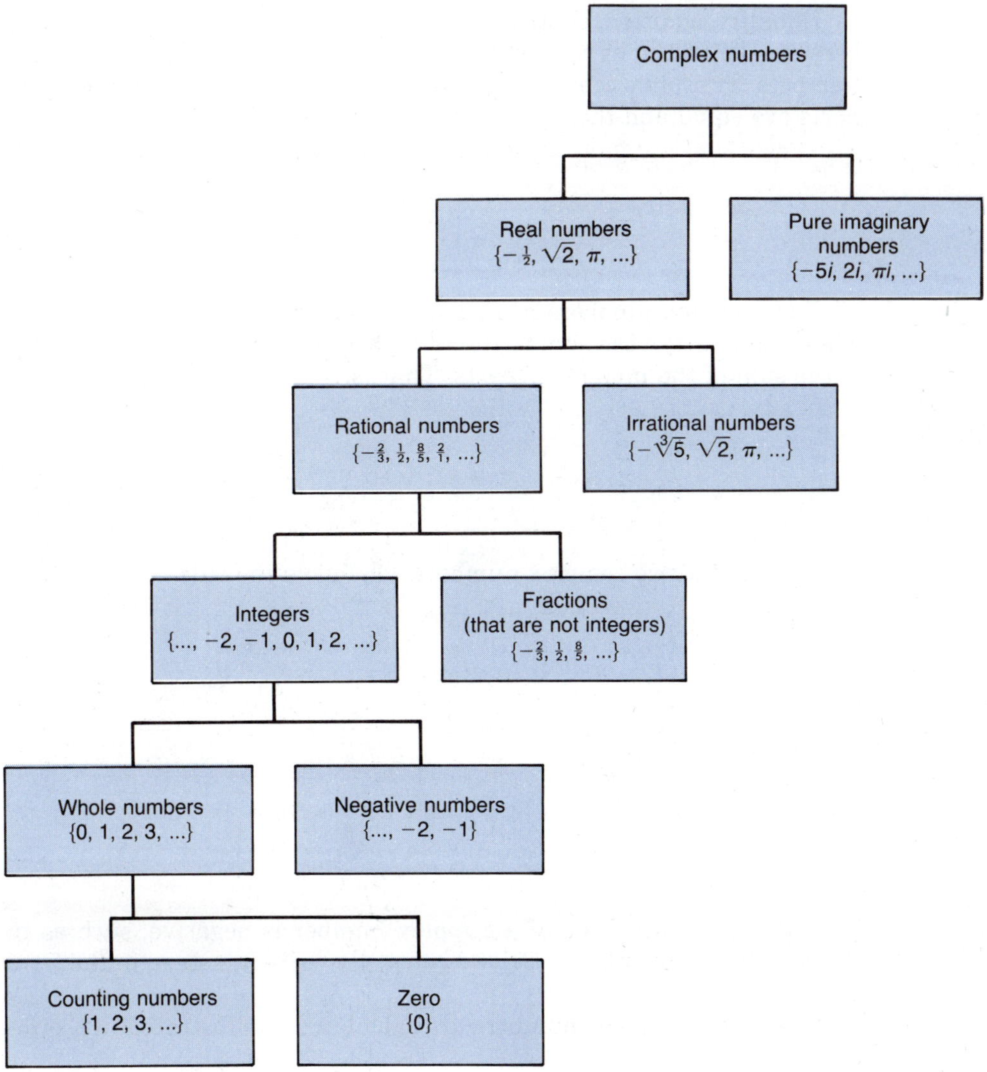

Figure 11

Complex Numbers

Complex numbers are numbers of the form $\mathbf{a + bi}$, where a and b are real numbers. The real number a is called the **real part** of the number $a + bi$; the real number b is called the **imaginary part** of $a + bi$.

For example, the complex number $-5 + 6i$ has the real part -5 and the imaginary part 6.

The complex number $a + 0i$ is usually written merely as a. This serves to remind us that the real numbers are a subset of the complex numbers. The complex number $0 + bi$ is usually written as bi. Sometimes the complex number bi is called a **pure imaginary number**.

Equality, addition, subtraction, and multiplication of complex numbers are defined so as to preserve the familiar rules of algebra for real numbers. Thus, two complex numbers are equal if and only if their real parts are equal and their imaginary parts are equal. That is,

$$a + bi = c + di \quad \text{if and only if} \quad a = c \text{ and } b = d \tag{1}$$

Two complex numbers are added by forming the complex number whose real part is the sum of the real parts and whose imaginary part is the sum of the imaginary parts. That is,

Sum of Complex Numbers

$$(a + bi) + (c + di) = (a + c) + (b + d)i \tag{2}$$

To subtract two complex numbers, we follow the rule

Difference of Complex Numbers

$$(a + bi) - (c + di) = (a - c) + (b - d)i \tag{3}$$

Example 1

(a) $(3 + 5i) + (-2 + 3i) = [3 + (-2)] + (5 + 3)i = 1 + 8i$

(b) $(6 + 4i) - (3 + 6i) = (6 - 3) + (4 - 6)i = 3 + (-2)i$ ■

If the imaginary part of a complex number is negative, such as in the complex number $3 + (-2)i$, we generally write it instead in the form $3 - 2i$.

Products of complex numbers are calculated as illustrated in Example 2.

Example 2 $(5 + 3i) \cdot (2 + 7i) \underset{\uparrow}{=} 5 \cdot (2 + 7i) + 3i(2 + 7i) \underset{\uparrow}{=} 10 + 35i + 6i + 21i^2$

Distributive property Distributive property

$$\underset{\uparrow}{=} 10 + 41i + 21(-1)$$

$i^2 = -1$

$$= -11 + 41i$$ ■

Based on the procedure of Example 2, we define the **product** of two complex numbers by the formula

Product of Complex Numbers

$$(a + bi) \cdot (c + di) = (ac - bd) + (ad + bc)i \qquad (4)$$

Do not bother to memorize formula (4). Instead, whenever it is necessary to multiply two complex numbers, follow the usual rules for multiplying two binomials, as in Example 2, remembering that $i^2 = -1$. For example,

$$(2i)(2i) = 4i^2 = -4 \qquad (2 + i)(1 - i) = 2 - 2i + i - i^2 = 3 - i$$

Algebraic properties for addition and multiplication, such as the commutative, associative, and distributive properties, and so on, hold for complex numbers. Of these, the property that every nonzero complex number has a multiplicative inverse or a reciprocal requires a closer look.

Conjugates

Conjugate If $z = a + bi$ is a complex number, then its **conjugate**, denoted by $\bar{z}$, is defined as

$$\bar{z} = \overline{a + bi} = a - bi$$

For example, $\overline{2 + 3i} = 2 - 3i$ and $\overline{-6 - 2i} = -6 + 2i$.

Example 3 If $z = 3 + 4i$, find $z\bar{z}$.

Solution Since $\bar{z} = 3 - 4i$, we have

$$z\bar{z} = (3 + 4i)(3 - 4i) = 9 + 12i - 12i - 16i^2 = 9 + 16 = 25$$ ■

The result obtained in Example 3 has an important generalization:

Theorem The product of a complex number and its conjugate is a nonnegative real number. Thus, if $z = a + bi$,

$$z\bar{z} = a^2 + b^2 \tag{5}$$

Proof If $z = a + bi$, then

$$z\bar{z} = (a + bi)(a - bi) = a^2 - (bi)^2 = a^2 - b^2i^2 = a^2 + b^2$$ ■

When a complex number is written in the form $a + bi$, where a and b are real numbers, we say it is in **standard form**. To express the reciprocal of a nonzero complex number z in standard form, multiply the numerator and denominator by its conjugate $\bar{z}$. Thus, if $z = a + bi$ is a nonzero complex number, then

$$\frac{1}{a + bi} = \frac{1}{z} = \frac{1}{z} \cdot \frac{\bar{z}}{\bar{z}} = \frac{\bar{z}}{z\bar{z}} = \frac{a - bi}{(a + bi)(a - bi)}$$

$$= \frac{a - bi}{a^2 + b^2}$$

↑ Use (5).

$$= \frac{a}{a^2 + b^2} - \frac{b}{a^2 + b^2}i$$

Example 4 Write $\frac{1}{3 + 4i}$ in standard form $a + bi$; that is, find the reciprocal of $3 + 4i$.

Solution The idea is to multiply the numerator and denominator by the conjugate of $3 + 4i$, namely, the complex number $3 - 4i$. The result is

$$\frac{1}{3 + 4i} = \frac{1}{3 + 4i} \cdot \frac{3 - 4i}{3 - 4i}$$

$$= \frac{3 - 4i}{9 + 16} = \frac{3}{25} - \frac{4}{25}i$$ ■

To express the quotient of two complex numbers in standard form, we multiply the numerator and denominator of the quotient by the conjugate of the denominator.

Example 5 Write each of the following in standard form:

(a) $\frac{1 + 4i}{5 - 12i}$ (b) $\frac{2 - 3i}{4 - 3i}$

Solution (a) $\dfrac{1+4i}{5-12i} = \dfrac{1+4i}{5-12i} \cdot \dfrac{5+12i}{5+12i} = \dfrac{5+20i+12i+48i^2}{25+144}$

$$= \frac{-43+32i}{169} = \frac{-43}{169} + \frac{32}{169}i$$

(b) $\dfrac{2-3i}{4-3i} = \dfrac{2-3i}{4-3i} \cdot \dfrac{4+3i}{4+3i} = \dfrac{8-12i+6i-9i^2}{16+9}$

$$= \frac{17-6i}{25} = \frac{17}{25} - \frac{6}{25}i$$ ■

Example 6 If $z = 2 - 3i$ and $w = 5 + 2i$, write each of the following expressions in standard form:

(a) $\dfrac{z}{w}$ (b) $\overline{z+w}$ (c) $z + \bar{z}$

Solution (a) $\dfrac{z}{w} = \dfrac{z \cdot \bar{w}}{w \cdot \bar{w}} = \dfrac{(2-3i)(5-2i)}{(5+2i)(5-2i)} = \dfrac{10-15i-4i+6i^2}{25+4}$

$$= \frac{4-19i}{29} = \frac{4}{29} - \frac{19}{29}i$$

(b) $\overline{z+w} = \overline{(2-3i)+(5+2i)} = \overline{7-i} = 7+i$

(c) $z + \bar{z} = (2-3i) + (2+3i) = 4$ ■

The conjugate of a complex number has certain general properties that we shall find useful later.

For a real number $a = a + 0i$, the conjugate is $\bar{a} = \overline{a+0i} = a - 0i = a$. That is:

Theorem The conjugate of a real number is the real number itself. ■

Other properties of the conjugate that are direct consequences of the definition are listed below. In each statement, z and w represent complex numbers.

Theorem The conjugate of the conjugate of a complex number is the complex number itself:

$$(\bar{\bar{z}}) = z \tag{6}$$

The conjugate of the sum of two complex numbers equals the sum of their conjugates:

$$\overline{z+w} = \bar{z} + \bar{w} \tag{7}$$

The conjugate of the product of two complex numbers equals the product of their conjugates:

$$\overline{z \cdot w} = \bar{z} \cdot \overline{w} \tag{8}$$

We leave the proofs of equations (6), (7), and (8) as exercises.

Powers of i

The **powers of** i follow a pattern that is useful to know:

$i^1 = i$	$i^5 = i^4 \cdot i = 1 \cdot i = i$
$i^2 = -1$	$i^6 = i^4 \cdot i^2 = -1$
$i^3 = i^2 \cdot i = -i$	$i^7 = i^4 \cdot i^3 = -i$
$i^4 = i^2 \cdot i^2 = (-1)(-1) = 1$	$i^8 = i^4 \cdot i^4 = 1$

And so on. Thus, the powers of i repeat with every fourth power.

Example 7 (a) $i^{27} = i^{24} \cdot i^3 = (i^4)^6 \cdot i^3 = 1^6 \cdot i^3 = -i$

(b) $i^{101} = i^{100} \cdot i^1 = (i^4)^{25} \cdot i = 1^{25} \cdot i = i$

Example 8 Write $(2 + i)^3$ in standard form.

Solution We use the special product formula for $(x + a)^3$, namely,

$$(x + a)^3 = x^3 + 3ax^2 + 3a^2x + a^3$$

Thus,

$$\begin{aligned}(2 + i)^3 &= 2^3 + 3 \cdot i \cdot 2^2 + 3 \cdot i^2 \cdot 2 + i^3\\ &= 8 + 12i + 6(-1) + (-i)\\ &= 2 + 11i\end{aligned}$$

EXERCISE 1.9

In Problems 1–44, write each expression in the standard form $a + bi$.

1. $(2 - 3i) + (6 + 8i)$ **2.** $(4 + 5i) + (-8 + 2i)$ **3.** $(-3 + 2i) - (4 - 4i)$

4. $(3 - 4i) - (-3 - 4i)$ **5.** $(2 - 5i) - (8 + 6i)$ **6.** $(-8 + 4i) - (2 - 2i)$

7. $3(2 - 6i)$ **8.** $-4(2 + 8i)$ **9.** $2i(2 - 3i)$

10. $3i(-3 + 4i)$ **11.** $(3 - 4i)(2 + i)$ **12.** $(5 + 3i)(2 - i)$

13. $(-6 + i)(-6 - i)$ **14.** $(-3 + i)(3 + i)$ **15.** $\dfrac{10}{3 - 4i}$

16. $\dfrac{13}{5 - 12i}$ **17.** $\dfrac{2 + i}{i}$ **18.** $\dfrac{2 - i}{-2i}$

19. $\dfrac{6 - i}{1 + i}$ **20.** $\dfrac{2 + 3i}{1 - i}$ **21.** $\left(\dfrac{1}{2} + \dfrac{\sqrt{3}}{2}i\right)^2$

22. $\left(\dfrac{\sqrt{3}}{2} - \dfrac{1}{2}i\right)^2$ **23.** $(1 + i)^2$ **24.** $(1 - i)^2$

25. $\dfrac{(2 - i)^2}{2 + i}$ **26.** $\dfrac{1 + 2i}{(1 - 2i)^2}$ **27.** $\dfrac{1}{(3 - i)(2 + i)}$

28. $\dfrac{3}{(2 + 3i)(2 - 3i)}$ **29.** i^{23} **30.** i^{14}

31. i^{-15} **32.** i^{-23} **33.** $i^6 - 5$

34. $4 + i^3$ **35.** $6i^3 - 4i^5$ **36.** $4i^3 - 2i^2 + 1$

37. $(1 + i)^3$ **38.** $(3i)^4 + 1$ **39.** $i^7(1 + i^2)$

40. $2i^4(1 + i^2)$ **41.** $\dfrac{i^4 + i^3 + i^2 + i + 1}{i + 1}$ **42.** $\dfrac{1 - i + i^2 - i^3 + i^4}{1 - i}$

43. $i^6 + i^4 + i^2 + 1$ **44.** $i^7 + i^5 + i^3 + i$

In Problems 45–52, $z = 3 - 4i$ and $w = 8 + 3i$. Write each expression in the standard form $a + bi$.

45. $z + \bar{z}$ **46.** $w - \bar{w}$ **47.** $z\bar{z}$ **48.** $\overline{z - w}$

49. $\overline{z \cdot w}$ **50.** $\bar{z} \cdot \bar{w}$ **51.** $\overline{\bar{z} + w}$ **52.** $\overline{z\bar{w}}$

53. Use $z = a + bi$ to show that $z + \bar{z} = 2a$ and that $z - \bar{z} = 2bi$.

54. Use $z = a + bi$ to show that $(\bar{\bar{z}}) = z$.

55. Use $z = a + bi$ and $w = c + di$ to show that $\overline{z + w} = \bar{z} + \bar{w}$.

56. Use $z = a + bi$ and $w = c + di$ to show that $\overline{z \cdot w} = \bar{z} \cdot \bar{w}$.

1.10 Geometry Topics

In this section we review some topics studied in geometry that we shall need for our study of algebra.

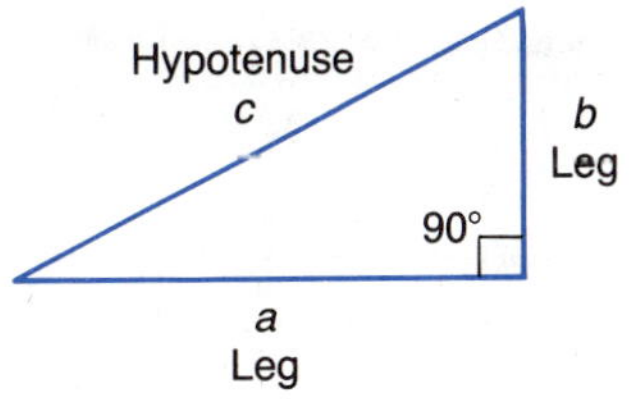

Figure 12

Pythagorean Theorem

The *Pythagorean Theorem* is a statement about *right triangles*. A **right triangle** is one that contains a **right angle**—that is, an angle of 90°. The side of the triangle opposite the 90° angle is called the **hypotenuse**; the remaining two sides are called **legs**. In Figure 12 we have used c to represent the length of the hypotenuse and a and b to represent the lengths of the legs. Notice the use of the symbol ⌐ to show the 90° angle. We now state the Pythagorean Theorem.

Theorem

Pythagorean Theorem

In a right triangle, the square of the length of the hypotenuse is equal to the sum of the squares of the lengths of the legs. That is, in the right triangle shown in Figure 12,

$$c^2 = a^2 + b^2 \tag{1}$$

■

We shall prove this result at the end of this section.

Example 1 In a right triangle, one leg is of length 4 and the other is of length 3. What is the length of the hypotenuse?

Solution Since the triangle is a right triangle, we use the Pythagorean Theorem with $a = 4$ and $b = 3$ to find the length c of the hypotenuse. Thus, from equation (1), we have

$$c^2 = a^2 + b^2$$
$$c^2 = 4^2 + 3^2 = 16 + 9 = 25$$
$$c = 5$$

■

The converse of the Pythagorean Theorem is also true.

Theorem

Converse of the Pythagorean Theorem

In a triangle, if the square of the length of one side equals the sum of the squares of the lengths of the other two sides, then the triangle is a right triangle. The 90° angle is opposite the longest side. ■

We shall prove this result at the end of this section.

Example 2 Show that a triangle whose sides are of lengths 5, 12, and 13 is a right triangle. Identify the hypotenuse.

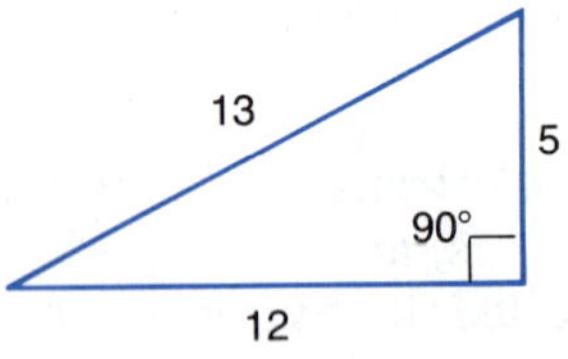

Figure 13

Solution We square the lengths of the sides:

$$25, \quad 144, \quad 169$$

Notice that the sum of the first two squares (25 and 144) equals the third square (169). Hence, the triangle is a right triangle. The longest side, 13, is the hypotenuse. See Figure 13. ■

Geometry Formulas

Certain formulas from geometry are useful in solving algebra problems. We list some of these formulas below.

For a rectangle of length l and width w:

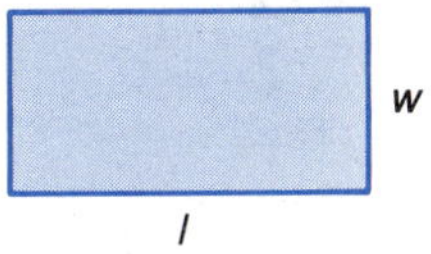

$$\text{Area} = lw \qquad \text{Perimeter} = 2l + 2w$$

For a triangle with base b and altitude h:

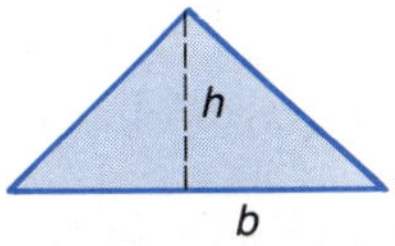

$$\text{Area} = \tfrac{1}{2}bh$$

For a circle of radius r:

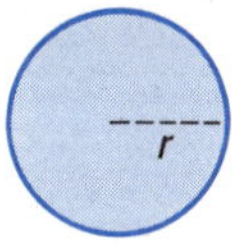

$$\text{Area} = \pi r^2 \qquad \text{Circumference} = 2\pi r$$

For a rectangular box of length l, width w, and height h:

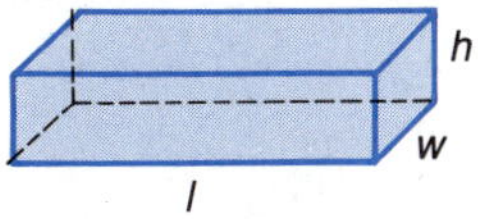

$$\text{Volume} = lwh$$

Proof of the Pythagorean Theorem

We begin with a square, each side of length $a + b$. In this square, we can form four right triangles, each having legs equal in length to a and b. See Figure 14. All of these triangles are **congruent** (two sides and

Figure 14

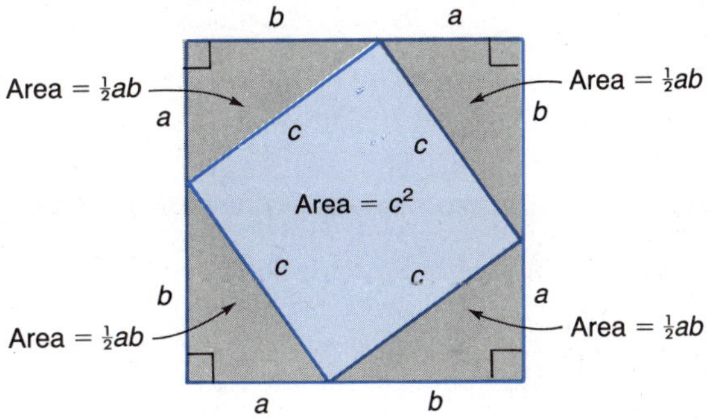

their included angle are equal). As a result, the hypotenuse of each is the same, say c, and the color shading in Figure 14 indicates a square with area equal to c^2. The area of the original square with side $a + b$ equals the sum of the areas of the four triangles (each of area $\frac{1}{2}ab$) plus the area of the square with side c. Thus,

$$(a + b)^2 = \tfrac{1}{2}ab + \tfrac{1}{2}ab + \tfrac{1}{2}ab + \tfrac{1}{2}ab + c^2$$
$$a^2 + 2ab + b^2 = 2ab + c^2$$
$$a^2 + b^2 = c^2$$

The proof is complete. ■

Proof of the Converse of the Pythagorean Theorem

We begin with two triangles: one a right triangle with legs a and b; the other a triangle with sides a, b, and c for which $c^2 = a^2 + b^2$ (see Figure 15). By the Pythagorean Theorem, the length x of the third side of the first triangle is

$$x^2 = a^2 + b^2$$

But $c^2 = a^2 + b^2$. Hence,

$$x^2 = c^2$$
$$x = c$$

The two triangles have the same sides and are therefore congruent; hence, corresponding angles are equal. Thus, the angle opposite side c of the second triangle equals 90°.

The proof is complete. ■

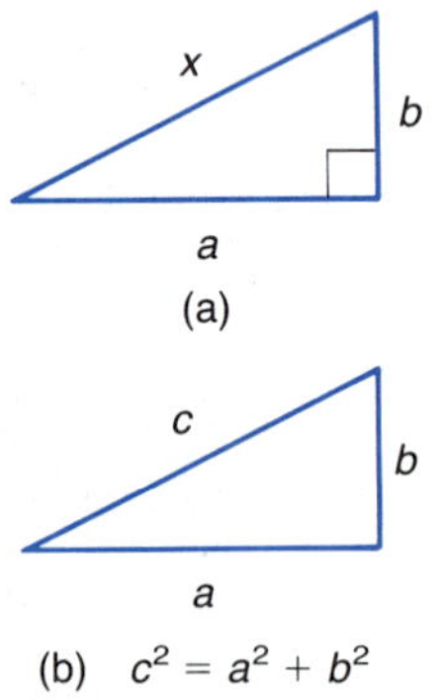

Figure 15

EXERCISE 1.10

In Problems 1–6, the lengths of the legs of a right triangle are given. Find the hypotenuse.

1. $a = 5, \quad b = 12$ **2.** $a = 6, \quad b = 8$ **3.** $a = 10, \quad b = 24$
4. $a = 4, \quad b = 3$ **5.** $a = 7, \quad b = 24$ **6.** $a = 14, \quad b = 48$

In Problems 7–14, the lengths of a triangle are given. Determine which are right triangles. For those that are, identify the hypotenuse.

7. 3, 4, 5 **8.** 6, 8, 10 **9.** 4, 5, 6 **10.** 2, 2, 3
11. 7, 24, 25 **12.** 10, 24, 26 **13.** 6, 4, 3 **14.** 5, 4, 7

C **15.** The tallest inhabited building is the Sears Tower in Chicago.* If the observation tower is 1454 feet above ground level, use the figure at the top of the next page to determine how far a person standing in the observation tower can see (with the aid of a telescope). Use 3960 miles for the radius of the Earth. [*Note:* 1 mile = 5280 feet]

Source: Guinness Book of World Records

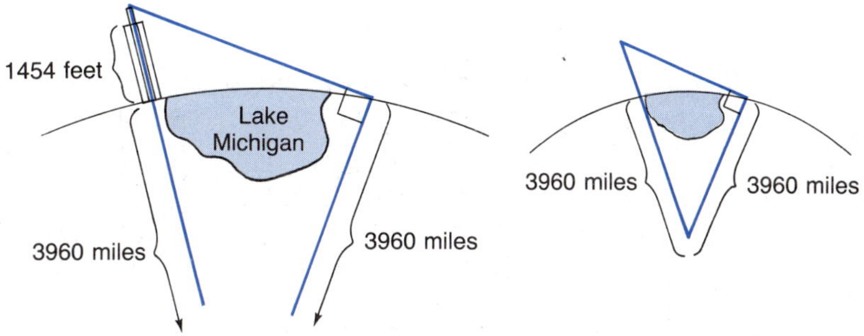

C **16.** The conning tower of the USS *Silversides*, a World War II submarine now permanently stationed in Muskegon, Michigan, is approximately 20 feet above sea level. How far can one see from the conning tower?

C **17.** A person who is 6 feet tall is standing on the beach in Fort Lauderdale, Florida, and looks out onto the Atlantic Ocean. Suddenly, a ship appears on the horizon. How far is the ship from shore?

C **18.** The deck of a destroyer is 100 feet above sea level. How far can a person see from the deck? How far can a person see from the bridge, which is 150 feet above sea level?

19. Suppose m and n are positive integers with $m > n$. If $a = m^2 - n^2$, $b = 2mn$, and $c = m^2 + n^2$, show that a, b, and c are the lengths of the sides of a right triangle. [This formula can be used to find the sides of a right triangle that are integers, such as 3, 4, 5; 5, 12, 13; and so on. Such triplets of integers are called *Pythagorean triples*.]

CHAPTER REVIEW

THINGS TO KNOW

Classification of numbers	Counting numbers	1, 2, 3 . . .
	Whole numbers	0, 1, 2, 3, . . .
	Integers	. . . , −2, −1, 0, 1, 2, . . .
	Rational numbers	Quotients of two integers (denominators not equal to 0); terminating or infinite, repeating decimals
	Irrational numbers	Infinite, nonrepeating decimals
	Real numbers	Rational or irrational numbers
	Complex numbers	Numbers of the form $a + bi$, where a, b are real numbers, $i^2 = -1$
Properties of real numbers	Commutative properties	$a + b = b + a, \quad a \cdot b = b \cdot a$
	Associative properties	$a + (b + c) = (a + b) + c$, $a \cdot (b \cdot c) = (a \cdot b) \cdot c$
	Distributive property	$a \cdot (b + c) = a \cdot b + a \cdot c$

Identity properties	$a + 0 = a, \quad a \cdot 1 = a$
Inverse properties	$a + (-a) = 0, \quad a \cdot \frac{1}{a} = 1$, where $a \neq 0$
Product law	If $ab = 0$, then $a = 0$ or $b = 0$ or both.

Absolute value: $|x| = x$ if $x \geq 0$, $\quad |x| = -x$ if $x < 0$

Exponents: $a^n = \underbrace{a \cdot a \cdot \cdots \cdot a}_{n \text{ factors}}$, n a positive integer

$a^0 = 1, a \neq 0$

$a^{-n} = \frac{1}{a^n}$, $a \neq 0$, n a positive integer

Laws of exponents: $a^m \cdot a^n = a^{m+n}, \quad (a^m)^n = a^{mn}, \quad (a \cdot b)^n = a^n \cdot b^n$

$\frac{a^m}{a^n} = a^{m-n} = \frac{1}{a^{n-m}}, \quad \left(\frac{a}{b}\right)^n = \frac{a^n}{b^n}$

Polynomial: Algebraic expression of the form

$a_nx^n + a_{n-1}x^{n-1} + \cdots + a_1x + a_0$, n a nonnegative integer

Special products/factoring formulas:

$(x - a)(x + a) = x^2 - a^2$

$(x + a)^2 = x^2 + 2ax + a^2, \quad (x - a)^2 = x^2 - 2ax + a^2$

$(x + a)(x + b) = x^2 + (a + b)x + ab$

$(ax + b)(cx + d) = acx^2 + (ad + bc)x + bd$

$(x + a)^3 = x^3 + 3ax^2 + 3a^2x + a^3, \quad (x - a)^3 = x^3 - 3ax^2 + 3a^2x - a^3$

$(x - a)(x^2 + ax + a^2) = x^3 - a^3, \quad (x + a)(x^2 - ax + a^2) = x^3 + a^3$

Radical: $\sqrt[n]{a} = b$ means $b^n = a$, where $b \geq 0$ if n is even

$a^{m/n} = \sqrt[n]{a^m} = (\sqrt[n]{a})^m$

Pythagorean Theorem: In a right triangle, the square of the length of the hypotenuse is equal to the sum of the squares of the lengths of the legs.

Converse of Pythagorean Theorem: In a triangle, if the square of the length of one side equals the sum of the squares of the lengths of the other two sides, then the triangle is a right triangle.

How To:

Add, subtract, multiply, divide, and factor polynomials

Add, subtract, multiply, divide, and simplify rational expressions

Rationalize denominators of quotients containing radicals

FILL-IN-THE-BLANK ITEMS

1. The __________ property of equality states that if $a = b$, then $b = a$.
2. The __________ properties state that $a + b = b + a$ and $ab = ba$.

3. The __________ property states that $a \cdot (b + c) = a \cdot b + a \cdot c$.
4. In the expression 2^4, the number 2 is called the __________, and 4 is called the __________.
5. The polynomial $5x^3 - 3x^2 + 2x - 4$ is of degree __________; the __________ coefficient is 5.
6. In the expression $\sqrt[3]{25}$, the number 3 is called the __________; the number 25 is called the __________.
7. In the complex number $5 + 2i$, the number 5 is called the __________ part; the number 2 is called the __________ part; and the number i is called the __________ __________.

REVIEW EXERCISES

In Problems 1–20, perform the indicated operations.

1. $3 - 4 \cdot 5 + 1$
2. $8 + 4 \cdot 2 - 1$
3. $\frac{3}{4} - \frac{7}{12}$
4. $\frac{9}{8} - \frac{3}{4}$
5. $\dfrac{\frac{15}{2} - \frac{1}{4}}{\frac{2}{3}}$
6. $\dfrac{\frac{9}{2} - \frac{3}{4}}{\frac{4}{3}}$
7. $5^2 - 3^3$
8. $2^3 + 3^2 - 1$
9. $\dfrac{2^{-3} \cdot 5^0}{4^2}$
10. $\dfrac{3^{-2} - 4^1}{2^{-2}}$
11. $(2\sqrt{5} - 2)(2\sqrt{5} + 2)$
12. $(2\sqrt{3} + \sqrt{2})(2\sqrt{3} - \sqrt{2})$
13. $\left(\frac{8}{9}\right)^{-2/3}$
14. $\left(\frac{4}{27}\right)^{-3/2}$
15. $(\sqrt[3]{2})^{-3}$
16. $(4\sqrt{32})^{-1/2}$
17. $|5 - 8^{1/3}|$
18. $|9^{1/2} - 27^{2/3}|$
19. $\sqrt{|3^2 - 5^2|}$
20. $\sqrt[3]{-|4^2 - 2^3|}$

In Problems 21–28, write each expression so that all exponents are positive. Assume $x > 0$ and $y > 0$.

21. $\dfrac{x^{-2}}{y^{-2}}$
22. $\left(\dfrac{x^{-1}}{y^{-3}}\right)^2$
23. $\dfrac{(x^2y)^{-4}}{(xy)^{-3}}$
24. $\dfrac{\left(\frac{x}{y}\right)^2}{\left(\frac{x}{y}\right)^{-1}}$
25. $(25x^{-4/3}y^{-2/3})^{3/2}$
26. $(16x^{-2/3}y^{4/3})^{-3/2}$
27. $\left(\dfrac{2x^{-1/2}}{y^{-3/4}}\right)^{-4}$
28. $\left(\dfrac{8x^{-3/2}}{y^{-3}}\right)^{-2/3}$

In Problems 29–38, perform the indicated operations. Express your answer as a polynomial.

29. $(2x - 3)(x^2 - 5x + 2)$
30. $(3x + 4)(2x^2 - 8x - 1)$
31. $4(3x^3 - 2x^2 + 1) - 3(x^3 + 4x^2 - 2x - 3)$
32. $8(1 - x^2 + x^3) - 4(1 + 2x^2 - 4x^4)$
33. $(2x - 5)(3x + 2)$
34. $(1 - 2x)(1 - 4x)$
35. $(x + 1)(x + 2)(x + 3)$
36. $(x^2 - 1)(x + 1)$
37. $(x - 2y)(x + 4y)$
38. $(x - y)^2(x + y)^2$

In Problems 39–52, factor each polynomial completely (over the integers). If the polynomial cannot be factored, say it is prime.

39. $x^2 + 5x - 14$
40. $x^2 + 15x + 14$
41. $6x^2 - 5x - 6$
42. $6x^2 + x - 2$
43. $3x^2 - 15x - 42$
44. $2x^3 + 18x^2 + 28x$

45. $8x^3 + 1$ **46.** $27x^3 - 8$ **47.** $2x^3 + 3x^2 - 2x - 3$

48. $2x^3 + 3x^2 + 2x + 3$ **49.** $25x^2 - 4$ **50.** $16x^2 - 1$

51. $9x^2 + 1$ **52.** $x^2 - x + 1$

In Problems 53–62, perform the indicated operation and simplify.

53. $\dfrac{2x^2 + 11x + 14}{x^2 - 4}$ **54.** $\dfrac{x^2 - 5x - 14}{4 - x^2}$

55. $\dfrac{9x^2 - 1}{x^2 - 9} \cdot \dfrac{3x - 9}{9x^2 + 6x + 1}$ **56.** $\dfrac{x^2 - 25}{x^3 - 4x^2 - 5x} \cdot \dfrac{x^2 + x}{1 - x^2}$

57. $\dfrac{x + 1}{x - 1} - \dfrac{x - 1}{x + 1}$ **58.** $\dfrac{x}{x + 1} - \dfrac{2x}{x + 2}$

59. $\dfrac{3x + 4}{x^2 - 4} - \dfrac{2x - 3}{x^2 + 4x + 4}$ **60.** $\dfrac{x^2}{2x^2 + 5x - 3} + \dfrac{x^2}{2x^2 - 5x + 2}$

61. $1 + \dfrac{1}{1 + \dfrac{1}{x}}$ **62.** $1 - \dfrac{1}{1 + \dfrac{1}{x}}$

In Problems 63–68, rationalize the denominator of each expression.

63. $\dfrac{2}{\sqrt{3}}$ **64.** $\dfrac{-1}{\sqrt{5}}$ **65.** $\dfrac{2}{1 - \sqrt{2}}$

66. $\dfrac{-4}{1 + \sqrt{3}}$ **67.** $\dfrac{1 + \sqrt{5}}{1 - \sqrt{5}}$ **68.** $\dfrac{4\sqrt{3} + 2}{2\sqrt{3} + 1}$

In Problems 69–72, write each expression as a single quotient in which only positive exponents and/or radicals appear.

69. $(2 + x^2)^{1/2} + x \cdot \frac{1}{2}(2 + x^2)^{-1/2} \cdot 2x$ **70.** $(x^2 + 4)^{2/3} + x \cdot \frac{2}{3}(x^2 + 4)^{-1/3} \cdot 2x$

71. $\dfrac{(x + 4)^{1/2} \cdot 2x - x^2 \cdot \frac{1}{2}(x + 4)^{-1/2}}{x + 4}$ **72.** $\dfrac{(x^2 + 4)^{1/2} \cdot 2x - x^2 \cdot \frac{1}{2}(x^2 + 4)^{-1/2} \cdot 2x}{x^2 + 4}$

In Problems 73–82, use the complex number system and write each expression in the standard form $a + bi$.

73. $(6 - 3i) - (2 + 4i)$ **74.** $(8 + 3i) + (-6 - 2i)$

75. $4(3 - i) + 3(-5 + 2i)$ **76.** $2(1 + i) - 3(2 - 3i)$

77. $\dfrac{3}{3 + i}$ **78.** $\dfrac{4}{2 - i}$ **79.** i^{68}

80. i^{21} **81.** $(2 + 3i)^3$ **82.** $(3 - 2i)^3$

2 Equations and Inequalities

The investigation of equations and their solutions has played a central role in algebra for many hundreds of years. In fact, as early as 200 BC, the Babylonians had a well-developed algebra that included a solution for quadratic equations (see the Historical Comment in Section 2.3). Of late, the study of inequalities (Sections 2.7 and 2.8) has become equally important. For example, the problem of optimizing airline scheduling leads to linear inequalities. As you go through this chapter, you will see evidence of the importance of the topics presented here for solving problems that occur in business, engineering, and the social and natural sciences.

We shall limit our discussion in this chapter to equations and inequalities containing a single variable. Later, we will study equations and inequalities containing more than one variable.

2.1 Equations

An **equation in one variable** is a statement in which two expressions, at least one containing the variable, are equal. The expressions are called the **sides** of the equation. Since an equation is a statement, it may or may not be true, depending on the value of the variable. Unless otherwise restricted, the admissible values of the variable are those in the domain of the variable. Those admissible values of the variable, if any, that result in a true statement are called **solutions**, or **roots**, of the equation. To **solve an equation** means to find all the solutions of the equation.

For example, the following are all equations in one variable, x:

$$x + 5 = 9 \qquad x^2 + 5x = 2x - 2 \qquad \frac{x^2 - 4}{x + 1} = 0 \qquad x^2 + 9 = 5$$

The first of these statements, $x + 5 = 9$, is true when $x = 4$ and is false for any other choice of x. Thus, 4 is a solution of the equation $x + 5 = 9$. We also say that 4 **satisfies** the equation $x + 5 = 9$, because, when x is replaced by 4, a true statement results.

Sometimes an equation will have more than one solution. For example, the equation

$$\frac{x^2 - 4}{x + 1} = 0$$

has either $x = -2$ or $x = 2$ as a solution.

Sometimes we will write the solutions of an equation in set notation. This set is called the **solution set** of the equation. For example, the solution set of the equation $x^2 - 9 = 0$ is $\{-3, 3\}$.

Unless indicated otherwise, we will limit ourselves to real solutions. Some equations have no real solution. For example, $x^2 + 9 = 5$ has no real solution, because there is no real number whose square when added to 9 equals 5.

An equation that is satisfied for every choice of the variable for which both sides are defined is called an **identity**. For example, the equation

$$3x + 5 = x + 3 + 2x + 2$$

is an identity, because this statement is true for any real number x.

Two or more equations that have precisely the same solutions are called **equivalent equations**. For example, all the following equations are equivalent, because each has only the solution $x = 5$:

$$\begin{aligned} 2x + 3 &= 13 \\ 2x &= 10 \\ x &= 5 \end{aligned}$$

These three equations illustrate one method for solving many types of equations: Replace the original equation by an equivalent one, and continue until an equation with an obvious solution, such as $x = 5$, is reached. The question, though, is: "How do I obtain an equivalent equation?" In general, there are five ways to do so.

Procedures That Result in Equivalent Equations

1. Interchange the two sides of the equation:

   Replace $3 = x$ by $x = 3$

2. Simplify the sides of the equation by combining like terms, eliminating parentheses, and so on:

   Replace $(x + 2) + 6 = 2x + (x + 1)$

   by $x + 8 = 3x + 1$

3. Add or subtract the same expression on both sides of the equation:

   Replace $3x - 5 = 4$

   by $(3x - 5) + 5 = 4 + 5$

4. Multiply or divide both sides of the equation by the same nonzero expression:

   Replace $\dfrac{3x}{x - 1} = \dfrac{6}{x - 1} \qquad x \neq 1$

   by $\dfrac{3x}{x - 1} \cdot (x - 1) = \dfrac{6}{x - 1} \cdot (x - 1)$

5. If the right-hand side of the equation is 0 and the left-hand side can be factored, then we may use the product law* and set each factor equal to 0:

   Replace $x(x - 3) = 0$

   by $x = 0$ or $x - 3 = 0$

Let's look at an elementary example to illustrate these ideas.

*Recall that the product law states that if $AB = 0$, then $A = 0$ or $B = 0$ or both equal 0.

Example 1 Solve the equation: $3x - 5 = 4$

Solution We replace the original equation by a succession of equivalent ones:

$$3x - 5 = 4$$

$$(3x - 5) + 5 = 4 + 5 \quad \text{Add 5 to both sides.}$$

$$3x = 9 \quad \text{Simplify.}$$

$$\frac{3x}{3} = \frac{9}{3} \quad \text{Divide both sides by 3.}$$

$$x = 3 \quad \text{Simplify.}$$

The last equation, $x = 3$, has the single solution 3. All the above equations are equivalent, so 3 is the only solution of the original equation $3x - 5 = 4$.

Check: It is a good practice to check the solution by replacing x by 3 in the original equation:

$$3x - 5 = 4$$

$$3(3) - 5 \stackrel{?}{=} 4$$

$$9 - 5 \stackrel{?}{=} 4$$

$$4 = 4$$

The solution checks. ■

Example 2 Solve the equation: $\dfrac{3x}{x-1} + 2 = \dfrac{3}{x-1}$

Solution First, we note that the domain of the variable is $\{x \mid x \neq 1\}$. Since the two quotients in the equation have the same denominator, $x - 1$, we can simplify by multiplying both sides by $x - 1$. The resulting equation is equivalent to the original one, since we are multiplying by $x - 1$, which is not 0 (remember, $x \neq 1$).

$$\frac{3x}{x-1} + 2 = \frac{3}{x-1}$$

$$\left(\frac{3x}{x-1} + 2\right) \cdot (x - 1) = \frac{3}{\cancel{x-1}} \cdot \cancel{(x-1)}$$

$$\left[\frac{3x}{\cancel{x-1}} \cdot \cancel{(x-1)}\right] + [2 \cdot (x - 1)] = 3$$

$$3x + (2x - 2) = 3$$

$$5x - 2 = 3$$

$$5x = 5$$

$$x = 1$$

The solution appears to be 1. But recall that $x = 1$ is not in the domain of the variable. Thus, the equation has no solution. ■

In the next example, we use the product law (procedure 5, listed in the box on page 87).

Example 3 Solve the equation: $(x + 1)(2x) = (x + 1)(2)$

Solution We begin by collecting all terms on the left side:

$$(x + 1)(2x) = (x + 1)(2)$$
$$(x + 1)(2x) - (x + 1)(2) = 0$$
$$(x + 1)(2x - 2) = 0 \qquad \text{Factor.}$$
$$x + 1 = 0 \quad \text{or} \quad 2x - 2 = 0 \qquad \text{Apply the product law.}$$
$$x = -1 \qquad\qquad 2x = 2$$
$$x = 1$$

The solution set is $\{-1, 1\}$. ■

Steps for Solving Equations

STEP 1: Determine the domain of the variable.
STEP 2: Simplify the equation by replacing the original equation by a succession of equivalent equations following the procedures listed earlier.
STEP 3: If the result of Step 2 is a product of factors equal to 0, use the product law and set each factor equal to 0 (procedure 5).

Linear Equations

Linear Equation in One Variable

A **linear equation in one variable** is an equation equivalent to one of the form

$$ax + b = 0$$

where a and b are real numbers and $a \neq 0$.

Sometimes, a linear equation is called a **first-degree equation**, because the left side is a polynomial in x of degree 1.

It is relatively easy to solve a linear equation:

$ax + b = 0$	
$(ax + b) - b = 0 - b$	Subtract b from both sides.
$ax = -b$	Simplify.
$\dfrac{ax}{a} = \dfrac{-b}{a}$	Divide both sides by a, $a \neq 0$.
$x = \dfrac{-b}{a}$	Simplify.

The linear equation $ax + b = 0$ has the single solution given by the formula $x = -b/a$.

Whenever it is possible to solve a linear equation in your head, do so. For example:

The solution of $2x = 8$ is $x = 4$.

The solution of $3x - 15 = 0$ is $x = 5$.

Often, though, some rearrangement is necessary.

Example 4 Solve the equation: $3(2x - 5) + 24 = 3x$

Solution

$3(2x - 5) + 24 = 3x$	
$6x - 15 + 24 = 3x$	Use the distributive property.
$6x + 9 = 3x$	Combine like terms.
$6x + 9 - 3x = 3x - 3x$	Subtract $3x$ from each side.
$3x + 9 = 0$	Simplify.
$x = -3$	Mental solution ■

Example 5 Solve the equation: $\frac{1}{2}(p + 5) - 3 = \frac{1}{3}(2p - 1)$

Solution To clear the equation of fractions, we multiply both sides by 6, the least common multiple of the denominators of the fractions $\frac{1}{2}$ and $\frac{1}{3}$:

$\frac{1}{2}(p + 5) - 3 = \frac{1}{3}(2p - 1)$	
$6\left[\frac{1}{2}(p + 5) - 3\right] = 6\left[\frac{1}{3}(2p - 1)\right]$	Multiply both sides by 6, the LCM of 2 and 3.
$3(p + 5) - 18 = 2(2p - 1)$	Use the distributive property on the left and the associative property on the right.
$3p + 15 - 18 = 4p - 2$	Use the distributive property.
$3p - 3 = 4p - 2$	Combine like terms.
$3p = 4p + 1$	Add 3 to each side.
$-p = 1$	Subtract $4p$ from each side.
$p = -1$	Multiply both sides by -1. ■

Equations That Lead to Linear Equations

The next two examples illustrate the solution of equations that are not linear but lead to linear equations upon simplification.

Example 6 Solve the equation: $(2y + 1)(y - 1) = (y + 5)(2y - 5)$

Solution

$$\begin{aligned}
(2y + 1)(y - 1) &= (y + 5)(2y - 5) \\
2y^2 - y - 1 &= 2y^2 + 5y - 25 && \text{Multiply, and combine like terms.} \\
-y - 1 &= 5y - 25 && \text{Subtract } 2y^2 \text{ from each side.} \\
-y &= 5y - 24 && \text{Add 1 to each side.} \\
-6y &= -24 && \text{Subtract } 5y \text{ from each side.} \\
y &= 4 && \text{Divide both sides by } -6.
\end{aligned}$$

■

Example 7 Solve the equation: $\dfrac{3}{x - 2} = \dfrac{1}{x - 1} + \dfrac{7}{(x - 1)(x - 2)}$

Solution First, we note that the domain of the variable is $\{x \mid x \neq 1, x \neq 2\}$. As we did in Example 5, we clear the equation of fractions by multiplying both sides by $(x - 1)(x - 2)$, the least common multiple of the denominators of the three fractions:

$$\begin{aligned}
\frac{3}{x - 2} &= \frac{1}{x - 1} + \frac{7}{(x - 1)(x - 2)} \\
(x - 1)\cancel{(x - 2)}\left(\frac{3}{\cancel{x - 2}}\right) &= (x - 1)(x - 2)\left[\frac{1}{x - 1} + \frac{7}{(x - 1)(x - 2)}\right] \\
(x - 1)(3) &= \cancel{(x - 1)}(x - 2)\frac{1}{\cancel{x - 1}} + \cancel{(x - 1)}\cancel{(x - 2)}\frac{7}{\cancel{(x - 1)}\cancel{(x - 2)}} \\
3x - 3 &= (x - 2) + 7 \\
3x - 3 &= x + 5 \\
2x &= 8 \\
x &= 4
\end{aligned}$$

■

Example 8 In the United States we measure temperature both in degrees Fahrenheit (°F) and in degrees Celsius (°C), which are related by the formula $C = \frac{5}{9}(F - 32)$. What are the Fahrenheit temperatures corresponding to Celsius temperatures of 0°, 10°, 20°, and 30°C?

Solution We could solve four equations for F by replacing C each time by 0, 10, 20, and 30. Instead, it is much easier and faster first to solve the equation $C = \frac{5}{9}(F - 32)$ for F and then substitute in the values of C:

$$C = \tfrac{5}{9}(F - 32)$$
$$9C = 5(F - 32) \quad \text{Multiply by 9.}$$
$$9C = 5F - 160 \quad \text{Use the distributive property.}$$
$$5F - 160 = 9C \quad \text{Interchange sides.}$$
$$5F = 9C + 160 \quad \text{Add 160 to each side.}$$
$$F = \tfrac{9}{5}C + 32 \quad \text{Divide both sides by 5.}$$

We can now do the required arithmetic:

$$0°\text{C}: \quad F = \tfrac{9}{5}(0) + 32 = 32°\text{F}$$
$$10°\text{C}: \quad F = \tfrac{9}{5}(10) + 32 = 50°\text{F}$$
$$20°\text{C}: \quad F = \tfrac{9}{5}(20) + 32 = 68°\text{F}$$
$$30°\text{C}: \quad F = \tfrac{9}{5}(30) + 32 = 86°\text{F}$$

■

Summary

Steps for Solving a Linear Equation

To solve a linear equation, follow these steps:

STEP 1: If necessary, clear the equation of fractions by multiplying both sides by the least common multiple (LCM) of the denominators of all the fractions.

STEP 2: Remove all parentheses and simplify.

STEP 3: Collect all terms containing the variable on one side and all other terms on the other side.

Historical Comment

■ The solution of equations is among the oldest of mathematical activities, and efforts to systematize this activity determined much of the shape of modern mathematics.

Consider the following problem, and its solution, using only words: Solve the problem of how many apples Jim has, given that

> "Bob's five apples and Jim's apples together make twelve apples" by thinking,
>
> "Jim's apples are all twelve apples less Bob's five apples" and then concluding,
>
> "Jim has seven apples."

The mental steps translated into algebra are

$$5 + x = 12$$
$$x = 12 - 5$$
$$x = 7$$

The solution of this problem is the earliest form of algebra. Such problems were solved exactly this way in Babylonia in 1800 BC. We know almost nothing of mathematical work before this date, although most authorities believe the sophistication of the earliest known texts indicates that a long period of previous development must have occurred. The method of writing out equations in words persisted for thousands of years, and although it now seems extremely cumbersome, it was used very effectively by many generations of mathematicians. The Arabs developed a good deal of the theory of cubic equations while writing out all the equations in words. About AD 1500, the tendency to abbreviate words in the written equations began to lead in the direction of modern notation; for example, the Latin word *et* (meaning *and*) developed into the plus sign, +. Although the occasional use of letters to represent variables dates back to AD 1200, the practice did not become common until about AD 1600. Development thereafter was rapid, and by 1635, algebraic notation did not differ essentially from what we use now. ■

EXERCISE 2.1

In Problems 1–8, mentally solve each equation.

1. $3x = 18$ **2.** $4x = -24$ **3.** $5x + 10 = 0$ **4.** $3x + 12 = 0$
5. $2x - 3 = 5$ **6.** $3x + 4 = -8$ **7.** $\frac{1}{3}x = \frac{5}{12}$ **8.** $\frac{2}{3}x = \frac{9}{2}$

In Problems 9–52, solve each equation.

9. $3x + 2 = x - 6$ **10.** $2x - 7 = 3x + 5$ **11.** $2t - 6 = 3 - t$
12. $5y + 6 = -18 - y$ **13.** $6 - x = 2x + 9$ **14.** $3 - 2x = 2 - x$
15. $3 + 2n = 5n + 7$ **16.** $3 - 2m = 3m + 1$ **17.** $2(3 + 2x) = 3(x - 4)$
18. $3(2 - x) = 2x - 1$ **19.** $8x - (2x + 1) = 3x - 10$ **20.** $5 - (2x - 1) = 10$
21. $\frac{3}{2}x + 2 = \frac{1}{2} - \frac{1}{2}x$ **22.** $\frac{1}{3}x = 2 - \frac{2}{3}x$ **23.** $\frac{1}{2}x - 4 = \frac{3}{4}x$
24. $1 - \frac{1}{2}x = 5$ **25.** $\frac{2}{3}p = \frac{1}{2}p + \frac{1}{3}$ **26.** $\frac{1}{2} - \frac{1}{3}p = \frac{4}{3}$
27. $0.9t = 0.4 + 0.1t$ **28.** $0.9t = 1 + t$ **29.** $\frac{x + 1}{3} + \frac{x + 2}{7} = 5$
30. $\frac{2x + 1}{3} + 16 = 3x$ **31.** $\frac{2}{y} + \frac{4}{y} = 3$ **32.** $\frac{4}{y} - 5 = \frac{5}{2y}$
33. $(x + 7)(x - 1) = (x + 1)^2$ **34.** $(x + 2)(x - 3) = (x - 3)^2$
35. $x(2x - 3) = (2x + 1)(x - 4)$ **36.** $x(1 + 2x) = (2x - 1)(x - 2)$
37. $z(z^2 + 1) = 3 + z^3$ **38.** $w(4 - w^2) = 8 - w^3$
39. $\frac{x}{x + 2} = \frac{1}{2}$ **40.** $\frac{3x}{x - 1} = 2$ **41.** $\frac{3}{2x - 3} = \frac{2}{x + 5}$
42. $\frac{-2}{x + 4} = \frac{-3}{x + 1}$ **43.** $(x + 2)(3x) = (x^2 + 2)(3)$ **44.** $(x + 5)(2x) = (x^2 + 5)(2)$
45. $\frac{6t + 7}{4t - 1} = \frac{3t + 8}{2t - 4}$ **46.** $\frac{8w + 5}{10w - 7} = \frac{4w - 3}{5w + 7}$

47. $\dfrac{2}{x-2} = \dfrac{3}{x+5} + \dfrac{10}{(x+5)(x-2)}$

48. $\dfrac{1}{2x+3} + \dfrac{1}{x-1} = \dfrac{1}{(2x+3)(x-1)}$

49. $\dfrac{2}{y+3} + \dfrac{3}{y-4} = \dfrac{5}{y+6}$

50. $\dfrac{5}{5z-11} + \dfrac{4}{2z-3} = \dfrac{-3}{5-z}$

51. $\dfrac{x}{x^2-1} - \dfrac{x+3}{x^2-x} = \dfrac{-3}{x^2+x}$

52. $\dfrac{x+1}{x^2+2x} - \dfrac{x+4}{x^2+x} = \dfrac{-3}{x^2+3x+2}$

In Problems 53–58, solve each equation. The letters a, b, and c are constants.

53. $ax - b = c, \quad a \neq 0$

54. $1 - ax = b, \quad a \neq 0$

55. $\dfrac{x}{a} + \dfrac{x}{b} = c, \quad a \neq 0, b \neq 0, a \neq -b$

56. $\dfrac{a}{x} + \dfrac{b}{x} = c, \quad c \neq 0$

57. $\dfrac{1}{x-a} + \dfrac{1}{x+a} = \dfrac{2}{x-1}$

58. $\dfrac{b+c}{x+a} = \dfrac{b-c}{x-a}, \quad c \neq 0, a \neq 0$

59. Find the number a for which $x = 4$ is a solution of the equation

$$x + 2a = 16 + ax - 6a$$

60. Find the number b for which $x = 2$ is a solution of the equation

$$x + 2b = x - 4 + 2bx$$

Problems 61–66 list some formulas that occur in applications. Solve each formula for the indicated variable.

61. $\dfrac{1}{R} = \dfrac{1}{R_1} + \dfrac{1}{R_2}$ for R (electricity)

62. $A = P(1 + rt)$ for r (finance)

63. $F = \dfrac{mv^2}{R}$ for R (mechanics)

64. $PV = nRT$ for T (chemistry)

65. $S = \dfrac{a}{1-r}$ for r (mathematics)

66. $v = -gt + v_0$ for t (mechanics)

67. Explain what is wrong in the following steps:

$$\begin{aligned} x &= 2 & (1)\\ 3x - 2x &= 2 & (2)\\ 3x &= 2x + 2 & (3)\\ x^2 + 3x &= x^2 + 2x + 2 & (4)\\ x^2 + 3x - 10 &= x^2 + 2x - 8 & (5)\\ (x-2)(x+5) &= (x-2)(x+4) & (6)\\ x + 5 &= x + 4 & (7)\\ 1 &= 0 & (8) \end{aligned}$$

68. Which of the following pairs of equations are equivalent?

(a) $x^2 = 9;\ x = 3$ (b) $x = \sqrt{9};\ x = 3$ (c) $(x-1)(x-2) = (x-1)^2;\ x - 2 = x - 1$

2.2 Setting Up Equations: Applications

The previous section provides the tools for solving equations. But, unfortunately, applied problems do not come in the form, "Solve the equation . . . ." Instead, they are narratives that supply information—hopefully, enough to answer the question that inevitably arises. Thus, to solve applied problems we must be able to translate the verbal description into the language of mathematics. We do this by using symbols (usually letters of the alphabet) to represent unknown quantities and then finding relationships (such as equations) that involve these symbols.

Let's look at a few examples that will help you translate certain words into mathematical symbols.

Example 1 (a) The area of a rectangle is the product of its length times its width.

Translation: If A is used to represent the area, l the length, and w the width, then $A = lw$.

(b) For uniform motion, the velocity of an object equals the distance traveled divided by the time required.

Translation: If v is the velocity, s the distance, and t the time, then $v = s/t$.

(c) The sum of two numbers is 50. If one number is known, how is the other found?

Translation: If N and M are the two numbers, then $N + M = 50$. Suppose M is the number we know. Then a formula for obtaining N, the one to be found, is $N = 50 - M$.

(d) Let x denote a number.

The number 5 times as large as x is $5x$.

The number 3 less than x is $x - 3$.

The number that exceeds x by 4 is $x + 4$.

The number that, when added to x, gives 5, is $5 - x$. ■

Mathematical equations that represent real situations should be consistent in terms of the units used. In Example 1(a), if l is measured in feet, then w also must be expressed in feet, and A will be expressed in square feet. In Example 1(b), if v is measured in miles per hour, then the distance s must be expressed in miles and the time t must be expressed in hours. It is a good practice to check units to be sure they are consistent and make sense.

Although each situation has its own unique features, we can provide an outline of the steps to follow in setting up applied problems.

Steps for Setting Up Applied Problems

STEP 1: Read the problem carefully, perhaps two or three times. Pay particular attention to the question being asked in order to identify what you are looking for.

STEP 2: Assign some letter (variable) to represent what you are looking for, and, if necessary, express any remaining unknown quantities in terms of this variable.

STEP 3: Make a list of all the known facts, and write down any relationships among them, especially any that involve the variable. These will usually take the form of an equation (or, later, an inequality) involving the variable. If possible, draw an appropriately labeled diagram to assist you. Sometimes, a table or chart helps.

STEP 4: Solve the equation for the variable, and then answer the question asked in the problem.

STEP 5: Check the answer with the facts in the problem. If it agrees, congratulations! If it does not agree, try again.

Let's look at an example.

Example 2 The sum of two numbers is 180. If one number is 30 less than the other, find the two numbers.

Solution

STEP 1: We are being asked to find two numbers.

STEP 2: We do not know either number, but we let x represent the larger number. Then $x - 30$ is the other number.

STEP 3: We set up a table:

ONE NUMBER	OTHER NUMBER	REASON
x	$x - 30$	One number is 30 less than the other.

It follows from the table and the fact that the sum of the two numbers is 180 that $x + (x - 30) = 180$.

STEP 4:

$$\begin{aligned} x + (x - 30) &= 180 \\ 2x - 30 &= 180 \\ 2x &= 210 \\ x &= 105 \end{aligned}$$

Thus, one number is 105 and the other is $105 - 30 = 75$.

STEP 5: The sum of 105 and 75 is 180, and 75 is 30 less than 105. ■

Example 3 A friend of yours grossed \$435 one week by working 52 hours. Her employer pays time-and-a-half for all hours worked in excess of 40 hours. What is your friend's regular hourly wage?

Solution STEP 1: We are looking for an hourly wage. Our answer will be in dollars per hour.

STEP 2: Let x represent the regular hourly wage; x is measured in dollars per hour.

STEP 3: We set up a table:

	HOURLY WAGE	SALARY
Regular hours, 40	x	$40x$
Overtime hours, 12	$1.5x$	$12(1.5x) = 18x$

The sum of regular salary plus overtime salary will equal \$435. Thus, from the table, $40x + 18x = 435$.

STEP 4:

$$40x + 18x = 435$$
$$58x = 435$$
$$x = 7.50$$

The regular hourly wage is \$7.50 per hour.

STEP 5: Forty hours yields a salary of $40(7.50) = \$300$, and 12 hours of overtime yields a salary of $12(1.5)(7.50) = \$135$, for a total of \$435. ■

Interest

The next example involves **interest**. Interest is money paid for the use of money. The total amount borrowed (whether by an individual from a bank in the form of a loan or by a bank from an individual in the form of a savings account) is called the **principal**. The **rate of interest**, expressed as a percent, is the amount charged for the use of the principal for a given period of time, usually on a yearly (that is, per annum) basis.

If a principal of P dollars is borrowed for a period of t years at a per annum interest rate r, expressed as a decimal, the interest I charged is

Simple Interest Formula

$$I = Prt \tag{1}$$

Interest charged according to formula (1) is called **simple interest**.

Example 4 If \$500 is borrowed for 6 months at the simple interest rate of 9% per annum, what is the interest charged on this loan?

Solution The rate of interest is given per annum, so the actual time the money is borrowed must be expressed in years. Thus, the interest charged would be the principal (\$500) times the rate of interest (9% = 0.09) times the time in years $\left(\frac{1}{2}\right)$:

$$\text{Interest charged} = I = Prt = (500)(0.09)\left(\tfrac{1}{2}\right) = \$22.50$$

■

Example 5 An investor with \$70,000 decides to place part of her money in corporate bonds paying 12% per year and the rest in a certificate of deposit paying 8% per year. If she wishes to obtain an overall return of 9% per year, how much should she place in each investment?

Solution STEP 1: The question is asking for two dollar amounts: the principal to invest in the corporate bonds and the principal to invest in the certificate of deposit.

STEP 2: We let x represent the amount (in dollars) to be invested in the bonds. Then $70{,}000 - x$ is the amount that will be invested in the certificate. (Do you see why?)

STEP 3: We set up a table:

	PRINCIPAL \$	RATE	TIME yr	INTEREST \$
Bonds	x	12% = 0.12	1	$0.12x$
Certificate	$70{,}000 - x$	8% = 0.08	1	$0.08(70{,}000 - x)$
Total	70,000	9% = 0.09	1	$0.09(70{,}000) = 6300$

Since the total interest from the investments is to equal $0.09(70{,}000) = 6300$, we must have the equation

$$0.12x + 0.08(70{,}000 - x) = 6300$$

(Note that the units are consistent—the unit is dollars on each side.)

STEP 4:

$$\begin{aligned} 0.12x + 5600 - 0.08x &= 6300 \\ 0.04x &= 700 \\ x &= 17{,}500 \end{aligned}$$

Thus, the investor should place \$17,500 in the bonds and \$70,000 − \$17,500 = \$52,500 in the certificate.

STEP 5: The interest on the bond after 1 year is 0.12(\$17,500) = \$2100; the interest on the certificate after 1 year is 0.08(\$52,500) = \$4200. The total annual interest is \$6300, the required amount.

■

Mixture Problems

The next example is a type usually referred to as a **mixture problem**.

Example 6 In a chemistry laboratory the concentration of one solution is 10% hydrochloric acid (HCl) and of a second solution is 60% HCl. How many milliliters of each should be mixed to obtain 50 milliliters of a 30% HCl solution?

Solution Let x represent the number of milliliters of the 10% solution. Then $50 - x$ equals the number of milliliters of the 60% solution. See Figure 1.

Figure 1

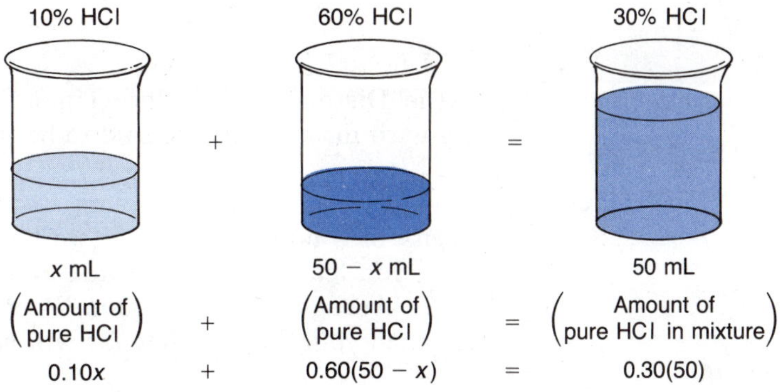

Based on the figure, we form the table:

	AMOUNT mL	CONCENTRATION OF HCl	AMOUNT OF PURE ACID mL
10% HCl	x	10% = 0.10	$0.10x$
60% HCl	$50 - x$	60% = 0.60	$0.60(50 - x)$
30% HCl	50	30% = 0.30	$0.30(50) = 15$

The amount of HCl in the 30% solution (15 milliliters) must equal the sum of the amounts of HCl found in the 10% solution and the 60% solution. Thus, we have the equation

$$0.10x + 0.60(50 - x) = 15$$
$$0.10x + 30 - 0.60x = 15$$
$$-0.50x = -15$$
$$x = 30 \text{ milliliters}$$

Thus, 30 milliliters of the 10% acid solution, when mixed with 20 milliliters of the 60% acid solution, yields 50 milliliters of a 30% acid solution.

Check: To check this answer, we note that there are $0.10(30) = 3$ milliliters of acid in the 10% solution and $0.60(20) = 12$ milliliters of acid in the 60% solution. The 50 milliliter mixture therefore contains 15 milliliters of acid, for an acid concentration of $\frac{15}{50} = 0.30 = 30\%$ acid solution. ■

Uniform Motion

The next example deals with moving objects.

If an object moves at an average velocity v, the distance s covered in time t is given by the formula

$$s = vt \tag{2}$$

That is, Distance = Velocity · Time. Objects that are moving in accordance with formula (2) are said to be in **uniform motion**.

Example 7 A friend of yours, who is a long-distance runner, runs at an average velocity of 8 miles per hour. Two hours after your friend leaves your house, you leave in your car and follow the same route as your friend. If your average velocity is 40 miles per hour, how long will it be before you reach your friend? How far will each of you be from your house?

Solution We use t to represent the time (in hours) it takes the car to catch up with the runner. When this occurs, the total time elapsed for the runner is $(t + 2)$ hours. Set up the following table:

	VELOCITY mi/hr	TIME hr	DISTANCE mi
Runner	8	$t + 2$	$8(t + 2)$
Car	40	t	$40t$

Since the distance traveled is the same, we are led to the equation:

$$\begin{aligned} 8(t + 2) &= 40t \\ 8t + 16 &= 40t \\ 32t &= 16 \\ t &= \tfrac{1}{2} \text{ hour} \end{aligned}$$

It will take you $\frac{1}{2}$ hour to reach your friend. Each of you will have gone 20 miles.

Check: In 2.5 hours, the runner travels a distance of $(2.5)(8) = 20$ miles. In $\frac{1}{2}$ hour, the car travels a distance of $\left(\frac{1}{2}\right)(40) = 20$ miles. ■

Constant Rate Jobs

Our final example in this section involves jobs that are performed at a **constant rate**. Our assumption is that if a job can be done in t units of time, then $1/t$ of the job is done in 1 unit of time. Let's look at an example.

Example 8 It takes Danny 4 hours to do a certain job, whereas it takes Mike 6 hours to do the same job. Assuming no gain or loss of efficiency, how long will it take them to do this job if they work together?

Solution In 1 hour, Danny does $\frac{1}{4}$ of the job, and in 1 hour, Mike does $\frac{1}{6}$ of the job. Let t be the time (in hours) it takes them to do the job together. In 1 hour, then, $1/t$ of the job is completed. We reason as follows:

$$\begin{pmatrix}\text{Part done by Danny} \\ \text{in 1 hour}\end{pmatrix} + \begin{pmatrix}\text{Part done by Mike} \\ \text{in 1 hour}\end{pmatrix} = \begin{pmatrix}\text{Part done together} \\ \text{in 1 hour}\end{pmatrix}$$

	HOURS TO DO JOB	PART OF JOB DONE IN 1 HOUR
Danny	4	$\frac{1}{4}$
Mike	6	$\frac{1}{6}$
Together	t	$\frac{1}{t}$

Now we set up the table in the margin. From the table,

$$\frac{1}{4} + \frac{1}{6} = \frac{1}{t}$$

$$\frac{3+2}{12} = \frac{1}{t}$$

$$t = \frac{12}{5}$$

Working together, the job can be done in $\frac{12}{5}$ hours, or 2 hours, 24 minutes. ■

EXERCISE 2.2

In Problems 1–8, translate each sentence into a mathematical equation. Be sure to identify the meaning of all symbols.

1. The area of a circle is the product of the number π times the square of the radius.

2. The circumference of a circle is the product of the number π times twice the radius.

3. The area of a square is the square of the length of a side.

4. The perimeter of a square is four times the length of a side.

5. Force equals the product of mass times acceleration.

6. Pressure is force per unit area.

7. Work equals force times distance.

8. Kinetic energy is one-half the product of the mass times the square of the velocity.

9. Find two consecutive integers whose sum is 83.

10. Find two consecutive even integers whose sum is 66.

11. Find two consecutive integers for which the difference of their squares is 27.

12. Find two consecutive odd integers for which the difference of their squares is 48.

13. The sum of two numbers is 90. Twice one of them equals the other less 15. Find the numbers.

14. The sum of two numbers is 35. Twice one of them less 10 is the other. Find the numbers.

15. A worker who is paid time-and-a-half for hours worked in excess of 40 hours had gross weekly wages of $442 for 48 hours worked. What is the regular hourly rate?

16. Colleen is paid time-and-a-half for hours worked in excess of 40 hours and double-time for hours worked on Sunday. If Colleen had gross weekly wages of $342 for working 50 hours, 4 of which were on Sunday, what is her regular hourly rate?

17. In an NFL football game, one team scored a total of 41 points, consisting of one safety (2 points) and two field goals (3 points each). The team missed 2 extra points (1 point each) after scoring touchdowns (6 points each). How many touchdowns did they get?

18. In a basketball game, one team scored a total of 70 points and made three times as many field goals (2 points each) as free throws (1 point each). How many field goals did they have?

19. The perimeter of a rectangle is 40 feet. Find its length and width if the length is 8 feet longer than the width.

20. The perimeter of a rectangle is 30 meters. Find its length and width if the length is twice the width.

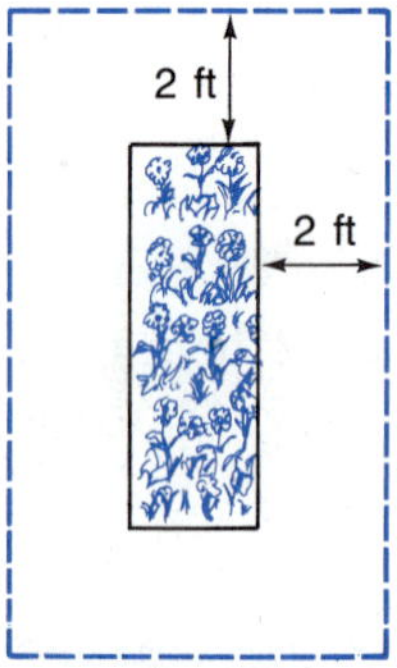

21. A gardener has 46 feet of fencing to be used to enclose a rectangular garden that has a border 2 feet wide surrounding it (see the figure). If the length of the garden is to be twice its width, what will be the dimensions of the garden?

22. A sugar molecule has twice as many atoms of hydrogen as it does oxygen and one more atom of carbon than oxygen. If a sugar molecule has a total of 45 atoms, how many are oxygen? How many are hydrogen?

23. A recent retiree requires $6000 per year in extra income. She has $50,000 to invest and can invest in B-rated bonds paying 15% per year or in a certificate paying 7% per year. How much money should be invested in each to realize exactly $6000 in interest per year?

24. After 2 years, the retiree referred to in Problem 23 finds she now will require $7000 per year. Assuming the remaining information is the same, how should the money be reinvested?

25. A bank loaned out $10,000, part of it at the rate of 8% per year and the rest at the rate of 18% per year. If the interest received from these loans totaled $1000, how much was loaned at 8%?

26. A loan officer at a bank has $100,000 to lend and is required to obtain an average return of 18% per year. If she can lend at the rate of 19% or the rate of 16%, how much can she lend at the 16% rate and still meet her requirement?

27. A nut store normally sells cashews for $4.00 per pound and peanuts for $1.50 per pound. But at the end of the month, the peanuts had not sold well, so in order to sell 60 pounds of peanuts, the manager decides to mix the 60 pounds of peanuts with some cashews and sell the mixture for $2.50 per pound. How many pounds of cashews should be mixed with the peanuts to ensure no change in the profit?

28. A candy store sells boxes of candy containing caramels and cremes. Each box sells for $9.50 and holds 30 pieces of candy (all pieces are the same size). If the caramels cost $0.25 to produce and the cremes cost $0.45 to produce, how many of each should be in a box for no profit or loss to occur?

29. For a certain experiment, a student requires 100 cubic centimeters of a solution that is 8% HCl. The storeroom has only solutions that are 15% HCl and 5% HCl. How many cubic centimeters of each available solution should be mixed to get 100 cubic centimeters of 8% HCl?

30. A coffee manufacturer wants to market a new blend of coffee that will cost $2.90 per pound by mixing two coffees that sell for $2.75 and $3 per pound, respectively. What amounts of each coffee should be blended to obtain the desired mixture? [*Hint:* Assume the total weight of the desired blend is 100 pounds.]

31. The manager of the Coral Theater wants to know whether the majority of its patrons are adults or children. During a week in July, 5200 tickets were sold and the receipts totaled $20,335. The adult admission is $4.75, and the children's admission is $2.50. How many adult patrons were there?

32. A wool suit, discounted by 30% for a clearance sale, has a price tag of $199. What was the suit's original price?

33. A builder of tract homes reduced the price of a model by 15%. If the new price is $93,500, what was its original price? How much can be saved by purchasing the model?

34. A car dealer, at a year-end clearance, reduces the list price of last year's models by 15%. If a certain four-door model has a discounted price of $8000, what was its list price? How much can be saved by purchasing last year's model?

35. A college book store marks up the price it pays the publisher for a book by 25%. If the selling price of the book is $35.00, how much did the book store pay for the book?

36. The suggested list price of a new car is $12,000. The dealer's cost is 85% of list. How much will you pay if the dealer is willing to accept $100 over cost for the car?

37. Mike can deliver his newspapers in 30 minutes. It takes Danny 20 minutes to do the same route. How long would it take them to deliver the newspapers if they work together?

38. A painter by himself can paint four rooms in 10 hours. If he hires a helper, they can do the same job together in 6 hours. If he lets the helper work alone, how long will it take for the helper to paint four rooms?

39. Going into the final exam, which will count as two tests, Colleen has test scores of 80, 83, 71, 61, and 95. What score does Colleen need on the final in order to have an average score of 80?

40. Going into the final exam, which will count as two-thirds of the final grade, Dan has test scores of 86, 80, 84, and 90. What score does Dan need on the final in order to earn a B, which requires an average score of 80? What does he need to earn an A, which requires an average of 90?

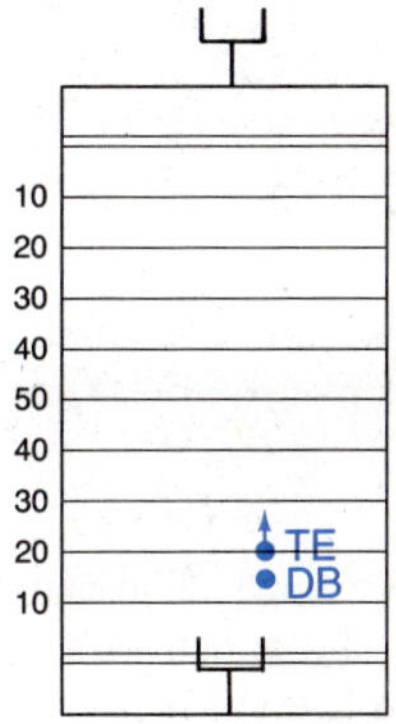

41. A tight end can run the 100 yard dash in 12 seconds. A defensive back can do it in 10 seconds. The tight end catches a pass at his own 20 yard line with the defensive back at the 15 yard line. (See the figure.) If no other players are nearby, at what yard line will the defensive back catch up to the tight end?

42. Debbie, an outside saleswoman, uses her car for both business and pleasure. Last year, she traveled 30,000 miles, using 900 gallons of gasoline. Her car gets 40 miles per gallon on the highway and 25 in the city. She can deduct all highway travel, but no city travel, on her taxes. How many miles should Debbie be allowed as a business expense?

43. How much water should be added to 1 gallon of pure antifreeze to obtain a solution that is 60% antifreeze?

44. The cooling system of a certain foreign-made car has a capacity of 15 liters. If the system is filled with a mixture that is 40% antifreeze, how much of this mixture should be drained and replaced by pure antifreeze so that the system is filled with a solution that is 60% antifreeze?

45. A motorboat can maintain a constant speed of 15 miles per hour relative to the water. The boat makes a trip upstream to a certain point in 20 minutes; the return trip takes 15 minutes. What is the speed of the current? (See the figure.)

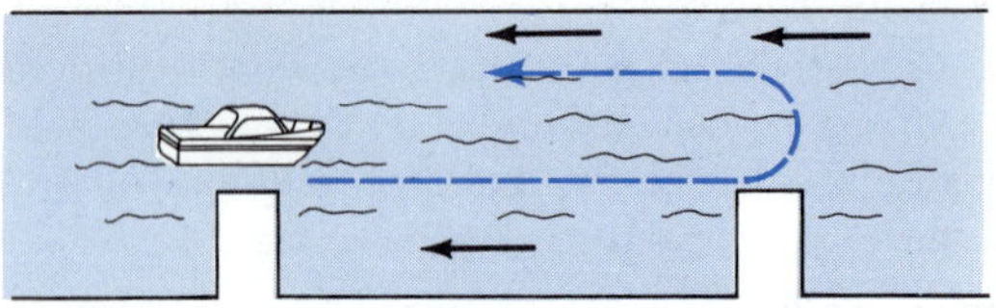

46. A car 15 feet in length overtakes a large semitrailer truck 40 feet in length. If the truck is traveling at a constant speed of 45 miles per hour, what constant speed should the car maintain to pass the truck within 550 feet? (5280 feet = 1 mile)

47. An oil tanker can be emptied by the main pump in 4 hours. An auxiliary pump can empty the tanker in 9 hours. If the main pump is started at 9 AM, when should the auxiliary pump be started so that the tanker is emptied by noon?

48. A 20 pound bag of Economy brand cement mix contains 25% cement and 75% sand. How much pure cement must be added to produce a cement mix that is 40% cement?

49. A bathroom tub will fill in 15 minutes with both faucets open and the stopper in place. With both faucets closed and the stopper removed, the tub will empty in 20 minutes. How long will it take for the tub to fill if both faucets are open and the stopper removed?

C 50. An air rescue plane averages 300 miles per hour in still air. It carries enough fuel for 5 hours of flying time. If, upon takeoff, it encounters a wind of 30 miles per hour, how far can it fly and return safely? (Assume the wind remains constant.)

51. Mike can run the mile in 6 minutes, and Dan can run the mile in 9 minutes. If Mike gives Dan a head start of 1 minute, how far from the start will Mike pass Dan? (See the figure.) How long does it take?

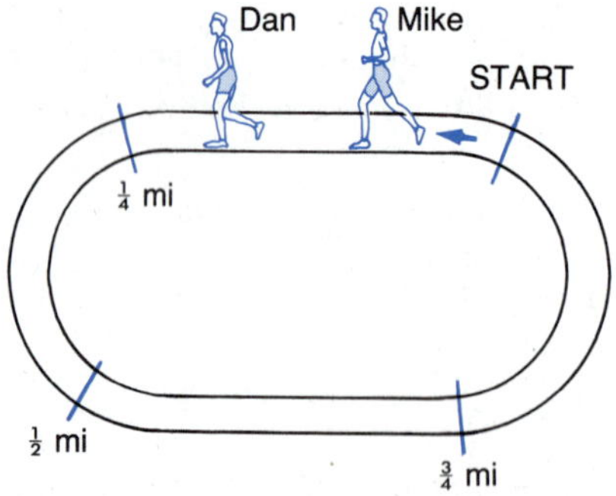

52. A motorboat heads upstream on a river that has a current of 3 miles per hour. The trip upstream takes 4 hours, while the return trip takes 2.5 hours. What is the speed of the motorboat? (Assume the motorboat maintains a constant speed relative to the water.)

53. Two cars enter the Florida Turnpike at Commercial Boulevard at 8:00 AM, each heading for Wildwood. One car's average speed is 10 miles per hour more than the other's. The faster car arrives at Wildwood at 11:00 AM, $\frac{1}{2}$ hour before the other car. What is the average speed of each car? How far did each travel?

54. A ship that is in danger of sinking radios the Coast Guard for assistance. When the rescue craft leaves the Coast Guard station, the ship is 60 miles away and heading directly toward the station. If the average speed of the ship is 10 miles per hour and the average speed of the rescue craft is 20 miles per hour, how long will it take for the rescue craft to reach the ship? (See the figure.)

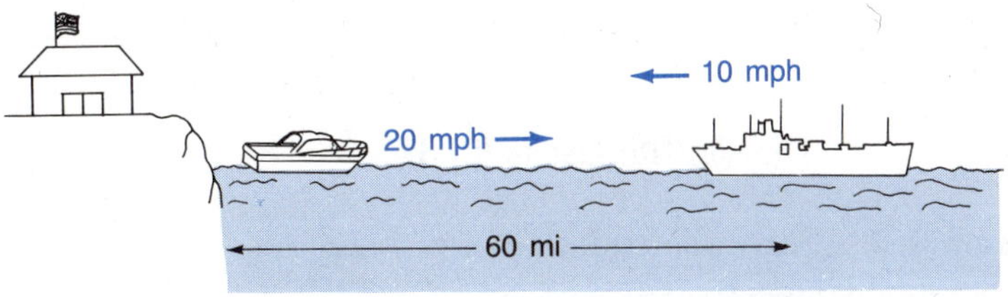

2.3 Quadratic Equations

Quadratic Equation A **quadratic equation** is an equation equivalent to one of the form

$$ax^2 + bx + c = 0 \qquad (1)$$

where a, b, and c are real numbers and $a \neq 0$.

A quadratic equation written in the form $ax^2 + bx + c = 0$ is said to be in **standard form**.

Sometimes, a quadratic equation is called a **second-degree equation**, because the left side is a polynomial of degree 2. We shall discuss three ways of solving quadratic equations: by factoring, by completing the square, and by use of the quadratic formula.

Factoring

When a quadratic equation is written in standard form, $ax^2 + bx + c = 0$, it may be possible to factor the expression on the left side as the product of two first-degree polynomials. Then, by setting each factor equal to 0 and solving the resulting linear equations, we obtain the solutions of the quadratic equation.

Let's look at an example.

Example 1 Solve the equation: $x^2 - 5x + 6 = 0$

Solution The equation is in the standard form specified in equation (1). The left side may be factored as

$$x^2 - 5x + 6 = 0$$
$$(x - 2)(x - 3) = 0$$

We set each factor equal to 0, and solve the resulting first-degree equations.

$$x - 2 = 0 \quad \text{or} \quad x - 3 = 0$$
$$x = 2 \qquad\qquad x = 3$$

The solution set is $\{2, 3\}$. ■

Example 2 Solve the equation: $x^2 = 12 - x$

Solution We put the equation in standard form by subtracting $12 - x$ from each side:

$$x^2 = 12 - x$$
$$x^2 + x - 12 = 0$$

The left side may now be factored as

$$(x + 4)(x - 3) = 0$$

so that

$$x + 4 = 0 \quad \text{or} \quad x - 3 = 0$$
$$x = -4 \qquad\qquad x = 3$$

The solution set is $\{-4, 3\}$. ■

Example 3 Solve the equation: $x^2 = 4x$

Solution We put the equation in standard form and factor:

$$x^2 = 4x$$
$$x^2 - 4x = 0$$
$$x(x - 4) = 0$$
$$x = 0 \quad \text{or} \quad x - 4 = 0$$
$$x = 4$$

Thus, the solution set is $\{0, 4\}$. ■

When the left side factors into two linear equations with the same solution, the quadratic equation is said to have a **repeated solution**. We also call this solution a **root of multiplicity 2**, or a **double root**.

Example 4 Solve the equation: $x^2 - 6x + 9 = 0$

Solution This equation is already in standard form, and the left side can be factored:

$$x^2 - 6x + 9 = 0$$
$$(x - 3)(x - 3) = 0$$

so that

$$x = 3 \quad \text{or} \quad x = 3$$

This equation has only the repeated solution 3. ■

Completing the Square

We begin with a preliminary result. Suppose we wish to solve the quadratic equation

$$x^2 = p \tag{2}$$

where $p \geq 0$ is a nonnegative number. We proceed as in the earlier examples:

$$x^2 - p = 0 \quad \text{Put in standard form.}$$
$$(x - \sqrt{p})(x + \sqrt{p}) = 0 \quad \text{Factor (over the real numbers).}$$
$$x = \sqrt{p} \quad \text{or} \quad x = -\sqrt{p} \quad \text{Solve.}$$

Thus, we have the following result:

If $x^2 = p$ and $p \geq 0$, then $x = \sqrt{p}$ or $x = -\sqrt{p}$. (3)

Note that if $p > 0$, the equation $x^2 = p$ has two solutions; namely, $x = \sqrt{p}$ and $x = -\sqrt{p}$. We usually abbreviate these solutions as $x = \pm\sqrt{p}$, read as "x equals plus or minus the square root of p." For example, the two solutions of the equation

$$x^2 = 4$$

are

$$x = \pm\sqrt{4}$$

and since $\sqrt{4} = 2$, we have

$$x = \pm 2$$

The solution set is $\{-2, 2\}$. Do not confuse the two solutions of the equation $x^2 = 4$ with value of the principal square root of 4, which is $\sqrt{4} = 2$. That is, the principal square root of a positive number is unique, while the equation $x^2 = p$, $p > 0$, has two solutions.

Example 5 Solve each equation.

(a) $x^2 = 5$ (b) $(x - 2)^2 = 16$

Solution (a) We use the result in equation (3) to get

$$x^2 = 5$$
$$x = \pm\sqrt{5}$$
$$x = \sqrt{5} \quad \text{or} \quad x = -\sqrt{5}$$

The solution set is $\{-\sqrt{5}, \sqrt{5}\}$.

(b) We use the result in equation (3) to get

$$(x - 2)^2 = 16$$
$$x - 2 = \pm\sqrt{16}$$
$$x - 2 = \sqrt{16} \quad \text{or} \quad x - 2 = -\sqrt{16}$$
$$x - 2 = 4 \qquad x - 2 = -4$$
$$x = 6 \qquad x = -2$$

The solution set is $\{-2, 6\}$. ■

We now introduce the method of **completing the square**. The idea behind this method is to "adjust" the left side of a quadratic equation, $ax^2 + bx + c = 0$, so that it becomes a perfect square—the square of a first-degree polynomial. For example, $x^2 + 6x + 9$ and $x^2 - 4x + 4$ are perfect squares because

$$x^2 + 6x + 9 = (x + 3)^2 \quad \text{and} \quad x^2 - 4x + 4 = (x - 2)^2$$

How do we "adjust" the left side? We do it by adding the appropriate number to the left side to create a perfect square. For example, to make $x^2 + 6x$ a perfect square, we add 9.

Let's look at several examples of completing the square when the coefficient of x^2 is 1:

START	ADD	RESULT
$x^2 + 4x$	4	$x^2 + 4x + 4 = (x + 2)^2$
$x^2 + 12x$	36	$x^2 + 12x + 36 = (x + 6)^2$
$x^2 - 6x$	9	$x^2 - 6x + 9 = (x - 3)^2$
$x^2 + x$	$\frac{1}{4}$	$x^2 + x + \frac{1}{4} = \left(x + \frac{1}{2}\right)^2$

Do you see the pattern? Provided the coefficient of x^2 *is* 1, we complete the square by adding the square of $\frac{1}{2}$ the coefficient of x:

START	ADD	RESULT
$x^2 + mx$	$\left(\frac{m}{2}\right)^2$	$x^2 + mx + \left(\frac{m}{2}\right)^2 = \left(x + \frac{m}{2}\right)^2$

The next example illustrates how the procedure of completing the square can be used to solve a quadratic equation.

Example 6 Solve by completing the square: $x^2 + 5x + 4 = 0$

Solution We always begin this procedure by rearranging the equation so that the constant is on the right side:

$$x^2 + 5x + 4 = 0$$
$$x^2 + 5x = -4$$

Since the coefficient of x^2 is 1, we can complete the square on the left side by adding $\left(\frac{1}{2} \cdot 5\right)^2 = \frac{25}{4}$. Of course, in an equation, whatever we add to the left side also must be added to the right side. Thus, we add $\frac{25}{4}$ to *both* sides:

$$x^2 + 5x + \frac{25}{4} = -4 + \frac{25}{4}$$
$$\left(x + \frac{5}{2}\right)^2 = \frac{9}{4}$$
$$x + \frac{5}{2} = \pm\sqrt{\frac{9}{4}}$$
$$x + \frac{5}{2} = \pm\frac{3}{2}$$
$$x = -\frac{5}{2} \pm \frac{3}{2}$$
$$x = -\frac{5}{2} + \frac{3}{2} = -1 \quad \text{or} \quad x = -\frac{5}{2} - \frac{3}{2} = -4$$

The solution set is $\{-4, -1\}$. ■

Example 7 Solve by completing the square: $2x^2 - 8x - 5 = 0$

Solution First, we rewrite the equation:

$$2x^2 - 8x - 5 = 0$$
$$2x^2 - 8x = 5$$

Next, we divide by 2 so that the coefficient of x^2 is 1. (This enables us to complete the square at the next step.)

$$x^2 - 4x = \frac{5}{2}$$

Finally, we complete the square by adding 4 to each side:

$$\begin{aligned} x^2 - 4x + 4 &= \tfrac{5}{2} + 4 \\ (x - 2)^2 &= \tfrac{13}{2} \\ x - 2 &= \pm\sqrt{\tfrac{13}{2}} \\ x &= 2 \pm \sqrt{\tfrac{13}{2}} \end{aligned}$$

We choose to leave our answer in this compact form. Thus, the solution set is $\{2 - \sqrt{\frac{13}{2}}, 2 + \sqrt{\frac{13}{2}}\}$.

Note: If we wanted an approximation, say, to two decimal places, of these solutions, we would use a calculator to get $\{-0.55, 4.55\}$. ■

The Quadratic Formula

We can use the method of completing the square to obtain a general formula for solving the quadratic equation

$$ax^2 + bx + c = 0 \qquad a \neq 0$$

As in Examples 6 and 7, we rearrange the terms as

$$ax^2 + bx = -c$$

Since $a \neq 0$, we can divide both sides by a to get

$$x^2 + \frac{b}{a}x = -\frac{c}{a}$$

Now the coefficient of x^2 is 1. To complete the square on the left side, add the square of $\frac{1}{2}$ the coefficient of x; that is, add

$$\left(\frac{1}{2} \cdot \frac{b}{a}\right)^2 = \frac{b^2}{4a^2}$$

to each side. Then,

$$\begin{aligned} x^2 + \frac{b}{a}x + \frac{b^2}{4a^2} &= \frac{b^2}{4a^2} - \frac{c}{a} \\ \left(x + \frac{b}{2a}\right)^2 &= \frac{b^2 - 4ac}{4a^2} \end{aligned} \qquad (4)$$

Provided $b^2 - 4ac \geq 0$, we now can apply the result in equation (3) to get

$$\begin{aligned} x + \frac{b}{2a} &= \pm\sqrt{\frac{b^2 - 4ac}{4a^2}} \\ x &= -\frac{b}{2a} \pm \frac{\sqrt{b^2 - 4ac}}{2a} = \frac{-b \pm \sqrt{b^2 - 4ac}}{2a} \end{aligned}$$

What if $b^2 - 4ac$ is negative? Then equation (4) states that the left expression (a real number squared) equals the right expression (a negative number). Since this occurrence is impossible for real numbers, we conclude that if $b^2 - 4ac < 0$, the quadratic equation has no *real* solution. (We discuss quadratic equations for which the quantity $b^2 - 4ac < 0$ in detail in the next section.)

We now state the *quadratic formula*.

Theorem Consider the quadratic equation

$$ax^2 + bx + c = 0 \qquad a \neq 0$$

If $b^2 - 4ac < 0$, this equation has no real solution.

If $b^2 - 4ac \geq 0$, the real solution(s) of this equation is (are) given by the **quadratic formula**:

Quadratic Formula

$$x = \frac{-b \pm \sqrt{b^2 - 4ac}}{2a} \tag{5}$$

■

The quantity $b^2 - 4ac$ is called the **discriminant** of the quadratic equation, because its value tells us whether the equation has real solutions. In fact, it also tells us how many solutions to expect.

Discriminant of a Quadratic Equation

For a quadratic equation $ax^2 + bx + c = 0$:

1. If $b^2 - 4ac > 0$, there are two unequal real solutions.
2. If $b^2 - 4ac = 0$, there is a repeated real solution—a root of multiplicity 2.
3. If $b^2 - 4ac < 0$, there is no real solution.

Thus, when asked to find the real solutions, if any, of a quadratic equation, always evaluate the discriminant first to see how many real solutions there are.

Example 8 Use the quadratic formula to find the real solutions, if any, of the equation: $3x^2 - 5x + 1 = 0$

Solution The equation is in standard form, so we compare it to $ax^2 + bx + c = 0$ to find a, b, and c:

$$\begin{aligned} 3x^2 - 5x + 1 &= 0 \\ ax^2 + bx + c &= 0 \end{aligned}$$

With $a = 3$, $b = -5$, and $c = 1$, we evaluate the discriminant $b^2 - 4ac$:

$$b^2 - 4ac = (-5)^2 - 4(3)(1) = 25 - 12 = 13$$

Since $b^2 - 4ac > 0$, there are two real solutions, which can be found using the quadratic formula:

$$x = \frac{-b \pm \sqrt{b^2 - 4ac}}{2a} = \frac{5 \pm \sqrt{13}}{6}$$

The solution set is $\{(5 - \sqrt{13})/6, (5 + \sqrt{13})/6\}$. ■

Example 9 Use the quadratic formula to find the real solutions, if any, of the equation: $\frac{25}{2}x^2 - 30x + 18 = 0$

Solution The equation is given in standard form. However, to simplify the arithmetic, we clear the fractions:

$$\begin{aligned} \tfrac{25}{2}x^2 - 30x + 18 &= 0 \\ 25x^2 - 60x + 36 &= 0 && \text{Clear fractions.} \\ ax^2 + bx + c &= 0 && \text{Compare to standard form.} \end{aligned}$$

With $a = 25$, $b = -60$, and $c = 36$, we evaluate the discriminant:

$$b^2 - 4ac = (-60)^2 - 4(25)(36) = 3600 - 3600 = 0$$

The equation has a repeated solution, which we find by using the quadratic formula:

$$x = \frac{-b \pm \sqrt{b^2 - 4ac}}{2a} = \frac{60 \pm \sqrt{0}}{50} = \frac{60}{50} = \frac{6}{5}$$

The repeated solution is $\frac{6}{5}$. ■

Example 10 Use the quadratic formula to find the real solutions, if any, of the equation: $3x^2 - 5x = -2$

Solution The equation, as given, is not in standard form.

$$\begin{aligned} 3x^2 - 5x &= -2 \\ 3x^2 - 5x + 2 &= 0 && \text{Put in standard form.} \\ ax^2 + bx + c &= 0 && \text{Compare to standard form.} \end{aligned}$$

With $a = 3$, $b = -5$, and $c = 2$, we find

$$b^2 - 4ac = (-5)^2 - 4(3)(2) = 25 - 24 = 1$$

The two real solutions are

$$x = \frac{-b \pm \sqrt{b^2 - 4ac}}{2a} = \frac{5 \pm \sqrt{1}}{6} = \frac{5 \pm 1}{6}$$

The solution set is $\left\{\frac{2}{3}, 1\right\}$. ■

Example 11 Use the quadratic formula to find the real solutions, if any, of the equation: $3x^2 + 2 = 4x$

Solution The equation, as given, is not in standard form.

A=3
B=4
c=2

$-4 \pm \sqrt{16 - 4(3)(2)}$ / 6

$$3x^2 + 2 = 4x$$

$$3x^2 - 4x + 2 = 0 \quad \text{Put in standard form.}$$

$$ax^2 + bx + c = 0 \quad \text{Compare to standard form.}$$

With $a = 3$, $b = -4$, and $c = 2$, we find

$$b^2 - 4ac = 16 - 24 = -8$$

Since $b^2 - 4ac < 0$, the equation has no real solution. ■

Sometimes a given equation can be transformed into a quadratic equation so that it can be solved using the quadratic formula.

Example 12 Find the real solutions, if any, of the equation: $9 + \frac{3}{x} - \frac{2}{x^2} = 0, x \neq 0$

Solution In its present form, the equation

$$9 + \frac{3}{x} - \frac{2}{x^2} = 0$$

is not a quadratic equation. However, it can be transformed into one by multiplying each side by x^2. The result is

$$9x^2 + 3x - 2 = 0$$

Although we multiplied each side by x^2, we know that $x^2 \neq 0$ (do you see why?), so this quadratic equation is equivalent to the original equation.

Using $a = 9$, $b = 3$, and $c = -2$, the discriminant is

$$b^2 - 4ac = 9 + 72 = 81$$

Since $b^2 - 4ac > 0$, the new equation has two real solutions:

$$x = \frac{-b \pm \sqrt{b^2 - 4ac}}{2a} = \frac{-3 \pm \sqrt{81}}{18} = \frac{-3 \pm 9}{18}$$

$$x = \frac{-3 + 9}{18} = \frac{6}{18} = \frac{1}{3} \quad \text{or} \quad x = \frac{-3 - 9}{18} = \frac{-12}{18} = \frac{-2}{3}$$

The solution set is $\{-\frac{2}{3}, \frac{1}{3}\}$. ■

Summary

Procedure for Solving a Quadratic Equation

To solve a quadratic equation, first put it in standard form:

$$ax^2 + bx + c = 0$$

Then:

STEP 1: Identify a, b, and c.

STEP 2: Evaluate the discriminant, $b^2 - 4ac$.

STEP 3: (a) If the discriminant is negative, the equation has no real solution.

(b) If the discriminant is nonnegative, then look to see whether the left side can be factored. If you can easily spot factors, use the factoring method to solve the equation. Otherwise, use the quadratic formula or the method of completing the square.

Applications

Many applied problems require the solution of a quadratic equation. Let's look at one that you will probably see again in a slightly different form if you study calculus.

Example 13 From each corner of a square piece of sheet metal, remove a square of side 9 centimeters. Turn up the edges to form an open box. If the box is to hold 144 cubic centimeters, what should be the dimensions of the piece of sheet metal?

Solution We use Figure 2 as a guide. We have labeled by x the length of a side of the square piece of sheet metal. The box will be of height 9 centimeters and its square base will have $x - 18$ as the length of a side. The volume (Length × Width × Height) of the box is therefore

$$9(x - 18)(x - 18) = 9(x - 18)^2$$

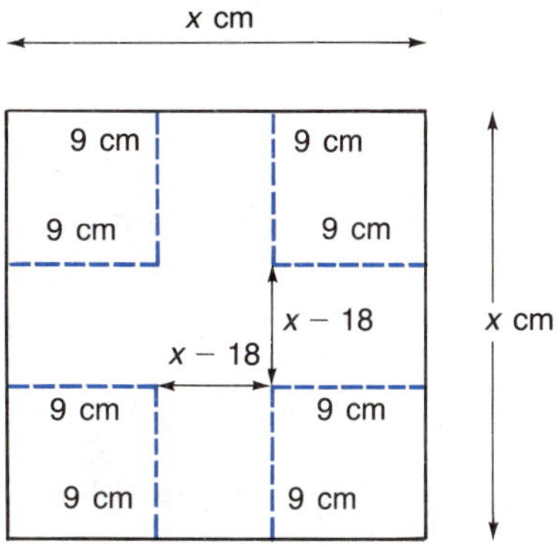

Figure 2

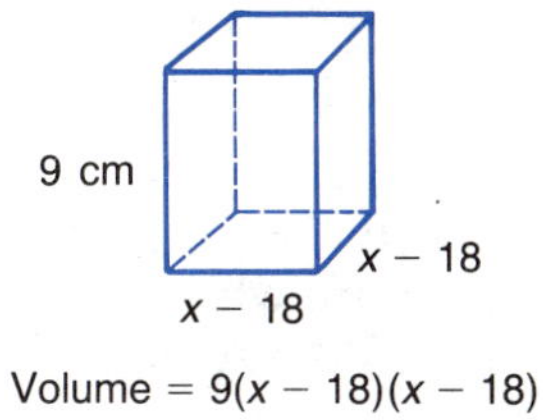

Since the volume of the box is to be 144 cubic centimeters, we have

$$\begin{aligned} 9(x-18)^2 &= 144 \\ (x-18)^2 &= 16 \\ x - 18 &= \pm 4 \\ x &= 18 \pm 4 \\ x &= 22 \qquad \text{or} \qquad x = 14 \end{aligned}$$

We discard the solution $x = 14$ (do you see why?) and conclude that the sheet metal should be 22 centimeters by 22 centimeters.

Check: If we begin with a piece of sheet metal 22 centimeters by 22 centimeters, cut out a 9 centimeter square from each corner, and fold up the edges, we get a box whose dimensions are 9 by 4 by 4, with volume $9 \times 4 \times 4 = 144$ cubic centimeters, as required. ■

Example 14 A piece of wire 8 feet in length is to be cut into two pieces. Each of these pieces will then be bent into a square. Where should the cut in the wire be made if the sum of the areas of these squares is to be 2 square feet?

Solution We use Figure 3 as a guide. We have labeled by x the length of one of the pieces of wire after it has been cut. The remaining piece will be of length $8 - x$. If each of the lengths is bent into a square, then one of the squares has a side of length $x/4$ and the other a side of length $(8 - x)/4$. Since the sum of the areas of these two squares is 2, we have the equation

$$\begin{aligned} \left(\frac{x}{4}\right)^2 + \left(\frac{8-x}{4}\right)^2 &= 2 \\ \frac{x^2}{16} + \frac{64 - 16x + x^2}{16} &= 2 \\ 2x^2 - 16x + 64 &= 32 \\ 2x^2 - 16x + 32 &= 0 && \text{Put in standard form.} \\ x^2 - 8x + 16 &= 0 && b^2 - 4ac = 64 - 64 = 0 \\ (x-4)^2 &= 0 \\ x &= 4 \end{aligned}$$

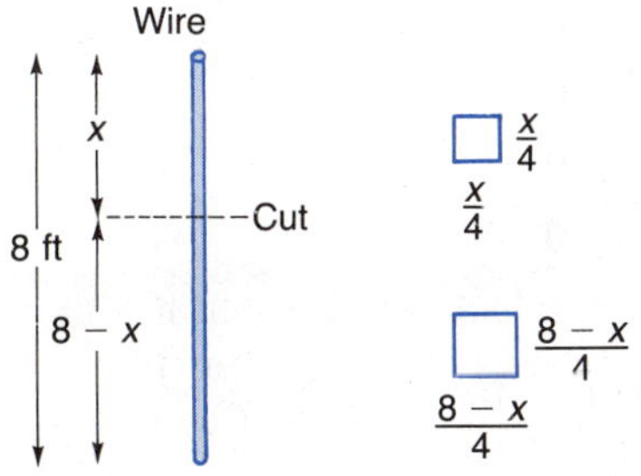

Figure 3

Since $x = 4$, $8 - x = 4$, and the original piece of wire should be cut into two pieces, each of length 4 feet.

Check: If the length of each piece of wire is 4 feet, then each piece can be formed into a square whose side is 1 foot. The area of each square is then 1 square foot, so the sum of the areas is 2 square feet, as required. ■

Example 15 A motorboat heads upstream a distance of 24 miles on a river whose current is running at 3 miles per hour. The trip up and back takes 6 hours. Assuming the motorboat maintained a constant speed relative to the water, what was its speed?

Solution We use v to represent the constant speed of the motorboat relative to the water. Then the true speed going upstream is $v - 3$ miles per hour, and the true speed going downstream is $v + 3$ miles per hour. Since Distance = Velocity × Time, then Time = Distance/Velocity. We set up the table below.

	VELOCITY mi/hr	DISTANCE mi	TIME = DISTANCE/VELOCITY hr
Upstream	$v - 3$	24	$\frac{24}{v - 3}$
Downstream	$v + 3$	24	$\frac{24}{v + 3}$

Since the total time up and back is 6 hours, we have

$$\frac{24}{v - 3} + \frac{24}{v + 3} = 6$$

$$\frac{24(v + 3) + 24(v - 3)}{(v - 3)(v + 3)} = 6$$

$$\frac{48v}{v^2 - 9} = 6$$

$$48v = 6(v^2 - 9)$$

$$6v^2 - 48v - 54 = 0$$

$$v^2 - 8v - 9 = 0$$

$$(v - 9)(v + 1) = 0$$

$$v = 9 \quad \text{or} \quad v = -1$$

We discard the solution $v = -1$ mile per hour, so the speed of the motorboat relative to the water is 9 miles per hour. ■

Historical Comment ■ Problems using quadratic equations are found in the oldest known mathematical literature. Babylonians and Egyptians were solving such problems before 1800 BC. Euclid solved quadratic equations geometrically in his *Data* (300 BC), and the Hindus and Arabs gave rules for solving any quadratic equation with real roots. Because negative numbers were not freely used before AD 1500, there were several different types of quadratic equations, each with its own rule. Thomas Harriot (1560–1621) introduced the method of factoring to obtain solutions, and François Vièta (1540–1603) introduced a method that is essentially completing the square.

Until modern times it was usual to neglect the negative roots (if there were any), and equations involving square roots of negative quantities were regarded as unsolvable until the 1500's.

Historical Problems

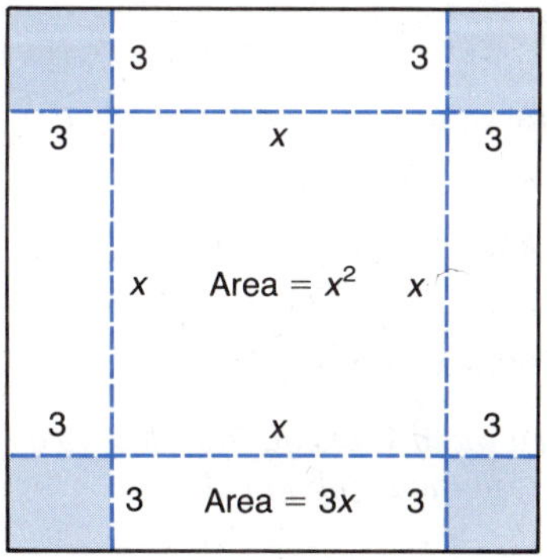

1. *One of al-Khowârizmî's solutions* We solve $x^2 + 12x = 85$ by drawing the square shown in the margin. The area of the unshaded part is $x^2 + 12x$. We then set this expression equal to 85 to get the equation $x^2 + 12x = 85$. If we add the four shaded squares, we will have a larger square of known area. Complete the solution.
2. *Vièta's method* We solve $x^2 + 12x - 85 = 0$ by letting $x = u + z$. Then

$$(u + z)^2 + 12(u + z) - 85 = 0$$
$$u^2 + (2z + 12)u + (z^2 + 12z - 85) = 0$$

Now select z so that $2z + 12 = 0$ and finish the solution.

3. *Another method to get the quadratic formula* Look at equation (4), page 110. Rewrite the right side as $(\sqrt{b^2 - 4ac}/2a)^2$ and then subtract it from each side. The right side is now 0 and the left side is a difference of two squares. If you factor this difference of two squares, you will easily be able to get the quadratic formula, and, moreover, the quadratic expression is factored, which is sometimes useful. ■

EXERCISE 2.3 ■

In Problems 1–20, solve each equation by factoring.

1. $x^2 = 4x$
2. $x^2 = -8x$
3. $x^2 - 9 = 0$
4. $x^2 - 1 = 0$
5. $z^2 + z - 12 = 0$
6. $v^2 + 7v + 12 = 0$
7. $2x^2 - 5x - 3 = 0$
8. $3x^2 + 5x + 2 = 0$
9. $6t^2 = 1 - t$
10. $6y^2 = 6 + 5y$
11. $x(x - 7) + 12 = 0$
12. $x(x + 1) = 12$
13. $4x^2 + 9 = 12x$
14. $25x^2 + 16 = 40x$
15. $6(p^2 - 1) = 5p$
16. $2(2u^2 - 4u) + 3 = 0$
17. $6x - 5 = \dfrac{6}{x}$
18. $x + \dfrac{12}{x} = 7$
19. $\dfrac{4(x - 2)}{x - 3} + \dfrac{3}{x} = \dfrac{-3}{x(x - 3)}$
20. $\dfrac{5}{x + 4} = 4 + \dfrac{3}{x - 2}$

In Problems 21–40, find the real solutions, if any, of each equation. Use the quadratic formula.

21. $x^2 - 4x + 2 = 0$ **22.** $x^2 + 4x + 2 = 0$ **23.** $x^2 - 5x - 1 = 0$

24. $x^2 + 5x + 3 = 0$ **25.** $2x^2 - 5x + 3 = 0$ **26.** $2x^2 + 5x + 3 = 0$

27. $4y^2 - y + 2 = 0$ **28.** $4t^2 + t + 1 = 0$ **29.** $4x^2 = 1 - 2x$

30. $2x^2 = 1 - 2x$ **31.** $4x^2 = 9x$ **32.** $5x = 4x^2$

33. $9t^2 - 6t + 1 = 0$ **34.** $4u^2 - 6u + 9 = 0$ **35.** $x^2 - 2x - 1 = 0$

36. $x^2 - 3x - 1 = 0$ **37.** $4 - \frac{1}{x} - \frac{2}{x^2} = 0$ **38.** $4 + \frac{1}{x} - \frac{1}{x^2} = 0$

39. $3x = 1 - \frac{1}{x}$ **40.** $x = 1 - \frac{4}{x}$

C *In Problems 41–48, find the real solutions, if any, of each equation. Use the quadratic formula, and express any solutions rounded off to two decimal places.*

41. $x^2 - 4x + 2 = 0$ **42.** $x^2 + 4x + 2 = 0$ **43.** $x^2 + \sqrt{3}x - 3 = 0$

44. $x^2 + \sqrt{2}x - 2 = 0$ **45.** $\pi x^2 - x - \pi = 0$ **46.** $\pi x^2 + \pi x - 2 = 0$

47. $x^2 + \pi x + \sqrt{29} = 0$ **48.** $4x^2 - 15\sqrt{2}x + 20 = 0$

In Problems 49–60, find the real solutions, if any, of each quadratic equation. Use any method.

49. $x^2 - 6 = 0$ **50.** $x^2 - 5 = 0$ **51.** $16x^2 - 8x + 1 = 0$

52. $9x^2 - 6x + 1 = 0$ **53.** $10x^2 - 19x - 15 = 0$ **54.** $6x^2 + 7x - 20 = 0$

55. $2 + z = 6z^2$ **56.** $2 = y + 6y^2$ **57.** $x^2 + \sqrt{2}x = \frac{1}{2}$

58. $\frac{1}{2}x^2 = \sqrt{2}x + 1$ **59.** $x^2 + x = 4$ **60.** $x^2 + x = 1$

In Problems 61–66, use the discriminant to determine whether each quadratic equation has two unequal real solutions, a repeated real solution, or no real solution, without solving the equation.

61. $x^2 - 5x + 7 = 0$ **62.** $x^2 + 5x + 7 = 0$ **63.** $9x^2 - 30x + 25 = 0$

64. $25x^2 - 20x + 4 = 0$ **65.** $3x^2 + 5x - 2 = 0$ **66.** $2x^2 - 3x - 4 = 0$

In Problems 67–72, what number should be added to complete the square of each expression?

67. $x^2 + 4x$ **68.** $x^2 - 2x$ **69.** $x^2 + \frac{1}{2}x$ **70.** $x^2 - \frac{1}{3}x$ **71.** $x^2 - \frac{2}{3}x$ **72.** $x^2 - \frac{2}{5}x$

In Problems 73–78, solve each equation by completing the square.

73. $x^2 + 4x - 21 = 0$ **74.** $x^2 - 6x = 13$ **75.** $x^2 - \frac{1}{2}x = \frac{3}{16}$

76. $x^2 + \frac{2}{3}x = \frac{1}{3}$ **77.** $3x^2 + x - \frac{1}{2} = 0$ **78.** $2x^2 - 3x = 1$

79. Find two consecutive odd integers whose product is 143.

80. Find two consecutive integers whose product is 306.

81. Find the dimensions of a rectangle whose perimeter is 30 meters and whose area is 56 square meters.

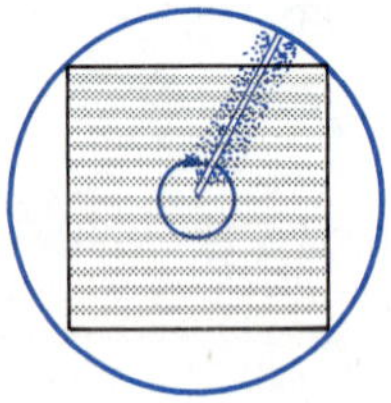

82. An adjustable water sprinkler that sprays water in a circular pattern is placed at the center of a square field whose area is 1250 square feet (see the figure). What is the shortest radius setting that can be used if the field is to be completely enclosed within the circle?

83. An open box is to be constructed from a square piece of sheet metal by removing a square of side 1 foot from each corner and turning up the edges. If the box is to hold 4 cubic feet, what should be the dimensions of the sheet metal?

84. Rework Problem 83 if the piece of sheet metal is a rectangle whose length is twice its width.

85. A ball is thrown vertically upward from the top of a building 96 feet tall with an initial velocity of 80 feet per second. The distance s (in feet) of the ball from the ground after t seconds is $s = 96 + 80t - 16t^2$.
(a) After how many seconds does the ball strike the ground?
(b) After how many seconds will the ball pass the top of the building on its way down?

C **86.** A 39 ounce can of Hills Bros.® coffee requires 188.5 square inches of aluminum. If its height is 7 inches, what is its radius? (The surface area A of a right circular cylinder is $A = 2\pi r^2 + 2\pi rh$, where r is the radius and h is the height.)

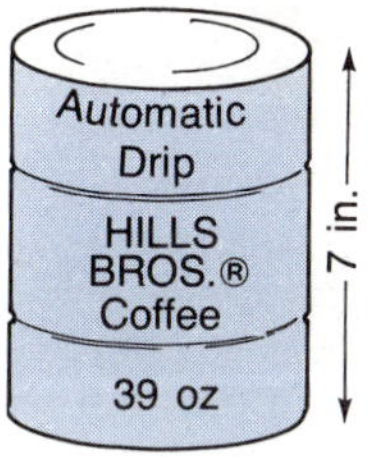

87. Mike and Dan, working together, can paint the exterior of a house in 6 days. Mike, by himself, can complete this job in 5 days less than Dan. How long will it take Mike to complete the job by himself?

88. Two pumps of different sizes, working together, can empty a fuel tank in 5 hours. The larger pump can empty this tank in 4 hours less than the smaller one. If the larger one is out of order, how long will it take the smaller one to do the job alone?

89. A company charges \$200 for each box of tools on orders of 150 or fewer boxes. If a customer orders x boxes in excess of 150, the cost for each box ordered is reduced by x dollars. If a customer's bill came to \$30,625, how many boxes were ordered?

90. A contractor orders 8 cubic yards of premixed cement, all of which is to be used to pour a patio that will be 4 inches thick. If the length of the patio is specified to be twice the width, what will be the patio dimensions? (1 cubic yard = 27 cubic feet)

C **91.** A landscaper, who just completed a rectangular flower garden measuring 6 feet by 10 feet, orders 1 cubic yard of premixed cement, all of which is to be used to create a border of uniform width around the garden. If the border is to have a depth of 3 inches, how wide will the border be?

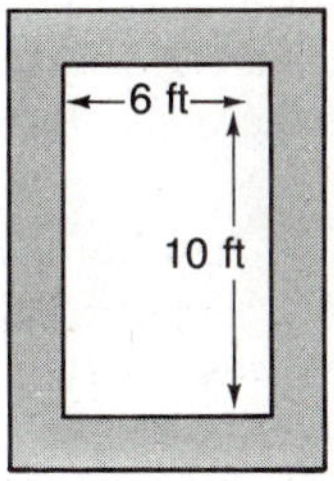

92. An object is propelled vertically upward with an initial velocity of 20 meters per second. The distance s (in meters) of the object from the ground after t seconds is $s = -4.9t^2 + 20t$.
(a) When will the object be 15 meters above the ground?
(b) When will it strike the ground?
(c) Will the object reach a height of 100 meters?
(d) What is the maximum height?

C **93.** A jumbo chocolate bar with a rectangular shape measures 12 centimeters in length, 7 centimeters in width, and 3 centimeters in thickness. Due to escalating costs of cocoa, management decides to reduce the volume of the bar by 10%. To accomplish this reduction, management decides the new bar should have the same 3 centimeter thickness, but the new length and width each should be reduced by the same amount. What should be the dimensions of the new candy bar?

C **94.** Rework Problem 93 if the reduction is to be 20%.

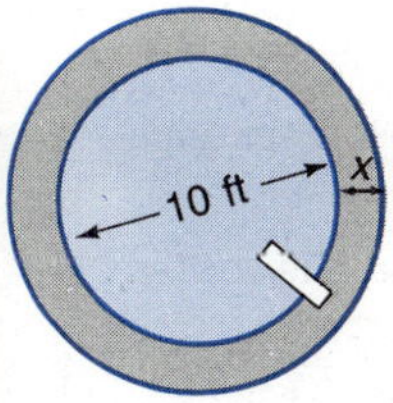

C **95.** A pool in the shape of a circle measures 10 feet across. One cubic yard of concrete is to be used to create a circular border of uniform width around the pool. If the border is to have a depth of 3 inches, how wide will the border be? (1 cubic yard = 27 cubic feet)

C **96.** Rework Problem 95 if the depth of the border is 4 inches.

97. A motorboat maintained a constant speed of 15 miles per hour relative to the water in going 10 miles upstream and then returning. The total time for the trip was 1.5 hours. Use this information to find the speed of the current.

98. The hypotenuse of a right triangle measures 13 centimeters. Find the lengths of the legs if their sum is 17 centimeters.

99. The diagonal of a rectangle measures 10 inches. If the length is 2 inches more than the width, find the dimensions of the rectangle.

100. An object is thrown down from the top of a building 1280 feet tall with an initial velocity of 32 feet per second. The distance s (in feet) of the object from the ground after t seconds is $s = 1280 - 32t - 16t^2$.
(a) When will the object strike ground?
(b) What is the height of the object after 4 seconds?

101. Show that the sum of the roots of a quadratic equation is $-b/a$.

102. Show that the product of the roots of a quadratic equation is c/a.

103. Find k such that the equation $kx^2 + x + k = 0$ has a repeated real solution.

104. Find k such that the equation $x^2 - kx + 4 = 0$ has a repeated real solution.

105. Show that the real solutions of the equation $ax^2 + bx + c = 0$ are the negatives of the real solutions of the equation $ax^2 - bx + c = 0$. Assume $b^2 - 4ac \geq 0$.

106. Show that the real solutions of the equation $ax^2 + bx + c = 0$ are the reciprocals of the real solutions of the equation $cx^2 + bx + a = 0$. Assume $b^2 - 4ac \geq 0$.

107. The sum of the consecutive integers 1, 2, 3, . . . , n is given by the formula $\frac{1}{2}n(n + 1)$. How many consecutive integers, starting with 1, must be added to get a sum of 666?

108. If a polygon of n sides has $\frac{1}{2}n(n - 3)$ diagonals, how many sides will a polygon with 65 diagonals have? Is there a polygon with 80 diagonals?

109. In going from Chicago to Atlanta, a car averages 45 miles per hour, and in going from Atlanta to Miami, it averages 55 miles per hour. If Atlanta is halfway between Chicago and Miami, what is the average speed from Chicago to Miami? [*Hint:* The answer is not 50 miles per hour—be careful!]

C **110.** On a recent flight from Phoenix to Kansas City, a distance of 919 nautical miles, the plane arrived 20 minutes early. On leaving the aircraft, I asked the captain, "What was our tail wind?" He replied, "I don't know, but our ground speed was 550 knots." What was the tail wind? (1 knot = 1 nautical mile per hour)

111. *Programming exercise* Develop code that will solve a quadratic equation:

```
{Enter the coefficient of x squared}    READ (a);
{Enter the coefficient of x}            READ (b);
{Enter the constant term}               READ (c);
IF b^2 - 4ac < 0
THEN {write no real solution}
ELSE IF b^2 - 4ac = 0
     THEN {write -b/2a is a double root}
     ELSE {write (-b + SQRT(b^2 - 4ac))/2a
          or (-b - SQRT(b^2 - 4ac))/2a
          is a solution};
```

2.4 Quadratic Equations with a Negative Discriminant

Quadratic equations with a negative discriminant have no real number solution. However, if we extend our number system to allow complex numbers (refer to Section 1.9), quadratic equations will always have a solution. Since the solution to a quadratic equation involves the square root of the discriminant, we begin with a discussion of square roots of negative numbers.

Square Roots of Negative Numbers

When we first introduced square roots, we made the observation that the equation

$$x^2 = N \qquad N > 0$$

has two solutions—one positive, the other negative. To avoid confusion, we agreed that the positive solution of the equation $x^2 = N$ would be called the **principal square root of** N, and we introduced the notation $\sqrt{N}$ to represent it. Thus, $\sqrt{4} = 2$, $\sqrt{9} = 3$, and so on.

With these ideas as background, we give the following definition.

Principal Square Root of $-N$

If N is a positive real number, we define the **principal square root of** $-N$, denoted by $\sqrt{-N}$, as

$$\sqrt{-N} = \sqrt{N}\,i$$

where i is the imaginary unit and $i^2 = -1$.

Example 1 (a) $\sqrt{-1} = \sqrt{1}\,i = i$ (b) $\sqrt{-4} = \sqrt{4}\,i = 2i$
(c) $\sqrt{-8} = \sqrt{8}\,i = 2\sqrt{2}\,i$ ■

Example 2 Solve each equation in the complex number system.

(a) $x^2 = 4$ (b) $x^2 = -9$

Solution (a)

$$x^2 = 4$$
$$x = \pm\sqrt{4} = \pm 2$$

The equation has two solutions, -2 and 2.

(b)

$$x^2 = -9$$
$$x = \pm\sqrt{-9} = \pm\sqrt{9}\,i = \pm 3i$$

The equation has two solutions, $-3i$ and $3i$. ■

Warning: When working with square roots of negative numbers, do not set the square root of a product equal to the product of the square roots (which can be done with positive numbers). To see why, look at this calculation: We know that $\sqrt{100} = 10$. However, it is also true that $100 = (-25)(-4)$, so

$$10 = \sqrt{100} = \sqrt{(-25)(-4)} \underset{\uparrow}{=} \sqrt{-25}\sqrt{-4}$$

Here is the error.

$$= (\sqrt{25}i)(\sqrt{4}i) = (5i)(2i) = 10i^2 = -10$$

The Quadratic Formula

Because we have defined the square root of a negative number, we now can restate the quadratic formula without restriction.

Theorem In the complex number system, the solutions of the quadratic equation $ax^2 + bx + c = 0$, where a, b, and c are real numbers and $a \neq 0$, are given by the formula

Quadratic Formula

$$x = \frac{-b \pm \sqrt{b^2 - 4ac}}{2a} \tag{1}$$

■

Example 3 Solve the equation $x^2 - 4x + 8 = 0$ in the complex number system.

Solution Here $a = 1$, $b = -4$, $c = 8$, and $b^2 - 4ac = 16 - 4(8) = -16$. Using equation (1), we find

$$x = \frac{4 \pm \sqrt{-16}}{2} = \frac{4 \pm \sqrt{16}i}{2} = \frac{4 \pm 4i}{2} = 2 \pm 2i$$

The equation has the solution set $\{2 - 2i, 2 + 2i\}$.

Check:

$2 + 2i$: $(2 + 2i)^2 - 4(2 + 2i) + 8 = 4 + 8i + 4i^2 - 8 - 8i + 8$
$= 4 - 4 = 0$

$2 - 2i$: $(2 - 2i)^2 - 4(2 - 2i) + 8 = 4 - 8i + 4i^2 - 8 + 8i + 8$
$= 4 - 4 = 0$ ■

The discriminant, $b^2 - 4ac$, of a quadratic equation still serves as a way to determine the character of the solutions.

Discriminant of a Quadratic Equation

In the complex number system, consider a quadratic equation $ax^2 + bx + c = 0$ with real coefficients.

1. If $b^2 - 4ac > 0$, the equation has two unequal real solutions.
2. If $b^2 - 4ac = 0$, the equation has a repeated real solution—a double root.
3. If $b^2 - 4ac < 0$, the equation has two complex solutions that are not real. The solutions are conjugates of each other.

The third conclusion in the box is a consequence of the fact that if $b^2 - 4ac = -N < 0$, then by the quadratic formula, the solutions are

$$x = \frac{-b + \sqrt{b^2 - 4ac}}{2a} = \frac{-b + \sqrt{-N}}{2a} = \frac{-b + \sqrt{N}i}{2a} = \frac{-b}{2a} + \frac{\sqrt{N}}{2a}i$$

and

$$x = \frac{-b - \sqrt{b^2 - 4ac}}{2a} = \frac{-b - \sqrt{-N}}{2a} = \frac{-b - \sqrt{N}i}{2a} = \frac{-b}{2a} - \frac{\sqrt{N}}{2a}i$$

which are conjugates of each other.

Example 4 Without solving, determine the character of the solution of each equation in the complex number system.

(a) $3x^2 + 4x + 5 = 0$ (b) $2x^2 + 4x + 1 = 0$

(c) $9x^2 - 6x + 1 = 0$

Solution (a) Here, $a = 3$, $b = 4$, and $c = 5$, so $b^2 - 4ac = 16 - 4(3)(5) = -44$. The solutions are complex numbers that are not real and are conjugates of each other.

(b) Here, $a = 2$, $b = 4$, and $c = 1$, so $b^2 - 4ac = 16 - 8 = 8$. The solutions are two unequal real numbers.

(c) Here, $a = 9$, $b = -6$, and $c = 1$, so $b^2 - 4ac = 36 - 4(9)(1) = 0$. The solution is a repeated real number—that is, a double root. ■

EXERCISE 2.4

In Problems 1–10, perform the indicated operations and express your answer in the form $a + bi$.

1. $\sqrt{-4}$ **2.** $\sqrt{-9}$ **3.** $\sqrt{-25}$ **4.** $\sqrt{-64}$

5. $\sqrt{(3 + 4i)(4i - 3)}$ **6.** $\sqrt{(4 + 3i)(3i - 4)}$ **7.** $(\sqrt{-16} + 1)^2$

8. $(\sqrt{-4} + 2)^2$ **9.** $(3 - \sqrt{-4})(2 + \sqrt{-9})$ **10.** $(8 + \sqrt{-4})(2 - \sqrt{-9})$

In Problems 11–30, solve each equation in the complex number system.

11. $x^2 + 4 = 0$ **12.** $x^2 - 4 = 0$ **13.** $x^2 - 16 = 0$

14. $x^2 + 25 = 0$ **15.** $x^2 - 6x + 13 = 0$ **16.** $x^2 + 4x + 8 = 0$

17. $x^2 - 6x + 10 = 0$ **18.** $x^2 - 2x + 5 = 0$ **19.** $8x^2 - 4x + 1 = 0$

20. $10x^2 + 6x + 1 = 0$ **21.** $5x^2 + 2x + 1 = 0$ **22.** $13x^2 + 6x + 1 = 0$

23. $x^2 + x + 1 = 0$ **24.** $x^2 - x + 1 = 0$ **25.** $x^3 - 8 = 0$

26. $x^3 + 27 = 0$ **27.** $x^4 - 16 = 0$ **28.** $x^4 - 1 = 0$

29. $x^4 + 13x^2 + 36 = 0$ **30.** $x^4 + 3x^2 - 4 = 0$

In Problems 31–36, without solving, determine the character of the solutions of each equation in the complex number system.

31. $3x^2 - 3x + 4 = 0$ **32.** $2x^2 - 4x + 1 = 0$ **33.** $2x^2 + 3x - 4 = 0$

34. $x^2 + 2x + 6 = 0$ **35.** $9x^2 - 12x + 4 = 0$ **36.** $4x^2 + 12x + 9 = 0$

37. $2 + 3i$ is a solution of a quadratic equation with real coefficients. Find the other solution.

38. $4 - i$ is a solution of a quadratic equation with real coefficients. Find the other solution.

2.5 Other Types of Equations

In this section we look at other types of equations, most of which can be solved using variations of the techniques already discussed. For the rest of this chapter, we shall be working in the real number system.

Equations Containing Radicals

When the variable in an equation occurs in a square root, cube root, and so on—that is, when it occurs in a radical—the equation is called a **radical equation**. Sometimes, a suitable operation will change a radical equation to one that is linear or quadratic. A commonly used procedure is to isolate the most complicated radical on one side of the equation and then eliminate it by raising each side to a power equal to the index of the radical. Care must be taken, however, because apparent solutions that are not, in fact, solutions of the original equation may result. These are called **extraneous solutions**. Thus, we need to check all answers when working with radical equations.

Example 1 Solve the equation: $\sqrt[3]{2x - 4} - 2 = 0$

Solution The equation contains a radical whose index is 3. We isolate it on the left side:

$$\sqrt[3]{2x - 4} - 2 = 0$$
$$\sqrt[3]{2x - 4} = 2$$

Now raise each side to the third power (the index of the radical is 3) and solve:

$$(\sqrt[3]{2x-4})^3 = 2^3$$
$$2x - 4 = 8$$
$$2x = 12$$
$$x = 6$$

Check: $\sqrt[3]{2(6)-4} - 2 = \sqrt[3]{12-4} - 2 = \sqrt[3]{8} - 2 = 2 - 2 = 0$

The solution is $x = 6$. ■

Sometimes, we need to raise each side to a power more than once in order to solve a radical equation.

Example 2 Solve the equation: $\sqrt{2x+3} - \sqrt{x+2} = 2$

Solution First, we choose to isolate the more complicated radical expression (in this case, $\sqrt{2x+3}$) on the left side:

$$\sqrt{2x+3} = \sqrt{x+2} + 2$$

Now square both sides (the index of the radical is 2):

$$(\sqrt{2x+3})^2 = (\sqrt{x+2} + 2)^2$$
$$2x + 3 = (\sqrt{x+2})^2 + 4\sqrt{x+2} + 4$$
$$2x + 3 = x + 2 + 4\sqrt{x+2} + 4$$

Because the equation still contains a radical, we combine like terms, isolate the remaining radical on the right side, and again square both sides:

$$x - 3 = 4\sqrt{x+2}$$
$$(x-3)^2 = 16(x+2)$$
$$x^2 - 6x + 9 = 16x + 32$$
$$x^2 - 22x - 23 = 0$$
$$(x-23)(x+1) = 0$$
$$x = 23 \quad \text{or} \quad x = -1$$

The original equation appears to have the solution set $\{-1, 23\}$. However, we have not yet checked.

Check: $\sqrt{2(23)+3} - \sqrt{23+2} = \sqrt{49} - \sqrt{25} = 7 - 5 = 2$

$\sqrt{2(-1)+3} - \sqrt{-1+2} = \sqrt{1} - \sqrt{1} = 1 - 1 = 0$

Thus, the equation has only one real solution, 23; the solution -1 is extraneous. ■

Note: In Example 2, since $x - 3 = 4\sqrt{x + 2}$, we know that in the real number system $x - 3 \geq 0$ or $x \geq 3$. As a result, when we found that $x = 23$ or $x = -1$, we could have discarded $x = -1$, leaving $x = 23$ as the only real solution.

Equations Quadratic in Form

The equation $x^4 + x^2 - 12 = 0$ is not quadratic in x, but it is quadratic in x^2. That is, if we let $u = x^2$, we get $u^2 + u - 12 = 0$, a quadratic equation. This equation can be solved for u and, in turn, by using $u = x^2$, we can find the solutions x of the original equation.

In general, if an appropriate substitution u transforms an equation into one of the form

$$au^2 + bu + c = 0 \qquad a \neq 0$$

then the original equation is called an **equation of the quadratic type**, or an **equation quadratic in form**.

The difficulty of solving such an equation lies in the determination that the equation is, in fact, quadratic in form. After you are told an equation is quadratic in form, it is easy enough to see it, but some practice is needed to enable you to see these on your own.

Example 3 Solve the equation: $x^4 + x^2 - 12 = 0$

Solution For the equation $x^4 + x^2 - 12 = 0$, we let $u = x^2$. Then $u^2 = x^4$, and the original equation,

$$x^4 + x^2 - 12 = 0$$

becomes

$$u^2 + u - 12 = 0 \qquad u = x^2,\ u^2 = x^4$$
$$(u + 4)(u - 3) = 0 \qquad \text{Factor.}$$
$$u = -4 \quad \text{or} \quad u = 3$$

But remember that we want to solve for x. Because $u = x^2$, we have

$$x^2 = -4 \quad \text{or} \quad x^2 = 3$$

The first of these has no real solution; the second has the solution set $\{-\sqrt{3}, \sqrt{3}\}$.

Check:

$$(\sqrt{3})^4 + (\sqrt{3})^2 - 12 = 9 + 3 - 12 = 0$$
$$(-\sqrt{3})^4 + (-\sqrt{3})^2 - 12 = 9 + 3 - 12 = 0$$

Thus, $\{-\sqrt{3}, \sqrt{3}\}$ is the solution set of the original equation. ■

Example 4 Solve the equation: $x + 2\sqrt{x} - 3 = 0$

Solution For the equation $x + 2\sqrt{x} - 3 = 0$, let $u = \sqrt{x}$. Then $u^2 = x$, and the original equation,

$$x + 2\sqrt{x} - 3 = 0$$

becomes

$$\begin{aligned} u^2 + 2u - 3 &= 0 && u = \sqrt{x},\ u^2 = x \\ (u + 3)(u - 1) &= 0 && \text{Factor.} \\ u = -3 \quad \text{or} \quad u &= 1 && \text{Solve.} \end{aligned}$$

Since $u = \sqrt{x}$, we have $\sqrt{x} = -3$ or $\sqrt{x} = 1$. The first of these, $\sqrt{x} = -3$, has no real solution, since the square root of a real number is never negative. The second one, $\sqrt{x} = 1$, has the solution $x = 1$.

Check: $1 + 2\sqrt{1} - 3 = 1 + 2 - 3 = 0$

Thus, $x = 1$ is the only solution of the original equation. ■

Example 5 Solve the equation: $\dfrac{1}{x^4} + \dfrac{11}{x^2} - 12 = 0$

Solution For this equation, we let $u = 1/x^2$. Then $u^2 = 1/x^4$, and the original equation,

$$\frac{1}{x^4} + \frac{11}{x^2} - 12 = 0$$

becomes

$$\begin{aligned} u^2 + 11u - 12 &= 0 && u = \frac{1}{x^2},\ u^2 = \frac{1}{x^4} \\ (u + 12)(u - 1) &= 0 \\ u = -12 \quad \text{or} \quad u &= 1 \\ \frac{1}{x^2} = -12 \quad \text{or} \quad \frac{1}{x^2} &= 1 \end{aligned}$$

The equation on the left has no real solution. We take reciprocals of each side of the right-hand equation to get

$$\begin{aligned} x^2 &= 1 \\ x &= \pm 1 \end{aligned}$$

Check: $\dfrac{1}{(\pm 1)^4} + \dfrac{11}{(\pm 1)^2} - 12 = \dfrac{1}{1} + \dfrac{11}{1} - 12 = 0$

The original equation has the solution set $\{-1, 1\}$. ■

The idea should now be clear. If an equation contains some expression and that same expression squared, make a substitution for the expression. You may get a quadratic equation.

Factorable Equations

We have already used factoring as a means of solving certain quadratic equations. This method also can be used to solve *any* equation that can be factored with 0 on one side of the equation. The solutions are then found by setting each factor equal to 0.

Example 6 Solve the equation: $x^3 = 25x$

Solution We first rearrange the equation to get 0 on the right side:

$$x^3 = 25x$$
$$x^3 - 25x = 0$$

We notice that x is a factor of each term on the left:

$$x(x^2 - 25) = 0$$

$$x = 0 \quad \text{or} \quad x^2 - 25 = 0 \qquad \text{Set each factor equal to 0.}$$

$$x = \pm 5 \qquad \text{Solve.}$$

The solution set is $\{-5, 0, 5\}$. ■

Example 7 Solve the equation: $x^3 - x^2 - 4x + 4 = 0$

Solution Do you recall the method of factoring by grouping from Chapter 1? We group the terms of $x^3 - x^2 - 4x + 4 = 0$ as follows:

$$(x^3 - x^2) - (4x - 4) = 0$$

Factor out x^2 from the first grouping and 4 from the second:

$$x^2(x - 1) - 4(x - 1) = 0$$

This reveals the common factor $(x - 1)$, so we have

$$(x^2 - 4)(x - 1) = 0$$

$$(x - 2)(x + 2)(x - 1) = 0 \qquad \text{Factor again.}$$

$$x - 2 = 0 \quad \text{or} \quad x + 2 = 0 \quad \text{or} \quad x - 1 = 0 \qquad \text{Set each factor equal to 0.}$$

$$x = 2 \qquad x = -2 \qquad x = 1 \qquad \text{Solve.}$$

The solution set is $\{-2, 1, 2\}$. ■

EXERCISE 2.5

In Problems 1–26, solve each equation.

1. $\sqrt{2t-1}=1$
2. $\sqrt{3t+4}=2$
3. $\sqrt{2t-1}=-1$
4. $\sqrt{3t+4}=-2$
5. $\sqrt[3]{1-2x}-3=0$
6. $\sqrt[3]{1-2x}-1=0$
7. $\sqrt{15-2x}=x$
8. $\sqrt{12-x}=x$
9. $x=2\sqrt{x-1}$
10. $x=2\sqrt{-x-1}$
11. $\sqrt{x^2-x-4}=x+2$
12. $\sqrt{3-x+x^2}=x-2$
13. $\sqrt[3]{y^3-2y+4}=y$
14. $\sqrt[3]{z^3+z+5}=z$
15. $\sqrt{2x+3}-\sqrt{x+1}=1$
16. $\sqrt{3x+7}+\sqrt{x+2}=1$
17. $\sqrt{3x+1}-\sqrt{x-1}=2$
18. $\sqrt{3x-5}-\sqrt{x+7}=2$
19. $\sqrt{3-2\sqrt{x}}=\sqrt{x}$
20. $\sqrt{10+3\sqrt{x}}=\sqrt{x}$
21. $(3x+1)^{1/2}=4$
22. $(3x-5)^{1/2}=2$
23. $(x-1)^{1/3}=2$
24. $(2x+1)^{1/3}=-2$
25. $(x^2+16)^{1/2}=5$
26. $(x^2-9)^{1/2}=4$

In Problems 27–50, solve each equation.

27. $x-4\sqrt{x}=0$
28. $x+8\sqrt{x}=0$
29. $x+\sqrt{x}=12$
30. $x+7\sqrt{x}+12=0$
31. $2(s+1)^2-5(s+1)=3$
32. $3(1-y)^2+5(1-y)+2=0$
33. $x^4-5x^2+4=0$
34. $x^4-10x^2+4=0$
35. $t^{1/2}-2t^{1/4}+1=0$
36. $z^{1/2}+2z^{1/4}+1=0$
37. $4x^{1/2}-9x^{1/4}+4=0$
38. $x^{1/2}-3x^{1/4}+2=0$
39. $\sqrt[4]{5x^2-6}=x$
40. $\sqrt[4]{4-5x^2}=x$
41. $x^2+3x+\sqrt{x^2+3x}=6$
42. $x^2-3x-\sqrt{x^2-3x}=2$
43. $\dfrac{1}{(x+1)^2}=\dfrac{1}{x+1}+2$
44. $\dfrac{1}{(x-1)^2}+\dfrac{1}{x-1}=12$
45. $3x^{-2}-7x^{-1}-6=0$
46. $2x^{-2}-3x^{-1}-4=0$
47. $2x^{2/3}-5x^{1/3}-3=0$
48. $3x^{4/3}+5x^{2/3}+2=0$
49. $\left(\dfrac{v}{v+1}\right)^2+\dfrac{2v}{v+1}=8$
50. $\left(\dfrac{y}{y-1}\right)^2=6\left(\dfrac{y}{y-1}\right)+7$

In Problems 51–62, solve each equation by factoring.

51. $x=8\sqrt{x}$
52. $x=6\sqrt{x}$
53. $x^{3/2}-2x^{1/2}=0$
54. $x^{3/4}-4x^{1/4}=0$
55. $x^3+x^2-12x=0$
56. $x^3-6x^2-7x=0$
57. $x^3+x^2+x+1=0$
58. $x^3+x^2-x-1=0$
59. $x^3-3x^2-4x+12=0$
60. $x^3-3x^2-x+3=0$
61. $t^6-t^4-t^2+1=0$
62. $y^6-4y^4-y^2+4=0$

C *In Problems 63–68, solve each equation. Use a calculator to express solutions correct to two decimal places.*

63. $x-4x^{1/2}+2=0$
64. $x^{2/3}+4x^{1/3}+2=0$
65. $x^4+\sqrt{3}x^2-3=0$
66. $x^4+\sqrt{2}x^2-2=0$
67. $\pi(1+t)^2=\pi+1+t$
68. $\pi(1+r)^2=2+\pi(1+r)$

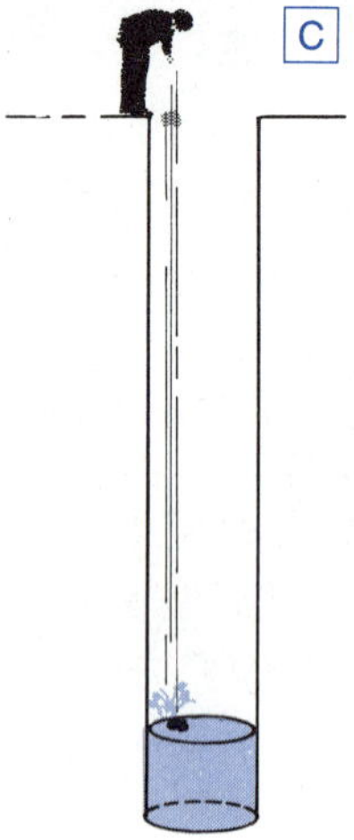

C **69.** The depth of a well can sometimes be found by dropping an object into the well and measuring the time elapsed until a sound is heard. If t_1 is the time (measured in seconds) it takes for the object to strike the bottom of the well, then t_1 will obey the equation $s = 16t_1^2$, where s is the distance (measured in feet). It follows that $t_1 = \sqrt{s}/4$. Suppose t_2 is the time it takes for the sound of impact to reach your ears. Because sound waves are known to travel at a speed of approximately 1100 feet per second, the time t_2 to travel the distance s will be $t_2 = s/1100$. Now $t_1 + t_2$ is the total time that elapses from the moment the object is dropped to the moment a sound is heard. Thus, we have the equation

$$\text{Total time elapsed} = \frac{\sqrt{s}}{4} + \frac{s}{1100}$$

Find the depth of a well if the total time elapsed from dropping a rock to hearing it hit bottom is 4 seconds. Express the answer correct to the nearest integer.

2.6 Inequalities

Properties of Inequalities

In working with inequalities, we will need to know certain properties they obey. In the properties listed below, a, b, and c are real numbers.

For any real number a, we have

Nonnegative Property

$$a^2 \geq 0$$

The **nonnegative property** is a consequence of the fact that the product of two positive real numbers is positive; the product of two negative real numbers is positive; and the product of 0 and 0 is 0. Thus, for any real number a, the value of a^2 is 0 or positive; that is, a^2 is nonnegative.

Transitive Property

If $a < b$ and $b < c$, then $a < c$.

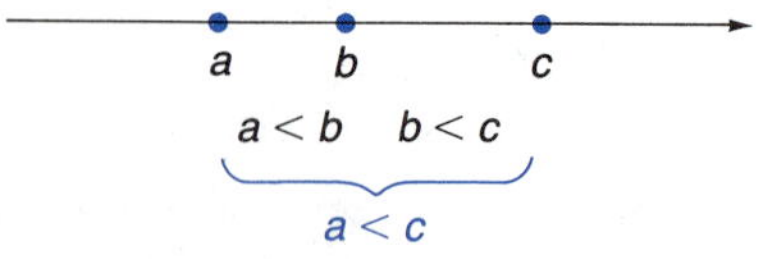

Figure 4

In Figure 4, we can see that geometrically the **transitive property** states that if a lies to the left of b and b lies to the left of c, then it follows that a must also lie to the left of c.

Addition Property

> If $a < b$, then $a + c < b + c$.

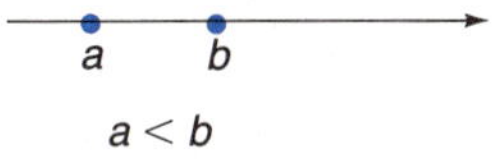

Figure 5

The **addition property** states that the sense, or direction, of an inequality remains unchanged if the same number is added to each side. In Figure 5, we see that a lies to the left of b. Now look at Figure 6(a). If c is positive, then $a + c$ and $b + c$ each lie c units to the right of a and b, respectively. Consequently, $a + c$ must lie to the left of $b + c$; that is, $a + c < b + c$. Figure 6(b) illustrates the situation if c is negative.

Figure 6

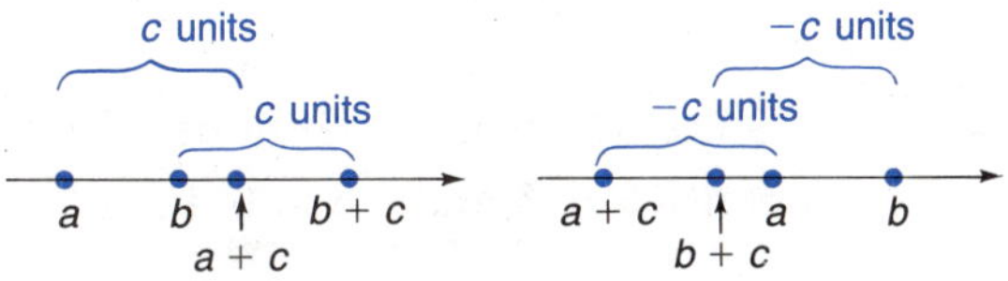

(a) If $a < b$ and $c > 0$, then $a + c < b + c$.

(b) If $a < b$ and $c < 0$, then $a + c < b + c$.

Example 1 (a) Since $3 < 5$, then $3 + 7 < 5 + 7$, or $10 < 12$.

(b) Since $4 < 10$, then $4 + (-4) < 10 + (-4)$, or $0 < 6$ or $6 > 0$.

(c) Since $-3 < 0$, then $-3 + 3 < 0 + 3$, or $0 < 3$ or $3 > 0$. ■

We'll use two examples to arrive at our next property.

Example 2 Express as an inequality the result of multiplying each side of the inequality $3 < 7$ by 2.

Solution We begin with

$$3 < 7$$

Multiplying each side by 2 yields the numbers 6 and 14, so we have

$$6 < 14$$ ■

Example 3 Express as an inequality the result of multiplying each side of the inequality $3 < 7$ by -4.

Solution We begin with

$$3 < 7$$

Multiplying each side by -4 yields the numbers -12 and -28, so we have

$$-12 > -28$$ ■

Examples 2 and 3 illustrate the following general properties of multiplication involving inequalities:

Multiplication Properties

(a) If $a < b$ and if $c > 0$, then $ac < bc$.
(b) If $a < b$ and if $c < 0$, then $ac > bc$.

The **multiplication properties** state that the sense, or direction, of an inequality *remains the same* if each side is multiplied by a *positive* real number, while the direction is *reversed* if each side is multiplied by a *negative* real number.

Example 4 (a) Since $3 < 5$ and $2 > 0$, then $3 \cdot 2 < 5 \cdot 2$, or $6 < 10$.
(b) Since $3 < 5$ and $-3 < 0$, then $3 \cdot (-3) > 5 \cdot (-3)$, or $-9 > -15$. ■

The **reciprocal property** states that the reciprocal of a positive real number is positive.

Reciprocal Property

If $a > 0$, then $\frac{1}{a} > 0$.

The transitive, addition, multiplication, and reciprocal properties may also be stated as follows:

Transitive Property

If $a > b$ and $b > c$, then $a > c$.

Addition Property

If $a > b$, then $a + c > b + c$.

Multiplication Properties

(a) If $a > b$ and if $c > 0$, then $ac > bc$.
(b) If $a > b$ and if $c < 0$, then $ac < bc$.

Reciprocal Property

If $a < 0$, then $\frac{1}{a} < 0$.

Finally, we have the **trichotomy property**, which states that either two numbers are equal or one of them is less than the other.

For any pair of numbers a and b,

Trichotomy Property

$$a < b \quad \text{or} \quad a = b \quad \text{or} \quad b < a$$

If $b = 0$, the trichotomy property states that for any real number a,

$$a < 0 \quad \text{or} \quad a = 0 \quad \text{or} \quad a > 0$$

That is, any real number is negative or 0 or positive, a fact we have already noted.

Graphing Inequalities

We shall find it useful in later work to graph inequalities on the real number line.

Example 5 (a) On the real number line, graph all numbers x for which $x > 4$.

(b) On the real number line, graph all numbers x for which $x \le 5$.

Solution (a) See Figure 7. Notice that we use a left parenthesis to indicate that the number 4 is *not* part of the graph.

Figure 7

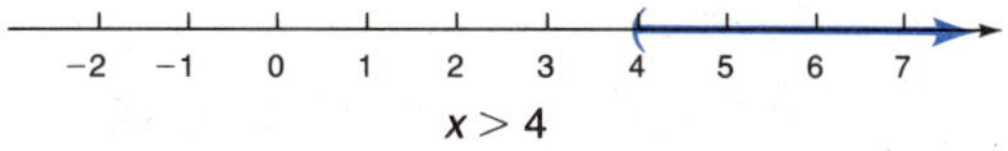

(b) See Figure 8. Notice that we use a right bracket to indicate that the number 5 *is* part of the graph.

Figure 8

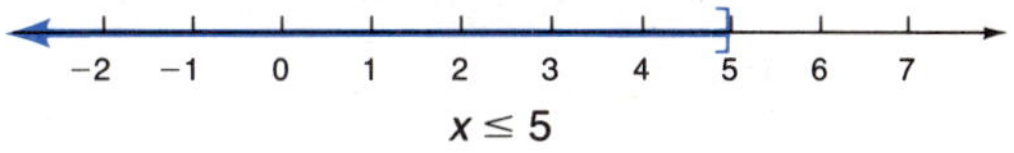

■

Inequalities are often combined.

Example 6 On the real number line, graph all numbers x for which $x > 4$ and $x < 6$.

Solution We first graph each inequality separately, as illustrated in Figure 9(a). Then it is easy to see that the numbers that belong to *both* the graph of $x > 4$ *and* the graph of $x < 6$ are graphed in Figure 9(b). For example,

5 is part of the graph because $5 > 4$ and $5 < 6$; 7 is not on the graph because 7 is not less than 6.

Figure 9

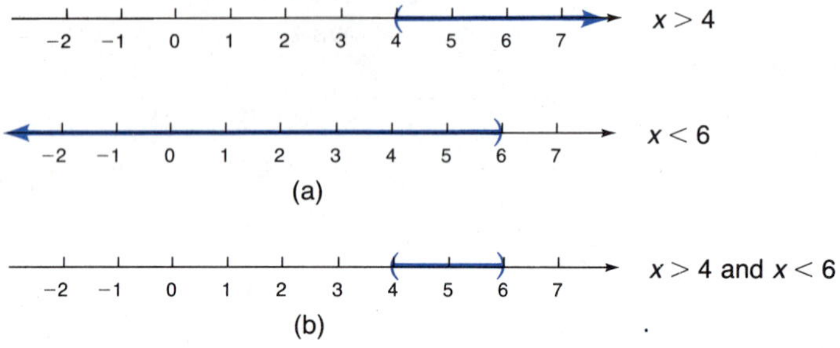

Example 7 On the real number line, graph all numbers x for which $x > 4$ or $x \leq -1$.

Solution See Figure 10(a), where each inequality is graphed separately. The numbers that belong to *either* the graph of $x > 4$ *or* the graph of $x \leq -1$ are graphed in Figure 10(b). For example, 5 is part of the graph because $5 > 4$; 2 is not on the graph because 2 is not greater than 4 nor is 2 less than or equal to -1.

Figure 10

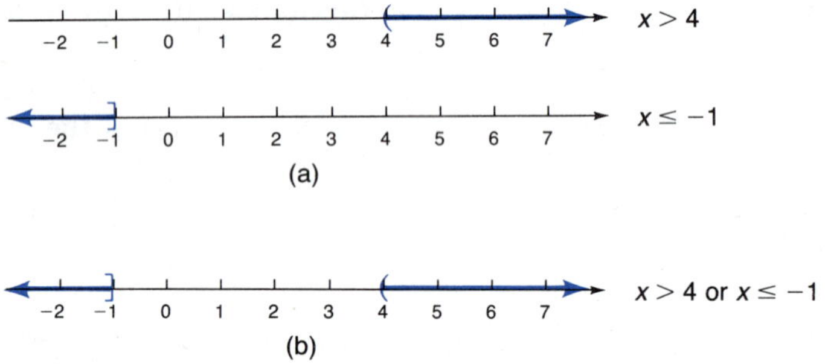

Example 8 On the real number line, graph all numbers x for which $x > 4$ and $x > 6$.

Solution Figure 11 shows the numbers for which *both* $x > 4$ *and* $x > 6$.

Figure 11

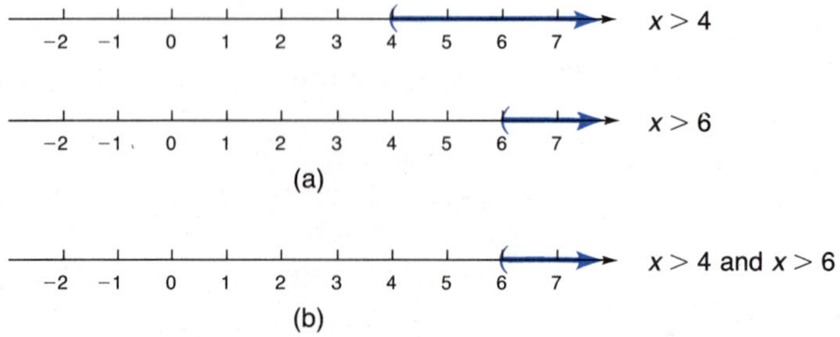

Example 9 On the real number line, graph all numbers x for which $x \geq 4$ or $x \geq 1$.

Solution Figure 12 shows the numbers for which *either* $x \geq 4$ *or* $x \geq 1$.

Figure 12

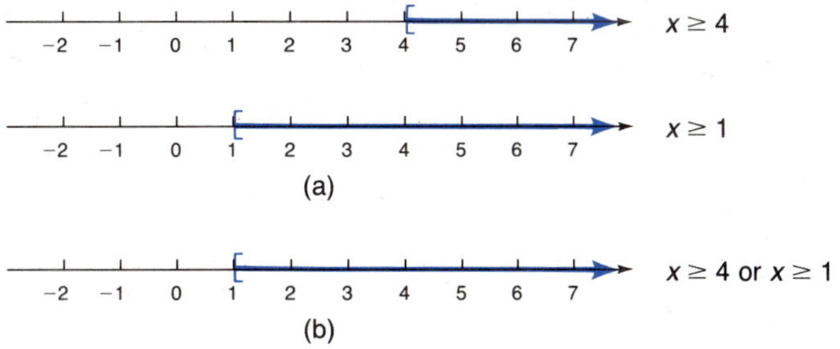

■

Example 10 On the real number line, graph all numbers x for which $x \leq 4$ and $x \geq 6$.

Solution Figure 13 illustrates that there are no numbers for which *both* $x \leq 4$ *and* $x \geq 6$.

Figure 13

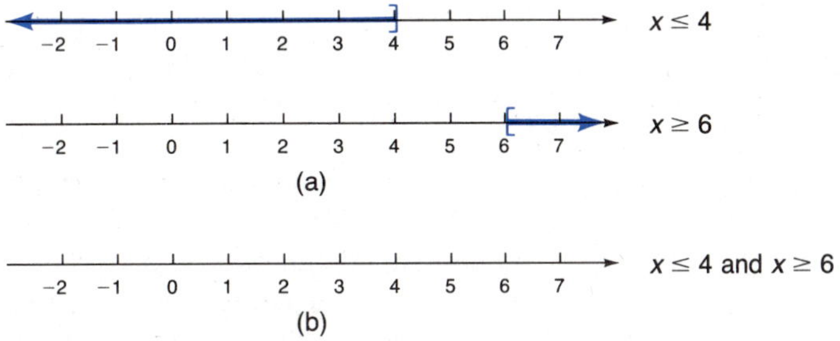

■

Suppose a and b are two real numbers and $a < b$. We shall use the notation $a < x < b$ to mean that x is a number *between* a and b. Thus, the expression $a < x < b$ is equivalent to the two inequalities $a < x$ and $x < b$. Similarly, the expression $a \leq x \leq b$ is equivalent to the two inequalities $a \leq x$ and $x \leq b$. The remaining two possibilities, $a \leq x < b$ and $a < x \leq b$, are defined analogously.

Although it is acceptable to write $3 \geq x \geq 2$, it is preferable to reverse the inequality symbols and write instead $2 \leq x \leq 3$, so that, as you read from left to right, the values go from smaller to larger.

A statement such as $2 \leq x \leq 1$ is false because there is no number x for which $2 \leq x$ and $x \leq 1$. Finally, we never mix inequality symbols, as in $2 \leq x \geq 3$.

The next example will give you some practice for future work involving inequalities.

Example 11 Choose a number x that is:

(a) Less than 2 (b) Between 2 and 4 (c) Larger than 4

Use it to determine whether the expression $(x - 2)(x - 4)$ is positive, negative, or 0.

Solution (a) Any number x less than 2 satisfies the inequality $x < 2$. We choose the number 1. Then the expression

$$(x - 2)(x - 4) = (1 - 2)(1 - 4) = (-1)(-3) = 3$$

is positive.

(b) Any number x between 2 and 4 satisfies $2 < x < 4$. We choose $x = 3$. Then the expression

$$(x - 2)(x - 4) = (3 - 2)(3 - 4) = (1)(-1) = -1$$

is negative.

(c) Any number x larger than 4 satisfies the inequality $x > 4$. We choose $x = 6$. Then the expression

$$(x - 2)(x - 4) = (6 - 2)(6 - 4) = (4)(2) = 8$$

is positive. ■

You may wonder what would happen if you picked numbers different from the ones we used in Example 11. Would the answers be different? Try some other choices and see. As it turns out, the answers are always the same. Let's see why.

For any choice of x less than 2, we have $x < 2$. As a result, both $x - 2$ and $x - 4$ are negative, making the product $(x - 2)(x - 4)$ positive. Similarly, for any choice of x between 2 and 4, we have $2 < x < 4$. Thus, $2 < x$ and $x < 4$ so that $x - 2$ is positive and $x - 4$ is negative, making the product $(x - 2)(x - 4)$ negative. Finally, for any choice of x greater than 4, we have $x > 4$. As a result, both $x - 2$ and $x - 4$ are positive, making the product $(x - 2)(x - 4)$ positive. Table 1 summarizes this discussion.

Table 1

	$x - 2$	$x - 4$	$(x - 2)(x - 4)$
$x < 2$	Negative	Negative	Positive
$2 < x < 4$	Positive	Negative	Negative
$x > 4$	Positive	Positive	Positive

Intervals

Let a and b represent two real numbers with $a < b$:

Closed Interval

A **closed interval**, denoted by **[a, b]**, consists of all real numbers x for which $a \le x \le b$.

Open Interval

An **open interval**, denoted by **(a, b)**, consists of all real numbers x for which $a < x < b$.

The **half-open**, or **half-closed**, **intervals** are **(a, b]**, consisting of all real numbers x for which $a < x \le b$, and **[a, b)**, consisting of all real numbers x for which $a \le x < b$.

In each of these definitions, a is called the **left endpoint** and b the **right endpoint** of the interval. Figure 14 illustrates each type of interval.

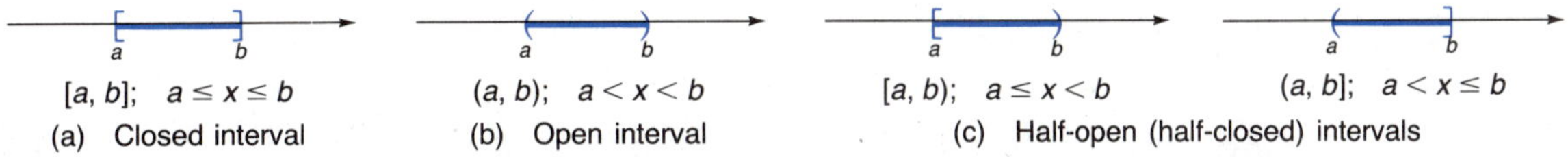

Figure 14

The symbol ∞ (read as "infinity") is not a real number but a notational device used to indicate unboundedness in the positive direction. The symbol $-\infty$ (read as "minus infinity") also is not a real number but a notational device used to indicate unboundedness in the negative direction. Using the symbols ∞ and $-\infty$, we can define four other kinds of intervals:

$[a, \infty)$	consists of all real numbers x for which $x \ge a$
(a, ∞)	consists of all real numbers x for which $x > a$
$(-\infty, a]$	consists of all real numbers x for which $x \le a$
$(-\infty, a)$	consists of all real numbers x for which $x < a$

Figure 15 illustrates these types of intervals.

Figure 15

Example 12 Write each inequality using interval notation.

(a) $1 \le x \le 3$ (b) $-4 < x < 0$ (c) $x > 5$ (d) $x \le 1$

Solution (a) $1 \leq x \leq 3$ describes all numbers x between 1 and 3, inclusive. In interval notation, we write $[1, 3]$.

(b) In interval notation, $-4 < x < 0$ is written $(-4, 0)$.

(c) $x > 5$ consists of all numbers x greater than 5. In interval notation, we write $(5, \infty)$.

(d) In interval notation, $x \leq 1$ is written $(-\infty, 1]$. ■

Example 13 Write each interval as an inequality involving x.

(a) $[1, 4)$ (b) $(2, \infty)$ (c) $[2, 3]$ (d) $(-\infty, -3]$

Solution (a) $[1, 4)$ consists of all numbers x for which $1 \leq x < 4$.

(b) $(2, \infty)$ consists of all numbers x for which $x > 2$.

(c) $[2, 3]$ consists of all numbers x for which $2 \leq x \leq 3$.

(d) $(-\infty, -3]$ consists of all numbers x for which $x \leq -3$. ■

EXERCISE 2.6

In Problems 1–10, graph the numbers x, if any, on the real number line.

1. $x \geq -1$
2. $x < 1$
3. $x \geq 4$ and $x < 6$
4. $x > 3$ and $x \leq 7$
5. $x \leq 0$ or $x < 6$
6. $x > 0$ or $x \geq 5$
7. $x \leq -2$ and $x > 1$
8. $x \geq 4$ and $x < -2$
9. $x \leq -2$ or $x > 1$
10. $x \geq 4$ or $x < -2$

In Problems 11–20, pick a number x that obeys the given inequality and use it to determine whether the given expression is positive, negative, or 0.

11. $3 < x < 5$; $(x - 5)(x - 3)$
12. $1 < x < 3$; $(x - 3)(x - 1)$
13. $-4 < x < 1$; $(x - 1)(x + 4)$
14. $-3 < x < -1$; $(x + 1)(x + 3)$
15. $x > 3$; $(x - 1)(x - 3)$
16. $x > -2$; $(x + 2)(x + 5)$
17. $x < 4$; $(x - 4)(x - 6)$
18. $x < 1$; $(x - 1)(x - 7)$
19. $x > 0$; $x(x + 4)$
20. $x < 0$; $x(x - 3)$

In Problems 21–28, write each inequality using interval notation, and illustrate each inequality using the real number line.

21. $0 \leq x \leq 4$
22. $-1 < x < 5$
23. $4 \leq x < 6$
24. $-2 < x \leq 0$
25. $x \geq 2$
26. $x \leq 2$
27. $x < -4$
28. $x > 1$

In Problems 29–36, write each interval as an inequality involving x, and illustrate each inequality using the real number line.

29. $[2, 5]$
30. $(1, 2)$
31. $(-3, -2)$
32. $[0, 1)$
33. $[4, \infty)$
34. $(-\infty, 2]$
35. $(-\infty, -1)$
36. $(-6, \infty)$

37. If $a \le b$ and $c > 0$, show that $ac \le bc$. [*Hint:* Since $a \le b$, it follows that $a - b \le 0$. Now multiply each side by c.]

38. If $a \le b$ and $c < 0$, show that $ac \ge bc$.

39. If $a < b$, show that $a < (a + b)/2 < b$. The number $(a + b)/2$ is called the *arithmetic mean* of a and b.

40. Refer to Problem 39. Show that the arithmetic mean of a and b is equidistant from a and b.

41. If $0 < a < b$, show that $a < \sqrt{ab} < b$. The number $\sqrt{ab}$ is called the *geometric mean* of a and b.

42. Refer to Problems 39 and 41. Show that the geometric mean of a and b is less than the arithmetic mean of a and b.

43. For $0 < a < b$, let h be defined by

$$\frac{1}{h} = \frac{1}{2}\left(\frac{1}{a} + \frac{1}{b}\right)$$

Show that $a < h < b$. The number h is called the *harmonic mean* of a and b.

44. Refer to Problems 39, 41, and 43. Show that the harmonic mean of a and b equals the quotient of the geometric mean squared divided by the arithmetic mean.

45. A young adult may be defined as someone older than 21, but less than 30 years of age. Express this statement using inequalities.

46. Middle-aged may be defined as being 40 or more and less than 50. Express this statement using inequalities.

47. Metropolitan Life Insurance Co. reported that an average 25-year-old male in 1987 could expect to live at least 48.4 more years, and an average 25-year-old female in 1987 could expect to live at least 54.7 more years.*

(a) To what age can an average 25-year-old male expect to live? Express your answer as an inequality.

(b) To what age can an average 25-year-old female expect to live? Express your answer as an inequality.

(c) Who can expect to live longer—a male or a female? By how many years?

2.7 Linear Inequalities

An **inequality in one variable** is a statement involving two expressions, at least one containing the variable, separated by one of the inequality symbols, $<$, $\le$, $>$, or $\ge$. To **solve an inequality** means to find all values of the variable for which the statement is true. These values are called **solutions** of the inequality.

For example, the following are all inequalities involving one variable, x:

$$x + 5 < 8 \qquad 2x - 3 \ge 4 \qquad x^2 - 1 \le 3 \qquad \frac{x + 1}{x - 2} > 0$$

**Source: Chicago Tribune*, June 26, 1988

Two inequalities having exactly the same solution set are called **equivalent inequalities**.

As with equations, one method for solving an inequality is to replace it by a series of equivalent inequalities, until an inequality with an obvious solution, such as $x < 3$, is obtained. We obtain equivalent inequalities by applying some of the same operations as those used to find equivalent equations. The addition property and the multiplication properties form the basis for the procedures listed below.

Procedures That Leave the Inequality Symbol Unchanged

1. Simplify both sides of the inequality by combining like terms and eliminating parentheses:

$$\text{Replace} \quad (x + 2) + 6 > 2x + (x + 1)$$
$$\text{by} \quad x + 8 > 3x + 1$$

2. Add or subtract the same expression on both sides of the inequality:

$$\text{Replace} \quad 3x - 5 < 4$$
$$\text{by} \quad (3x - 5) + 5 < 4 + 5$$

3. Multiply or divide both sides of the inequality by the same *positive* expression:

$$\text{Replace} \quad 4x > 16 \quad \text{by} \quad \frac{4x}{4} > \frac{16}{4}$$

Procedures That Reverse the Sense or Direction of the Inequality Symbol

1. Interchange the two sides of the inequality:

$$\text{Replace} \quad 3 < x \quad \text{by} \quad x > 3$$

2. Multiply or divide both sides of the inequality by the same *negative* expression:

$$\text{Replace} \quad -2x > 6 \quad \text{by} \quad \frac{-2x}{-2} < \frac{6}{-2}$$

A **linear inequality in one variable** is an inequality equivalent to one of the forms

$$ax + b < 0 \qquad ax + b > 0$$
$$ax + b \leq 0 \qquad ax + b \geq 0$$

where a and b are real numbers and $a \neq 0$.

The remainder of this section deals with solving linear inequalities. In the next section, we discuss the solution of other types of inequalities. As the examples that follow illustrate, we solve linear inequalities using many of the same steps we would use to solve a linear equation. In writing the solution of an inequality, we may use either set notation or interval notation, whichever is more convenient.

Example 1 Solve the inequality $3 - 2x < 5$, and draw a graph to illustrate the solution.

Solution

$$3 - 2x < 5$$

$$3 - 2x - 3 < 5 - 3 \quad \text{Subtract 3 from both sides.}$$

$$-2x < 2 \quad \text{Simplify.}$$

$$\frac{-2x}{-2} > \frac{2}{-2} \quad \text{Divide both sides by } -2\text{. (The sense of the inequality symbol is reversed.)}$$

$$x > -1 \quad \text{Simplify.}$$

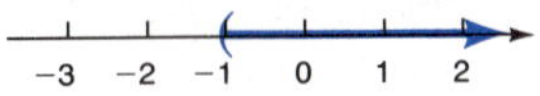

Figure 16
$x > -1$ or $(-1, \infty)$

The solution set is $\{x|x > -1\}$, or, using interval notation, all numbers in the interval $(-1, \infty)$. See Figure 16 for the graph. ■

Example 2 Solve the inequality $4x + 7 \geq 2x - 3$, and draw a graph to illustrate the solution.

Solution

$$4x + 7 \geq 2x - 3$$

$$4x + 7 - 7 \geq 2x - 3 - 7 \quad \text{Subtract 7 from both sides.}$$

$$4x \geq 2x - 10 \quad \text{Simplify.}$$

$$4x - 2x \geq 2x - 10 - 2x \quad \text{Subtract } 2x \text{ from both sides.}$$

$$2x \geq -10 \quad \text{Simplify.}$$

$$\frac{2x}{2} \geq \frac{-10}{2} \quad \text{Divide both sides by 2. (The sense of the inequality is unchanged.)}$$

$$x \geq -5 \quad \text{Simplify.}$$

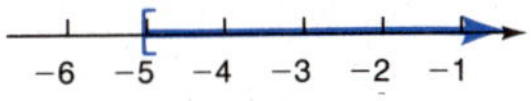

Figure 17
$x \geq -5$ or $[-5, \infty)$

The solution set is $\{x|x \geq -5\}$, or, using interval notation, all numbers in the interval $[-5, \infty)$.

See Figure 17 for the graph. ■

Example 3 Solve the inequality $-5 < 3x - 2 < 1$, and draw a graph to illustrate the solution.

Solution Recall that the inequality

$$-5 < 3x - 2 < 1$$

is equivalent to the two inequalities

$$-5 < 3x - 2 \qquad \text{and} \qquad 3x - 2 < 1$$

We will solve each of these inequalities separately. For the first inequality,

$$\begin{aligned}
-5 &< 3x - 2 \\
-5 + 2 &< 3x - 2 + 2 && \text{Add 2 to both sides.} \\
-3 &< 3x && \text{Simplify.} \\
\frac{-3}{3} &< \frac{3x}{3} && \text{Divide both sides by 3.} \\
-1 &< x && \text{Simplify.}
\end{aligned}$$

The second inequality is solved as follows:

$$\begin{aligned}
3x - 2 &< 1 \\
3x - 2 + 2 &< 1 + 2 && \text{Add 2 to both sides.} \\
3x &< 3 && \text{Simplify.} \\
\frac{3x}{3} &< \frac{3}{3} && \text{Divide both sides by 3.} \\
x &< 1 && \text{Simplify.}
\end{aligned}$$

The solution set of the original pair of inequalities consists of all x for which

$$-1 < x \qquad \text{and} \qquad x < 1$$

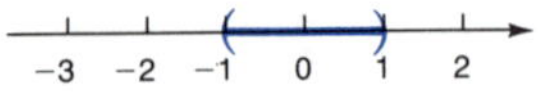

Figure 18
$-1 < x < 1$ or $(-1, 1)$

This may be written more compactly as $\{x \mid -1 < x < 1\}$. In interval notation, the solution is $(-1, 1)$. See Figure 18 for the graph. ■

We observe in the process above that the two inequalities we solved required exactly the same steps. A shortcut to solving the original inequality is to deal with the two inequalities at the same time as follows:

$$\begin{aligned}
-5 &< 3x - 2 < 1 \\
-5 + 2 &< 3x - 2 + 2 < 1 + 2 && \text{Add 2 to each part.} \\
-3 &< 3x < 3 && \text{Simplify.} \\
\frac{-3}{3} &< \frac{3x}{3} < \frac{3}{3} && \text{Divide each part by 3.} \\
-1 &< x < 1 && \text{Simplify.}
\end{aligned}$$

We use this shortcut in the next example.

Example 4 Solve the inequality:

$$-1 \le \frac{3 - 5x}{2} \le 9$$

Graph the solution set.

Solution

$$\begin{aligned}
-1 \le \quad \frac{3 - 5x}{2} \quad &\le 9 \\
2(-1) \le 2\left(\frac{3 - 5x}{2}\right) &\le 2(9) \qquad \text{Multiply each part by 2.} \\
-2 \le \quad 3 - 5x \quad &\le 18 \qquad \text{Simplify.} \\
-2 - 3 \le 3 - 5x - 3 &\le 18 - 3 \qquad \text{Subtract 3 from each part.} \\
-5 \le \quad -5x \quad &\le 15 \qquad \text{Simplify.} \\
\frac{-5}{-5} \ge \quad \frac{-5x}{-5} \quad &\ge \frac{15}{-5} \qquad \text{Divide each part by } -5 \text{ (change the sense of each inequality symbol).} \\
1 \ge \quad x \quad &\ge -3 \qquad \text{Simplify.} \\
-3 \le \quad x \quad &\le 1 \qquad \text{Reverse the order so the numbers get larger as you read from left to right.}
\end{aligned}$$

Figure 19
$-3 \le x \le 1$ or $[-3, 1]$

The solution set is $\{x \mid -3 \le x \le 1\}$; that is, all x in $[-3, 1]$. Figure 19 illustrates the graph. ■

Let's look at an applied problem involving linear inequalities.

Example 5 In electricity, Ohm's law states that $E = IR$, where E is the voltage (in volts), I is the current (in amperes), and R is the resistance (in ohms). An air-conditioning unit is rated at a resistance of 10 ohms. If the voltage varies from 110 to 120 volts, inclusive, what corresponding range of current will the air-conditioner draw?

Solution The voltage lies between 110 and 120, inclusive, so

$$\begin{aligned}
110 \le \quad E \quad &\le 120 \\
110 \le \quad IR \quad &\le 120 \qquad \text{Ohm's law, } E = IR \\
110 \le I(10) &\le 120 \qquad R = 10 \\
\frac{110}{10} \le \frac{I(10)}{10} &\le \frac{120}{10} \qquad \text{Divide each part by 10.} \\
11 \le \quad I \quad &\le 12 \qquad \text{Simplify.}
\end{aligned}$$

The air-conditioner will draw between 11 and 12 amperes of current, inclusive. ■

Historical Comment

■ Inequalities are a relatively new component of the algebra curriculum. They have been introduced in the last 25 years for two important reasons.

First, if approximations are made in a problem, inequalities allow calculation of how serious the error is likely to be. This use is important for practical applications of mathematics and also is critical for the understanding of calculus [in its modern version, due principally to Augustin Louis Cauchy (1789–1857) and Karl Weierstrass (1815–1897)].

Second, linear inequalities are the basis for solving a kind of problem that involves finding a maximum or minimum of a quantity, depending on variables that are subjected to certain constraints. For example, one might wish to ship several products by truck from several different factories, each factory having a limited supply of each product, and to get enough of the products to a central point within 3 days using the minimum amount of gas possible. Such problems, called *linear programming problems*, are discussed later in this text. ■

EXERCISE 2.7

In Problems 1–4, an inequality is given. Write the equivalent inequality obtained by:
(a) Adding −3 to each side of the given inequality.
(b) Subtracting 5 from each side of the given inequality.
(c) Multiplying each side of the given inequality by 3.
(d) Multiplying each side of the given inequality by −2.

1. $3 < 5$ **2.** $2 > 1$ **3.** $2x + 1 < 2$ **4.** $1 - 2x > 5$

In Problems 5–40, solve each inequality. Graph the solution set.

5. $x + 2 < 6$ **6.** $x - 5 < 2$ **7.** $1 - 2x \le 3$

8. $2 - 3x \le 5$ **9.** $3x - 7 > 2$ **10.** $2x + 5 > 1$

11. $3x - 1 \ge 3 + x$ **12.** $2x - 2 \ge 3 + x$ **13.** $-2(x + 3) < 6$

14. $-3(1 - x) < 9$ **15.** $4 - 3(1 - x) \le 3$ **16.** $8 - 4(2 - x) \le -2x$

17. $\frac{1}{2}(x - 4) > x + 8$ **18.** $3x + 4 > \frac{1}{3}(x - 2)$ **19.** $\frac{x}{2} \ge 1 - \frac{x}{4}$

20. $\frac{x}{3} \ge 2 + \frac{x}{6}$ **21.** $0 \le 2x - 6 \le 4$ **22.** $4 \le 2x + 2 \le 10$

23. $-6 \le 1 - 3x \le 2$ **24.** $-3 \le 2 - 2x \le 4$ **25.** $-3 < \frac{2x - 1}{4} < 0$

26. $0 < \frac{3x + 2}{2} < 4$ **27.** $1 < 1 - \frac{1}{2}x < 4$ **28.** $0 < 1 - \frac{1}{3}x < 1$

29. $(x + 2)(x - 3) > (x - 1)(x + 1)$ **30.** $(x - 1)(x + 1) > (x - 3)(x + 4)$

31. $x(4x + 3) \le (2x + 1)^2$ **32.** $x(9x - 5) \le (3x - 1)^2$ **33.** $\frac{1}{2} \le \frac{x + 1}{3} < \frac{3}{4}$

34. $\frac{1}{3} < \frac{x+1}{2} \leq \frac{2}{3}$ **35.** $(4x + 2)^{-1} < 0$ **36.** $(2x - 1)^{-1} > 0$

37. $0 < \frac{1}{x} < \frac{1}{3}$ **38.** $0 < \frac{1}{x} < \frac{1}{4}$ **39.** $0 < (2x - 4)^{-1} < \frac{1}{2}$

40. $0 < (3x + 6)^{-1} < \frac{1}{3}$

41. For what real numbers x is $\sqrt{3x - 6}$ defined?

42. For what real numbers x is $\sqrt{8 - 2x}$ defined?

43. For a certain ideal gas, the volume V (in cubic centimeters) equals 20 times the temperature T (in degrees Celsius). If the temperature varies from 80°C to 120°C, inclusive, what is the corresponding range of the volume of the gas?

44. An investor has \$1000 to invest for a period of 1 year. What range of per annum simple interest rates are needed to obtain interest that varies from \$90 to \$110, inclusive?

45. A real estate agent agrees to sell a large apartment complex according to the following commission schedule: \$45,000 plus 25% of the selling price in excess of \$900,000. Assuming the complex will sell at some price between \$900,000 and \$1,100,000, inclusive, over what range does the agent's commission vary? How does the commission vary as a percent of selling price?

46. A used car salesperson is paid a commission of \$25 plus 40% of the selling price in excess of owner's cost. The owner claims that used cars typically sell for at least owner's cost plus \$70 and at most owner's cost plus \$300. For each sale made, over what range can the salesperson expect the commission to vary?

C **47.** The percentage method of withholding for federal income tax (1989)* states that a single person claiming 0 dependents whose weekly wages are over \$378, but not over \$885, shall have \$53.55 plus 28% of the excess over \$378 withheld. Over what range does the amount withheld vary if the weekly wages vary from \$500 to \$550, inclusive?

C **48.** Rework Problem 47 if the weekly wages vary from \$475 to \$525, inclusive.

C **49.** Commonwealth Edison Company's summer charge for electricity is 12.255¢ per kilowatt-hour.† In addition, each monthly bill contains a customer charge of \$11.24. If last summer's bills ranged from a low of \$94.02 to a high of \$317.72, over what range did usage vary (in kilowatt-hours)?

C **50.** The Village of Oak Lawn charges homeowners \$17.76 per quarter year plus \$1.34 per 1000 gallons for water usage in excess of 12,000 gallons.‡ In 1989, one homeowner's quarterly bill ranged from a high of \$49.92 to a low of \$28.48. Over what range did water usage vary?

51. The markup over dealer's cost of a new car ranges from 12% to 18%. If the sticker price is \$8800, over what range will the dealer's cost vary?

52. A standard intelligence test has an average score of 100. According to statistical theory, of the people who take the test, the 2.5% with the highest scores will have scores of more than 1.95σ above the average, where σ (sigma, a number called the *standard deviation*) depends on the nature of the test. If $\sigma = 12$ for this test and there is (in principle) no upper limit to the score possible on the test, write the interval of possible test scores of the people in the top 2.5%.

**Source: Employer's Tax Guide*, Department of the Treasury, Internal Revenue Service, 1989
†*Source:* Commonwealth Edison Co., Chicago, Illinois, 1989
‡*Source:* Village of Oak Lawn, Illinois, 1989

53. In your Economics 101 class, you have scores of 70, 82, 85, and 89 on the first four of five tests. To get a grade of B, the average of the five test scores must be greater than or equal to 80 and less than 90. Solve an inequality to find the range of the score you need on the last test to get a B.

54. Repeat Problem 53 if the fifth test counts double.

55. A car that averages 25 miles per gallon has a tank that holds 20 gallons of gasoline. After a trip that covered at least 300 miles, the car ran out of gasoline. What is the range of the amount of gasoline (in gallons) that was in the tank at the start of the trip?

56. Repeat Problem 55 if the same car runs out of gasoline after a trip of no more than 250 miles.

2.8 Other Inequalities

In this section, we solve inequalities that contain polynomials of degree 2 and higher as well as some that contain rational expressions. Before we can solve such inequalities, we rearrange them so that the polynomial or rational expression is on the left side and 0 is on the right side. An example will show you why.

Example 1 Solve the inequality $x^2 + x - 12 > 0$, and graph the solution set.

Solution We factor the left side, obtaining

$$x^2 + x - 12 > 0$$
$$(x - 3)(x + 4) > 0$$

The product of two real numbers is positive either when both factors are positive or when both factors are negative.

BOTH POSITIVE	OR	BOTH NEGATIVE
$x - 3 > 0$ and $x + 4 > 0$		$x - 3 < 0$ and $x + 4 < 0$
$x > 3$ and $x > -4$		$x < 3$ and $x < -4$
The numbers x that are greater than 3 and at the same time greater than -4 are simply		The numbers x that are less than 3 and at the same time less than -4 are simply
$x > 3$	or	$x < -4$

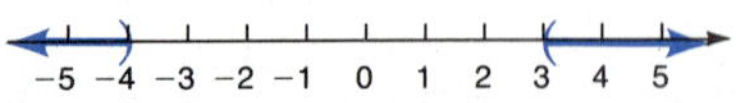

Figure 20
$x < -4$ or $x > 3$; $(-\infty, -4) \cup (3, \infty)$

The solution set is $\{x | x < -4 \text{ or } x > 3\}$. In interval notation, we write the solution as $(-\infty, -4) \cup (3, \infty)$. See Figure 20. ■

We also can obtain the solution to the inequality of Example 1 by another method. The left-hand side of the inequality is factored so that it becomes $(x - 3)(x + 4) > 0$, as before. We then construct a graph that uses the solutions to the equation

$$x^2 + x - 12 = (x - 3)(x + 4) = 0$$

namely, $x = 3$ and $x = -4$. These numbers, called **boundary points**, separate the real number line into three parts: $x < -4$, $-4 < x < 3$, and $x > 3$. See Figure 21(a).

Figure 21

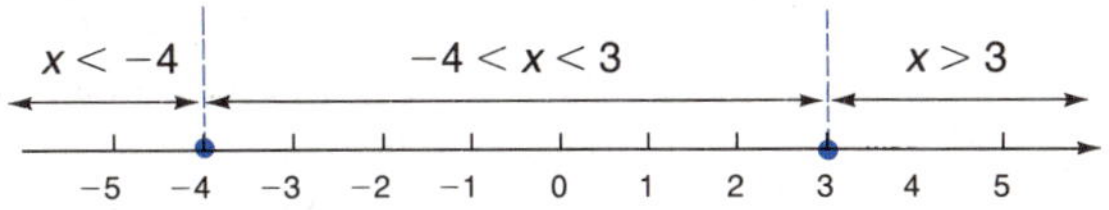

(a)

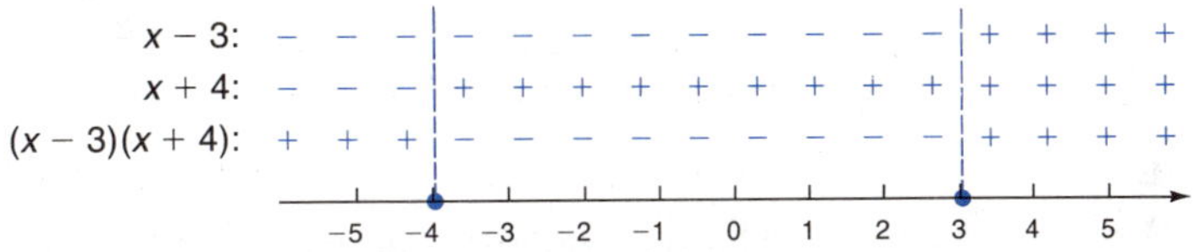

(b)

In the part of the line where $x < -4$, we deduce that both the quantities $(x - 3)$ and $(x + 4)$ are negative, so their product must be positive. See Figure 21(b). Therefore, $x < -4$ is a solution of the inequality. In the part of the line where $-4 < x < 3$, we deduce that $(x - 3)$ is negative and $(x + 4)$ is positive, so their product is negative. We conclude that the numbers between -4 and 3 are not solutions of the inequality. In the part of the line where $x > 3$, we deduce that both the quantities $(x - 3)$ and $(x + 4)$ are positive, so their product is positive. Hence, numbers greater than 3 are solutions of the inequality. Table 2 summarizes these results.

Table 2

	SIGN OF $x - 3$	SIGN OF $x + 4$	SIGN OF THE PRODUCT $(x - 3)(x + 4)$	CONCLUSION
$x < -4$	$-$	$-$	$+$	$x < -4$ is a solution
$-4 < x < 3$	$-$	$+$	$-$	$-4 < x < 3$ is not a solution
$x > 3$	$+$	$+$	$+$	$x > 3$ is a solution

The signs listed in Table 2 may be more easily obtained by using what we shall call a **test number**. For example, to determine the sign of $(x - 3)$ when $x < -4$, select any number that is less than -4, say, -5, and substitute it into the expression to be tested, to get $-5 - 3 = -8$. The negative result tells us that $x - 3$ is negative for all x for which $x < -4$.

As another example, in order to determine the sign of $(x + 4)$ when $-4 < x < 3$, we might select 0 as a test number (any number between

-4 and 3 will do). When $x = 0$, $(x + 4) = 4$. We conclude that $(x + 4)$ is positive for all x for which $-4 < x < 3$.

Example 2 Solve the inequality: $x^2 \le 4x + 12$

Solution First, we rearrange the inequality so that 0 is on the right side:

$$x^2 \le 4x + 12$$
$$x^2 - 4x - 12 \le 0$$
$$(x - 6)(x + 2) \le 0 \qquad \text{Factor.}$$

Next, we set the left side equal to 0 in order to locate the boundary points:

$$(x - 6)(x + 2) = 0$$

The boundary points (solutions of the equation) are -2 and 6, and they separate the real number line into three parts:

$$x < -2, \qquad -2 < x < 6, \qquad \text{and} \qquad x > 6$$

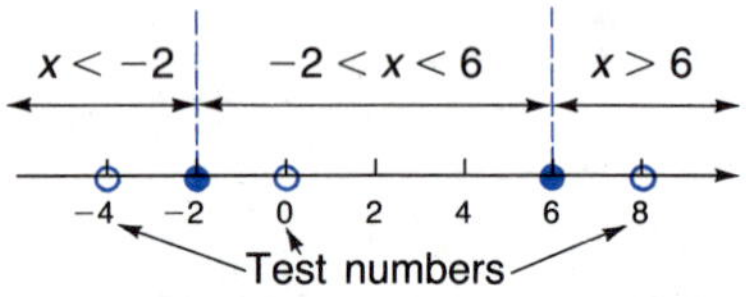

Figure 22

See Figure 22. Using the test numbers shown, we obtain the results shown in Table 3.

Table 3

	TEST NUMBER	$x - 6$	$x + 2$	$(x - 6)(x + 2)$
$x < -2$	-4	$-$	$-$	$+$
$-2 < x < 6$	0	$-$	$+$	$-$
$x > 6$	8	$+$	$+$	$+$

Since we want to know where the product $(x - 6)(x + 2)$ is negative, we conclude that the solutions are numbers x for which $-2 < x < 6$. However, because the original inequality is nonstrict, numbers x that satisfy the equation $x^2 = 4x + 12$ are also solutions of the inequality $x^2 \le 4x + 12$. Thus, we include the boundary points -2 and 6, and the solution set of the given inequality is $\{x \mid -2 \le x \le 6\}$; that is, all x in $[-2, 6]$. ■

The method used in Example 2 required that we find boundary points (solutions of an equation). When a polynomial equation has no real solutions, the polynomial is either always positive or always negative. For example, the equation

$$x^2 + 5x + 8 = 0$$

has no real solutions. (Do you see why? Its discriminant, $b^2 - 4ac = 25 - 32 = -7$, is negative.) The value of $x^2 + 5x + 8$ is therefore always positive or always negative. To see which is true, we test its value at some number (0 is the easiest). Because $0^2 + 5(0) + 8 = 8$ is positive, we conclude that $x^2 + 5x + 8 > 0$ for all x.

The boundary point method works well for inequalities containing polynomials of any degree, provided they can be factored.

Example 3 Solve the inequality: $x^3 - 5x^2 + 6x > 0$

Solution Because 0 is on the right-hand side, we proceed to factor. Because x is an obvious factor, we begin there:

$$x^3 - 5x^2 + 6x > 0$$
$$x(x^2 - 5x + 6) > 0$$
$$x(x - 2)(x - 3) > 0$$

The boundary points [solutions of the equation $x^3 - 5x^2 + 6x = x(x - 2)(x - 3) = 0$] are 0, 2, and 3, and they separate the real number line into four parts, namely,

$$x < 0, \qquad 0 < x < 2, \qquad 2 < x < 3, \qquad \text{and} \qquad x > 3$$

See Figure 23 and Table 4.

We want to know where the product $x(x - 2)(x - 3)$ is positive; thus, the solution set is $\{x \mid 0 < x < 2 \text{ or } x > 3\}$; that is, all x in $(0, 2) \cup (3, \infty)$.

Figure 23

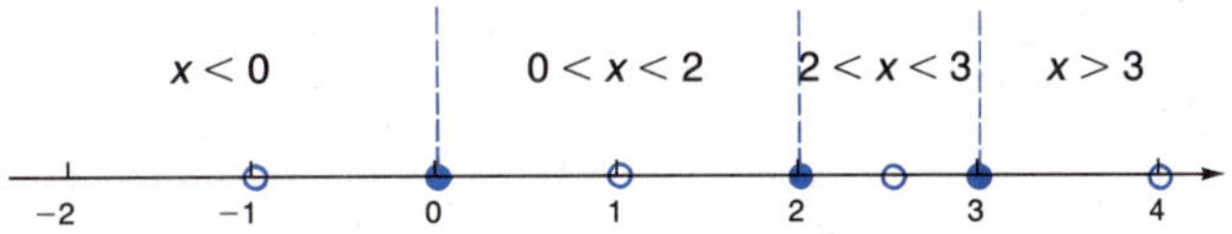

Table 4

	TEST NUMBER	x	$x - 2$	$x - 3$	$x(x - 2)(x - 3)$
$x < 0$	-1	$-$	$-$	$-$	$-$
$0 < x < 2$	1	$+$	$-$	$-$	$+$
$2 < x < 3$	2.5	$+$	$+$	$-$	$-$
$x > 3$	4	$+$	$+$	$+$	$+$

■

Let's use the boundary point method to solve a rational inequality.

Example 4 Solve the inequality: $\dfrac{4x + 5}{x + 2} \geq 3$

Solution We rearrange terms so that 0 is on the right side:

$$\frac{4x+5}{x+2} \geq 3$$

$$\frac{4x+5}{x+2} - 3 \geq 0$$

$$\frac{4x+5-3(x+2)}{x+2} \geq 0 \quad \text{Rewrite using } x+2 \text{ as the denominator.}$$

$$\frac{x-1}{x+2} \geq 0 \quad \text{Simplify.}$$

The sign of a rational expression depends on the sign of its numerator and the sign of its denominator. Thus, for a rational expression, we use as boundary points the numbers obtained by setting the numerator and the denominator equal to 0. For this example, the boundary points are 1 and -2. See Figure 24 and Table 5. The conclusions found in Figure 24 and Table 5 reveal the numbers x for which $(x-1)/(x+2)$ is positive. However, we want to know where the expression $(x-1)/(x+2)$ is positive or 0. Since $(x-1)/(x+2) = 0$ only if $x = 1$, we conclude that the solution set is $\{x \mid x < -2 \text{ or } x \geq 1\}$; that is, all x in $(-\infty, -2) \cup [1, \infty)$.

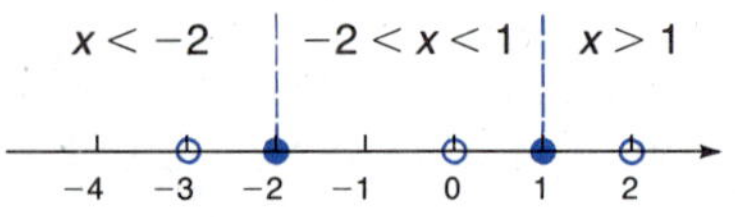

Figure 24

Table 5

	TEST NUMBER	$x-1$	$x+2$	$\frac{x-1}{x+2}$
$x < -2$	-3	$-$	$-$	$+$
$-2 < x < 1$	0	$-$	$+$	$-$
$x > 1$	2	$+$	$+$	$+$

■

In Example 4, you may wonder why we did not first multiply both sides of the inequality by $x + 2$ to clear the denominator. The reason is that we do not know whether $x + 2$ is positive or negative and, as a result, we do not know whether to reverse the sense of the inequality symbol after multiplying by $x + 2$. However, there is nothing to prevent us from multiplying both sides by $(x+2)^2$, which is always positive, since $x \neq -2$. (Do you see why?) Then

$$\frac{4x+5}{x+2} \geq 3 \qquad x \neq -2$$

$$\frac{4x+5}{x+2}(x+2)^2 \geq 3(x+2)^2$$

$$(4x+5)(x+2) \geq 3(x^2+4x+4)$$

$$4x^2+13x+10 \geq 3x^2+12x+12$$

$$x^2+x-2 \geq 0$$

$$(x+2)(x-1) \geq 0 \qquad x \neq -2$$

This last expression leads to the same solution set obtained in Example 4.

For our next example, we need the following equivalence property of inequalities.

If a and b are real numbers and $a \geq 0$, $b \geq 0$, then

$$a \leq b \quad \text{is equivalent to} \quad \sqrt{a} \leq \sqrt{b} \tag{1}$$

The proof of this property is left as an exercise.

Example 5 An object is dropped from the roof of a building 150 feet tall. After t seconds, the object will be $150 - 16t^2$ feet above the ground. During what interval of time will the object be between 54 and 118 feet above the ground?

Solution The inequality to be solved has the form

$$\begin{aligned} 54 \leq 150 - 16t^2 &\leq 118 \\ -96 \leq -16t^2 &\leq -32 \quad \text{Subtract 150 from each part.} \\ 6 \geq t^2 &\geq 2 \quad \text{Divide each part by } -16. \\ 2 \leq t^2 &\leq 6 \end{aligned}$$

For the situation described, the time t is assumed to be a nonnegative real number. With this restriction, by equation (1), the inequality

$$2 \leq t^2 \leq 6 \qquad t \geq 0$$

is equivalent to

$$\sqrt{2} \leq t \leq \sqrt{6}$$

which may be approximated by

$$1.41 \leq t \leq 2.45$$

For times ranging from approximately 1.41 seconds to 2.45 seconds, the object will be between 54 and 118 feet above the ground. ■

EXERCISE 2.8

In Problems 1–46, solve each inequality.

1. $(x - 3)(x + 1) < 0$ **2.** $(x - 1)(x + 2) < 0$ **3.** $x^2 - 4x > 0$

4. $x^2 + 8x > 0$ **5.** $x^2 - 9 < 0$ **6.** $x^2 - 1 < 0$

7. $x^2 + x > 12$
8. $x^2 + 7x < -12$
9. $2x^2 < 5x + 3$
10. $6x^2 < 6 + 5x$
11. $x(x - 7) > -12$
12. $x(x + 1) > 12$
13. $4x^2 + 9 < 6x$
14. $25x^2 + 16 < 40x$
15. $6(x^2 - 1) > 5x$
16. $2(2x^2 - 3x) > -9$
17. $(x - 1)(x^2 + x + 1) > 0$
18. $(x + 2)(x^2 - x + 1) > 0$
19. $(x - 1)(x - 2)(x - 3) < 0$
20. $(x + 1)(x + 2)(x + 3) < 0$
21. $x^3 - 2x^2 - 8x > 0$
22. $x^3 + 2x^2 - 8x > 0$
23. $x^3 > x$
24. $x^3 < 4x$
25. $x^3 > x^2$
26. $x^3 < 3x^2$
27. $x^4 > 1$
28. $x^3 > 1$
29. $\dfrac{x + 1}{x - 1} > 0$
30. $\dfrac{x - 3}{x + 1} > 0$
31. $\dfrac{(x - 1)(x + 1)}{x} < 0$
32. $\dfrac{(x - 3)(x + 2)}{x - 1} < 0$
33. $\dfrac{x - 2}{x^2 - 1} \geq 0$
34. $\dfrac{x + 5}{x^2 - 4} \geq 0$
35. $6x - 5 < \dfrac{6}{x}$
36. $x + \dfrac{12}{x} < 7$
37. $\dfrac{x + 4}{x - 2} \leq 1$
38. $\dfrac{x + 2}{x - 4} \geq 1$
39. $\dfrac{3x - 5}{x + 2} \leq 2$
40. $\dfrac{x - 4}{2x + 4} \geq 1$
41. $\dfrac{1}{x + 3} < \dfrac{2}{3x - 9}$
42. $\dfrac{5}{2x - 4} > \dfrac{3}{x + 1}$
43. $\dfrac{2x + 5}{x + 1} > \dfrac{x + 1}{x - 1}$
44. $\dfrac{1}{x + 2} > \dfrac{3}{x + 1}$
45. $\dfrac{x^2(3 + x)(x + 4)}{(x + 5)(x - 1)} > 0$
46. $\dfrac{x(x^2 + 1)(x - 2)}{(x - 1)(x + 1)} > 0$

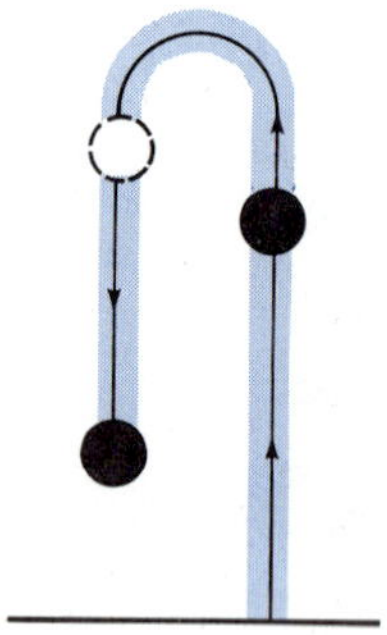

47. For what positive numbers will the cube of a number exceed 4 times its square?
48. For what positive numbers will the square of a number exceed twice the number?
49. For what real numbers x is $\sqrt{x^2 - 9}$ defined?
50. For what real numbers x is $\sqrt{(x - 1)/(x + 4)}$ defined?
51. A ball is thrown vertically upward with an initial velocity of 80 feet per second. The distance s (in feet) of the ball from the ground after t seconds is $s = 80t - 16t^2$. For what time interval is the ball more than 96 feet above the ground? (See the figure.)
52. Rework Problem 51 to find when the ball is less than 64 feet above the ground.
53. The monthly revenue achieved by selling x wristwatches is figured to be $x(40 - 0.2x)$ dollars. The wholesale cost of each watch is \$28. How many watches must be sold each month to achieve a profit (revenue − cost) of at least \$100?
54. The monthly revenue achieved by selling x boxes of candy is figured to be $x(5 - 0.05x)$ dollars. The wholesale cost of each box of candy is \$1.50. How many boxes must be sold each month to achieve a profit of at least \$60?
55. Prove that if a, b are real numbers and $a \geq 0$, $b \geq 0$, then

$$a \leq b \qquad \text{is equivalent to} \qquad \sqrt{a} \leq \sqrt{b}$$

[*Hint:* $b - a = (\sqrt{b} - \sqrt{a})(\sqrt{b} + \sqrt{a})$]
56. Find k such that the equation $x^2 + kx + 1 = 0$ has no real solution.
57. Find k such that the equation $kx^2 + 2x + 1 = 0$ has two distinct real solutions.

2.9 Equations and Inequalities Involving Absolute Value

The absolute value of a real number x has been defined as

$$|x| = \begin{cases} x & \text{if } x \geq 0 \\ -x & \text{if } x < 0 \end{cases}$$

Also, recall that geometrically the absolute value of x equals the distance from the origin to the point whose coordinate is x.

In this section, we shall discuss equations and inequalities involving absolute value. In solving such equations and inequalities, the properties listed below will prove useful.

Theorem

1. The absolute value of any real number a is always nonnegative; that is,

$$|a| \geq 0 \tag{1}$$

2. The absolute value of any real number a equals the principal square root of the number squared; that is,

$$|a| = \sqrt{a^2} \tag{2}$$

3. The absolute value of the product of two real numbers a and b equals the product of their absolute values; that is,

$$|ab| = |a| \cdot |b| \tag{3}$$

Proof Property (1) follows directly from the definition of absolute value. Property (2) is a special case of the fact that $\sqrt[n]{a^n} = |a|$ when n is even. Just let $n = 2$. Property (3) is proved by using property (2):

$$|ab| = \sqrt{(ab)^2} = \sqrt{a^2b^2} = \sqrt{a^2} \cdot \sqrt{b^2} = |a| \cdot |b| \quad \blacksquare$$

We shall now take up the problem of solving equations and inequalities that contain absolute values.

Because there are two points whose distance from the origin is 5 units, namely, -5 and 5, the equation $|x| = 5$ will have the solution set $\{-5, 5\}$. We are thus led to the following result:

Theorem If the absolute value of an expression equals some positive number a, then the expression itself equals either a or $-a$. Thus,

$$|u| = a \quad \text{is equivalent to} \quad u = a \quad \text{or} \quad u = -a \qquad (4)$$

■

Example 1 Solve the equation: $|x + 4| = 13$

Solution This follows the form of equation (4), where $u = x + 4$. Thus, there are two possibilities:

$$x + 4 = 13 \quad \text{or} \quad x + 4 = -13$$
$$x = 9 \qquad\qquad x = -17$$

Thus, the solution set is $\{-17, 9\}$. ■

Let's look at an inequality involving absolute value.

Example 2 Solve the inequality: $|x| < 4$

Solution We are looking for all points whose coordinate x is a distance less than 4 units from the origin. See Figure 25 for an illustration. Because any x between -4 and 4 satisfies the condition $|x| < 4$, the solution set consists of all numbers x for which $-4 < x < 4$; that is, all x in $(-4, 4)$.

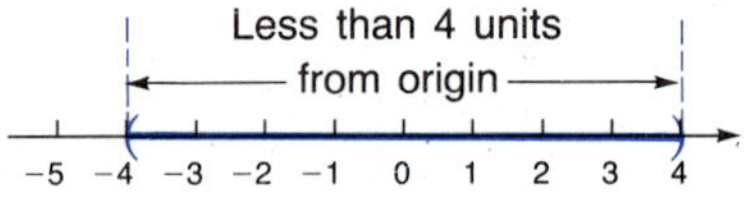

Figure 25
$-4 < x < 4$ or $(-4, 4)$

■

We are led to the following results:

Theorem If a is any positive number, then

$$|u| < a \quad \text{is equivalent to} \quad -a < u < a \qquad (5)$$
$$|u| \le a \quad \text{is equivalent to} \quad -a \le u \le a \qquad (6)$$

In other words, $|u| < a$ is equivalent to $-a < u$ and $u < a$. ■

Example 3 Solve the inequality $|2x + 4| \le 3$, and graph the solution set.

Solution

$\lvert 2x + 4\rvert \le 3$	This follows the form of equation (6); the expression $u = 2x + 4$ is inside the absolute value bars.
$-3 \le 2x + 4 \le 3$	Apply equation (6).
$-3 - 4 \le 2x + 4 - 4 \le 3 - 4$	Subtract 4 from each part.
$-7 \le 2x \le -1$	Simplify.
$\dfrac{-7}{2} \le \dfrac{2x}{2} \le \dfrac{-1}{2}$	Divide each part by 2.
$-\dfrac{7}{2} \le x \le -\dfrac{1}{2}$	Simplify.

The solution set is $\{x \mid -\frac{7}{2} \le x \le -\frac{1}{2}\}$; that is, all x in $\left[-\frac{7}{2}, -\frac{1}{2}\right]$. See Figure 26. ■

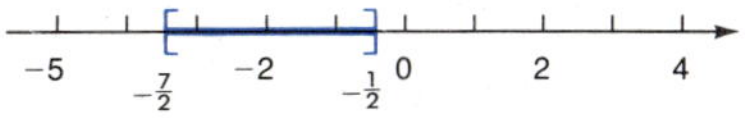

Figure 26
$-\frac{7}{2} \le x \le -\frac{1}{2}$ or $\left[-\frac{7}{2}, -\frac{1}{2}\right]$

Example 4 Solve the inequality $|1 - 4x| < 5$, and graph the solution set.

Solution

$\lvert 1 - 4x\rvert < 5$	This follows the form of equation (5); the expression $u = 1 - 4x$ is inside the absolute value bars.
$-5 < 1 - 4x < 5$	Apply equation (5).
$-5 - 1 < 1 - 4x - 1 < 5 - 1$	Subtract 1 from each part.
$-6 < -4x < 4$	Simplify.
$\dfrac{-6}{-4} > \dfrac{-4x}{-4} > \dfrac{4}{-4}$	Divide each part by -4, which reverses the sense of the inequality symbols.
$\dfrac{3}{2} > x > -1$	Simplify.
$-1 < x < \dfrac{3}{2}$	Rearrange the ordering.

The solution set is $\{x \mid -1 < x < \frac{3}{2}\}$; that is, all x in $\left(-1, \frac{3}{2}\right)$. See Figure 27. ■

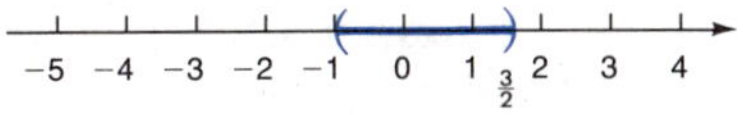

Figure 27
$-1 < x < \frac{3}{2}$ or $\left(-1, \frac{3}{2}\right)$

Example 5 Solve the inequality $|x| > 3$, and graph the solution set.

Solution We are looking for all points whose coordinate x is a distance greater than 3 units from the origin. Figure 28 illustrates the situation. We conclude that any x less than -3 or greater than 3 satisfies the condition $|x| > 3$. Consequently, the solution set consists of all numbers x for which $x < -3$ or $x > 3$; that is, all x in $(-\infty, -3) \cup (3, \infty)$. ■

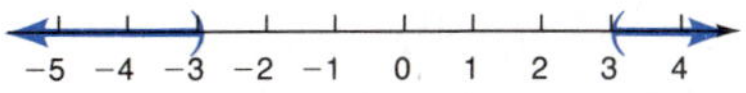

Figure 28
$x < -3$ or $x > 3$; $(-\infty, -3) \cup (3, \infty)$

This leads us to the following results:

Theorem If a is any positive number, then

$\lvert u\rvert > a$	is equivalent to	$u < -a$	or	$u > a$	(7)
$\lvert u\rvert \geq a$	is equivalent to	$u \leq -a$	or	$u \geq a$	(8)

■

Example 6 Solve the inequality $|2x - 5| > 3$, and graph the solution set.

Solution

$|2x - 5| > 3$ This follows the form of equation (7); the expression $u = 2x - 5$ is inside the absolute value bars.

$2x - 5 < -3$	or	$2x - 5 > 3$	Apply equation (7).
$2x - 5 + 5 < -3 + 5$	or	$2x - 5 + 5 > 3 + 5$	Add 5 to each part.
$2x < 2$	or	$2x > 8$	Simplify.
$\dfrac{2x}{2} < \dfrac{2}{2}$	or	$\dfrac{2x}{2} > \dfrac{8}{2}$	Divide each part by 2.
$x < 1$	or	$x > 4$	Simplify.

Figure 29
$x < 1$ or $x > 4$; $(-\infty, 1) \cup (4, \infty)$

The solution set is $\{x|x < 1 \text{ or } x > 4\}$; that is, all x in $(-\infty, 1) \cup (4, \infty)$. See Figure 29. ■

Warning: A common error to be avoided is to attempt to write the solution $x < 1$ or $x > 4$ as $1 > x > 4$, which is incorrect, since there are no numbers x for which $x < 1$ *and* $x > 4$. Another common error is to "mix" the symbols and write $1 < x > 4$, which, of course, makes no sense.

EXERCISE 2.9

In Problems 1–22, solve each equation.

1. $\lvert 2x\rvert = 6$	**2.** $\lvert 3x\rvert = 12$	**3.** $\lvert 2x + 3\rvert = 5$	**4.** $\lvert 3x - 1\rvert = 2$
5. $\lvert 1 - 4t\rvert = 5$	**6.** $\lvert 1 - 2z\rvert = 3$	**7.** $\lvert -2x\rvert = 8$	**8.** $\lvert -x\rvert = 1$
9. $\lvert -2\rvert x = 4$	**10.** $\lvert 3\rvert x = 9$	**11.** $\frac{2}{3}\lvert x\rvert = 8$	**12.** $\frac{3}{4}\lvert x\rvert = 9$
13. $\left\lvert \dfrac{x}{3} + \dfrac{2}{5}\right\rvert = 2$	**14.** $\left\lvert \dfrac{x}{2} - \dfrac{1}{3}\right\rvert = 1$	**15.** $\lvert u - 2\rvert = -\dfrac{1}{2}$	**16.** $\lvert 2 - v\rvert = -1$
17. $\lvert x^2 - 4\rvert = 0$	**18.** $\lvert x^2 - 9\rvert = 0$	**19.** $\lvert x^2 - 2x\rvert = 3$	**20.** $\lvert x^2 + x\rvert = 12$

21. $|x^2 + x - 1| = 1$ **22.** $|x^2 + 3x - 2| = 2$

In Problems 23–42, solve each inequality, and graph the solution set.

23. $|2x| < 6$ **24.** $|3x| < 12$ **25.** $|3x| > 12$ **26.** $|2x| > 6$

27. $|x - 2| < 1$ **28.** $|x + 4| < 2$ **29.** $|3t - 2| \le 4$ **30.** $|2u + 5| \le 7$

31. $|x - 1| \ge 2$ **32.** $|x + 3| \ge 2$ **33.** $|1 - 4x| < 5$ **34.** $|1 - 2x| < 3$

35. $|1 - 2x| > 3$ **36.** $|2 - 3x| > 1$ **37.** $|x + 2| > 0$ **38.** $|2 - x| > 0$

39. $|x + 2| > -3$ **40.** $|2 - x| > -2$ **41.** $|2x - 1| < 0.02$ **42.** $|3x - 2| < 0.02$

43. If $b \ne 0$, prove that $|a/b| = |a|/|b|$.

44. Show that $a \le |a|$.

45. Prove the *triangle inequality*: $|a + b| \le |a| + |b|$ [*Hint:* Expand $|a + b|^2 = (a + b)^2$, and use the result of Problem 44.]

46. Prove that $|a - b| \ge |a| - |b|$. [*Hint:* Apply the triangle inequality from Problem 45 to $|a| = |(a - b) + b|$.]

47. Express the fact that x differs from 2 by less than $\frac{1}{2}$ as an inequality involving an absolute value. Solve for x.

48. Express the fact that x differs from -1 by less than 1 as an inequality involving an absolute value. Solve for x.

49. Express the fact that x differs from -3 by more than 2 as an inequality involving an absolute value. Solve for x.

50. Express the fact that x differs from 2 by more than 3 as an inequality involving an absolute value. Solve for x.

51. "Normal" human body temperature is 98.6°F. If a temperature x that differs from normal by at least 1.5° is considered unhealthy, write the condition for an unhealthy temperature x as an inequality involving an absolute value, and solve for x.

52. In the United States, normal household voltage is 115 volts. However, it is not uncommon for actual voltage to differ from normal voltage by at most 5 volts. Express this situation as an inequality involving an absolute value. Use x as the actual voltage and solve for x.

53. If $a > 0$, show that the solution set of the inequality

$$x^2 < a$$

consists of all numbers x for which

$$-\sqrt{a} < x < \sqrt{a}$$

54. If $a > 0$, show that the solution set of the inequality

$$x^2 > a$$

consists of all numbers x for which

$$x > \sqrt{a} \quad \text{or} \quad x < -\sqrt{a}$$

In Problems 55–62, use the results found in Problems 53 and 54 to solve each inequality.

55. $x^2 < 1$ **56.** $x^2 < 4$ **57.** $x^2 \ge 9$ **58.** $x^2 \ge 1$

59. $x^2 \le 16$ **60.** $x^2 \le 9$ **61.** $x^2 > 4$ **62.** $x^2 > 16$

CHAPTER REVIEW

THINGS TO KNOW

Quadratic equation and quadratic formula	If $ax^2 + bx + c = 0$, $a \neq 0$, then $x = \dfrac{-b \pm \sqrt{b^2 - 4ac}}{2a}$.	
Discriminant	If $b^2 - 4ac > 0$, there are two distinct real solutions.	
	If $b^2 - 4ac = 0$, there is one repeated real solution.	
	If $b^2 - 4ac < 0$, there are two distinct complex solutions that are not real; the solutions are conjugates of each other.	
Inequality properties	Transitive property	If $a < b$ and $b < c$, then $a < c$.
	Addition property	If $a < b$, then $a + c < b + c$.
	Multiplication properties	(a) If $a < b$ and if $c > 0$, then $ac < bc$.
		(b) If $a < b$ and if $c < 0$, then $ac > bc$.
	Reciprocal property	If $a > 0$, then $\dfrac{1}{a} > 0$.
	Trichotomy property	$a < b$ or $a = b$ or $b < a$
Absolute value	If $\lvert u \rvert = a$, $a > 0$, then $u = -a$ or $u = a$.	
	If $\lvert u \rvert \le a$, $a > 0$, then $-a \le u \le a$.	
	If $\lvert u \rvert \ge a$, $a > 0$, then $u \le -a$ or $u \ge a$.	
Interval notation	$[a, b]$	$\{x \mid a \le x \le b\}$
	$[a, b)$	$\{x \mid a \le x < b\}$
	$(a, b]$	$\{x \mid a < x \le b\}$
	(a, b)	$\{x \mid a < x < b\}$
	$[a, \infty)$	$\{x \mid x \ge a\}$
	(a, ∞)	$\{x \mid x > a\}$
	$(-\infty, a]$	$\{x \mid x \le a\}$
	$(-\infty, a)$	$\{x \mid x < a\}$

How To:

Solve linear equations in one variable
Solve quadratic equations
Solve radical equations
Solve linear inequalities in one variable
Solve quadratic and rational inequalities
Solve equations and inequalities involving absolute value
Solve applied problems

FILL-IN-THE-BLANK ITEMS

1. Two equations (or inequalities) that have precisely the same solution set are called __________.
2. An equation that is satisfied for every choice of the variable for which both sides are meaningful is called a(n) __________.
3. To complete the square of the expression $x^2 + 5x$, you would __________ the number __________.
4. The quantity $b^2 - 4ac$ is called the __________ of a quadratic equation. If it is __________, the equation has no real solution.
5. If $a < 0$, then $|a| =$ __________.
6. When a quadratic equation has a repeated solution, it is called a(n) __________ root or a root of __________ __________.
7. When an apparent solution does not satisfy the original equation, it is called a(n) __________ solution.
8. If each side of an inequality is multiplied by a(n) __________ number, then the sense of the inequality symbol is reversed.

REVIEW EXERCISES

In Problems 1–38, find all real solutions, if any, of each equation. (Where they appear, a, b, m, and n are positive constants.)

1. $2 - \dfrac{x}{3} = 5$
2. $\dfrac{x}{4} - 2 = 4$
3. $-2(5 - 3x) + 8 = 4 + 5x$
4. $(6 - 3x) - 2(1 + x) = 6x$
5. $\dfrac{3x}{4} - \dfrac{x}{3} = \dfrac{1}{12}$
6. $\dfrac{4 - 2x}{3} + \dfrac{1}{6} = 2x$
7. $\dfrac{x}{x - 1} = \dfrac{5}{4}, \quad x \neq 1$
8. $\dfrac{4x - 5}{3 - 7x} = 4, \quad x \neq \frac{3}{7}$
9. $x(1 - x) = 6$
10. $x(1 + x) = 2$
11. $\dfrac{1}{2}\left(x - \dfrac{1}{3}\right) = \dfrac{3}{4} - \dfrac{x}{6}$
12. $\dfrac{1 - 3x}{4} = \dfrac{x + 6}{3} + \dfrac{1}{2}$
13. $(x - 1)(2x + 3) = 3$
14. $x(2 - x) = 3(x - 4)$
15. $2x + 3 = 4x^2$
16. $1 + 6x = 4x^2$
17. $\sqrt[3]{x^2 - 1} = 2$
18. $\sqrt{1 + x^3} = 3$
19. $x(x + 1) + 2 = 0$
20. $3x^2 - x + 1 = 0$
21. $x^4 - 5x^2 + 4 = 0$
22. $3x^4 + 4x^2 + 1 = 0$
23. $\sqrt{2x - 3} + x = 3$
24. $\sqrt{2x - 1} = x - 2$
25. $x^{3/2} + 2x^{1/2} = 0$
26. $x^{2/3} - x = 0$
27. $\sqrt{x + 1} + \sqrt{x - 1} = \sqrt{2x + 1}$
28. $\sqrt{2x - 1} - \sqrt{x - 5} = 3$
29. $2\sqrt[3]{x^2} - \sqrt[3]{x} = 1$
30. $4\sqrt[3]{x^2} = 1$
31. $x^{-6} - 7x^{-3} - 8 = 0$
32. $6x^{-1} - 5x^{-1/2} + 1 = 0$
33. $x^2 + m^2 = 2mx + (nx)^2$
34. $b^2x^2 + 2ax = x^2 + a^2$
35. $10a^2x^2 - 2abx - 36b^2 = 0$
36. $\dfrac{1}{x - m} + \dfrac{1}{x - n} = \dfrac{2}{x}, \quad x \neq 0, m, n$
37. $\sqrt{x^2 + 3x + 7} - \sqrt{x^2 - 3x + 9} + 2 = 0$
38. $\sqrt{x^2 + 3x + 7} - \sqrt{x^2 + 3x + 9} = 2$

In Problems 39–58, solve each inequality.

39. $\dfrac{2x-3}{5}+1 \le \dfrac{x}{2}$

40. $\dfrac{5-x}{3} \le 6x-1$

41. $-9 \le \dfrac{2x+3}{-4} \le 7$

42. $-4 < \dfrac{2x-2}{3} < 6$

43. $6 > \dfrac{3-3x}{12} > 2$

44. $6 > \dfrac{5-3x}{2} \ge -3$

45. $2x^2+5x-12<0$

46. $3x^2-2x-1 \ge 0$

47. $\dfrac{6}{x+2} \ge 1, \quad x \ne -2$

48. $\dfrac{-2}{1-3x} < -1, \quad x \ne \frac{1}{3}$

49. $\dfrac{2x-6}{1-x} < 2, \quad x \ne 1$

50. $\dfrac{3-2x}{2x+5} \ge 2, \quad x \ne -\frac{5}{2}$

51. $\dfrac{(x-2)(x-1)}{x-3} > 0, \quad x \ne 3$

52. $\dfrac{x+1}{x(x-5)} \le 0, \quad x \ne 0, 5$

53. $\dfrac{x^2-8x+12}{x^2-16} > 0, \quad x \ne -4, 4$

54. $\dfrac{x(x^2+x-2)}{x^2+9x+20} \le 0, \quad x \ne -5, -4$

55. $|3x+4| < \frac{1}{2}$

56. $|1-2x| < \frac{1}{3}$

57. $|2x-5| \ge 7$

58. $|3x+1| \ge 2$

In Problems 59–68, solve each equation in the complex number system.

59. $x^2+x+1=0$

60. $x^2-x+1=0$

61. $2x^2+x-2=0$

62. $3x^2-2x-1=0$

63. $x^2+3=x$

64. $2x^2+1=2x$

65. $x(1-x)=6$

66. $x(1+x)=2$

67. $x^4+2x^2-8=0$

68. $x^4-8x^2-9=0$

69. A flash of lightning is seen, and the resulting thunderclap is heard 3 seconds later. If the speed of sound averages 1100 feet per second, how far away is the storm?

70. The intensity I (in candlepower) of a certain light source obeys the equation $I = 900/x^2$, where x is the distance (in meters) from the light. Over what range of distances can an object be placed from this light source so that the range of intensity of light is from 1600 to 3600 candlepower, inclusive?

71. A search plane has a cruising speed of 250 miles per hour and carries enough fuel for at most 5 hours of flying. If there is a wind that averages 30 miles per hour and the direction of search is with the wind one way and against it the other, how far can the search plane travel?

72. If the search plane described in Problem 71 is able to add a supplementary fuel tank that allows for an additional 2 hours of flying, how much farther can the plane extend its search?

C **73.** A life raft, set adrift from a sinking ship 150 miles offshore, travels directly toward a Coast Guard station at the rate of 5 miles per hour. At the time the raft is set adrift, a rescue

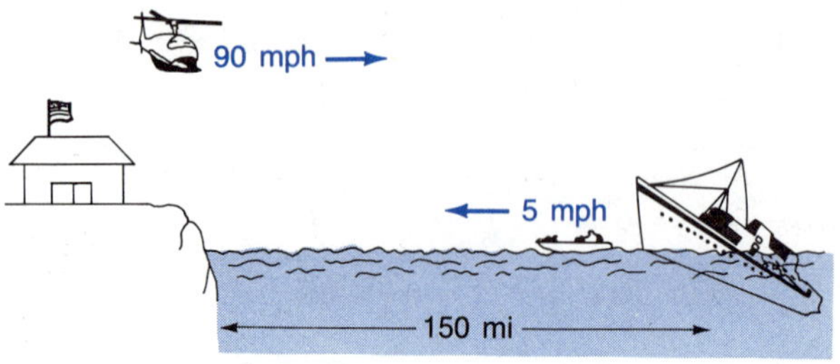

helicopter is dispatched from the Coast Guard station. If the helicopter's average speed is 90 miles per hour, how long will it take the helicopter to reach the life raft?

74. Two bees leave two locations 150 meters apart and fly, without stopping, back and forth between these two locations at average speeds of 3 meters per second and 5 meters per second, respectively. How long is it until the bees meet for the first time? How long is it until they meet for the second time?

C **75.** A man is walking at an average speed of 4 miles per hour alongside a railroad track. A freight train, going in the same direction at an average speed of 30 miles per hour, requires 5 seconds to pass the man. How long is the freight train? Give your answer in feet.

C **76.** One formula stating the relationship between the length l and width w of a rectangle of "pleasing proportion" is $l^2 = w(l + w)$. How should a 4 foot by 8 foot sheet of plasterboard be cut so that the result is a rectangle of "pleasing proportion" with a width of 4 feet?

C **77.** A group of 20 senior citizens can charter a bus for a 1 day excursion trip for $15 per person. The charter company agrees to reduce the price of each ticket by 10¢ for each additional passenger in excess of 20 who goes on the trip, up to a maximum of 44 passengers (the capacity of the bus). If the final bill from the charter company was $482.40, how many seniors went on the trip, and how much did each pay?

Graphs

3

The idea of using a system of rectangular coordinates dates back to ancient times, when such a system was used for surveying and city planning. Apollonius of Perga, in 200 BC, used a form of rectangular coordinates in his work on conics, although this use does not stand out as clearly as it does in modern treatments. Sporadic use of rectangular coordinates continued until the 1600's. By that time, algebra had developed sufficiently so that René Descartes (1596–1650) and Pierre de Fermat (1601–1665) could take the crucial step, which was the use of rectangular coordinates to translate geometry problems into algebra problems, and vice versa. This step was supremely important for two reasons. First, it allowed both geometers and algebraists to gain critical new insights into their subjects, which previously had been regarded as separate but now were seen to be connected in many important ways. Second, the insights gained made possible the development of calculus, which greatly enlarged the number of areas in which mathematics could be applied and made possible a much deeper understanding of these areas.

3.1 Rectangular Coordinates

We locate a point on the real number line by assigning it a single real number, called the *coordinate of the point*. For work in a two-dimensional plane, we locate points by using two numbers.

We begin with two real number lines located in the same plane: one horizontal and the other vertical. We call the horizontal line the **x-axis**, the vertical line the **y-axis**, and the point of intersection the **origin O**. We assign coordinates to every point on these number lines, as described earlier (Section 1.1) and shown in Figure 1, using a convenient scale. (The scales are usually, but not necessarily, the same on both axes.) Also, the origin O has a value of 0 on both the x-axis and the y-axis. We follow the usual convention that points on the x-axis to the right of O are associated with positive real numbers, and those to the left of O are associated with negative real numbers. Those on the y-axis above O are associated with positive real numbers, and those below O are associated with negative real numbers. In Figure 1, the x-axis and y-axis are labeled as x and y, respectively, and we have used an arrow at the end of each axis to denote the positive direction.

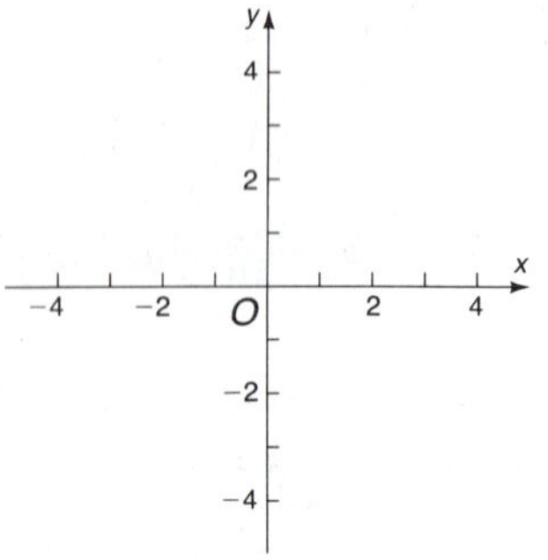

Figure 1

The coordinate system described here is called a **rectangular**, or **Cartesian* coordinate system**. The plane formed by the x-axis and y-axis is sometimes called the **xy-plane**, and the x-axis and y-axis are referred to as the **coordinate axes**.

Any point P in the xy-plane can then be located by using an **ordered pair (x, y)** of real numbers. Let x denote the signed distance of P from the y-axis (*signed* in the sense that if P is to the right of the y-axis, then $x > 0$, and if P is to the left of the y-axis, then $x < 0$); and let y denote the signed distance of P from the x-axis. The ordered pair (x, y), also called the **coordinates** of P, then gives us enough information to locate the point P in the plane.

For example, to locate the point whose coordinates are $(-3, 1)$, go 3 units along the x-axis to the left of O and then go straight up 1 unit. We **plot** this point by placing a dot at this location. See Figure 2, in which the points with coordinates $(-3, 1)$, $(-2, -3)$, $(3, -2)$ and $(3, 2)$, are plotted.

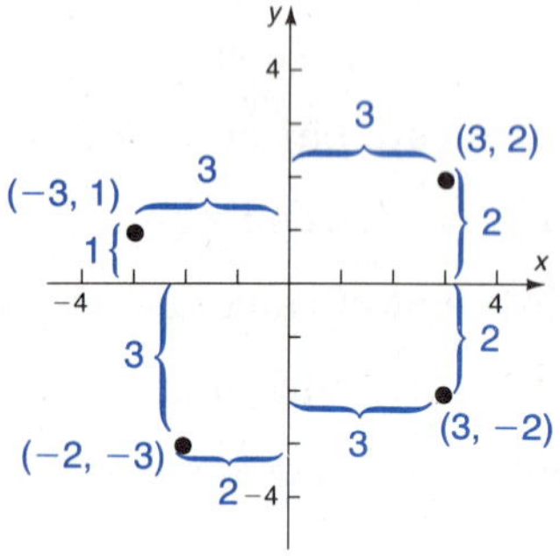

Figure 2

The origin has coordinates $(0, 0)$. Any point on the x-axis has coordinates of the form $(x, 0)$, and any point on the y-axis has coordinates of the form $(0, y)$.

If (x, y) are the coordinates of a point P, then x is called the **x-coordinate**, or **abscissa**, of P and y is the **y-coordinate**, or **ordinate**, of P. We identify the point P by its coordinates (x, y) by writing $P = (x, y)$. Usually, we will simply say "the point (x, y)" rather than "the point whose coordinates are (x, y)."

*Named after René Descartes (1596–1650), a French mathematician, philosopher, and theologian.

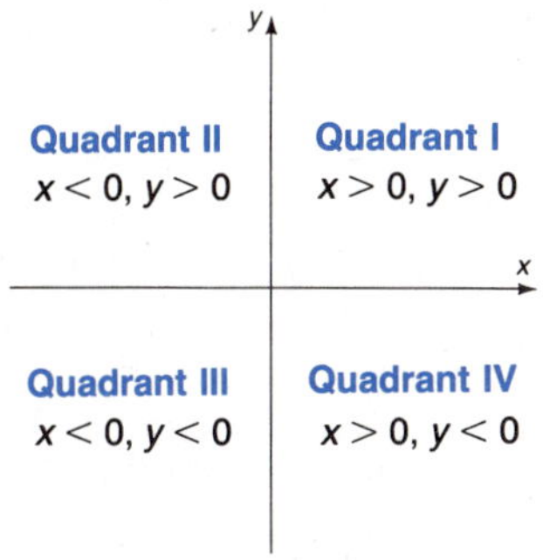

Figure 3

The coordinate axes divide the xy-plane into four sections, called **quadrants**, as shown in Figure 3. In quadrant I, both the x-coordinate and the y-coordinate of all points are positive; in quadrant II, x is negative and y is positive; in quadrant III, both x and y are negative; and in quadrant IV, x is positive and y is negative. Points on the coordinate axes belong to no quadrant.

Distance between Points

If the same units of measurement—such as feet, miles, and so on—are used for both the x-axis and the y-axis, then all distances in the xy-plane can be measured using this unit of measurement.

Example 1 Find the distance d between the points (1, 3) and (5, 6).

Solution First we plot the points (1, 3) and (5, 6), as shown in Figure 4(a). Then we draw a horizontal line from (1, 3) to (5, 3) and a vertical line from (5, 3) to (5, 6), forming a right triangle, as in Figure 4(b). One leg of the triangle is of length 4 and the other is of length 3. By the Pythagorean Theorem (see Section 1.10), the square of the distance d we seek is

$$d^2 = 4^2 + 3^2 = 16 + 9 = 25$$

so that

$$d = 5$$

Figure 4

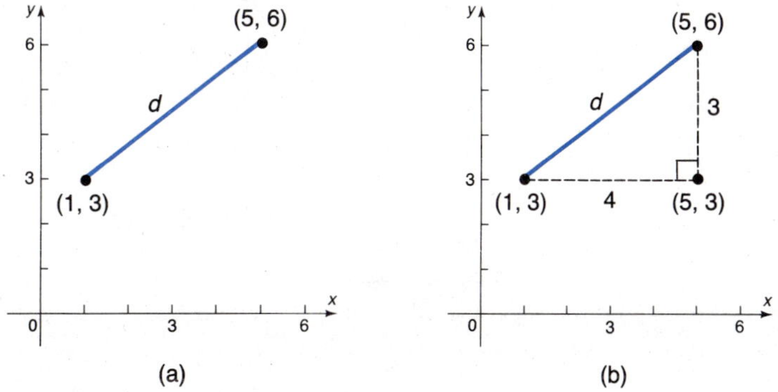

The *distance formula* provides a straightforward method for computing the distance between two points.

Theorem

The distance between two points $P_1 = (x_1, y_1)$ and $P_2 = (x_2, y_2)$, denoted by $d(P_1, P_2)$, is

Distance Formula

$$d(P_1, P_2) = \sqrt{(x_2 - x_1)^2 + (y_2 - y_1)^2} \tag{1}$$

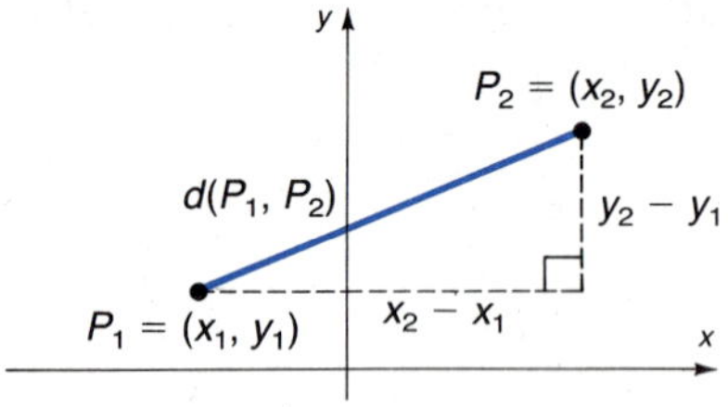

Figure 5

That is, to compute the distance between two points, find the difference of the x-coordinates, square it, and add this to the square of the difference of the y-coordinates. The square root of this sum is the distance. See Figure 5.

Proof of the Distance Formula

Let (x_1, y_1) denote the coordinates of point P_1, and let (x_2, y_2) denote the coordinates of point P_2. Assume the line joining P_1 and P_2 is neither horizontal nor vertical. Refer to Figure 6(a). The coordinates of P_3 are (x_2, y_1). The horizontal distance from P_1 to P_3 is the absolute value of the difference of the x-coordinates, namely, $|x_2 - x_1|$. The vertical distance from P_3 to P_2 is the absolute value of the difference of the y-coordinates, namely, $|y_2 - y_1|$. See Figure 6(b). The distance $d(P_1, P_2)$ that we seek is the length of the hypotenuse of the right triangle, so, by the Pythagorean Theorem, it follows that

$$\begin{aligned} [d(P_1, P_2)]^2 &= |x_2 - x_1|^2 + |y_2 - y_1|^2 \\ &= (x_2 - x_1)^2 + (y_2 - y_1)^2 \\ d(P_1, P_2) &= \sqrt{(x_2 - x_1)^2 + (y_2 - y_1)^2} \end{aligned}$$

Figure 6

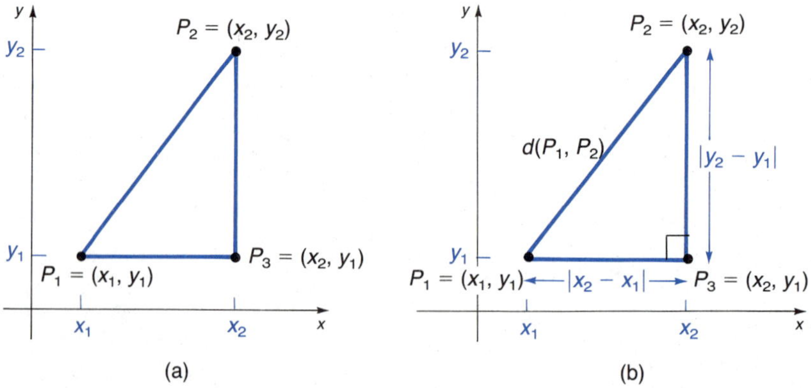

Now, if the line joining P_1 and P_2 is horizontal, then the y-coordinate of P_1 equals the y-coordinate of P_2; that is, $y_1 = y_2$. Refer to Figure 7(a). In this case, the distance formula (1) still works, because, for $y_1 = y_2$, it reduces to

$$d(P_1, P_2) = \sqrt{(x_2 - x_1)^2 + 0^2} = \sqrt{(x_2 - x_1)^2} = |x_2 - x_1|$$

Figure 7

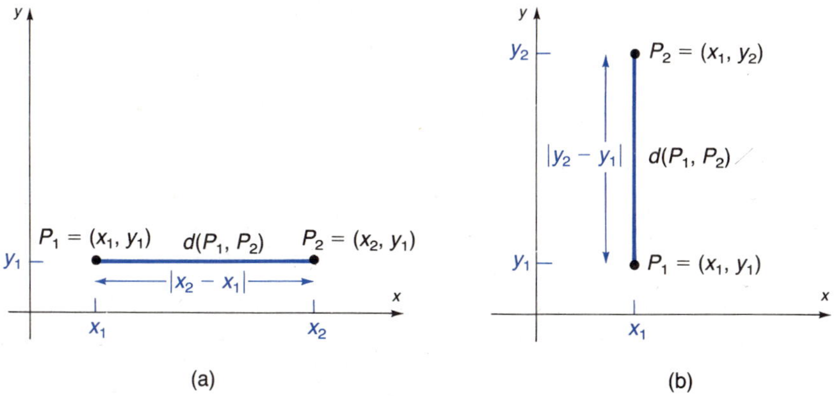

A similar argument holds if the line joining P_1 and P_2 is vertical. See Figure 7(b). Thus, the distance formula is valid in all cases. ■

Example 2 Find the distance d between the points $(-4, 5)$ and $(3, 2)$.

Solution Using the distance formula (1), the solution is obtained as follows:

$$d = \sqrt{[3 - (-4)]^2 + (2 - 5)^2} = \sqrt{7^2 + (-3)^2}$$
$$= \sqrt{49 + 9} = \sqrt{58} \approx 7.62$$ ■

The distance between two points $P_1 = (x_1, y_1)$ and $P_2 = (x_2, y_2)$ is never a negative number. Furthermore, the distance between two points is 0 only when the points are identical—that is, when $x_1 = x_2$ and $y_1 = y_2$. Also, because $(x_2 - x_1)^2 = (x_1 - x_2)^2$ and $(y_2 - y_1)^2 = (y_1 - y_2)^2$, it makes no difference whether the distance is computed from P_1 to P_2 or from P_2 to P_1—that is, $d(P_1, P_2) = d(P_2, P_1)$.

The introduction to this chapter mentioned that rectangular coordinates enable us to translate geometry problems into algebra problems, and vice versa. The next example shows how algebra (the distance formula) can be used to solve some geometry problems.

Example 3 Consider the three points $A = (-2, 1)$, $B = (2, 3)$, and $C = (3, 1)$.

(a) Plot each point and form the triangle ABC.

(b) Find the length of each side of the triangle.

(c) Verify that the triangle is a right triangle.

(d) Find the area of the triangle.

Solution (a) Points A, B, C, and triangle ABC are plotted in Figure 8 (next page).

(b) $d(A, B) = \sqrt{[2 - (-2)]^2 + (3 - 1)^2} = \sqrt{16 + 4} = \sqrt{20} = 2\sqrt{5}$

$d(B, C) = \sqrt{(3 - 2)^2 + (1 - 3)^2} = \sqrt{1 + 4} = \sqrt{5}$

$d(A, C) = \sqrt{[3 - (-2)]^2 + (1 - 1)^2} = \sqrt{5^2 + 0^2} = 5$

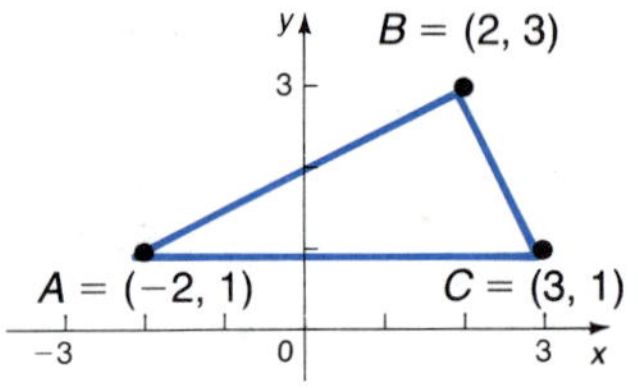

Figure 8

(c) To show that the triangle is a right triangle we need to show that the sum of the squares of the lengths of two of the sides equals the square of the length of the third side. (Why is this sufficient?) Looking at Figure 8, it seems reasonable to conjecture that the right angle is at vertex B. Thus, we shall check to see whether

$$[d(A, B)]^2 + [d(B, C)]^2 = [d(A, C)]^2$$

We find

$$\begin{aligned}[d(A, B)]^2 + [d(B, C)]^2 &= (2\sqrt{5})^2 + (\sqrt{5})^2 \\ &= 20 + 5 = 25 = [d(A, C)]^2\end{aligned}$$

so it follows from the converse of the Pythagorean Theorem that triangle ABC is a right triangle.

(d) Because the right angle is at B, the sides AB and BC form the base and altitude of the triangle. Its area is therefore

$$\text{Area} = \tfrac{1}{2}(\text{Base})(\text{Altitude}) = \tfrac{1}{2}(2\sqrt{5})(\sqrt{5}) = 5 \text{ square units}$$ ■

Midpoint Formula

We now derive a formula for the coordinates of the **midpoint of a line segment**. Let $P_1 = (x_1, y_1)$ and $P_2 = (x_2, y_2)$ be the endpoints of a line segment, and let $M = (x, y)$ be the point on the line segment that is the same distance from P_1 as it is from P_2. See Figure 9. The triangles P_1AM and MBP_2 are congruent.* [Do you see why? Angle AP_1M = Angle BMP_2,† Angle P_1MA = Angle MP_2B, and $d(P_1, M) = d(M, P_2)$ is given. Thus, we have Angle–Side–Angle.] Hence, corresponding sides are equal in length. That is,

$$\begin{aligned} x - x_1 &= x_2 - x & \quad\text{and}\quad y - y_1 &= y_2 - y \\ 2x &= x_1 + x_2 & 2y &= y_1 + y_2 \\ x &= \frac{x_1 + x_2}{2} & y &= \frac{y_1 + y_2}{2} \end{aligned}$$

Figure 9

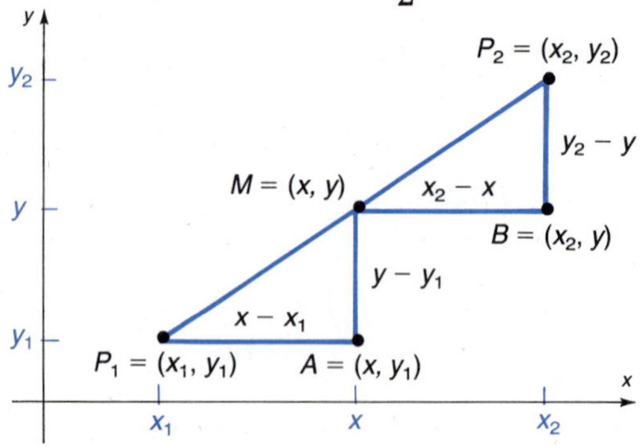

*Two triangles are congruent if their sides are the same length (SSS), or if two sides and the included angle are the same (SAS), or if two angles and the included side are the same (ASA).

†The transversal $\overline{P_1P_2}$ meets the parallel lines $\overline{P_1A}$ and $\overline{MB}$ in equal angles.

Theorem The midpoint (x, y) of the line segment from $P_1 = (x_1, y_1)$ to $P_2 = (x_2, y_2)$ is given by

Midpoint Formula

$$x = \frac{x_1 + x_2}{2} \quad \text{and} \quad y = \frac{y_1 + y_2}{2} \tag{2}$$

■

Thus, to find the midpoint of a line segment, we average the x-coordinates and the y-coordinates of the endpoints.

Example 4 Find the midpoint of the line segment from $P_1 = (-5, 3)$ to $P_2 = (3, 1)$. Plot the points P_1 and P_2, and their midpoint. Check your answer.

Solution We apply the midpoint formula (2) using $x_1 = -5$, $x_2 = 3$, $y_1 = 3$, and $y_2 = 1$. Then the coordinates (x, y) of the midpoint are

$$x = \frac{x_1 + x_2}{2} = \frac{-5 + 3}{2} = -1 \quad \text{and} \quad y = \frac{y_1 + y_2}{2} = \frac{3 + 1}{2} = 2$$

That is, $M = (-1, 2)$. See Figure 10.

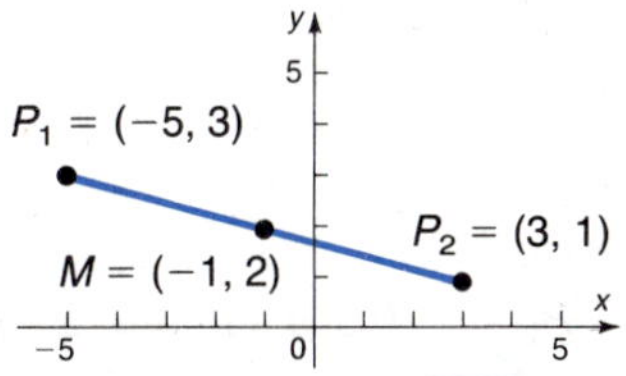

Figure 10

Check: Because M is the midpoint, we check the answer by verifying that $d(P_1, M) = d(M, P_2)$:

$$d(P_1, M) = \sqrt{[-1 - (-5)]^2 + (2 - 3)^2} = \sqrt{16 + 1} = \sqrt{17}$$
$$d(M, P_2) = \sqrt{[3 - (-1)]^2 + (1 - 2)^2} = \sqrt{16 + 1} = \sqrt{17}$$ ■

EXERCISE 3.1

In Problems 1 and 2, plot each point in the xy-plane. Tell in which quadrant or on what coordinate axis each point lies.

1. (a) $A = (-3, 2)$ (b) $B = (6, 0)$ (c) $C = (-2, -2)$
(d) $D = (6, 5)$ (e) $E = (0, -3)$ (f) $F = (6, -3)$

2. (a) $A = (1, 4)$ (b) $B = (-3, -4)$ (c) $C = (-3, 4)$
(d) $D = (4, 1)$ (e) $E = (0, 1)$ (f) $F = (-3, 0)$

3. Plot the points $(2, 0)$, $(2, -3)$, $(2, 4)$, $(2, 1)$, and $(2, -1)$. Describe the set of all points of the form $(2, y)$, where y is a real number.

4. Plot the points $(0, 3)$, $(1, 3)$, $(-2, 3)$, $(5, 3)$, and $(-4, 3)$. Describe the set of all points of the form $(x, 3)$, where x is a real number.

In Problems 5–14, find the distance $d(P_1, P_2)$ between the points P_1 and P_2.

5. $P_1 = (3, -4)$; $P_2 = (3, 1)$
6. $P_1 = (-1, 0)$; $P_2 = (2, 1)$
7. $P_1 = (-3, 2)$; $P_2 = (6, 0)$
8. $P_1 = (2, -3)$; $P_2 = (4, 2)$
9. $P_1 = (4, -3)$; $P_2 = (6, 1)$
10. $P_1 = (-4, -3)$; $P_2 = (2, 2)$

C 11. $P_1 = (-0.2, 0.3)$; $P_2 = (2.3, 1.1)$

C 12. $P_1 = (1.2, 2.3)$; $P_2 = (-0.3, 1.1)$

13. $P_1 = (a, b)$; $P_2 = (0, 0)$

14. $P_1 = (a, a)$; $P_2 = (0, 0)$

In Problems 15–20, plot each point and form the triangle ABC. Verify that the triangle is a right triangle. Find its area.

15. $A = (-2, 5)$; $B = (1, 3)$; $C = (-1, 0)$
16. $A = (-2, 5)$; $B = (12, 3)$; $C = (10, -11)$
17. $A = (-5, 3)$; $B = (6, 0)$; $C = (5, 5)$
18. $A = (-6, 3)$; $B = (3, -5)$; $C = (-1, 5)$
19. $A = (4, -3)$; $B = (0, -3)$; $C = (4, 2)$
20. $A = (4, -3)$; $B = (4, 1)$; $C = (2, 1)$

21. Find all points having an x-coordinate of 2 whose distance from the point $(-2, -1)$ is 5.
22. Find all points having a y-coordinate of -3 whose distance from the point $(1, 2)$ is 13.
23. Find all points on the x-axis that are 5 units from the point $(2, -3)$.
24. Find all points on the y-axis that are 5 units from the point $(-4, 4)$.

In Problems 25–34, find the midpoint of the line segment joining the points P_1 and P_2.

25. $P_1 = (3, -4)$; $P_2 = (3, 1)$
26. $P_1 = (-1, 0)$; $P_2 = (2, 1)$
27. $P_1 = (-3, 2)$; $P_2 = (6, 0)$
28. $P_1 = (2, -3)$; $P_2 = (4, 2)$
29. $P_1 = (4, -3)$; $P_2 = (6, 1)$
30. $P_1 = (-4, -3)$; $P_2 = (2, 2)$
31. $P_1 = (-0.2, 0.3)$; $P_2 = (2.3, 1.1)$
32. $P_1 = (1.2, 2.3)$; $P_2 = (-0.3, 1.1)$
33. $P_1 = (a, b)$; $P_2 = (0, 0)$
34. $P_1 = (a, a)$; $P_2 = (0, 0)$

35. The **medians** of a triangle are the line segments from each vertex to the midpoint of the opposite side. Find the lengths of the medians of the triangle with vertices at $(0, 0)$, $(0, 6)$, and $(8, 0)$.
36. An **equilateral triangle** is one in which all three sides are of equal length. If two vertices of an equilateral triangle are $(0, 4)$ and $(0, 0)$, find the third vertex. How many of these triangles are possible?

In Problems 37–40, find the length of each side of the triangle determined by the three points P_1, P_2, and P_3. State whether the triangle is an isosceles triangle, a right triangle, neither of these, or both. (An ***isosceles triangle*** *is one in which at least two of the sides are of equal length.)*

37. $P_1 = (2, 1)$; $P_2 = (-4, 1)$; $P_3 = (-4, -3)$
38. $P_1 = (-1, 4)$; $P_2 = (6, 2)$; $P_3 = (4, -5)$
39. $P_1 = (-2, -1)$; $P_2 = (0, 7)$; $P_3 = (3, 2)$
40. $P_1 = (7, 2)$; $P_2 = (-4, 0)$; $P_3 = (4, 6)$

Problems 42–46 use the result of Problem 41.

41. If r is a real number, prove that the point $P = (x, y)$ that divides the line segment from $P_1 = (x_1, y_1)$ to $P_2 = (x_2, y_2)$ in the ratio r—that is,

$$\frac{d(P_1, P)}{d(P_1, P_2)} = r$$

has coordinates

$$x = x_1 + r(x_2 - x_1) \qquad \text{and} \qquad y = y_1 + r(y_2 - y_1)$$

[*Hint:* Use similar triangles.*]

42. Verify that the midpoint of a line segment divides the line segment from $P_1 = (x_1, y_1)$ to $P_2 = (x_2, y_2)$ in the ratio $r = \frac{1}{2}$.

43. What point P divides the line segment from P_1 to P_2 in the ratio $r = 1$?

44. What point P divides the line segment from P_1 to P_2 in the ratio $r = 0$?

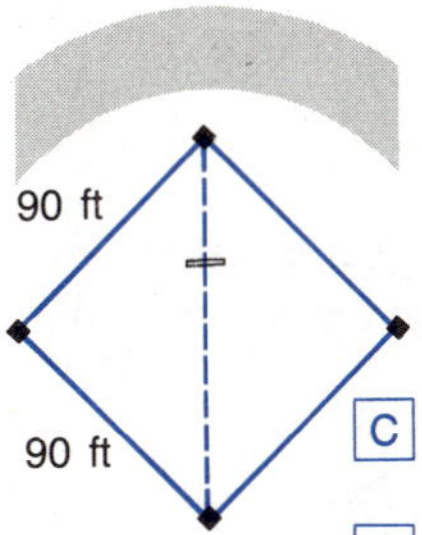

45. Find the point P on the line joining $P_1 = (1, 4)$ and $P_2 = (5, 6)$ that is twice as far from P_1 as P_2 is from P_1 and that lies on the same side of P_1 as P_2 does.

46. Find the point P on the line joining $P_1 = (0, 4)$ and $P_2 = (-1, 1)$ that is three times as far from P_1 as P_2 is from P_1.

C **47.** A major league baseball "diamond" is actually a square, 90 feet on a side (see the figure). What is the distance directly from home plate to second base (the diagonal of the square)?

C **48.** The layout of a Little League playing field is a square, 60 feet on a side.† How far is it directly from home plate to second base (the diagonal of the square)?

49. Find the midpoint of each diagonal of a square with side of length s. Draw the conclusion that the diagonals of a square intersect at their midpoints. [*Hint:* Use $(0, 0)$, $(0, s)$, $(s, 0)$, and (s, s) as the vertices of the square.]

50. Verify that the points $(0, 0)$, $(a, 0)$, and $(a/2, \sqrt{3}a/2)$ are the vertices of an equilateral triangle. Then show that the midpoints of the three sides are the vertices of a second equilateral triangle. (See Problem 36.)

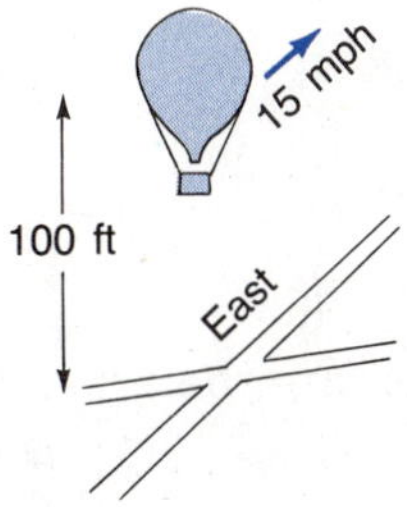

51. An automobile and a truck leave an intersection at the same time. The automobile heads east at an average speed of 40 miles per hour, while the truck heads south at an average speed of 30 miles per hour. Find an expression for their distance apart d (in miles) at the end of t hours.

52. A hot air balloon, headed due east at an average speed of 15 miles per hour and at a constant altitude of 100 feet, passes over an intersection (see the figure). Find an expression for its distance d (measured in feet) from the intersection t seconds later.

3.2 Graphs of Equations

In Chapter 2, we concerned ourselves mainly with equations involving a single real variable. In this section, we look at equations that contain two real variables.

Illustrations play an important role in helping us to visualize the relationships that exist between two variable quantities. For example, Figure 11 (at the top of the next page) shows the variation in the rate

*Two triangles are **similar** if they have the same angles. In similar triangles, corresponding sides are in proportion.

†*Source*: *Little League Baseball, Official Regulations and Playing Rules*, 1989

Figure 11
Variation in the rate of rotation of the Earth

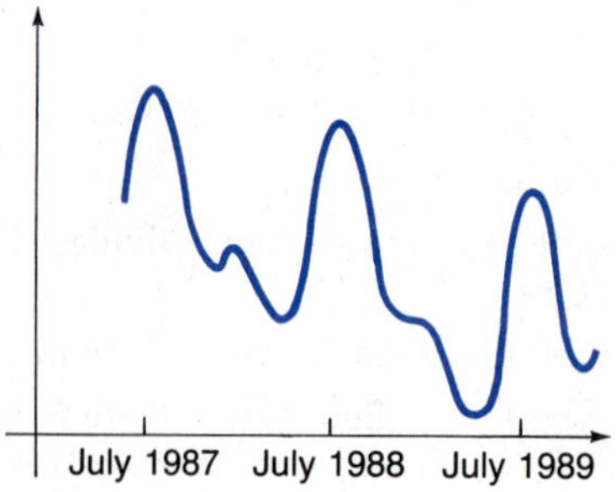

of rotation of the Earth. Such illustrations are usually referred to as *graphs*. The **graph of an equation** in two variables x and y consists of the set of points in the xy-plane whose coordinates (x, y) satisfy the equation.

Let's look at some examples.

Example 1 Graph the equation: $y = 2x + 5$

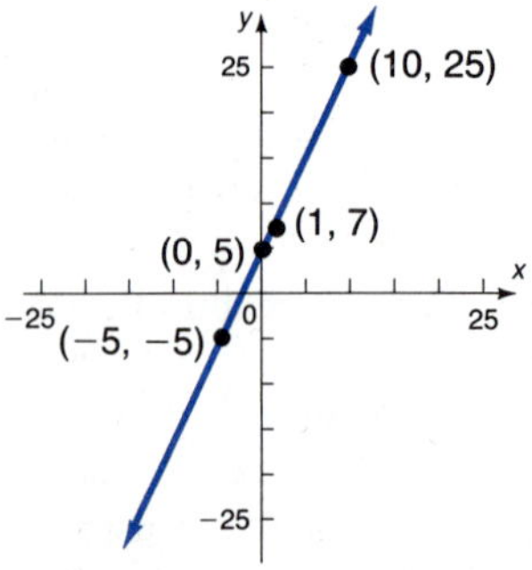

Figure 12
$y = 2x + 5$

Solution We want to find all points (x, y) that satisfy the equation. To locate some of these points (and thus get an idea of the pattern of the graph), we assign some numbers to x and find corresponding values for y:

IF	THEN	POINT ON GRAPH
$x = 0$	$y = 2(0) + 5 = 5$	$(0, 5)$
$x = 1$	$y = 2(1) + 5 = 7$	$(1, 7)$
$x = -5$	$y = 2(-5) + 5 = -5$	$(-5, -5)$
$x = 10$	$y = 2(10) + 5 = 25$	$(10, 25)$

By plotting these points and then connecting them, we obtain the graph of the equation (a straight line), as shown in Figure 12. ■

Example 2 Graph the equation: $y = x^2$

Table 1

x	$y = x^2$	(x, y)
−4	16	(−4, 16)
−3	9	(−3, 9)
−2	4	(−2, 4)
−1	1	(−1, 1)
0	0	(0, 0)
1	1	(1, 1)
2	4	(2, 4)
3	9	(3, 9)
4	16	(4, 16)

Solution Table 1 provides several points on the graph. In Figure 13 we plot these points and connect them with a smooth curve to obtain the graph (a *parabola*).

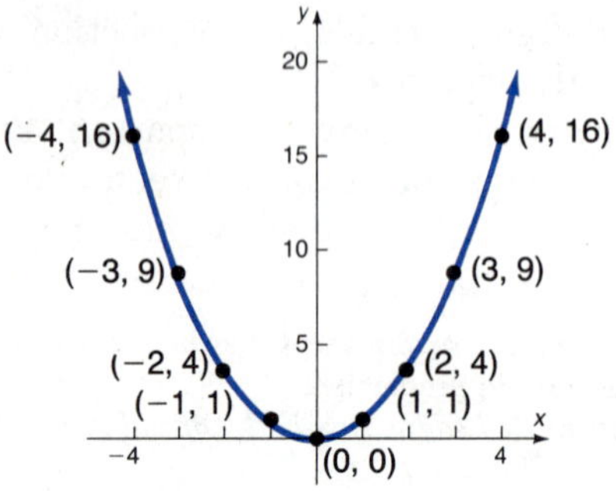

Figure 13
$y = x^2$

■

The graphs of the equations shown in Figures 12 and 13 are necessarily incomplete. For example, in Figure 12, the point (20, 45) is a part of the graph of $y = 2x + 5$, but it is not shown. Since the graph of $y = 2x + 5$ could be extended out as far as we please, we use arrows to indicate that the pattern shown continues. Thus, it is important when illustrating a graph to present enough of the graph so that any viewer of the illustration will "see" the rest of it as an obvious continuation of what is actually there. For the most part, we shall graph equations by plotting a sufficient number of points on the graph until a pattern becomes evident; then we connect these points with a smooth curve following the suggested pattern. Shortly, we shall investigate various techniques that will enable us to graph an equation without plotting so many points.

Example 3 Graph the equation: $y = x^3$

Solution We set up Table 2, listing several points on the graph. Figure 14 illustrates some of these points and the graph of $y = x^3$.

Table 2

x	$y = x^3$	(x, y)
−3	−27	(−3, −27)
−2	−8	(−2, −8)
−1	−1	(−1, −1)
0	0	(0, 0)
1	1	(1, 1)
2	8	(2, 8)
3	27	(3, 27)

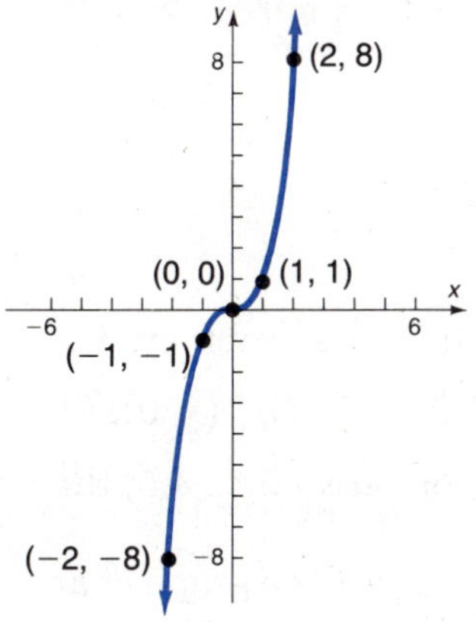

Figure 14
$y = x^3$

Example 4 Graph the equation: $x = y^2$

Solution We set up Table 3, listing several points on the graph. In this case, because of the form of the equation, we assign some numbers to y and find corresponding values of x. Figure 15 illustrates some of these points and the graph of $x = y^2$.

Table 3

y	$x = y^2$	(x, y)
−3	9	(9, −3)
−2	4	(4, −2)
−1	1	(1, −1)
0	0	(0, 0)
1	1	(1, 1)
2	4	(4, 2)
3	9	(9, 3)
4	16	(16, 4)

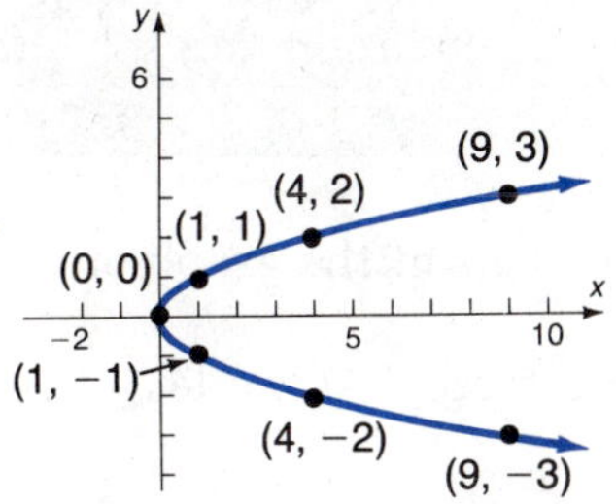

Figure 15
$x = y^2$

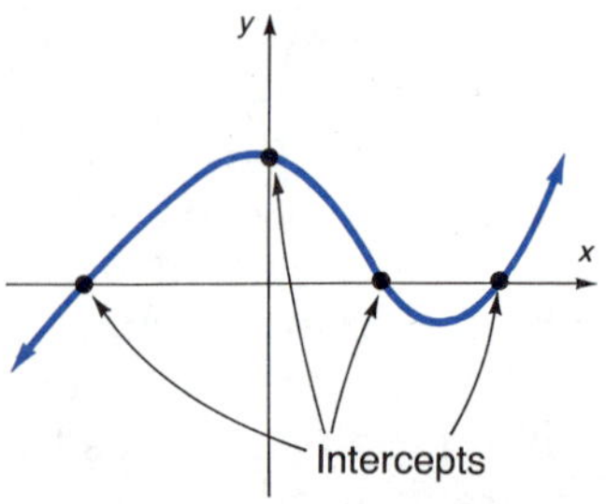

Figure 16

We said earlier that we would discuss techniques that reduce the number of points required to graph an equation. Two such techniques involve finding *intercepts* and checking for *symmetry*.

Intercepts

The points, if any, at which a graph intersects the coordinate axes are called the **intercepts**. See Figure 16. The x-coordinate of a point at which the graph crosses the x-axis is an **x-intercept**, and the y-coordinate of a point at which the graph crosses the y-axis is a **y-intercept**.

Example 5 Find the intercepts of the graph in Figure 17. What are its x-intercepts? What are its y-intercepts?

Figure 17

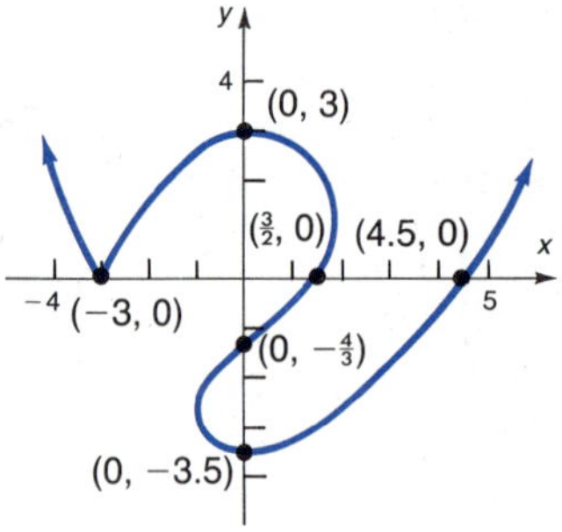

Solution The intercepts of the graph are the points

$$(-3, 0), \quad (0, 3), \quad \left(\tfrac{3}{2}, 0\right), \quad \left(0, -\tfrac{4}{3}\right), \quad (0, -3.5), \quad (4.5, 0)$$

The x-intercepts are $-3, \frac{3}{2}, 4.5$; the y-intercepts are $-3.5, -\frac{4}{3}, 3$. ■

The intercepts of the graph of an equation can be found by using the fact that points on the x-axis have y-coordinates equal to 0 and points on the y-axis have x-coordinates equal to 0.

Procedure for Finding Intercepts

1. To find the x-intercept(s), if any, of the graph of an equation, let $y = 0$ in the equation and solve for x.
2. To find the y-intercept(s), if any, of the graph of an equation, let $x = 0$ in the equation and solve for y.

Example 6 Find the intercepts of the graph of $y = x^2 - 4$.

Solution To find the x-intercept(s), we let $y = 0$ and obtain the equation

$$x^2 - 4 = 0$$

The equation has two solutions, -2 and 2. Thus, the x-intercepts are -2 and 2.

To find the y-intercept(s), we let $x = 0$ and obtain the equation

$$y = -4$$

Thus, the y-intercept is -4.

Since $x^2 \geq 0$ for all x, we deduce from the equation $y = x^2 - 4$ that $y \geq -4$ for all x. This information, plus the points from Table 4, provide enough information to graph $y = x^2 - 4$. See Figure 18.

Table 4

x	$y = x^2 - 4$	(x, y)
-3	5	$(-3, 5)$
-1	-3	$(-1, -3)$
1	-3	$(1, -3)$
3	5	$(3, 5)$

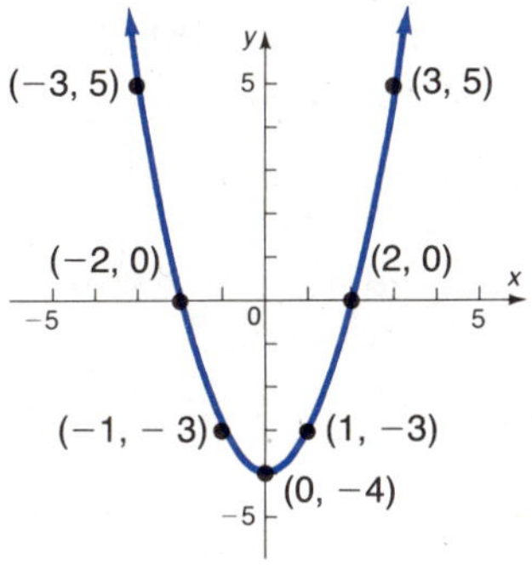

Figure 18
$y = x^2 - 4$

■

Symmetry

Another useful tool for graphing equations involves *symmetry*, particularly symmetry with respect to the x-axis, the y-axis, and the origin.

Symmetry with Respect to the x-Axis

A graph is said to be **symmetric with respect to the x-axis** if, for every point (x, y) on the graph, the point $(x, -y)$ is also on the graph.

Figure 19(a) illustrates the definition. Notice that when a graph is symmetric with respect to the x-axis, the part of the graph above the x-axis is a reflection of the part below it, and vice versa.

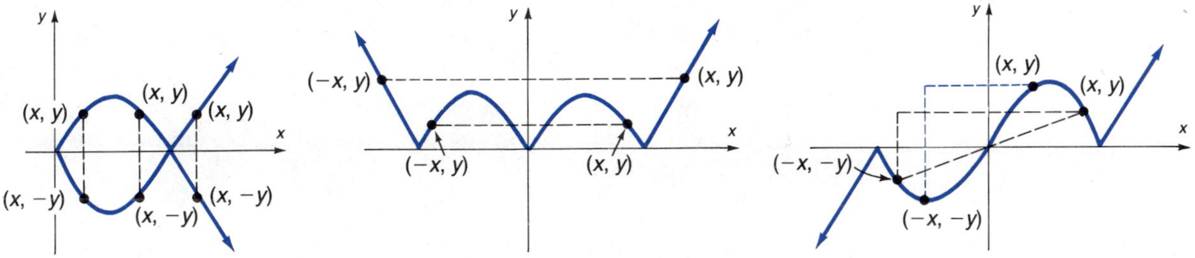

(a) Symmetric with respect to the x-axis

(b) Symmetric with respect to the y-axis

(c) Symmetric with respect to the origin

Figure 19

Symmetry with Respect to the y-Axis

A graph is said to be **symmetric with respect to the y-axis** if, for every point (x, y) on the graph, the point $(-x, y)$ is also on the graph.

Figure 19(b) illustrates the definition. Notice that when a graph is symmetric with respect to the y-axis, the part of the graph to the right of the y-axis is a reflection of the part to the left of it, and vice versa.

Symmetry with Respect to the Origin

A graph is said to be **symmetric with respect to the origin** if, for every point (x, y) on the graph, the point $(-x, -y)$ is also on the graph.

Figure 19(c) illustrates the definition. Notice that symmetry with respect to the origin may be viewed in two ways:

1. As a reflection about the y-axis, followed by a reflection about the x-axis.
2. As a projection along a line through the origin so that the distances from the origin are equal.

When the graph of an equation is symmetric, the number of points that you need to plot in order to see the pattern is reduced. For example, if the graph of an equation is symmetric with respect to the y-axis, then, once points to the right of the y-axis are plotted, an equal number of points on the graph can be obtained by reflecting them about the y-axis. Thus, before we graph an equation, we first want to determine whether it has any symmetry. The following tests are used for that purpose.

Tests for Symmetry

To test the graph of an equation for symmetry with respect to the:

x-axis Replace y by $-y$ in the equation. If an equivalent equation results, the graph of the equation is symmetric with respect to the x-axis.

y-axis Replace x by $-x$ in the equation. If an equivalent equation results, the graph of the equation is symmetric with respect to the y-axis.

Origin Replace x by $-x$ and y by $-y$ in the equation. If an equivalent equation results, the graph of the equation is symmetric with respect to the origin.

Let's look at an equation we have already graphed to see how these tests are used.

Example 7 (a) To test the graph of the equation $x = y^2$ for symmetry with respect to the x-axis, we replace y by $-y$ in the equation, as follows:

$$\begin{aligned} x &= y^2 && \text{Original equation} \\ x &= (-y)^2 && \text{Replace } y \text{ by } -y. \\ x &= y^2 && \text{Simplify.} \end{aligned}$$

When we replace y by $-y$, the result is the same equation. Thus, the graph is symmetric with respect to the x-axis.

(b) To test the graph of the equation $x = y^2$ for symmetry with respect to the y-axis, we replace x by $-x$ in the equation:

$$\begin{aligned} x &= y^2 && \text{Original equation} \\ -x &= y^2 && \text{Replace } x \text{ by } -x. \end{aligned}$$

Because we arrive at the equation $-x = y^2$, which is not equivalent to the original equation, we conclude that the graph is not symmetric with respect to the y-axis.

(c) To test for symmetry with respect to the origin, we replace x by $-x$ and y by $-y$:

$$\begin{aligned} x &= y^2 && \text{Original equation} \\ -x &= (-y)^2 && \text{Replace } x \text{ by } -x, \text{ and } y \text{ by } -y. \\ -x &= y^2 && \text{Simplify.} \end{aligned}$$

The resulting equation, $-x = y^2$, is not equivalent to the original equation. We conclude that the graph is not symmetric with respect to the origin. ■

Figure 20(a) illustrates the graph of $x = y^2$. In forming a table of points on the graph of $x = y^2$, we can restrict ourselves to points whose y-coordinates are positive. Once these are plotted and connected, a reflection about the x-axis (because of the symmetry) provides the rest of the graph.

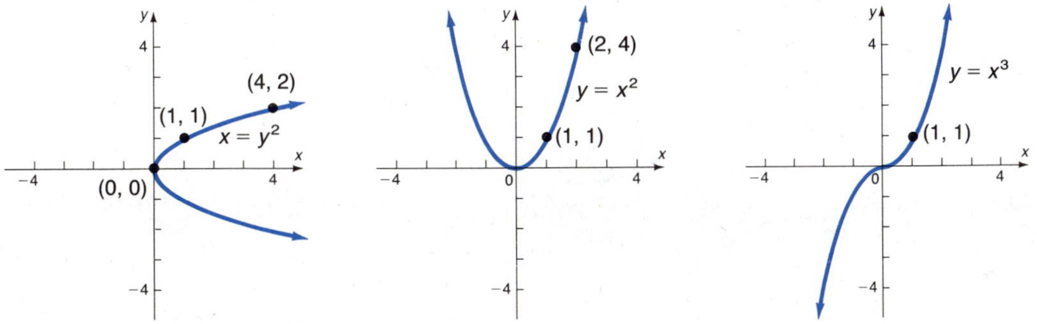

(a) Symmetry about the x-axis (b) Symmetry about the y-axis (c) Symmetry about the origin

Figure 20

Figures 20(b) and 20(c) illustrate two other equations we graphed earlier. Notice how the existence of symmetry reduces the number of points we need to plot.

Example 8 Graph the equation: $y = \dfrac{1}{x}$

Solution We check for intercepts first. If we let $x = 0$, we obtain a 0 denominator, which is not allowed. Hence, there is no y-intercept. If we let $y = 0$, we get the equation $1/x = 0$, which has no solution. Hence, there is no x-intercept. Thus, the graph of $y = 1/x$ does not cross the coordinate axes.

Next we check for symmetry:

x-Axis: Replacing y by $-y$ yields $-y = 1/x$, which is not equivalent to $y = 1/x$.

y-Axis: Replacing x by $-x$ yields $y = -1/x$, which is not equivalent to $y = 1/x$.

Origin: Replacing x by $-x$ and y by $-y$ yields $-y = -1/x$, which is equivalent to $y = 1/x$.

The graph is symmetric with respect to the origin.

Finally, we set up Table 5, listing several points on the graph. Because of the symmetry with respect to the origin, we use only positive values of x. From Table 5, we infer that if x is a large and positive number, then $y = 1/x$ is a positive number close to 0. We also infer that if x is a positive number close to 0, then $y = 1/x$ is a large and positive number. Armed with this information, we can graph the equation. Figure 21 illustrates some of these points and the graph of $y = 1/x$. Observe how the absence of intercepts and the existence of symmetry with respect to the origin were utilized.

Table 5

x	$y = 1/x$	(x, y)
$\frac{1}{10}$	10	$(\frac{1}{10}, 10)$
$\frac{1}{3}$	3	$(\frac{1}{3}, 3)$
$\frac{1}{2}$	2	$(\frac{1}{2}, 2)$
1	1	$(1, 1)$
2	$\frac{1}{2}$	$(2, \frac{1}{2})$
3	$\frac{1}{3}$	$(3, \frac{1}{3})$
10	$\frac{1}{10}$	$(10, \frac{1}{10})$

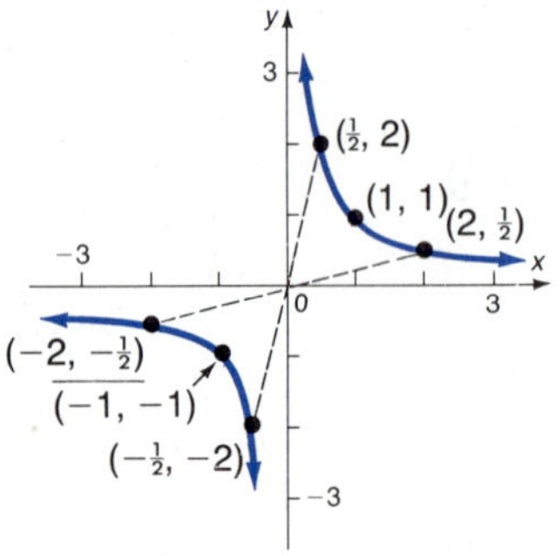

Figure 21
$y = \dfrac{1}{x}$

■

EXERCISE 3.2

In Problems 1–10, plot each point. Then plot the point that is symmetric to it with respect to:
(a) The x-axis (b) The y-axis (c) The origin

1. (3, 4) **2.** (5, 3) **3.** (−2, 1) **4.** (4, −2) **5.** (1, 1)
6. (−1, −1) **7.** (−3, −4) **8.** (4, 0) **9.** (0, −3) **10.** (−3, 0)

In Problems 11–34, graph each equation by plotting a sufficient number of points and connecting them with a smooth curve.

11. $y = 3x + 2$ **12.** $y = 2x - 3$ **13.** $y = -2x + 1$
14. $y = -3x + 1$ **15.** $3x - 2y + 6 = 0$ **16.** $2x - 3y + 6 = 0$
17. $y = 2x^2$ **18.** $y = 3x^2$ **19.** $y = -x^2$
20. $y = -x^2 + 3$ **21.** $y = x^2 + 3$ **22.** $y = x^2 - 3$
23. $y = x^3 - 1$ **24.** $y = x^3 - 8$ **25.** $x^2 = 4y$
26. $x^2 = y + 2$ **27.** $x^2 = y + 1$ **28.** $x^2 = -2y$
29. $y = \sqrt{x}$ **30.** $x = \sqrt{y}$ **31.** $y = \sqrt{x - 1}$
32. $x = \sqrt{y + 2}$ **33.** $y = \dfrac{1}{x - 2}$ **34.** $y = \dfrac{1}{x - 3}$

In Problems 35–38, use the graph below.

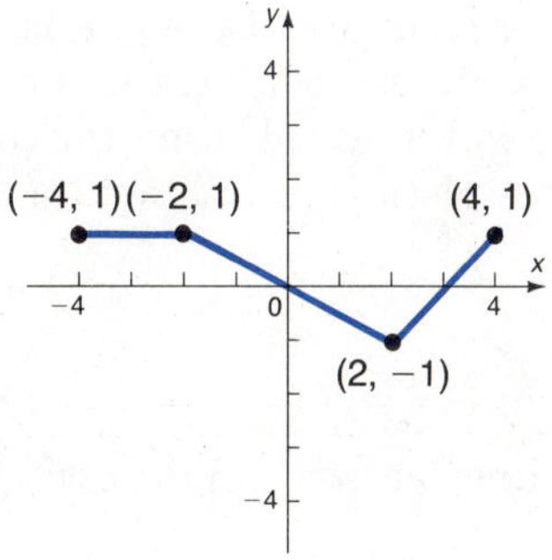

35. Extend the graph to make it symmetric with respect to the x-axis.

36. Extend the graph to make it symmetric with respect to the y-axis.

37. Extend the graph to make it symmetric with respect to the origin.

38. Extend the graph to make it symmetric with respect to the x-axis, y-axis, and origin.

In Problems 39–42, use the graph below.

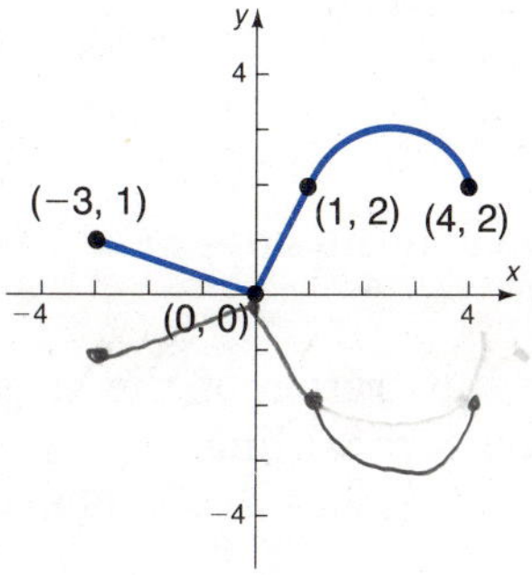

39. Extend the graph to make it symmetric with respect to the x-axis.

40. Extend the graph to make it symmetric with respect to the y-axis.

41. Extend the graph to make it symmetric with respect to the origin.

42. Extend the graph to make it symmetric with respect to the x-axis, y-axis, and origin.

In Problems 43–56, list the intercepts and test for symmetry.

43. $x^2 = y$ **44.** $y^2 = x$ **45.** $y = 3x$

46. $y = -5x$ **47.** $x^2 + y - 9 = 0$ **48.** $y^2 - x - 4 = 0$

49. $4x^2 + 9y^2 = 36$ **50.** $x^2 + 4y^2 = 4$ **51.** $y = x^3 - 27$

52. $y = x^4 - 1$ **53.** $y = x^2 - 3x - 4$ **54.** $y = x^2 + 4$

55. $y = \dfrac{x}{x^2 + 9}$ **56.** $y = \dfrac{x^2 - 4}{x}$

3.3 The Straight Line

In this section we study a certain type of equation that contains two variables, called a *linear equation*, and its graph, a *straight line*.

Slope of a Line

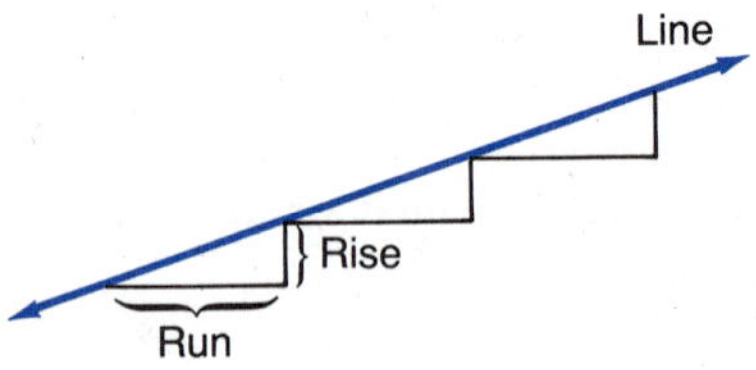

Figure 22

Consider the staircase illustrated in Figure 22. Each step contains exactly the same horizontal **run** and the same vertical **rise**. The ratio of the rise to the run, called the *slope*, is a numerical measure of the steepness of the staircase. For example, if the run is increased and the rise remains the same, the staircase becomes less steep. If the run is kept the same, but the rise is increased, the staircase becomes more steep. This important characteristic of a line is best defined using rectangular coordinates.

Slope of a Line

Let $P = (x_1, y_1)$ and $Q = (x_2, y_2)$ be two distinct points with $x_1 \neq x_2$. The **slope *m*** of the nonvertical line L containing P and Q is defined by the formula

$$m = \frac{y_2 - y_1}{x_2 - x_1} \qquad x_1 \neq x_2 \qquad (1)$$

If $x_1 = x_2$, L is a **vertical line** and the slope m of L is **undefined** (since this results in division by 0).

Figure 23(a) provides an illustration of the slope of a nonvertical line; Figure 23(b) illustrates a vertical line.

As Figure 23(a) illustrates, the slope m of a nonvertical line may be viewed as

$$m = \frac{y_2 - y_1}{x_2 - x_1} = \frac{\text{Rise}}{\text{Run}}$$

Figure 23

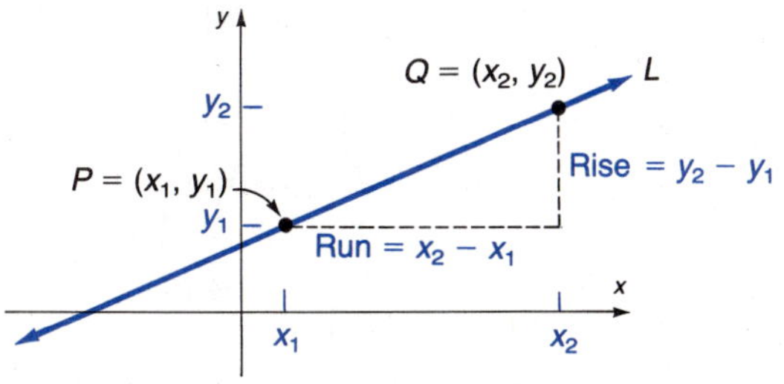

(a) Slope of L is $m = \dfrac{y_2 - y_1}{x_2 - x_1}$

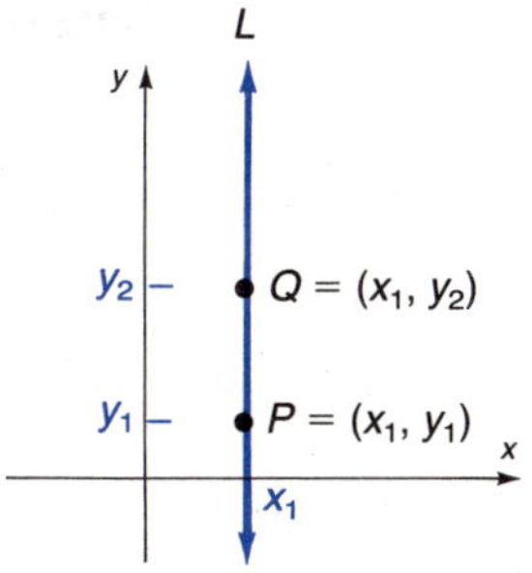

(b) Slope is undefined; L is vertical

We can also express the slope m of a nonvertical line as

$$m = \frac{y_2 - y_1}{x_2 - x_1} = \frac{\text{Change in } y}{\text{Change in } x}$$

That is, the slope m of a nonvertical line L is the ratio of the change in the y-coordinates from P to Q to the change in the x-coordinates from P to Q.

Two comments about computing the slope of a nonvertical line may prove helpful:

1. Any two distinct points on the line can be used to compute the slope of the line. (See Figure 24 for justification.)

Figure 24
Triangles ABC and PQR are similar (equal angles). Hence, ratios of corresponding sides are proportional. Thus:

Slope using P and Q $= \dfrac{y_2 - y_1}{x_2 - x_1}$

$=$ Slope using A and B $= \dfrac{d(B, C)}{d(A, C)}$

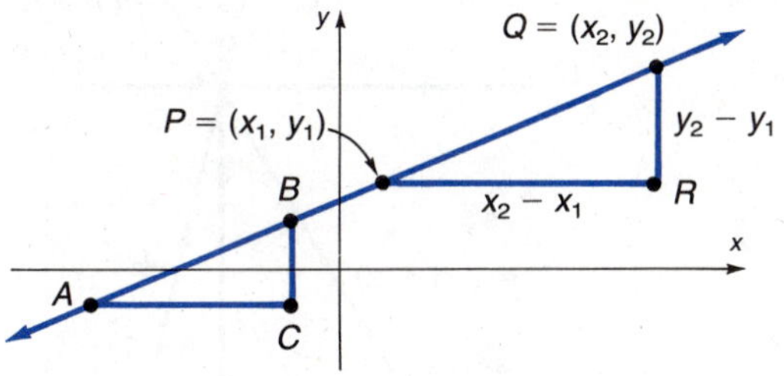

2. The slope of a line may be computed from $P = (x_1, y_1)$ to $Q = (x_2, y_2)$ or from Q to P, because

$$\frac{y_2 - y_1}{x_2 - x_1} = \frac{y_1 - y_2}{x_1 - x_2}$$

Example 1 The slope m of the line joining the points (1, 2) and (5, −3) may be computed as

$$m = \frac{-3 - 2}{5 - 1} = \frac{-5}{4} \qquad \text{or as} \qquad m = \frac{2 - (-3)}{1 - 5} = \frac{5}{-4} = \frac{-5}{4}$$ ■

To get a better idea of the meaning of the slope m of a line L, consider the following example.

Example 2 Compute the slopes of the lines L_1, L_2, L_3, and L_4 containing the following pairs of points. Graph all four lines on the same set of coordinate axes.

$$\begin{array}{llll} L_1: & P = (2, 3) & & Q_1 = (-1, -2) \\ L_2: & P = (2, 3) & & Q_2 = (3, -1) \\ L_3: & P = (2, 3) & & Q_3 = (5, 3) \\ L_4: & P = (2, 3) & & Q_4 = (2, 5) \end{array}$$

Solution Let m_1, m_2, m_3, and m_4 denote the slopes of the lines L_1, L_2, L_3, and L_4, respectively. Then

$$m_1 = \frac{-2 - 3}{-1 - 2} = \frac{-5}{-3} = \frac{5}{3} \qquad \text{A rise of 5 divided by a run of 3}$$

$$m_2 = \frac{-1 - 3}{3 - 2} = \frac{-4}{1} = -4$$

$$m_3 = \frac{3 - 3}{5 - 2} = \frac{0}{3} = 0$$

m_4 is undefined

The graphs of these lines are given in Figure 25.

Figure 25

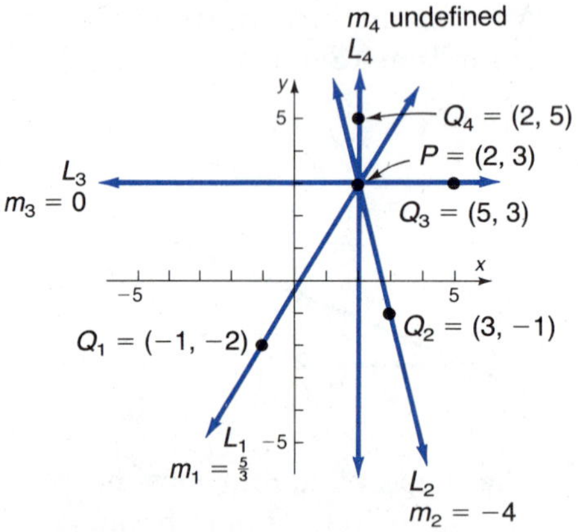

■

Figure 25 illustrates the following facts:

1. When the slope of a line is positive, the line slants upward from left to right (L_1).
2. When the slope of a line is negative, the line slants downward from left to right (L_2).
3. When the slope is 0, the line is horizontal (L_3).
4. When the slope is undefined, the line is vertical (L_4).

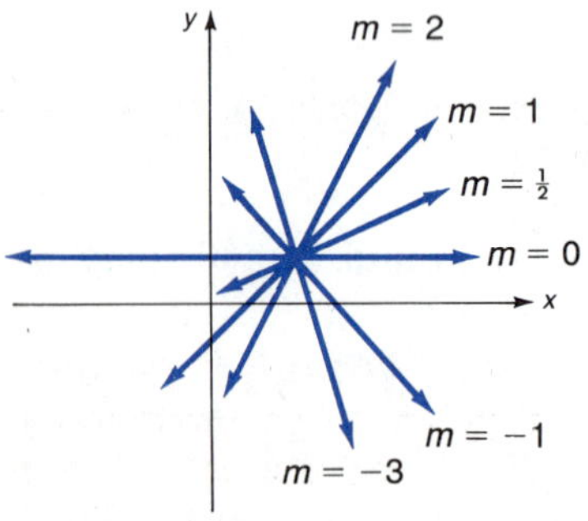

Figure 26

Figure 26 illustrates some additional facts about the slope of a line. Note that the closer the line is to the vertical position, the greater the magnitude of the slope.

The next example illustrates how the slope of a line can be used to graph the line.

Example 3 Draw a graph of the line that passes through the point (3, 2) and has a slope of:

(a) $\frac{3}{4}$ (b) $-\frac{4}{5}$

Solution (a) Slope = Rise/Run. The fact that the slope is $\frac{3}{4}$ means that for every horizontal movement (run) of 4 units to the right, there will be a vertical movement (rise) of 3 units. If we start at the given point (3, 2), and move 4 units to the right and 3 units up, we reach the point (7, 5). By drawing the line through this point and the point (3, 2), we have the graph. See Figure 27.

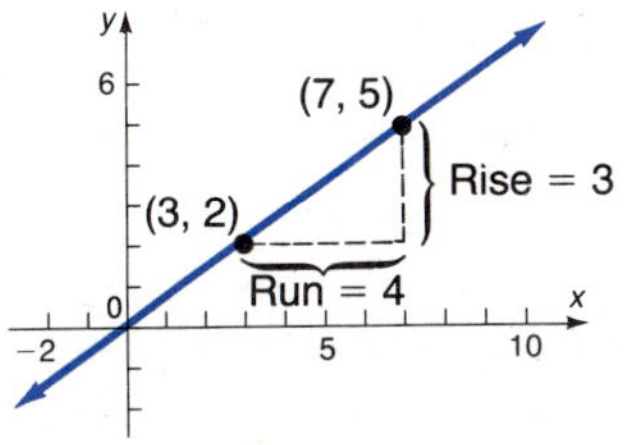

Figure 27
Slope = $\frac{3}{4}$

(b) The fact that the slope is

$$-\frac{4}{5} = \frac{-4}{5} = \frac{\text{Rise}}{\text{Run}}$$

means that for every horizontal movement of 5 units to the right, there will be a corresponding vertical movement of −4 units (a downward movement). If we start at the given point (3, 2), and move 5 units to the right and then 4 units down, we arrive at the point (8, −2). By drawing the line through these points, we have the graph. See Figure 28.

Alternatively, we can set

$$-\frac{4}{5} = \frac{4}{-5} = \frac{\text{Rise}}{\text{Run}}$$

so that for every horizontal movement of −5 units (a movement to the left), there will be a corresponding vertical movement of 4 units (upward). This approach brings us to the point (−2, 6), which is also on the graph shown in Figure 28. ■

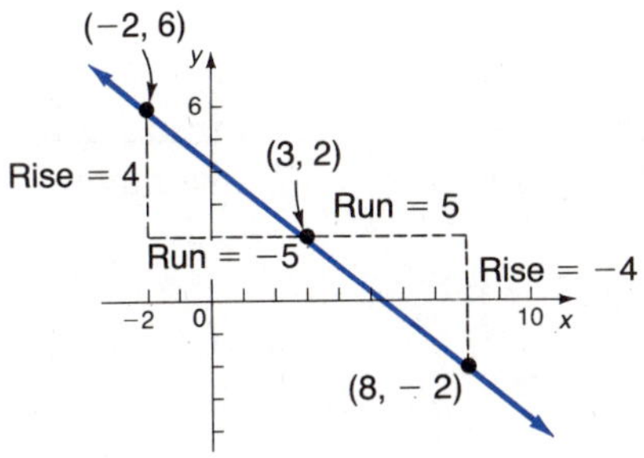

Figure 28
Slope = $-\frac{4}{5}$

Equations of Lines

Now that we have discussed the slope of a line, we are ready to derive equations of lines. As we shall see, there are several forms of the equation of a line. Let's start with an example.

Example 4 Graph the equation: $x = 3$

Solution We are looking for all points (x, y) in the plane for which $x = 3$. Thus, no matter what y-coordinate is used, the corresponding x-coordinate always equals 3. Consequently, the graph of the equation $x = 3$ is a vertical line with x-intercept 3 and undefined slope. See Figure 29.

Figure 29
$x = 3$

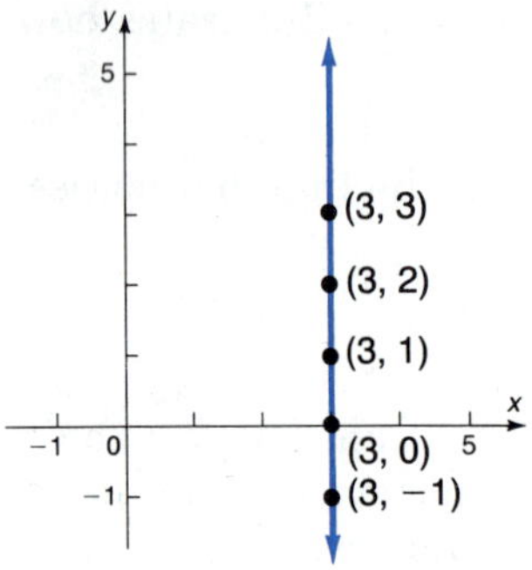

As suggested by Example 4, we have the following result:

Theorem A vertical line is given by an equation of the form

Equation of a Vertical Line

$$x = a$$

where a is the x-intercept.

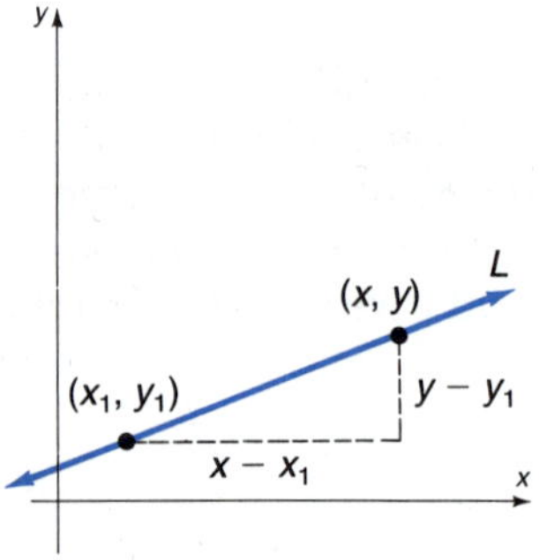

Figure 30

Now let L be a nonvertical line with slope m and containing the point (x_1, y_1). See Figure 30. For any other point (x, y) on L, we have

$$m = \frac{y - y_1}{x - x_1} \qquad \text{or} \qquad y - y_1 = m(x - x_1)$$

Theorem An equation of a nonvertical line of slope m that passes through the point (x_1, y_1) is

Point–Slope Form of an Equation of a Line

$$y - y_1 = m(x - x_1) \qquad (2)$$

Example 5 An equation of the line with slope 4 and passing through the point (1, 2) can be found by using the point–slope form with $m = 4$, $x_1 = 1$, and $y_1 = 2$:

$$y - y_1 = m(x - x_1)$$
$$y - 2 = 4(x - 1)$$
$$y = 4x - 2$$

See Figure 31.

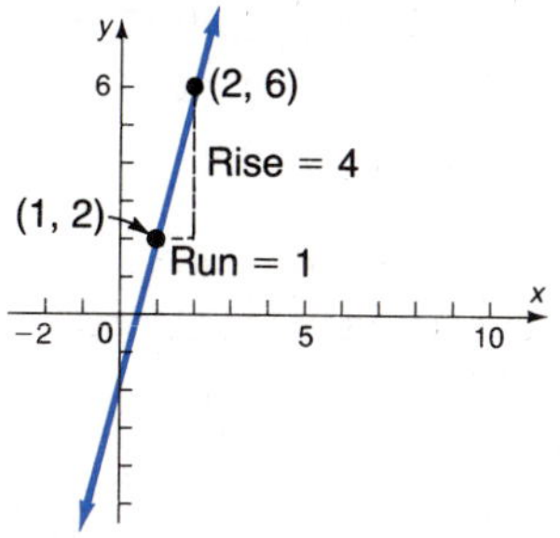

Figure 31
$y = 4x - 2$

Example 6 Find an equation of the horizontal line passing through the point (3, 2).

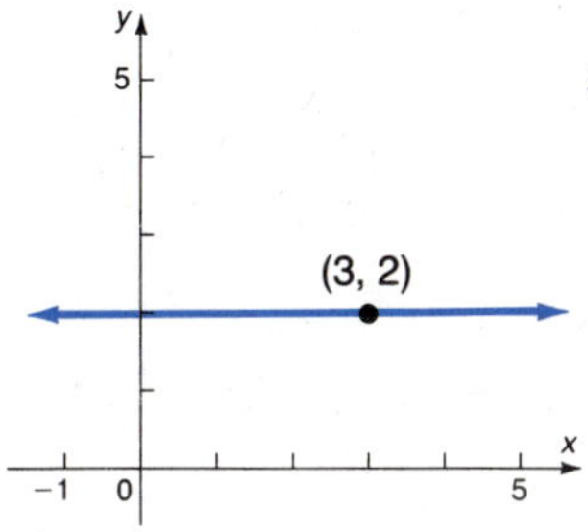

Figure 32
$y = 2$

Solution The slope of a horizontal line is 0. To get an equation, we use the point–slope form with $m = 0$, $x_1 = 3$, and $y_1 = 2$:

$$y - y_1 = m(x - x_1)$$
$$y - 2 = 0 \cdot (x - 3)$$
$$y - 2 = 0$$
$$y = 2$$

See Figure 32 for the graph. ■

As suggested by Example 6, we have the following result:

Theorem A horizontal line is given by an equation of the form

Equation of a Horizontal Line

$$y = b$$

where b is the y-intercept. ■

Example 7 Find an equation of the line L passing through the points (2, 3) and (−4, 5). Graph the line L.

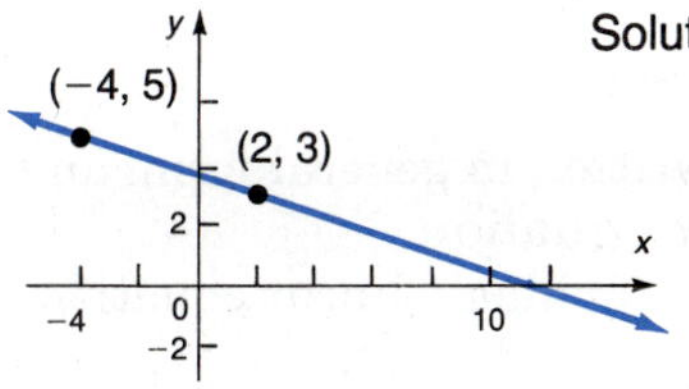

Figure 33
$y - 3 = -\frac{1}{3}(x - 2)$

Solution Since two points are given, we first compute the slope of the line:

$$m = \frac{5 - 3}{-4 - 2} = \frac{2}{-6} = \frac{-1}{3}$$

We use the point (2, 3) and the fact that the slope $m = -\frac{1}{3}$ to get the point–slope form of the equation of the line:

$$y - 3 = -\tfrac{1}{3}(x - 2)$$

See Figure 33 for the graph. ■

In the solution in Example 7, we could have used the other point, $(-4, 5)$, instead of the point $(2, 3)$. The equation that results, although it looks different, is equivalent to the equation we obtained in the example. (Try it for yourself.)

Another form of the equation of the line in Example 7 can be obtained by multiplying both sides of the point–slope equation by 3 and collecting terms:

$$y - 3 = -\tfrac{1}{3}(x - 2)$$
$$3(y - 3) = 3\left(-\tfrac{1}{3}\right)(x - 2) \quad \text{Multiply by 3.}$$
$$3y - 9 = -1(x - 2)$$
$$3y - 9 = -x + 2$$
$$x + 3y - 11 = 0$$

The equation of a line L is in **general form** when it is written as

General Form of an Equation of a Line

$$Ax + By + C = 0 \qquad (3)$$

where A, B, and C are three real numbers and A and B are not both 0.

Every line has an equation that is equivalent to an equation written in general form. For example, a vertical line whose equation is

$$x = a$$

can be written in the general form

$$1 \cdot x + 0 \cdot y - a = 0 \quad A = 1, B = 0, C = -a$$

A horizontal line whose equation is

$$y = b$$

can be written in the general form

$$0 \cdot x + 1 \cdot y - b = 0 \quad A = 0, B = 1, C = -b$$

Lines that are neither vertical nor horizontal have general equations of the form

$$Ax + By + C = 0 \quad A \neq 0 \text{ and } B \neq 0$$

Because the equation of every line can be written in general form, any equation equivalent to (3) is called a **linear equation**.

The next example illustrates one way of graphing a linear equation.

Example 8 Find the intercepts of the line $2x + 3y - 6 = 0$. Graph this line.

Solution To find the point at which the graph crosses the x-axis—that is, to find the x-intercept—we need to find the number x for which $y = 0$. Thus, we let $y = 0$ to get

$$2x + 3(0) - 6 = 0$$
$$2x - 6 = 0$$
$$x = 3$$

The x-intercept is 3. To find the y-intercept, we let $x = 0$ and solve for y:

$$2(0) + 3y - 6 = 0$$
$$3y - 6 = 0$$
$$y = 2$$

The y-intercept is 2.

We now know two points on the line: (3, 0) and (0, 2). Because two points determine a unique line, we do not need further information to graph the line. See Figure 34. ■

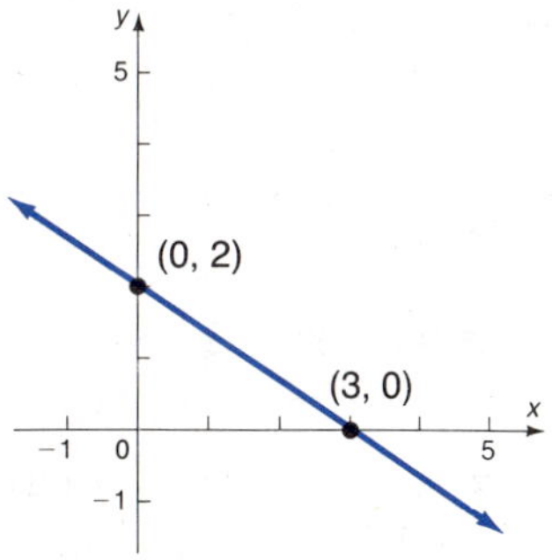

Figure 34
$2x + 3y - 6 = 0$

Another useful equation of a line is obtained when the slope m and y-intercept b are known. In this event, we know both the slope m of the line and a point $(0, b)$ on the line; thus, we may use the point–slope form, equation (2), to obtain the following equation:

$$y - b = m(x - 0) \qquad \text{or} \qquad y = mx + b$$

Theorem An equation of a line L with slope m and y-intercept b is

Slope–Intercept Form of an Equation of a Line

$$y = mx + b \tag{4}$$

■

When the equation of a line is written in slope–intercept form, it is easy to find the slope m and y-intercept b of the line. For example, suppose the equation of a line is

$$y = -2x + 3$$

Compare it to $y = mx + b$:

$$\begin{array}{ccc} y = & -2x & + \; 3 \\ & \uparrow & \uparrow \\ y = & mx & + \; b \end{array}$$

The slope of this line is -2 and its y-intercept is 3.

Let's look at another example.

Example 9 Find the slope m and y-intercept b of the line $2x + 4y - 8 = 0$. Graph the line.

Solution To obtain the slope and y-intercept, we transform the equation into its slope–intercept form. Thus, we need to solve for y:

$$\begin{aligned} 2x + 4y - 8 &= 0 \\ 4y &= -2x + 8 \\ y &= -\tfrac{1}{2}x + 2 \end{aligned}$$

The coefficient of x, $-\frac{1}{2}$, is the slope, and the y-intercept is 2. We can graph the line in two ways:

1. Use the fact that the y-intercept is 2 and the slope is $-\frac{1}{2}$. Then, starting at the point (0, 2), go to the right 2 units and then down 1 unit to the point (2, 1).

Or:

2. Locate the intercepts. Because the y-intercept is 2, we know one intercept is (0, 2). To obtain the x-intercept, let $y = 0$ and solve for x. When $y = 0$, we have

$$\begin{aligned} 2x + 4 \cdot 0 - 8 &= 0 \\ 2x - 8 &= 0 \\ x &= 4 \end{aligned}$$

Thus, the intercepts are (4, 0) and (0, 2).

See Figure 35. ■

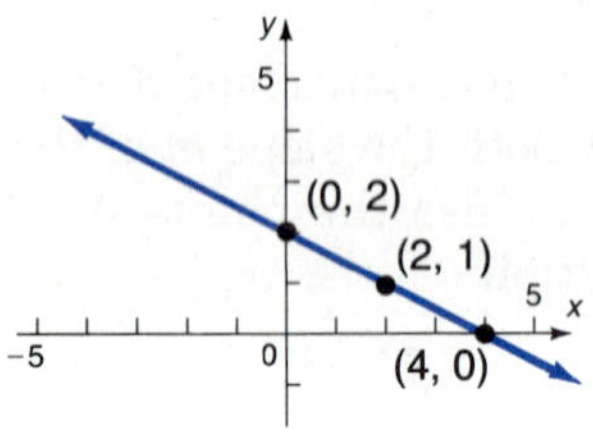

Figure 35
$2x + 4y - 8 = 0$

The next example illustrates a typical situation that requires the use of linear equations.

Example 10 The National Car Rental Company has determined that the average cost of operating a vehicle is \$0.41 per mile. Write an equation that relates the average cost C, in dollars, of operating a car and the number x of miles it has been driven.

Solution If x is the number of miles the car has been driven, then the average cost C, in dollars, is $0.41x$. Thus, an equation relating C and x is

$$C = 0.41x \qquad x \geq 0$$

The average cost per mile, \$0.41, is the slope of the line $C = 0.41x$. In other words, the cost increases by \$0.41 for each additional mile driven. See Figure 36.

Figure 36
$C = 0.41x$

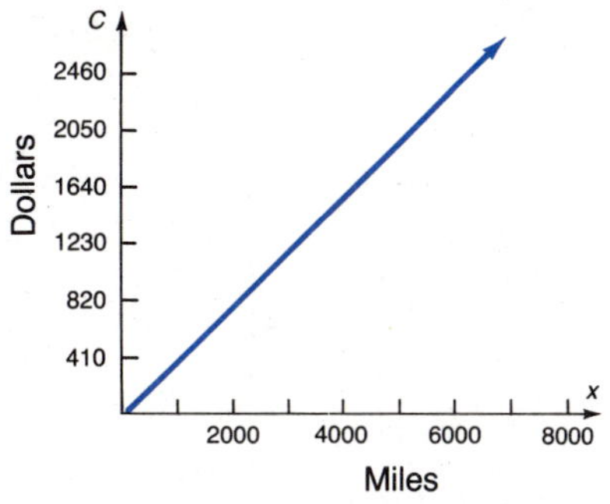

EXERCISE 3.3

In Problems 1–10, plot each pair of points and determine the slope of the line containing them. Graph the line.

1. $(2, 3)$; $(1, 0)$ **2.** $(1, 2)$; $(3, 4)$ **3.** $(-2, 3)$; $(2, 1)$

4. $(-1, 1)$; $(2, 3)$ **5.** $(-3, -1)$; $(2, -1)$ **6.** $(4, 2)$; $(-5, 2)$

7. $(-1, 2)$; $(-1, -2)$ **8.** $(2, 0)$; $(2, 2)$ C **9.** $(\sqrt{2}, 3)$; $(1, \sqrt{3})$

C **10.** $(-2\sqrt{2}, 0)$; $(4, \sqrt{5})$

In Problems 11–18, graph the line passing through the point P and having slope m.

11. $P = (1, 2)$; $m = 2$ **12.** $P = (2, 1)$; $m = 3$ **13.** $P = (2, 4)$; $m = -3$

14. $P = (1, 3)$; $m = -2$ **15.** $P = (-1, 3)$; $m = 0$ **16.** $P = (2, -4)$; $m = 0$

17. $P = (0, 3)$; slope undefined **18.** $P = (-2, 0)$; slope undefined

In Problems 19–30, find a general equation for the line with the given properties.

19. Slope = 2; passing through $(-2, 3)$ **20.** Slope = 3; passing through $(4, -3)$

21. Slope = $-\frac{2}{3}$; passing through $(1, -1)$ **22.** Slope = $\frac{1}{2}$; passing through $(3, 1)$

23. Passing through $(1, 3)$ and $(-1, 2)$ **24.** Passing through $(-3, 4)$ and $(2, 5)$

25. Slope = -3; y-intercept = 3 **26.** Slope = -2; y-intercept = -2

27. x-intercept = 2; y-intercept = -1 **28.** x-intercept = -4; y-intercept = 4

29. Slope undefined; passing through $(1, 4)$ **30.** Slope undefined; passing through $(2, 1)$

In Problems 31–42, find the slope and y-intercept of each line. Graph the line.

31. $y = 2x + 3$ **32.** $y = -3x + 4$ **33.** $\frac{1}{2}y = x - 1$ **34.** $\frac{1}{3}x + y = 2$

35. $2x - 3y = 6$ **36.** $3x + 2y = 6$ **37.** $x + y = 1$ **38.** $x - y = 2$

39. $x = -4$ **40.** $y = -1$ **41.** $2y - 3x = 0$ **42.** $x + y = 0$

In Problems 43–46, find the general equation of each line.

43.

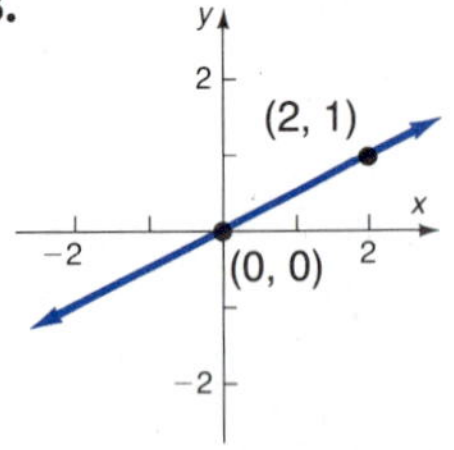

44.

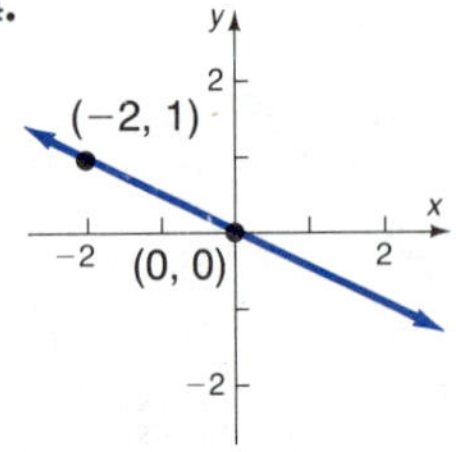

45.

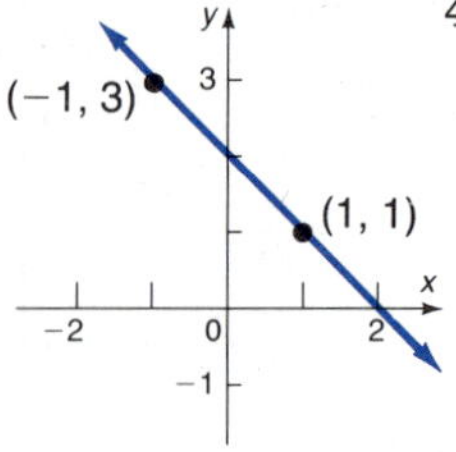

46.

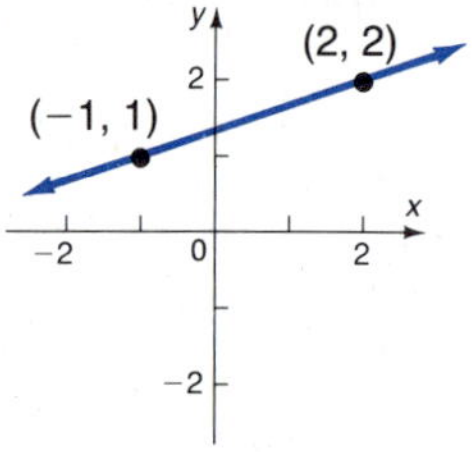

47. Find the general equation of the x-axis.

48. Find the general equation of the y-axis.

49. The relationship between Celsius (°C) and Fahrenheit (°F) degrees for measuring temperature is linear. Find an equation relating °C and °F if 0°C corresponds to 32°F and 100°C corresponds to 212°F. Use the equation to find the Celsius measure of 70°F.

50. The Kelvin (K) scale for measuring temperature is obtained by adding 273 to the Celsius temperature.
(a) Write an equation relating K and °C.
(b) Write an equation relating K and °F (see Problem 49).

51. Each Sunday, a newspaper agency sells x copies of a certain newspaper for \$1.00 per copy. The cost to the agency of each newspaper is \$0.50. The agency pays a fixed cost for storage, delivery, and so on, of \$100 per Sunday. Write an equation that relates the profit P, in dollars, to the number x of copies sold. Graph this equation.

52. Repeat Problem 51 if the cost to the agency is \$0.45 per copy and the fixed cost is \$125 per Sunday.

53. Show that an equation for a line with nonzero x- and y-intercepts can be written as

$$\frac{x}{a} + \frac{y}{b} = 1$$

where a is the x-intercept and b is the y-intercept. This is called the **intercept form** of the equation of a line.

3.4 Parallel and Perpendicular Lines; Circles

Parallel and Perpendicular Lines

When two lines (in the plane) have no points in common, they are said to be **parallel**. Look at Figure 37. There we have drawn two lines and have constructed two right triangles by drawing sides parallel to the

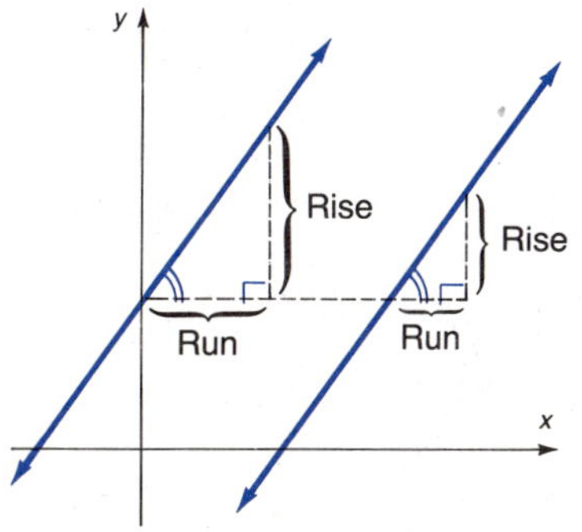

Figure 37
The lines are parallel if and only if their slopes are equal.

coordinate axes. These lines are parallel if and only if the right triangles are similar. (Do you see why? Two angles are equal.) But the triangles are similar if and only if the ratios of corresponding sides are equal. This suggests the following result:

Theorem Two nonvertical lines are parallel if and only if their slopes are equal. ■

The use of the words "if and only if" in the above theorem means that there are actually two statements being made:

If two nonvertical lines are parallel, then their slopes are equal.

If two nonvertical lines have equal slopes, then they are parallel.

Example 1 Show that the lines given by the equations below are parallel:

$$L: \quad 2x + 3y - 6 = 0 \qquad M: \quad 4x + 6y = 0$$

Solution To determine whether these lines have equal slopes, we write each equation in slope–intercept form:

$$\begin{aligned} L: \quad 2x + 3y - 6 &= 0 \\ 3y &= -2x + 6 \\ y &= -\tfrac{2}{3}x + 2 \\ \text{Slope} &= -\tfrac{2}{3} \end{aligned} \qquad \begin{aligned} M: \quad 4x + 6y &= 0 \\ 6y &= -4x \\ y &= -\tfrac{2}{3}x \\ \text{Slope} &= -\tfrac{2}{3} \end{aligned}$$

Because each line has slope $-\frac{2}{3}$, the lines are parallel. See Figure 38. ■

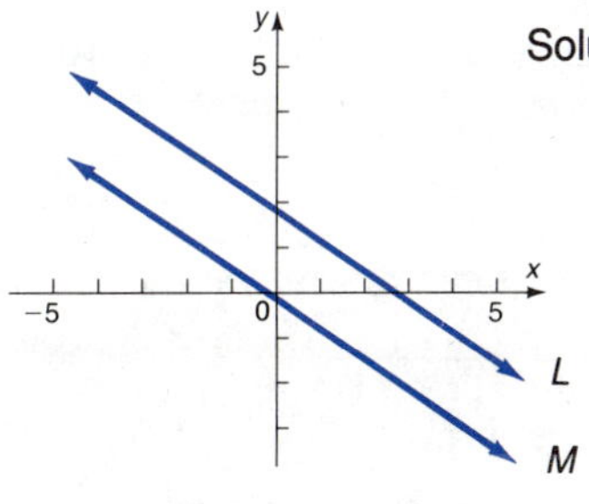

Figure 38
Parallel lines

When two lines intersect at a right angle (90°), they are said to be **perpendicular**. See Figure 39.

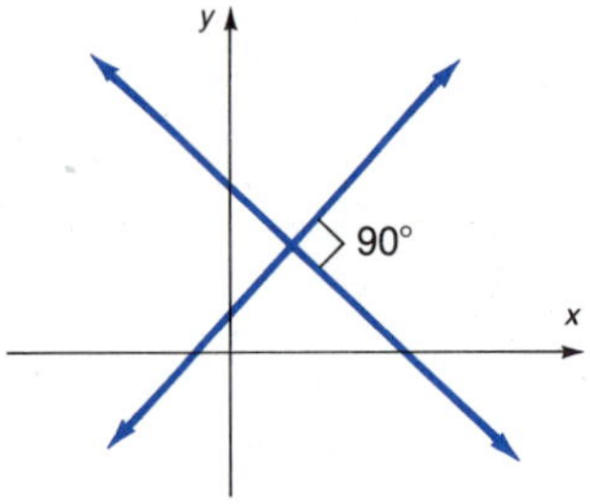

Figure 39
Perpendicular lines

The following result gives a condition, in terms of their slopes, for two lines to be perpendicular:

Theorem Two nonvertical lines are perpendicular if and only if the product of their slopes is -1. ■

Here, we shall prove only the statement:

If two nonvertical lines are perpendicular, then the product of their slopes is -1.

In Problem 57, you are asked to prove the second part of the theorem; that is:

If two nonvertical lines have slopes whose product is -1, then the lines are perpendicular.

Proof Let m_1 and m_2 denote the slopes of the two lines. There is no loss in generality (that is, neither the angle nor the slopes are affected) if we situate the lines so that they meet at the origin. See Figure 40. The point $A = (1, m_2)$ is on the line having slope m_2, and the point $B = (1, m_1)$ is on the line having slope m_1. (Do you see why this must be true?)

Suppose the lines are perpendicular. Then triangle OAB is a right triangle. As a result of the Pythagorean Theorem, it follows that

$$[d(O, A)]^2 + [d(O, B)]^2 = [d(A, B)]^2 \quad (1)$$

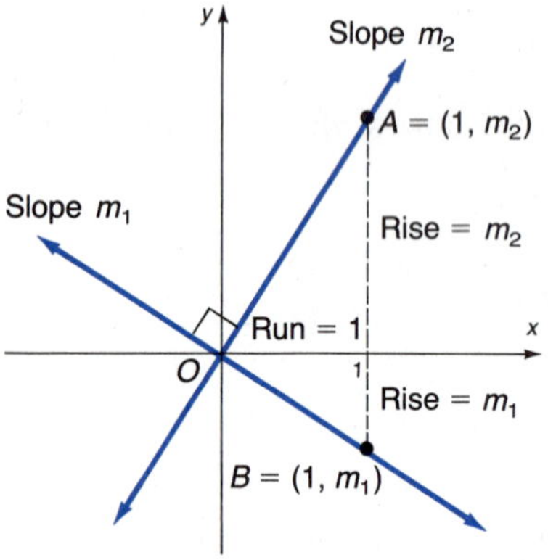

Figure 40

By the distance formula, we can write each of these distances as

$$[d(O, A)]^2 = (1 - 0)^2 + (m_2 - 0)^2 = 1 + m_2^2$$
$$[d(O, B)]^2 = (1 - 0)^2 + (m_1 - 0)^2 = 1 + m_1^2$$
$$[d(A, B)]^2 = (1 - 1)^2 + (m_2 - m_1)^2 = m_2^2 - 2m_1m_2 + m_1^2$$

Using these facts in equation (1), we get

$$(1 + m_2^2) + (1 + m_1^2) = m_2^2 - 2m_1m_2 + m_1^2$$

which, upon simplification, can be written as

$$m_1m_2 = -1$$

Thus, if the lines are perpendicular, the product of their slopes is -1. ■

You may find it easier to remember the condition for two nonvertical lines to be perpendicular by observing that the equality $m_1m_2 = -1$ means that m_1 and m_2 are negative reciprocals of each other; that is, either $m_1 = -1/m_2$ or $m_2 = -1/m_1$.

Example 2 If a line has slope $\frac{3}{2}$, any line having slope $-\frac{2}{3}$ is perpendicular to it. ■

Example 3 Find the general form of the equation of a line passing through the point $(1, -2)$ and perpendicular to the line $x + 3y - 6 = 0$.

Solution We first write the equation of the given line in slope–intercept form to find its slope:

$$\begin{aligned} x + 3y - 6 &= 0 \\ 3y &= -x + 6 \\ y &= -\tfrac{1}{3}x + 2 \end{aligned}$$

The given line has slope $-\frac{1}{3}$. Any line perpendicular to this line will have slope 3. Because we require the point $(1, -2)$ to be on this line with slope 3, we use the point–slope form of the equation of a line:

$$\begin{aligned} y - (-2) &= 3(x - 1) \\ y + 2 &= 3(x - 1) \end{aligned}$$

This equation is equivalent to the general form

$$3x - y - 5 = 0$$

■

Circles

One of the advantages of a coordinate system is that it enables us to translate a geometric statement into an algebraic statement, and vice versa. Consider, for example, the following geometric statement that defines a circle.

Circle A **circle** is a set of points in the xy-plane that are a fixed distance r from a fixed point (h, k). The fixed distance r is called the **radius**, and the fixed point (h, k) is called the **center** of the circle.

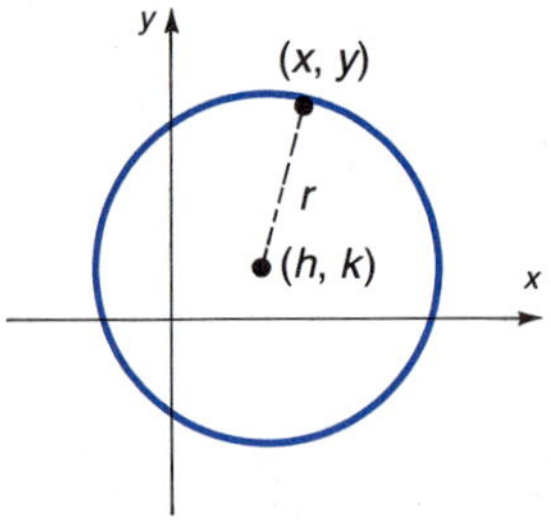

Figure 41

Figure 41 shows the graph of a circle. Is there an equation having this graph? If so, what is the equation? To find the equation, we let (x, y) represent the coordinates of any point on a circle with radius r and center (h, k). Then the distance between the points (x, y) and (h, k) must always equal r. That is, by the distance formula,

$$\sqrt{(x - h)^2 + (y - k)^2} = r$$

or, equivalently,

$$(x - h)^2 + (y - k)^2 = r^2$$

The **standard form of an equation of a circle** with radius r and center (h, k) is

Standard Form of an Equation of a Circle

$$(x - h)^2 + (y - k)^2 = r^2 \qquad (2)$$

Conversely, by reversing the steps, we conclude: The graph of any equation of the form of equation (2) is that of a circle with radius r and center (h, k).

Example 4 Graph the equation: $(x + 3)^2 + (y - 2)^2 = 16$

Solution By comparing the given equation to the standard form of the equation of a circle, we conclude that the graph of the given equation is a circle. Moreover, the comparison yields information about the circle:

$$(x - h)^2 + (y - k)^2 = r^2$$
$$(x + 3)^2 + (y - 2)^2 = 16$$

We see that $h = -3$, $k = 2$, and $r = 4$. Hence, the circle has center $(-3, 2)$ and a radius of 4 units. To graph this circle, we first plot the center $(-3, 2)$. Since the radius is 4, we can locate four points on the circle by going out 4 units to the left, to the right, up, and down from the center. These four points can then be used as guides to obtain the graph. See Figure 42. ■

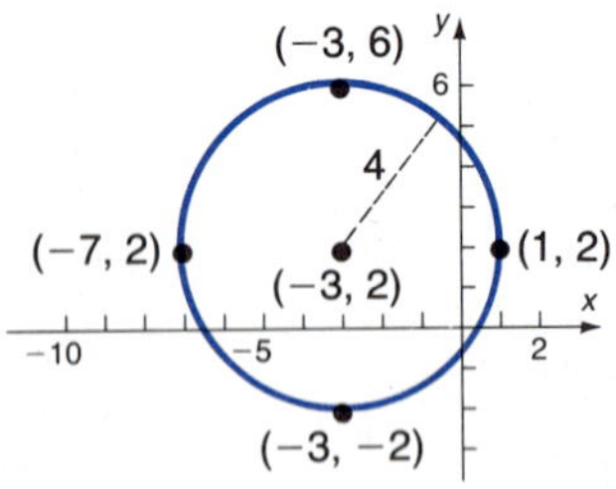

Figure 42

Example 5 Write the standard form of an equation of the circle with radius 3 and center $(1, -2)$.

Solution Using the form of equation (2) and substituting the values $r = 3$, $h = 1$, and $k = -2$, we have

$$(x - h)^2 + (y - k)^2 = r^2$$
$$(x - 1)^2 + (y + 2)^2 = 9$$ ■

The standard form of an equation of a circle of radius r with center at the origin (0, 0) is

$$x^2 + y^2 = r^2$$

If the radius $r = 1$, the circle whose center is at the origin is called the **unit circle** and has the equation

$$x^2 + y^2 = 1$$

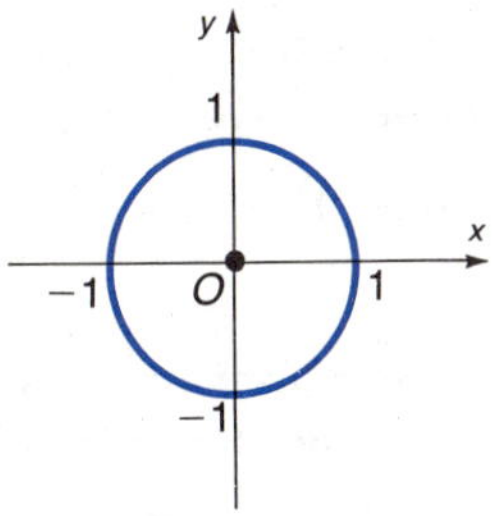

Figure 43
Unit circle $x^2 + y^2 = 1$

See Figure 43.

If we eliminate the parentheses from the standard form of the equation of the circle obtained in Example 5, we get

$$(x - 1)^2 + (y + 2)^2 = 9$$
$$x^2 - 2x + 1 + y^2 + 4y + 4 = 9$$

which we find, upon simplifying, is equivalent to

$$x^2 + y^2 - 2x + 4y - 4 = 0$$

By completing the squares on both the x- and y-terms, it can be shown that any equation of the form

$$x^2 + y^2 + ax + by + c = 0$$

has a graph that is a circle, or a point, or has no graph at all. For example, the graph of the equation $x^2 + y^2 = 0$ is the single point (0, 0). The equation $x^2 + y^2 + 5 = 0$, or $x^2 + y^2 = -5$, has no graph, because sums of squares of real numbers are never negative. When its graph is a circle, the equation

General Form of an Equation of a Circle

$$x^2 + y^2 + ax + by + c = 0$$

is referred to as the **general form of an equation of a circle**.

The next example shows how to transform an equation in the general form to an equivalent equation in standard form. As we said earlier, the idea is to use the method of completing the square on both the x- and y-terms.

Example 6 Graph the equation: $x^2 + y^2 + 4x - 6y + 12 = 0$

Solution We rearrange the equation as follows:

$$(x^2 + 4x) + (y^2 - 6y) = -12$$

Next, we complete the square of each expression in parentheses. Remember that any number added on the left also must be added on the right:

$$(x^2 + 4x + 4) + (y^2 - 6y + 9) = -12 + 4 + 9$$
$$(x + 2)^2 + (y - 3)^2 = 1$$

We recognize this equation as the standard form of the equation of a circle with radius 1 and center $(-2, 3)$. Figure 44 illustrates the graph.

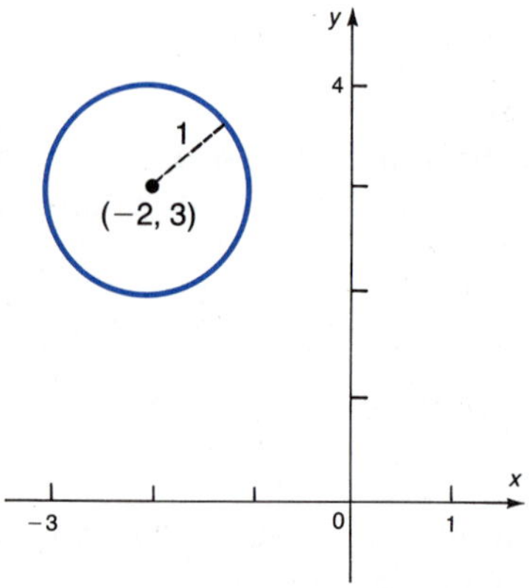

Figure 44
$x^2 + y^2 + 4x - 6y + 12 = 0$

Example 7 Find the general equation of the circle whose center is $(1, -2)$ and whose graph contains the point $(4, 2)$.

Solution To find the equation of a circle, we need to know its center and its radius. Here, we know that the center is $(1, -2)$. Since the point $(4, 2)$ is on the graph, the radius r will equal the distance from $(4, 2)$ to the center $(1, -2)$. See Figure 45. Thus,

$$r = \sqrt{(4 - 1)^2 + [2 - (-2)]^2} = \sqrt{9 + 16} = 5$$

The standard form of the equation of the circle is

$$(x - 1)^2 + (y + 2)^2 = 25$$

Eliminating the parentheses and rearranging terms, we get the general equation

$$x^2 + y^2 - 2x + 4y - 20 = 0$$

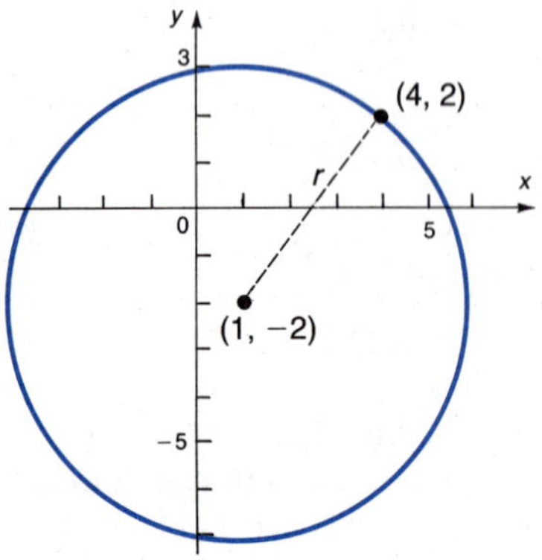

Figure 45
$x^2 + y^2 - 2x + 4y - 20 = 0$

Overview

The preceding discussion about lines and circles dealt with two main types of problems that can be generalized as follows:

1. Given an equation, classify it and graph it.
2. Given a graph, or information about a graph, find its equation.

This text deals mainly with the first type of problem. We shall study various equations, classify them, and graph them. The second type of problem is usually more difficult to solve than the first. Although we shall tackle such problems when it is practical to do so, we leave problems such as "fitting" equations to graphs to more advanced courses.

EXERCISE 3.4

In Problems 1–10, the equation of a line L is given. Find the slope of a line that is:
(a) Parallel to L *(b) Perpendicular to L*

1. $y = 2x$
2. $y = -3x$
3. $y = -\frac{1}{2}x + 2$
4. $y = \frac{2}{3}x - 1$
5. $2x - 4y + 5 = 0$
6. $3x + y = 4$
7. $3x + 5y - 10 = 0$
8. $4x - 3y + 7 = 0$
9. $x = 4$
10. $y = 5$

In Problems 11–22, find a general equation for the line with the given properties.

11. Parallel to the line $y = 3x$; passing through $(-1, 2)$
12. Parallel to the line $y = -2x$; passing through $(-1, 2)$
13. Parallel to the line $2x - y + 2 = 0$; passing through $(0, 0)$
14. Parallel to the line $x - 2y + 5 = 0$; passing through $(0, 0)$
15. Parallel to the line $x = 5$; passing through $(4, 2)$
16. Parallel to the line $y = 5$; passing through $(4, 2)$

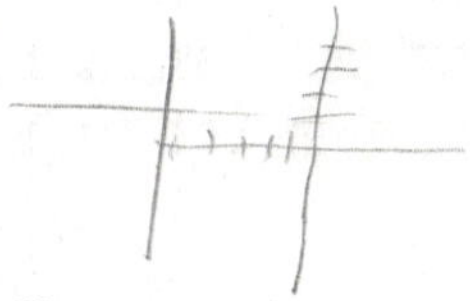

17. Perpendicular to the line $y = \frac{1}{2}x + 4$; passing through $(1, -2)$
18. Perpendicular to the line $y = 2x - 3$; passing through $(1, -2)$
19. Perpendicular to the line $2x + y - 2 = 0$; passing through $(-3, 0)$
20. Perpendicular to the line $x - 2y + 5 = 0$; passing through $(0, 4)$
21. Perpendicular to the line $x = 5$; passing through $(3, 4)$
22. Perpendicular to the line $y = 5$; passing through $(3, 4)$

In Problems 23–30, write the standard form of the equation and the general form of the equation of each circle of radius r and center (h, k).

23. $r = 1$; $(h, k) = (1, -1)$
24. $r = 2$; $(h, k) = (-2, 1)$
25. $r = 2$; $(h, k) = (0, 2)$
26. $r = 3$; $(h, k) = (1, 0)$
27. $r = 5$; $(h, k) = (4, -3)$
28. $r = 4$; $(h, k) = (2, -3)$
29. $r = 2$; $(h, k) = (0, 0)$
30. $r = 3$; $(h, k) = (0, 0)$

In Problems 31–40, find the center (h, k) and radius r of each circle.

31. $x^2 + y^2 = 4$

32. $x^2 + (y - 1)^2 = 1$

33. $(x - 3)^2 + y^2 = 4$

34. $(x + 1)^2 + (y - 1)^2 = 2$

35. $x^2 + y^2 + 4x - 4y - 1 = 0$

36. $x^2 + y^2 - 6x + 2y + 9 = 0$

37. $x^2 + y^2 - x + 2y + 1 = 0$

38. $x^2 + y^2 + x + y - \frac{1}{2} = 0$

39. $2x^2 + 2y^2 - 12x + 8y - 24 = 0$

40. $2x^2 + 2y^2 + 8x + 7 = 0$

In Problems 41–46, find the general form of the equation of each circle.

41. Center at the origin and containing the point $(2, -3)$

42. Center $(1, 0)$ and containing the point $(-3, 1)$

43. Center $(2, 3)$ and tangent to the x-axis

44. Center $(-3, 1)$ and tangent to the y-axis

45. With endpoints of a diameter at $(1, 4)$ and $(-3, 2)$

46. With endpoints of a diameter at $(4, 3)$ and $(0, 1)$

47. Use slopes to show that the triangle whose vertices are $(-2, 5)$, $(1, 3)$, and $(-1, 0)$ is a right triangle.

48. Use slopes to show that the quadrilateral whose vertices are $(1, -1)$, $(4, 1)$, $(2, 2)$, and $(5, 4)$ is a parallelogram.

49. Use slopes to show that the quadrilateral whose vertices are $(-1, 0)$, $(2, 3)$, $(1, -2)$, and $(4, 1)$ is a rectangle.

50. Use slopes to show that the quadrilateral whose vertices are $(0, 0)$, $(1, 3)$, $(4, 2)$, and $(3, -1)$ is a square.

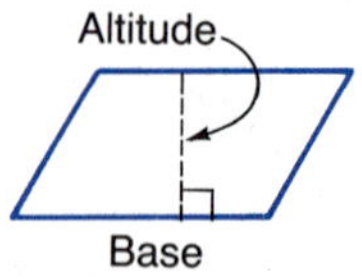

In Problems 51–56, find the area of the parallelogram with the given vertices. [*Hint:* *The area of a parallelogram is the product of the base times the altitude to that base. See the figure.*]

51. $(0, 0)$; $(4, 0)$; $(1, 3)$; $(5, 3)$

52. $(0, 0)$; $(0, 2)$; $(3, 1)$; $(3, -1)$

53. $(2, 1)$; $(4, 2)$; $(2, 3)$; $(4, 4)$

54. $(1, 2)$; $(2, 0)$; $(3, 0)$; $(2, 2)$

55. $(1, 1)$; $(3, 2)$; $(2, -1)$; $(4, 0)$

56. $(1, 3)$; $(0, 1)$; $(3, 2)$; $(2, 0)$

57. Prove that if two nonvertical lines have slopes whose product is -1, then the lines are perpendicular. [*Hint:* Use the converse of the Pythagorean Theorem.]

C **58.** The Earth is represented on a map of a portion of the solar system so that its surface is the circle with equation $x^2 + y^2 + 2x + 4y - 4091 = 0$. A satellite circles 0.6 unit above the Earth with the center of its circular orbit at the center of the Earth. Find the equation for the orbit of the satellite on this map.

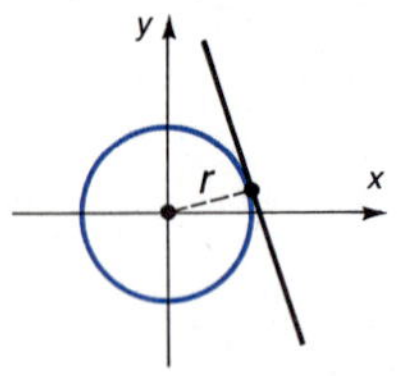

59. The **tangent line** to a circle may be defined as the line that intersects the circle in a single point, called the **point of tangency** (see the figure). If the equation of the circle is $x^2 + y^2 = r^2$ and the equation of the tangent line is $y = mx + b$, show that:

(a) $r^2(1 + m^2) = b^2$ [*Hint:* The quadratic equation $x^2 + (mx + b)^2 = r^2$ has exactly one solution.]

(b) The point of tangency is $(-r^2m/b, r^2/b)$.

(c) The tangent line is perpendicular to the line containing the center of the circle and the point of tangency.

60. The Greek method for finding the equation of the tangent line to a circle used the fact that at any point on a circle the line containing the radius and the tangent line are perpendicular (see Problem 59). Use this method to find an equation of the tangent line to the circle $x^2 + y^2 = 9$ at the point $(1, 2\sqrt{2})$.

61. Use the Greek method described in Problem 60 to find an equation of the tangent line to the circle $x^2 + y^2 - 4x + 6y + 4 = 0$ at the point $(3, 2\sqrt{2} - 3)$.

62. Refer to Problem 59. The line $x - 2y + 4 = 0$ is tangent to a circle at $(0, 2)$. The line $y = 2x - 7$ is tangent to the same circle at $(3, -1)$. Find the center of the circle.

63. Find an equation of the line containing the centers of the two circles

$$x^2 + y^2 - 4x + 6y + 4 = 0 \quad \text{and} \quad x^2 + y^2 + 6x + 4y + 9 = 0$$

64. Show that the line containing the points (a, b) and (b, a) is perpendicular to the line $y = x$. Also show that the midpoint of (a, b) and (b, a) lies on the line $y = x$.

65. The equation $2x - y + C = 0$ defines a **family of lines**, one line for each value of C. On one set of coordinate axes, graph the members of the family when $C = -4$, $C = 0$, and $C = 2$. Can you draw a conclusion from the graph about each member of the family?

66. Rework Problem 65 for the family of lines $Cx + y + 4 = 0$.

3.5 Variation

Relationships that exist between quantities are often expressed using language such as the following:

> Force is proportional to acceleration.
>
> For an ideal gas held at a constant temperature, pressure and volume are inversely proportional.
>
> The force of attraction between two heavenly bodies is inversely proportional to the square of the distance between them.
>
> Revenue is directly proportional to sales.

Each of these statements illustrates the idea of **variation**, or how one quantity varies in relation to another quantity. Quantities may vary *directly*, *inversely*, or *jointly*.

Direct Variation

Let x and y denote two quantities. Then y **varies directly** as x, or y is **directly proportional to** x, if there is a nonzero number k such that

$$y = kx$$

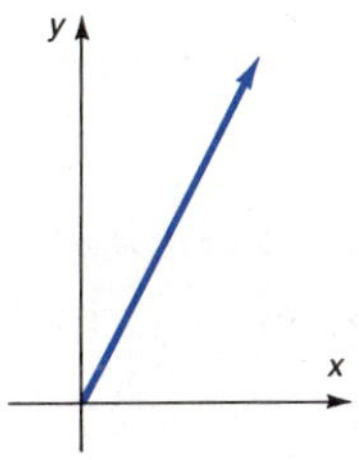

Figure 46
$y = kx$, $k > 0$, $x \geq 0$

The number k is called the **constant of proportionality**.

The graph in Figure 46 illustrates the relationship between y and x if y varies directly as x and $k > 0$, $x \geq 0$. Note that the constant of proportionality is, in fact, the slope of the line.

If we know that two quantities vary directly, then knowing the value of each quantity in one instance enables us to write a formula that is true in all cases.

Example 1 For a certain gas enclosed in a container of fixed volume, the pressure P (in pounds per square inch) varies directly as the temperature T (in degrees Celsius). If the pressure is found to be 10 pounds per square inch at a temperature of 30°C, find a formula that relates pressure P to temperature T. Then find the pressure P when $T = 120$°C.

Solution Because P varies directly as T, we know that

$$P = kT$$

for some constant k. Because $P = 10$ when $T = 30$,

$$10 = k(30)$$
$$k = \tfrac{1}{3}$$

Thus, in all cases,

$$P = \tfrac{1}{3}T$$

In particular, when $T = 120$°C, we find

$$P = \tfrac{1}{3}(120) = 40 \text{ pounds per square inch}$$

Figure 47 illustrates the relationship between the pressure P and the temperature T.

Figure 47

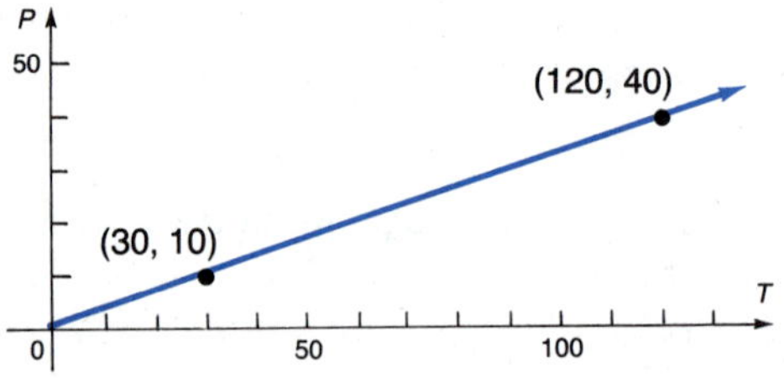

Inverse Variation

Let x and y denote two quantities. Then y **varies inversely** as x, or y is **inversely proportional to** x, if there is a nonzero constant k such that

$$y = \frac{k}{x}$$

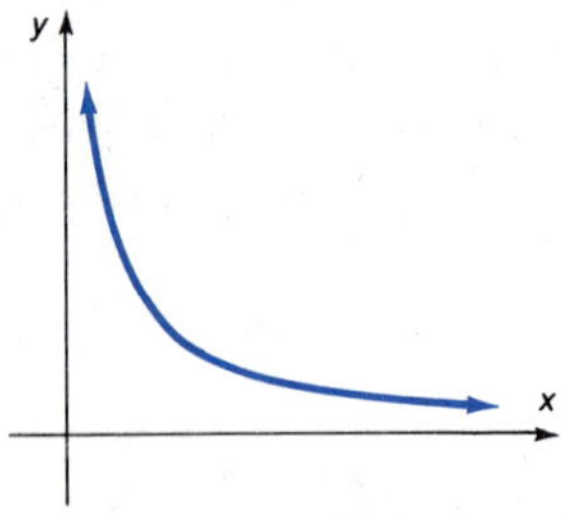

Figure 48
$y = \frac{k}{x}$; $k > 0, x > 0$

The graph in Figure 48 illustrates the relationship between y and x if y varies inversely as x and $k > 0$, $x > 0$.

Example 2 The weight W that can be safely supported by a 2 inch by 4 inch piece of lumber varies inversely as its length l. Experiments indicate that the maximum weight a 10 foot pine 2 by 4 can support is 500 pounds. Write a general formula relating the safe weight W (in pounds) to length l (in feet). Find the maximum weight W that can be safely supported by a length of 25 feet.

Solution Because W varies inversely as l, we know that

$$W = \frac{k}{l}$$

for some constant k. Because $W = 500$ when $l = 10$, we have

$$500 = \frac{k}{10}$$
$$k = 5000$$

Thus, in all cases,

$$W = \frac{5000}{l}$$

In particular, the maximum weight W that can be safely supported by a piece of pine 25 feet in length is

$$W = \frac{5000}{25} = 200 \text{ pounds}$$

Figure 49 illustrates the relationship between the weight W and the length l. ■

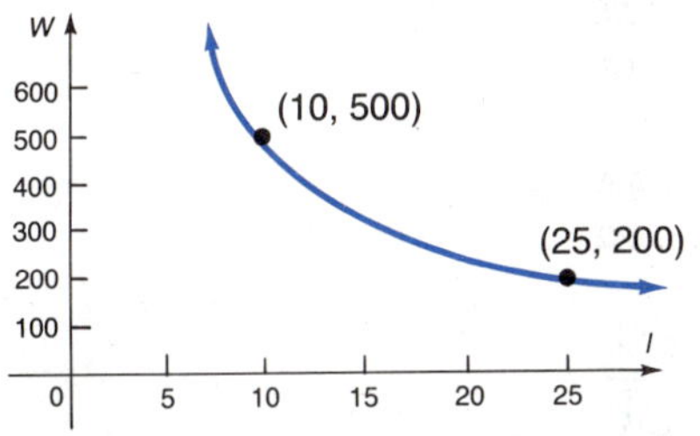

Figure 49

In direct or inverse variation, the quantities that vary may be raised to powers. For example, the square of the period T of a planet varies directly as the cube of its mean distance a from the Sun. That is, $T^2 = ka^3$, where k is the constant of proportionality.

Joint Variation and Combined Variation

When a variable quantity Q is proportional to the product of two or more other variables, then we say that Q **varies jointly** as (or with) these quantities. Finally, combinations of direct and/or inverse variation may occur. This is usually referred to as **combined variation**.

Let's look at an example.

Example 3 The loss of heat through a wall varies jointly as the area of the wall and the difference between the inside and outside temperatures, and varies inversely as the thickness of the wall. Write an equation that relates these quantities.

Solution We begin by assigning symbols to represent the quantities:

L = Heat loss $\quad$ T = Temperature difference

A = Area of wall $\quad$ d = Thickness of wall

Then,

$$L = k\frac{AT}{d}$$

where k is the constant of proportionality. ■

Example 4 The force F of the wind on a plane surface positioned at a right angle to the direction of the wind varies jointly as the area A of the surface and the square of the speed v of the wind. A wind of 30 miles per hour blowing on a window measuring 4 feet by 5 feet has a force of 150 pounds. What is the force on a window measuring 3 feet by 4 feet caused by a wind of 50 miles per hour?

Solution Since F varies jointly as A and v^2, we have

$$F = kAv^2$$

where k is the constant of proportionality. We are told that $F = 150$ when $v = 30$ and $A = 4 \cdot 5 = 20$. Thus,

$$150 = k(20)(900)$$
$$k = \tfrac{1}{120}$$

The general formula is therefore

$$F = \tfrac{1}{120}Av^2$$

For a wind of 50 miles per hour blowing on a window whose area is $A = 3 \cdot 4 = 12$ square feet, the force F is

$$F = \tfrac{1}{120}(12)(2500) = 250 \text{ pounds}$$ ■

EXERCISE 3.5

In Problems 1–12, write a general formula to describe each variation.

1. y varies directly as x; $y = 5$ when $x = 10$
2. v varies directly as t; $v = 16$ when $t = 1$
3. A varies directly as x^2; $A = 4\pi$ when $x = 2$
4. V varies directly as x^3; $V = 36\pi$ when $x = 3$
5. F varies inversely as d^2; $F = 10$ when $d = 5$
6. y varies inversely as $\sqrt{x}$; $y = 4$ when $x = 9$

7. z varies directly as the sum of the squares of x and y; $z = 5$ when $x = 3$ and $y = 4$
8. T varies jointly as the cube root of x and the square of d; $T = 18$ when $x = 8$ and $d = 3$
9. M varies directly as the square of d and inversely as the square root of x; $M = 24$ when $x = 9$ and $d = 4$
10. z varies directly as the sum of the cube of x and the square of y; $z = 1$ when $x = 2$ and $y = 3$
11. The square of T varies directly as the cube of a and inversely as the square of d; $T = 2$ when $a = 1$ and $d = 4$
12. The cube of z varies directly as the sum of the squares of x and y; $z = 2$ when $x = 4$ and $y = 9$

In Problems 13–20, write an equation that relates the quantities.

13. The volume V of a sphere varies directly as the cube of its radius r. The constant of proportionality is $4\pi/3$.
14. The square of the hypotenuse c of a right triangle varies directly as the sum of the squares of the legs a and b. The constant of proportionality is 1.
15. The area A of a triangle varies jointly as the lengths of the base b and the height h. The constant of proportionality is $\frac{1}{2}$.

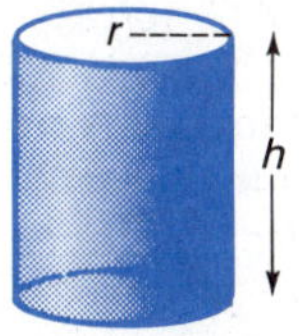

16. The perimeter p of a rectangle varies directly as the sum of the lengths of its sides l and w. The constant of proportionality is 2.
17. The volume V of a right circular cylinder varies jointly as the square of its radius r and its height h. The constant of proportionality is π. (See the figure.)

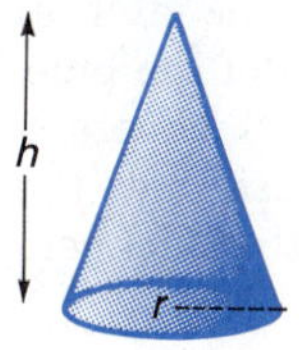

18. The volume V of a right circular cone varies jointly as the square of its radius r and its height h. The constant of proportionality is $\pi/3$. (See the figure.)
19. The force F (in newtons) of attraction between two bodies varies jointly as their masses m and M (in kilograms) and inversely as the square of the distance d (in meters) between them. The constant of proportionality is $G = 6.67 \times 10^{-11}$.
20. The *period* of a pendulum is the time required for one oscillation; the pendulum is usually referred to as *simple* when the angle made to the vertical is less than 5°. The period T of a simple pendulum (in seconds) varies directly as the square root of its length l (in feet). The constant of proportionality is $2\pi/\sqrt{32}$.
21. The distance s an object falls is directly proportional to the square of the time t of the fall. If an object falls 16 feet in 1 second, how far will it fall in 3 seconds? How long will it take an object to fall 64 feet?
22. The velocity v of a falling object is directly proportional to the time t of the fall. If, after 2 seconds, the velocity of the object is 64 feet per second, what will its velocity be after 3 seconds?

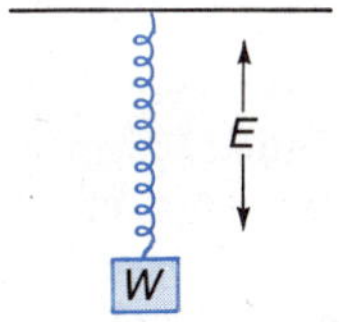

23. The elongation E of a spring balance varies directly as the applied weight W (see the figure). If $E = 3$ when $W = 20$, find E when $W = 15$.
24. The rate of vibration of a string under constant tension varies inversely as the length of the string. If a string is 48 inches long and vibrates 256 times per second, what is the length of a string that vibrates 576 times per second?

C **25.** The weight of a body above the surface of the Earth varies inversely as the square of the distance from the center of the Earth. If a certain body weighs 55 pounds when it is 4×10^3 miles from the center of the Earth, how much will it weigh when it is 4.4×10^3 miles from the center?

C **26.** The force exerted by the wind on a plane surface varies jointly as the area of the surface and the square of the velocity of the wind. If the force on an area of 20 square feet is 11 pounds when the wind velocity is 22 miles per hour, find the force on a surface area of 47.125 square feet when the wind velocity is 36.5 miles per hour.

C **27.** The horsepower that a shaft can safely transmit varies jointly as its speed (in revolutions per minute, rpm) and the cube of its diameter. If a shaft of a certain material 3 inches in diameter can transmit 36 horsepower at 75 rpm, what diameter must the shaft have in order to transmit 45 horsepower at 125 rpm?

C **28.** The weight of a body varies inversely as the square of its distance from the center of the Earth. Assuming that the radius of the Earth is 4000 miles, how much would a man weigh at an altitude of 1 mile above the Earth's surface if he weighs 180 pounds on the Earth's surface?

29. The kinetic energy K of a moving object varies jointly as its mass m and the square of its velocity v. If an object weighing 25 pounds and moving with a velocity of 100 feet per second has a kinetic energy of 400 foot-pounds, find its kinetic energy when the velocity is 150 feet per second.

C **30.** The electrical resistance of a wire varies directly as the length of the wire and inversely as the square of the diameter of the wire. If a wire 432 feet long and 4 millimeters in diameter has a resistance of 1.24 ohms, find the length of a wire of the same material whose resistance is 1.44 ohms and whose diameter is 3 millimeters.

31. The stress in the material of a pipe subject to internal pressure varies jointly as the internal pressure and the internal diameter of the pipe and inversely as the thickness of the pipe. The stress is 100 pounds per square inch when the diameter is 5 inches, the thickness is 0.75 inch, and the internal pressure is 25 pounds per square inch. Find the stress when the internal pressure is 40 pounds per square inch, if the diameter is 8 inches and the thickness is 0.50 inch.

C **32.** The maximum safe load for a horizontal rectangular beam varies jointly as the width of the beam and the square of the thickness of the beam and inversely as its length. If an 8 foot beam will support up to 750 pounds when the beam is 4 inches wide and 2 inches thick, what is the maximum safe load in a similar beam 10 feet long, 6 inches wide, and 2 inches thick?

33. The resistance (in ohms) of a circular conductor varies directly as the length of the conductor and inversely as the square of the radius of the conductor. If 50 feet of a wire with a radius of 6×10^{-3} inch has a resistance of 10 ohms, what would be the resistance of 100 feet of the same wire if the radius is increased to 7×10^{-3} inch?

34. The volume V of an ideal gas varies directly with the temperature T and inversely with the pressure P. Write an equation relating V, T, and P using k as the constant of proportionality. If a cylinder contains oxygen at a temperature of 20°C and a pressure of 15 atmospheres in a volume of 100 liters, what is the constant of proportionality k? If a piston is lowered into the cylinder, decreasing the volume occupied by the gas to 80 liters and raising the temperature to 25°C, what is the gas pressure?

Problems 36–40 use the result obtained in Problem 35.

35. The speed v required of a satellite to maintain a near-Earth circular orbit is directly proportional to the square root of the distance r of the satellite from the center of the Earth.* The constant of proportionality is $\sqrt{g}$, where g is the acceleration of gravity for Earth. Write an equation that shows the relationship between v and r. (The radius of the Earth is approximately 4000 miles.)

C **36.** What speed is required to maintain a communications satellite in a circular orbit 500 miles above the Earth's surface?

C **37.** Find the speed of a satellite that moves in a circular orbit 100 miles above the Earth's surface.

C **38.** Find the distance of a satellite from the surface of the Earth as it moves around the Earth in a circular orbit at a constant speed of 18,630 miles per hour.

C **39.** A weather satellite orbits the Earth in a circle every 1.5 hours. How high is it above the Earth?

C **40.** Some communications satellites remain stationary above a fixed point on the equator of the Earth's surface. How high are such satellites, and what is their common speed? (Assume the Earth turns once every 24 hours.)

Problems 42–46 use the result obtained in Problem 41.

41. The force F (in newtons) required to maintain an object in a circular path varies jointly with the mass m (in kilograms) of the object and the square of its speed v (in meters per second) and inversely with the radius r (in meters) of the circular path. The constant of proportionality is 1. Write an equation relating F, m, v, and r.

C **42.** A motorcycle that weighs 150 kilograms is driven at a constant speed of 120 kilometers per hour on a circular track with a radius of 100 meters. To keep the motorcycle from skidding, what frictional force must be exerted by the tires on the track?

43. If the speed of the motorcycle described in Problem 42 is increased by 10%, by how much is the frictional force of the tires increased?

44. If the radius of the track described in Problem 42 is cut in half, how much slower should the motorcycle be driven to maintain the same frictional force?

45. A girl is spinning a bucket of water in a horizontal plane at the end of a rope of length L (see the figure). If she triples the speed of the bucket, how many times as hard must she pull on the rope?

46. If the girl in Problem 45 doubles the length of the rope and maintains the same speed for the bucket, will she have to pull on the rope more or less? How much?

*Near-Earth orbits are at least 100 miles above the Earth's surface (out of the Earth's atmosphere) and up to an altitude of approximately 15,000 miles. The effect of the gravitational attraction of other bodies is ignored. Although the acceleration of gravity at such altitudes is somewhat less than $g \approx 32$ feet per second per second $\approx$ 79,036 miles per hour per hour, we shall ignore this discrepancy in our calculations.

CHAPTER REVIEW

THINGS TO KNOW

Formulas

Distance formula	$d = \sqrt{(x_2 - x_1)^2 + (y_2 - y_1)^2}$
Midpoint formula	$x = \frac{x_1 + x_2}{2}, \quad y = \frac{y_1 + y_2}{2}$
Slope	$m = \frac{y_2 - y_1}{x_2 - x_1}$, if $x_1 \neq x_2$; undefined if $x_1 = x_2$
Parallel lines	Equal slopes ($m_1 = m_2$)
Perpendicular lines	Product of slopes is -1 ($m_1 \cdot m_2 = -1$)
Direct variation	$y = kx$
Inverse variation	$y = \frac{k}{x}$

Equations

Vertical line	$x = a$
Horizontal line	$y = b$
Point–slope form of an equation of a line	$y - y_1 = m(x - x_1)$; m is the slope of the line, (x_1, y_1) is a point on the line
General form of an equation of a line	$Ax + By + C = 0$, A, B not both 0
Slope–intercept form of an equation of a line	$y = mx + b$; m is the slope of the line, b is the y-intercept
Standard form of an equation of a circle	$(x - h)^2 + (y - k)^2 = r^2$; r is the radius of the circle, (h, k) is the center of the circle
General form of an equation of a circle	$x^2 + y^2 + ax + by + c = 0$
Equation of a unit circle	$x^2 + y^2 = 1$

How To:

Use the distance formula
Graph equations by plotting points
Find the intercepts of a graph
Test an equation for symmetry
Find the slope and intercepts of a line, given the equation
Graph lines
Obtain the equation of a line
Obtain the equation of a circle
Find the center and radius of a circle, given the equation
Graph circles

FILL-IN-THE-BLANK ITEMS

1. If (x, y) are the coordinates of a point P in the xy-plane, then x is called the ________ of P and y is the ________ of P.
2. If three points P, Q, and R all lie on a line and if $d(P, Q) = d(Q, R)$, then Q is called the ________ of the line segment from P to R.
3. If for every point (x, y) on a graph, the point $(-x, y)$ is also on the graph, then the graph is symmetric with respect to the ________.
4. The set of points in the xy-plane that are a fixed distance from a fixed point is called a(n) ________. The fixed distance is called the ________; the fixed point is called the ________.
5. The slope of a vertical line is ________; the slope of a horizontal line is ________.
6. Two nonvertical lines have slopes m_1 and m_2, respectively. The lines are parallel if ________; the lines are perpendicular if ________.
7. If z varies jointly as x^2 and y^3 and inversely as $\sqrt{t}$, then $z =$ ________, where k is the constant of proportionality.

REVIEW EXERCISES

In Problems 1–10, find a general equation of the line having the given characteristics.

1. Slope $= -2$; passing through $(2, -1)$
2. Slope $= 0$; passing through $(-3, 4)$
3. Slope undefined; passing through $(-3, 4)$
4. x-intercept $= 2$; passing through $(4, -5)$
5. y-intercept $= -2$; passing through $(5, -3)$
6. Passing through $(3, -4)$ and $(2, 1)$
7. Parallel to the line $2x - 3y + 4 = 0$; passing through $(-5, 3)$
8. Parallel to the line $x + y - 2 = 0$; passing through $(1, -3)$
9. Perpendicular to the line $x + y - 2 = 0$; passing through $(1, -3)$
10. Perpendicular to the line $3x - y + 4 = 0$; passing through $(-2, 2)$

In Problems 11–16, graph each line, labeling the x-intercept and y-intercept.

11. $4x - 5y + 20 = 0$
12. $3x + 4y - 12 = 0$
13. $\frac{1}{2}x - \frac{1}{3}y + \frac{1}{6} = 0$
14. $-\frac{3}{4}x + \frac{1}{2}y = 0$
15. $\sqrt{2}x + \sqrt{3}y = \sqrt{6}$
16. $\frac{x}{3} + \frac{y}{4} = 1$

In Problems 17–20, find the center and radius of each circle.

17. $x^2 + y^2 - 2x + 4y - 4 = 0$
18. $x^2 + y^2 + 4x - 4y - 1 = 0$
19. $3x^2 + 3y^2 - 6x + 12y = 0$
20. $2x^2 + 2y^2 - 4x = 0$

21. Find the slope of the line containing the points (1, 4) and (−3, 2). What is the distance between these points? What is the midpoint between them?

22. Find the slope of the line containing the points (0, 5) and (6, −3). What is the distance between these points? What is the midpoint between them?

In Problems 23–30, graph each equation by plotting a sufficient number of points and connecting them with a smooth curve.

23. $2x - 3y - 6 = 0$ 24. $3x + 2y + 6 = 0$ 25. $y = 2x^2 + 1$

26. $y = -2x^2 + 1$ 27. $9x = y^2$ 28. $x = -4y^2$

29. $y = \sqrt{2 - x}$ 30. $y = \dfrac{1}{x + 2}$

In Problems 31–38, list the intercepts and test for symmetry.

31. $2x = 3y^2$ 32. $y = 5x$ 33. $4x^2 + y^2 = 1$

34. $x^2 - 9y^2 = 9$ 35. $y = x^4 + 2x^2 + 1$ 36. $y = x^3 - x$

37. $x^2 + x + y^2 + 2y = 0$ 38. $x^2 + 4x + y^2 - 2y = 0$

C 39. The area of an equilateral triangle varies directly as the square of the length of a side. If the area of the equilateral triangle whose sides are of length 1 centimeter is $\sqrt{3}/4$, find the length s of each side of an equilateral triangle whose area A is 16 square centimeters.

40. In a vibrating string, the pitch varies directly as the square root of the tension of the string. If a certain string vibrates 300 times per second under a tension of 9 pounds, find the tension required to cause the string to vibrate 400 times per second.

C 41. Kepler's third law of planetary motion states that the square of the period T of revolution of a planet is proportional to the cube of its mean distance from the Sun. If the mean distance of the Earth from the Sun is 93 million miles, what is the mean distance a of the planet Mercury from the Sun, given that Mercury has a "year" of 88 days?

C 42. Use Problem 41 to find the mean distance of the planet Jupiter from the Sun, given that Jupiter circles the Sun every $5\sqrt{5}$ years.

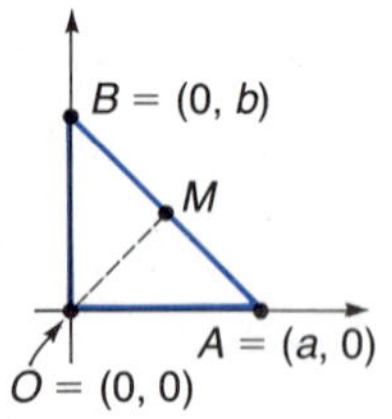

43. Show that the midpoint of the hypotenuse of a right triangle is the same distance from each of the three vertices. [*Hint:* Use the figure in the margin.]

44. Show that the line joining the midpoints of two sides of a triangle is parallel to the third side.

45. Show that the points $A = (3, 4)$, $B = (1, 1)$, and $C = (-2, 3)$ are the vertices of an isosceles triangle.

46. Show that the points $A = (-2, 0)$, $B = (-4, 4)$, and $C = (8, 5)$ are the vertices of a right triangle in two ways:

(a) By using the converse of the Pythagorean Theorem
(b) By using the slopes of the lines joining the vertices

47. Show that the points $A = (2, 5)$, $B = (6, 1)$, and $C = (8, -1)$ lie on a straight line by using slopes.

48. Show that the points $A = (1, 5)$, $B = (2, 4)$, and $C = (-3, 5)$ lie on a circle with center $(-1, 2)$. What is the radius of this circle?

49. The endpoints of the diameter of a circle are (−3, 2) and (5, −6). Find the center and radius of the circle. Write the general equation of this circle.

50. Find two numbers y such that the distance from (−3, 2) to (5, y) is 10.

4 Functions and Their Graphs

Perhaps the most central idea in mathematics is the notion of a *function*. This important chapter deals with what a function is, how to graph functions, and how to perform operations on functions.

The word *function* apparently was introduced by René Descartes in 1637. For him, a function simply meant any positive integral power of a variable x. Gottfried Wilhelm von Leibniz (1646–1716), who always emphasized the geometric side of mathematics, used the word function to denote any quantity associated with a curve, such as the coordinates of a point on the curve. Leonhard Euler (1707–1783) employed the word to mean any equation or formula involving variables and constants. His idea of a function is similar to the one most often used today in courses that precede calculus. Later, the use of functions in investigating heat flow equations led to a very broad definition, due to Lejeune Dirichlet (1805–1859), which describes a function as a rule or correspondence between two sets. It is his definition that we use here.

4.1 Functions

In many applications, a correspondence often exists between two sets of numbers. For example, the revenue R resulting from the sale of x items selling for \$10 each is $R = 10x$ dollars. If we know how many items have been sold, then we can calculate the revenue by using the rule $R = 10x$. This rule is an example of a *function*.

As another example, if an object is dropped from a height of 64 feet above the ground, the distance s (in feet) of the object from the ground after t seconds is given (approximately) by the formula $s = 64 - 16t^2$. When $t = 0$ seconds, the object is $s = 64$ feet above the ground. After 1 second, the object is $s = 64 - 16(1)^2 = 48$ feet above the ground. After 2 seconds, the object strikes the ground. The formula $s = 64 - 16t^2$ provides a way of finding the distance s when the time t $(0 \le t \le 2)$ is prescribed. There is a correspondence between each time t in the interval $0 \le t \le 2$ and the distance s. We say that the distance s is a *function* of the time t because:

1. There is a correspondence between the set of times and the set of distances.
2. There is exactly one distance s obtained for a prescribed time t in the interval $0 \le t \le 2$.

Let's now look at the definition of a function.

Definition of a Function

Function

Let X and Y be two nonempty sets of real numbers.* A **function** from X into Y is a rule or a correspondence that associates with each element of X a unique element of Y. The set X is called the **domain** of the function. For each element x in X, the corresponding element y in Y is called the **value** of the function at x, or the **image** of x. The set of all images of the elements of the domain is called the **range** of the function.

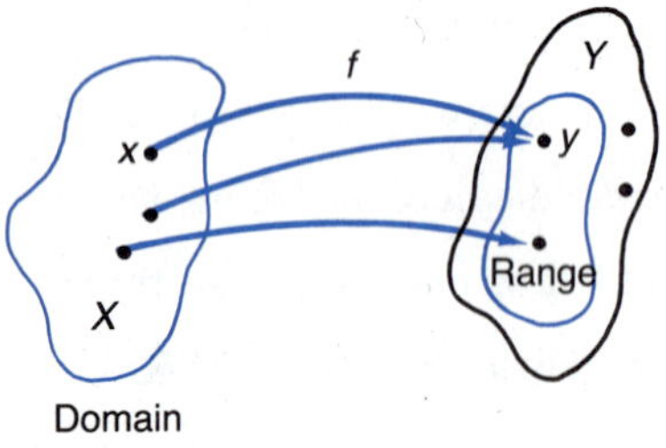

Figure 1

Refer to Figure 1. Since there may be some elements in Y that are not the image of some x in X, it follows that the range of a function may be a subset of Y.

The rule (or correspondence) referred to in the definition of a function is most often given as an equation in two variables, usually denoted x and y.

*The two sets X and Y also can be sets of complex numbers, and then we have defined a complex function. In the broad definition (due to Lejeune Dirichlet), X and Y can be any two sets.

Example 1 Consider the function defined by the equation

$$y = 2x - 5 \qquad 1 \le x \le 6$$

The domain $1 \le x \le 6$ specifies that the number x is restricted to the real numbers from 1 to 6, inclusive. The rule $y = 2x - 5$ specifies that the number x is to be multipled by 2 and then 5 is to be subtracted from the result to get y. For example, the value of the function at $x = \frac{3}{2}$ (that is, the image of $x = \frac{3}{2}$) is $y = 2 \cdot \frac{3}{2} - 5 = -2$. ■

Functions are often denoted by letters such as f, F, g, G, and so on. If f is a function, then for each number x in its domain, the corresponding image in the range is designated by the symbol $f(x)$, read as "f of x." We refer to $f(x)$ as the **value of f at the number x**. Thus, $f(x)$ is the number that results when x is given and the rule for f is applied; $f(x)$ does *not* mean "f times x." For example, the function given in Example 1 may be written as $f(x) = 2x - 5$, $1 \le x \le 6$.

In general, when the rule that defines a function f is given by an equation in x and y, we say the function f is given **implicitly**. If it is possible to solve the equation for y in terms of x, then we write $y = f(x)$ and say the function is given **explicitly**. In fact, we usually write "the function $y = f(x)$" when we mean "the function f defined by the equation $y = f(x)$." Although this usage is not entirely correct, it is rather common and should not cause any confusion. For example:

IMPLICIT FORM	EXPLICIT FORM
$3x + y = 5$	$y = f(x) = -3x + 5$
$x^2 - y = 6$	$y = f(x) = x^2 - 6$
$xy = 4$	$y = f(x) = 4/x$

We list below a summary of some important facts to remember about a function f.

Summary of Important Facts about Functions

1. $f(x)$ is the image of x, or the value of f at x, when the rule f is applied to an x in the domain.
2. To each x in the domain of f, there is one and only one image $f(x)$ in the range.
3. f is the symbol we use to denote the function. It is symbolic of a domain and a rule we use to get from an x in the domain to $f(x)$ in the range.

Domain of a Function

Usually, the domain of a function f is not specified; instead, only a rule or equation defining the function is given. In such cases, we automatically assume that the domain of f is the largest set of real numbers for which the rule makes sense or, more precisely, for which the value $f(x)$ is a real number. The range will then consist of all the images of the numbers in the domain.

Example 2 The functions f, g, F, and G are defined by:

(a) $f(x) = x^2$ (b) $g(x) = 1/x$ (c) $F(x) = \sqrt{x}$ (d) $G(x) = 3$

Find the domain and range of each function.

Solution (a) The operation of squaring a number can be performed on any real number x. Therefore, the domain of f is the set of all real numbers. What is the result when real numbers are squared? We obtain nonnegative real numbers. The range of f is therefore the set of nonnegative real numbers.

(b) We can divide 1 by any nonzero real number. Hence, the domain of g is all real numbers except 0. The range of g is also the set of nonzero real numbers.

(c) Because the square root of a negative number is not defined, the domain of F consists of all nonnegative real numbers. Because the square root of a nonnegative number is itself nonnegative, the range of F is the set of nonnegative real numbers.

(d) No matter what real number x is chosen, the image is 3. Therefore, the domain of G is the set of all real numbers, and the range of G is the single number 3.

Figure 2 illustrates these functions for selected choices of x.

Figure 2

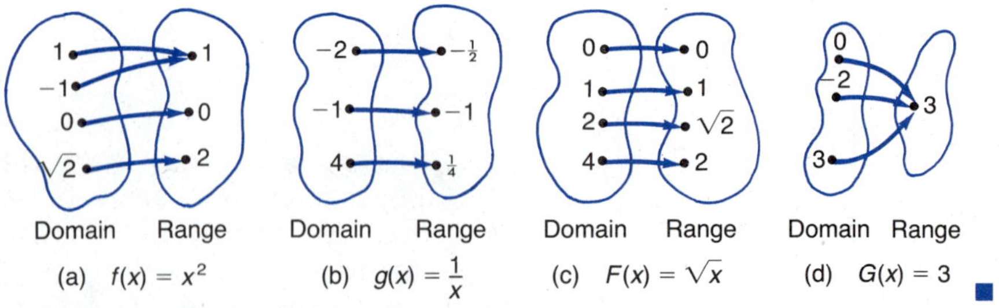

It is often difficult to find the range of a function. Therefore, we shall usually be content to find just the domain of a function when only the rule for the function is given. We shall express the domain of a function using interval notation, set notation, or words, whichever is most convenient.

Example 3 Find the domain of each of the following functions:

(a) $f(x) = \dfrac{3x}{x^2 - 4}$ (b) $g(x) = \sqrt{4 - 3x}$

Solution (a) The rule f tells us to divide $3x$ by $x^2 - 4$. Division by 0 is not allowed; therefore, the denominator $x^2 - 4$ can never be 0. Thus, x can never equal 2 or -2, and the domain of the function f is $\{x \mid x \neq -2, x \neq 2\}$.

(b) The rule g tells us to take the square root of $4 - 3x$. But only nonnegative numbers have real square roots. Hence, we require that

$$4 - 3x \geq 0$$
$$-3x \geq -4$$
$$x \leq \tfrac{4}{3}$$

The domain of g is $\{x \mid x \leq \frac{4}{3}\}$ or $\left(-\infty, \frac{4}{3}\right]$. ■

If x is in the domain of a function f, we shall say that **f is defined at x**, or **$f(x)$ exists**. If x is not in the domain of f, we say that **f is not defined at x**, or **$f(x)$ does not exist**. For example, if $f(x) = x/(x^2 - 1)$, then $f(0)$ exists, but $f(1)$ and $f(-1)$ do not exist. (Do you see why?)

When we use functions in applications, the domain may be restricted by physical or geometric considerations. For example, the domain of the function f defined by $f(x) = x^2$ is the set of all real numbers. However, if f is used as the rule for obtaining the area of a square when the length x of a side is known, then we must restrict the domain of f to the positive real numbers, since the length of a side never can be 0 or negative.

Independent Variable; Dependent Variable

Consider a function $y = f(x)$. The number x that appears here is called the **independent variable**, because it can be assigned any of the permissible numbers from the domain. The number y is called the **dependent variable**, because its value depends on the number x.

Any symbol can be used to represent the independent and dependent variables. For example, if f is the *cube function*, then f can be defined by $f(x) = x^3$ or $f(t) = t^3$ or $f(z) = z^3$. All three rules are identical—each tells us to cube the independent variable. In practice, the symbols used for the independent and dependent variables are based on common usage.

Example 4 (a) The cost per square foot to build a house is \$110. Express the cost C as a function of x, the number of square feet.

(b) Express the area of a circle as a function of its radius.

Solution (a) The cost C of building a house containing x square feet is $110x$ dollars. A function expressing this relationship is

$$C(x) = 110x$$

where x is the independent variable and C is the dependent variable. In this setting, the domain is $\{x|x > 0\}$.

(b) We know that the formula for the area A of a circle of radius r is $A = \pi r^2$. If we use r to represent the independent variable and A to represent the dependent variable, the function expressing this relationship is

$$A(r) = \pi r^2$$

In this setting, the domain is $\{r|r > 0\}$. ■

Function Notation

The independent variable x of a function $y = f(x)$ is sometimes called the **argument** of the function f. Thinking of the independent variable x as an argument sometimes can make it easier to apply the rule of the function. For example, suppose f is the function defined by $f(x) = x^3$. Here, x is the argument of the function f. Thus, the function $f(x) = x^3$ is the rule that tells us to cube the argument. Then $f(2)$ means to cube the argument 2; $f(-1)$ means to cube -1; $f(a)$ means to cube the number a; $f(x + h)$ means to cube the quantity $(x + h)$. In actually evaluating $f(2)$, for example, we would replace x by 2 and compute the result: $f(2) = 2^3 = 8$.

Example 5 For the function G defined by $G(x) = 2x^2 - 3x$, evaluate:

(a) $G(3)$ (b) $G(-1)$ (c) $G(-x)$ (d) $-G(x)$
(e) $G(h)$ (f) $G(x + h)$ (g) $G(x) + G(h)$

Solution (a) We replace x by 3 in the rule for G to get

$$G(3) = 2(3)^2 - 3(3) = 18 - 9 = 9$$

(b) $G(-1) = 2(-1)^2 - 3(-1) = 2 + 3 = 5$
(c) $G(-x) = 2(-x)^2 - 3(-x) = 2x^2 + 3x$
(d) $-G(x) = -(2x^2 - 3x) = -2x^2 + 3x$
(e) $G(h) = 2(h)^2 - 3(h) = 2h^2 - 3h$
(f) $G(x + h) = 2(x + h)^2 - 3(x + h)$
$= 2(x^2 + 2xh + h^2) - 3x - 3h$
$= 2x^2 + 4xh + 2h^2 - 3x - 3h$

Notice the use of parentheses here.

(g) $G(x) + G(h) = 2x^2 - 3x + 2h^2 - 3h$ ■

Example 5 illustrates certain uses of **function notation**. Another important use of function notation is to find the **difference quotient** of f:

Difference Quotient

$$\frac{f(x+h)-f(x)}{h} \qquad h \neq 0 \qquad (1)$$

This expression is used often in calculus.

Example 6 Find the difference quotient (1) of the function f defined by

$$f(x) = 2x^2 - x + 1$$

Solution It is helpful to proceed in steps.

STEP 1: First, we calculate $f(x+h)$:

$$\begin{aligned} f(x+h) &= 2(x+h)^2 - (x+h) + 1 \\ &= 2(x^2 + 2xh + h^2) - x - h + 1 \\ &= 2x^2 + 4xh + 2h^2 - x - h + 1 \end{aligned}$$

STEP 2: Now, subtract $f(x)$ from this result:

$$f(x+h) - f(x) = 2x^2 + 4xh + 2h^2 - x - h + 1 - \underbrace{(2x^2 - x + 1)}$$

Be careful to subtract the quantity $f(x)$.

$$\begin{aligned} &= 2x^2 + 4xh + 2h^2 - x - h + 1 - 2x^2 + x - 1 \\ &= 4xh + 2h^2 - h \end{aligned}$$

STEP 3: Now, divide by h:

$$\frac{f(x+h)-f(x)}{h} = \frac{4xh + 2h^2 - h}{h}$$

$$= \frac{\not{h}(4x + 2h - 1)}{\not{h}}$$

↑ Factor numerator.

$$= 4x + 2h - 1$$

↑ Cancellation property ■

Other Notations

Sometimes, **arrow notation** is used to signify a function. Suppose f is a function defined by the equation $y = f(x)$. In arrow notation this would be written as

$$f : x \to y \qquad \text{or} \qquad f : x \to f(x)$$

We read this notation as: "f maps x onto y" or "f maps x onto $f(x)$."

Another notation for functions, called *ordered-pair notation*, is discussed in the next section.

Calculators

Most calculators have special keys that enable you to find the value of many functions. On your calculator, you should be able to find the square function, $f(x) = x^2$; the square root function, $f(x) = \sqrt{x}$; the reciprocal function, $f(x) = 1/x$; and many others that will be discussed later in this book (such as $\ln x$, $\log x$, and so on). When you enter x and then press one of these function keys, you get the value of that function at x. Try it with the functions listed in Example 7.

C **Example 7** (a) $f(x) = x^2$; $f(1.234) \approx 1.522756$

(b) $g(x) = 1/x$; $g(1.234) \approx 0.8103728$

(c) $F(x) = \sqrt{x}$; $F(1.234) \approx 1.1108555$ ■

Summary

We list here some of the important vocabulary introduced in this section, with a brief description of each term.

Function	A rule or correspondence between two sets of real numbers so that each number x in the first set, the domain, has corresponding to it exactly one number y in the second set.
	The range is the set of y values of the function for the x values in the domain.
	A function f may be defined implicitly, by an equation involving x and y; or explicitly, by writing $y = f(x)$.
Function notation	$y = f(x)$
	f is a symbol for the rule that defines the function.
	x is the argument, or independent variable.
	y is the dependent variable.
	$f(x)$ is the value of the function at x, or the image of x.

Unspecified domain

If a function f is defined by an equation and no domain is specified, then the domain will be taken to be the largest set of real numbers for which the rule defines a real number.

EXERCISE 4.1

In Problems 1–8, find the following for each function:

(a) $f(0)$ (b) $f(1)$ (c) $f(-1)$ (d) $f(2)$

1. $f(x) = -3x^2 + 2x - 4$ **2.** $f(x) = 2x^2 + x - 1$ **3.** $f(x) = \dfrac{x}{x^2 + 1}$

4. $f(x) = \dfrac{x^2 - 1}{x + 4}$ **5.** $f(x) = |x| + 4$ **6.** $f(x) = \sqrt{x^2 + x}$

7. $f(x) = \dfrac{2x + 1}{3x - 5}$ **8.** $f(x) = 1 - \dfrac{1}{(x + 2)^2}$

In Problems 9–18, find the following for each function:

(a) $f(-x)$ (b) $-f(x)$ (c) $f(2x)$ (d) $f(x - 3)$ (e) $f(1/x)$ (f) $1/f(x)$

9. $f(x) = 2x + 3$ **10.** $f(x) = 4 - x$ **11.** $f(x) = 2x^2 - 4$

12. $f(x) = x^3 + 1$ **13.** $f(x) = x^3 - 3x$ **14.** $f(x) = x^2 + x$

15. $f(x) = \dfrac{x}{x^2 + 1}$ **16.** $f(x) = \dfrac{x^2}{x^2 + 1}$ **17.** $f(x) = |x|$

18. $f(x) = \dfrac{1}{x}$

In Problems 19–32, find the domain of each function.

19. $f(x) = 2x + 1$ **20.** $f(x) = 3x^2 - 2$ **21.** $f(x) = \dfrac{x}{x^2 + 1}$

22. $f(x) = \dfrac{x^2}{x^2 + 1}$ **23.** $g(x) = \dfrac{x}{x^2 - 1}$ **24.** $h(x) = \dfrac{x}{x - 1}$

25. $F(x) = \dfrac{x - 2}{x^3 + x}$ **26.** $G(x) = \dfrac{x + 4}{x^3 - 4x}$ **27.** $h(x) = \sqrt{3x - 12}$

28. $G(x) = \sqrt{1 - x}$ **29.** $f(x) = \sqrt{x^2 - 9}$ **30.** $f(x) = \dfrac{1}{\sqrt{x^2 - 4}}$

31. $p(x) = \sqrt{\dfrac{x - 2}{x - 1}}$ **32.** $q(x) = \sqrt{x^2 - x - 2}$

In Problems 33–42, find the difference quotient,

$$\frac{f(x + h) - f(x)}{h} \qquad h \neq 0$$

for each function. Be sure to simplify.

33. $f(x) = 3$ **34.** $f(x) = 2x$ **35.** $f(x) = 1 - 3x$

36. $f(x) = x^2 + 1$ **37.** $f(x) = 3x^2 - 2x$ **38.** $f(x) = 4x - 2x^2$

39. $f(x) = x^3 - x$ **40.** $f(x) = x^3 + x$ **41.** $f(x) = \frac{1}{x}$

42. $f(x) = \frac{1}{x^2}$

43. For the function $f(x) = \sqrt{x}$, show that

$$\frac{f(x+h) - f(x)}{h} = \frac{1}{\sqrt{x+h} + \sqrt{x}} \qquad h \neq 0$$

[*Hint:* Multiply the numerator and denominator by $\sqrt{x+h} + \sqrt{x}$.]

44. For the function $f(x) = \sqrt{x+3}$, show that

$$\frac{f(x+h) - f(x)}{h} = \frac{1}{\sqrt{x+3+h} + \sqrt{x+3}} \qquad h \neq 0$$

45. If $f(x) = 2x^3 + Ax^2 + 4x - 5$ and $f(2) = 3$, what is the value of A?

46. If $f(x) = 3x^2 - Bx + 4$ and $f(-1) = 10$, what is the value of B?

47. If $f(x) = (3x + 8)/(2x - A)$ and $f(0) = 2$, what is the value of A?

48. If $f(x) = (2x - B)/(3x + 4)$ and $f(2) = \frac{1}{2}$, what is the value of B?

49. If $f(x) = (2x - A)/(x - 3)$ and $f(4) = 0$, what is the value of A? Where is f not defined?

50. If $f(x) = (x - B)/(x - A)$, $f(2) = 0$, and $f(1)$ is undefined, what are the values of A and B?

C **51.** If a rock falls from a height of 20 meters on Earth, the height H (in meters) after x seconds is approximately

$$H(x) = 20 - 4.9x^2$$

(a) What is the height of the rock when $x = 1$ second? $x = 1.1$ seconds? $x = 1.2$ seconds? $x = 1.3$ seconds?
(b) When does the rock strike the ground?

C **52.** If a rock falls from a height of 20 meters on the planet Jupiter, its height H (in meters) after x seconds is approximately

$$H(x) = 20 - 13x^2$$

(a) What is the height of the rock when $x = 1$ second? $x = 1.1$ seconds? $x = 1.2$ seconds?
(b) When does the rock strike the ground?

53. Express the area A of a rectangle as a function of the length x if the length is twice the width of the rectangle.

54. Express the area A of an isosceles right triangle as a function of the length x of one of the two equal sides.

55. Express the gross salary G of a person who earns \$5 per hour as a function of the number x of hours worked.

56. A commissioned salesperson earns \$100 base pay plus \$10 per item sold. Express the gross salary G as a function of the number x of items sold.

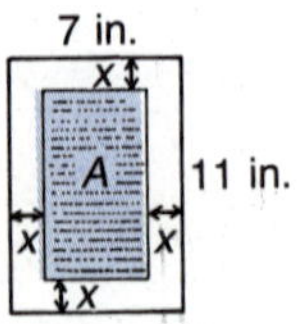

57. A page with dimensions of 11 inches by 7 inches has a border of uniform width x surrounding the printed matter of the page, as shown in the figure. Write a formula for the area A of the printed part of the page as a function of the width x of the border. Give the domain and range of A.

C **58.** An airplane crosses the Atlantic Ocean (3000 miles) with an airspeed of 500 miles per hour. The cost C (in dollars) per passenger is

$$C(x) = 100 + \frac{x}{10} + \frac{36{,}000}{x}$$

where x is the ground speed (airspeed $\pm$ wind).
(a) What is the cost per passenger for quiescent (no wind) conditions?
(b) What is the cost per passenger with a head wind of 50 miles per hour?
(c) What is the cost per passenger with a tail wind of 100 miles per hour?
(d) What is the cost per passenger with a head wind of 100 miles per hour?

C **59.** The period T (in seconds) of a simple pendulum is a function of its length l (in feet) defined by the equation

$$T(l) = 2\pi\sqrt{\frac{l}{g}}$$

where $g \approx 32.2$ feet per second per second is the acceleration of gravity. Use a calculator to determine the period of a pendulum whose length is 1 foot. By how much does the period increase if the length is increased to 2 feet?

C **60.** If an object weighs m pounds at sea level, then its weight W (in pounds) at a height of h miles above sea level is given approximately by

$$W(h) = m\left(\frac{4000}{4000 + h}\right)^2$$

If a woman weighs 120 pounds at sea level, how much will she weigh on Pike's Peak, which is 14,110 feet above sea level?

61. A function f has the property that $f(a + b) = f(a) + f(b)$ for all real numbers a and b. Which of the following functions also have this property?
(a) $h(x) = 2x$ (b) $g(x) = x^2$ (c) $F(x) = 5x - 2$ (d) $G(x) = 1/x$

62. Let f and g be two functions defined on the same interval $[a, b]$. Suppose we define two functions $\min(f, g)$ and $\max(f, g)$ as follows:

$$\min(f, g)(x) = \begin{cases} f(x) & \text{if } f(x) \le g(x) \\ g(x) & \text{if } f(x) > g(x) \end{cases} \qquad \max(f, g)(x) = \begin{cases} g(x) & \text{if } f(x) \le g(x) \\ f(x) & \text{if } f(x) > g(x) \end{cases}$$

Show that

$$\min(f, g)(x) = \frac{f(x) + g(x)}{2} - \frac{|f(x) - g(x)|}{2}$$

Develop a similar formula for $\max(f, g)$.

4.2 The Graph of a Function

In applications, a graph often demonstrates more clearly the relationship between two variables than, say, an equation or table would. For example, Figure 3 (on the next page) shows OPEC crude oil production in millions of barrels per day (vertical axis) from January 1981 through August 1983 (horizontal axis). It is easy to see from the graph that

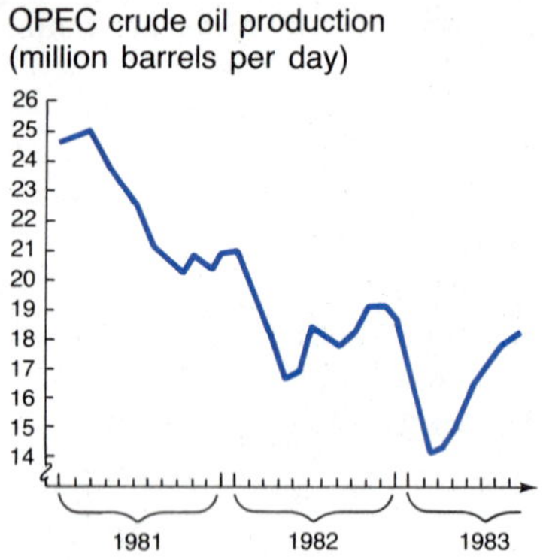

Figure 3

production was falling during 1981 and that production was rising in 1983. It is also easy to see that production was at the lowest level some time at the beginning of 1983. Equations and tables, on the other hand, usually require some calculations and interpretation before this kind of information can be "seen."

Look again at Figure 3. The graph shows that, for each time on the horizontal axis, there is only one production amount on the vertical axis. Thus, the graph represents a function, although the exact rule for getting from time to production amount is not given.

When the rule that defines a function f is given by an equation in x and y, the **graph of *f*** is the graph of the equation, namely, the set of points (x, y) in the xy-plane that satisfies the equation.

Not every collection of points in the xy-plane represents the graph of a function. Remember, for a function f, each number x in the domain of f has one and only one image $f(x)$. Thus, the graph of a function f cannot contain two points with the same x-coordinate and different y-coordinates. Therefore, the graph of a function must satisfy the following **vertical-line test**:

Theorem
Vertical-Line Test

A set of points in the xy-plane is the graph of a function if and only if a vertical line intersects the graph in at most one point. ■

In other words, if any vertical line intersects a graph at more than one point, the graph is not the graph of a function.

Example 1 Which of the graphs in Figure 4 are graphs of functions?

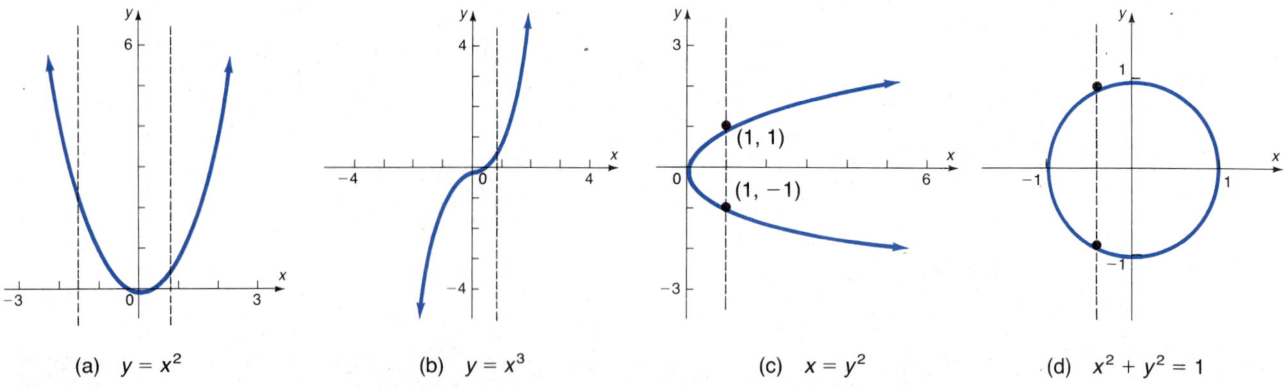

Figure 4

Solution The graphs in Figures 4(a) and 4(b) are graphs of functions, because a vertical line intersects each graph in at most one point. The graphs in Figures 4(c) and 4(d) are not graphs of functions, because some vertical line intersects each graph in more than one point. ■

Ordered Pairs

The preceding discussion provides an alternative way to think of a function. We may consider a function f as a set of **ordered pairs** (x, y) or $(x, f(x))$, in which no two pairs have the same first element. The set of all first elements is the domain, and the set of all second elements is the range of the function. Thus, there is associated with each element x in the domain a unique element y in the range. An example is the set of all ordered pairs (x, y) such that $y = x^2$. Some of the pairs in this set are

$$(2, 2^2) = (2, 4) \qquad (0, 0^2) = (0, 0)$$
$$(-2, (-2)^2) = (-2, 4) \qquad \left(\tfrac{1}{2}, \left(\tfrac{1}{2}\right)^2\right) = \left(\tfrac{1}{2}, \tfrac{1}{4}\right)$$

In this set, no two pairs have the same *first* element (although there are pairs that have the same *second* element). This set is the *square function*, which associates with each real number x the number x^2. Look again at Figure 4(a).

On the other hand, the ordered pairs (x, y) for which $y^2 = x$ do not represent a function, because there are ordered pairs with the same first element but different second elements. For example, $(1, 1)$ and $(1, -1)$ are ordered pairs obeying the relationship $y^2 = x$ with the same first element but different second elements. Look again at Figure 4(c).

The next example illustrates how to determine the domain and range of a function if its graph is given.

Example 2 Let f be a function whose graph is given in Figure 5. Some points on the graph are labeled.

(a) What is the value of the function when $x = -6$, $x = -4$, $x = 0$, and $x = 6$?

(b) What is the domain of f?

(c) What is the range of f?

(d) List the intercepts. (Recall that these are the points, if any, where the graph crosses the coordinate axes.)

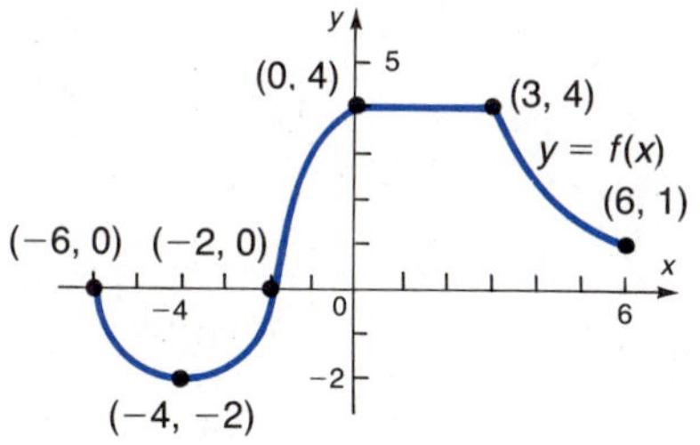

Figure 5

Solution (a) Since $(-6, 0)$ is on the graph of f, the y-coordinate 0 must be the value of f at the x-coordinate -6; that is, $f(-6) = 0$. In a similar way, we find that when $x = -4$, then $y = -2$, or $f(-4) = -2$; when $x = 0$, then $y = 4$, or $f(0) = 4$; and when $x = 6$, then $y = 1$, or $f(6) = 1$.

(b) To determine the domain of f, we notice that the points on the graph of f all have x-coordinates between -6 and 6, inclusive; and, for each number x between -6 and 6, there is a point $(x, f(x))$ on the graph. Thus, the domain of f is $\{x \mid -6 \le x \le 6\}$, or $[-6, 6]$.

(c) The points on the graph all have y-coordinates between -2 and 4, inclusive; and, for each such number y, there is at least one number x in the domain. Hence, the range of f is $\{y | -2 \le y \le 4\}$, or $[-2, 4]$.

(d) The intercepts are $(-6, 0)$, $(-2, 0)$, and $(0, 4)$. ■

Increasing and Decreasing Functions

Consider again the graph given in Figure 5. If you look from left to right along the graph of this function, you will notice that parts of the graph are rising, parts are falling, and parts are horizontal. In such cases, the function is described as *increasing*, *decreasing*, and *constant*, respectively. More precise definitions follow.

Increasing Function A function f is (strictly) **increasing** on an interval I if, for any choice of x_1 and x_2 in I, with $x_1 < x_2$, we have $f(x_1) < f(x_2)$.

Decreasing Function A function f is (strictly) **decreasing** on an interval I if, for any choice of x_1 and x_2 in I, with $x_1 < x_2$, we have $f(x_1) > f(x_2)$.

Constant Function A function f is **constant** on an interval I if, for all choices of x in I, the values $f(x)$ are equal.

Thus, the graph of an increasing function goes up from left to right; the graph of a decreasing function goes down from left to right; and the graph of a constant function remains at a fixed height. Figure 6 illustrates the definitions.

Figure 6

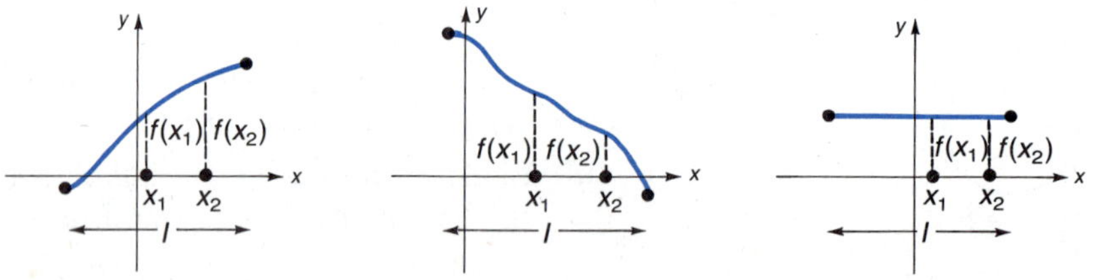

(a) For $x_1 < x_2$ in I, $f(x_1) < f(x_2)$; f is increasing

(b) For $x_1 < x_2$ in I, $f(x_1) > f(x_2)$; f is decreasing

(c) Values of f are equal; f is constant

To answer the question of where a function is increasing, where it is decreasing, and where it is constant, we use inequalities involving the independent variable x or intervals of x-coordinates.

Example 3 Where is the function in Figure 5 (page 221) increasing? Where is it decreasing? Where is it constant?

Solution The graph is rising for $-4 \le x \le 0$; that is, the function is increasing on the interval $[-4, 0]$. It is decreasing for $-6 \le x \le -4$ and for $3 \le x \le 6$; that is, on the intervals $[-6, -4]$ and $[3, 6]$. It is constant for $0 \le x \le 3$; that is, on the interval $[0, 3]$. ■

Even and Odd Functions

Even Function A function f is **even** if for every number x in its domain, the number $-x$ is also in the domain and

$$f(-x) = f(x)$$

Odd Function A function f is **odd** if for every number x in its domain, the number $-x$ is also in the domain and

$$f(-x) = -f(x)$$

Refer to Section 3.2, where the tests for symmetry are listed. The following results are then evident:

Theorem A function is even if and only if its graph is symmetric with respect to the y-axis.

A function is odd if and only if its graph is symmetric with respect to the origin. ■

Example 4 Determine whether each graph given in Figure 7 is the graph of an even function, an odd function, or a function that is neither even nor odd.

Figure 7

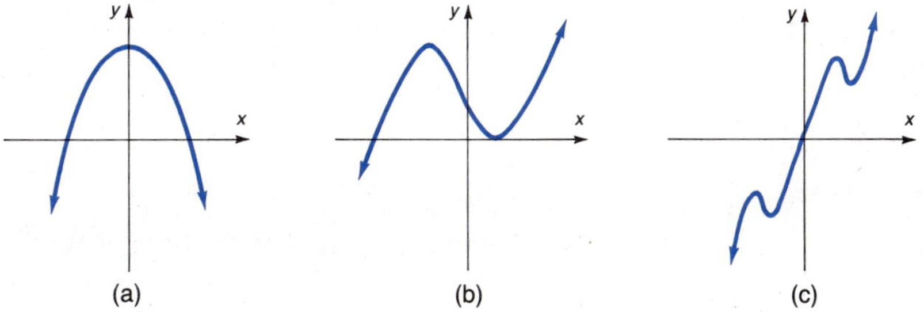

Solution The graph in Figure 7(a) is that of an even function, because the graph is symmetric with respect to the y-axis. The function whose graph is given in Figure 7(b) is neither even nor odd, because the graph is neither symmetric with respect to the y-axis nor symmetric with respect to the origin. The function whose graph is given in Figure 7(c) is odd, because its graph is symmetric with respect to the origin. ■

Example 5 Determine whether each of the following functions is even, odd, or neither:

(a) $f(x) = x^2 - 5$ (b) $g(x) = x^3 - 1$

(c) $h(x) = 5x^3 - x$ (d) $F(x) = |x|$

Without graphing, determine whether the graph is symmetric with respect to the y-axis or with respect to the origin.

Solution (a) We replace x by $-x$ in $f(x) = x^2 - 5$. Then,

$$f(-x) = (-x)^2 - 5 = x^2 - 5$$

Since $f(-x) = f(x)$, we conclude that f is an even function, and the graph is symmetric with respect to the y-axis.

(b) We replace x by $-x$. Then,

$$g(-x) = (-x)^3 - 1 = -x^3 - 1$$

Since $g(-x) \neq g(x)$ and $g(-x) \neq -g(x)$, we conclude that g is neither even nor odd. The graph is not symmetric with respect to the y-axis nor with respect to the origin.

(c) We replace x by $-x$ in $h(x) = 5x^3 - x$. Then,

$$h(-x) = 5(-x)^3 - (-x) = -5x^3 + x$$

Since $h(-x) = -h(x)$, h is an odd function, and the graph of h is symmetric with respect to the origin.

(d) We replace x by $-x$ in $F(x) = |x|$. Then,

$$F(-x) = |-x| = |x|$$

Since $F(-x) = F(x)$, F is an even function, and the graph of F is symmetric with respect to the y-axis. ■

Important Functions

We now give names to some of the functions we have encountered.

Linear Function

$f(x) = mx + b$ m and b are real numbers

The domain of the **linear function** f consists of all real numbers. The graph of this function is a nonvertical straight line with slope m and y-intercept b. A linear function is increasing if $m > 0$, decreasing if $m < 0$, and constant if $m = 0$.

Constant Function

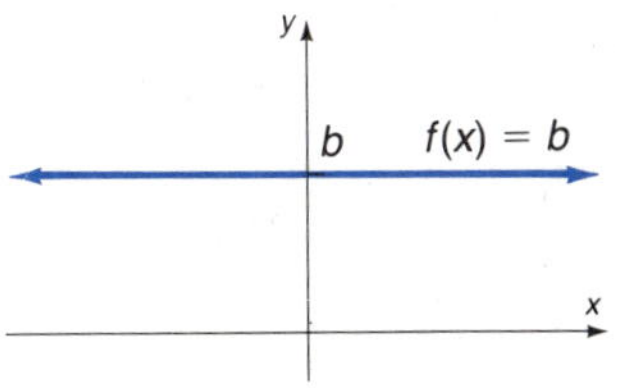

Figure 8

$$f(x) = b \qquad b \text{ a real number}$$

A **constant function** is a special linear function ($m = 0$). Its domain is the set of all real numbers; its range is the set consisting of a single number b. Its graph is a horizontal line whose y-intercept is b. The constant function is an even function whose graph is constant over its domain. See Figure 8.

Identity Function

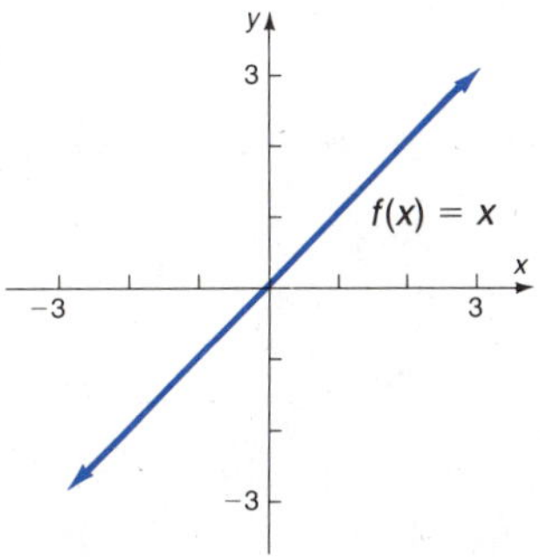

Figure 9

$$f(x) = x$$

The **identity function** is also a special linear function. Its domain and its range are the set of all real numbers. Its graph is a line whose slope is $m = 1$ and whose y-intercept is 0. The line consists of all points for which the x-coordinate equals the y-coordinate. The identity function is an odd function that is increasing over its domain. See Figure 9. Note that the graph bisects quadrants I and III.

Square Function

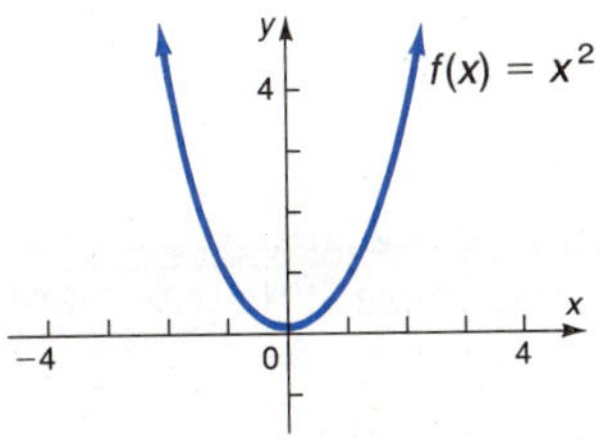

Figure 10

$$f(x) = x^2$$

The domain of the **square function** f is the set of all real numbers; its range is the set of nonnegative real numbers. The graph of this function is a parabola, whose intercept is at $(0, 0)$. The square function is an even function that is decreasing on the interval $(-\infty, 0]$ and increasing on the interval $[0, \infty)$. See Figure 10.

Cube Function

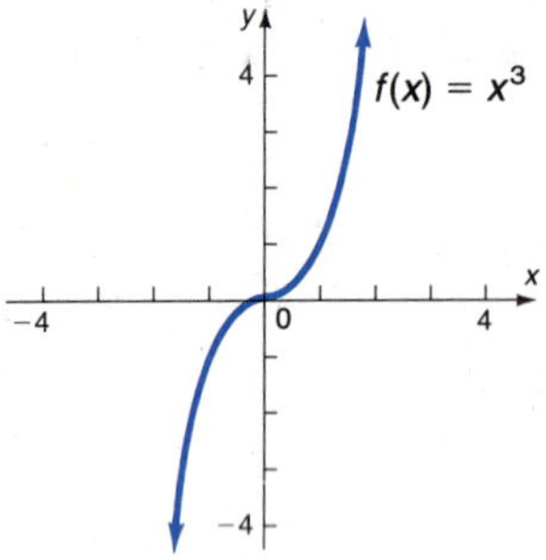

Figure 11

$$f(x) = x^3$$

The domain and range of the **cube function** are the set of all real numbers. The intercept of the graph is at (0, 0). The cube function is odd and is increasing on the interval $(-\infty, \infty)$. See Figure 11.

Square Root Function

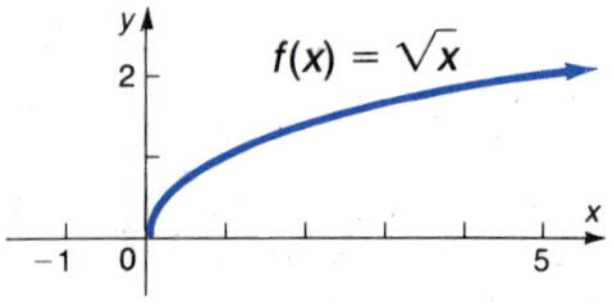

Figure 12

$$f(x) = \sqrt{x}$$

The domain and range of the **square root function** are the set of nonnegative real numbers. The intercept of the graph is at (0, 0). The square root function is neither even nor odd and is increasing on the interval $[0, \infty)$. See Figure 12.

Reciprocal Function

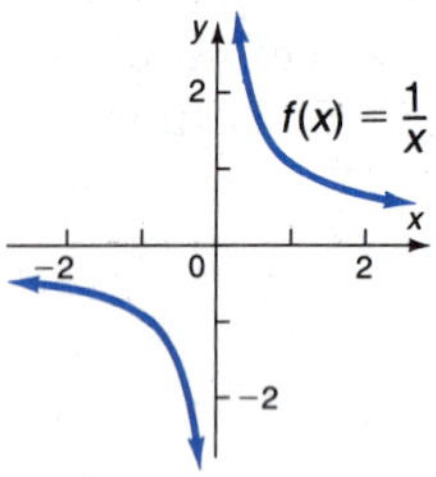

Figure 13

$$f(x) = \frac{1}{x}$$

The domain and range of the **reciprocal function** are the set of all nonzero real numbers. The graph has no intercepts. The reciprocal function is decreasing on the intervals $(-\infty, 0)$ and $(0, \infty)$, and is an odd function. See Figure 13.

Absolute Value Function

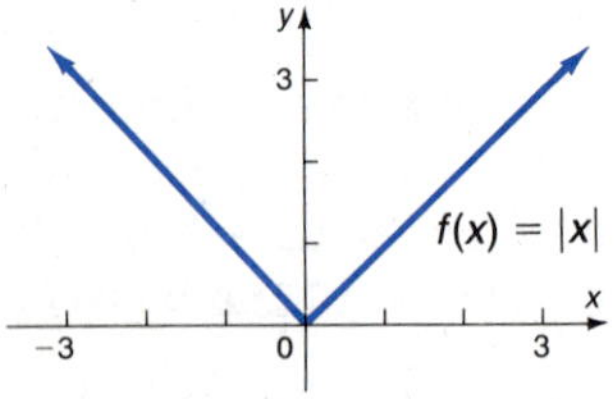

Figure 14

$$f(x) = |x|$$

The domain of the **absolute value function** is the set of all real numbers; its range is the set of nonnegative real numbers. The intercept of the graph is at (0, 0). If $x \geq 0$, then $f(x) = x$ and the graph of f is part of the line $y = x$; if $x < 0$, then $f(x) = -x$ and the graph of f is part of the line $y = -x$. The absolute value function is an even function; it is decreasing on the interval $(-\infty, 0]$ and increasing on the interval $[0, \infty)$. See Figure 14.

The symbol $[\![x]\!]$, read as "**bracket *x*,**" stands for the largest integer less than or equal to x. For example,

$$[\![1]\!] = 1 \qquad [\![2.5]\!] = 2 \qquad [\![\tfrac{1}{2}]\!] = 0 \qquad [\![-\tfrac{3}{4}]\!] = -1 \qquad [\![\pi]\!] = 3$$

This type of correspondence occurs frequently enough in mathematics that we give it a name.

Greatest-Integer Function

$$f(x) = [\![x]\!] = \text{Greatest integer less than or equal to } x$$

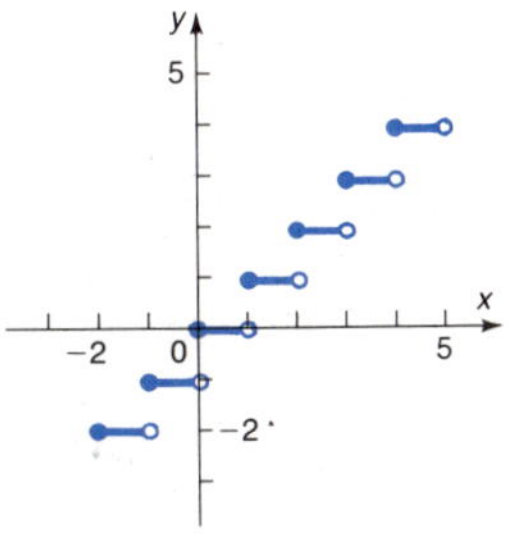

Figure 15

The domain of the **greatest-integer function** is the set of all real numbers; its range is the set of integers. The y-intercept of the graph is at 0. The x-intercepts lie in the interval $[0, 1)$. The greatest-integer function is neither even nor odd. It is constant on every interval of the form $[k, k + 1)$, for k an integer. In Figure 15, we use a solid dot to indicate, for example, that at $x = 1$, the value of f is $f(1) = 1$; we use an open circle to illustrate that the function does not assume the value 0 at $x = 1$.

From the graph of the greatest-integer function, we can see why it is also called a **step function**. At $x = 0$, $x = \pm 1$, $x = \pm 2$, and so on, this function exhibits what is called a *discontinuity*; that is, at integer values, the graph suddenly "steps" from one value to another without taking on any of the intermediate values. For example, to the immediate left of $x = 3$, the y-coordinates are 2, and to the immediate right of $x = 3$, the y-coordinates are 3.

The functions we have discussed so far are basic. Whenever you encounter one of them, you should see a mental picture of its graph. For example, if you encounter the function $f(x) = x^2$, you should see in your mind's eye a picture like Figure 10.

Piecewise-Defined Functions

Sometimes, a function is defined by a rule consisting of two or more equations. The choice of which equation to use depends on the value of the independent variable x. For example, the absolute value function $f(x) = |x|$ is actually defined by two equations: $f(x) = x$ if $x \geq 0$ and $f(x) = -x$ if $x < 0$. For convenience, we generally combine these equations into one expression as

$$f(x) = |x| = \begin{cases} x & \text{if } x \geq 0 \\ -x & \text{if } x < 0 \end{cases}$$

When functions are defined by more than one equation, they are called **piecewise-defined** functions.

Let's look at another example of a piecewise-defined function.

Example 6 For the following function f:

$$f(x) = \begin{cases} -x + 1 & \text{if } -1 \le x < 1 \\ 2 & \text{if } x = 1 \\ x^2 & \text{if } x > 1 \end{cases}$$

(a) Find $f(0)$, $f(1)$, and $f(2)$. (b) Determine the domain of f.
(c) Graph f.
(d) Use the graph to find the range of f.

Solution (a) To find $f(0)$, we observe that when $x = 0$, the equation for f is given by $f(x) = -x + 1$. So, we have

$$f(0) = -0 + 1 = 1$$

When $x = 1$, the equation for f is $f(x) = 2$. Thus,

$$f(1) = 2$$

When $x = 2$, the equation for f is $f(x) = x^2$. So,

$$f(2) = 2^2 = 4$$

(b) To find the domain of f, we look at its definition. We conclude that the domain of f is $\{x|x \ge -1\}$, or $[-1, \infty)$.

(c) To graph f, we graph "each piece." Thus, we first graph the line $y = -x + 1$, and keep only the part for which $-1 \le x < 1$. Then we plot the point $(1, 2)$, because when $x = 1$, $f(x) = 2$. Finally, we graph the parabola $y = x^2$, and keep only the part for which $x > 1$. See Figure 16.

(d) From the graph, we conclude that the range of f is $\{y|y > 0\}$, or $(0, \infty)$. ■

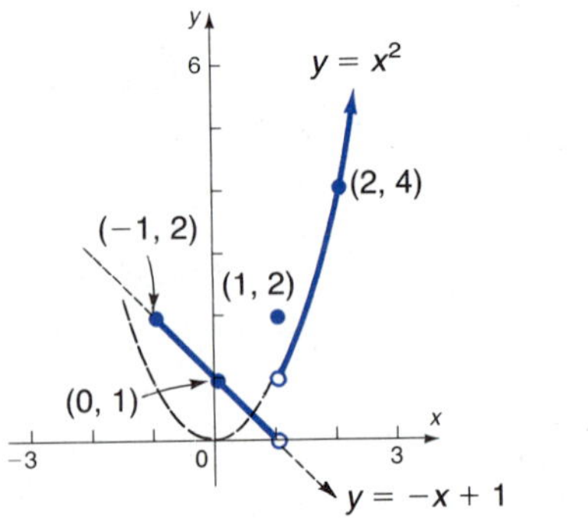

Figure 16
$y = f(x)$

EXERCISE 4.2

In Problems 1–12, use the graph of the function f given below.

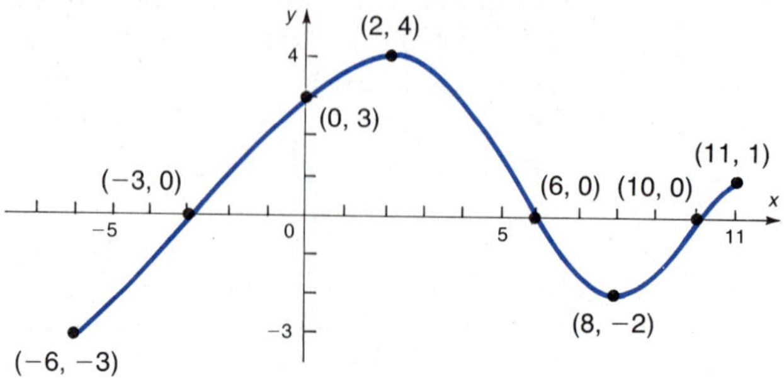

1. Find $f(0)$ and $f(2)$.
2. Find $f(8)$ and $f(-3)$.
3. Is $f(2)$ positive or negative?
4. Is $f(8)$ positive or negative?
5. For what numbers x is $f(x) = 0$?
6. For what numbers x is $f(x) > 0$?
7. What is the domain of f?
8. What is the range of f?
9. Where is f increasing?
10. Where is f decreasing?
11. How often does the line $y = \frac{1}{2}$ intersect the graph?
12. How often does the line $y = 3$ intersect the graph?

In Problems 13–16, answer the questions about the given function.

13. $f(x) = \dfrac{x + 2}{x - 6}$
    (a) Is the point (3, 14) on the graph of f? (b) If $x = 4$, what is $f(x)$?
    (c) If $f(x) = 2$, what is x? (d) What is the domain of f?

14. $f(x) = \dfrac{x^2 + 2}{x + 4}$
    (a) Is the point $\left(1, \frac{3}{5}\right)$ on the graph of f? (b) If $x = 0$, what is $f(x)$?
    (c) If $f(x) = \frac{1}{2}$, what is x? (d) What is the domain of f?

15. $f(x) = \dfrac{2x^2}{x^4 + 1}$
    (a) Is the point $(-1, 1)$ on the graph of f? (b) If $x = 2$, what is $f(x)$?
    (c) If $f(x) = 1$, what is x? (d) What is the domain of f?

16. $f(x) = \dfrac{2x}{x - 2}$
    (a) Is the point $\left(\frac{1}{2}, -\frac{2}{3}\right)$ on the graph of f? (b) If $x = 4$, what is $f(x)$?
    (c) If $f(x) = 1$, what is x? (d) What is the domain of f?

In Problems 17–30, determine whether the graph is that of a function by using the vertical-line test. If it is, use the graph to find:

(a) Its domain and range

(b) The intervals on which it is increasing, decreasing, or constant

(c) Whether it is even, odd, or neither

(d) The intercepts, if any

17.

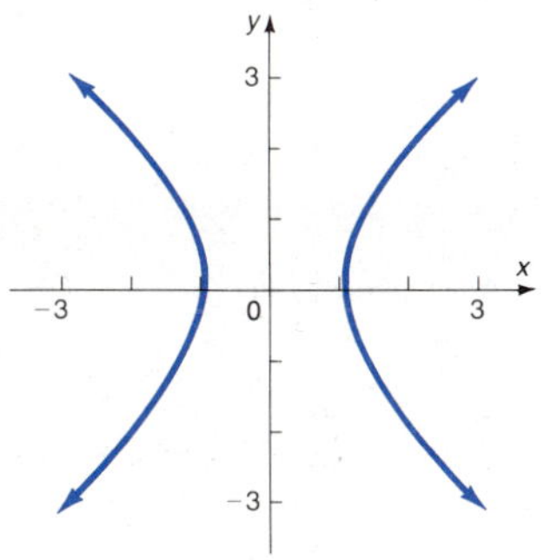

18.

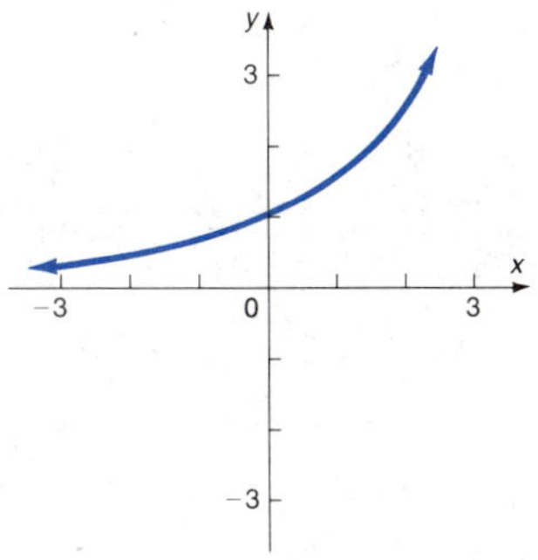

19.

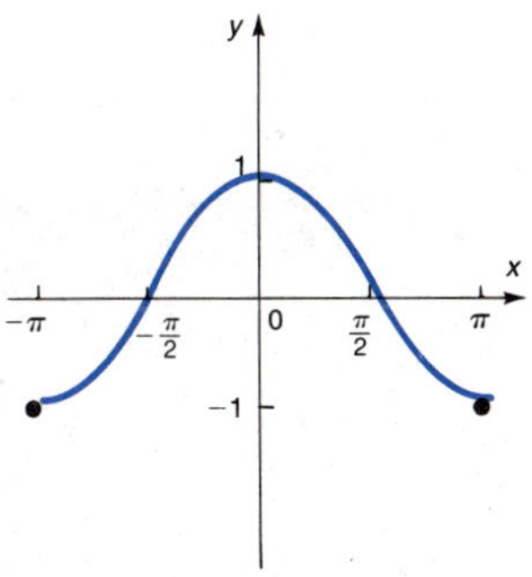

20.

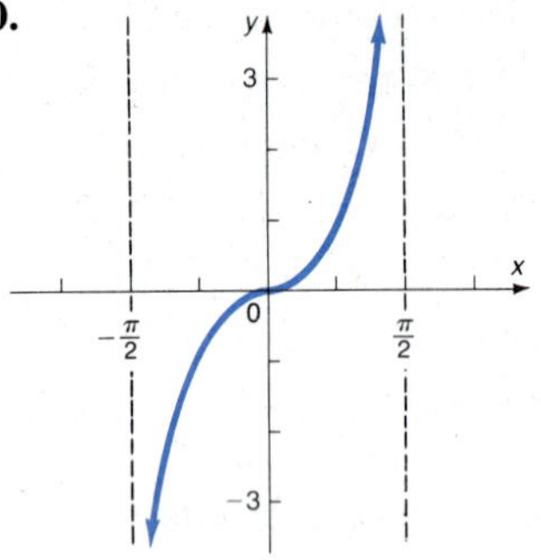

21.

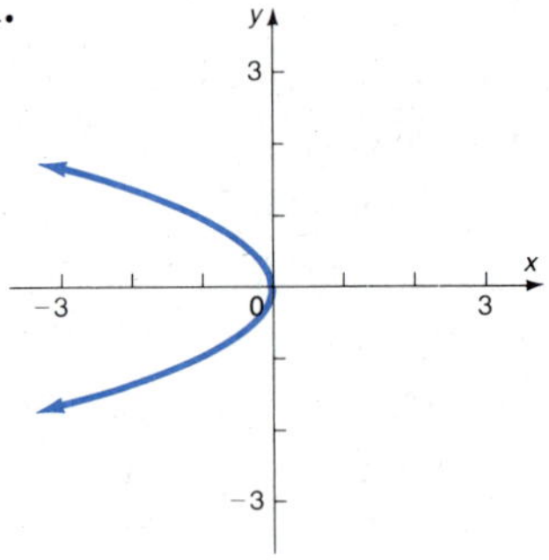

22.

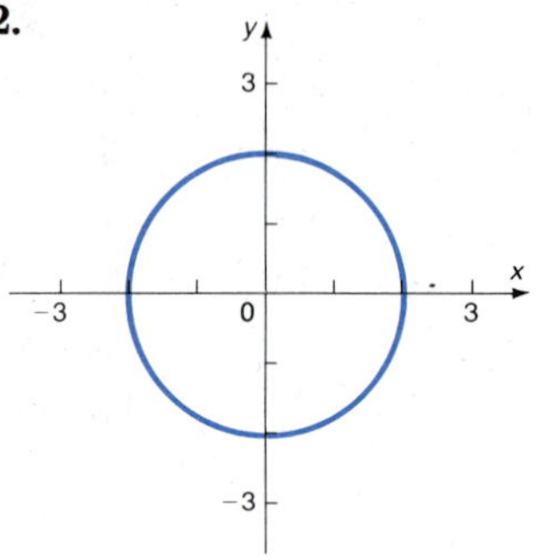

23.

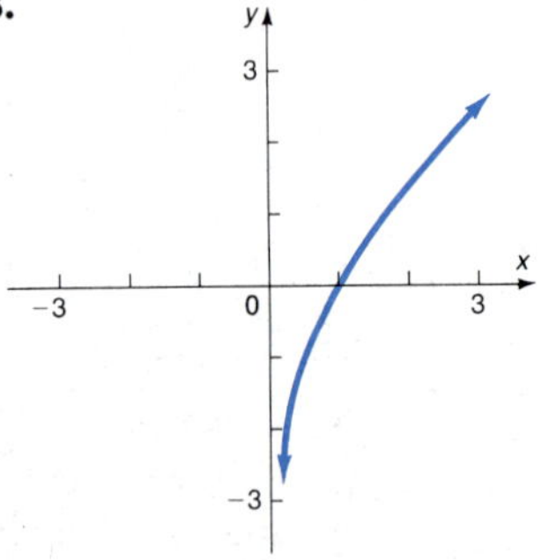

24.

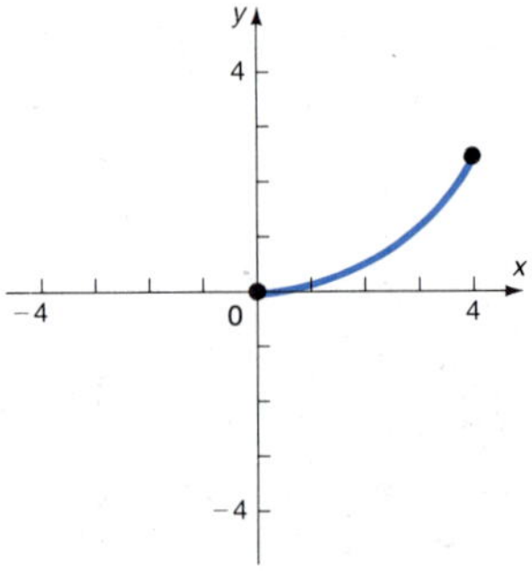

25.

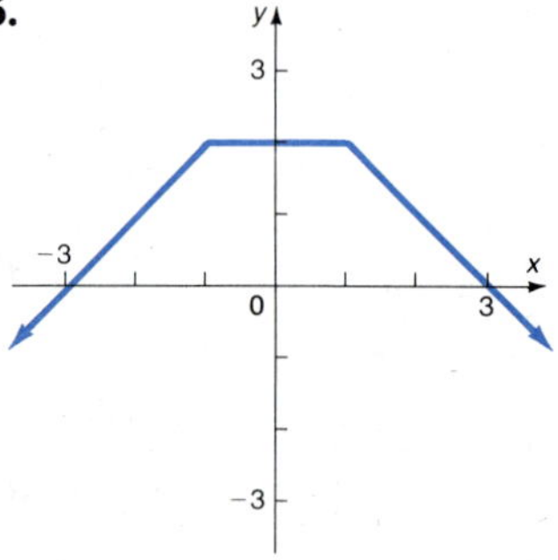

26.

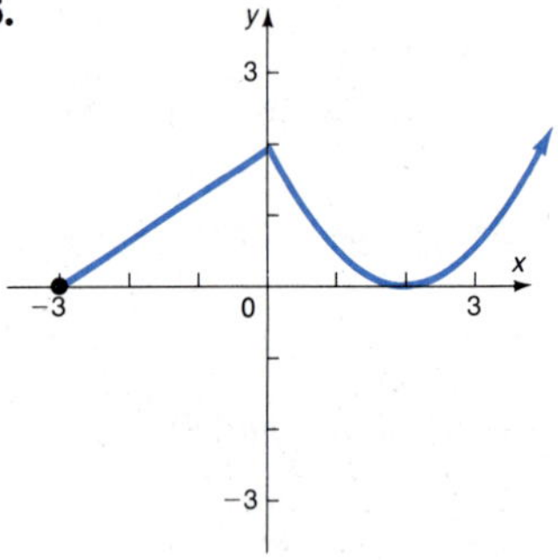

27.

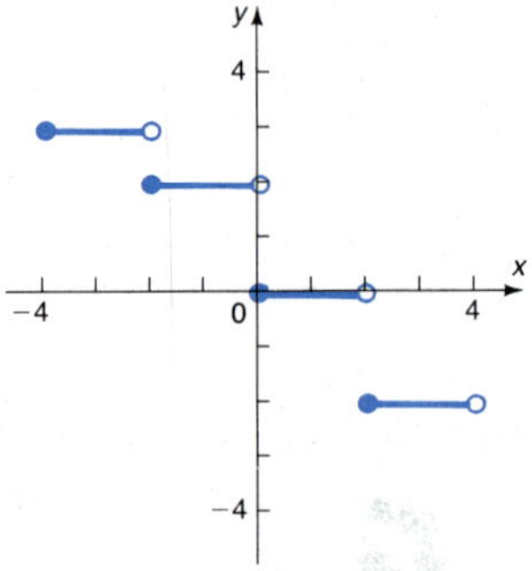

28.

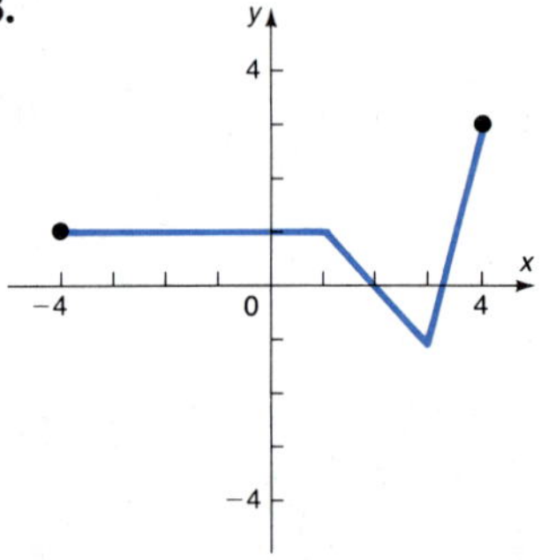

29.

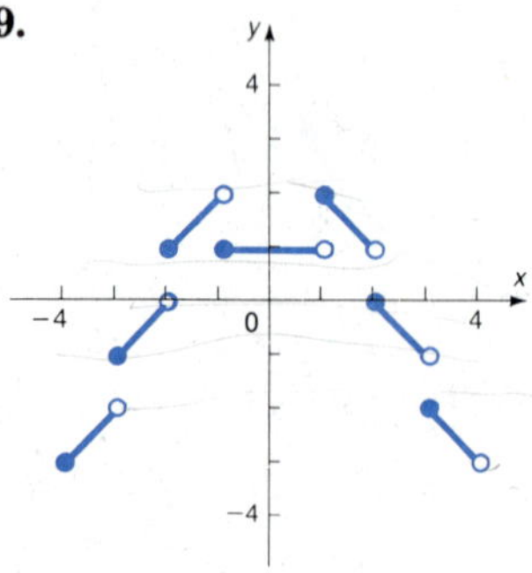

30.

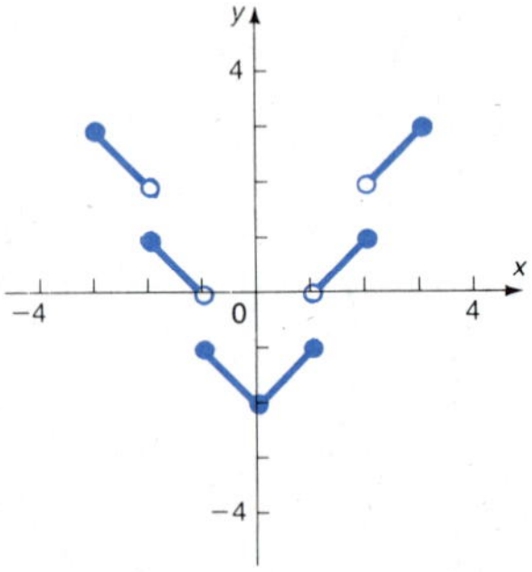

In Problems 31–42, tell whether each function is even, odd, or neither without drawing a graph.

31. $f(x) = 2x^3$ **32.** $f(x) = x^4 - x^2$ **33.** $g(x) = 2x^2 - 5$ **34.** $h(x) = 3x^3 + 2$

35. $F(x) = \sqrt[3]{x}$ **36.** $G(x) = \sqrt{x}$ **37.** $f(x) = x + |x|$ **38.** $f(x) = \sqrt[3]{2x^2 + 1}$

39. $g(x) = \dfrac{1}{x^2}$ **40.** $h(x) = \dfrac{x}{x^2 - 1}$ **41.** $h(x) = \dfrac{x^3}{3x^2 - 9}$ **42.** $F(x) = \dfrac{x}{|x|}$

43. How many x-intercepts can a function defined on an interval have if it is increasing on that interval? Explain.

44. How many y-intercepts can a function have? Explain.

In Problems 45–68:

(a) *Find the domain of each function.* (b) *Locate any intercepts.*

(c) *Graph each function.* (d) *Based on the graph, find the range.*

45. $f(x) = 3x - 3$ **46.** $f(x) = 4 - 2x$ **47.** $g(x) = x^2 - 4$ **48.** $g(x) = x^2 + 4$

49. $h(x) = -x^2$ **50.** $F(x) = 2x^2$ **51.** $f(x) = \sqrt{x - 2}$ **52.** $g(x) = \sqrt{x} + 2$

53. $h(x) = \sqrt{2 - x}$ **54.** $F(x) = -\sqrt{x}$ **55.** $f(x) = |x| + 3$ **56.** $g(x) = |x + 3|$

57. $h(x) = -|x|$ **58.** $F(x) = |3 - x|$

59. $f(x) = \begin{cases} 2x & \text{if } x \neq 0 \\ 0 & \text{if } x = 0 \end{cases}$ **60.** $f(x) = \begin{cases} 3x & \text{if } x \neq 0 \\ 4 & \text{if } x = 0 \end{cases}$

61. $f(x) = \begin{cases} 1 + x & \text{if } x < 0 \\ x^2 & \text{if } x \geq 0 \end{cases}$ **62.** $f(x) = \begin{cases} 1/x & \text{if } x < 0 \\ \sqrt{x} & \text{if } x \geq 0 \end{cases}$

63. $f(x) = \begin{cases} |x| & \text{if } -2 \leq x < 0 \\ 1 & \text{if } x = 0 \\ x^3 & \text{if } x > 0 \end{cases}$ **64.** $f(x) = \begin{cases} 3 + x & \text{if } -3 \leq x < 0 \\ 3 & \text{if } x = 0 \\ \sqrt{x} & \text{if } x > 0 \end{cases}$

65. $g(x) = \begin{cases} 1 & \text{if } x \text{ is an integer} \\ -1 & \text{if } x \text{ is not an integer} \end{cases}$ **66.** $g(x) = \begin{cases} x & \text{if } x \geq 1 \\ 1 & \text{if } x < 1 \end{cases}$

67. $h(x) = 2[\![x]\!]$ **68.** $f(x) = [\![2x]\!]$

In Problems 69–76, tell whether the set of ordered pairs (x, y) defined by each equation is a function.

69. $y = x^2 + 2x$ **70.** $y = x^3 - 3x$ **71.** $y = \dfrac{2}{x}$ **72.** $y = \dfrac{3}{x} - 3$

73. $y^2 = 1 - x^2$ **74.** $y = \pm\sqrt{1 - 2x}$ **75.** $x^2 + y = 1$ **76.** $x + 2y^2 = 1$

In Problems 77–80, the graph of a piecewise-defined function is given. Write a definition for each function.

77.

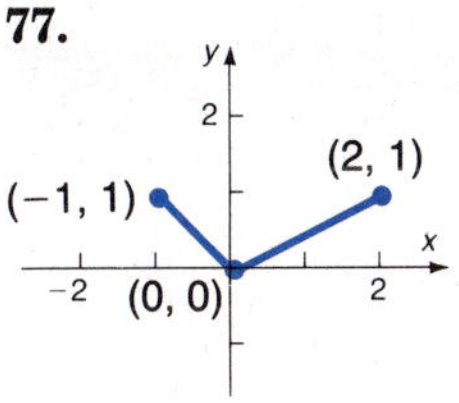

78.

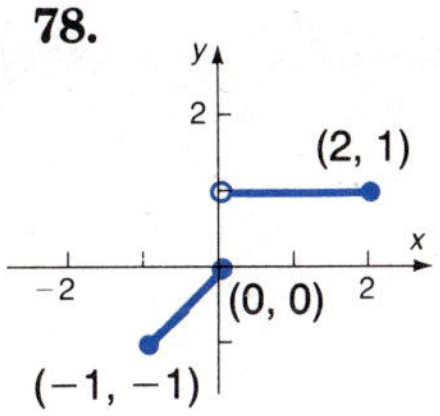

79.

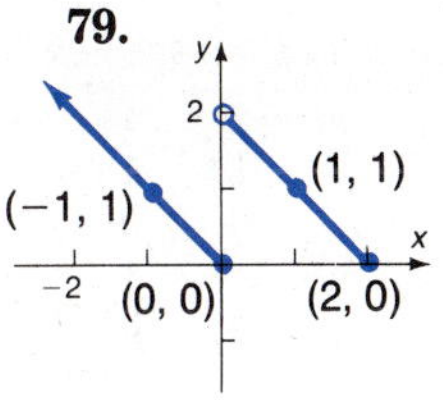

80.

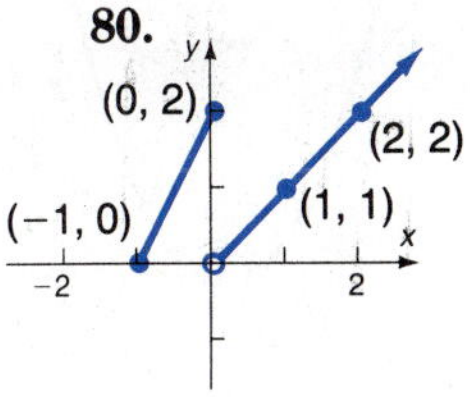

In Problems 81 and 82, decide whether each function is even. Give a reason.

81. $f(x) = \begin{cases} x^2 + 4 & \text{if } x \neq 2 \\ 6 & \text{if } x = 2 \end{cases}$ **82.** $f(x) = \begin{cases} x^2 + 4 & \text{if } x \neq 2 \\ 5 & \text{if } x = 2 \end{cases}$

83. Let f denote any function with the property that whenever x is in its domain, then so is $-x$. Define the functions $E(x)$ and $O(x)$ to be

$$E(x) = \tfrac{1}{2}[f(x) + f(-x)] \qquad O(x) = \tfrac{1}{2}[f(x) - f(-x)]$$

(a) Show that $E(x)$ is an even function. (b) Show that $O(x)$ is an odd function.
(c) Show that $f(x) = E(x) + O(x)$.
(d) Draw the conclusion that any such function f can be written as the sum of an even function and an odd function.

4.3 Graphing Techniques*

At this stage, if you were asked to graph any of the functions defined by $y = x^2$, $y = x^3$, $y = x$, $y = \sqrt{x}$, $y = |x|$, or $y = 1/x$, your response should be, "Yes, I recognize these functions and know the general shapes of their graphs." (If this is not your answer, review the previous section and Figures 9–14.)

Sometimes, we are asked to graph a function that is "almost" like one we already know how to graph. In this section, we look at some of these functions and develop techniques for graphing them.

Vertical Shifts

If a real number c is added to the right side of a function $y = f(x)$, the graph of the new function $y = f(x) + c$ is the graph of f **shifted vertically** up (if $c > 0$) or down (if $c < 0$). Let's look at an example.

Example 1 Use the graph of $f(x) = x^2$ to obtain the graph of $g(x) = x^2 + 3$.

Solution We begin by obtaining some points on the graphs of f and g. For example, when $x = 0$, then $y = f(0) = 0$ and $y = g(0) = 3$. When $x = 1$, then $y = f(1) = 1$ and $y = g(1) = 4$. Table 1 lists these and a few other points on each graph. We conclude that the graph of g is identical to that of f, except that it is shifted vertically up 3 units. See Figure 17.

Table 1

x	$y = f(x) = x^2$	$y = g(x) = x^2 + 3$
−2	4	7
−1	1	4
0	0	3
1	1	4
2	4	7

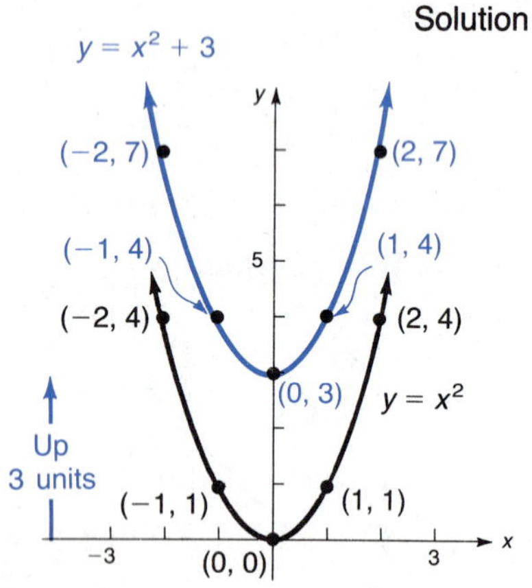

Figure 17

*[C] Refer to Graphics Calculator Supplement: Activity I.

Example 2 Use the graph of $f(x) = x^2$ to obtain the graph of $h(x) = x^2 - 4$.

Solution Table 2 lists some points on the graphs of f and h. The graph of h is identical to that of f, except that it is shifted down 4 units. See Figure 18.

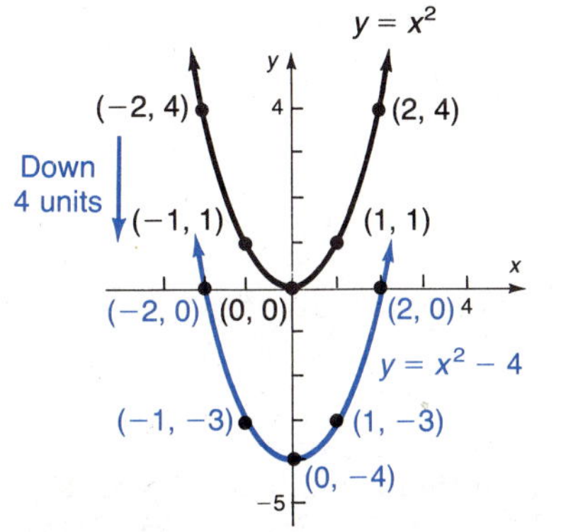

Figure 18

Table 2

x	$y = f(x) = x^2$	$y = h(x) = x^2 - 4$
−2	4	0
−1	1	−3
0	0	−4
1	1	−3
2	4	0

■

Horizontal Shifts

If a real number c is added to the argument x of a function f, the graph of the new function $g(x) = f(x + c)$ is the graph of f **shifted horizontally** left (if $c > 0$) or right (if $c < 0$). Let's look at an example.

Example 3 Use the graph of $f(x) = x^2$ to obtain the graph of $g(x) = (x - 2)^2$.

Solution The function $g(x) = (x - 2)^2$ is basically a square function. Table 3 lists some points on the graphs of f and g. Note that when $f(x) = 0$, then $x = 0$, and when $g(x) = 0$, then $x = 2$. Also, when $f(x) = 4$, then $x = -2$ or 2, and when $g(x) = 4$, then $x = 0$ or 4. We conclude that the graph of g is identical to that of f, except that it is shifted 2 units to the right. See Figure 19.

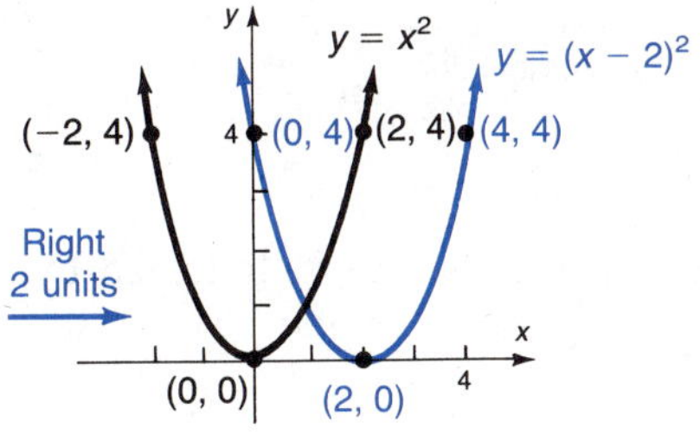

Figure 19

Table 3

x	$y = f(x) = x^2$	$y = g(x) = (x - 2)^2$
−2	4	16
0	0	4
2	4	0
4	16	4

■

Example 4 Use the graph of $f(x) = x^2$ to obtain the graph of $h(x) = (x + 4)^2$.

Solution Again, the function $h(x) = (x + 4)^2$ is basically a square function. Thus, its graph is the same as that of f, except it is shifted 4 units to the left. (Do you see why?) See Figure 20 (at the top of the next page).

Figure 20

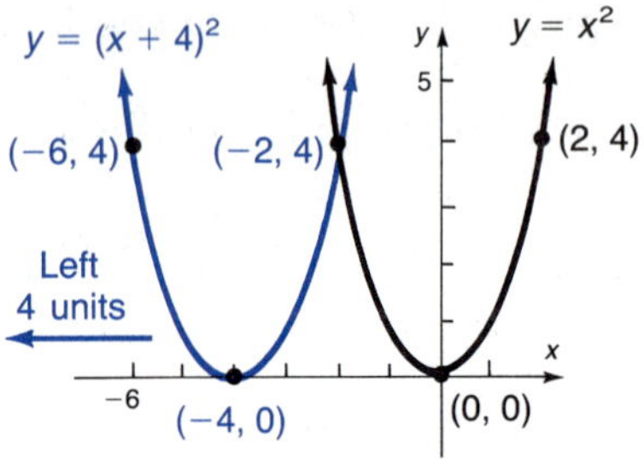

Vertical and horizontal shifts are sometimes combined.

Example 5 Graph the function: $f(x) = (x - 1)^3 + 3$

Solution We graph f in steps. First, we note that the rule for f is basically a cube function. Thus, we begin with the graph of $y = x^3$. See Figure 21(a). Next, to get the graph of $y = (x - 1)^3$, we shift the graph of $y = x^3$ horizontally 1 unit to the right. See Figure 21(b). Finally, to get the graph of $y = (x - 1)^3 + 3$, we shift the graph of $y = (x - 1)^3$ vertically up 3 units. See Figure 21(c). Note the three points that have been plotted on each graph. Using key points such as these can be helpful in keeping track of just what is taking place.

Figure 21

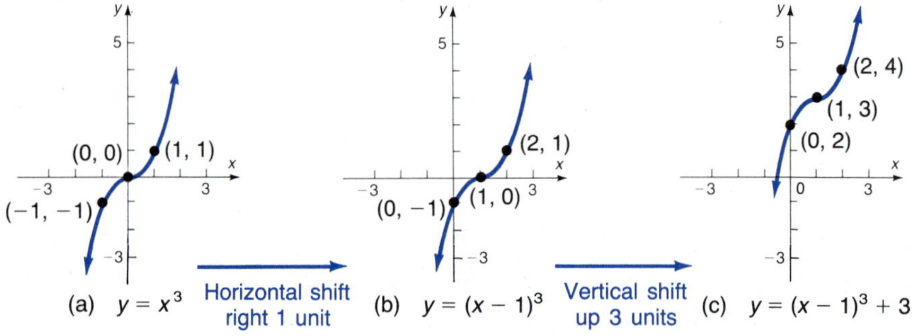

In Example 5, if the vertical shift had been done first, followed by the horizontal shift, the result would have been the same. (Try it for yourself.)

Compressions and Stretches

When the right side of a function $y = f(x)$ is multiplied by a positive number k, the graph of the new function $y = kf(x)$ is a vertically "compressed" (if $0 < k < 1$) or "stretched" (if $k > 1$) version of the graph of $y = f(x)$. The next two examples will clarify this idea.

Example 6 Use the graph of $f(x) = |x|$ to obtain the graph of $g(x) = 3|x|$.

Solution To see the relationship between the graphs of f and g, we form Table 4, listing points on each graph. For each x, the y-coordinate of a point on the graph of g is 3 times as large as the corresponding y-coordinate on the graph of f. The graph of $f(x) = |x|$ is vertically "stretched" [for example, from (1, 1) to (1, 3)] to obtain the graph of $g(x) = 3|x|$. See Figure 22.

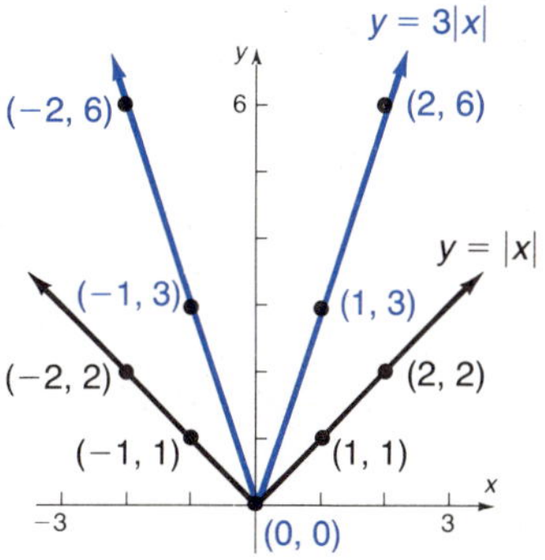

Figure 22

Table 4

x	$y = f(x) = \|x\|$	$y = g(x) = 3\|x\|$
−2	2	6
−1	1	3
0	0	0
1	1	3
2	2	6

■

Example 7 Use the graph $f(x) = |x|$ to obtain the graph of $h(x) = \frac{1}{2}|x|$.

Solution For each x, the y-coordinate of a point on the graph of h is $\frac{1}{2}$ as large as the corresponding y-coordinate on the graph of f. The graph of $f(x) = |x|$ is vertically "compressed" [for example, from (2, 2) to (2, 1)] to obtain the graph of $h(x) = \frac{1}{2}|x|$. See Table 5 and Figure 23.

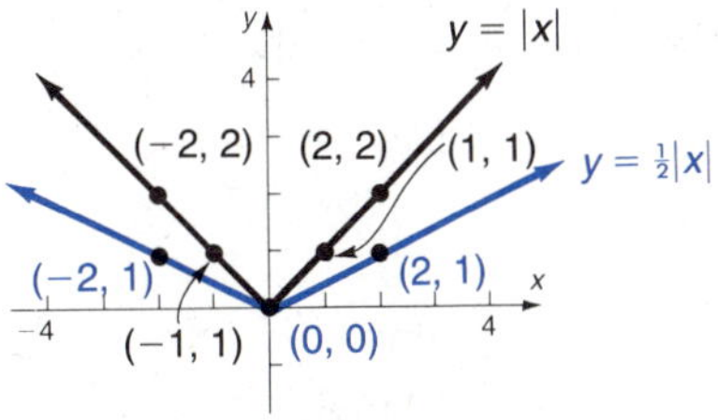

Figure 23

Table 5

x	$y = f(x) = \|x\|$	$y = h(x) = \frac{1}{2}\|x\|$
−2	2	1
−1	1	$\frac{1}{2}$
0	0	0
1	1	$\frac{1}{2}$
2	2	1

■

Compare Figures 22 and 23. Notice how the graph of $f(x) = |x|$ has been vertically "stretched" to obtain the graph of $g(x) = 3|x|$, while it has been vertically "compressed" to obtain the graph of $h(x) = \frac{1}{2}|x|$. To put it another way, for a given x, the graph of $g(x) = 3|x|$ has larger y values than the graph of $f(x) = |x|$, while the graph of $h(x) = \frac{1}{2}|x|$ has smaller y values than that of f.

If the argument x of a function $y = f(x)$ is multiplied by a positive number k, the graph of the new function $y = f(kx)$ is also a "compressed" or "stretched" version of the graph of $y = f(x)$, but, in this case, it occurs horizontally rather than vertically. To see why, we look at the following example.

Example 8 Use the graph of $f(x) = \sqrt{x}$ to obtain the graph of $g(x) = \sqrt{2x}$.

Solution The graph of f is familiar to us, and is shown in Figure 24. Now, we list some points on the graphs of f and g in Table 6. As the table indicates, the graph of g increases more rapidly than that of f. That is, the graph of g is compressed horizontally toward the y-axis. See Figure 24.

Table 6

x	$y = f(x) = \sqrt{x}$	$y = g(x) = \sqrt{2x}$
0	0	0
1	1	$\sqrt{2}$
2	$\sqrt{2}$	2
4	2	$2\sqrt{2}$

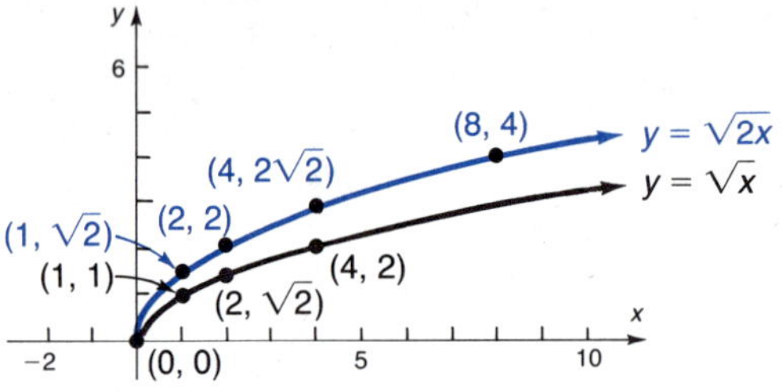

Figure 24

Note that since $\sqrt{2x} = \sqrt{2}\sqrt{x}$, the graph of $g(x) = \sqrt{2x}$ also can be viewed as a vertical "stretch" of the graph of $f(x) = \sqrt{x}$. ■

Reflection about the x-Axis

When the right side of the equation $y = f(x)$ is multiplied by -1, the graph of the new function $y = -f(x)$ is the **reflection** about the x-axis of the graph of the function $y = f(x)$.

Example 9 Graph the function: $f(x) = -x^2$

Solution We begin with the graph of $y = x^2$, as shown in Figure 25. For each point (x, y) on the graph of $y = x^2$, the point $(x, -y)$ is on the graph of $y = -x^2$, as indicated in Table 7. Thus, we can draw the graph of $y = -x^2$ by reflecting the graph of $y = x^2$ about the x-axis. See Figure 25.

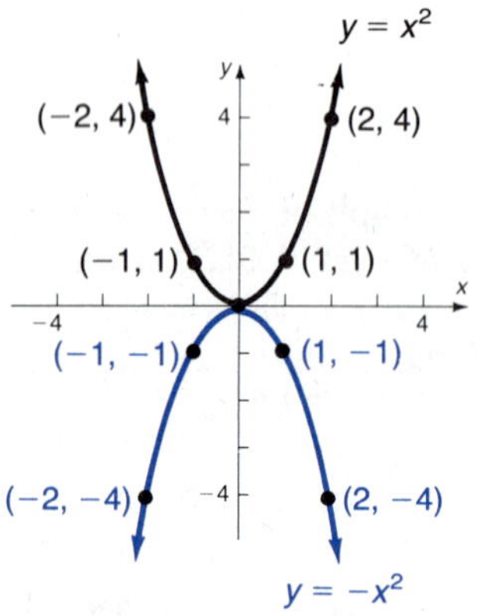

Figure 25

Table 7

x	$y = x^2$	$y = -x^2$
−2	4	−4
−1	1	−1
0	0	0
1	1	−1
2	4	−4

■

Reflection about the y-Axis

When the graph of the function $y = f(x)$ is known, the graph of the new function $y = f(-x)$ is the reflection about the y-axis of the graph of the function $y = f(x)$.

Example 10 Graph the function: $f(x) = \sqrt{-x}$

Solution First, notice that the domain of f consists of all real numbers x for which $-x \geq 0$, or for which $x \leq 0$. To get the graph of $f(x) = \sqrt{-x}$, we begin with the graph of $y = \sqrt{x}$, as shown in Figure 26. For each point (x, y) on the graph of $y = \sqrt{x}$, the point $(-x, y)$ is on the graph of $y = \sqrt{-x}$. Thus, we get the graph of $y = \sqrt{-x}$ by reflecting the graph of $y = \sqrt{x}$ about the y-axis. See Figure 26.

Figure 26

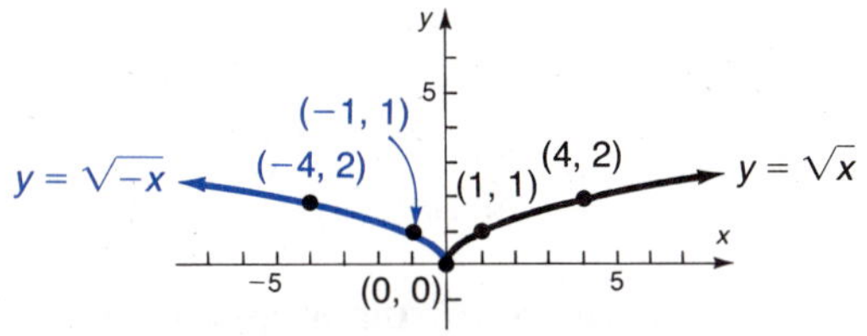

■

The examples that follow combine some of the procedures outlined in this section to get the required graph.

Example 11 Graph the function: $f(x) = \dfrac{3}{x-2} + 1$

Solution We use the following steps to obtain the graph of f:

STEP 1: $y = \dfrac{1}{x}$ Reciprocal function

STEP 2: $y = \dfrac{3}{x}$ Vertical "stretch" of the graph of $y = \dfrac{1}{x}$

STEP 3: $y = \dfrac{3}{x-2}$ Horizontal shift to the right 2 units

STEP 4: $y = \dfrac{3}{x-2} + 1$ Vertical shift up 1 unit

See Figure 27.

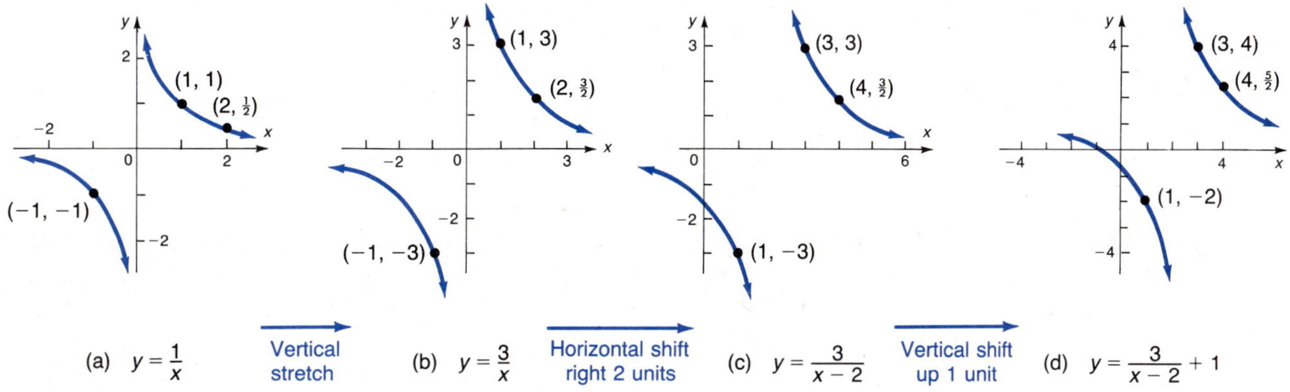

Figure 27

■

There are other orderings of the steps shown in Example 11 that would also result in the graph of f. For example, try this one:

STEP 1: $y = \dfrac{1}{x}$ Reciprocal function

STEP 2: $y = \dfrac{1}{x-2}$ Horizontal shift to the right 2 units

STEP 3: $y = \dfrac{3}{x-2}$ Vertical "stretch" of the graph of $y = \dfrac{1}{x-2}$

STEP 4: $y = \dfrac{3}{x-2} + 1$ Vertical shift up 1 unit

Example 12 Graph the function: $f(x) = \sqrt{1-x} + 2$

Solution We use the following steps to get the graph of $y = \sqrt{1-x} + 2$:

STEP 1: $y = \sqrt{x}$ Square root function

STEP 2: $y = \sqrt{-x}$ Reflect about y-axis

STEP 3: $y = \sqrt{1-x}$ Replace x by $x - 1$ or $-x$ by $-(x-1) = 1 - x$; horizontal shift right 1 unit

STEP 4: $y = \sqrt{1-x} + 2$ Vertical shift up 2 units

See Figure 28.

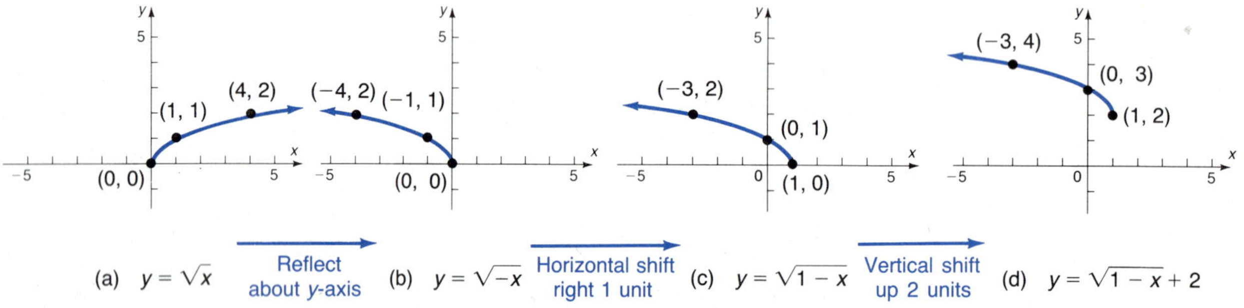

Figure 28

■

A different ordering of the steps used to solve Example 12 is given below:

STEP 1: $y = \sqrt{x}$ Square root function

STEP 2: $y = \sqrt{x+1}$ Replace x by $x + 1$; horizontal shift left 1 unit

STEP 3: $y = \sqrt{-x+1} = \sqrt{1-x}$ Replace x by $-x$; reflect about y-axis

STEP 4: $y = \sqrt{1-x} + 2$ Vertical shift up 2 units

Summary of Graphing Techniques

Table 8 summarizes the graphing procedures we have discussed.

Table 8

TO GRAPH:	DRAW THE GRAPH OF f AND:
Vertical shifts	
$y = f(x) + c, \quad c > 0$	Raise the graph of f c units.
$y = f(x) - c, \quad c > 0$	Lower the graph of f c units.
Horizontal shifts	
$y = f(x + c), \quad c > 0$	Shift the graph of f to the left c units.
$y = f(x - c), \quad c > 0$	Shift the graph of f to the right c units.
Compressing or stretching	
$y = kf(x), \quad k > 0$	Compress or stretch the graph of f by a factor of k.
$y = f(kx), \quad k > 0$	
Reflection about the x-axis	
$y = -f(x)$	Reflect the graph of f about the x-axis.
Reflection about the y-axis	
$y = f(-x)$	Reflect the graph of f about the y-axis.

EXERCISE 4.3

In Problems 1–30, graph each function using the techniques of shifting, compressing, stretching, and/or reflecting.

1. $f(x) = x^2 - 1$
2. $f(x) = x^2 + 4$
3. $g(x) = x^3 + 1$
4. $g(x) = x^3 - 1$
5. $h(x) = \sqrt{x - 2}$
6. $h(x) = \sqrt{x + 1}$
7. $f(x) = (x - 1)^3$
8. $f(x) = (x + 2)^3$
9. $g(x) = 4\sqrt{x}$
10. $g(x) = \frac{1}{2}\sqrt{x}$
11. $h(x) = \dfrac{1}{2x}$
12. $h(x) = \dfrac{4}{x}$
13. $f(x) = -|x|$
14. $f(x) = -\sqrt{x}$
15. $g(x) = -\dfrac{1}{x}$
16. $g(x) = -x^3$
17. $h(x) = [\![-x]\!]$
18. $h(x) = \dfrac{1}{-x}$
19. $f(x) = (x + 1)^2 - 3$
20. $f(x) = (x - 2)^2 + 1$
21. $g(x) = \sqrt{x - 2} + 1$
22. $g(x) = |x + 1| - 3$
23. $h(x) = \sqrt{-x} - 2$
24. $h(x) = \dfrac{4}{x} + 2$
25. $f(x) = (x + 1)^3 - 1$
26. $f(x) = 4\sqrt{x - 1}$
27. $g(x) = 2|1 - x|$
28. $g(x) = 4\sqrt{2 - x}$
29. $h(x) = 2[\![x - 1]\!]$
30. $h(x) = -x^3 + 2$

In Problems 31–36, the graph of a function f is illustrated. Use the graph of f as the first step toward graphing each of the following functions:

(a) $F(x) = f(x) + 3$ (b) $G(x) = f(x + 2)$ (c) $P(x) = -f(x)$

(d) $Q(x) = \frac{1}{2}f(x)$ (e) $g(x) = f(-x)$ (f) $h(x) = 3f(x)$

31.

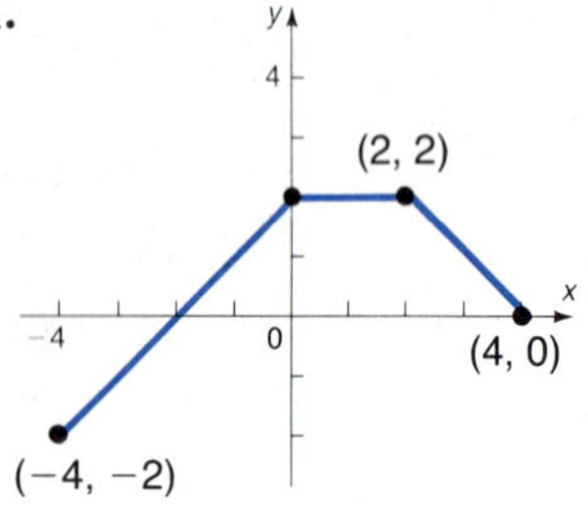

32.

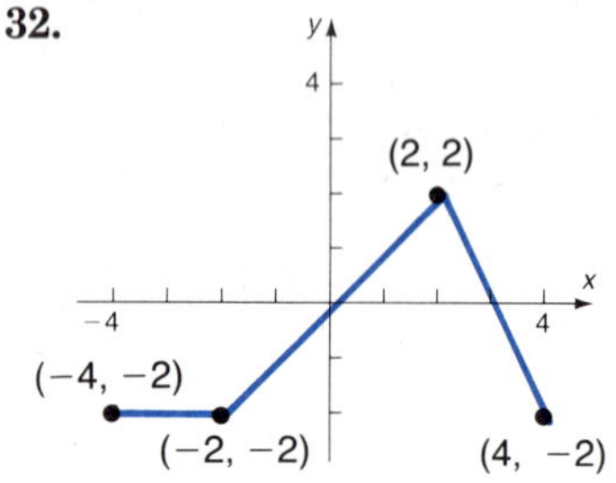

33.

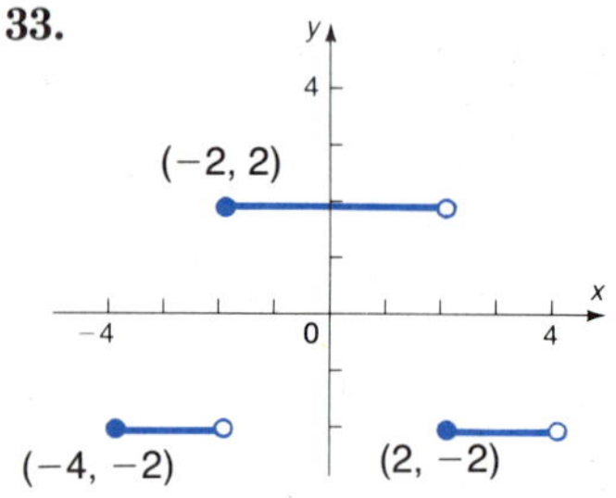

34.

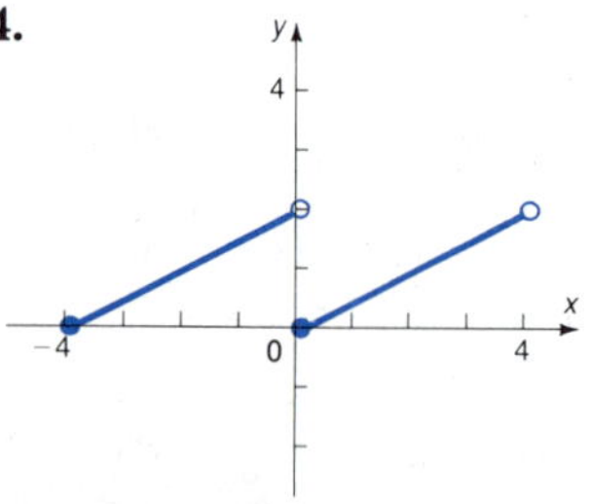

35.

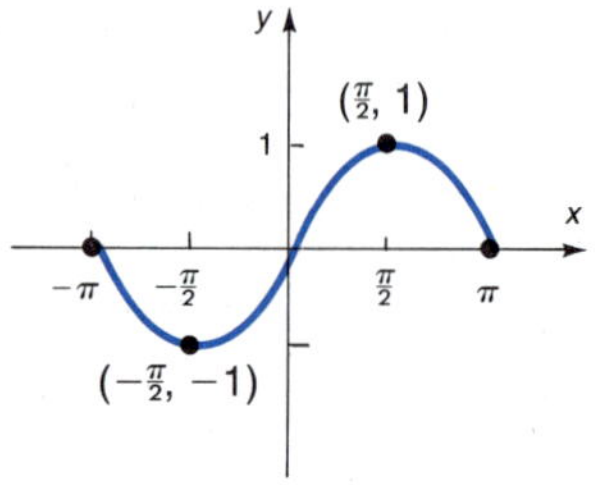

36.

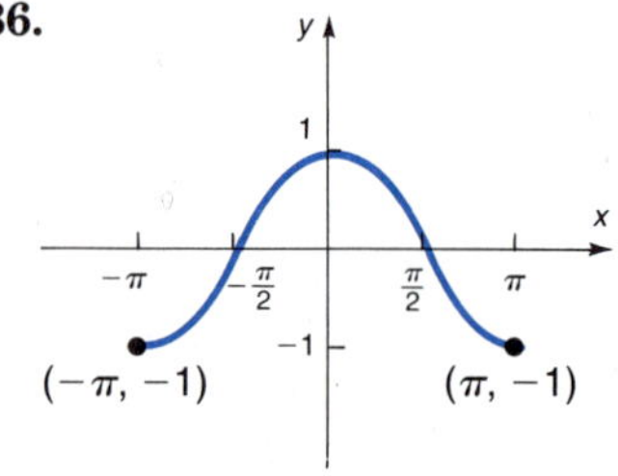

In Problems 37–42, complete the square of each quadratic expression. Then graph each function using the technique of shifting.

37. $f(x) = x^2 + 2x$ **38.** $f(x) = x^2 - 6x$ **39.** $f(x) = x^2 - 8x + 1$

40. $f(x) = x^2 + 4x + 2$ **41.** $f(x) = x^2 + x + 1$ **42.** $f(x) = x^2 - x + 1$

43. The equation $y = (x - c)^2$ defines a *family of parabolas*, one parabola for each value of c. On one set of coordinate axes, graph the members of the family for $c = 0$, $c = 3$, $c = -2$.

44. Repeat Problem 43 for the family of parabolas $y = x^2 + c$.

45. The relationship between Celsius (°C) and Fahrenheit (°F) for measuring temperature is given by the equation

$$F = \tfrac{9}{5}C + 32$$

The relationship between Celsius (°C) and Kelvin (K) is $K = C + 273$. Graph the equation $F = \frac{9}{5}C + 32$, using degrees Fahrenheit on the y-axis and degrees Celsius on the x-axis. Use the techniques introduced in this section to obtain the graph showing the relationship between Kelvin and Fahrenheit temperatures.

C **46.** The period T (in seconds) of a simple pendulum is a function of its length l (in feet) defined by the equation

$$T = 2\pi\sqrt{\frac{l}{g}}$$

where $g \approx 32.2$ feet per second per second is the acceleration of gravity. Graph this function using T on the y-axis and l on the x-axis. On the same coordinate axes, graph the following functions:

(a) $T = 2\pi\sqrt{\dfrac{l+2}{g}}$ (b) $T = 2\pi\sqrt{\dfrac{4l}{g}}$

Discuss how changes in the length l affect the period T of the pendulum.

47. The graph of a function f is illustrated at the right.
(a) Draw the entire graph of $y = |f(x)|$.
(b) Draw the entire graph of $y = f(|x|)$.

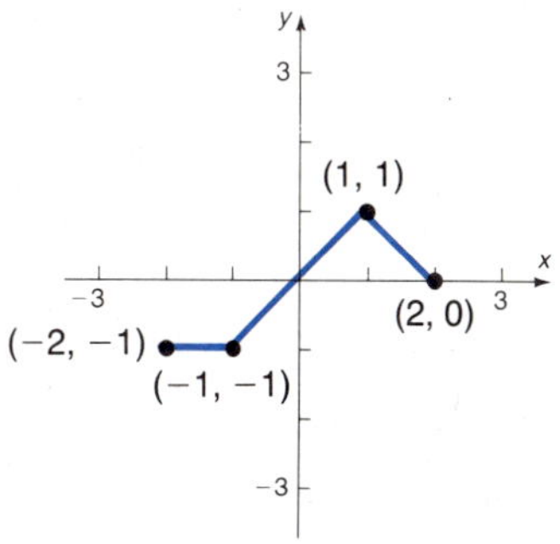

4.4 Operations on Functions; Composite Functions

In this section, we introduce some operations on functions. We shall see that functions, like numbers, can be added, subtracted, multiplied, and divided.

If f and g are functions:

Sum Function

Their **sum $f + g$** is defined by

$$(f+g)(x) = f(x) + g(x)$$

Difference Function

Their **difference $f - g$** is defined by

$$(f-g)(x) = f(x) - g(x)$$

Product Function

Their **product $f \cdot g$** is defined by

$$(f \cdot g)(x) = f(x) \cdot g(x)$$

Quotient Function

Their **quotient f/g** is defined by

$$\left(\frac{f}{g}\right)(x) = \frac{f(x)}{g(x)}$$

In each case, the domain of the resulting function consists of the numbers x that are common to the domains of f and g, but the numbers x for which $g(x) = 0$ must be excluded from the domain of the quotient f/g.

Thus, the sum function, $f + g$, is defined as the sum of the values of the functions f and g, and so on.

Example 1 Let f and g be two functions defined as

$$f(x) = \sqrt{x + 2} \quad \text{and} \quad g(x) = \sqrt{x - 3}$$

Find the following, and determine the domain in each case:

(a) $(f + g)(x)$ (b) $(f - g)(x)$ (c) $(f \cdot g)(x)$ (d) $(f/g)(x)$

Solution (a) $(f + g)(x) = f(x) + g(x) = \sqrt{x + 2} + \sqrt{x - 3}$

(b) $(f - g)(x) = f(x) - g(x) = \sqrt{x + 2} - \sqrt{x - 3}$

(c) $(f \cdot g)(x) = f(x) \cdot g(x) = (\sqrt{x + 2})(\sqrt{x - 3}) = \sqrt{(x + 2)(x - 3)}$

(d) $\left(\frac{f}{g}\right)(x) = \frac{f(x)}{g(x)} = \frac{\sqrt{x + 2}}{\sqrt{x - 3}} = \sqrt{\frac{x + 2}{x - 3}}$

The domain of f consists of all numbers x for which $x \geq -2$; the domain of g consists of all numbers x for which $x \geq 3$. The numbers x common to both these domains are those for which $x \geq 3$. As a result, the numbers x for which $x \geq 3$ comprise the domain of the sum function $f + g$, the difference function $f - g$, and the product function $f \cdot g$. For the quotient function f/g, we must exclude from this set the number 3, because the denominator, g, has the value 0 when $x = 3$. Thus, the domain of f/g consists of all x for which $x > 3$. ■

It is sometimes helpful to view a complicated function as the sum, difference, product, or quotient of simpler functions. For example:

$F(x) = x^2 + \sqrt{x}$ is the sum of $f(x) = x^2$ and $g(x) = \sqrt{x}$.

$H(x) = (x^2 - 1)/(x^2 + 1)$ is the quotient of $f(x) = x^2 - 1$ and $g(x) = x^2 + 1$.

One use of this view of functions is to obtain a graph. The next example illustrates this graphing technique when the function to be graphed is the sum of two simpler functions. In this instance, the method used is called **adding y-coordinates**.

Example 2 Graph the function: $F(x) = x + \sqrt{x}$

Solution First, we notice that the domain of F is $x \geq 0$. Next, we graph the two functions $f(x) = x$ and $g(x) = \sqrt{x}$ for $x \geq 0$. See Figures 29(a) and 29(b). To plot a point $(x, F(x))$ on the graph of F, we select a nonnegative number x and add the y-coordinates $f(x)$ and $g(x)$ to get the y-coordinate $F(x) = f(x) + g(x)$. For example, when $x = 1$, then $f(1) = 1$, $g(1) = 1$, and $F(1) = f(1) + g(1) = 1 + 1 = 2$. When $x = 4$, then $f(4) = 4$, $g(4) = 2$, and $F(4) = f(4) + g(4) = 4 + 2 = 6$; and so on. See Table 9. Figure 29(c) illustrates the graph of F.

Table 9

x	$f(x) = x$	$g(x) = \sqrt{x}$	$F(x) = x + \sqrt{x}$
0	0	0	0
1	1	1	2
4	4	2	6

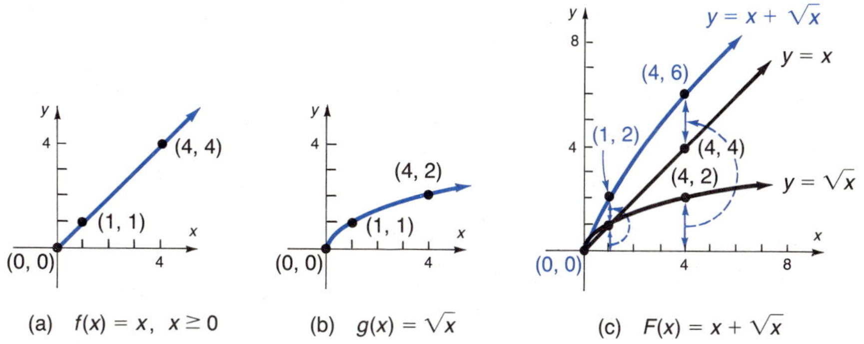

Figure 29

Composite Functions

Consider the function $y = (2x + 3)^2$. If we write $y = f(u) = u^2$ and $u = g(x) = 2x + 3$, then, by a substitution process, we can obtain the original function, namely, $y = f(u) = f(g(x)) = (2x + 3)^2$. This process is called **composition**. In general, suppose that f and g are two functions, and suppose that x is a number in the domain of g. By evaluating g at x, we get $g(x)$. If $g(x)$ is in the domain of f, then we may evaluate f at $g(x)$ and thereby obtain the expression $f(g(x))$. If we do this for all x such that x is in the domain of g and $g(x)$ is in the domain of f, the resulting correspondence from x to $f(g(x))$ is called a *composite function*. See Figure 30.

Figure 30

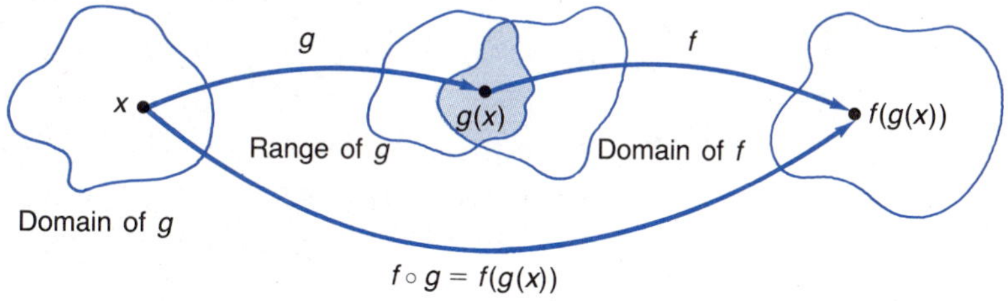

Composite Function

Given the two functions f and g, the **composite function**, denoted by $f \circ g$ (read as "f composed with g"), is defined by

$$(f \circ g)(x) = f(g(x))$$

where the domain of $f \circ g$ is the set of all numbers x in the domain of g such that $g(x)$ is in the domain of f.

Figure 31 (on the next page) provides a second illustration of the definition. Notice that the "inside" function g in $f(g(x))$ is done first.

Figure 31
$f \circ g$

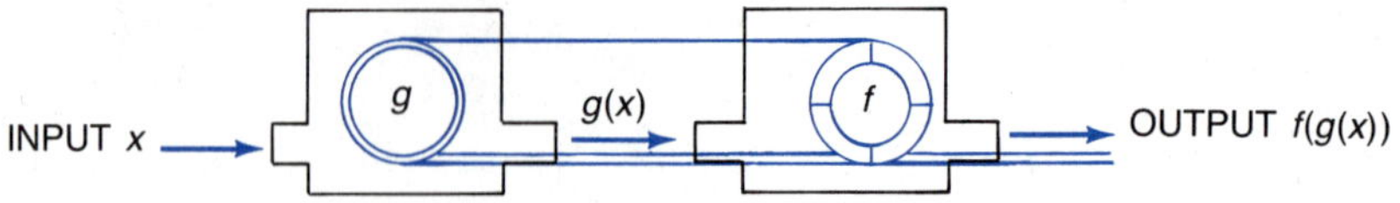

Let's look at some examples.

Example 3 Suppose $f(x) = 2x^2 - 3$ and $g(x) = 4x$. Find:

(a) $(f \circ g)(1)$ (b) $(g \circ f)(1)$ (c) $(f \circ f)(-2)$ (d) $(g \circ g)(-1)$

Solution (a) $(f \circ g)(1) = f(g(1)) = f(4) = 2 \cdot 16 - 3 = 29$

$g(x) = 4x$ $\quad f(x) = 2x^2 - 3$
$g(1) = 4$

(b) $(g \circ f)(1) = g(f(1)) = g(-1) = 4 \cdot (-1) = -4$

$f(x) = 2x^2 - 3$ $\quad g(x) = 4x$
$f(1) = -1$

(c) $(f \circ f)(-2) = f(f(-2)) = f(5) = 2 \cdot 25 - 3 = 47$

$f(-2) = 5$

(d) $(g \circ g)(-1) = g(g(-1)) = g(-4) = 4 \cdot (-4) = -16$

$g(-1) = -4$ ■

Example 4 Suppose $f(x) = \sqrt{x}$ and $g(x) = x^3 - 1$. Find the following composite functions, and then find the domain of each composite function:

(a) $f \circ g$ (b) $g \circ f$ (c) $f \circ f$ (d) $g \circ g$

Solution (a)
$$(f \circ g)(x) = f(g(x)) = f(x^3 - 1) = \sqrt{x^3 - 1}$$

The domain of $f \circ g$ is the interval $[1, \infty)$, which is found by determining those x in the domain of g for which $x^3 - 1 \geq 0$.

(b)
$$(g \circ f)(x) = g(f(x)) = g(\sqrt{x}) = (\sqrt{x})^3 - 1 = x^{3/2} - 1$$

The domain of $g \circ f$ is $[0, \infty)$.

(c)
$$(f \circ f)(x) = f(f(x)) = f(\sqrt{x}) = \sqrt{\sqrt{x}} = \sqrt[4]{x}$$

The domain of $f \circ f$ is $[0, \infty)$.

(d)
$$(g \circ g)(x) = g(g(x)) = g(x^3 - 1) = (x^3 - 1)^3 - 1$$

The domain of $g \circ g$ is the set of all real numbers. ■

Examples 4(a) and 4(b) illustrate that, in general, $f \circ g \neq g \circ f$. However, sometimes $f \circ g$ does equal $g \circ f$, as shown in the next example.

Example 5 If $f(x) = 3x - 4$ and $g(x) = \frac{1}{3}(x + 4)$, show that $(f \circ g)(x) = (g \circ f)(x) = x$ for every x.

Solution

$$(f \circ g)(x) = f(g(x))$$
$$= f\left(\frac{x+4}{3}\right) \qquad g(x) = \frac{1}{3}(x + 4) = \frac{x+4}{3}$$
$$= 3\left(\frac{x+4}{3}\right) - 4 \qquad \text{Substitute } g(x) \text{ into the rule for } f,\ f(x) = 3x - 4.$$
$$= x + 4 - 4 = x$$

$$(g \circ f)(x) = g(f(x))$$
$$= g(3x - 4) \qquad f(x) = 3x - 4$$
$$= \tfrac{1}{3}[(3x - 4) + 4] \qquad \text{Substitute } f(x) \text{ into the rule for } g,\ g(x) = \tfrac{1}{3}(x + 4).$$
$$= \tfrac{1}{3}(3x) = x$$

Thus, $(f \circ g)(x) = (g \circ f)(x) = x$. ■

In the next section, we shall see that there is an important relationship between functions f and g for which $(f \circ g)(x) = (g \circ f)(x) = x$.

Calculus Application

Some techniques in calculus require that we be able to determine the components of a composite function. For example, the function $H(x) = \sqrt{x + 1}$ is the composition of the functions f and g, where $f(x) = \sqrt{x}$ and $g(x) = x + 1$, because $H(x) = (f \circ g)(x) = f(g(x)) = f(x + 1) = \sqrt{x + 1}$.

Example 6 Find functions f and g such that $f \circ g = H$ if $H(x) = (x^2 + 1)^{50}$.

Solution The function H takes $x^2 + 1$ and raises it to the power 50. A natural choice (there are others) for f is, therefore, to raise x to the power 50. The choice of g is to square x and add 1. In other words, we let $f(x) = x^{50}$ and $g(x) = x^2 + 1$. Then,

$$(f \circ g)(x) = f(g(x)) = f(x^2 + 1) = (x^2 + 1)^{50} = H(x)$$

See Figure 32. ■

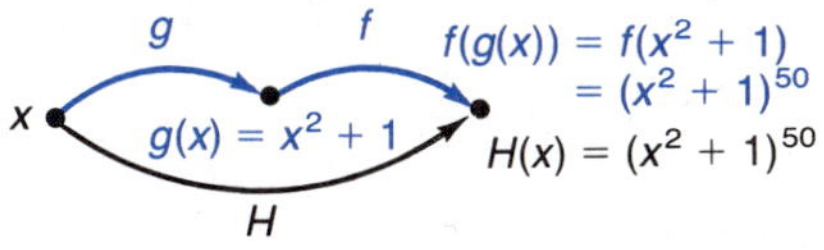

Figure 32

Other functions f and g may be found for which $f \circ g = H$ in Example 6. For example, if $f(x) = x^2$ and $g(x) = (x^2 + 1)^{25}$, then

$$(f \circ g)(x) = f(g(x)) = f((x^2 + 1)^{25}) = [(x^2 + 1)^{25}]^2 = (x^2 + 1)^{50}$$

Thus, although the functions f and g found as a solution to Example 6 are not unique, there is usually a "natural" selection for f and g that comes to mind first.

This natural selection usually will enable you to use your calculator most efficiently. Let's look again at Example 6. To calculate the value of H at, say, 2, we proceed as follows:

	Enter 2	Press x^2	Press +	Enter 1
Display:	2	4		1
	Press =	Press y^x	Enter 50	Press =
Display:	5		50	8.8818 E34

Example 7 Find functions f and g such that $f \circ g = H$ if $H(x) = 1/(x + 1)$.

Solution If we let $f(x) = 1/x$ and $g(x) = x + 1$, we find that

$$(f \circ g)(x) = f(g(x)) = f(x + 1) = \frac{1}{x + 1} = H(x)$$ ■

EXERCISE 4.4

In Problems 1–10, for the given functions f and g, find the functions below, and state the domain of each:

(a) $f + g$ (b) $f - g$ (c) $f \cdot g$ (d) f/g

1. $f(x) = 3x - 4$; $g(x) = 2x + 3$

2. $f(x) = 2x - 1$; $g(x) = 3x + 2$

3. $f(x) = x - 1$; $g(x) = 2x^2$

4. $f(x) = 2x^2 + 3$; $g(x) = 4x^3 + 1$

5. $f(x) = \sqrt{x}$; $g(x) = 3x - 5$

6. $f(x) = |x|$; $g(x) = x$

7. $f(x) = 1 + \dfrac{1}{x}$; $g(x) = \dfrac{1}{x}$

8. $f(x) = 2x^2 - x$; $g(x) = 2x^2 + x$

9. $f(x) = \dfrac{2x + 3}{3x - 2}$; $g(x) = \dfrac{x}{3x - 2}$

10. $f(x) = \sqrt{x - 1}$; $g(x) = \dfrac{1}{x}$

11. Given $f(x) = 3x + 1$ and $(f + g)(x) = 6 - \frac{1}{2}x$, find the function g.

12. Given $f(x) = 1/x$ and $(f/g)(x) = (x + 1)/(x^2 - x)$, find the function g.

In Problems 13–16, use the method of adding y-coordinates to graph each function on the interval [0, 2].

13. $f(x) = |x| + x^2$ **14.** $f(x) = |x| + \sqrt{x}$ **15.** $f(x) = x^3 + x$ **16.** $f(x) = x^3 + x^2$

In Problems 17–26, for the given functions f and g, find:

(a) $(f \circ g)(4)$ (b) $(g \circ f)(2)$ (c) $(f \circ f)(1)$ (d) $(g \circ g)(0)$

17. $f(x) = 2x;\ \ g(x) = 3x^2 + 1$

18. $f(x) = 3x + 2;\ \ g(x) = 2x^2 - 1$

19. $f(x) = 4x^2 - 3;\ \ g(x) = 3 - \frac{1}{2}x^2$

20. $f(x) = 2x^2;\ \ g(x) = 1 - 3x^2$

21. $f(x) = \sqrt{x};\ \ g(x) = 2x$

22. $f(x) = \sqrt{x+1};\ \ g(x) = 3x$

23. $f(x) = |x|;\ \ g(x) = \dfrac{1}{x^2 + 1}$

24. $f(x) = |x - 2|;\ \ g(x) = \dfrac{3}{x^2 + 2}$

25. $f(x) = \dfrac{3}{x^2 + 1};\ \ g(x) = \sqrt{x}$

26. $f(x) = x^3;\ \ g(x) = \dfrac{2}{x^2 + 1}$

In Problems 27–40, for the given functions f and g, find:

(a) $f \circ g$ (b) $g \circ f$ (c) $f \circ f$ (d) $g \circ g$

27. $f(x) = 2x + 1;\ \ g(x) = 3x$

28. $f(x) = -x;\ \ g(x) = 2x - 3$

29. $f(x) = 3x + 1;\ \ g(x) = x^2$

30. $f(x) = \sqrt{x+1};\ \ g(x) = x + 4$

31. $f(x) = \sqrt{x};\ \ g(x) = x^2 - 1$

32. $f(x) = \sqrt{x+1};\ \ g(x) = \dfrac{1}{x^2}$

33. $f(x) = \dfrac{x-1}{x+1};\ \ g(x) = \dfrac{1}{x}$

34. $f(x) = x + \dfrac{1}{x};\ \ g(x) = x^2$

35. $f(x) = x^2;\ \ g(x) = \sqrt{x}$

36. $f(x) = 2x + 4;\ \ g(x) = \frac{1}{2}x - 2$

37. $f(x) = \dfrac{1}{2x+3};\ \ g(x) = 2x + 3$

38. $f(x) = \dfrac{x+1}{x-1};\ \ g(x) = \dfrac{x-1}{x+1}$

39. $f(x) = ax + b;\ \ g(x) = cx + d$

40. $f(x) = \dfrac{ax+b}{cx+d};\ \ g(x) = mx$

In Problems 41–48, show that $(f \circ g)(x) = (g \circ f)(x) = x$.

41. $f(x) = 3x;\ \ g(x) = \frac{1}{3}x$

42. $f(x) = 2x;\ \ g(x) = \frac{1}{2}x$

43. $f(x) = x^3;\ \ g(x) = \sqrt[3]{x}$

44. $f(x) = x + 5;\ \ g(x) = x - 5$

45. $f(x) = 2x - 6;\ \ g(x) = \frac{1}{2}(x + 6)$

46. $f(x) = 4 - 3x;\ \ g(x) = \frac{1}{3}(4 - x)$

47. $f(x) = ax + b;\ \ g(x) = \dfrac{1}{a}(x - b),\ a \neq 0$

48. $f(x) = \dfrac{1}{x};\ \ g(x) = \dfrac{1}{x}$

49. If $f(x) = 2x^3 - 3x^2 + 4x - 1$ and $g(x) = 2$, find $(f \circ g)(x)$ and $(g \circ f)(x)$.

50. If $f(x) = x/(x - 1)$, find $(f \circ f)(x)$.

In Problems 51–54, use $f(x) = x^2$, $g(x) = \sqrt{x} + 2$ *and* $h(x) = 1 - 3x$ *to find the indicated composite function.*

51. $f \circ (g \circ h)$

52. $(f \circ g) \circ h$

53. $(f + g) \circ h$

54. $(f \circ h) + (g \circ h)$

In Problems 55–62, let $f(x) = x^2$, $g(x) = 3x$, *and* $h(x) = \sqrt{x} + 1$. *Express each function as a composite of f, g, and/or h.*

55. $F(x) = 9x^2$

56. $G(x) = 3x^2$

57. $H(x) = |x| + 1$

58. $p(x) = 3\sqrt{x} + 3$

59. $q(x) = x + 2\sqrt{x} + 1$

60. $R(x) = 9x$

61. $P(x) = x^4$

62. $Q(x) = \sqrt{\sqrt{x} + 1} + 1$

In Problems 63–70, find functions f and g so that $f \circ g = H$.

63. $H(x) = (2x + 5)^3$

64. $H(x) = (1 - x^2)^{3/2}$

65. $H(x) = \sqrt{x^2 + x + 1}$

66. $H(x) = \dfrac{1}{1 + x^2}$

67. $H(x) = \left(1 - \dfrac{1}{x^2}\right)^2$

68. $H(x) = |2x^2 + 3|$

69. $H(x) = [\![x^2 + 1]\!]$ **70.** $H(x) = (4 - x^2)^{-4}$

71. The surface area S (in square meters) of a hot air balloon is given by

$$S(r) = 4\pi r^2$$

where r is the radius of the balloon (in meters). If the radius r is increasing with time t (in seconds) according to the formula $r(t) = \frac{2}{3}t^3$, $t \geq 0$, find the surface area S of the balloon as a function of the time t.

72. The volume V (in cubic meters) of the hot air balloon described in Problem 71 is given by $V(r) = \frac{4}{3}\pi r^3$. If the radius r is the same function of t as in Problem 71, find the volume V as a function of the time t.

73. The number N of cars produced at a certain factory in 1 day after t hours of operation is given by $N(t) = 100t - 5t^2$, $0 \leq t \leq 10$. If the cost C (in dollars) of producing x cars is $C(x) = 5000 + 6000x$, find the cost C as a function of the time t of operation of the factory.

4.5 Inverse Functions

We begin the discussion of inverse functions with the idea of a *one-to-one function.*

One-to-One Functions

One-to-One Function

A function f is said to be **one-to-one** if, for any choice of numbers x_1 and x_2, $x_1 \neq x_2$, in the domain of f, then $f(x_1) \neq f(x_2)$.

For a function $y = f(x)$, two distinct ordered pairs (x_1, y_2) and (x_2, y_2) always have different first elements $(x_1 \neq x_2)$. If f is also one-to-one, then the ordered pairs (x_1, y_1) and (x_2, y_2) must also have different second elements $(y_1 \neq y_2)$. In other words, if f is a one-to-one function, then for each x in the domain of f, there is exactly one y in the range, and no y in the range is the image of more than one x in the domain. See Figure 33.

Figure 33

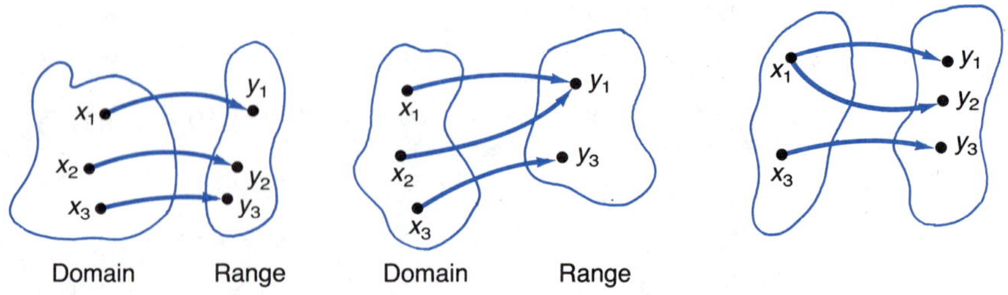

(a) One-to-one function: Each x in the domain has one and only one image in the range

(b) Not a one-to-one function: y_1 is the image of both x_1 and x_2

(c) Not a function: x_1 has two images, y_1 and y_2

Example 1 (a) The function $f(x) = |x|$ is not one-to-one because, for example, the distinct elements 2 and -2 of the domain have the same value 2 in the range. That is, because $f(2) = f(-2)$, the ordered pairs (2, 2) and $(-2, 2)$ have the same second element.

(b) The function $g(x) = 3x$ is one-to-one because all the ordered pairs (x, y), where $y = 3x$, have different second elements. ■

If the graph of a function f is known, there is a simple test, called the **horizontal-line test**, to determine whether f is one-to-one.

Theorem If horizontal lines intersect the graph of a function f in at most one point, then f is one-to-one. ■

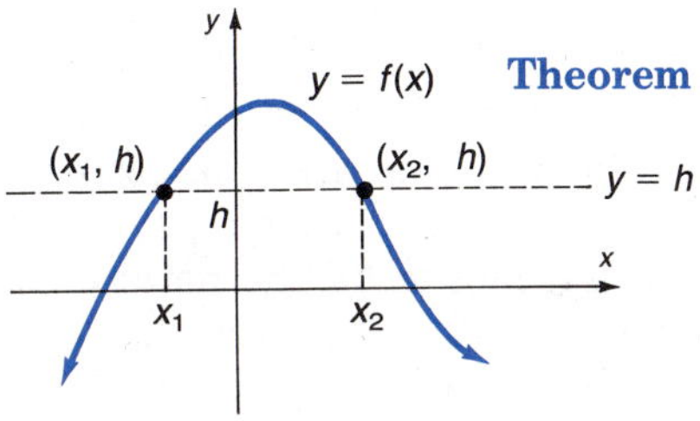

Figure 34
$x_1 \neq x_2$, but $f(x_1) = f(x_2) = h$; f is not a one-to-one function

The reason this test works can be seen in Figure 34, where the horizontal line $y = h$ intersects the graph at two distinct points, (x_1, h) and (x_2, h), with the same second element. Thus, f is not one-to-one.

Example 2 For each given function, use the graph to determine whether the function is one-to-one.

(a) $f(x) = x^2$ (b) $g(x) = x^3$

Solution (a) Figure 35 illustrates the horizontal-line test for $f(x) = x^2$. The horizontal line $y = 1$ meets the graph of f twice, at (1, 1) and at $(-1, 1)$, so f is not one-to-one.

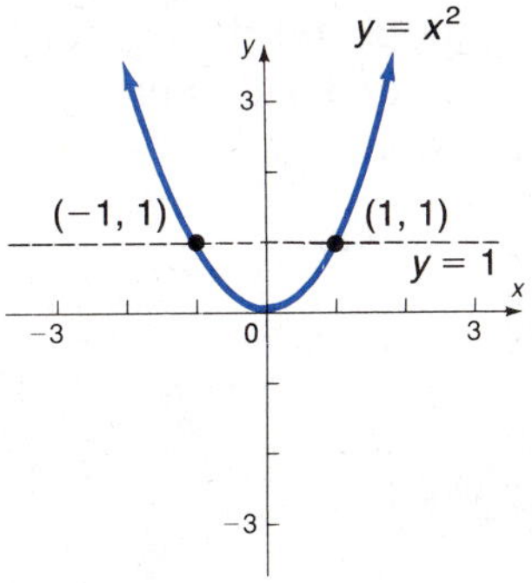

Figure 35
A horizontal line intersects the graph twice; thus, f is not one-to-one

(b) Figure 36 illustrates the horizontal-line test for $g(x) = x^3$. Because each horizontal line will intersect the graph of g exactly once, it follows that g is one-to-one.

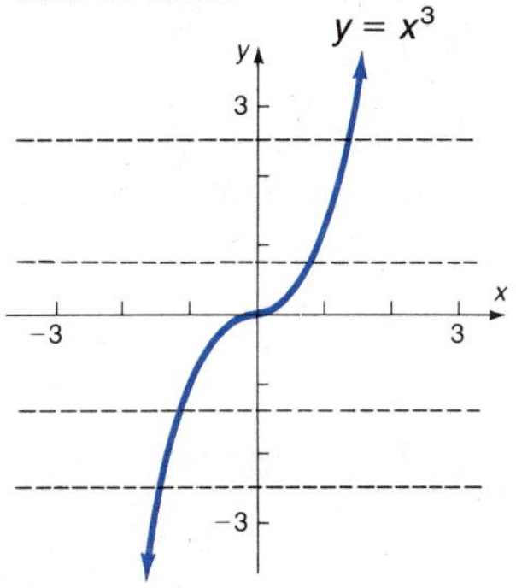

Figure 36
Horizontal lines intersect the graph exactly once; thus, g is one-to-one

■

Let's look more closely at the one-to-one function $g(x) = x^3$. This function is an increasing function. Because an increasing (or decreasing) function will always have different y values for unequal x values, it follows that a function that is increasing (or decreasing) on its domain is also a one-to-one function.

Theorem An increasing (decreasing) function is a one-to-one function. ■

Inverse of a Function

Let f be a one-to-one function defined by $y = f(x)$. Then for each x in its domain, there is exactly one y in its range; furthermore, to each y in the range there corresponds exactly one x in the domain. The correspondence from the range of f onto the domain of f is, therefore, also a function, which we call the *inverse of f* and symbolize by the notation f^{-1}.

Inverse of f

Let f denote a one-to-one function $y = f(x)$. The **inverse of f**, denoted by $\boldsymbol{f^{-1}}$, is a function such that $f^{-1}(f(x)) = x$ for every x in the domain of f and $f(f^{-1}(x)) = x$ for every x in the domain of f^{-1}.

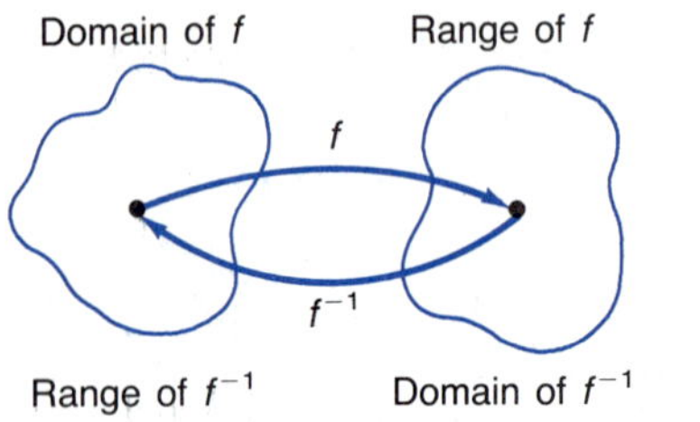

Figure 37

Warning: Be careful! The -1 used in f^{-1} is not an exponent. Thus, f^{-1} does *not* mean the reciprocal of f; it means the inverse of f.

Figure 37 illustrates the definition.

Two facts are now apparent about a function f and its inverse f^{-1}.

Domain of f = Range of f^{-1}	Range of f = Domain of f^{-1}

Look again at Figure 37 to visualize the relationship. If we start with x, apply f, and then apply f^{-1}, we get x back again. If we start with x, apply f^{-1}, and then apply f, we get the number x back again. To put it simply, whatever f does, f^{-1} undoes, and vice versa:

$$\boxed{\text{Input } x} \xrightarrow{\text{Apply } f} \boxed{f(x)} \xrightarrow{\text{Apply } f^{-1}} \boxed{f^{-1}(f(x)) = x}$$

$$\boxed{\text{Input } x} \xrightarrow{\text{Apply } f^{-1}} \boxed{f^{-1}(x)} \xrightarrow{\text{Apply } f} \boxed{f(f^{-1}(x)) = x}$$

In other words,

$$f^{-1}(f(x)) = x \quad \text{and} \quad f(f^{-1})(x)) = x$$

The above conditions can be used to verify that a function is, in fact, the inverse of f, as Example 3 demonstrates.

Example 3 (a) We verify that the inverse of $g(x) = x^3$ is $g^{-1}(x) = \sqrt[3]{x}$ by showing that

$$g^{-1}(g(x)) = g^{-1}(x^3) = \sqrt[3]{x^3} = x$$

and

$$g(g^{-1}(x)) = g(\sqrt[3]{x}) = (\sqrt[3]{x})^3 = x$$

(b) We verify that the inverse of $h(x) = 3x$ is $h^{-1}(x) = \frac{1}{3}x$ by showing that

$$h^{-1}(h(x)) = h^{-1}(3x) = \tfrac{1}{3}(3x) = x$$

and

$$h(h^{-1}(x)) = h\left(\tfrac{1}{3}x\right) = 3\left(\tfrac{1}{3}x\right) = x$$

(c) We verify that the inverse of $f(x) = 2x + 3$ is $f^{-1}(x) = \frac{1}{2}(x - 3)$ by showing that

$$f^{-1}(f(x)) = f^{-1}(2x + 3) = \tfrac{1}{2}[(2x + 3) - 3] = \tfrac{1}{2}(2x) = x$$

and

$$f(f^{-1}(x)) = f\left(\tfrac{1}{2}(x - 3)\right) = 2\left[\tfrac{1}{2}(x - 3)\right] + 3 = (x - 3) + 3 = x \quad ■$$

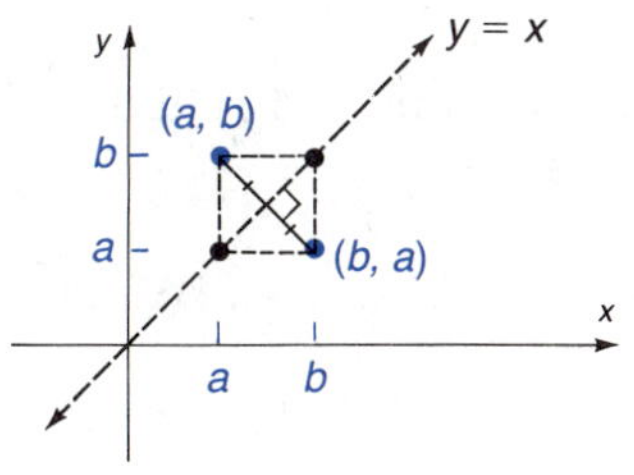

Figure 38

Geometric Interpretation

Suppose (a, b) is a point on the graph of the one-to-one function f defined by $y = f(x)$. Then $b = f(a)$. This means that $a = f^{-1}(b)$, so (b, a) is a point on the graph of the inverse function f^{-1}. The relationship between the point (a, b) on f and the point (b, a) on f^{-1} is shown in Figure 38. Since the line joining (a, b) and (b, a) is perpendicular to the line $y = x$ and is bisected by the line $y = x$, it follows that the point (b, a) on f^{-1} is the reflection about the line $y = x$ of the point (a, b) on f.

Theorem The graph of a function f and the graph of its inverse f^{-1} are symmetric with respect to the line $y = x$. ■

Figure 39 (on the next page) illustrates this result. Notice that once the graph of f is known, the graph of f^{-1} may be obtained by folding the paper along the line $y = x$.

Figure 39

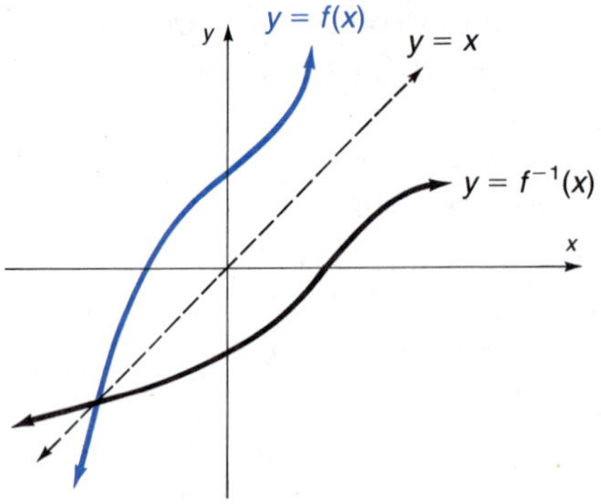

Finding the Inverse Function

The above result tells us more. It says that we can obtain f^{-1} by interchanging the roles of x and y in f. Look again at Figure 39. That is, if f is defined by the equation

$$y = f(x)$$

then f^{-1} is defined by the equation

$$x = f(y)$$

Be careful! The equation $x = f(y)$ defines f^{-1} implicitly. If we can solve this equation for y, we will have the explicit form of f^{-1}, namely,

$$y = f^{-1}(x)$$

Let's use this procedure to find the inverse of $f(x) = 2x + 3$.

Example 4 Find the inverse of $f(x) = 2x + 3$. Also find the domain and range of f and f^{-1}. Graph f and f^{-1} on the same coordinate axes.

Solution In the equation $y = 2x + 3$, interchange the variables x and y. The result,

$$x = 2y + 3$$

is an equation that defines the inverse f^{-1} implicitly. Solving for y, we obtain

$$2y + 3 = x$$
$$2y = x - 3$$
$$y = \tfrac{1}{2}(x - 3)$$

The explicit form of the inverse f^{-1} is therefore

$$f^{-1}(x) = \tfrac{1}{2}(x - 3)$$

which we verified in Example 3(c).

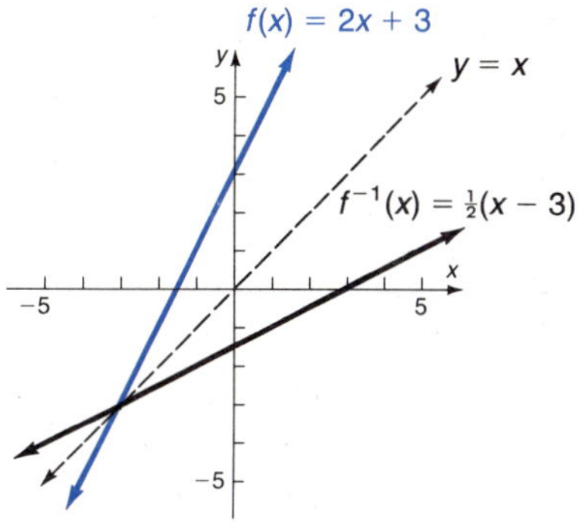

Figure 40

Then we find

$$\text{Domain } f = \text{Range } f^{-1} = (-\infty, \infty)$$
$$\text{Range } f = \text{Domain } f^{-1} = (-\infty, \infty)$$

The graphs of $f(x) = 2x + 3$ and its inverse $f^{-1}(x) = \frac{1}{2}(x - 3)$ are shown in Figure 40. Note the symmetry of the graphs with respect to the line $y = x$. ■

We outline below the steps to follow for finding the inverse of a function:

Procedure for Finding the Inverse Function

STEP 1: Determine whether the function $y = f(x)$ is one-to-one. If f is not one-to-one, then f has no inverse. If f is one-to-one, then f has an inverse, f^{-1}.

STEP 2: In $y = f(x)$, interchange the variables x and y to obtain

$$x = f(y)$$

This equation defines the inverse function f^{-1} implicitly.

STEP 3: If possible, solve the implicit equation for y in terms of x to obtain the explicit form of f^{-1}, namely,

$$y = f^{-1}(x)$$

STEP 4: Verify the result by showing that

$$f^{-1}(f(x)) = x \quad \text{and} \quad f(f^{-1}(x)) = x$$

We have said that if a function is not one-to-one, then it will have no inverse. Sometimes, though, an appropriate restriction on the domain of such a function will yield a new function that is one-to-one. Let's look at an example of this common practice.

Example 5 Find the inverse of $y = f(x) = x^2$ if $x \geq 0$.

Solution The function $f(x) = x^2$ is not one-to-one. [Refer to Example 2(a).] However, if we restrict f to only that part of its domain for which $x \geq 0$, as indicated, we have a new function that is increasing and therefore is one-to-one.

We follow the steps given above to find f^{-1}:

STEP 1: The function defined by $y = x^2$, $x \geq 0$, is one-to-one and, hence, has an inverse, f^{-1}.

STEP 2: In the equation $y = x^2$, $x \geq 0$, interchange the variables x and y. The result is

$$x = y^2 \qquad y \geq 0$$

This equation defines (implicitly) the inverse function.

STEP 3: We solve for y to get the explicit form of the inverse. Since $y \geq 0$, only one solution for y is obtained, namely,

$$y = \sqrt{x}$$

so that $f^{-1}(x) = \sqrt{x}$.

STEP 4: *Check:* $f^{-1}(f(x)) = f^{-1}(x^2) = \sqrt{x^2} = |x| = x$, since $x \geq 0$

$$f(f^{-1}(x)) = f(\sqrt{x}) = (\sqrt{x})^2 = x$$

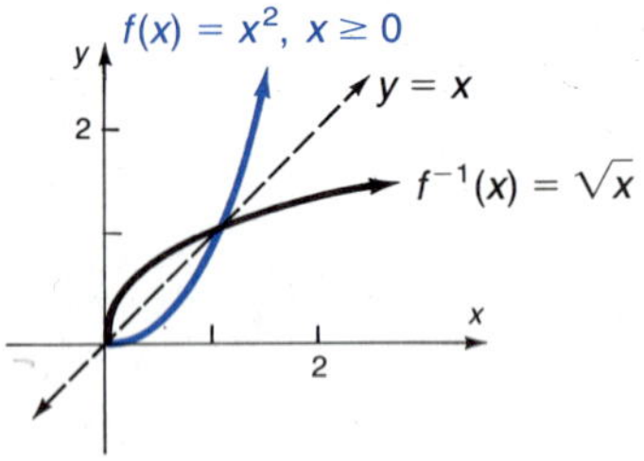

Figure 41

Figure 41 illustrates the graphs of $f(x) = x^2$, $x \geq 0$, and $f^{-1}(x) = \sqrt{x}$. ■

Calculators

We noted earlier that many calculators have keys that allow you to find the value of a function. These same calculators usually have a key labeled [inv] or [inverse] that enables you to calculate the value of the inverse function. (If the actual inverse is present as a function key, such as [$\sqrt{x}$] and [x^2], the inverse key is usually disengaged for such functions.)

Try the following experiment if you have a calculator with an inverse key:

Enter	[1.234]		Press [$\sqrt{x}$]		Press [x^2]		
Display:		1.234		1.1108555		1.234	

Enter	[1.234]		Press [ln x]		Press [inv]		Press [ln x]	
Display:		1.234		0.2102609		0.2102609		1.234

Summary

1. If a function f is one-to-one, then it has an inverse f^{-1}.
2. Domain f = Range f^{-1}; Range f = Domain f^{-1}
3. To verify that f^{-1} is the inverse of f, show that $f^{-1}(f(x)) = x$ and $f(f^{-1}(x)) = x$.
4. The graphs of f and f^{-1} are symmetric with respect to the line $y = x$.

EXERCISE 4.5

In Problems 1–6, the graph of a function f is given. Use the horizontal-line test to determine whether f is one-to-one.

1.

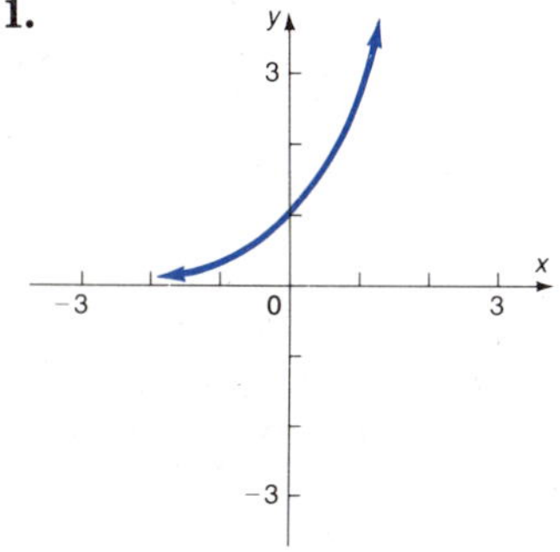

2.

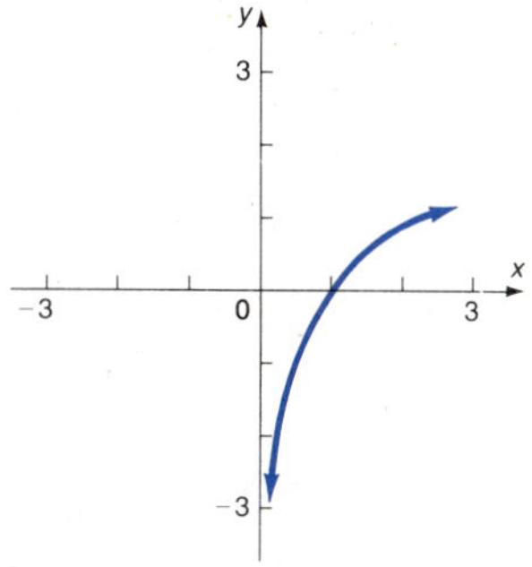

3.

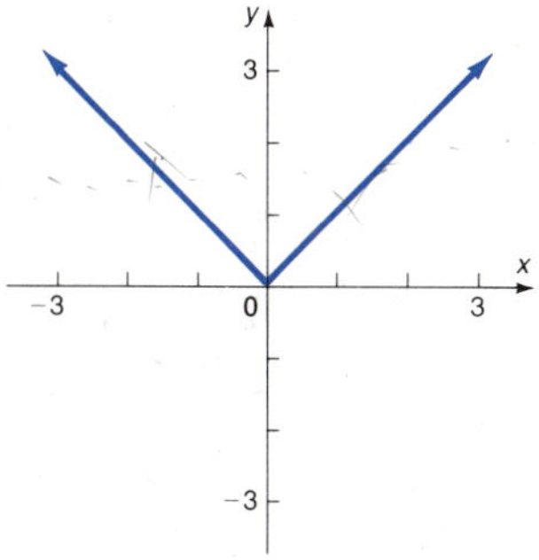

4.

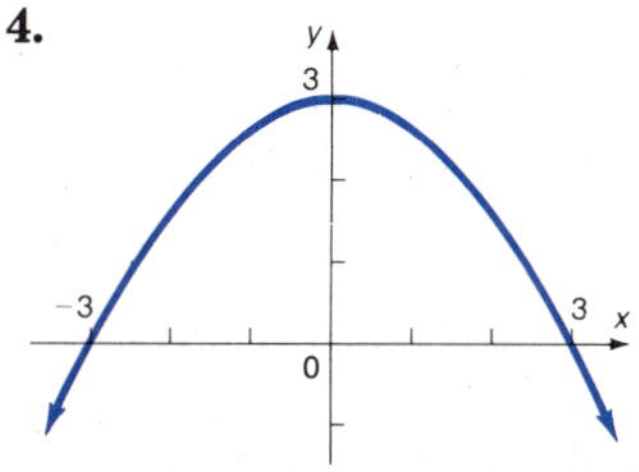

5.

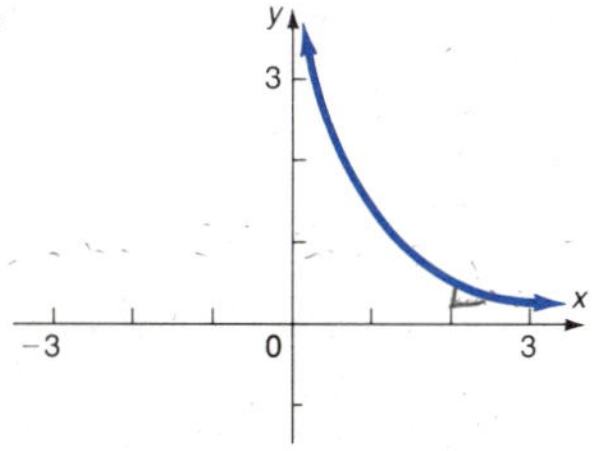

6.

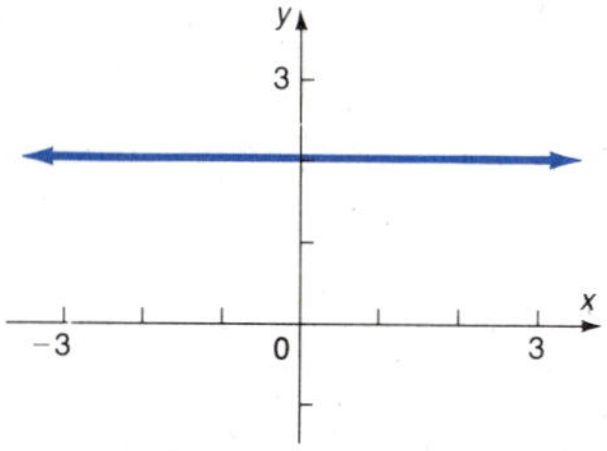

In Problems 7–12, the graph of a one-to-one function f is given. Find the graph of the inverse function f^{-1}. For convenience (and as a hint) the graph of $y = x$ is also given.

7.

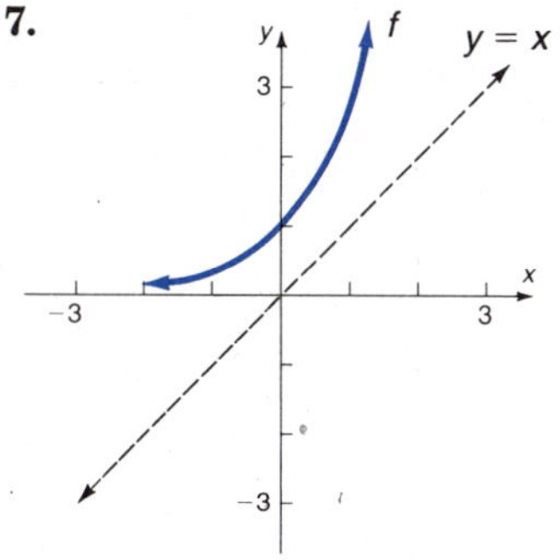

8.

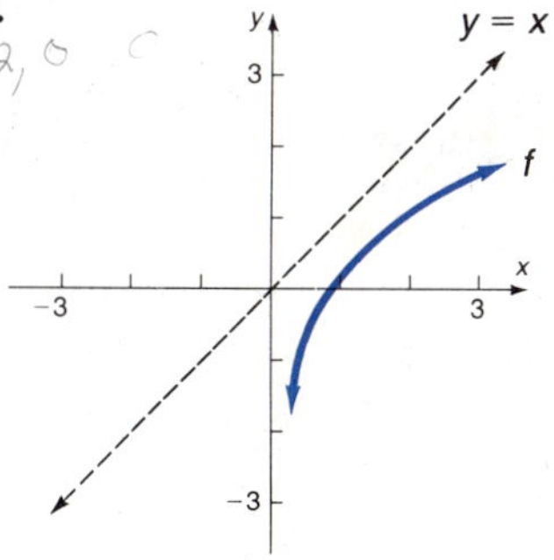

9.

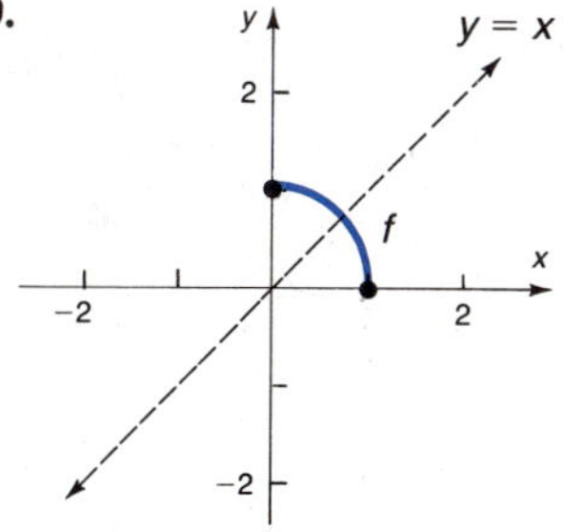

10.

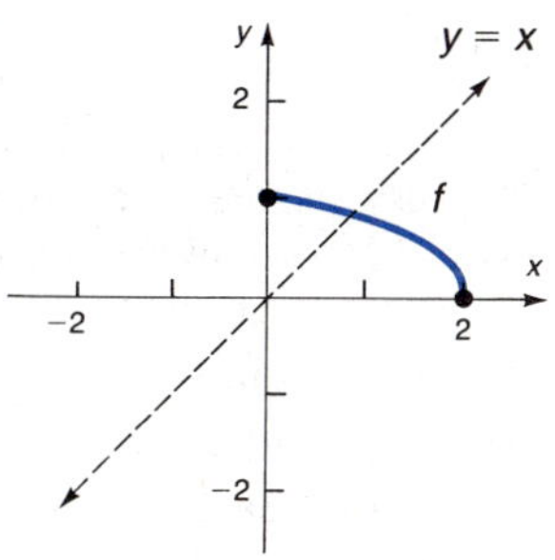

11.

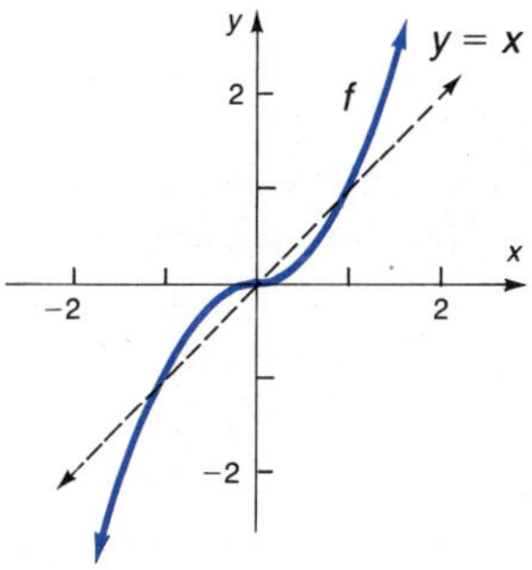

12.

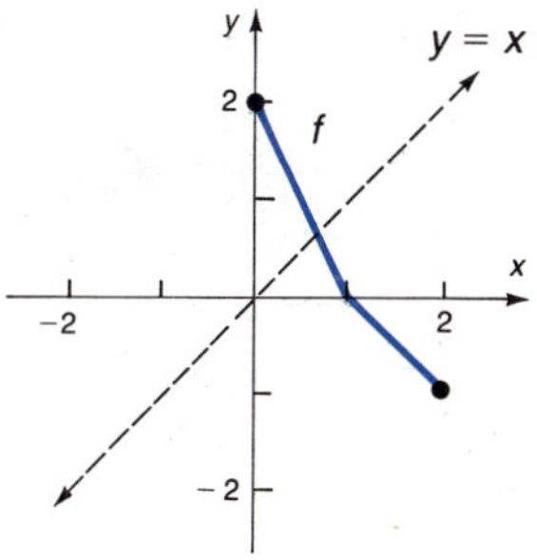

In Problems 13–22, verify that the functions f and g are inverses of each other by showing that $f(g(x)) = x$ and $g(f(x)) = x$.

13. $f(x) = 3x - 4;\ g(x) = \frac{1}{3}(x + 4)$

14. $f(x) = 1 - 2x;\ g(x) = -\frac{1}{2}(x - 1)$

15. $f(x) = 4x - 8;\ g(x) = \frac{x}{4} + 2$

16. $f(x) = 2x + 6;\ g(x) = \frac{1}{2}x - 3$

17. $f(x) = x^3 - 8;\ g(x) = \sqrt[3]{x + 8}$

18. $f(x) = (x - 2)^2,\ x \geq 2;\ g(x) = \sqrt{x} + 2,\ x \geq 0$

19. $f(x) = \frac{1}{x};\ g(x) = \frac{1}{x}$

20. $f(x) = x;\ g(x) = x$

21. $f(x) = \frac{2x + 3}{x + 4};\ g(x) = \frac{4x - 3}{2 - x}$

22. $f(x) = \frac{x - 5}{2x + 3};\ g(x) = \frac{3x + 5}{1 - 2x}$

In Problems 23–34, the function f is one-to-one. Find its inverse, and verify your answer. State the domain and range of f and f^{-1}. Graph f and f^{-1} on the same coordinate axes.

23. $f(x) = 2x$

24. $f(x) = -3x$

25. $f(x) = 4x + 2$

26. $f(x) = 1 - 3x$

27. $f(x) = x^3 - 1$

28. $f(x) = x^3 + 1$

29. $f(x) = x^2 + 4,\ x \geq 0$

30. $f(x) = x^2 + 9,\ x \geq 0$

31. $f(x) = \frac{4}{x}$

32. $f(x) = -\frac{3}{x}$

33. $f(x) = \frac{1}{x - 2}$

34. $f(x) = \frac{4}{x + 2}$

In Problems 35–46, the function f is one-to-one. Find its inverse, and verify your answer. State the domain and range of f and f^{-1}.

35. $f(x) = \frac{1}{3 + x}$

36. $f(x) = \frac{1}{1 - x}$

37. $f(x) = (x + 2)^2,\ x \geq -2$

38. $f(x) = (x - 1)^2,\ x \geq 1$

39. $f(x) = \frac{2x}{x - 1}$

40. $f(x) = \frac{3x + 1}{x}$

41. $f(x) = \frac{3x + 4}{2x - 3}$

42. $f(x) = \frac{2x - 3}{x + 4}$

43. $f(x) = \frac{2x + 3}{x + 2}$

44. $f(x) = \frac{-3x - 4}{x - 2}$

45. $f(x) = 2\sqrt[3]{x}$

46. $f(x) = \frac{4}{\sqrt{x}}$

47. Find the inverse of the linear function $f(x) = mx + b,\ m \neq 0$.

48. Find the inverse of the function $f(x) = \sqrt{r^2 - x^2},\ 0 \leq x \leq r$.

49. Can an even function be one-to-one? Explain.

50. Is every odd function one-to-one? Explain.

51. A function f has an inverse. If the graph of f lies in quadrant I, in which quadrant does the graph of f^{-1} lie?

52. A function f has an inverse. If the graph of f lies in quadrant II, in which quadrant does the graph of f^{-1} lie?

53. The function $f(x) = |x|$ is not one-to-one. Find a suitable restriction on the domain of f so that the new function that results is one-to-one. Then find the inverse of f.

54. The function $f(x) = x^4$ is not one-to-one. Find a suitable restriction on the domain of f so that the new function that results is one-to-one. Then find the inverse of f.

55. To convert from x degrees Celsius to y degrees Fahrenheit, we use the formula $y = f(x) = \frac{9}{5}x + 32$. To convert from x degrees Fahrenheit to y degrees Celsius, we use the formula $y = g(x) = \frac{5}{9}(x - 32)$. Show that f and g are inverse functions.

56. The demand for corn obeys the equation $p(x) = 300 - 50x$, where p is the price per bushel (in dollars) and x is the number of bushels produced, in millions. Express the production amount x as a function of the price p.

57. The period T (in seconds) of a simple pendulum is a function of its length l (in feet), given by $T(l) = 2\pi\sqrt{l/g}$, where $g \approx 32.2$ feet per second per second is the acceleration of gravity. Express the length l as a function of the period T.

58. Give an example of a function whose domain is the set of real numbers and that is neither increasing nor decreasing on its domain, but is one-to-one. [*Hint:* Use a piecewise-defined function.]

59. Given

$$f(x) = \frac{ax + b}{cx + d}$$

find $f^{-1}(x)$. If $c \neq 0$, under what conditions on a, b, c, and d is $f = f^{-1}$?

60. We said earlier that finding the range of a function f is not easy. However, if f is one-to-one, we can find its range by finding the domain of the inverse function f^{-1}. Use this technique to find the range of each of the following one-to-one functions:

(a) $f(x) = \dfrac{2x + 5}{x - 3}$ (b) $g(x) = 4 - \dfrac{2}{x}$ (c) $F(x) = \dfrac{3}{4 - x}$

4.6 Constructing Functions: Applications

The techniques we used in Sections 2.2 and 2.3 to set up equations also can be used to construct functions. Below, we repeat the steps for setting up applied problems from Section 2.2.

Steps for Setting Up Applied Problems

STEP 1: Read the problem carefully, perhaps two or three times. Pay particular attention to the question being asked in order to identify what you are looking for.

STEP 2: Assign some letter (variable) to represent what you are looking for, and, if necessary, express any remaining unknown quantities in terms of this variable.

STEP 3: Make a list of all the known facts, and write down any relationships among them, especially any that involve the variables. These will usually take the form of an equation (or an inequality) involving the variable. If possible, draw an appropriately labeled diagram to assist you. Sometimes, a table or chart helps.

STEP 4: Solve the equation (or inequality) for the variable, and then answer the question asked in the problem.

STEP 5: Check the answer with the facts in the problem. If it agrees, congratulations! If it does not agree, try again.

Example 1 In economics, revenue R is defined as the amount of money derived from the sale of a product and is equal to the unit selling price p of the product times the number x of units actually sold. That is,

$$R = xp$$

Usually, p and x are related—as one increases, the other decreases.

Suppose p and x are related by the following **demand equation**:

$$p = -\tfrac{1}{10}x + 20 \qquad 0 \le x \le 200$$

Express the revenue R as a function of the number x of units sold.

Solution Since $R = xp$ and $p = -\frac{1}{10}x + 20$, it follows that

$$R(x) = xp = x\left(-\tfrac{1}{10}x + 20\right) = -\tfrac{1}{10}x^2 + 20x$$ ■

Example 2 Let $P = (x, y)$ be a point on the graph of $y = x^2 - 1$.

(a) Express the distance d from P to the origin O as a function of x.

(b) What is d if $x = 0$? (c) What is d if $x = 1$?

(d) What is d if $x = \sqrt{2}/2$?

Solution (a) Figure 42 illustrates the graph. The distance d from P to O is

$$d = \sqrt{x^2 + y^2}$$

Since P is a point on the graph of $y = x^2 - 1$, we have

$$d(x) = \sqrt{x^2 + (x^2 - 1)^2} = \sqrt{x^4 - x^2 + 1}$$

Thus, we have expressed the distance d as a function of x.

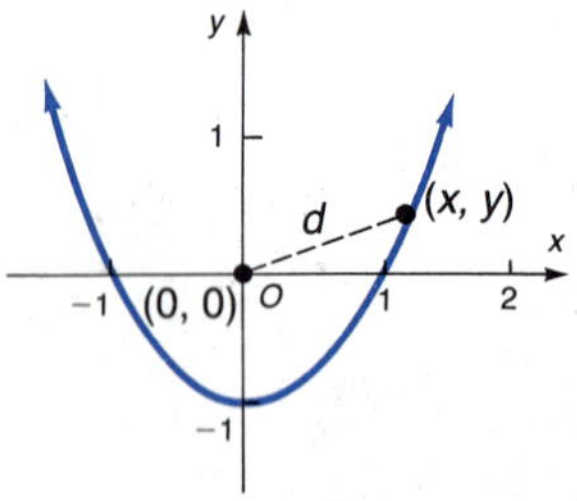

Figure 42
$y = x^2 - 1$

(b) If $x = 0$, the distance d is

$$d(0) = \sqrt{1} = 1$$

(c) If $x = 1$, the distance d is

$$d(1) = \sqrt{1 - 1 + 1} = 1$$

(d) If $x = \sqrt{2}/2$, the distance d is

$$d\left(\frac{\sqrt{2}}{2}\right) = \sqrt{\left(\frac{\sqrt{2}}{2}\right)^4 - \left(\frac{\sqrt{2}}{2}\right)^2 + 1} = \sqrt{\frac{1}{4} - \frac{1}{2} + 1} = \frac{\sqrt{3}}{2}$$ ■

Example 3 A rectangular swimming pool 20 meters long and 10 meters wide is 4 meters deep at one end and 1 meter deep at the other. Figure 43

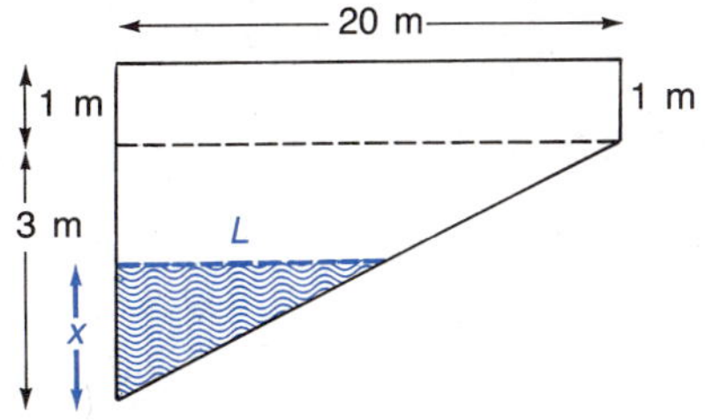

Figure 43

illustrates a cross-sectional view of the pool. Water is being pumped into the pool at the deep end.

(a) Find a function that expresses the volume V of water in the pool as a function of the height x of the water at the deep end.

(b) Find the volume when the height is 1 meter.

(c) Find the volume when the height is 2 meters.

Solution (a) Let L denote the distance (in meters) measured at water level from the deep end to the short end. Notice that L and x form the sides of a triangle that is similar to the triangle whose sides are 20 meters by 3 meters. Thus, L and x are related by the equation

$$\frac{L}{x} = \frac{20}{3} \quad \text{or} \quad L = \frac{20x}{3} \qquad 0 \le x \le 3$$

The volume V of water in the pool at any time is

$$V = \begin{pmatrix}\text{Cross-sectional}\\ \text{triangular area}\end{pmatrix}(\text{Width}) = \left(\tfrac{1}{2}Lx\right)(10) \text{ cubic meters}$$

Since $L = 20x/3$, we have

$$V(x) = \left(\frac{1}{2} \cdot \frac{20x}{3} \cdot x\right)(10) = \frac{100}{3}x^2 \text{ cubic meters}$$

(b) The volume V when the height x is 1 meter is

$$V(1) = \frac{100}{3} \cdot 1^2 = 33.3 \text{ cubic meters}$$

(c) The volume V when the height x is 2 meters is

$$V(2) = \frac{100}{3} \cdot 2^2 = \frac{400}{3} = 133.3 \text{ cubic meters}$$ ■

Example 4 Consider an isosceles triangle of fixed perimeter p.

(a) If x equals the length of one of the two equal sides, express the area A as a function of x.

(b) What is the domain of A?

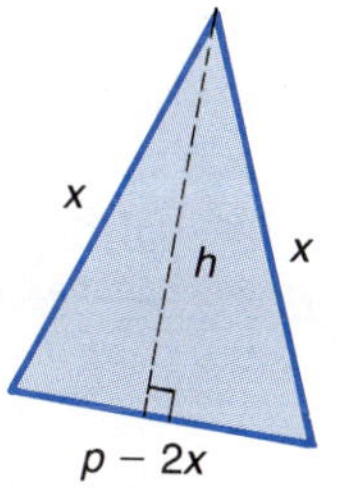

Figure 44

Solution (a) Look at Figure 44. Since the equal sides are of length x, the third side must be of length $p - 2x$. (Do you see why?) We know that the area A is

$$A = \tfrac{1}{2}(\text{Base})(\text{Height})$$

To find the height h, we drop the perpendicular to the base of length $p - 2x$ and use the fact that the perpendicular bisects the base. Then, by the Pythagorean Theorem, we have

$$h^2 = x^2 - \left(\frac{p - 2x}{2}\right)^2 = x^2 - \frac{1}{4}(p^2 - 4px + 4x^2)$$
$$= px - \frac{1}{4}p^2 = \frac{4px - p^2}{4}$$
$$h = \sqrt{\frac{4px - p^2}{4}} = \frac{\sqrt{p}}{2}\sqrt{4x - p}$$

The area A is given by

$$A = \frac{1}{2} \cdot (p - 2x)\frac{\sqrt{p}}{2}\sqrt{4x - p} = \frac{\sqrt{p}}{4}(p - 2x)\sqrt{4x - p}$$

(b) The domain of A is found as follows. Because of the expression $\sqrt{4x - p}$, we require that

$$4x - p > 0$$
$$x > \frac{p}{4}$$

Since $p - 2x$ is a side of the triangle, we also require that

$$p - 2x > 0$$
$$-2x > -p$$
$$x < \frac{p}{2}$$

Thus, the domain of A is $p/4 < x < p/2$, or $(p/4, p/2)$, and we state the function as

$$A(x) = \frac{\sqrt{p}}{4}(p - 2x)\sqrt{4x - p} \qquad \frac{p}{4} < x < \frac{p}{2}$$ ■

Example 5 An island is 2 miles from the nearest point P on a straight shoreline. A town is 12 miles down the shore from P.

(a) If a person can row a boat at an average speed of 5 miles per hour and the same person can walk 8 miles per hour, express the time T it takes to go from the island to town as a function of the distance x from P to where the person lands the boat.

(b) Compute the time if $x = 4$ miles and if $x = 8$ miles.

Solution (a) Figure 45 illustrates the situation. The distance d_1 from the island to the landing point satisfies the equation

$$d_1^2 = 4 + x^2 \tag{1}$$

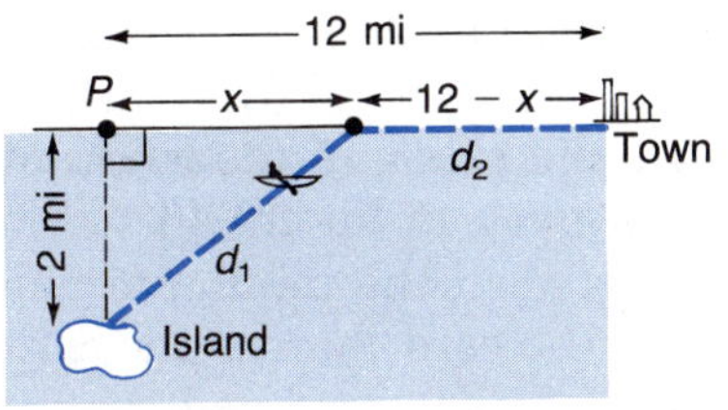

Figure 45

Since the average speed of the boat is 5 miles per hour, the time t_1 it takes to cover the distance d_1 is

$$d_1 = 5t_1$$

Thus, from equation (1),

$$t_1 = \frac{d_1}{5} = \frac{\sqrt{4 + x^2}}{5}$$

The distance d_2 from the landing point to town is $12 - x$, and the time t_2 it takes to cover this distance is

$$d_2 = 8t_2$$

Thus,

$$t_2 = \frac{d_2}{8} = \frac{12 - x}{8}$$

The total time T of the trip is $t_1 + t_2$. Thus,

$$T(x) = \frac{\sqrt{4 + x^2}}{5} + \frac{12 - x}{8}$$

(b) If $x = 4$ miles, the time is

$$T(4) = \frac{\sqrt{20}}{5} + \frac{8}{8} \approx 1.89 \text{ hours}$$

If $x = 8$ miles, the time is

$$T(8) = \frac{\sqrt{68}}{5} + \frac{4}{8} \approx 2.15 \text{ hours}$$

■

Example 6 Holders of credit cards issued by banks, department stores, oil companies, and so on, receive bills each month that state minimum amounts that must be paid by a certain due date. The minimum due depends on the total amount owed. One such credit card company uses the following rules: For a bill of less than \$10, the entire amount is due. For a bill of at least \$10 but less than \$500, the minimum due is \$10. There is a minimum of \$30 due on a bill of at least \$500 but less than \$1000, a minimum of \$50 due on a bill of at least \$1000 but less than \$1500, and a minimum of \$70 due on bills of \$1500 or more. The function f that describes the minimum payment due on a bill of x dollars is

$$f(x) = \begin{cases} x & \text{if } 0 \le x < 10 \\ 10 & \text{if } 10 \le x < 500 \\ 30 & \text{if } 500 \le x < 1000 \\ 50 & \text{if } 1000 \le x < 1500 \\ 70 & \text{if } 1500 \le x \end{cases}$$

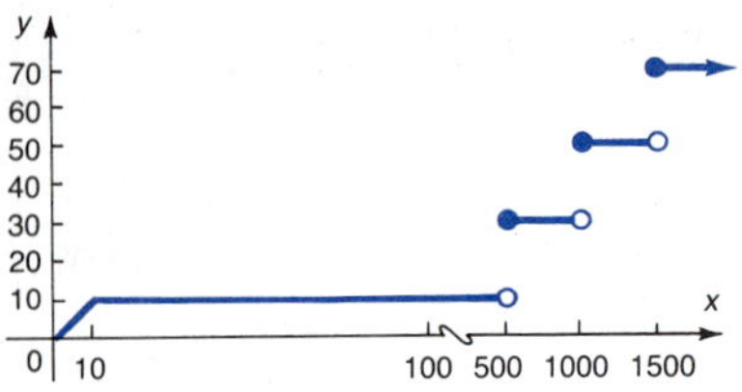

Figure 46

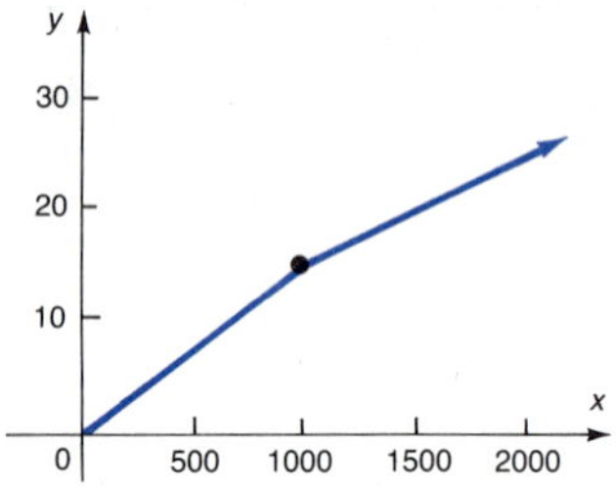

Figure 47

To graph this function f, we proceed as follows: For $0 \le x < 10$, draw the graph of $y = x$; for $10 \le x < 500$, draw the graph of the constant function $y = 10$; for $500 \le x < 1000$, draw the graph of the constant function $y = 30$; and so on. The graph of f is given in Figure 46.

The card holder may pay any amount between the minimum due and the total owed. The organization issuing the card charges the card holder interest of 1.5% per month for the first \$1000 owed and 1% per month on any unpaid balance over \$1000. Thus, if $g(x)$ is the amount of interest charged per month on a balance of x, then $g(x) = 0.015x$ for $0 \le x \le 1000$. The amount of the unpaid balance above \$1000 is $x - 1000$. If the balance due is $x > 1000$, then the interest is $0.015(1000) + 0.01(x - 1000) = 15 + 0.01x - 10 = 5 + 0.01x$, so

$$g(x) = \begin{cases} 0.015x & \text{if } 0 \le x \le 1000 \\ 5 + 0.01x & \text{if } x > 1000 \end{cases}$$

See Figure 47. ■

Example 7 A company that manufactures aluminum cans requires a cylindrical container with a capacity of 500 cubic centimeters $\left(\frac{1}{2}\text{ liter}\right)$. The top and bottom of the can will be made of a special aluminum alloy that costs \$0.05 per square centimeter. The sides of the can are to be made of material that costs \$0.02 per square centimeter. Express the cost of materials for the can as a function of the radius r of the can.

Figure 48

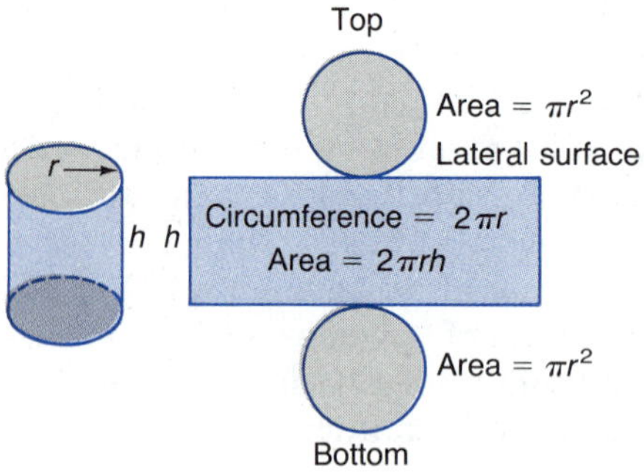

Solution Figure 48 illustrates the situation. Notice that the material required to produce a cylindrical can of height h and radius r consists of a rectangle of area $2\pi rh$ and two circles, each of area πr^2. The total cost C (in cents) of manufacturing the can is therefore

$$\begin{aligned} C &= \text{Cost of top and bottom} + \text{Cost of side} \\ &= \underbrace{2(\pi r^2)}_{\substack{\text{Total area}\\ \text{of top and}\\ \text{bottom}}} \quad \underbrace{(5¢)}_{\substack{\text{Cost/unit}\\ \text{area}}} + \underbrace{(2\pi rh)}_{\substack{\text{Total}\\ \text{area of}\\ \text{side}}} \quad \underbrace{(2¢)}_{\substack{\text{Cost/unit}\\ \text{area}}} \\ &= 10\pi r^2 + 4\pi rh \end{aligned}$$

But we have the additional restriction that the height h and radius r must be chosen so that the volume V of the can is 500 cubic centimeters. Since $V = \pi r^2 h$, we have

$$500 = \pi r^2 h \qquad \text{or} \qquad h = \frac{500}{\pi r^2}$$

Thus, the cost C, in cents, as a function of the radius r is

$$C(r) = 10\pi r^2 + 4\pi r\left(\frac{500}{\pi r^2}\right) = 10\pi r^2 + \frac{2000}{r}$$ ■

Example 8 A manufacturer of children's playpens makes a square model that can be opened at one corner and attached at right angles to a wall or, perhaps, the side of a house. If each side is 3 feet in length, the open configuration doubles the available area in which the child can play from 9 feet to 18 square feet. See Figure 49(a).

Now, suppose we place hinges at the outer corners to allow for a configuration like the one shown in Figure 49(b).

(a) Express the area A of this configuration as a function of the distance x between the two parallel sides.

(b) Find the domain of A. (c) Find A if $x = 5$.

(d) Find A if $x = 5.8$.*

Solution (a) Refer to Figure 49(b). The area A we seek consists of the area of a rectangle (with width x and height 3) and the area of an isosceles triangle (with base x and two equal sides of length 3). The height h of the triangle may be found using the Pythagorean Theorem:

$$h^2 = 3^2 - \left(\frac{x}{2}\right)^2 = 9 - \frac{x^2}{4} = \frac{36 - x^2}{4}$$

$$h = \tfrac{1}{2}\sqrt{36 - x^2}$$

The area A enclosed by the playpen is

$$A = \text{Area of rectangle} + \text{Area of triangle} = 3x + \tfrac{1}{2}x\left(\tfrac{1}{2}\sqrt{36 - x^2}\right)$$

$$A(x) = 3x + \frac{x\sqrt{36 - x^2}}{4}$$

Thus, we have expressed the area A as a function of x.

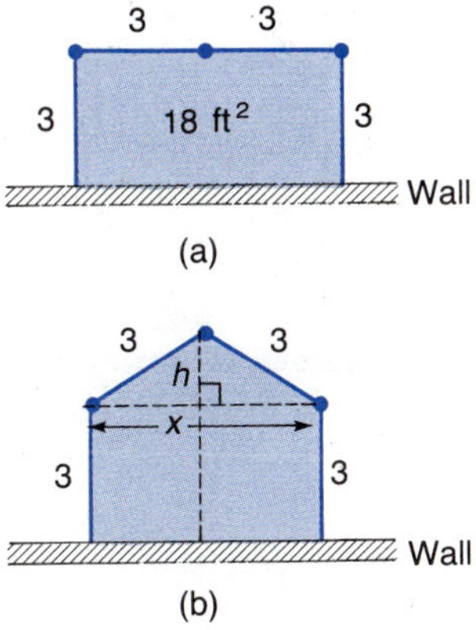

Figure 49

*Adapted from *Proceedings, Summer Conference for College Teachers on Applied Mathematics* (University of Missouri, Rolla), 1971.

(b) To find the domain of A, we note first that $x > 0$, since x is a distance. Also, the expression under the radical must be nonnegative, so

$$36 - x^2 \geq 0$$
$$x^2 \leq 36$$
$$-6 \leq x \leq 6$$

Combining these restrictions, we find that the domain of A is $0 < x \leq 6$, or $(0, 6]$.

(c) If $x = 5$, the area is

$$A(5) = 3(5) + \frac{5}{4}\sqrt{36 - (5)^2} \approx 19.15 \text{ square feet}$$

(d) If $x = 5.8$, the area is

$$A(5.8) = 3(5.8) + \frac{5.8}{4}\sqrt{36 - (5.8)^2} \approx 19.63 \text{ square feet}$$

(Note that both of these values of x provide more area than the 18 square foot rectangular configuration.) ■

Note: Once you know some calculus, you will be able to show that the value of x that provides the most area is $x \approx 5.58$ feet.

EXERCISE 4.6

1. The volume V of a right circular cylinder of height h and radius r is $V = \pi r^2 h$. If the height is twice the radius, express the volume V as a function of r.
2. The volume V of a right circular cone is $V = \frac{1}{3}\pi r^2 h$. If the height is twice the radius, express the volume V as a function of r.
3. The price p and the quantity x sold of a certain product obey the demand equation

$$p = -\tfrac{1}{5}x + 100 \qquad 0 \leq x \leq 500$$

Express the revenue R as a function of x. (Remember, $R = xp$.)

4. The price p and the quantity x sold of a certain product obey the demand equation

$$p = -\tfrac{1}{4}x + 100 \qquad 0 \leq x \leq 400$$

Express the revenue R as a function of x.

5. The price p and the quantity x sold of a certain product obey the demand equation

$$x = -20p + 100 \qquad 0 \leq p \leq 5$$

Express the revenue R as a function of x.

6. The price p and the quantity x sold of a certain product obey the demand equation

$$x = -5p + 500 \qquad 0 \leq p \leq 100$$

Express the revenue R as a function of x.

7. A farmer has available 100 yards of fencing and wishes to enclose a rectangular area. Express the area A of the rectangle as a function of the width x of the rectangle. What is the domain of A?

8. A farmer has 1000 feet of fencing available to enclose a rectangular field. One side of the field lies along a river, so only three sides require fencing. Express the area A of the rectangle as a function of x, where x is the length of the side parallel to the river.

9. A wire of length x is bent into the shape of a circle.
   (a) Express the circumference of the circle as a function of x.
   (b) Express the area of the circle as a function of x.

10. A wire of length x is bent into the shape of a square.
   (a) Express the perimeter of the square as a function of x.
   (b) Express the area of the square as a function of x.

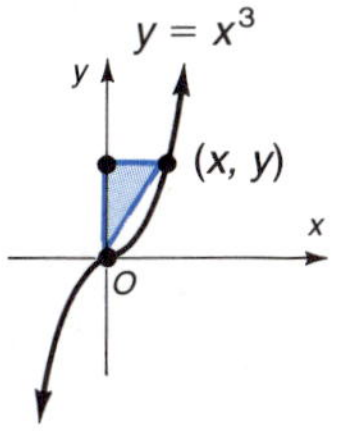

11. A triangle has one vertex on the graph of $y = x^3$, $x > 0$, another at the origin, and the third on the positive y-axis, as shown in the figure. Express the area A of the triangle as a function of x.

12. A triangle has one vertex on the graph of $y = 9 - x^2$, $x > 0$, another at the origin, and the third on the positive x-axis. Express the area A of the triangle as a function of x.

13. Let $P = (x, y)$ be a point on the graph of $y = x^2 - 4$.
   (a) Express the distance d from P to the origin as a function of x.
   (b) What is d if $x = 0$? (c) What is d if $x = 1$?

14. Let $P = (x, y)$ be a point on the graph of $y = x^2 - 4$.
   (a) Express the distance d from P to the point $(0, -1)$ as a function of x.
   (b) What is d if $x = 0$? (c) What is d if $x = -1$?

15. Let $P = (x, y)$ be a point on the graph of $y = \sqrt{x}$. Express the distance d from P to the point $(1, 0)$ as a function of x.

16. Let $P = (x, y)$ be a point on the graph of $y = 1/x$. Express the distance d from P to the origin as a function of x.

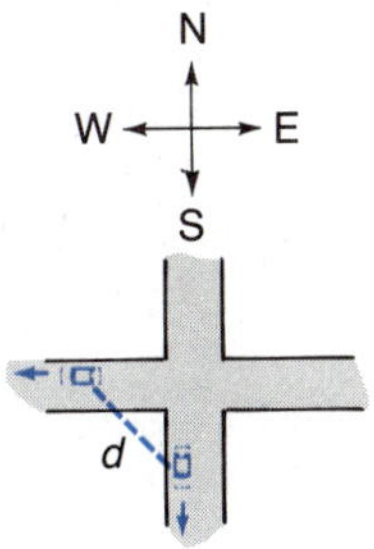

17. Two cars leave an intersection at the same time. One is headed south at a constant speed of 25 miles per hour; the other is headed west at a constant speed of 40 miles per hour (see the figure). Express the distance d between the cars as a function of the time t. [*Hint:* At $t = 0$, the cars leave the intersection.]

18. Two cars are approaching an intersection. One is 2 miles south of the intersection and is moving at a constant speed of 30 miles per hour. At the same time, the other car is 3 miles east of the intersection and is moving at a constant speed of 40 miles per hour. Express the distance d between the cars as a function of time t. [*Hint:* At $t = 0$, the cars are 2 miles south and 3 miles east of the intersection, respectively.]

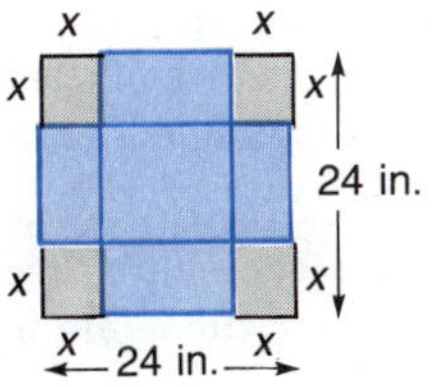

19. An open box with a square base is to be made from a square piece of cardboard 24 inches on a side by cutting out a square from each corner and turning up the sides (see the figure). Express the volume V of the box as a function of the length x of the side of the square cut from each corner.

20. An open box with a square base is required to have a volume of 10 cubic feet. Express the amount A of material used to make such a box as a function of the length x of a side of the square base.

21. A closed box with a square base is required to have a volume of 10 cubic feet. Express the amount A of material used to make such a box as a function of the length x of a side of the square base.

22. The volume V of a sphere of radius r is $V = \frac{4}{3}\pi r^3$; the surface area S of this sphere is $S = 4\pi r^2$. Express the volume V as a function of the surface area S. If the surface area doubles, how does the volume change?

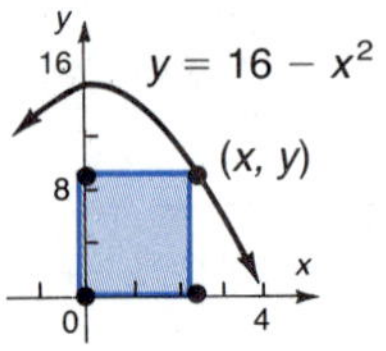

23. A rectangle has one corner on the graph of $y = 16 - x^2$, another at the origin, a third on the positive y-axis, and the fourth on the positive x-axis (see the figure). Express the area A of the rectangle as a function of x. What is the domain of A?

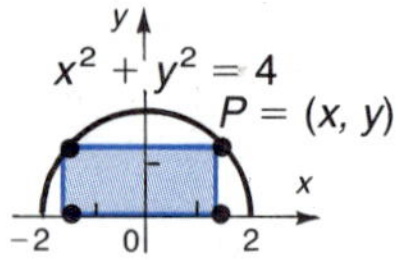

24. A rectangle is inscribed in a semicircle of radius 2 (see the figure). Let $P = (x, y)$ be the point in quadrant I that is a vertex of the rectangle and is on the circle.
(a) Express the area A of the rectangle as a function of x.
(b) Express the perimeter p of the rectangle as a function of x.

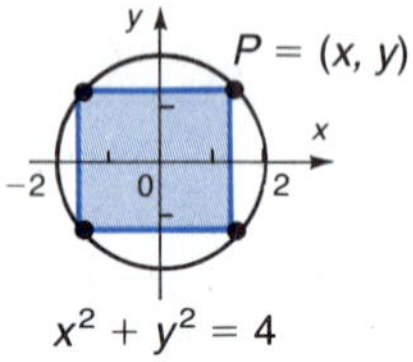

25. A rectangle is inscribed in a circle of radius 2 (see the figure). Let $P = (x, y)$ be the point in quadrant I that is a vertex of the rectangle and is on the circle.
(a) Express the area A of the rectangle as a function of x.
(b) Express the perimeter p of the rectangle as a function of x.

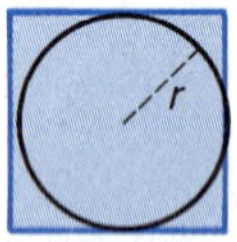

26. A circle of radius r is inscribed in a square (see the figure).
(a) Express the area A of the square as a function of the radius r of the circle.
(b) Express the perimeter p of the square as a function of r.

27. A can in the shape of a right circular cylinder is required to have a volume of 500 cubic centimeters. The top and bottom are made of material that costs 6¢ per square centimeter, while the sides are made of material that costs 4¢ per square centimeter. Express the total cost C of the material as a function of the radius r of the cylinder. (Refer to Figure 48.)

28. A steel drum in the shape of a right circular cylinder is required to have a volume of 100 cubic feet.
(a) Express the amount A of material required to make the drum as a function of the radius r of the cylinder.
(b) How much material is required if the can is of radius 3 feet?
(c) How much material is required for a can of radius 4 feet?
(d) How much material is required for a can of radius 5 feet?

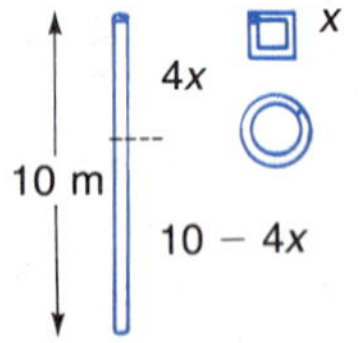

29. A wire 10 meters long is to be cut into two pieces. One piece will be shaped as a square, and the other piece will be shaped as a circle (see the figure). Express the total area A enclosed by the pieces of wire as a function of the length x of a side of the square. What is the domain of A?

30. A wire 10 meters long is to be cut into two pieces. One piece will be shaped as an equilateral triangle, and the other piece will be shaped as a circle. Express the total area A enclosed by the pieces of wire as a function of the length x of a side of the equilateral triangle. What is the domain of A?

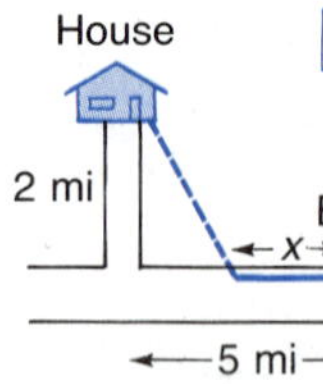

C

31. A cable TV company is asked to provide service to a customer whose house is located 2 miles from the road along which the cable is buried. The nearest connection box for the cable is located 5 miles down the road (see the figure).
(a) If the installation cost is \$10 per mile along the road and \$14 per mile off the road, express the total cost C of installation as a function of the distance x (in miles) from the connection box to the point where the cable installation turns off the road.

(b) What is the domain of C?

(c) Compute the cost for $x = 1$, $x = 2$, $x = 3$, and $x = 4$.

32. An island is 3 miles from the nearest point P on a straight shoreline. A town is located 20 miles down the shore from P. (Refer to Figure 45 for a similar situation.)

(a) If a person has a boat that averages 12 miles per hour and the same person can walk 5 miles per hour, express the time T it takes to go from the island to town as a function of x, where x is the distance from P to where the person lands the boat.

(b) Compute the time T if $x = 8$ miles and if $x = 12$ miles.

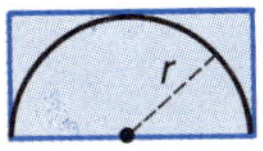

33. A semicircle of radius r is inscribed in a rectangle so that the diameter of the semicircle is the length of the rectangle (see the figure).

(a) Express the area A of the rectangle as a function of the radius r of the semicircle.

(b) Express the perimeter p of the rectangle as a function of r.

34. An equilaterial triangle is inscribed in a circle of radius r. Express the circumference C of the circle as a function of the length x of a side of the triangle. [*Hint:* First show that $r^2 = x^2/3$.]

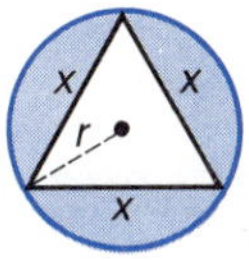

35. An equilateral triangle is inscribed in a circle of radius r. Express the area A within the circle, but outside the triangle, as a function of the length x of a side of the triangle.

36. A trucking company transports goods between Chicago and New York, a distance of 960 miles. The company's policy is to charge, for each pound, \$0.50 per mile for the first 100 miles, \$0.40 per mile for the next 300 miles, \$0.25 per mile for the next 400 miles, and no charge for the remaining 160 miles.

(a) Graph the relationship between the cost of transportation in dollars and mileage over the entire 960 mile route.

(b) Find the cost as a function of mileage for hauls between 100 and 400 miles from Chicago.

(c) Find the cost as a function of mileage for hauls between 400 and 800 miles from Chicago.

37. An economy car rented in California from National Car Rental® on a weekly basis costs \$95 per week.* Extra days cost \$24 per day until the daily rate exceeds the weekly rate, in which case the weekly rate applies. Find the cost C of renting an economy car as a piecewise-defined function of the number x of days used, where $7 \le x \le 14$. Graph this function. [*Note:* Any part of a day counts as a full day.]

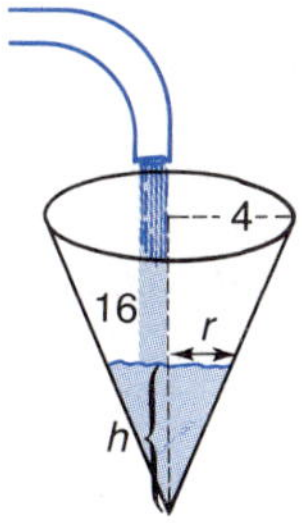

38. Rework Problem 37 for a luxury car, which costs \$219 on a weekly basis with extra days at \$45 per day.

39. Water is poured into a container in the shape of a right circular cone with radius 4 feet and height 16 feet (see the figure). Express the volume V of water in the cone as a function of the height h of the water. [*Hint:* The volume V of a cone of radius r and height h is $V = \frac{1}{3}\pi r^2 h$.]

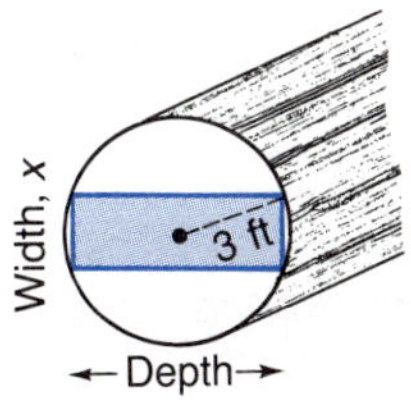

40. The strength of a rectangular wooden beam is proportional to the product of the width and the cube of its depth (see the figure). If the beam is to be cut from a log in the shape of a cylinder of radius 3 feet, express the strength S of the beam as a function of the width x. What is the domain of S?

41. Two 1988 Tax Rate Schedules are given at the top of the next page. If x equals the amount on Form 1040, line 37, and y equals the tax due, construct a function f for each schedule. Assume $0 < x \le 89{,}560$ for Schedule X and $0 < x \le 149{,}250$ for Schedule Y-1.

**Source:* National Car Rental®, 1989

1988 Tax Rate Schedules

SCHEDULE X—USE IF YOUR FILING STATUS IS SINGLE				SCHEDULE Y-1—USE IF YOUR FILING STATUS IS MARRIED FILING JOINTLY OR QUALIFYING WIDOW(ER)			
If the amount on Form 1040, line 37, is: *Over—*	*But not over—*	Enter on Form 1040, line 38	*of the amount over—*	If the amount on Form 1040, line 37, is: *Over—*	*But not over—*	Enter on Form 1040, line 38	*of the amount over—*
\$0	\$17,850	------- 15%	\$0	\$0	\$29,750	------- 15%	\$0
17,850	43,150	\$2,677.50 + 28%	17,850	29,710	71,900	\$4,462.50 + 28%	29,750
43,150	89,560	9,761.50 + 33%	43,150	71,900	149,250	16,264.50 + 33%	71,900
89,560	-------	Use Worksheet below to figure your tax.		149,250	-------	Use Worksheet below to figure your tax.	

CHAPTER REVIEW

THINGS TO KNOW

Function	A rule or correspondence between two sets of real numbers so that each number x in the first set, the domain, has corresponding to it exactly one number y in the second set. The range is the set of y values of the function for the x values in the domain.
	x is the independent variable; y is the dependent variable.
	A function f may be defined implicitly, by an equation involving x and y; or explicitly, by writing $y = f(x)$.
Function notation	$y = f(x)$
	f is a symbol for the function or rule that defines the function.
	x is the argument, or independent variable.
	y is the dependent variable.
	$f(x)$ is the value of the function at x, or the image of x.
Domain	If unspecified, the domain of a function f is the largest set of real numbers for which the rule defines a real number.
Vertical-line test	A set of points in the plane is the graph of a function if and only if a vertical line intersects the graph in at most one point.
Even function f	$f(-x) = f(x)$ for every x in the domain ($-x$ also must be in the domain).
Odd function f	$f(-x) = -f(x)$ for every x in the domain ($-x$ also must be in the domain).
One-to-one function f	If $x_1 \neq x_2$, then $f(x_1) \neq f(x_2)$ for any choice of x_1 and x_2 in the domain.
Horizontal-line test	If horizontal lines intersect the graph of a function f in at most one point, then f is one-to-one.

Inverse function f^{-1} of f	Domain of f = Range of f^{-1}; Range of f = Domain of f^{-1} $f^{-1}(f(x)) = x$ and $f(f^{-1}(x)) = x$ Graphs of f and f^{-1} are symmetric with respect to the line $y = x$.

Important Functions

Linear function	$f(x) = mx + b$	Graph is a straight line with slope m and y-intercept b.
Constant function	$f(x) = b$	Graph is a horizontal line with y-intercept b (see Figure 8).
Identity function	$f(x) = x$	Graph is a straight line with slope 1 and y-intercept 0 (see Figure 9).
Square function	$f(x) = x^2$	Graph is a parabola with intercept at (0, 0) (see Figure 10).
Cube function	$f(x) = x^3$	See Figure 11.
Square root function	$f(x) = \sqrt{x}$	See Figure 12.
Reciprocal function	$f(x) = 1/x$	See Figure 13.
Absolute value function	$f(x) = \lvert x \rvert$	See Figure 14.

How To:

Find the domain and range of a function from its graph

Determine whether a function is even or odd without graphing it

Graph certain functions by shifting, compressing, stretching, and/or reflecting (see Table 8)

Find the composite of two functions

Find the inverse of certain one-to-one functions (see the procedure given on page 253)

Graph f^{-1} given the graph of f

Construct functions in applications, including piecewise-defined functions

FILL-IN-THE-BLANK ITEMS

1. If f is a function defined by the equation $y = f(x)$, then x is called the ________ variable and y is the ________ variable.
2. A set of points in the xy-plane is the graph of a function if and only if no ________ line contains more than one point of the set.
3. A(n) ________ function f is one for which $f(-x) = f(x)$ for every x in the domain of f; a(n) ________ function f is one for which $f(-x) = -f(x)$ for every x in the domain of f.
4. Suppose the graph of a function f is known. Then the graph of $y = f(x - 2)$ may be obtained by a(n) ________ shift of the graph of f to the ________ a distance of 2 units.

5. If $f(x) = x + 1$ and $g(x) = x^3$, then __________ $= (x + 1)^3$.
6. If every horizontal line intersects the graph of a function f at no more than one point, then f is a(n) __________ function.
7. If f^{-1} denotes the inverse of a function f, then the graphs of f and f^{-1} are symmetric with respect to the line __________.

REVIEW EXERCISES

1. Given that f is a linear function, $f(4) = -2$, and $f(1) = 4$, write the equation that defines f.
2. Given that g is a linear function with slope $= -2$ and $g(-2) = 2$, write the equation that defines g.
3. A function f is defined by the equation below. If $f(1) = 4$, find A.

$$f(x) = \frac{Ax + 5}{6x - 2}$$

4. A function g is defined by the equation below. If $g(-1) = 0$, find A.

$$g(x) = \frac{A}{x} + \frac{8}{x^2}$$

5. (a) Tell which of the graphs below are graphs of functions.
(b) Tell which of the graphs below are graphs of one-to-one functions.

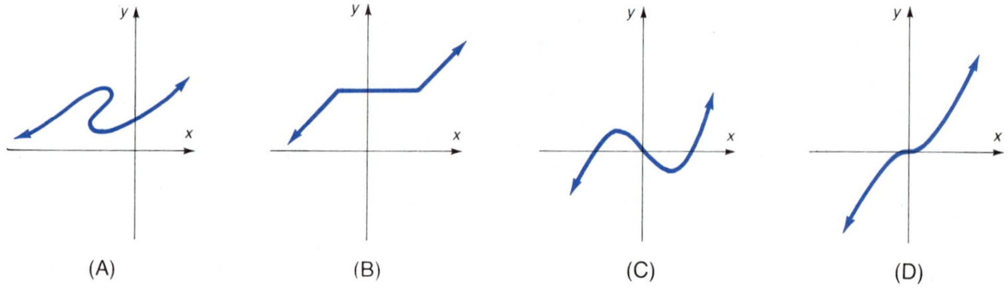

6. Use the graph of the function f shown below to find:
(a) The domain and range of f
(b) The intervals on which f is increasing
(c) The intervals on which f is constant
(d) The intercepts of f

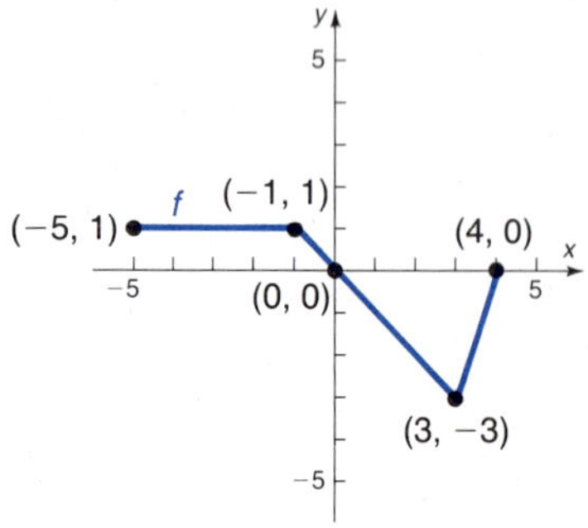

In Problems 7–12, find the following for each function:

(*a*) $f(-x)$ (*b*) $-f(x)$ (*c*) $f(x + 2)$ (*d*) $f(x - 2)$

7. $f(x) = \dfrac{x}{x^2 - 4}$ **8.** $f(x) = \dfrac{x^2}{x - 2}$ **9.** $f(x) = \sqrt{x^2 - 4}$

10. $f(x) = |x^2 - 4|$ **11.** $f(x) = \dfrac{x^2 - 4}{x^2}$ **12.** $f(x) = \dfrac{x^3}{x^2 - 4}$

In Problems 13–18, determine whether the given function is even, odd, or neither without drawing a graph.

13. $f(x) = x^3 - x$ **14.** $g(x) = \dfrac{1 + x^2}{1 + x^4}$ **15.** $h(x) = \dfrac{1}{x^4} + \dfrac{1}{x^2} + 1$

16. $F(x) = \sqrt{1 - x^3}$ **17.** $G(x) = 1 - x + x^3$ **18.** $H(x) = 1 + x + x^2$

In Problems 19–30, find the domain of each function.

19. $f(x) = \dfrac{x}{x^2 - 4}$ **20.** $f(x) = \dfrac{x^2}{x - 2}$ **21.** $f(x) = \sqrt{2 - x}$

22. $f(x) = \sqrt{x + 2}$ **23.** $h(x) = \dfrac{\sqrt{x}}{|x|}$ **24.** $g(x) = \dfrac{|x|}{x}$

25. $f(x) = \dfrac{x}{x^2 + 2x - 3}$ **26.** $F(x) = \dfrac{1}{x^2 - 3x - 4}$

27. $G(x) = \begin{cases} |x| & \text{if } -1 \le x \le 1 \\ 1/x & \text{if } x > 1 \end{cases}$ **28.** $H(x) = \begin{cases} 1/x & \text{if } 0 < x < 4 \\ x - 4 & \text{if } 4 \le x \le 8 \end{cases}$

29. $f(x) = \begin{cases} 1/(x - 2) & \text{if } x > 2 \\ 0 & \text{if } x = 2 \\ 3 & \text{if } 0 \le x < 2 \end{cases}$ **30.** $g(x) = \begin{cases} |1 - x| & \text{if } x < 1 \\ 3 & \text{if } x = 1 \\ x & \text{if } 1 < x \le 3 \end{cases}$

In Problems 31–50:
(*a*) *Find the domain of each function.* (*b*) *Locate any intercepts.*
(*c*) *Graph each function.* (*d*) *Based on the graph, find the range.*

31. $F(x) = |x| - 4$ **32.** $f(x) = |x| + 4$ **33.** $g(x) = -|x|$

34. $g(x) = \frac{1}{2}|x|$ **35.** $h(x) = \sqrt{x - 1}$ **36.** $h(x) = \sqrt{x} - 1$

37. $f(x) = \sqrt{1 - x}$ **38.** $f(x) = -\sqrt{x}$

39. $F(x) = \begin{cases} x^2 + 4 & \text{if } x < 0 \\ 4 - x^2 & \text{if } x \ge 0 \end{cases}$ **40.** $H(x) = \begin{cases} |1 - x| & \text{if } 0 \le x \le 2 \\ |x - 1| & \text{if } x > 2 \end{cases}$

41. $h(x) = (x - 1)^2 + 2$ **42.** $h(x) = (x + 2)^2 - 3$

43. $g(x) = (x - 1)^3 + 1$ **44.** $g(x) = (x + 2)^3 - 8$

45. $f(x) = \begin{cases} 2\sqrt{x} & \text{if } x \ge 4 \\ x & \text{if } 0 < x < 4 \end{cases}$ **46.** $f(x) = \begin{cases} 3|x| & \text{if } x < 0 \\ \sqrt{1 - x} & \text{if } 0 \le x \le 1 \end{cases}$

47. $g(x) = \dfrac{1}{x - 1} + 1$ **48.** $g(x) = \dfrac{1}{x + 2} - 2$

49. $h(x) = [\![-x]\!]$ **50.** $h(x) = -[\![x]\!]$

In Problems 51–56, the function f is one-to-one. Find the inverse of each function, and verify your answer. Find the domain and range of f and f^{-1}.

51. $f(x) = \dfrac{2x + 3}{x - 2}$ **52.** $f(x) = \dfrac{2 - x}{1 + x}$ **53.** $f(x) = \dfrac{1}{x - 1}$

54. $f(x) = \sqrt{x - 2}$ **55.** $f(x) = \dfrac{3}{x^{1/3}}$ **56.** $f(x) = x^{1/3} + 1$

In Problems 57–62, for the given functions f and g, find:

(a) $(f \circ g)(2)$ (b) $(g \circ f)(-2)$ (c) $(f \circ f)(4)$ (d) $(g \circ g)(-1)$

57. $f(x) = 3x - 5$; $g(x) = 1 - 2x^2$

58. $f(x) = 4 - x$; $g(x) = 1 + x^2$

59. $f(x) = \sqrt{x + 2}$; $g(x) = 2x^2 + 1$

60. $f(x) = 1 - 3x^2$; $g(x) = \sqrt{4 - x}$

61. $f(x) = \dfrac{1}{x^2 + 4}$; $g(x) = 3x - 2$

62. $f(x) = \dfrac{2}{1 + 2x^2}$; $g(x) = 3x$

In Problems 63–68, find $f \circ g$, $g \circ f$, $f \circ f$, and $g \circ g$ for each pair of functions.

63. $f(x) = \dfrac{2 - x}{x}$; $g(x) = 3x + 2$

64. $f(x) = \dfrac{x}{x + 1}$; $g(x) = \dfrac{x}{x - 1}$

65. $f(x) = 3x^2 + x + 1$; $g(x) = |3x|$

66. $f(x) = \sqrt{3x}$; $g(x) = 1 + x + x^2$

67. $f(x) = \dfrac{x + 1}{x - 1}$; $g(x) = \dfrac{1}{x}$

68. $f(x) = \sqrt{x^2 - 3}$; $g(x) = \sqrt{3 - x^2}$

69. For the graph of the function f shown below:
(a) Draw the graph of $y = f(-x)$.
(b) Draw the graph of $y = -f(x)$.
(c) Draw the graph of $y = f(x + 2)$.
(d) Draw the graph of $y = f(x) + 2$.
(e) Draw the graph of $y = f(2 - x)$.
(f) Draw the graph of f^{-1}.

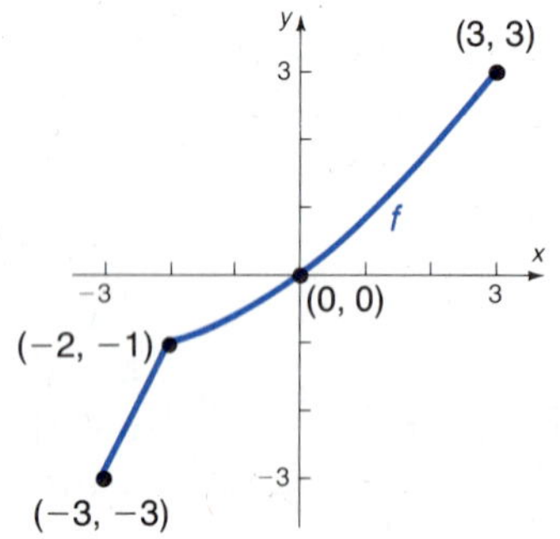

70. Repeat Problem 69 for the graph of the function g shown below.

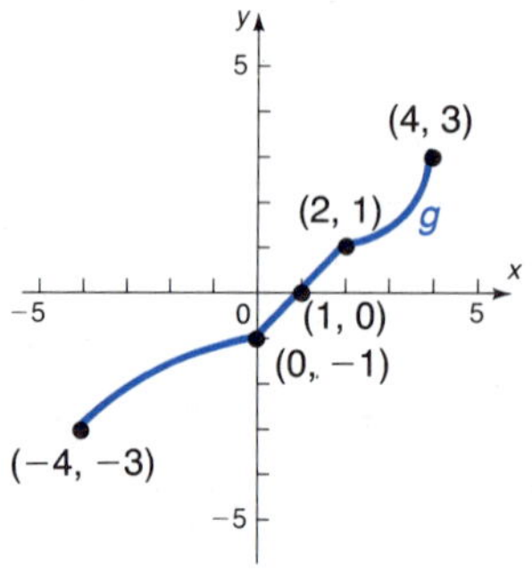

71. The temperature T of the air is approximately a linear function of the altitude h for altitudes within 6 miles of the surface of the Earth. If the surface temperature is 30°C and the temperature at 10,000 meters is 5°C, find the equation of the function $T = T(h)$.

72. The speed v (in feet per second) of a car is a linear function of the time t (in seconds) for $10 \le t \le 30$. If after each second, the speed of the car has increased by 5 feet per second, and if after 20 seconds, the speed is 80 feet per second, how fast is the car going after 30 seconds? Find the equation of the function $v = v(t)$.

5

Polynomial and Rational Functions

In Chapters 3 and 4, we graphed linear functions $f(x) = ax + b$, $a \neq 0$; the square function $f(x) = x^2$; and the cube function $f(x) = x^3$. Each of these functions belongs to the class of functions called *polynomial functions*, which we discuss further in this chapter. We will also discuss *rational functions*, which are ratios of polynomial functions. In this chapter, we place special emphasis on the graphs of polynomial and rational functions. This emphasis will demonstrate the importance of evaluating polynomials (Section 5.4) and solving polynomial equations (Sections 5.5 and 5.6). Section 5.7 deals with polynomials having coefficients that are complex numbers.

5.1 Quadratic Functions*

A **quadratic function** is a function of the form

$$f(x) = ax^2 + bx + c \qquad (1)$$

where a, b, and c are real numbers and $a \neq 0$. The domain of a quadratic function consists of all real numbers.

Many applications require a knowledge of quadratic functions. For example, suppose the equation that relates the number x of units sold and the price p per unit is given by

$$x = 15{,}000 - 750p$$

Then the revenue R derived from selling x units at the price p per unit is

$$R = xp = (15{,}000 - 750p)p = -750p^2 + 15{,}000p$$

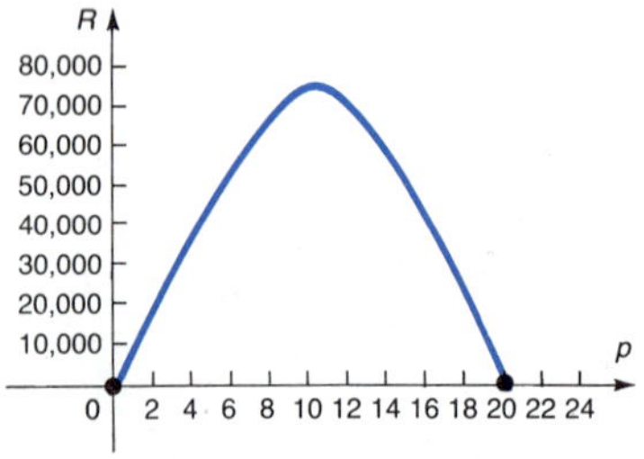

Figure 1
Graph of a revenue function:
$R = -750p^2 + 15{,}000p$

Figure 1 illustrates the graph of this revenue function, whose domain is $0 \leq p \leq 20$, since both x and p must be nonnegative.

A second situation in which a quadratic function appears involves the motion of a projectile. Based on Newton's Second Law of Motion (force equals mass times acceleration, $F = ma$), it can be shown that, ignoring air resistance, the path of a projectile propelled upward at an inclination to the horizontal is the graph of a quadratic function. See Figure 2 for an illustration.

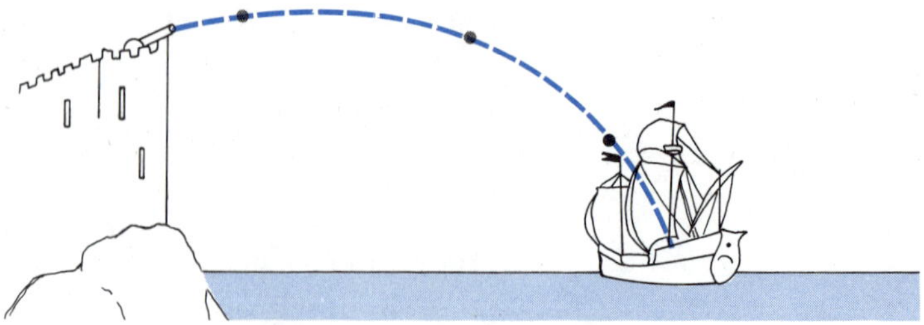

Figure 2
Path of a cannonball

* C Refer to Graphics Calculator Supplement: Activity II.

Graphing Quadratic Functions

We already know how to graph quadratic functions. For example, based on the discussion in Section 4.3, we know how to graph quadratic functions of the form $f(x) = ax^2$, $a \neq 0$. Figure 3 illustrates the graphs of $f(x) = ax^2$ for $a = 1$, $a = 3$, and $a = \frac{1}{2}$, drawn on the same set of coordinate axes. Observe that the choice of a larger value of a in $f(x) = ax^2$ results in a "thinner," or "narrower," graph.

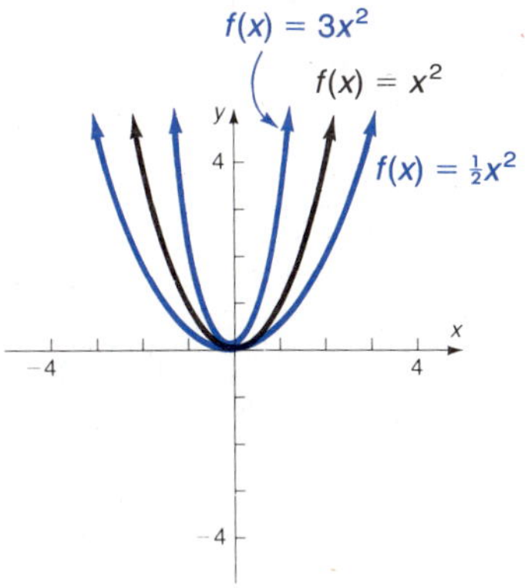

Figure 3

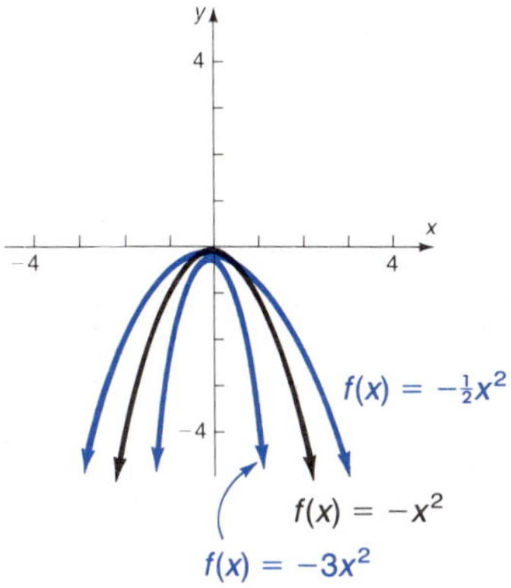

Figure 4

Of course, the graphs of $f(x) = ax^2$ for $a < 0$ are merely the reflection about the x-axis of the corresponding graphs of $f(x) = |a|x^2$. See Figure 4.

The graphs in Figures 3 and 4 are typical of the graphs of all quadratic functions, which we call **parabolas**.* Refer to Figure 5, where two parabolas are pictured. The one on the left **opens up** and has a lowest point; the one on the right **opens down** and has a highest point. The lowest or highest point of a parabola is called the **vertex**. The dashed vertical line passing through the vertex in each parabola in Figure 5 is called the **axis of symmetry** (usually abbreviated to **axis**) of the parabola. Because the parabola is symmetric about its axis, the axis of symmetry of a parabola can be used to advantage in graphing the parabola.

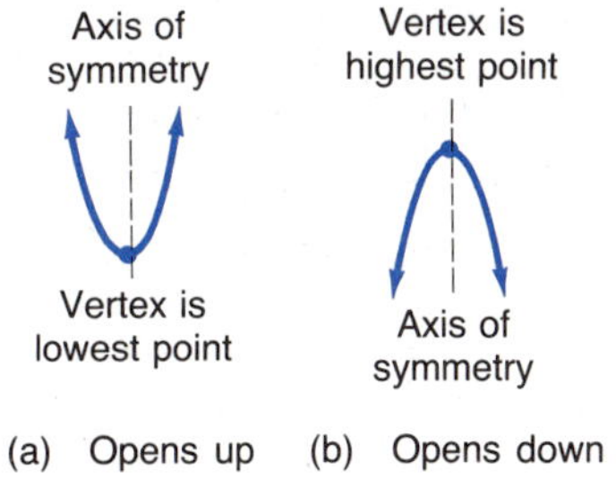

Figure 5
Graphs of a quadratic function, $f(x) = ax^2 + bx + c$, $a \neq 0$

The parabolas in Figure 5 are the graphs of a quadratic function $f(x) = ax^2 + bx + c$, $a \neq 0$. Notice that the coordinate axes are not included in the figure. Depending on the values of a, b, and c, the axes could be placed anywhere. The important fact is that, except possibly for compression or stretching, the shape of the graph of a quadratic function will look like one of the parabolas in Figure 5.

In the following example, we graph a quadratic function $f(x) = ax^2 + bx + c$, $a \neq 0$, using techniques from Section 4.3. In so doing, we

*We shall study parabolas using a geometric definition in Section 11.2.

shall complete the square and write the function f in the form $f(x) = a(x - h)^2 + k$.

Example 1 Graph the function: $f(x) = 2x^2 + 8x + 5$

Solution We begin by completing the square on the right side:

$$\begin{aligned} f(x) &= 2x^2 + 8x + 5 \\ &= 2(x^2 + 4x) + 5 && \text{Factor out the 2 from } 2x^2 + 8x. \\ &= 2(x^2 + 4x + 4) + 5 - 8 && \text{Complete the square of } 2(x^2 + 4x). \text{ Notice that the factor of 2 requires that 8 be added and subtracted.} \\ &= 2(x + 2)^2 - 3 \end{aligned}$$

The graph of f can be obtained in three stages, as shown in Figure 6. Now compare this graph to the graph in Figure 5(a). The graph of $f(x) = 2x^2 + 8x + 5$ is a parabola that opens up and has its vertex (lowest point) at $(-2, -3)$. Its axis of symmetry is the line $x = -2$.

Figure 6

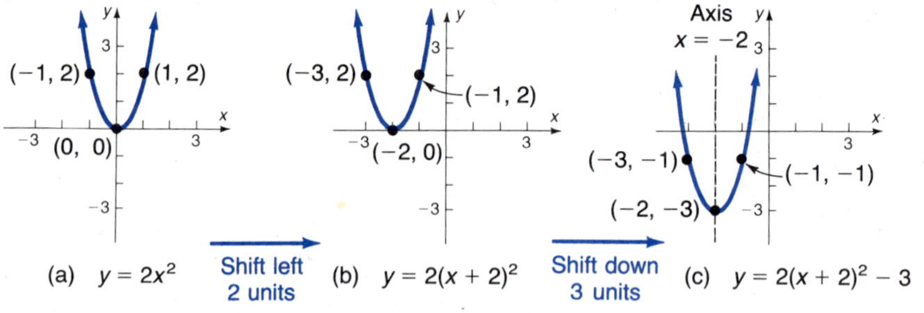

The method used in Example 1 can be used to graph any quadratic function $f(x) = ax^2 + bx + c$, $a \neq 0$, as follows:

$$\begin{aligned} f(x) &= ax^2 + bx + c \\ &= a\left(x^2 + \frac{b}{a}x\right) + c && \text{Factor out } a \text{ from } ax^2 + bx. \\ &= a\left(x^2 + \frac{b}{a}x + \frac{b^2}{4a^2}\right) + c - a\left(\frac{b^2}{4a^2}\right) && \text{Complete the square by adding and subtracting } a(b^2/4a^2). \text{ Look closely at this step!} \\ &= a\left(x + \frac{b}{2a}\right)^2 + c - \frac{b^2}{4a} \\ &= a\left(x + \frac{b}{2a}\right)^2 + \frac{4ac - b^2}{4a} \end{aligned}$$

If we let $h = -b/2a$ and let $k = (4ac - b^2)/4a$, this last equation can be rewritten in the form

$$f(x) = a(x - h)^2 + k$$

The graph of f is the parabola $y = ax^2$ shifted horizontally h units and vertically k units. As a result, the vertex is at (h, k), and the graph opens up if $a > 0$ and down if $a < 0$. The axis is the vertical line $x = h$.

In almost every case, it is easier to obtain the vertex of a quadratic function f by remembering that its x-coordinate is $h = -b/2a$. The y-coordinate can then be found by evaluating f at $-b/2a$, instead of by calculating $k = (4ac - b^2)/4a$.

These results are summarized below:

Characteristics of the Graph of a Quadratic Function

$$f(x) = ax^2 + bx + c$$

$$\text{Vertex} = \left(\frac{-b}{2a}, f\left(\frac{-b}{2a}\right)\right) \qquad \text{Axis: The line } x = \frac{-b}{2a} \qquad (2)$$

Parabola opens up if $a > 0$. Parabola opens down if $a < 0$.

Example 2 Without graphing, locate the vertex and axis of the parabola defined by $f(x) = -3x^2 + 6x + 1$. Does it open up or down?

Solution For this quadratic function, $a = -3$, $b = 6$, and $c = 1$. The x-coordinate of the vertex is

$$\frac{-b}{2a} = \frac{-6}{-6} = 1$$

The y-coordinate of the vertex is therefore

$$f\left(\frac{-b}{2a}\right) = f(1) = -3 + 6 + 1 = 4$$

The vertex is located at the point $(1, 4)$. The axis of symmetry is the line $x = 1$. Finally, because $a = -3 < 0$, the parabola opens down. ■

The information we gathered in Example 2, together with the location of the intercepts, usually provides enough information to sketch the graph of $f(x) = ax^2 + bx + c$, $a \neq 0$. The y-intercept is the value of f at $x = 0$, namely, $f(0) = c$. The x-intercepts, if there are any, are found by solving the equation

$$f(x) = ax^2 + bx + c = 0$$

This equation has two, one, or no real solutions, according as the discriminant $b^2 - 4ac$ is positive, 0, or negative. Thus, it has corresponding x-intercepts, as follows:

The x-Intercepts of a Quadratic Function

1. If the discriminant $b^2 - 4ac > 0$, the graph of $f(x) = ax^2 + bx + c$ has two distinct x-intercepts, and so will cross the x-axis in two places.
2. If the discriminant $b^2 - 4ac = 0$, the graph of $f(x) = ax^2 + bx + c$ has one x-intercept, and is tangent to the x-axis at its vertex.
3. If the discriminant $b^2 - 4ac < 0$, the graph of $f(x) = ax^2 + bx + c$ has no x-intercept, and so will not cross the x-axis.

Figure 7 illustrates these possibilities for parabolas that open up.

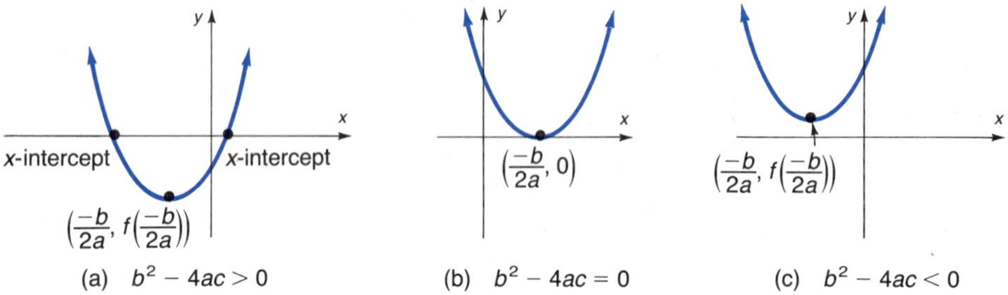

Figure 7
$f(x) = ax^2 + bx + c,\ a > 0$

Example 3 Use the information from Example 2 and the location of intercepts to graph $f(x) = -3x^2 + 6x + 1$.

Solution The y-intercept is found by letting $x = 0$. Thus, the y-intercept is $f(0) = 1$. The x-intercepts are found by letting $f(x) = 0$. This results in the equation

$$-3x^2 + 6x + 1 = 0$$

The discriminant $b^2 - 4ac = (6)^2 - 4(-3)(1) = 36 + 12 = 48 > 0$, so the equation has two real solutions and the graph has two x-intercepts.

Using the quadratic formula, we find

$$x = \frac{-b + \sqrt{b^2 - 4ac}}{2a} = \frac{-6 + \sqrt{48}}{-6} = \frac{-6 + 4\sqrt{3}}{-6} \approx -0.15$$

and

$$x = \frac{-b - \sqrt{b^2 - 4ac}}{2a} = \frac{-6 - \sqrt{48}}{-6} = \frac{-6 - 4\sqrt{3}}{-6} \approx 2.15$$

The x-intercepts are approximately -0.15 and 2.15.

The graph is illustrated in Figure 8. Notice how we used the y-intercept and the axis of symmetry, $x = 1$, to obtain the additional point $(2, 1)$ on the graph. ■

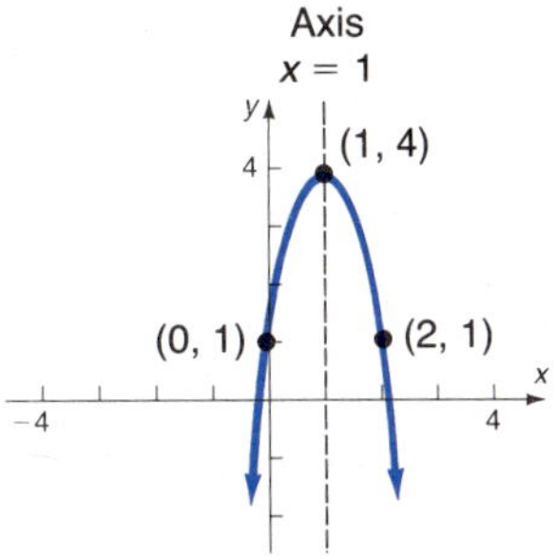

Figure 8
$f(x) = -3x^2 + 6x + 1$

If the graph of a quadratic function has one x-intercept or none, it may be necessary to plot two additional points to obtain the graph.

Example 4 Graph $f(x) = x^2 - 6x + 9$ by determining whether its graph opens up or down and by finding its vertex, axis of symmetry, y-intercept, and x-intercepts, if any.

Solution For $f(x) = x^2 - 6x + 9$, we have $a = 1$, $b = -6$, and $c = 9$. Since $a = 1 > 0$, the parabola opens up. The x-coordinate of the vertex is

$$\frac{-b}{2a} = \frac{-(-6)}{2 \cdot 1} = 3$$

The y-coordinate of the vertex is

$$f(3) = 9 - 6 \cdot 3 + 9 = 0$$

So the vertex is at $(3, 0)$. The axis of symmetry is the line $x = 3$. The y-intercept is $f(0) = 9$. Since the vertex $(3, 0)$ lies on the x-axis, the graph will touch the x-axis at the x-intercept. By using the axis of symmetry and the y-intercept at $(0, 9)$, we can locate the point $(6, 9)$ on the graph. See Figure 9. ■

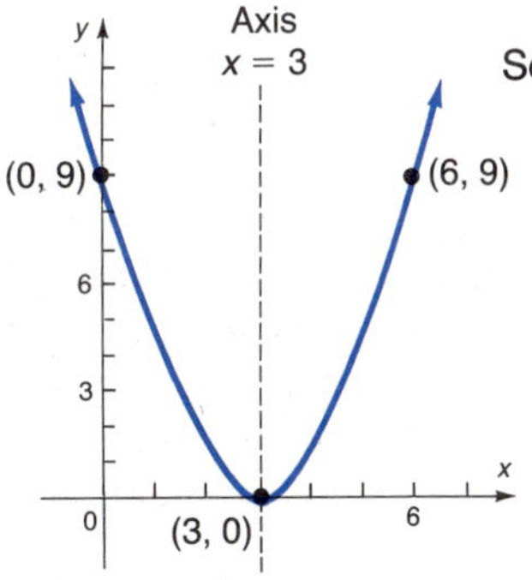

Figure 9
$f(x) = x^2 - 6x + 9$

Example 5 Graph $f(x) = 2x^2 + x + 1$ by determining whether its graph opens up or down and by finding its vertex, axis of symmetry, y-intercept, and x-intercepts, if any.

Solution For $f(x) = 2x^2 + x + 1$, we have $a = 2$, $b = 1$, and $c = 1$. Since $a = 2 > 0$, the parabola opens up. The x-coordinate of the vertex is

$$\frac{-b}{2a} = -\frac{1}{4}$$

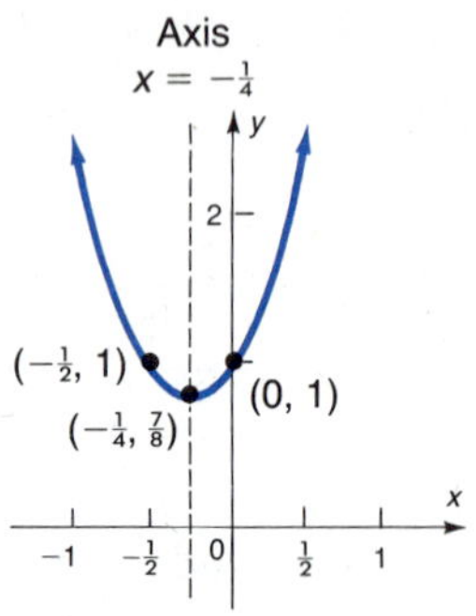

Figure 10
$f(x) = 2x^2 + x + 1$

The y-coordinate of the vertex is

$$f\left(-\tfrac{1}{4}\right) = 2\left(\tfrac{1}{16}\right) + \left(-\tfrac{1}{4}\right) + 1 = \tfrac{7}{8}$$

So the vertex is at $\left(-\frac{1}{4}, \frac{7}{8}\right)$. The axis of symmetry is the line $x = -\frac{1}{4}$. The y-intercept is $f(0) = 1$. The x-intercept(s), if any, obey the equation

$$2x^2 + x + 1 = 0$$

Since the discriminant $b^2 - 4ac = 1 - 8 = -7 < 0$, this equation has no real solution, and so the graph has no x-intercepts. We use the point (0, 1) and the axis of symmetry $x = -\frac{1}{4}$ to locate the point $\left(-\frac{1}{2}, 1\right)$ on the graph. See Figure 10. ■

We now know two ways to graph a quadratic function:

1. Complete the square and apply shifting techniques (Example 1).
2. Use the results given in display (2) to find the vertex and axis of symmetry and to determine whether the graph opens up or down. Then locate the y-intercept and the x-intercepts, if there are any (Examples 2–5).

Applications

We have already seen that the graph of a quadratic function $f(x) = ax^2 + bx + c$ is a parabola with vertex at $(-b/2a, f(-b/2a))$. This vertex is the highest point on the graph if $a < 0$ and the lowest point on the graph if $a > 0$. If the vertex is the highest point ($a < 0$), then $f(-b/2a)$ is the **maximum value** of f. If the vertex is the lowest point ($a > 0$), then $f(-b/2a)$ is the **minimum value** of f. These ideas give rise to many applications.

Example 6 The manufacturer of a digital watch has found that when the watches are sold at a price of p dollars per unit, the revenue R (in dollars) as a function of the price p is

$$R(p) = -750p^2 + 15{,}000p$$

What unit price should be established in order to maximize revenue? If this price is charged, what is the maximum revenue?

Solution The revenue R is

$$\begin{aligned} R(p) &= -750p^2 + 15{,}000p \\ &= ap^2 + bp + c \end{aligned}$$

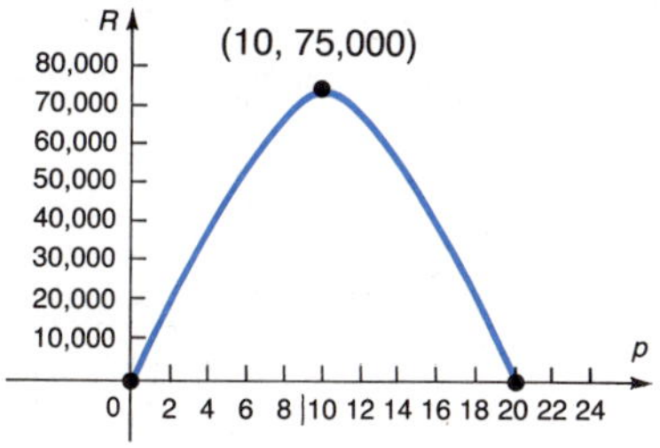

Figure 11
$R(p) = -750p^2 + 15{,}000p$

The function R is a quadratic function with $a = -750$, $b = 15{,}000$, and $c = 0$. Because $a < 0$, the vertex is the highest point of the parabola. The revenue R is therefore a maximum when

$$p = \frac{-b}{2a} = \frac{-15{,}000}{2(-750)} = \frac{-15{,}000}{-1500} = \$10$$

The maximum revenue R is

$$R(10) = -750(10)^2 + 15{,}000(10) = \$75{,}000$$

See Figure 11 for an illustration. ■

Example 7 A company charges \$200 for each box of tools on orders of 150 or fewer boxes. The cost to the buyer on every box is reduced by \$1 for each box ordered in excess of 150. For what size order is revenue maximum? What is the maximum revenue?

Solution For an order of exactly 150 boxes, the company's revenue is

$$\$200(150) = \$30{,}000$$

For an order of 160 boxes (which is 10 in excess of 150), the charge per box is $200 - 10(1) = 190$, and the revenue is

$$\$190(160) = \$30{,}400$$

To solve the problem, let x denote the number of boxes sold. The revenue R is

$$\begin{aligned} R &= (\text{Number of boxes})(\text{Charge per box}) \\ &= x(\text{Charge per box}) \end{aligned}$$

If $x \geq 150$, the charge per box is

$$200 - 1\begin{pmatrix}\text{Number of boxes} \\ \text{in excess of 150}\end{pmatrix} = 200 - 1(x - 150) = 350 - x$$

Hence, the revenue R is

$$R = x(350 - x) = -x^2 + 350x \qquad x \geq 150$$

which is a quadratic function with $a = -1$, $b = 350$, and $c = 0$. Because $a < 0$, the vertex is the highest point. The revenue R is therefore a maximum when

$$x = \frac{-b}{2a} = \frac{-350}{-2} = 175$$

The maximum revenue is

$$R(175) = 175(350 - 175) = \$30{,}625$$

The company would want to set 175 as the maximum number of boxes a person could purchase on this plan, because revenue to the company starts to decrease for orders in excess of 175. ■

Example 8 A cruise ship leaves the Port of Miami heading due east at a constant speed of 5 knots (1 knot = 1 nautical mile per hour). At 5:00 PM, the cruise ship is 5 nautical miles due south of a cabin cruiser that is moving south at a constant speed of 10 knots. At what time are the two ships closest?

Solution We begin with an illustration depicting the relative position of each ship at 5:00 PM. See Figure 12(a). After a time t (in hours) has passed, the cruise ship has moved east $5t$ nautical miles, and the cabin cruiser has moved south $10t$ nautical miles. Figure 12(b) illustrates the relative position of each ship after t hours. Figure 12(c) shows a right triangle extracted from Figure 12(b). By the Pythagorean Theorem, the square of the distance d between the ships after time t is

$$\begin{aligned} d^2 &= (5 - 10t)^2 + (5t)^2 \\ &= 125t^2 - 100t + 25 \end{aligned}$$

Figure 12

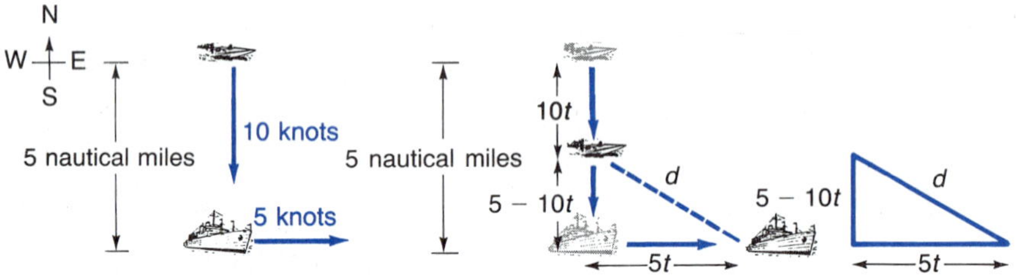

(a) Position at 5:00 PM (b) Position at time t (c)

Now, the distance d is a minimum when d^2 is a minimum. Because d^2 is a quadratic function of t, it follows that d^2, and hence d, is a minimum when

$$t = \frac{-b}{2a} = \frac{100}{2(125)} = \frac{2}{5} \text{ hour}$$

Thus, the ships are closest after $\frac{2}{5}(60) = 24$ minutes—that is, at 5:24 PM. ■

In a suspensión bridge, the main cables are of parabolic shape because if the total weight of a bridge is uniformly distributed along its length, the only cable shape that will bear the load evenly is that of a parabola.

Example 9 A suspension bridge with weight uniformly distributed along its length has twin towers that extend 100 meters above the road surface and are 400 meters apart. The cables are parabolic in shape and touch the road surface at the center of the bridge. Find the height of the cables at a point 100 meters from the center. (The road is assumed to be level.)

Solution We begin by choosing the placement of the coordinate axes so that the x-axis coincides with the road surface and the origin coincides with the center of the bridge. As a result, the twin towers will be vertical (height 100 meters) and located 200 meters from the center. Also, the cable, which has the shape of a parabola, will extend from the towers, open up, and have its vertex at (0, 0). As illustrated in Figure 13, the choice of placement of the axes enables us to identify the equation of the parabola as $y = ax^2$, $a > 0$. We also can see that the points (200, 100) and (−200, 100) are on the graph.

Figure 13

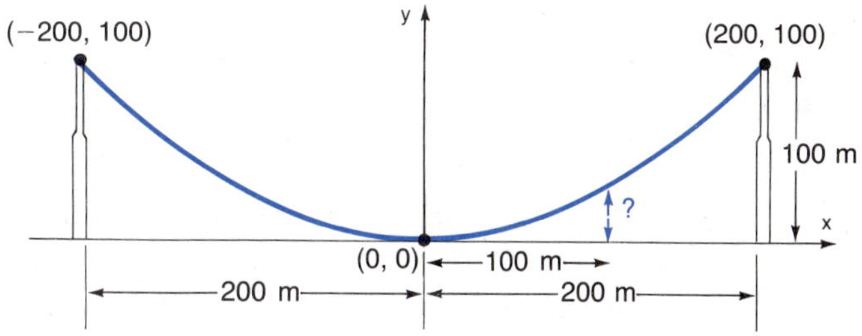

Based on these facts, we can find the value of a in $y = ax^2$:

$$y = ax^2$$
$$100 = a(200)^2$$
$$a = \frac{1}{400}$$

The equation of the parabola is therefore

$$y = \frac{1}{400}x^2$$

The height of the cable when $x = 100$ is

$$y = \frac{1}{400}(100)^2 = 25$$

Thus, the cable is 25 meters high at a distance of 100 meters from the center of the bridge. ■

EXERCISE 5.1

In Problems 1–20, graph the function f by starting with the graph of $y = x^2$ and using shifting, compressing, stretching, and/or reflection.

1. $f(x) = \frac{1}{4}x^2$
2. $f(x) = 2x^2$
3. $f(x) = \frac{1}{4}x^2 - 2$
4. $f(x) = 2x^2 - 3$
5. $f(x) = \frac{1}{4}x^2 + 2$
6. $f(x) = 2x^2 + 4$
7. $f(x) = x^2 + 1$
8. $f(x) = x^2 - 2$
9. $f(x) = -x^2 + 1$
10. $f(x) = -x^2 - 2$
11. $f(x) = \frac{1}{3}x^2 + 1$
12. $f(x) = -2x^2 - 2$
13. $f(x) = x^2 + 4x + 2$
14. $f(x) = x^2 - 6x - 1$
15. $f(x) = 2x^2 - 4x + 1$
16. $f(x) = 3x^2 + 6x$
17. $f(x) = -x^2 - 2x$
18. $f(x) = -2x^2 + 6x + 2$
19. $f(x) = \frac{1}{2}x^2 + x - 1$
20. $f(x) = \frac{2}{3}x^2 + \frac{4}{3}x - 1$

In Problems 21–34, graph each quadratic function by determining whether its graph opens up or down and by finding its vertex, axis of symmetry, y-intercept, and x-intercepts, if any.

21. $f(x) = x^2 + 2x - 3$
22. $f(x) = x^2 - 2x - 8$
23. $f(x) = -x^2 - 3x + 4$
24. $f(x) = -x^2 + x + 2$
25. $f(x) = x^2 + 2x + 1$
26. $f(x) = -x^2 + 4x - 4$
27. $f(x) = 2x^2 - x + 2$
28. $f(x) = 4x^2 - 2x + 1$
29. $f(x) = -2x^2 + 2x - 3$
30. $f(x) = -3x^2 + 3x - 2$
C **31.** $f(x) = 3x^2 - 6x + 2$
C **32.** $f(x) = 2x^2 - 5x + 3$
C **33.** $f(x) = -4x^2 - 6x + 2$
C **34.** $f(x) = 3x^2 - 8x + 2$

In Problems 35–40, determine whether the given quadratic function has a maximum value or a minimum value, and then find the value.

35. $f(x) = 6x^2 + 12x - 3$
36. $f(x) = 4x^2 - 8x + 2$
37. $f(x) = -x^2 + 10x - 4$
38. $f(x) = -2x^2 + 8x + 3$
39. $f(x) = -3x^2 + 12x + 1$
40. $f(x) = 4x^2 - 4x$

41. On one set of coordinate axes, graph the family of parabolas $f(x) = x^2 + 2x + c$, for $c = -3$, $c = 0$, and $c = 1$. Can you describe the characteristics of a member of this family?

42. Repeat Problem 41 for the family of parabolas $f(x) = x^2 + cx$.

43. Find two positive numbers whose sum is 30 and whose product is a maximum.

44. Find two numbers whose difference is 50 and whose product is a minimum.

45. Suppose the manufacturer of a gas clothes dryer has found that when the unit price is p dollars, the revenue R (in dollars) is

$$R = -4p^2 + 4000p$$

What unit price should be established for the dryer to maximize revenue? What is the maximum revenue?

46. A tractor company has found that the revenue from sales of heavy-duty tractors is a function of the unit price p it charges. If the revenue R is

$$R = -\frac{1}{2}p^2 + 1900p$$

what unit price p should be charged to maximize revenue? What is the maximum revenue?

47. What is the largest rectangular area that can be enclosed within 100 feet of fencing? What are the dimensions of the rectangle?

48. What are the dimensions of a rectangle of a fixed perimeter P that result in the largest area?

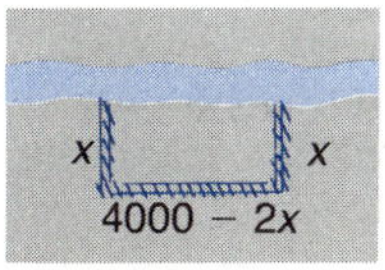

49. A farmer with 4000 meters of fencing wants to enclose a rectangular plot that borders on a river. If the farmer does not fence the side along the river, what is the largest area that can be enclosed? See the figure.

50. A farmer with 3000 meters of fencing wants to enclose a rectangular plot that borders on a straight highway. If the farmer does not fence the side along the highway, what is the largest area that can be enclosed?

51. A farmer with 30,000 meters of fencing wants to enclose a rectangular field and then divide it into two plots with a fence parallel to one of the sides. (See the figure.) What is the largest area that can be enclosed?

52. Repeat Problem 51 if the farmer wants to divide the rectangular field into three plots with two fences parallel to one of the sides.

53. A charter flight club charges its members $200 per year. But, for each new member in excess of 60, the charge for every member is reduced by $2. What number of members leads to a maximum revenue?

54. A car rental agency has 24 identical cars. The owner of the agency finds that all the cars can be rented at a price of $10 per day. However, for each $1 increase in rental, one of the cars is not rented. What should be charged to maximize income?

55. An aircraft carrier maintains a constant speed of 10 knots heading due north. At 4:00 PM, the ship's radar detects a destroyer 100 nautical miles due east of the carrier. If the destroyer is heading due west at 20 knots, when will the two ships be closest? (1 knot = 1 nautical mile per hour)

C **56.** An air traffic controller sees two aircraft flying at the same altitude on his screen. One, a Piper Cub, is headed due west at 150 miles per hour. The other, a Lear jet, is 5 miles due north of the Piper and is headed due south at 400 miles per hour. How close will the two aircraft come to each other?

57. A suspension bridge with weight uniformly distributed along its length has twin towers that extend 75 meters above the road surface and are 300 meters apart. The cables are parabolic in shape and are suspended from the tops of the towers. The cables touch the road surface at the center of the bridge. Find the height of the cables at a point 80 meters from the center. (Assume the road is level.)

58. A parabolic arch has a span of 120 feet and a maximum height of 25 feet. Choose suitable rectangular coordinate axes and find the equation of the parabola. Then calculate the height of the arch at points 10 feet, 20 feet, and 40 feet from the center.

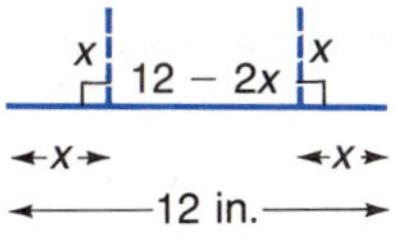

59. A rain gutter is to be made of aluminum sheets that are 12 inches wide by turning up the edges 90°. What depth will provide maximum cross-sectional area and hence allow the most water to flow?

60. A projectile is fired at an inclination of 45° to the horizontal with an initial velocity of v_0 feet per second. If the starting point is the origin, the x-axis is horizontal, and the y-axis is vertical, then the height y (in feet) after a horizontal distance x has been traversed is approximately

$$y = -\frac{32}{v_0^2}x^2 + x$$

Find the maximum height in terms of the initial velocity v_0. If the initial velocity is doubled, what happens to the maximum height? Assuming the ground is flat, how far from the starting point will the projectile land if the initial velocity is 64 feet per second?

C **61.** A Norman window has the shape of a rectangle surmounted by a semicircle of diameter equal to the width of the rectangle (see the figure). If the perimeter of the window is 20 feet, what dimensions will admit the most light (maximize the area)? [*Hint:* Circumference of circle = $2\pi r$; Area of circle = πr^2, where r is the radius of the circle.]

C **62.** A track and field playing area is in the shape of a rectangle with semicircles at each end (see the figure). The inside perimeter of the track is to be 1500 meters. What should the dimensions of the rectangle be so that the area of the rectangle is a maximum?

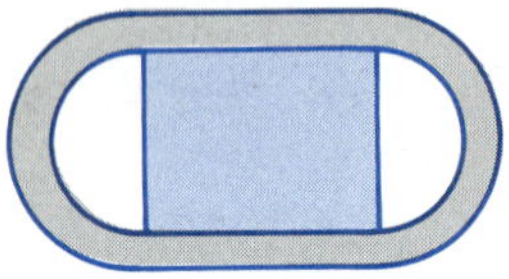

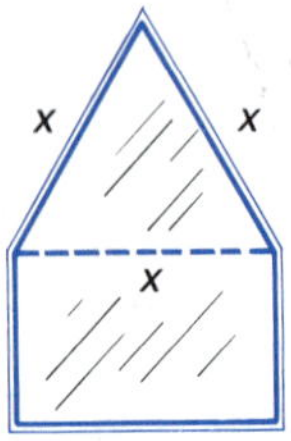

C **63.** A special window has the shape of a rectangle surmounted by an equilateral triangle (see the figure). If the perimeter of the window is 16 feet, what dimensions will admit the most light? [*Hint:* Area of an equilateral triangle = $(\sqrt{3}/4)x^2$, where x is the length of a side of the triangle.]

C **64.** At 4 PM a cruise ship leaves the Port of Miami heading due east at a constant speed of 15 knots. At the same time, a pleasure boat located 100 nautical miles northeast of the Port of Miami is headed due south at a constant speed of 12 knots. When are the two ships closest? How close do they get to each other? (Express your answer in nautical miles; 1 knot = 1 nautical mile per hour.)

65. The graph of the function $f(x) = ax^2 + bx + c$ has vertex at $x = 0$ and passes through the points (0, 2) and (1, 8). Find a, b, and c.

66. The graph of the function $f(x) = ax^2 + bx + c$ has vertex at $x = 1$ and passes through the points (0, 1) and (−1, −8). Find a, b, and c.

67. A self-catalytic chemical reaction results in the formation of a compound that causes the formation ratio to increase. If the reaction rate V is given by

$$V(x) = kx(a - x) \qquad 0 \le x \le a$$

where k is a positive constant, a is the initial amount of the compound, and x is the variable amount of the compound, for what value of x is the reaction rate a maximum?

68. A rectangle has one vertex on the line $y = 10 - x$, $x > 0$, another at the origin, one on the positive x-axis, and one on the positive y-axis. Find the largest area A that can be enclosed by the rectangle.

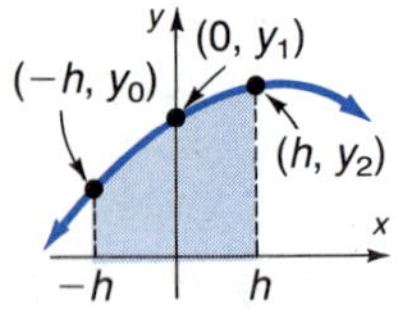

69. The figure shows the graph of $y = ax^2 + bx + c$. Suppose the points $(-h, y_0)$, $(0, y_1)$, and (h, y_2) are on the graph. It can be shown that the area enclosed by the parabola, the x-axis, and the lines $x = -h$ and $x = h$ is

$$\text{Area} = \frac{h}{3}(2ah^2 + 6c)$$

Show that this area also may be given by

$$\text{Area} = \frac{h}{3}(y_0 + 4y_1 + y_2)$$

5.2 Polynomial Functions*

Polynomial Function

A **polynomial function** is a function of the form

$$f(x) = a_n x^n + a_{n-1} x^{n-1} + \cdots + a_1 x + a_0 \qquad (1)$$

where $a_n, a_{n-1}, \ldots, a_1, a_0$ are real numbers and n is a nonnegative integer. The domain consists of all real numbers.

Thus, a polynomial function is a function whose rule is given by a polynomial in one variable (refer to Section 1.4). The **degree** of a polynomial function is the degree of the polynomial in one variable.

Example 1 Determine which of the following are polynomial functions. For those that are, state the degree; for those that are not, tell why not.

(a) $f(x) = 2 - 3x^4$ (b) $g(x) = \sqrt{x}$ (c) $h(x) = \dfrac{x^2 - 2}{x^3 - 1}$

(d) $F(x) = 0$ (e) $G(x) = 8$

Solution
(a) f is a polynomial function of degree 4.
(b) g is not a polynomial function. The variable x is raised to the $\frac{1}{2}$ power, which is not a nonnegative integer.
(c) h is not a polynomial function. It is the ratio of two polynomials, and the polynomial in the denominator is of positive degree.
(d) F is the zero polynomial function; it is not assigned a degree.
(e) G is a nonzero constant function, a polynomial function of degree 0. ■

We have already discussed in detail polynomial functions of degrees 0, 1, and 2. See Table 1 for a summary of the characteristics of the graphs of these polynomial functions.

Table 1

DEGREE	FORM	NAME	GRAPH
No degree	$f(x) = 0$	Zero function	The x-axis
0	$f(x) = a_0, \quad a_0 \neq 0$	Constant function	Horizontal line with y-intercept a_0
1	$f(x) = a_1 x + a_0, \quad a_1 \neq 0$	Linear function	Nonvertical, nonhorizontal line with slope a_1 and y-intercept a_0
2	$f(x) = a_2 x^2 + a_1 x + a_0, \quad a_2 \neq 0$	Quadratic function	Parabola: graph opens up if $a_2 > 0$; graph opens down if $a_2 < 0$

* C Refer to Graphics Calculator Supplement: Activity II.

Power Functions

First, we consider a special kind of polynomial function called a *power function*.

Power Function of Degree n

A **power function of degree n** is a function of the form

$$f(x) = ax^n \tag{2}$$

where a is a real number, $a \neq 0$, and $n > 0$ is an integer.

The graph of a power function of degree 1, $f(x) = ax$, is a straight line, with slope a, that passes through the origin. The graph of a power function of degree 2, $f(x) = ax^2$, is a parabola, with vertex at the origin, that opens up if $a > 0$ and down if $a < 0$.

If we know how to graph a power function of the form $f(x) = x^n$, then a compression or stretch and, perhaps, a reflection about the x-axis will enable us to obtain the graph of $g(x) = ax^n$. Consequently, we shall concentrate on graphing power functions of the form $f(x) = x^n$.

We begin with power functions of even degree of the form $f(x) = x^n$, $n \geq 2$ and n even. The domain of f is the set of all real numbers, and the range is the set of nonnegative real numbers. Such a power function is an even function (do you see why?), and hence its graph is symmetric with respect to the y-axis. Its graph always contains the origin and the points $(-1, 1)$ and $(1, 1)$.

If $n = 2$, the graph is the familiar parabola $y = x^2$ that opens up, with vertex at the origin. If $n \geq 4$, the graph of $f(x) = x^n$, n even, will be closer to the x-axis than the parabola $y = x^2$ if $-1 < x < 1$ and farther from the x-axis than the parabola $y = x^2$ if $x < -1$ or if $x > 1$. Figure 14(a) illustrates this conclusion. Figure 14(b) shows the graphs of $y = x^4$ and $y = x^8$ for further comparison.

From Figure 14, we can see that as n increases, the graph of $f(x) = x^n$, $n \geq 2$ and n even, tends to flatten out near the origin and to increase very rapidly when x is far from 0. For large n, it may appear that the graph coincides with the x-axis near the origin, but it does not; the graph actually touches the x-axis only at the origin (see Table 2). Also, for large n, it may appear that for $x < -1$ or for $x > 1$, the graph is vertical, but it is not; it is only increasing very rapidly in those intervals. If the graphs were enlarged many times, these distinctions would be clear.

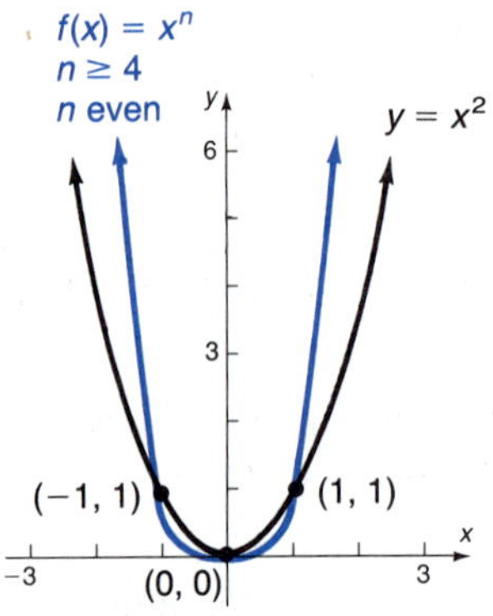

(a)

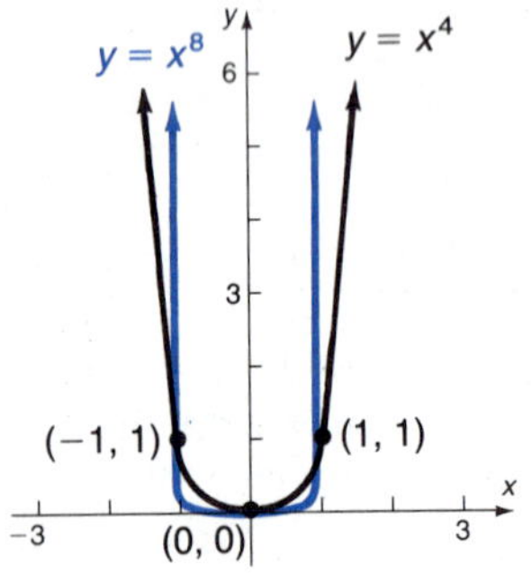

(b)

Figure 14

Table 2

	$x = 0.1$	$x = 0.3$	$x = 0.5$
$f(x) = x^8$	10^{-8}	0.0000656	0.0039063
$f(x) = x^{20}$	10^{-20}	$3.487 \cdot 10^{-11}$	0.000001
$f(x) = x^{40}$	10^{-40}	$1.216 \cdot 10^{-21}$	$9.095 \cdot 10^{-13}$

Now, we consider power functions of odd degree of the form $f(x) = x^n$, $n \geq 3$ and n odd. The domain and range of f are the set of real numbers. Such a power function is an odd function (do you see why?), and hence its graph is symmetric with respect to the origin. Its graph always contains the origin and the points $(-1, -1)$ and $(1, 1)$.

The graph of $f(x) = x^n$ when $n = 3$ has been shown several times and is repeated in Figure 15. If $n \geq 5$, the graph of $f(x) = x^n$, n odd, will be closer to the x-axis than that of $y = x^3$ if $-1 < x < 1$ and farther from the x-axis than that of $y = x^3$ if $x < -1$ or if $x > 1$. Figure 15 also illustrates this conclusion. Figure 16 shows the graph of $y = x^5$ and the graph of $y = x^9$ for further comparison.

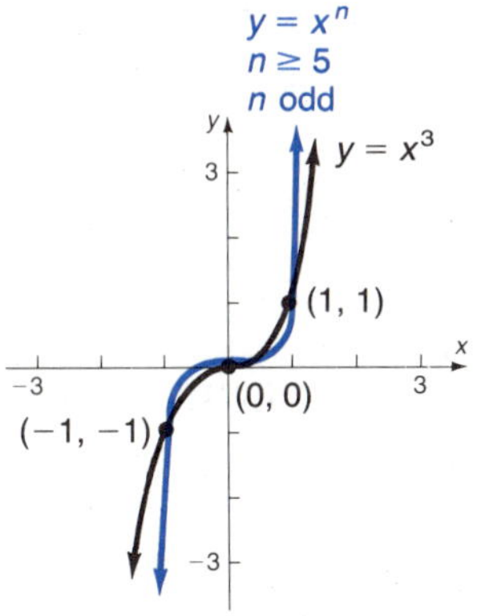

Figure 15

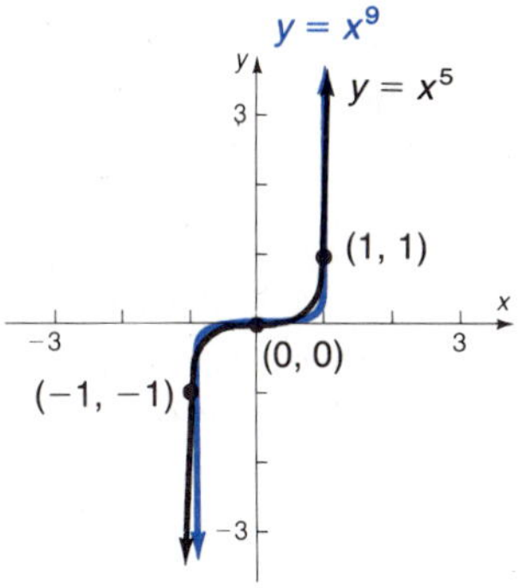

Figure 16

From Figures 15 and 16, we can see that as n increases, the graph of $f(x) = x^n$, $n \geq 3$ and n odd, tends to flatten out near the origin and to become nearly vertical when x is far from 0.

The methods of shifting, compression, stretching, and reflection studied in Section 4.3, when used in conjunction with the facts just presented, enable us to graph a variety of polynomials.

Example 2 Graph: $f(x) = 1 - x^5$

Solution Figure 17 shows the required stages.

Figure 17

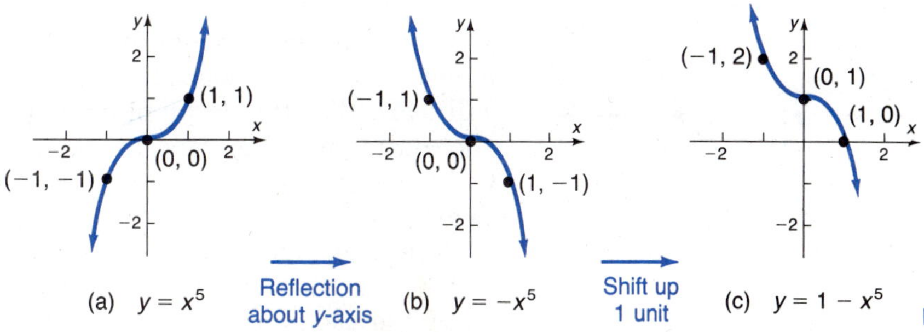

Example 3 Graph: $f(x) = \frac{1}{2}(x - 1)^4$

Solution Figure 18 shows the required stages.

Figure 18

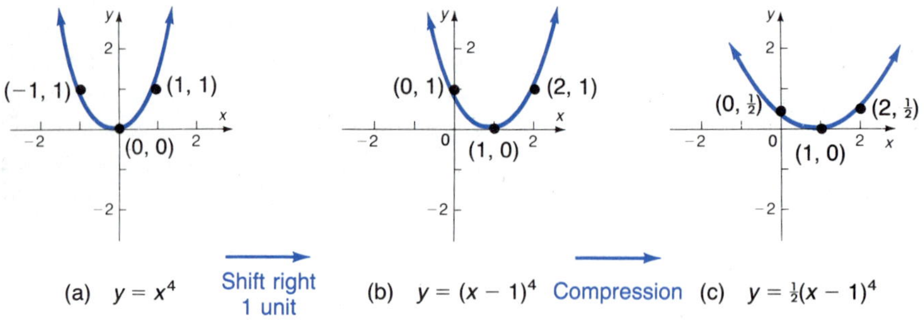

Example 4 Graph: $f(x) = (x + 1)^4 - 3$

Solution Figure 19 shows the required stages.

Figure 19

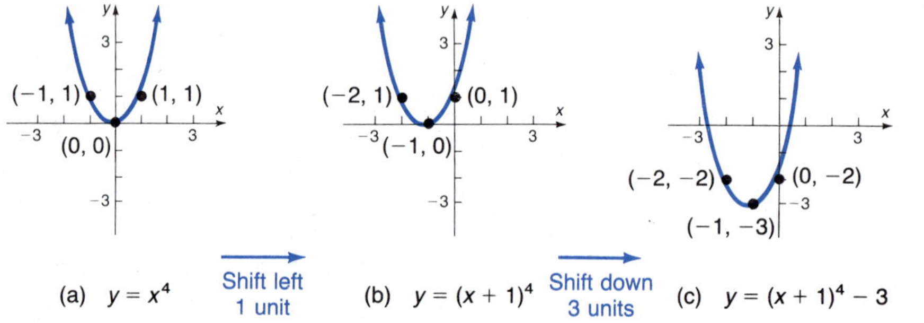

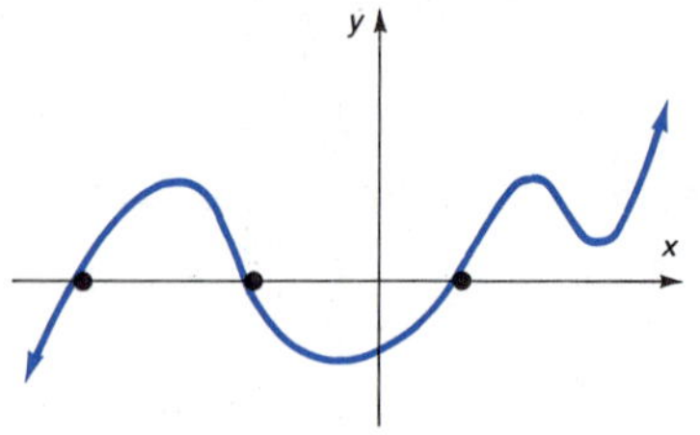

(a) Graph of a polynomial function: smooth, continuous

Figure 20(a)

Graphing Other Polynomials

To graph most polynomial functions of degree 3 or higher requires techniques beyond the scope of this text. If you take a course in calculus you will learn that the graph of every polynomial function is both smooth and continuous, like the graph illustrated in Figure 20(a). Such graphs will never contain sharp corners or gaps like the graph illustrated in Figure 20(b).

Notice in Figure 20(a) that the x-intercepts of the graph divide the x-axis into intervals on which the graph is either above the x-axis or below the x-axis. We will make use of this characteristic shortly.

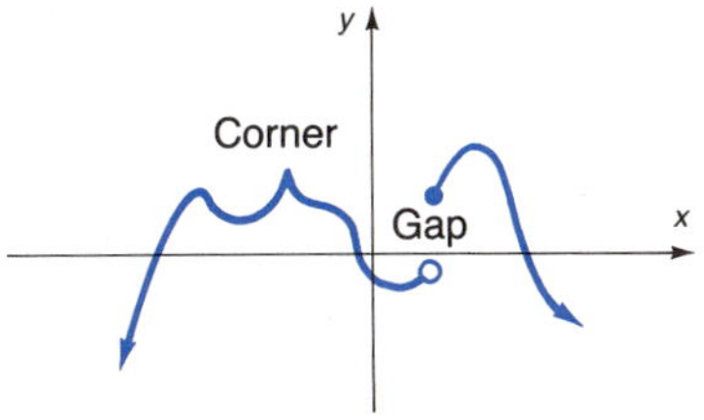

(b) Cannot be the graph of a polynomial function

Figure 20(b)

If a polynomial function f is factored completely, it is easy to solve the equation $f(x) = 0$ and locate the x-intercepts of the graph. For example, if $f(x) = (x - 1)^2(x + 3)$, then the solutions of the equation

$$f(x) = (x - 1)^2(x + 3) = 0$$

are easily identified as 1 and -3. In general, if f is a polynomial function and r is a real number for which $f(r) = 0$, then r is called a (real) **zero of f**, or **root of f**. Thus, the real zeros of a polynomial function are the x-intercepts of its graph. Also, if $x - r$ is a factor of a polynomial f, then $f(r) = 0$ and so r is a zero of f. If the same factor $x - r$ occurs more than once, then r is called a **repeated**, or **multiple**, **zero of f**. More precisely, we have the following definition.

Zero of Multiplicity m

If $(x - r)^m$ is a factor of a polynomial f and $(x - r)^{m+1}$ is not a factor of f, then r is called a **zero of multiplicity m of f**.

Example 5 For the polynomial

$$f(x) = 5(x - 2)(x + 3)^2\left(x - \tfrac{1}{2}\right)^4$$

2 is a zero of multiplicity 1

-3 is a zero of multiplicity 2

$\frac{1}{2}$ is a zero of multiplicity 4 ■

Suppose it is possible to factor completely a polynomial function and, as a result, locate all the x-intercepts of its graph (the real zeros of the function). As mentioned earlier, these x-intercepts then divide the x-axis into open intervals, and on each such interval, the graph of the polynomial will be either above or below the x-axis. Let's look at an example.

Example 6 (a) Find the x- and y-intercepts of the polynomial

$$f(x) = x^2(x - 4)(x + 1)$$

(b) Use the x-intercepts to find the intervals on which the graph of f is above the x-axis and the intervals on which the graph of f is below the x-axis.

Solution (a) The y-intercept is $f(0) = 0$. The x-intercepts satisfy the equation

$$f(x) = x^2(x - 4)(x + 1) = 0$$

So

$$\begin{array}{ccccc} x^2 = 0 & \text{or} & x - 4 = 0 & \text{or} & x + 1 = 0 \\ x = 0 & & x = 4 & & x = -1 \end{array}$$

The x-intercepts are -1, 0, and 4.

(b) The three x-intercepts divide the x-axis into four intervals: $x < -1$, $-1 < x < 0$, $0 < x < 4$, and $x > 4$. See Figure 21:

Figure 21

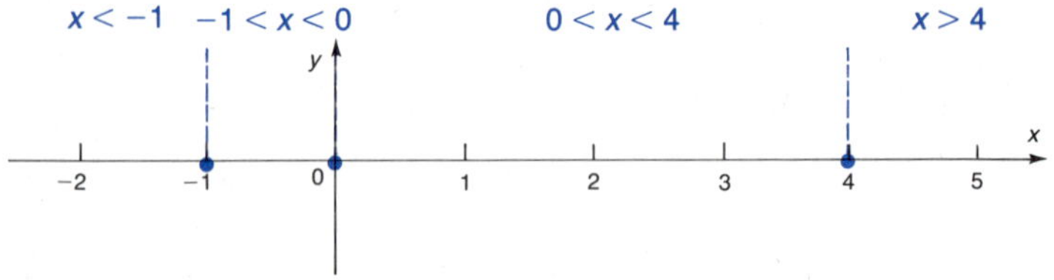

Since the graph of f crosses (or touches) the x-axis only at $x = -1$, $x = 0$, and $x = 4$, it follows that the graph of f is either above or below the x-axis in each of these intervals. To determine which is the case, we select a test number in each interval (refer to Section 2.8). Next, we evaluate the function f at each test number to obtain the sign of f on the interval. The sign of f, in turn, tells us whether the graph is above or below the x-axis. Table 3 illustrates this process:

Table 3

INTERVAL	TEST NUMBER	VALUE OF x^2	VALUE OF $x + 1$	VALUE OF $x - 4$	VALUE OF $f(x)$	SIGN OF $f(x)$	GRAPH OF f
$x < -1$	-2	4	-1	-6	24	+	Above x-axis
$-1 < x < 0$	$-\frac{1}{2}$	$\frac{1}{4}$	$\frac{1}{2}$	$-\frac{9}{2}$	$-\frac{9}{16}$	−	Below x-axis
$0 < x < 4$	1	1	2	-3	-6	−	Below x-axis
$x > 4$	5	25	6	1	150	+	Above x-axis

Thus, the graph of f is above the x-axis for $x < -1$ and $x > 4$, and the graph is below the x-axis for $-1 < x < 0$ and $0 < x < 4$. ■

What more can we determine about the graph of the function f in Example 6? From Table 3, we know the graph will cross the x-axis at $x = -1$ and at $x = 4$; it just touches the x-axis at $x = 0$, since the graph is below the x-axis on both sides of $(0, 0)$. The test numbers we used to fill in Table 3 provide four additional points on the graph: $(-2, 24)$, $\left(-\frac{1}{2}, -\frac{9}{16}\right)$, $(1, -6)$, and $(5, 150)$. Figure 22 illustrates what we know so far. Notice how we scaled the y-axis to accommodate the points we know are on the graph.

We are still missing information about just how low or high the graph actually goes on each interval. Also, we cannot be sure of the general shape of the graph. Calculus provides the tools needed to determine that the low points are $(-0.7, -0.7)$ and $(2.9, -36.1)$, and the graph in Figure 23 is the graph of f.

Figure 22

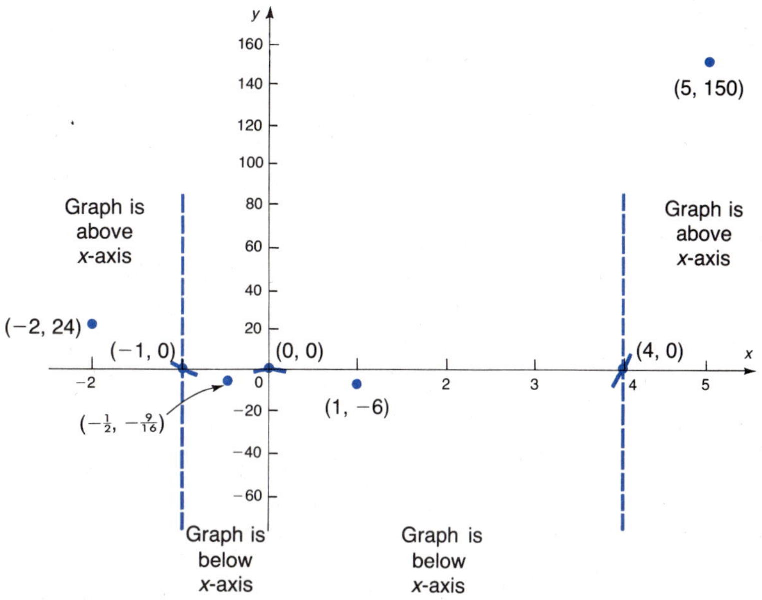

Figure 23
$f(x) = x^2(x - 4)(x + 1)$

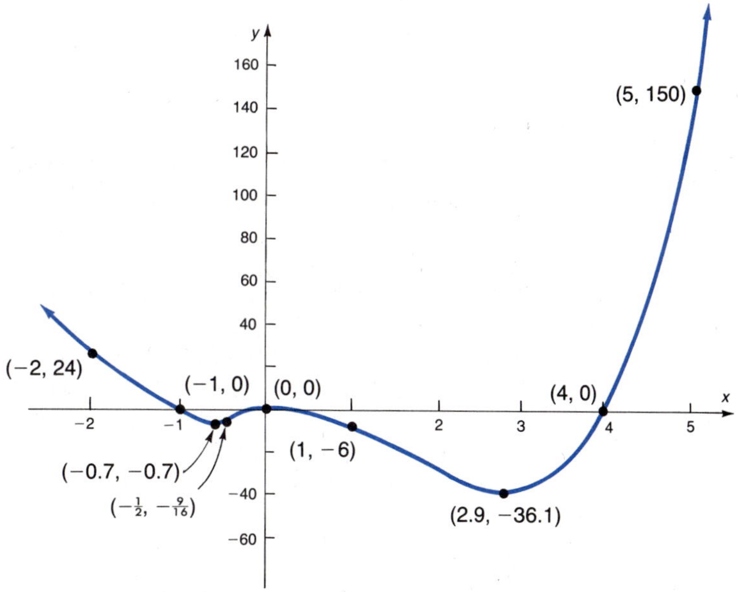

Notice the points $(-0.7, -0.7)$, $(0, 0)$, and $(2.9, -36.1)$ in Figure 23. At these points, the graph changes direction, or turns, so we refer to such points as **turning points**. Needless to say, you will not be asked to identify the turning points of a graph. For now, all that will be expected is that you locate intercepts, determine on what intervals the graph is above and below the x-axis, plot a few additional points on the graph, and connect them with a smooth curve.

Before presenting another example, it is interesting to note some other features of the graph of a polynomial function. In calculus it is shown that if f is a polynomial function of degree n, then the number of turning points of the graph of f cannot exceed $n - 1$. Notice that the graph of $f(x) = x^2(x - 4)(x + 1)$ shown in Figure 23 has exactly three turning points, and the degree of the polynomial function is 4.

Also, the function $f(x) = x^2(x - 4)(x + 1)$ has the number 0 as a zero of multiplicity 2, and the numbers -1 and 4 as zeros of multiplicity 1. An examination of Table 3 reveals that the sign of $f(x)$ does not change from one side to the other side of 0 and does change from one side to the other side of -1 and 4. This suggests the following result:

r Is a Zero of Even Multiplicity	Sign of $f(x)$ does not change from one side to the other side of r.	Graph **touches** x-axis at r.
r Is a Zero of Odd Multiplicity	Sign of $f(x)$ changes from one side to the other side of r.	Graph **crosses** x-axis at r.

One More Comment: For large values of x, the corresponding values of a polynomial function

$$f(x) = a_n x^n + a_{n-1}x^{n-1} + \cdots + a_1 x + a_0 \qquad a_n \neq 0$$

behave like the values of the power function $g(x) = a_n x^n$. As an illustration, compare the values of $f(x) = 2x^5 + 50x^4 + 100x^3$ and $g(x) = 2x^5$ at $x = 10{,}000$ and at $x = 100{,}000$ (use your calculator). For example, the graph of $f(x) = x^2(x - 4)(x + 1)$ in Figure 23 behaves like the graph of $g(x) = x^4$ for x very large. Therefore, the graph of f will rise very rapidly as x gets larger in either the positive or the negative direction.

The box below summarizes some features of the graph of a polynomial function:

Summary: Graph of a Polynomial Function $f(x) = a_n x^n + a_{n-1}x^{n-1} + \cdots + a_1 x + a_0$, $a_n \neq 0$

Degree of the polynomial f: n

Maximum number of turning points: $n - 1$

At zero of even multiplicity: graph of f touches x-axis

At zero of odd multiplicity: graph of f crosses x-axis

For large x, graph of f behaves like graph of $y = a_n x^n$.

Example 7 (a) Find the x- and y-intercepts of the polynomial $f(x) = 6x^2(x + 4)$.

(b) Determine whether the graph of f crosses or touches the x-axis at each x-intercept.

(c) Find the power function that approximates f for large values of x.

(d) Determine the maximum number of turning points on the graph of f.

Solution (a) The y-intercept is $f(0) = 0$. The x-intercepts are 0 and -4.

(b) The graph of f crosses the x-axis at $(-4, 0)$ and touches the x-axis at $(0, 0)$.

(c) The power function $y = 6x^3$ approximates f for large values of x.

(d) The maximum number of turning points is 2. ■

Example 8 (a) Find the x- and y-intercepts of the polynomial $f(x) = x^3 + x^2 - 12x$.

(b) Use the x-intercepts to find the intervals on which the graph of f is above the x-axis and the intervals on which the graph of f is below the x-axis.

(c) Obtain several other points on the graph and connect them with a smooth curve.

Solution (a) The y-intercept is $f(0) = 0$. To find the x-intercepts, if any, we factor f:

$$f(x) = x^3 + x^2 - 12x = x(x^2 + x - 12) = x(x + 4)(x - 3)$$

Solving the equation $f(x) = x(x + 4)(x - 3) = 0$, we find that the x-intercepts, or zeros of f, are -4, 0, and 3. Since each is of multiplicity 1, the graph of f will cross the x-axis at each of these x-intercepts.

(b) The three x-intercepts divide the x-axis into four intervals: $x < -4$, $-4 < x < 0$, $0 < x < 3$, and $x > 3$. To determine the sign of $f(x)$ on each interval, we construct Table 4:

Table 4

INTERVAL	TEST NUMBER	VALUE OF $f(x)$	SIGN OF $f(x)$	GRAPH OF f
$x < -4$	-5	-40	$-$	Below x-axis
$-4 < x < 0$	-2	20	$+$	Above x-axis
$0 < x < 3$	1	-10	$-$	Below x-axis
$x > 3$	4	32	$+$	Above x-axis

(c) The graph of f for large values of x will behave like $y = x^3$. Also, the graph will contain at most two turning points. Using the points

$(-5, -40)$, $(-2, 20)$, $(1, -10)$, and $(4, 32)$ from Table 4, we arrive at the graph of $f(x) = x^3 + x^2 - 12x$ shown in Figure 24. [Again, the turning points $(1.7, -12.6)$ amd $(-2.4, 21)$ were obtained using calculus.]

Figure 24
$f(x) = x^3 + x^2 - 12x$

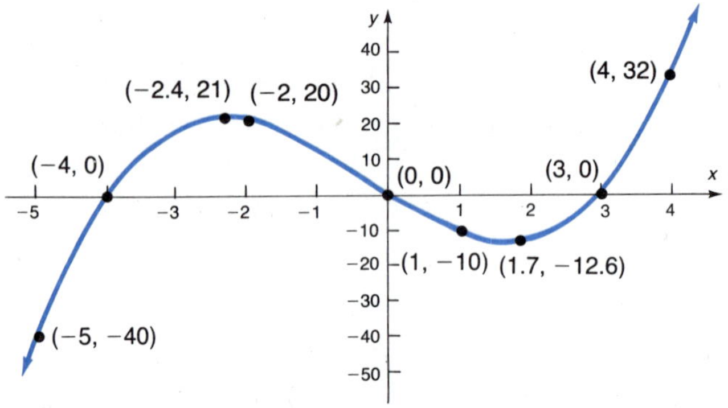

Examples 6, 7, and 8 point out the importance of being able to factor polynomials in order to obtain information about their graphs. The ability to evaluate a polynomial at a given number is equally important. Later in this chapter, we discuss various techniques for factoring and evaluating polynomials.

Summary

To sketch the graph of a polynomial function $y = f(x)$, follow these steps:

Steps for Graphing a Polynomial

STEP 1: (a) Find the x-intercepts, if any, by solving the equation $f(x) = 0$.
(b) Find the y-intercept by letting $x = 0$ and finding the value of $f(0)$.

STEP 2: Use the x-intercepts to find the intervals on which the graph of f is above and below the x-axis. During this step, several additional points on the graph are located.

STEP 3: Connect the points with a smooth curve.

EXERCISE 5.2

In Problems 1–10, determine which functions are polynomial functions. For those that are, state the degree. For those that are not, tell why not.

1. $f(x) = 2x - x^3$

2. $f(x) = 3x^2 - 4x^4$

3. $g(x) = \dfrac{1 - x^2}{2}$

4. $h(x) = 3 - \frac{1}{2}x$

5. $f(x) = 1 - \dfrac{1}{x}$

6. $f(x) = x(x - 1)$

7. $g(x) = x^{3/2} - x^2 + 2$

8. $h(x) = \sqrt{x}(\sqrt{x} - 1)$

9. $F(x) = 5x^4 - \pi x^3 + \frac{1}{2}$

10. $F(x) = \dfrac{x^2 - 5}{x^3}$

In Problems 11–18, use the graph of $y = x^4$ to graph each function.

11. $f(x) = (x - 1)^4$

12. $f(x) = x^4 + 3$

13. $f(x) = \frac{1}{2}x^4$

14. $f(x) = -x^4$

15. $f(x) = 2(x + 1)^4 + 1$

16. $f(x) = 3 - (x + 2)^4$

17. $f(x) = -\frac{1}{2}(x - 2)^4 - 1$

18. $f(x) = 1 - 2(x + 1)^4$

In Problems 19–28, for each polynomial function, list each real zero and its multiplicity.

19. $f(x) = 3(x - 4)(x + 5)^2$

20. $f(x) = 4(x + 1)(x - 3)^3$

21. $f(x) = 4(x^2 + 1)(x - 2)^3$

22. $f(x) = 2(x - 3)(x + 4)^3$

23. $f(x) = -2\left(x + \frac{1}{2}\right)^2(x^2 + 4)^2$

24. $f(x) = \left(x - \frac{1}{3}\right)^2(x - 1)^3$

25. $f(x) = (x - 5)^3(x + 4)^2$

26. $f(x) = (x + \sqrt{3})^2(x - 2)^4$

27. $f(x) = 3(x^2 + 4)(x^2 + 9)^2$

28. $f(x) = -2(x^2 + 1)^3$

In Problems 29–38, for each polynomial function f:

(a) Find the x- and y-intercepts of f.

(b) Determine whether the graph of f crosses or touches the x-axis at each x-intercept.

(c) Find the power function that approximates f for large values of x.

(d) Determine the maximum number of turning points on the graph of f.

29. $f(x) = (x - 1)^2$

30. $f(x) = (x - 2)^3$

31. $f(x) = x^2(x - 1)$

32. $f(x) = x(x + 2)^2$

33. $f(x) = 6x^3(x + 4)$

34. $f(x) = 5x(x - 1)^3$

35. $f(x) = (x - 1)(x - 2)(x - 3)$

36. $f(x) = x(x + 1)(x + 2)$

37. $f(x) = -4x^2(x + 2)$

38. $f(x) = -\frac{1}{2}x^3(x + 4)$

In Problems 39–56:

(a) Find the x- and y-intercepts of each polynomial function f.

(b) Determine whether the graph of f touches or crosses the x-axis at each x-intercept.

(c) Find the intervals on which the graph of f is above the x-axis and those on which it is below the x-axis.

(d) Obtain several other points on the graph and connect them with a smooth curve.

39. $f(x) = x(x - 2)(x + 4)$

40. $f(x) = x(x + 4)(x - 3)$

41. $f(x) = x^2(x - 3)$

42. $f(x) = x(x + 2)^2$

43. $f(x) = 4x - x^3$

44. $f(x) = x - x^3$

45. $f(x) = x^3 - 2x^2 - 3x$

46. $f(x) = x^3 + 2x^2 - 3x$

47. $f(x) = x^2(x - 2)(x + 2)$

48. $f(x) = x^2(x - 3)(x + 4)$

49. $f(x) = x^2(x - 2)^2$

50. $f(x) = x^3(x - 3)$

51. $f(x) = x^2(x - 3)(x + 1)$

52. $f(x) = x^2(x - 3)(x - 1)$

53. $f(x) = x(x + 2)(x - 4)(x - 6)$

54. $f(x) = x(x - 2)(x + 2)(x + 4)$

55. $f(x) = x^2(x - 2)(x^2 + 3)$

56. $f(x) = x^2(x^2 + 1)(x + 4)$

5.3 Rational Functions*

Ratios of integers are called *rational numbers*. Similarly, ratios of polynomial functions are called *rational functions*.

Rational Function

A **rational function** is a function of the form

$$R(x) = \frac{p(x)}{q(x)}$$

where p and q are polynomial functions and q is not the zero polynomial. The domain consists of all real numbers except those for which the denominator q is 0.

Example 1 (a) The domain of $R(x) = \dfrac{2x^2 - 4}{x + 5}$ consists of all real numbers x except -5.

(b) The domain of $R(x) = \dfrac{1}{x^2 - 4}$ consists of all real numbers x except -2 and 2.

(c) The domain of $R(x) = \dfrac{x^3}{x^2 + 1}$ consists of all real numbers.

(d) The domain of $R(x) = \dfrac{-x^2 + 2}{3}$ consists of all real numbers.

(e) The domain of $R(x) = \dfrac{x^2 - 1}{x - 1}$ consists of all real numbers x except 1. ■

* [C] Refer to Graphics Calculator Supplement: Activity II.

Notice in Example 1(e) that

$$R(x) = \frac{x^2 - 1}{x - 1} \quad \text{and} \quad f(x) = x + 1$$

are not equal, since the domain of R is $\{x|x \neq 1\}$ and the domain of f is all real numbers.

If $R(x) = p(x)/q(x)$ is a rational function and if p and q have no common factors, then the rational function R is said to be in **lowest terms**. We shall assume throughout this section that all rational functions are in lowest terms.

For a rational function $R(x) = p(x)/q(x)$ in lowest terms, the zeros, if any, of the numerator are the x-intercepts of the graph of R and so will play a major role in graphing R. The zeros of the denominator of R (that is, the numbers x, if any, for which $q(x) = 0$), although not in the domain of R, also play a major role in the graph of R. We will discuss this role shortly.

Example 2 Graph: $H(x) = \dfrac{1}{x^2}$

Solution The domain of $H(x) = 1/x^2$ consists of all real numbers x except 0. Thus, the graph has no y-intercept, because x can never equal 0. The graph has no x-intercept, because the equation $H(x) = 0$ has no solution. Therefore, the graph of H will not cross the coordinate axes. Because

$$H(-x) = \frac{1}{(-x)^2} = \frac{1}{x^2} = H(x)$$

H is an even function, so its graph is symmetric with respect to the y-axis. Table 5 shows the behavior of $H(x) = 1/x^2$ for selected positive numbers x (we will use symmetry to obtain the graph of H when $x < 0$). From Table 5, it is apparent that as the values of x approach (get closer to) 0, the values of $H(x)$ become larger and larger positive numbers. When this happens, we say that H is **unbounded in the positive direction**. We symbolize this by writing $H \to \infty$ (read as "H **approaches infinity**"). Look again at Table 5. As $x \to \infty$, the values of $H(x)$ approach 0. Figure 25 illustrates the graph.

Table 5

x	$H(x) = 1/x^2$
$\frac{1}{100}$	10,000
$\frac{1}{10}$	100
$\frac{1}{2}$	4
1	1
2	$\frac{1}{4}$
10	$\frac{1}{100}$
100	$\frac{1}{10,000}$

Figure 25
$H(x) = \dfrac{1}{x^2}$

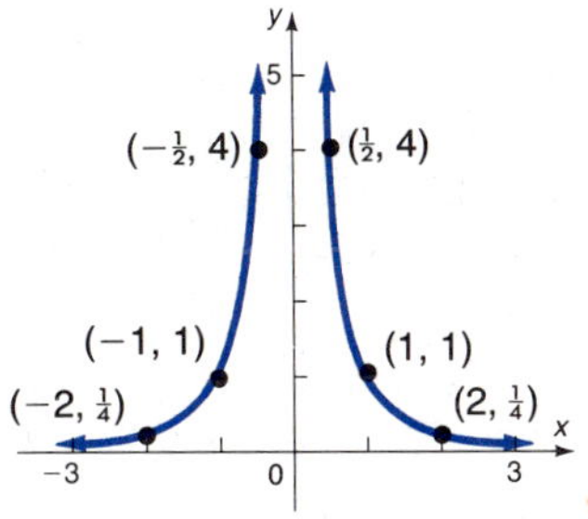

■

Sometimes the techniques of shifting, compressing, stretching, and reflection can be used to graph a rational function.

Example 3 Graph the rational function: $R(x) = \dfrac{1}{(x-2)^2} + 1$

Solution First, we take note of the fact that the domain of R consists of all real numbers except $x = 2$. To graph R, we start with the graph of $y = 1/x^2$. See Figure 26 for the steps.

Figure 26

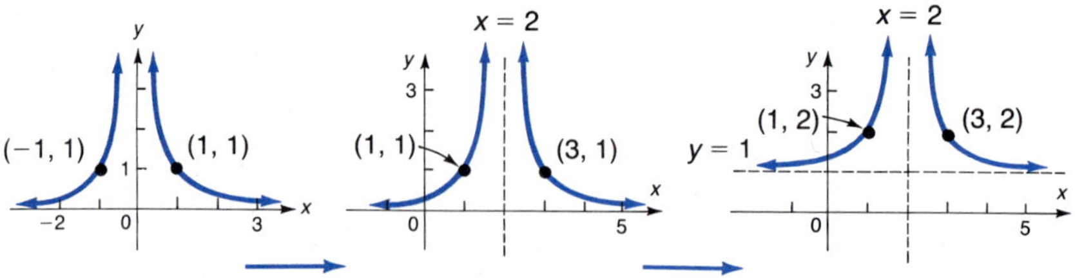

(a) $y = \dfrac{1}{x^2}$ → Shift right 2 units → (b) $y = \dfrac{1}{(x-2)^2}$ → Shift up 1 unit → (c) $y = \dfrac{1}{(x-2)^2} + 1$

■

Asymptotes

In Figure 26(c), notice that as the values of x become more negative—that is, as x becomes **unbounded in the negative direction** ($x \to -\infty$, read as "x **approaches negative infinity**")—the values $R(x)$ approach 1. In fact, we can conclude the following from Figure 26(c):

1. As $x \to -\infty$, the values $R(x)$ approach 1.
2. As x approaches 2, the values $R(x) \to \infty$.
3. As $x \to \infty$, the values $R(x)$ approach 1.

This behavior of the graph is depicted by the dashed vertical line $x = 2$ and the dashed horizontal line $y = 1$. These lines are called *asymptotes* of the graph, which we define as follows:

Let R denote a function:

Horizontal Asymptote

If, as $x \to -\infty$ or as $x \to \infty$, the values of $R(x)$ approach some fixed number L, then the line $y = L$ is a **horizontal asymptote** of the graph of R.

Vertical Asymptote

If, as x approaches some number c, the values $|R(x)| \to \infty$, then the line $x = c$ is a **vertical asymptote** of the graph of R.

Even though asymptotes of a function are not part of the graph of the function, they provide information about the way the graph looks. Figure 27 illustrates some of the possibilities.

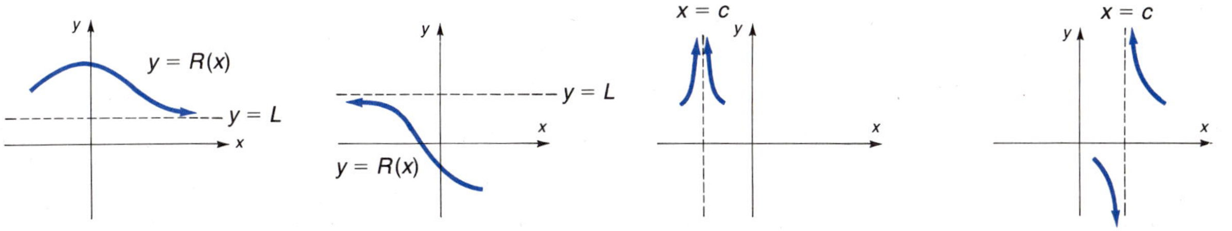

(a) As $x \to \infty$, the values of $R(x)$ approach L; $y = L$ is a horizontal asymptote

(b) As $x \to -\infty$, the values of $R(x)$ approach L; $y = L$ is a horizontal asymptote

(c) As x approaches c, the values $|R(x)| \to \infty$; $x = c$ is a vertical asymptote

(d) As x approaches c, the values $|R(x)| \to \infty$; $x = c$ is a vertical asymptote

Figure 27

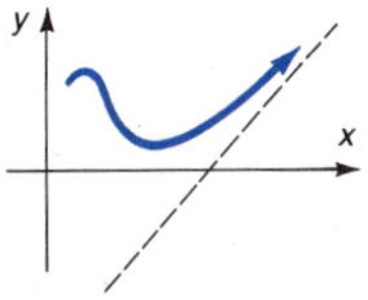

Figure 28
Oblique asymptote

Thus, an asymptote is a line that a certain part of the graph of a function gets closer and closer to, but never touches. However, other parts of the graph of the function may intersect a nonvertical asymptote. The graph of the function will never intersect a vertical asymptote. Notice that a horizontal asymptote, when it occurs, describes a certain behavior of the graph as $x \to \infty$ or as $x \to -\infty$, while a vertical asymptote, when it occurs, describes a certain behavior of the graph when x is close to some number c.

If an asymptote is neither horizontal nor vertical, it is called **oblique**. Figure 28 shows an oblique asymptote.

Finding Asymptotes

The vertical asymptotes, if any, of a rational function $R(x) = p(x)/q(x)$ are found by factoring the denominator $q(x)$. Suppose $x - r$ is a factor of the denominator. Now, as x approaches r, symbolized as $x \to r$, the values of $x - r$ approach 0, causing the ratio to become unbounded—that is, causing $|R(x)| \to \infty$. Based on the definition, we conclude that the line $x = r$ is a vertical asymptote.

Theorem
Locating Vertical Asymptotes

A rational function $R(x) = p(x)/q(x)$ in lowest terms will have a vertical asymptote $x = r$ if $x - r$ is a factor of the denominator q. ■

Warning: If a rational function is not in lowest terms, an application of this theorem may result in an incorrect listing of vertical asymptotes.

Example 4 Find the vertical asymptotes, if any, of the graph of each rational function.

(a) $R(x) = \dfrac{x}{x^2 - 4}$ (b) $F(x) = \dfrac{x + 3}{x - 1}$ (c) $H(x) = \dfrac{x^2}{x^2 + 1}$

Solution (a) The factors of the denominator are $x^2 - 4 = (x + 2)(x - 2)$. Hence, the lines $x = -2$ and $x = 2$ are the vertical asymptotes of the graph of R.

(b) The only factor of the denominator is $x - 1$. Hence, the line $x = 1$ is the only vertical asymptote of the graph of F.

(c) The denominator has no factors of the form $x - r$. Hence, the graph of H has no vertical asymptotes. ■

As Example 4 points out, rational functions can have no vertical asymptotes, one vertical asymptote, or more than one vertical asymptote. However, the graph of a rational function will never intersect any of its vertical asymptotes. (Do you know why?)

The procedure for finding horizontal and oblique asymptotes is somewhat more involved. To find such asymptotes, we need to know how the values of a function behave as $x \to -\infty$ or as $x \to \infty$.

If a rational function $R(x)$ is **proper**—that is, if the degree of the numerator is less than the degree of the denominator—then as $x \to -\infty$ or as $x \to \infty$, the values of $R(x)$ approach 0. Consequently, the line $y = 0$ (the x-axis) is the only horizontal asymptote of the graph.

If a rational function $R(x)$ is **improper**—that is, if the degree of the numerator is greater than or equal to the degree of the denominator—we must use long division to write the rational function as the sum of a polynomial $f(x)$ plus a proper rational function $r(x)$. That is, we write

$$R(x) = \frac{p(x)}{q(x)} = f(x) + r(x)$$

where $f(x)$ is a polynomial and $r(x)$ is a proper rational function. Since $r(x)$ is proper, then $r(x) \to 0$ as $x \to -\infty$ or as $x \to \infty$. Thus,

$$R(x) = \frac{p(x)}{q(x)} \to f(x) \qquad \text{as} \qquad x \to -\infty \text{ or as } x \to \infty$$

The possibilities are listed below:

1. If (Degree of p) = (Degree of q), then $f(x) = b$, a constant, and the line $y = b$ is a horizontal asymptote of the graph of R.
2. If (Degree of p) = (Degree of q) + 1, then $f(x) = ax + b$, $a \neq 0$, and the line $y = ax + b$ is an oblique asymptote of the graph of R.
3. If (Degree of p) > (Degree of q) + 1, then the graph of R approaches the graph of f, and there are no horizontal or oblique asymptotes.

The following examples demonstrate the conclusions drawn above.

Example 5 Find the horizontal or oblique asymptotes, if any, of the graph of

$$R(x) = \frac{x - 12}{4x^2 + x + 1}$$

Solution The rational function R is proper, since the degree of the numerator (1) is less than the degree of the denominator (2). We conclude that the line $y = 0$ is a horizontal asymptote of the graph of R.

To see why this is so, we divide the numerator and denominator of R by x^2 (the highest power of x in the denominator) and let $x \to -\infty$ or $x \to \infty$. Then,

$$R(x) = \frac{x - 12}{4x^2 + x + 1} = \frac{\dfrac{x}{x^2} - \dfrac{12}{x^2}}{\dfrac{4x^2}{x^2} + \dfrac{x}{x^2} + \dfrac{1}{x^2}}$$

$$= \frac{\dfrac{1}{x} - \dfrac{12}{x^2}}{4 + \dfrac{1}{x} + \dfrac{1}{x^2}} \to \frac{0}{4} = 0 \qquad \text{as } x \to -\infty \text{ or as } x \to \infty$$

Consequently, the line $y = 0$ is a horizontal asymptote of the graph of R. ■

Example 6 Find the horizontal or oblique asymptotes, if any, of the graph of

$$H(x) = \frac{3x^4 - x^2}{x^3 - x^2 + 1}$$

Solution The rational function H is improper, since the degree of the numerator (4) is larger than the degree of the denominator (3). Since the difference is 1, the graph of H will have an oblique asymptote. To find it, we use long division:

$$\begin{array}{r l}
 & 3x + 3 \\
x^3 - x^2 + 1 \,\big) & 3x^4 \qquad\qquad - x^2 \\
 & \underline{3x^4 - 3x^3 \qquad\quad + 3x} \\
 & \qquad 3x^3 - x^2 - 3x \\
 & \qquad \underline{3x^3 - 3x^2 \qquad + 3} \\
 & \qquad\qquad 2x^2 - 3x - 3
\end{array}$$

Thus,

$$H(x) = \frac{3x^4 - x^2}{x^3 - x^2 + 1} = 3x + 3 + \frac{2x^2 - 3x - 3}{x^3 - x^2 + 1}$$

Then, as $x \to -\infty$ or as $x \to \infty$, we have $H(x) \to 3x + 3$, since the proper rational function approaches 0:

$$\frac{2x^2 - 3x - 3}{x^3 - x^2 + 1} = \frac{\dfrac{2}{x} - \dfrac{3}{x^2} - \dfrac{3}{x^3}}{1 + \dfrac{1}{x} + \dfrac{1}{x^3}} \to \frac{0}{1} = 0$$

↑ Divide by x^3

We conclude that the graph of the rational function H has the oblique asymptote $y = 3x + 3$. ■

Example 7 Find the horizontal or oblique asymptotes, if any, of the graph of

$$R(x) = \frac{8x^2 - x + 2}{4x^2 - 1}$$

Solution The rational function R is improper, since the degree of the numerator (2) equals the degree of the denominator (2). In this case, the graph of R has a horizontal asymptote. To find it, we use long division:

$$\begin{array}{r} 2 \\ 4x^2 - 1 \overline{)\,8x^2 - x + 2} \\ \underline{8x^2 - 2} \\ - x + 4 \end{array}$$

Thus,

$$R(x) = \frac{8x^2 - x + 2}{4x^2 - 1} = 2 + \frac{-x + 4}{4x^2 - 1}$$

Then, as $x \to -\infty$ or as $x \to \infty$, we have $R(x) \to 2$, since the proper rational function approaches 0:

$$\frac{-x + 4}{4x^2 - 1} \to 0$$

We conclude that $y = 2$ is a horizontal asymptote of the graph. ■

In Example 7, note that the quotient 2 obtained by long division is the quotient of the leading coefficients of the numerator polynomial and the denominator polynomial $\left(\frac{8}{4}\right)$. This means we can avoid the long division process for rational functions whose numerator and denominator *are of the same degree* and conclude that the quotient of the leading coefficients will give us the horizontal asymptote.

Example 8 Find the horizontal or oblique asymptotes, if any, of the graph of

$$G(x) = \frac{2x^5 - x^3 + 2}{x^3 - 1}$$

Solution The rational function G is improper, since the degree of the numerator (5) is larger than the degree of the denominator (3). Since the difference is 2, the graph of G has no horizontal and no oblique asymptotes, and falls into category 3 listed earlier on page 302.

We proceed with long division to find the polynomial f:

$$\begin{array}{r} 2x^2 - 1 \\ x^3 - 1 \overline{)\, 2x^5 - x^3 + 2} \\ \underline{2x^5 - 2x^2 } \\ -x^3 + 2x^2 + 2 \\ \underline{-x^3 + 1} \\ 2x^2 + 1 \end{array}$$

Thus,

$$G(x) = \frac{2x^5 - x^3 + 2}{x^3 - 1} = 2x^2 - 1 + \frac{2x^2 + 1}{x^3 - 1}$$

Then, as $x \to -\infty$ or as $x \to \infty$, we have $G(x) \to 2x^2 - 1$. We conclude that, for large values of x, the graph of G approaches the graph of $y = 2x^2 - 1$. ■

We summarize below the procedure for finding horizontal and oblique asymptotes.

Finding Horizontal and Oblique Asymptotes

Consider the rational function

$$R(x) = \frac{p(x)}{q(x)} = \frac{a_n x^n + a_{n-1}x^{n-1} + \cdots + a_1 x + a_0}{b_m x^m + b_{m-1}x^{m-1} + \cdots + b_1 x + b_0}$$

in which the degree of the numerator is n and the degree of the denominator is m.

1. If $n < m$, the line $y = 0$ (the x-axis) is a horizontal asymptote of the graph of R.
2. If $n = m$, then the line $y = a_n/b_m$ is a horizontal asymptote of the graph of R.
3. If $n = m + 1$ (that is, if the degree of the numerator is 1 more than the degree of the denominator), the graph of $R(x)$ will have an oblique asymptote which is found using long division.
4. If $n > m + 1$, the graph of $R(x)$ has no horizontal and no oblique asymptote.

Note: The graph of a rational function either has one horizontal or one oblique asymptote or else has no horizontal and no oblique asymptote.

Now we are ready to graph rational functions.

Graphing Rational Functions

We commented earlier that calculus provides the tools required to graph a polynomial function accurately. The same holds true for rational functions. However, we can gather together quite a bit of information about their graphs to get an idea of the general shape and position of the graph.

In the examples that follow, we will discuss the graph of a rational function $R(x) = p(x)/q(x)$ by applying the following steps:

Steps for Graphing a Rational Function

STEP 1: Locate the intercepts, if any, of the graph. The x-intercepts, if any, of $R(x) = p(x)/q(x)$ satisfy the equation $p(x) = 0$. The y-intercept, if there is one, is $R(0)$.

STEP 2: Test for symmetry. Replace x by $-x$ in $R(x)$. If $R(-x) = R(x)$, there is symmetry with respect to the y-axis; if $R(-x) = -R(x)$, there is symmetry with respect to the origin.

STEP 3: Locate the vertical asymptotes, if any, by factoring the denominator $q(x)$ of $R(x)$ and identifying its zeros.

STEP 4: Locate the horizontal or oblique asymptotes, if any, using the procedure given on page 305. Determine points, if any, at which the graph of R intersects these asymptotes.

STEP 5: Determine where the graph is above the x-axis and where the graph is below the x-axis.

STEP 6: Plot some additional points and sketch the graph.

Example 9 Discuss the graph of the rational function: $R(x) = \dfrac{x - 1}{x^2 - 4}$

Solution First, we factor both the numerator and the denominator of R:

$$R(x) = \frac{x - 1}{(x + 2)(x - 2)}$$

STEP 1: We locate the x-intercepts by finding the zeros of the numerator. By inspection, 1 is the only x-intercept. The y-intercept is $R(0) = \frac{1}{4}$.

STEP 2: Because

$$R(-x) = \frac{-x - 1}{x^2 - 4}$$

we conclude that R is neither even nor odd. Thus, no symmetry is present.

STEP 3: We locate the vertical asymptotes by factoring the denominator: $x^2 - 4 = (x + 2)(x - 2)$. The graph of R thus has two vertical asymptotes: the lines $x = -2$ and $x = 2$.

STEP 4: The degree of the numerator is less than the degree of the denominator, so the line $y = 0$ (the x-axis) is a horizontal asymptote of the graph. To determine if the graph of R intersects the horizontal asymptote, we solve the equation $R(x) = 0$. The only solution is $x = 1$, so the graph of R intersects the horizontal asymptote at $(1, 0)$.

STEP 5: The zero of the numerator, 1, and the zeros of the denominator, -2 and 2, divide the x-axis into four intervals: $x < -2$, $-2 < x < 1$, $1 < x < 2$, and $x > 2$. Now we construct Table 6.

Table 6

INTERVAL	TEST NUMBER	VALUE OF R	GRAPH OF R
$x < -2$	-3	$R(-3) = -0.8$	Below x-axis
$-2 < x < 1$	0	$R(0) = \frac{1}{4}$	Above x-axis
$1 < x < 2$	$\frac{3}{2}$	$R\left(\frac{3}{2}\right) = -\frac{2}{7}$	Below x-axis
$x > 2$	3	$R(3) = 0.4$	Above x-axis

STEP 6: Now we are ready to put all the information together to sketch the graph. In Figure 29(a) we have plotted the points found in Table 6. Since the x-axis is a horizontal asymptote and the graph lies below the x-axis for $x < -2$, we can sketch a portion of the graph by placing a small arrow to the far left and under the x-axis. Since the line $x = -2$ is a vertical asymptote and the graph lies below the x-axis for $x < -2$, we continue the sketch with an arrow placed well below the x-axis and approaching the line $x = -2$ on the left. Similar explanations account for the positions of the other portions of the graph. In particular, note how we use the facts that the graph lies above the x-axis for $-2 < x < 1$, below the x-axis for $1 < x < 2$, and $(1, 0)$ is an x-intercept to draw the conclusion that the graph crosses the x-axis at $(1, 0)$. Figure 29(b) shows the complete sketch.

Figure 29
$R(x) = \dfrac{x - 1}{x^2 - 4}$

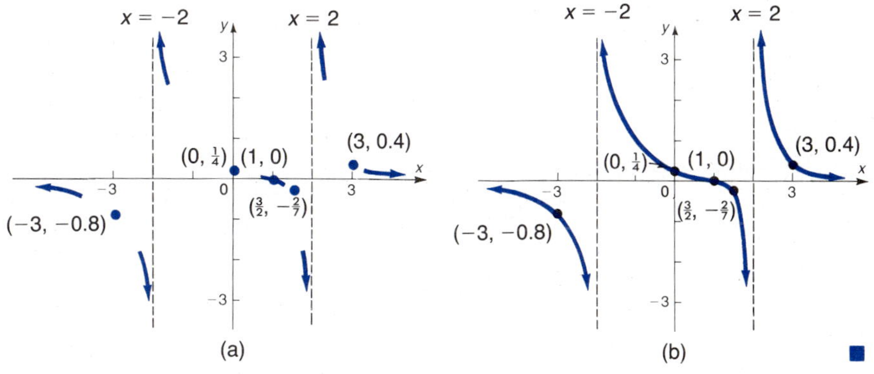

■

Example 10 Discuss the graph of the rational function: $R(x) = \dfrac{x^2 - 1}{x}$

Solution

STEP 1: The graph has two x-intercepts: -1 and 1. There is no y-intercept.

STEP 2: Since $R(-x) = -R(x)$, the function is odd and the graph is symmetric with respect to the origin.

STEP 3: The graph of $R(x)$ has the line $x = 0$ (the y-axis) as a vertical asymptote.

STEP 4: The degree of the numerator is 1 more than the degree of the denominator, so the graph of $R(x)$ has an oblique asymptote. To find it, we use long division:

$$\begin{array}{r} x \phantom{{}-1} \\ x\overline{)x^2 - 1} \\ \underline{x^2 \phantom{{}-1}} \\ -1 \end{array}$$

Thus,

$$R(x) = \frac{x^2 - 1}{x} = x + \frac{-1}{x}$$

The line $y = x$ is an oblique asymptote of the graph. To determine whether the graph of R intersects the asymptote $y = x$, we solve the equation $R(x) = x$:

$$R(x) = \frac{x^2 - 1}{x} = x$$
$$x^2 - 1 = x^2$$
$$-1 = 0 \quad \text{Impossible}$$

We conclude that the equation $(x^2 - 1)/x = x$ has no solution, so the graph of $R(x)$ does not intersect the line $y = x$.

STEP 5: The zeros of the numerator are -1 and 1; the denominator has the zero 0. Thus, we divide the x-axis into four intervals: $x < -1$, $-1 < x < 0$, $0 < x < 1$, and $x > 1$. Now we construct Table 7.

Table 7

INTERVAL	TEST NUMBER	VALUE OF R	GRAPH OF R
$x < -1$	-2	$R(-2) = -\frac{3}{2}$	Below x-axis
$-1 < x < 0$	$-\frac{1}{2}$	$R\left(-\frac{1}{2}\right) = \frac{3}{2}$	Above x-axis
$0 < x < 1$	$\frac{1}{2}$	$R\left(\frac{1}{2}\right) = -\frac{3}{2}$	Below x-axis
$x > 1$	2	$R(2) = \frac{3}{2}$	Above x-axis

STEP 6: Figure 30(a) shows a partial graph using the facts we have gathered. The complete graph is given in Figure 30(b).

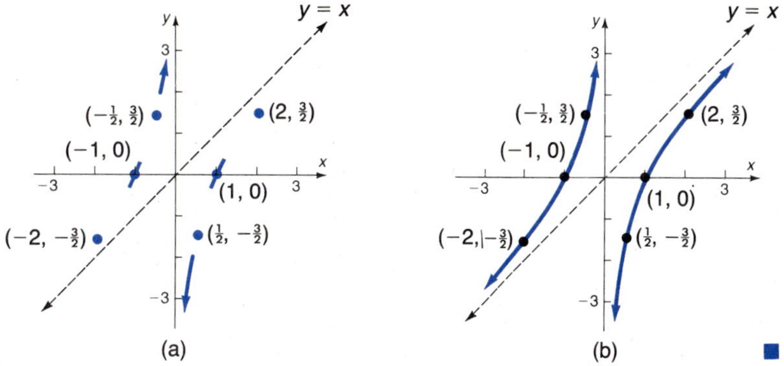

Figure 30
$R(x) = \dfrac{x^2 - 1}{x}$

Example 11 Discuss the graph of the rational function: $R(x) = \dfrac{x^4 + 1}{x^2}$

Solution

STEP 1: The graph has no x-intercepts and no y-intercepts.

STEP 2: Since $R(-x) = R(x)$, the function is even and the graph is symmetric with respect to the y-axis.

STEP 3: The graph of $R(x)$ has the line $x = 0$ (the y-axis) as a vertical asymptote.

STEP 4: The degree of the numerator exceeds the degree of the denominator by 2, so the graph has no horizontal and no oblique asymptotes. However, since

$$R(x) = \frac{x^4 + 1}{x^2} = x^2 + \frac{1}{x^2}$$

the graph of R approaches the graph of $y = x^2$ as $x \to -\infty$ and as $x \to \infty$.

STEP 5: The numerator has no zeros, and the denominator has one zero at 0. Thus, we divide the x-axis into two intervals: $x < 0$ and $x > 0$. Now we set up Table 8:

Table 8

INTERVAL	TEST NUMBER	VALUE OF R	GRAPH OF R
$x < 0$	-1	$R(-1) = 2$	Above x-axis
$x > 0$	1	$R(1) = 2$	Above x-axis

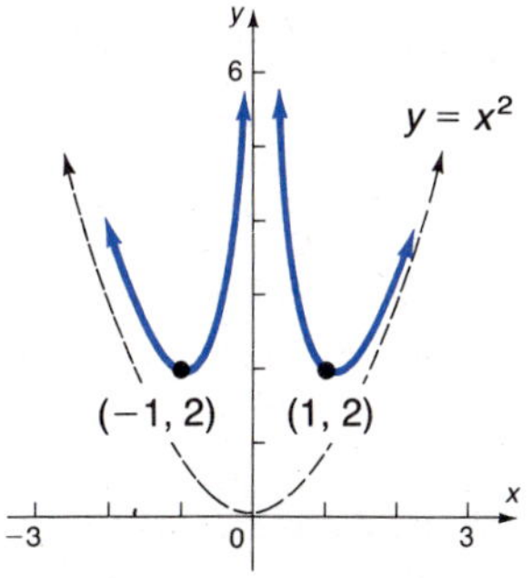

Figure 31
$R(x) = \dfrac{x^4 + 1}{x^2}$

STEP 6: Figure 31 shows the graph. ■

Example 12 Discuss the graph of the rational function: $R(x) = \dfrac{3x^2 - 3x}{x^2 + x - 12}$

Solution We factor R to get

$$R(x) = \frac{3x(x-1)}{(x+4)(x-3)}$$

STEP 1: The graph has two x-intercepts: 0 and 1. The y-intercept is $R(0) = 0$.

STEP 2: No symmetry is present.

STEP 3: The graph of R has two vertical asymptotes: $x = -4$ and $x = 3$.

STEP 4: Since the degree of the numerator equals the degree of the denominator, the graph has a horizontal asymptote. To find it, we form the quotient of the leading coefficient of the numerator (3) and the leading coefficient of the denominator (1). Thus, the graph of R has the horizontal asymptote $y = 3$. To find out whether the graph of R intersects the asymptote, we solve the equation $R(x) = 3$:

$$R(x) = \frac{3x^2 - 3x}{x^2 + x - 12} = 3$$

$$3x^2 - 3x = 3x^2 + 3x - 36$$

$$-6x = -36$$

$$x = 6$$

Thus, the graph intersects the line $y = 3$ only at $x = 6$, and $(6, 3)$ is a point on the graph of R.

STEP 5: The zeros of the numerator, 0 and 1, and the zeros of the denominator, -4 and 3, divide the x-axis into five intervals: $x < -4$, $-4 < x < 0$, $0 < x < 1$, $1 < x < 3$, and $x > 3$. Now we can construct Table 9:

Table 9

INTERVAL	TEST NUMBER	VALUE OF R	GRAPH OF R
$x < -4$	-5	$R(-5) = 11.25$	Above x-axis
$-4 < x < 0$	-1	$R(-1) = -\frac{1}{2}$	Below x-axis
$0 < x < 1$	$\frac{1}{2}$	$R(\frac{1}{2}) = \frac{1}{15}$	Above x-axis
$1 < x < 3$	2	$R(2) = -1$	Below x-axis
$x > 3$	4	$R(4) = 4.5$	Above x-axis

STEP 6: Figure 32(a) shows a partial graph. Notice that we have not yet used the fact that the line $y = 3$ is a horizontal asymptote,

because we do not know yet whether the graph of R crosses or touches the line $y = 3$ at $(6, 3)$. To see whether the graph, in fact, crosses or touches the line $y = 3$, we plot an additional point to the right of $(6, 3)$. We use $x = 7$ to find $R(7) = \frac{63}{22} < 3$. Thus, the graph crosses $y = 3$ at $x = 6$. Because $y = 3$ is an asymptote of the graph, the graph approaches the line $y = 3$ from above as $x \to -\infty$ and approaches the line $y = 3$ from below as $x \to \infty$. See Figure 32(b). The completed graph is shown in Figure 32(c).

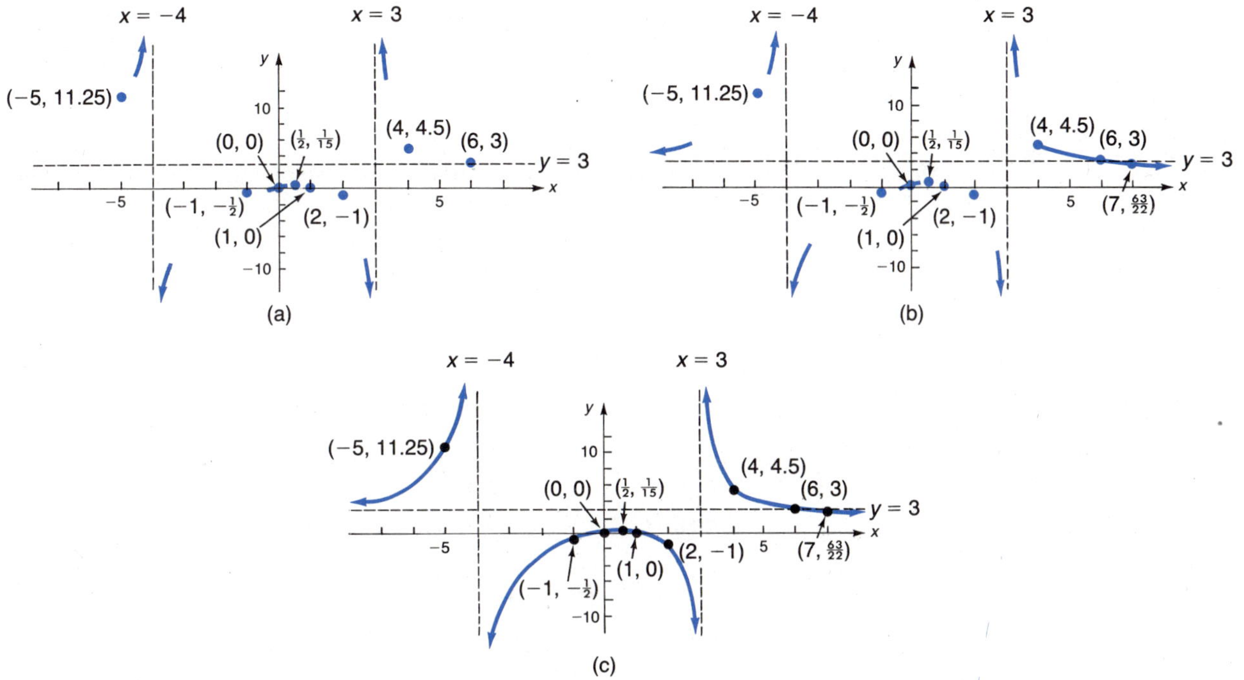

Figure 32
$R(x) = \dfrac{3x^2 - 3x}{x^2 + x - 12}$

■

EXERCISE 5.3

In Problems 1–10, find the domain of each rational function.

1. $R(x) = \dfrac{x}{x - 2}$ **2.** $R(x) = \dfrac{5x^2}{3 - x}$ **3.** $H(x) = \dfrac{-4x^2}{(x - 2)(x + 1)}$

4. $G(x) = \dfrac{1}{(x + 2)(4 - x)}$ **5.** $F(x) = \dfrac{3x(x - 1)}{2x^2 - 5x - 3}$ **6.** $Q(x) = \dfrac{-x(1 - x)}{3x^2 + 5x - 2}$

7. $R(x) = \dfrac{x}{x^3 - 8}$ **8.** $R(x) = \dfrac{x}{x^4 - 1}$ **9.** $H(x) = \dfrac{3x^2 + x}{x^2 + 4}$

10. $G(x) = \dfrac{x - 3}{x^4 + 1}$

In Problems 11–20, graph each rational function using the methods of shifting, compression, stretching, and reflection, or the Steps 1–6 on page 306.

11. $R(x) = \dfrac{1}{(x - 1)^2}$ **12.** $R(x) = \dfrac{3}{x}$ **13.** $H(x) = \dfrac{-2}{x + 1}$

14. $G(x) = \dfrac{2}{(x + 2)^2}$ **15.** $R(x) = \dfrac{1}{x^2 + 4x + 4}$ **16.** $R(x) = \dfrac{1}{x - 1} + 1$

17. $F(x) = 1 - \dfrac{1}{x}$ **18.** $Q(x) = 1 + \dfrac{1}{x}$ **19.** $R(x) = \dfrac{x^2 - 4}{x^2}$

20. $R(x) = \dfrac{x - 4}{x}$

In Problems 21–30, find the vertical, horizontal, and oblique asymptotes, if any, of each rational function. Do not graph.

21. $R(x) = \dfrac{x}{x + 1}$ **22.** $R(x) = \dfrac{2x + 3}{x - 4}$ **23.** $H(x) = \dfrac{x^4 + 2x^2 + 1}{3x^2 - x + 1}$

24. $G(x) = \dfrac{-x^2 + 1}{x + 5}$ **25.** $T(x) = \dfrac{x^3}{x^4 - 1}$ **26.** $P(x) = \dfrac{4x^5}{x^3 - 1}$

27. $Q(x) = \dfrac{5 - x^2}{3x^4}$ **28.** $F(x) = \dfrac{-2x^2 + 1}{2x^3 + 4x^2}$ **29.** $R(x) = \dfrac{3x^4 + 1}{x^3 + 5x}$

30. $R(x) = \dfrac{6x^2 + x - 12}{3x^2 + 5x - 2}$

In Problems 31–56, follow Steps 1–6 on page 306 to graph each rational function.

31. $R(x) = \dfrac{x + 1}{x(x + 4)}$ **32.** $R(x) = \dfrac{x}{(x - 1)(x + 2)}$ **33.** $R(x) = \dfrac{3x + 3}{2x + 4}$

34. $R(x) = \dfrac{2x + 4}{x - 1}$ **35.** $R(x) = \dfrac{3}{x^2 - 4}$ **36.** $R(x) = \dfrac{6}{x^2 - x - 6}$

37. $P(x) = \dfrac{x^4 + x^2 + 1}{x^2 - 1}$ **38.** $Q(x) = \dfrac{x^4 - 1}{x^2 - 4}$ **39.** $H(x) = \dfrac{x^3 - 1}{x^2 - 9}$

40. $G(x) = \dfrac{x^3 + 1}{x^2 + 2x}$ **41.** $R(x) = \dfrac{x^2}{x^2 + x - 6}$ **42.** $R(x) = \dfrac{x^2 + x - 12}{x^2 - 4}$

43. $G(x) = \dfrac{x}{x^2 - 4}$ **44.** $G(x) = \dfrac{3x}{x^2 - 1}$ **45.** $R(x) = \dfrac{3}{(x - 1)(x^2 - 4)}$

46. $R(x) = \dfrac{-4}{(x + 1)(x^2 - 9)}$ **47.** $H(x) = 4\dfrac{x^2 - 1}{x^4 - 16}$ **48.** $H(x) = \dfrac{x^2 + 1}{x^4 - 1}$

49. $F(x) = \dfrac{x^2 - 3x - 4}{x + 2}$ **50.** $F(x) = \dfrac{x^2 + 3x + 2}{x - 1}$ **51.** $R(x) = \dfrac{x^2 + x - 12}{x - 4}$

52. $R(x) = \dfrac{x^2 - x - 12}{x + 5}$ **53.** $F(x) = \dfrac{x^2 + x - 12}{x + 2}$ **54.** $G(x) = \dfrac{x^2 - x - 12}{x + 1}$

55. $R(x) = \dfrac{x(x-1)^2}{(x+3)^3}$ **56.** $R(x) = \dfrac{(x-1)(x+2)(x-3)}{x(x-4)^2}$

57. If the graph of a rational function R has the vertical asymptote $x = 4$, then the factor $x - 4$ must be present in the denominator of R. Explain why.

58. If the graph of a rational function R has the horizontal asymptote $y = 2$, then the degree of the numerator of R equals the degree of the denominator of R. Explain why.

5.4 Synthetic Division*

Long Division

The process of long division for dividing one polynomial by another is familiar to you. However, since synthetic division is based on long division, we begin with a detailed example.

Example 1

$$
\begin{array}{rrrrrrl}
 & x^2 & -\,2x & & -\,3 & & \leftarrow \text{Quotient} \\ \cline{2-6}
\underset{\text{Divisor}}{x^2 - x + 1} \big) & x^4 & -\,3x^3 & & +\,2x & -\,5 & \leftarrow \text{Dividend (note space left for missing } x^2\text{-term)} \\
 & x^4 & -\,x^3 & +\,x^2 & & & \\ \cline{2-4}
\text{Subtract} & & -2x^3 & -\,x^2 & +\,2x & -\,5 & \\
 & & -2x^3 & +\,2x^2 & -\,2x & & \\ \cline{3-6}
\text{Subtract} & & & -3x^2 & +\,4x & -\,5 & \\
 & & & -3x^2 & +\,3x & -\,3 & \\ \cline{4-6}
\text{Subtract} & & & & x & -\,2 & \leftarrow \text{Remainder}
\end{array}
$$

Thus,

$$\frac{x^4 - 3x^3 + 2x - 5}{x^2 - x + 1} = x^2 - 2x - 3 + \frac{x - 2}{x^2 - x + 1}$$ ■

The process of long division carried out in Example 1 yielded the quotient $x^2 - 2x - 3$ and the remainder $x - 2$ when the dividend $x^4 - 3x^3 + 2x - 5$ was divided by the divisor $x^2 - x + 1$. The process ends when, after subtracting, we obtain either the zero polynomial or a polynomial whose degree is less than the degree of the divisor.

The work we did in Example 1 may be checked by verifying that

$$(\text{Divisor})(\text{Quotient}) + \text{Remainder} = \text{Dividend}$$

* C Refer to Graphics Calculator Supplement: Activity III.

Thus,

$$\begin{aligned}&(\text{Divisor})(\text{Quotient}) + \text{Remainder}\\ &\quad = (x^2 - x + 1)(x^2 - 2x - 3) + x - 2\\ &\quad = x^4 - 2x^3 - 3x^2 - x^3 + 2x^2 + 3x + x^2 - 2x - 3 + x - 2\\ &\quad = x^4 - 3x^3 + 2x - 5 = \text{Dividend}\end{aligned}$$

This checking routine is the basis for a famous theorem called the **division algorithm*** **for polynomials**, which we now state without proof.

Theorem

Division Algorithm for Polynomials

If $f(x)$ and $g(x)$ denote polynomial functions and if $g(x)$ is not the zero polynomial, then there are unique polynomial functions $q(x)$ and $r(x)$ such that

$$\frac{f(x)}{g(x)} = q(x) + \frac{r(x)}{g(x)} \quad \text{or} \quad f(x) = g(x)q(x) + r(x) \qquad (1)$$

where $r(x)$ is either the zero polynomial or a polynomial of degree less than that of $g(x)$. ■

In equation (1), $f(x)$ is the **dividend**, $g(x)$ is the **divisor**, $q(x)$ is the **quotient**, and $r(x)$ is the **remainder**.

If the divisor $g(x)$ is a first-degree polynomial of the form

$$g(x) = x - c \qquad c \text{ a real number}$$

then the remainder $r(x)$ is either the zero polynomial or a polynomial of degree 0. Thus, for such divisors, the remainder is some number, say, R, and we may write

$$f(x) = (x - c)q(x) + R \qquad (2)$$

This equation is an identity in x and is true for all real numbers x. In particular, it is true when $x = c$. Thus, if $x = c$, then equation (2) becomes

$$\begin{aligned}f(c) &= (c - c)q(c) + R\\ f(c) &= R\end{aligned}$$

and equation (2) takes the form

*A systematic process in which certain steps are repeated a finite number of times is called an **algorithm**. Thus, long division is an algorithm.

$$f(x) = (x - c)q(x) + f(c) \qquad (3)$$

We have now proved the following result, called the **Remainder Theorem**:

Theorem
Remainder Theorem

Let f be a polynomial function. If $f(x)$ is divided by $x - c$, then the remainder is $f(c)$. ■

Example 2 Find the remainder if $f(x) = x^3 - 4x^2 + 2x - 5$ is divided by:

(a) $x - 3$ (b) $x + 2$

Solution

(a) We could use long division. However, it is much easier here to use the Remainder Theorem, which says the remainder is

$$f(3) = (3)^3 - 4(3)^2 + 2(3) - 5 = 27 - 36 + 6 - 5 = -8$$

(b) To find the remainder when $f(x)$ is divided by $x + 2 = x - (-2)$, we evaluate

$$f(-2) = (-2)^3 - 4(-2)^2 + 2(-2) - 5 = -8 - 16 - 4 - 5 = -33$$

Thus, the remainder is -33. ■

An important and useful consequence of the Remainder Theorem is the **Factor Theorem**.

Theorem
Factor Theorem

Let f be a polynomial function. Then $x - c$ is a factor of $f(x)$ if and only if $f(c) = 0$. ■

The Factor Theorem actually consists of two separate statements:

1. If $f(c) = 0$, then $x - c$ is a factor of $f(x)$.
2. If $x - c$ is a factor of $f(x)$, then $f(c) = 0$.

Thus, the proof requires two parts.

Proof

1. Suppose $f(c) = 0$. Then, by equation (3), we have

$$f(x) = (x - c)q(x)$$

for some polynomial $q(x)$. That is, $x - c$ is a factor of $f(x)$.
2. Suppose $x - c$ is a factor of $f(x)$. Then there is a polynomial function q such that

$$f(x) = (x - c)q(x)$$

Replacing x by c, we find

$$f(c) = (c - c)q(c) = 0 \cdot q(c) = 0$$

This completes the proof. ■

One use of the Factor Theorem is to determine whether a polynomial has a particular factor.

Example 3 Use the Factor Theorem to determine whether the function $f(x) = 2x^3 - x^2 + 2x - 3$ has the factor:

(a) $x - 1$ (b) $x + 3$

Solution (a) Because $x - 1$ is of the form $x - c$ with $c = 1$, we find the value of $f(1)$:

$$f(1) = 2(1)^3 - (1)^2 + 2(1) - 3 = 2 - 1 + 2 - 3 = 0$$

By the Factor Theorem, $x - 1$ is a factor of $f(x)$.

(b) To test the factor $x + 3$, we first need to write it in the form $x - c$. Since $x + 3 = x - (-3)$, we find the value of $f(-3)$:

$$f(-3) = 2(-3)^3 - (-3)^2 + 2(-3) - 3 = -54 - 9 - 6 - 3 = -72$$

Because $f(-3) \neq 0$, we conclude from the Factor Theorem that $x - (-3) = x + 3$ is not a factor of $f(x)$. ■

Synthetic Division

To find the quotient as well as the remainder when a polynomial function f of degree 1 or higher is divided by $g(x) = x - c$, a shortened version of long division, called **synthetic division**, makes the task simpler.

To see how synthetic division works, we divide the polynomial $f(x) = 2x^3 - x^2 + 3$ by $g(x) = x - 3$. First, in long division, we have

$$\begin{array}{rrrrr}
 & 2x^2 & + 5x & + 15 & \\
\hline
x - 3 \,) & 2x^3 & - x^2 & & + 3 \\
 & 2x^3 & - 6x^2 & & \\
\hline
 & & 5x^2 & & \\
 & & 5x^2 & - 15x & \\
\hline
 & & & 15x & + 3 \\
 & & & 15x & - 45 \\
\hline
 & & & & 48
\end{array}$$

The process of synthetic division arises from rewriting the long division in a more compact form, using simpler notation. For example, in the long division above, the circled terms are not really necessary because they are identical to the terms directly above them. With these terms removed, we have

$$
\begin{array}{rrrrr}
 & 2x^2 & +5x & +15 & \\ \cline{2-5}
x-3) & 2x^3 & -x^2 & & +3\\
 & & -6x^2 & & \\ \cline{2-3}
 & & 5x^2 & & \\
 & & & -15x & \\ \cline{3-4}
 & & & 15x & \\
 & & & & -45\\ \cline{4-5}
 & & & & 48
\end{array}
$$

Most of the x's that appear in this process also can be removed, provided we are careful about positioning each coefficient. In this regard, we will need to use 0 as the coefficient of x in the dividend, because that power of x is missing. Now we have

$$
\begin{array}{rrrrr}
 & 2x^2 & +5x & +15 & \\ \cline{2-5}
x-3) & 2 & -1 & 0 & 3\\
 & & -6 & & \\ \cline{2-3}
 & & \textcircled{5} & & \\
 & & & -15 & \\ \cline{3-5}
 & & & \textcircled{15} & \\
 & & & & -45\\ \cline{4-5}
 & & & & \textcircled{48}
\end{array}
$$

We can make this display more compact by moving the lines up until the circled numbers align horizontally:

$$
\begin{array}{rrrrrl}
 & 2x^2 & +5x & +15 & & \text{Row 1}\\ \cline{2-5}
x-3) & 2 & -1 & 0 & 3 & \text{Row 2}\\
 & & -6 & -15 & -45 & \text{Row 3}\\ \cline{2-5}
 & \bigcirc & 5 & 15 & 48 & \text{Row 4}
\end{array}
$$

Now, if we place the leading coefficient of the quotient (2) in the circled position, the first three numbers in Row 4 are precisely the coefficients of the quotient, and the last number in Row 4 is the remainder. Thus, Row 1 is not really needed, so we can compress the process to three rows, where the bottom row contains the coefficients of both the quotient and the remainder:

$$
\begin{array}{rrrrrl}
x-3) & 2 & -1 & 0 & 3 & \text{Row 1}\\
 & & -6 & -15 & -45 & \text{Row 2 (subtract)}\\ \cline{2-5}
 & 2 & 5 & 15 & 48 & \text{Row 3}
\end{array}
$$

Recall that the entries in Row 3 are obtained by subtracting the entries in Row 2 from those in Row 1. Rather than subtracting the entries in Row 2 we can change the sign of each entry and add. With this modification, our display will look like this:

$x - 3\overline{)2}$	-1	0	3	Row 1
	6	15	45	Row 2 (add)
2	5	15	48	Row 3

Notice that the entries in Row 2 are 3 times the prior entries in Row 3. Our last modification to the display replaces the $x - 3$ by 3. The entries in Row 3 give the quotient and the remainder as shown below.

$3\overline{)2}$	-1	0	3	Row 1
	6	15	45	Row 2 (add)
2	5	15	48	Row 3

(Each Row 2 entry is ×3 the prior Row 3 entry.)

Quotient: $2x^2 + 5x + 15$ Remainder: $R = 48$

Let's go through another example step by step.

Example 4 Use synthetic division to find the quotient and remainder when

$$f(x) = 3x^4 + 8x^2 - 7x + 4 \quad \text{is divided by} \quad g(x) = x - 1$$

Solution STEP 1: Write the dividend in descending powers of x. Then copy the coefficients, remembering to insert a 0 for any missing powers of x:

3	0	8	-7	4	Row 1

STEP 2: Insert the usual division symbol. Since the divisor is $x - 1$, we insert 1 to the left of the division symbol:

$1\overline{)3}$	0	8	-7	4	Row 1

STEP 3: Bring the 3 down two rows, and enter it in Row 3:

$1\overline{)3}$	0	8	-7	4	Row 1
↓					Row 2
3					Row 3

STEP 4: Multiply the latest entry in Row 3 by 1 and place the result in Row 2, but one column over to the right:

$1\overline{)3}$	0	8	-7	4	Row 1
	3				Row 2
3					Row 3

STEP 5: Add the entry in Row 2 to the entry above it in Row 1, and enter the sum in Row 3:

1)	3	0	8	−7	4	Row 1
		3				Row 2
	3	3				Row 3

STEP 6: Repeat Steps 4 and 5 until no more entries are available in Row 1:

1)	3	0	8	−7	4	Row 1
		3	3	11	4	Row 2 (add)
	3	3	11	4	8	Row 3

STEP 7: The final entry in Row 3, an 8, is the remainder; the other entries in Row 3 (3, 3, 11, and 4) are the coefficients (in descending order) of a polynomial whose degree is 1 less than that of the dividend; this is the quotient. Thus,

$$\text{Quotient} = 3x^3 + 3x^2 + 11x + 4 \qquad \text{Remainder} = 8$$

Check:

$$\begin{aligned}
&(\text{Divisor})(\text{Quotient}) + \text{Remainder} \\
&\quad = (x - 1)(3x^3 + 3x^2 + 11x + 4) + 8 \\
&\quad = 3x^4 + 3x^3 + 11x^2 + 4x - 3x^3 - 3x^2 - 11x - 4 + 8 \\
&\quad = 3x^4 + 8x^2 - 7x + 4 = \text{Dividend}
\end{aligned}$$

■

Let's do an example in which all seven steps are combined.

Example 5 Use synthetic division to show that $g(x) = x + 3$ is a factor of

$$f(x) = 2x^5 + 5x^4 - 2x^3 + 2x^2 - 2x + 3$$

Solution The divisor is $x + 3 = x - (-3)$, so the Row 3 entries will be multiplied by -3, entered in Row 2, and added to Row 1:

−3)	2	5	−2	2	−2	3	Row 1
		−6	3	−3	3	−3	Row 2
	2	−1	1	−1	1	0	Row 3

Because the remainder is 0, it follows that $f(-3) = 0$. Hence, by the Factor Theorem, $x - (-3) = x + 3$ is a factor of $f(x)$. ■

One important use of synthetic division is to find the value of a polynomial.

Example 6 Use synthetic division to find the value of $f(x) = -3x^4 + 2x^3 - x + 1$ at $x = -2$; that is, find $f(-2)$.

Solution The Remainder Theorem tells us that the value of a polynomial function at c equals the remainder when the polynomial is divided by $x - c$. This remainder is the final entry of the third row in the process of synthetic division. We want $f(-2)$, so we divide by $x - (-2)$:

$$\begin{array}{r|rrrrr} -2 & -3 & 2 & 0 & -1 & 1 \\ & & 6 & -16 & 32 & -62 \\ \hline & -3 & 8 & -16 & 31 & -61 \end{array}$$

The quotient is $q(x) = -3x^3 + 8x^2 - 16x + 31$; the remainder is $R = -61$. Because the remainder was found to be -61, it follows from the Remainder Theorem that $f(-2) = -61$. ■

As Example 6 illustrates, we can use the process of synthetic division to find the value of a polynomial function at a number c as an alternative to merely substituting c for x. Compare the work required in Example 6 with the arithmetic involved in substituting:

$$\begin{aligned} f(-2) &= -3(-2)^4 + 2(-2)^3 - (-2) + 1 \\ &= -3(16) + 2(-8) + 2 + 1 \\ &= -48 - 16 + 2 + 1 = -61 \end{aligned}$$

As you can see, finding $f(-2)$ may be easier using synthetic division.

Sometimes, neither substitution nor synthetic division avoids the need for messy calculations. Consider the problem of evaluating $f(x) = 3x^5 - 5x^4 + 0.2x^3 - 1.5x^2 + 2x - 6$ at $x = 1.2$. Here, a third method—using the *nested form* of a polynomial—is more helpful.

Nested Form of a Polynomial

Consider the polynomial $f(x) = 3x^3 - 5x^2 + 2x - 7$. We can factor $f(x)$ as follows:

$$\begin{aligned} f(x) &= 3x^3 - 5x^2 + 2x - 7 \\ &= (3x^3 - 5x^2 + 2x) - 7 && \text{Group terms containing } x. \\ &= (3x^2 - 5x + 2)x - 7 && \text{Factor } x. \\ &= [(3x^2 - 5x) + 2]x - 7 && \text{Regroup.} \\ &= [(3x - 5)x + 2]x - 7 && \text{Factor } x \text{ from parentheses.} \end{aligned}$$

Notice that this form of the polynomial contains only linear expressions. A polynomial function written in this way is said to be in **nested form**.

Let's look at some other examples.

Example 7 Write each polynomial in nested form:

(a) $f(x) = 2x^2 - 3x + 5$ (b) $f(x) = 5x^3 - 6x^2 + 2$

(c) $f(x) = -5x^4 + 3x^3 - 2x^2 + 10x - 8$

Solution (a) We proceed in steps as follows:

$$f(x) = (2x^2 - 3x) + 5 = (2x - 3)x + 5$$

The expression $(2x - 3)x + 5$ is the nested form of $2x^2 - 3x + 5$.

(b) $$f(x) = (5x^3 - 6x^2) + 2 = (5x^2 - 6x)x + 2 = [(5x - 6)x]x + 2$$

The expression $[(5x - 6)x]x + 2$ is the nested form of the polynomial $5x^3 - 6x^2 + 2$.

(c)
$$\begin{aligned} f(x) &= (-5x^4 + 3x^3 - 2x^2 + 10x) - 8 \\ &= (-5x^3 + 3x^2 - 2x + 10)x - 8 \\ &= [(-5x^2 + 3x - 2)x + 10]x - 8 \\ &= \{[(-5x + 3)x - 2]x + 10\}x - 8 \end{aligned}$$ ■

The advantage of evaluating a polynomial in nested form is that this method avoids the need to raise a number to a power, which on a calculator or computer can cause serious round-off errors. Further, computers can perform the operation of addition much faster than the operation of multiplication, and the nested form requires fewer multiplications than the ordinary form of a polynomial. In Example 7(b), to evaluate $f(x) = 5x^3 - 6x^2 + 2$ in its ordinary form requires five multiplications and two additions:

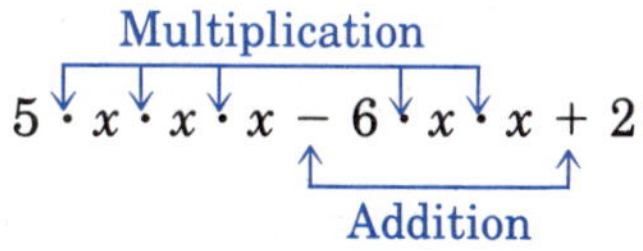

In nested form, it requires three multiplications and two additions:

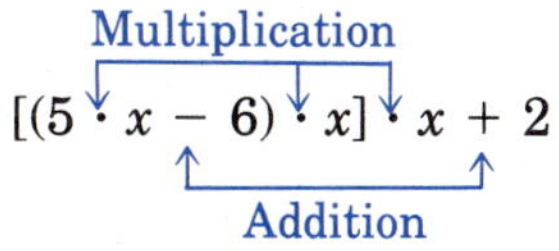

Thus, to avoid errors and to speed up calculations, many computers evaluate polynomials by using the nested form.

C **Example 8** Use the nested form and a calculator to evaluate the following polynomial at $x = 1.3$:

$$f(x) = 0.5x^3 - 1.2x^2 + 5.1x - 6.2$$

Solution We write f in nested form as

$$f(x) = [(0.5x - 1.2)x + 5.1]x - 6.2$$

We start inside the parentheses by multiplying 0.5 by $x = 1.3$. Then subtract 1.2. Multiply the result by $x = 1.3$ and add 5.1. Multiply this result by $x = 1.3$ and subtract 6.2. The value is

$$f(1.3) = -0.4995$$

On a calculator, you would proceed as follows:

	Enter [0.5]	Press [×]	Enter [1.3]	Press [−]
Display:	0.5		1.3	0.65
	Enter [1.2]	Press [=]	Press [×]	Enter [1.3]
Display:	1.2	−0.55		1.3
	Press [+]	Enter [5.1]	Press [=]	
Display:	−0.715	5.1	4.385	
	Press: [×]	Enter [1.3]	Press [−]	
Display:		1.3	5.7005	
	Enter [6.2]	Press [=]		
Display:	6.2	−0.4995		

Notice that no memory key was used in this process. ■

Summary

We now have three ways to find the value of a polynomial function $f(x)$ at a number c:

1. Replace x by the number c to find $f(c)$.
2. Use synthetic division to find $f(c)$.
3. Write $f(x)$ in nested form and use a calculator to find $f(c)$.

EXERCISE 5.4

In Problems 1–12, use long division to find the quotient $q(x)$ and the remainder $r(x)$ when $f(x)$ is divided by $g(x)$. Check your work by verifying that (Divisor)(Quotient) + Remainder = Dividend.

1. $f(x) = 4x^3 - 2x^2 - x + 1$; $g(x) = x^2 + x + 1$
2. $f(x) = -5x^3 + 2x^2 - x + 2$; $g(x) = x^2 - x - 1$
3. $f(x) = -2x^4 + x^3 - 3x^2 + 2$; $g(x) = x + 3$
4. $f(x) = 4x^3 - 3x^2 + 2$; $g(x) = x - 3$
5. $f(x) = 3x^5 - 5x^4 + 2x - 4$; $g(x) = x^3 + 2$
6. $f(x) = 4x^4 - 2x^3 + x^2 - 1$; $g(x) = x^2 - 2$
7. $f(x) = 1 - x^5$; $g(x) = x^2 - 1$
8. $f(x) = 1 - x^4$; $g(x) = x^2 + 1$
9. $f(x) = x^4 - c^4$; $g(x) = x - c$
10. $f(x) = x^4 + c^4$; $g(x) = x + c$
11. $f(x) = 5x^3 + 3x + 3$; $g(x) = 2x^2 + 1$
12. $f(x) = 4x^4 + 5x - 6$; $g(x) = 4x - 3$

In Problems 13–22, use synthetic division to find the quotient $q(x)$ and remainder R when $f(x)$ is divided by $g(x)$.

13. $f(x) = 3x^3 + 2x^2 - x + 3$; $g(x) = x - 3$
14. $f(x) = -4x^3 + 2x^2 - x + 1$; $g(x) = x + 2$
15. $f(x) = x^5 - 4x^3 + x$; $g(x) = x + 3$
16. $f(x) = x^4 + x^2 + 2$; $g(x) = x - 2$
17. $f(x) = 4x^6 - 3x^4 + x^2 + 5$; $g(x) = x - 1$
18. $f(x) = x^5 + 5x^3 - 10$; $g(x) = x + 1$
C **19.** $f(x) = 0.1x^3 + 0.2x$; $g(x) = x + 1.1$
C **20.** $f(x) = 0.1x^2 - 0.2$; $g(x) = x + 2.1$
21. $f(x) = x^5 - 1$; $g(x) = x - 1$
22. $f(x) = x^5 + 1$; $g(x) = x + 1$

In Problems 23–32, use synthetic division to determine whether $x - c$ is a factor of $f(x)$.

23. $f(x) = 3x^3 + 2x^2 - 3x + 3$; $c = 2$
24. $f(x) = -4x^3 + 3x^2 + 2$; $c = -3$
25. $f(x) = 3x^4 - 6x^3 - 5x + 10$; $c = 2$
26. $f(x) = 4x^4 - 15x^2 - 4$; $c = 2$
27. $f(x) = 3x^6 + 82x^3 + 27$; $c = -3$
28. $f(x) = 2x^6 - 18x^4 + x^2 - 9$; $c = -3$
29. $f(x) = 4x^6 - 64x^4 + x^2 - 15$; $c = -4$
30. $f(x) = x^6 - 16x^4 + x^2 - 16$; $c = -4$
31. $f(x) = 2x^4 - x^3 + 2x - 1$; $c = \frac{1}{2}$
32. $f(x) = 3x^4 + x^3 - 3x + 1$; $c = -\frac{1}{3}$

In Problems 33–38, use synthetic division to find $f(c)$.

33. $f(x) = 3x^4 - 2x^2 + 1$; $c = 2$
34. $f(x) = -2x^3 + 2x^2 + 3$; $c = -2$
35. $f(x) = 4x^5 - 3x^3 + 2x - 1$; $c = -1$
36. $f(x) = -3x^4 + 3x^3 - 2x^2 + 5$; $c = -1$
37. $f(x) = 9x^{17} - 8x^{10} + 9x^8 + 5$; $c = 1$
38. $f(x) = 10x^{15} + 4x^{12} - 2x^5 + x^2$; $c = -1$

In Problems 39–48, write each polynomial in nested form.

39. $f(x) = 3x^3 + 2x^2 - 3x + 3$
40. $f(x) = -4x^3 + 3x^2 + 2$
41. $f(x) = 3x^4 - 6x^3 - 5x + 10$
42. $f(x) = 4x^4 - 15x^2 - 4$
43. $f(x) = 3x^6 - 82x^3 + 27$
44. $f(x) = 2x^6 - 18x^4 + x^2 - 9$
45. $f(x) = 4x^6 - 64x^4 + x^2 - 15$
46. $f(x) = x^6 - 16x^4 + x^2 - 16$
47. $f(x) = 2x^4 - x^3 + 2x - 1$
48. $f(x) = 3x^4 + x^3 - 3x + 1$

C *In Problems 49–58, use the nested form and a calculator to evaluate each polynomial at $x = 1.2$. Avoid using any memory key.*

49. $f(x) = 3x^3 + 2x^2 - 3x + 3$

50. $f(x) = -4x^3 + 3x^2 + 2$

51. $f(x) = 3x^4 - 6x^3 - 5x + 10$

52. $f(x) = 4x^4 - 15x^2 - 4$

53. $f(x) = 3x^6 - 82x^3 + 27$

54. $f(x) = 2x^6 - 18x^4 + x^2 - 9$

55. $f(x) = 4x^6 - 64x^4 + x^2 - 15$

56. $f(x) = x^6 - 16x^4 + x^2 - 16$

57. $f(x) = 2x^4 - x^3 + 2x - 1$

58. $f(x) = 3x^4 + x^3 - 3x + 1$

59. Find k such that $f(x) = x^3 - kx^2 + kx + 2$ has the factor $x - 2$.

60. Find k such that $f(x) = x^4 - kx^3 + kx^2 + 1$ has the factor $x + 2$.

61. What is the remainder when $f(x) = 2x^{20} - 8x^{10} + x - 2$ is divided by $x - 1$?

62. What is the remainder when $f(x) = -3x^{17} + x^9 - x^5 + 2x$ is divided by $x + 1$?

63. Use the Factor Theorem to prove that $x - c$ is a factor of $x^n - c^n$ for any positive integer n.

64. Use the Factor Theorem to prove that $x + c$ is a factor of $x^n + c^n$ if $n \geq 1$ is an odd integer.

C **65.** An IBM-AT microcomputer finds powers by multiplication. Suppose each multiplication of two numbers requires 33,333 nanoseconds and each addition or subtraction requires 500 nanoseconds. (Note that 1 nanosecond $= 10^{-9}$ second.) If all other times are disregarded, how long will it take the computer to find the value of $f(x) = 2x^3 - 6x^2 + 4x - 10$ at $x = 2.013$ by:
(a) Replacing x by 2.013 in the expression for $f(x)$?
(b) Replacing x by 2.013 in the nested form for $f(x)$?

C **66.** Using the microcomputer described in Problem 65, how long would it take by method (a) and by method (b) to find the value of $f(x) = ax^3 + bx^2 + cx + d$ for 5000 values of x?

67. *Programming exercise* Write a program that simulates synthetic division and divides a polynomial by $x - c$. Your input should consist of the coefficients of the polynomial, in order from highest to lowest, followed by the number c. Your output should consist of the numbers that would appear in the third row of the process of synthetic division.

68. *Programming exercise* Write a program that will evaluate a polynomial by using the nested form. Your input should consist of the coefficients of the polynomial, in order from highest to lowest, followed by the number at which the polynomial is to be evaluated. Test your program on the polynomials given in Problems 49–58.

5.5 The Zeros of a Polynomial Function

The (real) zeros of a polynomial function f are the (real) solutions, if any, of the equation $f(x) = 0$. They are also the x-intercepts of the graph of f. For polynomial and rational functions, we have seen the importance of locating the zeros for graphing. In most cases, however, the zeros of a polynomial function are difficult to find. There are no nice formulas like the quadratic formula available to us for polynomials of degree higher than 2. Although formulas do exist for solving any third- or fourth-degree polynomial equation, they are somewhat complicated. (If you are interested in learning about them, consult a book on the theory of equations.) It has been proved that no general formulas exist for polynomial equations of degree 5 or higher. In this section, we shall learn some ways of detecting information about the character of the zeros, which, in turn, may help us find them or, at least, isolate them.

Our first theorem concerns the number of zeros a polynomial function may have. In counting the zeros of a polynomial, we count each zero as many times as its multiplicity.

Theorem

Number of Zeros

A polynomial function cannot have more zeros than its degree.

Proof The proof is based on the Factor Theorem. If r is a zero of a polynomial function f, then $f(r) = 0$ and, hence, $x - r$ is a factor of $f(x)$. Thus, each zero corresponds to a factor of degree 1. Because f cannot have more first-degree factors than its degree, the result follows. ■

The next theorem, called **Descartes' Rule of Signs**, provides information about the number and location of the zeros of a polynomial function, so we know where to look for zeros. Descartes' Rule of Signs assumes that the polynomial is written in descending powers of x and requires that we count the number of variations in sign of the coefficients of $f(x)$ and $f(-x)$.

For example, the polynomial function below has two variations in the signs of coefficients:

$$\begin{aligned} f(x) &= -3x^7 + 4x^4 + 3x^2 - 2x - 1 \\ &= \underbrace{-3x^7 + 0x^6 + 0x^5 + 4x^4}_{-\text{ to }+} + 0x^3 + \underbrace{3x^2 - 2x}_{+\text{ to }-} - 1 \end{aligned}$$

Notice that we ignored the zero coefficients in $0x^6$, $0x^5$, and $0x^3$ in counting the number of variations in sign of $f(x)$. Replacing x by $-x$, we get

$$f(-x) = 3x^7 + 4x^4 + 3x^2 + \underbrace{2x - 1}_{+\text{ to }-}$$

which has one variation in sign.

Theorem

Descartes' Rule of Signs

Let f denote a polynomial function.

The number of positive zeros of f either equals the number of variations in sign of the coefficients of $f(x)$ or else equals that number less some even integer.

The number of negative zeros of f either equals the number of variations in sign of the coefficients of $f(-x)$ or else equals that number less some even integer. ■

We shall not prove Descartes' Rule of Signs. Let's see how it is used.

Example 1 Discuss the zeros of: $f(x) = 3x^6 - 4x^4 + 3x^3 + 2x^2 - x - 3$

Solution There are at most six zeros, because the equation is of degree 6. Since there are three variations in sign of the coefficients of $f(x)$, by Descartes' Rule of Signs we expect either three or one positive zero(s). To continue, we look at $f(-x)$:

$$f(-x) = 3x^6 - 4x^4 - 3x^3 + 2x^2 + x - 3$$

There are three variations in sign, so we expect either three or one negative zero(s). ■

Although we have not actually found the zeros, we know something about the number of zeros and how many might be positive or negative. The next result, which you are asked to prove in Exercise 5.5 (Problem 72), is called the **Rational Zeros Theorem**. It provides information about the rational zeros of a polynomial with integer coefficients.

Theorem

Rational Zeros Theorem

Let f be a polynomial function of degree 1 or higher of the form

$$f(x) = a_n x^n + a_{n-1}x^{n-1} + \cdots + a_1 x + a_0 \qquad a_n \neq 0,\ a_0 \neq 0$$

where each coefficient is an integer. If p/q, in lowest terms, is a rational zero of f, then p must be a factor of a_0 and q must be a factor of a_n. ■

Example 2 List the potential rational zeros of

$$f(x) = 3x^5 - 2x^4 - 15x^3 + 10x^2 + 12x - 8$$

Solution Because f has integer coefficients, we may use the Rational Zeros Theorem. First, we list all the integers p that are factors of $a_0 = -8$ and all the integers q that are factors of $a_5 = 3$:

$$p: \quad \pm 1, \pm 2, \pm 4, \pm 8$$
$$q: \quad \pm 1, \pm 3$$

Now we form all possible ratios p/q:

$$\frac{p}{q}: \quad \pm 1, \pm 2, \pm 4, \pm 8, \pm\frac{1}{3}, \pm\frac{2}{3}, \pm\frac{4}{3}, \pm\frac{8}{3}$$

If f has a rational zero, it will be found in this list, which contains 16 possibilities. ■

Be sure you understand what the Rational Zeros Theorem says: For a polynomial with integer coefficients, *if* there is a rational zero, it is one of those listed. There may not be any rational zeros. Synthetic division may be used to test each potential rational zero to determine whether it is indeed a zero. To make the work easier, the integers are usually tested first. Let's continue this example.

Example 3 Continue working with Example 2 to find the zeros of

$$f(x) = 3x^5 - 2x^4 - 15x^3 + 10x^2 + 12x - 8$$

Solution We gather all the information we can about the zeros:

STEP 1: There are at most five zeros.

STEP 2: By Descartes' Rule of Signs, there are three or one positive zero(s). Also, because

$$f(-x) = -3x^5 - 2x^4 + 15x^3 + 10x^2 - 12x - 8$$

there are two or no negative zeros.

STEP 3: Now we use our list of potential rational zeros from Example 2: $\pm 1, \pm 2, \pm 4, \pm 8, \pm \frac{1}{3}, \pm \frac{2}{3}, \pm \frac{4}{3}, \pm \frac{8}{3}$. We can test the potential rational zero 1 using synthetic division:

$$\begin{array}{r|rrrrrr} 1 & 3 & -2 & -15 & 10 & 12 & -8 \\ & & 3 & 1 & -14 & -4 & 8 \\ \hline & 3 & 1 & -14 & -4 & 8 & 0 \end{array}$$

The remainder is 0. Thus, 1 is a zero and $x - 1$ is a factor. The entries in the bottom row of this synthetic division can be used to factor f:

$$\begin{aligned} f(x) &= 3x^5 - 2x^4 - 15x^3 + 10x^2 + 12x - 8 \\ &= (x - 1)(3x^4 + x^3 - 14x^2 - 4x + 8) \end{aligned}$$

Let

$$q_1(x) = 3x^4 + x^3 - 14x^2 - 4x + 8$$

If r is a zero of q_1, then $q_1(r) = 0$. Since $f(x) = (x - 1)q_1(x)$, it follows that $f(r) = 0$ also. In other words, every zero of q_1 is also a zero of f. Because of this fact, we call the equation $q_1(x) = 0$ a **depressed equation** of f. Since the degree of the depressed equation is lower than that of the original equation, it is usually easier to find the zeros of the depressed equation than to continue working with the original equation.

STEP 4: By Descartes' Rule of Signs, q_1 has two or no positive zeros. Also, because

$$q_1(-x) = 3x^4 - x^3 - 14x^2 + 4x + 8$$

q_1 has two or no negative zeros.

STEP 5: The potential rational zeros of q_1 are the same as those listed earlier for f. We choose to test 1 again because it may be a repeated root:

$$\begin{array}{r|rrrrr} 1 & 3 & 1 & -14 & -4 & 8 \\ & & 3 & 4 & -10 & -14 \\ \hline & 3 & 4 & -10 & -14 & -6 \end{array}$$

The remainder tells us that 1 is not a zero of q_1. Now we test -1:

$$\begin{array}{r|rrrrr} -1 & 3 & 1 & -14 & -4 & 8 \\ & & -3 & 2 & 12 & -8 \\ \hline & 3 & -2 & -12 & 8 & 0 \end{array}$$

We find that -1 is a zero of q_1 and therefore $x - (-1) = x + 1$ is a factor of q_1. Thus, we have

$$f(x) = (x - 1)(x + 1)(3x^3 - 2x^2 - 12x + 8)$$

STEP 6: We work now with the depressed equation

$$q_2(x) = 3x^3 - 2x^2 - 12x + 8 = 0$$

By Descartes' Rule of Signs, q_2 has two or no positive zeros. Also, because

$$q_2(-x) = -3x^3 - 2x^2 + 12x + 8$$

q_2 has one negative zero. The list of potential rational zeros of q_2 is the same as that of f. However, because 1 was not a zero of q_1, it cannot be a zero of q_2. Also, the fact that -1 is a zero of q_1 does not mean it cannot also be a zero of q_2 (that is, it could be a repeated root of q_1). We know there is one negative zero (which, of course, may not be rational), so we test -1 once more to determine whether it is a root of q_2:

$$\begin{array}{r|rrrr} -1 & 3 & -2 & -12 & 8 \\ & & -3 & 5 & 7 \\ \hline & 3 & -5 & -7 & 15 \end{array}$$

It is not. Next, we choose to test -2:

$$\begin{array}{r|rrrr} -2 & 3 & -2 & -12 & 8 \\ & & -6 & 16 & -8 \\ \hline & 3 & -8 & 4 & 0 \end{array}$$

We find that -2 is a zero, so that $x - (-2) = x + 2$ is a factor. Thus, we have

$$f(x) = (x - 1)(x + 1)(x + 2)(3x^2 - 8x + 4) \qquad (1)$$

STEP 7: The new depressed equation of f, $q_3(x) = 3x^2 - 8x + 4 = 0$, is a quadratic equation with a discriminant of $b^2 - 4ac = (-8)^2 - 4(3)(4) = 16$. Therefore, $q_3(x)$ has two real solutions, and, in this case, we find them by factoring:

$$3x^2 - 8x + 4 = 0$$
$$(3x - 2)(x - 2) = 0$$
$$3x - 2 = 0 \quad \text{or} \quad x - 2 = 0$$
$$x = \tfrac{2}{3} \qquad\qquad x = 2$$

The zeros of f are $-2, -1, \frac{2}{3}, 1$, and 2. ■

The procedure outlined in Example 3 for finding the zeros of a polynomial also can be used to solve polynomial equations.

Example 4 Solve the equation: $x^5 - 5x^4 + 12x^3 - 24x^2 + 32x - 16 = 0$

Solution The solutions of this equation are the zeros of the polynomial function

$$f(x) = x^5 - 5x^4 + 12x^3 - 24x^2 + 32x - 16$$

STEP 1: There are at most five real solutions.

STEP 2: By Descartes' Rule of Signs, there are five, three, or one positive solution(s). Because

$$f(-x) = -x^5 - 5x^4 - 12x^3 - 24x^2 - 32x - 16$$

there are no negative solutions.

STEP 3: Because $a_5 = 1$ and there are no negative solutions, the potential rational solutions are the positive integers 1, 2, 4, 8, and 16. We test the potential rational solution 1 first, using synthetic division:

$$\begin{array}{r|rrrrrr} 1 & 1 & -5 & 12 & -24 & 32 & -16 \\ & & 1 & -4 & 8 & -16 & 16 \\ \hline & 1 & -4 & 8 & -16 & 16 & 0 \end{array}$$

Thus, 1 is a solution. The remaining solutions satisfy the depressed equation

$$x^4 - 4x^3 + 8x^2 - 16x + 16 = 0$$

STEP 4: The potential rational solutions are still 1, 2, 4, 8, and 16. We test 1 first, since it may be a repeated solution:

$$\begin{array}{r|rrrrr} 1 & 1 & -4 & 8 & -16 & 16 \\ & & 1 & -3 & 5 & -11 \\ \hline & 1 & -3 & 5 & -11 & 5 \end{array}$$

Thus, 1 is not a solution of the depressed equation. We try 2 next:

$$\begin{array}{r|rrrrr} 2 & 1 & -4 & 8 & -16 & 16 \\ & & 2 & -4 & 8 & -16 \\ \hline & 1 & -2 & 4 & -8 & 0 \end{array}$$

Thus, 2 is a solution. The remaining solutions satisfy the depressed equation

$$x^3 - 2x^2 + 4x - 8 = 0$$

STEP 5: The potential rational solutions are now 1, 2, 4, and 8. We know 1 is not a solution (why?), so we start with 2:

$$\begin{array}{r|rrrr} 2 & 1 & -2 & 4 & -8 \\ & & 2 & 0 & 8 \\ \hline & 1 & 0 & 4 & 0 \end{array}$$

Thus, 2 is a solution. The remaining solutions satisfy the depressed equation

$$x^2 + 4 = 0$$

which has no real solutions.

STEP 6: Thus, the real solutions are 1 and 2 (the latter being a repeated solution). ■

In Example 3, we found the zeros of the polynomial function $f(x)$. In so doing, we also factored f over the real numbers. Starting at equation (1), we found the factored form of f over the real numbers to be

$$\begin{aligned} f(x) &= 3x^5 - 2x^4 - 15x^3 + 10x^2 + 12x - 8 \\ &= (x - 1)(x + 1)(x + 2)(3x^2 - 8x + 4) \\ &= (x - 1)(x + 1)(x + 2)(3x - 2)(x - 2) \\ &= 3(x - 1)(x + 1)(x + 2)\left(x - \tfrac{2}{3}\right)(x - 2) \end{aligned}$$

Example 5 Use the factored form of $f(x) = 3x^5 - 2x^4 - 15x^3 + 10x^2 + 12x - 8$ to graph f.

Solution Using the zeros of f and the factored form of f, we construct Table 10. A calculator and the nested form of f were used to evaluate f at the test numbers (except $x = 0$).

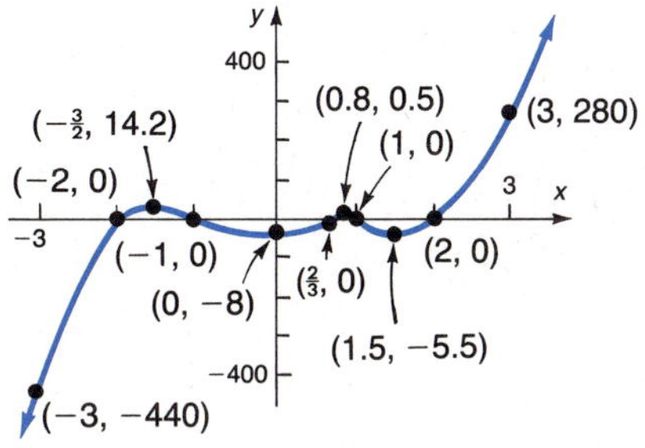

Figure 33

Table 10

INTERVAL	TEST NUMBER	VALUE OF f	GRAPH OF f
$x < -2$	-3	$f(-3) = -440$	Below x-axis
$-2 < x < -1$	$-\frac{3}{2}$	$f(-\frac{3}{2}) \approx 14.2$	Above x-axis
$-1 < x < \frac{2}{3}$	0	$f(0) = -8$	Below x-axis
$\frac{2}{3} < x < 1$	0.8	$f(0.8) \approx 0.5$	Above x-axis
$1 < x < 2$	$\frac{3}{2}$	$f(\frac{3}{2}) \approx -5.5$	Below x-axis
$x > 2$	3	$f(3) = 280$	Above x-axis

The graph of f has at most four turning points. For large values of x, the graph will behave like the graph of $y = 3x^5$. Figure 33 illustrates the graph of f. ■

Example 6 Use Descartes' Rule of Signs and the Rational Zeros Theorem to find the real zeros of the polynomial function

$$g(x) = x^5 - x^4 - x^3 + x^2 - 2x + 2$$

Use the zeros to factor g over the real numbers. Then graph g.

Solution

STEP 1: There are at most five zeros.

STEP 2: There are four, two, or no positive zeros. Also, because

$$g(-x) = -x^5 - x^4 + x^3 + x^2 + 2x + 2$$

there is one negative zero.

STEP 3: The potential rational zeros of g are ± 1, ± 2. We test 1:

$$\begin{array}{r|rrrrrr} 1 & 1 & -1 & -1 & 1 & -2 & 2 \\ & & 1 & 0 & -1 & 0 & -2 \\ \hline & 1 & 0 & -1 & 0 & -2 & 0 \end{array}$$

Thus, 1 is a zero, so $x - 1$ is a factor and

$$g(x) = (x - 1)(x^4 - x^2 - 2)$$

STEP 4: The depressed equation $q_1(x) = x^4 - x^2 - 2 = 0$ is quadratic in form and can be factored as follows:

$$x^4 - x^2 - 2 = 0$$
$$(x^2 - 2)(x^2 + 1) = 0$$
$$x^2 - 2 = 0 \quad \text{or} \quad x^2 + 1 = 0$$
$$x = \pm\sqrt{2}$$

Because $x^2 + 1 = 0$ has no real solution, the depressed equation has only the two solutions $\sqrt{2}$ and $-\sqrt{2}$.

Thus, the zeros of g are $-\sqrt{2}$, 1, and $\sqrt{2}$. The factored form of g over the real numbers is

$$\begin{aligned} g(x) &= x^5 - x^4 - x^3 + x^2 - 2x + 2 \\ &= (x-1)(x^4 - x^2 + 2) \\ &= (x-1)(x^2-2)(x^2+1) \\ &= (x-1)(x-\sqrt{2})(x+\sqrt{2})(x^2+1) \end{aligned}$$

Now we construct Table 11:

Table 11

INTERVAL	TEST NUMBER	VALUE OF g	GRAPH OF g
$x < -\sqrt{2}$	-2	$g(-2) = -30$	Below x-axis
$-\sqrt{2} < x < 1$	-1	$g(-1) = 4$	Above x-axis
$1 < x < \sqrt{2}$	1.1	$g(1.1) \approx -0.17$	Below x-axis
$x > \sqrt{2}$	2	$g(2) = 10$	Above x-axis

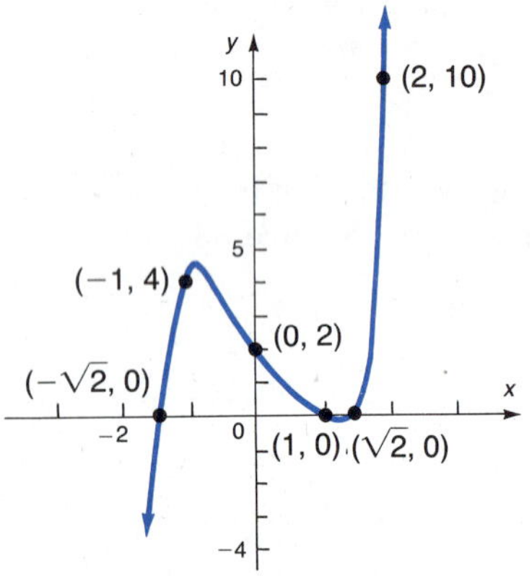

Figure 34
$g(x) = x^5 - x^4 - x^3 + x^2 - 2x + 2$

The graph of g has at most four turning points. For large values of x, the graph will behave like the graph of $y = x^5$. Figure 34 illustrates the graph of g. ■

In Example 6, the quadratic factor $x^2 + 1$ that appears in the factored form of $g(x)$ is called *irreducible*, because the polynomial $x^2 + 1$ cannot be factored over the real numbers. In general, we say that a quadratic factor $ax^2 + bx + c$ is **irreducible** if it cannot be factored over the real numbers—that is, if it is prime over the real numbers.

Refer back to Example 3, where the factored form of $f(x)$ was found to be

$$f(x) = 3(x-1)(x+1)(x+2)\left(x - \tfrac{2}{3}\right)(x-2)$$

The factored form of the polynomial g from Example 6 is

$$g(x) = (x-1)(x-\sqrt{2})(x+\sqrt{2})(x^2+1)$$

Note that the polynomial function f has five real zeros, and its factored form contains five linear factors. The polynomial function g has three real zeros, and its factored form contains three linear factors and one irreducible quadratic factor. The next result tells us what to expect when we factor a polynomial.

Theorem Every polynomial function (with real coefficients) can be uniquely factored into a product of linear factors and/or irreducible quadratic factors. ■

We shall prove this result in Section 5.7, and, in fact, we shall draw several additional conclusions about the zeros of a polynomial function. For now, though, we summarize the results obtained thus far.

Summary

To obtain information about the (real) zeros of a polynomial function, follow these steps:

Procedure for Finding the Zeros of a Polynomial Function

STEP 1: Use the degree of the polynomial to determine the maximum number of zeros.

STEP 2: Use Descartes' Rule of Signs to determine the possible number of positive zeros and negative zeros.

STEP 3: (a) If the polynomial has integer coefficients, use the Rational Zeros Theorem to identify those rational numbers that potentially can be zeros.
(b) Use synthetic division to test each potential rational zero.
(c) Each time a zero (and thus a factor) is found, repeat Steps 2 and 3 on the depressed equation.

STEP 4: In attempting to find the zeros, remember to use (if possible) the factoring techniques you already know (special products, factoring by grouping, and so on).

If these procedures fail to locate all the zeros, you may have to be satisfied with "estimating" or "approximating" the zeros—the subject of the next section.

Historical Comment

■ Formulas for the solution of third- and fourth-degree polynomial equations exist, and, while not very practical, they do have an interesting history.

In the 1500's in Italy, mathematical contests were a popular pastime, and persons possessing methods for solving problems kept them secret. (Solutions that were published were already common knowledge.) Nicolo of Brescia (1499–1557), commonly referred to as Tartaglia ("the stammerer"), had the secret for solving cubic (third-degree) equations, which gave him a decided advantage in the contests. Girolamo Cardano (1501–1576) found out that Tartaglia had the secret, and, being interested in cubics, he requested it from Tartaglia. The reluctant Tartaglia hesitated for some time, but finally, swearing Cardano to secrecy with midnight oaths by candlelight, told him the secret. Cardano then published the solution in his book *Ars Magna* (1545), giving Tartaglia the credit but rather compromising the secrecy. Tartaglia exploded into bitter recriminations, and each wrote pamphlets that reflected on the other's mathematics, moral character, and ancestry.

The quartic (fourth-degree) equation was solved by Cardano's student Lodovico Ferrari, and this solution also was included, with credit and this time with permission, in the *Ars Magna*.

Attempts were made to solve the fifth-degree equation in similar ways, all of which failed. In the early 1800's, P. Ruffini, Niels Abel, and Evariste Galois all found ways to show that it is not possible to solve fifth-degree equations by formula, but the proofs required the introduction of new methods. Galois' methods eventually developed into a large part of modern algebra. ■

EXERCISE 5.5

In Problems 1–12, use Descartes' Rule of Signs to determine how many positive and how many negative zeros each polynomial function may have. Do not attempt to find the zeros.

1. $f(x) = -4x^7 + x^3 - 1$
2. $f(x) = 5x^4 - 2x^2 + x - 2$
3. $f(x) = 2x^6 - 3x^2 - x + 1$
4. $f(x) = -3x^5 + 4x^4 + 2$
5. $f(x) = 3x^3 - 2x^2 + x + 2$
6. $f(x) = -x^3 - x^2 + x + 1$
7. $f(x) = -x^4 + x^2 - 1$
8. $f(x) = x^4 + 5x^3 - 2$
9. $f(x) = x^5 + x^4 + x^2 + x + 1$
10. $f(x) = x^5 - x^4 + x^3 - x^2 + x - 1$
11. $f(x) = x^6 - 1$
12. $f(x) = x^6 + 1$

In Problems 13–24, list the potential rational zeros of each polynomial function. Do not attempt to find the zeros.

13. $f(x) = x^4 - 3x^3 + x^2 - x + 1$
14. $f(x) = x^5 - x^4 + 2x^2 + 2$
15. $f(x) = x^5 - 6x^2 + 9x - 3$
16. $f(x) = 2x^5 - x^4 - x^2 + 1$
17. $f(x) = -2x^3 - x^2 + x + 1$
18. $f(x) = 3x^4 - x^2 + 2$
19. $f(x) = 3x^4 - x^2 + 2$
20. $f(x) = -4x^3 + x^2 + x + 2$
21. $f(x) = 2x^5 - x^3 + 2x^2 + 4$
22. $f(x) = 3x^5 - x^2 + 2x + 3$
23. $f(x) = 6x^4 + 2x^3 - x^2 + 2$
24. $f(x) = -6x^3 - x^2 + x + 3$

In Problems 25–36, use Descartes' Rule of Signs and the Rational Zeros Theorem to find all the real zeros of each polynomial function. Use the zeros to factor f over the real numbers.

25. $f(x) = x^3 + 2x^2 - 5x - 6$
26. $f(x) = x^3 + 8x^2 + 11x - 20$
27. $f(x) = 2x^3 - x^2 + 2x - 1$
28. $f(x) = 2x^3 + x^2 + 2x + 1$
29. $f(x) = x^4 + x^2 - 2$
30. $f(x) = x^4 - 3x^2 - 4$
31. $f(x) = 4x^4 + 7x^2 - 2$
32. $f(x) = 4x^4 + 15x^2 - 4$

33. $f(x) = x^4 + x^3 - 3x^2 - x + 2$

34. $f(x) = x^4 - x^3 - 6x^2 + 4x + 8$

35. $f(x) = 4x^5 - 8x^4 - x + 2$

36. $f(x) = 4x^5 + 12x^4 - x - 3$

In Problems 37–46, solve each equation in the real number system.

37. $x^4 - x^3 + 2x^2 - 4x - 8 = 0$

38. $2x^3 + 3x^2 + 2x + 3 = 0$

39. $3x^3 + 4x^2 - 7x + 2 = 0$

40. $2x^3 - 3x^2 - 3x - 5 = 0$

41. $3x^3 - x^2 - 15x + 5 = 0$

42. $2x^3 - 11x^2 + 10x + 8 = 0$

43. $x^4 + 4x^3 + 2x^2 - x + 6 = 0$

44. $x^4 - 2x^3 + 10x^2 - 18x + 9 = 0$

45. $x^3 - \frac{2}{3}x^2 + \frac{8}{3}x + 1 = 0$

46. $x^3 + \frac{3}{2}x^2 + 3x - 2 = 0$

In Problems 47–56, find the intercepts of each polynomial function $f(x)$. Find the intervals x for which the graph of f is above the x-axis and below the x-axis. Obtain several other points on the graph, and connect them with a smooth curve. [Hint: Use the factored form of f (see Problems 27–36).]

47. $f(x) = 2x^3 - x^2 + 2x - 1$

48. $f(x) = 2x^3 + x^2 + 2x + 1$

49. $f(x) = x^4 + x^2 - 2$

50. $f(x) = x^4 - 3x^2 - 4$

51. $f(x) = 4x^4 + 7x^2 - 2$

52. $f(x) = 4x^4 + 15x^2 - 4$

53. $f(x) = x^4 + x^3 - 3x^2 - x + 2$

54. $f(x) = x^4 - x^3 - 6x^2 + 4x + 8$

55. $f(x) = 4x^5 - 8x^4 - x + 2$

56. $f(x) = 4x^5 + 12x^4 - x - 3$

In Problems 57–64, solve each equation in the complex number system.

57. $x^4 + 3x^2 - 4 = 0$

58. $x^4 + 5x^2 - 36 = 0$

59. $x^4 + x^3 - x - 1 = 0$

60. $x^4 - x^3 + x - 1 = 0$

61. $x^4 + 3x^3 - x^2 - 12x - 12 = 0$

62. $x^4 - 3x^3 - 5x^2 + 27x - 36 = 0$

63. $x^5 - x^4 + 2x^3 - 2x^2 + x - 1 = 0$

64. $x^5 + x^4 + x^3 + x^2 - 2x - 2 = 0$

65. Is $\frac{1}{3}$ a zero of $f(x) = 2x^3 + 3x^2 - 6x + 7$?

66. Is $\frac{1}{3}$ a zero of $f(x) = 4x^3 - 5x^2 - 3x + 1$?

67. Is $\frac{3}{5}$ a zero of $f(x) = 2x^6 - 5x^4 + x^3 - x + 1$?

68. Is $\frac{2}{3}$ a zero of $f(x) = x^7 + 6x^5 - x^4 + x + 2$?

69. What is the length of the edge of a cube if, after a slice 1 inch thick is cut from one side, the volume remaining is 294 cubic inches?

70. What is the length of the edge of a cube if its volume could be doubled by an increase of 6 centimeters in one edge, an increase of 12 centimeters in a second edge, and a decrease of 4 centimeters in the third edge?

71. Let $f(x)$ be a polynomial function whose coefficients are integers. Suppose that r is a (real) zero of f and that the leading coefficient of f is 1. Use the Rational Zeros Theorem to show that r is either an integer or an irrational number.

72. Prove the Rational Zeros Theorem. [*Hint:* Let p/q, where p and q have no common factors except 1 and -1, be a solution of the polynomial $f(x) = a_nx^n + a_{n-1}x^{n-1} + \cdots + a_1x + a_0$, whose coefficients are all integers. Show that $a_np^n + a_{n-1}p^{n-1}q + \cdots + a_1pq^{n-1} + a_0q^n = 0$. Now, because p is a factor of the first n terms of this equation, p also must be a factor of the term a_0q^n. Since p is not a factor of q (why?), p must be a factor of a_0. Similarly, q must be a factor of a_n.]

Problems 73–81 develop the Tartaglia–Cardano solution of the cubic equation and show why it is not altogether practical.

73. Show that the general cubic equation, $y^3 + by^2 + cy + d = 0$, can be transformed into an equation of the form $x^3 + px + q = 0$ by using the substitution $y = x - b/3$.

74. In the equation $x^3 + px + q = 0$, replace x by $H + K$. Let $3HK = -p$, and show that $H^3 + K^3 = -q$. [*Hint:* $3H^2K + 3HK^2 = 3HKx$]

75. Based on Problem 74, we have the two equations

$$3HK = -p \qquad \text{and} \qquad H^3 + K^3 = -q$$

Solve for K in $3HK = -p$ and substitute into $H^3 + K^3 = -q$. Then show that

$$H = \sqrt[3]{\frac{-q}{2} + \sqrt{\frac{q^2}{4} + \frac{p^3}{27}}}$$

[*Hint:* Look for an equation that is quadratic in form.]

76. Use the solution for H from Problem 75 and the equation $H^3 + K^3 = -q$ to show that

$$K = \sqrt[3]{\frac{-q}{2} - \sqrt{\frac{q^2}{4} + \frac{p^3}{27}}}$$

77. Use the results from Problems 74–76 to show that the solution of $x^3 + px + q = 0$ is

$$x = \sqrt[3]{\frac{-q}{2} + \sqrt{\frac{q^2}{4} + \frac{p^3}{27}}} + \sqrt[3]{\frac{-q}{2} - \sqrt{\frac{q^2}{4} + \frac{p^3}{27}}}$$

78. Use the result of Problem 77 to solve the equation: $x^3 - 6x - 9 = 0$

C **79.** Use the result of Problem 77 to solve the equation: $x^3 + 3x - 14 = 0$

80. Use the methods of this chapter to solve the equation: $x^3 + 3x - 14 = 0$

81. *Requires complex numbers* Show that the formula derived in Problem 77 leads to the cube roots of a complex number when applied to the equation $x^3 - 6x + 4 = 0$. Use the methods of this chapter to solve the equation.

5.6 Approximating the Real Zeros of a Polynomial Function*

Sometimes the procedures we discussed in Section 5.5 yield limited information about the zeros of a polynomial. Let's look at an example.

Example 1 Discuss the zeros of: $f(x) = x^5 - x^3 - 1$

Solution

STEP 1: f has at most five zeros.

STEP 2: f has one positive zero, and because $f(-x) = -x^5 + x^3 - 1$, f has two or no negative zeros.

* C Refer to Graphics Calculator Supplement: Activity II.

STEP 3: The potential rational zeros are ± 1, neither of which is an actual zero. We conclude that f has one positive irrational zero and perhaps two negative irrational zeros. ■

To obtain more information about the zeros of the polynomial of Example 1, we need some additional results.

Upper and Lower Bounds

The search for the zeros of a polynomial function can be reduced somewhat if upper and lower bounds to the zeros can be found. A number M is an **upper bound** to the zeros of a polynomial f if no zero of f exceeds M. The number m is a **lower bound** if no zero of f is less than m.

Thus, if m is a lower bound and M is an upper bound to the zeros of a polynomial f, then

$$m \leq \text{Any zero of } f \leq M$$

One immediate advantage of knowing the values of a lower bound m and an upper bound M is that for polynomials with integer coefficients, it may allow you to eliminate some potential rational zeros—namely, any that lie outside the interval $[m, M]$. The next result tells us how to locate lower and upper bounds.

Theorem
Bounds on Zeros

Let f denote a polynomial function whose leading coefficient is positive.

If $M > 0$ is a real number and if the third row in the process of synthetic division of f by $x - M$ contains only numbers that are positive or 0, then M is an upper bound to the zeros of f.

If $m \leq 0$ is a real number and if the third row in the process of synthetic division of f by $x - m$ contains numbers that are alternately positive (or 0) and negative (or 0), then m is a lower bound to the zeros of f.

Proof (Outline) We shall give only an outline of the proof of the first part of the theorem. Suppose M is a positive real number and the third row in the process of synthetic division of the polynomial f by $x - M$ contains only numbers that are positive or 0. Then there is a quotient q and a remainder R so that

$$f(x) = (x - M)q(x) + R$$

where the coefficients of $q(x)$ are positive or 0 and the remainder $R \geq 0$. Then, for any $x > M$, we must have $x - M > 0$, $q(x) > 0$, and $R \geq 0$, so that $f(x) > 0$. That is, there is no zero of f larger than M. ■

Example 2 Find upper and lower bounds to the zeros of: $f(x) = x^5 - x^3 - 1$

Solution To get an upper bound to the zeros, the usual practice is to start with 1, and continue with 2, 3, . . . , until the third row of the process of synthetic division yields only numbers that are positive or 0. Thus, we start with 1:

1)	1	0	−1	0	0	−1
		1	1	0	0	0
	1	1	0	0	0	−1

2)	1	0	−1	0	0	−1
		2	4	6	12	24
	1	2	3	6	12	23

The third row has only positive numbers; thus, 2 is an upper bound.

To get a lower bound to the zeros, we start with −1, and continue with −2, −3, . . . , until the third row of the process of synthetic division yields numbers that alternate in sign:

−1)	1	0	−1	0	0	−1
		−1	1	0	0	0
	1	−1	0	0	0	−1
			Count as positive	Count as negative	Count as positive	

Because the entries alternate in sign, −1 is a lower bound. Thus, the zeros of f lie between −1 and 2. ■

Note: In determining lower bounds, a 0 in the bottom row following a nonzero entry may be counted as positive or negative, as needed. If the next entry is also a 0, it must be counted opposite to the way the preceding 0 was counted (refer to Example 2).

In Example 2, we found that the zeros of $f(x) = x^5 - x^3 - 1$ lie in the interval $[-1, 2]$. However, remember that in finding the lower bound −1 and the upper bound 2, we tested only integers. Were we to test other positive numbers less than 2 and other negative numbers greater than −1, we might be able to "fine-tune" the bounds and find a smaller interval containing the zeros of f. However, the effort required to do this is usually not worth it, since more efficient methods are available. We look at one such method next.

Intermediate Value Theorem

The next result, called the **Intermediate Value Theorem**, is based on the fact that the graph of a polynomial function is continuous—that is, contains no "jumps" or "gaps."

Theorem

Intermediate Value Theorem

Let f denote a polynomial function. If $a < b$ and if $f(a)$ and $f(b)$ are of opposite sign, then there is at least one zero of f between a and b. ■

Although the proof of this result requires advanced methods in calculus, it is easy to "see" why the result is true. Look at Figure 35.

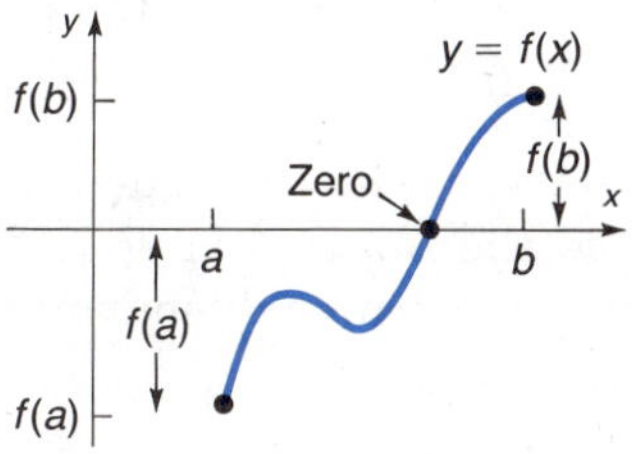

Figure 35
If $f(a) < 0$ and $f(b) > 0$, there is a zero between a and b

Example 3 Show that $f(x) = x^5 - x^3 - 1$ has a zero between 1 and 2.

Solution We know from Example 1 that f has exactly one positive zero. Now, look back at the solution to Example 2, where we used synthetic division to divide f by $x - 1$ and then by $x - 2$. There, we see that

$$f(1) = -1 \qquad \text{and} \qquad f(2) = 23$$

Because $f(1) < 0$ and $f(2) > 0$, it follows from the Intermediate Value Theorem that f has a zero between 1 and 2. ■

Note that the zero we now know to lie between 1 and 2 is irrational, because we found in Example 1 that the only possible rational zeros are -1 and 1.

We can use the Intermediate Value Theorem to get a better approximation of the zero of a function f as follows:

Approximating the Zeros of a Polynomial Function

STEP 1: Find two consecutive integers a and $a + 1$ such that f has a zero between them.
STEP 2: Divide the interval $[a, a + 1]$ into ten equal subintervals.
STEP 3: Evaluate f at each endpoint of the subintervals until the Intermediate Value Theorem applies; that interval then contains a zero.
STEP 4: Repeat the process starting at Step 2 until the desired accuracy is achieved.

C **Example 4** Find the positive zero of $f(x) = x^5 - x^3 - 1$ to within 0.01.

Solution From Example 3 we know the positive zero is between 1 and 2. We divide the interval [1, 2] into ten equal subintervals: [1, 1.1], [1.1, 1.2], [1.2, 1.3], [1.3, 1.4], [1.4, 1.5], [1.5, 1.6], [1.6, 1.7], [1.7, 1.8], [1.8, 1.9], [1.9, 2]. Now, we find the value of f at each endpoint until the Intermediate Value Theorem applies. The easiest method is to write $f(x)$ in nested form and use a calculator. Thus, we write

$$\begin{aligned} f(x) &= x^5 - x^3 - 1 \\ &= (x^2 - 1) \cdot x \cdot x \cdot x - 1 = (x \cdot x - 1) \cdot x \cdot x \cdot x - 1 \end{aligned}$$

$$f(1.0) = -1 \qquad f(1.2) = -0.23968$$
$$f(1.1) = -0.72049 \qquad f(1.3) = 0.51593$$

We can stop here and conclude that the zero is between 1.2 and 1.3. Now we divide the interval [1.2, 1.3] into ten equal subintervals and proceed to evaluate f at each endpoint:

$$f(1.20) = -0.23968 \qquad f(1.23) \approx -0.0455613$$
$$f(1.21) \approx -0.1778185 \qquad f(1.24) \approx 0.025001$$
$$f(1.22) \approx -0.1131398$$

We conclude that the zero lies between 1.23 and 1.24, and so we have approximated it to within 0.01. ■

There are many other numerical techniques for approximating the zeros of a polynomial. The one outlined in Example 4 (a variation of the *bisection method*) has the advantages that it will always work, that it can be programmed rather easily on a computer, and each time it is used another decimal place of accuracy is achieved. See Problem 23 for the bisection method.

EXERCISE 5.6

In Problems 1–6, use the Intermediate Value Theorem to show that each polynomial function has a zero in the given interval.

1. $f(x) = 8x^4 - 2x^2 + 5x - 1$; [0, 1]

2. $f(x) = x^4 + 8x^3 - x^2 + 2$; [−1, 0]

3. $f(x) = 2x^3 + 6x^2 - 8x + 2$; [−5, −4]

4. $f(x) = 3x^3 - 10x + 9$; [−3, −2]

C **5.** $f(x) = x^5 - x^4 + 7x^3 - 7x^2 - 18x + 18$; [1.4, 1.5]

C **6.** $f(x) = x^5 - 3x^4 - 2x^3 + 6x^2 + x + 2$; [1.7, 1.8]

In Problems 7–12, find integer-valued upper and lower bounds to the zeros of each polynomial function.

7. $f(x) = 2x^3 + x^2 - 1$

8. $f(x) = 3x^3 - 2x^2 + x + 4$

9. $f(x) = x^3 + 5x^2 + 11x + 11$

10. $f(x) = 2x^3 + x^2 + 11x - 6$

11. $f(x) = x^4 + 3x^3 - 5x^2 + 9$

12. $f(x) = 4x^4 - 12x^3 + 27x^2 - 54x + 81$

C *In Problems 13–16, each polynomial function has exactly one positive root. Use the method of Example 4 to approximate the zero to within 0.01.*

13. $f(x) = x^3 + x^2 + x - 4$

14. $f(x) = 2x^4 + x^2 - 1$

15. $f(x) = 2x^4 - 3x^3 - 4x^2 - 8$

16. $f(x) = 3x^3 - 2x^2 - 20$

In Problems 17–20, each equation has a solution r in the interval indicated. Use the method of Example 4 to approximate this solution to within 0.01.

17. $8x^4 - 2x^2 + 5x - 1 = 0; \quad 0 \le r \le 1$

18. $x^4 + 8x^3 - x^2 + 2 = 0; \quad -1 \le r \le 0$

19. $2x^3 + 6x^2 - 8x + 2 = 0; \quad -5 \le r \le -4$

20. $3x^3 - 10x + 9 = 0; \quad -3 \le r \le -2$

21. *Programming exercise* Write a computer program that will estimate the positive zero of a polynomial function to any desired degree of accuracy. Input should consist of the coefficients of the polynomial, in order from highest to lowest, followed by the degree N of accuracy wanted (that is, the zero is to be estimated to within 10^{-N}), followed by two consecutive integers between which the zero lies. Output will consist of two decimal numbers between which the zero lies. The program should contain a subroutine that writes the polynomial in nested form.

22. *Programming exercise* Modify the program in Problem 21 to include a subroutine that will locate the two consecutive integers between which the zero lies.

23. *The bisection method for approximating zeros of a function f* We begin with two consecutive integers, a and $a + 1$, such that $f(a)$ and $f(a + 1)$ are of opposite sign. Evaluate f at the midpoint m_1 of a and $a + 1$. If $f(m_1) = 0$, then m_1 is the zero of f, and we are finished. Otherwise, $f(m_1)$ is of opposite sign to either $f(a)$ or $f(a + 1)$. Suppose it is $f(a)$ and $f(m_1)$ that are of opposite sign. Now evaluate f at the midpoint m_2 of a and m_1. Repeat this process until the desired degree of accuracy is obtained. Note that each iteration places the zero in an interval whose length is half that of the previous interval. Use the bisection method to solve Problems 13–20.

5.7 Complex Polynomials; Fundamental Theorem of Algebra

A variable z in the complex number system is referred to as a **complex variable**. A **complex polynomial function $f(z)$** of degree n is a complex function of the form

$$f(z) = a_n z^n + a_{n-1} z^{n-1} + \cdots + a_1 z + a_0 \qquad (1)$$

where $a_n, a_{n-1}, \ldots, a_1, a_0$ are complex numbers, $a_n \neq 0$, and n is a nonnegative integer. Here, a_n is called the **leading coefficient** of f. A complex number r is called a (complex) **zero** of a complex function f if $f(r) = 0$.

In Chapter 2, we discovered that some quadratic equations have no real solutions, but that in the complex number system, every quadratic equation has a solution—either real or complex. The next result, proved by Karl Friedrich Gauss (1777–1855) when he was 22 years of age,* gives an extension to complex polynomials. In fact, this result is so important and useful, it has become known as the **Fundamental Theorem of Algebra**.

Theorem

Fundamental Theorem of Algebra

Every complex polynomial function $f(z)$ of degree $n \geq 1$ has at least one complex zero. ■

We shall not prove this result, as the proof is beyond the scope of this book. However, using the Fundamental Theorem of Algebra and the Factor Theorem, we can prove the following result:

Theorem Every complex polynomial function $f(z)$ of degree $n \geq 1$ can be factored into n linear factors (not necessarily distinct) of the form

$$f(z) = a_n(z - r_1)(z - r_2) \cdot \cdots \cdot (z - r_n) \tag{2}$$

where $a_n, r_1, r_2, \ldots, r_n$ are complex numbers.

Proof Let

$$f(z) = a_n z^n + a_{n-1}z^{n-1} + \cdots + a_1 z + a_0$$

By the Fundamental Theorem of Algebra, f has at least one zero, say, r_1. Then, by the Factor Theorem, $z - r_1$ is a factor, and

$$f(z) = (z - r_1)q_1(z)$$

where $q_1(z)$ is a complex polynomial of degree $n - 1$ whose leading coefficient is a_n. Again, by the Fundamental Theorem of Algebra, the complex polynomial $q_1(z)$ has at least one zero, say, r_2. By the Factor Theorem, $q_1(z)$ has the factor $z - r_2$, so that

$$q_1(z) = (z - r_2)q_2(z)$$

*In all, Gauss gave four different proofs of this theorem, the first one in 1799 being the subject of his doctoral dissertation.

where $q_2(z)$ is a complex polynomial of degree $n - 2$ whose leading coefficient is a_n. Consequently,

$$f(z) = (z - r_1)(z - r_2)q_2(z)$$

Repeating this argument n times, we finally arrive at

$$f(z) = (z - r_1)(z - r_2) \cdot \cdots \cdot (z - r_n)q_n(z)$$

where $q_n(z)$ is a complex polynomial of degree $n - n = 0$ whose leading coefficient is a_n. Thus, $q_n(z) = a_n z^0 = a_n$, and so,

$$f(z) = a_n(z - r_1)(z - r_2) \cdot \cdots \cdot (z - r_n)$$ ■

Complex Polynomials with Real Coefficients

We can use the Fundamental Theorem of Algebra to obtain valuable information about the zeros of complex polynomials whose coefficients are real numbers.

Theorem
Conjugate Pairs

Let $f(z)$ be a complex polynomial whose coefficients are real numbers. If $r = a + bi$ is a zero of f, then the complex conjugate $\bar{r} = a - bi$ is also a zero of f. ■

In other words, for complex polynomials whose coefficients are real numbers, the zeros occur in conjugate pairs.

Proof Let

$$f(z) = a_n z^n + a_{n-1}z^{n-1} + \cdots + a_1 z + a_0$$

where $a_n, a_{n-1}, \ldots, a_1, a_0$ are real numbers and $a_n \neq 0$. If r is a zero of f, then $f(r) = 0$, so that

$$a_n r^n + a_{n-1}r^{n-1} + \cdots + a_1 r + a_0 = 0$$

We take the conjugate of both sides to get

$$\overline{a_n r^n + a_{n-1}r^{n-1} + \cdots + a_1 r + a_0} = \bar{0}$$

$$\overline{a_n r^n} + \overline{a_{n-1}r^{n-1}} + \cdots + \overline{a_1 r} + \overline{a_0} = \bar{0}$$ The conjugate of a sum equals the sum of the conjugates (see Section 1.9).

$$\overline{a_n}(\bar{r})^n + \overline{a_{n-1}}(\bar{r})^{n-1} + \cdots + \overline{a_1}\bar{r} + \overline{a_0} = \bar{0}$$ The conjugate of a product equals the product of the conjugates.

$$a_n(\bar{r})^n + a_{n-1}(\bar{r})^{n-1} + \cdots + a_1\bar{r} + a_0 = 0$$ The conjugate of a real number equals the real number.

This last equation states that $f(\bar{r}) = 0$; that is, $\bar{r}$ is a zero of f. ■

The value of this result should be clear. Once we know that, say, $3 + 4i$ is a zero of a polynomial with real coefficients, then we know that $3 - 4i$ is also a zero. This result has an important corollary.

Corollary A complex polynomial $f(z)$ of odd degree with real coefficients has at least one real zero.

Proof Because complex zeros occur as conjugate pairs in a complex polynomial with real coefficients, there will always be an even number of zeros that are not real numbers. Consequently, since f is of odd degree, one of its zeros has to be a real number. ■

For example, the polynomial $f(z) = z^5 - 3z^4 + 4z^3 - 5$ has at least one zero that is a real number, since f is of degree 5 (odd) and has real coefficients.

Now we can prove the theorem we conjectured earlier in Section 5.5.

Theorem Every polynomial function with real coefficients can be uniquely factored over the real numbers into a product of linear factors and/or irreducible quadratic factors.

Proof Every complex polynomial $f(z)$ of degree n has exactly n zeros and can be factored into a product of n linear factors. If its coefficients are real, then those zeros that are complex numbers will always occur as conjugate pairs. As a result, if $r = a + bi$ is a complex zero, then so is $\bar{r} = a - bi$. Consequently, when the linear factors $z - r$ and $z - \bar{r}$ of $f(z)$ are multiplied, we have

$$(z - r)(z - \bar{r}) = z^2 - (r + \bar{r})z + r\bar{r} = z^2 - 2az + a^2 + b^2$$

This second-degree polynomial has real coefficients and is irreducible (over the real numbers). Thus, the factors of f are either linear or irreducible quadratic factors. ■

Example 1 A polynomial $f(z)$ of degree 5 whose coefficients are real numbers has the zeros 1, $5i$, and $1 + i$. Find the remaining two zeros.

Solution Since complex zeros appear as conjugate pairs, it follows that $-5i$, the conjugate of $5i$, and $1 - i$, the conjugate of $1 + i$, are the two remaining zeros. ■

Polynomials with Complex Coefficients

The division algorithm for polynomials (see Section 5.4) is true for polynomials with complex coefficients. As a result, the Remainder Theorem and Factor Theorem are also true. In fact, the process of synthetic division also works for polynomials with complex coefficients.

Example 2 Use synthetic division and the Factor Theorem to show that $1 + 2i$ is a zero of

$$f(z) = (1 + i)z^2 + (2 - i)z + (3 - 4i)$$

Solution We use synthetic division and divide $f(z)$ by $z - (1 + 2i)$:

$$\begin{array}{r|rrr} 1+2i & 1+i & 2-i & 3-4i \\ & & -1+3i & -3+4i \\ \hline & 1+i & 1+2i & 0 \end{array}$$

Thus, $z - (1 + 2i)$ is a factor, and $1 + 2i$ is a zero. ■

Of course, we could have shown that $1 + 2i$ is a zero of the polynomial $f(z)$ in Example 2 by using substitution as follows:

$$\begin{aligned} f(1 + 2i) &= (1 + i)(1 + 2i)^2 + (2 - i)(1 + 2i) + (3 - 4i) \\ &= -7 + i + 4 + 3i + 3 - 4i = 0 \end{aligned}$$

Based on equation (2), a complex polynomial $f(z)$ of degree n has n linear factors. These n linear factors do not have to be distinct; some may be repeated more than once. When a linear factor $z - r$ appears exactly m times in the factored form of $f(z)$, then r is called a **zero of multiplicity *m*** of f. Thus, if a zero of multiplicity m is counted m times, it follows that a complex polynomial $f(z)$ of degree $n \geq 1$ has exactly n zeros.

Example 3 The complex polynomial function

$$f(z) = (2 + i)(z - 5)^3(z + i)^2[z - (3 + i)]^4(z - i)$$

of degree 10 has $2 + i$ as leading coefficient. Its zeros are listed below.

5:	Multiplicity	3
$-i$:	Multiplicity	2
$3 + i$:	Multiplicity	4
i:	Multiplicity	1
Degree:		10

■

Example 4 Form a polynomial $f(z)$ with complex coefficients of degree 3 and with the following zeros:

$1 + i$: Multiplicity 1

$-i$: Multiplicity 2

Solution Since $1 + i$ is a zero of multiplicity 1 and $-i$ is a zero of multiplicity 2, then $z - (1 + i)$ and $(z + i)^2$ are factors of f. Thus, $f(z)$ is of the form

$$\begin{aligned} f(z) &= [z - (1 + i)](z + i)^2 \\ &= [z - (1 + i)](z^2 + 2iz - 1) \\ &= z^3 + (-1 + i)z^2 + (1 - 2i)z + 1 + i \end{aligned}$$

Although there are other polynomials with complex coefficients having the three required zeros, the only ones of degree 3 will be $f(z)$ or $kf(z)$, where $k \neq 0$ is some complex number. ■

EXERCISE 5.7

In Problems 1–10, information is given about a complex polynomial $f(z)$ whose coefficients are real numbers. Find the remaining zeros of f.

1. Degree 3; zeros: $2, 1 - i$
2. Degree 3; zeros: $1, 2 + i$
3. Degree 4; zeros: $i, 1 + i$
4. Degree 4; zeros: $1, 2, 2 + i$
5. Degree 5; zeros: $1, i, 2i$
6. Degree 5; zeros: $0, 1, 2, i$
7. Degree 4; zeros: $i, 2, -2$
8. Degree 4; zeros: $2 - i, -i$
9. Degree 6; zeros: $2, 2 + i, -3 - i, 0$
10. Degree 6; zeros: $i, 3 - 2i, -2 + i$

In Problems 11 and 12, tell why the facts given are contradictory.

11. $f(z)$ is a complex polynomial of degree 3 whose coefficients are real numbers; its zeros are $1 + i$, $1 - i$, and $2 + i$.
12. $f(z)$ is a complex polynomial of degree 3 whose coefficients are real numbers; its zeros are 2, i, and $1 + i$.
13. $f(z)$ is a complex polynomial of degree 4 whose coefficients are real numbers; three of its zeros are 2, $1 + 2i$, and $1 - 2i$. Explain why the remaining zero must be a real number.
14. $f(z)$ is a complex polynomial of degree 4 whose coefficients are real numbers; two of its zeros are -3 and $4 - i$. Explain why one of the remaining zeros must be a real number. Write down one of the missing zeros.
15. Find all the zeros of $f(z) = z^3 - 1$.
16. Find all the zeros of $f(z) = z^4 - 1$.

In Problems 17–22, evaluate each complex polynomial function $f(z)$ at $z = 1 + i$.

17. $f(z) = iz - 2$
18. $f(z) = 2z - i$
19. $f(z) = 3z^2 - z$
20. $f(z) = (4 + i)z^2 + 5 - 2i$
21. $f(z) = z^3 + iz - 1 + i$
22. $f(z) = iz^3 - 2z^2 + 1$

In Problems 23–28, use synthetic division to find the value of $f(r)$.

23. $f(z) = 5z^5 - iz^4 + 2; \quad r = 1 + i$
24. $f(z) = iz^4 + (2 + i)z^2 - z; \quad r = 1 - i$
25. $f(z) = (1 + i)z^4 - z^3 + iz; \quad r = 2 - i$
26. $f(z) = 2iz^3 + 8z^2 - 4iz + 1; \quad r = 2 + i$
27. $f(z) = iz^5 + iz^3 + iz; \quad r = 1 + 2i$
28. $f(z) = z^4 + z^2 + 1; \quad r = 1 - 2i$

In Problems 29–34, form a polynomial $f(z)$ with complex coefficients having the given degree and zeros.

29. Degree 3; zeros: $1 + 2i$, multiplicity 1; 3, multiplicity 2

30. Degree 3; zeros: $-i$, multiplicity 1; $1 + 2i$, multiplicity 2

31. Degree 3; zeros: 2, multiplicity 1; $-i$, multiplicity 1; $1 + i$, multiplicity 1

32. Degree 3; zeros: i, multiplicity 1; $4 - i$, multiplicity 1; $2 + i$, multiplicity 1

33. Degree 4; zeros: 3, multiplicity 2; $-i$, multiplicity 2

34. Degree 4; zeros: 1, multiplicity 3; $1 + i$, multiplicity 1

CHAPTER REVIEW

THINGS TO KNOW

Quadratic function	$f(x) = ax^2 + bx + c,\ a \neq 0$	Vertex: $(-b/2a, f(-b/2a))$ Axis: The line $x = -b/2a$ Parabola opens up if $a > 0$. Parabola opens down if $a < 0$.
Polynomial function	$f(x) = a_n x^n + a_{n-1}x^{n-1} + \cdots + a_1 x + a_0,\ a_n \neq 0$	At most $n - 1$ turning points; behaves like $y = a_n x^n$ for large x (see Steps 1–3, page 296).
Power function	$f(x) = x^n$, n even	Even function: Passes through $(-1, 1)$, $(0, 0)$, $(1, 1)$ Opens up
	$f(x) = x^n$, n odd	Odd function: Passes through $(-1, -1)$, $(0, 0)$, $(1, 1)$ Increasing
Rational function	$R(x) = \dfrac{p(x)}{q(x)}$, p, q are polynomial functions in lowest terms	See Steps 1–6 on page 306.
Zeros of a polynomial f	Numbers for which $f(x) = 0$; these are the x-intercepts of the graph of f.	
Remainder Theorem	If a polynomial $f(x)$ is divided by $x - c$, then the remainder is $f(c)$.	
Factor Theorem	$x - c$ is a factor of a polynomial $f(x)$ if and only if $f(c) = 0$.	
Descartes' Rule of Signs	Let f denote a polynomial function. The number of positive zeros of f either equals the number of variations in sign of the coefficients of $f(x)$	

or else equals that number less some even integer. The number of negative zeros of f either equals the number of variations in sign of the coefficients of $f(-x)$ or else equals that number less some even integer.

Rational Zeros Theorem — Let f be a polynomial function of degree 1 or higher of the form

$$f(x) = a_n x^n + a_{n-1}x^{n-1} + \cdots + a_1 x + a_0,\ a_n \neq 0,\ a_0 \neq 0$$

where each coefficient is an integer. If p/q, in lowest terms, is a rational zero of f, then p must be a factor of a_0 and q must be a factor of a_n.

Intermediate Value Theorem — If $a < b$ and if $f(a)$ and $f(b)$ are of opposite sign, then there is at least one zero of f between a and b.

Fundamental Theorem of Algebra — Every complex polynomial function $f(z)$ of degree $n \geq 1$ has at least one complex zero.

Conjugate Pairs Theorem — Let $f(z)$ be a complex polynomial whose coefficients are real numbers. If $r = a + bi$ is a zero of f, then the complex conjugate $\bar{r} = a - bi$ is also a zero of f.

How To:

Graph quadratic functions

Graph polynomial functions (see Steps 1–3, page 296)

Graph rational functions (see Steps 1–6, page 306)

Use synthetic division to divide a polynomial by $x - c$

Write a polynomial in nested form

Find the zeros of a polynomial by using Descartes' Rule of Signs, the Rational Zeros Theorem, and depressed equations

Solve polynomial equations using Descartes' Rule of Signs, the Rational Zeros Theorem, and depressed equations

Approximate the zeros of a polynomial

FILL-IN-THE-BLANK ITEMS

1. The graph of a quadratic function is called a(n) __________. Its lowest or highest point is called the __________.
2. In the process of long division,

$$\text{(Divisor)(Quotient)} + \_\_\_\_\_\_\_\_\_\_ = \_\_\_\_\_\_\_\_\_\_$$

3. When a polynomial function f is divided by $x - c$, the remainder is __________.
4. A polynomial function f has the factor $x - c$ if and only if __________.
5. A number r for which $f(r) = 0$ is called a(n) __________ of the function f.
6. The polynomial function $f(x) = x^5 - 2x^3 + x^2 + x - 1$ has either __________ or __________ positive zeros; it has __________ or __________ negative zeros.
7. The possible rational zeros of $f(x) = 2x^5 - x^3 + x^2 - x + 1$ are __________.

8. The line ________ is a horizontal asymptote of: $R(x) = \dfrac{x^3 - 1}{x^3 + 1}$

9. The line ________ is a vertical asymptote of: $R(x) = \dfrac{x^3 - 1}{x^3 + 1}$

10. If $3 + 4i$ is a zero of a polynomial of degree 5 with real coefficients, then so is ________.

REVIEW EXERCISES

In Problems 1–10, graph each quadratic function by determining whether its graph opens up or down, and by finding its vertex, axis of symmetry, y-intercept, and x-intercepts, if any.

1. $f(x) = (x - 2)^2 + 2$
2. $f(x) = (x + 1)^2 - 4$
3. $f(x) = \frac{1}{4}x^2 - 16$
4. $f(x) = -\frac{1}{2}x^2 + 2$
5. $f(x) = -4x^2 + 4x$
6. $f(x) = 9x^2 - 6x + 3$
7. $f(x) = \frac{9}{2}x^2 + 3x + 1$
8. $f(x) = -x^2 + x + \frac{1}{2}$
9. C $f(x) = 3x^2 + 4x - 1$
10. C $f(x) = -2x^2 - x + 4$

In Problems 11–16, graph each function using the techniques of shifting, compressing, stretching, and reflection.

11. $f(x) = (x + 2)^3$
12. $f(x) = -x^3 + 3$
13. $f(x) = -(x - 1)^4$
14. $f(x) = (x - 1)^4 - 2$
15. $f(x) = (x - 1)^4 + 2$
16. $f(x) = (1 - x)^3$

In Problems 17–22, determine whether the given quadratic function has a maximum value or a minimum value, and then find the value.

17. $f(x) = 2x^2 - 4x + 3$
18. $f(x) = 3x^2 + 6x + 4$
19. $f(x) = -x^2 + 8x - 4$
20. $f(x) = -x^2 - 10x - 3$
21. $f(x) = -3x^2 + 12x + 4$
22. $f(x) = -2x^2 + 4$

In Problems 23–30:

(a) *Find the x- and y-intercepts of each polynomial function f.*

(b) *Determine whether the graph of f touches or crosses the x-axis at each x-intercept.*

(c) *Find the intervals on which the graph is above and below the x-axis.*

(d) *Obtain several other points on the graph and connect them with a smoooth curve.*

23. $f(x) = x(x + 2)(x + 4)$
24. $f(x) = x(x - 2)(x - 4)$
25. $f(x) = (x - 2)^2(x + 4)$
26. $f(x) = (x - 2)(x + 4)^2$
27. $f(x) = x^3 - 4x^2$
28. $f(x) = x^3 + 4x$
29. $f(x) = (x - 1)^2(x + 3)(x + 1)$
30. $f(x) = (x - 4)(x + 2)^2(x - 2)$

In Problems 31–40, discuss each rational function following the six steps outlined in Section 5.3.

31. $R(x) = \dfrac{2x - 6}{x}$
32. $R(x) = \dfrac{4 - x}{x}$
33. $H(x) = \dfrac{x + 2}{x(x - 2)}$
34. $H(x) = \dfrac{x}{x^2 - 1}$
35. $R(x) = \dfrac{x^2}{(x - 1)^2}$
36. $R(x) = \dfrac{(x - 3)^2}{x^2}$
37. $F(x) = \dfrac{x^3}{x^2 - 4}$
38. $F(x) = \dfrac{3x^3}{(x - 1)^2}$
39. $R(x) = \dfrac{2x^4}{(x - 1)^2}$
40. $R(x) = \dfrac{x^4}{x^2 - 9}$

In Problems 41–44, use synthetic division to find the quotient $q(x)$ and remainder R when $f(x)$ is divided by $g(x)$.

41. $f(x) = 8x^3 - 2x^2 + x - 4;\ \ g(x) = x - 1$

42. $f(x) = 2x^3 + 4x^2 - 6x + 5;\ \ g(x) = x - 2$

43. $f(x) = x^4 - 2x^3 + x - 1;\ \ g(x) = x + 2$

44. $f(x) = x^4 - x^2 + 3x;\ \ g(x) = x + 1$

45. Find the value of $f(x) = 12x^6 - 8x^4 + 1$ at $x = 4$.

46. Find the value of $f(x) = -16x^3 + 18x^2 - x + 2$ at $x = -2$.

In Problems 47 and 48 use Descartes' Rule of Signs to determine how many positive and negative zeros each polynomial function may have. Do not attempt to find the zeros.

47. $f(x) = 12x^8 - x^7 + 6x^4 - x^3 + x - 3$

48. $f(x) = -6x^5 + x^4 + 2x^3 - x + 1$

49. List all the potential rational zeros of: $f(x) = 12x^8 - x^7 + 6x^4 - x^3 + x - 3$

50. List all the potential rational zeros of: $f(x) = -6x^5 + x^4 + 2x^3 - x + 1$

In Problems 51–56, use Descartes' Rule of Signs and the Rational Zeros Theorem to find all the real zeros of each polynomial function. Use the zeros to factor f over the real numbers.

51. $f(x) = x^3 - 3x^2 - 6x + 8$

52. $f(x) = x^3 - x^2 - 10x - 8$

53. $f(x) = 4x^3 + 4x^2 - 7x + 2$

54. $f(x) = 4x^3 - 4x^2 - 7x - 2$

55. $f(x) = x^4 - 4x^3 + 9x^2 - 20x + 20$

56. $f(x) = x^4 + 6x^3 + 11x^2 + 12x + 18$

In Problems 57–60, solve each equation in the real number system.

57. $2x^4 + 2x^3 - 11x^2 + x - 6 = 0$

58. $3x^4 + 3x^3 - 17x^2 + x - 6 = 0$

59. $2x^4 + 7x^3 + x^2 - 7x - 3 = 0$

60. $2x^4 + 7x^3 - 5x^2 - 28x - 12 = 0$

In Problems 61–70, find the intercepts of each polynomial $f(x)$. Find the numbers x for which the graph of f is above and below the x-axis. Obtain several other points on the graph and connect them with a smooth curve.

61. $f(x) = x^3 - 3x^2 - 6x + 8$

62. $f(x) = x^3 - x^2 - 10x - 8$

63. $f(x) = 4x^3 + 4x^2 - 7x + 2$

64. $f(x) = 4x^3 - 4x^2 - 7x - 2$

65. $f(x) = x^4 - 4x^3 + 9x^2 - 20x + 20$

66. $f(x) = x^4 + 6x^3 + 11x^2 + 12x + 18$

67. $f(x) = 2x^4 + 2x^3 - 11x^2 + x - 6$

68. $f(x) = 3x^4 + 3x^3 - 17x^2 + x - 6$

69. $f(x) = 2x^4 + 7x^3 + x^2 - 7x - 3$

70. $f(x) = 2x^4 + 7x^3 - 5x^2 - 28x - 12$

In Problems 71–74, use the Intermediate Value Theorem to show that each polynomial has a zero in the given interval.

71. $f(x) = 3x^3 - x - 1;\ \ [0, 1]$

72. $f(x) = 2x^3 - x^2 - 3;\ \ [1, 2]$

73. $f(x) = 8x^4 - 4x^3 - 2x - 1;\ \ [0, 1]$

74. $f(x) = 3x^4 + 4x^3 - 8x - 2;\ \ [1, 2]$

In Problems 75–78, find integer-valued upper and lower bounds to the zeros of each polynomial function.

75. $f(x) = 2x^3 - x^2 - 4x + 2$

76. $f(x) = 2x^3 + x^2 - 10x - 5$

77. $f(x) = 2x^3 - 7x^2 - 10x + 35$

78. $f(x) = 3x^3 - 7x^2 - 6x + 14$

C *In Problems 78–82, each polynomial has exactly one positive zero. Approximate the zero to within 0.01.*

79. $f(x) = x^3 - x - 2$

80. $f(x) = 2x^3 - x^2 - 3$

81. $f(x) = 8x^4 - 4x^3 - 2x - 1$

82. $f(x) = 3x^4 + 4x^3 - 8x - 2$

In Problems 83–86, information is given about a complex polynomial $f(z)$ whose coefficients are real numbers. Find the remaining zeros of f.

83. Degree 3; zeros: $1 + i$, 1

84. Degree 3; zeros: $3 - 4i$, 2

85. Degree 4; zeros: i, $1 + i$

86. Degree 4; zeros: 1, 2, $1 + i$

In Problems 87–90, form a polynomial $f(z)$ with complex coefficients having the given degree and zeros.

87. Degree 4; zeros: 1, multiplicity 2; i, multiplicity 1; 2, multiplicity 1

88. Degree 4; zeros: i, multiplicity 2; 3, multiplicity 2

89. Degree 3; zeros: $1 + i$, 2, 3, each of multiplicity 1

90. Degree 3; zeros: 1, $1 + i$, $1 + 2i$, each of multiplicity 1

91. Find the quotient and remainder if $x^4 + 2x^3 - 7x^2 - 8x + 12$ is divided by $(x - 2)(x - 1)$.

92. Find the quotient and remainder if $x^4 + 2x^3 - 4x^2 - 5x - 6$ is divided by $(x - 2)(x + 3)$.

In Problems 93–96, solve each equation in the complex number system.

93. $x^3 - x^2 - 8x + 12 = 0$

94. $x^3 - 3x^2 - 4x + 12 = 0$

95. $3x^4 - 4x^3 + 4x^2 - 4x + 1 = 0$

96. $x^4 + 4x^3 + 2x^2 - 8x - 8 = 0$

C *In Problems 97–100, write each polynomial in nested form. Then use a calculator to evaluate each polynomial at $x = 1.5$. Avoid using any memory key.*

97. $f(x) = 8x^3 - 2x^2 + x - 4$

98. $f(x) = 2x^3 + 4x^2 - 6x + 5$

99. $f(x) = x^4 - 2x^3 + x - 1$

100. $f(x) = x^4 - x^2 + 3x$

101. Find integer-valued upper and lower bounds to the zeros of

$$f(x) = 4x^5 - 3x^4 + 8x^2 + x + 2$$

102. Find integer-valued upper and lower bounds to the zeros of

$$f(x) = 8x^6 - x^4 + 6x^2 + 24x + 15$$

103. Find the point on the line $y = x$ that is closest to the point (3, 1). [*Hint:* Find the minimum value of the function $f(x) = d^2$, where d is the distance from (3, 1) to a point on the line.]

104. Find the point on the line $y = x + 1$ that is closest to the point (4, 1).

105. A horizontal bridge is in the shape of a parabolic arch. Given the information shown in the figure, what is the height h of the arch 2 feet from shore?

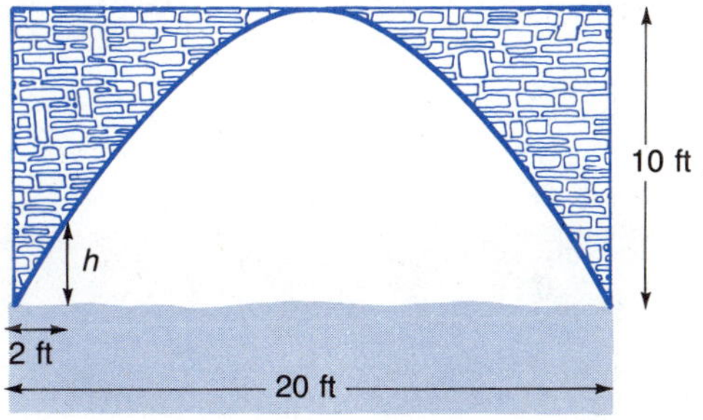

106. Find the length and width of a rectangle whose perimeter is 20 feet and whose area is 16 square feet.

Exponential and Logarithmic Functions

6

Until now, our study of functions has concentrated primarily on polynomial and rational functions. These functions belong to the class of **algebraic functions**—that is, functions that can be expressed in terms of sums, differences, products, quotients, powers, or roots of polynomials. Functions that are not algebraic are termed **transcendental** (they transcend, or go beyond, algebraic functions).

In this chapter, we study two transcendental functions: the *exponential* and *logarithmic functions*. These functions occur frequently in a wide variety of applications.

6.1 Exponential Functions*

In Chapter 1, we gave a definition for raising a real number a to a rational power. Based on that discussion, we gave meaning to expressions of the form

$$a^r$$

where the base a is a positive real number and the exponent r is a rational number.

But what is the meaning of a^x, where the base a is a positive real number and the exponent x is an irrational number? Although a rigorous definition requires methods discussed in calculus, the basis for the definition is easy to follow: Select a rational number r that is formed by truncating (removing) all but a finite number of digits from the irrational number x. Then it is reasonable to expect that

$$a^x \approx a^r$$

For example, take the irrational number $\pi = 3.14159\ldots$. Then, an approximation to a^π is

$$a^\pi \approx a^{3.14}$$

where the digits after the hundredths position have been removed from the value for π. A better approximation would be

$$a^\pi \approx a^{3.14159}$$

where the digits after the hundred-thousandths position have been removed. Continuing in this way, we can obtain approximations to a^π to any desired degree of accuracy.

Most scientific calculators have a $\boxed{y^x}$ key for working with exponents. To use this key, first enter the base y, then press the $\boxed{y^x}$ key, enter x, and press the $\boxed{=}$ key.

C **Example 1** Using a calculator with a $\boxed{y^x}$ key, evaluate:

(a) $2^{1.4}$ (b) $2^{1.41}$ (c) $2^{1.414}$ (d) $2^{1.4142}$ (e) $2^{\sqrt{2}}$

Solution (a) $2^{1.4} \approx 2.6390158$ (b) $2^{1.41} \approx 2.6573716$
(c) $2^{1.414} \approx 2.6647497$ (d) $2^{1.4142} \approx 2.6651191$
(e) $2^{\sqrt{2}} \approx 2.6651441$ ■

It can be shown that the familiar laws of rational exponents hold for real exponents.

* C Refer to Graphics Calculator Supplement: Activity IV.

Theorem If s, t, a, and b are real numbers with $a > 0$ and $b > 0$, then

Laws of Exponents

$$a^s \cdot a^t = a^{s+t} \qquad (a^s)^t = a^{st} \qquad (ab)^s = a^s \cdot b^s$$
$$1^s = 1 \qquad a^{-s} = \frac{1}{a^s} = \left(\frac{1}{a}\right)^s \qquad a^0 = 1 \tag{1}$$

■

We are now ready for the following definition.

Exponential Function

An **exponential function** is a function of the form

$$f(x) = a^x$$

where a is a positive real number and $a \neq 1$. The domain of f is the set of all real numbers.

We exclude the base $a = 1$, because this function is simply the constant function $f(x) = 1^x = 1$. We also need to exclude bases that are negative, because, otherwise, we would have to exclude many values of x from the domain, such as $x = \frac{1}{2}$, $x = \frac{3}{4}$, and so on. [Recall that $(-2)^{1/2}$, $(-3)^{3/4}$, and so on, are not defined in the system of real numbers.]

Graphs of Exponential Functions

First, we graph the exponential function $y = 2^x$.

Example 2 Graph the exponential function: $f(x) = 2^x$

Solution The domain of $f(x) = 2^x$ consists of all real numbers. We begin by locating some points on the graph of $f(x) = 2^x$, as listed in Table 1 (page 356).

Since $2^x > 0$ for all x, the range of f is $(0, \infty)$. From this, we conclude that the graph has no x-intercepts, and, in fact, the graph will lie above the x-axis. As Table 1 indicates, the y-intercept is 1. Table 1 also indicates that as $x \to -\infty$, the value of $f(x) = 2^x$ gets closer and closer to 0. Thus, the x-axis is a horizontal asymptote to the graph as $x \to -\infty$. Look again at Table 1. As $x \to \infty$, $f(x) = 2^x$ grows very quickly, causing the graph of $f(x) = 2^x$ to rise very rapidly. Thus, it is apparent that f is an increasing function and, hence, is one-to-one. Using all this information, we plot some of the points from Table 1 and connect them with a smooth curve, as shown in Figure 1.

Table 1

x	$f(x) = 2^x$
-10	$2^{-10} \approx 0.00098$
-3	$2^{-3} = \frac{1}{8}$
-2	$2^{-2} = \frac{1}{4}$
-1	$2^{-1} = \frac{1}{2}$
0	$2^0 = 1$
1	$2^1 = 2$
2	$2^2 = 4$
3	$2^3 = 8$
10	$2^{10} = 1024$

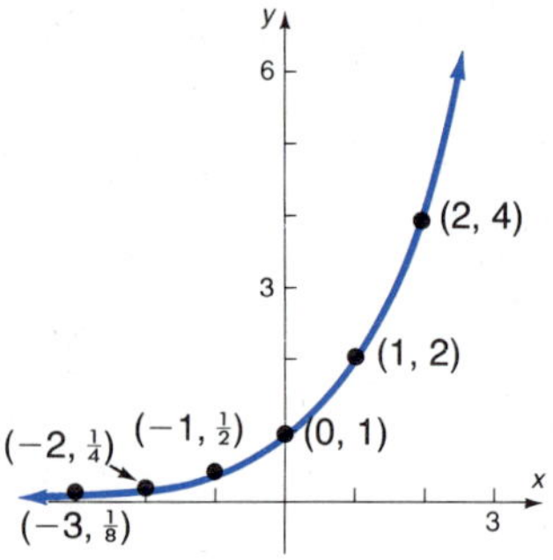

Figure 1
$y = 2^x$

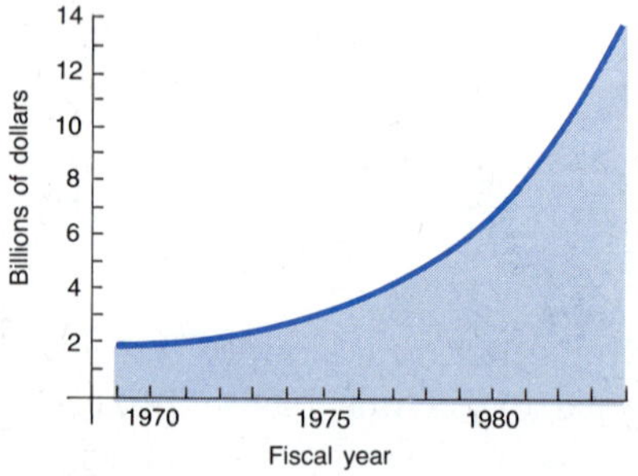

Figure 2
Medicare payments to doctors for care of elderly
Source: Office of Management and Budget

As we shall see, graphs that look like the one in Figure 1 occur very frequently in a variety of situations. For example, look at the graph in Figure 2, which illustrates the amount of money paid to doctors in the Medicare program. Researchers might conclude from this graph that Medicare payments "behave exponentially"; that is, the graph exhibits "rapid, or exponential, growth." We shall have more to say about situations that lead to exponential growth later in this chapter. For now, we continue to seek properties of the exponential functions.

The graph of $f(x) = 2^x$ in Figure 1 is typical of all exponential functions that have a base larger than 1. Such functions are increasing functions and hence are one-to-one. Their graphs lie above the x-axis, pass through the point (0, 1), and thereafter rise rapidly as $x \to \infty$. As $x \to -\infty$, the x-axis is a horizontal asymptote. There are no vertical asymptotes. Finally, the graphs are smooth, with no corners or gaps. Figure 3 illustrates the graphs of two more exponential functions whose bases are larger than 1. Notice that for the larger base, the graph is steeper when $x > 0$ and is closer to the x-axis when $x < 0$.

The box summarizes the information we have about $f(x) = a^x$, $a > 1$:

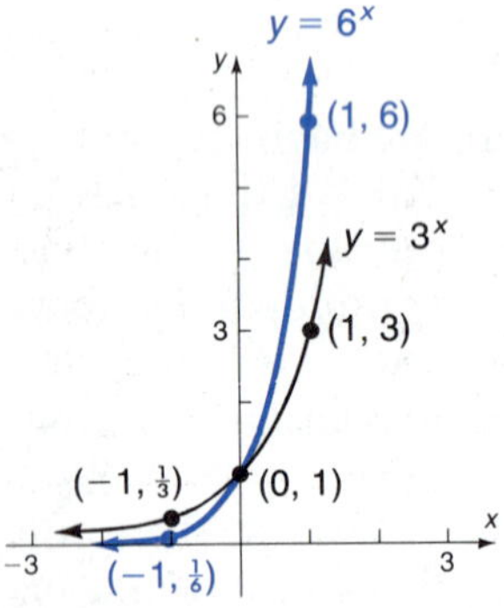

Figure 3

$f(x) = a^x \qquad a > 1$

Domain: $(-\infty, \infty)$ Range: $(0, \infty)$

x-intercepts: None y-intercept: 1

Horizontal asymptote: x-axis, as $x \to -\infty$

$f(x)$ is an increasing function; $f(x)$ is one-to-one

Now, what happens when $0 < a < 1$? Example 3 will help us find out.

Example 3 Graph the exponential function: $f(x) = \left(\frac{1}{2}\right)^x$

Solution The domain of $f(x) = \left(\frac{1}{2}\right)^x$ consists of all real numbers. As before, we locate some points on the graph, as listed in Table 2. Since $\left(\frac{1}{2}\right)^x > 0$ for all x, the range of f is $(0, \infty)$. Thus, the graph lies above the x-axis and so has no x-intercepts. The y-intercept is 1. As $x \to -\infty$, $f(x) = \left(\frac{1}{2}\right)^x$ grows very quickly. As $x \to \infty$, the values of $f(x)$ approach 0. Thus, the x-axis ($y = 0$) is a horizontal asymptote as $x \to \infty$. It is apparent that f is a decreasing function and, hence, is one-to-one. Figure 4 illustrates the graph.

Table 2

x	$f(x) = \left(\frac{1}{2}\right)^x$
-10	$\left(\frac{1}{2}\right)^{-10} = 1024$
-3	$\left(\frac{1}{2}\right)^{-3} = 8$
-2	$\left(\frac{1}{2}\right)^{-2} = 4$
-1	$\left(\frac{1}{2}\right)^{-1} = 2$
0	$\left(\frac{1}{2}\right)^{0} = 1$
1	$\left(\frac{1}{2}\right)^{1} = \frac{1}{2}$
2	$\left(\frac{1}{2}\right)^{2} = \frac{1}{4}$
3	$\left(\frac{1}{2}\right)^{3} = \frac{1}{8}$
10	$\left(\frac{1}{2}\right)^{10} \approx 0.00098$

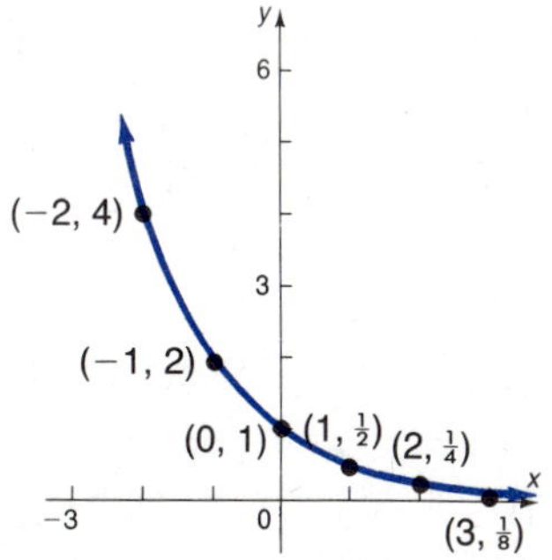

Figure 4
$y = \left(\frac{1}{2}\right)^x$

■

Note that we could have obtained the graph of $y = \left(\frac{1}{2}\right)^x$ from the graph of $y = 2^x$. Since $\left(\frac{1}{2}\right)^x = 2^{-x}$, the graph of $y = 2^{-x} = \left(\frac{1}{2}\right)^x$ is a reflection about the y-axis of the graph of $y = 2^x$. Compare Figures 1 and 4.

The graph of $f(x) = \left(\frac{1}{2}\right)^x$ in Figure 4 is typical of all exponential functions that have a base between 0 and 1. Such functions are decreasing, one-to-one functions. Their graphs lie above the x-axis and pass through the point $(0, 1)$. The graphs rise rapidly as $x \to -\infty$. As $x \to \infty$, the x-axis is a horizontal asymptote. There are no vertical asymptotes. Finally, the graphs are smooth, with no corners or gaps. Figure 5 illustrates the graphs of two more exponential functions whose bases are between 0 and 1. Notice that the choice of a base closer to 0 results in a graph that is steeper when $x < 0$ and closer to the x-axis when $x > 0$.

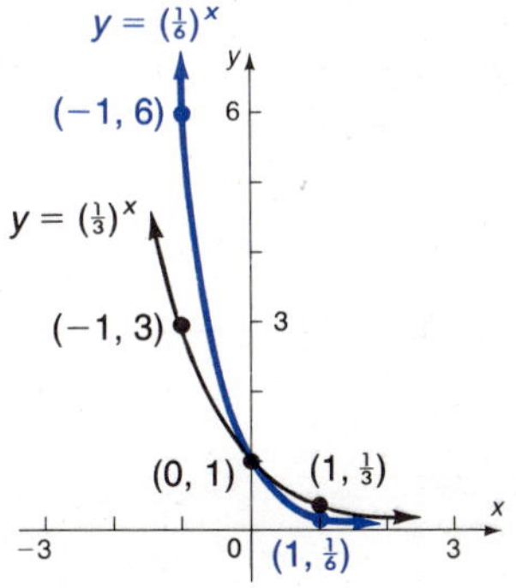

Figure 5

The box summarizes the information we have about $f(x) = a^x$, $0 < a < 1$:

$f(x) = a^x$	$0 < a < 1$
Domain: $(-\infty, \infty)$	Range: $(0, \infty)$
x-intercepts: None	y-intercept: 1
Horizontal asymptote:	x-axis, as $x \to \infty$
$f(x)$ is a decreasing function;	$f(x)$ is one-to-one

The techniques of shifting, compression, stretching, and reflection may be used to graph many exponential functions.

Example 4 Graph: $f(x) = 2^{-x} - 3$

Solution Figure 6 shows the various steps.

Figure 6

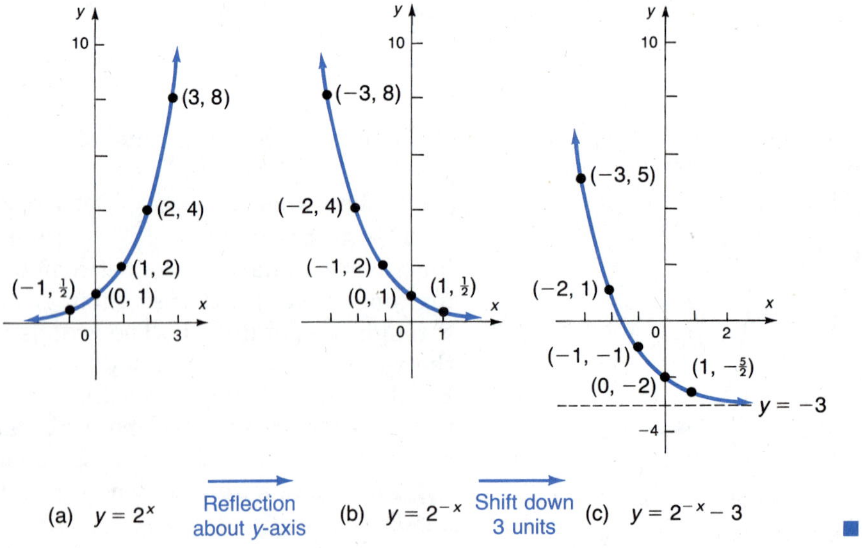

Example 5 Graph: $f(x) = -(2^{x-3})$

Solution Figure 7 shows the various steps.

Figure 7

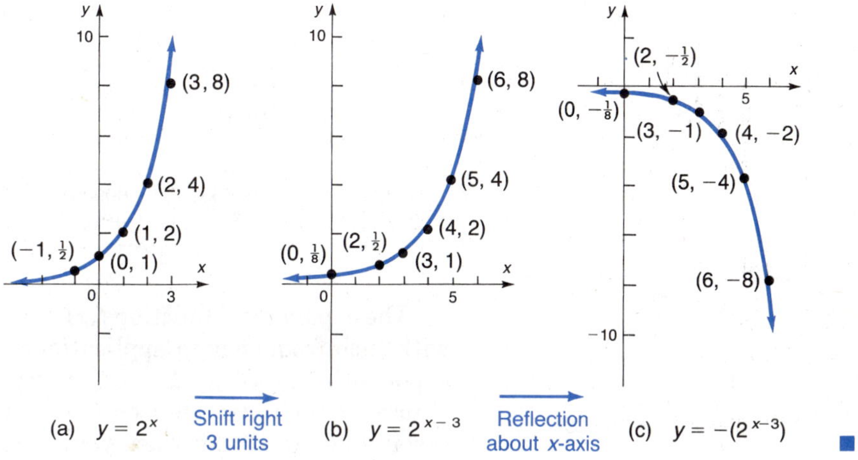

The Base *e*

As we shall see shortly, many problems that occur in nature require the use of an exponential function whose base is a certain irrational number, symbolized by the letter *e*.

Let's look now at one way of arriving at this important number *e*.

The **number *e*** is defined as the number that the expression

The Number *e*

$$\left(1 + \frac{1}{n}\right)^n \qquad (2)$$

approaches as $n \to \infty$.

Table 3 (page 360) illustrates what happens to the defining expression (2) as n takes on increasingly large values. The last number in the last column in the table is correct to nine decimal places and is the same as the entry given for *e* on your calculator (if expressed correct to nine decimal places).

Table 3

n	$\frac{1}{n}$	$1 + \frac{1}{n}$	$\left(1 + \frac{1}{n}\right)^n$
1	1	2	2
2	0.5	1.5	2.25
5	0.2	1.2	2.48832
10	0.1	1.1	2.59374246
100	0.01	1.01	2.704813829
1,000	0.001	1.001	2.716923932
10,000	0.0001	1.0001	2.718145926
100,000	0.00001	1.00001	2.718268237
1,000,000	0.000001	1.000001	2.718280469
1,000,000,000	10^{-9}	$1 + 10^{-9}$	2.718281828

The exponential function $f(x) = e^x$, whose base is the number e, occurs with such frequency in applications that it is usually referred to as *the* exponential function. Indeed, many calculators have the key e^x or exp(x), which may be used to evaluate the exponential function for a given value of x.* Now use your calculator to find e^x (or refer to Table I in the back of this book) for $x = -2$, $x = -1$, $x = 0$, $x = 1$, and $x = 2$, as we have done to create Table 4 (after rounding). The graph of the exponential function $f(x) = e^x$ is given in Figure 8. Since $2 < e < 3$, the graph of $y = e^x$ lies between the graphs of $y = 2^x$ and $y = 3^x$. (Refer to Figures 1 and 3.)

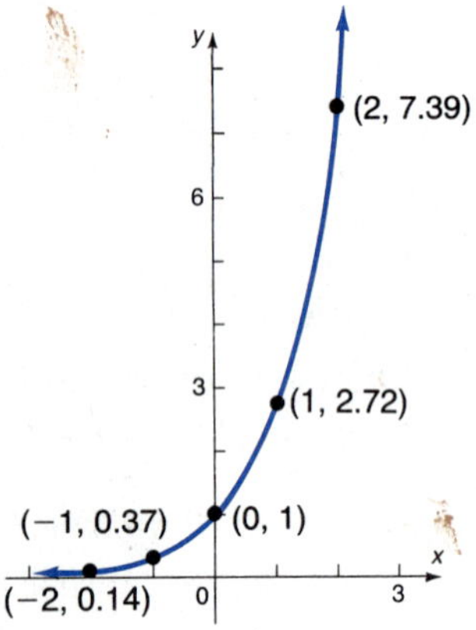

Figure 8
$y = e^x$

Table 4

x	e^x
−2	0.14
−1	0.37
0	1
2	7.39

There are many applications involving the exponential function. Let's look at one.

*If your calculator does not have this key but does have an inv key and an ln x key, you can display the number e as follows:

Enter 1 Press inv Press ln x

Display: 1 → 2.7182818

The reason this works will become clear in Section 6.3.

C **Example 6** Suppose the percentage R of people who respond to a newspaper advertisement for a new product and purchase the item advertised after t days is found using the formula

$$R = 0.5 - e^{-0.3t}$$

(a) What percentage has responded after 5 days? After 10 days?

(b) What is the highest percentage of people expected to respond?

Solution (a) When $t = 5$, we have

$$R = 0.5 - e^{(-0.3)(5)} = 0.5 - e^{-1.5}$$

We use a calculator to evaluate this expression:

Enter	1.5	Press ±	Press e^x
Display:	1.5	−1.5	0.2231302
	Press ±	Press +	Enter 0.5
Display:	−0.2231302		0.5
	Press =		
Display:	0.27687698		

Thus, about 28% will have responded after 5 days. When $t = 10$, we have

$$R = 0.5 - e^{(-0.3)(10)} = 0.5 - e^{-3} \approx 0.45021$$

About 45% will have responded after 10 days.

(b) As time passes, more people are expected to respond. The highest percentage expected to respond is therefore found for the value of R as $t \to \infty$. Since $e^{-0.3t} \to 0$ as $t \to \infty$, it follows that the highest percentage expected is 50%. ■

Exponential Equations

Equations that involve terms of the form a^x, $a > 0$, $a \neq 1$, are often referred to as **exponential equations**. Such equations sometimes can be solved by appropriately applying the laws of exponents and the following fact:

$$\text{If} \quad a^u = a^v, \quad \text{then} \quad u = v. \tag{3}$$

The result (3) is a consequence of the fact that exponential functions are one-to-one.

Example 7 Solve the equation: $3^{x+1} = 81$

Solution In order to apply result (3), each side of the equality must be written with the same base. Thus, we need to express 81 as a power of 3:

$$3^{x+1} = 81$$
$$3^{x+1} = 3^4$$

Now apply (3) to get

$$x + 1 = 4$$
$$x = 3$$

■

Example 8 Solve the equation: $e^{-x^2} = (e^x)^2 \cdot \frac{1}{e^3}$

Solution We use some laws of exponents first to get the same base e on each side:

$$e^{-x^2} = (e^x)^2 \cdot \frac{1}{e^3}$$
$$e^{-x^2} = e^{2x} \cdot e^{-3}$$
$$e^{-x^2} = e^{2x-3}$$

Now apply (3) to get

$$-x^2 = 2x - 3$$
$$x^2 + 2x - 3 = 0$$
$$(x + 3)(x - 1) = 0$$
$$x = -3 \quad \text{or} \quad x = 1$$

■

Example 9 Solve the equation: $4^x - 2^x - 12 = 0$

Solution We note that $4^x = (2^2)^x = 2^{2x} = (2^x)^2$, so the equation is actually quadratic in form, and we can rewrite it as

$$(2^x)^2 - 2^x - 12 = 0$$

Now, we can factor as usual:

$$(2^x - 4)(2^x + 3) = 0$$

$$2^x - 4 = 0 \quad \text{or} \quad 2^x + 3 = 0$$

$$2^x = 4 \qquad\qquad 2^x = -3$$

The equation on the left has the solution $x = 2$, since $2^x = 4 = 2^2$; the equation on the right has no solution, since $2^x > 0$ for all x. ■

We hasten to point out that most exponential equations are not as easily solved as those in Examples 7, 8, and 9. Consider, for example, the equation $x + e^x = 2$. Approximate solutions to such equations are usually found by numerical methods studied in calculus. Other types of exponential equations that can be solved using algebra are discussed in Section 6.4.

EXERCISE 6.1

C *In Problems 1–6, approximate each number using a calculator.*

1. (a) $3^{2.2}$ (b) $3^{2.23}$ (c) $3^{2.236}$ (d) $3^{\sqrt{5}}$
2. (a) $5^{1.7}$ (b) $5^{1.73}$ (c) $5^{1.732}$ (d) $5^{\sqrt{3}}$
3. (a) $2^{3.14}$ (b) $2^{3.141}$ (c) $2^{3.1415}$ (d) 2^{π}
4. (a) $2^{2.7}$ (b) $2^{2.71}$ (c) $2^{2.718}$ (d) 2^{e}
5. (a) $3.1^{2.7}$ (b) $3.14^{2.71}$ (c) $3.141^{2.718}$ (d) π^{e}
6. (a) $2.7^{3.1}$ (b) $2.71^{3.14}$ (c) $2.718^{3.141}$ (d) e^{π}

C *In Problems 7 and 8, graph each function by carefully preparing a table listing points on the graph.*

7. $f(x) = (\sqrt{2})^x$ 8. $g(x) = \pi^x$

In Problems 9–18, use the results of Problems 7 and 8, along with the techniques of shifting, compression, stretching, and reflection, to graph each function.

9. $f(x) = (\sqrt{2})^x + 2$ 10. $g(x) = \pi^x + 2$ 11. $f(x) = (\sqrt{2})^{x+2}$ 12. $g(x) = \pi^{x+2}$
13. $f(x) = (\sqrt{2})^{-x}$ 14. $g(x) = \pi^{-x}$ 15. $f(x) = (2\sqrt{2})^x$ 16. $g(x) = (\pi^2)^x$
17. $f(x) = (\sqrt{2}/2)^x$ 18. $g(x) = \pi^{-x}$

In Problems 19–22, use the graph of $y = e^x$ (Figure 8), along with the techniques of shifting, compression, stretching, and reflection, to graph each function.

19. $y = e^{-x}$ 20. $y = -e^x$ 21. $y = e^{x+2}$ 22. $y = e^x - 1$

In Problems 23–34, solve each equation.

23. $2^{2x+1} = 4$ 24. $5^{1-2x} = \frac{1}{5}$ 25. $3^{x^3} = 9^x$ 26. $4^{x^2} = 2^x$
27. $8^{x^2-2x} = \frac{1}{2}$ 28. $9^{-x} = \frac{1}{3}$ 29. $2^x \cdot 8^{-x} = 4^x$ 30. $\left(\frac{1}{2}\right)^{1-x} = 4$

31. $2^{2x} - 2^x - 12 = 0$

32. $3^{2x} + 3^x - 2 = 0$

33. $3^{2x} + 3^{x+1} - 4 = 0$

34. $4^x - 2^x = 0$

C **35.** The percentage R of viewers who respond to a television commercial for a new product after t days is found by using the formula

$$R = 0.7 - e^{-0.2t}$$

(a) What percentage is expected to respond after 10 days? After 20 days?
(b) What is the highest percentage of viewers expected to respond?

C **36.** Rework Problem 35 if the formula is changed to: $R = 0.7 - e^{-0.4t}$

C **37.** Rework Problem 35 if the formula is changed to: $R = 0.9 - e^{-0.2t}$

C **38.** Rework Problem 35 if the formula is changed to: $R = 0.9 - e^{-0.1t}$

C **39.** The equation governing the amount of current I (in amperes) after time t (in seconds) in a single RL circuit consisting of a resistance R (in ohms), an inductance L (in henrys), and an electromotive force E (in volts) is

$$I = \frac{E}{R}(1 - e^{-(R/L)t})$$

(a) If $E = 120$ volts, $R = 10$ ohms, and $L = 5$ henrys, how much current is available after 0.01 second? After 0.5 second?
(b) What is the maximum current?
(c) Graph this function, measuring I along the y-axis and t along the x-axis.

C **40.** Rework Problem 39 if $E = 120$ volts, $R = 5$ ohms, and $L = 10$ henrys.

C **41.** The annual profit P of a company due to the sales of a particular item after it has been on the market x years is determined to be

$$P = \$100{,}000 - \$60{,}000\left(\tfrac{1}{2}\right)^x$$

(a) How much profit is earned after 5 years? After 10 years?
(b) What is the maximum profit the company expects from this item?
(c) Graph this function, measuring P along the y-axis and x along the x-axis.

C **42.** The demand for a new product increases rapidly at first, and then levels off. The percentage P of actual purchases of this product after it has been on the market t months is

$$P = 90 - 80\left(\tfrac{1}{4}\right)^t$$

(a) What is the percentage of purchases of the product after 5 months? After 10 months?
(b) What is the maximum percentage of purchases of the product?

C **43.** If $f(x) = e^x$, compute: $f(3) - f(2)$

C **44.** If $f(x) = e^{-x}$, compute: $\dfrac{f(5) - f(2)}{3}$

Problems 45 and 46 provide definitions for two other transcendental functions.

45. The **hyperbolic sine function**, designated by $\sinh x$, is defined as

$$\sinh x = \tfrac{1}{2}(e^x - e^{-x})$$

(a) Show that $f(x) = \sinh x$ is an odd function.
(b) Graph $y = e^x$ and $y = e^{-x}$ on the same set of coordinate axes, and use the method of subtracting y-coordinates to obtain a graph of $f(x) = \sinh x$.

46. The **hyperbolic cosine function**, designated by cosh x, is defined as

$$\cosh x = \tfrac{1}{2}(e^x + e^{-x})$$

(a) Show that $f(x) = \cosh x$ is an even function.
(b) Graph $y = e^x$ and $y = e^{-x}$ on the same set of coordinate axes, and use the method of adding y-coordinates to obtain a graph of $f(x) = \cosh x$.
(c) Refer to Problem 45. Show that, for every x,

$$(\cosh x)^2 - (\sinh x)^2 = 1$$

47. If $f(x) = a^x$, show that: $\dfrac{f(x+h) - f(x)}{h} = a^x\left(\dfrac{a^h - 1}{h}\right)$

48. If $f(x) = a^x$, show that: $f(A + B) = f(A) \cdot f(B)$

49. If $f(x) = a^x$, show that: $f(-x) = \dfrac{1}{f(x)}$

50. If $f(x) = a^x$, show that: $f(\alpha x) = [f(x)]^\alpha$

C **51.** *Historical problem* Pierre de Fermat (1601–1665) conjectured that the function

$$f(x) = 2^{(2^x)} + 1$$

for $x = 1, 2, 3, \ldots$, would always have a value equal to a prime number. But Leonhard Euler (1707–1783) showed that this formula fails for $x = 5$. Determine the prime numbers produced by f for $x = 1, 2, 3, 4$. Then show that $f(5) = 641 \times 6{,}700{,}417$, which is not prime.

C **52.** Compute the value of

$$2 + \frac{1}{2!} + \frac{1}{3!} + \cdots + \frac{1}{n!}$$

for $n = 4, 6, 8$, and 10. Compare each result with e. [*Hint:* $1! = 1$, $2! = 2 \cdot 1$, $3! = 3 \cdot 2 \cdot 1$, $n! = n(n-1) \cdot \cdots \cdot (3)(2)(1)$]

6.2 Compound Interest

When the interest due at the end of a payment period is added to the principal so that the interest computed at the end of the next payment period is based on this new principal amount (old principal + interest), the interest is said to have been **compounded**. Thus, **compound interest** is interest paid on previously earned interest.

Example 1 A savings and loan association pays interest of 8% per annum compounded quarterly on a certain savings plan. If \$1000 is deposited in such a plan and the interest is left to accumulate, how much is in the account after 1 year?

Solution We use the simple interest formula, $I = Prt$ (see Section 2.2). After the first quarter of a year, the interest earned is

$$I = Prt = (\$1000)(0.08)\left(\tfrac{1}{4}\right) = \$20$$

The new principal is $P + I = \$1000 + \$20 = \$1020$. At the end of the second quarter, the interest on this principal is

$$I = (\$1020)(0.08)\left(\tfrac{1}{4}\right) = \$20.40$$

At the end of the third quarter, the interest on the new principal of $\$1020 + \$20.40 = \$1040.40$ is

$$I = (\$1040.40)(0.08)\left(\tfrac{1}{4}\right) \approx \$20.81$$

Finally, after the fourth quarter, the interest is

$$I = (\$1061.21)(0.08)\left(\tfrac{1}{4}\right) \approx \$21.22$$

Thus, after 1 year, the account contains \$1082.43. ■

The pattern of the calculations performed in Example 1 leads to a general formula for compound interest. To fix our ideas, let P represent the principal to be invested at a per annum interest rate r, which is compounded n times per year. (For computing purposes, r is expressed as a decimal.) The interest earned after each compounding period is the principal times r/n. Thus, the amount A after one compounding period is

$$A = P + P\left(\frac{r}{n}\right) = P\left(1 + \frac{r}{n}\right)$$

After two compounding periods, the amount A, based on the new principal $P(1 + r/n)$, is

$$A = \underbrace{P\left(1 + \frac{r}{n}\right)}_{\text{New principal}} + \underbrace{P\left(1 + \frac{r}{n}\right)\left(\frac{r}{n}\right)}_{\text{Interest on new principal}} = P\left(1 + \frac{r}{n}\right)\left(1 + \frac{r}{n}\right) = P\left(1 + \frac{r}{n}\right)^2$$

After three compounding periods,

$$A = P\left(1 + \frac{r}{n}\right)^2 + P\left(1 + \frac{r}{n}\right)^2\left(\frac{r}{n}\right) = P\left(1 + \frac{r}{n}\right)^2\left(1 + \frac{r}{n}\right) = P\left(1 + \frac{r}{n}\right)^3$$

Continuing in this way, after n compounding periods (1 year),

$$A = P\left(1 + \frac{r}{n}\right)^n$$

Because t years will contain $n \cdot t$ compounding periods, after t years we have

$$A = P\left(1 + \frac{r}{n}\right)^{nt}$$

Theorem The amount A after t years due to a principal P invested at an annual interest rate r compounded n times per year is

Compound Interest Formula

$$A = P\left(1 + \frac{r}{n}\right)^{nt} \tag{1}$$

■

C **Example 2** Investing \$1000 at an annual rate of 10% compounded annually, quarterly, monthly, and daily will yield the following amounts after 1 year:

Annual compounding: $A = P(1 + r)$
$= (\$1000)(1 + 0.10) = \$1100.00$

Quarterly compounding: $A = P\left(1 + \frac{r}{4}\right)^4$
$= (\$1000)(1 + 0.025)^4 \approx \$1103.81$

Monthly compounding: $A = P\left(1 + \frac{r}{12}\right)^{12}$
$= (\$1000)(1 + 0.00833)^{12} \approx \$1104.71$

Daily compounding: $A = P\left(1 + \frac{r}{365}\right)^{365}$
$= (\$1000)(1 + 0.000274)^{365} \approx \$1105.16$

■

From Example 2, we can see that the effect of compounding more frequently is that the amount after 1 year is higher: \$1000 compounded 4 times a year at 10% results in \$1103.81; \$1000 compounded 12 times a year at 10% results in \$1104.71; and \$1000 compounded 365 times a year at 10% results in \$1105.16. This leads to the following question: What would happen to the amount after 1 year if the number of times the interest is compounded were increased without bound?

Let's find the answer. Suppose P is the principal, r is the per annum interest rate, and n is the number of times the interest is compounded each year. The amount after 1 year is

$$A = P\left(1 + \frac{r}{n}\right)^n$$

Now suppose the number n of times the interest is compounded per year gets larger and larger; that is, suppose $n \to \infty$. Table 5 compares $(1 + r/n)^n$, for large values of n, to e^r for $r = 0.05$, $r = 0.10$, $r = 0.15$, and $r = 1$. The larger n gets, the closer $(1 + r/n)^n$ gets to e^r (a fact that is proved in many calculus books). Thus, no matter how frequent the compounding, the amount after 1 year has the definite ceiling Pe^r.

Table 5

	$\left(1+\frac{r}{n}\right)^n$			
	$n = 100$	$n = 1000$	$n = 10{,}000$	e^r
$r = 0.05$	1.0512579	1.05127	1.051271	1.0512711
$r = 0.10$	1.1051157	1.1051654	1.1051703	1.1051709
$r = 0.15$	1.1617037	1.1618212	1.1618329	1.1618342
$r = 1$	2.7048138	2.7169239	2.7181459	2.7182818

When interest is compounded so that the amount after 1 year is Pe^r, we say the interest is **compounded continuously**.

Theorem
Continuous Compounding

The amount A after t years due to a principal P invested at an annual interest rate r compounded continuously is

$$A = Pe^{rt} \tag{2}$$

■

C **Example 3** The amount A that results from investing a principal P of \$1000 at an annual rate r of 10% compounded continuously for a time t of 1 year is

$$A = \$1000e^{0.10} = (\$1000)(1.10517) = \$1105.17$$ ■

The **effective rate of interest** is the equivalent annual simple rate of interest that would yield the same amount as compounding after 1 year. For example, based on Example 3, a principal of \$1000 will result in \$1105.17 at a rate of 10% compounded continuously. To get this same amount using a simple rate of interest would require that interest of \$1105.17 − \$1000.00 = \$105.17 be earned on the principal. Since \$105.17 is 10.517% of \$1000, a simple rate of interest of 10.517% is needed to equal 10% compounded continuously. Thus, the effective rate of interest of 10% compounded continuously is 10.517%.

Based on the results of Examples 2 and 3, we find the following comparisons:

	ANNUAL RATE	EFFECTIVE RATE
Annual compounding	10%	10%
Quarterly compounding	10%	10.381%
Monthly compounding	10%	10.471%
Daily compounding	10%	10.516%
Continuous compounding	10%	10.517%

C **Example 4** On January 2, 1990, \$2000 is placed in an Individual Retirement Account (IRA) that will pay interest of 10% per annum compounded continuously. What will the IRA be worth on January 1, 2010?

Solution The amount A after 20 years is

$$A = Pe^{rt} = \$2000e^{(0.10)(20)} = \$14,778.11$$ ■

When people engaged in finance speak of the "time value of money," they are usually referring to the **present value** of money. The present value of A dollars to be received at a future date is the principal you would need to invest now so that it would grow to A dollars in the specified time period. Thus, the present value of money to be received at a future date is always less than the amount to be received, since the amount to be received will equal the present value (money invested now) *plus* the interest accrued over the time period.

We use the compound interest formula (1) to get a formula for present value. If P is the present value of A dollars to be received after t years at a per annum interest rate r compounded n times per year, then by formula (1),

$$A = P\left(1 + \frac{r}{n}\right)^{nt}$$

To solve for P, we divide both sides by $(1 + r/n)^{nt}$, and the result is

$$\frac{A}{(1 + r/n)^{nt}} = P \quad \text{or} \quad P = A\left(1 + \frac{r}{n}\right)^{-nt}$$

Theorem

Present Value Formulas

The present value P of A dollars to be received after t years assuming a per annum interest rate r compounded n times per year is

$$P = A\left(1 + \frac{r}{n}\right)^{-nt} \tag{3}$$

If the interest is compounded continuously, then

$$P = Ae^{-rt} \tag{4}$$

You are asked to derive formula (4) in Problem 31.

Example 5 A zero-coupon (noninterest-bearing) bond can be redeemed in 10 years for \$1000. How much should you be willing to pay for it now if you want a return of:

(a) 8% compounded monthly? (b) 7% compounded continuously?

Solution (a) We are seeking the present value of \$1000. Thus, we use formula (3) with $A = \$1000$, $n = 12$, $r = 0.08$, and $t = 10$:

$$\begin{aligned} P &= A\left(1 + \frac{r}{n}\right)^{-nt} \\ &= \$1000\left(1 + \frac{0.08}{12}\right)^{-12(10)} \\ &\approx \$450.52 \end{aligned}$$

(b) Here, we use formula (4) with $A = \$1000$, $r = 0.07$, and $t = 10$:

$$\begin{aligned} P &= Ae^{-rt} \\ &= \$1000e^{-(0.07)(10)} \\ &\approx \$496.59 \end{aligned}$$

EXERCISE 6.2

C *In Problems 1–10, find the amount that results from each investment.*

1. \$100 invested at 8% compounded monthly after a period of 2 years
2. \$50 invested at 6% compounded daily after a period of 3 years
3. \$500 invested at 10% compounded quarterly after a period of $2\frac{1}{2}$ years
4. \$300 invested at 12% compounded monthly after a period of $1\frac{1}{2}$ years
5. \$600 invested at 5% compounded daily after a period of 3 years
6. \$700 invested at 9% compounded quarterly after a period of 2 years
7. \$10 invested at 11% compounded continuously after a period of 2 years
8. \$40 invested at 7% compounded continuously after a period of 3 years
9. \$100 invested at 10% compounded continuously after a period of $2\frac{1}{4}$ years
10. \$100 invested at 12% compounded continuously after a period of $3\frac{3}{4}$ years

C *In Problems 11–20, find the principal needed now to get each amount; that is, find the present value.*

11. To get \$100 after 2 years at 8% compounded monthly
12. To get \$75 after 3 years at 10% compounded quarterly
13. To get \$1000 after $2\frac{1}{2}$ years at 6% compounded daily
14. To get \$800 after $3\frac{1}{2}$ years at 7% compounded monthly
15. To get \$600 after 2 years at 12% compounded quarterly
16. To get \$300 after 4 years at 14% compounded daily
17. To get \$80 after $3\frac{1}{4}$ years at 9% compounded continuously
18. To get \$800 after $2\frac{1}{2}$ years at 8% compounded continuously
19. To get \$400 after 1 year at 10% compounded continuously
20. To get \$1000 after 1 year at 12% compounded continuously

C 21. A zero-coupon bond can be redeemed in 20 years for \$10,000. How much should you be willing to pay for it now if you want a return of:
(a) 10% compounded monthly? (b) 12% compounded continuously?

C 22. Rework Problem 21 if the bond can be redeemed in 10 years for \$5000.

C 23. You invest \$2000 in a bond trust that pays 9% interest compounded semiannually. Your friend invests \$2000 in a Certificate of Deposit (CD) that pays $8\frac{1}{2}$% compounded continuously. Who has more money after 20 years, you or your friend?

C 24. Rework Problem 23 if the bond trust pays 10% interest compounded semiannually whereas the CD pays only 8% compounded continuously.

C 25. Suppose you have access to an investment that will pay 10% interest compounded continuously. Which is better: To be given \$1000 now so that you can take advantage of this investment opportunity, or to be given \$1325 after 3 years?

C 26. Rework Problem 25 if the 10% interest is compounded annually.

C 27. You have just purchased a house for \$150,000, with the seller holding a second mortgage of \$50,000. You promise to pay the seller \$50,000 plus all accrued interest 5 years from now. The seller offers you three interest options on the second mortgage:
(a) Simple interest at 12% per annum (b) $11\frac{1}{2}$% interest compounded monthly
(c) $11\frac{1}{4}$% interest compounded continuously
Which option is best; that is, which one results in the least interest on the loan?

C 28. Refer to Problem 27. If you promise to pay the seller \$50,000 plus all accrued interest 3 years from now, which option is best?

C 29. Financial Federal Savings Bank of Illinois advertised the following investment options:

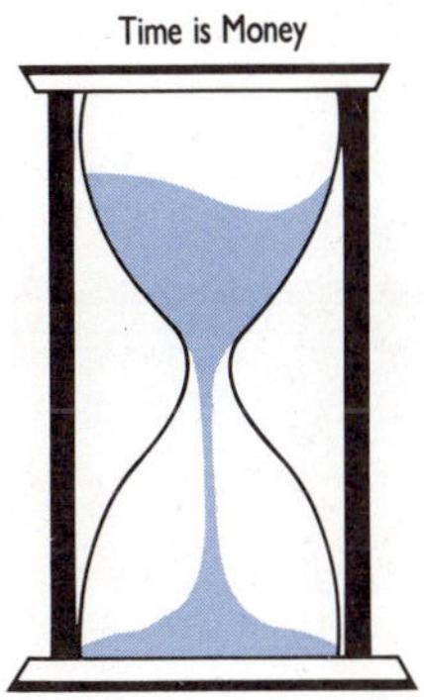

Fixed-Term Certificate Investment Options

MATURITY TERM	RATE	YIELD
91 days	8.50%	8.77%
6 months	9.15%	9.47%
1 year	9.50%	9.84%
$1\frac{1}{2}$ years	9.80%	10.17%
$2\frac{1}{2}$ years	10.15%	10.54%
3 years	10.35%	10.76%
Passbook savings account, no maturity restrictions:		
Regular savings	5.50%	5.65%

For the 1 year fixed-term certificate of 9.50%, determine whether the yield of 9.84% is an effective rate for 9.50% compounded continuously, daily (365 days), monthly, or quarterly. Assume a principal deposit of $100.00.

C **30.** Rework Problem 29 for the regular savings account of 5.50% showing a yield of 5.65%.

31. Derive the formula $P = Ae^{-rt}$ for the present value P of A dollars if the interest rate r is compounded continuously.

6.3 Logarithmic Functions*

Recall that a one-to-one function $y = f(x)$ has an inverse that is defined (implicitly) by the equation $x = f(y)$. In particular, the exponential function $y = f(x) = a^x$, $a > 0$, $a \neq 1$, is one-to-one and, hence, has an inverse that is defined implicitly by the equation

$$x = a^y \qquad a > 0, a \neq 1$$

This inverse is so important that it is given a name, the *logarithmic function.*

Logarithmic Function

The **logarithmic function to the base *a***, where $a > 0$ and $a \neq 1$, is denoted by $y = \log_a x$ (read as "y is the logarithm to the base a of x") and is defined by

$$y = \log_a x \quad \text{if and only if} \quad x = a^y$$

For a function f and its inverse f^{-1}, we know that

$$\text{Domain } f^{-1} = \text{Range } f \quad \text{and} \quad \text{Range } f^{-1} = \text{Domain } f$$

Consequently, it follows that:

Domain of logarithmic function = Range of exponential function $= (0, \infty)$

Range of logarithmic function = Domain of exponential function $= (-\infty, \infty)$

*C Refer to Graphics Calculator Supplement: Activity IV.

The box below summarizes some properties of the logarithmic function:

$y = \log_a x$	(Defining equation: $x = a^y$)
Domain: $0 < x < \infty$	Range: $-\infty < y < \infty$

Remember that the domain of a logarithmic function consists of the *positive* real numbers.

Example 1 Find the domain of each logarithmic function:

(a) $F(x) = \log_2(1 - x)$ (b) $g(x) = \log_5\left(\dfrac{1 + x}{1 - x}\right)$ (c) $h(x) = \log_{1/2}|x|$

Solution (a) The domain of F consists of all x for which $(1 - x) > 0$; that is, all $x < 1$, or $(-\infty, 1)$.

(b) The domain of g is restricted to

$$\frac{1 + x}{1 - x} > 0$$

Solving this inequality, we find that the domain of g consists of all x between -1 and 1; that is, $-1 < x < 1$, or $(-1, 1)$.

(c) Since $|x| > 0$ provided $x \neq 0$, the domain of h consists of all nonzero real numbers. ■

Since exponential functions and logarithmic functions are inverses of each other, the graph of a logarithmic function $y = \log_a x$ is the reflection about the line $y = x$ of the graph of the exponential function $y = a^x$, as shown in Figure 9.

Figure 9

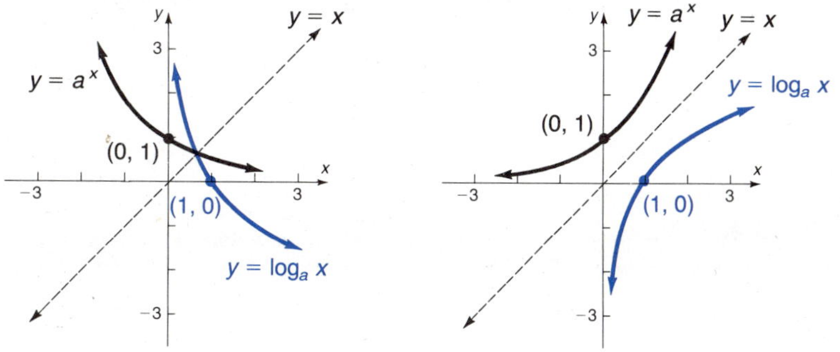

(a) $0 < a < 1$ (b) $a > 1$

Some other facts about a logarithmic function $f(x) = \log_a x$ are now apparent:

1. The x-intercept of the graph is 1. There is no y-intercept.
2. The y-axis is a vertical asymptote of the graph.
3. A logarithmic function is increasing if $a > 1$ and decreasing if $0 < a < 1$.
4. The graph is smooth, with no corners or gaps.

If the base of a logarithmic function is the number e, then we have the **natural logarithm function**. This function occurs so frequently in applications that it is given a special symbol, **ln** (from the Latin, *logarithmus naturalis*). Thus,

$$y = \ln x \qquad \text{if and only if} \qquad x = e^y$$

Table 6 shows some values of the function $f(x) = \ln x$, and Figure 10 illustrates its graph.

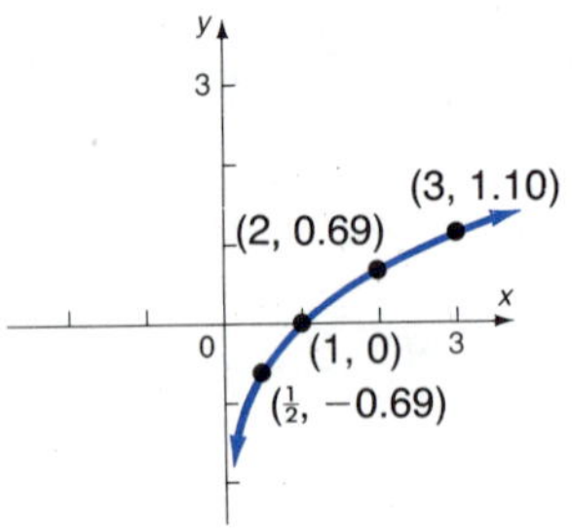

Figure 10
$y = \ln x$

Table 6

x	$\ln x$
$\frac{1}{2}$	−0.69
1	0
2	0.69
3	1.10

Example 2 Graph $y = -\ln x$ by starting with the graph of $y = \ln x$ given in Figure 10.

Solution The graph of $y = -\ln x$ is obtained by a reflection about the x-axis of the graph of $y = \ln x$. See Figure 11.

Figure 11
$y = -\ln x$

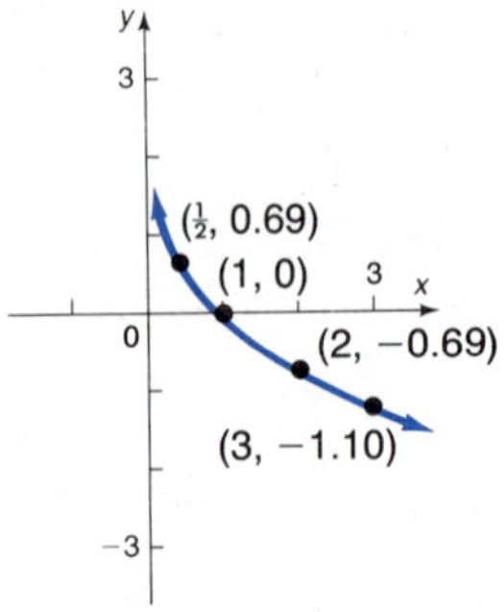

■

Example 3 Graph: $y = \ln(x + 2)$

Solution The domain consists of all x for which

$$x + 2 > 0 \qquad \text{or} \qquad x > -2$$

The graph is obtained by applying a horizontal shift to the left 2 units, as shown in Figure 12. Notice that the line $x = -2$ is a vertical asymptote.

Figure 12
$y = \ln(x + 2)$

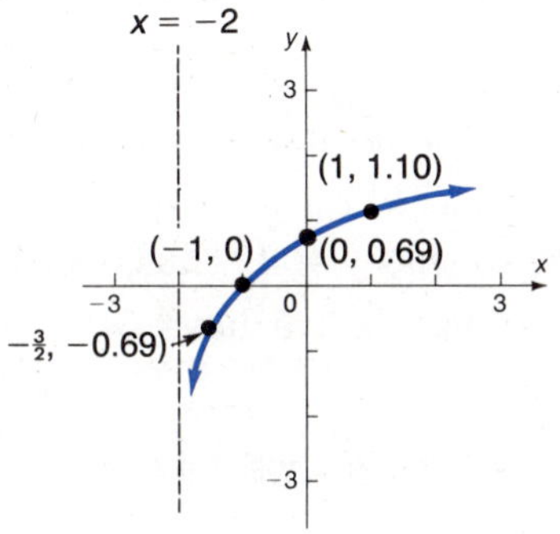

■

Example 4 Graph: $y = \ln(1 - x)$

Solution The domain consists of all x for which

$$1 - x > 0 \qquad \text{or} \qquad x < 1$$

To obtain the graph of $y = \ln(1 - x)$, we use the following steps:

$$\begin{aligned} y &= \ln x \\ &= \ln(-x) && \text{Reflection about } y\text{-axis} \\ &= \ln[-(x - 1)] = \ln(1 - x) && \text{Replace } x \text{ by } x - 1\text{; shift right 1 unit.} \end{aligned}$$

See Figure 13.

Figure 13

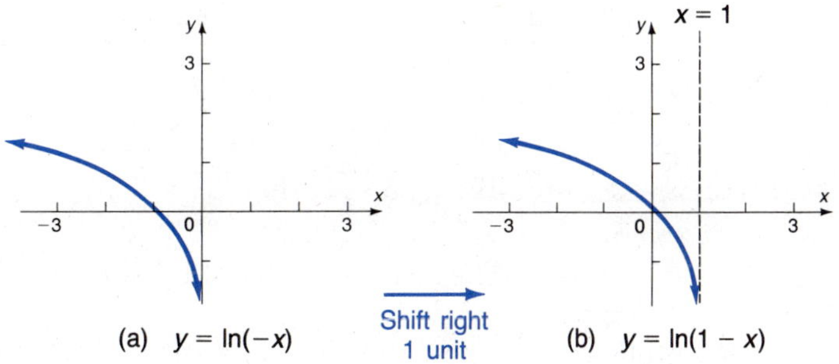

■

Evaluating Logarithmic Functions

To find the value of a logarithm, we may use the equivalent exponential equation.

Example 5 Find the value of:

(a) $\ln 1$ (b) $\log_2 8$ (c) $\log_3 \frac{1}{3}$ (d) $\log_5 25$

Solution (a) Let $y = \ln 1$. Remember, the base of a natural logarithm is understood to be e, so the equivalent exponential equation is

$$e^y = 1$$

Since $1 = e^0$, we find $e^y = e^0$, so that $y = 0$. Thus, $\ln 1 = 0$.

(b) For $y = \log_2 8$, we have the equivalent exponential equation $2^y = 8 = 2^3$, so that $y = 3$. Thus, $\log_2 8 = 3$.

(c) For $y = \log_3 \frac{1}{3}$, we have $3^y = \frac{1}{3} = 3^{-1}$, so that $y = -1$. Thus, $\log_3 \frac{1}{3} = -1$.

(d) For $y = \log_5 25$, we have $5^y = 25 = 5^2$, so that $y = 2$. Thus, $\log_5 25 = 2$. ■

Example 6 Change each exponential expression to an equivalent expression involving a logarithm.

(a) $1.2^3 = m$ (b) $e^b = 9$ (c) $a^4 = 24$

Solution We use the fact that $y = \log_a x$ and $x = a^y$, $a > 0$, $a \neq 1$, are equivalent.

(a) If $1.2^3 = m$, then $3 = \log_{1.2} m$.

(b) If $e^b = 9$, then $b = \log_e 9 = \ln 9$.

(c) If $a^4 = 24$, then $4 = \log_a 24$. ■

Example 7 Change each logarithmic expression to an equivalent expression involving an exponent.

(a) $\log_a 4 = 5$ (b) $\ln b = -3$ (c) $\log_3 5 = c$

Solution (a) If $\log_a 4 = 5$, then $a^5 = 4$.

(b) If $\ln b = -3$, then $e^{-3} = b$.

(c) If $\log_3 5 = c$, then $3^c = 5$. ■

The next example establishes some general results about logarithms.

Example 8 (a) Show that $\log_a 1 = 0$. (b) Show that $\log_a a = 1$.

Solution (a) This fact was established when we graphed $y = \log_a x$ (see Figure 9). Algebraically, for $y = \log_a 1$, we have $a^y = 1 = a^0$, so that $y = 0$.

(b) For $y = \log_a a$, we have $a^y = a = a^1$, so that $y = 1$. ■

$$\log_a 1 = 0 \qquad \log_a a = 1$$

EXERCISE 6.3

In Problems 1–10, find the domain of each function.

1. $f(x) = \ln(3 - x)$ **2.** $g(x) = \ln(x^2 - 1)$ **3.** $F(x) = \log_2 x^2$

4. $H(x) = \log_5 x^3$ **5.** $h(x) = \log_{1/2}(x^2 - x - 6)$ **6.** $G(x) = \log_{1/2}\left(\frac{1}{x}\right)$

7. $f(x) = \frac{1}{\ln x}$ **8.** $g(x) = \ln(x - 5)$ **9.** $g(x) = \log_5\left(\frac{x + 1}{x}\right)$

10. $h(x) = \log_3\left(\frac{x^2}{x - 1}\right)$

In Problems 11–22, change each logarithmic expression to an equivalent expression involving an exponent.

11. $\log_2 8 = 3$ **12.** $\log_3\left(\frac{1}{9}\right) = -2$ **13.** $\log_a 3 = 6$

14. $\log_b 4 = 2$ **15.** $\log_3 2 = x$ **16.** $\log_2 6 = x$

17. $\log_2 M = 1.3$ **18.** $\log_3 N = 2.1$ **19.** $\log_{\sqrt{2}} \pi = x$

20. $\log_\pi x = \frac{1}{2}$ **21.** $\ln 4 = x$ **22.** $\ln x = 4$

In Problems 23–34, change each exponential expression to an equivalent expression involving a logarithm.

23. $9 = 3^2$ **24.** $16 = 4^2$ **25.** $a^2 = 1.6$ **26.** $a^3 = 2.1$

27. $1.1^2 = M$ **28.** $2.2^3 = N$ **29.** $2^x = 7.2$ **30.** $3^x = 4.6$

31. $x^{\sqrt{2}} = \pi$ **32.** $x^\pi = e$ **33.** $e^x = 8$ **34.** $e^{2.2} = M$

In Problems 35–46, evaluate each logarithm without using a calculator or a table.

35. $\log_2 1$ **36.** $\log_8 8$ **37.** $\log_5 25$ **38.** $\log_3\left(\frac{1}{9}\right)$

39. $\log_{1/2} 16$ **40.** $\log_{1/3} 9$ **41.** $\log_{10} \sqrt{10}$ **42.** $\log_5 \sqrt[3]{25}$

43. $\log_{\sqrt{2}} 4$ **44.** $\log_{\sqrt{3}} 9$ **45.** $\ln \sqrt{e}$ **46.** $\ln e^3$

3

36. $\log_8 8 = x$

$8^x = 8$

$x = 1$

47. Find a such that the graph of $f(x) = \log_a x$ contains the point $(2, 2)$.

48. Find a such that the graph of $f(x) = \log_a x$ contains the point $\left(\frac{1}{2}, -4\right)$.

C *In Problems 49 and 50, graph each function by carefully preparing a table listing points on the graph and then plotting them.*

49. $f(x) = \log_2 x$ **50.** $g(x) = \log_5 x$

In Problems 51–64, graph each function.

51. $f(x) = \log_2(x + 4)$ **52.** $g(x) = \log_5(x - 3)$ **53.** $f(x) = \log_2(-x)$

54. $g(x) = \log_5(-x)$ **55.** $f(x) = \log_2(3 - x)$ **56.** $g(x) = \log_5(4 - x)$

57. $f(x) = \log_2\left(\frac{1}{x}\right)$ **58.** $g(x) = \log_5\left(\frac{1}{x}\right)$ **59.** $f(x) = \log_2 4x$

60. $g(x) = \log_5 25x$ **61.** $f(x) = \log_2 x^2$ **62.** $g(x) = \log_5 x^2$

63. $f(x) = \log_{1/2} x$ **64.** $g(x) = \log_{1/5} x$

6.4 Properties of Logarithms

Logarithms have some very useful properties that can be derived directly from the definition and the laws of exponents.

Theorem

Properties of Logarithms

In the properties listed below, M and a are positive real numbers, with $a \neq 1$, and r is any real number.

The number $\log_a M$ is the exponent to which a must be raised to obtain M. That is,

$$a^{\log_a M} = M \qquad (1)$$

The logarithm to the base a of a raised to a power equals that power. That is,

$$\log_a a^r = r \qquad (2)$$

Proof of Property (1)

Let $x = \log_a M$. Change this logarithmic expression to the equivalent exponential expression:

$$a^x = M$$

Now replace x by $\log_a M$, to get

$$a^{\log_a M} = M$$

Proof of Property (2)

Let $x = a^r$. Change this exponential expression to the equivalent logarithmic expression

$$\log_a x = r$$

Now replace x by a^r, to get

$$\log_a a^r = r$$ ■

Example 1

(a) $\sqrt{2}^{\log_{\sqrt{2}} \pi} = \pi$ (b) $\log_{0.2} 0.2^{-\sqrt{2}} = -\sqrt{2}$ (c) $\ln e^{kt} = kt$ ■

Other useful properties of logarithms are listed below.

Theorem

In the following properties, M, N, and a are positive real numbers, with $a \neq 1$, and r is any real number:

Log of a Product Equals the Sum of the Logs

$$\log_a MN = \log_a M + \log_a N \qquad (3)$$

The Log of a Quotient Equals the Difference of the Logs

$$\log_a\left(\frac{M}{N}\right) = \log_a M - \log_a N \qquad (4)$$

$$\log_a\left(\frac{1}{N}\right) = -\log_a N \qquad (5)$$

$$\log_a M^r = r \log_a M \qquad (6)$$

■

We shall derive properties (3) and (6) and leave the derivations of properties (4) and (5) as exercises (see Problems 99 and 100).

Proof of Property (3)

Let $A = \log_a M$ and let $B = \log_a N$. These expressions are equivalent to the exponential expressions

$$a^A = M \qquad \text{and} \qquad a^B = N$$

Now,

$$\begin{aligned} \log_a MN = \log_a a^A a^B &= \log_a a^{A+B} && \text{Law of exponents} \\ &= A + B && \text{Property (2) of logarithms} \\ &= \log_a M + \log_a N \end{aligned}$$

Proof of Property (6) Let $A = \log_a M$. This expression is equivalent to

$$a^A = M$$

Now,

$$\begin{aligned} \log_a M^r = \log_a(a^A)^r &= \log_a a^{rA} && \text{Law of exponents} \\ &= rA && \text{Property (2) of logarithms} \\ &= r \log_a M \end{aligned}$$

■

Logarithms can be used to transform products into sums, quotients into differences, and powers into factors. Such transformations prove useful in certain types of calculus problems.

Example 2 Write $\log_a(x\sqrt{x^2 + 1})$ as a sum of logarithms. Express all powers as factors.

Solution

$$\begin{aligned} \log_a(x\sqrt{x^2 + 1}) &= \log_a x + \log_a \sqrt{x^2 + 1} && \text{Property (3)} \\ &= \log_a x + \log_a(x^2 + 1)^{1/2} \\ &= \log_a x + \tfrac{1}{2} \log_a(x^2 + 1) && \text{Property (6)} \end{aligned}$$

■

Example 3 Write

$$\log_a \frac{x^2}{(x - 1)^3}$$

as a difference of logarithms. Express all powers as factors.

Solution

$$\log_a \frac{x^2}{(x - 1)^3} \underset{\substack{\uparrow \\ \text{Property (4)}}}{=} \log_a x^2 - \log_a(x - 1)^3 \underset{\substack{\uparrow \\ \text{Property (6)}}}{=} 2 \log_a x - 3 \log_a(x - 1)$$

■

Example 4 Write

$$\log_a \frac{x^3\sqrt{x^2 + 1}}{(x + 1)^4}$$

as a sum and difference of logarithms. Express all powers as factors.

Solution

$$\begin{aligned} \log_a \frac{x^3\sqrt{x^2 + 1}}{(x + 1)^4} &= \log_a(x^3\sqrt{x^2 + 1}) - \log_a(x + 1)^4 \\ &= \log_a x^3 + \log_a \sqrt{x^2 + 1} - \log_a(x + 1)^4 \\ &= \log_a x^3 + \log_a(x^2 + 1)^{1/2} - \log_a(x + 1)^4 \\ &= 3 \log_a x + \tfrac{1}{2} \log_a(x^2 + 1) - 4 \log_a(x + 1) \end{aligned}$$

■

Another use of properties (3)–(6) is to write sums and/or differences of logarithms with the same base as a single logarithm.

Example 5 Write each of the following as a single logarithm:

(a) $\log_a 7 + 4 \log_a 3$ (b) $\frac{2}{3} \log_a 8 - \log_a (3^4 - 8)$

(c) $\log_a x + \log_a 9 + \log_a(x^2 + 1) - \log_a 5$

Solution (a) $\log_a 7 + 4 \log_a 3 = \log_a 7 + \log_a 3^4$ Property (6)

$= \log_a 7 + \log_a 81$

$= \log_a(7 \cdot 81)$ Property (3)

$= \log_a 567$

(b) $\frac{2}{3} \log_a 8 - \log_a(3^4 - 8) = \log_a 8^{2/3} - \log_a(81 - 8)$ Property (6)

$= \log_a 4 - \log_a 73$

$= \log_a\left(\frac{4}{73}\right)$ Property (4)

(c) $\log_a x + \log_a 9 + \log_a(x^2 + 1) - \log_a 5$

$= \log_a 9x + \log_a(x^2 + 1) - \log_a 5$

$= \log_a[9x(x^2 + 1)] - \log_a 5$

$= \log_a\left[\frac{9x(x^2 + 1)}{5}\right]$ ■

Warning: A common error made by some students is to express the logarithm of a sum as the sum of logarithms:

Not correct: $\log_a(M + N) = \log_a M + \log_a N$

Correct; property (3): $\log_a MN = \log_a M + \log_a N$

Another common error is to express the difference of logarithms as the quotient of logarithms:

Not correct: $\log_a M - \log_a N = \dfrac{\log_a M}{\log_a N}$

Correct; property (4): $\log_a M - \log_a N = \log_a\left(\dfrac{M}{N}\right)$

There remain two other properties of logarithms we need to know. They are a consequence of the fact that the logarithmic function $y = \log_a x$ is one-to-one.

Theorem In the following properties, M, N and a are positive real numbers, with $a \neq 1$:

$$\text{If} \quad M = N, \quad \text{then} \quad \log_a M = \log_a N. \tag{7}$$

$$\text{If} \quad \log_a M = \log_a N, \quad \text{then} \quad M = N. \tag{8}$$

■

Properties (7) and (8) are useful for solving *logarithmic equations.*

Logarithmic Equations

Equations that contain terms of the form $\log_a x$, where a is a positive real number, with $a \neq 1$, are often called **logarithmic equations**.

Let's see how we can use the properties of logarithms to solve logarithmic equations.

Example 6 Solve: $2 \log_5 x = \log_5 9$

Solution

$$2 \log_5 x = \log_5 9$$

$$\log_5 x^2 = \log_5 9 \qquad \text{Property (6)}$$

$$x^2 = 9 \qquad \text{Property (8)}$$

$x = 3$ or ~~$x = -3$~~ Recall that logarithms of negative numbers are not defined, so, in the expression $2 \log_5 x$, x must be positive. Therefore, -3 is extraneous and we discard it.

The equation has only one solution, 3. ■

Example 7 Solve: $\log_3(4x - 7) = 2$

Solution We change the expression to exponential form to solve:

$$\log_3(4x - 7) = 2$$

$$4x - 7 = 3^2$$

$$4x - 7 = 9$$

$$4x = 16$$

$$x = 4$$

■

Example 8 Solve: $\log_5 x + \log_5(2x - 3) = 1$

Solution We need to express the left side as a single logarithm. Then, we will change the expression to exponential form.

$$\log_5 x + \log_5(2x - 3) = 1$$

$$\log_5[x(2x - 3)] = 1 \qquad \text{Property (3)}$$

$$x(2x - 3) = 5^1 = 5$$

$$2x^2 - 3x - 5 = 0$$

$$(2x - 5)(x + 1) = 0$$

$$x = \tfrac{5}{2} \quad \text{or} \quad \cancel{x = -1}$$

The solution $x = -1$ is extraneous, because x must be larger than $\frac{3}{2}$. (Do you see why?)

The equation has only one solution, $\frac{5}{2}$. ■

Care must be taken when solving logarithmic equations. Be sure to check each apparent solution in the original equation and discard any that are extraneous. Remember that in the expression $\log_a M$, a and M are positive and $a \neq 1$.

Common Logarithms; Natural Logarithms

Sometimes, **common logarithms**, that is, logarithms to the base 10, are used to facilitate arithmetic computations. The next example illustrates this application.

Example 9 Given that $\log_{10} 2 \approx 0.3010$ and $\log_{10} 3 \approx 0.4771$,* compute:

(a) $\log_{10} 4$ (b) $\log_{10} 6$ (c) $\log_{10} 200$ (d) $\log_{10} 15$

Solution

(a) $\log_{10} 4 = \log_{10} 2^2 = 2 \log_{10} 2 \approx 2(0.3010) = 0.6020$

(b) $\log_{10} 6 = \log_{10}(2 \cdot 3) = \log_{10} 2 + \log_{10} 3 \approx 0.3010 + 0.4771 = 0.7781$

(c) $\log_{10} 200 = \log_{10}(2 \cdot 10^2) = \log_{10} 2 + \log_{10} 10^2 \approx 0.3010 + 2 = 2.3010$

(d) $\log_{10} 15 = \log_{10}\left(\frac{30}{2}\right) = \log_{10} 30 - \log_{10} 2 = \log_{10}(3 \cdot 10) - \log_{10} 2$
$= \log_{10} 3 + \log_{10} 10 - \log_{10} 2 \approx 0.4771 + 1 - 0.3010$
$= 1.1761$ ■

The widespread use of calculators has made this particular application of common logarithms less important than it once was. However, as mentioned earlier, natural logarithms, that is, logarithms to the base $e = 2.718\ldots$, remain very important because they arise frequently in the study of natural phenomena.

*These values are found in Table II in the back of the book.

Common logarithms are usually abbreviated by writing **log**, with the base understood to be 10, just as natural logarithms are abbreviated by **ln**, with the base understood to be e.

Most calculators have both [log x] and [ln x] keys to calculate the common logarithm and natural logarithm of a number. For your convenience, Tables II and III in the back of this book give values of $\log x$ and $\ln x$ for selected numbers x. To calculate logarithms having a base other than 10 or e, we employ the **change-of-base formula**.

Theorem If $a \neq 1$, $b \neq 1$, and M are positive real numbers, then

Change-of-Base Formula

$$\log_a M = \frac{\log_b M}{\log_b a} \tag{9}$$

Proof We derive this formula as follows: Let $y = \log_a M$. Then $a^y = M$, so that

$$\begin{aligned} \log_b a^y &= \log_b M && \text{Property (7)} \\ y \log_b a &= \log_b M && \text{Property (6)} \\ \log_a M \cdot \log_b a &= \log_b M && y = \log_a M \\ \log_a M &= \frac{\log_b M}{\log_b a} \end{aligned}$$

■

In practice, the change-of-base formula uses either $b = 10$ or $b = e$. Thus,

$$\log_a M = \frac{\log M}{\log a} \quad \text{and} \quad \log_a M = \frac{\ln M}{\ln a} \tag{10}$$

C **Example 10** Calculate:

(a) $\log_5 89$ (b) $\log_{\sqrt{2}} \sqrt{5}$

Solution (a) $\log_5 89 = \dfrac{\log 89}{\log 5} \approx \dfrac{1.94939}{0.69897} = 2.7889$

$$\log_5 89 = \frac{\ln 89}{\ln 5} \approx \frac{4.4886}{1.6094} = 2.7889$$

(b) $\log_{\sqrt{2}} \sqrt{5} = \dfrac{\log \sqrt{5}}{\log \sqrt{2}} = \dfrac{\frac{1}{2}\log 5}{\frac{1}{2}\log 2} \approx \dfrac{0.69897}{0.30103} = 2.3219$

$$\log_{\sqrt{2}} \sqrt{5} = \frac{\ln \sqrt{5}}{\ln \sqrt{2}} = \frac{\frac{1}{2}\ln 5}{\frac{1}{2}\ln 2} \approx \frac{1.6094}{0.6931} = 2.3219$$

■

Logarithms are very useful for solving certain types of exponential equations. Let's see how.

C **Example 11** Solve for x: $2^x = 5$

Solution We write the exponential equation as the equivalent logarithmic equation:

$$2^x = 5$$

$$x = \log_2 5 = \frac{\log 5}{\log 2} \approx 2.3219$$

↑ Change-of-base formula (10)

Alternatively, we can solve the equation $2^x = 5$ by taking the natural logarithm (or common logarithm) of each side. Taking the natural logarithm,

$$\begin{aligned} 2^x &= 5 \\ \ln 2^x &= \ln 5 \\ x \ln 2 &= \ln 5 \\ x &= \frac{\ln 5}{\ln 2} \approx 2.3219 \end{aligned}$$

■

C **Example 12** Solve for x: $5^{x-2} = 3^{3x+2}$

Solution Because the bases are different, we cannot use the method shown in Section 6.1. However, if we take the natural logarithm of each side and apply appropriate properties of logarithms, the result is an equation in x that we can solve.

$$\begin{aligned} 5^{x-2} &= 3^{3x+2} \\ \ln 5^{x-2} &= \ln 3^{3x+2} && \text{Property (7)} \\ (x-2)\ln 5 &= (3x+2)\ln 3 && \text{Property (6)} \\ (\ln 5)x - 2\ln 5 &= (3\ln 3)x + 2\ln 3 \\ (\ln 5 - 3\ln 3)x &= 2\ln 3 + 2\ln 5 \\ x &= \frac{2(\ln 3 + \ln 5)}{\ln 5 - 3\ln 3} \approx \frac{2(2.7081)}{-1.6864} = -3.2116 \end{aligned}$$

■

Historical Comment ■ Logarithms were invented about 1590 by John Napier (1550–1617) and Jobst Bürgi (1552–1632), working independently. Napier, whose work had the greater influence, was a Scottish lord, a secretive man whose neighbors were inclined to believe him to be in league with the devil. His approach to logarithms was quite different from ours; it was based on the relationship between arithmetic and geometric series (see the chapter on induction and sequences), and not on the inverse function relationship of logarithms to exponential functions (described in Section 6.3). Napier's tables, published in 1614, listed what would now be called *natural logarithms* of sines, and were rather difficult to use. A London professor, Henry Briggs, became interested in the tables and visited Napier. In their conversations, they developed the idea of common logarithms, and Briggs then converted Napier's tables into tables of common logarithms, which were published in 1617. Their importance for calculation was immediately recognized, and, by 1650, they were being printed as far away as China. They remained an important calculation tool until the advent of the inexpensive handheld calculator about 1972, which has decreased their calculational, but not their theoretical, importance.

A side effect of the invention of logarithms was the popularization of the decimal system of notation for real numbers. ■

EXERCISE 6.4

In Problems 1–12, suppose $\ln 2 = a$ *and* $\ln 3 = b$. *Write each expression in terms of a and b.*

1. $\ln 6$ **2.** $\ln \frac{2}{3}$ **3.** $\ln 1.5$ **4.** $\ln 0.5$

5. $\ln 2e$ **6.** $\ln\left(\frac{3}{e}\right)$ **7.** $\ln 12$ **8.** $\ln 24$

9. $\ln \sqrt[5]{18}$ **10.** $\ln \sqrt[4]{48}$ **11.** $\log_2 3$ **12.** $\log_3 2$

In Problems 13–22, write each expression as a sum and/or difference of logarithms. Express powers as factors.

13. $\ln(x^2\sqrt{1-x})$ **14.** $\ln(x\sqrt{1+x^2})$ **15.** $\log_2\left(\frac{x^3}{x-3}\right)$

16. $\log_5\left(\frac{\sqrt[3]{x^2+1}}{x^2-1}\right)$ **17.** $\log\left[\frac{x(x+2)}{(x+3)^2}\right]$ **18.** $\log \frac{x^3\sqrt{x+1}}{(x-2)^2}$

19. $\ln\left[\frac{x^2-x-2}{(x+4)^2}\right]^{1/3}$ **20.** $\ln\left[\frac{(x-4)^2}{x^2-1}\right]^{2/3}$ **21.** $\ln \frac{5x\sqrt{1-3x}}{(x-4)^3}$

22. $\ln\left[\frac{5x^2\sqrt[3]{1-x}}{4(x+1)^2}\right]$

In Problems 23–32, write each expression as a single logarithm.

23. $3 \log_5 u + 4 \log_5 v$ **24.** $\log_3 u^2 - \log_3 v$

25. $\log_{1/2} \sqrt{x} - \log_{1/2} x^3$ **26.** $\log_2\left(\frac{1}{x}\right) + \log_2\left(\frac{1}{x^2}\right)$

27. $\ln\left(\frac{x}{x-1}\right) + \ln\left(\frac{x+1}{x}\right) - \ln(x^2-1)$ **28.** $\log\left(\frac{x^2+2x-3}{x^2-4}\right) - \log\left(\frac{x^2+7x+6}{x+2}\right)$

29. $8 \log_2 \sqrt{3x-2} - \log_2\left(\frac{4}{x}\right) + \log_2 4$ **30.** $21 \log_3 \sqrt[3]{x} + \log_3 9x^2 - \log_5 25$

31. $2 \log_a 5x^3 - \frac{1}{2}\log_a(2x+3)$ **32.** $\frac{1}{3}\log(x^3+1) + \frac{1}{2}\log(x^2+1)$

In Problems 33–52, solve each equation.

33. $\frac{1}{2}\log_3 x = 2$

34. $2\log_4 x = 3$

35. $3\log_2(x-1) + \log_2 4 = 5$

36. $2\log_3(x+4) - \log_3 9 = 2$

37. $\log_{10} x + \log_{10}(x+15) = 2$

38. $\log_4 x + \log_4(x-3) = 1$

39. $\log_x 4 = 2$

40. $\log_x\left(\frac{1}{8}\right) = 3$

41. $\log_3(x-1)^2 = 2$

42. $\log_2(x+4)^3 = 6$

43. $\log_{1/2}(3x+1)^{1/3} = -2$

44. $\log_{1/3}(1-2x)^{1/2} = -1$

45. $\log_a(x-1) - \log_a(x+6) = \log_a(x-2) - \log_a(x+3)$

46. $\log_a x + \log_a(x-2) = \log_a(x+4)$

47. $\log_{1/3}(x^2+x) - \log_{1/3}(x^2-x) = -1$

48. $\log_4(x^2-9) - \log_4(x+3) = 3$

49. $\log_2 8^x = -3$

50. $\log_3 3^x = -1$

51. $\log_2(x^2+1) - \log_4 x^2 = 1$
[*Hint:* Change $\log_4 x^2$ to base 2.]

52. $\log_2(3x+2) - \log_4 x = 3$

53. Show that: $\log_a(x + \sqrt{x^2-1}) + \log_a(x - \sqrt{x^2-1}) = 0$

54. Show that: $\log_a(\sqrt{x} + \sqrt{x-1}) + \log_a(\sqrt{x} - \sqrt{x-1}) = 0$

C *In Problems 55–62, use a calculator that has either a* log x *key or an* ln x *key to evaluate each logarithm.*

55. $\log_3 21$

56. $\log_5 18$

57. $\log_{1/3} 71$

58. $\log_{1/2} 15$

59. $\log_{\sqrt{2}} 7$

60. $\log_{\sqrt{5}} 8$

61. $\log_\pi e$

62. $\log_\pi \sqrt{2}$

C *In Problems 63–74, solve for x using a calculator that has either a* log x *key or an* ln x *key.*

63. $2^x = 10$

64. $3^x = 14$

65. $8^{-x} = 1.2$

66. $2^{-x} = 1.5$

67. $3^{1-2x} = 4^x$

68. $2^{x+1} = 5^{1-2x}$

69. $\left(\frac{3}{5}\right)^x = 7^{1-x}$

70. $\left(\frac{4}{3}\right)^{1-x} = 5^x$

71. $1.2^x = (0.5)^{-x}$

72. $(0.3)^{1+x} = 1.7^{2x-1}$

73. $\pi^{1-x} = e^x$

74. $e^{x+3} = \pi^x$

75. Find the domain of $f(x) = \log_a x^2$ and the domain of $g(x) = 2\log_a x$. Since $\log_a x^2 = 2\log_a x$, how do you reconcile the fact that the domains are not equal?

76. If $f(x) = \log_a x$, show that: $\dfrac{f(x+h) - f(x)}{h} = \log_a\left(1 + \dfrac{h}{x}\right)^{1/h}, \quad h \neq 0$

77. If $f(x) = \log_a x$, show that: $-f(x) = \log_{1/a} x$

78. If $f(x) = \log_a x$, show that: $f(1/x) = -f(x)$

79. If $f(x) = \log_a x$, show that: $f(AB) = f(A) + f(B)$

80. If $f(x) = \log_a x$, show that: $f(x^\alpha) = \alpha f(x)$.

In Problems 81–90, express y as a function of x. The constant C is a positive number.

81. $\ln y = \ln x + \ln C$

82. $\ln y = \ln(x + C)$

83. $\ln y = \ln x + \ln(x+1) + \ln C$

84. $\ln y = 2\ln x - \ln(x+1) + \ln C$

85. $\ln y = 3x + \ln C$

86. $\ln y = -2x + \ln C$

87. $\ln(y-3) = -4x + \ln C$

88. $\ln(y+4) = 5x + \ln C$

89. $3 \ln y = \frac{1}{2} \ln(2x + 1) - \frac{1}{3} \ln(x + 4) + \ln C$

90. $2 \ln y = -\frac{1}{2} \ln x + \frac{1}{3} \ln(x^2 + 1) + \ln C$

91. Find the value of $\log_2 3 \cdot \log_3 4 \cdot \log_4 5 \cdot \log_5 6 \cdot \log_6 7 \cdot \log_7 8$.

92. Find the value of $\log_2 4 \cdot \log_4 6 \cdot \log_6 8$.

93. Find the value of $\log_2 3 \cdot \log_3 4 \cdot \cdots \cdot \log_n(n + 1) \cdot \log_{n+1} 2$.

94. Find the value of $\log_2 2 \cdot \log_2 4 \cdot \cdots \cdot \log_2 2^n$.

C 95. How long does it take for an investment to double in value if it is invested at 8% per annum compounded monthly? Compounded continuously?

C 96. How long does it take for an investment to double in value if it is invested at 10% per annum compounded monthly? Compounded continuously?

C 97. If you have \$100 to invest at 8% per annum compounded monthly, how long will it be before the amount is \$150? If the compounding is continuous, how long will it be?

C 98. If you have \$100 to invest at 10% per annum compounded monthly, how long will it be before the amount is \$175? If the compounding is continuous, how long will it be?

99. Show that $\log_a(M/N) = \log_a M - \log_a N$, where a, M, and N are positive real numbers, with $a \neq 1$.

100. Show that $\log_a(1/N) = -\log_a N$, where a and N are positive real numbers, with $a \neq 1$.

6.5 Applications

Common logarithms often appear in the measurement of quantities, because they provide a way to scale down positive numbers that vary from very small to very large. For example, if a certain quantity can take on values from $0.0000000001 = 10^{-10}$ to $10{,}000{,}000{,}000 = 10^{10}$, the common logarithms of such numbers would be between -10 and 10.

Our first application utilizes a logarithmic scale to measure the loudness of a sound. Physicists define the **intensity of a sound wave** as the amount of energy the wave transmits through a given area. For example, the least intense sound that a human ear can detect at a frequency of 100 hertz is about 10^{-12} watt per square meter. The **loudness** $L(x)$, measured in **decibels** (named in honor of Alexander Graham Bell), of a sound of intensity x (measured in watts per square meter) is defined as

Loudness

$$L(x) = 10 \log \frac{x}{I_0} \tag{1}$$

where $I_0 = 10^{-12}$ watt per square meter is the least intense sound that a human ear can detect. If we let $x = I_0$ in equation (1), we get

$$L(I_0) = 10 \log \frac{I_0}{I_0} = 10 \log 1 = 0$$

Thus, at the threshold of human hearing, the loudness is 0 decibels. Figure 14 gives the loudness of some common sounds.

DECIBELS		
140	Shotgun blast, jet 100 feet away at takeoff	Pain
130	Motor test chamber	Human ear pain threshold
120	Firecrackers, severe thunder, pneumatic jackhammer, hockey crowd	Uncomfortably loud
110	Amplified rock music	
100	Textile loom, subway train, elevated train, farm tractor, power lawn mower, newspaper press	Loud
90	Heavy city traffic, noisy factory	
80	Diesel truck going 40 mph 50 feet away, crowded restaurant, garbage disposal, average factory, vacuum cleaner	Moderately loud
70	Passenger car going 50 mph 50 feet away	
60	Quiet typewriter, singing birds, window air-conditioner, quiet automobile	Quiet
50	Normal conversation, average office	
40	Household refrigerator, quiet office	Very quiet
30	Average home, dripping faucet, whisper 5 feet away	
20	Light rainfall, rustle of leaves	Average person's threshold of hearing
10	Whisper across room	Just audible
0		Threshold for acute hearing

Figure 14
Loudness of common sounds (in decibels)

Note that a decibel is not a linear unit like the meter. For example, a noise level of 10 decibels is 10 times as great as a noise level of 0 decibels. [If $L(x) = 10$, then $x = 10I_0$.] A noise level of 20 decibels is 100 times as great as a noise level of 0 decibels. [If $L(x) = 20$, then $x = 100I_0$.] A noise level of 30 decibels is 1000 times as great as a noise level of 0 decibels, and so on.

C **Example 1** Use Figure 14 to find the intensity of the sound of a dripping faucet.

Solution From Figure 14, we see that the loudness of the sound of dripping water is 30 decibels. Thus, by equation (1), its intensity x may be found as follows:

$$30 = 10 \log\left(\frac{x}{I_0}\right)$$

$$3 = \log\left(\frac{x}{I_0}\right) \quad \text{Divide by 10.}$$

$$\frac{x}{I_0} = 10^3 \quad \text{Write in exponential form.}$$

$$x = 1000I_0$$

where $I_0 = 10^{-12}$ watt per square meter. Thus, the intensity of the sound of a dripping faucet is 1000 times as great as a noise level of 0 decibels; that is, such a sound has an intensity of $1000 \cdot 10^{-12} = 10^{-9}$ watt per square meter. ■

Example 2 Use Figure 14 to determine the loudness of a subway train if it is known that this sound is 10 times as intense as the sound due to heavy city traffic.

Solution The sound due to heavy city traffic has a loudness of 90 decibels. Its intensity, therefore, is the value of x in the equation

$$90 = 10 \log\left(\frac{x}{I_0}\right)$$

A sound 10 times as intense as x has loudness $L(10x)$. Thus, the loudness of the subway train is

$$\begin{aligned} L(10x) &= 10 \log\left(\frac{10x}{I_0}\right) && \text{Replace } x \text{ by } 10x. \\ &= 10\left[\log 10 + \log\left(\frac{x}{I_0}\right)\right] && \text{Log of quotient} = \text{Sum of logs} \\ &= 10 \log 10 + 10 \log\left(\frac{x}{I_0}\right) && \log 10 = 1 \\ &= 10 + 90 = 100 \text{ decibels} \end{aligned}$$

■

Our second application uses a logarithmic scale to measure the magnitude of an earthquake.

The **Richter scale*** is one way of converting seismographic readings into numbers that provide an easy reference for measuring the magnitude M of an earthquake. All earthquakes are compared to a so-called **zero-level earthquake** whose seismographic reading measures 0.001 millimeter at a distance of 100 kilometers from the epicenter. An earthquake whose seismographic reading measures x millimeters has **magnitude** $M(x)$ given by

Magnitude of an Earthquake

$$M(x) = \log\left(\frac{x}{x_0}\right) \qquad (2)$$

where $x_0 = 10^{-3}$ is the reading of a zero-level earthquake the same distance from its epicenter.

*Named after the American scientist, C. F. Richter, who devised it in 1935.

Example 3 What is the magnitude of an earthquake whose seismographic reading is 0.1 millimeter at a distance of 100 kilometers from its epicenter?

Solution If $x = 0.1$, the magnitude $M(x)$ of this earthquake is

$$M(0.1) = \log\left(\frac{x}{x_0}\right) = \log\left(\frac{0.1}{0.001}\right) = \log\left(\frac{10^{-1}}{10^{-3}}\right) = \log 10^2 = 2$$

This earthquake thus measures 2.0 on the Richter scale. ■

Based on formula (2), we define the **intensity of an earthquake** as the ratio of x to x_0. For example, the intensity of the earthquake described in Example 3 is $\frac{0.1}{0.001} = 10^2 = 100$. That is, it is 100 times as intense as a zero-level earthquake.

Example 4 The devastating San Francisco earthquake of 1906 measured 8.9 on the Richter scale. How did the intensity of that earthquake compare to the Papua New Guinea earthquake of 1988, which measured 6.7 on the Richter scale?

Solution Let x_1 and x_2 denote the seismographic readings, respectively, of the 1906 San Francisco earthquake and the Papua New Guinea earthquake. Then, based on formula (2),

$$8.9 = \log\left(\frac{x_1}{x_0}\right) \qquad 6.7 = \log\left(\frac{x_2}{x_0}\right)$$

Consequently,

$$\frac{x_1}{x_0} = 10^{8.9} \qquad \frac{x_2}{x_0} = 10^{6.7}$$

The 1906 San Francisco earthquake was $10^{8.9}$ times as intense as a zero-level earthquake. The Papua New Guinea earthquake was $10^{6.7}$ times as intense as a zero-level earthquake. Thus,

$$\frac{x_1}{x_2} = \frac{10^{8.9}x_0}{10^{6.7}x_0} = 10^{2.2} \approx 158$$

$$x_1 \approx 158x_2$$

Hence, the San Francisco earthquake was 158 times as intense as the Papua New Guinea earthquake. ■

Example 4 demonstrates that the relative intensity of two earthquakes can be found by raising 10 to a power equal to the difference of their readings on the Richter scale.

EXERCISE 6.5

1. Find the loudness of a dishwasher that operates at an intensity of 10^{-5} watt per square meter. Express your answer in decibels.
2. Find the loudness of a diesel engine that operates at an intensity of 10^{-3} watt per square meter. Express your answer in decibels.
3. [C] With engines at full throttle, a Boeing 727 jetliner produces noise at an intensity of 0.15 watt per square meter. Find the loudness of the engines in decibels.
4. [C] A whisper produces noise at an intensity of $10^{-9.8}$ watt per square meter. What is the loudness of a whisper in decibels?
5. For humans, the threshold of pain due to sound averages 130 decibels. What is the intensity of such a sound in watts per square meter?
6. [C] If one sound is 50 times as intense as another, what is the difference in the loudness of the two sounds? Express your answer in decibels.
7. Find the magnitude of an earthquake whose seismograph reading is 10.0 millimeters at a distance of 100 kilometers from its epicenter.
8. [C] Find the magnitude of an earthquake whose seismograph reading is 1210 millimeters at a distance of 100 kilometers from its epicenter.
9. [C] The Mexico City earthquake of 1978 registered 7.85 on the Richter scale. What would a seismograph 100 kilometers from the epicenter have measured for this earthquake? How does this earthquake compare in intensity to the 1906 San Francisco earthquake, which registered 8.9 on the Richter scale?
10. Two earthquakes differ by 1.0 when measured on the Richter scale. How would the seismographic readings differ at a distance of 100 kilometers from the epicenter? How do their intensities compare?

In Problems 11 and 12, use the following result: If x is the atmospheric pressure (measured in millimeters of mercury), then the formula for the altitude $h(x)$ (measured in meters above sea level) is

$$h(x) = (30T + 8000)\log\left(\frac{P_0}{x}\right)$$

where T is the temperature (in degrees Celsius) and P_0 is the atmospheric pressure at sea level, which is approximately 760 millimeters of mercury.

11. [C] At what height is an aircraft whose instruments record an outside temperature of 0°C and a barometric pressure of 300 millimeters of mercury?
12. [C] What is the atmospheric pressure (in millimeters of mercury) on Mt. Everest, which has an altitude of approximately 8900 meters, if the air temperature is 5°C?

6.6 Growth and Decay

Many natural phenomena have been found to follow the law that an amount A varies with time t according to

$$A = A_0e^{kt} \qquad (1)$$

where A_0 is the original amount ($t = 0$) and $k \neq 0$ is a constant.

If $k > 0$, then equation (1) states that the amount A is increasing over time; if $k < 0$, the amount A is decreasing over time. In either case, when an amount A varies over time according to equation (1), it is said to follow the **exponential law** or the **law of uninhibited growth** ($k > 0$) **or decay** ($k < 0$).

For example, we saw in Section 6.2 that continuously compounded interest follows the law of uninhibited growth. In this section we shall look at three additional phenomena that follow the exponential law.

Biology

Mitosis, or division of cells, is a universal process in the growth of living organisms such as amebas, plants, human skin cells, and many others. (Human nerve cells are one of the few exceptions.) Based on an ideal situation in which no cells die and no by-products are produced, the number of cells present at a given time follows the law of uninhibited growth. Actually, however, after enough time has passed, growth at an exponential rate will cease due to the influence of factors such as lack of living space, dwindling food supply, and so on. The law of uninhibited growth accurately reflects only the early stages of the mitosis process.

The mitosis process begins with a culture containing N_0 cells. Each cell in the culture grows for a certain period of time and then divides into two identical cells. We assume that the time needed for each cell to divide in two is constant and does not change as the number of cells increases. These cells then grow, and eventually each divides in two; and so on. A formula that gives the number N of cells in the culture after a time t has passed (in the early stages of growth) is

Uninhibited Growth of Cells

$$N(t) = N_0e^{kt} \qquad (2)$$

where k is a positive constant.

C **Example 1** A colony of bacteria increases according to the law of uninhibited growth. If the number of bacteria doubles in 3 hours, how long will it take for the size of the colony to triple?

Solution Using formula (2), the number N of cells at a time t is

$$N(t) = N_0e^{kt}$$

where N_0 is the initial number of bacteria present and k is a positive number. We first seek the number k. The number of cells doubles in 3 hours; thus, we have

$$N(3) = 2N_0$$

But $N(3) = N_0e^{k(3)}$, so

$$\begin{aligned} N_0e^{k(3)} &= 2N_0 \\ e^{3k} &= 2 \\ 3k &= \ln 2 \\ k &= \tfrac{1}{3}\ln 2 \approx \tfrac{1}{3}(0.6931) = 0.2310 \end{aligned}$$

Write the equivalent logarithmic equation.

Formula (2) for this growth process is therefore

$$N(t) = N_0e^{0.2310t}$$

The time t needed for the size of the colony to triple requires that $N = 3N_0$. Thus, we substitute $3N_0$ for N to get

$$\begin{aligned} 3N_0 &= N_0e^{0.2310t} \\ 3 &= e^{0.2310t} \\ 0.2310t &= \ln 3 \\ t &= \frac{1}{0.2310}\ln 3 \approx \frac{1.0986}{0.2310} = 4.756 \text{ hours} \end{aligned}$$

It will take about 4.756 hours for the size of the colony to triple. ■

Radioactive Decay

Radioactive materials follow the law of uninhibited decay. Thus, the amount A of a radioactive material present at time t is given by the formula

Uninhibited Radioactive Decay

$$A = A_0e^{kt} \qquad (3)$$

where A_0 is the original amount of radioactive material and k is a negative number.

All radioactive substances have a specific **half-life**, which is the time required for half of the particular substance to decay. In **carbon dating**, we use the fact that all living organisms contain two kinds of carbon, carbon-12 (a stable carbon) and carbon-14 (a radioactive carbon, with a half-life of 5600 years). While an organism is living, the ratio of carbon-12 to carbon-14 is constant. But when an organism dies, the original amount of carbon-12 present remains unchanged, whereas the amount of carbon-14 begins to decrease. This change in the amount of carbon-14 present relative to the amount of carbon-12 present makes it possible to calculate when an organism died.

Example 2 Traces of burned wood found along with ancient stone tools in an archaeological dig in Chile were found to contain approximately 1.67% of the original amount of carbon-14. If the half-life of carbon-14 is 5600 years, approximately when was the tree cut and burned?

Solution Using equation (3), the amount A of carbon-14 present at time t is

$$A = A_0 e^{kt}$$

where A_0 is the original amount of carbon-14 present and k is a negative number. We first seek the number k. To find it, we use the fact that after 5600 years, half of the original amount of carbon-14 remains. Thus,

$$\tfrac{1}{2}A_0 = A_0 e^{k(5600)}$$

$$\tfrac{1}{2} = e^{5600k}$$

$$5600k = \ln \tfrac{1}{2}$$

Write the equivalent logarithmic equation.

$$k = \tfrac{1}{5600} \ln \tfrac{1}{2} \approx -0.000124$$

Formula (3) therefore becomes

$$A = A_0 e^{-0.000124t}$$

If the amount A of carbon-14 now present is 1.67% of the original amount, it follows that

$$0.0167A_0 = A_0 e^{-0.000124t}$$

$$0.0167 = e^{-0.000124t}$$

$$-0.000124t = \ln 0.0167$$

$$t = \frac{1}{-0.000124} \ln 0.0167 \approx 33{,}000 \text{ years}$$

The tree was cut and burned about 33,000 years ago. Some archaeologists use this conclusion to argue that humans lived in the Americas 33,000 years ago, much earlier than is generally accepted. ■

Newton's Law of Cooling

Newton's Law of Cooling* states that the temperature of a heated object decreases exponentially over time toward the temperature of the surrounding medium. That is, the temperature u of a heated object at a given time t satisfies the equation

*Named after Sir Isaac Newton (1642–1727), one of the cofounders of calculus.

Newton's Law of Cooling

$$u = T + (u_0 - T)e^{kt} \tag{4}$$

where T is the constant temperature of the surrounding medium, u_0 is the initial temperature of the heated object, and k is a negative number.

Example 3 An object is heated to 100°C and is then allowed to cool in a room whose air temperature is 30°C. If the temperature of the object is 80°C after 5 minutes, when will its temperature be 50°C?

Solution Using equation (4) with $T = 30$ and $u_0 = 100$, the temperature (in degrees Celsius) of the object at time t (in minutes) is

$$u = 30 + (100 - 30)e^{kt} = 30 + 70e^{kt} \tag{5}$$

where k is a negative number. To find k, we use the fact that $u = 80$ when $t = 5$. Then,

$$\begin{aligned} 80 &= 30 + 70e^{k(5)} \\ 50 &= 70e^{5k} \\ e^{5k} &= \tfrac{50}{70} \\ 5k &= \ln \tfrac{5}{7} \\ k &= \tfrac{1}{5} \ln \tfrac{5}{7} \approx -0.0673 \end{aligned}$$

Formula (5) therefore becomes

$$u = 30 + 70e^{-0.0673t}$$

Now, we want to find t when $u = 50$°C, so

$$\begin{aligned} 50 &= 30 + 70e^{-0.0673t} \\ 20 &= 70e^{-0.0673t} \\ e^{-0.0673t} &= \tfrac{20}{70} \\ -0.0673t &= \ln \tfrac{2}{7} \\ t &= \tfrac{1}{-0.0673} \ln \tfrac{2}{7} \approx 18.6 \text{ minutes} \end{aligned}$$

Thus, the temperature of the object will be 50°C after about 18.6 minutes. ■

C EXERCISE 6.6

1. The size P of a certain insect population at time t (in days) obeys the equation $P = 500e^{0.02t}$. After how many days will the population reach 1000? 2000?
2. The number N of bacteria present in a culture at time t (in hours) obeys the equation $N = 1000e^{0.01t}$. After how many hours will the population equal 1500? 2000?

3. Strontium-90 is a radioactive material that decays according to the law $A = A_0e^{-0.0244t}$, where A_0 is the initial amount present and A is the amount present at time t (in years). What is the half-life of strontium-90?

4. Iodine-131 is a radioactive material that decays according to the law $A = A_0e^{-0.087t}$, where A_0 is the initial amount present and A is the amount present at time t (in days). What is the half-life of iodine-131?

5. Use the information in Problem 3 to determine how long it takes for 100 grams of strontium-90 to decay to 10 grams.

6. Use the information in Problem 4 to determine out how long it takes for 100 grams of iodine-131 to decay to 10 grams.

7. The population of a colony of mosquitoes obeys the law of uninhibited growth. If there are 1000 mosquitoes initially, and there are 1800 after 1 day, what is the size of the colony after 3 days? How long is it until there are 10,000 mosquitoes?

8. A culture of bacteria obeys the law of uninhibited growth. If there are 500 bacteria present initially, and there are 800 after 1 hour, how many will be present in the culture after 5 hours? How long is it until there are 20,000 bacteria?

9. The population of a southern city follows the exponential law. If the population doubled in size over an 18 month period and the current population is 10,000, what will the population be 2 years from now?

10. The population of a midwestern city follows the exponential law. If the population decreased from 900,000 to 800,000 from 1988 to 1990, what will the population be in 1992?

11. The half-life of radium is 1690 years. If 10 grams are present now, how much will be present in 50 years?

12. The half-life of radioactive potassium is 1.3 billion years. If 10 grams are present now, how much will be present in 100 years? In 1000 years?

13. A piece of charcoal is found to contain 30% of the carbon-14 it originally had. When did the tree from which the charcoal came die? Use 5600 years as the half-life of carbon-14.

14. A fossilized leaf contains 70% of its normal amount of carbon-14. How old is the fossil?

15. A pizza baked at 450°F is removed from the oven at 5:00 PM into a room that is a constant 70°F. After 5 minutes, the pizza is at 300°F. At what time can you begin eating the pizza if you want its temperature to be 135°F?

16. A thermometer reading 72°F is placed in a refrigerator where the temperature is a constant 38°F. If the thermometer reads 60°F after 2 minutes, what will it read after 7 minutes? How long will it take before the thermometer reads 39°F?

17. A thermometer reading 8°C is brought into a room with a constant temperature of 35°C. If the thermometer reads 15°C after 3 minutes, what will it read after being in the room for 5 minutes? For 10 minutes? [*Hint:* You need to construct a formula similar to equation (4).]

18. A frozen steak has a temperature of 28°F. It is placed in a room with a constant temperature of 70°F. After 10 minutes, the temperature of the steak has risen to 35°F. What will the temperature of the steak be after 30 minutes? How long will it take the steak to thaw to a temperature of 45°F? [See the hint given for Problem 17.]

19. Salt (NaCl) decomposes in water into sodium (Na^+) and chloride (Cl^-) ions according to the law of uninhibited decay. If the initial amount of salt is 25 kilograms and, after 10 hours, 15 kilograms of salt are left, how much salt is left after 1 day? How long does it take until $\frac{1}{2}$ kilogram of salt is left?

20. The voltage of a certain condenser decreases over time according to the law of uninhibited decay. If the initial voltage is 40 volts, and 2 seconds later it is 10 volts, what is the voltage after 5 seconds?

21. The concentration of alcohol in a person's blood is measurable. Recent medical research suggests that the risk R (given as a percentage) of having a car accident obeys the law of uninhibited growth,

$$R = 1.5e^{ka}$$

where a is the variable concentration of alcohol in the blood and k is a constant. Suppose a concentration of alcohol in the blood of 0.11 results in a 10% risk ($R = 10$) of an accident. What is the risk, then, if the concentration is 0.17? What concentration of alcohol corresponds to a risk of 100%? If anyone whose risk of having an accident is 20% or more should not drive, what concentration of alcohol in the blood should be used to test a person's ability to continue driving?

22. Rework Problem 21 if the law changes to $R = 1.0e^{ka}$.

23. The equation governing the amount of current I (in amperes) after time t (in seconds) in a simple RL circuit consisting of a resistance R (in ohms), an inductance L (in henrys), and an electromotive force E (in volts) is

$$I = \frac{E}{R}[1 - e^{-(R/L)t}]$$

If $E = 12$ volts, $R = 10$ ohms, and $L = 5$ henrys, how long does it take to obtain a current of 0.5 ampere? Of 1.0 ampere? Graph this function, measuring I along the y-axis and t along the x-axis.

24. Rework Problem 23 if $E = 12$ volts, $R = 5$ ohms, and $L = 10$ henrys.

25. Psychologists sometimes use the function

$$L(t) = A(1 - e^{-kt})$$

to measure the amount L learned at time t. The number A represents the amount to be learned, and the number k measures the rate of learning. Suppose a student has an amount A of 200 vocabulary words to learn. A psychologist determines that the student learned 20 vocabulary words after 5 minutes. Determine the rate of learning k. Approximately how many words will the student have learned after 10 minutes? After 15 minutes? How long does it take for the student to learn 180 words?

26. Rework Problem 25 for a student who has to learn 200 vocabulary words but learns only 15 words after 5 minutes.

27. After the fallout from Chernobyl, the hay in Austria was contaminated by iodine-131 (see Problem 4). If it is alright to feed the hay to cows when 10% of the iodine-131 remains, how long do the farmers need to wait to use this hay?

28. The hotel Bora-Bora is having a pig roast. At noon, the chef put the pig in a large earthen oven. The pig's original temperature was 75°F. At 2:00 PM, the chef checked the pig's temperature and was upset because it had reached only 100°F. If the oven's temperature remains a constant 325°F, at what time may the hotel serve its guests, assuming that pork is done when it reaches 175°F?

CHAPTER REVIEW

THINGS TO KNOW

Properties of the exponential function	$f(x) = a^x$, $a > 1$	Domain: $(-\infty, \infty)$; Range: $(0, \infty)$; x-intercepts: none; y-intercept: 1; horizontal asymptote: x-axis as $x \to -\infty$; increasing; one-to-one See Figure 3 for a typical graph.
	$f(x) = a^x$, $0 < a < 1$	Domain: $(-\infty, \infty)$; Range: $(0, \infty)$; x-intercepts: none, y-intercept: 1; horizontal asymptote: x-axis as $x \to \infty$; decreasing; one-to-one See Figure 5 for a typical graph.
Properties of the logarithmic function	$f(x) = \log_a x$, $a > 1$ ($y = \log_a x$ means $x = a^y$)	Domain: $(0, \infty)$; Range: $(-\infty, \infty)$; x-intercept: 1; y-intercept: none; vertical asymptote: y-axis; increasing; one-to-one See Figure 9(b) for a typical graph.
	$f(x) = \log_a x$, $0 < a < 1$ ($y = \log_a x$ means $x = a^y$)	Domain: $(0, \infty)$; Range: $(-\infty, \infty)$; x-intercept: 1; y-intercept: none; vertical asymptote: y-axis; decreasing; one-to-one See Figure 9(a) for a typical graph.
Number e	Value approached by the expression $\left(1 + \frac{1}{n}\right)^n$ as $n \to \infty$	
Natural logarithm	$y = \ln x$ means $x = e^y$	
Properties of logarithms	$\log_a 1 = 0$; $\log_a a = 1$ $a^{\log_a M} = M$; $\log_a a^r = r$ $\log_a MN = \log_a M + \log_a N$ $\log_a\left(\frac{M}{N}\right) = \log_a M - \log_a N$ $\log_a\left(\frac{1}{N}\right) = -\log_a N$ $\log_a M^r = r \log_a M$	
Change-of-base formula	$\log_a M = \frac{\log_b M}{\log_b a}$	

Formulas

Compound interest	$A = P\left(1 + \frac{r}{n}\right)^{nt}$
Continuous compounding	$A = Pe^{rt}$
Present value	$P = A\left(1 + \frac{r}{n}\right)^{-nt}$ or $P = Ae^{-rt}$
Growth and decay	$A = A_0e^{kt}$

How To:

Solve certain exponential equations
Solve certain logarithmic equations
Solve problems involving intensity of sound and intensity of earthquakes
Solve problems involving growth and decay

FILL-IN-THE-BLANK ITEMS

1. The graph of every exponential function $f(x) = a^x$, $a > 0$, $a \neq 1$, passes through the point ________.
2. If the graph of an exponential function $f(x) = a^x$, $a > 0$, $a \neq 1$, is decreasing, then its base must be less than ________.
3. If $3^x = 3^4$, then $x =$ ________.
4. The logarithm of a product equals the ________ of the logarithms.
5. For every base, the logarithm of ________ equals 0.
6. If $\log_8 M = \log_5 7/\log_5 8$, then $M =$ ________.
7. The domain of the logarithmic function $f(x) = \log_a x$ consists of ________.
8. The graph of every logarithmic function $f(x) = \log_a x$, $a > 0$, $a \neq 1$, passes through the point ________.
9. If the graph of a logarithmic function $f(x) = \log_a x$, $a > 0$, $a \neq 1$, is increasing, then its base must be larger than ________.
10. If $\log_3 x = \log_3 7$, then $x =$ ________.

REVIEW EXERCISES

In Problems 1–6, evaluate each expression.

1. $\log_2\left(\frac{1}{8}\right)$ **2.** $\log_3 81$ **3.** $\ln e^{\sqrt{2}}$ **4.** $e^{\ln 0.1}$ **5.** $2^{\log_2 0.4}$ **6.** $\log_2 2^{\sqrt{3}}$

In Problems 7–12, write each expression as a single logarithm.

7. $3 \log_4 x^2 + \frac{1}{2} \log_4 \sqrt{x}$ **8.** $-2 \log_3\left(\frac{1}{x}\right) + \frac{1}{3} \log_3 \sqrt{x}$

9. $\ln\left(\frac{x-1}{x}\right) + \ln\left(\frac{x}{x+1}\right) - \ln(x^2 - 1)$

10. $\log(x^2 - 9) - \log(x^2 + 7x + 12)$

11. $2 \log 2 + 3 \log x - \frac{1}{2}[\log(x + 3) + \log(x - 2)]$

12. $\frac{1}{2} \ln(x^2 + 1) - 4 \ln \frac{1}{2} - \frac{1}{2}[\ln(x - 4) + \ln x]$

In Problems 13–20, find y as a function of x. The constant C is a positive number.

13. $\ln y = 2x^2 + \ln C$

14. $\ln(y - 3) = \ln 2x^2 + \ln C$

15. $\frac{1}{2} \ln y = 3x^2 + \ln C$

16. $\ln 2y = \ln(x + 1) + \ln(x + 2) + \ln C$

17. $\ln(y - 3) + \ln(y + 3) = x + C$

18. $\ln(y - 1) + \ln(y + 1) = -x + C$

19. $e^{y+C} = x^2 + 4$

20. $e^{3y-C} = (x + 4)^2$

In Problems 21–30, graph each function. Begin each problem either with the graph of $y = e^x$ or with the graph of $y = \ln x$.

21. $f(x) = e^{-x}$

22. $f(x) = \ln(-x)$

23. $f(x) = 1 - e^x$

24. $f(x) = 3 + \ln x$

25. $f(x) = 3e^x$

26. $f(x) = \frac{1}{2} \ln x$

27. $f(x) = e^{|x|}$

28. $f(x) = \ln|x|$

29. $f(x) = 3 - e^{-x}$

30. $f(x) = 4 - \ln(-x)$

In Problems 31–50, solve each equation.

31. $4^{1-2x} = 2$

32. $8^{6+3x} = 4$

33. $3^{x^2+x} = \sqrt{3}$

34. $4^{x-x^2} = \frac{1}{2}$

35. $\log_x 64 = -3$

36. $\log_{\sqrt{2}} x = -6$

37. $\log_{\sqrt{3}}(9\sqrt{3}) = x$

38. $\log_x 3 = \frac{1}{5}$

C **39.** $5^x = 3^{x+2}$

C **40.** $5^{x+2} = 7^{x-2}$

41. $9^{2x} = 27^{3x-4}$

42. $25^{2x} = 5^{x^2-12}$

43. $8 = 4^{x^2} \cdot 2^{5x}$

44. $2^x \cdot 5 = 10^x$

45. $\log_6(x + 3) + \log_6(x + 4) = 1$

46. $\log_{10}(7x - 12) = 2 \log_{10} x$

C **47.** $e^{1-x} = 5$

C **48.** $e^{1-2x} = 4$

C **49.** $2^{3x} = 3^{2x+1}$

C **50.** $2^{x^3} = 3^{x^2}$

C **51.** First Colonial Bankshares Corporation advertised the following IRA investment plans. Assuming continuous compounding, what was the annual rate of interest they offered?

Target IRA Plans

FOR EACH $5000 MATURITY VALUE DESIRED

Deposit:	At a term of:
$620.17	20 years
$1045.02	15 years
$1760.92	10 years
$2967.26	5 years

C **52.** See Problem 51. First Colonial Bankshares claims that $4000 invested today will have a value of over $32,000 in 20 years. Use the answer found in Problem 51 to find the actual value of $4000 in 20 years. Assume continuous compounding.

C **53.** Find the loudness of a garbage disposal unit that operates at an intensity of 10^{-4} watt per square meter. Express your answer in decibels.

C **54.** On September 9, 1985, the western suburbs of Chicago experienced a mild earthquake that registered 3.0 on the Richter scale. How did this earthquake compare in intensity to the great San Francisco earthquake of 1906, which registered 8.9 on the Richter scale?

C **55.** The bones of a prehistoric man found in the desert of New Mexico contain approximately 5% of the original amount of carbon-14. If the half-life of carbon-14 is 5600 years, approximately how long ago did the man die?

C **56.** A skillet is removed from an oven whose temperature is 450°F and placed in a room whose temperature is 70°F. After 5 minutes, the temperature of the skillet is 400°F. How long will it be until its temperature is 150°F?

7 Trigonometric Functions

Trigonometry was developed by Greek astronomers who regarded the sky as the inside of a sphere, so it was natural that triangles on a sphere were investigated early (by Menelaus of Alexandria about AD 100) and that triangles in the plane were studied much later. The first book containing a systematic treatment of plane and spherical trigonometry was written by the Persian astronomer Nasîr ed-dîn (about AD 1250).

Regiomontanus (1436–1476) is the man most responsible for moving trigonometry from astronomy into mathematics. His work was improved by Copernicus (1473–1543) and Copernicus's student Rhaeticus (1514–1576). Rhaeticus's book was the first to define the six trigonometric functions as ratios of sides of triangles, although he did not give the functions their present names. Credit for this is due to Thomas Fincke (1583), but Fincke's notation was by no means universally accepted at the time. The notation was finally stabilized by the textbooks of Leonhard Euler (1707–1783).

Trigonometry has since evolved from its use by surveyors, navigators, and engineers to present applications involving ocean tides, the rise and fall of food supplies in certain ecologies, brain wave patterns, and many other phenomena.

7.1 Angles and Their Measure

A **ray**, or **half-line**, is that portion of a line that starts at a point V on the line and extends indefinitely in one direction. The starting point V of a ray is called its **vertex**. See Figure 1.

Figure 1

If two rays are drawn with a common vertex, they form an **angle**. We call one of the rays of an angle the **initial side** and the other the **terminal side**. The angle that is formed is identified by showing the direction of rotation from the initial side to the terminal side. If the rotation is in the counterclockwise direction, the angle is **positive**; if the rotation is clockwise, the angle is **negative**. See Figure 2. Lowercase Greek letters, such as α (alpha), β (beta), γ (gamma), θ (theta), and so on, will be used to denote angles. Thus, in Figure 2, we have used α to denote the positive angle and β to denote the negative angle.

Figure 2

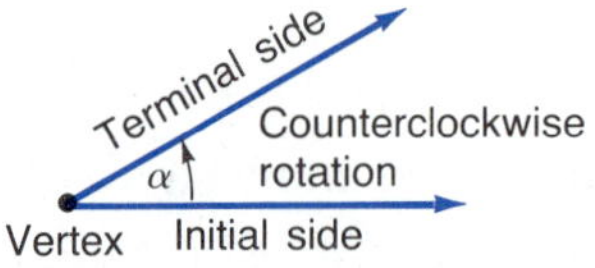

(a) Positive angle

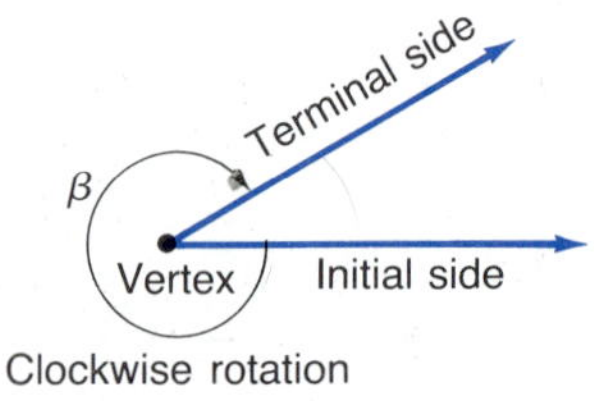

(b) Negative angle

An angle θ is said to be in **standard position** if its vertex is at the origin of a rectangular coordinate system and its initial side coincides with the positive x-axis. See Figure 3.

Figure 3

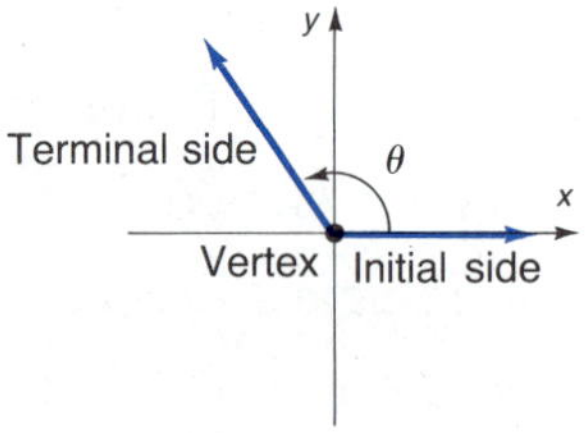

(a) θ in standard position; θ positive

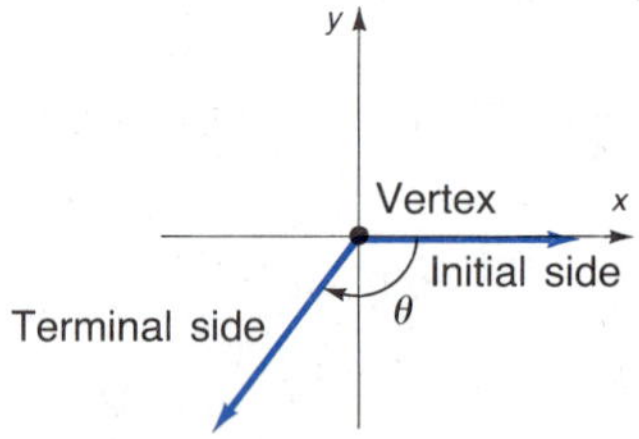

(b) θ in standard position; θ negative

When an angle θ is in standard position, the terminal side either will lie in a quadrant, in which case we say ***θ* lies in that quadrant**, or it will lie on the x-axis or the y-axis, in which case we say θ is a **quadrantal angle**. For example, the angle θ in Figure 4(a) lies in quadrant II; the angle θ in Figure 4(b) lies in quadrant IV; and the angle θ in Figure 4(c) is a quadrantal angle.

Figure 4

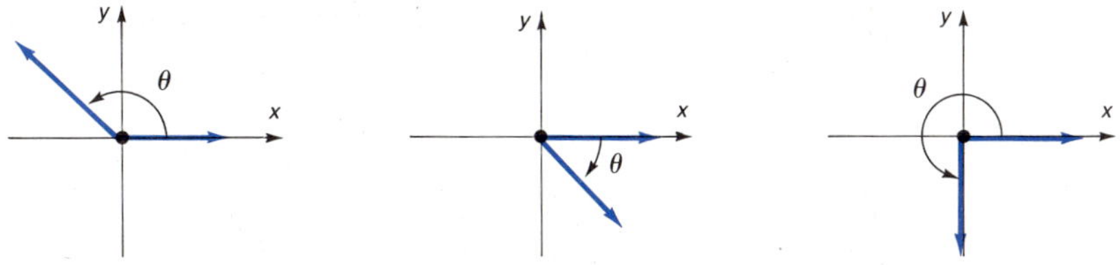

(a) θ lies in quadrant II (b) θ lies in quadrant IV (c) θ is a quadrantal angle

We measure angles by determining the amount of rotation needed for the initial side to become coincident with the terminal side. There are two commonly used measures for angles: *degrees* and *radians*.

Degrees

The angle formed by rotating the initial side exactly once in the counterclockwise direction until it coincides with itself (1 revolution) is said to measure 360 degrees, abbreviated 360°. Thus **one degree, 1°,** is $\frac{1}{360}$ revolution. A **right angle** is an angle of 90°, or $\frac{1}{4}$ revolution; a **straight angle** is an angle of 180°, or $\frac{1}{2}$ revolution. See Figure 5. As Figure 5(b) shows, it is customary to indicate a right angle by using the symbol ∟.

Figure 5

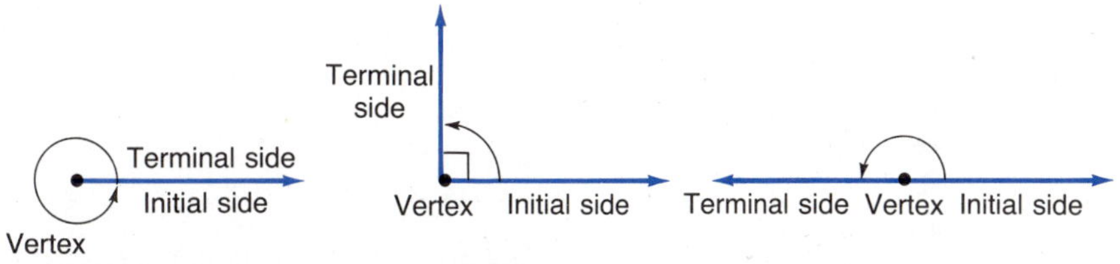

(a) 1 revolution counterclockwise, 360° (b) $\frac{1}{4}$ revolution counterclockwise, 90° (c) $\frac{1}{2}$ revolution counterclockwise, 180°

Example 1 Draw each angle:

(a) 45° (b) −90° (c) 225° (d) 405°

Solution (a) An angle of 45° is $\frac{1}{2}$ of a right angle. See Figure 6.

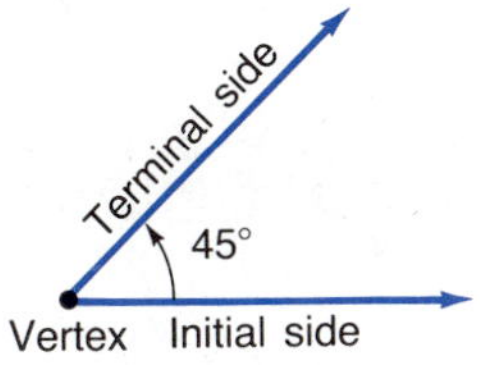

Figure 6

(b) An angle of −90° is $\frac{1}{4}$ revolution in the clockwise direction. See Figure 7.

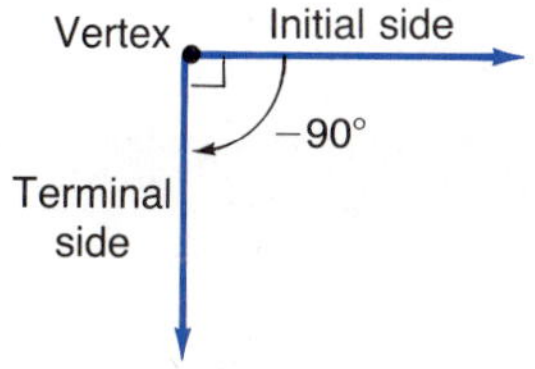

Figure 7

(c) An angle of 225° consists of a rotation through 180° followed by a rotation through 45°. See Figure 8.

(d) An angle of 405° consists of 1 revolution (360°) followed by a rotation through 45°. See Figure 9.

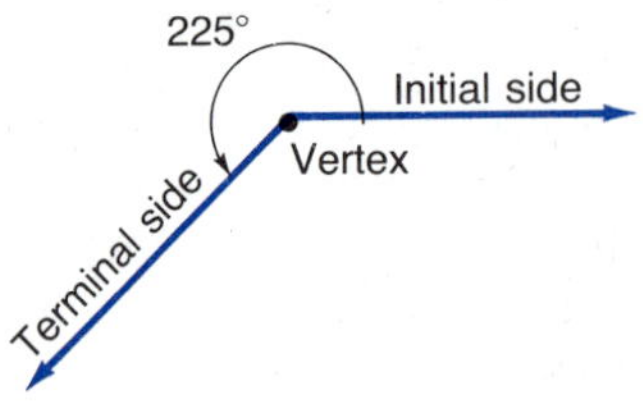

Figure 8

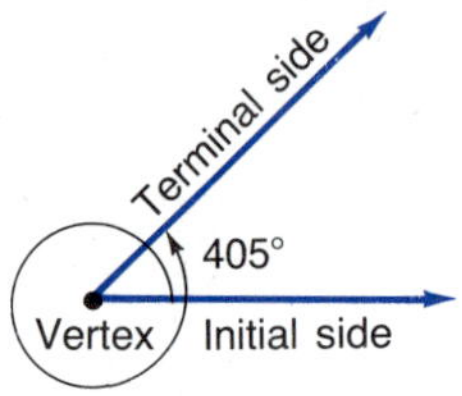

Figure 9 ■

Although subdivisions of a degree may be obtained by using decimals, we also may use the notion of *minutes* and *seconds*. **One minute,** denoted by **1′**, is defined as $\frac{1}{60}$ degree. **One second**, denoted by **1″**, is defined as $\frac{1}{60}$ minute or, equivalently, $\frac{1}{3600}$ degree. An angle of, say, 30 degrees, 40 minutes, 10 seconds is written compactly as 30°40′10″. To summarize:

$$1 \text{ counterclockwise revolution} = 360^\circ$$
$$60' = 1^\circ \qquad 60'' = 1' \tag{1}$$

Because calculators use decimals, it is important to be able to convert from the degree, minute, second notation (D°M′S″) to a decimal form, and vice versa. Check your calculator; it may have a special key that does the conversion for you. If you do not have this key on your calculator, you can perform the conversion as described below.

Before getting started, though, you must set the calculator to receive degrees, because there are two common ways to measure angles. Many calculators show which mode is currently in use by displaying *deg* for degree mode or *rad* for radian mode. (We will define radians shortly.) Usually, a key is used to change from one mode to another. Check your instruction manual to find out how your particular calculator works.

Now let's see how to convert from the degree, minute, second notation (D°M′S″) to a decimal form, and vice versa, by looking at some examples: 15°30′ = 15.5°, because $30' = \frac{1}{2}^\circ = 0.5^\circ$, and 32.25° = 32°15′, because $0.25^\circ = \frac{1}{4}^\circ = 15'$. For most conversions, a calculator will be helpful.

C **Example 2** (a) Convert 50°6′21″ to a decimal in degrees.

(b) Convert 21.256° to the D°M′S″ form. Round off your answer to the nearest second.

Solution (a) Because $1' = \frac{1}{60}^\circ$ and $1'' = \frac{1}{60}' = \left(\frac{1}{60} \cdot \frac{1}{60}\right)^\circ$, we convert as follows:

$$\begin{aligned} 50^\circ 6' 21'' &= \left(50 + 6 \cdot \tfrac{1}{60} + 21 \cdot \tfrac{1}{60} \cdot \tfrac{1}{60}\right)^\circ \\ &\approx (50 + 0.1 + 0.005833)^\circ \\ &= 50.105833^\circ \end{aligned}$$

(b) We start with the decimal part of 21.256°, namely, 0.256°:

$$0.256^\circ = (0.256)(1^\circ) \underset{\substack{\uparrow \\ 1^\circ = 60'}}{=} (0.256)(60') = 15.36'$$

Now we work with the decimal part of 15.36′, namely, 0.36′:

$$0.36' = (0.36)(1') = (0.36)(60'') = 21.6'' \approx 22''$$

Thus,

$$\begin{aligned} 21.256^\circ &= 21^\circ + 0.256^\circ = 21^\circ + 15.36' = 21^\circ + 15' + 0.36' \\ &= 21^\circ + 15' + 21.6'' \approx 21^\circ 15' 22'' \end{aligned}$$

■

In many applications, such as describing the exact location of a star or the precise position of a boat at sea, angles measured in degrees, minutes, and even seconds are used. For calculation purposes, these are transformed to decimal form. In many other applications, especially those in calculus, angles are measured using *radians*.

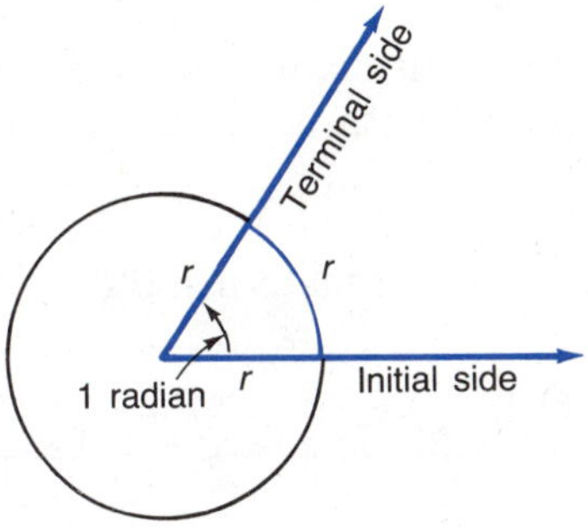

Figure 10

Radians

Consider a circle of radius r. Construct an angle whose vertex is at the center of this circle, called a **central angle**, and whose rays subtend an arc on the circle whose length equals r. See Figure 10. The measure of such an angle is **1 radian**.

Now consider a circle and two central angles, θ and θ_1. Suppose these central angles subtend arcs of lengths s and s_1, respectively, as shown in Figure 11. From geometry, we know that the ratio of the measures of the angles equals the ratio of the corresponding lengths of the arcs subtended by those angles; that is,

$$\frac{\theta}{\theta_1} = \frac{s}{s_1} \tag{2}$$

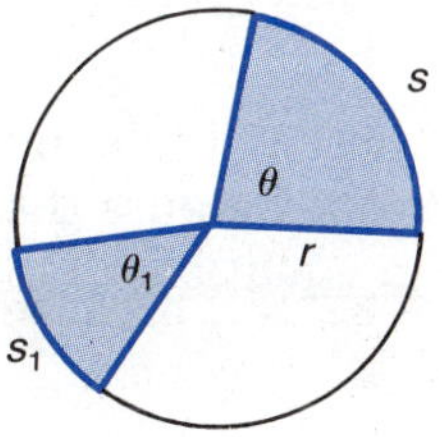

Figure 11
$\dfrac{\theta}{\theta_1} = \dfrac{s}{s_1}$

Suppose θ and θ_1 are measured in radians. If $\theta_1 = 1$ radian, then by definition, $s_1 = r$, the radius of the circle. Formula (2) then reduces to

$$\frac{\theta}{1} = \frac{s}{r} \quad \text{or} \quad s = r\theta \tag{3}$$

Theorem

Arc Length

For a circle of radius r, a central angle of θ radians subtends an arc whose length s is

$$s = r\theta \tag{4}$$

■

Note: Formulas must be consistent with regard to the units used. In equation (4), we write

$$s = r\theta$$

To see the units, however, we must go back to equation (3) and write

$$\frac{\theta \text{ radians}}{1 \text{ radian}} = \frac{s \text{ length units}}{r \text{ length units}}$$

$$s \text{ length units} = (r \text{ length units})\frac{\theta \text{ radians}}{1 \text{ radian}}$$

Since the radians "cancel," we are left with

$$s \text{ length units} = (r \text{ length units})\theta \qquad s = r\theta$$

where θ appears to be "dimensionless" but, in fact, is measured in radians. Thus, in using the formula $s = r\theta$, the dimension of radians for θ is usually omitted, and any convenient unit of length (such as inches, meters, and so on) may be used for s and r.

Example 3 Find the length of the arc of a circle of radius $r = 2$ meters subtended by a central angle of 0.25 radian.

Solution We use equation (4) with $r = 2$ meters and $\theta = 0.25$. The length s of the arc is

$$s = r\theta = 2(0.25) = 0.5 \text{ meter}$$

■

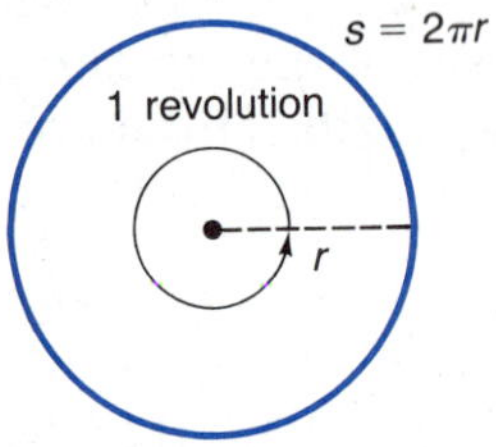

Figure 12
1 revolution = 2π radians

Relationship between Degrees and Radians

Consider a circle of radius r. A central angle of 1 revolution will subtend an arc equal to the circumference of the circle (Figure 12). Because the circumference of a circle equals $2\pi r$, we use $s = 2\pi r$ in equation (4) to find that, for an angle θ of 1 revolution,

$$s = r\theta$$

$$2\pi r = r\theta$$

$$\theta = 2\pi \text{ radians}$$

Thus,

$$1 \text{ revolution} = 2\pi \text{ radians} \tag{5}$$

so that

$$360^\circ = 2\pi \text{ radians}$$

or

$$180^\circ = \pi \text{ radians} \tag{6}$$

Based on equation (6), we have the following two conversion formulas:

$$1 \text{ degree} = \frac{\pi}{180} \text{ radian} \qquad 1 \text{ radian} = \frac{180}{\pi} \text{ degrees}$$

Example 4 Convert each angle in degrees to radians:

(a) 60° (b) 150° (c) -45° (d) 90°

Solution (a) $60^\circ = 60 \cdot 1 \text{ degree} = 60 \cdot \frac{\pi}{180} \text{ radian} = \frac{\pi}{3} \text{ radians}$

(b) $150^\circ = 150 \cdot \frac{\pi}{180} \text{ radian} = \frac{5\pi}{6} \text{ radians}$

(c) $-45^\circ = -45 \cdot \frac{\pi}{180} \text{ radian} = -\frac{\pi}{4} \text{ radian}$

(d) $90^\circ = 90 \cdot \frac{\pi}{180} \text{ radian} = \frac{\pi}{2} \text{ radians}$ ■

Example 4 illustrates that angles that are fractions of a revolution are expressed in radian measure as fractional multiples of π, rather than as decimals. Thus, a right angle, as in Example 4(d), is left in the form $\pi/2$ radians, which is exact, rather than using the approximation $\pi/2 \approx 3.1416/2 = 1.5708$ radians.

Example 5 Convert each angle in radians to degrees.

(a) $\frac{\pi}{6}$ radian (b) $\frac{3\pi}{2}$ radians (c) $-\frac{3\pi}{4}$ radians (d) $\frac{7\pi}{3}$ radians

Solution (a) $\frac{\pi}{6}$ radian $= \frac{\pi}{6} \cdot 1$ radian $= \frac{\pi}{6} \cdot \frac{180}{\pi}$ degrees $= 30°$

(b) $\frac{3\pi}{2}$ radians $= \frac{3\pi}{2} \cdot \frac{180}{\pi}$ degrees $= 270°$

(c) $-\frac{3\pi}{4}$ radians $= -\frac{3\pi}{4} \cdot \frac{180}{\pi}$ degrees $= -135°$

(d) $\frac{7\pi}{3}$ radians $= \frac{7\pi}{3} \cdot \frac{180}{\pi}$ degrees $= 420°$ ■

Table 1 lists the degree and radian measures of some commonly encountered angles. You should learn to feel equally comfortable using degree or radian measure for these angles.

Table 1

DEGREES	0°	30°	45°	60°	90°	120°	135°	150°	180°
RADIANS	0	$\frac{\pi}{6}$	$\frac{\pi}{4}$	$\frac{\pi}{3}$	$\frac{\pi}{2}$	$\frac{2\pi}{3}$	$\frac{3\pi}{4}$	$\frac{5\pi}{6}$	π
DEGREES	210°	225°	240°	270°	300°	315°	330°	360°	
RADIANS	$\frac{7\pi}{6}$	$\frac{5\pi}{4}$	$\frac{4\pi}{3}$	$\frac{3\pi}{2}$	$\frac{5\pi}{3}$	$\frac{7\pi}{4}$	$\frac{11\pi}{6}$	2π	

Example 6 Find the length of the arc of a circle of radius $r = 3$ feet subtended by a central angle of 30°.

Solution We use equation (4), but first we must convert the central angle of 30° to radians. Since $30° = \pi/6$ radian, we use $\theta = \pi/6$ and $r = 3$ feet in equation (4). The length of the arc is

$$s = r\theta = 3 \cdot \frac{\pi}{6} = \frac{\pi}{2} \approx \frac{3.14}{2} = 1.57 \text{ feet}$$ ■

When an angle is measured in degrees, the degree symbol always will be shown. However, when an angle is measured in radians, we will follow the usual practice and omit the word *radians*. Thus, if the measure of an angle is given as $\pi/6$, it is understood to mean $\pi/6$ radian.

Circular Motion

We have already defined the average speed of an object as the distance traveled divided by the elapsed time. Suppose the motion of an object is along a circle of radius r. If s is the distance traveled in time t along this circle, then the **linear speed** v of the object is defined as

$$v = \frac{s}{t} \tag{7}$$

As this object travels along the circle, suppose θ (measured in radians) is the central angle swept out in time t. Then the **angular speed** ω (the Greek letter omega) of this object is the angle (measured in radians) swept out divided by the elapsed time; that is,

$$\omega = \frac{\theta}{t} \tag{8}$$

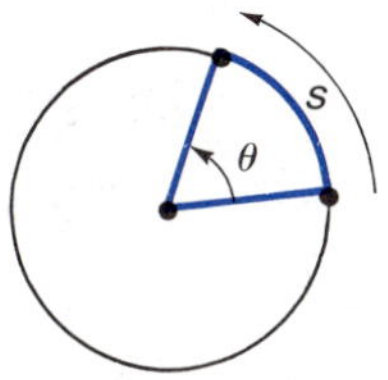

Figure 13
$v = \frac{s}{t}$, $\omega = \frac{\theta}{t}$

See Figure 13.

Angular speed is the way the speed of a phonograph record is described. For example, a 45 rpm (revolutions per minute) record is one that rotates at an angular speed of

$$\frac{45 \text{ revolutions}}{\text{Minute}} = \frac{45 \text{ revolutions}}{\text{Minute}} \cdot \frac{2\pi \text{ radians}}{\text{Revolution}} = \frac{90\pi \text{ radians}}{\text{Minute}}$$

There is an important relationship between linear speed and angular speed. In the formula $s = r\theta$, divide each side by t:

$$\frac{s}{t} = r\frac{\theta}{t}$$

Then, using equations (7) and (8), we obtain

$$v = r\omega \tag{9}$$

When using equation (9), remember that $v = s/t$ (the linear speed) has the dimensions of length per unit of time (such as feet per second, miles per hour, etc.); r (the radius of the circular motion) has the same length dimension as s; and ω (the angular speed) has the dimensions of radians per unit of time. As noted earlier, we leave the radian dimension off the numerical value of the angular speed ω so that both sides of the equation will be dimensionally consistent (with "length per unit of time"). If the angular speed is given in terms of *revolutions* per unit of time (as is often the case), be sure to convert it to *radians* per unit of time before attempting to use equation (9).

Example 7 Find the linear speed of a needle playing a $33\frac{1}{3}$ rpm record when the needle is 3 inches from the spindle (center of the record).

Solution Look at Figure 14. The needle is traveling along a circle of radius r = 3 inches. The angular speed ω of the record is

$$\omega = \frac{33\frac{1}{3}\text{ revolutions}}{\text{Minute}} = \frac{100\text{ revolutions}}{3\text{ minutes}} \cdot \frac{2\pi\text{ radians}}{\text{Revolution}} = \frac{200\pi\text{ radians}}{3\text{ minutes}}$$

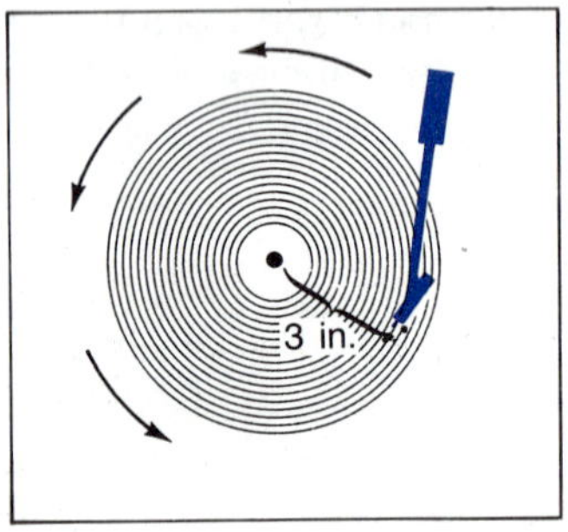

Figure 14

From equation (9), the linear speed v of the needle is

$$v = r\omega = 3\text{ inches} \cdot \frac{200\pi\text{ radians}}{3\text{ minutes}} = \frac{200\pi\text{ inches}}{\text{Minute}} \approx \frac{628\text{ inches}}{\text{Minute}}$$ ■

EXERCISE 7.1

In Problems 1–12, draw each angle.

1. $30°$ **2.** $60°$ **3.** $135°$ **4.** $-120°$ **5.** $450°$
6. $540°$ **7.** $3\pi/4$ **8.** $4\pi/3$ **9.** $-\pi/6$ **10.** $-2\pi/3$
11. $16\pi/3$ **12.** $21\pi/4$

In Problems 13–22, convert each angle in degrees to radians. Express your answer as a multiple of π.

13. $30°$ **14.** $120°$ **15.** $240°$ **16.** $330°$ **17.** $-60°$
18. $-30°$ **19.** $50°$ **20.** $40°$ **21.** $225°$ **22.** $-135°$

In Problems 23–32, convert each angle in radians to degrees.

23. $\pi/3$ **24.** $5\pi/6$ **25.** $-5\pi/4$ **26.** $-2\pi/3$ **27.** $7\pi/2$
28. $9\pi/4$ **29.** $\pi/12$ **30.** $5\pi/12$ **31.** $2\pi/3$ **32.** $5\pi/4$

In Problems 33–40, s denotes the length of arc of a circle of radius r subtended by the central angle θ. Find the missing quantity.

33. $r = 10$ meters, $\theta = \frac{1}{2}$ radian, $s = ?$
34. $r = 6$ feet, $\theta = 2$ radians, $s = ?$

35. $\theta = \frac{1}{3}$ radian, $s = 2$ feet, $r = ?$

36. $\theta = \frac{1}{4}$ radian, $s = 6$ centimeters, $r = ?$

37. $r = 5$ miles, $s = 3$ miles, $\theta = ?$

38. $r = 6$ meters, $s = 8$ meters, $\theta = ?$

39. $r = 2$ inches, $\theta = 30°$, $s = ?$

40. $r = 3$ meters, $\theta = 120°$, $s = ?$

C *In Problems 41–48, convert each angle in degrees to radians. Express your answer in decimal form, rounded to four decimal places.*

41. 17° **42.** 73° **43.** −40° **44.** −51°

45. 125° **46.** 200° **47.** 340° **48.** 350°

C *In Problems 49–56, convert each angle in radians to degrees. Express your answer in decimal form, rounded to one decimal place.*

49. 3.14 **50.** π **51.** 10.25 **52.** 0.75

53. 2 **54.** 3 **55.** 6.32 **56.** $\sqrt{2}$

C *In Problems 57–62, convert each angle to a decimal in degrees. Round off your answer to four decimal places.*

57. 40°10′25″ **58.** 61°42′21″ **59.** 1°2′3″

60. 73°40′40″ **61.** 9°9′9″ **62.** 98°22′45″

C *In Problems 63–68, convert each angle to* D°M′S″ *form. Round off your answer to the nearest second.*

63. 40.32° **64.** 61.24° **65.** 18.255°

66. 29.411° **67.** 19.99° **68.** 44.01°

69. The minute hand of a clock is 6 inches long. How far does the tip of the minute hand move in 15 minutes? How far does it move in 25 minutes?

70. A pendulum clock swings through an angle of 20° each second. If the pendulum is 40 inches long, how far does its tip move each second?

71. An object is traveling around a circle with a radius of 5 centimeters. If in 20 seconds a central angle of $\frac{1}{3}$ radian is swept out, what is the angular speed of the object? What is its linear speed?

72. An object is traveling around a circle with a radius of 2 meters. If in 20 seconds the object travels 5 meters, what is its angular speed? What is its linear speed?

C **73.** The diameter of each wheel of a bicycle is 26 inches. If you are traveling at a speed of 35 miles per hour on this bicycle, through how many revolutions per minute are the wheels turning?

C **74.** Rework Problem 73 if the diameter of each wheel is 24 inches.

C **75.** The windshield wiper of a car is 18 inches long. How many inches will the tip of the wiper trace out in $\frac{1}{3}$ revolution?

C **76.** The windshield wiper of a car is 18 inches long. If it takes 1 second to trace out $\frac{1}{3}$ revolution, how fast is the tip of the wiper moving?

C **77.** The mean distance of the Moon from the Earth is 2.39×10^5 miles. Assuming that the orbit of the Moon around the Earth is circular and that 1 revolution takes 27.3 days, find the linear speed of the Moon. Express your answer in miles per hour.

C **78.** The mean distance of the Earth from the Sun is 9.29×10^7 miles. Assuming that the orbit of the Earth around the Sun is circular and that 1 revolution takes 365 days, find the linear speed of the Earth. Express your answer in miles per hour.

79. Two pulleys, one with radius 2 inches and the other with radius 8 inches, are connected by a belt. (See the figure.) If the 2 inch pulley is caused to rotate at 3 revolutions per minute, determine the revolutions per minute of the 8 inch pulley. [*Hint:* The linear speeds of the pulleys—that is, the speed of the belt—are the same.]

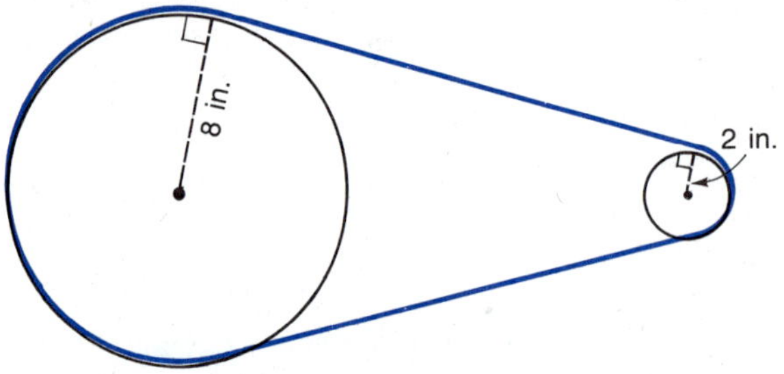

80. Two pulleys, one with radius r_1 and the other with radius r_2, are connected by a belt. The pulley with radius r_1 rotates at ω_1 revolutions per minute, whereas the pulley with radius r_2 rotates at ω_2 revolutions per minute. Show that $r_1/r_2 = \omega_2/\omega_1$.

C **81.** To approximate the speed of the current of a river, a circular paddle wheel with radius 4 feet is lowered into the water. If the current causes the wheel to rotate at a speed of 10 revolutions per minute, what is the speed of the current? Express your answer in miles per hour.

C **82.** A spin balancer rotates the wheel of a car at 480 revolutions per minute. If the diameter of the wheel is 26 inches, what road speed is being tested? Express your answer in miles per hour. At how many revolutions per minute should the balance be set to test a road speed of 80 miles per hour?

C **83.** A **nautical mile** equals the length of arc subtended by a central angle of 1 minute on a great circle* on the surface of the Earth. (See the figure.) If the radius of the Earth is taken as 3960 miles, express 1 nautical mile in terms of ordinary, or **statute**, miles (5280 feet).

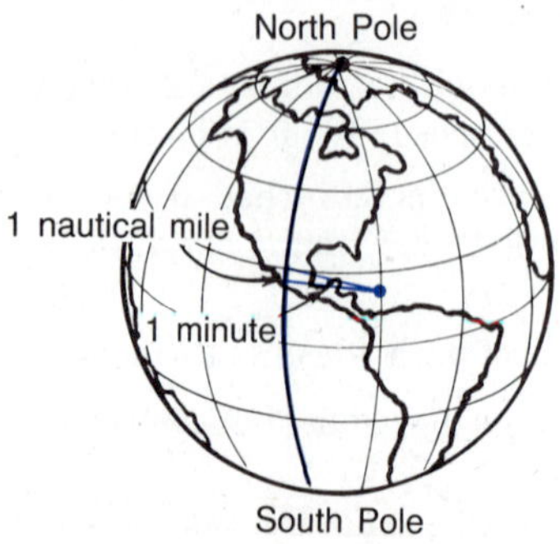

*Any circle drawn on the surface of the Earth that divides the Earth into two equal hemispheres.

84. Naples, Florida, is approximately 90 nautical miles due west of Ft. Lauderdale. How much sooner would a person in Ft. Lauderdale first see the rising Sun than a person in Naples? [*Hint:* Consult the figure. When a person at Q sees the first rays of the Sun, a person at P is still in the dark. The person at P sees the first rays after the Earth has rotated so that P is at the location Q. Now use the fact that in 24 hours, a length of arc of $2\pi(3960)$ miles is subtended.]

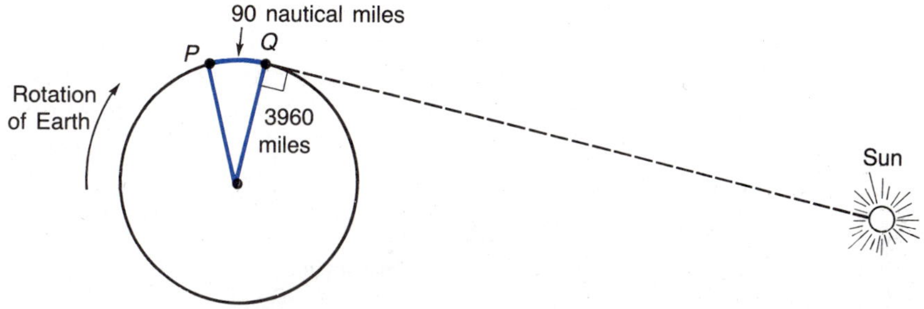

7.2 Right Triangle Trigonometry

A triangle in which one angle is a right angle (90°) is called a **right triangle**. Recall that the side opposite the right angle is called the **hypotenuse**, and the remaining two sides are called the **legs** of the triangle. In Figure 15 we have labeled the hypotenuse as c, to indicate its length is c units, and, in a like manner, we have labeled the legs as a and b. Because the triangle is a right triangle, the Pythagorean Theorem tells us that

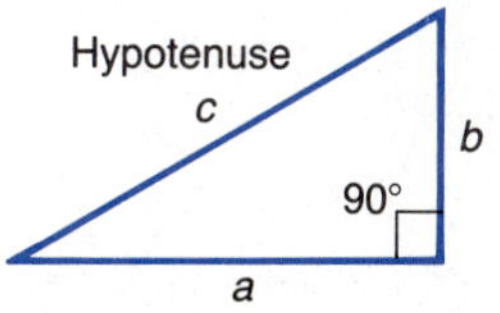

Figure 15

$$c^2 = a^2 + b^2$$

Now, suppose θ is an **acute angle**; that is, $0° < \theta < 90°$ (if θ is measured in degrees) and $0 < \theta < \pi/2$ (if θ is measured in radians). See Figure 16(a). Using this acute angle θ, we can form a right triangle, like the one illustrated in Figure 16(b), with hypotenuse of length c and legs of lengths a and b. Using the three sides of this triangle, we can form exactly six ratios, namely,

$$\frac{b}{c}, \quad \frac{a}{c}, \quad \frac{b}{a}, \quad \frac{c}{b}, \quad \frac{c}{a}, \quad \frac{a}{b}$$

Figure 16

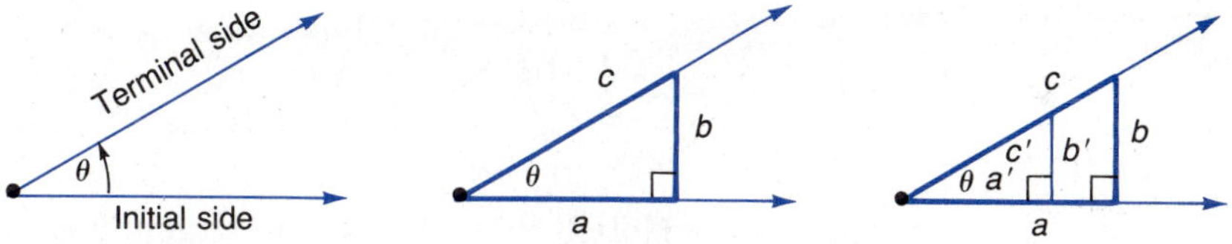

(a) Acute angle (b) Right triangle (c) Similar triangles

In fact, these ratios depend only on the size of the angle θ and not on the triangle formed. To see why, look at Figure 16(c). Any two right triangles formed using the angle θ will be similar, and hence, corresponding ratios will be equal. As a result,

$$\frac{b}{c}=\frac{b'}{c'},\quad \frac{a}{c}=\frac{a'}{c'},\quad \frac{b}{a}=\frac{b'}{a'},\quad \frac{c}{b}=\frac{c'}{b'},\quad \frac{c}{a}=\frac{c'}{a'},\quad \frac{a}{b}=\frac{a'}{b'}$$

Because the ratios depend only on the angle θ and not on the triangle itself, we give each ratio a name that involves θ: sine of θ, cosine of θ, tangent of θ, cosecant of θ, secant of θ, cotangent of θ.

Trigonometric Functions of Acute Angles

The six ratios of a right triangle are called **trigonometric functions** of acute angles and are defined as follows:

FUNCTION NAME	ABBREVIATION	VALUE
sine of θ	$\sin\theta$	$\frac{b}{c}$
cosine of θ	$\cos\theta$	$\frac{a}{c}$
tangent of θ	$\tan\theta$	$\frac{b}{a}$
cosecant of θ	$\csc\theta$	$\frac{c}{b}$
secant of θ	$\sec\theta$	$\frac{c}{a}$
cotangent of θ	$\cot\theta$	$\frac{a}{b}$

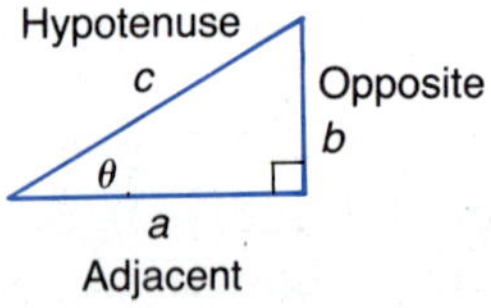

Figure 17

As an aid to remembering these definitions, it may be helpful to refer to the lengths of the sides of the triangle by the names *hypotenuse* (c), *opposite* (b), and *adjacent* (a). See Figure 17. In terms of these names, we have the following ratios:

$$\sin\theta=\frac{\text{Opposite}}{\text{Hypotenuse}}=\frac{b}{c}\qquad \cos\theta=\frac{\text{Adjacent}}{\text{Hypotenuse}}=\frac{a}{c}$$
$$\tan\theta=\frac{\text{Opposite}}{\text{Adjacent}}=\frac{b}{a}\qquad \csc\theta=\frac{\text{Hypotenuse}}{\text{Opposite}}=\frac{c}{b}\qquad (1)$$
$$\sec\theta=\frac{\text{Hypotenuse}}{\text{Adjacent}}=\frac{c}{a}\qquad \cot\theta=\frac{\text{Adjacent}}{\text{Opposite}}=\frac{a}{b}$$

Notice that each of the trigonometric functions of the acute angle θ is positive.

Example 1 Find the value of each of the six trigonometric functions of the angle θ in Figure 18.

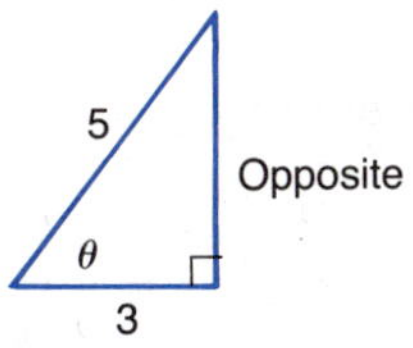

Figure 18

Solution We see in Figure 18 that the two given sides of the triangle are

$$c = \text{Hypotenuse} = 5 \qquad a = \text{Adjacent} = 3$$

To find the length of the opposite side, we use the Pythagorean Theorem:

$$(\text{Adjacent})^2 + (\text{Opposite})^2 = (\text{Hypotenuse})^2$$
$$3^2 + (\text{Opposite})^2 = 5^2$$
$$(\text{Opposite})^2 = 25 - 9 = 16$$
$$\text{Opposite} = 4$$

Now that we know the lengths of the three sides, we use the ratios in (1) to find the value of each of the six trigonometric functions:

$$\sin\theta = \frac{\text{Opposite}}{\text{Hypotenuse}} = \frac{4}{5} \qquad \cos\theta = \frac{\text{Adjacent}}{\text{Hypotenuse}} = \frac{3}{5} \qquad \tan\theta = \frac{\text{Opposite}}{\text{Adjacent}} = \frac{4}{3}$$
$$\csc\theta = \frac{\text{Hypotenuse}}{\text{Opposite}} = \frac{5}{4} \qquad \sec\theta = \frac{\text{Hypotenuse}}{\text{Adjacent}} = \frac{5}{3} \qquad \cot\theta = \frac{\text{Adjacent}}{\text{Opposite}} = \frac{3}{4}$$

■

Fundamental Identities

You should have observed some relationships that exist among the six trigonometric functions of acute angles. For example, the **reciprocal identities** are

Reciprocal Identities

$$\csc\theta = \frac{1}{\sin\theta} \qquad \sec\theta = \frac{1}{\cos\theta} \qquad \cot\theta = \frac{1}{\tan\theta} \qquad (2)$$

Two other fundamental identities that are easy to see are

$$\tan\theta = \frac{\sin\theta}{\cos\theta} \qquad \cot\theta = \frac{\cos\theta}{\sin\theta} \qquad (3)$$

If $\sin\theta$ and $\cos\theta$ are known, formulas (2) and (3) make it easy to find the values of the remaining trigonometric functions.

Example 2 Given $\sin\theta = 1/\sqrt{5}$ and $\cos\theta = 2/\sqrt{5}$, find the value of each of the four remaining trigonometric functions of θ.

Solution Based on formula (3), we have

$$\tan \theta = \frac{\sin \theta}{\cos \theta} = \frac{1/\sqrt{5}}{2/\sqrt{5}} = \frac{1}{2}$$

Then we use the reciprocal identities from formula (2) to get

$$\csc \theta = \frac{1}{\sin \theta} = \frac{1}{1/\sqrt{5}} = \sqrt{5} \qquad \sec \theta = \frac{1}{\cos \theta} = \frac{1}{2/\sqrt{5}} = \frac{\sqrt{5}}{2} \qquad \cot \theta = \frac{1}{\tan \theta} = \frac{1}{\frac{1}{2}} = 2$$

■

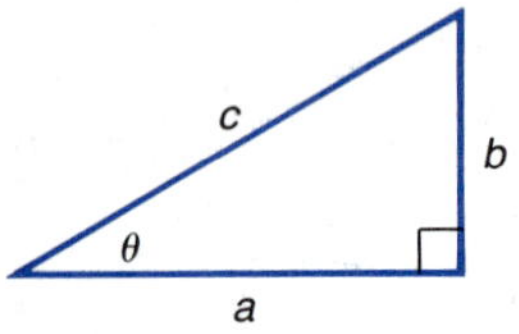

Figure 19
$a^2 + b^2 = c^2$

Refer now to the triangle in Figure 19. The Pythagoream Thoerem states that

$$b^2 + a^2 = c^2$$

Dividing each side by c^2, we get

$$\frac{b^2}{c^2} + \frac{a^2}{c^2} = 1 \quad \text{or} \quad \left(\frac{b}{c}\right)^2 + \left(\frac{a}{c}\right)^2 = 1$$

In terms of trigonometric functions of the angle θ, this equation states that

$$(\sin \theta)^2 + (\cos \theta)^2 = 1 \tag{4}$$

Equation (4) is, in fact, an identity, since the equation is true for any acute angle θ.

It is customary to write $\sin^2 \theta$ instead of $(\sin \theta)^2$, $\cos^2 \theta$ instead of $(\cos \theta)^2$, and so on. With this notation, we can rewrite equation (4) as

$$\sin^2 \theta + \cos^2 \theta = 1 \tag{5}$$

Another identity can be obtained from equation (5) by dividing each side by $\cos^2 \theta$:

$$\frac{\sin^2 \theta}{\cos^2 \theta} + 1 = \frac{1}{\cos^2 \theta}$$

Now use formulas (2) and (3) to get

$$\tan^2 \theta + 1 = \sec^2 \theta \tag{6}$$

Similarly, by dividing each side of equation (5) by $\sin^2 \theta$, we get

$$1 + \cot^2 \theta = \csc^2 \theta \qquad (7)$$

Collectively, the identities in equations (5), (6), and (7) are referred to as the **Pythagorean identities**.

Let's pause here to summarize the fundamental identities.

Fundamental Identities

$$\tan \theta = \frac{\sin \theta}{\cos \theta} \qquad \cot \theta = \frac{\cos \theta}{\sin \theta}$$

$$\cot \theta = \frac{1}{\tan \theta} \qquad \sec \theta = \frac{1}{\cos \theta} \qquad \csc \theta = \frac{1}{\sin \theta}$$

$$\sin^2 \theta + \cos^2 \theta = 1 \qquad \tan^2 \theta + 1 = \sec^2 \theta \qquad 1 + \cot^2 \theta = \csc^2 \theta$$

Once the value of one trigonometric function is known, it is possible to find the value of each of the remaining five trigonometric functions.

Example 3 Given that $\sin \theta = \frac{1}{3}$ and θ is an acute angle, find the value of each of the remaining five trigonometric functions of θ.

We solve this problem in two ways: the first way uses the definition of the trigonometric functions; the second method uses the fundamental identities.

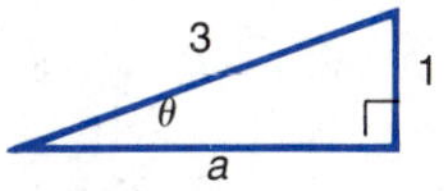

Figure 20

Solution 1 We draw a right triangle with acute angle θ, opposite side of length 1, and hypotenuse of length 3 (because $\sin \theta = \frac{1}{3} = b/c$). See Figure 20. The adjacent side a can be found by using the Pythagorean Theorem:

$$a^2 + 1 = 9$$
$$a^2 = 8$$
$$a = 2\sqrt{2}$$

Now the definitions given in equation (1) can be used to find the value of each of the remaining five trigonometric functions. (Refer back to the method used in Example 1.)

$$\cos \theta = \frac{2\sqrt{2}}{3} \qquad \tan \theta = \frac{1}{2\sqrt{2}} = \frac{\sqrt{2}}{4}$$

$$\csc \theta = \frac{3}{1} = 3 \qquad \sec \theta = \frac{3}{2\sqrt{2}} = \frac{3\sqrt{2}}{4} \qquad \cot \theta = \frac{2\sqrt{2}}{1} = 2\sqrt{2}$$

■

Solution 2 We begin by seeking $\cos\theta$, which can be found by using equation (5):

$$\sin^2\theta + \cos^2\theta = 1$$
$$\tfrac{1}{9} + \cos^2\theta = 1$$
$$\cos^2\theta = 1 - \tfrac{1}{9} = \tfrac{8}{9}$$

Because $\cos\theta > 0$ for an acute angle θ, we have

$$\cos\theta = \sqrt{\frac{8}{9}} = \frac{2\sqrt{2}}{3}$$

Now we know that $\sin\theta = \frac{1}{3}$ and $\cos\theta = 2\sqrt{2}/3$, so we can proceed as we did in Example 2:

$$\tan\theta = \frac{\sin\theta}{\cos\theta} = \frac{\frac{1}{3}}{2\sqrt{2}/3} = \frac{1}{2\sqrt{2}} = \frac{\sqrt{2}}{4} \qquad \cot\theta = \frac{1}{\tan\theta} = \frac{1}{1/2\sqrt{2}} = 2\sqrt{2}$$

$$\sec\theta = \frac{1}{\cos\theta} = \frac{1}{2\sqrt{2}/3} = \frac{3}{2\sqrt{2}} = \frac{3\sqrt{2}}{4} \qquad \csc\theta = \frac{1}{\sin\theta} = \frac{1}{\frac{1}{3}} = 3$$

■

Finding the Values of the Trigonometric Functions When One Is Known

Given the value of one trigonometric function, the value of each of the remaining five trigonometric functions can be found in either of two ways:

METHOD 1

STEP 1: Draw a right triangle showing the angle θ.
STEP 2: Two of the sides can then be assigned values based on the given trigonometric function.
STEP 3: Find the length of the third side by using the Pythagorean Theorem.
STEP 4: Use the definitions in equation (1) to find the value of each of the remaining trigonometric functions.

METHOD 2

Use appropriately selected identities to find the value of each of the remaining trigonometric functions.

Complementary Angles; Cofunctions

Two acute angles are called **complementary** if their sum is a right angle. Because the sum of the angles of any triangle is 180°, it follows that, for a right triangle, the two acute angles are complementary.

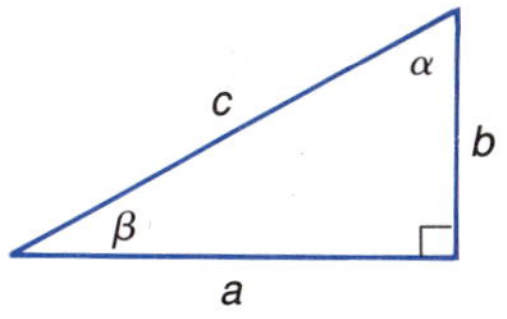

Figure 21

Refer now to Figure 21; we have labeled the angle opposite side b as β and the angle opposite side a as α. Notice that side a is adjacent to angle β and is opposite angle α. Similarly, side b is opposite angle β and is adjacent to angle α. As a result,

$$\sin\beta = \frac{b}{c} = \cos\alpha \qquad \cos\beta = \frac{a}{c} = \sin\alpha \qquad \tan\beta = \frac{b}{a} = \cot\alpha$$
$$\csc\beta = \frac{c}{b} = \sec\alpha \qquad \sec\beta = \frac{c}{a} = \csc\alpha \qquad \cot\beta = \frac{a}{b} = \tan\alpha \tag{8}$$

Because of these relationships, the functions sine and cosine, tangent and cotangent, and secant and cosecant are called **cofunctions** of each other. The identities (8) may be expressed in words as follows:

Theorem Cofunctions of complementary angles are equal. ■

Examples of this theorem are given below:

Complementary angles: $\sin 30° = \cos 60°$ — Cofunctions

Complementary angles: $\tan 40° = \cot 50°$ — Cofunctions

Complementary angles: $\sec 80° = \csc 10°$ — Cofunctions

If an angle θ is measured in degrees, we shall use the degree symbol when writing a trigonometric function of θ, as, for example, in $\sin 30°$, $\tan 45°$, etc. If an angle θ is measured in radians, then no symbol is used when writing a trigonometric function of θ, as, for example, in $\cos\pi$, $\sec\pi/3$, etc.

If θ is an acute angle measured in degrees, the angle $90° - \theta$ (or $\pi/2 - \theta$, if θ is in radians) is the angle complementary to θ. Table 2 restates the above theorem on cofunctions.

Table 2

θ (DEGREES)	θ (RADIANS)
$\sin\theta = \cos(90° - \theta)$	$\sin\theta = \cos(\pi/2 - \theta)$
$\cos\theta = \sin(90° - \theta)$	$\cos\theta = \sin(\pi/2 - \theta)$
$\tan\theta = \cot(90° - \theta)$	$\tan\theta = \cot(\pi/2 - \theta)$
$\csc\theta = \sec(90° - \theta)$	$\csc\theta = \sec(\pi/2 - \theta)$
$\sec\theta = \csc(90° - \theta)$	$\sec\theta = \csc(\pi/2 - \theta)$
$\cot\theta = \tan(90° - \theta)$	$\cot\theta = \tan(\pi/2 - \theta)$

Example 4 (a) $\sin 62° = \cos(90° - 62°) = \cos 28°$

(b) $\tan\frac{\pi}{12} = \cot\left(\frac{\pi}{2} - \frac{\pi}{12}\right) = \cot\frac{5\pi}{12}$

(c) $\cos\frac{\pi}{4} = \sin\left(\frac{\pi}{2} - \frac{\pi}{4}\right) = \sin\frac{\pi}{4}$ (d) $\csc\frac{\pi}{6} = \sec\left(\frac{\pi}{2} - \frac{\pi}{6}\right) = \sec\frac{\pi}{3}$ ■

Historical Comment

■ The name *sine* for the sine function is due to a medieval confusion. The name comes from the Sanskrit word *jīva* (meaning chord), first used in India by Aryabhata the Elder (AD 510). He really meant half-chord, but abbreviated it. This was brought into Arabic as *jība*, which was meaningless. Because the proper Arabic word *jaib* would be written the same way (short vowels are not written out in Arabic), *jība* was pronounced as *jaib*, which meant bosom or hollow, and *jaib* remains as the Arabic word for sine to this day. Scholars translating the Arabic works into Latin found that the word *sinus* also meant bosom or hollow, and from *sinus* we get the word *sine*.

The name *tangent*, due to Thomas Fincke (1583), can be understood by looking at Figure 22. The line segment $\overline{DC}$ is tangent to the circle at C. If $d(O, B) = d(O, C) = 1$, then the length of the line segment $\overline{DC}$ is

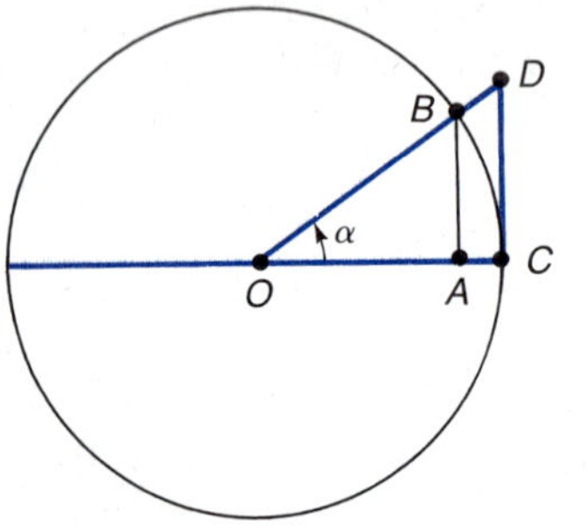

Figure 22

$$d(D, C) = \frac{d(D, C)}{1} = \frac{d(D, C)}{d(O, C)} = \tan \alpha$$

The old name for the tangent is *umbra versa* (meaning turned shadow), referring to the use of the tangent in solving height problems with shadows.

The names of the cofunctions came about as follows. If α and β are complementary angles, then $\cos \alpha = \sin \beta$. Because β is the complement of α, it was natural to write the cosine of α as *sin co* α. Probably for reasons involving ease of pronunciation, the *co* migrated to the front, and then cosine received a three-letter abbreviation to match sin, sec, and tan. The two other cofunctions were similarly treated, except that the long forms *cotan* and *cosec* survive to this day in some countries. ■

EXERCISE 7.2

In Problems 1–10, find the value of each of the six trigonometric functions of the angle θ in each figure.

1.
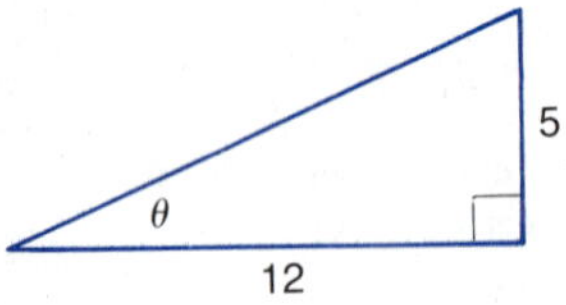

2.
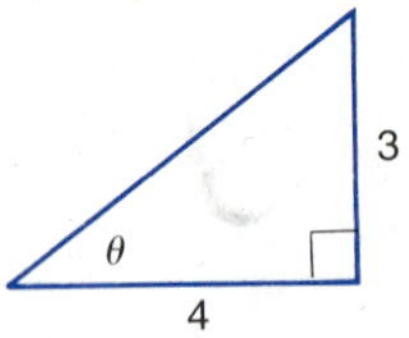

3.
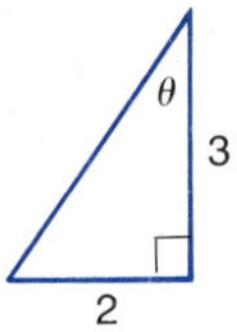

4.
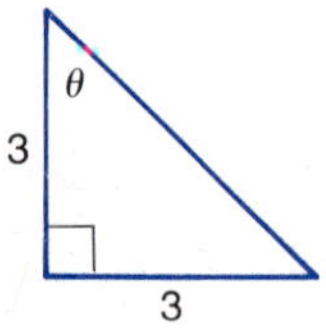

5.
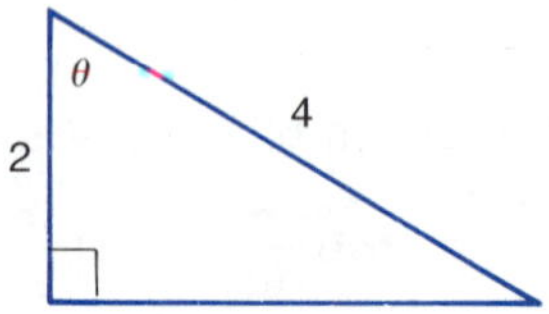

6.
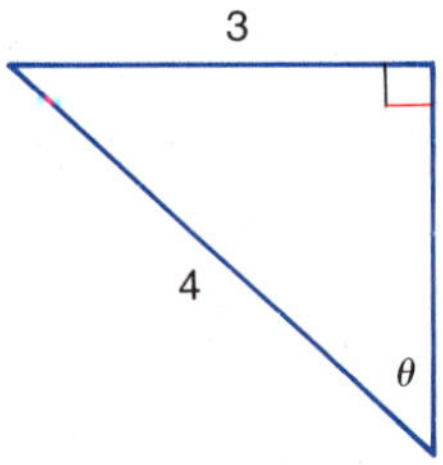

7.

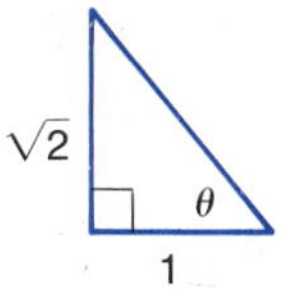

8.

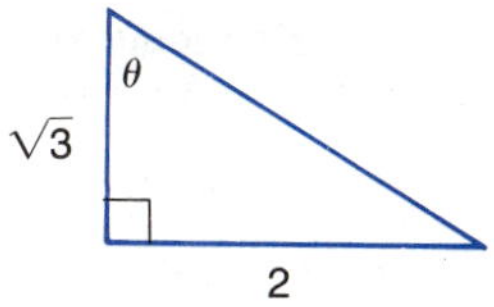

9.

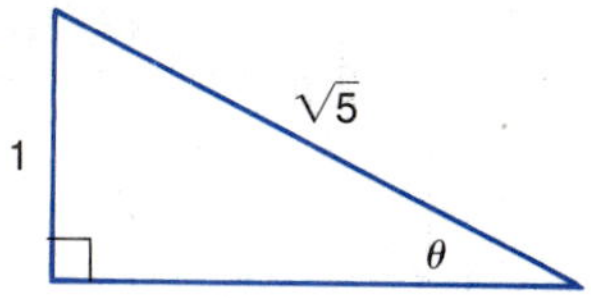

10.

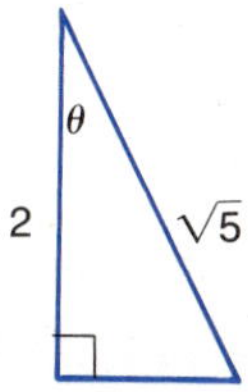

In Problems 11–14, find the value of each of the four remaining trigonometric functions of the acute angle θ.

11. $\sin\theta = \frac{1}{2}$, $\cos\theta = \sqrt{3}/2$

12. $\sin\theta = \sqrt{3}/2$, $\cos\theta = \frac{1}{2}$

13. $\sin\theta = \frac{1}{3}$, $\cos\theta = 2\sqrt{2}/3$

14. $\sin\theta = \frac{2}{3}$, $\cos\theta = \sqrt{5}/3$

In Problems 15–24, find the value of each of the remaining five trigonometric functions of the acute angle θ.

15. $\sin\theta = \sqrt{2}/2$
16. $\cos\theta = \sqrt{2}/2$
17. $\cos\theta = \frac{1}{3}$
18. $\sin\theta = \sqrt{3}/4$
19. $\tan\theta = \frac{1}{2}$
20. $\cot\theta = \frac{1}{2}$
21. $\sec\theta = 3$
22. $\csc\theta = 5$
23. $\tan\theta = \sqrt{2}$
24. $\sec\theta = \frac{5}{3}$

In Problems 25–40, find the value of each expression.

25. $\sin^2 40° + \cos^2 40°$
26. $\sec^2 18° - \tan^2 18°$
27. $\sin 80° \csc 80°$
28. $\tan 10° \cot 10°$
29. $\tan 40° - \dfrac{\sin 40°}{\cos 40°}$
30. $\cot 20° - \dfrac{\cos 20°}{\sin 20°}$
31. $\sin 38° - \cos 52°$
32. $\tan 12° - \cot 78°$
33. $\dfrac{\cos 10°}{\sin 80°}$
34. $\dfrac{\cos 40°}{\sin 50°}$
35. $1 - \cos^2 20° - \cos^2 70°$
36. $1 + \tan^2 5° - \csc^2 85°$
37. $\tan 20° - \dfrac{\cos 70°}{\cos 20°}$
38. $\cot 40° - \dfrac{\sin 50°}{\sin 40°}$
39. $\cos 35° \sin 55° + \cos 55° \sin 35°$
40. $\sec 35° \csc 55° - \tan 35° \cot 55°$

41. Given $\sin 30° = \frac{1}{2}$, use trigonometric identities to find the value of:

(a) $\cos 60°$ (b) $\cos^2 30°$ (c) $\csc \dfrac{\pi}{6}$ (d) $\sec \dfrac{\pi}{3}$

42. Given $\sin 60° = \sqrt{3}/2$, use trigonometric identities to find the value of:

(a) $\cos 30°$ (b) $\cos^2 60°$ (c) $\sec \dfrac{\pi}{6}$ (d) $\csc \dfrac{\pi}{3}$

43. Given $\tan\theta = 4$, use trigonometric identities to find the value of:

(a) $\sec^2\theta$ (b) $\cot\theta$ (c) $\cot\left(\dfrac{\pi}{2} - \theta\right)$ (d) $\csc^2\theta$

44. Given $\sec\theta = 3$, use trigonometric identities to find the value of:
(a) $\cos\theta$ (b) $\tan^2\theta$ (c) $\csc(90° - \theta)$ (d) $\sin^2\theta$

45. Given $\csc\theta = 4$, use trigonometric identities to find the value of:
(a) $\sin\theta$ (b) $\cot^2\theta$ (c) $\sec(90° - \theta)$ (d) $\sec^2\theta$

46. Given $\cot\theta = 2$, use trigonometric identities to find the value of:
(a) $\tan\theta$ (b) $\csc^2\theta$ (c) $\tan\left(\frac{\pi}{2} - \theta\right)$ (d) $\sec^2\theta$

C **47.** Given the approximation $\sin 38° \approx 0.6157$, use trigonometric identities to find the approximate value of:
(a) $\cos 38°$ (b) $\tan 38°$ (c) $\cot 38°$ (d) $\sec 38°$
(e) $\csc 38°$ (f) $\sin 52°$ (g) $\cos 52°$ (h) $\tan 52°$

C **48.** Given the approximation $\cos 21° \approx 0.9336$, use trigonometric identities to find the approximate value of:
(a) $\sin 21°$ (b) $\tan 21°$ (c) $\cot 21°$ (d) $\sec 21°$
(e) $\csc 21°$ (f) $\sin 69°$ (g) $\cos 69°$ (h) $\tan 69°$

49. If $\sin\theta = 0.3$, find the value of: $\sin\theta + \cos\left(\frac{\pi}{2} - \theta\right)$

50. If $\tan\theta = 4$, find the value of: $\tan\theta + \tan\left(\frac{\pi}{2} - \theta\right)$

51. Find an acute angle θ that satisfies the equation: $\sin\theta = \cos(2\theta + 30°)$

52. Find an acute angle θ that satisfies the equation: $\tan\theta = \cot(\theta + 45°)$

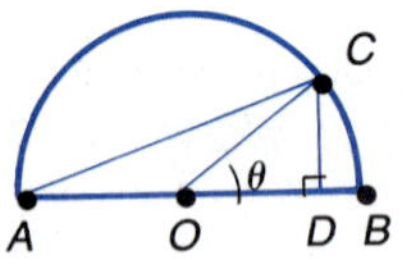

53. Suppose the angle θ is a central angle of a circle of radius 1 (see the figure). Show that:
(a) Angle $OAC = \theta/2$ (b) $|CD| = \sin\theta$ and $|OD| = \cos\theta$ (c) $\tan\dfrac{\theta}{2} = \dfrac{\sin\theta}{1 + \cos\theta}$

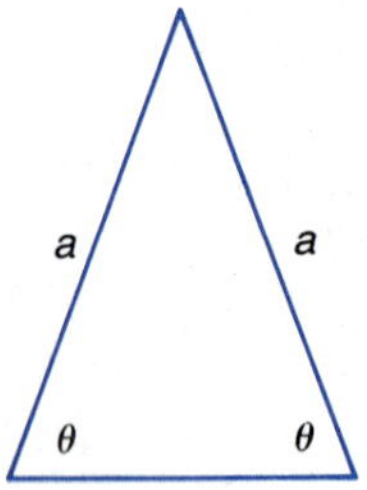

54. Show that the area of an isosceles triangle is $A = a^2 \sin\theta\cos\theta$, where a is the length of one of the two equal sides and θ is the measure of one of the two equal angles (see the figure).

55. Let $n > 0$ be any real number, and let θ be any angle for which $0 < \theta < \pi/(1 + n)$. Then we can construct a triangle with the angles θ and $n\theta$ and included side of length 1 (do you see why?) and place it on the unit circle as illustrated. Now, drop the perpendicular from C to $D = (x, 0)$, and show that

$$x = \frac{\tan n\theta}{\tan\theta + \tan n\theta}$$

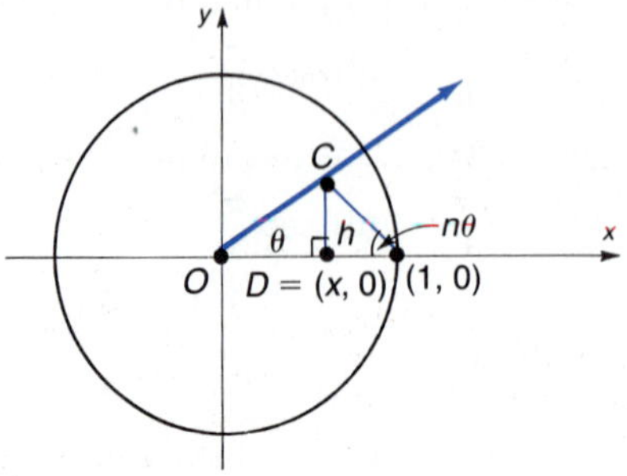

56. Refer to the figure below. The smaller circle, whose radius is a, is tangent to the larger circle, whose radius is b. The ray OA contains a diameter of each circle, and the ray OB is tangent to each circle. Show that

$$\cos\theta = \frac{\sqrt{ab}}{\dfrac{a+b}{2}}$$

(that is, $\cos\theta$ equals the ratio of the geometric mean of a and b to the arithmetic mean of a and b). [*Hint:* First show that $\sin\theta = (b-a)/(b+a)$.]

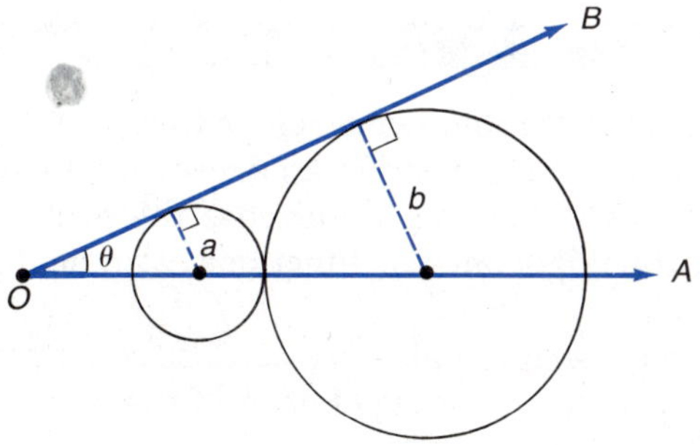

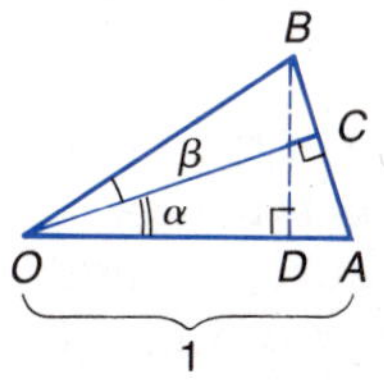

57. Refer to the figure in the margin. If $|OA| = 1$, show that:
(a) Area $\triangle OAC = \frac{1}{2}\sin\alpha\cos\alpha$ (b) Area $\triangle OCB = \frac{1}{2}|OB|^2\sin\beta\cos\beta$
(c) Area $\triangle OAB = \frac{1}{2}|OB|\sin(\alpha+\beta)$ (d) $|OB| = \dfrac{\cos\alpha}{\cos\beta}$
(e) $\sin(\alpha+\beta) = \sin\alpha\cos\beta + \cos\alpha\sin\beta$
[*Hint:* Area $\triangle OAB$ = Area $\triangle OAC$ + Area $\triangle OCB$]

58. Refer to the figure below, where a unit circle is drawn. The line DB is tangent to the circle.
(a) Express the area of $\triangle OBC$ in terms of $\sin\theta$ and $\cos\theta$.
(b) Express the area of $\triangle OBD$ in terms of $\sin\theta$ and $\cos\theta$.
(c) The area of the sector of the circle $\widehat{OBC}$ is $\frac{1}{2}\theta$, where θ is measured in radians. Use the results of parts (a) and (b) and the fact that

$$\text{Area } \triangle OBC < \text{Area } \widehat{OBC} < \text{Area } \triangle OBD$$

to show that

$$1 < \frac{\theta}{\sin\theta} < \frac{1}{\cos\theta}$$

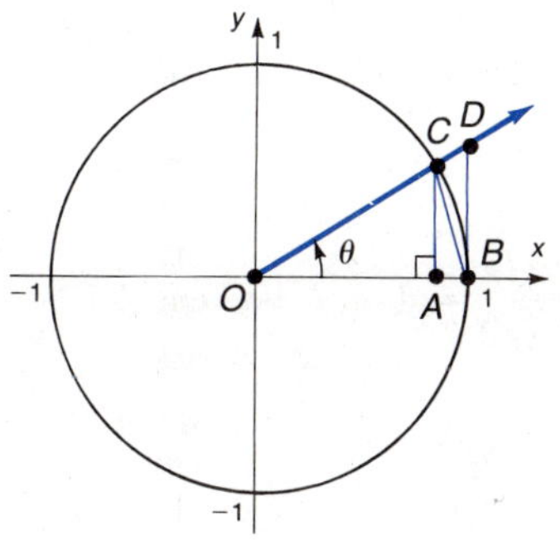

59. If $\cos \alpha = \tan \beta$ and $\cos \beta = \tan \alpha$, where α and β are acute angles, show that

$$\sin \alpha = \sin \beta = \sqrt{\frac{3 - \sqrt{5}}{2}}$$

60. Explain why, if θ is an acute angle, then $\sec \theta > 1$.

61. Explain why, if θ is an acute angle, then $0 < \sin \theta < 1$.

7.3 Computing the Values of Trigonometric Functions of Given Angles

In the previous section, we developed ways to find the value of each of the trigonometric functions of an acute angle when one of the functions is known. In this section, we discuss the problem of finding the value of each of the trigonometric functions of an acute angle when the angle is given.

For three special angles, we can use some results from plane geometry to find the exact value of each of the six trigonometric functions.

45°, 30°, and 60° Angles

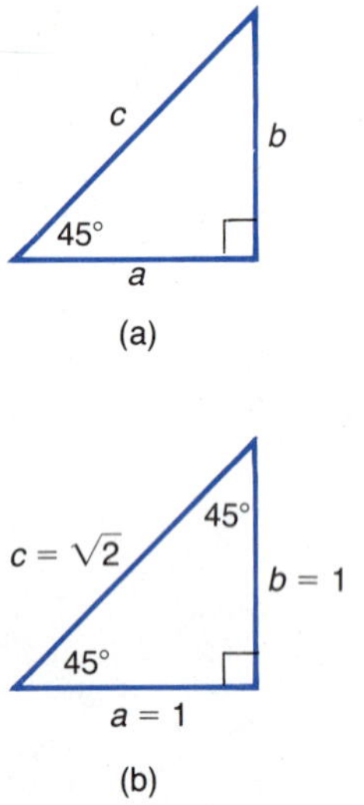

Figure 23

Consider the right triangle in Figure 23(a), in which one of the angles is 45°. It follows that the other acute angle is also 45°, and hence, the triangle is isosceles. As a result, side a and side b are equal in length. Because the values of the trigonometric functions of an angle depend only on the angle and not on the size of the triangle, we may choose to use the triangle for which

$$a = b = 1$$

Then, by the Pythagorean Theorem,

$$c^2 = a^2 + b^2 = 1 + 1 = 2$$
$$c = \sqrt{2}$$

As a result, we have the triangle in Figure 23(b), from which we find

$$\sin 45° = \frac{b}{c} = \frac{1}{\sqrt{2}} = \frac{\sqrt{2}}{2} \qquad \cos 45° = \frac{a}{c} = \frac{1}{\sqrt{2}} = \frac{\sqrt{2}}{2}$$

Thus,

$$\tan 45° = \frac{\sin 45°}{\cos 45°} = \frac{\sqrt{2}/2}{\sqrt{2}/2} = 1 \qquad \cot 45° = \frac{1}{\tan 45°} = \frac{1}{1} = 1$$

$$\sec 45° = \frac{1}{\cos 45°} = \frac{1}{1/\sqrt{2}} = \sqrt{2} \qquad \csc 45° = \frac{1}{\sin 45°} = \frac{1}{1/\sqrt{2}} = \sqrt{2}$$

Now, consider a right triangle in which one of the angles is 30°. It then follows that the other angle is 60°. Figure 24(a) illustrates such a triangle with hypotenuse of length 2. Our problem is to determine a and b.

Figure 24

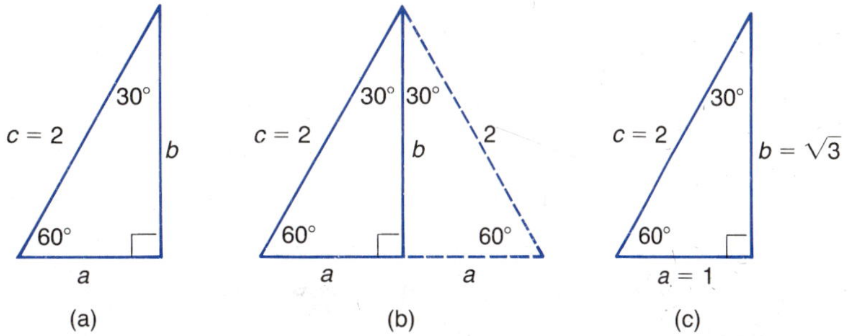

We begin by placing next to the triangle in Figure 24(a) another triangle congruent to the first, as shown in Figure 24(b). Notice that we now have a triangle whose angles are each 60°. This triangle is therefore equilateral, so each side is of length 2. In particular, the base is $2a = 2$, and so $a = 1$. By the Pythagorean Theorem, b satisfies the equation $a^2 + b^2 = c^2$, so we have

$$\begin{aligned} a^2 + b^2 &= c^2 \\ 1^2 + b^2 &= 2^2 \\ b^2 &= 4 - 1 = 3 \\ b &= \sqrt{3} \end{aligned}$$

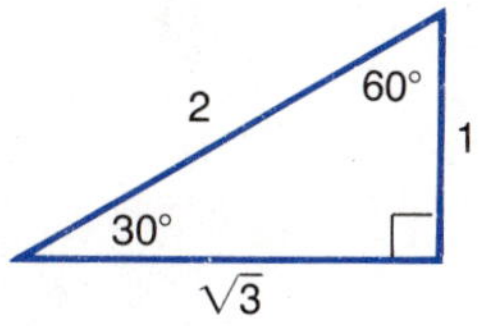

Figure 25

Using the triangle in Figure 25 and the fact that 30° and 60° are complementary angles, we find

$$\sin 30° = \frac{\text{Opposite}}{\text{Hypotenuse}} = \frac{1}{2} \qquad \cos 60° = \frac{1}{2}$$

$$\cos 30° = \frac{\text{Adjacent}}{\text{Hypotenuse}} = \frac{\sqrt{3}}{2} \qquad \sin 60° = \frac{\sqrt{3}}{2}$$

$$\tan 30° = \frac{\sin 30°}{\cos 30°} = \frac{\frac{1}{2}}{\sqrt{3}/2} = \frac{1}{\sqrt{3}} = \frac{\sqrt{3}}{3} \qquad \cot 60° = \frac{\sqrt{3}}{3}$$

$$\csc 30° = \frac{1}{\sin 30°} = \frac{1}{\frac{1}{2}} = 2 \qquad \sec 60° = 2$$

$$\sec 30° = \frac{1}{\cos 30°} = \frac{1}{\sqrt{3}/2} = \frac{2}{\sqrt{3}} = \frac{2\sqrt{3}}{3} \qquad \csc 60° = \frac{2\sqrt{3}}{3}$$

$$\cot 30° = \frac{1}{\tan 30°} = \frac{1}{\sqrt{3}/3} = \frac{3}{\sqrt{3}} = \sqrt{3} \qquad \tan 60° = \sqrt{3}$$

Table 3 summarizes the information just derived for the angles 30°, 45°, and 60°. Until you memorize the entries in Table 3, you should draw the appropriate triangle to determine the values given in the table.

Table 3

θ (RADIANS)	θ (DEGREES)	$\sin\theta$	$\cos\theta$	$\tan\theta$	$\csc\theta$	$\sec\theta$	$\cot\theta$
$\pi/6$	30°	$\frac{1}{2}$	$\sqrt{3}/2$	$\sqrt{3}/3$	2	$2\sqrt{3}/3$	$\sqrt{3}$
$\pi/4$	45°	$\sqrt{2}/2$	$\sqrt{2}/2$	1	$\sqrt{2}$	$\sqrt{2}$	1
$\pi/3$	60°	$\sqrt{3}/2$	$\frac{1}{2}$	$\sqrt{3}$	$2\sqrt{3}/3$	2	$\sqrt{3}/3$

Example 1 Find the value of each expression:

(a) $\sin 45° \cos 30°$ (b) $\tan\dfrac{\pi}{4} - \sin\dfrac{\pi}{3}$

Solution (a) $\sin 45° \cos 30° = \dfrac{\sqrt{2}}{2}\cdot\dfrac{\sqrt{3}}{2} = \dfrac{\sqrt{6}}{4}$

(b) $\tan\dfrac{\pi}{4} - \sin\dfrac{\pi}{3} = 1 - \dfrac{\sqrt{3}}{2} = \dfrac{2-\sqrt{3}}{2}$ ■

The values of the trigonometric functions for the angles 30°, 45°, and 60° are relatively easy to calculate, because the triangles that contain such angles have "nice" geometric features. To find the values of the trigonometric functions of most other acute angles, we will need either a calculator or a table of values of trigonometric functions.

Using a Calculator to Find Values of Trigonometric Functions

Before getting started, you must first decide whether to enter the angle in the calculator using radians or degrees and then set the calculator to the correct mode. If your calculator does not display the mode, you can determine the current mode by entering 30 and then pressing the key marked [sin]. If you are in the degree mode, the display will show [0.5] ($\sin 30° = 0.5$). If you are in the radian mode, the display will show [−0.9880316]. Most calculators have a key that allows you to change from one mode to the other. (Check your instruction manual to find out how your calculator handles degrees and radians.)

C **Example 2** Use a calculator to find the approximate value of:

(a) $\cos 48°$ (b) $\csc 21°$ (c) $\tan\dfrac{\pi}{12}$

Solution (a) First, we set the mode to receive degrees.

Enter:	48	Press:	cos*
Display:	48		0.6691306

Thus, $$\cos 48° \approx 0.6691306 \approx 0.6691$$

rounded to four decimal places.

(b) Most calculators do not have a csc key. The manufacturers assume the user knows some trigonometry. Thus, to find the value of csc 21°, we use the fact that csc 21° = 1/(sin 21°) and proceed as follows:

Enter:	21	Press:	sin
Display:	21		0.3583679
Press:	1/x		
Display:	2.7904281		

Thus, $$\csc 21° \approx 2.7904281 \approx 2.7904$$

rounded to four decimal places.

(c) Set the mode to receive radians. Then, to find tan(π/12),

Enter:	π	Press:	÷	Enter:	12
Display:	3.1415927				12
Press:	=	Press:	tan		
Display:	0.2617994		0.2679492		

Thus, $$\tan \frac{\pi}{12} \approx 0.2679492 \approx 0.2679$$

rounded to four decimal places. ■

*On some calculators, you may have to press a key labeled 2nd before pressing cos. For example, on the TI57, the cosine function is a second function associated with the $\sqrt{x}$ key, appearing as cos / $\sqrt{x}$ on the keyboard.

Tables of Values of Trigonometric Functions (Optional)

Before calculators were readily available, tables containing values of the trigonometric functions were used. Such a table is included in the back of this book (Table IV). A portion of Table IV is reproduced here. This particular table is called a four-place table, because values are rounded to four decimal places.

Table IV Trigonometric Functions

DEGREES	RADIANS	sin	cos	tan	cot		
35	0.6109	0.5736	0.8192	0.7002	1.4280	0.9599	55
36	0.6283	0.5878	0.8090	0.7265	1.3764	0.9425	54
37	0.6458	0.6018	0.7986	0.7536	1.3270	0.9250	53
38	0.6632	0.6157	0.7880	0.7813	1.2799	0.9076	52
39	0.6807	0.6293	0.7771	0.8098	1.2349	0.8901	51
40	0.6981	0.6428	0.7660	0.8391	1.1918	0.8727	50
41	0.7156	0.6561	0.7547	0.8693	1.1504	0.8552	49
42	0.7330	0.6691	0.7431	0.9004	1.1106	0.8378	48
43	0.7505	0.6820	0.7314	0.9325	1.0724	0.8203	47
44	0.7679	0.6947	0.7193	0.9657	1.0355	0.8029	46
45	0.7854	0.7071	0.7071	1.0000	1.0000	0.7854	45
		cos	sin	cot	tan	RADIANS	DEGREES

To find the approximate value of a trigonometric function of an angle between 0° and 45° (or between 0 and $\pi/4 \approx 0.7854$), read down the left column, stopping at the desired angle, and then read across, stopping under the desired function (located in the top row). As you can see, the table is quite efficient and utilizes the fact that cofunctions of complementary angles are equal. Thus, to obtain a value of a trigonometric function of an angle between 45° and 90°, read from the right column (these angles are complementary to the angles in the left column) and locate the desired function using the entries in the bottom row (these are the cofunctions of those in the top row).

Example 3 Use Table IV to find the approximate value of:

(a) cos 36° (b) tan 41° (c) sin 51°

Solution (a) Locate 36° in the left column. Read across to the column labeled cos in the top row. You should see 0.8090. Thus,

$$\cos 36° \approx 0.8090$$

(b) Locate 41° in the left column. Read across to the column labeled tan in the top row. Then

$$\tan 41° \approx 0.8693$$

(c) Locate 51° in the right column. Read across to the column labeled sin in the bottom row. Then

$$\sin 51° \approx 0.7771$$

EXERCISE 7.3

1. Write down the value of each of the six trigonometric functions of 45°.
2. Write down the value of each of the six trigonometric functions of 30°.
3. Write down the value of each of the six trigonometric functions of 60°.

In Problems 4–14, evaluate the given expression. Do not use a calculator or a table.

4. $2 \sin 45° - 4 \sin 30°$ **5.** $6 \tan 45° + 8 \cos 60°$ **6.** $\sin 30° \cdot \tan 60°$

7. $\sec \frac{\pi}{4} - 2 \csc \frac{\pi}{3}$ **8.** $\tan \frac{\pi}{4} + \cot \frac{\pi}{4}$ **9.** $\sec^2 \frac{\pi}{6} - 4$

10. $4 + \tan^2 \frac{\pi}{3}$ **11.** $\sin^2 40° + \cos^2 40°$ **12.** $\sec^2 18° - \tan^2 18°$

13. $1 - \cos^2 20° - \cos^2 70°$ **14.** $1 + \tan^2 5° - \csc^2 85°$

In Problems 15–24, find the value of each expression if $\theta = 60°$.

15. $\sin \theta$ **16.** $\cos \theta$ **17.** $\sin \frac{\theta}{2}$ **18.** $\cos \frac{\theta}{2}$ **19.** $(\sin \theta)^2$

20. $(\cos \theta)^2$ **21.** $2 \sin \theta$ **22.** $2 \cos \theta$ **23.** $\frac{\sin \theta}{2}$ **24.** $\frac{\cos \theta}{2}$

C *In Problems 25–48, use a calculator to find the approximate value of each expression rounded to four decimal places.*

25. $\sin 28°$ **26.** $\cos 14°$ **27.** $\tan 21°$ **28.** $\sin 15°$ **29.** $\sec 41°$

30. $\csc 55°$ **31.** $\cot 70°$ **32.** $\tan 80°$ **33.** $\sin \frac{\pi}{10}$ **34.** $\cos \frac{\pi}{8}$

35. $\tan \frac{5\pi}{12}$ **36.** $\sin \frac{3\pi}{10}$ **37.** $\sec \frac{\pi}{12}$ **38.** $\csc \frac{5\pi}{13}$ **39.** $\cot \frac{\pi}{18}$

40. $\sin \frac{\pi}{18}$ **41.** $\sin 1$ **42.** $\tan 1$ **43.** $\sin 1°$ **44.** $\tan 1°$

45. $\cos 21.5°$ **46.** $\cos 35.2°$ **47.** $\tan 0.3$ **48.** $\tan 0.1$

In Problems 49–64, use Table IV in the back of the book to find the approximate value of each expression.

49. $\sin 28°$ **50.** $\cos 14°$ **51.** $\tan 21°$ **52.** $\sin 15°$ **53.** $\cos 41°$

54. $\cot 55°$ **55.** $\cot 70°$ **56.** $\tan 80°$ **57.** $\sin 0.6981$ **58.** $\cos 0.3491$

59. $\tan 0.1745$ **60.** $\sin 0.4363$ **61.** $\cos 1.0472$ **62.** $\cot 0.8727$ **63.** $\cot 0.6109$

64. $\tan 1.2217$

The path of a projectile fired at an inclination θ to the horizontal with initial speed v_0 is a parabola (see the figure). The range R of the projectile—that is, the horizontal distance the projectile travels—is found by using the formula

$$R = \frac{2v_0^2 \sin\theta \cos\theta}{g}$$

where $g \approx 32.2$ feet per second per second ≈ 9.8 meters per second per second is the acceleration due to gravity. The maximum height H of the projectile is

$$H = \frac{v_0^2 \sin^2\theta}{2g}$$

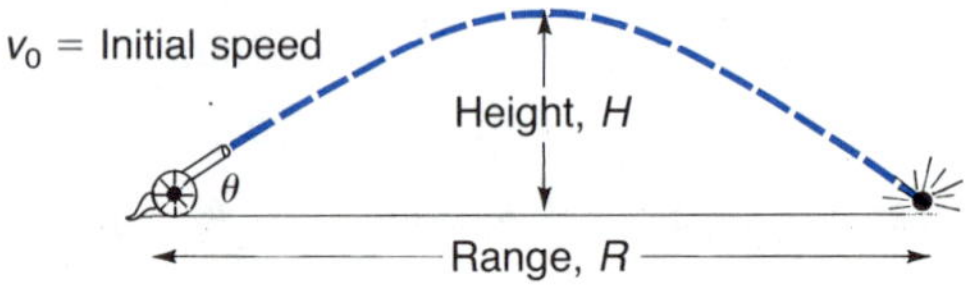

In Problems 65–68, find the range R and maximum height H.

C **65.** The projectile is fired at an angle of 45° to the horizontal with initial speed 100 feet per second.

C **66.** The projectile is fired at an angle of 30° to the horizontal with initial speed 150 meters per second.

C **67.** The projectile is fired at an angle of 25° to the horizontal with initial speed 500 meters per second.

C **68.** The projectile is fired at an angle of 50° to the horizontal with initial speed 200 feet per second.

C **69.** If friction is ignored, the time t (in seconds) required for a block to slide down an inclined plane (see the figure) is given by the formula

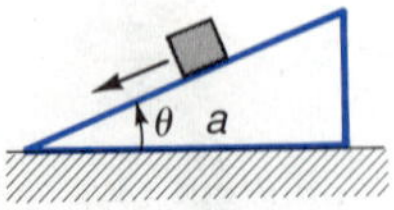

$$t = \sqrt{\frac{2a}{g \sin\theta \cos\theta}}$$

where a is the length (in feet) of the base and $g \approx 32$ feet per second per second is the acceleration of gravity. How long does it take a block to slide down an inclined plane with base $a = 10$ feet when:

(a) $\theta = 30°$? (b) $\theta = 45°$? (c) $\theta = 60°$?

C **70.** In a certain piston engine, the distance x (in meters) from the center of the drive shaft to the head of the piston is given by

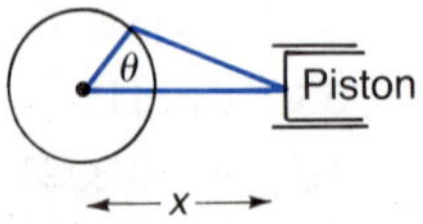

$$x = \cos\theta + \sqrt{16 + 0.5(2\cos^2\theta - 1)}$$

where θ is the angle between the crank and the path of the piston head (see the figure). Find x when $\theta = 30°$ and when $\theta = 45°$.

C **71.** Use a calculator set in radian mode to complete the table below. What can you conclude about the ratio $(\sin\theta)/\theta$ as θ approaches 0?

θ	0.5	0.4	0.2	0.1	0.01	0.001	0.0001	0.00001
$\sin\theta$								
$\dfrac{\sin\theta}{\theta}$								

C **72.** Use a calculator set in radian mode to complete the table below. What can you conclude about the ratio $(\cos\theta - 1)/\theta$ as θ approaches 0?

θ	0.5	0.4	0.2	0.1	0.01	0.001	0.0001	0.00001
$\cos\theta - 1$								
$\dfrac{\cos\theta - 1}{\theta}$								

73. Find the value of: $\tan 1° \cdot \tan 2° \cdot \tan 3° \cdot \cdots \cdot \tan 89°$

74. Find the value of: $\cot 1° \cdot \cot 2° \cdot \cot 3° \cdot \cdots \cdot \cot 89°$

75. Find the value of: $\cos 1° \cdot \cos 2° \cdot \cdots \cdot \cos 45° \cdot \csc 46° \cdot \cdots \cdot \csc 89°$

76. Find the value of: $\sin 1° \cdot \sin 2° \cdot \cdots \cdot \sin 45° \cdot \sec 46° \cdot \cdots \cdot \sec 89°$

7.4 Applications

One common use for trigonometry is to measure heights and distances that are either awkward or impossible to measure by ordinary means.

Example 1 A surveyor can measure the width of a river by setting up a transit* at a point C on one side of the river and taking a sighting of a point A on the other side. Refer to Figure 26. After turning through an angle of 90° at C, a distance of 200 meters is walked off to point B. Using the transit at B, the angle β is measured and found to be 20°. What is the width of the river?

Solution We seek the length of side b. We know a and β, so we use the relationship

$$\tan\beta = \frac{b}{a}$$

to get

$$\tan 20° = \frac{b}{200}$$

$$b = 200\tan 20° \approx 200(0.3640) = 72.8 \text{ meters}$$

Thus, the width of the river is approximately 72.8 meters. ■

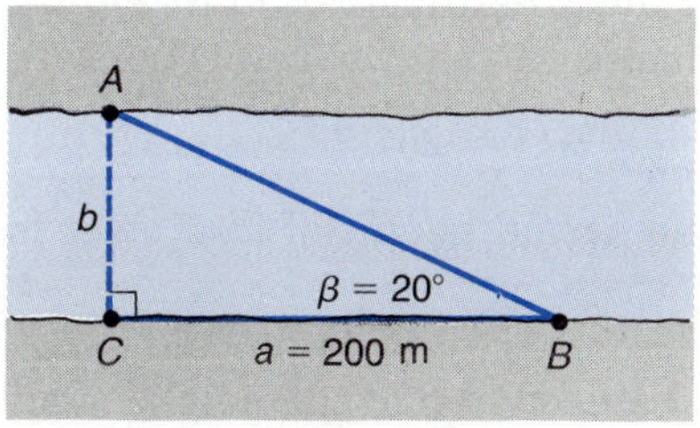

Figure 26

*An instrument used in surveying to measure angles.

Vertical heights may be measured by using what is commonly referred to as the **angle of elevation**, defined as the acute angle measured from the horizontal to a line-of-sight observation of the object. See Figure 27.

Figure 27

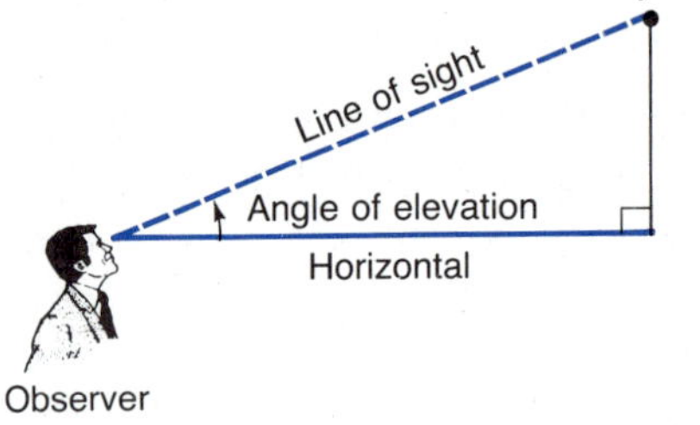

Example 2 To determine the height of a radio transmission tower, a surveyor walks off a distance of 300 meters from the base of the tower. See Figure 28(a). The angle of elevation is then measured and found to be 40°. If the transit is 2 meters off the ground when the sighting is taken, how high is the radio tower?

Figure 28

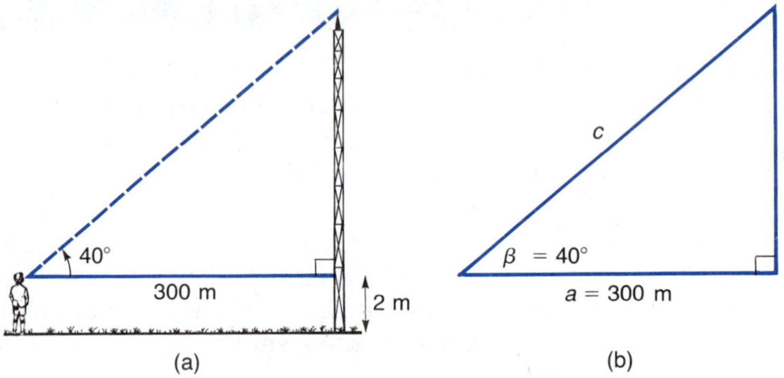

Solution Figure 28(b) shows a triangle that replicates the illustration in Figure 28(a). To find the length b, we use the fact that $\tan \beta = b/a$. Then,

$$b = a \tan \beta = 300 \tan 40° \approx 300(0.8391) = 251.73 \text{ meters}$$

Because the transit is 2 meters high, the actual height of the tower is approximately 253.73 meters. ■

The idea behind Example 2 also can be used to find the height of an object with a base that is not accessible to the horizontal.

C **Example 3** Adorning the top of the Board of Trade building in Chicago is a statue of the Greek goddess Ceres, goddess of wheat. From street level, two observations are taken 400 feet from the center of the building. The angle of elevation to the base of the statue is found to be 45°; the angle of elevation to the top of the statue is 47.2°. See Figure 29(a). What is the height of the statue?

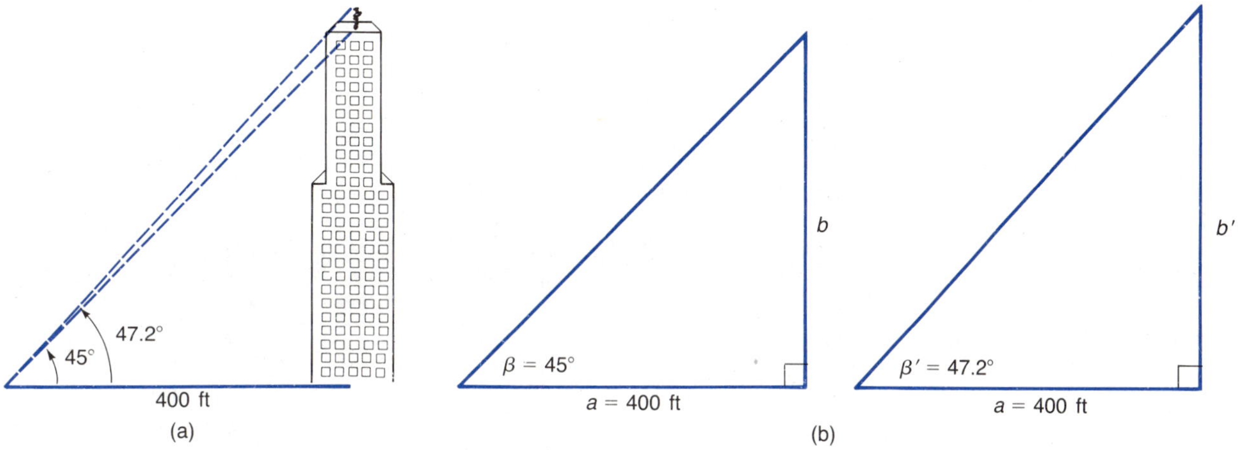

Figure 29

Solution Figure 29(b) shows two triangles that replicate Figure 29(a). The height of the statue of Ceres will be $b' - b$. To find b and b', we refer to Figure 29(b):

$$\tan 45° = \frac{b}{400} \qquad \tan 47.2° = \frac{b'}{400}$$

$$b = 400 \tan 45° = 400 \qquad b' = 400 \tan 47.2° \approx 432$$

The height of the statue is therefore $b' - b \approx 32$ feet. ■

When it is not possible to walk off a distance from the base of the object whose height we seek, a more imaginative solution is required.

Example 4 To measure the height of a mountain, a surveyor takes two sightings of the peak at a distance 900 meters apart on a direct line to the mountain.* See Figure 30(a), page 436. The first observation results in an angle of elevation of 47°, whereas the second results in an angle of elevation of 35°. If the transit is 2 meters high, what is the height h of the mountain?

*For simplicity, we assume these two sightings are at the same level.

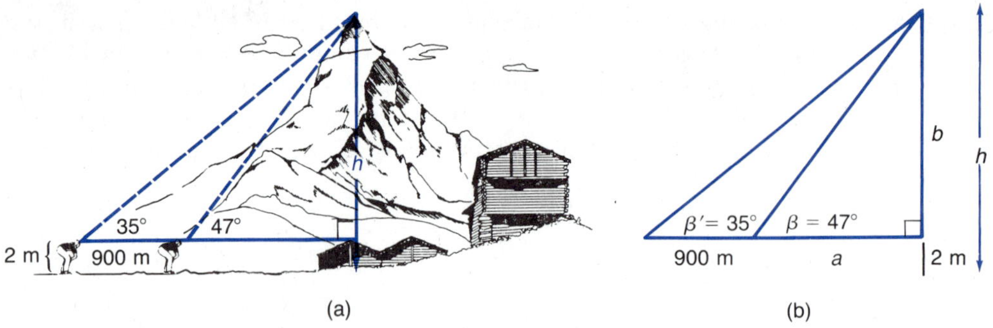

Figure 30

Solution Figure 30(b) shows two triangles that replicate the illustration in Figure 30(a). From the two right triangles shown, we find

$$\cot \beta' = \frac{a + 900}{b} \qquad \cot \beta = \frac{a}{b}$$

$$\cot 35° = \frac{a + 900}{b} \qquad \cot 47° = \frac{a}{b}$$

This is a system of two equations involving two variables, a and b. Because we seek b, we choose to solve the right-hand equation for a and substitute the result, $a = b \cot 47°$, in the left-hand equation. The result is

$$\cot 35° = \frac{b \cot 47° + 900}{b}$$

$$b \cot 35° = b \cot 47° + 900$$

$$b(\cot 35° - \cot 47°) = 900$$

$$b = \frac{900}{\cot 35° - \cot 47°} \approx 1816$$

↑ Use a table or a calculator.

The height of the peak from ground level is therefore approximately $1816 + 2 = 1818$ meters. ■

EXERCISE 7.4

In Problems 1–10, use the right triangle shown in the margin. Then, using the given information, compute the indicated quantity. Use either Table IV or a calculator.

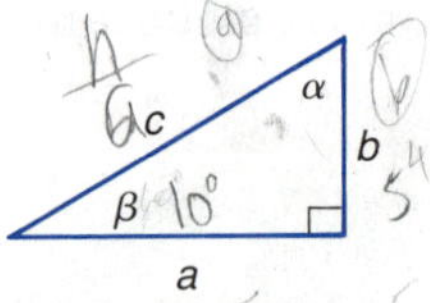

1. $b = 5$, $\beta = 10°$; find a and c
2. $b = 4$, $\beta = 20°$; find a and c
3. $a = 6$, $\beta = 40°$; find b and c
4. $a = 7$, $\beta = 50°$; find b and c
5. $b = 4$, $\alpha = 10°$; find a and c
6. $b = 6$, $\alpha = 20°$; find a and c
7. $a = 5$, $\alpha = 25°$; find b and c
8. $a = 6$, $\alpha = 40°$; find b and c
9. $c = 9$, $\beta = 20°$; find b and a
10. $c = 10$, $\alpha = 40°$; find b and a

11. A right triangle has a hypotenuse of length 3 inches. If one angle is 35°, find the length of each leg.

12. A right triangle has a hypotenuse of length 2 centimeters. If one angle is 40°, find the length of each leg.

13. A right triangle contains a 25° angle. If one leg is of length 5 inches, what is the length of the hypotenuse? [*Hint:* Two answers are possible.]

14. A right triangle contains an angle of $\pi/10$ radian. If one leg is of length 2 meters, what is the length of the hypotenuse? [*Hint:* Two answers are possible.]

15. Find the distance from A to C across the gorge illustrated in the figure.

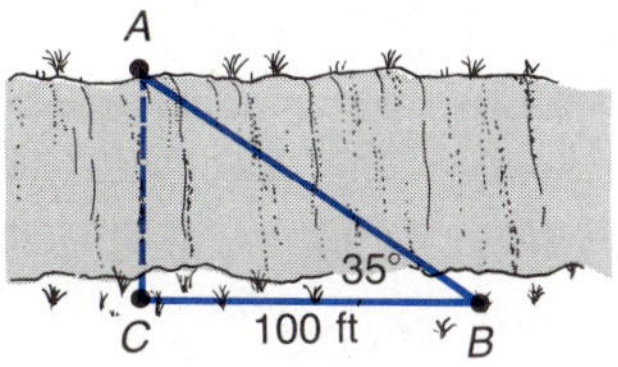

16. Find the distance from A to C across the pond illustrated in the figure.

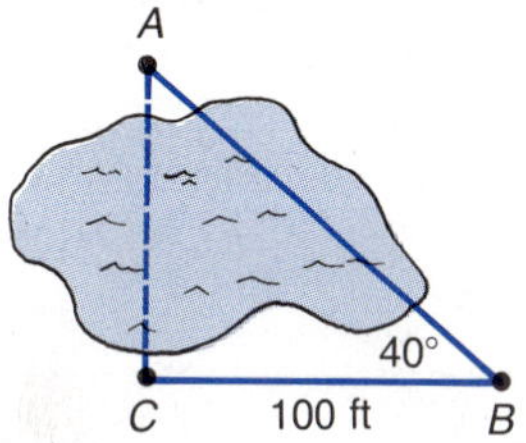

17. The tallest tower built before the era of television masts,* the Eiffel Tower was completed on March 31, 1889. Find the height of the Eiffel Tower (before a television mast was added to the top) using the information given in the illustration.

18. A ship, offshore from a vertical cliff known to be 100 feet in height, takes a sighting of the top of the cliff. If the angle of elevation is found to be 25°, how far offshore is the ship?

19. At 2 PM, a tower casts a shadow whose length is 10 meters. A meter stick simultaneously casts a shadow of length $\frac{1}{2}$ meter. How tall is the tower?

20. A street light causes a shadow 8 feet long to be cast by a girl who is 5 feet tall and is standing 10 feet from the base of the light. How tall is the street light?

21. A state trooper is hidden 30 feet from a highway. One second after a truck passes, the angle between the highway and the line of observation from the patrol car to the truck measures 15° (see the figure). How fast is the truck traveling? (Express the answer in miles per hour.)

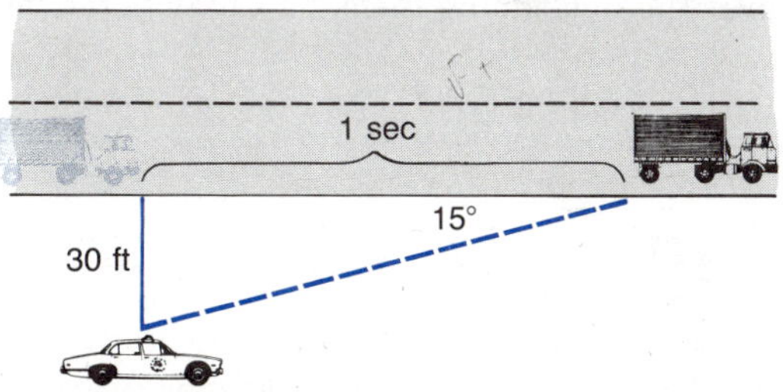

22. Rework Problem 21 if the angle is 20°.

23. Suppose you are headed toward a plateau 50 meters high. If the angle of elevation to the top of the plateau is 20°, how far are you from the base of the plateau?

*Source: *Guinness Book of World Records*

24. A ship is just offshore New York City. A sighting is taken of the Statue of Liberty, which is about 305 feet tall. If the angle of elevation to the top of the statue is 20°, how far is the ship from the base of the statue?

25. A 22 foot extension ladder leaning against a building makes a 70° angle with the ground. How far up the building does the ladder touch?

26. To measure the height of a building, two sightings are taken a distance of 50 feet apart. If the first angle of elevation is 40° and the second one is 32°, what is the height of the building?

27. One of the original Seven Wonders of the World, the Great Pyramid of Cheops was built about 2580 BC. Its original height was 480 feet 11 inches, but due to the loss of its topmost stones, it is now shorter.* Find the current height of the Great Pyramid, using the information given in the illustration.

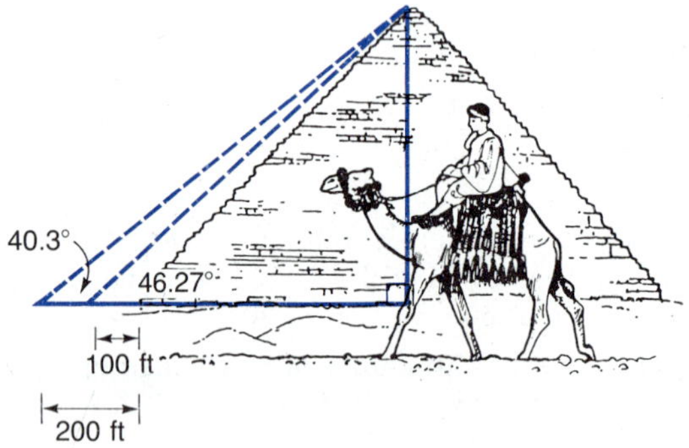

28. To measure the height of Lincoln's caricature on Mt. Rushmore, two sightings 800 feet from the base of the mountain are taken. If the angle of elevation to the bottom of Lincoln's face is 32° and the angle of elevation to the top is 35°, what is the height of Lincoln's face?

29. Two observers simultaneously measure the angle of elevation of a helicopter. One angle is measured as 25°, the other as 40° (see the figure). If the observers are 100 feet apart and the helicopter lies over the line joining them, how high is the helicopter?

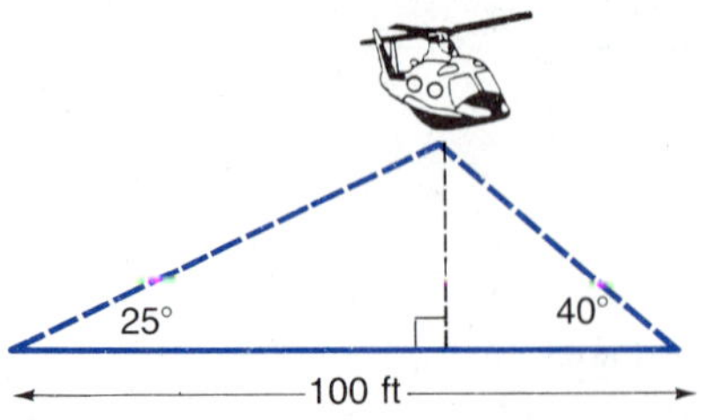

*Source: *Guinness Book of World Records**

30. A blimp, suspended in the air at a height of 500 feet, lies directly over a line from Soldier Field to the Adler Planetarium on Lake Michigan (see the figure). If the angle of depression from the blimp to the stadium is 32° and from the blimp to the planetarium is 23°, find the distance between Soldier Field and the Adler Planetarium.

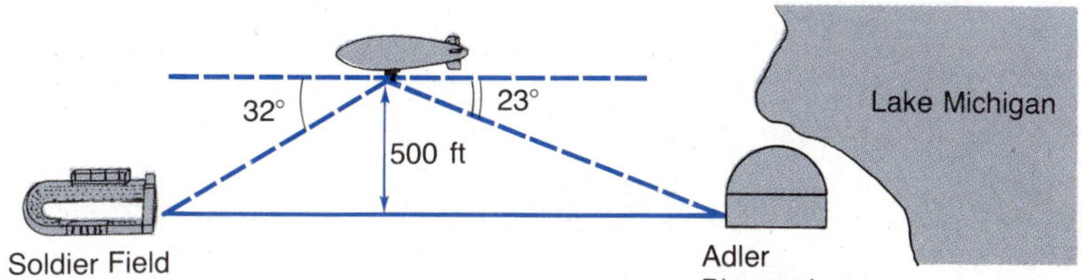

31. A radio transmission tower is 200 feet high. How long should a guy wire be if it is to be attached to the tower 10 feet from the top and is to make an angle of 21° with the ground?

32. A guy wire 80 feet long is attached to the top of a radio transmission tower, making an angle of 25° with the ground. How high is the tower?

33. The angle of elevation of the Washington Monument is 35.1° at the instant it casts a shadow 789 feet long. Use this information to calculate the height of the monument.

34. A straight trail with a uniform inclination of 17° leads from a hotel at an elevation of 9000 feet to a mountain lake at an elevation of 11,200 feet. What is the length of the trail?

7.5 Trigonometric Functions of General Angles

In order to extend the definitions of the trigonometric functions of acute angles to include angles that are not acute, we employ a rectangular coordinate system.

Definition of the Six Trigonometric Functions of a General Angle

Let θ be any angle in standard position, and let (a, b) denote the coordinates of any point except the origin $(0, 0)$ on the terminal side of θ. If $r = \sqrt{a^2 + b^2}$ denotes the distance from $(0, 0)$ to (a, b), then the six trigonometric functions of θ are defined as the ratios

$$\sin\theta = \frac{b}{r} \qquad \cos\theta = \frac{a}{r} \qquad \tan\theta = \frac{b}{a}$$

$$\csc\theta = \frac{r}{b} \qquad \sec\theta = \frac{r}{a} \qquad \cot\theta = \frac{a}{b}$$

provided no denominator equals 0. if a denominator equals 0, that trigonometric function of the angle θ is not defined.

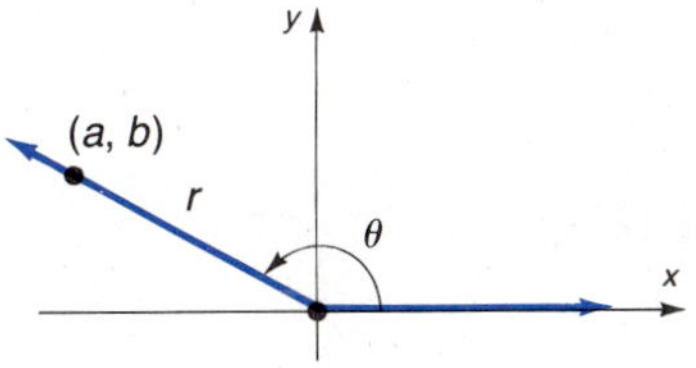

Figure 31

Figure 31 illustrates the definition.

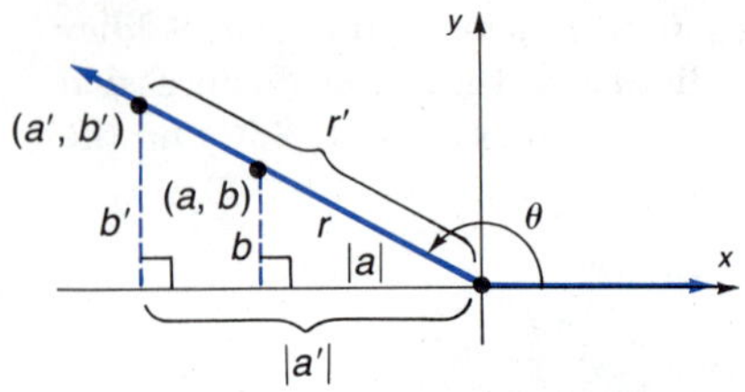

Figure 32

By constructing similar triangles, you should be convinced that the values of the six trigonometric functions of an angle θ do not depend on the selection of the point (a, b) on the terminal side of θ, but rather depend only on the angle θ itself. See Figure 32 for an illustration when θ lies in quadrant II. Since the triangles are similar, the ratio b/r equals the ratio b'/r', the common value being $\sin\theta$. Also, the ratio $|a|/r$ equals the ratio $|a'|/r'$, so that $a/r = a'/r'$, the common value being $\cos\theta$. And so on.

Also, observe that if θ is acute, these definitions reduce to the right triangle definitions given in Section 7.2, as illustrated in Figure 33.

Figure 33

$\sin\theta = \dfrac{b}{r} = \dfrac{\text{Opposite}}{\text{Hypotenuse}}$;

$\cos\theta = \dfrac{a}{r} = \dfrac{\text{Adjacent}}{\text{Hypotenuse}}$

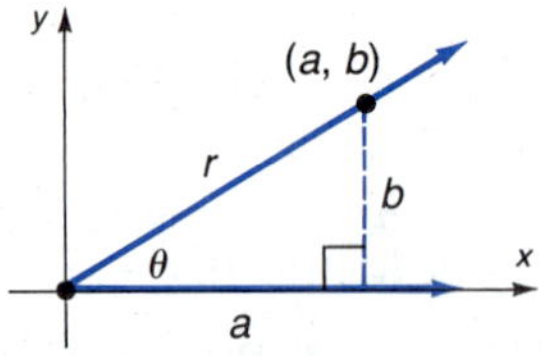

Example 1 Find the value of each of the six trigonometric functions of a positive angle θ if $(4, -3)$ is a point on its terminal side.

Solution Figure 34 illustrates the situation. For the point $(a, b) = (4, -3)$, we have $a = 4$ and $b = -3$. Then $r = \sqrt{a^2 + b^2} = \sqrt{16 + 9} = 5$. Thus,

$$\sin\theta = \frac{b}{r} = -\frac{3}{5} \qquad \cos\theta = \frac{a}{r} = \frac{4}{5} \qquad \tan\theta = \frac{b}{a} = -\frac{3}{4}$$

$$\csc\theta = \frac{r}{b} = -\frac{5}{3} \qquad \sec\theta = \frac{r}{a} = \frac{5}{4} \qquad \cot\theta = \frac{a}{b} = -\frac{4}{3}$$

Figure 34

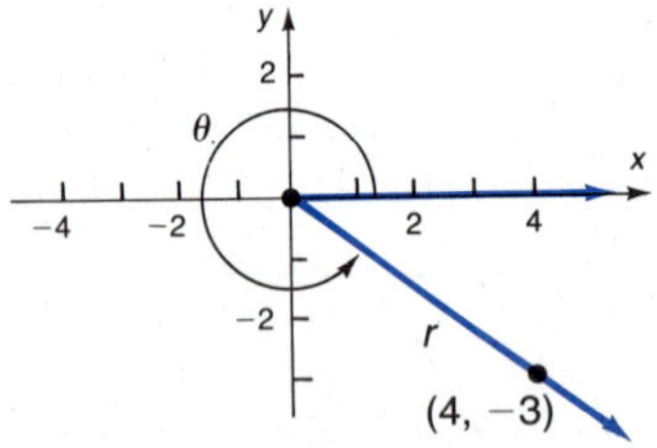

■

In the next example, we find the value of each of the six trigonometric functions at the quadrantal angles 0, $\pi/2$, π, and $3\pi/2$.

Example 2 Find the value of each of the trigonometric functions at:

(a) $\theta = 0 = 0°$ (b) $\theta = \pi/2 = 90°$ (c) $\theta = \pi = 180°$

(d) $\theta = 3\pi/2 = 270°$

Solution (a) The point $P = (1, 0)$ is on the terminal side of $\theta = 0 = 0°$ and is a distance of 1 unit from the origin. See Figure 35. Thus,

$$\sin 0 = \sin 0° = \frac{0}{1} = 0$$

$$\cos 0 = \cos 0° = \frac{1}{1} = 1$$

$$\tan 0 = \tan 0° = \frac{0}{1} = 0$$

$$\sec 0 = \sec 0° = \frac{1}{1} = 1$$

Since the y-coordinate of P is 0, csc 0 and cot 0 are not defined.

(b) The point $P = (0, 1)$ is on the terminal side of $\theta = \pi/2 = 90°$ and is a distance of 1 unit from the origin. See Figure 36. Thus,

$$\sin \frac{\pi}{2} = \sin 90° = \frac{1}{1} = 1$$

$$\cos \frac{\pi}{2} = \cos 90° = \frac{0}{1} = 0$$

$$\csc \frac{\pi}{2} = \csc 90° = \frac{1}{1} = 1$$

$$\cot \frac{\pi}{2} = \cot 90° = \frac{0}{1} = 0$$

Since the x-coordinate of P is 0, $\tan \pi/2$ and $\sec \pi/2$ are not defined.

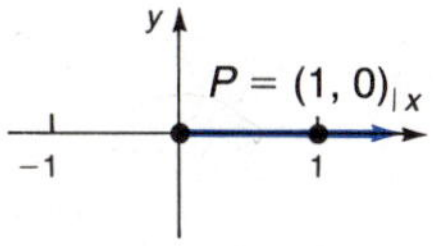

Figure 35
$\theta = 0 = 0°$

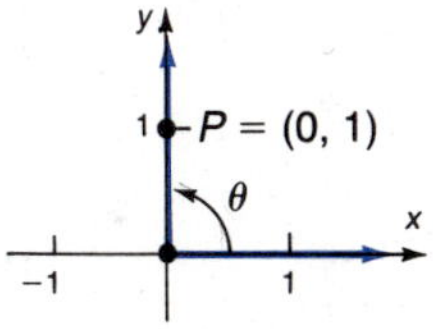

Figure 36
$\theta = \pi/2 = 90°$

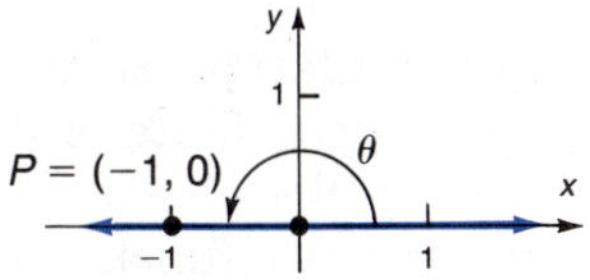

Figure 37
$\theta = \pi = 180°$

(c) The point $P = (-1, 0)$ is on the terminal side of $\theta = \pi = 180°$ and is a distance of 1 unit from the origin. See Figure 37. Thus,

$$\sin \pi = \sin 180° = \frac{0}{1} = 0$$

$$\cos \pi = \cos 180° = \frac{-1}{1} = -1$$

$$\tan \pi = \tan 180° = \frac{0}{1} = 0$$

$$\sec \pi = \sec 180° = \frac{1}{-1} = -1$$

Since the y-coordinate of P is 0, $\csc \pi$ and $\cot \pi$ are not defined.

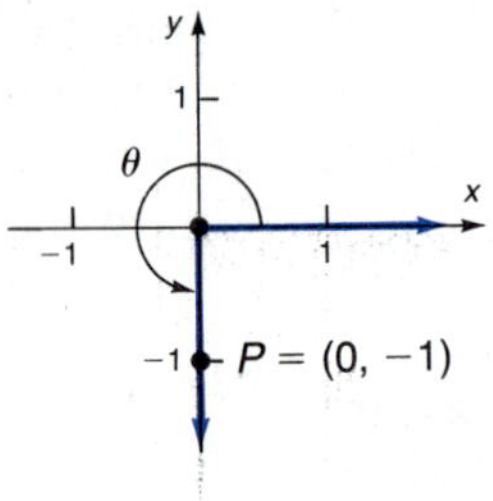

Figure 38
$\theta = 3\pi/2 = 270°$

(d) The point $P = (0, -1)$ is on the terminal side of $\theta = 3\pi/2 = 270°$ and is a distance of 1 unit from the origin. See Figure 38. Thus,

$$\sin\frac{3\pi}{2} = \sin 270° = \frac{-1}{1} = -1$$

$$\cos\frac{3\pi}{2} = \cos 270° = \frac{0}{1} = 0$$

$$\csc\frac{3\pi}{2} = \csc 270° = \frac{1}{-1} = -1$$

$$\cot\frac{3\pi}{2} = \cot 270° = \frac{0}{-1} = 0$$

Since the x-coordinate of P is 0, $\tan 3\pi/2$ and $\sec 3\pi/2$ are not defined. ■

Table 4 summarizes the values of the trigonometric functions found in Example 2.

Table 4

θ (RADIANS)	θ (DEGREES)	$\sin\theta$	$\cos\theta$	$\tan\theta$	$\csc\theta$	$\sec\theta$	$\cot\theta$
0	0°	0	1	0	Not defined	1	Not defined
$\pi/2$	90°	1	0	Not defined	1	Not defined	0
π	180°	0	−1	0	Not defined	−1	Not defined
$3\pi/2$	270°	−1	0	Not defined	−1	Not defined	0

The Signs of the Trigonometric Functions

If θ is not a quadrantal angle, then it will lie in a particular quadrant. In such a case, the signs of the x-coordinate and the y-coordinate of a point (a, b) on the terminal side of θ are known. Because $r = \sqrt{a^2 + b^2} > 0$, it follows that the signs of the trigonometric functions of an angle θ can be found if we know in which quadrant θ lies.

For example, if θ lies in quadrant II, then a point (a, b) on the terminal side of θ has a negative x-coordinate and a positive y-coordinate. Thus,

$$\sin\theta = \frac{b}{r} > 0 \qquad \cos\theta = \frac{a}{r} < 0 \qquad \tan\theta = \frac{b}{a} < 0$$

$$\csc\theta = \frac{r}{b} > 0 \qquad \sec\theta = \frac{r}{a} < 0 \qquad \cot\theta = \frac{a}{b} < 0$$

Table 5 lists the signs of the six trigonometric functions for each quadrant. See also Figure 39.

Table 5

QUADRANT OF θ	$\sin\ \theta$, $\csc\ \theta$	$\cos\ \theta$, $\sec\ \theta$	$\tan\ \theta$, $\cot\ \theta$
I	Positive	Positive	Positive
II	Positive	Negative	Negative
III	Negative	Negative	Positive
IV	Negative	Positive	Negative

Figure 39

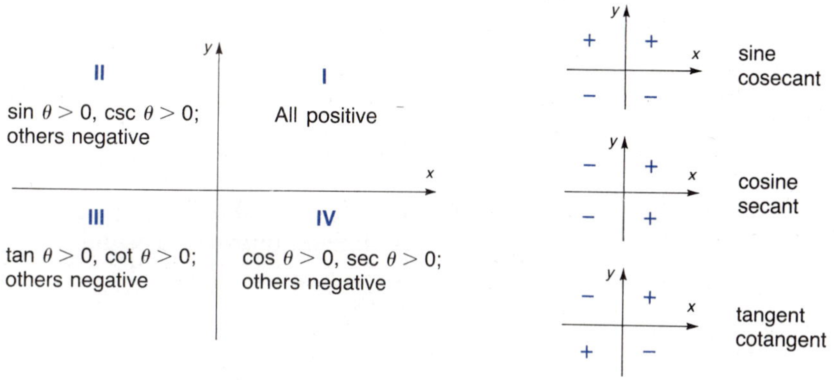

(a) (b)

Example 3 If $\sin\ \theta < 0$ and $\cos\ \theta < 0$, name the quadrant in which the point P corresponding to the angle θ lies.

Solution Let $P = (a, b)$ be a point on the terminal side of the angle θ a distance r from the origin. Then $\sin\ \theta = b/r < 0$ and $\cos\ \theta = a/r < 0$. Thus, $b < 0$ and $a < 0$, so that $P = (a, b)$ must lie in quadrant III. ■

Coterminal Angles

Coterminal Angles Two angles in standard position are said to be **coterminal** if they have the same terminal side.

See Figure 40.

Figure 40

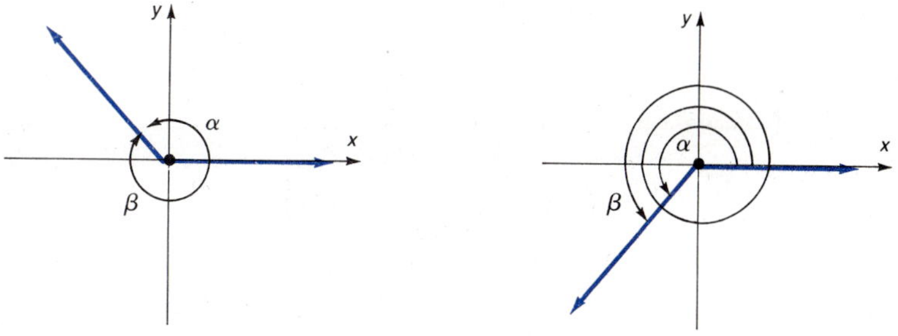

(a) α and β are coterminal (b) α and β are coterminal

In general, if θ is an angle measured in degrees, then $\theta \pm 360°k$, where k is any integer, is an angle coterminal with θ. If θ is measured in radians, then $\theta \pm 2\pi k$, where k is any integer, is an angle coterminal with θ.

Because coterminal angles have the same terminal side, it follows that the values of the trigonometric functions of coterminal angles are equal. We use this fact in the next example.

Example 4 Find the value of each of the following trigonometric functions:

(a) $\sin 390°$ (b) $\cos 420°$ (c) $\tan \dfrac{9\pi}{4}$ (d) $\sec\left(-\dfrac{7\pi}{4}\right)$

(e) $\csc(-270°)$

Solution (a) It is best to sketch the angle first. See Figure 41. The angle 390° is coterminal with 30°. Thus,

$$\sin 390° = \sin(360° + 30°) = \sin 30° = \tfrac{1}{2}$$

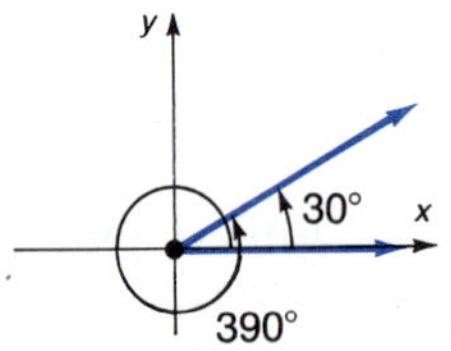

Figure 41

(b) See Figure 42. The angle 420° is coterminal with 60°. Thus,

$$\cos 420° = \cos(360° + 60°) = \cos 60° = \tfrac{1}{2}$$

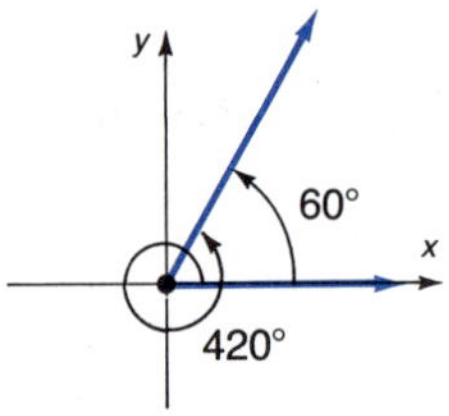

Figure 42

(c) See Figure 43. The angle $9\pi/4$ is coterminal with $\pi/4$. Thus,

$$\tan \frac{9\pi}{4} = \tan\left(2\pi + \frac{\pi}{4}\right) = \tan \frac{\pi}{4} = 1$$

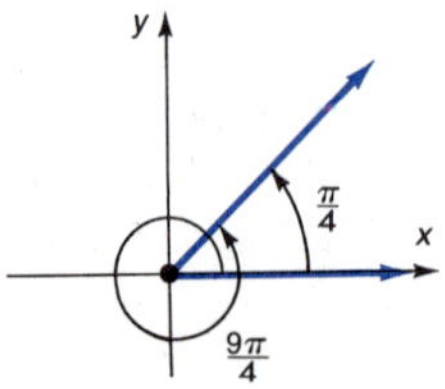

Figure 43

(d) See Figure 44. The angle $-7\pi/4$ is coterminal with $\pi/4$. Thus,

$$\sec\left(-\frac{7\pi}{4}\right) = \sec\left(-2\pi + \frac{\pi}{4}\right) = \sec \frac{\pi}{4} = \sqrt{2}$$

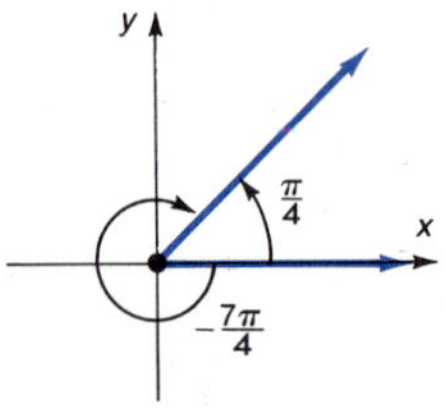

Figure 44

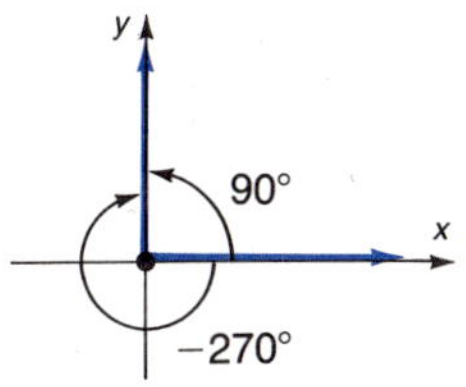

Figure 45

(e) See Figure 45. The angle $-270°$ is coterminal with $90°$. Thus,

$$\begin{aligned}\csc(-270°) &= \csc(-360° + 90°)\\ &= \csc 90° = 1\end{aligned}$$

■

As Example 4 illustrates, the value of a trigonometric function of any angle is equal to the value of the same trigonometric function of an angle θ where $0° \le \theta < 360°$ (or $0 \le \theta < 2\pi$). Moreover, because the angles θ and $\theta \pm 360°k$ (or $\theta \pm 2\pi k$), where k is any integer, are coterminal, and because the values of the trigonometric functions are equal for coterminal angles, it follows that

$$\begin{aligned}
\sin(\theta \pm 360°k) &= \sin\theta & \sin(\theta \pm 2\pi k) &= \sin\theta\\
\cos(\theta \pm 360°k) &= \cos\theta & \cos(\theta \pm 2\pi k) &= \cos\theta\\
\tan(\theta \pm 360°k) &= \tan\theta & \tan(\theta \pm 2\pi k) &= \tan\theta\\
\csc(\theta \pm 360°k) &= \csc\theta & \csc(\theta \pm 2\pi k) &= \csc\theta\\
\sec(\theta \pm 360°k) &= \sec\theta & \sec(\theta \pm 2\pi k) &= \sec\theta\\
\cot(\theta \pm 360°k) &= \cot\theta & \cot(\theta \pm 2\pi k) &= \cot\theta
\end{aligned} \tag{1}$$

These formulas show that the values of the trigonometric functions repeat themselves every $360°$ (or 2π radians).

Reference Angle

Once we know in which quadrant an angle lies, we know the sign of each value of the trigonometric functions of that angle. The use of a certain reference angle may help us to evaluate the trigonometric functions of such an angle.

Let θ denote a nonacute angle that lies in a quadrant. The acute angle formed by the terminal side of θ and either the positive x-axis or the negative x-axis is called the **reference angle** for θ. Figure 46 illustrates the reference angle for some general angles θ. Note that a reference angle is always an acute angle, namely, an angle whose measure is between 0° and 90°.

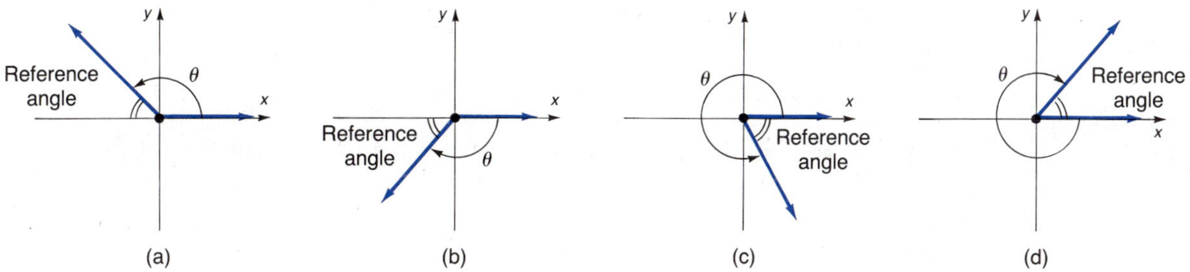

Figure 46

Although formulas can be given for calculating reference angles, usually it is easier to find the reference angle for a given angle by making a quick sketch of the angle.

Example 5 Find the reference angle for each of the following angles:

(a) $150°$ (b) $-45°$ (c) $9\pi/4$ (d) $-5\pi/6$

Solution (a) Refer to Figure 47. The reference angle for $150°$ is $30°$.

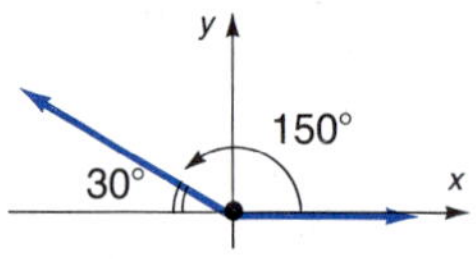

Figure 47

(b) Refer to Figure 48. The reference angle for $-45°$ is $45°$.

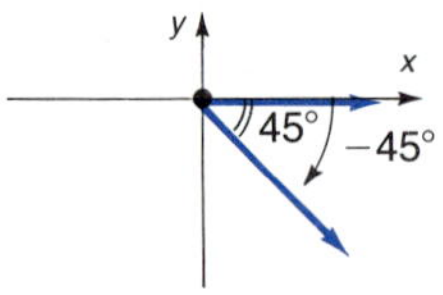

Figure 48

(c) Refer to Figure 49. The reference angle for $9\pi/4$ is $\pi/4$.

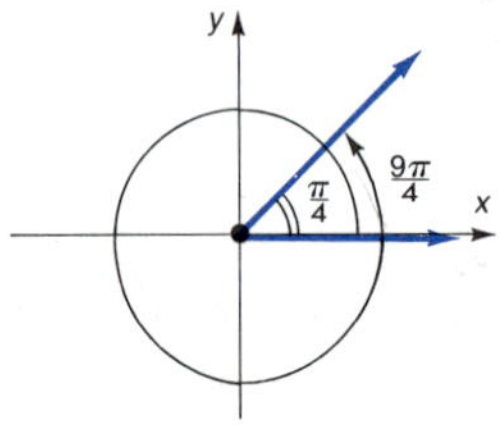

Figure 49

(d) Refer to Figure 50. The reference angle for $-5\pi/6$ is $\pi/6$.

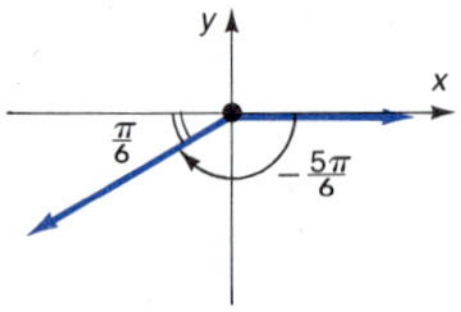

Figure 50 ■

The advantage of using reference angles is that, except for the correct sign, the values of the trigonometric functions of a general angle θ equal the values of the trigonometric functions of its reference angle.

Theorem
Reference Angles

If θ is an angle that lies in a quadrant and if α is its reference angle, then

$$\sin\theta = \pm\sin\alpha \qquad \cos\theta = \pm\cos\alpha \qquad \tan\theta = \pm\tan\alpha$$
$$\csc\theta = \pm\csc\alpha \qquad \sec\theta = \pm\sec\alpha \qquad \cot\theta = \pm\cot\alpha \tag{2}$$

where the + or − sign depends on the quadrant in which θ lies. ■

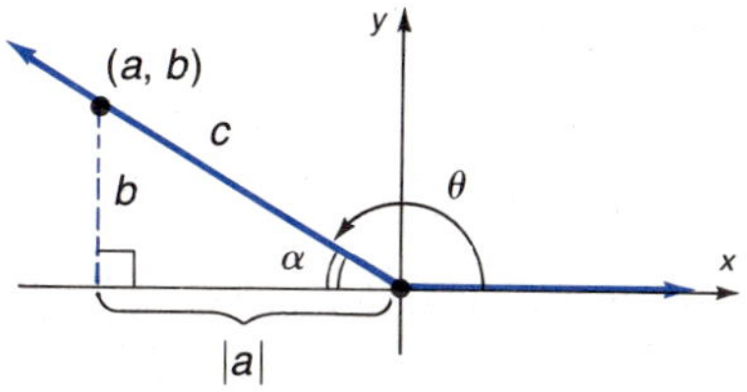

Figure 51
$\sin\theta = b/c$, $\sin\alpha = b/c$;
$\cos\theta = a/c$, $\cos\alpha = |a|/c$

For example, suppose θ lies in quadrant II and α is its reference angle. See Figure 51. If (a, b) is a point on the terminal side of θ and if $c = \sqrt{a^2 + b^2}$, we have

$$\sin\theta = \frac{b}{c} = \sin\alpha \qquad \cos\theta = \frac{a}{c} = \frac{-|a|}{c} = -\cos\alpha$$

and so on.

The next example illustrates how the theorem on reference angles is used.

Example 6 Find the value of each of the following trigonometric functions using reference angles:

(a) $\sin 135°$ (b) $\cos 240°$ (c) $\cos\dfrac{5\pi}{6}$ (d) $\tan\left(-\dfrac{\pi}{3}\right)$

Solution (a) Refer to Figure 52. The angle 135° is in quadrant II, where the sine function is positive. The reference angle for 135° is 45°. Thus,

$$\sin 135° = \sin 45° = \frac{\sqrt{2}}{2}$$

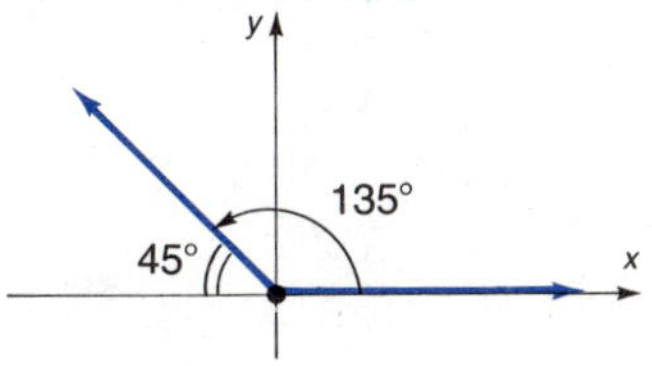

Figure 52

(b) Refer to Figure 53. The angle 240° is in quadrant III, where the cosine function is negative. The reference angle for 240° is 60°. Thus,

$$\cos 240° = -\cos 60° = -\tfrac{1}{2}$$

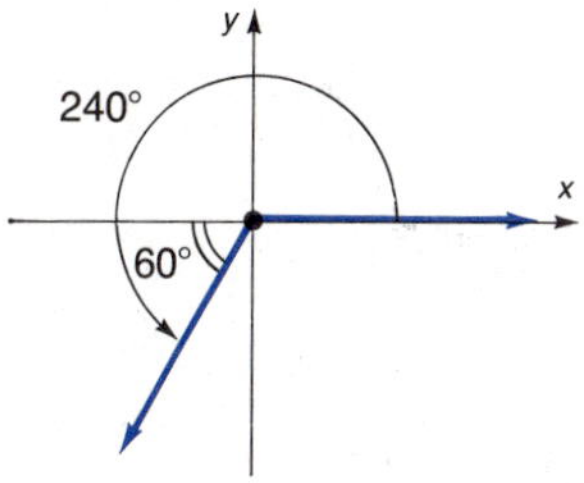

Figure 53

(c) Refer to Figure 54. The angle $5\pi/6$ is in quadrant II, where the cosine function is negative. The reference angle for $5\pi/6$ is $\pi/6$. Thus,

$$\cos\frac{5\pi}{6} = -\cos\frac{\pi}{6} = -\frac{\sqrt{3}}{2}$$

Figure 54

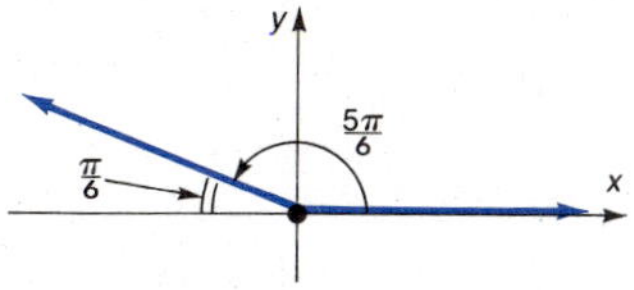

(d) Refer to Figure 55. The angle $-\pi/3$ is in quadrant IV, where the tangent function is negative. The reference angle for $-\pi/3$ is $\pi/3$. Thus,

$$\tan\left(-\frac{\pi}{3}\right) = -\tan\frac{\pi}{3} = -\sqrt{3}$$

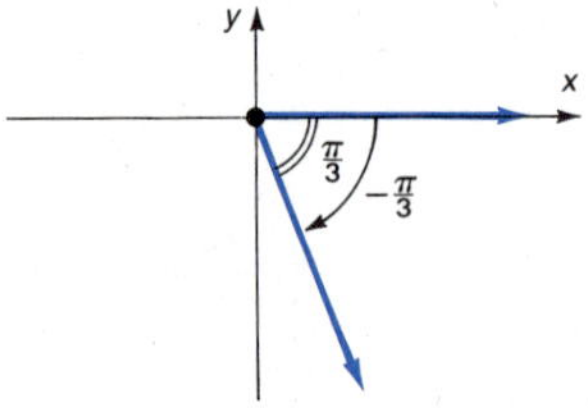

Figure 55

Example 7 Given that $\cos\theta = -\frac{2}{3}$, $\pi/2 < \theta < \pi$, find the value of each of the remaining trigonometric functions.

Solution The angle θ lies in quadrant II, so we know that $\sin\theta$ and $\csc\theta$ are positive, whereas the other trigonometric functions are negative. If α is the reference angle for θ, then $\cos\alpha = \frac{2}{3}$. The values of the remaining trigonometric functions of the angle α can be found by drawing the appropriate triangle. We use Figure 56 to obtain

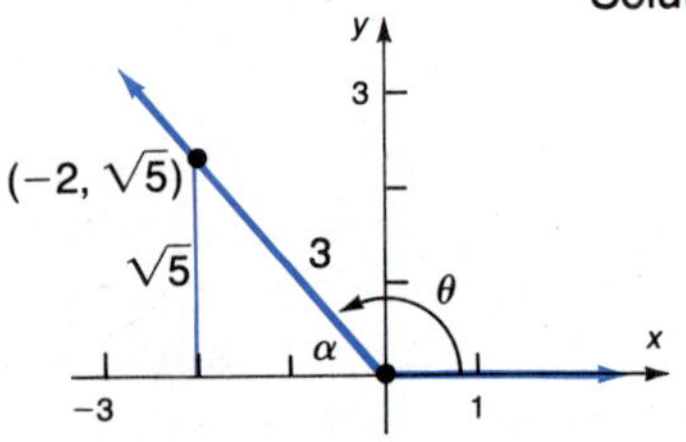

Figure 56

$$\sin\alpha = \frac{\sqrt{5}}{3} \qquad \cos\alpha = \frac{2}{3} \qquad \tan\alpha = \frac{\sqrt{5}}{2}$$

$$\csc\alpha = \frac{3}{\sqrt{5}} = \frac{3\sqrt{5}}{5} \qquad \sec\alpha = \frac{3}{2} \qquad \cot\alpha = \frac{2}{\sqrt{5}} = \frac{2\sqrt{5}}{5}$$

Now, we assign the appropriate sign to each of these values to find the values of the trigonometric functions of θ:

$$\sin\theta = \frac{\sqrt{5}}{3} \qquad \cos\theta = -\frac{2}{3} \qquad \tan\theta = -\frac{\sqrt{5}}{2}$$

$$\csc\theta = \frac{3\sqrt{5}}{5} \qquad \sec\theta = -\frac{3}{2} \qquad \cot\theta = -\frac{2\sqrt{5}}{5}$$

Example 8 If $\tan\theta = -4$ and $\sin\theta < 0$, find the value of each of the remaining trigonometric functions of θ.

Solution Since $\tan\theta = -4 < 0$ and $\sin\theta < 0$, it follows that θ lies in quadrant IV. If α is the reference angle for θ, then $\tan\alpha = 4$. See Figure 57. Thus,

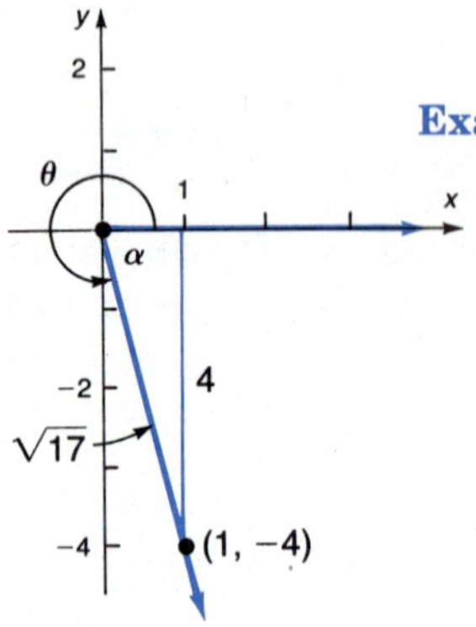

Figure 57

$$\sin\alpha = \frac{4}{\sqrt{17}} = \frac{4\sqrt{17}}{17} \qquad \cos\alpha = \frac{1}{\sqrt{17}} = \frac{\sqrt{17}}{17} \qquad \tan\alpha = \frac{4}{1} = 4$$

$$\csc\alpha = \frac{\sqrt{17}}{4} \qquad \sec\alpha = \frac{\sqrt{17}}{1} = \sqrt{17} \qquad \cot\alpha = \frac{1}{4}$$

We assign the appropriate sign to each of these to obtain the values of the trigonometric functions of θ:

$$\sin\theta = \frac{-4\sqrt{17}}{17} \qquad \cos\theta = \frac{\sqrt{17}}{17} \qquad \tan\theta = -4$$

$$\csc\theta = \frac{-\sqrt{17}}{4} \qquad \sec\theta = \sqrt{17} \qquad \cot\theta = -\frac{1}{4}$$

Summary

To find the values of the trigonometric functions of a general angle:

STEP 1: If the angle is not between 0° and 360° (0 and 2π radians), replace it by an angle θ, $0° \le \theta < 360°$ ($0 \le \theta < 2\pi$), that is coterminal with it.

STEP 2: If the angle θ is a quadrantal angle, pick a point on its terminal side and apply the definitions of the trigonometric functions.

STEP 3: If the angle θ lies in a quadrant, determine the correct signs of the trigonometric functions in that quadrant and determine the reference angle α for θ. Now express each trigonometric function of θ in terms of the same value (except for the sign) of the trigonometric function of α, an acute angle.

EXERCISE 7.5

In Problems 1–8, a point on the terminal side of an angle θ is given. Find the value of each of the six trigonometric functions of θ.

1. $(-3, 4)$ **2.** $(5, -12)$ **3.** $(2, -3)$ **4.** $(-1, -2)$
5. $(-2, -2)$ **6.** $(1, -1)$ **7.** $(-3, -2)$ **8.** $(2, 2)$

In Problems 9–16, name the quadrant in which the point P corresponding to the angle θ lies.

9. $\sin\theta > 0$, $\cos\theta < 0$ **10.** $\sin\theta < 0$, $\cos\theta > 0$ **11.** $\sin\theta < 0$, $\tan\theta < 0$
12. $\cos\theta > 0$, $\tan\theta > 0$ **13.** $\cos\theta > 0$, $\tan\theta < 0$ **14.** $\cos\theta < 0$, $\tan\theta > 0$
15. $\sec\theta < 0$, $\sin\theta > 0$ **16.** $\csc\theta > 0$, $\cos\theta < 0$

In Problems 17–32, find the reference angle of each angle.

17. $-30°$ **18.** $60°$ **19.** $120°$ **20.** $300°$ **21.** $210°$ **22.** $330°$
23. $5\pi/4$ **24.** $5\pi/6$ **25.** $8\pi/3$ **26.** $7\pi/4$ **27.** $-135°$ **28.** $-240°$
29. $-2\pi/3$ **30.** $-7\pi/6$ **31.** $420°$ **32.** $480°$

In Problems 33–68, find the value of each expression.

33. $\sin 405°$ **34.** $\cos 420°$ **35.** $\tan 405°$ **36.** $\sin 390°$
37. $\csc 450°$ **38.** $\sec 540°$ **39.** $\cot 390°$ **40.** $\sec 420°$
41. $\cos \frac{33\pi}{4}$ **42.** $\sin \frac{9\pi}{4}$ **43.** $\tan 21\pi$ **44.** $\csc \frac{9\pi}{2}$
45. $\sec \frac{17\pi}{4}$ **46.** $\cot \frac{17\pi}{4}$ **47.** $\tan \frac{19\pi}{6}$ **48.** $\sec \frac{25\pi}{6}$
49. $\sin 150°$ **50.** $\cos 210°$ **51.** $\cos 315°$ **52.** $\sin 120°$
53. $\sec 240°$ **54.** $\csc 300°$ **55.** $\cot 330°$ **56.** $\tan 225°$
57. $\sin \frac{3\pi}{4}$ **58.** $\cos \frac{2\pi}{3}$ **59.** $\cot \frac{7\pi}{6}$ **60.** $\csc \frac{7\pi}{4}$
61. $\cos(-60°)$ **62.** $\tan(-120°)$ **63.** $\sin\left(-\frac{2\pi}{3}\right)$ **64.** $\cot\left(-\frac{\pi}{6}\right)$
65. $\tan \frac{14\pi}{3}$ **66.** $\sec \frac{11\pi}{4}$ **67.** $\csc(-315°)$ **68.** $\sec(-225°)$

In Problems 69–84, find the value of each of the remaining trigonometric functions of θ.

69. $\sin\theta = \frac{12}{13}$, $90° < \theta < 180°$ **70.** $\cos\theta = \frac{3}{5}$, $270° < \theta < 360°$
71. $\cos\theta = -\frac{4}{5}$, $\pi < \theta < 3\pi/2$ **72.** $\sin\theta = -\frac{5}{13}$, $\pi < \theta < 3\pi/2$
73. $\sin\theta = \frac{5}{13}$, $\cos\theta < 0$ **74.** $\cos\theta = \frac{4}{5}$, $\sin\theta < 0$
75. $\cos\theta = -\frac{1}{3}$, $\csc\theta > 0$ **76.** $\sin\theta = -\frac{2}{3}$, $\sec\theta > 0$
77. $\sin\theta = \frac{2}{3}$, $\tan\theta < 0$ **78.** $\cos\theta = -\frac{1}{4}$, $\tan\theta > 0$
79. $\sec\theta = 2$, $\sin\theta < 0$ **80.** $\csc\theta = 3$, $\cot\theta < 0$
81. $\tan\theta = \frac{3}{4}$, $\sin\theta < 0$ **82.** $\cot\theta = \frac{4}{3}$, $\cos\theta < 0$
83. $\tan\theta = -\frac{1}{3}$, $\sin\theta > 0$ **84.** $\sec\theta = -2$, $\tan\theta > 0$

85. Find the value of: $\sin 45° + \sin 135° + \sin 225° + \sin 315°$
86. Find the value of: $\tan 60° + \tan 150°$
87. If $\sin\theta = 0.1$, find $\sin(\theta + \pi)$.
88. If $\cos\theta = 0.3$, find $\cos(\theta + \pi)$.
89. If $\tan\theta = 3$, find $\tan(\theta + \pi)$.
90. If $\cot\theta = -2$, find $\cot(\theta + \pi)$.
91. If $\sin\theta = \frac{1}{5}$, find $\csc\theta$.
92. If $\cos\theta = \frac{2}{3}$, find $\sec\theta$.
93. Find the value of: $\sin 1° + \sin 2° + \sin 3° + \cdots + \sin 358° + \sin 359°$
94. Find the value of: $\cos 1° + \cos 2° + \cos 3° + \cdots + \cos 358° + \cos 359°$

7.6 Trigonometric Functions: Unit Circle Approach

There are two widely accepted approaches to the development of the trigonometric functions: One uses right triangles; the other uses circles, especially the unit circle. In this section, we introduce trigonometric functions using the unit circle. We shall show that this approach leads to the definition given earlier.

The Unit Circle

1485 $8\frac{1}{4}$ 45 4665

Recall that the **unit circle** is a circle whose radius is 1 and whose center is at the origin of a rectangular coordinate system. Because the radius r of the unit circle is 1, we see from the formula $s = r\theta$ that a central angle of θ radians subtends an arc whose length s is

$$s = \theta$$

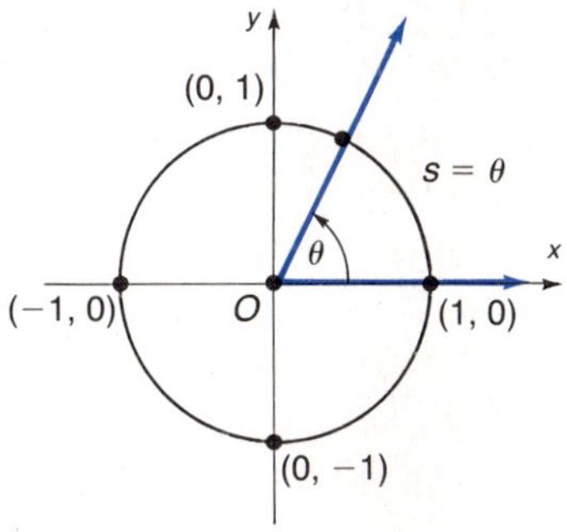

Figure 58
Unit circle: $x^2 + y^2 = 1$

See Figure 58. Thus, on the unit circle, the length measure of the arc s equals the radian measure of the central angle θ. In other words, on the unit circle, the real number used to measure an angle θ in radians corresponds exactly with the real number used to measure the length of the arc subtended by that angle.

For example, suppose $r = 1$ foot. Then if $s = 3$ feet, $\theta = 3$ radians; if $\theta = 8.2$ radians, then $s = 8.2$ feet; and so on.

Now, let t be any real number and let θ be the angle, in standard position, equal to t radians. Let P be the point on the unit circle that is also on the terminal side of θ. See Figure 59(a). If $t \geq 0$, this point P is reached by moving *counterclockwise* along the unit circle, starting at (1, 0), for a length of arc equal to t units. See Figure 59(b). If $t < 0$, this point P is reached by moving *clockwise* along the unit circle, starting at (1, 0), for a length of arc equal to $|t|$ units. See Figure 59(c).

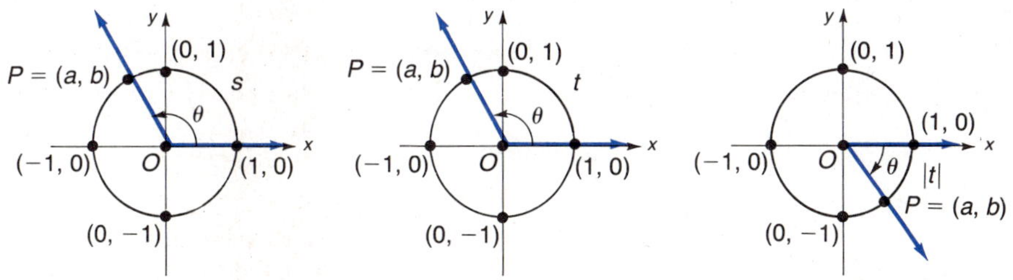

(a) $\theta = t$ radians $s = |t|$ units

(b) $\theta = t$ radians; length of arc from (1, 0) to P is t units, $t \geq 0$

(c) $\theta = t$ radians; length of arc from (1, 0) to P is $|t|$ units, $t < 0$

Figure 59

Thus, to each real number t, there corresponds a unique point $P = (a, b)$ on the unit circle. This is the important idea here. No matter what real number t is chosen, there corresponds a unique point P on the unit circle. We use the coordinates of the point $P = (a, b)$ on the unit circle corresponding to the real number t to define the six trigonometric functions of the real number t.

Trigonometric Functions

Let t be a real number and let $P = (a, b)$ be the point on the unit circle that corresponds to t.

Sine Function

The **sine function** associates with t the y-coordinate of P and is denoted by

$$\sin t = b$$

Cosine Function

The **cosine function** associates with t the x-coordinate of P and is denoted by

$$\cos t = a$$

Tangent Function

If $a \neq 0$, the **tangent function** is defined as

$$\tan t = \frac{b}{a}$$

Cosecant Function

If $b \neq 0$, the **cosecant function** is defined as

$$\csc t = \frac{1}{b}$$

Secant Function

If $a \neq 0$, the **secant function** is defined as

$$\sec t = \frac{1}{a}$$

Cotangent Function

If $b \neq 0$, the **cotangent function** is defined as

$$\cot t = \frac{a}{b}$$

Because we use the unit circle in these definitions of the trigonometric functions, they are also sometimes referred to as **circular functions**.

Example 1

Let t be a real number and let $P = \left(-\frac{1}{2}, \sqrt{3}/2\right)$ be the point on the unit circle that corresponds to t. See Figure 60. Then

$$\sin t = \frac{\sqrt{3}}{2} \qquad \cos t = -\frac{1}{2} \qquad \tan t = \frac{\sqrt{3}/2}{-\frac{1}{2}} = -\sqrt{3}$$

$$\csc t = \frac{1}{\sqrt{3}/2} = \frac{2\sqrt{3}}{3} \qquad \sec t = \frac{1}{-\frac{1}{2}} = -2 \qquad \cot t = \frac{-\frac{1}{2}}{\sqrt{3}/2} = -\frac{\sqrt{3}}{3}$$

■

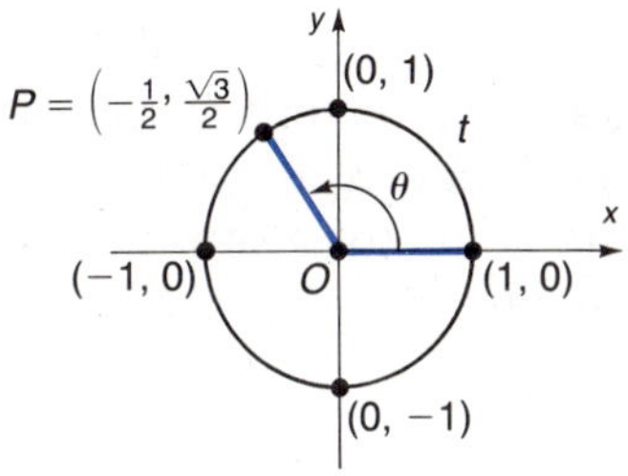

Figure 60
$\theta = t$ radians

Trigonometric Functions of Angles

Let P be the point on the unit circle corresponding to the real number t. Then the angle θ, in standard position and measured in radians, whose terminal side is the ray from the origin through P is

$$\theta = t \text{ radians}$$

See Figure 61. Thus, on the unit circle, the measure of the angle θ in radians equals the value of the real number t. As a result, we can say that

$$\underset{\text{Real number}}{\sin \underset{\uparrow}{t}} = \sin \underset{\substack{\uparrow \\ \theta = t \text{ radians}}}{\theta}$$

and so on. We now can define the trigonometric functions of the angle θ.

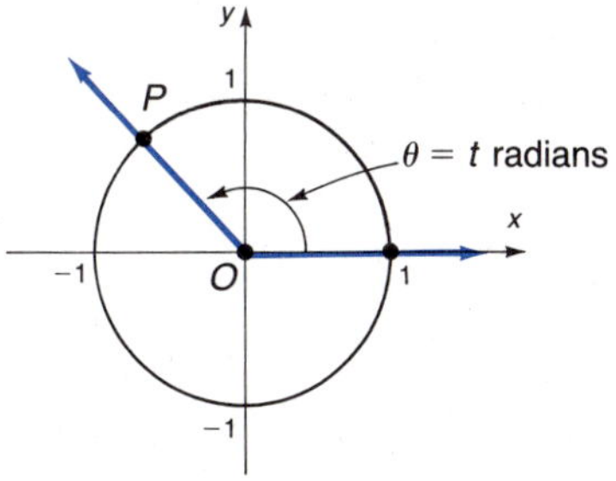

Figure 61

If $\theta = t$ radians, the six trigonometric functions of the angle θ are defined as

$$\sin \theta = \sin t \qquad \cos \theta = \cos t \qquad \tan \theta = \tan t$$
$$\csc \theta = \csc t \qquad \sec \theta = \sec t \qquad \cot \theta = \cot t$$

Even though the distinction between trigonometric functions of real numbers and trigonometric functions of angles is important, it is customary to refer to trigonometric functions of real numbers and trigonometric functions of angles collectively as *the trigonometric functions*. We shall follow this practice from now on.

Since the values of the trigonometric functions of an angle θ are determined by the coordinates of the point $P = (a, b)$ on the unit circle corresponding to θ, the units used to measure the angle θ are irrelevant. For example, it does not matter whether we write $\theta = \pi/2$ radians or $\theta = 90°$. The point on the unit circle corresponding to this angle is $P = (0, 1)$. Hence,

$$\sin\frac{\pi}{2} = \sin 90° = 1 \qquad \text{and} \qquad \cos\frac{\pi}{2} = \cos 90° = 0$$

To find the exact value of a trigonometric function of an angle θ requires that we locate the corresponding point P on the unit circle. In fact, though, any circle whose center is at the origin can be used.

Let θ be any nonquadrantal angle placed in standard position. Let $P^* = (a^*, b^*)$ be the point where the terminal side of θ intersects the unit circle. See Figure 62.

Figure 62

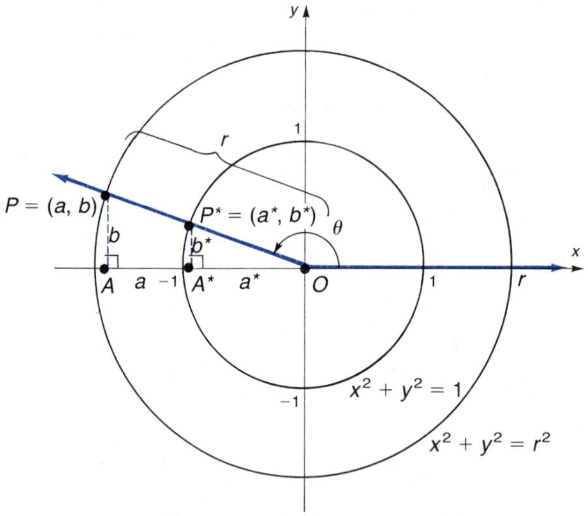

Let $P = (a, b)$ be any point on the terminal side of θ. Suppose r is the distance from the origin to P. Then P is on the circle $x^2 + y^2 = r^2$. Refer again to Figure 62. Notice that the triangles OA^*P^* and OAP are similar; thus, ratios of corresponding sides are equal:

$$\frac{b^*}{1} = \frac{b}{r} \qquad \frac{a^*}{1} = \frac{a}{r} \qquad \frac{b^*}{a^*} = \frac{b}{a}$$

$$\frac{1}{b^*} = \frac{r}{b} \qquad \frac{1}{a^*} = \frac{r}{a} \qquad \frac{a^*}{b^*} = \frac{a}{b}$$

These results lead us to formulate the following theorem:

Theorem For an angle θ in standard position, let $P = (a, b)$ be any point on the terminal side of θ. Let r equal the distance from the origin to P. Then

$$\sin\theta = \frac{b}{r} \qquad \cos\theta = \frac{a}{r} \qquad \tan\theta = \frac{b}{a},\quad a \neq 0$$

$$\csc\theta = \frac{r}{b},\quad b \neq 0 \qquad \sec\theta = \frac{r}{a},\quad a \neq 0 \qquad \cot\theta = \frac{a}{b},\quad b \neq 0$$

■

This result coincides with the definition given in Section 7.5 for the six trigonometric functions of a general angle θ.

Example 2 Given that $\sin\theta = \frac{1}{3}$ and $\cos\theta < 0$, find the value of each of the remaining five trigonometric functions.

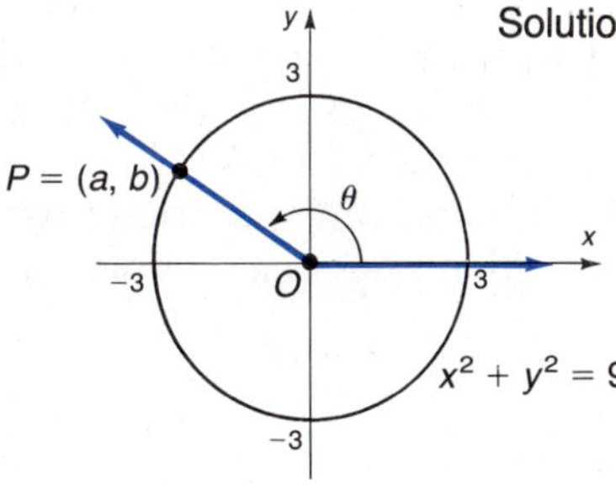

Figure 63

Solution Suppose $P = (a, b)$ is a point on the terminal side of θ that lies a distance of $r = 3$ units from the origin. See Figure 63. (Do you see why we chose 3 units? Notice that $\sin\theta = \frac{1}{3} = b/r$. The choice $r = 3$ will make our calculations easy.) With this choice, $b = 1$ and $r = 3$. Since $\cos\theta = a/r < 0$, it follows that $a < 0$. Thus, we have

$$\begin{aligned} a^2 + b^2 &= r^2 \qquad b = 1,\ r = 3,\ a < 0 \\ a^2 + 1^2 &= 3^2 \\ a^2 &= 8 \\ a &= -2\sqrt{2} \end{aligned}$$

Thus,

$$\cos\theta = \frac{a}{r} = \frac{-2\sqrt{2}}{3} \qquad \tan\theta = \frac{b}{a} = \frac{1}{-2\sqrt{2}} = \frac{-\sqrt{2}}{4}$$

$$\csc\theta = \frac{r}{b} = \frac{3}{1} = 3 \qquad \sec\theta = \frac{r}{a} = \frac{3}{-2\sqrt{2}} = \frac{-3\sqrt{2}}{4} \qquad \cot\theta = \frac{a}{b} = \frac{-2\sqrt{2}}{1} = -2\sqrt{2}$$

■

You should compare the method used above with the two methods used earlier to solve the same problem. See Example 3, Section 7.2.

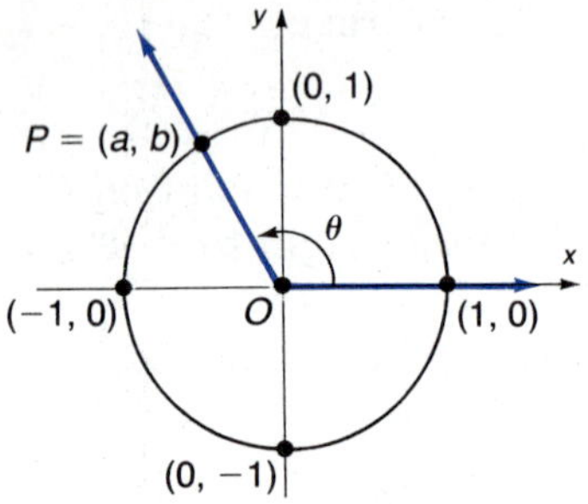

Figure 64

Domain and Range of the Trigonometric Functions

Let θ be an angle in standard position, and let $P = (a, b)$ be a point on the terminal side of θ. Suppose, for convenience, that P also lies on the unit circle. See Figure 64. Then, by definition,

$$\sin\theta = b \qquad \cos\theta = a \qquad \tan\theta = \frac{b}{a},\quad a \neq 0$$
$$\csc\theta = \frac{1}{b},\quad b \neq 0 \qquad \sec\theta = \frac{1}{a},\quad a \neq 0 \qquad \cot\theta = \frac{a}{b},\quad b \neq 0$$

For $\sin\theta$ and $\cos\theta$, θ can be any angle, so it follows that the domain of the sine function and cosine function is the set of all real numbers.

> The domain of the sine function is all real numbers.
>
> The domain of the cosine function is all real numbers.

Notice that for the tangent function and secant function the x-coordinate of P cannot be 0. On the unit circle, there are two such points, namely, (0, 1) and (0, −1). These two points correspond to the angles $\pi/2$ (90°) and $3\pi/2$ (270°) or, more generally, to any angle that is an odd multiple of $\pi/2$ (90°) such as $\pi/2$ (90°), $3\pi/2$ (270°), $5\pi/2$ (450°), $-\pi/2$ (−90°), $-3\pi/2$ (−270°), and so on. Such angles must therefore be excluded from the domain of the tangent function and secant function.

> The domain of the tangent function is all real numbers, except odd multiples of $\pi/2$ (90°).
>
> The domain of the secant function is all real numbers, except odd multiples of $\pi/2$ (90°).

Notice that for the cotangent function and cosecant function the y-coordinate of P cannot be 0. On the unit circle, there are two such points, namely, (1, 0) and (−1, 0). These two points correspond to the angles 0 (0°) and π (180°) or, more generally, to any angle that is an integral multiple of π (180°) such as 0 (0°), π (180°), 2π (360°), 3π (540°), $-\pi$ (−180°), and so on. Such angles must therefore be excluded from the domain of the cotangent function and cosecant function.

The domain of the cotangent function is all real numbers, except integral multiples of π (180°).

The domain of the cosecant function is all real numbers, except integral multiples of π (180°).

Next, we determine the range of each of the six trigonometric functions. Refer again to Figure 64. Let $P = (a, b)$ be the point on the unit circle that corresponds to the angle θ. It follows that $-1 \le a \le 1$ and $-1 \le b \le 1$. Consequently, since $\sin\theta = b$ and $\cos\theta = a$, we have

$$-1 \le \sin\theta \le 1 \qquad \text{and} \qquad -1 \le \cos\theta \le 1$$

Thus, the range of both the sine function and the cosine function consists of all real numbers between -1 and 1, inclusive. In terms of absolute value notation, we have $|\sin\theta| \le 1$ and $|\cos\theta| \le 1$.

Similarly, if θ is not a multiple of π (180°), then $\csc\theta = 1/b$. Since $b = \sin\theta$ and $|b| = |\sin\theta| \le 1$, it follows that $|\csc\theta| = 1/|b| \ge 1$. Thus, the range of the cosecant function consists of all real numbers less than or equal to -1 or greater than or equal to 1. That is,

$$\csc\theta \le -1 \qquad \text{or} \qquad \csc\theta \ge 1$$

If θ is not an odd multiple of $\pi/2$ (90°), then, by definition, $\sec\theta = 1/a$. Since $a = \cos\theta$ and $|a| = |\cos\theta| \le 1$, it follows that $|\sec\theta| = 1/|a| \ge 1$. Thus, the range of the secant function consists of all real numbers less than or equal to -1 or greater than or equal to 1.

$$\sec\theta \le -1 \qquad \text{or} \qquad \sec\theta \ge 1$$

The range of both the tangent function and the cotangent function consists of all real numbers. That is,

$$-\infty < \tan\theta < \infty \qquad \text{and} \qquad -\infty < \cot\theta < \infty$$

You are asked to prove this in Problems 35 and 36.

Table 6 summarizes these results.

Table 6

FUNCTION	SYMBOL	DOMAIN	RANGE
sine	$f(\theta) = \sin \theta$	All real numbers	All real numbers from -1 to 1, inclusive
cosine	$f(\theta) = \cos \theta$	All real numbers	All real numbers from -1 to 1, inclusive
tangent	$f(\theta) = \tan \theta$	All real numbers, except odd multiples of $\pi/2$ (90°)	All real numbers
cosecant	$f(\theta) = \csc \theta$	All real numbers, except integral multiples of π (180°)	All real numbers greater than or equal to 1 or less than or equal to -1
secant	$f(\theta) = \sec \theta$	All real numbers, except odd multiples of $\pi/2$ (90°)	All real numbers greater than or equal to 1 or less than or equal to -1
cotangent	$f(\theta) = \cot \theta$	All real numbers, except integral multiples of π (180°)	All real numbers

Period of the Trigonometric Functions

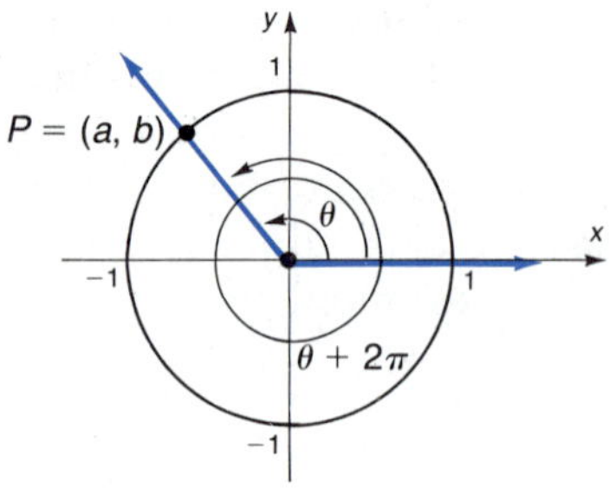

Figure 65

Suppose that for a given angle θ, measured in radians, we know the corresponding point $P = (a, b)$ on the unit circle. Now add 2π to θ. The point on the unit circle corresponding to $\theta + 2\pi$ is identical to the point P on the unit circle corresponding to θ. See Figure 65. Thus, the values of the trigonometric functions of $\theta + 2\pi$ are equal to the values of the corresponding trigonometric functions of θ.

For example,

$$\sin \frac{\pi}{4} = \frac{\sqrt{2}}{2} \quad \text{and} \quad \sin\left(\frac{\pi}{4} + 2\pi\right) = \sin \frac{9\pi}{4} = \frac{\sqrt{2}}{2}$$

$$\cos \frac{\pi}{4} = \frac{\sqrt{2}}{2} \quad \text{and} \quad \cos\left(\frac{\pi}{4} + 2\pi\right) = \cos \frac{9\pi}{4} = \frac{\sqrt{2}}{2}$$

If we add (or subtract) integral multiples of 2π to θ, the trigonometric values remain unchanged. That is, for all θ,

$$\sin(\theta + 2\pi k) = \sin \theta \qquad \cos(\theta + 2\pi k) = \cos \theta \qquad (1)$$

where k is any integer. Functions that exhibit this kind of behavior are called *periodic functions*.

Periodic Function

A function f is called **periodic** if there is a positive number p such that whenever θ is in the domain of f, so is $\theta + p$, and

$$f(\theta + p) = f(\theta)$$

Period

If there is a smallest such number p, this smallest value is called the **(fundamental) period** of f.

Thus, based on equation (1), the sine and cosine functions are periodic. In fact, the sine and cosine functions have period 2π. You are asked to prove this fact in Problems 37 and 38. The secant and cosecant functions are also periodic with period 2π; the tangent and cotangent functions are periodic with period π. You are asked to prove these statements in Problems 39–42.

These facts are summarized below:

Periodic Properties

$\sin(\theta + 2\pi) = \sin\theta$	$\cos(\theta + 2\pi) = \cos\theta$	$\tan(\theta + \pi) = \tan\theta$
$\csc(\theta + 2\pi) = \csc\theta$	$\sec(\theta + 2\pi) = \sec\theta$	$\cot(\theta + \pi) = \cot\theta$

Because the sine, cosine, secant, and cosecant functions have period 2π, once we know their values for $0 \le \theta < 2\pi$, we know all their values; similarly, since the tangent and cotangent functions have period π, once we know their values for $0 \le \theta < \pi$, we know all their values.

The periodic properties of the trigonometric functions will be very helpful to us when we study their graphs in the next chapter.

Even–Odd Properties

Recall that a function f is even if $f(-\theta) = f(\theta)$ for all θ in the domain of f; a function f is odd if $f(-\theta) = -f(\theta)$ for all θ in the domain of f. We will now show that the trigonometric functions sine, tangent, cotangent, and cosecant are odd functions, whereas the functions cosine and secant are even functions.

Theorem

Even–Odd Properties

$\sin(-\theta) = -\sin\theta$	$\cos(-\theta) = \cos\theta$	$\tan(-\theta) = -\tan\theta$
$\csc(-\theta) = -\csc\theta$	$\sec(-\theta) = \sec\theta$	$\cot(-\theta) = -\cot\theta$

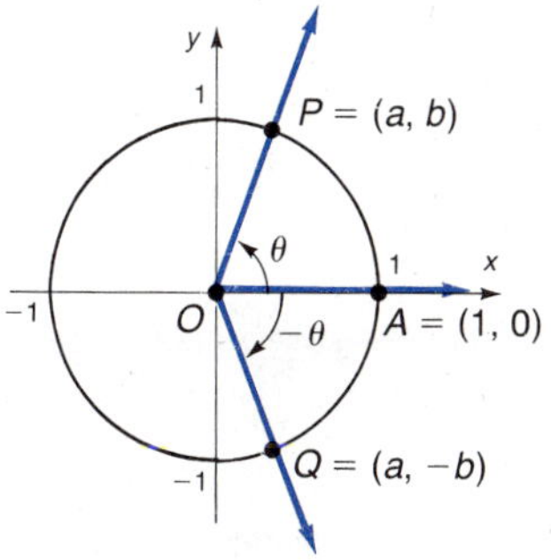

Figure 66

Proof

Let $P = (a, b)$ be the point on the terminal side of the angle θ that is on the unit circle. (See Figure 66.) The point Q on the terminal side of the angle $-\theta$ that is on the unit circle will have coordinates $(a, -b)$. Using the definition for the trigonometric functions, we have

$$\sin(-\theta) = -b = -\sin\theta \qquad \cos(-\theta) = a = \cos\theta$$

Now, using these results and some of the fundamental identities, we have

$$\tan(-\theta) = \frac{\sin(-\theta)}{\cos(-\theta)} = \frac{-\sin\theta}{\cos\theta} = -\tan\theta$$

$$\cot(-\theta) = \frac{1}{\tan(-\theta)} = \frac{1}{-\tan\theta} = -\cot\theta$$

$$\sec(-\theta) = \frac{1}{\cos(-\theta)} = \frac{1}{\cos\theta} = \sec\theta$$

$$\csc(-\theta) = \frac{1}{\sin(-\theta)} = \frac{1}{-\sin\theta} = -\csc\theta$$

EXERCISE 7.6

In Problems 1–8, t is a real number and P is the point on the unit circle that corresponds to t. Find the value of each of the six trigonometric functions of t.

1. $P = \left(\sqrt{3}/2, \frac{1}{2}\right)$ **2.** $P = (\sqrt{2}/2, \sqrt{2}/2)$ **3.** $P = \left(-\frac{1}{2}, -\sqrt{3}/2\right)$

4. $P = (\sqrt{2}/2, -\sqrt{2}/2)$ **5.** $P = \left(-\frac{3}{5}, \frac{4}{5}\right)$ **6.** $P = \left(-\frac{12}{13}, \frac{5}{13}\right)$

7. $P = (\sqrt{5}/5, -2\sqrt{5}/5)$ **8.** $P = (-2\sqrt{5}/5, -\sqrt{5}/5)$

In Problems 9–16, find the value of each of the remaining five trigonometric functions of θ. Use the method shown in Example 2.

9. $\sin\theta = \frac{2}{3}, \quad \pi/2 < \theta < \pi$

10. $\cos\theta = \frac{1}{3}, \quad 3\pi/2 < \theta < 2\pi$

11. $\tan\theta = \frac{1}{2}, \quad \pi < \theta < 3\pi/2$

12. $\sec\theta = 3, \quad 0 < \theta < \pi/2$

13. $\sin\theta = -\frac{1}{4}, \quad \cos\theta > 0$

14. $\cos\theta = -\frac{3}{4}, \quad \sin\theta < 0$

15. $\sec\theta = 3, \quad \tan\theta < 0$

16. $\csc\theta = -4, \quad \tan\theta > 0$

17. What is the domain of the sine function?

18. What is the domain of the cosine function?

19. For what numbers θ is $f(\theta) = \tan\theta$ not defined?

20. For what numbers θ is $f(\theta) = \cot\theta$ not defined?

21. For what numbers θ is $f(\theta) = \sec\theta$ not defined?

22. For what numbers θ is $f(\theta) = \csc\theta$ not defined?

23. What is the range of the sine function?

24. What is the range of the cosine function?

25. What is the range of the tangent function?

26. What is the range of the cotangent function?

27. What is the range of the secant function?

28. What is the range of the cosecant function?

29. Is the sine function even, odd, or neither? Is its graph symmetric? With respect to what?

30. Is the cosine function even, odd, or neither? Is its graph symmetric? With respect to what?

31. Is the tangent function even, odd, or neither? Is its graph symmetric? With respect to what?

32. Is the cotangent function even, odd, or neither? Is its graph symmetric? With respect to what?

33. Is the secant function even, odd, or neither? Is its graph symmetric? With respect to what?

34. Is the cosecant function even, odd, or neither? Is its graph symmetric? With respect to what?

35. Show that the range of the tangent function is the set of all real numbers.

36. Show that the range of the cotangent function is the set of all real numbers.

37. Show that the period of $f(\theta) = \sin\theta$ is 2π. [*Hint:* Assume $0 < p < 2\pi$ exists so that $\sin(\theta + p) = \sin\theta$ for all θ. Let $\theta = 0$ to find p. Then let $\theta = \pi/2$ to obtain a contradiction.]

38. Show that the period of $f(\theta) = \cos\theta$ is 2π.

39. Show that the period of $f(\theta) = \sec\theta$ is 2π.

40. Show that the period of $f(\theta) = \csc\theta$ is 2π.

41. Show that the period of $f(\theta) = \tan\theta$ is π.

42. Show that the period of $f(\theta) = \cot\theta$ is π.

43. If θ $(0 < \theta < \pi)$ is the angle between a horizontal ray directed to the right (say, the positive x-axis) and a nonhorizontal, nonvertical line L, show that the slope m of L equals $\tan\theta$. The angle θ is called the **inclination** of L. [*Hint:* See the illustration, where we have drawn the line L^* parallel to L and passing through the origin. Use the fact that L^* intersects the unit circle at the point $(\cos\theta, \sin\theta)$.]

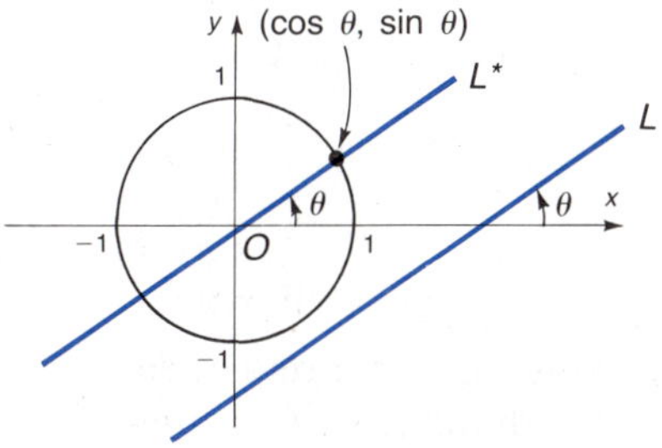

CHAPTER REVIEW

THINGS TO KNOW

Definitions

Standard position of an angle	Vertex is at the origin; initial side is along the positive x-axis
Degree (1°)	$1° = \frac{1}{360}$ revolution
Radian (1)	The measure of a central angle whose rays subtend an arc whose length is the radius of the circle
Acute angle	An angle θ whose measure is $0° < \theta < 90°$ (or $0 < \theta < \pi/2$)

Trigonometric functions	$P = (a, b)$ is the point on the terminal side of θ a distance r from the origin; $\sin\theta = \frac{b}{r}$; $\cos\theta = \frac{a}{r}$; $\tan\theta = \frac{b}{a}, a \neq 0$; $\cot\theta = \frac{a}{b}, b \neq 0$; $\csc\theta = \frac{r}{b}, b \neq 0$; $\sec\theta = \frac{r}{a}, a \neq 0$
Complementary angles	Two acute angles whose sum is 90° ($\pi/2$)
Cofunction	The following pairs of functions are cofunctions of each other: sine and cosine; tangent and cotangent; secant and cosecant
Reference angle of θ	The acute angle formed by the terminal side of θ and either the positive or negative x-axis
Unit circle	Center at origin; radius = 1 Equation: $x^2 + y^2 = 1$
Periodic function	$f(\theta + p) = f(\theta)$, for all θ, $p > 0$, where the smallest such p is the fundamental period

Formulas

$$1 \text{ revolution} = 360° = 2\pi \text{ radians}$$

$s = r\theta$	θ is measured in radians; s is the length of arc subtended by the central angle θ of the circle of radius r.
$v = r\omega$	v is the linear speed along the circle of radius r; ω is the angular speed (measured in radians per unit time)

Table of Values

θ (RADIANS)	θ (DEGREES)	$\sin\theta$	$\cos\theta$	$\tan\theta$	$\csc\theta$	$\sec\theta$	$\cot\theta$
0	0°	0	1	0	Not defined	1	Not defined
$\pi/6$	30°	$\frac{1}{2}$	$\sqrt{3}/2$	$\sqrt{3}/3$	2	$2\sqrt{3}/3$	$\sqrt{3}$
$\pi/4$	45°	$\sqrt{2}/2$	$\sqrt{2}/2$	1	$\sqrt{2}$	$\sqrt{2}$	1
$\pi/3$	60°	$\sqrt{3}/2$	$\frac{1}{2}$	$\sqrt{3}$	$2\sqrt{3}/3$	2	$\sqrt{3}/3$
$\pi/2$	90°	1	0	Not defined	1	Not defined	0
π	180°	0	−1	0	Not defined	−1	Not defined
$3\pi/2$	270°	−1	0	Not defined	−1	Not defined	0

Properties of the Trigonometric Functions

$f(\theta) = \sin\theta$	Domain: all real numbers; Range: all real numbers from -1 to 1, inclusive Periodic: period $= 2\pi$ (360°) Odd function
$f(\theta) = \cos\theta$	Domain: all real numbers Range: all real numbers from -1 to 1, inclusive Periodic: period $= 2\pi$ (360°) Even function
$f(\theta) = \tan\theta$	Domain: all real numbers, except odd multiples of $\pi/2$ (90°) Range: all real numbers Periodic: period $= \pi$ (180°) Odd function
$f(\theta) = \csc\theta$	Domain: all real numbers, except integral multiples of π (180°) Range: all real numbers greater than or equal to 1 or less than or equal to -1 Periodic: period $= 2\pi$ (360°) Odd function
$f(\theta) = \sec\theta$	Domain: all real numbers, except odd multiples of $\pi/2$ (90°) Range: all real numbers greater than or equal to 1 or less than or equal to -1 Periodic: period $= 2\pi$ (360°) Even function
$f(\theta) = \cot\theta$	Domain: all real numbers, except integral multiples of π (180°) Range: all real numbers Periodic: period $= \pi$ (180°) Odd function

Identities

$$\tan\theta = \frac{\sin\theta}{\cos\theta}, \quad \cot\theta = \frac{\cos\theta}{\sin\theta}$$

$$\cot\theta = \frac{1}{\tan\theta}, \quad \sec\theta = \frac{1}{\cos\theta}, \quad \csc\theta = \frac{1}{\sin\theta}$$

$$\sin^2\theta + \cos^2\theta = 1, \quad \tan^2\theta + 1 = \sec^2\theta, \quad 1 + \cot^2\theta = \csc^2\theta$$

How To:

Convert an angle from radian measure to degree measure
Convert an angle from degree measure to radian measure
Find the value of each of the remaining trigonometric functions if the value of one function and the quadrant of the angle are given
Use the theorem on cofunctions of complementary angles
Use reference angles to find the value of a trigonometric function
Use a calculator (or a table) to find the value of a trigonometric function
Use right triangle trigonometry to solve applied problems

FILL-IN-THE-BLANK ITEMS

1. Two rays drawn with a common vertex form a(n) ________. One of the rays is called the ________ ________; the other is called the ________.
2. In the formula $s = r\theta$ for measuring the length s of arc along a circle of radius r, the angle θ must be measured in ________.
3. 180 degrees = ________ radians.
4. Two acute angles whose sum is a right angle are called ________.
5. The sine and ________ functions are cofunctions.
6. An angle is in ________ ________ if its vertex is at the origin and its initial side coincides with the positive x-axis.
7. The reference angle of 135° is ________.
8. The sine, cosine, cosecant, and secant functions have period ________; the tangent and cotangent functions have period ________.

REVIEW EXERCISES

In Problems 1–4, convert each angle in degrees to radians. Express your answer as a multiple of π.

1. 135° **2.** 210° **3.** 18° **4.** 15°

In Problems 5–8, convert each angle in radians to degrees.

5. $3\pi/4$ **6.** $2\pi/3$ **7.** $-5\pi/2$ **8.** $-3\pi/2$

In Problems 9–30, find the value of each expression. Do not use a calculator or a table.

9. $\tan\frac{\pi}{4} - \sin\frac{\pi}{6}$ **10.** $\cos\frac{\pi}{3} + \sin\frac{\pi}{2}$ **11.** $3\sin 45° - 4\tan\frac{\pi}{6}$

12. $4\cos 60° + 3\tan\frac{\pi}{3}$ **13.** $6\cos\frac{3\pi}{4} + 2\tan\left(-\frac{\pi}{3}\right)$ **14.** $3\sin\frac{2\pi}{3} - 4\cos\frac{5\pi}{2}$

15. $\sec\left(-\frac{\pi}{3}\right) - \cot\left(-\frac{5\pi}{4}\right)$ **16.** $4\csc\frac{3\pi}{4} - \cot\left(-\frac{\pi}{4}\right)$ **17.** $\tan\pi + \sin\pi$

18. $\cos\frac{\pi}{2} - \sec\left(-\frac{\pi}{2}\right)$

19. $\cos 180° - \tan(-45°)$

20. $\sin 270° + \cos(-180°)$

21. $\sin^2 20° + \dfrac{1}{\sec^2 20°}$

22. $\dfrac{1}{\cos^2 40°} - \dfrac{1}{\cot^2 40°}$

23. $\sec 50° \cos 50°$

24. $\tan 10° \cot 10°$

25. $\dfrac{\sin 50°}{\cos 40°}$

26. $\dfrac{\tan 20°}{\cot 70°}$

27. $\dfrac{\sin(-40°)}{\cos 50°}$

28. $\tan(-20°) \cot 20°$

29. $\sin 400° \sec(-50°)$

30. $\cot 200° \cot(-70°)$

In Problems 31–46, find the value of each of the remaining trigonometric functions.

31. $\sin\theta = -\frac{4}{5}, \quad \cos\theta > 0$

32. $\cos\theta = -\frac{3}{5}, \quad \sin\theta < 0$

33. $\tan\theta = \frac{12}{5}, \quad \sin\theta < 0$

34. $\cot\theta = \frac{12}{5}, \quad \cos\theta < 0$

35. $\sec\theta = -\frac{5}{4}, \quad \tan\theta < 0$

36. $\csc\theta = -\frac{5}{3}, \quad \cot\theta < 0$

37. $\sin\theta = \frac{12}{13}, \quad \theta$ in quadrant II

38. $\cos\theta = -\frac{3}{5}, \quad \theta$ in quadrant III

39. $\sin\theta = -\frac{5}{13}, \quad 3\pi/2 < \theta < 2\pi$

40. $\cos\theta = \frac{12}{13}, \quad 3\pi/2 < \theta < 2\pi$

41. $\tan\theta = \frac{1}{3}, \quad 180° < \theta < 270°$

42. $\tan\theta = -\frac{2}{3}, \quad 90° < \theta < 180°$

43. $\sec\theta = 3, \quad 3\pi/2 < \theta < 2\pi$

44. $\csc\theta = -4, \quad \pi < \theta < 3\pi/2$

45. $\cot\theta = -2, \quad \pi/2 < \theta < \pi$

46. $\tan\theta = -2, \quad 3\pi/2 < \theta < 2\pi$

47. Find the length of arc subtended by a central angle of 30° on a circle of radius 2 feet.

48. The minute hand of a clock is 8 inches long. How far does the tip of the minute hand move in 30 minutes? How far does it move in 20 minutes?

C **49.** A race car is driven around a circular track at a constant speed of 180 miles per hour. If the diameter of the track is $\frac{1}{2}$ mile, what is the angular speed of the car? Express your answer in revolutions per hour (which is equivalent to laps per hour).

C **50.** Repeat Problem 49 if the car goes only 150 miles per hour.

C **51.** From a stationary hot air balloon 500 feet above the ground, two sightings of a lake are made (see the figure). How long is the lake?

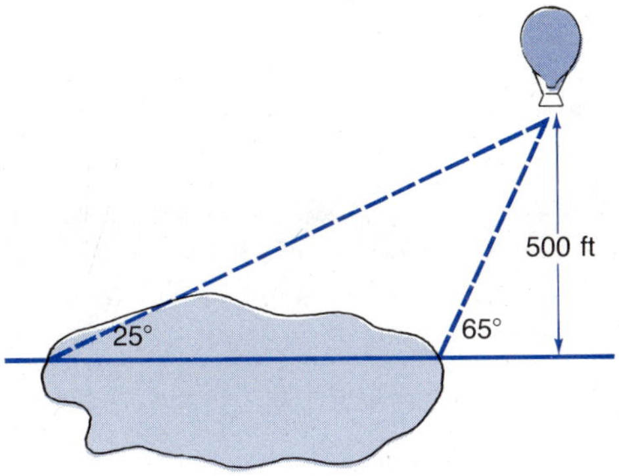

C **52.** From a glider 200 feet above the ground, two sightings of a stationary object directly in front are taken 1 minute apart (see the figure). What is the speed of the glider?

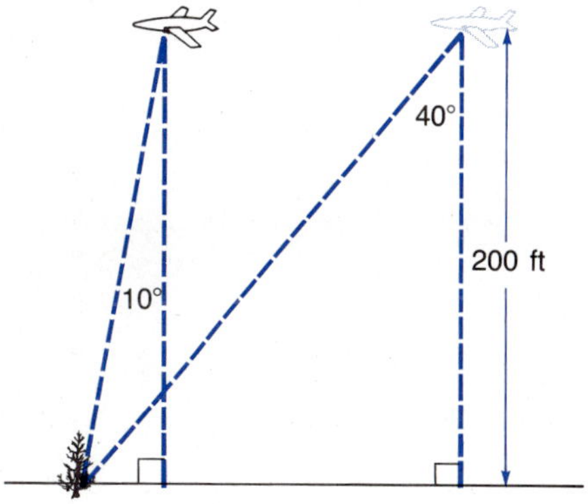

C **53.** Find the distance from A to C across the river illustrated in the figure.

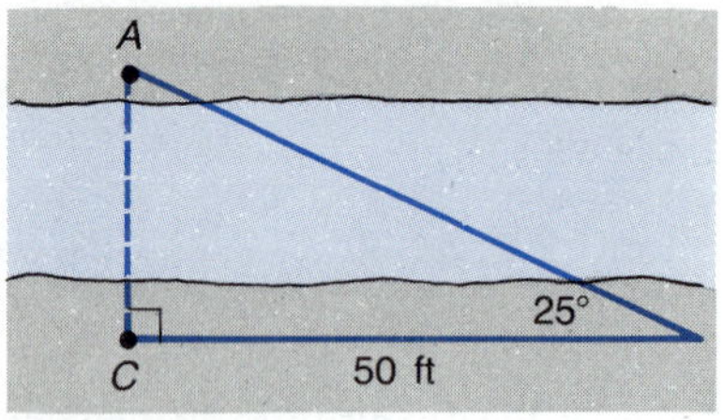

C **54.** Find the height of the building shown in the figure.

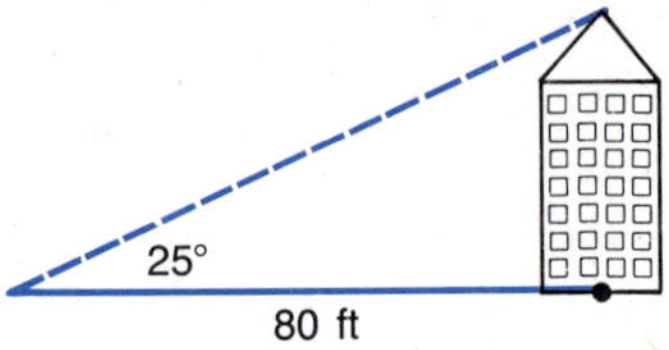

C **55.** The Sears Tower in Chicago is 1454 feet tall and is situated about a mile inland from the shore of Lake Michigan, as indicated in the figure. An observer in a pleasure boat on the lake directly in front of the Sears Tower looks at the top of the tower and measures the angle of elevation as 5°. How far offshore is the boat?

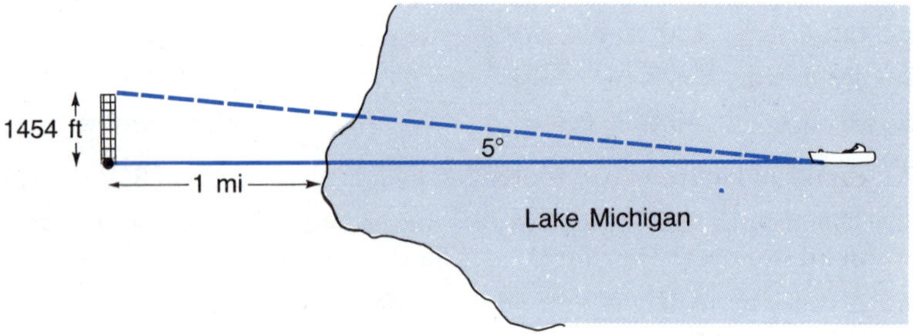

Graphs of Trigonometric Functions

8

In this chapter, we graph the six trigonometric functions. Then we introduce the inverse trigonometric functions and show how to graph them.

8.1 Graphs of the Six Trigonometric Functions*

In the preceding chapter, we defined the trigonometric functions $f(\theta) = \sin \theta$, $f(\theta) = \cos \theta$, and so on. In this chapter, we shall use the traditional symbols x to represent the independent variable (or argument) and y for the dependent variable (or value at x) for each function. Thus, we write the six trigonometric functions as

$y = \sin x$	$y = \cos x$	$y = \tan x$
$y = \csc x$	$y = \sec x$	$y = \cot x$

Our purpose in this section is to graph each of these functions. Unless indicated otherwise, we shall use radian measure throughout for the independent variable x.

The Graph of $y = \sin x$

Since the sine function has period 2π, we need to graph $y = \sin x$ only on the interval $[0, 2\pi]$. The remainder of the graph will consist of repetitions of this portion of the graph.

We begin by constructing Table 1, which lists some points on the graph of $y = \sin x$, $0 \le x \le 2\pi$. As the table shows, the graph of $y = \sin x$, $0 \le x \le 2\pi$, begins at the origin. As x increases from 0 to $\pi/2$, the value of $y = \sin x$ increases from 0 to 1; as x increases from $\pi/2$ to π to $3\pi/2$, the value of y decreases from 1 to 0 to -1; as x increases from $3\pi/2$ to 2π, the value of y increases from -1 to 0. If we plot the points listed in Table 1 and connect them with a smooth curve, we obtain the graph shown in Figure 1.

Table 1

x	$y = \sin x$	(x, y)
0	0	$(0, 0)$
$\pi/6$	$\frac{1}{2}$	$(\pi/6, \frac{1}{2})$
$\pi/2$	1	$(\pi/2, 1)$
$5\pi/6$	$\frac{1}{2}$	$(5\pi/6, \frac{1}{2})$
π	0	$(\pi, 0)$
$7\pi/6$	$-\frac{1}{2}$	$(7\pi/6, -\frac{1}{2})$
$3\pi/2$	-1	$(3\pi/2, -1)$
$11\pi/6$	$-\frac{1}{2}$	$(11\pi/6, -\frac{1}{2})$
2π	0	$(2\pi, 0)$

* C Refer to Graphics Calculator Supplement: Activity V.

Figure 1
$y = \sin x,\ 0 \le x \le 2\pi$

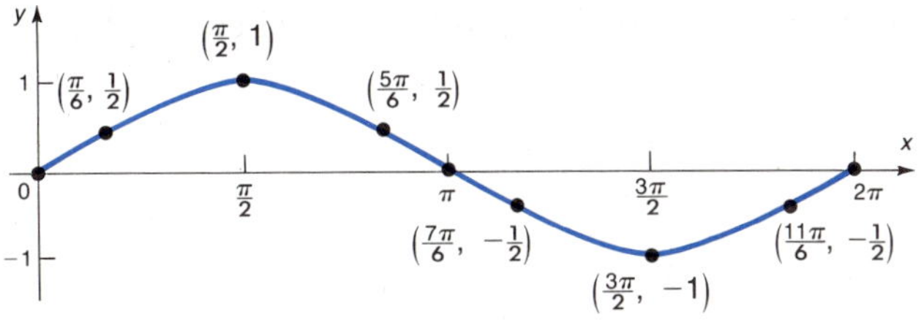

The graph in Figure 1 is one period of the graph of $y = \sin x$. To obtain a more complete graph of $y = \sin x$, we repeat this period in each direction, as shown in Figure 2.

Figure 2
$y = \sin x,\ -\infty < x < \infty$

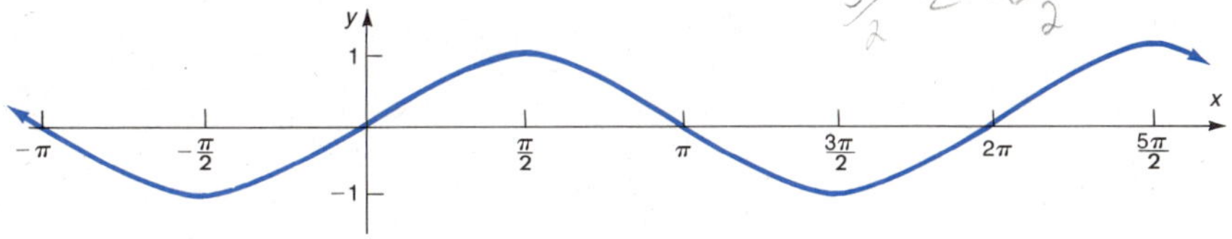

The graph of $y = \sin x$ illustrates some of the facts we already know about the sine function:

Characteristics of the Sine Function

1. The domain is the set of all real numbers.
2. The range consists of all real numbers from -1 to 1, inclusive.
3. The sine function is an odd function, as the symmetry of the graph with respect to the origin indicates.
4. The sine function is periodic, with period 2π.
5. The x-intercepts are $\ldots, -2\pi, -\pi, 0, \pi, 2\pi, 3\pi, \ldots$; the y-intercept is 0.
6. The maximum value is 1 and occurs at $x = \ldots, -3\pi/2, \pi/2, 5\pi/2, 9\pi/2, \ldots$; the minimum value is -1 and occurs at $x = \ldots, -\pi/2, 3\pi/2, 7\pi/2, 11\pi/2, \ldots$.

The graphing techniques introduced in Chapter 4 may be used to graph functions that are variations of the sine function.

Example 1 Use the graph of $y = \sin x$ to graph: $y = -\sin x + 2$

Solution Figure 3 illustrates the steps.

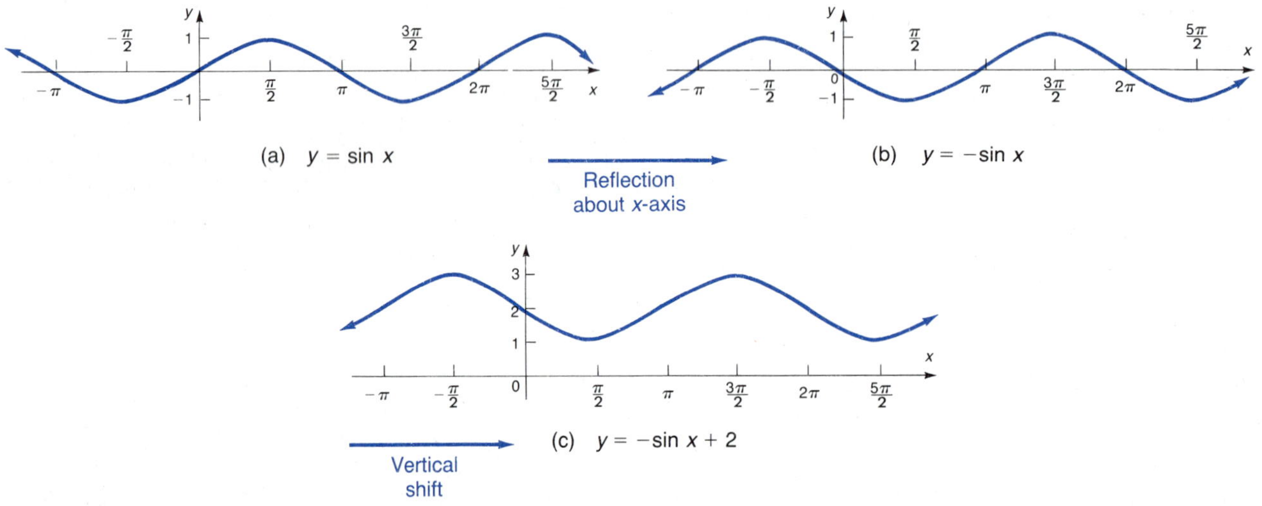

Figure 3

Example 2 Use the graph of $y = \sin x$ to graph: $y = \sin\left(x - \frac{\pi}{4}\right)$

Solution Figure 4 illustrates the steps.

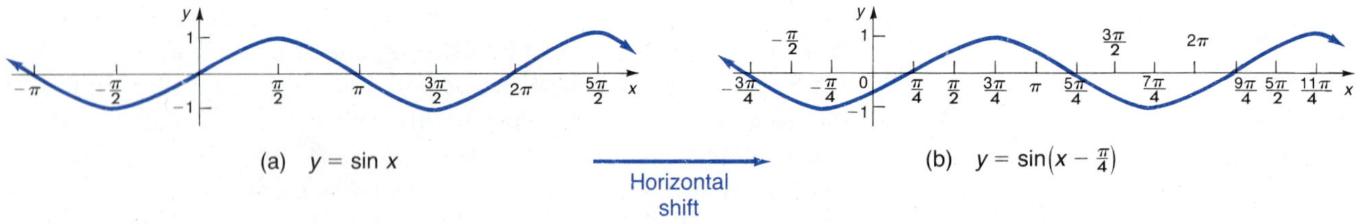

Figure 4

The Graph of $y = \cos x$

The cosine function also has period 2π. Thus, we proceed as we did with the sine function by constructing Table 2, which lists some points on the graph of $y = \cos x$, $0 \le x \le 2\pi$. As the table shows, the graph of $y = \cos x$, $0 \le x \le 2\pi$, begins at the point $(0, 1)$. As x increases from 0 to $\pi/2$ to π, the value of y decreases from 1 to 0 to -1; as x increases from π to $3\pi/2$ to 2π, the value of y increases from -1 to 0 to 1. As before, we plot the points in Table 2 to get one period of the graph of $y = \cos x$. See Figure 5.

Table 2

x	$y = \cos x$	(x, y)
0	1	(0, 1)
$\pi/3$	$\frac{1}{2}$	$(\pi/3, \frac{1}{2})$
$\pi/2$	0	$(\pi/2, 0)$
$2\pi/3$	$-\frac{1}{2}$	$(2\pi/3, -\frac{1}{2})$
π	-1	$(\pi, -1)$
$4\pi/3$	$-\frac{1}{2}$	$(4\pi/3, -\frac{1}{2})$
$3\pi/2$	0	$(3\pi/2, 0)$
$5\pi/3$	$\frac{1}{2}$	$(5\pi/3, \frac{1}{2})$
2π	1	$(2\pi, 1)$

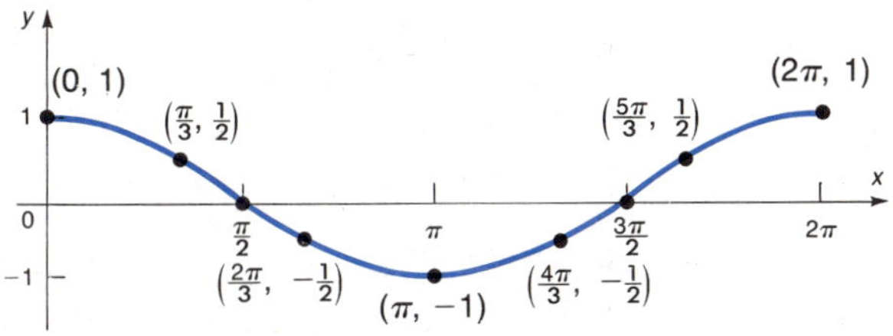

Figure 5
$y = \cos x,\ 0 \le x \le 2\pi$

A more complete graph of $y = \cos x$ is obtained by repeating this period in each direction, as shown in Figure 6.

Figure 6
$y = \cos x,\ -\infty < x < \infty$

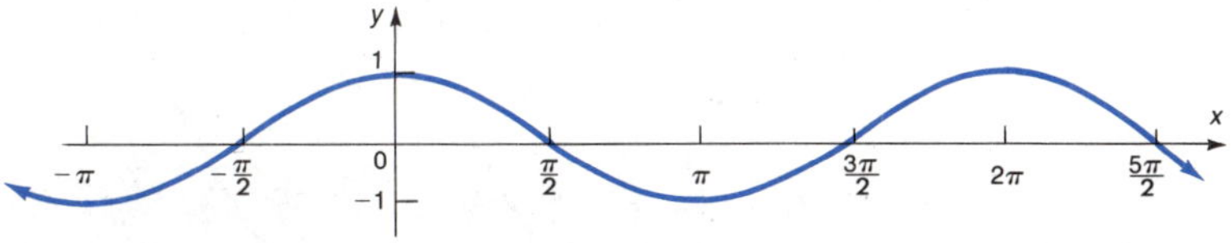

The graph of $y = \cos x$ illustrates some of the facts we already know about the cosine function:

Characteristics of the Cosine Function

1. The domain is the set of all real numbers
2. The range consists of all real numbers from -1 to 1, inclusive.
3. The cosine function is an even function, as the symmetry of the graph with respect to the y-axis indicates.
4. The cosine function is periodic, with period 2π.
5. The x-intercepts are $\ldots, -3\pi/2, -\pi/2, \pi/2, 3\pi/2, 5\pi/2, \ldots$; the y-intercept is 1.
6. The maximum value is 1 and occurs at $x = \ldots, -2\pi, 0, 2\pi, 4\pi, 6\pi, \ldots$; the minimum value is -1 and occurs at $x = \ldots, -\pi, \pi, 3\pi, 5\pi, \ldots$.

Again, the graphing techniques from Chapter 4 may be used to graph variations of the cosine function.

Example 3 Use the graph of $y = \cos x$ to graph: $y = 2\cos x$

Solution Figure 7 illustrates the graph, which is a vertical stretch of the graph of $y = \cos x$.

Figure 7
$y = 2\cos x$

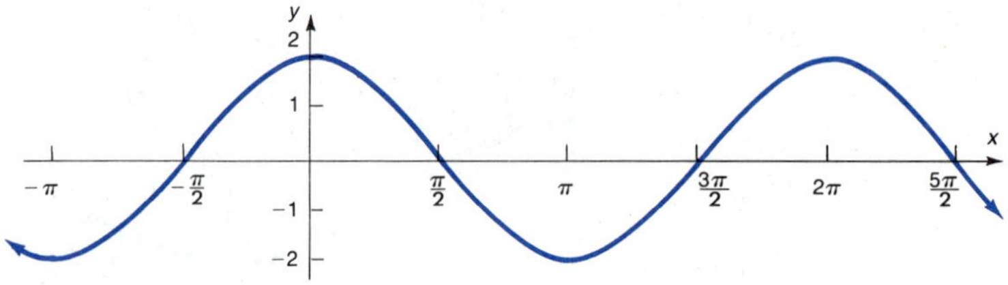

Example 4 Use the graph of $y = \cos x$ to graph: $y = \cos 3x$

Solution Figure 8 illustrates the graph, which is a horizontal compression of the graph of $y = \cos x$. Notice that, due to this compression, the period of $y = \cos 3x$ is $2\pi/3$, whereas the period of $y = \cos x$ is 2π. We shall comment more about this in the next section.

Figure 8
$y = \cos 3x$

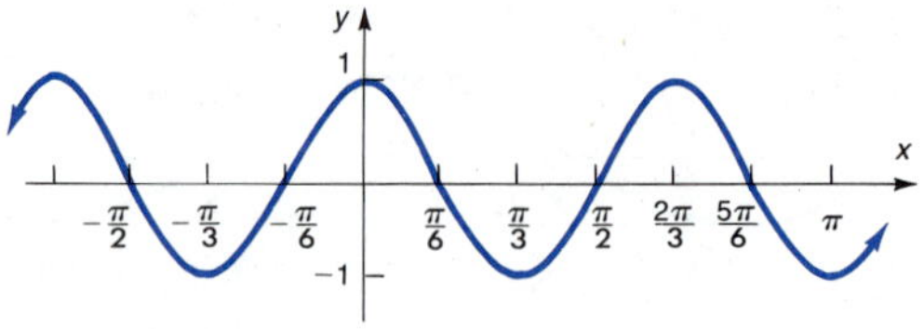

The Graph of $y = \tan x$

Because the tangent function has period π, we only need to determine the graph over some interval of length π. The rest of the graph will consist of repetitions of that graph. Because the tangent function is not defined at $\ldots, -3\pi/2, -\pi/2, \pi/2, 3\pi/2, \ldots$, we shall concentrate on the interval $(-\pi/2, \pi/2)$, of length π, and construct Table 3, which lists

Table 3

x	$y = \tan x$	(x, y)
$-\pi/3$	$-\sqrt{3} \approx -1.73$	$(-\pi/3, -\sqrt{3})$
$-\pi/4$	-1	$(-\pi/4, -1)$
$-\pi/6$	$-\sqrt{3}/3 \approx -0.58$	$(-\pi/6, -\sqrt{3}/3)$
0	0	$(0, 0)$
$\pi/6$	$\sqrt{3}/3 \approx 0.58$	$(\pi/6, \sqrt{3}/3)$
$\pi/4$	1	$(\pi/4, 1)$
$\pi/3$	$\sqrt{3} \approx 1.73$	$(\pi/3, \sqrt{3})$

some points on the graph of $y = \tan x$, $-\pi/2 < x < \pi/2$. As before, we plot the points in the table and connect them with a smooth curve. See Figure 9 for a partial graph of $y = \tan x$, where $-\pi/3 \le x \le \pi/3$.

Figure 9
$y = \tan x,\ -\pi/3 \le x \le \pi/3$

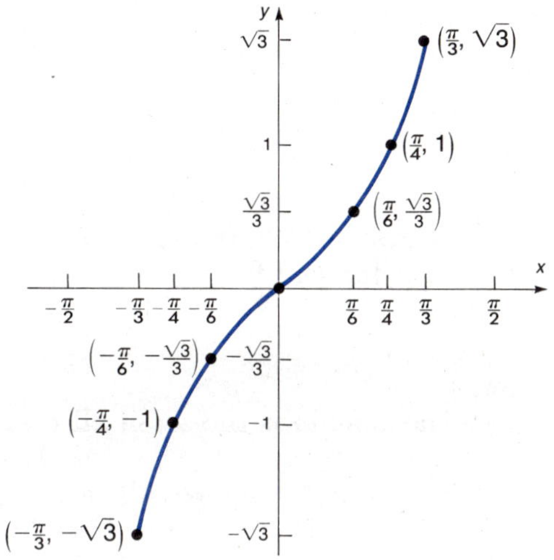

To complete the graph of $y = \tan x$, we need to investigate the behavior of the function as x approaches $-\pi/2$ and $\pi/2$. We must be careful, though, because $y = \tan x$ is not defined at these numbers. To determine this behavior, we use the identity

$$\tan x = \frac{\sin x}{\cos x}$$

If x is close to $\pi/2$ but remains less than $\pi/2$, then $\sin x$ will be close to 1 and $\cos x$ will be positive and close to 0. (Refer back to the graphs of the sine function and the cosine function.) Hence, the ratio $(\sin x)/(\cos x)$ will be positive and large. In fact, the closer x gets to $\pi/2$, the closer $\sin x$ gets to 1 and $\cos x$ gets to 0, so that $\tan x$ approaches ∞. In other words, the vertical line $x = \pi/2$ is a vertical asymptote to the graph of $y = \tan x$.

If x is close to $-\pi/2$ but remains greater than $-\pi/2$, then $\sin x$ will be close to -1 and $\cos x$ will be positive and close to 0. Hence, the ratio $(\sin x)/(\cos x)$ approaches $-\infty$. In other words, the vertical line $x = -\pi/2$ is also a vertical asymptote to the graph.

With these observations, we can complete one period of the graph, and we obtain the graph of $y = \tan x$ by repeating this period, as shown in Figure 10.

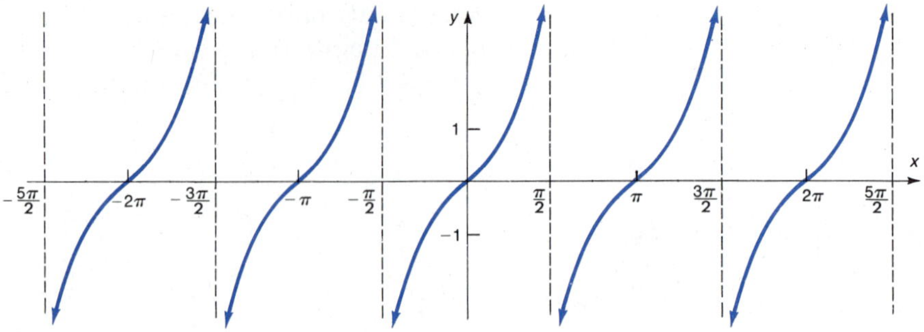

Figure 10
$y = \tan x$, $-\infty < x < \infty$, x not equal to odd multiples of $\pi/2$

The graph of $y = \tan x$ illustrates some of the facts we already know about the tangent function:

Characteristics of the Tangent Function

1. The domain is the set of all real numbers, except odd multiples of $\pi/2$.
2. The range consists of all real numbers.
3. The tangent function is an odd function, as the symmetry of the graph with respect to the origin indicates.
4. The tangent function is periodic, with period π.
5. The x-intercepts are $\ldots, -2\pi, -\pi, 0, \pi, 2\pi, 3\pi, \ldots$; the y-intercept is 0.
6. Vertical asymptotes occur at $x = \ldots, -3\pi/2, -\pi/2, \pi/2, 3\pi/2, \ldots$.

Example 5 Graph: $y = \tan\left(x + \frac{\pi}{4}\right)$

Solution We start with the graph of $y = \tan x$ and shift it horizontally to the left $\pi/4$ unit. See Figure 11.

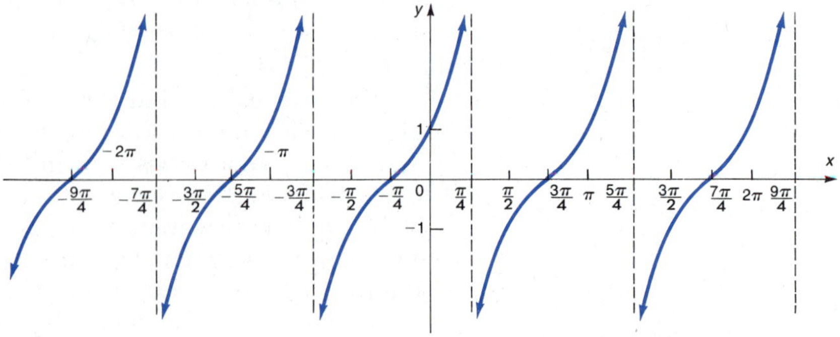

Figure 11
$y = \tan\left(x + \frac{\pi}{4}\right)$

■

The Graphs of $y = \csc x$, $y = \sec x$, and $y = \cot x$

The cosecant and secant functions, sometimes referred to as **reciprocal functions**, are graphed by making use of the reciprocal identities

$$\csc x = \frac{1}{\sin x} \qquad \text{and} \qquad \sec x = \frac{1}{\cos x}$$

For example, the value of the cosecant function $y = \csc x$ at a given number x equals the reciprocal of the corresponding value of the sine function, provided the value of the sine function is not 0. If the value of $\sin x$ is 0 then, at such numbers x, the cosecant function is not defined. In fact, the graph of the cosecant function has vertical asymptotes at integral multiples of π. See Figure 12 for the graph.

Figure 12
$y = \csc x$, $-\infty < x < \infty$, x not equal to integral multiples of π, $|y| \geq 1$

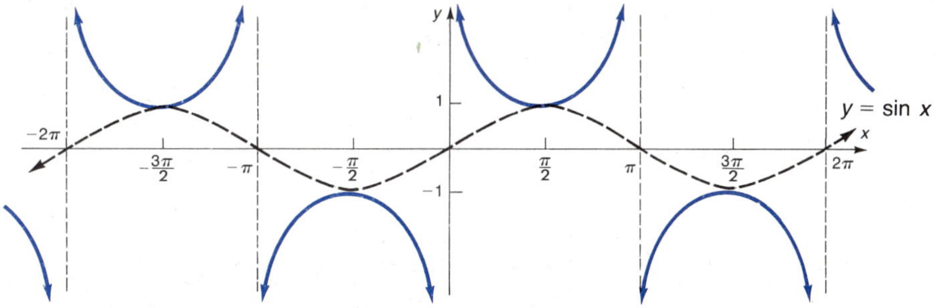

Using the idea of reciprocals, we can similarly obtain the graph of $y = \sec x$. See Figure 13.

Figure 13
$y = \sec x$, $-\infty < x < \infty$, x not equal to odd multiples of $\pi/2$, $|y| \geq 1$

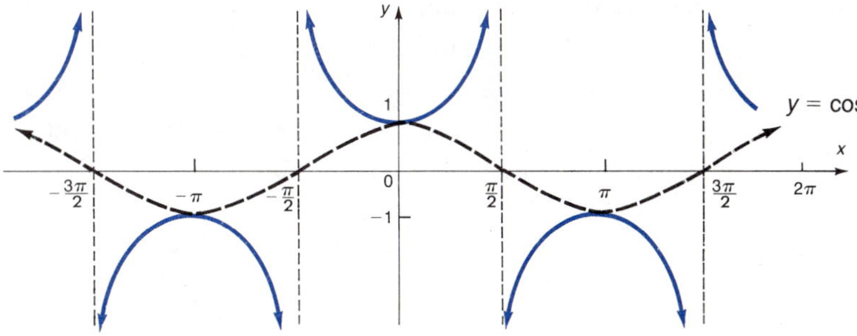

We obtain the graph of $y = \cot x$ as we did the graph of $y = \tan x$. The period of $y = \cot x$ is π. Because the cotangent function is not defined for integral multiples of π, we shall concentrate on the interval $(0, \pi)$. Table 4 lists some points on the graph of $y = \cot x$, $0 < x < \pi$. As x approaches 0, but remains greater than 0, the value of $\cos x$ will be close to 1 and the value of $\sin x$ will be positive and close to 0. Hence, the ratio $(\cos x)/(\sin x) = \cot x$ will be positive and large, so that as x approaches 0, $\cot x$ approaches ∞. Similarly, as x approaches but remains less than π, the value of $\cos x$ will be close to -1 and the value of $\sin x$ will be positive and close to 0. Hence, the ratio $(\cos x)/(\sin x) = \cot x$ will be negative and will approach $-\infty$ as x approaches π. Figure 14 shows the graph.

Table 4

x	$y = \cot x$	(x, y)
$\pi/6$	$\sqrt{3}$	$(\pi/6, \sqrt{3})$
$\pi/4$	1	$(\pi/4, 1)$
$\pi/3$	$\sqrt{3}/3$	$(\pi/3, \sqrt{3}/3)$
$\pi/2$	0	$(\pi/2, 0)$
$2\pi/3$	$-\sqrt{3}/3$	$(2\pi/3, -\sqrt{3}/3$
$3\pi/4$	-1	$(3\pi/4, -1)$
$5\pi/6$	$-\sqrt{3}$	$(5\pi/6, -\sqrt{3})$

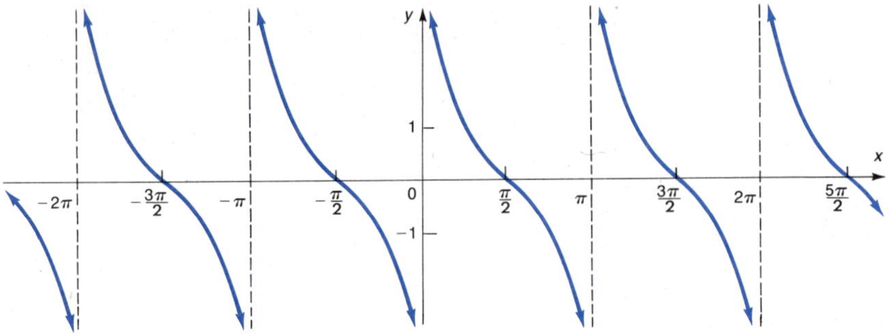

Figure 14
$y = \cot x$, $-\infty < x < \infty$, x not equal to integral multiples of π, $-\infty < y < \infty$

EXERCISE 8.1

In Problems 1–20, refer to the graphs of $y = \sin x$, $y = \cos x$, $y = \tan x$, $y = \csc x$, $y = \sec x$, and $y = \cot x$ in this section to answer each question, if necessary.

1. What is the y-intercept of $y = \sin x$?
2. What is the y-intercept of $y = \cos x$?
3. What is the y-intercept of $y = \tan x$?
4. What is the y-intercept of $y = \cot x$?
5. For what numbers x, $-\pi \le x \le \pi$, is the graph of $y = \sin x$ increasing?
6. For what numbers x, $-\pi \le x \le \pi$, is the graph of $y = \cos x$ decreasing?
7. Which of the trigonometric functions have graphs that are symmetric with respect to the y-axis?
8. Which of the trigonometric functions have graphs that are symmetric with respect to the origin?
9. What is the largest value of $y = \sin x$?
10. What is the smallest value of $y = \cos x$?
11. For what numbers x, $0 \le x \le 2\pi$, does $\sin x = 0$?
12. For what numbers x, $0 \le x \le 2\pi$, does $\cos x = 0$?
13. For what numbers x, $-2\pi \le x \le 2\pi$, does the graph of $y = \sec x$ have vertical asymptotes?
14. For what numbers x, $-2\pi \le x \le 2\pi$, does the graph of $y = \csc x$ have vertical asymptotes?
15. For what numbers x, $-2\pi \le x \le 2\pi$, does the graph of $y = \tan x$ have vertical asymptotes?

16. For what numbers x, $-2\pi \le x \le 2\pi$, does the graph of $y = \cot x$ have vertical asymptotes?

17. For what numbers x, $-2\pi \le x \le 2\pi$, does $\sec x = 1$? What about $\sec x = -1$?

18. For what numbers x, $-2\pi \le x \le 2\pi$, does $\csc x = 1$? What about $\csc x = -1$?

19. For what numbers x, $-2\pi \le x \le 2\pi$, does $\sin x = 1$? What about $\sin x = -1$?

20. For what numbers x, $-2\pi \le x \le 2\pi$, does $\cos x = 1$? What about $\cos x = -1$?

In Problems 21–50, graph each function. Show at least one period.

21. $y = 3 \sin x$	**22.** $y = 4 \cos x$	**23.** $y = \cos\left(x + \frac{\pi}{4}\right)$
24. $y = \sin(x - \pi)$	**25.** $y = \tan x - 1$	**26.** $y = \cot x + 1$
27. $y = -2 \sin x$	**28.** $y = -3 \cos x$	**29.** $y = \sin \pi x$
30. $y = \cos \frac{\pi}{2}x$	**31.** $y = 2 \sec x$	**32.** $y = 3 \csc x$
33. $y = -\sec x$	**34.** $y = -\cot x$	**35.** $y = \sec\left(x - \frac{\pi}{2}\right)$
36. $y = \csc(x - \pi)$	**37.** $y = \tan(x - \pi)$	**38.** $y = \cot(x - \pi)$
39. $y = 3 \sin 2x$	**40.** $y = 4 \cos 2x$	**41.** $y = 3 \tan 2x$
42. $y = 4 \tan \frac{1}{2}x$	**43.** $y = \sec 2x$	**44.** $y = \csc \frac{1}{2}x$
45. $y = \cot \pi x$	**46.** $y = \cot 2x$	**47.** $y = -3 \tan 4x$
48. $y = -3 \tan 2x$	**49.** $y = 2 \sec \frac{1}{2}x$	**50.** $y = 2 \sec 3x$

51. Graph

$$y = \tan x \quad \text{and} \quad y = -\cot\left(x + \frac{\pi}{2}\right)$$

Do you think that $\tan x = -\cot\left(x + \frac{\pi}{2}\right)$?

52. Graph

$$y = \sin x \quad \text{and} \quad y = \cos\left(x - \frac{\pi}{2}\right)$$

Do you think that $\sin x = \cos\left(x - \frac{\pi}{2}\right)$?

8.2 Sinusoidal Graphs

The graph of $y = \cos x$, when compared to the graph of $y = \sin x$, suggests that the graph of $y = \sin x$ is the same as the graph of $y = \cos x$ after a horizontal shift of $\pi/2$ units to the right. That is,

$$\sin x = \cos\left(x - \frac{\pi}{2}\right)$$

(We shall prove this fact in the next chapter.) Because of this similarity, the graphs of sine functions and cosine functions are referred to as **sinusoidal graphs**. Let's look at some general characteristics of such graphs.

Refer to Figure 7 (in the preceding section), where the graph of $y = 2 \cos x$ is shown. We see that the values of $y = 2 \cos x$ lie between -2 and 2, inclusive. In general, the values of the functions $y = A \sin x$ and $y = A \cos x$, where $A \neq 0$, will always satisfy the inequalities

$$-|A| \leq A \sin x \leq |A| \qquad \text{and} \qquad -|A| \leq A \cos x \leq |A|$$

respectively. The number $|A|$ is called the **amplitude** of $y = A \sin x$ or $y = A \cos x$. See Figure 15(a).

(a) $y = A \sin x$, $A > 0$
Period $= 2\pi$

(b) $y = A \sin \omega x$, $A > 0$, $\omega > 0$
Period $= \frac{2\pi}{\omega}$

Figure 15

Now, refer to Figure 8 (in the preceding section), where the graph of $y = \cos 3x$ is shown. We see that the period of this function is $2\pi/3$. In general, if $\omega > 0$, the functions $y = \sin \omega x$ and $y = \cos \omega x$ will have period $T = 2\pi/\omega$. To see why, recall that the graph of $y = \sin \omega x$ is obtained from the graph of $y = \sin x$ by performing an appropriate horizontal compression or stretch. This horizontal compression replaces the interval $[0, 2\pi]$, which contains one period of the graph of $y = \sin x$, by the interval $[0, 2\pi/\omega]$, which contains one period of the graph of $y = \sin \omega x$. Thus, the period of the functions $y = \sin \omega x$ and $y = \cos \omega x$, $\omega > 0$, is $2\pi/\omega$. See Figure 15(b).

If $\omega < 0$ in $y = \sin \omega x$ or $y = \cos \omega x$, we use the facts that

$$\sin \omega x = -\sin(-\omega x) \qquad \text{and} \qquad \cos \omega x = \cos(-\omega x)$$

to obtain an equivalent form in which the coefficient of x is positive.

Theorem If $\omega > 0$, the amplitude and period of $y = A \sin \omega x$ and $y = A \cos \omega x$ are given by

$$\text{Amplitude} = |A| \qquad \text{Period} = T = \frac{2\pi}{\omega} \tag{1}$$

■

Example 1 Determine the amplitude and period of $y = 3 \sin 4x$, and graph the function.

Solution Comparing $y = 3 \sin 4x$ to $y = A \sin \omega x$, we find that $A = 3$ and $\omega = 4$. Thus, from equation (1), the amplitude is $|A| = 3$ and the period is $T = 2\pi/\omega = 2\pi/4 = \pi/2$. We can use this information to graph $y = 3 \sin 4x$ by beginning as shown in Figure 16(a). Notice that the amplitude is used to scale the y-axis, and the period is used to scale the x-axis. (This will usually result in a different scale for each axis.) Now we fill in the graph of the sine function. See Figure 16(b).

Figure 16

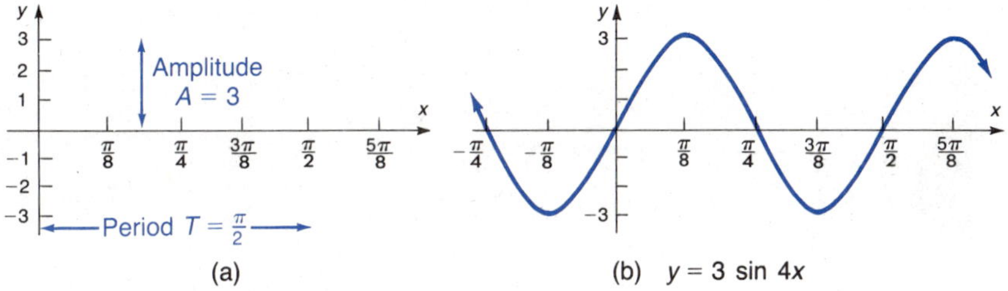

Example 2 Determine the amplitude and period of $y = -4 \cos \pi x$, and graph the function.

Solution Comparing $y = -4 \cos \pi x$ with $y = A \cos \omega x$, we find that $A = -4$ and $\omega = \pi$. Thus, the amplitude is $|A| = |-4| = 4$, and the period is $T = 2\pi/\omega = 2\pi/\pi = 2$. We use the amplitude to scale the y-axis, the period to scale the x-axis, and fill in the graph of the cosine function, thus obtaining the graph of $y = 4 \cos \pi x$ shown in Figure 17(a). Now, since we want the graph of $y = -4 \cos \pi x$, we reflect the graph of $y = 4 \cos \pi x$ about the x-axis, as shown in Figure 17(b).

Figure 17

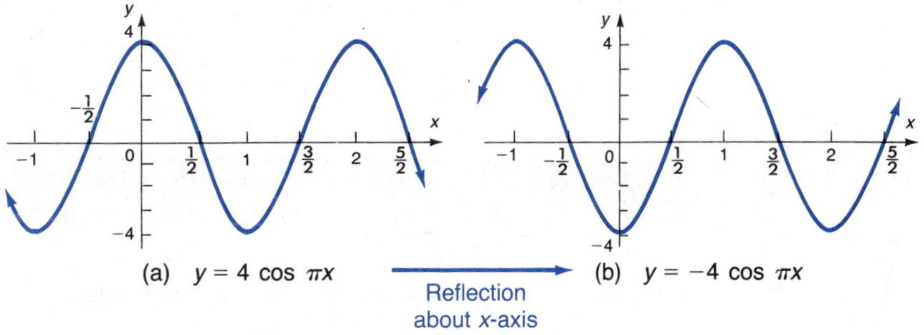

Example 3 Determine the amplitude and period of $y = 2 \sin(-\pi x)$, and graph the function.

Solution Since the sine function is odd, we use the equivalent form

$$y = -2 \sin \pi x$$

Comparing $y = -2 \sin \pi x$ to $y = A \sin \omega x$, we find that $A = -2$ and $\omega = \pi$. Thus, the amplitude is $|A| = 2$, and the period is $T = 2\pi/\omega = 2\pi/\pi = 2$. Again, we use the amplitude to scale the y-axis, the period to scale the x-axis, and fill in the graph of the sine function. Figure 18(a) shows the resulting graph of $y = 2 \sin \pi x$. To obtain the graph of $y = -2 \sin \pi x$, we reflect the graph of $y = 2 \sin \pi x$ about the x-axis, as shown in Figure 18(b).

Figure 18

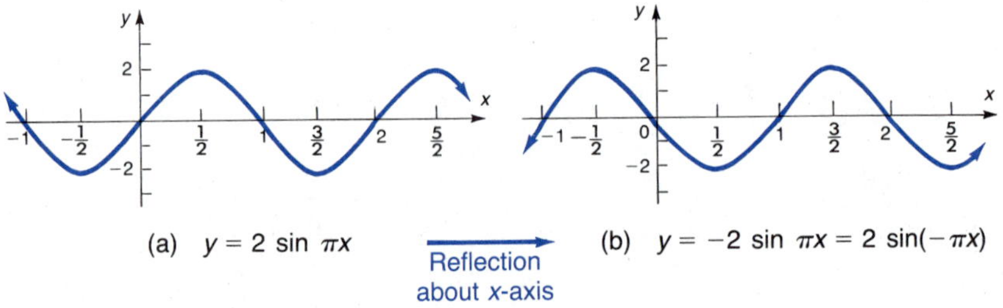

(a) $y = 2 \sin \pi x$ (b) $y = -2 \sin \pi x = 2 \sin(-\pi x)$

We also can use the ideas of amplitude and period to identify a function when its graph is given.

Example 4 Find an equation for the graph shown in Figure 19.

Figure 19

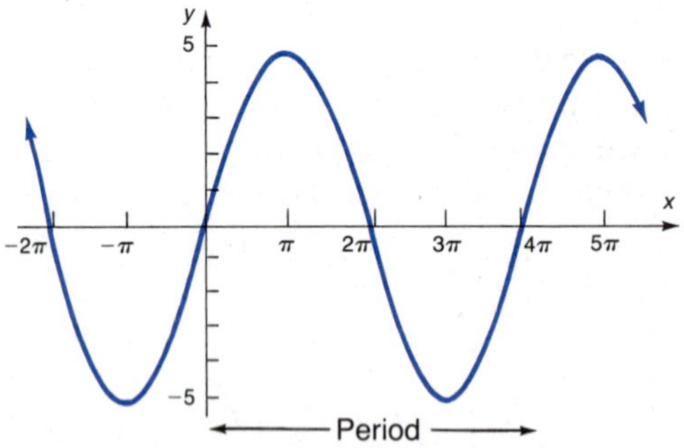

Solution This graph can be viewed as the graph of a sine function with amplitude $A = 5$.* The period T is observed to be 4π. Thus, by equation (1),

*The equation also could be viewed as a cosine function with a horizontal shift, but viewing it as a sine function is easier.

$$T = \frac{2\pi}{\omega}$$

$$4\pi = \frac{2\pi}{\omega}$$

$$\omega = \frac{2\pi}{4\pi} = \frac{1}{2}$$

A sine function whose graph is given in Figure 19 is

$$y = A \sin \omega x = 5 \sin \frac{x}{2}$$

■

Example 5 Find an equation for the graph shown in Figure 20.

Figure 20

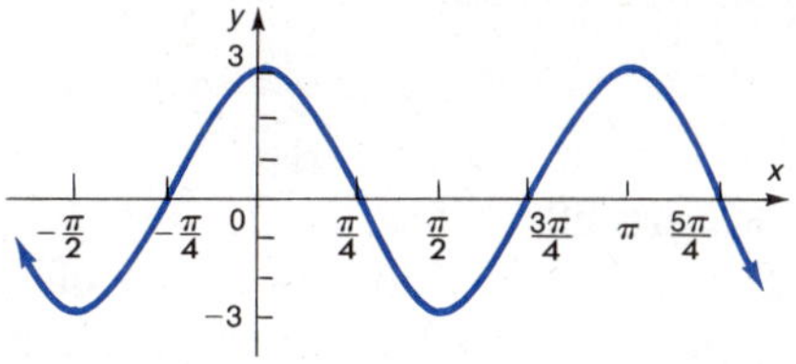

Solution From the graph we conclude that it is easiest to view the equation as a cosine function with amplitude $A = 3$ and period $T = \pi$. Thus, $2\pi/\omega = \pi$, so $\omega = 2$. A cosine function whose graph is given in Figure 20 is

$$y = A \cos \omega x = 3 \cos 2x$$

■

Example 6 Find an equation for the graph shown in Figure 21.

Figure 21

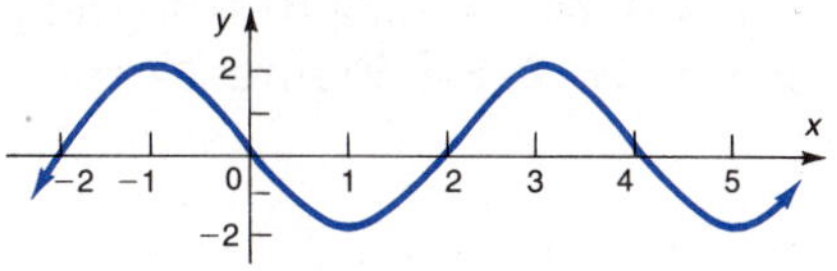

Solution The graph is sinusoidal, with amplitude $A = 2$. The period is 4, so that $2\pi/\omega = 4$, or $\omega = \pi/2$. Since the graph passes through the origin, it is easiest to view the equation as a sine function, but notice that the graph is actually the reflection of a sine function about the x-axis (since the graph is decreasing near the origin). Thus, we have

$$y = -A \sin \omega x = -2 \sin \frac{\pi}{2}x \qquad A > 0$$

■

Phase Shift

We have seen that the graph of $y = A \sin \omega x$, $\omega > 0$, has amplitude $|A|$ and period $T = 2\pi/\omega$. Thus, one period can be drawn as x varies from 0 to $2\pi/\omega$ or, equivalently, as ωx varies from 0 to 2π. See Figure 22.

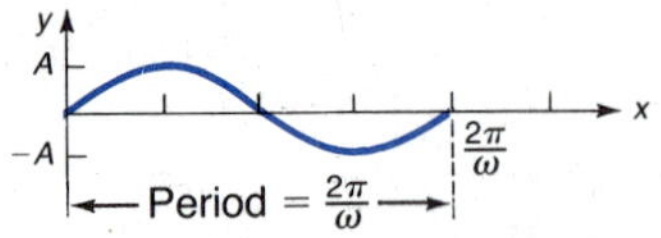

Figure 22
One period of $y = A \sin \omega x$, $A > 0$, $\omega > 0$

We now want to discuss the graph of

$$y = A \sin(\omega x - \phi)$$

where $\omega > 0$ and ϕ (the Greek letter phi) are real numbers. The graph will be a sine curve of amplitude $|A|$. As $\omega x - \phi$ varies from 0 to 2π, one period will be traced out. This period will begin when

$$\omega x - \phi = 0 \qquad \text{or} \qquad x = \frac{\phi}{\omega}$$

and will end when

$$\omega x - \phi = 2\pi \qquad \text{or} \qquad x = \frac{2\pi}{\omega} + \frac{\phi}{\omega}$$

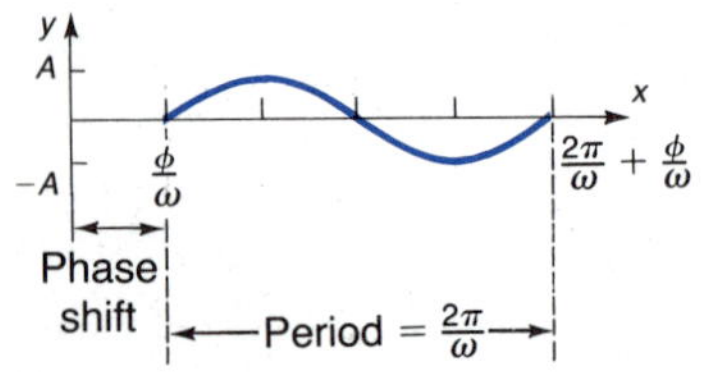

Figure 23
One period of $y = A \sin(\omega x - \phi)$, $A > 0$, $\omega > 0$, $\phi > 0$

See Figure 23.

Thus, we see that the graph of $y = A \sin(\omega x - \phi)$ is the same as the graph of $y = A \sin \omega x$, except that it has been shifted ϕ/ω units (to the right if $\phi > 0$ and to the left if $\phi < 0$). This number ϕ/ω is called the **phase shift** of the graph of $y = A \sin(\omega x - \phi)$.

These results also apply to the graph of $y = A \cos(\omega x - \phi)$.

Example 7 Find the amplitude, period, and phase shift of $y = 3 \sin(2x - \pi)$, and graph the function.

Solution Comparing $y = 3 \sin(2x - \pi)$ to $y = A \sin(\omega x - \phi)$, we find that $A = 3$, $\omega = 2$, and $\phi = \pi$. The graph is a sine curve with amplitude $A = 3$ and period $T = 2\pi/\omega = 2\pi/2 = \pi$. One period of the sine curve begins at $2x - \pi = 0$, or $x = \pi/2$ (this is the phase shift) and ends at $2x - \pi = 2\pi$, or $x = 3\pi/2$. See Figure 24.

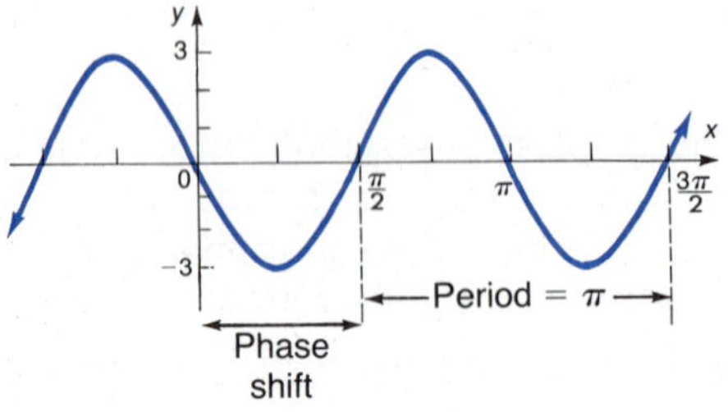

Figure 24
$y = 3 \sin(2x - \pi)$

■

Example 8 Find the amplitude, period, and phase shift of $y = 3 \cos(4x + 2\pi)$, and graph the function.

Solution Comparing $y = 3\cos(4x + 2\pi)$ to $y = A\cos(\omega x - \phi)$, we see that $A = 3$, $\omega = 4$, and $\phi = -2\pi$. The graph is a cosine curve with amplitude $A = 3$ and period $T = 2\pi/\omega = 2\pi/4 = \pi/2$. One period of the cosine curve begins at $4x + 2\pi = 0$, or $x = -\pi/2$ (the phase shift) and ends at $4x + 2\pi = 2\pi$, or $x = 0$. See Figure 25.

Figure 25
$y = 3\cos(4x + 2\pi)$

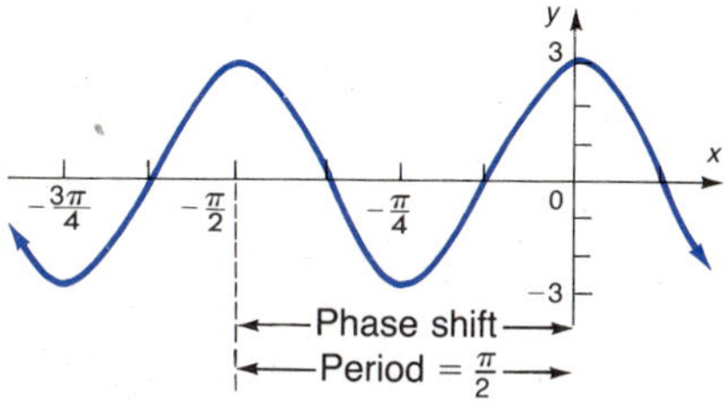

EXERCISE 8.2

In Problems 1–10, determine the amplitude and period of each function without graphing.

1. $y = 2\sin x$ **2.** $y = 3\cos x$ **3.** $y = -4\cos 2x$ **4.** $y = -\sin \frac{1}{2}x$

5. $y = 6\sin \pi x$ **6.** $y = -3\cos 3x$ **7.** $y = -\frac{1}{2}\cos \frac{3}{2}x$ **8.** $y = \frac{4}{3}\sin \frac{2}{3}x$

9. $y = \frac{5}{3}\sin\left(-\frac{2\pi}{3}x\right)$ **10.** $y = \frac{9}{5}\cos\left(-\frac{3\pi}{2}x\right)$

In Problems 11–20, match the given function to one of the graphs (A)–(J).

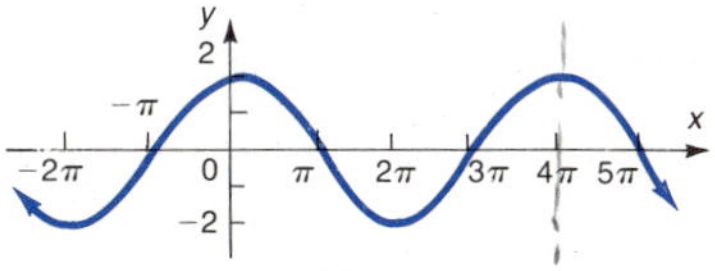

(A)

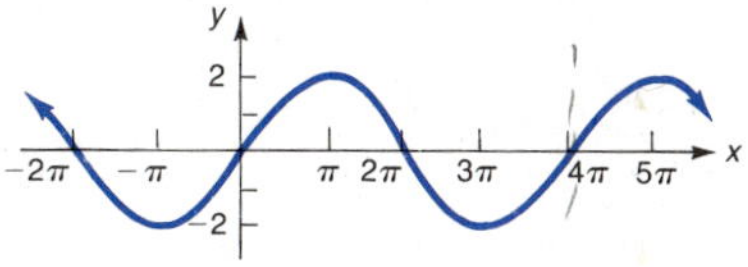

(B)

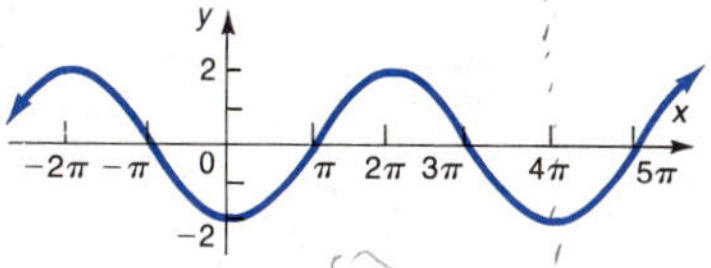

(C)

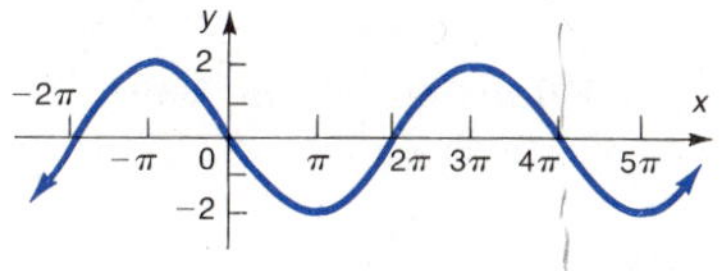

(D)

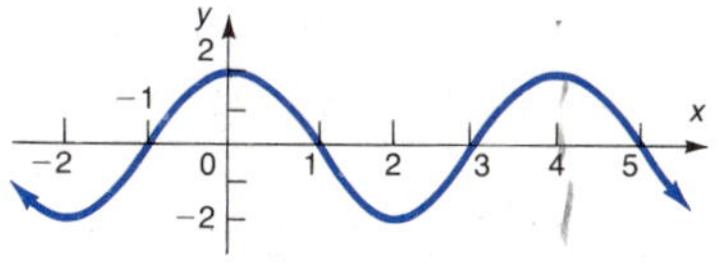

(E)

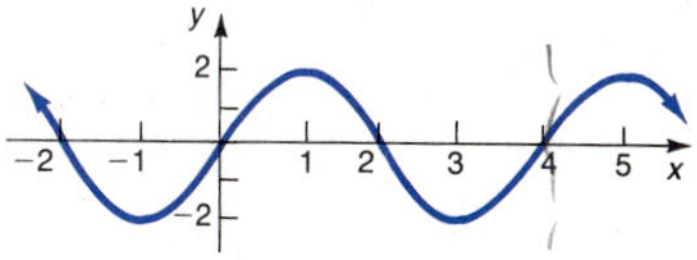

(F)

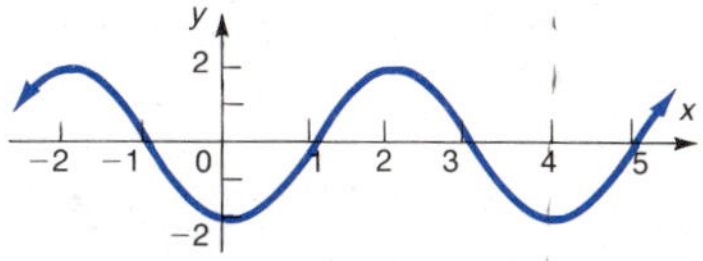

(G)

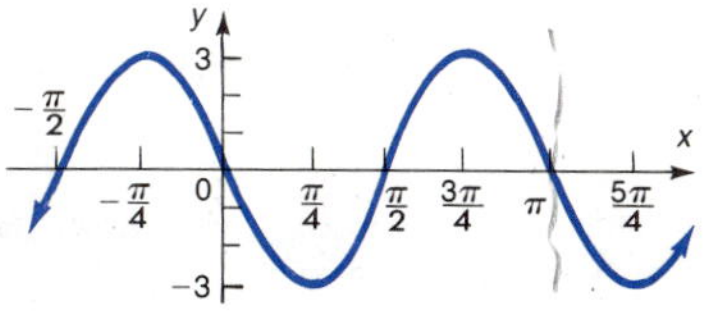

(H)

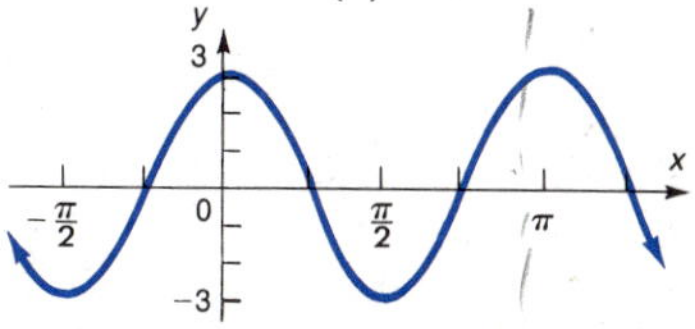

(I)

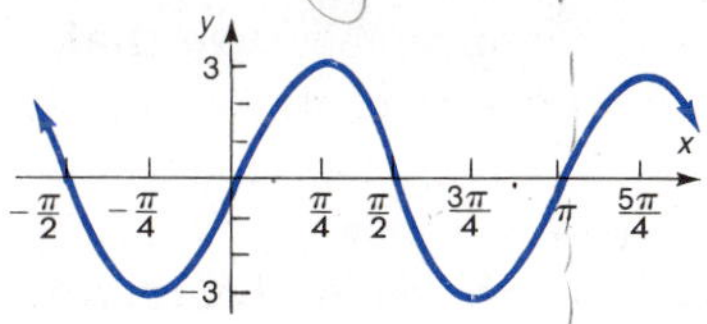

(J)

11. $y = 2\sin \frac{\pi}{2}x$ **12.** $y = 2\cos \frac{\pi}{2}x$ **13.** $y = 2\cos \frac{1}{2}x$

14. $y = 3\cos 2x$ **15.** $y = -3\sin 2x$ **16.** $y = 2\sin \frac{1}{2}x$

17. $y = -2\cos \frac{1}{2}x$ **18.** $y = -2\cos \frac{\pi}{2}x$ **19.** $y = 3\sin 2x$

20. $y = -2\sin \frac{1}{2}x$

In Problems 21–30, graph each function.

21. $y = 5 \sin 4x$ **22.** $y = 4 \cos 6x$ **23.** $y = 5 \cos \pi x$ **24.** $y = 2 \sin \pi x$

25. $y = -2 \cos 2\pi x$ **26.** $y = -5 \cos 2\pi x$ **27.** $y = -4 \sin \frac{1}{2}x$ **28.** $y = -2 \cos \frac{1}{2}x$

29. $y = \frac{3}{2} \sin\left(-\frac{2}{3}x\right)$ **30.** $y = \frac{4}{3} \cos\left(-\frac{1}{3}x\right)$

In Problems 31–40, find a function whose graph is given.

31.

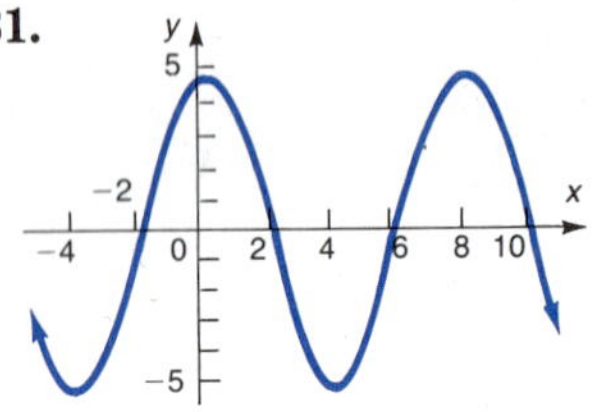

32.

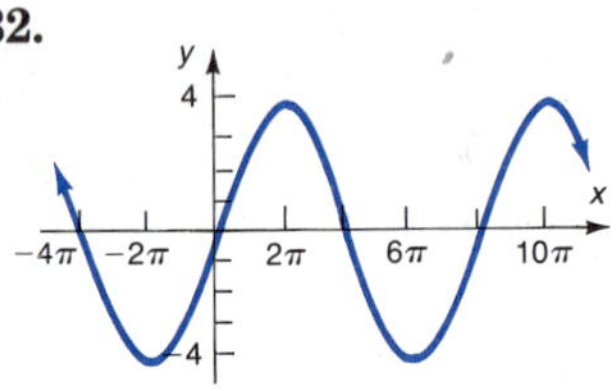

33.

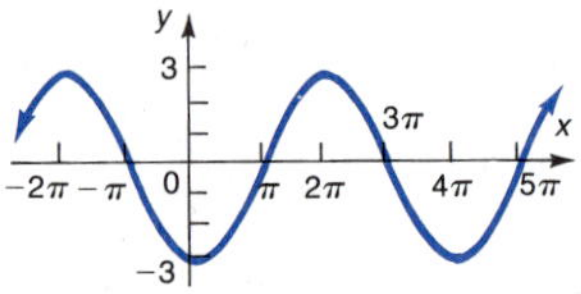

34.

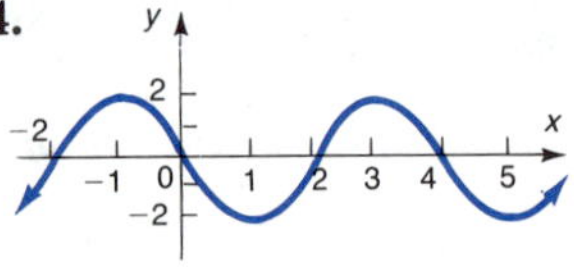

35.

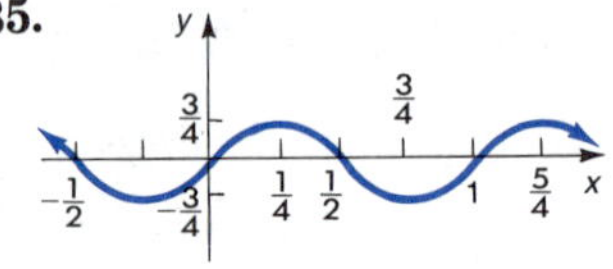

36.

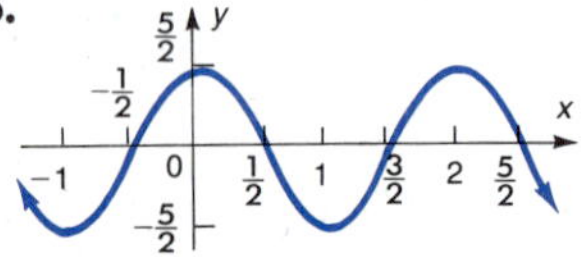

37.

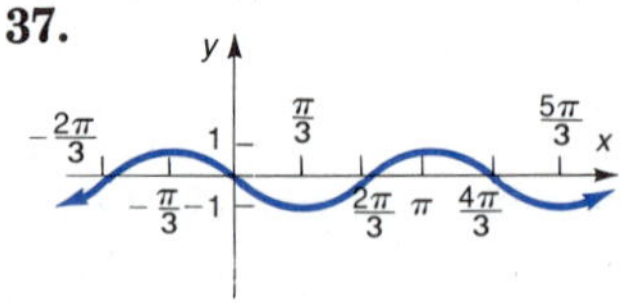

38.

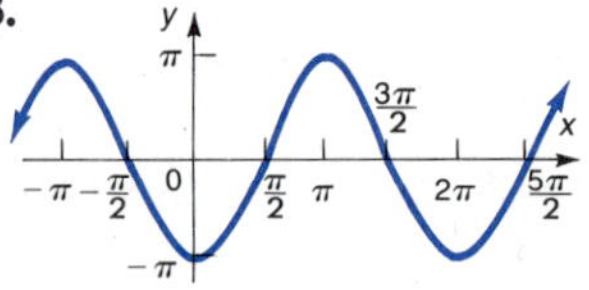

39.

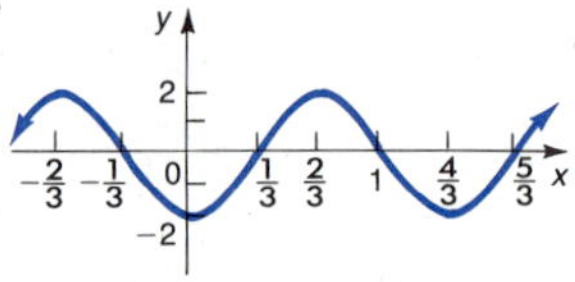

40.

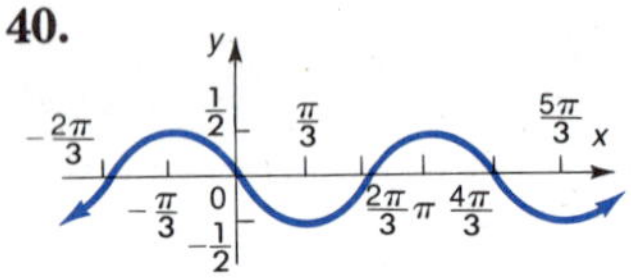

In Problems 41–46, find the amplitude, period, and phase shift of each function. Graph one period.

41. $y = 4 \sin(2x - \pi)$ **42.** $y = 3 \sin(3x - \pi)$ **43.** $y = 2 \cos\left(3x + \dfrac{\pi}{2}\right)$

44. $y = 3 \cos(2x + \pi)$ **45.** $y = -3 \sin\left(2x + \dfrac{\pi}{2}\right)$ **46.** $y = -2 \cos\left(2x - \dfrac{\pi}{2}\right)$

8.3 The Inverse Trigonometric Functions

In Section 4.5 we discussed inverse functions, and we noted that if a function is one-to-one, it will have an inverse. We also observed that if a function is not one-to-one, it may be possible to restrict its domain in some suitable manner such that the restricted function is one-to-one. In this section, we use these ideas to define inverse trigonometric functions. (You may wish to review Section 4.5 at this time.) We begin with the inverse of the sine function.

The Inverse Sine Function

In Figure 26 we reproduce the graph of $y = \sin x$. Because every horizontal line $y = b$, where b is between -1 and 1, intersects the graph of $y = \sin x$ infinitely many times, it follows from the horizontal-line test that the function $y = \sin x$ is not one-to-one.

Figure 26
$y = \sin x$, $-\infty < x < \infty$, $-1 \leq y \leq 1$

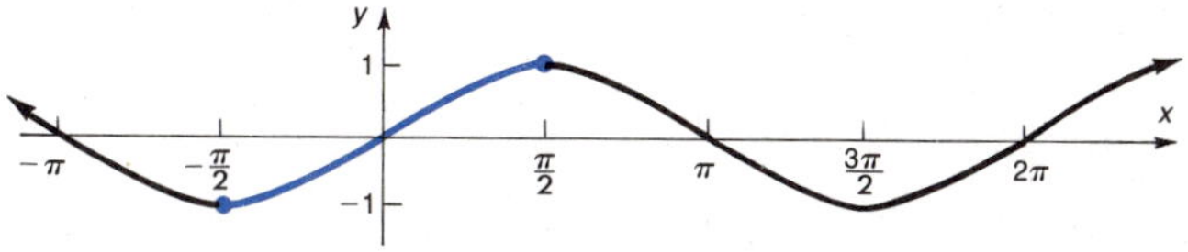

However, if we restrict the domain of $y = \sin x$ to the interval $[-\pi/2, \pi/2]$, the restricted function

$$y = \sin x \qquad -\pi/2 \leq x \leq \pi/2$$

is one-to-one, and hence, will have an inverse.* See Figure 27.

Figure 27
$y = \sin x$, $-\pi/2 \leq x \leq \pi/2$,
$-1 \leq y \leq 1$

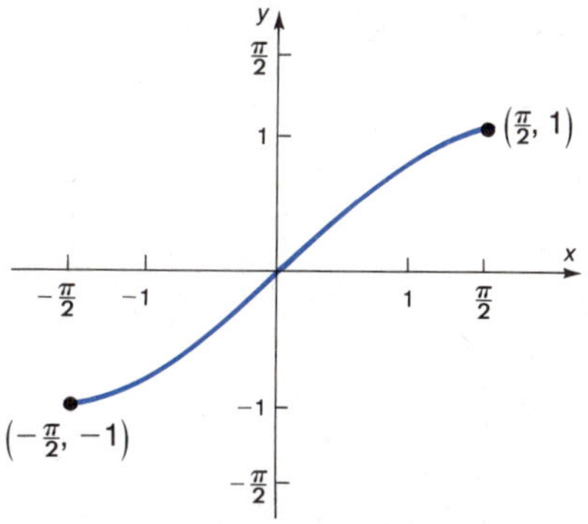

The inverse function is called the **inverse sine** of x and is symbolized by $y = \sin^{-1} x$. Thus,

Inverse Sine Function

$$y = \sin^{-1} x \quad \text{means} \quad x = \sin y \tag{1}$$
$$\text{where} \quad -\pi/2 \leq y \leq \pi/2 \quad \text{and} \quad -1 \leq x \leq 1$$

Because $y = \sin^{-1} x$ means $x = \sin y$, we read $y = \sin^{-1} x$ as "y is the angle whose sine equals x." Alternatively, we can say "y is the inverse sine of x." Be careful about the notation used. The superscript -1 that

*Although there are many other ways to restrict the domain and obtain a one-to-one function, mathematicians have agreed on a consistent use of the interval $[-\pi/2, \pi/2]$ in order to define the inverse of $y = \sin x$.

appears in $y = \sin^{-1} x$ is not an exponent, but is reminiscent of the symbolism f^{-1} used to denote the inverse of a function f. (To avoid this notation, some books use the notation $y = \arcsin x$ instead of $y = \sin^{-1} x$.)

Based on the general discussion of functions and their inverses (Section 4.5), we have the following results:

$$\sin^{-1}(\sin u) = u \quad \text{where} \quad -\pi/2 \leq u \leq \pi/2$$
$$\sin(\sin^{-1} v) = v \quad \text{where} \quad -1 \leq v \leq 1$$

Let's examine the function $y = \sin^{-1} x$. Its domain is $-1 \leq x \leq 1$, and its range is $-\pi/2 \leq y \leq \pi/2$. Its graph can be obtained by reflecting the restricted portion of the graph of $y = \sin x$ about the line $y = x$, as shown in Figure 28.

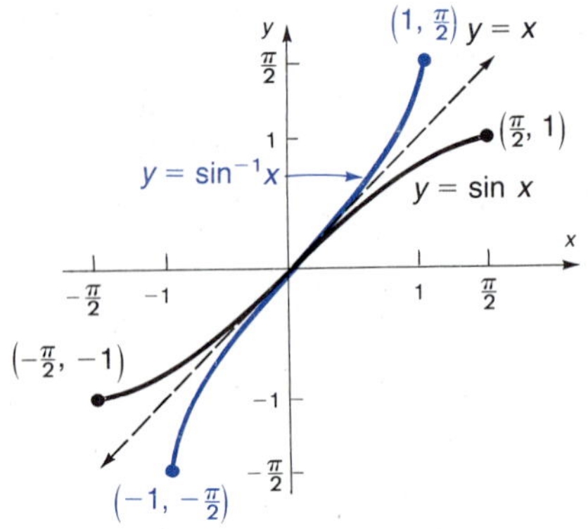

Figure 28
$y = \sin x, \ -\pi/2 \leq x \leq \pi/2$
$y = \sin^{-1} x, \ -1 \leq x \leq 1$

For some numbers x it is possible to find the exact value of $y = \sin^{-1} x$.

Example 1 Find the exact value of $\sin^{-1} 1$.

Solution Let $\theta = \sin^{-1} 1$. Then, we seek the angle θ, $-\pi/2 \leq \theta \leq \pi/2$, whose sine equals 1:

$$\theta = \sin^{-1} 1 \qquad -\pi/2 \leq \theta \leq \pi/2$$
$$\sin \theta = 1 \qquad -\pi/2 \leq \theta \leq \pi/2 \quad \text{By definition}$$

From Figure 29, we see that the only angle θ within the interval $[-\pi/2, \pi/2]$ whose sine is 1 is $\pi/2$. (Note that $\sin(5\pi/2)$ also equals 1, but $5\pi/2$ lies outside the interval $[-\pi/2, \pi/2]$, and hence, is not admissible.)

Figure 29

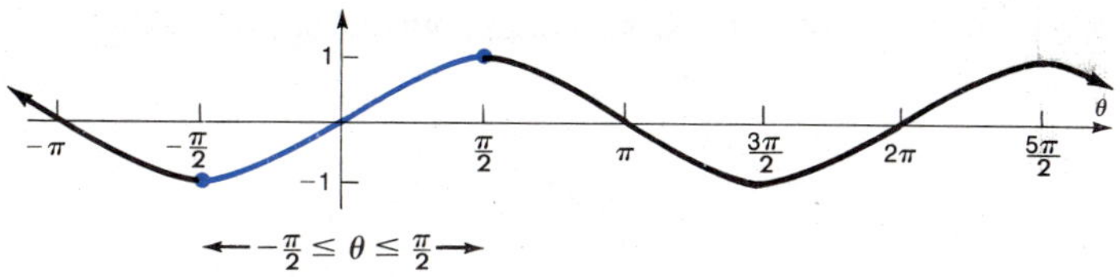

We conclude that

$$\theta = \frac{\pi}{2}$$

so that

$$\sin^{-1} 1 = \frac{\pi}{2}$$ ■

Example 2 Find the exact value of $\sin^{-1}(\sqrt{3}/2)$.

Solution Let $\theta = \sin^{-1}(\sqrt{3}/2)$. Then, we seek the angle θ, $-\pi/2 \le \theta \le \pi/2$, whose sine equals $\sqrt{3}/2$:

$$\theta = \sin^{-1}\frac{\sqrt{3}}{2} \qquad -\pi/2 \le \theta \le \pi/2$$

$$\sin\theta = \frac{\sqrt{3}}{2} \qquad -\pi/2 \le \theta \le \pi/2 \quad \text{By definition}$$

From Figure 30, we see that the only angle θ within the interval $[-\pi/2, \pi/2]$ whose sine is $\sqrt{3}/2$ is $\pi/3$. (Note that $\sin(2\pi/3)$ also equals $\sqrt{3}/2$, but $2\pi/3$ lies outside the interval $[-\pi/2, \pi/2]$, and hence, is not admissible.)

Figure 30

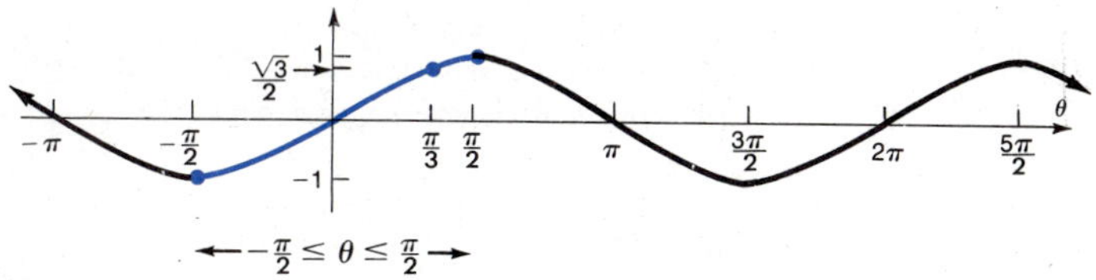

We conclude that

$$\theta = \frac{\pi}{3}$$

so that

$$\sin^{-1}\frac{\sqrt{3}}{2} = \frac{\pi}{3}$$ ■

Example 3 Find the exact value of $\sin^{-1}(-\frac{1}{2})$.

Solution Let $\theta = \sin^{-1}(-\frac{1}{2})$. Then, we seek the angle θ, $-\pi/2 \le \theta \le \pi/2$, whose sine equals $-\frac{1}{2}$:

$$\theta = \sin^{-1}\left(-\tfrac{1}{2}\right) \qquad -\pi/2 \le \theta \le \pi/2$$
$$\sin\theta = -\tfrac{1}{2} \qquad -\pi/2 \le \theta \le \pi/2$$

(Refer to Figure 30, if necessary.) The only angle within the interval $[-\pi/2, \pi/2]$ whose sine is $-\frac{1}{2}$ is $-\pi/6$. Thus,

$$\theta = -\frac{\pi}{6}$$

so that

$$\sin^{-1}\left(-\frac{1}{2}\right) = -\frac{\pi}{6}$$ ■

For most numbers x, the value $y = \sin^{-1} x$ must be approximated either on a calculator or by reading Table IV in the back of the book in reverse. Some calculators have a single key labeled [sin⁻¹] or [arcsin]. For others, you need to press the [inv] key, followed by the [sin] key, to evaluate the inverse sine function. In the following example, our calculations were done using a TI57.

C **Example 4** Find the approximate value of:

(a) $\sin^{-1}\frac{1}{3}$ (b) $\sin^{-1}(-\frac{1}{4})$

Solution Because we want the angle measured in radians, we first set the mode of the calculator to radians.

(a) Enter: [1] [÷] [3] [=]

Display: 1 3 0.3333333

Press: [inv] [sin]

Display: 0.3398369

Thus, $\sin^{-1}\frac{1}{3} \approx 0.3398369$.

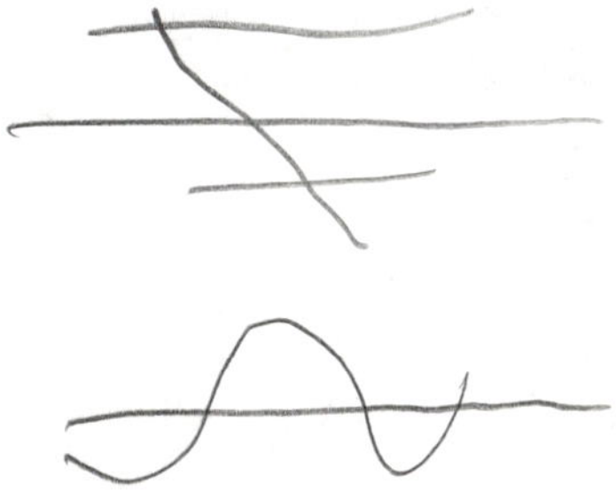

(b) Enter: [0.25] Press: [+/−]

Display: [0.25] [−0.25]

Press: [inv] [sin]

Display: [−0.2526803]

Thus, $\sin^{-1}\left(-\frac{1}{4}\right) \approx -0.2526803$. ■

The Inverse Cosine Function

In Figure 31 we reproduce the graph of $y = \cos x$. Because every horizontal line $y = b$, where b is between -1 and 1, intersects the graph of $y = \cos x$ infinitely many times, it follows that the cosine function is not one-to-one.

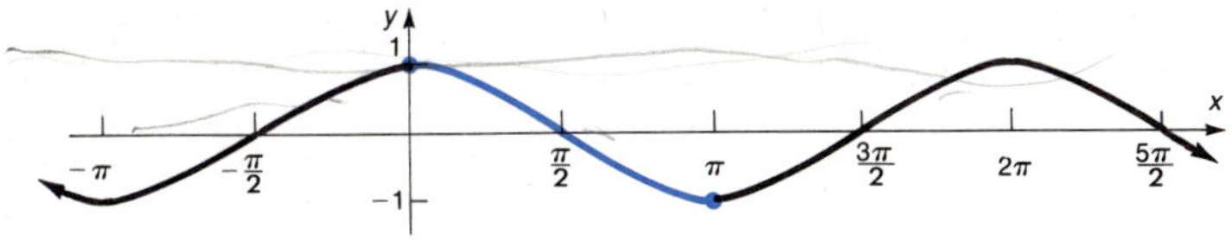

Figure 31
$y = \cos x$, $-\infty < x < \infty$, $-1 \le y \le 1$

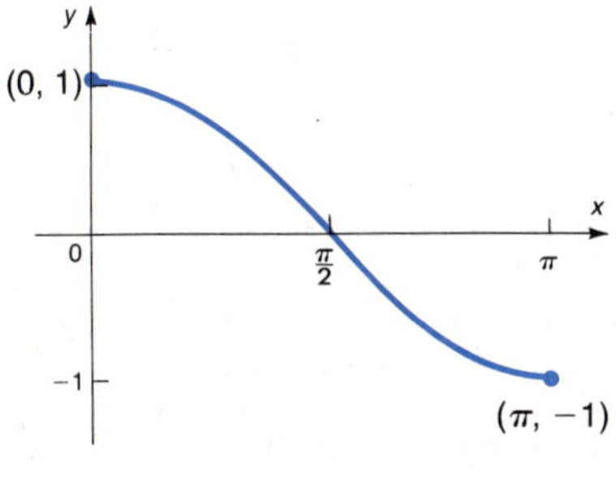

Figure 32
$y = \cos x$, $0 \le x \le \pi$, $-1 \le y \le 1$

However, if we restrict the domain of $y = \cos x$ to the interval $[0, \pi]$, the restricted function

$$y = \cos x \qquad 0 \le x \le \pi$$

is one-to-one, and hence, will have an inverse.* See Figure 32.

The inverse function is called the **inverse cosine** of x and is symbolized by $y = \cos^{-1} x$ (or by $y = \arccos x$). Thus,

Inverse Cosine Function

$$y = \cos^{-1} x \quad \text{means} \quad x = \cos y \qquad (2)$$
$$\text{where} \quad 0 \le y \le \pi \quad \text{and} \quad -1 \le x \le 1$$

Here, y is the angle whose cosine is x. The domain of the function $y = \cos^{-1} x$ is $-1 \le x \le 1$, and its range is $0 \le y \le \pi$. The graph of $y = \cos^{-1} x$ can be obtained by reflecting the restricted portion of the graph of $y = \cos x$ about the line $y = x$, as shown in Figure 33.

*This is the generally accepted restriction to define the inverse.

Figure 33
$y = \cos x,\ 0 \le x \le \pi$
$y = \cos^{-1} x,\ -1 \le x \le 1$

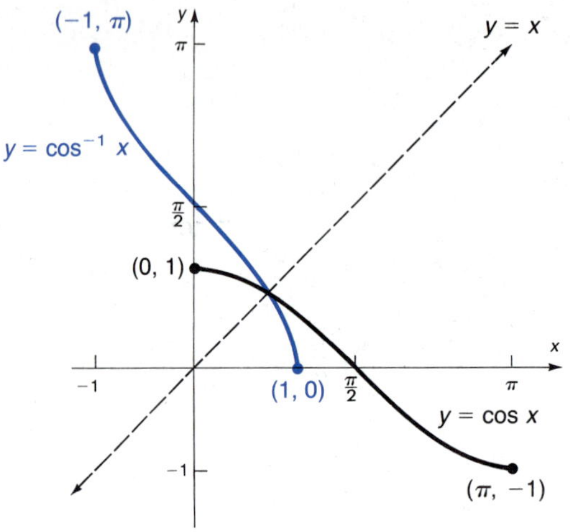

In general,

$$\cos^{-1}(\cos u) = u \qquad \text{where} \quad 0 \le u \le \pi$$
$$\cos(\cos^{-1} v) = v \qquad \text{where} \quad -1 \le v \le 1$$

Example 5 Find the exact value of $\cos^{-1} 0$.

Solution Let $\theta = \cos^{-1} 0$. Then, we seek the angle θ, $0 \le \theta \le \pi$, whose cosine equals 0:

$$\theta = \cos^{-1} 0 \qquad 0 \le \theta \le \pi$$
$$\cos \theta = 0 \qquad 0 \le \theta \le \pi$$

From Figure 34, we see that the only angle θ within the interval $[0, \pi]$ whose cosine is 0 is $\pi/2$. (Note that $\cos(3\pi/2)$ also equals 0, but $3\pi/2$ lies outside the interval $[0, \pi]$, and hence, is not admissible.)

Figure 34

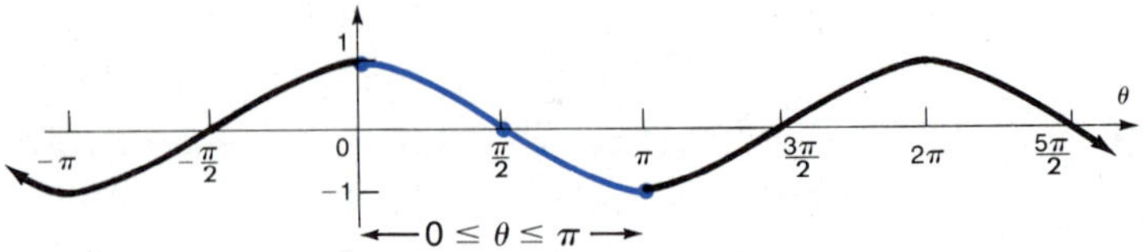

We conclude that

$$\theta = \frac{\pi}{2}$$

so that

$$\cos^{-1} 0 = \frac{\pi}{2}$$ ■

Example 6 Find the exact value of $\cos^{-1}(\sqrt{2}/2)$.

Solution Let $\theta = \cos^{-1}(\sqrt{2}/2)$. Then, we seek the angle θ, $0 \le \theta \le \pi$, whose cosine equals $\sqrt{2}/2$:

$$\theta = \cos^{-1}\frac{\sqrt{2}}{2} \qquad 0 \le \theta \le \pi$$

$$\cos\theta = \frac{\sqrt{2}}{2} \qquad 0 \le \theta \le \pi$$

From Figure 35, we see that the only angle θ within the interval $[0, \pi]$ whose cosine is $\sqrt{2}/2$ is $\pi/4$.

Figure 35

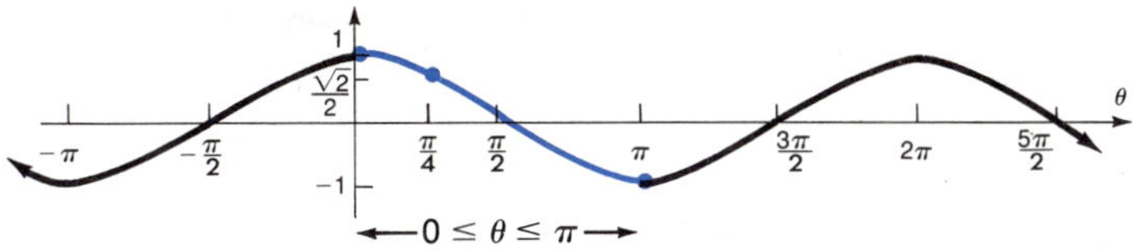

We conclude that

$$\theta = \frac{\pi}{4}$$

so that

$$\cos^{-1}\frac{\sqrt{2}}{2} = \frac{\pi}{4}$$ ■

Example 7 Find the exact value of $\cos^{-1}\left(-\frac{1}{2}\right)$.

Solution Let $\theta = \cos^{-1}\left(-\frac{1}{2}\right)$. Then, we seek the angle θ, $0 \le \theta \le \pi$, whose cosine equals $-\frac{1}{2}$:

$$\theta = \cos^{-1}\left(-\tfrac{1}{2}\right) \qquad 0 \le \theta \le \pi$$

$$\cos\theta = -\tfrac{1}{2} \qquad 0 \le \theta \le \pi$$

(Refer to Figure 35, if necessary.) The only angle within the interval $[0, \pi]$ whose cosine is $-\frac{1}{2}$ is $2\pi/3$. Thus,

$$\theta = \frac{2\pi}{3}$$

so that

$$\cos^{-1}\left(-\frac{1}{2}\right) = \frac{2\pi}{3}$$ ■

The Inverse Tangent Function

In Figure 36, we reproduce the graph of $y = \tan x$. Because every horizontal line intersects the graph infinitely many times, it follows that the tangent function is not one-to-one.

Figure 36
$y = \tan x$, $-\infty < x < \infty$, x not equal to odd multiples of $\pi/2$, $-\infty < y < \infty$

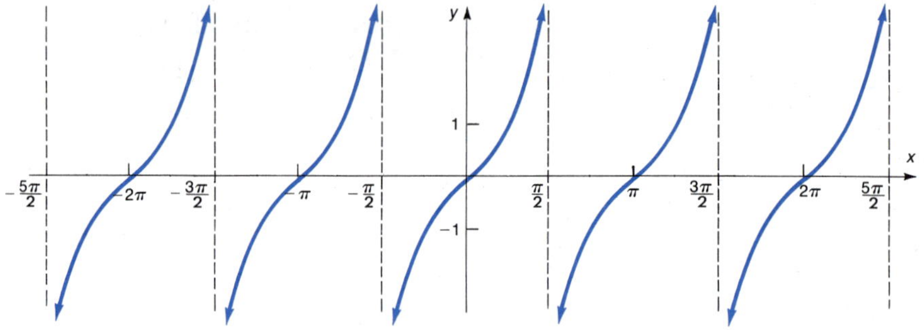

However, if we restrict the domain of $y = \tan x$ to the interval $(-\pi/2, \pi/2)$,* the restricted function

$$y = \tan x \qquad -\pi/2 < x < \pi/2$$

is one-to-one, and hence, has an inverse. See Figure 37.

Figure 37
$y = \tan x$, $-\pi/2 < x < \pi/2$, $-\infty < y < \infty$

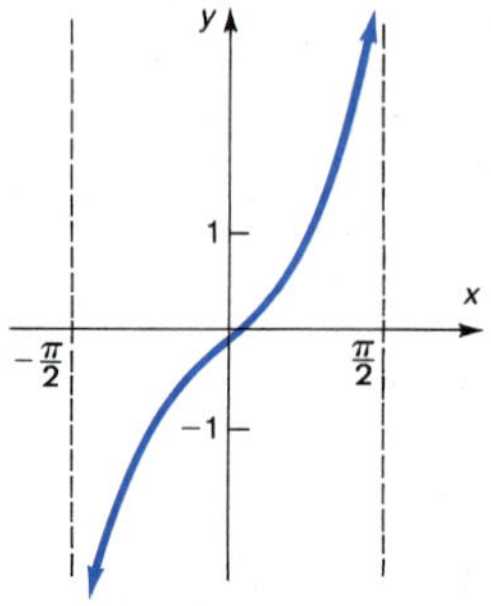

The inverse function is called the **inverse tangent** of x and is symbolized by $y = \tan^{-1} x$ (or by $y = \arctan x$). Thus,

Inverse Tangent Function

$$y = \tan^{-1} x \quad \text{means} \quad x = \tan y \tag{3}$$
$$\text{where} \quad -\pi/2 < y < \pi/2 \quad \text{and} \quad -\infty < x < \infty$$

Here, y is the angle whose tangent is x. The domain of the function $y = \tan^{-1} x$ is $-\infty < x < \infty$, and its range is $-\pi/2 < y < \pi/2$. The graph of $y = \tan^{-1} x$ can be obtained by reflecting the restricted portion of the graph of $y = \tan x$ about the line $y = x$, as shown in Figure 38.

*This is the generally accepted restriction.

Figure 38

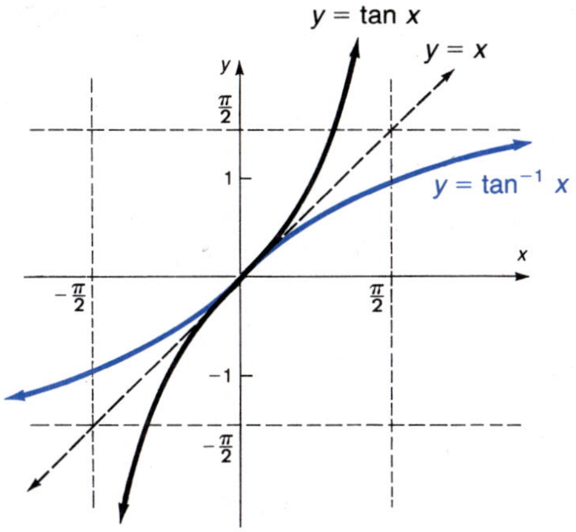

In general,

$$\tan^{-1}(\tan u) = u \quad \text{where} \quad -\pi/2 < u < \pi/2$$
$$\tan(\tan^{-1} v) = v \quad \text{where} \quad -\infty < v < \infty$$

Example 8 Find the exact value of $\tan^{-1} 1$.

Solution Let $\theta = \tan^{-1} 1$. Then, we seek the angle θ, $-\pi/2 < \theta < \pi/2$, whose tangent equals 1:

$$\theta = \tan^{-1} 1 \qquad -\pi/2 < \theta < \pi/2$$
$$\tan \theta = 1 \qquad -\pi/2 < \theta < \pi/2$$

Refer to Figure 38. The only angle θ within the interval $(-\pi/2, \pi/2)$ whose tangent is 1 is $\pi/4$.

We conclude that

$$\theta = \frac{\pi}{4}$$

so that

$$\tan^{-1} 1 = \frac{\pi}{4}$$

■

Example 9 Find the exact value of $\tan^{-1}(-\sqrt{3})$.

Solution Let $\theta = \tan^{-1}(-\sqrt{3})$. Then, we seek the angle θ, $-\pi/2 < \theta < \pi/2$, whose tangent equals $-\sqrt{3}$:

$$\theta = \tan^{-1}(-\sqrt{3}) \qquad -\pi/2 < \theta < \pi/2$$
$$\tan\theta = -\sqrt{3} \qquad -\pi/2 < \theta < \pi/2$$

Refer again to Figure 38. The only angle θ within the interval $(-\pi/2, \pi/2)$ whose tangent is $-\sqrt{3}$ is $-\pi/3$. Thus,

$$\theta = -\frac{\pi}{3}$$

so that

$$\tan^{-1}(-\sqrt{3}) = -\frac{\pi}{3}$$ ■

Example 10 Find the exact value of $\sin[\cos^{-1}(\sqrt{3}/2)]$.

Solution We first find the angle θ, $0 \le \theta \le \pi$, whose cosine equals $\sqrt{3}/2$:

$$\cos\theta = \frac{\sqrt{3}}{2} \qquad 0 \le \theta \le \pi$$
$$\theta = \frac{\pi}{6}$$

Now,

$$\sin\left(\cos^{-1}\frac{\sqrt{3}}{2}\right) = \sin\theta = \sin\frac{\pi}{6} = \frac{1}{2}$$ ■

Example 11 Find the exact value of $\cos[\tan^{-1}(-1)]$.

Solution We first find the angle θ, $-\pi/2 < \theta < \pi/2$, whose tangent equals -1:

$$\tan\theta = -1 \qquad -\pi/2 < \theta < \pi/2$$
$$\theta = -\frac{\pi}{4}$$

Now,

$$\cos[\tan^{-1}(-1)] = \cos\theta = \cos\left(-\frac{\pi}{4}\right) = \frac{\sqrt{2}}{2}$$ ■

Example 12 Find the exact value of $\sec\left(\sin^{-1}\frac{1}{2}\right)$.

Solution We first find the angle θ, $-\pi/2 \le \theta \le \pi/2$, whose sine equals $\frac{1}{2}$:

$$\sin\theta = \frac{1}{2} \qquad -\pi/2 \le \theta \le \pi/2$$

$$\theta = \frac{\pi}{6}$$

Now,

$$\sec\left(\sin^{-1}\frac{1}{2}\right) = \sec\theta = \sec\frac{\pi}{6} = \frac{2}{\sqrt{3}} = \frac{2\sqrt{3}}{3}$$

■

It is not necessary to be able to find the angle in order to solve problems like those given in Examples 10–12.

Example 13 Find the exact value of $\sin\left(\tan^{-1}\frac{1}{2}\right)$.

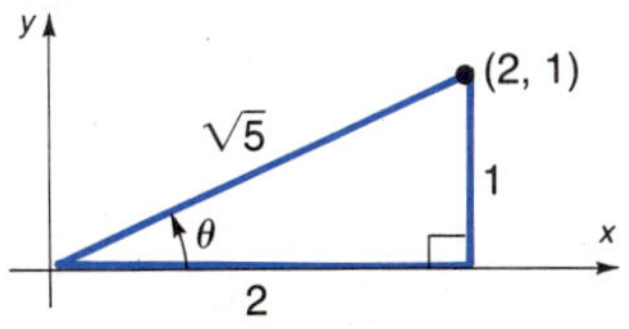

Figure 39
$\tan\theta = \frac{1}{2}$

Solution Let $\theta = \tan^{-1}\frac{1}{2}$. Then $\tan\theta = \frac{1}{2}$, where $-\pi/2 < \theta < \pi/2$. Because $\tan\theta > 0$, it follows that $0 < \theta < \pi/2$. Now, in Figure 39, we draw a triangle in the appropriate quadrant depicting $\tan\theta = \frac{1}{2}$. The hypotenuse of this triangle is easily found to be of length $\sqrt{5}$. Hence, the sine of θ is

$$\sin\left(\tan^{-1}\frac{1}{2}\right) = \sin\theta = \frac{1}{\sqrt{5}} = \frac{\sqrt{5}}{5}$$

■

Example 14 Find the exact value of $\tan\left(\cos^{-1}\frac{1}{3}\right)$.

Solution Let $\theta = \cos^{-1}\frac{1}{3}$. Then $\cos\theta = \frac{1}{3}$, where $0 \le \theta \le \pi$. Because $\cos\theta > 0$, it follows that $0 \le \theta \le \pi/2$. Look at the triangle in Figure 40. The opposite side to angle θ is $\sqrt{8} = 2\sqrt{2}$. Thus,

$$\tan\left(\cos^{-1}\tfrac{1}{3}\right) = \tan\theta = 2\sqrt{2}$$

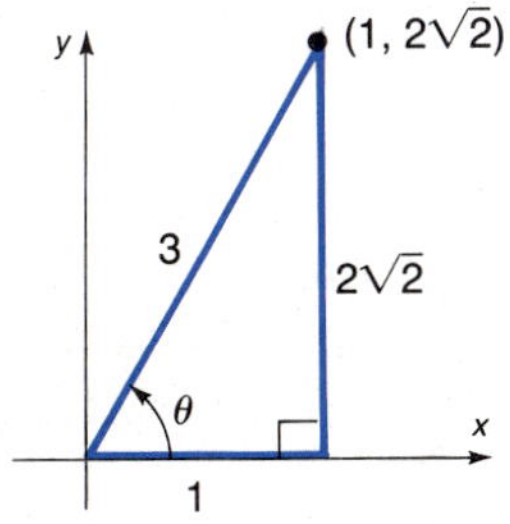

Figure 40
$\cos\theta = \frac{1}{3}$

■

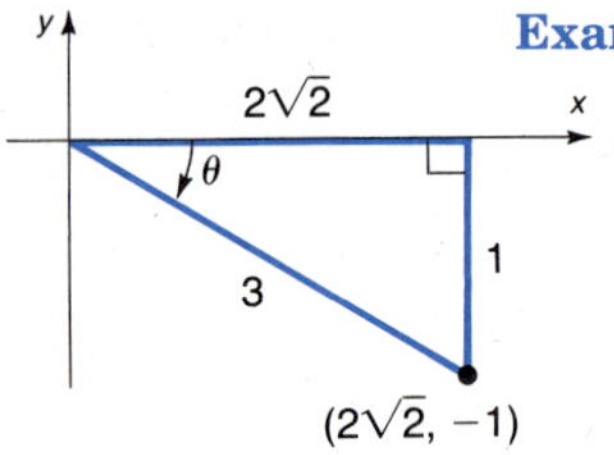

Figure 41
$\sin\theta = -\frac{1}{3}$

Example 15 Find the exact value of $\cos\left[\sin^{-1}\left(-\frac{1}{3}\right)\right]$.

Solution Let $\theta = \sin^{-1}\left(-\frac{1}{3}\right)$. Then $\sin\theta = -\frac{1}{3}$ and $-\pi/2 \le \theta \le \pi/2$. Because $\sin\theta < 0$, it follows that $-\pi/2 \le \theta \le 0$. Based on Figure 41, we conclude that

$$\cos\left[\sin^{-1}\left(-\frac{1}{3}\right)\right] = \cos\theta = \frac{2\sqrt{2}}{3}$$ ■

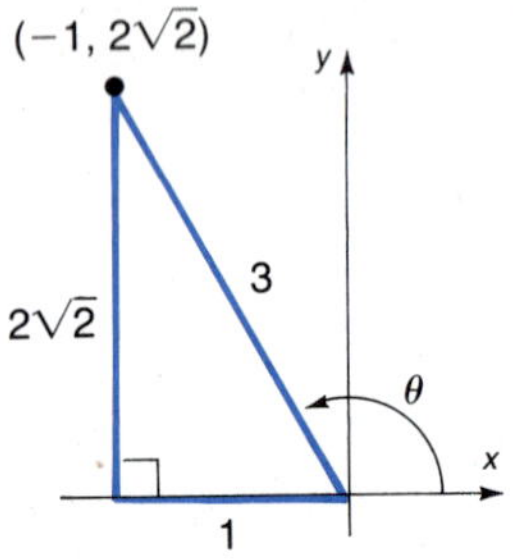

Figure 42
$\cos\theta = -\frac{1}{3}$

Example 16 Find the exact value of $\tan\left[\cos^{-1}\left(-\frac{1}{3}\right)\right]$.

Solution Let $\theta = \cos^{-1}\left(-\frac{1}{3}\right)$. Then $\cos\theta = -\frac{1}{3}$ and $0 \le \theta \le \pi$. Because $\cos\theta < 0$, it follows that $\pi/2 \le \theta \le \pi$. Based on Figure 42, we conclude that

$$\tan\left[\cos^{-1}\left(-\tfrac{1}{3}\right)\right] = \tan\theta = -2\sqrt{2}$$ ■

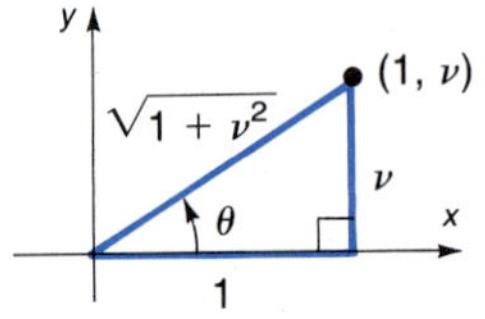

Figure 43

Example 17 Show that: $\sin(\tan^{-1}\nu) = \dfrac{\nu}{\sqrt{1+\nu^2}}$

Solution Let $\theta = \tan^{-1}\nu$ so that $\tan\theta = \nu$, $-\pi/2 < \theta < \pi/2$. There are two possibilities: either $-\pi/2 < \theta < 0$ or $0 \le \theta < \pi/2$. If $0 \le \theta < \pi/2$, then $\tan\theta = \nu \ge 0$. Based on Figure 43, we conclude that

$$\sin(\tan^{-1}\nu) = \sin\theta = \frac{\nu}{\sqrt{1+\nu^2}}$$

If $-\pi/2 < \theta < 0$, then $\tan\theta = \nu < 0$. Based on Figure 44, we conclude that

$$\sin(\tan^{-1}\nu) = \sin\theta = \frac{\nu}{\sqrt{1+\nu^2}}$$ ■

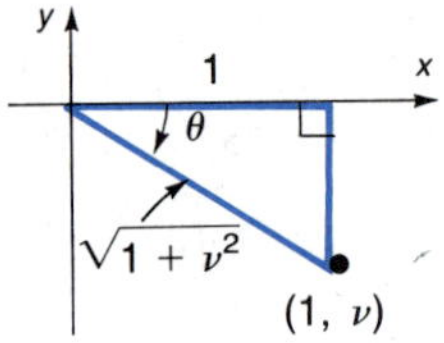

Figure 44

An alternative solution to Example 17 uses the fundamental identities. If $\theta = \tan^{-1}\nu$, then $\tan\theta = \nu$ and $-\pi/2 < \theta < \pi/2$, as before. Thus,

$$\sin(\tan^{-1}\nu) = \sin\theta = \cos\theta\tan\theta = \frac{\tan\theta}{\sec\theta} = \frac{\tan\theta}{\sqrt{1+\tan^2\theta}} = \frac{\nu}{\sqrt{1+\nu^2}}$$

$\tan\theta = \dfrac{\sin\theta}{\cos\theta}$ (for the step $\sin\theta = \cos\theta\tan\theta$)

$\sec^2\theta = 1 + \tan^2\theta$, $\sec\theta > 0$ (for the step $\dfrac{\tan\theta}{\sec\theta} = \dfrac{\tan\theta}{\sqrt{1+\tan^2\theta}}$)

The Remaining Inverse Trigonometric Functions

The inverse cotangent, inverse secant, and inverse cosecant functions are defined below.*

$$y = \cot^{-1} x \quad \text{means} \quad x = \cot y \tag{4}$$
$$\text{where} \quad 0 < y < \pi \quad \text{and} \quad -\infty < x < \infty$$

$$y = \sec^{-1} x \quad \text{means} \quad x = \sec x \tag{5}$$
$$\text{where} \quad 0 \le y \le \pi, \quad y \ne \pi/2 \quad \text{and} \quad |x| \ge 1$$

$$y = \csc^{-1} x \quad \text{means} \quad x = \csc y \tag{6}$$
$$\text{where} \quad -\pi/2 \le y \le \pi/2, \quad y \ne 0, \quad \text{and} \quad |x| \ge 1$$

Most calculators do not have keys for evaluating these inverse trigonometric functions. The easiest way to evaluate them is to convert to an inverse trigonometric function whose range is the same as the one to be evaluated. In this regard, notice that $y = \cot^{-1} x$ and $y = \sec^{-1} x$ (except where undefined) each have the same range as $y = \cos^{-1} x$, while $y = \csc^{-1} x$ (except where undefined) has the same range as $y = \sin^{-1} x$.

C **Example 18** Use a calculator to approximate to two decimal places:

(a) $\sec^{-1} 3$ (b) $\csc^{-1}(-4)$ (c) $\cot^{-1} \frac{1}{2}$ (d) $\cot^{-1}(-2)$

Solution First, set your calculator to radian mode.

(a) Let $\theta = \sec^{-1} 3$. Then $\sec \theta = 3$ and $0 \le \theta \le \pi$, $\theta \ne \pi/2$. Thus, $\cos \theta = \frac{1}{3}$ and

$$\sec^{-1} 3 = \theta = \cos^{-1} \tfrac{1}{3} \approx 1.23$$

Use a calculator.

(b) Let $\theta = \csc^{-1}(-4)$. Then $\csc \theta = -4$, $-\pi/2 \le \theta \le \pi/2$, $\theta \ne 0$. Thus, $\sin \theta = -\frac{1}{4}$ and

$$\csc^{-1}(-4) = \theta = \sin^{-1}\left(-\tfrac{1}{4}\right) \approx -0.25$$

(c) Let $\theta = \cot^{-1} \frac{1}{2}$. Then $\cot \theta = \frac{1}{2}$, $0 < \theta < \pi$. Thus, $\cos \theta = 1/\sqrt{5}$ and $0 < \theta < \pi/2$ (see Figure 45), and

$$\cot^{-1} \frac{1}{2} = \theta = \cos^{-1} \frac{1}{\sqrt{5}} \approx 1.11$$

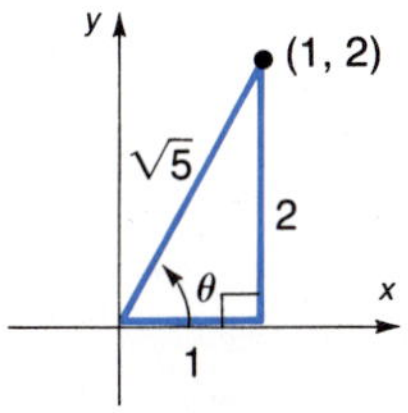

Figure 45
$\cot \theta = \frac{1}{2}$, $0 < \theta < \pi$

*The definitions of $\sec^{-1} x$ and $\csc^{-1} x$ may vary from text to text. If you encounter these functions in another course, be sure to check the definition carefully.

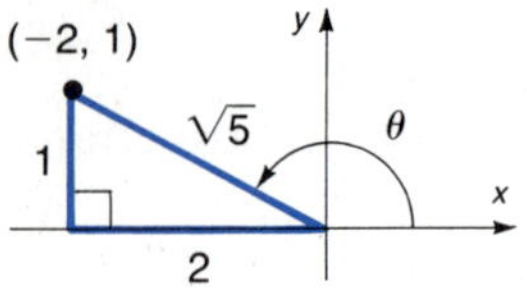

Figure 46
$\cot\theta = -2,\ 0 < \theta < \pi$

(d) Let $\theta = \cot^{-1}(-2)$. Then, $\cot\theta = -2$, $0 < \theta < \pi$. Thus, $\cos\theta = -2/\sqrt{5}$ and $\pi/2 < \theta < \pi$ (see Figure 46), and

$$\cot^{-1}(-2) = \theta = \cos^{-1}\left(-\frac{2}{\sqrt{5}}\right) \approx 2.68$$ ■

EXERCISE 8.3

In Problems 1–12, find the exact value of each expression.

1. $\sin^{-1} 0$ **2.** $\cos^{-1} 1$ **3.** $\sin^{-1}(-1)$ **4.** $\cos^{-1}(-1)$

5. $\tan^{-1} 0$ **6.** $\tan^{-1}(-1)$ **7.** $\sin^{-1}\frac{\sqrt{2}}{2}$ **8.** $\tan^{-1}\frac{\sqrt{3}}{3}$

9. $\tan^{-1}\sqrt{3}$ **10.** $\sin^{-1}\left(-\frac{\sqrt{3}}{2}\right)$ **11.** $\cos^{-1}\left(-\frac{\sqrt{3}}{2}\right)$ **12.** $\sin^{-1}\left(-\frac{\sqrt{2}}{2}\right)$

C *In Problems 13–24, use a calculator to find the approximate value of each expression to four decimal places.*

13. $\sin^{-1} 0.1$ **14.** $\cos^{-1} 0.6$ **15.** $\tan^{-1} 5$ **16.** $\tan^{-1} 0.2$

17. $\cos^{-1}\frac{7}{8}$ **18.** $\sin^{-1}\frac{1}{8}$ **19.** $\tan^{-1}(-0.4)$ **20.** $\tan^{-1}(-3)$

21. $\sin^{-1}(-0.12)$ **22.** $\cos^{-1}(-0.44)$ **23.** $\cos^{-1}\frac{\sqrt{2}}{3}$ **24.** $\sin^{-1}\frac{\sqrt{3}}{5}$

In Problems 25–46, find the exact value of each expression.

25. $\cos\left(\sin^{-1}\frac{\sqrt{2}}{2}\right)$ **26.** $\sin\left(\cos^{-1}\frac{1}{2}\right)$ **27.** $\tan\left[\cos^{-1}\left(-\frac{\sqrt{3}}{2}\right)\right]$

28. $\tan\left[\sin^{-1}\left(-\frac{1}{2}\right)\right]$ **29.** $\sec\left(\cos^{-1}\frac{1}{2}\right)$ **30.** $\cot\left[\sin^{-1}\left(-\frac{1}{2}\right)\right]$

31. $\csc(\tan^{-1} 1)$ **32.** $\sec(\tan^{-1}\sqrt{3})$ **33.** $\sin[\tan^{-1}(-1)]$

34. $\cos\left[\sin^{-1}\left(-\frac{\sqrt{3}}{2}\right)\right]$ **35.** $\sec\left[\sin^{-1}\left(-\frac{1}{2}\right)\right]$ **36.** $\csc\left[\cos^{-1}\left(-\frac{\sqrt{3}}{2}\right)\right]$

37. $\tan\left(\sin^{-1}\frac{1}{3}\right)$ **38.** $\tan\left(\cos^{-1}\frac{1}{3}\right)$ **39.** $\sec\left(\tan^{-1}\frac{1}{2}\right)$

40. $\cos\left(\sin^{-1}\frac{\sqrt{2}}{3}\right)$ **41.** $\cot\left[\sin^{-1}\left(-\frac{\sqrt{2}}{3}\right)\right]$ **42.** $\csc[\tan^{-1}(-2)]$

43. $\sin[\tan^{-1}(-3)]$ **44.** $\cot\left[\cos^{-1}\left(-\frac{\sqrt{3}}{3}\right)\right]$ **45.** $\sec\left(\sin^{-1}\frac{2\sqrt{5}}{5}\right)$

46. $\csc\left(\tan^{-1}\frac{1}{2}\right)$

C *In Problems 47–56, use a calculator to approximate the value of each expression to four decimal places.*

47. $\sin^{-1}(\tan 0.5)$ **48.** $\cos^{-1}(\tan 0.4)$ **49.** $\tan^{-1}(\sin 0.1)$ **50.** $\tan^{-1}(\cos 0.2)$

51. $\cos^{-1}(\sin 1)$ **52.** $\tan^{-1}(\cos 1)$ **53.** $\sin^{-1}\left(\tan \frac{\pi}{8}\right)$ **54.** $\cos^{-1}\left(\sin \frac{\pi}{8}\right)$

55. $\tan^{-1}\left(\sin \frac{\pi}{8}\right)$ **56.** $\tan^{-1}\left(\cos \frac{\pi}{8}\right)$

57. Show that $\sec(\tan^{-1} \nu) = \sqrt{1 + \nu^2}$.

58. Show that $\tan(\sin^{-1} \nu) = \nu/\sqrt{1 - \nu^2}$.

59. Show that $\tan(\cos^{-1} \nu) = \sqrt{1 - \nu^2}/\nu$.

60. Show that $\sin(\cos^{-1} \nu) = \sqrt{1 - \nu^2}$.

61. Show that $\cos(\sin^{-1} \nu) = \sqrt{1 - \nu^2}$.

62. Show that $\cos(\tan^{-1} \nu) = 1/\sqrt{1 + \nu^2}$.

63. Show that $\sin^{-1} \nu + \cos^{-1} \nu = \pi/2$.

64. Show that $\tan^{-1} \nu + \cot^{-1} \nu = \pi/2$.

C *In Problems 65–72, use a calculator to approximate each expression to two decimal places.*

65. $\sec^{-1} 4$ **66.** $\csc^{-1} 5$ **67.** $\cot^{-1} 2$ **68.** $\sec^{-1}(-3)$

69. $\csc^{-1}(-3)$ **70.** $\cot^{-1}\left(-\frac{1}{2}\right)$ **71.** $\cot^{-1}(-\sqrt{5})$ **72.** $\cot^{-1}(-8.1)$

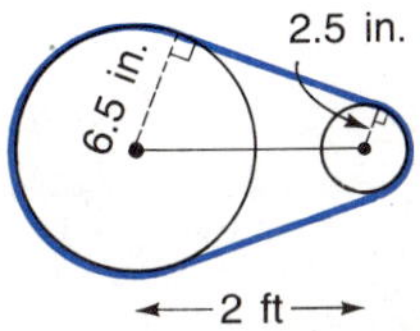

73. C The drive wheel of an engine is 13 inches in diameter, and the pulley on the rotary pump is 5 inches in diameter. If the shafts of the drive wheel and the pulley are 2 feet apart, what length of belt is required to join them as shown in the figure?

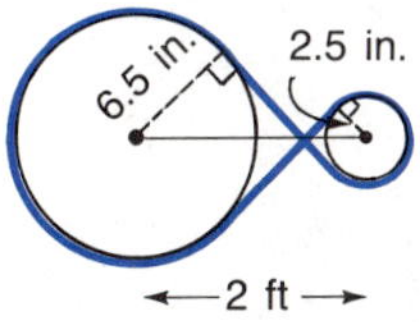

74. C Rework Problem 73 if the belt is crossed, as shown in the figure in the margin.

75. For what numbers x does $\sin(\sin^{-1} x) = x$?

76. For what numbers x does $\sin^{-1}(\sin x) = x$?

77. Draw the graph of $y = \cot^{-1} x$.

78. Draw the graph of $y = \sec^{-1} x$.

79. Draw the graph of $y = \csc^{-1} x$.

C **80.** Cadillac Mountain, elevation 1530 feet, is located in Acadia National Park, Maine, and is the highest peak on the east coast of the United States. It is said that a person standing on the summit will be the first person in the United States to see the rays of the rising Sun. How much sooner would a person atop Cadillac Mountain see the first rays than a person standing below, at sea level? [*Hint:* Consult the figure. When the person at D sees the first rays of the Sun, the person at P does not. The person at P sees the first rays of the Sun only after the Earth has rotated so that P is at location Q. Compute the length of arc s subtended by the central angle θ. Then use the fact that in 24 hours, a length of $2\pi(3960)$ miles is subtended, and find the time it takes to subtend the length s.]

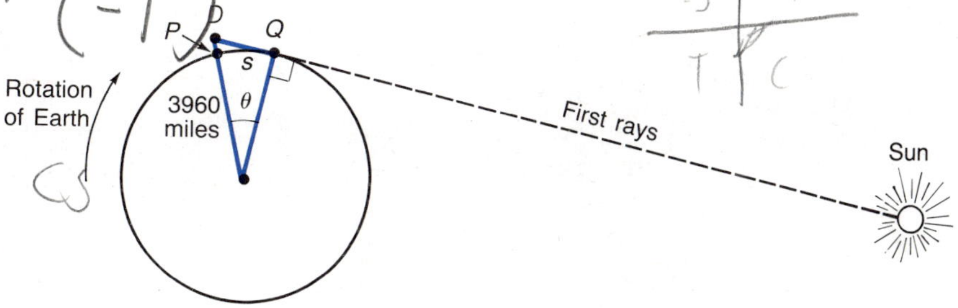

CHAPTER REVIEW

THINGS TO KNOW

The graphs of the six trigonometric functions

$y = \sin x$
$-\infty < x < \infty$
$-1 \le y \le 1$

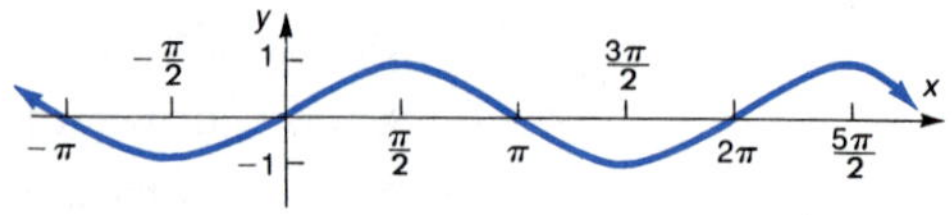

$y = \cos x$
$-\infty < x < \infty$
$-1 \le y \le 1$

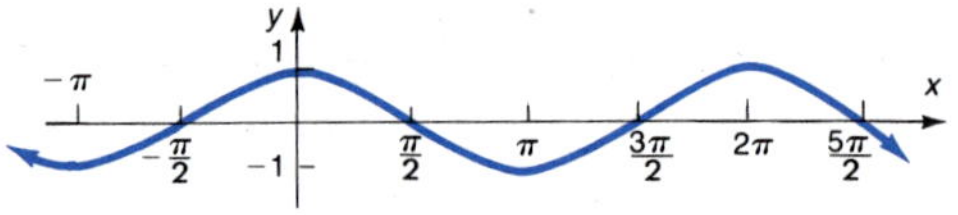

$y = \tan x$
$-\infty < x < \infty$, x not equal to odd multiples of $\pi/2$,
$-\infty < y < \infty$

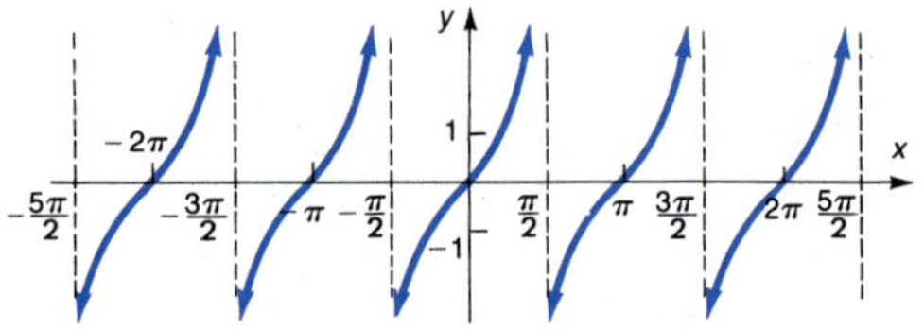

$y = \csc x$
$-\infty < x < \infty$, x not equal to integral multiples of π,
$|y| \ge 1$

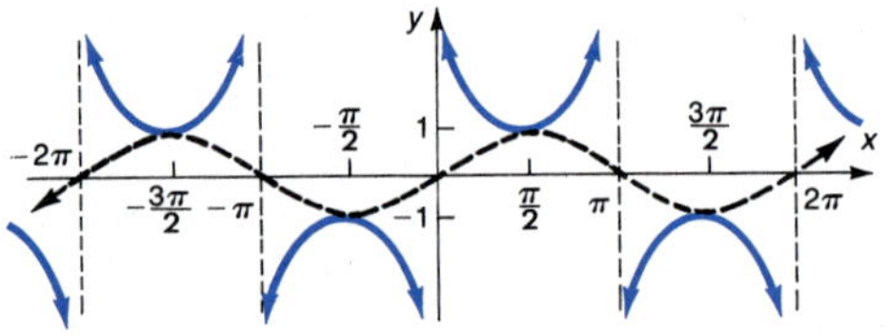

$y = \sec x$
$-\infty < x < \infty$, x not equal to odd multiples of $\pi/2$,
$|y| \ge 1$

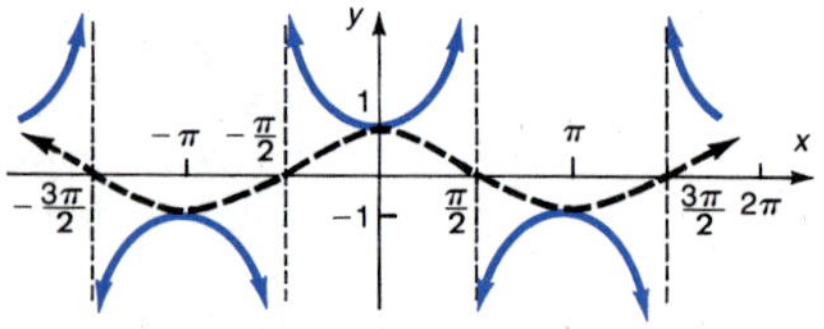

$y = \cot x$
$-\infty < x < \infty$, x not equal to integral multiples of π,
$-\infty < y < \infty$

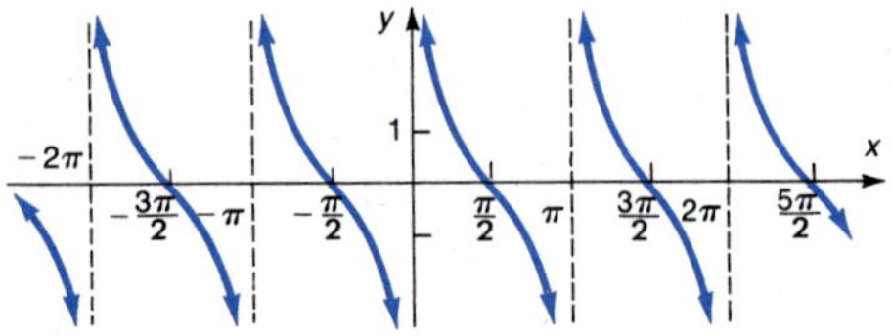

Sinusoidal graphs	$y = A \sin \omega x,\ \omega > 0$ $y = A \cos \omega x,\ \omega > 0$ $y = A \sin(\omega x - \phi)$ $y = A \cos(\omega x - \phi)$	Period $= 2\pi/\omega$ Amplitude $= \lvert A \rvert$ Phase shift $= \phi/\omega$
Definitions of the six inverse trigonometric functions	$y = \sin^{-1} x$ means $x = \sin y$ where $-1 \le x \le 1,\ -\pi/2 \le y \le \pi/2$ $y = \cos^{-1} x$ means $x = \cos y$ where $-1 \le x \le 1,\ 0 \le y \le \pi$ $y = \tan^{-1} x$ means $x = \tan y$ where $-\infty < x < \infty,\ -\pi/2 < y < \pi/2$ $y = \cot^{-1} x$ means $x = \cot y$ where $-\infty < x < \infty,\ 0 < y < \pi$ $y = \sec^{-1} x$ means $x = \sec y$ where $\lvert x \rvert \ge 1,\ 0 \le y \le \pi,\ y \ne \pi/2$ $y = \csc^{-1} x$ means $x = \csc y$ where $\lvert x \rvert \ge 1,\ -\pi/2 \le y \le \pi/2,\ y \ne 0$	

How To:

Graph the trigonometric functions, including variations

Find the period and amplitude of a sinusoidal function and use them to graph the function

Find a function whose sinusoidal graph is given

Find the value of certain inverse trigonometric functions

Use a calculator to find the approximate values of inverse trigonometric functions

FILL-IN-THE-BLANK ITEMS

1. The function below has amplitude 3 and period 2:

$$y = \_\_\_\_\_\_ \sin \_\_\_\_\_\_ x$$

2. The function $y = 3 \sin 6x$ has amplitude ______ and period ______.
3. The function $y = \sin^{-1} x$ has domain ______ and range ______.
4. The value of $\sin^{-1}[\cos(\pi/2)]$ is ______.

REVIEW EXERCISES

In Problems 1–4, determine the amplitude and period of each function without graphing.

1. $y = 4 \cos x$ **2.** $y = \sin 2x$ **3.** $y = -8 \sin \frac{\pi}{2}x$ **4.** $y = -2 \cos 3\pi x$

In Problems 5–12, find the amplitude, period, and phase shift of each function. Graph one period.

5. $y = 4 \sin 3x$ **6.** $y = 2 \cos \frac{1}{3}x$ **7.** $y = -2 \sin\left(\frac{\pi}{2}x + \frac{1}{2}\right)$

8. $y = -6 \sin(2\pi x - 2)$ **9.** $y = \frac{1}{2} \sin\left(\frac{3}{2}x - \pi\right)$ **10.** $y = \frac{3}{2} \cos(6x + 3\pi)$

11. $y = -\frac{2}{3} \cos(\pi x - 6)$ **12.** $y = -7 \sin\left(\frac{\pi}{3}x + \frac{4}{3}\right)$

In Problems 13–16, find a function whose graph is given.

13.

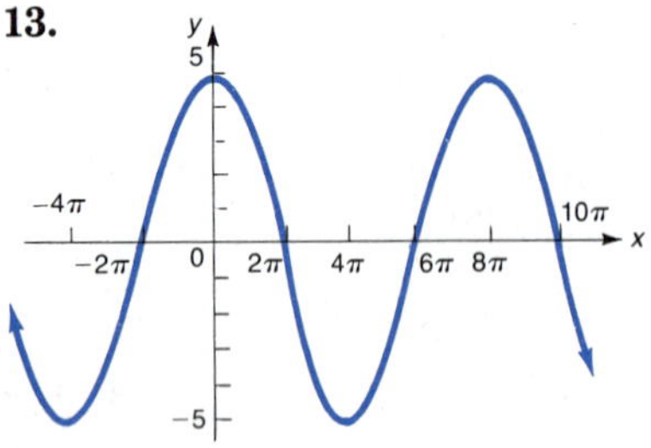

14.

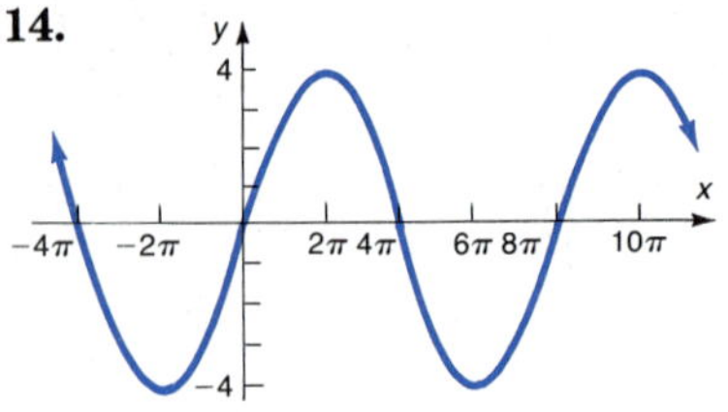

15.

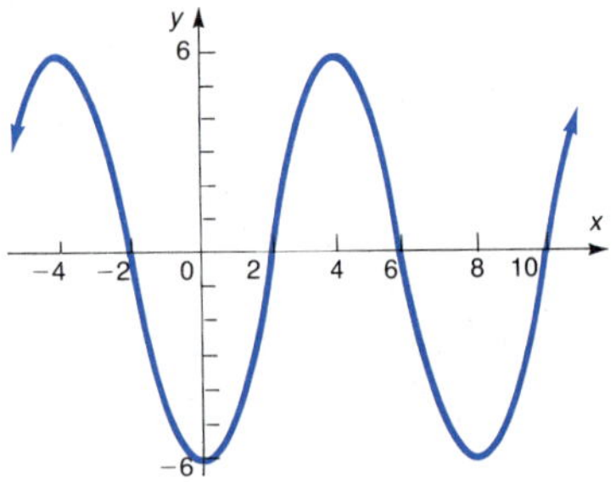

16.

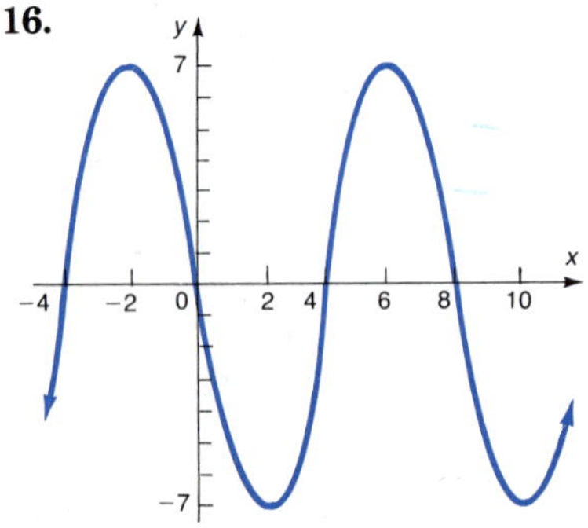

In Problems 17–20, graph each function. Each graph should contain at least one period.

17. $y = \tan(x + \pi)$

18. $y = -\tan\left(x - \dfrac{\pi}{2}\right)$

19. $y = -2 \tan 3x$

20. $y = 4 \tan 2x$

In Problems 21–36, find the exact value of each expression.

21. $\sin^{-1} 1$

22. $\cos^{-1} 0$

23. $\tan^{-1} 1$

24. $\sin^{-1}\left(-\frac{1}{2}\right)$

25. $\cos^{-1}\left(-\dfrac{\sqrt{3}}{2}\right)$

26. $\tan^{-1}(-\sqrt{3})$

27. $\sin\left(\cos^{-1} \dfrac{\sqrt{2}}{2}\right)$

28. $\cos(\sin^{-1} 0)$

29. $\tan\left[\sin^{-1}\left(-\dfrac{\sqrt{3}}{2}\right)\right]$

30. $\tan\left[\cos^{-1}\left(-\frac{1}{2}\right)\right]$

31. $\sec\left(\tan^{-1} \dfrac{\sqrt{3}}{3}\right)$

32. $\csc\left(\sin^{-1} \dfrac{\sqrt{3}}{2}\right)$

33. $\sin\left(\tan^{-1} \frac{3}{4}\right)$

34. $\cos\left(\sin^{-1} \frac{3}{5}\right)$

35. $\tan\left[\sin^{-1}\left(-\frac{4}{5}\right)\right]$

36. $\tan\left[\cos^{-1}\left(-\frac{3}{5}\right)\right]$

Analytic Trigonometry

9

In this chapter we continue the derivation of identities involving the trigonometric functions. These identities play an important role in calculus, the physical and life sciences, and economics, where they are used to simplify complicated expressions.

The last section of this chapter provides some techniques for solving equations that contain trigonometric functions.

9.1 Basic Trigonometric Identities

First, we review a fundamental definition:

Two functions f and g are said to be **identically equal** if they have the same domain and if

$$f(x) = g(x) \qquad \text{for every } x \text{ in the domain}$$

Such an equation is referred to as an **identity**. An equation that is not an identity is called a **conditional equation**.

For example, the expressions below are identities:

$$(x + 1)^2 = x^2 + 2x + 1 \qquad \sin^2 x + \cos^2 x = 1 \qquad \csc x = \frac{1}{\sin x}$$

The following expressions are conditional equations:

$2x + 5 = 0$	True only if $x = -\frac{5}{2}$
$\sin x = 0$	True only if $x = k\pi$, k an integer
$\sin x = \cos x$	True only if $x = \frac{\pi}{4} + 2k\pi$ or $x = \frac{5\pi}{4} + 2k\pi$, k an integer

The box below summarizes the trigonometric identities we have established thus far:

$$\tan\theta = \frac{\sin\theta}{\cos\theta} \qquad \cot\theta = \frac{\cos\theta}{\sin\theta}$$

Reciprocal Identities

$$\csc\theta = \frac{1}{\sin\theta} \qquad \sec\theta = \frac{1}{\cos\theta} \qquad \cot\theta = \frac{1}{\tan\theta}$$

Pythagorean Identities

$$\sin^2\theta + \cos^2\theta = 1 \qquad \tan^2\theta + 1 = \sec^2\theta$$
$$1 + \cot^2\theta = \csc^2\theta$$

Even–Odd Identities

$$\sin(-\theta) = -\sin\theta \qquad \cos(-\theta) = \cos\theta \qquad \tan(-\theta) = -\tan\theta$$
$$\csc(-\theta) = -\csc\theta \qquad \sec(-\theta) = \sec\theta \qquad \cot(-\theta) = -\cot\theta$$

This list of identities comprises what we shall refer to as the **basic trigonometric identities**. These identities should not merely be memorized, but should be *known* (just as you know your name rather than have it memorized). In fact, the use made of a basic identity is often a minor variation of the form listed above. For example, we might want to use $\sin^2\theta = 1 - \cos^2\theta$ instead of $\sin^2\theta + \cos^2\theta = 1$. For this reason, among others, you need to know these relationships and be quite comfortable with variations of them.

In the examples that follow, the directions will read "Establish the identity . . . ." As you will see, this is accomplished by starting with one side of the given equation (usually the one containing the more complicated expression) and, using appropriate basic identities and algebraic manipulations, arriving at the other side. The selection of an appropriate basic identity to obtain the desired result is learned only through experience and lots of practice.

Example 1 Establish the identity: $\sec\theta \cdot \sin\theta = \tan\theta$

Solution We start with the left side, because it contains the more complicated expression, and apply a reciprocal identity:

$$\sec\theta \cdot \sin\theta = \frac{1}{\cos\theta} \cdot \sin\theta = \frac{\sin\theta}{\cos\theta} = \tan\theta$$

Having arrived at the right side, the identity is established. ■

Example 2 Establish the identity: $\sin^2(-\theta) + \cos^2(-\theta) = 1$

Solution We begin with the left side and apply even–odd identities:

$$\begin{aligned}\sin^2(-\theta) + \cos^2(-\theta) &= [\sin(-\theta)]^2 + [\cos(-\theta)]^2\\ &= (-\sin\theta)^2 + (\cos\theta)^2\\ &= (\sin\theta)^2 + (\cos\theta)^2\\ &= 1\end{aligned}$$

■

Example 3 Establish the identity: $\dfrac{\sin^2(-\theta) - \cos^2(-\theta)}{\sin(-\theta) - \cos(-\theta)} = \cos\theta - \sin\theta$

Solution We begin with two observations: The left side appears to contain the more complicated expression. Also, the left side contains expressions with the argument $-\theta$, whereas the right side contains expressions with the argument θ. We decide, therefore, to start with the left side and apply even–odd identities:

$$\begin{aligned}\frac{\sin^2(-\theta) - \cos^2(-\theta)}{\sin(-\theta) - \cos(-\theta)} &= \frac{[\sin(-\theta)]^2 - [\cos(-\theta)]^2}{\sin(-\theta) - \cos(-\theta)}\\ &= \frac{(-\sin\theta)^2 - (\cos\theta)^2}{-\sin\theta - \cos\theta} && \text{Even–odd identities}\\ &= \frac{(\sin\theta)^2 - (\cos\theta)^2}{-\sin\theta - \cos\theta} && \text{Simplify.}\\ &= \frac{(\sin\theta - \cos\theta)\cancel{(\sin\theta + \cos\theta)}}{\cancel{-(\sin\theta + \cos\theta)}} && \text{Factor.}\\ &= \cos\theta - \sin\theta && \text{Cancel and simplify.}\end{aligned}$$

■

Example 4 Establish the identity: $\dfrac{1 + \tan\theta}{1 + \cot\theta} = \tan\theta$

Solution

$$\frac{1 + \tan\theta}{1 + \cot\theta} = \frac{1 + \tan\theta}{1 + \dfrac{1}{\tan\theta}} = \frac{1 + \tan\theta}{\dfrac{\tan\theta + 1}{\tan\theta}} = \frac{\tan\theta\cancel{(1 + \tan\theta)}}{\cancel{\tan\theta + 1}} = \tan\theta$$ ■

When sums or differences of quotients appear, it is usually best to rewrite them as a single quotient, especially if the other side of the identity consists of only one term.

Example 5 Establish the identity: $\dfrac{\sin\theta}{1 + \cos\theta} + \dfrac{1 + \cos\theta}{\sin\theta} = 2\csc\theta$

Solution The left side is more complicated, so we start with it and proceed to add:

$$\begin{aligned}\frac{\sin\theta}{1 + \cos\theta} + \frac{1 + \cos\theta}{\sin\theta} &= \frac{\sin^2\theta + (1 + \cos\theta)^2}{(1 + \cos\theta)(\sin\theta)}\\ &= \frac{\sin^2\theta + 1 + 2\cos\theta + \cos^2\theta}{(1 + \cos\theta)(\sin\theta)}\\ &= \frac{(\sin^2\theta + \cos^2\theta) + 1 + 2\cos\theta}{(1 + \cos\theta)(\sin\theta)}\\ &= \frac{2 + 2\cos\theta}{(1 + \cos\theta)(\sin\theta)}\\ &= \frac{2\cancel{(1 + \cos\theta)}}{\cancel{(1 + \cos\theta)}(\sin\theta)}\\ &= \frac{2}{\sin\theta}\\ &= 2\csc\theta\end{aligned}$$

■

Example 6 Establish the identity: $\dfrac{1}{\cos\theta} - \dfrac{\cos\theta}{1+\sin\theta} = \tan\theta$

Solution

$$\begin{aligned}\frac{1}{\cos\theta} - \frac{\cos\theta}{1+\sin\theta} &= \frac{1+\sin\theta-\cos^2\theta}{\cos\theta(1+\sin\theta)}\\ &= \frac{\sin\theta+(1-\cos^2\theta)}{\cos\theta(1+\sin\theta)} \quad 1-\cos^2\theta=\sin^2\theta\\ &= \frac{\sin\theta+\sin^2\theta}{\cos\theta(1+\sin\theta)}\\ &= \frac{\sin\theta(1+\sin\theta)}{\cos\theta(1+\sin\theta)}\\ &= \tan\theta\end{aligned}$$

■

Sometimes, multiplying the numerator and denominator by an appropriate factor will result in a simplification.

Example 7 Establish the identity: $\dfrac{1-\sin\theta}{\cos\theta} = \dfrac{\cos\theta}{1+\sin\theta}$

Solution We start with the left side and multiply the numerator and denominator by $1 + \sin\theta$. (Alternatively, we could multiply the right side by $1 - \sin\theta$.)

$$\begin{aligned}\frac{1-\sin\theta}{\cos\theta} &= \frac{1-\sin\theta}{\cos\theta}\cdot\frac{1+\sin\theta}{1+\sin\theta}\\ &= \frac{1-\sin^2\theta}{\cos\theta(1+\sin\theta)}\\ &= \frac{\cos^2\theta}{\cos\theta(1+\sin\theta)}\\ &= \frac{\cos\theta}{1+\sin\theta}\end{aligned}$$

■

Although a lot of practice is the only real way to learn how to establish identities, the following guidelines should prove helpful:

Guidelines for Establishing Identities

1. It is almost always preferable to start with the side containing the more complicated expression.
2. Rewrite sums or differences of quotients as a single quotient.
3. Always keep your goal in mind. As you manipulate one side of the expression, you must keep in mind the form of the expression on the other side.

Warning: Be careful not to handle identities to be established as if they were equations. You *cannot* establish an identity by such methods as adding the same expression to each side and obtaining a true statement. This practice is not allowed, because the original statement is precisely the one you are trying to establish. You do not know until it has been established that it is, in fact, true.

EXERCISE 9.1

In Problems 1–80, establish each identity.

1. $\csc\theta \cdot \cos\theta = \cot\theta$
2. $\csc\theta \cdot \tan\theta = \sec\theta$
3. $1 + \tan^2(-\theta) = \sec^2\theta$
4. $1 + \cot^2(-\theta) = \csc^2\theta$
5. $\cos\theta(\tan\theta + \cot\theta) = \csc\theta$
6. $\sin\theta(\cot\theta + \tan\theta) = \sec\theta$
7. $\tan\theta\cot\theta - \cos^2\theta = \sin^2\theta$
8. $\sin\theta\csc\theta - \cos^2\theta = \sin^2\theta$
9. $(\sec\theta - 1)(\sec\theta + 1) = \tan^2\theta$
10. $(\csc\theta - 1)(\csc\theta + 1) = \cot^2\theta$
11. $(\sec\theta + \tan\theta)(\sec\theta - \tan\theta) = 1$
12. $(\csc\theta + \cot\theta)(\csc\theta - \cot\theta) = 1$
13. $\sin^2\theta(1 + \cot^2\theta) = 1$
14. $(1 - \sin^2\theta)(1 + \tan^2\theta) = 1$
15. $(\sin\theta + \cos\theta)^2 + (\sin\theta - \cos\theta)^2 = 2$
16. $\tan^2\theta\cos^2\theta + \cot^2\theta\sin^2\theta = 1$
17. $\sec^4\theta - \sec^2\theta = \tan^4\theta + \tan^2\theta$
18. $\csc^4\theta - \csc^2\theta = \cot^4\theta + \cot^2\theta$
19. $\sec\theta - \tan\theta = \dfrac{\cos\theta}{1 + \sin\theta}$
20. $\csc\theta - \cot\theta = \dfrac{\sin\theta}{1 + \cos\theta}$
21. $3\sin^2\theta + 4\cos^2\theta = 3 + \cos^2\theta$
22. $9\sec^2\theta - 5\tan^2\theta = 5 + 4\sec^2\theta$
23. $1 - \dfrac{\cos^2\theta}{1 + \sin\theta} = \sin\theta$
24. $1 - \dfrac{\sin^2\theta}{1 - \cos\theta} = -\cos\theta$
25. $\dfrac{1 + \tan\theta}{1 - \tan\theta} = \dfrac{\cot\theta + 1}{\cot\theta - 1}$
26. $\dfrac{\csc\theta - 1}{\csc\theta + 1} = \dfrac{1 - \sin\theta}{1 + \sin\theta}$
27. $\dfrac{\sec\theta}{\csc\theta} + \dfrac{\sin\theta}{\cos\theta} = 2\tan\theta$
28. $\dfrac{\csc\theta - 1}{\cot\theta} = \dfrac{\cot\theta}{\csc\theta + 1}$
29. $\dfrac{1 + \sin\theta}{1 - \sin\theta} = \dfrac{\csc\theta + 1}{\csc\theta - 1}$
30. $\dfrac{\cos\theta + 1}{\cos\theta - 1} = \dfrac{1 + \sec\theta}{1 - \sec\theta}$
31. $\dfrac{1 - \sin\theta}{\cos\theta} + \dfrac{\cos\theta}{1 - \sin\theta} = 2\sec\theta$
32. $\dfrac{\cos\theta}{1 + \sin\theta} + \dfrac{1 + \sin\theta}{\cos\theta} = 2\sec\theta$
33. $\dfrac{\sin\theta}{\sin\theta - \cos\theta} = \dfrac{1}{1 - \cot\theta}$
34. $1 - \dfrac{\sin^2\theta}{1 + \cos\theta} = \cos\theta$
35. $\dfrac{1 - \sin\theta}{1 + \sin\theta} = (\sec\theta - \tan\theta)^2$
36. $\dfrac{1 - \cos\theta}{1 + \cos\theta} = (\csc\theta - \cot\theta)^2$
37. $\dfrac{\cos\theta}{1 - \tan\theta} + \dfrac{\sin\theta}{1 - \cot\theta} = \sin\theta + \cos\theta$
38. $\dfrac{\cot\theta}{1 - \tan\theta} + \dfrac{\tan\theta}{1 - \cot\theta} = 1 + \tan\theta + \cot\theta$
39. $\tan\theta + \dfrac{\cos\theta}{1 + \sin\theta} = \sec\theta$
40. $\dfrac{\sin\theta\cos\theta}{\cos^2\theta - \sin^2\theta} = \dfrac{\tan\theta}{1 - \tan^2\theta}$

41. $\dfrac{\tan\theta + \sec\theta - 1}{\tan\theta - \sec\theta + 1} = \tan\theta + \sec\theta$

42. $\dfrac{\sin\theta - \cos\theta + 1}{\sin\theta + \cos\theta - 1} = \dfrac{\sin\theta + 1}{\cos\theta}$

43. $\dfrac{\tan\theta - \cot\theta}{\tan\theta + \cot\theta} = \sin^2\theta - \cos^2\theta$

44. $\dfrac{\sec\theta - \cos\theta}{\sec\theta + \cos\theta} = \dfrac{\sin^2\theta}{1 + \cos^2\theta}$

45. $\dfrac{\tan\theta - \cot\theta}{\tan\theta + \cot\theta} = 2\sin^2\theta - 1$

46. $\dfrac{\tan\theta - \cot\theta}{\tan\theta + \cot\theta} = 1 - 2\cos^2\theta$

47. $\dfrac{\sec\theta + \tan\theta}{\cot\theta + \cos\theta} = \tan\theta\sec\theta$

48. $\dfrac{\sec\theta}{1 + \sec\theta} = \dfrac{1 - \cos\theta}{\sin^2\theta}$

49. $\dfrac{1 - \tan^2\theta}{1 + \tan^2\theta} = 2\cos^2\theta - 1$

50. $\dfrac{1 - \cot^2\theta}{1 + \cot^2\theta} = 1 - 2\cos^2\theta$

51. $\dfrac{\sec\theta - \csc\theta}{\sec\theta\csc\theta} = \sin\theta - \cos\theta$

52. $\dfrac{\sin^2\theta - \tan\theta}{\cos^2\theta - \cot\theta} = \tan^2\theta$

53. $\sec\theta - \cos\theta = \sin\theta\tan\theta$

54. $\tan\theta + \cot\theta = \sec\theta\csc\theta$

55. $\dfrac{1}{1 - \sin\theta} + \dfrac{1}{1 + \sin\theta} = 2\sec^2\theta$

56. $\dfrac{1 + \sin\theta}{1 - \sin\theta} - \dfrac{1 - \sin\theta}{1 + \sin\theta} = 4\tan\theta\sec\theta$

57. $\dfrac{\sec\theta}{1 - \sin\theta} = \dfrac{1 + \sin\theta}{\cos^3\theta}$

58. $\dfrac{1 - \sin\theta}{1 + \sin\theta} = (\sec\theta - \tan\theta)^2$

59. $\dfrac{(\sec\theta - \tan\theta)^2 + 1}{\csc\theta(\sec\theta - \tan\theta)} = 2\tan\theta$

60. $\dfrac{\sec^2\theta - \tan^2\theta + \tan\theta}{\sec\theta} = \sin\theta + \cos\theta$

61. $\dfrac{\sin\theta + \cos\theta}{\cos\theta} - \dfrac{\sin\theta - \cos\theta}{\sin\theta} = \sec\theta\csc\theta$

62. $\dfrac{\sin\theta + \cos\theta}{\sin\theta} - \dfrac{\cos\theta - \sin\theta}{\cos\theta} = \sec\theta\csc\theta$

63. $\dfrac{\sin^3\theta + \cos^3\theta}{\sin\theta + \cos\theta} = 1 - \sin\theta\cos\theta$

64. $\dfrac{\sin^3\theta + \cos^3\theta}{1 - 2\cos^2\theta} = \dfrac{\sec\theta - \sin\theta}{\tan\theta - 1}$

65. $\dfrac{\cos^2\theta - \sin^2\theta}{1 - \tan^2\theta} = \cos^2\theta$

66. $\dfrac{\cos\theta + \sin\theta - \sin^3\theta}{\sin\theta} = \cot\theta + \cos^2\theta$

67. $\dfrac{(2\cos^2\theta - 1)^2}{\cos^4\theta - \sin^4\theta} = 1 - 2\sin^2\theta$

68. $\dfrac{1 - 2\cos^2\theta}{\sin\theta\cos\theta} = \tan\theta - \cot\theta$

69. $\dfrac{1 + \sin\theta + \cos\theta}{1 + \sin\theta - \cos\theta} = \dfrac{1 + \cos\theta}{\sin\theta}$

70. $\dfrac{1 + \cos\theta + \sin\theta}{1 + \cos\theta - \sin\theta} = \sec\theta + \tan\theta$

71. $(a\sin\theta + b\cos\theta)^2 + (a\cos\theta - b\sin\theta)^2 = a^2 + b^2$

72. $(2a\sin\theta\cos\theta)^2 + a^2(\cos^2\theta - \sin^2\theta)^2 = a^2$

73. $\dfrac{\tan\alpha + \tan\beta}{\cot\alpha + \cot\beta} = \tan\alpha\tan\beta$

74. $(\tan\alpha + \tan\beta)(1 - \cot\alpha\cot\beta) + (\cot\alpha + \cot\beta)(1 - \tan\alpha\tan\beta) = 0$

75. $(\sin\alpha + \cos\beta)^2 + (\cos\beta + \sin\alpha)(\cos\beta - \sin\alpha) = 2\cos\beta(\sin\alpha + \cos\beta)$

76. $(\sin\alpha - \cos\beta)^2 + (\cos\beta + \sin\alpha)(\cos\beta - \sin\alpha) = -2\cos\beta(\sin\alpha - \cos\beta)$

77. $\ln|\sec\theta| = -\ln|\cos\theta|$

78. $\ln|\tan\theta| = \ln|\sin\theta| - \ln|\cos\theta|$

79. $\ln|1 + \cos\theta| + \ln|1 - \cos\theta| = 2\ln|\sin\theta|$

80. $\ln|\sec\theta + \tan\theta| + \ln|\sec\theta - \tan\theta| = 0$

9.2 Sum and Difference Formulas

In this section, we continue our derivation of trigonometric identities by obtaining formulas that involve the sum or difference of two angles, such as $\cos(\alpha + \beta)$, $\cos(\alpha - \beta)$, $\sin(\alpha + \beta)$, and so on. These formulas are referred to as the **sum and difference formulas**. We begin with the formulas for $\cos(\alpha + \beta)$ and $\cos(\alpha - \beta)$.

Theorem

Sum and Difference Formulas for Cosines

$$\cos(\alpha + \beta) = \cos\alpha\cos\beta - \sin\alpha\sin\beta \qquad (1)$$
$$\cos(\alpha - \beta) = \cos\alpha\cos\beta + \sin\alpha\sin\beta \qquad (2)$$

In words, formula (1) states that the cosine of the sum of two angles equals the cosine of the first times the cosine of the second minus the sine of the first times the sine of the second.

Proof

We shall prove formula (2) first. Although this formula is true for all numbers α and β, we shall assume in our proof that $0 < \beta < \alpha < 2\pi$. We begin with a circle with center at the origin (0, 0) and radius of 1 unit (the unit circle), and we place the angles α and β in standard position, as shown in Figure 1(a). The point $P_1 = (x_1, y_1)$ lies on the terminal side of β, and the point $P_2 = (x_2, y_2)$ lies on the terminal side of α.

Now, place the angle $\alpha - \beta$ in standard position, as shown in Figure 1(b), where the point A has coordinates (1, 0) and the point $P_3 = (x_3, y_3)$ is on the terminal side of the angle $\alpha - \beta$.

Looking at triangle OP_1P_2 in Figure 1(a) and triangle OAP_3 in Figure 1(b), we see that these triangles are congruent. (Do you see why? Two sides and the included angle, $\alpha - \beta$, are equal.) Hence, the unknown sides of these triangles must be equal; that is,

$$d(A, P_3) = d(P_1, P_2)$$

Using the distance formula, we find that

$$\sqrt{(x_3 - 1)^2 + y_3^2} = \sqrt{(x_2 - x_1)^2 + (y_2 - y_1)^2}$$
$$(x_3 - 1)^2 + y_3^2 = (x_2 - x_1)^2 + (y_2 - y_1)^2 \quad \text{Square each side.}$$
$$x_3^2 - 2x_3 + 1 + y_3^2 = x_2^2 - 2x_1x_2 + x_1^2 + y_2^2 - 2y_1y_2 + y_1^2 \qquad (3)$$

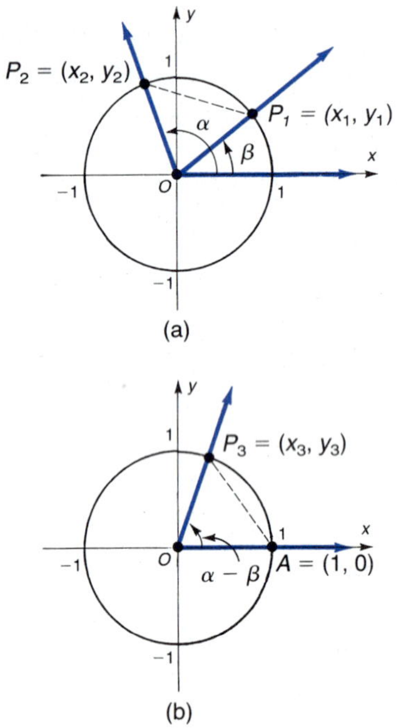

Figure 1

Since $P_1 = (x_1, y_1)$, $P_2 = (x_2, y_2)$, and $P_3 = (x_3, y_3)$ are points on the unit circle $x^2 + y^2 = 1$, it follows that

$$x_1^2 + y_1^2 = 1 \qquad x_2^2 + y_2^2 = 1 \qquad x_3^2 + y_3^2 = 1$$

Consequently, equation (3) simplifies to

$$\begin{aligned} x_3^2 + y_3^2 - 2x_3 + 1 &= (x_2^2 + y_2^2) + (x_1^2 + y_1^2) - 2x_1x_2 - 2y_1y_2 \\ 2 - 2x_3 &= 2 - 2x_1x_2 - 2y_1y_2 \\ x_3 &= x_1x_2 + y_1y_2 \end{aligned} \tag{4}$$

But $P_1 = (x_1, y_1)$ is on the terminal side of angle β and is a distance of 1 unit from the origin. Thus,

$$\sin \beta = \frac{y_1}{1} = y_1 \qquad \cos \beta = \frac{x_1}{1} = x_1 \tag{5}$$

Similarly,

$$\sin \alpha = \frac{y_2}{1} = y_2 \qquad \cos \alpha = \frac{x_2}{1} = x_2 \qquad \cos(\alpha - \beta) = \frac{x_3}{1} = x_3 \tag{6}$$

Using equations (5) and (6) in equation (4), we get

$$\cos(\alpha - \beta) = \cos \alpha \cos \beta + \sin \alpha \sin \beta$$

which is formula (2).

The proof of formula (1) follows from formula (2). We use the fact that $\alpha + \beta = \alpha - (-\beta)$. Then,

$$\begin{aligned} \cos(\alpha + \beta) &= \cos[\alpha - (-\beta)] \\ &= \cos \alpha \cos(-\beta) + \sin \alpha \sin(-\beta) && \text{Use formula (2).} \\ &= \cos \alpha \cos \beta - \sin \alpha \sin \beta && \text{Even–odd identities} \end{aligned}$$

■

One use of formulas (1) and (2) is to obtain the exact value of the cosine of an angle that can be expressed as the sum or difference of angles whose sine and cosine are known exactly.

Example 1 Find the exact value of $\cos 75°$.

Solution Since $75° = 45° + 30°$, we use formula (2) to obtain

$$\cos 75° = \cos(45° + 30°) \underset{\text{Formula (1)}}{\uparrow =} \cos 45° \cos 30° - \sin 45° \sin 30°$$

$$= \frac{\sqrt{2}}{2} \cdot \frac{\sqrt{3}}{2} - \frac{\sqrt{2}}{2} \cdot \frac{1}{2} = \frac{1}{4}(\sqrt{6} - \sqrt{2})$$

■

Example 2 Find the exact value of $\cos(\pi/12)$.

Solution

$$\cos\frac{\pi}{12} = \cos\left(\frac{3\pi}{12} - \frac{2\pi}{12}\right) = \cos\left(\frac{\pi}{4} - \frac{\pi}{6}\right)$$

$$= \cos\frac{\pi}{4}\cos\frac{\pi}{6} + \sin\frac{\pi}{4}\sin\frac{\pi}{6} \quad \text{Use formula (2).}$$

$$= \frac{\sqrt{2}}{2}\cdot\frac{\sqrt{3}}{2} + \frac{\sqrt{2}}{2}\cdot\frac{1}{2} = \frac{1}{4}(\sqrt{6} + \sqrt{2})$$

■

Another use of formulas (1) and (2) is to establish other identities. One important pair of identities is given below:

$$\cos\left(\frac{\pi}{2} - \theta\right) = \sin\theta \qquad (7a)$$

$$\sin\left(\frac{\pi}{2} - \theta\right) = \cos\theta \qquad (7b)$$

Proof To prove formula (7a), we use the formula for $\cos(\alpha - \beta)$ with $\alpha = \pi/2$ and $\beta = \theta$:

$$\cos\left(\frac{\pi}{2} - \theta\right) = \cos\frac{\pi}{2}\cos\theta + \sin\frac{\pi}{2}\sin\theta$$

$$= 0\cdot\cos\theta + 1\cdot\sin\theta$$

$$= \sin\theta$$

To prove formula (7b), we make use of the identity (7a) just established:

$$\sin\left(\frac{\pi}{2} - \theta\right) \underset{\uparrow}{=} \cos\left[\frac{\pi}{2} - \left(\frac{\pi}{2} - \theta\right)\right] = \cos\theta$$

Use (7a).

■

Formulas (7a) and (7b) should look familiar. They are the basis for the theorem stated in Chapter 7: Cofunctions of complementary angles are equal.

Furthermore, since $\cos(\pi/2 - \theta) = \cos(\theta - \pi/2)$, it follows from formula (7a) that $\cos(\theta - \pi/2) = \sin\theta$. Thus, the graphs of $y = \sin x$ and $y = \cos(x - \pi/2)$ are identical, a fact we used earlier in Section 8.2.

Formulas for $\sin(\alpha + \beta)$ and $\sin(\alpha - \beta)$

Having established the identities in formulas (7a) and (7b), we now can derive the sum and difference formulas for $\sin(\alpha + \beta)$ and $\sin(\alpha - \beta)$.

Proof

$$\sin(\alpha + \beta) = \cos\left[\frac{\pi}{2} - (\alpha + \beta)\right] \qquad \text{Formula (7a)}$$
$$= \cos\left[\left(\frac{\pi}{2} - \alpha\right) - \beta\right]$$
$$= \cos\left(\frac{\pi}{2} - \alpha\right)\cos\beta + \sin\left(\frac{\pi}{2} - \alpha\right)\sin\beta \qquad \text{Formula (2)}$$
$$= \sin\alpha\cos\beta + \cos\alpha\sin\beta \qquad \text{Formulas (7a) and (7b)}$$

$$\sin(\alpha - \beta) = \sin[\alpha + (-\beta)]$$
$$= \sin\alpha\cos(-\beta) + \cos\alpha\sin(-\beta)$$
$$= \sin\alpha\cos\beta + \cos\alpha(-\sin\beta) \qquad \text{Even–odd identities}$$
$$= \sin\alpha\cos\beta - \cos\alpha\sin\beta$$

■

Thus,

Theorem

Sum and Difference Formulas for Sines

$$\sin(\alpha + \beta) = \sin\alpha\cos\beta + \cos\alpha\sin\beta \qquad (8)$$
$$\sin(\alpha - \beta) = \sin\alpha\cos\beta - \cos\alpha\sin\beta \qquad (9)$$

■

In words, formula (8) states that the sine of the sum of two angles equals the sine of the first times the cosine of the second plus the cosine of the first times the sine of the second.

Example 3 Find the exact value of $\sin(7\pi/12)$.

Solution

$$\sin\frac{7\pi}{12} = \sin\left(\frac{3\pi}{12} + \frac{4\pi}{12}\right) = \sin\left(\frac{\pi}{4} + \frac{\pi}{3}\right)$$
$$= \sin\frac{\pi}{4}\cos\frac{\pi}{3} + \cos\frac{\pi}{4}\sin\frac{\pi}{3} \qquad \text{Formula (8)}$$
$$= \frac{\sqrt{2}}{2}\cdot\frac{1}{2} + \frac{\sqrt{2}}{2}\cdot\frac{\sqrt{3}}{2} = \frac{1}{4}(\sqrt{2} + \sqrt{6})$$

■

Example 4 Find the exact value of $\sin 165°$.

Solution

$$\sin 165° = \sin(225° - 60°)$$
$$= \sin 225°\cos 60° - \cos 225°\sin 60° \qquad \text{Formula (9)}$$
$$= \frac{-\sqrt{2}}{2}\cdot\frac{1}{2} - \frac{-\sqrt{2}}{2}\cdot\frac{\sqrt{3}}{2} = \frac{1}{4}(\sqrt{6} - \sqrt{2})$$

■

Example 5 Find the exact value of $\cos 80° \cos 20° + \sin 80° \sin 20°$.

Solution The form of the expression $\cos 80° \cos 20° + \sin 80° \sin 20°$ is that of the right side of the formula for $\cos(\alpha - \beta)$ with $\alpha = 80°$ and $\beta = 20°$. Thus,

$$\cos 80° \cos 20° + \sin 80° \sin 20° = \cos(80° - 20°) = \cos 60° = \tfrac{1}{2} \quad \blacksquare$$

Example 6 Find the exact value of $\sin \frac{\pi}{9} \cos \frac{\pi}{18} + \cos \frac{\pi}{9} \sin \frac{\pi}{18}$.

Solution We observe that the form of the given expression is that of the right side of the formula for $\sin(\alpha + \beta)$ with $\alpha = \pi/9$ and $\beta = \pi/18$. Thus,

$$\sin \frac{\pi}{9} \cos \frac{\pi}{18} + \cos \frac{\pi}{9} \sin \frac{\pi}{18} = \sin\left(\frac{\pi}{9} + \frac{\pi}{18}\right) = \sin \frac{3\pi}{18} = \sin \frac{\pi}{6} = \frac{1}{2} \quad \blacksquare$$

Example 7 If it is known that $\sin \alpha = \frac{4}{5}$, $\pi/2 < \alpha < \pi$, and that $\sin \beta = -2/\sqrt{5} = -2\sqrt{5}/5$, $\pi < \beta < 3\pi/2$, find the exact value of:

(a) $\cos \alpha$ (b) $\cos \beta$ (c) $\cos(\alpha + \beta)$ (d) $\sin(\alpha + \beta)$

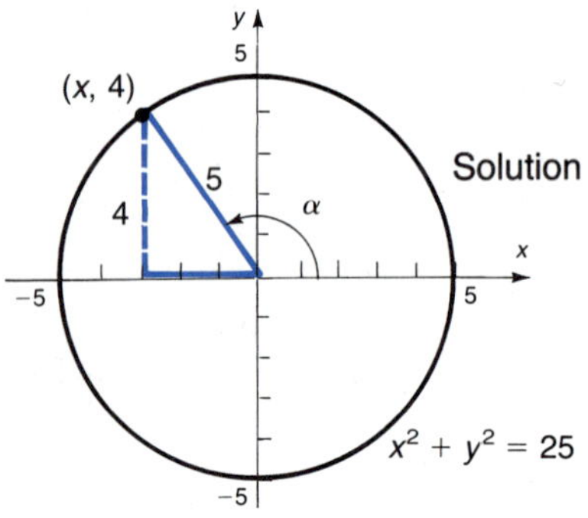

Figure 2
Given $\sin \alpha = \frac{4}{5}$, $\pi/2 < \alpha < \pi$

Solution (a) See Figure 2. Notice that $y = 4$ and $r = 5$, so that

$$x^2 + 4^2 = 5^2 \qquad x < 0$$
$$x^2 = 25 - 16 = 9$$
$$x = -3$$

Thus,

$$\cos \alpha = \frac{x}{r} = -\frac{3}{5}$$

(b) See Figure 3. Notice that $y = -2$ and $r = \sqrt{5}$, so that

$$x^2 + (-2)^2 = (\sqrt{5})^2 \qquad x < 0$$
$$x^2 = 5 - 4 = 1$$
$$x = -1$$

Thus,

$$\cos \beta = \frac{x}{r} = -\frac{1}{\sqrt{5}} = -\frac{\sqrt{5}}{5}$$

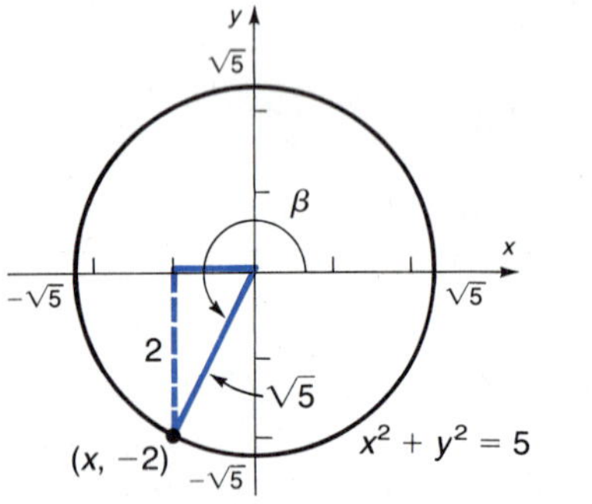

Figure 3
Given $\sin \beta = -2/\sqrt{5}$, $\pi < \beta < 3\pi/2$

(c) Using the results found in parts (a) and (b) and formula (1), we have

$$\cos(\alpha + \beta) = \cos \alpha \cos \beta - \sin \alpha \sin \beta$$
$$= -\frac{3}{5}\left(-\frac{\sqrt{5}}{5}\right) - \frac{4}{5}\left(-\frac{2\sqrt{5}}{5}\right) = \frac{11\sqrt{5}}{25}$$

(d)
$$\sin(\alpha + \beta) = \sin\alpha\cos\beta + \cos\alpha\sin\beta$$
$$= \frac{4}{5}\left(-\frac{\sqrt{5}}{5}\right) + \left(-\frac{3}{5}\right)\left(-\frac{2\sqrt{5}}{5}\right) = \frac{2\sqrt{5}}{25}$$
■

Formulas for $\tan(\alpha + \beta)$ and $\tan(\alpha - \beta)$

We use the identity $\tan\theta = (\sin\theta)/(\cos\theta)$ and the sum formulas for $\sin(\alpha + \beta)$ and $\cos(\alpha + \beta)$ to derive a formula for $\tan(\alpha + \beta)$.

Proof
$$\tan(\alpha + \beta) = \frac{\sin(\alpha + \beta)}{\cos(\alpha + \beta)} = \frac{\sin\alpha\cos\beta + \cos\alpha\sin\beta}{\cos\alpha\cos\beta - \sin\alpha\sin\beta}$$

Now we divide the numerator and denominator by $\cos\alpha\cos\beta$:

$$\tan(\alpha + \beta) = \frac{\dfrac{\sin\alpha\cos\beta + \cos\alpha\sin\beta}{\cos\alpha\cos\beta}}{\dfrac{\cos\alpha\cos\beta - \sin\alpha\sin\beta}{\cos\alpha\cos\beta}} = \frac{\dfrac{\sin\alpha\cancel{\cos\beta}}{\cos\alpha\cancel{\cos\beta}} + \dfrac{\cancel{\cos\alpha}\sin\beta}{\cancel{\cos\alpha}\cos\beta}}{\dfrac{\cancel{\cos\alpha\cos\beta}}{\cancel{\cos\alpha\cos\beta}} - \dfrac{\sin\alpha\sin\beta}{\cos\alpha\cos\beta}}$$
$$= \frac{\dfrac{\sin\alpha}{\cos\alpha} + \dfrac{\sin\beta}{\cos\beta}}{1 - \dfrac{\sin\alpha\sin\beta}{\cos\alpha\cos\beta}} = \frac{\tan\alpha + \tan\beta}{1 - \tan\alpha\tan\beta}$$

We use the sum formula for $\tan(\alpha + \beta)$ to get the difference formula:

$$\tan(\alpha - \beta) = \tan[\alpha + (-\beta)] = \frac{\tan\alpha + \tan(-\beta)}{1 - \tan\alpha\tan(-\beta)} = \frac{\tan\alpha - \tan\beta}{1 + \tan\alpha\tan\beta}$$
■

Thus, we have proved the following results:

Theorem

Sum and Difference Formulas for Tangents

$$\tan(\alpha + \beta) = \frac{\tan\alpha + \tan\beta}{1 - \tan\alpha\tan\beta} \quad (10)$$
$$\tan(\alpha - \beta) = \frac{\tan\alpha - \tan\beta}{1 + \tan\alpha\tan\beta} \quad (11)$$
■

In words, formula (10) states that the tangent of the sum of two angles equals the tangent of the first plus the tangent of the second divided by 1 minus their product.

Example 8 Prove the identity: $\tan(\theta + \pi) = \tan\theta$

Solution $$\tan(\theta + \pi) = \frac{\tan\theta + \tan\pi}{1 - \tan\theta\tan\pi} = \frac{\tan\theta + 0}{1 - \tan\theta \cdot 0} = \tan\theta$$ ■

The result obtained in Example 8 verifies that the tangent function is periodic with period π, a fact we mentioned earlier.

Warning: Be careful when using formulas (10) and (11). These formulas can be used only for angles α and β for which $\tan\alpha$ and $\tan\beta$ are defined, namely, all angles except odd multiples of $\pi/2$.

Example 9 Prove the identity: $\tan\left(\theta + \frac{\pi}{2}\right) = -\cot\theta$

Solution We cannot use formula (10), since $\tan(\pi/2)$ is not defined. Instead, we proceed as follows:

$$\tan\left(\theta + \frac{\pi}{2}\right) = \frac{\sin\left(\theta + \frac{\pi}{2}\right)}{\cos\left(\theta + \frac{\pi}{2}\right)} = \frac{\sin\theta\cos\frac{\pi}{2} + \cos\theta\sin\frac{\pi}{2}}{\cos\theta\cos\frac{\pi}{2} - \sin\theta\sin\frac{\pi}{2}}$$

$$= \frac{(\sin\theta)(0) + (\cos\theta)(1)}{(\cos\theta)(0) - (\sin\theta)(1)} = \frac{\cos\theta}{-\sin\theta} = -\cot\theta$$ ■

Example 10 Find the exact value of $\sin\left(\cos^{-1}\frac{1}{2} + \sin^{-1}\frac{3}{5}\right)$.

Solution Let $\alpha = \cos^{-1}\frac{1}{2}$ and $\beta = \sin^{-1}\frac{3}{5}$. Then

$$\cos\alpha = \tfrac{1}{2}, \quad 0 \le \alpha \le \pi \qquad \text{and} \qquad \sin\beta = \tfrac{3}{5}, \quad -\pi/2 \le \beta \le \pi/2$$

Based on Figure 4, we obtain $\sin\alpha = \sqrt{3}/2$ and $\cos\beta = \frac{4}{5}$.

Figure 4

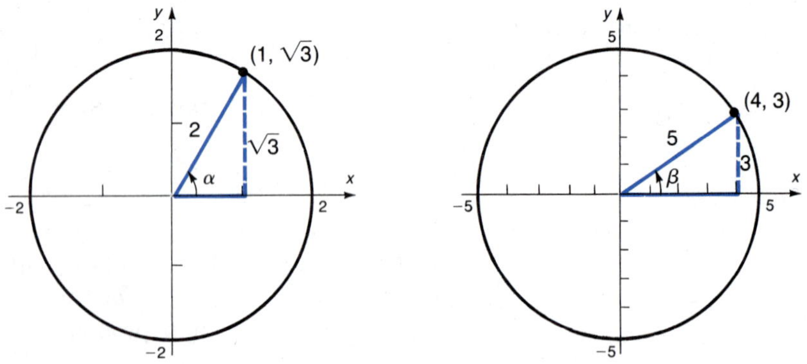

(a) $\cos\alpha = \frac{1}{2}, 0 \le \alpha \le \pi$ (b) $\sin\beta = \frac{3}{5}, -\pi/2 \le \beta \le \pi/2$

Thus,

$$\sin\left(\cos^{-1}\frac{1}{2} + \sin^{-1}\frac{3}{5}\right) = \sin(\alpha + \beta) = \sin\alpha\cos\beta + \cos\alpha\sin\beta$$
$$= \frac{\sqrt{3}}{2}\cdot\frac{4}{5} + \frac{1}{2}\cdot\frac{3}{5} = \frac{4\sqrt{3}+3}{10}$$

Summary

The box below summarizes the sum and difference formulas:

Sum and Difference Formulas

$$\cos(\alpha + \beta) = \cos\alpha\cos\beta - \sin\alpha\sin\beta$$
$$\cos(\alpha - \beta) = \cos\alpha\cos\beta + \sin\alpha\sin\beta$$
$$\sin(\alpha + \beta) = \sin\alpha\cos\beta + \cos\alpha\sin\beta$$
$$\sin(\alpha - \beta) = \sin\alpha\cos\beta - \cos\alpha\sin\beta$$
$$\tan(\alpha + \beta) = \frac{\tan\alpha + \tan\beta}{1 - \tan\alpha\tan\beta}$$
$$\tan(\alpha - \beta) = \frac{\tan\alpha - \tan\beta}{1 + \tan\alpha\tan\beta}$$

EXERCISE 9.2

In Problems 1–12, find the exact value of each trigonometric function.

1. $\sin\frac{5\pi}{12}$ **2.** $\sin\frac{\pi}{12}$ **3.** $\cos\frac{7\pi}{12}$ **4.** $\tan\frac{7\pi}{12}$

5. $\cos 165°$ **6.** $\sin 105°$ **7.** $\tan 15°$ **8.** $\tan 195°$

9. $\sin\frac{17\pi}{12}$ **10.** $\tan\frac{19\pi}{12}$ **11.** $\sec\left(-\frac{\pi}{12}\right)$ **12.** $\cot\left(-\frac{5\pi}{12}\right)$

In Problems 13–22, find the exact value of each expression.

13. $\sin 20°\cos 10° + \cos 20°\sin 10°$ **14.** $\sin 20°\cos 80° - \cos 20°\sin 80°$

15. $\cos 70°\cos 20° - \sin 70°\sin 20°$ **16.** $\cos 40°\cos 10° + \sin 40°\sin 10°$

17. $\frac{\tan 20° + \tan 25°}{1 - \tan 20°\tan 25°}$ **18.** $\frac{\tan 40° - \tan 10°}{1 + \tan 40°\tan 10°}$

19. $\sin\frac{\pi}{12}\cos\frac{7\pi}{12} - \cos\frac{\pi}{12}\sin\frac{7\pi}{12}$ **20.** $\cos\frac{5\pi}{12}\cos\frac{7\pi}{12} - \sin\frac{5\pi}{12}\sin\frac{7\pi}{12}$

21. $\sin\frac{\pi}{12}\cos\frac{5\pi}{12} - \sin\frac{5\pi}{12}\cos\frac{\pi}{12}$ **22.** $\sin\frac{\pi}{18}\cos\frac{5\pi}{18} + \cos\frac{\pi}{18}\sin\frac{5\pi}{18}$

In Problems 23–28, find the exact value of each of the following under the given conditions:

(*a*) $\sin(\alpha + \beta)$ (*b*) $\cos(\alpha + \beta)$ (*c*) $\sin(\alpha - \beta)$ (*d*) $\tan(\alpha - \beta)$

23. $\sin \alpha = \frac{3}{5}$, $0 < \alpha < \pi/2$; $\cos \beta = 2/\sqrt{5}$, $-\pi/2 < \beta < 0$

24. $\cos \alpha = 1/\sqrt{5}$, $0 < \alpha < \pi/2$; $\sin \beta = -\frac{4}{5}$, $-\pi/2 < \beta < 0$

25. $\tan \alpha = -\frac{4}{3}$, $\pi/2 < \alpha < \pi$; $\cos \beta = \frac{1}{2}$, $0 < \beta < \pi/2$

26. $\tan \alpha = \frac{5}{12}$, $\pi < \alpha < 3\pi/2$; $\sin \beta = -\frac{1}{2}$, $\pi < \beta < 3\pi/2$

27. $\sin \alpha = \frac{5}{13}$, $-3\pi/2 < \alpha < -\pi$; $\tan \beta = -\sqrt{3}$, $\pi/2 < \beta < \pi$

28. $\cos \alpha = \frac{1}{2}$, $-\pi/2 < \alpha < 0$; $\sin \beta = \frac{1}{3}$, $0 < \beta < \pi/2$

In Problems 29–54, establish each identity.

29. $\sin\left(\frac{\pi}{2} + \theta\right) = \cos \theta$

30. $\cos\left(\frac{\pi}{2} + \theta\right) = -\sin \theta$

31. $\sin(\pi - \theta) = \sin \theta$

32. $\cos(\pi - \theta) = -\cos \theta$

33. $\sin(\pi + \theta) = -\sin \theta$

34. $\cos(\pi + \theta) = -\cos \theta$

35. $\tan(\pi - \theta) = -\tan \theta$

36. $\tan(2\pi - \theta) = -\tan \theta$

37. $\sin\left(\frac{3\pi}{2} + \theta\right) = -\cos \theta$

38. $\cos\left(\frac{3\pi}{2} + \theta\right) = \sin \theta$

39. $\sin(\alpha + \beta) + \sin(\alpha - \beta) = 2 \sin \alpha \cos \beta$

40. $\cos(\alpha + \beta) + \cos(\alpha - \beta) = 2 \cos \alpha \cos \beta$

41. $\frac{\sin(\alpha + \beta)}{\sin \alpha \cos \beta} = 1 + \cot \alpha \tan \beta$

42. $\frac{\sin(\alpha + \beta)}{\cos \alpha \cos \beta} = \tan \alpha + \tan \beta$

43. $\frac{\cos(\alpha + \beta)}{\cos \alpha \cos \beta} = 1 - \tan \alpha \tan \beta$

44. $\frac{\cos(\alpha - \beta)}{\sin \alpha \cos \beta} = \cot \alpha + \tan \beta$

45. $\frac{\sin(\alpha + \beta)}{\sin(\alpha - \beta)} = \frac{\tan \alpha + \tan \beta}{\tan \alpha - \tan \beta}$

46. $\frac{\cos(\alpha + \beta)}{\cos(\alpha - \beta)} = \frac{1 - \tan \alpha \tan \beta}{1 + \tan \alpha \tan \beta}$

47. $\cot(\alpha + \beta) = \frac{\cot \alpha \cot \beta - 1}{\cot \beta + \cot \alpha}$

48. $\cot(\alpha - \beta) = \frac{\cot \alpha \cot \beta + 1}{\cot \beta - \cot \alpha}$

49. $\sec(\alpha + \beta) = \frac{\csc \alpha \csc \beta}{\cot \alpha \cot \beta - 1}$

50. $\sec(\alpha - \beta) = \frac{\sec \alpha \sec \beta}{1 + \tan \alpha \tan \beta}$

51. $\sin(\alpha - \beta) \sin(\alpha + \beta) = \sin^2 \alpha - \sin^2 \beta$

52. $\cos(\alpha - \beta) \cos(\alpha + \beta) = \cos^2 \alpha - \sin^2 \beta$

53. $\sin(\theta + k\pi) = (-1)^k \cdot \sin \theta$, k any integer

54. $\cos(\theta + k\pi) = (-1)^k \cdot \cos \theta$, k any integer

In Problems 55–68, find the exact value for each expression.

55. $\sin\left(\sin^{-1} \frac{1}{2} + \cos^{-1} 0\right)$

56. $\sin\left(\sin^{-1} \frac{\sqrt{3}}{2} + \cos^{-1} 1\right)$

57. $\cos\left(\sin^{-1} 0 - \cos^{-1} \frac{\sqrt{3}}{2}\right)$

58. $\cos\left[\sin^{-1}(-1) + \cos^{-1} \frac{1}{2}\right]$

59. $\tan(\sin^{-1} 1 + \tan^{-1} 1)$

60. $\tan\left(\cos^{-1} \frac{1}{2} - \tan^{-1} 0\right)$

61. $\sin\left[\sin^{-1} \frac{3}{5} - \cos^{-1}\left(-\frac{4}{5}\right)\right]$

62. $\sin\left[\sin^{-1}\left(-\frac{4}{5}\right) - \tan^{-1} \frac{3}{4}\right]$

63. $\cos\left(\tan^{-1} \frac{4}{3} + \cos^{-1} \frac{5}{13}\right)$

64. $\sin\left[\tan^{-1} \frac{5}{12} - \sin^{-1}\left(-\frac{3}{5}\right)\right]$

65. $\sec\left(\sin^{-1} \frac{5}{13} - \tan^{-1} \frac{3}{4}\right)$

66. $\sec\left(\tan^{-1} \frac{4}{3} + \cot^{-1} \frac{5}{12}\right)$

67. $\cot\left(\sec^{-1}\frac{5}{3}+\frac{\pi}{6}\right)$ **68.** $\cos\left(\frac{\pi}{4}-\csc^{-1}\frac{5}{3}\right)$

69. Show that the difference quotient for $f(x) = \sin x$ is given by

$$\frac{f(x+h)-f(x)}{h} = \frac{\sin(x+h)-\sin x}{h} = \cos x \cdot \frac{\sin h}{h} - \sin x \cdot \frac{1-\cos h}{h}$$

70. Show that the difference quotient for $f(x) = \cos x$ is given by

$$\frac{f(x+h)-f(x)}{h} = \frac{\cos(x+h)-\cos x}{h} = -\sin x \cdot \frac{\sin h}{h} - \cos x \cdot \frac{1-\cos h}{h}$$

71. Show that $\sin(\sin^{-1} u + \cos^{-1} u) = 1$.

72. Show that $\cos(\sin^{-1} u + \cos^{-1} u) = 0$.

73. Explain why formula (11) cannot be used to show that

$$\tan\left(\frac{\pi}{2}-\theta\right) = \cot\theta$$

Establish this identity by using formulas (7a) and (7b).

74. If $\tan\alpha = x + 1$ and $\tan\beta = x - 1$, show that $2\cot(\alpha - \beta) = x^2$.

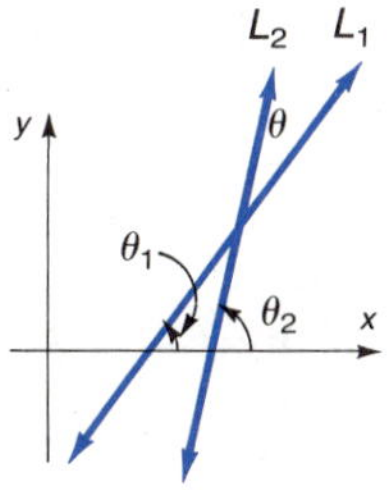

75. Let L_1 and L_2 denote two nonvertical intersecting lines, and let θ denote the acute angle between L_1 and L_2 (see the figure). Show that

$$\tan\theta = \frac{m_2 - m_1}{1 + m_1 m_2}$$

where m_1 and m_2 are the slopes of L_1 and L_2, respectively. [*Hint:* Use the facts that $\tan\theta_1 = m_1$ and $\tan\theta_2 = m_2$.]

76. If $\alpha + \beta + \gamma = 180°$ and $\cot\theta = \cot\alpha + \cot\beta + \cot\gamma$, $0 < \theta < 90°$, show that

$$\sin^3\theta = \sin(\alpha - \theta)\sin(\beta - \theta)\sin(\gamma - \theta)$$

9.3 Double-Angle and Half-Angle Formulas

In this section we derive formulas for $\sin 2\theta$, $\cos 2\theta$, $\sin\frac{1}{2}\theta$, and $\cos\frac{1}{2}\theta$ in terms of $\sin\theta$ and $\cos\theta$.

Double-Angle Formulas

In the sum formulas for $\sin(\alpha + \beta)$ and $\cos(\alpha + \beta)$, let $\alpha = \beta = \theta$. Then,

$$\sin(\alpha+\beta) = \sin\alpha\cos\beta + \cos\alpha\sin\beta$$
$$\sin(\theta+\theta) = \sin\theta\cos\theta + \cos\theta\sin\theta$$
$$\sin 2\theta = 2\sin\theta\cos\theta \qquad (1)$$

and

$$\cos(\alpha+\beta) = \cos\alpha\cos\beta - \sin\alpha\sin\beta$$
$$\cos(\theta+\theta) = \cos\theta\cos\theta - \sin\theta\sin\theta$$
$$\cos 2\theta = \cos^2\theta - \sin^2\theta \qquad (2)$$

An application of the Pythagorean identity $\sin^2 \theta + \cos^2 \theta = 1$ results in two other ways to write formula (2) for $\cos 2\theta$:

$$\cos 2\theta = \cos^2 \theta - \sin^2 \theta = (1 - \sin^2 \theta) - \sin^2 \theta = 1 - 2 \sin^2 \theta$$

and

$$\cos 2\theta = \cos^2 \theta - \sin^2 \theta = \cos^2 \theta - (1 - \cos^2 \theta) = 2 \cos^2 \theta - 1$$

Thus, we have established the following **double-angle formulas**:

Theorem

Double-Angle Formulas

$$\sin 2\theta = 2 \sin \theta \cos \theta \qquad (3)$$
$$\cos 2\theta = \cos^2 \theta - \sin^2 \theta \qquad (4a)$$
$$\cos 2\theta = 1 - 2 \sin^2 \theta \qquad (4b)$$
$$\cos 2\theta = 2 \cos^2 \theta - 1 \qquad (4c)$$

■

Example 1 If $\sin \theta = \frac{3}{5}$, $\pi/2 < \theta < \pi$, find the exact value of:

(a) $\sin 2\theta$ (b) $\cos 2\theta$

Solution (a) Because $\sin 2\theta = 2 \sin \theta \cos \theta$ and we already know $\sin \theta = \frac{3}{5}$, we only need to find $\cos \theta$. Since $\pi/2 < \theta < \pi$, it follows from Figure 5 that

$$x^2 + 3^2 = 5^2 \qquad x < 0$$
$$x^2 = 25 - 9 = 16$$
$$x = -4$$

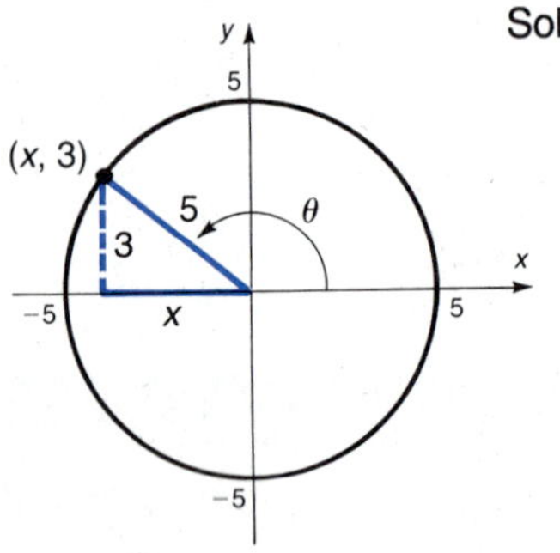

Figure 5

Thus, $\cos \theta = -\frac{4}{5}$. Now we use formula (3) to obtain

$$\sin 2\theta = 2 \sin \theta \cos \theta = 2\left(\frac{3}{5}\right)\left(-\frac{4}{5}\right) = -\frac{24}{25}$$

(b) Because we are given $\sin \theta = \frac{3}{5}$, it is easiest to use formula (4b) to get $\cos 2\theta$:

$$\cos 2\theta = 1 - 2 \sin^2 \theta = 1 - 2\left(\frac{9}{25}\right) = 1 - \frac{18}{25} = \frac{7}{25}$$ ■

Warning: In finding $\cos 2\theta$ in Example 1(b), we chose to use a version of the double-angle formula—namely, formula (4b). Note that we are unable to use the Pythagorean identity $\cos 2\theta = \pm\sqrt{1 - \sin^2 2\theta}$, with $\sin 2\theta = -\frac{24}{25}$, because we have no way of knowing which sign to choose.

Example 2 (a) Develop a formula for $\tan 2\theta$ in terms of $\tan \theta$.

(b) Develop a formula for $\sin 3\theta$ in terms of $\sin \theta$ and $\cos \theta$.

Solution (a) In the sum formula for $\tan(\alpha + \beta)$, let $\alpha = \beta = \theta$. Then,

$$\tan(\alpha + \beta) = \frac{\tan \alpha + \tan \beta}{1 - \tan \alpha \tan \beta}$$

$$\tan(\theta + \theta) = \frac{\tan \theta + \tan \theta}{1 - \tan \theta \tan \theta}$$

$$\tan 2\theta = \frac{2 \tan \theta}{1 - \tan^2 \theta} \qquad (5)$$

(b) To get a formula for $\sin 3\theta$, we use the sum formula and write 3θ as $2\theta + \theta$.

$$\sin 3\theta = \sin(2\theta + \theta) = \sin 2\theta \cos \theta + \cos 2\theta \sin \theta$$

Now use the double-angle formulas to get

$$\begin{aligned}\sin 3\theta &= (2 \sin \theta \cos \theta)(\cos \theta) + (\cos^2 \theta - \sin^2 \theta)(\sin \theta)\\ &= 2 \sin \theta \cos^2 \theta + \sin \theta \cos^2 \theta - \sin^3 \theta\\ &= 3 \sin \theta \cos^2 \theta - \sin^3 \theta\end{aligned}$$

■

The formula obtained in Example 2(b) also can be written as

$$\begin{aligned}\sin 3\theta = 3 \sin \theta \cos^2 \theta - \sin^3 \theta &= 3 \sin \theta(1 - \sin^2 \theta) - \sin^3 \theta\\ &= 3 \sin \theta - 4 \sin^3 \theta\end{aligned}$$

That is, $\sin 3\theta$ is a third-degree polynomial in the variable $\sin \theta$. In fact, $\sin n\theta$, n a positive odd integer, always can be written as a polynomial of degree n in the variable $\sin \theta$.*

Other Variations of the Double-Angle Formulas

By rearranging the double-angle formulas (4b) and (4c), we obtain other formulas that we will use a little later in this section.

If we solve formula (4b) for $\sin^2 \theta$, we get

$$\cos 2\theta = 1 - 2 \sin^2 \theta$$

$$2 \sin^2 \theta = 1 - \cos 2\theta$$

$$\sin^2 \theta = \frac{1 - \cos 2\theta}{2} \qquad (6)$$

*Due to the work done by P. L. Tchebycheff, these polynomials are sometimes called *Tchebycheff polynomials*.

Similarly, we can solve for $\cos^2 \theta$ in formula (4c):

$$\cos 2\theta = 2\cos^2 \theta - 1$$
$$2\cos^2 \theta = 1 + \cos 2\theta$$

$$\cos^2 \theta = \frac{1 + \cos 2\theta}{2} \qquad (7)$$

Formulas (6) and (7) can be used to develop a formula for $\tan^2 \theta$:

$$\tan^2 \theta = \frac{\sin^2 \theta}{\cos^2 \theta} = \frac{\dfrac{1 - \cos 2\theta}{2}}{\dfrac{1 + \cos 2\theta}{2}}$$

$$\tan^2 \theta = \frac{1 - \cos 2\theta}{1 + \cos 2\theta} \qquad (8)$$

Formulas (6)–(8) do not have to be memorized since their derivations are so straightforward.

The next example illustrates a problem that arises in calculus requiring the use of formula (7).

Example 3 Write an equivalent expression for $\cos^4 \theta$ that does not involve any powers of sine or cosine greater than 1.

Solution The idea here is to apply formula (7) twice:

$$\begin{aligned}
\cos^4 \theta = (\cos^2 \theta)^2 &= \left(\frac{1 + \cos 2\theta}{2}\right)^2 && \text{Formula (7)} \\
&= \frac{1}{4}(1 + 2\cos 2\theta + \cos^2 2\theta) \\
&= \frac{1}{4} + \frac{1}{2}\cos 2\theta + \frac{1}{4}\cos^2 2\theta \\
&= \frac{1}{4} + \frac{1}{2}\cos 2\theta + \frac{1}{4}\left[\frac{1 + \cos 2(2\theta)}{2}\right] && \text{Formula (7)} \\
&= \frac{1}{4} + \frac{1}{2}\cos 2\theta + \frac{1}{8}(1 + \cos 4\theta) \\
&= \frac{3}{8} + \frac{1}{2}\cos 2\theta + \frac{1}{8}\cos 4\theta
\end{aligned}$$

■

Half-Angle Formulas

Another important use of formulas (6)–(8) is to prove the **half-angle formulas**. In formulas (6)–(8), let $\theta = \alpha/2$. Then:

$$\sin^2 \frac{\alpha}{2} = \frac{1 - \cos \alpha}{2} \qquad \cos^2 \frac{\alpha}{2} = \frac{1 + \cos \alpha}{2}$$
$$\tan^2 \frac{\alpha}{2} = \frac{1 - \cos \alpha}{1 + \cos \alpha} \tag{9}$$

If we solve for the trigonometric functions on the left sides of equations (9), we obtain the half-angle formulas:

Theorem

Half-Angle Formulas

$$\sin \frac{\alpha}{2} = \pm\sqrt{\frac{1 - \cos \alpha}{2}} \tag{10a}$$

$$\cos \frac{\alpha}{2} = \pm\sqrt{\frac{1 + \cos \alpha}{2}} \tag{10b}$$

$$\tan \frac{\alpha}{2} = \pm\sqrt{\frac{1 - \cos \alpha}{1 + \cos \alpha}} \tag{10c}$$

where the + or − sign is determined by the quadrant of the angle $\alpha/2$. ■

We use the half-angle formulas in the next example.

Example 4 Find the exact value of:

(a) $\cos 15°$ (b) $\sin(-15°)$

Solution (a) Because $15° = 30°/2$, we can use the half-angle formula for $\cos(\alpha/2)$ with $\alpha = 30°$. Also, because $15°$ is in quadrant I, $\cos 15° > 0$, and we choose the + sign in using formula (10b):

$$\cos 15° = \cos \frac{30°}{2} = \sqrt{\frac{1 + \cos 30°}{2}}$$
$$= \sqrt{\frac{1 + \sqrt{3}/2}{2}} = \sqrt{\frac{2 + \sqrt{3}}{4}} = \frac{\sqrt{2 + \sqrt{3}}}{2}$$

(b) We use the fact that $\sin(-15°) = -\sin 15°$ and then apply formula (10a):

$$\sin(-15°) = -\sin\frac{30°}{2} = -\sqrt{\frac{1-\cos 30°}{2}}$$

$$= -\sqrt{\frac{1-\sqrt{3}/2}{2}} = -\sqrt{\frac{2-\sqrt{3}}{4}} = -\frac{\sqrt{2-\sqrt{3}}}{2}$$

■

It is interesting to compare the answer found in Example 4(a) with the answer to Example 2 of Section 9.2. There, we calculated

$$\cos\frac{\pi}{12} = \cos 15° = \frac{1}{4}(\sqrt{6}+\sqrt{2})$$

Based on these results, we conclude that

$$\frac{1}{4}(\sqrt{6}+\sqrt{2}) \quad \text{and} \quad \frac{\sqrt{2+\sqrt{3}}}{2}$$

are equal. (You can verify this by squaring each expression.) Thus, two very different looking, yet correct, answers can be obtained, depending on the approach taken to solve a problem.

Example 5 If $\cos\alpha = -\frac{3}{5}$, $\pi < \alpha < 3\pi/2$, find the exact value of:

(a) $\sin\frac{\alpha}{2}$ (b) $\cos\frac{\alpha}{2}$ (c) $\tan\frac{\alpha}{2}$

Solution First, we observe that if $\pi < \alpha < 3\pi/2$, then $\pi/2 < \alpha/2 < 3\pi/4$. As a result, $\alpha/2$ lies in quadrant II.

(a) Because $\alpha/2$ lies in quadrant II, $\sin(\alpha/2) > 0$. Thus, we use the + sign in formula (10a) to get

$$\sin\frac{\alpha}{2} = \sqrt{\frac{1-\cos\alpha}{2}} = \sqrt{\frac{1-\left(-\frac{3}{5}\right)}{2}}$$

$$= \sqrt{\frac{\frac{8}{5}}{2}} = \sqrt{\frac{4}{5}} = \frac{2}{\sqrt{5}} = \frac{2\sqrt{5}}{5}$$

(b) Because $\alpha/2$ lies in quadrant II, $\cos(\alpha/2) < 0$. Thus, we use the − sign in formula (10b) to get

$$\cos\frac{\alpha}{2} = -\sqrt{\frac{1+\cos\alpha}{2}} = -\sqrt{\frac{1+\left(-\frac{3}{5}\right)}{2}}$$

$$= -\sqrt{\frac{\frac{2}{5}}{2}} = -\frac{1}{\sqrt{5}} = -\frac{\sqrt{5}}{5}$$

(c) Because $\alpha/2$ lies in quadrant II, $\tan(\alpha/2) < 0$. Thus, we use the $-$ sign in formula (10c) to get

$$\tan\frac{\alpha}{2} = -\sqrt{\frac{1 - \cos\alpha}{1 + \cos\alpha}} = -\sqrt{\frac{1 - \left(-\frac{3}{5}\right)}{1 + \left(-\frac{3}{5}\right)}} = -\sqrt{\frac{\frac{8}{5}}{\frac{2}{5}}} = -2$$

Another way to solve Example 5(c) is to use the solutions found in parts (a) and (b):

$$\tan\frac{\alpha}{2} = \frac{\sin(\alpha/2)}{\cos(\alpha/2)} = \frac{2\sqrt{2}/5}{-\sqrt{5}/5} = -2$$

The next example illustrates a problem that arises in calculus.

Example 6 If $z = \tan(\alpha/2)$, show that:

(a) $\sin\alpha = \dfrac{2z}{1 + z^2}$ (b) $\cos\alpha = \dfrac{1 - z^2}{1 + z^2}$

Solution (a)

$$\frac{2z}{1 + z^2} = \frac{2\tan(\alpha/2)}{1 + \tan^2(\alpha/2)} = \frac{2\tan(\alpha/2)}{\sec^2(\alpha/2)} = \frac{2 \cdot \dfrac{\sin(\alpha/2)}{\cos(\alpha/2)}}{\dfrac{1}{\cos^2(\alpha/2)}}$$

$$= \frac{2\sin(\alpha/2)}{\cos(\alpha/2)} \cdot \cos^2\frac{\alpha}{2} = 2\sin\frac{\alpha}{2}\cos\frac{\alpha}{2}$$

$$= \sin 2\left(\frac{\alpha}{2}\right) = \sin\alpha$$

↑ Double-angle formula (3)

(b)

$$\frac{1 - z^2}{1 + z^2} = \frac{1 - \tan^2(\alpha/2)}{1 + \tan^2(\alpha/2)} = \frac{1 - \dfrac{\sin^2(\alpha/2)}{\cos^2(\alpha/2)}}{1 + \dfrac{\sin^2(\alpha/2)}{\cos^2(\alpha/2)}}$$

$$= \frac{\cos^2(\alpha/2) - \sin^2(\alpha/2)}{\cos^2(\alpha/2) + \sin^2(\alpha/2)} = \frac{\cos 2(\alpha/2)}{1} = \cos\alpha$$

↑ Double-angle formula (4a) and Pythagorean identity

EXERCISE 9.3

In Problems 1–10, use the information given about the angle θ to find the exact value of:

(a) $\sin 2\theta$ (b) $\cos 2\theta$ (c) $\sin\frac{1}{2}\theta$ (d) $\cos\frac{1}{2}\theta$

1. $\sin\theta = \frac{3}{5}$, $0 < \theta < \pi/2$

2. $\cos\theta = \frac{3}{5}$, $0 < \theta < \pi/2$

3. $\tan\theta = \frac{4}{3}$, $\pi < \theta < 3\pi/2$

4. $\tan\theta = \frac{1}{2}$, $\pi < \theta < 3\pi/2$

5. $\cos\theta = -\sqrt{2}/\sqrt{3}, \quad \pi/2 < \theta < \pi$

6. $\sin\theta = -1/\sqrt{3}, \quad 3\pi/2 < \theta < 2\pi$

7. $\sec\theta = 3, \quad \sin\theta > 0$

8. $\csc\theta = -\sqrt{5}, \quad \cos\theta < 0$

9. $\cot\theta = -2, \quad \sec\theta < 0$ 0

10. $\sec\theta = 2, \quad \csc\theta < 0$

In Problems 11–20, use the half-angle formulas to find the exact value of each trigonometric function.

11. $\sin 22.5°$

12. $\cos 22.5°$

13. $\tan\dfrac{7\pi}{8}$

14. $\tan\dfrac{9\pi}{8}$

15. $\cos 165°$

16. $\sin 195°$

17. $\sec\dfrac{15\pi}{8}$

18. $\csc\dfrac{7\pi}{8}$

19. $\sin\left(-\dfrac{\pi}{8}\right)$

20. $\cos\left(-\dfrac{3\pi}{8}\right)$

21. Show that $\sin^4\theta = \frac{3}{8} - \frac{1}{2}\cos 2\theta + \frac{1}{8}\cos 4\theta$.

22. Develop a formula for $\cos 3\theta$ as a third-degree polynomial in the variable $\cos\theta$.

23. Show that $\sin 4\theta = (\cos\theta)(4\sin\theta - 8\sin^3\theta)$.

24. Develop a formula for $\cos 4\theta$ as a fourth-degree polynomial in the variable $\cos\theta$.

25. Find an expression for $\sin 5\theta$ as a fifth-degree polynomial in the variable $\sin\theta$.

26. Find an expression for $\cos 5\theta$ as a fifth-degree polynomial in the variable $\cos\theta$.

27. Show that: $\tan\dfrac{\theta}{2} = \dfrac{1-\cos\theta}{\sin\theta}$

28. Show that: $\tan\dfrac{\theta}{2} = \dfrac{\sin\theta}{1+\cos\theta}$

In Problems 29–48, establish each identity.

29. $\cos^4\theta - \sin^4\theta = \cos 2\theta$

30. $\dfrac{\cot\theta - \tan\theta}{\cot\theta + \tan\theta} = \cos 2\theta$

31. $\cot 2\theta = \dfrac{\cot^2\theta - 1}{2\cot\theta}$

32. $\cot 2\theta = \frac{1}{2}(\cot\theta - \tan\theta)$

33. $\sec 2\theta = \dfrac{\sec^2\theta}{2 - \sec^2\theta}$

34. $\csc 2\theta = \frac{1}{2}\sec\theta\csc\theta$

35. $\cos^2 2\theta - \sin^2 2\theta = \cos 4\theta$

36. $(4\sin\theta\cos\theta)(1 - 2\sin^2\theta) = \sin 4\theta$

37. $\dfrac{\cos 2\theta}{1+\sin 2\theta} = \dfrac{\cot\theta - 1}{\cot\theta + 1}$

38. $\sin^2\theta\cos^2\theta = \frac{1}{8}(1 - \cos 4\theta)$

39. $\sec^2\dfrac{\theta}{2} = \dfrac{2}{1+\cos\theta}$

40. $\csc^2\dfrac{\theta}{2} = \dfrac{2}{1-\cos\theta}$

41. $\cot^2\dfrac{\theta}{2} = \dfrac{\sec\theta + 1}{\sec\theta - 1}$

42. $\tan\dfrac{\theta}{2} = \csc\theta - \cot\theta$

43. $\cos\theta = \dfrac{1 - \tan^2(\theta/2)}{1 + \tan^2(\theta/2)}$

44. $1 - \frac{1}{2}\sin 2\theta = \dfrac{\sin^3\theta + \cos^3\theta}{\sin\theta + \cos\theta}$

45. $\dfrac{\sin 3\theta}{\sin\theta} - \dfrac{\cos 3\theta}{\cos\theta} = 2$

46. $\dfrac{\cos\theta + \sin\theta}{\cos\theta - \sin\theta} - \dfrac{\cos\theta - \sin\theta}{\cos\theta + \sin\theta} = 2\tan 2\theta$

47. $\tan 3\theta = \dfrac{3\tan\theta - \tan^3\theta}{1 - 3\tan^2\theta}$

48. $\tan\theta + \tan(\theta + 120°) + \tan(\theta + 240°) = 3\tan 3\theta$

49. Graph $f(x) = \sin^2 x = (1 - \cos 2x)/2$ for $0 \le x \le 2\pi$ by using the ideas of shifting, compression, and so on.

50. Repeat Problem 49 for $g(x) = \cos^2 x$.

51. Use the fact that

$$\cos\frac{\pi}{12} = \frac{1}{4}(\sqrt{6} + \sqrt{2})$$

to find $\sin(\pi/24)$ and $\cos(\pi/24)$.

52. Show that

$$\sin\frac{\pi}{8} = \frac{\sqrt{2 - \sqrt{2}}}{2}$$

and use it to find $\sin(\pi/16)$ and $\cos(\pi/16)$.

In Problems 53–66, find the exact value of each expression.

53. $\sin\left(2\sin^{-1}\frac{1}{2}\right)$ **54.** $\sin[2\sin^{-1}(\sqrt{3}/2)]$ **55.** $\cos\left(2\sin^{-1}\frac{3}{5}\right)$

56. $\cos\left(2\cos^{-1}\frac{4}{5}\right)$ **57.** $\tan\left[2\cos^{-1}\left(-\frac{3}{5}\right)\right]$ **58.** $\tan\left(2\tan^{-1}\frac{3}{4}\right)$

59. $\sin\left(2\cos^{-1}\frac{4}{5}\right)$ **60.** $\cos\left[2\tan^{-1}\left(-\frac{4}{3}\right)\right]$ **61.** $\sin^2\left(\frac{1}{2}\cos^{-1}\frac{3}{5}\right)$

62. $\cos^2\left(\frac{1}{2}\sin^{-1}\frac{3}{5}\right)$ **63.** $\sec\left(2\tan^{-1}\frac{3}{4}\right)$ **64.** $\csc\left[2\sin^{-1}\left(-\frac{3}{5}\right)\right]$

65. $\cot^2\left(\frac{1}{2}\tan^{-1}\frac{4}{3}\right)$ **66.** $\cot^2\left(\frac{1}{2}\cos^{-1}\frac{5}{13}\right)$

67. Show that: $\sin^3\theta + \sin^3(\theta + 120°) + \sin^3(\theta + 240°) = -\frac{3}{4}\sin 3\theta$

68. If $\tan\theta = a\tan(\theta/3)$, express $\tan(\theta/3)$ in terms of a.

In Problems 69 and 70, establish each identity.

69. $\ln|\sin\theta| = \frac{1}{2}(\ln|1 - \cos 2\theta| - \ln 2)$ **70.** $\ln|\cos\theta| = \frac{1}{2}(\ln|1 + \cos 2\theta| - \ln 2)$

9.4 Product-to-Sum and Sum-to-Product Formulas

The sum and difference formulas can be used to derive formulas for writing the products of sines and/or cosines as sums or differences. These identities are usually called the **product-to-sum formulas.**

Theorem

Product-to-Sum Formulas

$$\sin\alpha\sin\beta = \tfrac{1}{2}[\cos(\alpha - \beta) - \cos(\alpha + \beta)] \tag{1}$$

$$\cos\alpha\cos\beta = \tfrac{1}{2}[\cos(\alpha - \beta) + \cos(\alpha + \beta)] \tag{2}$$

$$\sin\alpha\cos\beta = \tfrac{1}{2}[\sin(\alpha + \beta) + \sin(\alpha - \beta)] \tag{3}$$

These formulas do not have to be memorized. Instead, you should remember how they are derived. Then, when you want to use them, either look them up or derive them, as needed.

To derive formulas (1) and (2), write down the sum and difference formulas for the cosine:

$$\cos(\alpha - \beta) = \cos\alpha\cos\beta + \sin\alpha\sin\beta \tag{4}$$

$$\cos(\alpha + \beta) = \cos\alpha\cos\beta - \sin\alpha\sin\beta \tag{5}$$

Subtract equation (5) from equation (4) to get

$$\cos(\alpha - \beta) - \cos(\alpha + \beta) = 2\sin\alpha\sin\beta$$

from which

$$\sin \alpha \sin \beta = \tfrac{1}{2}[\cos(\alpha - \beta) - \cos(\alpha + \beta)]$$

Now, add equations (4) and (5) to get

$$\cos(\alpha - \beta) + \cos(\alpha + \beta) = 2 \cos \alpha \cos \beta$$

from which

$$\cos \alpha \cos \beta = \tfrac{1}{2}[\cos(\alpha - \beta) + \cos(\alpha + \beta)]$$

To derive product-to-sum formula (3), use the sum and difference formulas for sine in a similar way. (You are asked to do this in Problem 39 at the end of this section.)

Example 1 Express each of the following products as a sum containing only sines or cosines:

(a) $\sin 6\theta \sin 4\theta$ (b) $\cos 3\theta \cos \theta$ (c) $\sin 3\theta \cos 5\theta$

Solution (a) We use formula (1) to get

$$\begin{aligned}\sin 6\theta \sin 4\theta &= \tfrac{1}{2}[\cos(6\theta - 4\theta) - \cos(6\theta + 4\theta)]\\ &= \tfrac{1}{2}(\cos 2\theta - \cos 10\theta)\end{aligned}$$

(b) We use formula (2) to get

$$\begin{aligned}\cos 3\theta \cos \theta &= \tfrac{1}{2}[\cos(3\theta - \theta) + \cos(3\theta + \theta)]\\ &= \tfrac{1}{2}(\cos 2\theta + \cos 4\theta)\end{aligned}$$

(c) We use formula (3) to get

$$\begin{aligned}\sin 3\theta \cos 5\theta &= \tfrac{1}{2}[\sin(3\theta + 5\theta) + \sin(3\theta - 5\theta)]\\ &= \tfrac{1}{2}[\sin 8\theta + \sin(-2\theta)] = \tfrac{1}{2}(\sin 8\theta - \sin 2\theta)\end{aligned}$$

■

The **sum-to-product formulas** are given below.

Theorem

Sum-to-Product Formulas

$$\sin \alpha + \sin \beta = 2 \sin \frac{\alpha + \beta}{2} \cos \frac{\alpha - \beta}{2} \tag{6}$$

$$\sin \alpha - \sin \beta = 2 \sin \frac{\alpha - \beta}{2} \cos \frac{\alpha + \beta}{2} \tag{7}$$

$$\cos \alpha + \cos \beta = 2 \cos \frac{\alpha + \beta}{2} \cos \frac{\alpha - \beta}{2} \tag{8}$$

$$\cos \alpha - \cos \beta = -2 \sin \frac{\alpha + \beta}{2} \sin \frac{\alpha - \beta}{2} \tag{9}$$

■

We shall derive formula (6) and leave the derivations of formulas (7)–(9) as exercises (see Problems 40–42).

Proof

$$2 \sin \frac{\alpha+\beta}{2} \cos \frac{\alpha-\beta}{2} = 2 \cdot \frac{1}{2}\left[\sin\left(\frac{\alpha+\beta}{2}+\frac{\alpha-\beta}{2}\right)+\sin\left(\frac{\alpha+\beta}{2}-\frac{\alpha-\beta}{2}\right)\right]$$

↑ Product-to-sum formula (3)

$$= \sin \frac{2\alpha}{2} + \sin \frac{2\beta}{2} = \sin \alpha + \sin \beta$$ ■

Example 2 Express each sum or difference as a product of sines and/or cosines:

(a) $\sin 5\theta - \sin 3\theta$ (b) $\cos 3\theta + \cos 2\theta$

Solution (a) We use formula (7) to get

$$\sin 5\theta - \sin 3\theta = 2 \sin \frac{5\theta - 3\theta}{2} \cos \frac{5\theta + 3\theta}{2} = 2 \sin \theta \cos 4\theta$$

(b) $$\cos 3\theta + \cos 2\theta = 2 \cos \frac{3\theta + 2\theta}{2} \cos \frac{3\theta - 2\theta}{2} = 2 \cos \frac{5\theta}{2} \cos \frac{\theta}{2}$$ Formula (8) ■

EXERCISE 9.4

In Problems 1–10, express each product as a sum containing only sines or cosines

1. $\sin 4\theta \sin 2\theta$ **2.** $\cos 4\theta \cos 2\theta$ **3.** $\sin 4\theta \cos 2\theta$ **4.** $\sin 3\theta \sin 5\theta$

5. $\cos 3\theta \cos 5\theta$ **6.** $\sin 4\theta \cos 6\theta$ **7.** $\sin \theta \sin 2\theta$ **8.** $\cos 3\theta \cos 4\theta$

9. $\sin \frac{3\theta}{2} \cos \frac{\theta}{2}$ **10.** $\sin \frac{\theta}{2} \cos \frac{5\theta}{2}$

In Problems 11–18, express each sum or difference as a product of sines and/or cosines.

11. $\sin 4\theta - \sin 2\theta$ **12.** $\sin 4\theta + \sin 2\theta$ **13.** $\cos 2\theta + \cos 4\theta$ **14.** $\cos 5\theta - \cos 3\theta$

15. $\sin \theta + \sin 3\theta$ **16.** $\cos \theta + \cos 3\theta$ **17.** $\cos \frac{\theta}{2} - \cos \frac{3\theta}{2}$ **18.** $\sin \frac{\theta}{2} - \sin \frac{3\theta}{2}$

In Problems 19–36, establish each identity.

19. $\frac{\sin \theta + \sin 3\theta}{2 \sin 2\theta} = \cos \theta$ **20.** $\frac{\cos \theta + \cos 3\theta}{2 \cos 2\theta} = \cos \theta$

21. $\frac{\sin 4\theta + \sin 2\theta}{\cos 4\theta + \cos 2\theta} = \tan 3\theta$ **22.** $\frac{\cos \theta - \cos 3\theta}{\sin 3\theta - \sin \theta} = \tan 2\theta$

23. $\frac{\cos \theta - \cos 3\theta}{\sin \theta + \sin 3\theta} = \tan \theta$ **24.** $\frac{\cos \theta - \cos 5\theta}{\sin \theta + \sin 5\theta} = \tan 2\theta$

25. $\sin\theta(\sin\theta + \sin 3\theta) = \cos\theta(\cos\theta - \cos 3\theta)$

26. $\sin\theta(\sin 3\theta + \sin 5\theta) = \cos\theta(\cos 3\theta - \cos 5\theta)$

27. $\dfrac{\sin 4\theta + \sin 8\theta}{\cos 4\theta + \cos 8\theta} = \tan 6\theta$

28. $\dfrac{\sin 4\theta - \sin 8\theta}{\cos 4\theta - \cos 8\theta} = -\cot 6\theta$

29. $\dfrac{\sin 4\theta + \sin 8\theta}{\sin 4\theta - \sin 8\theta} = -\dfrac{\tan 6\theta}{\tan 2\theta}$

30. $\dfrac{\cos 4\theta - \cos 8\theta}{\cos 4\theta + \cos 8\theta} = \tan 2\theta \tan 6\theta$

31. $\dfrac{\sin\alpha + \sin\beta}{\sin\alpha - \sin\beta} = \tan\dfrac{\alpha+\beta}{2}\cot\dfrac{\alpha-\beta}{2}$

32. $\dfrac{\cos\alpha + \cos\beta}{\cos\alpha - \cos\beta} = -\cot\dfrac{\alpha+\beta}{2}\cot\dfrac{\alpha-\beta}{2}$

33. $\dfrac{\sin\alpha + \sin\beta}{\cos\alpha + \cos\beta} = \tan\dfrac{\alpha+\beta}{2}$

34. $\dfrac{\sin\alpha - \sin\beta}{\cos\alpha - \cos\beta} = -\cot\dfrac{\alpha+\beta}{2}$

35. $1 + \cos 2\theta + \cos 4\theta + \cos 6\theta = 4\cos\theta\cos 2\theta\cos 3\theta$

36. $1 - \cos 2\theta + \cos 4\theta - \cos 6\theta = 4\sin\theta\cos 2\theta\sin 3\theta$

37. If $\alpha + \beta + \gamma = \pi$, show that: $\sin 2\alpha + \sin 2\beta + \sin 2\gamma = 4\sin\alpha\sin\beta\sin\gamma$

38. If $\alpha + \beta + \gamma = \pi$, show that: $\tan\alpha + \tan\beta + \tan\gamma = \tan\alpha\tan\beta\tan\gamma$

39. Derive formula (3).

40. Derive formula (7).

41. Derive formula (8).

42. Derive formula (9).

9.5 Trigonometric Equations

The previous sections of this chapter were devoted to trigonometric identities—that is, equations involving trigonometric functions that are satisfied by every value in the domain of the variable. In this section, we discuss **trigonometric equations**—that is, equations involving trigonometric functions that are satisfied only by some values of the variable (or, possibly, are not satisfied by any values of the variable). The values that satisfy the equation are called **solutions** of the equation.

Example 1 Determine whether $\theta = \pi/4$ is a solution of the equation $\sin\theta = \frac{1}{2}$. Is $\theta = \pi/6$ a solution?

Solution Replace θ by $\pi/4$ in the given equation. The result is

$$\sin\frac{\pi}{4} = \frac{\sqrt{2}}{2} \neq \frac{1}{2}$$

We conclude that $\pi/4$ is not a solution.

Next, replace θ by $\pi/6$ in the equation. The result is

$$\sin\frac{\pi}{6} = \frac{1}{2}$$

Thus, $\pi/6$ is a solution of the given equation. ■

The equation given in Example 1 has other solutions besides $\theta = \pi/6$. For example, $\theta = 5\pi/6$ is also a solution, as is $\theta = 13\pi/6$. (You should check this for yourself.) In fact, the equation has an infinite number of solutions due to the periodicity of the sine function, as can be seen in Figure 6.

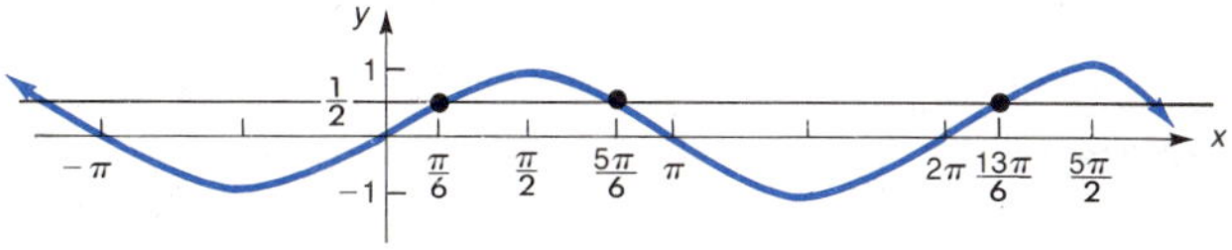

Figure 6
$y = \sin x$

Unless otherwise indicated, in solving trigonometric equations, we need to find *all* solutions. As the next example illustrates, finding all the solutions can be accomplished by first finding solutions over an interval whose length equals the period of the function and then adding multiples of that period to the solutions found. Let's look at some examples.

Example 2 Solve the equation: $\cos\theta = \frac{1}{2}$

Solution The period of the cosine function is 2π. In the interval $[0, 2\pi)$, there are two angles θ for which $\cos\theta = \frac{1}{2}$, namely, $\theta = \pi/3$ and $\theta = 5\pi/3$. See Figure 7. Because the cosine function has period 2π, all the solutions of $\cos\theta = \frac{1}{2}$ may be given by

$$\theta = \frac{\pi}{3} + 2k\pi \qquad \text{or} \qquad \theta = \frac{5\pi}{3} + 2k\pi \qquad k \text{ any integer}$$ ■

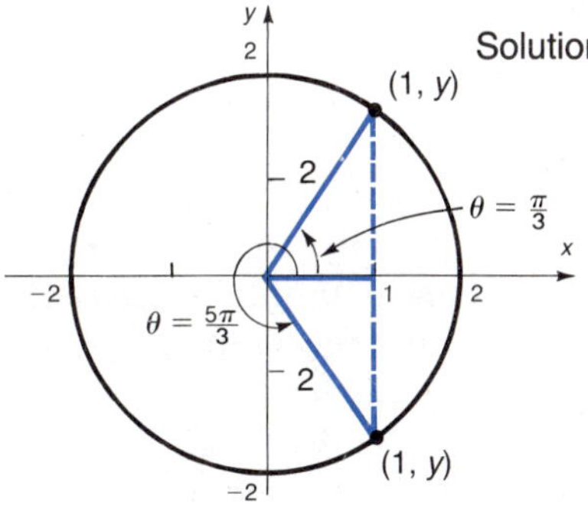

Figure 7
$\cos\theta = \frac{1}{2}$

In most of our work, we shall be interested only in finding solutions of trigonometric equations for $0 \le \theta < 2\pi$.

Example 3 Solve the equation: $\sin 2\theta = 1,\ 0 \le \theta < 2\pi$

Solution The period of the sine function is 2π. In the interval $[0, 2\pi)$, the sine function has the value 1 only at $\pi/2$. Because the argument is 2θ in the given equation, we have

$$2\theta = \frac{\pi}{2} + 2k\pi \qquad k \text{ any integer}$$

$$\theta = \frac{\pi}{4} + k\pi$$

In the interval $[0, 2\pi)$, the solutions of $\sin 2\theta = 1$ are $\pi/4$ $(k = 0)$ and $\pi/4 + \pi = 3\pi/4$ $(k = 1)$. ■

Warning: In solving a trigonometric equation for θ, $0 \le \theta \le 2\pi$, in which the argument is not θ (as in Example 3), you must write down all the solutions first, and then list those that are in the interval $[0, 2\pi)$. Otherwise, solutions may be lost. For example, in solving $\sin 2\theta = 1$, if you merely write the solution $2\theta = \pi/2$, you will find only $\theta = \pi/4$ and miss the solution $\theta = 3\pi/4$.

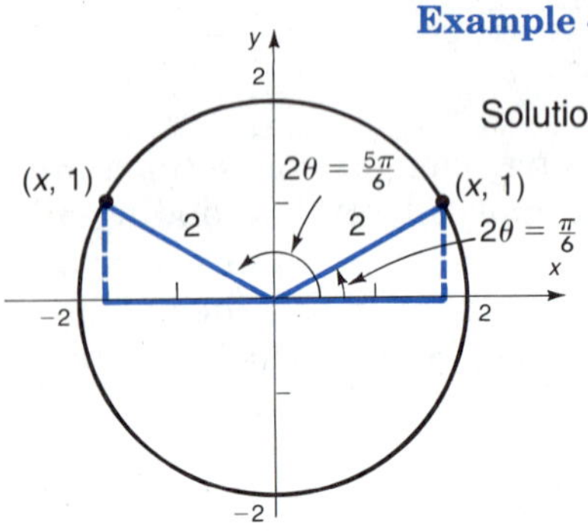

Figure 8
$\sin 2\theta = \frac{1}{2}$, $0 \le \theta < 2\pi$

Example 4 Solve the equation: $\sin 2\theta = \frac{1}{2}$, $0 \le \theta < 2\pi$

Solution The period of the sine function is 2π. In the interval $[0, 2\pi)$, the sine function has the value $\frac{1}{2}$ at $\pi/6$ and $5\pi/6$. See Figure 8. Consequently, because the argument is 2θ in the equation $\sin 2\theta = \frac{1}{2}$, we have

$$2\theta = \frac{\pi}{6} + 2k\pi \quad \text{or} \quad 2\theta = \frac{5\pi}{6} + 2k\pi \qquad k \text{ any integer}$$

$$\theta = \frac{\pi}{12} + k\pi \qquad\qquad \theta = \frac{5\pi}{12} + k\pi$$

In the interval $[0, 2\pi)$, the solutions of $\sin 2\theta = \frac{1}{2}$ are $\pi/12$, $\pi/12 + \pi = 13\pi/12$, $5\pi/12$, and $5\pi/12 + \pi = 17\pi/12$. ■

Example 5 Solve the equation: $\tan \frac{\theta}{4} = 1$, $0 \le \theta < 2\pi$

Solution The period of the tangent function is π. In the interval $[0, \pi)$, the tangent function has the value 1 only at $\pi/4$. Because the argument is $\theta/4$ in the given equation, we have

$$\frac{\theta}{4} = \frac{\pi}{4} + k\pi \qquad k \text{ any integer}$$

$$\theta = \pi + 4k\pi$$

In the interval $[0, 2\pi)$, $\theta = \pi$ is the only solution. ■

C **Example 6** Use a calculator to solve the equation: $\sin \theta = 0.3$, $0 \le \theta < 2\pi$

Solution To solve $\sin \theta = 0.3$ on a calculator, we first choose the mode. If we set it to radians, we find

Enter:	0.3	Press:	inv sin
Display:	0.3		0.3046927

The angle 0.3046927 radian is the angle $-\pi/2 \le \theta \le \pi/2$ for which $\sin \theta = 0.3$. Another angle for which $\sin \theta = 0.3$ is $\pi - 0.3046927$. See

Michelle 369-7145

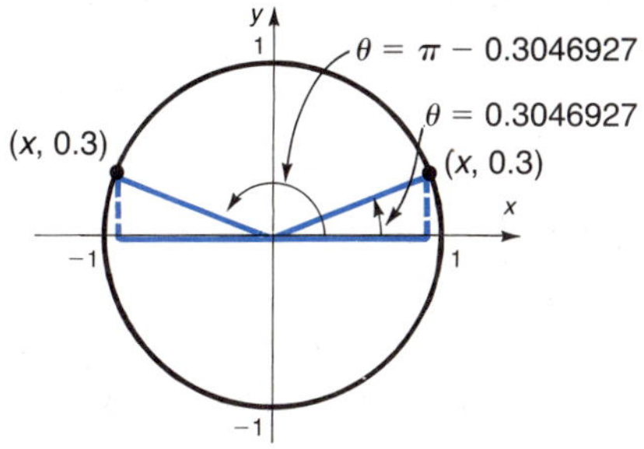

Figure 9
$\sin\theta = 0.3$

Figure 9. The angle $\pi - 0.3046927$ is the angle in quadrant II where $\sin\theta = 0.3$. Thus, the solutions for $\sin\theta = 0.3$, $0 \le \theta < 2\pi$, are

$$\theta = 0.3046927 \quad \text{or} \quad \pi - 0.3046927 \approx 2.8369$$ ■

Example 6 illustrates that caution must be exercised when solving trigonometric equations on a calculator. Remember that the calculator supplies an angle only within the restrictions of the definition of the inverse trigonometric function. To find the remaining solutions, you must identify other quadrants, if any, in which the angle may be located.

Many trigonometric equations can be solved by applying techniques we already know, such as applying the quadratic formula (if the equation is a second-degree polynomial), factoring, and so on.

Example 7 Solve the equation: $2\sin^2\theta - 3\sin\theta + 1 = 0$, $0 \le \theta < 2\pi$

Solution The equation we wish to solve is a quadratic equation (in $\sin\theta$) that can be factored:

$$2\sin^2\theta - 3\sin\theta + 1 = 0 \qquad 2x^2 - 3x + 1 = 0,\ x = \sin\theta$$
$$(2\sin\theta - 1)(\sin\theta - 1) = 0 \qquad (2x - 1)(x - 1) = 0$$

$$2\sin\theta - 1 = 0 \quad \text{or} \quad \sin\theta - 1 = 0$$
$$\sin\theta = \tfrac{1}{2} \qquad\qquad \sin\theta = 1$$

Thus,

$$\theta = \frac{\pi}{6} \qquad \theta = \frac{5\pi}{6} \qquad\qquad \theta = \frac{\pi}{2}$$ ■

When a trigonometric equation contains more than one trigonometric function, identities sometimes can be used to obtain an equivalent equation that contains only one trigonometric function.

Example 8 Solve the equation: $3\cos\theta + 3 = 2\sin^2\theta$, $0 \le \theta < 2\pi$

Solution The equation in its present form contains sines and cosines. However, a form of the Pythagorean identity can be used to transform the equation into an equivalent expression containing only cosines:

$$3\cos\theta + 3 = 2\sin^2\theta$$
$$3\cos\theta + 3 = 2(1 - \cos^2\theta) \qquad \sin^2\theta = 1 - \cos^2\theta$$
$$3\cos\theta + 3 = 2 - 2\cos^2\theta$$
$$2\cos^2\theta + 3\cos\theta + 1 = 0 \qquad \text{Quadratic in } \cos\theta$$
$$(2\cos\theta + 1)(\cos\theta + 1) = 0 \qquad \text{Factor}$$

$$2\cos\theta + 1 = 0 \quad \text{or} \quad \cos\theta + 1 = 0$$
$$\cos\theta = -\tfrac{1}{2} \qquad\qquad \cos\theta = -1$$

Thus,

$$\theta = \frac{2\pi}{3} \qquad \theta = \frac{4\pi}{3} \qquad \theta = \pi$$ ■

Example 9 Solve the equation: $\cos 2\theta + 3 = 5 \cos \theta$, $0 \le \theta < 2\pi$

Solution First, we observe that the given equation contains two different trigonometric functions, the functions $\cos \theta$ and $\cos 2\theta$. However, we can use the double-angle formula $\cos 2\theta = 2 \cos^2 \theta - 1$ to obtain an equivalent equation containing only $\cos \theta$:

$$\begin{aligned} \cos 2\theta + 3 &= 5 \cos \theta \\ (2 \cos^2 \theta - 1) + 3 &= 5 \cos \theta \\ 2 \cos^2 \theta - 5 \cos \theta + 2 &= 0 \\ (\cos \theta - 2)(2 \cos \theta - 1) &= 0 \\ \cos \theta = 2 \quad \text{or} \quad \cos \theta &= \tfrac{1}{2} \end{aligned}$$

For any angle θ, $-1 \le \cos \theta \le 1$; thus, the equation $\cos \theta = 2$ has no solution. The solutions of $\cos \theta = \frac{1}{2}$ are

$$\theta = \frac{\pi}{3} \qquad \theta = \frac{5\pi}{3}$$ ■

Example 10 Solve the equation: $\cos^2 \theta + \sin \theta = 2$, $0 \le \theta < 2\pi$

Solution We use a form of the Pythagorean identity:

$$\begin{aligned} \cos^2 \theta + \sin \theta &= 2 \\ (1 - \sin^2 \theta) + \sin \theta &= 2 \\ \sin^2 \theta - \sin \theta + 1 &= 0 \end{aligned}$$

This is a quadratic equation in $\sin \theta$. The discriminant is $b^2 - 4ac = 1 - 4 = -3 < 0$. Therefore, the equation has no real solution. ■

Example 11 Solve the equation: $\sin \theta \cos \theta = -\frac{1}{2}$, $0 \le \theta < 2\pi$

Solution The left side of the given equation is in the form of the double-angle formula $2 \sin \theta \cos \theta = \sin 2\theta$, except for a factor of 2. Thus, we multiply each side by 2:

$$\begin{aligned} \sin \theta \cos \theta &= -\tfrac{1}{2} \\ 2 \sin \theta \cos \theta &= -1 \\ \sin 2\theta &= -1 \end{aligned}$$

The argument here is 2θ. Thus, we need to write all the solutions of this equation and then list those that are in the interval $[0, 2\pi)$.

$$2\theta = \frac{3\pi}{2} + 2k\pi \qquad k \text{ any integer}$$

$$\theta = \frac{3\pi}{4} + k\pi$$

The solutions in the interval $[0, 2\pi)$ are

$$\theta = \frac{3\pi}{4} \qquad \theta = \frac{7\pi}{4}$$

■

Sometimes it is necessary to square both sides of an equation in order to obtain expressions that allow use of identities. Remember, however, that in squaring both sides, extraneous solutions may be introduced. As a result, apparent solutions must be checked.

Example 12 Solve the equation: $\sin\theta + \cos\theta = 1$, $0 \le \theta < 2\pi$

Solution Attempts to use available identities do not lead to equations that are easy to solve. (Try it yourself.) Thus, given the form of this equation, we decide to square each side:

$$\sin\theta + \cos\theta = 1$$

$$(\sin\theta + \cos\theta)^2 = 1$$

$$\sin^2\theta + 2\sin\theta\cos\theta + \cos^2\theta = 1$$

$$2\sin\theta\cos\theta = 0 \qquad \sin^2\theta + \cos^2\theta = 1$$

$$\sin\theta\cos\theta = 0$$

Thus,

$$\sin\theta = 0 \qquad \text{or} \qquad \cos\theta = 0$$

and the apparent solutions are

$$\theta = 0 \qquad \theta = \pi \qquad \theta = \frac{\pi}{2} \qquad \theta = \frac{3\pi}{2}$$

We must check these apparent solutions:

$\theta = 0$: $\sin 0 + \cos 0 = 0 + 1 = 1$ A solution

$\theta = \pi$: $\sin\pi + \cos\pi = 0 + (-1) = -1$ Not a solution

$\theta = \frac{\pi}{2}$: $\sin\frac{\pi}{2} + \cos\frac{\pi}{2} = 1 + 0 = 1$ A solution

$\theta = \frac{3\pi}{2}$: $\sin\frac{3\pi}{2} + \cos\frac{3\pi}{2} = -1 + 0 = -1$ Not a solution

Thus, $\theta = 3\pi/2$ and $\theta = \pi$ are extraneous. The actual solutions are $\theta = 0$ and $\theta = \pi/2$. ■

We can solve the equation given in Example 12 another way.

Solution We start with the equation

$$\sin \theta + \cos \theta = 1$$

and divide each side by $\sqrt{2}$. (The reason for this choice will become apparent shortly.) Then,

$$\frac{1}{\sqrt{2}} \sin \theta + \frac{1}{\sqrt{2}} \cos \theta = \frac{1}{\sqrt{2}}$$

The left side now resembles the formula for the sine of the sum of two angles, one of which is θ. The other angle is unknown (call it ϕ). Then,

$$\sin(\theta + \phi) = \sin \theta \cos \phi + \cos \theta \sin \phi = \frac{1}{\sqrt{2}} \tag{1}$$

where

$$\cos \phi = \frac{1}{\sqrt{2}} \qquad \sin \phi = \frac{1}{\sqrt{2}} \qquad 0 \le \phi < 2\pi$$

The angle ϕ is therefore $\pi/4$. As a result, equation (1) becomes

$$\sin\left(\theta + \frac{\pi}{4}\right) = \frac{1}{\sqrt{2}}$$

We solve this equation to get

$$\theta + \frac{\pi}{4} = \frac{\pi}{4} \qquad \text{or} \qquad \theta + \frac{\pi}{4} = \frac{3\pi}{4}$$

$$\theta = 0 \qquad\qquad \theta = \frac{\pi}{2}$$

These solutions agree with the solutions found earlier. ■

This second method of solution can be used to solve any linear equation in the variables $\sin \theta$ and $\cos \theta$.

Example 13 Solve

$$a \sin \theta + b \cos \theta = c \qquad 0 \le \theta < 2\pi \tag{2}$$

where a, b, and c are constants and either $a \ne 0$ or $b \ne 0$.

Solution We divide each side of equation (2) by $\sqrt{a^2 + b^2}$. Then,

$$\frac{a}{\sqrt{a^2 + b^2}} \sin \theta + \frac{b}{\sqrt{a^2 + b^2}} \cos \theta = \frac{c}{\sqrt{a^2 + b^2}} \tag{3}$$

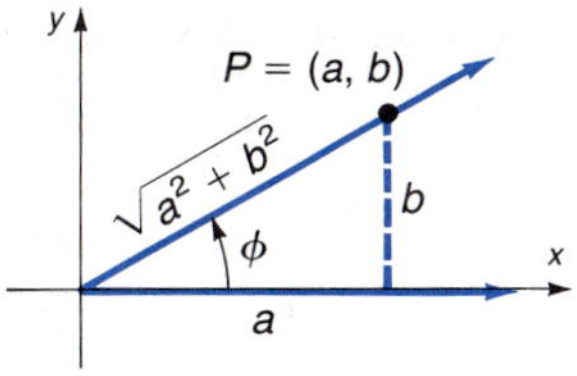

Figure 10

There is a unique angle ϕ, $0 \le \phi < 2\pi$, for which

$$\cos \phi = \frac{a}{\sqrt{a^2 + b^2}} \quad \text{and} \quad \sin \phi = \frac{b}{\sqrt{a^2 + b^2}} \tag{4}$$

(see Figure 10). Thus, equation (3) may be written as

$$\sin \theta \cos \phi + \cos \theta \sin \phi = \frac{c}{\sqrt{a^2 + b^2}}$$

or, equivalently,

$$\sin(\theta + \phi) = \frac{c}{\sqrt{a^2 + b^2}} \tag{5}$$

where ϕ satisfies equations (4).

If $|c| > \sqrt{a^2 + b^2}$, then $\sin(\theta + \phi) > 1$ or $\sin(\theta + \phi) < -1$, and equation (5) has no solution.

If $|c| \le \sqrt{a^2 + b^2}$, then the solutions of equation (5) are

$$\theta + \phi = \sin^{-1} \frac{c}{\sqrt{a^2 + b^2}} \quad \text{or} \quad \theta + \phi = \pi - \sin^{-1} \frac{c}{\sqrt{a^2 + b^2}}$$

Because the angle ϕ is determined by equations (4), these are the solutions to equation (2). ■

EXERCISE 9.5

In Problems 1–12, solve each equation on the interval $0 \le \theta < 2\pi$.

1. $\sin \theta = \frac{1}{2}$ **2.** $\tan \theta = 1$ **3.** $\tan \theta = -\frac{1}{\sqrt{3}}$ **4.** $\cos \theta = -\frac{\sqrt{3}}{2}$

5. $\cos \theta = 0$ **6.** $\sin \theta = \frac{\sqrt{2}}{2}$ **7.** $\sin 3\theta = -1$ **8.** $\tan \frac{\theta}{2} = \sqrt{3}$

9. $\cos\left(2\theta - \frac{\pi}{2}\right) = -1$ **10.** $\sin\left(3\theta + \frac{\pi}{18}\right) = 1$ **11.** $\sec \frac{3\theta}{2} = -2$ **12.** $\cot \frac{2\theta}{3} = -\sqrt{3}$

C *In Problems 13–20, solve each equation on the interval $0 \le \theta < 2\pi$.*

13. $\sin \theta = 0.4$ **14.** $\cos \theta = 0.6$ **15.** $\tan \theta = 5$ **16.** $\cot \theta = 2$

17. $\cos \theta = -0.9$ **18.** $\sin \theta = -0.2$ **19.** $\sec \theta = -4$ **20.** $\csc \theta = -3$

In Problems 21–46, solve each equation on the interval $0 \le \theta < 2\pi$.

21. $\cos \theta = \sin \theta$ **22.** $\cos \theta + \sin \theta = 0$ **23.** $\tan \theta = 2 \sin \theta$

24. $\sin 2\theta = \cos \theta$ **25.** $\sin \theta = \csc \theta$ **26.** $\tan \theta = \cot \theta$

27. $\cos 2\theta = \cos \theta$ **28.** $\sin 2\theta \sin \theta = \cos \theta$ **29.** $\sin 2\theta + \sin 4\theta = 0$

30. $\cos 2\theta + \cos 4\theta = 0$ **31.** $\cos 4\theta - \cos 6\theta = 0$ **32.** $\sin 4\theta - \sin 6\theta = 0$

33. $2 \sin^2 \theta + \sin \theta - 1 = 0$ **34.** $2 \cos^2 \theta - \cos \theta - 1 = 0$ **35.** $1 + \sin \theta = 2 \cos^2 \theta$

36. $\sin^2 \theta = 2 \cos \theta + 2$ **37.** $\tan^2 \theta = \frac{3}{2} \sec \theta$ **38.** $\csc^2 \theta = \cot \theta + 1$

39. $3 - \sin \theta = \cos 2\theta$ **40.** $\cos 2\theta + 5 \cos \theta + 3 = 0$ **41.** $\sec^2 \theta + \tan \theta = 0$

42. $\sec \theta = \tan \theta + \cot \theta$ **43.** $\sin \theta - \sqrt{3} \cos \theta = 1$ **44.** $\sqrt{3} \sin \theta + \cos \theta = 1$

45. $\tan 2\theta + 2 \sin \theta = 0$ **46.** $\tan 2\theta + 2 \cos \theta = 0$

The following discussion of ***Snell's Law of Refraction*** *(named after Willebrod Snell, 1591–1626) is needed for Problems 47–53: Light, sound, and other waves travel at different speeds, depending on the media (air, water, wood, and so on) through which they pass. Suppose light travels from a point A in one medium, where its speed is v_1, to a point B in another medium, where its speed is v_2. Refer to the figure, where the angle θ_1 is called the* ***angle of incidence*** *and the angle θ_2 is the* ***angle of refraction****. Snell's Law,* which can be proved using calculus, states that*

$$\frac{\sin \theta_1}{\sin \theta_2} = \frac{v_1}{v_2}$$

The ratio v_1/v_2 is called the ***index of refraction****. Some values are given in the table below.*

Some Indices of Refraction

MEDIUM	INDEX OF REFRACTION
Water	1.33
Ethyl alcohol	1.36
Carbon bisulfide	1.63
Air (1 atm and 20°C)	1.0003
Methylene iodide	1.74
Fused quartz	1.46
Glass, crown	1.52
Glass, dense flint	1.66
Sodium chloride	1.53

For light of wavelength 589 nanometers, measured with respect to a vacuum. The index with respect to air is negligibly different in most cases.

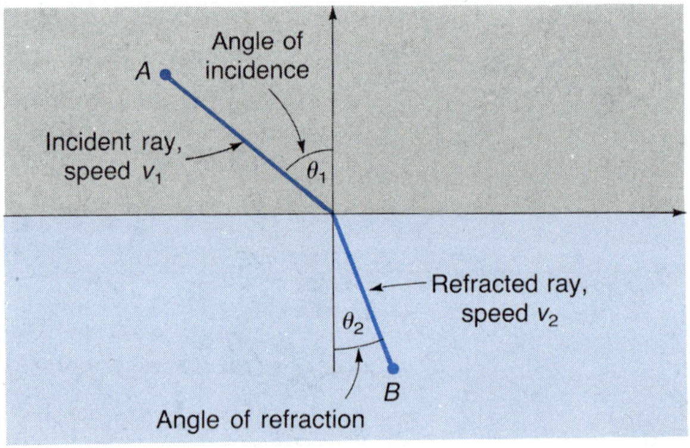

C **47.** The index of refraction of light in passing from a vacuum into water is 1.33. If the angle of incidence is 40°, determine the angle of refraction.

C **48.** The index of refraction of light in passing from a vacuum into dense glass is 1.66. If the angle of incidence is 50°, determine the angle of refraction.

C **49.** Ptolemy, who founded the city of Alexandria in Egypt toward the end of the first century AD, gave the measured values in the table in the margin for the angle of incidence θ_1 and the angle of refraction θ_2 for a light beam passing from air into water. Do these values agree with Snell's Law? If so, what index of refraction results? (These data are interesting as the oldest recorded physical measurements.)†

θ_1	θ_2
10°	7°45′
20°	15°30′
30°	22°30′
40°	29°0′
50°	35°0′
60°	40°30′
70°	45°30′
80°	50°0′

C **50.** The speed of yellow sodium light (wavelength 589 nanometers) in a certain liquid is measured to be 1.92×10^8 meters per second. What is the index of refraction of this liquid, with respect to air, for sodium light?†

*Because this law was also deduced by René Descartes, in France it is known as Descartes' Law.

†Adapted from Halliday and Resnick, *Physics, Parts 1 & 2*, 3rd ed., New York: Wiley, 1978 p. 953.

C **51.** A beam of light of wavelength 589 nanometers traveling in air makes an angle of incidence of 40° upon a slab of transparent material, and the refracted beam makes an angle of refraction of 26°. Find the index of refraction of the material.*

52. A light ray of wavelength 589 nanometers (produced by a sodium lamp) traveling through air makes an angle of incidence of 30° on a smooth, flat slab of crown glass. Find the angle of refraction.*

53. A light beam passes from one medium to another through a thick slab of material whose index of refraction is n_2. Show that the emerging beam is parallel to the incident beam.*

CHAPTER REVIEW

THINGS TO KNOW

Formulas

Sum and difference formulas

$$\cos(\alpha + \beta) = \cos\alpha\cos\beta - \sin\alpha\sin\beta$$
$$\cos(\alpha - \beta) = \cos\alpha\cos\beta + \sin\alpha\sin\beta$$
$$\sin(\alpha + \beta) = \sin\alpha\cos\beta + \cos\alpha\sin\beta$$
$$\sin(\alpha - \beta) = \sin\alpha\cos\beta - \cos\alpha\sin\beta$$
$$\tan(\alpha + \beta) = \frac{\tan\alpha + \tan\beta}{1 - \tan\alpha\tan\beta}$$
$$\tan(\alpha - \beta) = \frac{\tan\alpha - \tan\beta}{1 + \tan\alpha\tan\beta}$$

Double-angle formulas

$$\sin 2\theta = 2\sin\theta\cos\theta$$
$$\cos 2\theta = \cos^2\theta - \sin^2\theta$$
$$\cos 2\theta = 1 - 2\sin^2\theta$$
$$\cos 2\theta = 2\cos^2\theta - 1$$
$$\tan 2\theta = \frac{2\tan\theta}{1 - \tan^2\theta}$$

Half-angle formulas

$$\sin^2\frac{\alpha}{2} = \frac{1 - \cos\alpha}{2}$$
$$\cos^2\frac{\alpha}{2} = \frac{1 + \cos\alpha}{2}$$
$$\tan^2\frac{\alpha}{2} = \frac{1 - \cos\alpha}{1 + \cos\alpha}$$

$$\sin\frac{\alpha}{2} = \pm\sqrt{\frac{1 - \cos\alpha}{2}}$$
$$\cos\frac{\alpha}{2} = \pm\sqrt{\frac{1 + \cos\alpha}{2}}$$
$$\tan\frac{\alpha}{2} = \pm\sqrt{\frac{1 - \cos\alpha}{1 + \cos\alpha}}$$

where the + or − sign is determined by the quadrant of the angle $\alpha/2$

*Adapted from Serway, *Physics*, 3rd ed., Saunders, p. 805.

Product-to-sum formulas

$$\sin\alpha\sin\beta = \tfrac{1}{2}[\cos(\alpha-\beta) - \cos(\alpha+\beta)]$$
$$\cos\alpha\cos\beta = \tfrac{1}{2}[\cos(\alpha-\beta) + \cos(\alpha+\beta)]$$
$$\sin\alpha\cos\beta = \tfrac{1}{2}[\sin(\alpha+\beta) + \sin(\alpha-\beta)]$$

Sum-to-product formulas

$$\sin\alpha + \sin\beta = 2\sin\frac{\alpha+\beta}{2}\cos\frac{\alpha-\beta}{2}$$
$$\sin\alpha - \sin\beta = 2\sin\frac{\alpha-\beta}{2}\cos\frac{\alpha+\beta}{2}$$
$$\cos\alpha + \cos\beta = 2\cos\frac{\alpha+\beta}{2}\cos\frac{\alpha-\beta}{2}$$
$$\cos\alpha - \cos\beta = -2\sin\frac{\alpha+\beta}{2}\sin\frac{\alpha-\beta}{2}$$

How To:

Establish identities
Solve a trigonometric equation

FILL-IN-THE-BLANK ITEMS

1. Suppose f and g are two functions with the same domain. If $f(x) = g(x)$ for every x in the domain, the equation is called a(n) ________. Otherwise, it is called a(n) ________ equation.
2. $\cos(\alpha+\beta) = \cos\alpha\cos\beta$ ________ $\sin\alpha\sin\beta$
3. $\sin(\alpha+\beta) = \sin\alpha\cos\beta$ ________ $\cos\alpha\sin\beta$
4. $\cos 2\theta = \cos^2\theta -$ ________ $=$ ________ $- 1 = 1 -$ ________
5. $\sin^2\frac{\alpha}{2} = \frac{\text{________}}{2}$

REVIEW EXERCISES

In Problems 1–32, establish each identity.

1. $\tan\theta\cot\theta - \sin^2\theta = \cos^2\theta$
2. $\sin\theta\csc\theta - \sin^2\theta = \cos^2\theta$
3. $\cos^2\theta(1 + \tan^2\theta) = 1$
4. $(1 - \cos^2\theta)(1 + \cot^2\theta) = 1$
5. $4\cos^2\theta + 3\sin^2\theta = 3 + \cos^2\theta$
6. $4\sin^2\theta + 2\cos^2\theta = 4 - 2\cos^2\theta$
7. $\dfrac{1-\cos\theta}{\sin\theta} + \dfrac{\sin\theta}{1-\cos\theta} = 2\csc\theta$
8. $\dfrac{\sin\theta}{1+\cos\theta} + \dfrac{1+\cos\theta}{\sin\theta} = 2\csc\theta$
9. $\dfrac{\cos\theta}{\cos\theta - \sin\theta} = \dfrac{1}{1-\tan\theta}$
10. $1 - \dfrac{\cos^2\theta}{1+\sin\theta} = \sin\theta$
11. $\dfrac{\csc\theta}{1+\csc\theta} = \dfrac{1-\sin\theta}{\cos^2\theta}$
12. $\dfrac{1+\sec\theta}{\sec\theta} = \dfrac{\sin^2\theta}{1-\cos\theta}$
13. $\csc\theta - \sin\theta = \cos\theta\cot\theta$
14. $\dfrac{\csc\theta}{1-\cos\theta} = \dfrac{1+\cos\theta}{\sin^3\theta}$

15. $\dfrac{1 - \sin\theta}{\sec\theta} = \dfrac{\cos^3\theta}{1 + \sin\theta}$

16. $\dfrac{1 - \cos\theta}{1 + \cos\theta} = (\csc\theta - \cot\theta)^2$

17. $\dfrac{1 - 2\sin^2\theta}{\sin\theta\cos\theta} = \cot\theta - \tan\theta$

18. $\dfrac{(2\sin^2\theta - 1)^2}{\sin^4\theta - \cos^4\theta} = 1 - 2\cos^2\theta$

19. $\dfrac{\cos(\alpha + \beta)}{\cos\alpha\sin\beta} = \cot\beta - \tan\alpha$

20. $\dfrac{\sin(\alpha - \beta)}{\sin\alpha\cos\beta} = 1 - \cot\alpha\tan\beta$

21. $\dfrac{\cos(\alpha - \beta)}{\cos\alpha\cos\beta} = 1 + \tan\alpha\tan\beta$

22. $\dfrac{\cos(\alpha + \beta)}{\sin\alpha\cos\beta} = \cot\alpha - \tan\beta$

23. $(1 + \cos\theta)\left(\tan\dfrac{\theta}{2}\right) = \sin\theta$

24. $\sin\theta\tan\dfrac{\theta}{2} = 1 - \cos\theta$

25. $2\cot\theta\cot 2\theta = \cot^2\theta - 1$

26. $2\sin 2\theta(1 - 2\sin^2\theta) = \sin 4\theta$

27. $1 - 8\sin^2\theta\cos^2\theta = \cos 4\theta$

28. $\dfrac{\sin 3\theta\cos\theta - \sin\theta\cos 3\theta}{\sin 2\theta} = 1$

29. $\dfrac{\sin 2\theta + \sin 4\theta}{\cos 2\theta + \cos 4\theta} = \tan 3\theta$

30. $\dfrac{\sin 2\theta + \sin 4\theta}{\sin 2\theta - \sin 4\theta} + \dfrac{\tan 3\theta}{\tan\theta} = 0$

31. $\dfrac{\cos 2\theta - \cos 4\theta}{\cos 2\theta + \cos 4\theta} - \tan\theta\tan 3\theta = 0$

32. $\cos 2\theta - \cos 10\theta = (\tan 4\theta)(\sin 2\theta + \sin 10\theta)$

In Problems 33–40, find the exact value of each expression.

33. $\sin 165°$

34. $\tan 105°$

35. $\cos\dfrac{5\pi}{12}$

36. $\sin\left(-\dfrac{\pi}{12}\right)$

37. $\cos 80°\cos 20° + \sin 80°\sin 20°$

38. $\sin 70°\cos 40° - \cos 70°\sin 40°$

39. $\tan\dfrac{\pi}{8}$

40. $\sin\dfrac{5\pi}{8}$

In Problems 41–50, use the information given about the angles α and β to find the exact value of:

(a) $\sin(\alpha + \beta)$ (b) $\cos(\alpha + \beta)$ (c) $\sin(\alpha - \beta)$ (d) $\tan(\alpha + \beta)$

(e) $\sin 2\alpha$ (f) $\cos 2\beta$ (g) $\sin\dfrac{\beta}{2}$ (h) $\cos\dfrac{\alpha}{2}$

41. $\sin\alpha = \frac{4}{5}$, $0 < \alpha < \pi/2$; $\sin\beta = \frac{5}{13}$, $\pi/2 < \beta < \pi$
42. $\cos\alpha = \frac{4}{5}$, $0 < \alpha < \pi/2$; $\cos\beta = \frac{5}{13}$, $-\pi/2 < \beta < 0$
43. $\sin\alpha = -\frac{3}{5}$, $\pi < \alpha < 3\pi/2$; $\cos\beta = \frac{12}{13}$, $3\pi/2 < \beta < 2\pi$
44. $\sin\alpha = -\frac{4}{5}$, $-\pi/2 < \alpha < 0$; $\cos\beta = -\frac{5}{13}$, $\pi/2 < \beta < \pi$
45. $\tan\alpha = \frac{3}{4}$, $\pi < \alpha < 3\pi/2$; $\tan\beta = \frac{12}{5}$, $0 < \beta < \pi/2$
46. $\tan\alpha = -\frac{4}{3}$, $\pi/2 < \alpha < \pi$; $\cot\beta = \frac{12}{5}$, $\pi < \beta < 3\pi/2$
47. $\sec\alpha = 2$, $-\pi/2 < \alpha < 0$; $\sec\beta = 3$, $3\pi/2 < \beta < 2\pi$
48. $\csc\alpha = 2$, $\pi/2 < \alpha < \pi$; $\sec\beta = -3$, $\pi/2 < \beta < \pi$
49. $\sin\alpha = -\frac{2}{3}$, $\pi < \alpha < 3\pi/2$; $\cos\beta = -\frac{2}{3}$, $\pi < \beta < 3\pi/2$
50. $\tan\alpha = -2$, $\pi/2 < \alpha < \pi$; $\cot\beta = -2$, $\pi/2 < \beta < \pi$

In Problems 51–70, solve each equation on the interval $0 \le \theta < 2\pi$.

51. $\cos\theta = \frac{1}{2}$

52. $\sin\theta = -\sqrt{3}/2$

53. $\cos\theta = -\sqrt{2}/2$

54. $\tan\theta = -\sqrt{3}$

55. $\sin 2\theta = -1$

56. $\cos 2\theta = 0$

57. $\tan 2\theta = 0$

58. $\sin 3\theta = 1$

C 59. $\sin \theta = 0.9$

C 60. $\tan \theta = 25$

61. $\sin \theta = \tan \theta$

62. $\cos \theta = \sec \theta$

63. $\sin \theta + \sin 2\theta = 0$

64. $\cos 2\theta = \sin \theta$

65. $\sin 2\theta - \cos \theta - 2 \sin \theta + 1 = 0$

66. $\sin 2\theta - \sin \theta - 2 \cos \theta + 1 = 0$

67. $2 \sin^2 \theta - 3 \sin \theta + 1 = 0$

68. $2 \cos^2 \theta + \cos \theta - 1 = 0$

69. $\sin \theta - \cos \theta = 1$

70. $\sin \theta + 2 \cos \theta = 1$

Additional Applications of Trigonometry

10

In Chapter 7, we used the trigonometric functions to solve right triangles—that is, triangles with a 90° angle. In this chapter, we will use the trigonometric functions to solve oblique triangles—that is, triangles that do not have a 90° angle. To solve such triangles, we shall develop the Law of Sines (in Section 10.1) and the Law of Cosines (in Section 10.2). In addition to finding the sides and angles of such triangles, we shall derive formulas for finding their areas (in Section 10.3).

The final sections of this chapter deal with polar coordinates (an alternative to rectangular coordinates for plotting points), graphing in polar coordinates, and finding roots of complex numbers (DeMoivre's Theorem).

10.1 The Law of Sines

If none of the angles of a triangle is a right angle, the triangle is called **oblique**. Thus, an oblique triangle will have either three acute angles or two acute angles and one obtuse angle (an angle between 90° and 180°). See Figure 1.

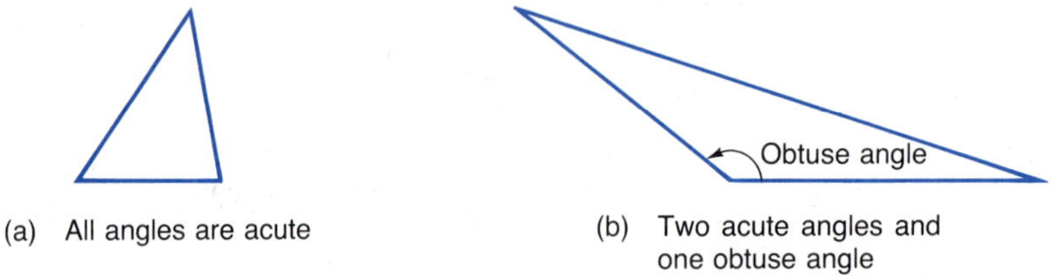

Figure 1
Oblique triangles

In the discussion that follows, we shall always label an oblique triangle so that side a is opposite angle α, side b is opposite angle β, and side c is opposite angle γ, as shown in Figure 2.

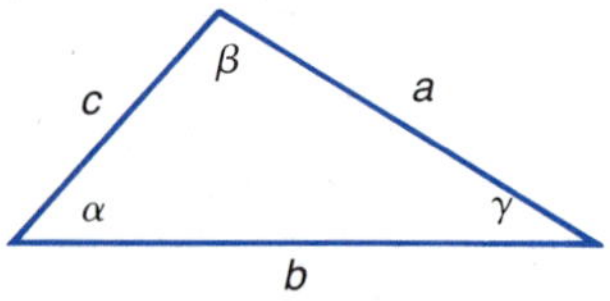

Figure 2

To **solve an oblique triangle** means to find the lengths of its sides and the measurements of its angles. To do this, we shall need to know the length of one side along with two other facts: either two angles, or the other two sides, or one angle and one other side.* Thus, there are four possibilities to consider:

CASE 1: One side and two angles are known (SAA).
CASE 2: Two sides and the angle opposite one of them are known (SSA).
CASE 3: Two sides and the included angle are known (SAS).
CASE 4: Three sides are known (SSS).

Figure 3 illustrates the four cases.

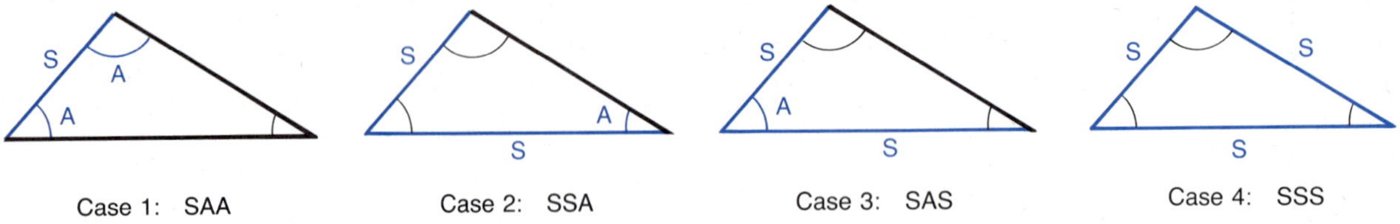

Figure 3

The **Law of Sines** is used to solve triangles for which Case 1 or 2 holds.

*Recall from plane geometry the fact that knowing three angles of a triangle determines a family of *similar triangles*—that is, triangles that have the same shape but different sizes.

Theorem

Law of Sines

For a triangle with sides a, b, c and opposite angles α, β, γ, respectively,

$$\frac{\sin \alpha}{a} = \frac{\sin \beta}{b} = \frac{\sin \gamma}{c} \tag{1}$$

Proof

To prove the Law of Sines, we construct an altitude of length h from one of the vertices of such a triangle. Figure 4(a) shows h for a triangle with three acute angles, and Figure 4(b) shows h for a triangle with an obtuse angle. In each case, the altitude is drawn from the vertex at β. Using either illustration, we have

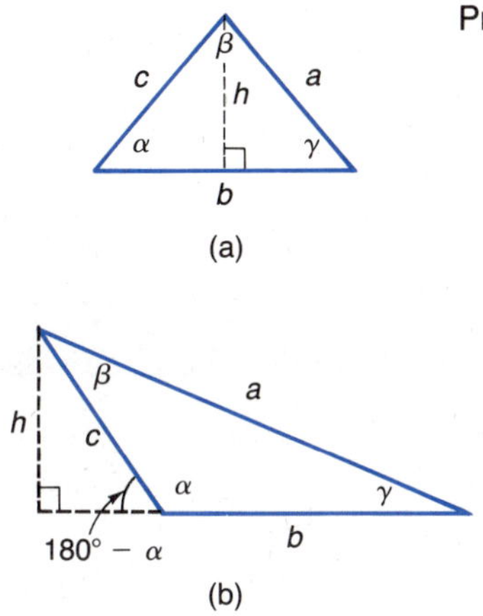

Figure 4

$$\sin \gamma = \frac{h}{a}$$

from which

$$h = a \sin \gamma \tag{2}$$

From Figure 4(a), it also follows that

$$\sin \alpha = \frac{h}{c}$$

from which

$$h = c \sin \alpha \tag{3}$$

From Figure 4(b), it follows that

$$\sin \alpha = \sin(180° - \alpha) = \frac{h}{c}$$

which again gives

$$h = c \sin \alpha$$

Thus, whether the triangle has three acute angles or has two acute angles and one obtuse angle, equations (2) and (3) hold. As a result, we may equate the expressions for h in equations (2) and (3) to get

$$a \sin \gamma = c \sin \alpha$$

from which

$$\frac{\sin \alpha}{a} = \frac{\sin \gamma}{c} \tag{4}$$

In a similar manner, by constructing the altitude h' from the vertex of angle α as shown in Figure 5, it is easy to show that

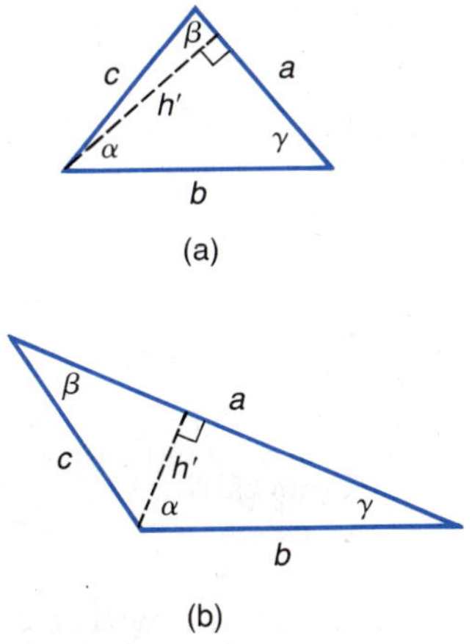

Figure 5

$$\sin \beta = \frac{h'}{c} \quad \text{and} \quad \sin \gamma = \frac{h'}{b}$$

Thus,

$$h' = c \sin \beta = b \sin \gamma$$

and

$$\frac{\sin \beta}{b} = \frac{\sin \gamma}{c} \tag{5}$$

When equations (4) and (5) are combined, we have equation (1), the Law of Sines. ■

In applying the Law of Sines to solve triangles, we use the fact that the sum of the angles of any triangle equals 180°; that is,

$$\alpha + \beta + \gamma = 180° \tag{6}$$

Our first two examples show how to solve a triangle when one side and two angles are known (Case 1: SAA).

Example 1 Solve the triangle: $\alpha = 40°$, $\beta = 60°$, $a = 4$

Solution Figure 6 shows the triangle we want to solve. The third angle γ is easily found using equation (6):

$$\alpha + \beta + \gamma = 180°$$
$$40° + 60° + \gamma = 180°$$
$$\gamma = 80°$$

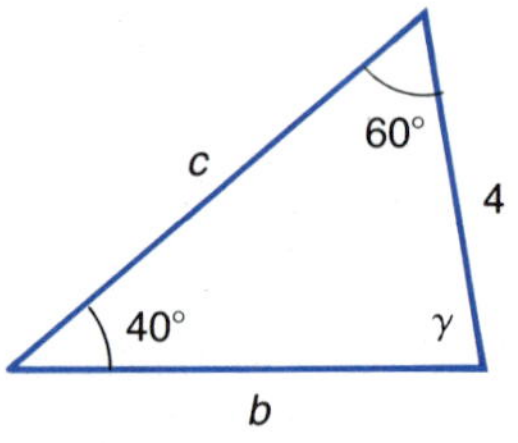

Figure 6

Now we use the Law of Sines (twice) to find the unknown sides b and c:

$$\frac{\sin \alpha}{a} = \frac{\sin \beta}{b} \qquad \frac{\sin \alpha}{a} = \frac{\sin \gamma}{c}$$

Because $a = 4$, $\alpha = 40°$, $\beta = 60°$, and $\gamma = 80°$, we have

$$\frac{\sin 40°}{4} = \frac{\sin 60°}{b} \qquad \frac{\sin 40°}{4} = \frac{\sin 80°}{c}$$

Thus,

$$b = \frac{4 \sin 60°}{\sin 40°} \approx 5.3892 \qquad c = \frac{4 \sin 80°}{\sin 40°} \approx 6.1284$$

↑ From table or calculator (for $\sin 40°$ in each)

■

Notice that in Example 1, we found b and c by working with the given side a. This is better than finding b first and working with a rounded value of b to find c.

Example 2 Solve the triangle: $\alpha = 35°, \beta = 15°, c = 5$

Solution Figure 7 illustrates the triangle we want to solve. Because we know two angles ($\alpha = 35°$ and $\beta = 15°$), it is easy to find the third angle using equation (6):

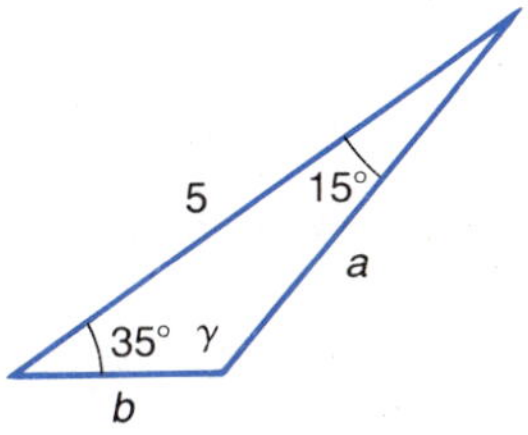

Figure 7

$$\alpha + \beta + \gamma = 180°$$
$$35° + 15° + \gamma = 180°$$
$$\gamma = 130°$$

Now we know the three angles and one side ($c = 5$) of the triangle. To find the remaining two sides a and b, we use the Law of Sines (twice):

$$\frac{\sin \alpha}{a} = \frac{\sin \gamma}{c} \qquad \frac{\sin \beta}{b} = \frac{\sin \gamma}{c}$$
$$\frac{\sin 35°}{a} = \frac{\sin 130°}{5} \qquad \frac{\sin 15°}{b} = \frac{\sin 130°}{5}$$
$$a = \frac{5 \sin 35°}{\sin 130°} \approx 3.7438 \qquad b = \frac{5 \sin 15°}{\sin 130°} \approx 1.6893$$ ■

Note: In subsequent examples and in the exercises that follow, unless otherwise indicated, we shall measure angles in degrees and round off to one decimal place; we shall round off all sides to four decimal places.

Ambiguous Case

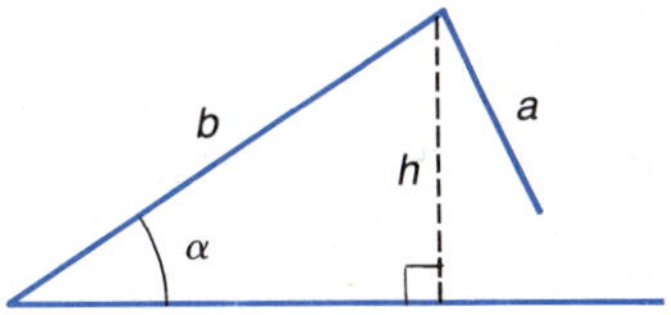

Figure 8

Case 2 (SSA), which applies to triangles where two sides and the angle opposite one of them are known, is referred to as the **ambiguous case**, because the known information may result in one triangle, two triangles, or no triangle at all. Suppose we are given sides a and b and angle α, as illustrated in Figure 8. The key to determining the possible triangles, if any, that may be formed from the given information, lies primarily with the height h and the fact that $h = b \sin \alpha$.

No Triangle: If $a < b \sin \alpha = h$, then clearly side a is not sufficiently long to form a triangle. See Figure 9.

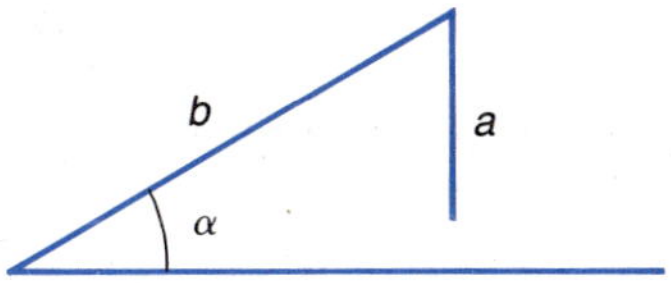

Figure 9
$a < b \sin \alpha$

One Right Triangle: If $a = b \sin \alpha = h$, then side a is just long enough to form a right triangle. See Figure 10.

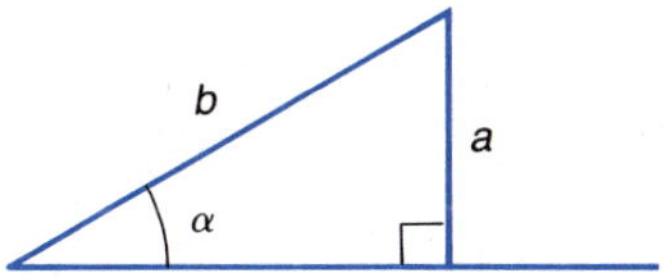

Figure 10
$a = b \sin \alpha$

Two Triangles: If $a < b$ and $h = b \sin \alpha < a$, then two distinct triangles can be formed from the given information. See Figure 11.

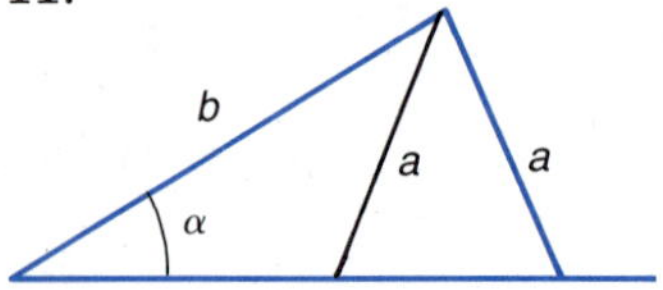

Figure 11
$b \sin \alpha < a$ and $a < b$

One Triangle: If $a \geq b$, then only one triangle can be formed. See Figure 12.

Figure 12
$a \geq b$

Fortunately, we do not have to rely on an illustration to draw the correct conclusion in the ambiguous case. The Law of Sines will lead us to the correct determination. Let's see how.

Example 3 Solve the triangle: $a = 3,\ b = 2,\ \alpha = 40°$

Solution Because $a = 3$, $b = 2$, and $\alpha = 40°$ are known, we use the Law of Sines to find the angle β:

$$\frac{\sin \alpha}{a} = \frac{\sin \beta}{b}$$

Then,

$$\frac{\sin 40°}{3} = \frac{\sin \beta}{2}$$

$$\sin \beta = \frac{2 \sin 40°}{3} \approx 0.4285$$

There are two angles β, $0° < \beta < 180°$, for which $\sin \beta \approx 0.4285$, namely,

$$\beta \approx 25.4° \quad \text{and} \quad \beta \approx 154.6°$$

The second possibility is ruled out, because $\alpha = 40°$, making $\alpha + \beta \approx 194.6° > 180°$. Now, using $\beta \approx 25.4°$, we find

$$\gamma = 180° - \alpha - \beta \approx 180° - 40° - 25.4° = 114.6°$$

The third side c may now be determined using the Law of Sines:

$$\frac{\sin \gamma}{c} = \frac{\sin \alpha}{a}$$

$$\frac{\sin 114.6°}{c} = \frac{\sin 40°}{3}$$

$$c = \frac{3 \sin 114.6°}{\sin 40°} \approx 4.2436$$

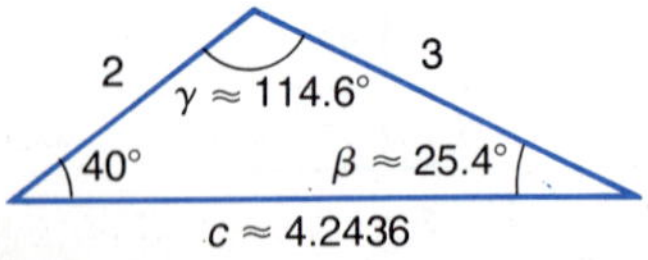

Figure 13

Figure 13 illustrates the solved triangle. ■

Example 4 Solve the triangle: $a = 6$, $b = 8$, $\alpha = 35°$

Solution Because $a = 6$, $b = 8$, and $\alpha = 35°$ are known, we use the Law of Sines to find the angle β:

$$\frac{\sin \alpha}{a} = \frac{\sin \beta}{b}$$

Then,

$$\frac{\sin 35°}{6} = \frac{\sin \beta}{8}$$

$$\sin \beta = \frac{8 \sin 35°}{6} \approx 0.7648$$

$$\beta_1 \approx 49.9° \quad \text{or} \quad \beta_2 \approx 130.1°$$

For both possibilities we have $\alpha + \beta < 180°$. Hence, there are two triangles—one containing the angle $\beta = \beta_1 \approx 49.9°$ and the other containing the angle $\beta = \beta_2 \approx 130.1°$. The third angle γ is either

$$\gamma_1 = 180° - \alpha - \beta_1 \approx 95.1° \quad \text{or} \quad \gamma_2 = 180° - \alpha - \beta_2 \approx 14.9°$$

↑ $\alpha = 35°$, $\beta_1 \approx 49.9°$ ↑ $\alpha = 35°$, $\beta_2 \approx 130.1°$

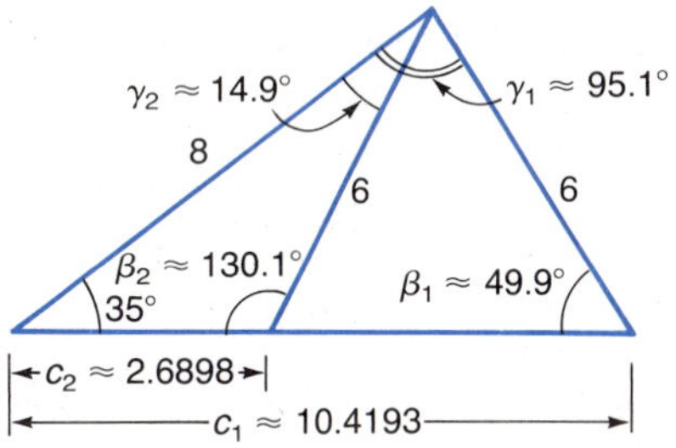

Figure 14

The third side c obeys the Law of Sines, so we have

$$\frac{\sin \gamma}{c} = \frac{\sin \alpha}{a}$$

$$\frac{\sin 95.1°}{c_1} = \frac{\sin 35°}{6} \quad \text{or} \quad \frac{\sin 14.9°}{c_2} = \frac{\sin 35°}{6}$$

$$c_1 = \frac{6 \sin 95.1°}{\sin 35°} \approx 10.4193 \qquad c_2 = \frac{6 \sin 14.9°}{\sin 35°} \approx 2.6898$$

The two solved triangles are illustrated in Figure 14. ■

Example 5 Solve the triangle: $a = 2$, $c = 1$, $\gamma = 50°$

Solution Because $a = 2$, $c = 1$, and $\gamma = 50°$ are known, we use the Law of Sines to find the angle α:

$$\frac{\sin \alpha}{a} = \frac{\sin \gamma}{c}$$

$$\frac{\sin \alpha}{2} = \frac{\sin 50°}{1}$$

$$\sin \alpha = 2 \sin 50° \approx 1.5321$$

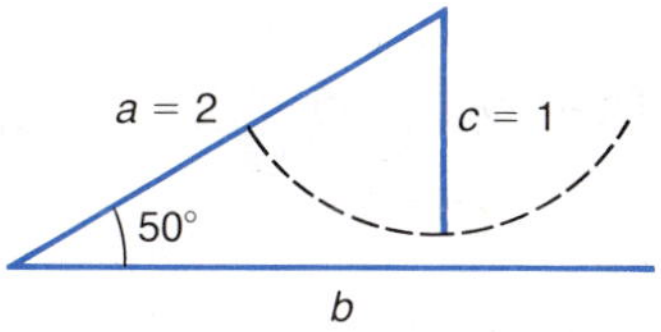

Figure 15

There is no angle α for which $\sin \alpha > 1$. Hence, there can be no triangle with the given measurements. Figure 15 illustrates the measurements given. Notice that no matter how we attempt to position side c, it will never intersect side b to form a triangle. ■

Applied Problems

The Law of Sines is particularly useful for solving certain applied problems.

C **Example 6** Coast Guard Station Zebra is located 120 miles due west of Station X-Ray. A ship at sea sends an SOS call that is received by each station. The call to Station Zebra indicates the location of the ship is 40° east of north; the call to Station X-Ray indicates the location of the ship is 30° west of north.

(a) How far is each station from the ship?

(b) If a helicopter capable of flying 200 miles per hour is dispatched from the nearest station to the ship, how long will it take to reach the ship?

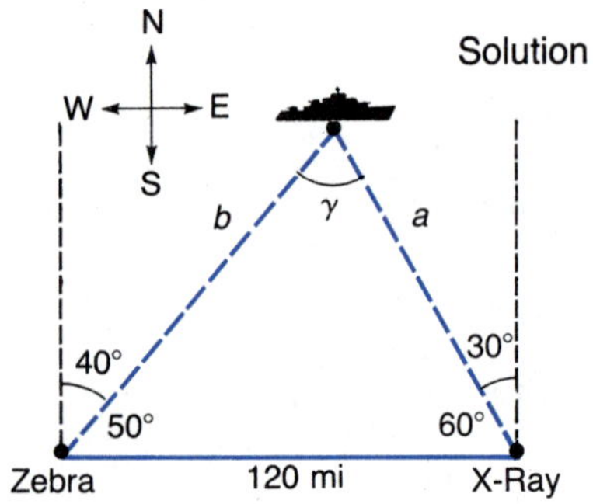

Figure 16

Solution (a) Figure 16 illustrates the situation. The angle γ is found to be

$$\gamma = 180° - 50° - 60° = 70°$$

The Law of Sines can now be used to find the two distances a and b we seek:

$$\frac{\sin 50°}{a} = \frac{\sin 70°}{120}$$

$$a = \frac{120 \sin 50°}{\sin 70°} \approx 97.8 \text{ miles}$$

$$\frac{\sin 60°}{b} = \frac{\sin 70°}{120}$$

$$b = \frac{120 \sin 60°}{\sin 70°} \approx 110.6 \text{ miles}$$

Thus, Station Zebra is 110.6 miles from the ship, and Station X-Ray is 97.8 miles from the ship.

(b) The time t needed for the helicopter to reach the ship from Station X-Ray is found by using the formula

$$(\text{Velocity}, v)(\text{Time}, t) = \text{Distance}$$

Then,

$$t = \frac{a}{v} = \frac{97.8}{200} \approx 0.49 \text{ hour} \approx 29 \text{ minutes}$$ ■

EXERCISE 10.1

In Problems 1–8, solve each triangle.

1.

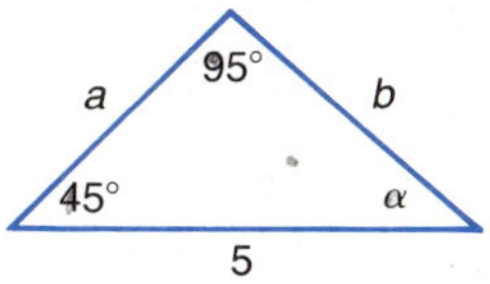

2.

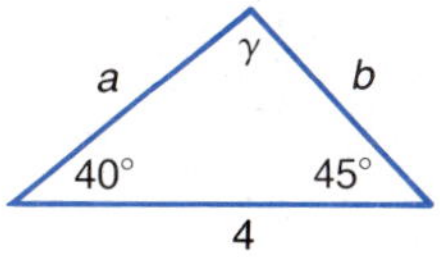

3.

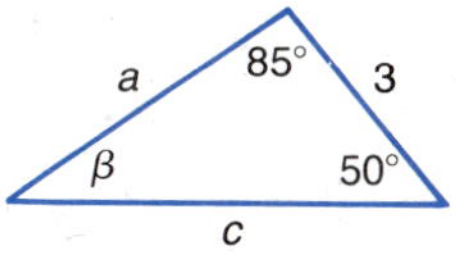

4.

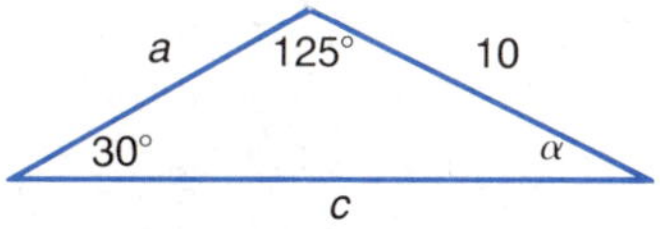

5.

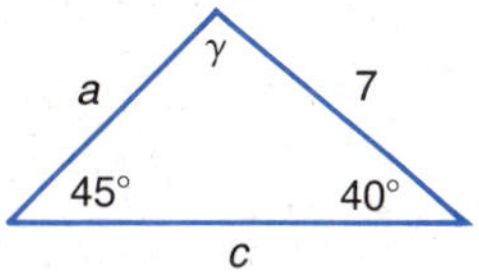

6.

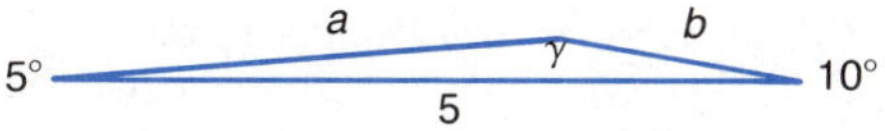

7.

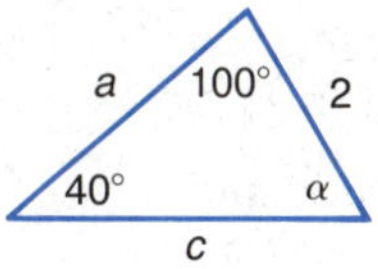

8.

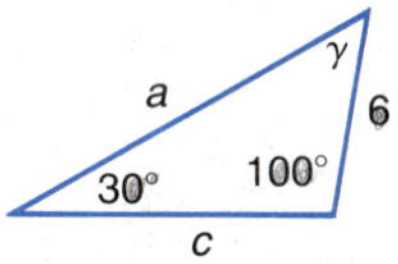

In Problems 9–16, solve each triangle.

9. $\alpha = 40°, \beta = 20°, a = 2$

10. $\alpha = 50°, \gamma = 20°, a = 3$

11. $\beta = 70°, \gamma = 10°, b = 5$

12. $\alpha = 70°, \beta = 60°, c = 4$

13. $\alpha = 110°, \gamma = 30°, c = 3$

14. $\beta = 10°, \gamma = 100°, b = 2$

15. $\alpha = 40°, \beta = 40°, c = 2$

16. $\beta = 20°, \gamma = 70°, a = 1$

In Problems 17–28, two sides and an angle are given. Determine whether the given information results in one triangle, two triangles, or no triangle at all. Solve any triangle(s) that results.

17. $a = 3, b = 2, \alpha = 50°$

18. $b = 4, c = 3, \beta = 40°$

19. $b = 5, c = 3, \beta = 100°$

20. $a = 2, c = 1, \alpha = 120°$

21. $a = 4, b = 5, \alpha = 60°$

22. $b = 2, c = 3, \beta = 40°$

23. $b = 4, c = 6, \beta = 20°$

24. $a = 3, b = 7, \alpha = 70°$

25. $a = 2, c = 1, \gamma = 100°$

26. $b = 4, c = 5, \beta = 95°$

27. $a = 2, c = 1, \gamma = 25°$

28. $b = 4, c = 5, \beta = 40°$

29. Coast Guard Station Able is located 150 miles due south of Station Baker. A ship at sea sends an SOS call that is received by each station. The call to Station Able indicates the ship is located 30° east of north; the call to Station Baker indicates the ship is located 35° east of south.

(a) How far is each station from the ship?

(b) If a helicopter capable of flying 200 miles per hour is dispatched from the nearest station to the ship, how long will it take to reach the ship?

30. Consult the figure. To find the distance from the house at A to the house at B, a surveyor measures the angle BAC to be 40°, then walks off a distance of 100 feet to C, and measures the angle ACB to be 50°. What is the distance from A to B?

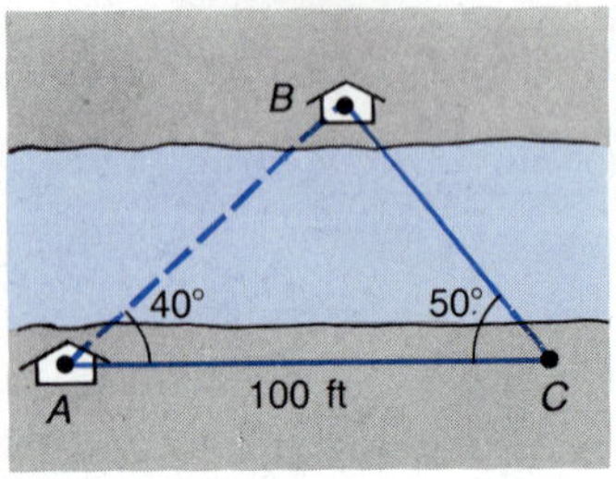

31. Consult the figure. To find the length of the span of a proposed ski lift from A to B, a surveyor measures the angle DAB to be 25°, then walks off a distance of 100 feet to C, and measures the angle ACB to be 15°. What is the distance from A to B?

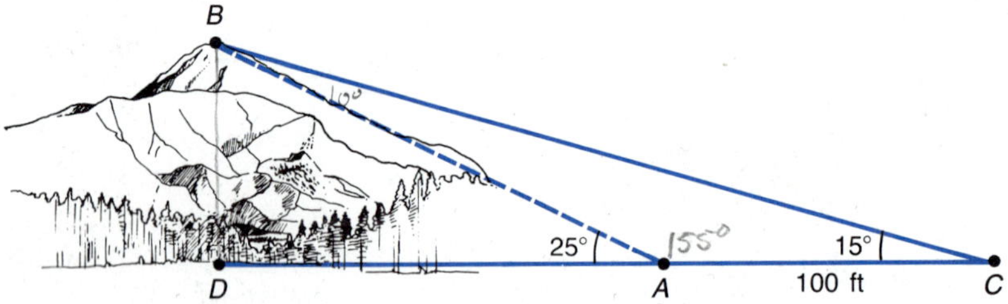

32. Use the illustration in Problem 31 to find the height BD of the mountain at B.

33. An aircraft is spotted by two observers who are 1000 feet apart. As the airplane passes over the line joining them, each observer takes a sighting of the angle of elevation to the plane, as indicated in the figure. How high is the airplane?

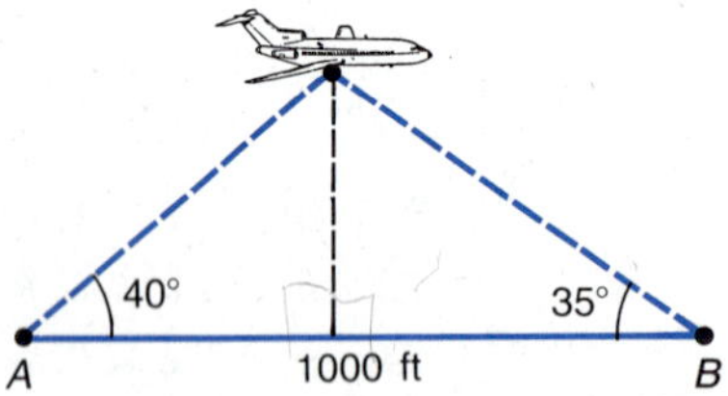

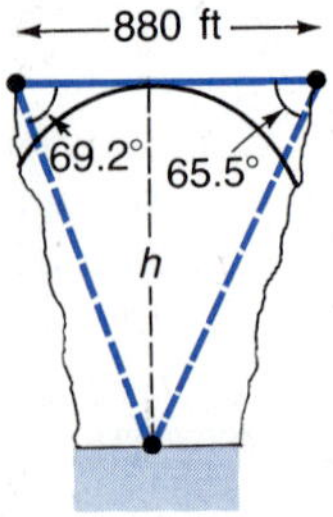

34. The highest bridge in the world is the bridge over the Royal Gorge of the Arkansas River in Colorado.* Sightings to the same point at water level directly under the bridge are taken from each side of the 880-foot-long bridge, as indicated in the figure. How high is the bridge?

*Source: *Guinness Book of World Records*

35. An airplane flies from city A to city B, a distance of 150 miles, then turns through an angle of 40°, and heads toward city C, as shown in the figure.
(a) If the distance between cities A and C is 300 miles, how far is it from city B to city C?
(b) Through what angle should the pilot turn at city C to return to city A?

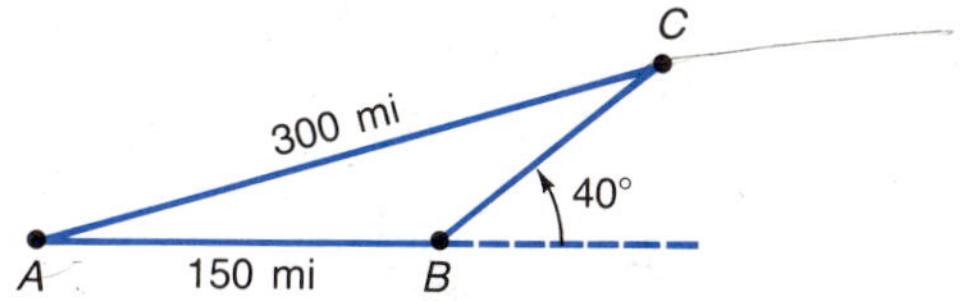

36. In attempting to fly from city A to city B, an aircraft followed a course that was 10° in error, as indicated in the figure. After flying a distance of 50 miles, the pilot corrected the course by turning at point C and flying 70 miles further. If the constant speed of the aircraft was 250 miles per hour, how much time was lost due to the error?

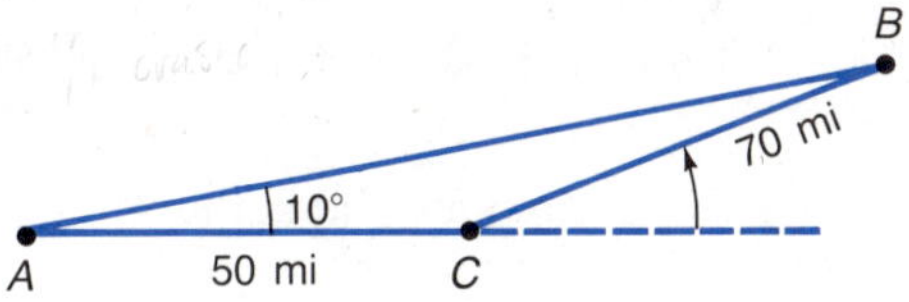

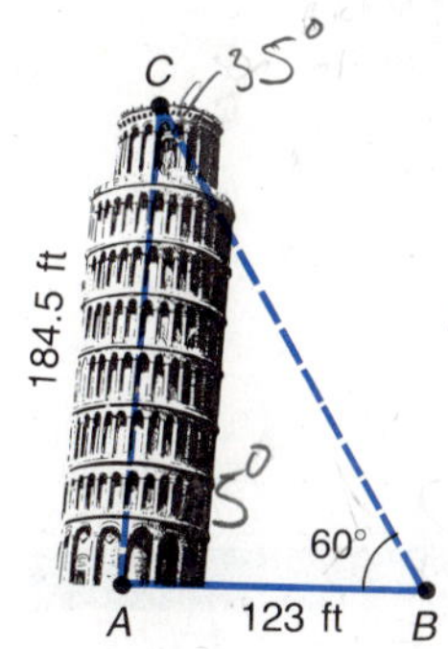

37. The famous Leaning Tower of Pisa was originally 184.5 feet high.* After walking 123 feet from the base of the tower, the angle of elevation to the top of the tower is found to be 60°. Find the angle CAB indicated in the figure. Also, find the perpendicular distance from C to AB.

C

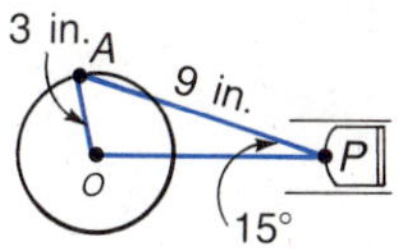

38. On a certain automobile, the crankshaft is 3 inches long and the connecting rod is 9 inches long (see the figure). At the time when the angle OPA is 15°, how far is the piston (P) from the center (O) of the crankshaft?

39. For any triangle, **Mollweide's Formula** (named after Karl Mollweide, 1774–1825) states that

$$\frac{a+b}{c} = \frac{\cos \frac{1}{2}(\alpha - \beta)}{\sin \frac{1}{2}\gamma}$$

Derive it. [*Hint:* Use the Law of Sines and then a sum-to-product formula.] Notice that this formula involves all six parts of a triangle. As a result, it is sometimes used to check the solution of a triangle.

*Making their annual report on the fragile 7-century-old bell tower, scientists in Pisa, Italy, said the Leaning Tower of Pisa had increased its famous lean by 1 millimeter, or 0.04 inch. That is about the annual average, although the tilting had slowed to about half that much in the previous 2 years. (*Source:* United Press International, June 29, 1986.)

40. Another form of Mollweide's Formula is

$$\frac{a-b}{c} = \frac{\sin \frac{1}{2}(\alpha - \beta)}{\cos \frac{1}{2}\gamma}$$

Derive it.

41. For any triangle, derive the formula

$$a = b \cos \gamma + c \cos \beta$$

[*Hint:* Use the fact that $\sin \alpha = \sin(180° - \beta - \gamma)$.]

42. For any triangle, derive the **Law of Tangents**, namely,

$$\frac{a-b}{a+b} = \frac{\tan \frac{1}{2}(\alpha - \beta)}{\tan \frac{1}{2}(\alpha + \beta)}$$

[*Hint:* Use Mollweide's Formula.]

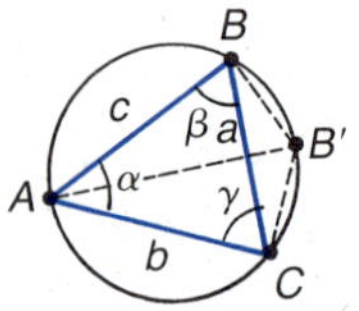

43. Show that

$$\frac{\sin \alpha}{a} = \frac{\sin \beta}{b} = \frac{\sin \gamma}{c} = \frac{1}{2r}$$

where r is the radius of the circle circumscribing the triangle ABC whose sides are a, b, c, as shown in the figure in the margin. [*Hint:* Draw the diameter AB. Then β = Angle ABC = Angle $AB'C$ and Angle $ACB' = 90°$.]

10.2 The Law of Cosines

In Section 10.1, we used the Law of Sines to solve Case 1 (SAA) and Case 2 (SSA) of an oblique triangle. In this section, we derive the Law of Cosines and use it to solve the remaining Cases 3 and 4.

CASE 3: Two sides and the included angle are known (SAS).
CASE 4: Three sides are known (SSS).

Theorem

Law of Cosines

For a triangle with sides a, b, c and opposite angles α, β, γ, respectively,

$$c^2 = a^2 + b^2 - 2ab \cos \gamma \quad (1)$$
$$b^2 = a^2 + c^2 - 2ac \cos \beta \quad (2)$$
$$a^2 = b^2 + c^2 - 2bc \cos \alpha \quad (3)$$

Proof We shall prove only formula (1) here. Formulas (2) and (3) may be proved using the same argument.

We begin by strategically placing a triangle on a rectangular coordinate system, so that the vertex of angle γ is at the origin and side b lies along the positive x-axis. Regardless of whether γ is acute, as in Figure 17(a), or obtuse, as in Figure 17(b), the vertex B has coordinates $(a \cos \gamma, a \sin \gamma)$. Vertex A has coordinates $(b, 0)$.

Figure 17

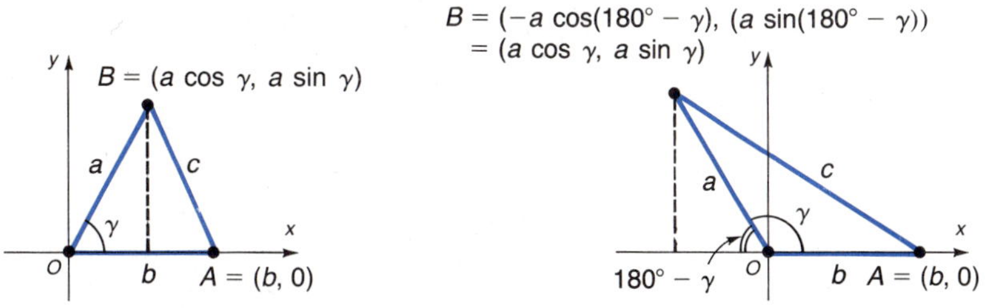

(a) Angle γ is acute (b) Angle γ is obtuse

We can now use the distance formula to compute c^2:

$$\begin{aligned} c^2 &= (b - a \cos \gamma)^2 + (0 - a \sin \gamma)^2 \\ &= b^2 - 2ab \cos \gamma + a^2 \cos^2 \gamma + a^2 \sin^2 \gamma \\ &= a^2(\cos^2 \gamma + \sin^2 \gamma) + b^2 - 2ab \cos \gamma \\ &= a^2 + b^2 - 2ab \cos \gamma \end{aligned}$$ ■

Each of formulas (1), (2), and (3) may be stated in words as follows:

Theorem

Law of Cosines

The square of one side of a triangle equals the sum of the squares of the other two sides minus twice their product times the cosine of their included angle. ■

Observe that if the triangle is a right triangle (so that, say, $\gamma = 90°$), then formula (1) becomes the familiar Pythagorean Theorem: $c^2 = a^2 + b^2$. Thus, the Pythagorean Theorem is a special case of the Law of Cosines.

Let's see how to use the Law of Cosines to solve Case 3 (SAS), which applies to triangles where two sides and the included angle are known.

Example 1 Solve the triangle: $a = 2$, $b = 3$, $\gamma = 60°$

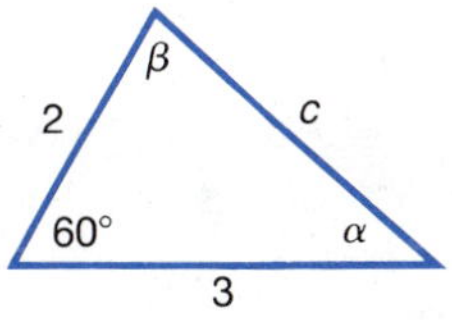

Figure 18

Solution See Figure 18. The Law of Cosines makes it easy to find the third side, c:

$$\begin{aligned} c^2 &= a^2 + b^2 - 2ab \cos \gamma \\ &= 4 + 9 - 2 \cdot 2 \cdot 3 \cdot \cos 60° \\ &= 13 - (12 \cdot \tfrac{1}{2}) = 7 \\ c &= \sqrt{7} \end{aligned}$$

Side c is of length $\sqrt{7}$. To find the angles α and β, we may use either the Law of Sines or the Law of Cosines. We choose formulas (2) and (3) of the Law of Cosines.

For α:

$$a^2 = b^2 + c^2 - 2bc \cos \alpha$$
$$2bc \cos \alpha = b^2 + c^2 - a^2$$
$$\cos \alpha = \frac{b^2 + c^2 - a^2}{2bc} = \frac{9 + 7 - 4}{2 \cdot 3\sqrt{7}} = \frac{12}{6\sqrt{7}} = \frac{2\sqrt{7}}{7}$$
$$\alpha \approx 40.9°$$

For β:

$$b^2 = a^2 + c^2 - 2ac \cos \beta$$
$$\cos \beta = \frac{a^2 + c^2 - b^2}{2ac} = \frac{4 + 7 - 9}{4\sqrt{7}} = \frac{1}{2\sqrt{7}} = \frac{\sqrt{7}}{14}$$
$$\beta \approx 79.1°$$

Notice that $\alpha + \beta + \gamma = 60° + 40.9° + 79.1° = 180°$, as required. ■

The next example illustrates how the Law of Cosines is used when three sides of a triangle are known, Case 4 (SSS).

Example 2 Solve the triangle: $a = 4$, $b = 3$, $c = 6$

Solution See Figure 19. To find the angles α, β, and γ, we proceed as we did in the latter part of the solution to Example 1.

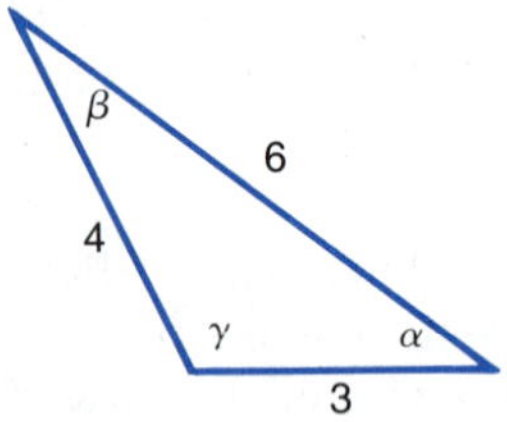

Figure 19

For α:

$$\cos \alpha = \frac{b^2 + c^2 - a^2}{2bc} = \frac{9 + 36 - 16}{2 \cdot 3 \cdot 6} = \frac{29}{36}$$
$$\alpha \approx 36.3°$$

For β:

$$\cos \beta = \frac{a^2 + c^2 - b^2}{2ac} = \frac{16 + 36 - 9}{2 \cdot 4 \cdot 6} = \frac{43}{48}$$
$$\beta \approx 26.4°$$

Since we know α and β,

$$\gamma = 180° - \alpha - \beta \approx 180° - 36.3° - 26.4° \approx 117.3°$$ ■

The Law of Cosines provides a very practical way to measure distances that cannot be measured directly due to inaccessibility or obstacles.

Example 3 Two homes are located on opposite sides of a small hill. See Figure 20. To measure the distance between them, a surveyor walks a distance of 50 feet from house A to point C, uses a transit to measure the angle ACB, which is found to be 80°, and then walks to house B, a distance of 60 feet. How far apart are the houses?

Figure 20

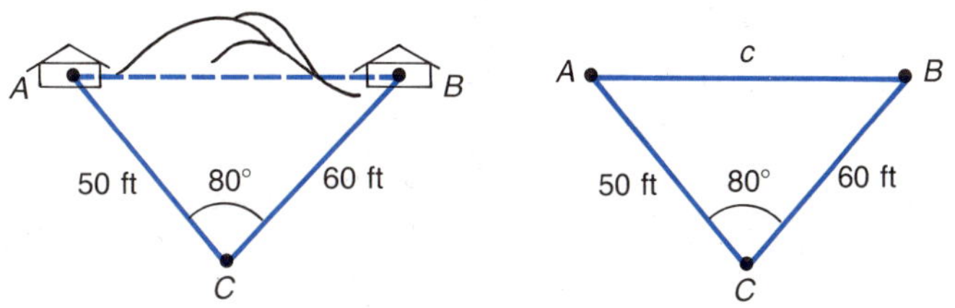

Solution The distance c we seek is the third side of a triangle in which the other two sides and their included angle are known. Consequently, we use the Law of Cosines:

$$\begin{aligned} c^2 &= a^2 + b^2 - 2ab\cos\gamma \\ &= (50)^2 + (60)^2 - 2 \cdot 50 \cdot 60 \cos 80° \\ &\approx 2500 + 3600 - 1042 = 5058 \\ c &\approx 71 \text{ feet} \end{aligned}$$

The houses are about 71 feet apart. ■

Historical Comment ■ The Law of Sines was known vaguely long before it was explicitly stated by Nasîr ed-dîn (about AD 1250). Ptolemy (about AD 150) was aware of it in a form using a chord function instead of the sine function. But it was first clearly stated in Europe by Regiomontanus, writing in 1464.

The Law of Cosines appears first in Euclid's *Elements* (Book II), but in a well-disguised form in which squares built on the sides of triangles are added and a rectangle representing the cosine term is subtracted. It was thus known to all mathematicians because of their familiarity with Euclid's work. An early modern form of the Law of Cosines—that for finding the angle when the sides are known—was stated by François Vièta (in 1593).

The Law of Tangents (see Problem 42 of Exercise 10.1) has become obsolete. In the past it was used in place of the Law of Cosines, because the Law of Cosines was very inconvenient for calculation with logarithms or slide rules. Mixing of addition and multiplication is now quite easy on a calculator, however, and the Law of Tangents has been shelved along with the slide rule. ■

EXERCISE 10.2

In Problems 1–8, solve each triangle.

1.

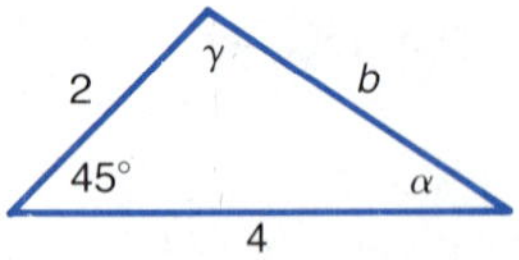

2.

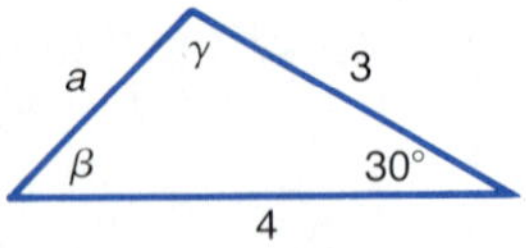

3.

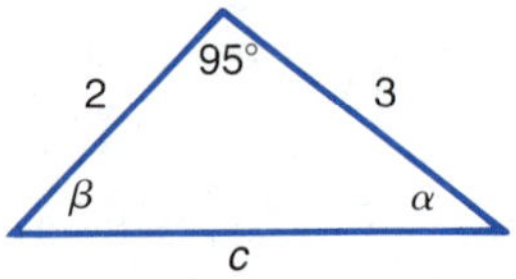

4.

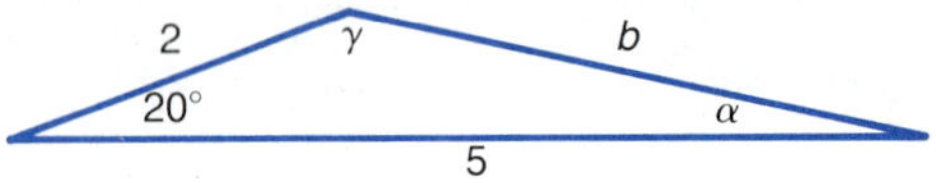

5.

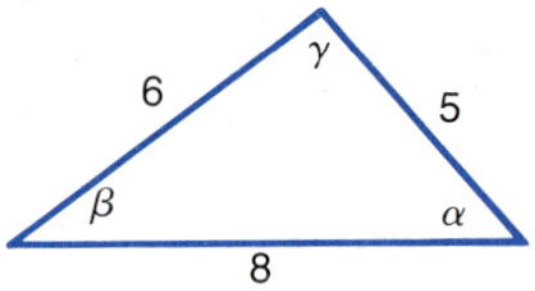

6.

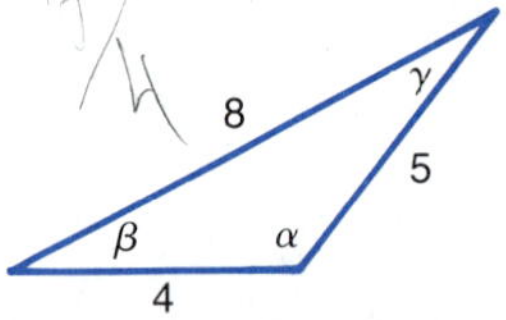

7.

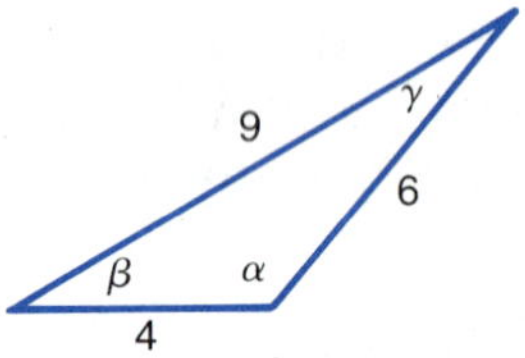

8.

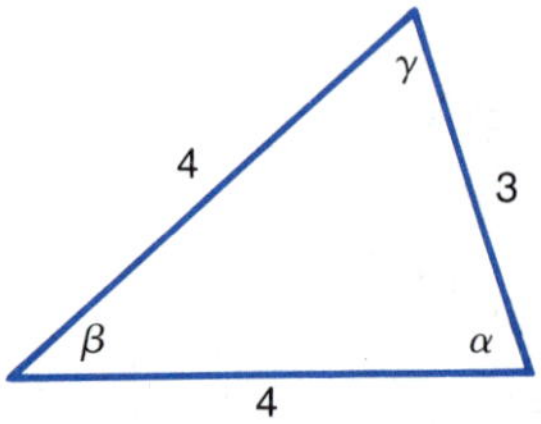

In Problems 9–24, solve each triangle.

9. $a = 3, b = 4, \gamma = 40°$ **10.** $a = 2, c = 1, \beta = 10°$ **11.** $b = 1, c = 3, \alpha = 80°$

12. $a = 6, b = 4, \gamma = 60°$ **13.** $a = 3, c = 2, \beta = 110°$ **14.** $b = 4, c = 1, \alpha = 120°$

15. $a = 2, b = 2, \gamma = 50°$ **16.** $a = 3, c = 2, \beta = 90°$ **17.** $a = 12, b = 13, c = 5$

18. $a = 4, b = 5, c = 3$ **19.** $a = 2, b = 2, c = 2$ **20.** $a = 3, b = 3, c = 2$

21. $a = 5, b = 8, c = 9$ **22.** $a = 4, b = 3, c = 6$ **23.** $a = 10, b = 8, c = 5$

24. $a = 9, b = 7, c = 10$

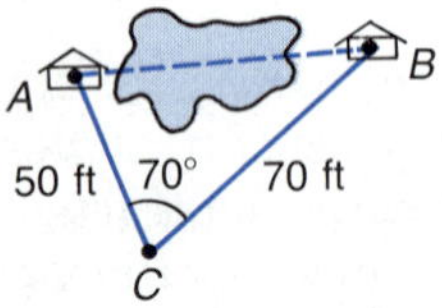

25. Consult the figure. To find the distance from the house at A to the house at B, a surveyor measures the angle ACB, which is found to be 70°, then walks off the distance to each house, 50 feet and 70 feet, respectively. How far apart are the houses?

26. An airplane flies from city A to city B, a distance of 150 miles, then turns through an angle of 50° and flies to city C, a distance of 100 miles (see the figure).
(a) How far is it from city A to city C?
(b) Through what angle should the pilot turn at city C to return to city A?

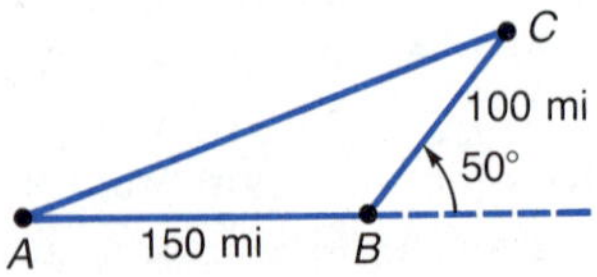

27. In attempting to fly from city A to city B, a distance of 330 miles, a pilot inadvertantly took a course that was 10° in error, as indicated in the figure.
    (a) If the aircraft maintains an average speed of 220 miles per hour and if the error in direction is discovered after 15 minutes, through what angle should the pilot turn to head toward city B?
    (b) What new average speed should the pilot maintain so that the total time of the trip is 90 minutes?

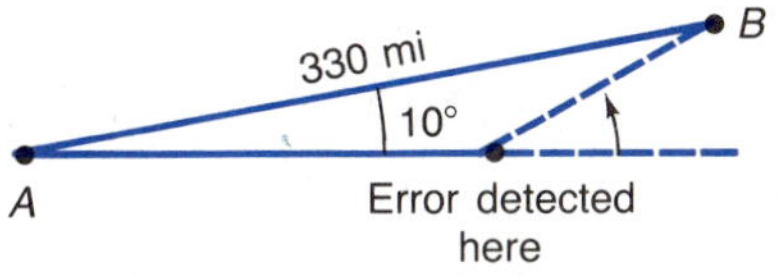

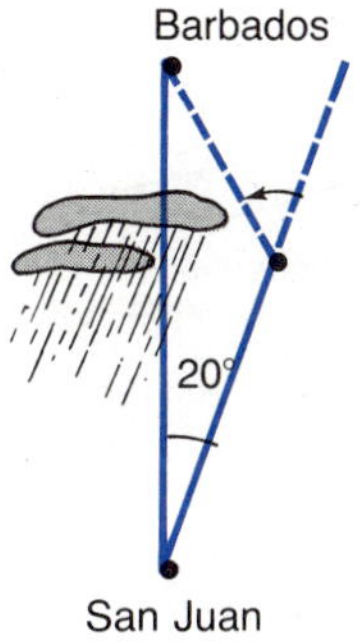

28. A cruise ship maintains an average speed of 15 knots in going from San Juan, Puerto Rico, to Barbados, West Indies, a distance of 600 nautical miles. To avoid a tropical storm, the captain heads out of San Juan in a direction of 20° off a direct heading to Barbados. The captain maintains the 15 knot speed for 10 hours, after which time the path to Barbados becomes clear of storms.
    (a) Through what angle should the captain turn to head directly to Barbados?
    (b) How long will it be before the ship reaches Barbados if the same 15 knot speed is maintained?

29. A Major League baseball diamond is actually a square 90 feet on a side. The pitching rubber is located 60.5 feet from home plate on a line joining home plate and second base.
    (a) How far is it from the pitching rubber to first base?
    (b) How far is it from the pitching rubber to second base?
    (c) If a pitcher faces home plate, through what angle does he need to turn to face first base?

30. According to Little League baseball official regulations, a diamond is a square 60 feet on a side. The pitching rubber is located 46 feet from home plate on a line joining home plate and second base.
    (a) How far is it from the pitching rubber to first base?
    (b) How far is it from the pitching rubber to second base?
    (c) If a pitcher faces home plate, through what angle does he need to turn to face first base?

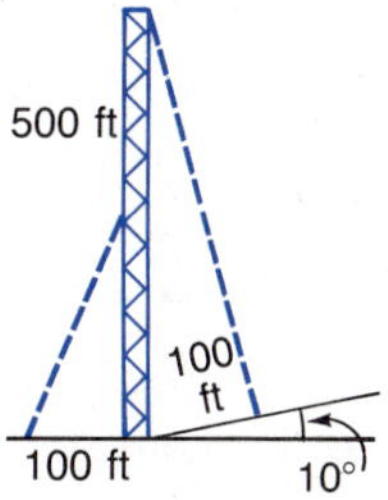

31. The height of a radio tower is 500 feet, and the ground on one side of the tower slopes upward at an angle of 10° (see the figure).
    (a) How long should a guy wire be if it is to connect at the top of the tower and be secured at a point on the slope 100 feet from the base of the tower?
    (b) How long should a second guy wire be if it is to connect to the middle of the tower and be secured at a point 100 feet from the base?

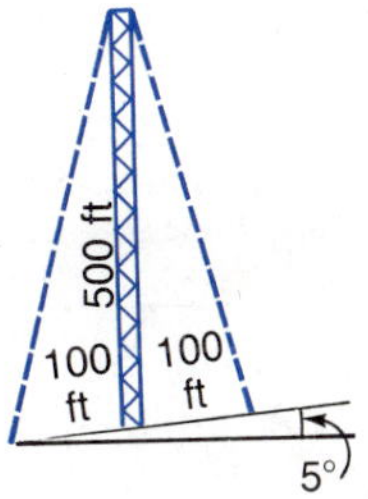

32. A radio tower 500 feet high is located on the side of a hill with an inclination to the horizontal of 5° (see the figure). How long should two guy wires be if they are to connect at the top of the tower and be secured at two points 100 feet directly above and directly below the base of the tower?

33. The distance from home plate to dead center in Wrigley Field is 400 feet (see the figure). How far is it from dead center to third base?

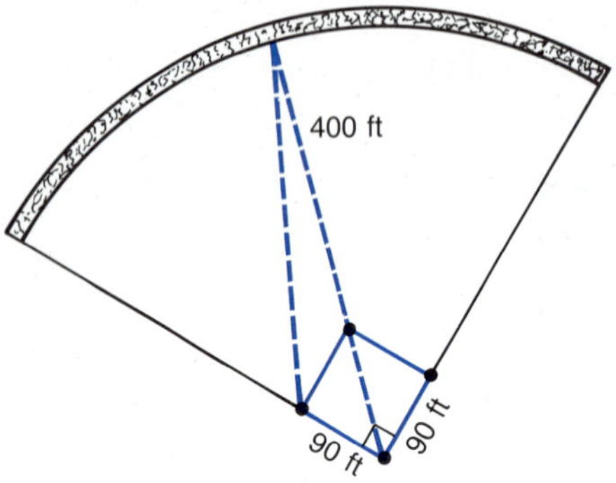

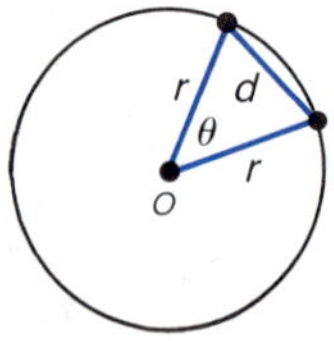

34. Show that the length d of a chord of a circle of radius r is given by the formula

$$d = 2r \sin \frac{\theta}{2}$$

where θ is the central angle formed by the radii to the ends of the chord (see the figure). Use this result to derive the fact that $\sin \theta < \theta$, where $\theta > 0$ is measured in radians.

35. For any triangle, show that

$$\cos \frac{\gamma}{2} = \sqrt{\frac{s(s-c)}{ab}}$$

where $s = \frac{1}{2}(a + b + c)$. [*Hint:* Use a half-angle formula and the Law of Cosines.]

36. For any triangle, show that

$$\sin \frac{\gamma}{2} = \sqrt{\frac{(s-a)(s-b)}{ab}}$$

where $s = \frac{1}{2}(a + b + c)$.

37. Use the Law of Cosines to prove the identity

$$\frac{\cos \alpha}{a} + \frac{\cos \beta}{b} + \frac{\cos \gamma}{c} = \frac{a^2 + b^2 + c^2}{2abc}$$

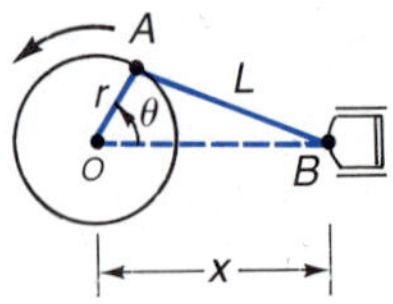

38. Rod OA (see the figure) rotates about the fixed point O so that point A travels on a circle of radius r. Connected to point A is another rod AB of length $L > r$, and point B is connected to a piston. Show that the distance x between point O and point B is given by

$$x = r \cos \theta + \sqrt{r^2 \cos^2 \theta + L^2 - r^2}$$

where θ is the angle of rotation of rod OA.*

*Adapted from Mizrahi and Sullivan, *Calculus and Analytic Geometry*, 3rd ed., Wadsworth, 1990, p. 68.

10.3 The Area of a Triangle

In this section, we shall derive several formulas for calculating the area A of a triangle. The most familiar of these is the following:

Theorem The area A of a triangle is

$$A = \tfrac{1}{2}bh \qquad (1)$$

where b is the base and h is an altitude drawn to that base.

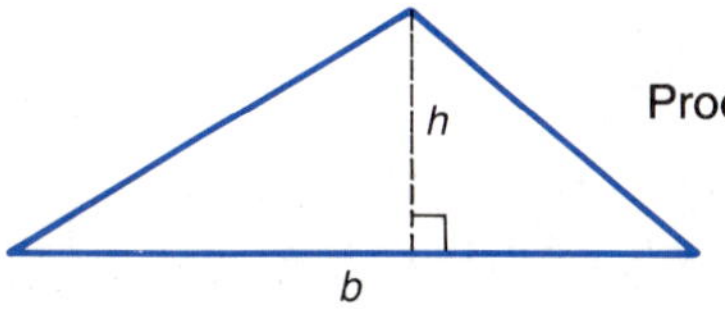

Figure 21
$A = \frac{1}{2}bh$

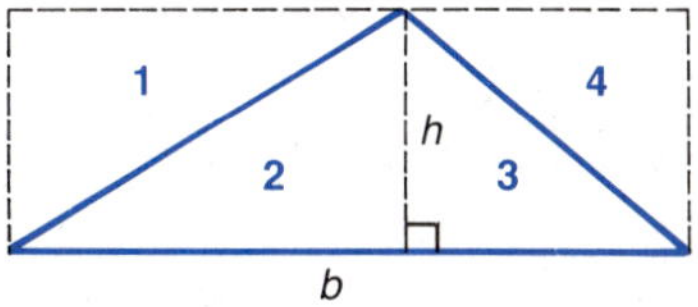

Figure 22

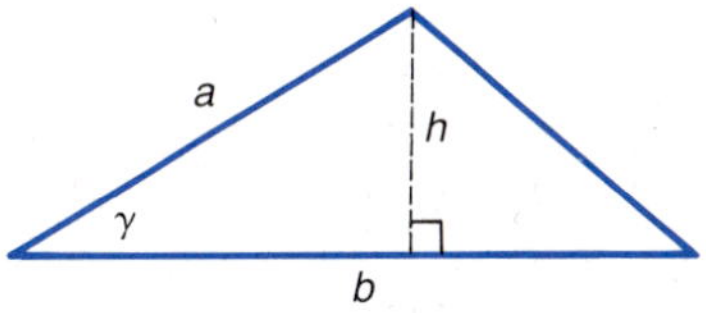

Figure 23
$h = a \sin \gamma$

Proof The derivation of this formula is rather easy once a rectangle of base b and height h is constructed around the triangle. See Figures 21 and 22.

Triangles 1 and 2 in Figure 22 are equal in area, as are triangles 3 and 4. Consequently, the area of the triangle with base b and altitude h is exactly half the area of the rectangle, which is bh. ■

If the base b and altitude h to that base are known, then we can easily find the area of such a triangle using formula (1). Usually, though, the information required to use formula (1) is not given. Suppose, for example, we know two sides a and b and the included angle γ (see Figure 23). Then the altitude h can be found by noting that

$$\frac{h}{a} = \sin \gamma$$

so that

$$h = a \sin \gamma$$

Using this fact in formula (1) produces

$$A = \tfrac{1}{2}bh = \tfrac{1}{2}b(a \sin \gamma) = \tfrac{1}{2}ab \sin \gamma$$

Thus, we have the formula

$$A = \tfrac{1}{2}ab \sin \gamma \qquad (2)$$

By dropping altitudes from the other two vertices of the triangle, we obtain the following corresponding formulas:

$$A = \tfrac{1}{2}bc \sin \alpha \qquad (3)$$
$$A = \tfrac{1}{2}ac \sin \beta \qquad (4)$$

It is easiest to remember these formulas using the following wording:

Theorem The area A of a triangle equals one-half the product of two of its sides times the sine of their included angle. ■

Example 1 Find the area A of the triangle for which $a = 8$, $b = 6$, and $\gamma = 30°$.

Solution We use formula (2) to get

$$A = \tfrac{1}{2}ab \sin \gamma = \tfrac{1}{2} \cdot 8 \cdot 6 \sin 30° = 12$$ ■

If the three sides of a triangle are known, another formula, called **Heron's Formula** (named after Heron of Alexandria),* can be used to find the area of a triangle.

Theorem The area A of a triangle with sides a, b, and c is

Heron's Formula

$$A = \sqrt{s(s-a)(s-b)(s-c)} \qquad (5)$$

where $s = \tfrac{1}{2}(a + b + c)$.

Proof The proof we shall give uses the Law of Cosines and is quite different from the proof given by Heron.

From the Law of Cosines,

$$c^2 = a^2 + b^2 - 2ab \cos \gamma$$

and the two half-angle formulas,

$$\cos^2 \frac{\gamma}{2} = \frac{1 + \cos \gamma}{2} \qquad \sin^2 \frac{\gamma}{2} = \frac{1 - \cos \gamma}{2}$$

we find

$$\cos^2 \frac{\gamma}{2} = \frac{1 + \cos \gamma}{2} = \frac{1 + \dfrac{a^2 + b^2 - c^2}{2ab}}{2}$$

$$= \frac{a^2 + 2ab + b^2 - c^2}{4ab} = \frac{(a + b)^2 - c^2}{4ab}$$

$$= \frac{(a + b - c)(a + b + c)}{4ab} = \frac{2(s - c) \cdot 2s}{4ab} = \frac{s(s - c)}{ab} \qquad (6)$$

$$a + b - c = a + b + c - 2c = 2s - 2c$$

*[C] Refer to Graphics Calculator Supplement: Activity IX.

Similarly,

$$\sin^2 \frac{\gamma}{2} = \frac{(s-a)(s-b)}{ab} \tag{7}$$

Now, we use formula (2) for the area:

$$\begin{aligned} A &= \frac{1}{2}ab \sin \gamma \\ &= \frac{1}{2}ab \cdot 2 \sin \frac{\gamma}{2} \cos \frac{\gamma}{2} && \sin \gamma = \sin 2\left(\frac{\gamma}{2}\right) = 2 \sin \frac{\gamma}{2} \cos \frac{\gamma}{2} \\ &= ab\sqrt{\frac{(s-a)(s-b)}{ab}}\sqrt{\frac{s(s-c)}{ab}} && \text{Use equations (6) and (7).} \\ &= \sqrt{s(s-a)(s-b)(s-c)} \end{aligned}$$

■

Example 2 Find the area of a triangle whose sides are 4, 5, and 7.

Solution We let $a = 4$, $b = 5$, and $c = 7$. Then,

$$s = \tfrac{1}{2}(a + b + c) = \tfrac{1}{2}(4 + 5 + 7) = 8$$

Heron's Formula then gives the area A as

$$A = \sqrt{s(s-a)(s-b)(s-c)} = \sqrt{8 \cdot 4 \cdot 3 \cdot 1} = \sqrt{96} = 4\sqrt{6}$$

■

Historical Comment

■ Heron's Formula (also known as *Hero's Formula*) is due to Heron of Alexandria (about AD 75), who had, besides his mathematical talents, a good deal of engineering skill. In various temples his mechanical devices produced effects that seemed supernatural, and visitors presumably were thus influenced to generosity. Heron's book *Metrica*, on making such devices, has survived, and was discovered in 1896 in the city of Constantinople.

Heron's Formula for the area of a triangle caused some mild discomfort in Greek mathematics, because a product with two factors was an area and with three factors was a volume, but four factors seemed contradictory in Heron's time.

Karl Mollweide (1774–1875), a mathematician and astronomer, discovered the formulas named for him (see Problems 39–41 of Exercise 10.1). These formulas are not too important in themselves but often simplify the derivation of other formulas, as demonstrated in Historical Problems 1 and 2, below.

Historical Problems

■ **1.** This derivation of Heron's formula uses Mollweide's Formula.

(a) Show that

$$\frac{s}{c} = \frac{\cos(\alpha/2)\cos(\beta/2)}{\sin(\gamma/2)}$$

where $s = \frac{1}{2}(a + b + c)$. *Hint:* Use Mollweide's Formula (see Problem 39 in Exercise 10.1) and add 1 to both sides. Then use the fact that

$$\sin\frac{\gamma}{2} = \sin\frac{180^\circ - (\alpha + \beta)}{2} = \cos\frac{\alpha + \beta}{2}$$

(b) Similarly, show that

$$\frac{s}{a} = \frac{\cos(\beta/2)\cos(\gamma/2)}{\sin(\alpha/2)} \qquad \text{and} \qquad \frac{s}{b} = \frac{\cos(\alpha/2)\cos(\gamma/2)}{\sin(\beta/2)}$$

(c) Use the results of parts (a) and (b) to show that

$$\frac{s - a}{a} = \frac{\sin(\beta/2)\sin(\gamma/2)}{\sin(\alpha/2)} \qquad \frac{s - b}{b} = \frac{\sin(\alpha/2)\sin(\gamma/2)}{\sin(\beta/2)}$$

$$\frac{s - c}{c} = \frac{\sin(\alpha/2)\sin(\beta/2)}{\sin(\gamma/2)}$$

(d) Now form the product

$$\frac{s}{c} \cdot \frac{s - a}{a} \cdot \frac{s - b}{b} \cdot \frac{s - c}{c}$$

After cancellations, multiply each side by $cabc$, use the double-angle formulas, and use the fact that $A = \frac{1}{2}bc \sin\alpha = \frac{1}{2}ac \sin\beta$. Heron's Formula will then follow.

2. We again use Mollweide's Formula to derive some other interesting formulas. (These were derived in another way in Problems 35 and 36 of Exercise 10.2.)

(a) Add 1 to both sides of the second form of Mollweide's Formula (see Problem 40 in Exercise 10.1) and simplify to obtain

$$\frac{s - b}{c} = \frac{\sin(\alpha/2)\cos(\beta/2)}{\cos(\gamma/2)}$$

(b) Similarly, show that

$$\frac{s - c}{b} = \frac{\sin(\alpha/2)\cos(\gamma/2)}{\cos(\beta/2)}$$

(c) Use the results of part (b) and Problem 1(b) and (c) to show that

$$\cos^2\frac{\gamma}{2} = \frac{s(s - c)}{ab} \qquad \text{and} \qquad \sin^2\frac{\gamma}{2} = \frac{(s - a)(s - b)}{ab}$$

(d) Show that

$$\tan^2\frac{\gamma}{2} = \frac{(s - a)(s - b)}{s(s - c)}$$

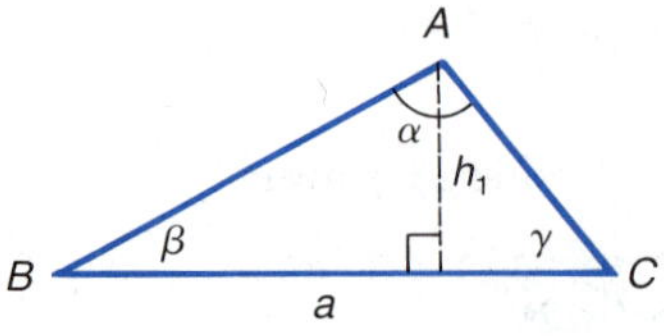

3. (a) If h_1, h_2, and h_3 are the altitudes dropped from A, B, and C, respectively, in a triangle (see the figure), show that

$$\frac{1}{h_1} + \frac{1}{h_2} + \frac{1}{h_3} = \frac{s}{K}$$

where K is the area of the triangle and $s = \frac{1}{2}(a + b + c)$. [*Hint:* $h_1 = 2K/a$]

(b) Show that a formula for the altitude h from a vertex to the opposite side a of a triangle is

$$h = \frac{a \sin \beta \sin \gamma}{\sin \alpha}$$

4. *Inscribed circle* The lines that bisect each angle of a triangle meet in a single point O, and the perpendicular distance r from O to each of the sides of the triangle is the same. The circle with center at O and radius r is called the *inscribed circle* of the triangle (see the figure).

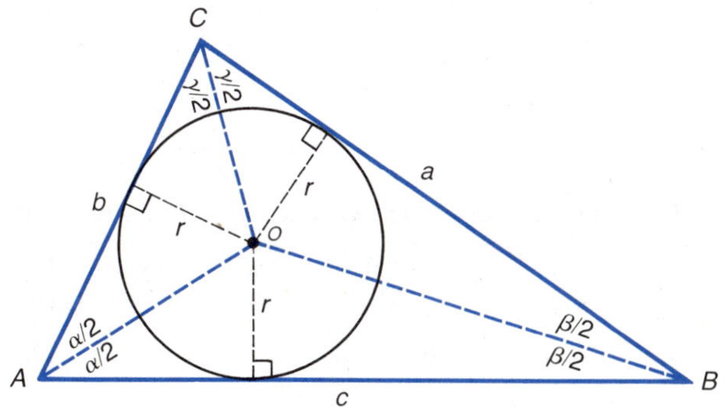

(a) Apply Problem 3(b) above to triangle OAB to show that

$$r = \frac{c \sin(\alpha/2) \sin(\beta/2)}{\cos(\gamma/2)}$$

(b) Use the results of part (a) and Problem 1(c) to show that

$$r = (s - c) \tan \frac{\gamma}{2}$$

(c) Show that

$$\cot \frac{\alpha}{2} + \cot \frac{\beta}{2} + \cot \frac{\gamma}{2} = \frac{s}{r}$$

(d) Show that the area K of triangle ABC is $K = rs$. Then show that

$$r = \sqrt{\frac{(s-a)(s-b)(s-c)}{s}}$$

where $s = \frac{1}{2}(a + b + c)$. ■

C EXERCISE 10.3

In Problems 1–8, find the area of each triangle. Round off answers to four decimal places.

1.

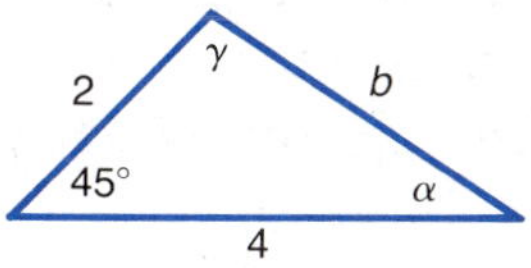

2.

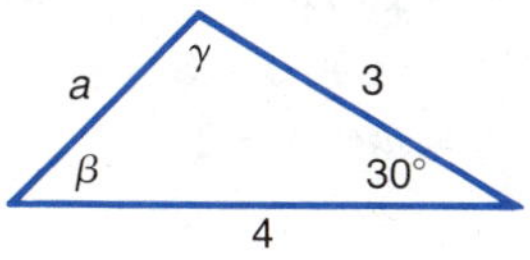

3.

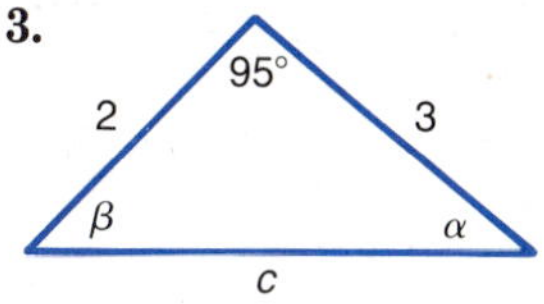

4.

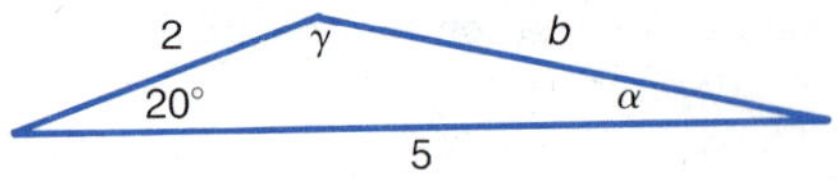

5.

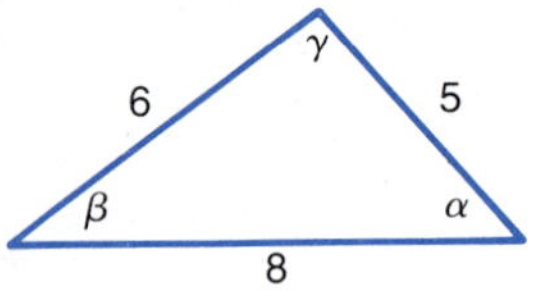

6.

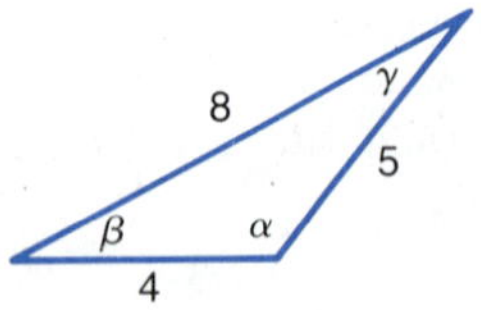

7.

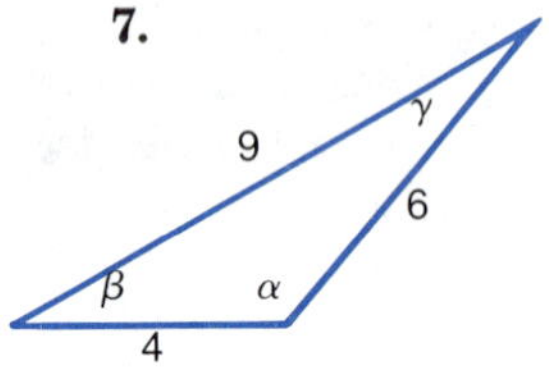

8.

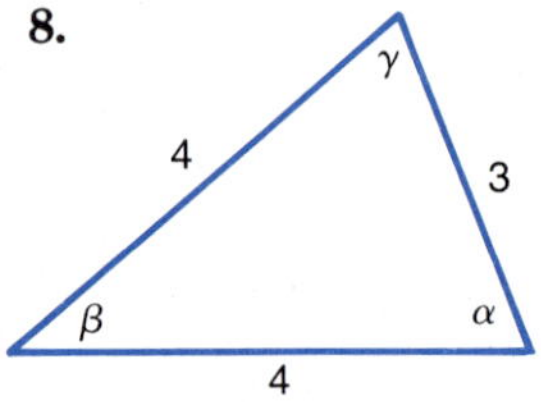

In Problems 9–24, find the area of each triangle. Round off answers to four decimal places.

9. $a = 3, b = 4, \gamma = 40°$

10. $a = 2, c = 1, \beta = 10°$

11. $b = 1, c = 3, \alpha = 80°$

12. $a = 6, b = 4, \gamma = 60°$

13. $a = 3, c = 2, \beta = 110°$

14. $b = 4, c = 1, \alpha = 120°$

15. $a = 2, b = 2, \gamma = 50°$

16. $a = 3, c = 2, \beta = 90°$

17. $a = 12, b = 13, c = 5$

18. $a = 4, b = 5, c = 3$

19. $a = 2, b = 2, c = 2$

20. $a = 3, b = 3, c = 2$

21. $a = 5, b = 8, c = 9$

22. $a = 4, b = 3, c = 6$

23. $a = 10, b = 8, c = 5$

24. $a = 9, b = 7, c = 10$

25. The dimensions of a triangular lot are 100 feet by 50 feet by 75 feet. If the price of such land is \$3 per square foot, how much does the lot cost?

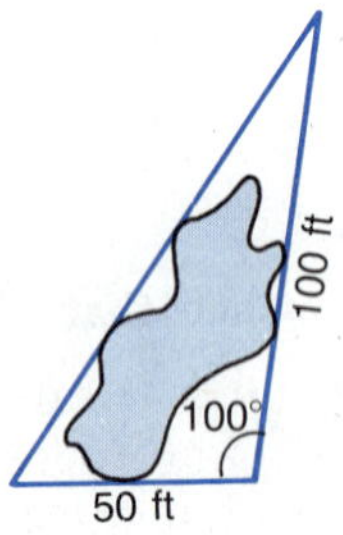

26. Refer to the figure in the margin. Use the area of the triangle to approximate the area of the lake.

27. Prove that the area A of a triangle is given by the formula

$$A = \frac{a^2 \sin \beta \sin \gamma}{2 \sin \alpha}$$

28. Prove the two other forms of the formula given in Problem 27,

$$A = \frac{b^2 \sin \alpha \sin \gamma}{2 \sin \beta} \quad \text{and} \quad A = \frac{c^2 \sin \alpha \sin \beta}{2 \sin \gamma}$$

In Problems 29–36, use the results of Problem 27 or 28 to find the area of each triangle. Round off answers to four decimal places.

29. $\alpha = 40°, \beta = 20°, a = 2$

30. $\alpha = 50°, \gamma = 20°, a = 3$

31. $\beta = 70°, \gamma = 10°, b = 5$

32. $\alpha = 70°, \beta = 60°, c = 4$

33. $\alpha = 110°, \gamma = 30°, c = 3$

34. $\beta = 10°, \gamma = 100°, b = 2$

35. $\alpha = 40°, \beta = 40°, c = 2$

36. $\beta = 20°, \gamma = 70°, a = 1$

10.4 Polar Coordinates*

So far, we have always used a system of rectangular coordinates to plot points in the plane. Now we are ready to describe another system called *polar coordinates*. As we shall soon see, in many instances, polar coordinates offer certain advantages over rectangular coordinates.

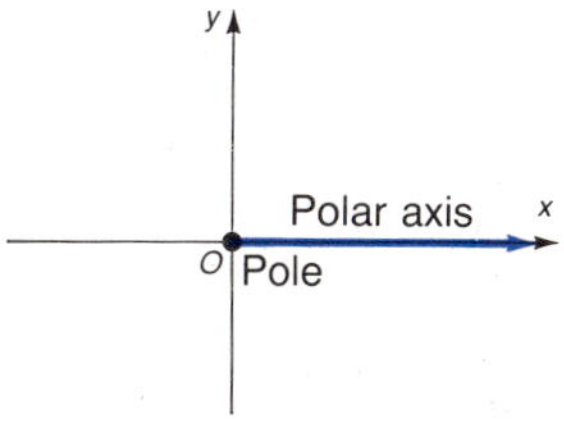

Figure 24

In a rectangular coordinate system, you will recall, a point in the plane is represented by an ordered pair of numbers (x, y), where x and y equal the signed distance of the point from the y-axis and x-axis, respectively. In a polar coordinate system, we select a point, called the **pole**, and then a ray with vertex at the pole, called the **polar axis**. Comparing the rectangular and polar coordinate systems, we see (in Figure 24) that the origin in rectangular coordinates coincides with the pole in polar coordinates, and the positive x-axis in rectangular coordinates coincides with the polar axis in polar coordinates.

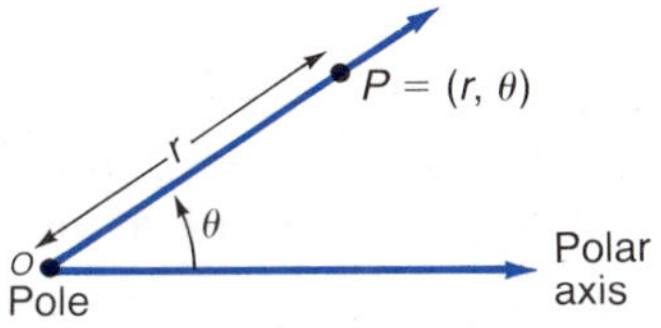

Figure 25

A point P in a polar coordinate system is represented by an ordered pair of numbers (r, θ). The number r is the distance of the point from the pole, and θ is an angle (in degrees or radians) formed by the polar axis and a ray from the origin through the point. We call the ordered pair (r, θ) the **polar coordinates** of the point. See Figure 25.

As an example, suppose the polar coordinates of a point P are $(2, \pi/4)$. We locate P by first drawing an angle of $\pi/4$ radian, placing its vertex at the pole and its initial side along the polar axis. Then we go out a distance of 2 units along the terminal side of the angle to reach the point P. See Figure 26.

Figure 26

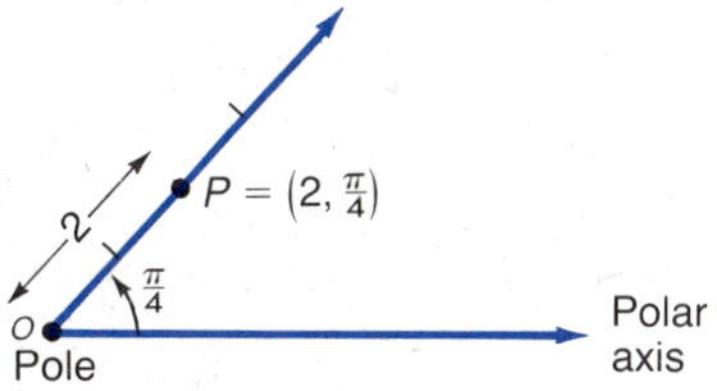

Recall that an angle measured counterclockwise is positive, whereas one measured clockwise is negative. This convention has some interesting consequences relating to polar coordinates. Let's see what these consequences are.

Example 1 Consider again a point P with polar coordinates $(2, \pi/4)$, as shown in Figure 27(a). Because $\pi/4$, $9\pi/4$, and $-7\pi/4$ all have the same terminal side, we also could have located this point P by using the polar coordinates $(2, 9\pi/4)$ or $(2, -7\pi/4)$, as shown in Figures 27(b) and (c).

* [C] Refer to Graphics Calculator Supplement: Activity VI.

Figure 27

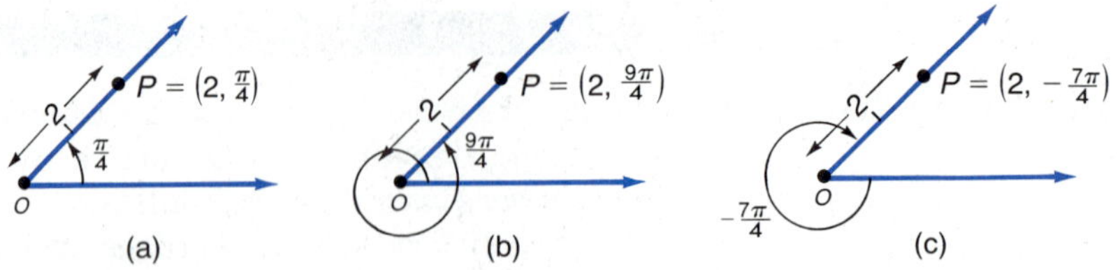

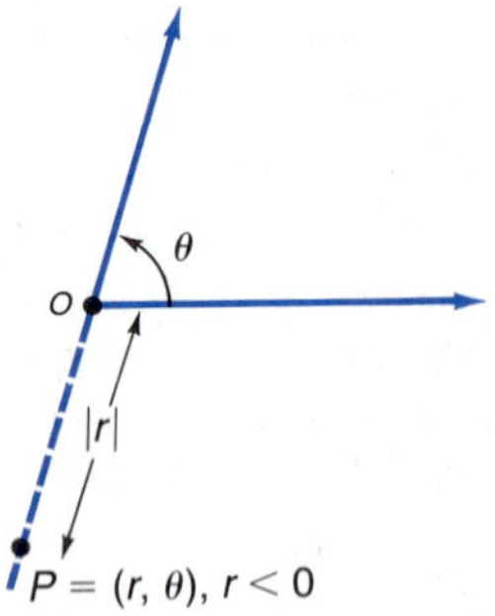

Figure 28

In using polar coordinates (r, θ), it is possible for the first entry r to be negative. When this happens, we follow the convention that the location of the point, instead of being on the terminal side of θ, is on the ray from the pole extending in the direction *opposite* the terminal side of θ at a distance $|r|$ from the pole. See Figure 28 for an illustration.

Example 2 Consider again the point P with polar coordinates $(2, \pi/4)$, as shown in Figure 29(a). This same point P can be assigned the polar coordinates $(-2, 5\pi/4)$, as indicated in Figure 29(b). To locate the point $(-2, 5\pi/4)$, we use the ray in the opposite direction of $5\pi/4$ and go out 2 units along that ray to find the point P.

Figure 29

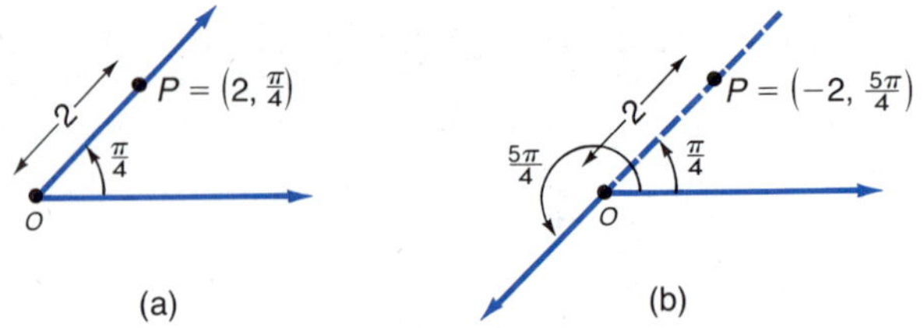

These examples show a major difference between rectangular coordinates and polar coordinates. In the former, each point has exactly one pair of rectangular coordinates; in the latter, a point can have infinitely many pairs of polar coordinates.

Summary

> A point with polar coordinates (r, θ) also can be represented by any of the following:
>
> $(r, \theta + 2k\pi)$ or $(-r, \theta + \pi + 2k\pi)$ k any integer
>
> The polar coordinates of the pole are $(0, \theta)$, where θ can be any angle.

Example 3 Plot the points with the following polar coordinates:

(a) $(3, 5\pi/3)$ (b) $(2, -\pi/4)$ (c) $(3, 0)$ (d) $(-2, \pi/4)$

Solution Figure 30 shows the points.

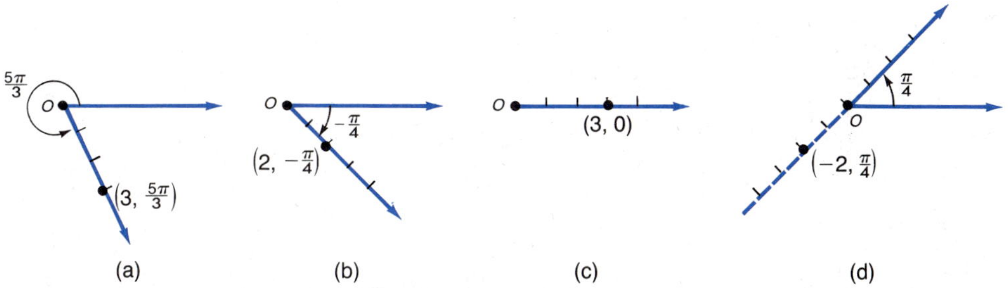

Figure 30

■

Example 4 Plot the point P with polar coordinates $(3, \pi/6)$, and find other polar coordinates (r, θ) of this same point for which:

(a) $r > 0, \quad 2\pi \le \theta \le 4\pi$ (b) $r < 0, \quad 0 \le \theta \le 2\pi$

(c) $r > 0, \quad -2\pi \le \theta \le 0$

Solution The point $(3, \pi/6)$ is plotted in Figure 31.

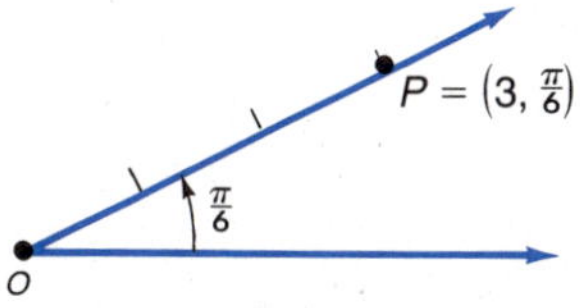

Figure 31

(a) We add 1 revolution (2π radians) to the angle $\pi/6$ to get $P = (3, \pi/6 + 2\pi) = (3, 13\pi/6)$. See Figure 32.

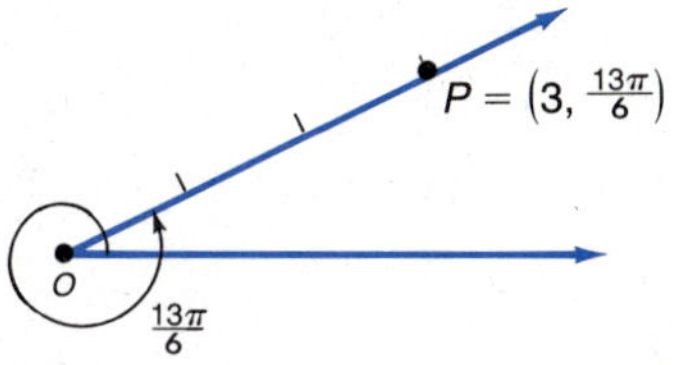

Figure 32

(b) We add $\frac{1}{2}$ revolution (π radians) to the angle and replace r by $-r$ to get $P = (-3, \pi/6 + \pi) = (-3, 7\pi/6)$. See Figure 33.

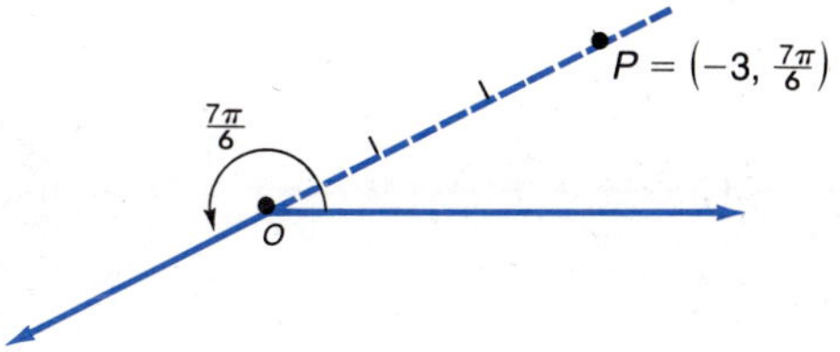

Figure 33

(c) We subtract 2π from the angle $\pi/6$ to get $P = (3, \pi/6 - 2\pi) = (3, -11\pi/6)$. See Figure 34.

Figure 34

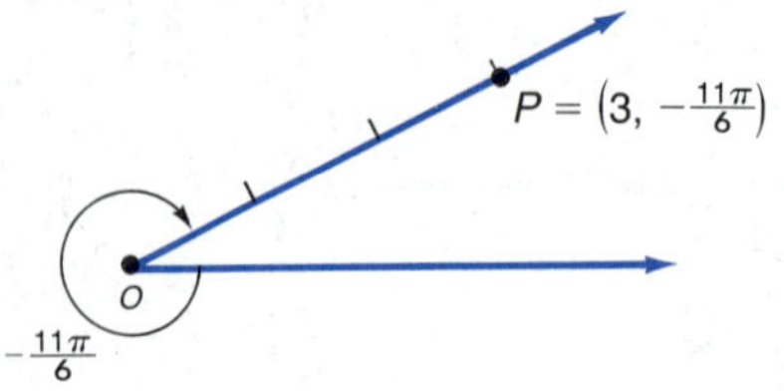

Conversion between Polar Coordinates and Rectangular Coordinates

It is sometimes convenient and, indeed, necessary to be able to convert coordinates or equations in rectangular form to polar form, and vice versa. To do this, we recall that the origin in rectangular coordinates is the pole in polar coordinates and that the positive x-axis in rectangular coordinates is the polar axis in polar coordinates.

Theorem

Conversion from Polar Coordinates to Rectangular Coordinates

If P is a point with polar coordinates (r, θ), the rectangular coordinates (x, y) of P are given by

$$x = r\cos\theta \qquad y = r\sin\theta \tag{1}$$

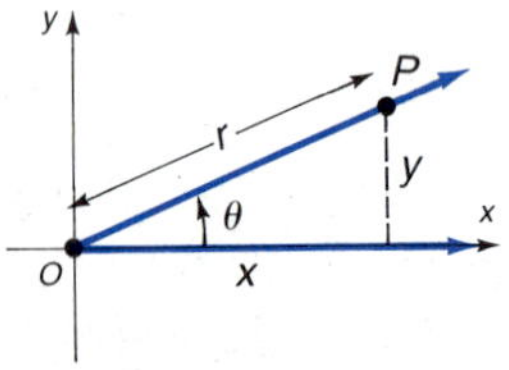

Figure 35

Proof Suppose P has the polar coordinates (r, θ). We seek the rectangular coordinates (x, y) of P. Refer to Figure 35.

If $r = 0$, then, regardless of θ, the point P is the pole, for which the rectangular coordinates are $(0, 0)$. Thus, formula (1) is valid for $r = 0$.

If $r > 0$, the point P is on the terminal side of θ and $r = d(O, P)$. Thus,

$$\cos\theta = \frac{x}{r} \qquad \sin\theta = \frac{y}{r}$$

so that

$$x = r\cos\theta \qquad y = r\sin\theta$$

If $r < 0$, then the point $P = (r, \theta)$ can be represented as $(-r, \pi + \theta)$, where $-r > 0$. Thus,

$$\cos(\pi + \theta) = -\cos\theta = \frac{x}{-r} \qquad \sin(\pi + \theta) = -\sin\theta = \frac{y}{-r}$$

so that

$$x = r\cos\theta \qquad y = r\sin\theta$$

Example 5 Find the rectangular coordinates of the points with the following polar coordinates:

(a) $(6, \pi/6)$ (b) $(-2, 5\pi/4)$ (c) $(-4, -\pi/4)$

Solution We use formula (1), namely, $x = r\cos\theta$ and $y = r\sin\theta$.

(a)
$$x = r\cos\theta = 6\cos\frac{\pi}{6} = 6\cdot\frac{\sqrt{3}}{2} = 3\sqrt{3}$$
$$y = r\sin\theta = 6\sin\frac{\pi}{6} = 6\cdot\frac{1}{2} = 3$$

The rectangular coordinates of the point $(6, \pi/6)$ are $(3\sqrt{3}, 3)$.

(b)
$$x = r\cos\theta = -2\cos\frac{5\pi}{4} = -2\left(-\frac{\sqrt{2}}{2}\right) = \sqrt{2}$$
$$y = r\sin\theta = -2\sin\frac{5\pi}{4} = -2\left(-\frac{\sqrt{2}}{2}\right) = \sqrt{2}$$

The rectangular coordinates of the point $(-2, 5\pi/4)$ are $(\sqrt{2}, \sqrt{2})$.

(c)
$$x = r\cos\theta = -4\cos\left(-\frac{\pi}{4}\right) = -4\cdot\frac{\sqrt{2}}{2} = -2\sqrt{2}$$
$$y = r\sin\theta = -4\sin\left(-\frac{\pi}{4}\right) = -4\left(-\frac{\sqrt{2}}{2}\right) = 2\sqrt{2}$$

The rectangular coordinates of the given point $(-4, -\pi/4)$ are $(-2\sqrt{2}, 2\sqrt{2})$. ■

Now, suppose P has the rectangular coordinates (x, y). We seek a representation (r, θ) for P in polar coordinates.

If $x = 0$, then the point $P = (x, y) = (0, y)$ lies on the y-axis a distance $r = |y|$ from the origin (pole). We can locate P using polar coordinates (r, θ), where $r = |y|$, by choosing $\theta = \pi/2$ or $= -\pi/2$, according as $y > 0$ or $y < 0$. See Figure 36.

Figure 36

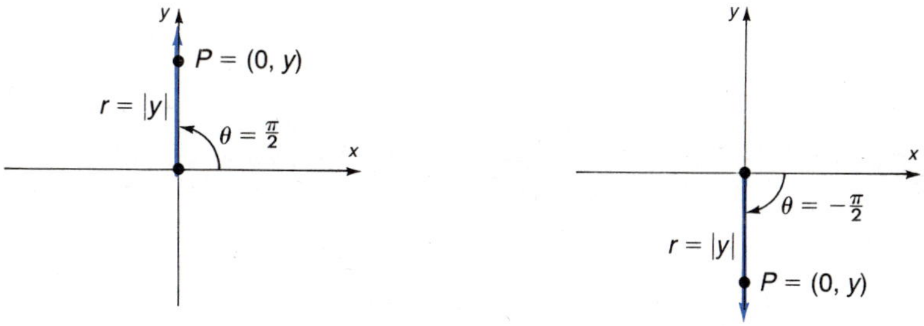

(a) $x = 0, y > 0$
$\theta = \frac{\pi}{2}, r = |y|$

(b) $x = 0, y < 0$
$\theta = -\frac{\pi}{2}, r = |y|$

Suppose $x \neq 0$. From formula (1), we have $x/r = \cos\theta$ and $y/r = \sin\theta$. Then

$$\frac{x^2}{r^2} + \frac{y^2}{r^2} = \cos^2\theta + \sin^2\theta = 1 \qquad \text{and} \qquad \tan\theta = \frac{\sin\theta}{\cos\theta} = \frac{y/r}{x/r}$$

$$r^2 = x^2 + y^2 \qquad \text{and} \qquad \tan\theta = \frac{y}{x} \qquad x \neq 0 \tag{2}$$

from which

$$r = \pm\sqrt{x^2 + y^2} \qquad \text{and} \qquad \theta = \tan^{-1}\frac{y}{x} \qquad -\pi/2 < \theta < \pi/2$$

Since the values of $\theta = \tan^{-1}(y/x)$ lie between $-\pi/2$ and $\pi/2$, the terminal side of θ will lie either in quadrant I or in quadrant IV.

If the point $P = (x, y)$ also lies in quadrant I or in quadrant IV, then P is on the terminal side of θ and so we can locate P by using the polar coordinates (r, θ), where $r > 0$. In this situation, the polar coordinates (r, θ) of P are $r = \sqrt{x^2 + y^2}$ and $\theta = \tan^{-1}(y/x)$. See Figure 37.

Figure 37

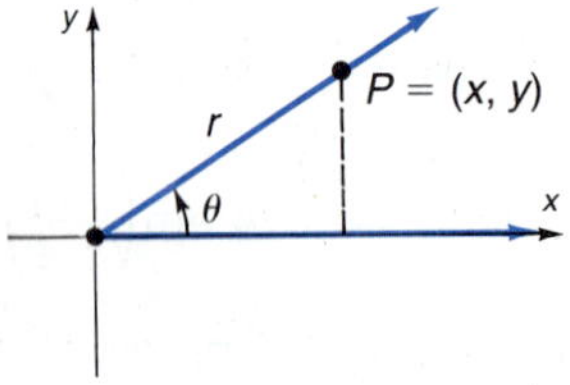

(a) P in quadrant I
$\theta = \tan^{-1}\frac{y}{x}$
$r = \sqrt{x^2 + y^2}$

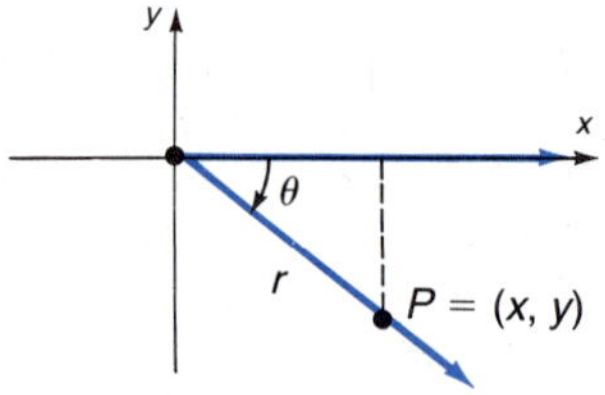

(b) P in quadrant IV
$\theta = \tan^{-1}\frac{y}{x}$
$r = \sqrt{x^2 + y^2}$

If the point $P = (x, y)$ lies in quadrant II or in quadrant III, then P does not lie on the terminal side of θ. To locate P by using the polar coordinates (r, θ), we must have $r < 0$. In this situation, the polar coordinates (r, θ) of P are $r = -\sqrt{x^2 + y^2}$ and $\theta = \tan^{-1}(y/x)$. See Figure 38.

Figure 38

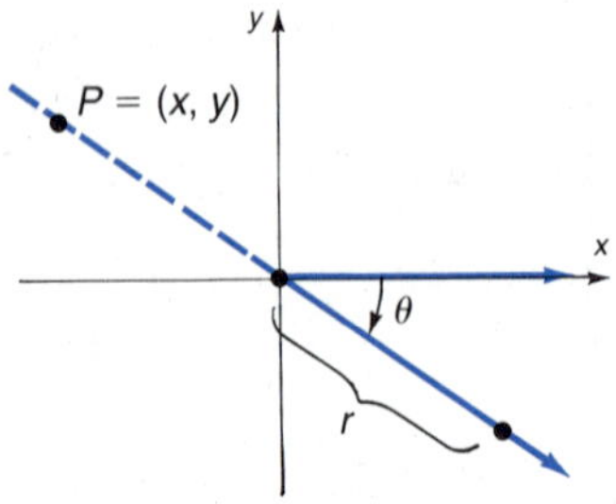

(a) P in quadrant II
$\theta = \tan^{-1}\frac{y}{x}$
$r = -\sqrt{x^2 + y^2}$

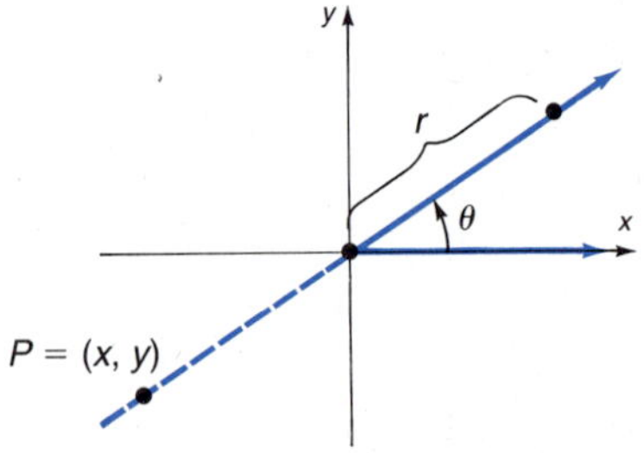

(b) P in quadrant III
$\theta = \tan^{-1}\frac{y}{x}$
$r = -\sqrt{x^2 + y^2}$

Thus, we have the following theorem:

Theorem

Conversion from Rectangular Coordinates to Polar Coordinates

If P is a point with rectangular coordinates (x, y), a corresponding set of polar coordinates (r, θ) for P is found as follows:

If $x = 0$, then $r = |y|$ and $\theta = \pi/2$ or $-\pi/2$ according as $y > 0$ or $y < 0$.

If $x \neq 0$ and P is in quadrant I or IV, then $r = \sqrt{x^2 + y^2}$ and $\theta = \tan^{-1}(y/x)$.

If $x \neq 0$ and P is in quadrant II or III, then $r = -\sqrt{x^2 + y^2}$ and $\theta = \tan^{-1}(y/x)$. ■

Warning: Care must be taken in using the above results. The value of θ in $\theta = \tan^{-1}(y/x)$, $x \neq 0$, will lie in the interval $-\pi/2 < \theta < \pi/2$. The quadrant location of the point P will determine the proper sign to choose for r, where $r = \pm\sqrt{x^2 + y^2}$.

Example 6 Find polar coordinates of a point P whose rectangular coordinates are $(2, -2)$.

Solution See Figure 39. The point $P = (2, -2)$ lies in quadrant IV, so

$$r = \sqrt{x^2 + y^2} = \sqrt{8} = 2\sqrt{2}$$

$$\theta = \tan^{-1}\frac{y}{x} = \tan^{-1}\left(\frac{-2}{2}\right) = \tan^{-1}(-1)$$

Since $\theta = \tan^{-1}(-1) = -\pi/4$, a set of polar coordinates for P is $(2\sqrt{2}, -\pi/4)$. Other possible representations include $(2\sqrt{2}, 7\pi/4)$, $(-2\sqrt{2}, 3\pi/4)$, and so on. ■

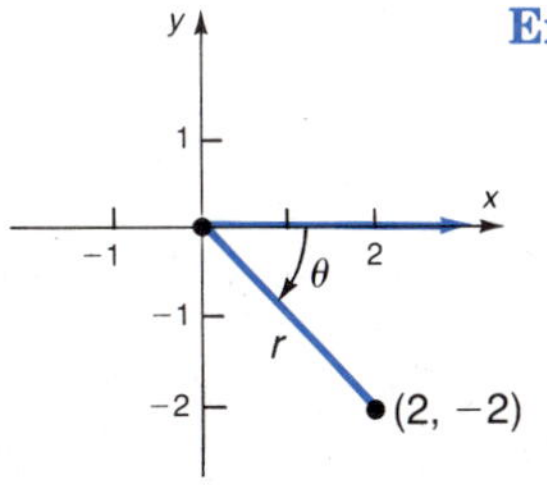

Figure 39

$r = 2\sqrt{2}$, $\theta = \tan^{-1}(-1) = \dfrac{-\pi}{4}$

Example 7 Find polar coordinates of a point P whose rectangular coordinates are $(-1, -\sqrt{3})$.

Solution See Figure 40. The point $P = (-1, -\sqrt{3})$ lies in quadrant III, so

$$r = -\sqrt{x^2 + y^2} = -\sqrt{4} = -2$$

$$\theta = \tan^{-1}\frac{y}{x} = \tan^{-1}\sqrt{3}$$

Since $\theta = \tan^{-1}\sqrt{3} = \pi/3$, a set of polar coordinates is $(-2, \pi/3)$. Other possible representations include $(2, 4\pi/3)$, $(2, -2\pi/3)$, and so on. ■

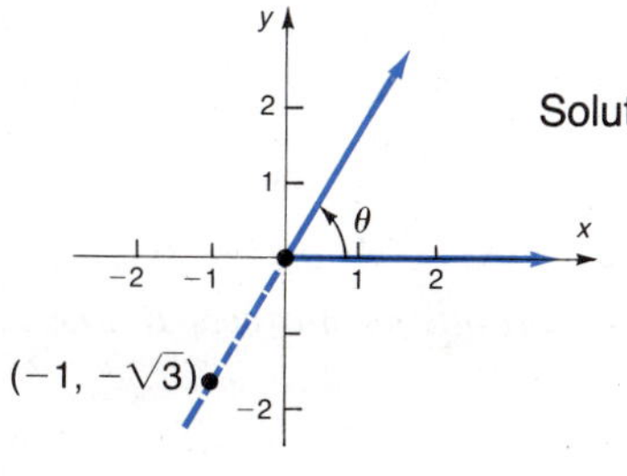

Figure 40

$r = -2$, $\theta = \tan^{-1}\sqrt{3} = \dfrac{\pi}{3}$

The conversion formulas may also be used to transform equations.

Example 8 Transform the equation $r = 4 \sin \theta$ from polar coordinates to rectangular coordinates, and identify the graph.

Solution If we multiply each side by r, it will be easier to apply formulas (1) and (2):

$$r = 4 \sin \theta$$
$$r^2 = 4r \sin \theta \quad \text{Multiply each side by } r.$$
$$x^2 + y^2 = 4y \quad \text{Apply formulas (1) and (2).}$$

This is the equation of a circle:

$$x^2 + (y^2 - 4y) = 0$$
$$x^2 + (y^2 - 4y + 4) = 4$$
$$x^2 + (y - 2)^2 = 4$$

Its center is at (0, 2), and its radius is 2. ■

Example 9 Transform the equation $4xy = 9$ from rectangular coordinates to polar coordinates.

Solution We use formula (1):

$$4xy = 9$$
$$4(r \cos \theta)(r \sin \theta) = 9 \quad \text{Formula (1)}$$
$$4r^2 \cos \theta \sin \theta = 9$$
$$2r^2 \sin 2\theta = 9 \quad \text{Double-angle formula}$$
■

EXERCISE 10.4

In Problems 1–12, plot each point given in polar coordinates.

1. $(3, 90°)$ **2.** $(4, 270°)$ **3.** $(-2, 0)$ **4.** $(-3, \pi)$
5. $(6, \pi/6)$ **6.** $(5, 5\pi/3)$ **7.** $(-2, 135°)$ **8.** $(-3, 120°)$
9. $(-1, -\pi/3)$ **10.** $(-3, -3\pi/4)$ **11.** $(-2, -\pi)$ **12.** $(-3, -\pi/2)$

In Problems 13–20, plot each point given in polar coordinates, and find other polar coordinates (r, θ) of the point for which:

(a) $r > 0, \quad -2\pi \le \theta < 0$ (b) $r < 0, \quad 0 \le \theta < 2\pi$ (c) $r > 0, \quad 2\pi \le \theta < 4\pi$

13. $(5, 2\pi/3)$ **14.** $(4, 3\pi/4)$ **15.** $(-2, 3\pi)$ **16.** $(-3, 4\pi)$
17. $(1, \pi/2)$ **18.** $(2, \pi)$ **19.** $(-3, -\pi/4)$ **20.** $(-2, -2\pi/3)$

In Problems 21–32, polar coordinates of a point are given. Find the rectangular coordinates of each point.

21. $(3, \pi/2)$ **22.** $(4, 3\pi/2)$ **23.** $(-2, 0)$ **24.** $(-3, \pi)$
25. $(6, 150°)$ **26.** $(5, 300°)$ **27.** $(-2, 3\pi/4)$ **28.** $(-3, 2\pi/3)$
29. $(-1, -\pi/3)$ **30.** $(-3, -3\pi/4)$ **31.** $(-2, -180°)$ **32.** $(-3, -90°)$

In Problems 33–40, the rectangular coordinates of a point are given. Find polar coordinates for each point.

33. $(3, 0)$ **34.** $(0, 2)$ **35.** $(-1, 0)$ **36.** $(0, -2)$
37. $(1, -1)$ **38.** $(-3, 3)$ **39.** $(\sqrt{3}, 1)$ **40.** $(-2, -2\sqrt{3})$

In Problems 41–48, the letters x and y represent rectangular coordinates. Write each equation using polar coordinates (r, θ).

41. $2x^2 + 2y^2 = 3$ **42.** $x^2 + y^2 = x$ **43.** $x^2 = 4y$ **44.** $y^2 = 2x$
45. $2xy = 1$ **46.** $4x^2y = 1$ **47.** $x = 4$ **48.** $y = -3$

In Problems 49–56, the letters r and θ represent polar coordinates. Write each equation using rectangular coordinates (x, y).

49. $r = \cos\theta$ **50.** $r = \sin\theta + 1$ **51.** $r^2 = \cos\theta$ **52.** $r = \sin\theta - \cos\theta$
53. $r = 2$ **54.** $r = 4$ **55.** $r = \dfrac{4}{1 - \cos\theta}$ **56.** $r = \dfrac{3}{3 - \cos\theta}$

57. Show that the formula for the distance d between two points $P_1 = (r_1, \theta_1)$ and $P_2 = (r_2, \theta_2)$ is

$$d = \sqrt{r_1^2 + r_2^2 - 2r_1r_2\cos(\theta_2 - \theta_1)}$$

10.5 Polar Equations and Graphs

Polar Equation An equation whose variables are polar coordinates is called a **polar equation**. The **graph of a polar equation** is the set of all points whose polar coordinates satisfy the equation.

Just as a rectangular grid may be used to plot points given by rectangular coordinates, as in Figure 41(a), we can use a grid consisting of concentric circles (with centers at the pole) and rays (with vertices at the pole) to plot points given by polar coordinates, as shown in Figure 41(b). We shall use such **polar grids** to graph polar equations.

One method we can use to graph a polar equation is to convert the equation to rectangular coordinates. In the discussion that follows, (x, y) represents the rectangular coordinates of a point P and (r, θ) represents polar coordinates of the point P.

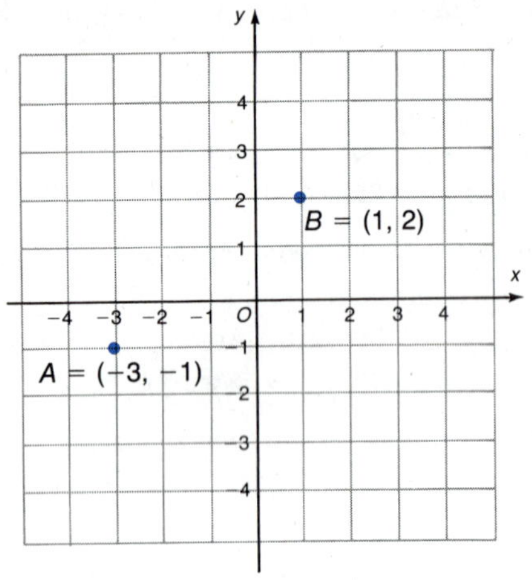

(a) Rectangular grid

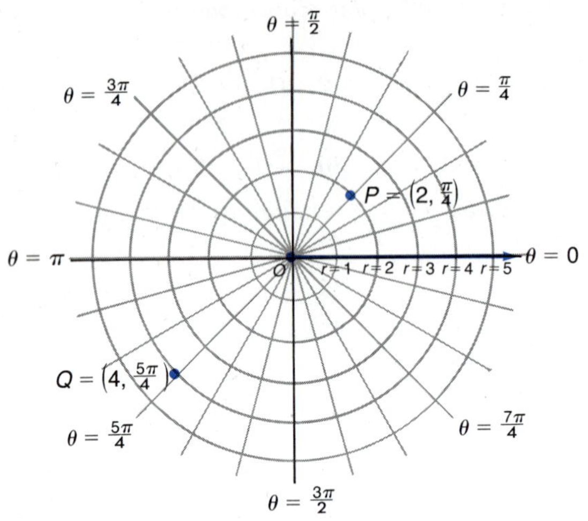

(b) Polar grid

Figure 41

Example 1 Identify and graph the equation: $r = 3$

Solution We convert the polar equation to a rectangular equation:

$$r = 3$$
$$r^2 = 9$$
$$x^2 + y^2 = 9$$

Thus, the graph of $r = 3$ is a circle, with center at the pole and radius 3. See Figure 42.

Figure 42
$r = 3$ or $x^2 + y^2 = 9$

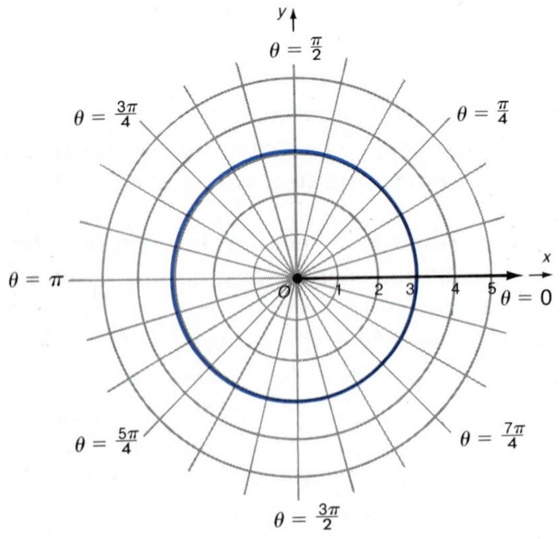

■

Example 2 Identify and graph the equation: $\theta = \pi/4$

Solution We convert the polar equation to a rectangular equation:

$$\theta = \frac{\pi}{4}$$

$$\tan\theta = \tan\frac{\pi}{4} = 1$$

$$\frac{y}{x} = 1$$

$$y = x$$

The graph of $\theta = \pi/4$ is a line passing through the pole making an angle of $\pi/4$ with the polar axis. See Figure 43.

Figure 43
$\theta = \frac{\pi}{4}$ or $y = x$

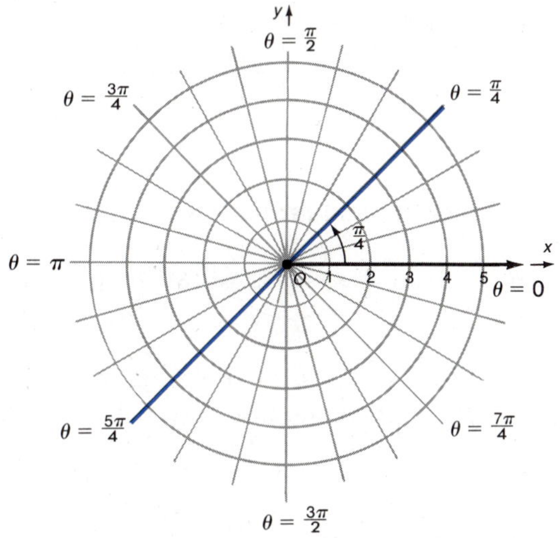

Example 3 Identify and graph the equation: $\theta = -\pi/6$

Solution We convert the polar equation to a rectangular equation:

$$\theta = -\frac{\pi}{6}$$

$$\tan\theta = \tan\left(-\frac{\pi}{6}\right) = -\frac{\sqrt{3}}{3}$$

$$\frac{y}{x} = -\frac{\sqrt{3}}{3}$$

$$y = -\frac{\sqrt{3}}{3}x$$

The graph of $\theta = -\pi/6$ is a line passing through the pole making an angle of $-\pi/6$ with the polar axis. See Figure 44.

Figure 44
$\theta = -\frac{\pi}{6}$ or $y = -\frac{\sqrt{3}}{3}x$

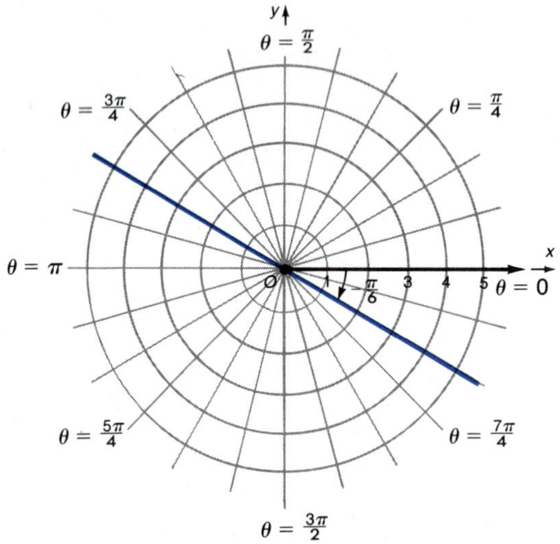

■

Example 4 Identify and graph the equation: $r \sin \theta = 2$

Solution Since $y = r \sin \theta$, we can simply write the equation as

$$y = 2$$

We conclude that the graph of $r \sin \theta = 2$ is a horizontal line 2 units above the pole. See Figure 45.

Figure 45
$r \sin \theta = 2$ or $y = 2$

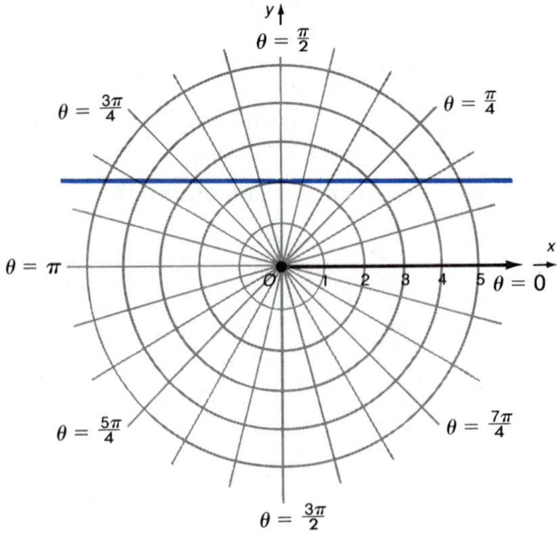

■

Example 5 Identify and graph the equation: $r \cos \theta = -3$

Solution Since $x = r \cos \theta$, we can simply write the equation as

$$x = -3$$

We conclude that the graph of $r \cos \theta = -3$ is a vertical line 3 units to the left of the pole. See Figure 46.

Figure 46
$r \cos \theta = -3$ or $x = -3$

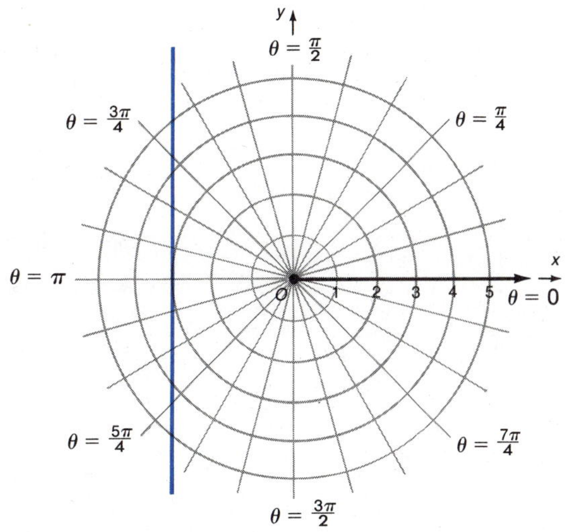

Based on Examples 4 and 5, we are led to the following results. (The proofs are left as exercises.)

Theorem If a is a nonzero real number:

The graph of the equation

$$r \sin \theta = a$$

is a horizontal line a units above the pole if $a > 0$ and $|a|$ units below the pole if $a < 0$.

The graph of the equation

$$r \cos \theta = a$$

is a vertical line a units to the right of the pole if $a > 0$ and $|a|$ units to the left of the pole if $a < 0$.

Example 6 Identify and graph the equation: $r = 4 \sin \theta$

Solution To transform the equation to rectangular coordinates, we multiply each side by r:

$$r^2 = 4r \sin \theta$$

Now we use the facts that $r^2 = x^2 + y^2$ and $y = r \sin \theta$. Then,

$$\begin{aligned} x^2 + y^2 &= 4y \\ x^2 + (y^2 - 4y) &= 0 \\ x^2 + (y - 2)^2 &= 4 \end{aligned}$$

This is the equation of a circle with center at (0, 2) in rectangular coordinates and radius 2. See Figure 47.

Figure 47
$r = 4 \sin \theta$ or $x^2 + (y - 2)^2 = 4$

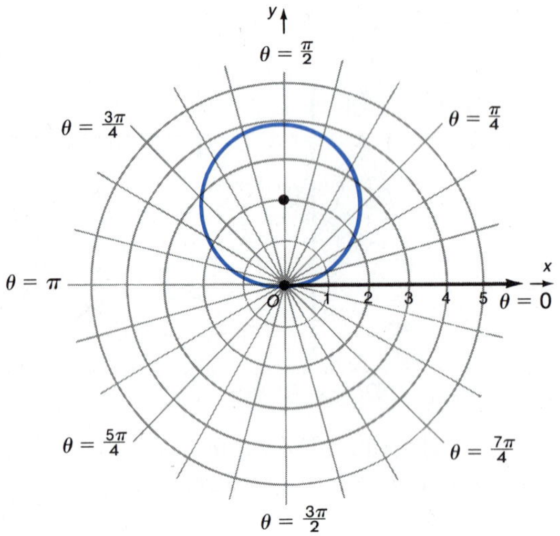

Example 7 Identify and graph the equation: $r = -2 \cos \theta$

Solution We proceed as in Example 6:

$$\begin{aligned} r^2 &= -2r \cos \theta \\ x^2 + y^2 &= -2x \\ x^2 + 2x + y^2 &= 0 \\ (x + 1)^2 + y^2 &= 1 \end{aligned}$$

This is the equation of a circle with center at $(-1, 0)$ in rectangular coordinates and radius 1. See Figure 48.

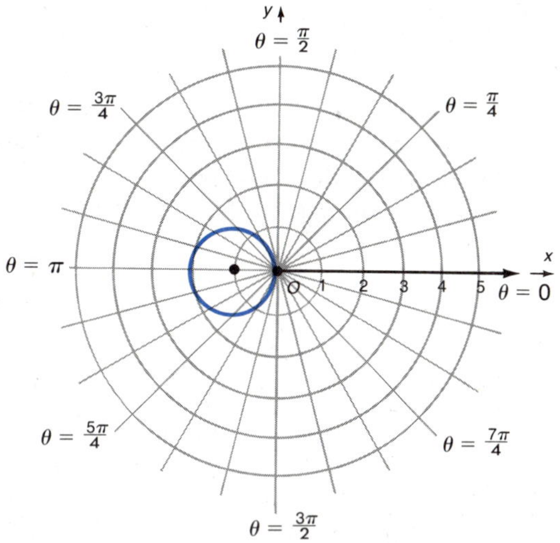

Figure 48
$r = -2 \cos \theta$ or $(x + 1)^2 + y^2 = 1$

Based on Examples 6 and 7, we are led to the following results. (The proofs are left as exercises.)

Theorem Let a be a positive real number. Then:

	EQUATION	DESCRIPTION
(a)	$r = 2a \sin \theta$	Circle: radius a; center at $(0, a)$ in rectangular coordinates
(b)	$r = -2a \sin \theta$	Circle: radius a; center at $(0, -a)$ in rectangular coordinates
(c)	$r = 2a \cos \theta$	Circle: radius a; center at $(a, 0)$ in rectangular coordinates
(d)	$r = -2a \cos \theta$	Circle: radius a; center at $(-a, 0)$ in rectangular coordinates

The method of converting a polar equation to an identifiable rectangular equation in order to graph it is not always helpful, nor is it always necessary. Usually, we set up a table that lists several points on the graph. By checking for symmetry, it may be possible to reduce the number of points needed to draw the graph.

Symmetry

In polar coordinates the points (r, θ) and $(r, -\theta)$ are symmetric with respect to the polar axis (x-axis). See Figure 49(a). The points (r, θ) and $(r, \pi - \theta)$ are symmetric with respect to the line $\theta = \pi/2$ (y-axis). See Figure 49(b). The points (r, θ) and $(-r, \theta)$ are symmetric with respect to the pole (origin). See Figure 49(c).

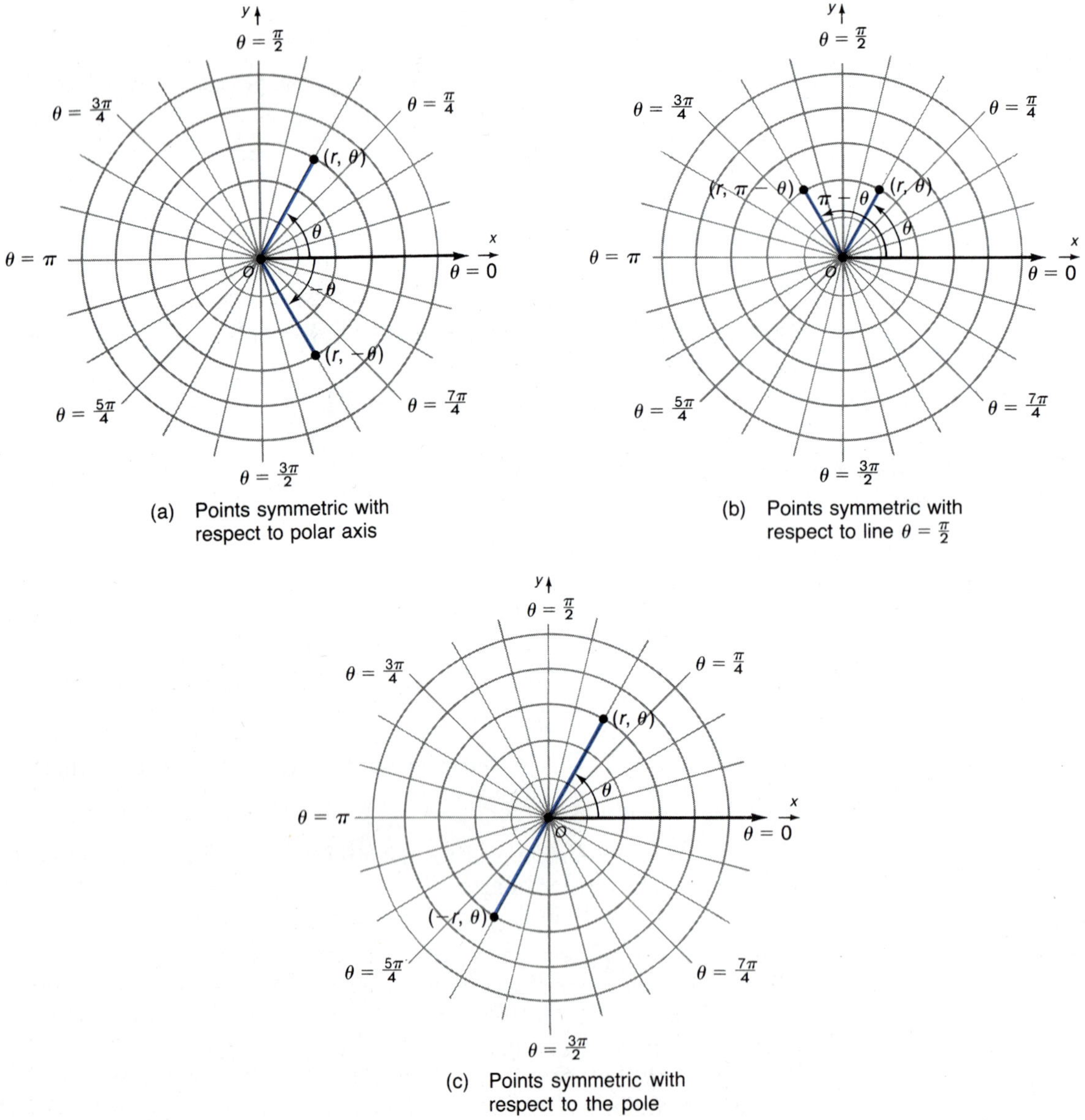

Figure 49

The following tests are a consequence of these observations.

Theorem

Tests for Symmetry

Symmetry with Respect to the Polar Axis (x-Axis): In a polar equation, replace θ by $-\theta$. If an equivalent equation results, the graph is symmetric with respect to the polar axis.

Symmetry with Respect to the Line $\theta = \pi/2$ (y-Axis): In a polar equation, replace θ by $\pi - \theta$. If an equivalent equation results, the graph is symmetric with respect to the line $\theta = \pi/2$.

Symmetry with Respect to the Pole (Origin): In a polar equation, replace r by $-r$. If an equivalent equation results, the graph is symmetric with respect to the pole. ■

The three tests for symmetry given above are *sufficient* conditions for symmetry, but they are not *necessary* conditions; that is, an equation may fail the above tests and still have a graph that is symmetric with respect to the polar axis, the line $\theta = \pi/2$, or the pole. For example, the graph of $r = \sin 2\theta$ turns out to be symmetric with respect to the polar axis, the line $\theta = \pi/2$, and the pole, but all three tests given above fail.

Example 8 Graph the equation: $r = 1 - \sin\theta$

Solution We check for symmetry first:

Polar axis: Replace θ by $-\theta$. The result is

$$r = 1 - \sin(-\theta) = 1 + \sin\theta$$

The test fails, so the graph may or may not be symmetric with respect to the polar axis.

The Line $\theta = \pi/2$: Replace θ by $\pi - \theta$. The result is

$$\begin{aligned} r = 1 - \sin(\pi - \theta) &= 1 - (\sin\pi\cos\theta - \cos\pi\sin\theta) \\ &= 1 - [0 \cdot \cos\theta - (-1)\sin\theta] = 1 - \sin\theta \end{aligned}$$

Thus, the graph is symmetric with respect to the line $\theta = \pi/2$.

The Pole: Replace r by $-r$. Then the result is $-r = 1 - \sin\theta$, so $r = -1 + \sin\theta$. The test fails, so the graph may or may not be symmetric with respect to the pole.

Next, we identify points on the graph by assigning values to the angle θ and calculating the corresponding values of r. Due to the symmetry with respect to the line $\theta = \pi/2$, we only need to assign values to θ from $-\pi/2$ to $\pi/2$, as given in Table 1.

Now we plot the points (r, θ) from Table 1 and trace out the graph, beginning at the point $(2, -\pi/2)$ and ending at the point $(0, \pi/2)$. Then we reflect this portion of the graph about the line $\theta = \pi/2$ (y-axis) to obtain the complete graph. See Figure 50.

Table 1

θ	$r = 1 - \sin\theta$
$-\pi/2$	$1 + 1 = 2$
$-\pi/3$	$1 + \sqrt{3}/2 \approx 1.87$
$-\pi/6$	$1 + \frac{1}{2} = \frac{3}{2}$
0	1
$\pi/6$	$1 - \frac{1}{2} = \frac{1}{2}$
$\pi/3$	$1 - \sqrt{3}/2 \approx 0.13$
$\pi/2$	0

Figure 50
$r = 1 - \sin\theta$

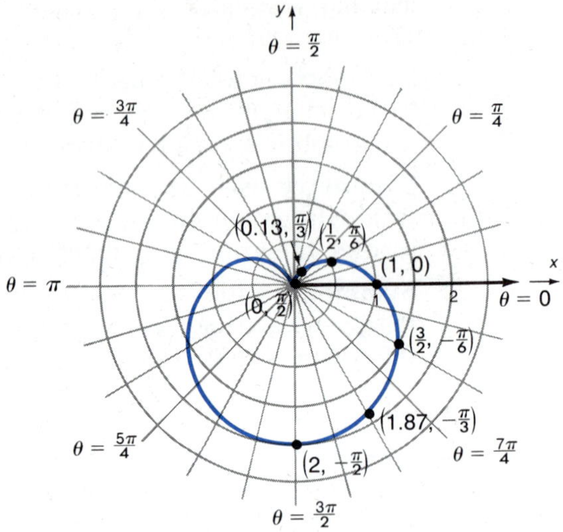

The curve in Figure 50 is an example of a **cardioid** (a heart-shaped curve). Cardioids are characterized by equations of the form

$$r = a(1 + \cos\theta) \qquad r = a(1 + \sin\theta)$$
$$r = a(1 - \cos\theta) \qquad r = a(1 - \sin\theta)$$

where $a > 0$. The graph of a cardioid contains the pole.

Example 9 Graph the equation: $r = 3 + 2\cos\theta$

Solution We check for symmetry first:

Polar Axis: Replace θ by $-\theta$. The result is

$$r = 3 + 2\cos(-\theta) = 3 + 2\cos\theta$$

Thus, the graph is symmetric with respect to the polar axis.

The Line $\theta = \pi/2$: Replace θ by $\pi - \theta$. The result is

$$r = 3 + 2\cos(\pi - \theta) = 3 + 2(\cos\pi\cos\theta + \sin\pi\sin\theta)$$
$$= 3 - 2\cos\theta$$

The test fails, so the graph may or may not be symmetric with respect to the line $\theta = \pi/2$.

The Pole: Replace r by $-r$. The test fails, so the graph may or may not be symmetric with respect to the pole.

Next, we identify points on the graph by assigning values to the angle θ and calculating the corresponding values of r. Due to the symmetry with respect to the polar axis, we only need to assign values to θ from 0 to π, as given in Table 2.

Table 2

θ	$r = 3 + 2\cos\theta$
0	5
$\pi/6$	$3 + \sqrt{3} \approx 4.73$
$\pi/3$	4
$\pi/2$	3
$2\pi/3$	2
$5\pi/6$	$3 - \sqrt{3} \approx 1.27$
π	1

Now we plot the points (r, θ) from Table 2 and trace out the graph, beginning at the point $(5, 0)$ and ending at the point $(\pi, 1)$. Then we reflect this portion of the graph about the pole (x-axis) to obtain the complete graph. See Figure 51.

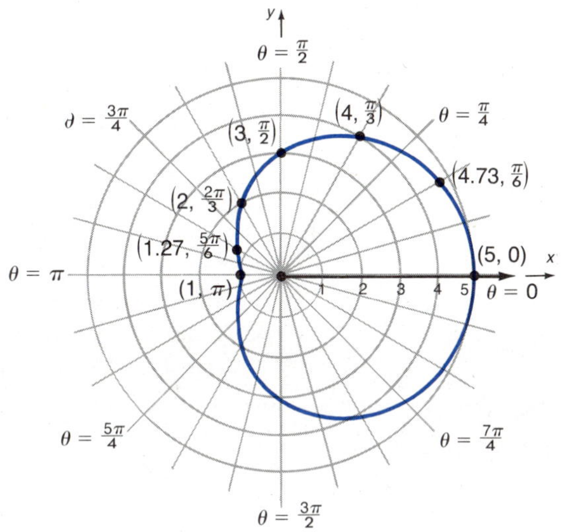

Figure 51
$r = 3 + 2\cos\theta$

■

The curve in Figure 51 is an example of a **limaçon** without an inner loop. Such limaçons are characterized by equations of the form

$$r = a + b\cos\theta \qquad r = a + b\sin\theta$$
$$r = a - b\cos\theta \qquad r = a - b\sin\theta$$

where $a > 0$, $b > 0$, and $a > b$. The graph of a limaçon without an inner loop does not contain the pole.

Example 10 Graph the equation: $r = 1 + 2\cos\theta$

Solution First, we check for symmetry:

Polar Axis: Replace θ by $-\theta$. The result is

$$r = 1 + 2\cos(-\theta) = 1 + 2\cos\theta$$

Thus, the graph is symmetric with respect to the polar axis.

The Line $\theta = \pi/2$: Replace θ by $\pi - \theta$. The result is

$$r = 1 + 2\cos(\pi - \theta) = 1 + 2(\cos\pi\cos\theta + \sin\pi\sin\theta)$$
$$= 1 - 2\cos\theta$$

The test fails, so the graph may or may not be symmetric with respect to the line $\theta = \pi/2$.

The Pole: Replace r by $-r$. The test fails, so the graph may or may not be symmetric with respect to the pole.

Table 3

θ	$r = 1 + 2\cos\theta$
0	3
$\pi/6$	$1 + \sqrt{3} \approx 2.73$
$\pi/3$	2
$\pi/2$	1
$2\pi/3$	0
$5\pi/6$	$1 - \sqrt{3} \approx -0.73$
π	-1

Next, we identify points on the graph of $r = 1 + 2\cos\theta$ by assigning values to the angle θ and calculating the corresponding values of r. Due to the symmetry with respect to the polar axis, we only need to assign values to θ from 0 to π, as given in Table 3.

Now we plot the points (r, θ) from Table 3, beginning at (3, 0) and ending at $(-1, \pi)$. See Figure 52(a). Finally, we reflect this portion of the graph about the polar axis (x-axis) to obtain the complete graph. See Figure 52(b).

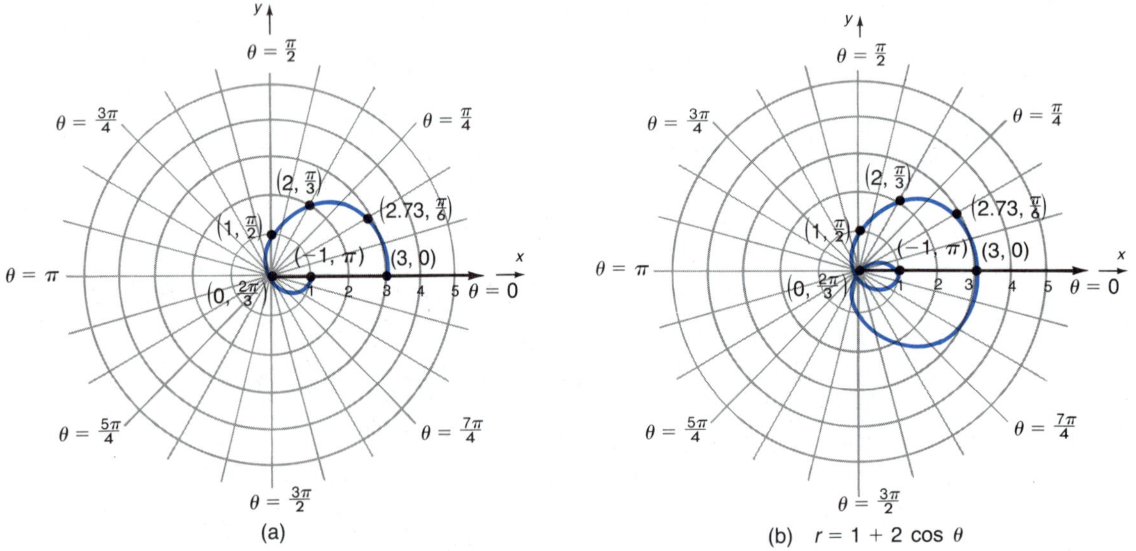

Figure 52

The curve in Figure 52(b) is an example of a limaçon with an inner loop. Such limaçons are characterized by equations of the form

$$r = a + b\cos\theta \qquad r = a + b\sin\theta$$
$$r = a - b\cos\theta \qquad r = a - b\sin\theta$$

where $a > 0$, $b > 0$, and $a < b$. The graph of a limaçon with an inner loop will pass through the pole twice.

Example 11 Graph the equation: $r = 2\cos 2\theta$

Solution We check for symmetry:

Polar Axis: If we replace θ by $-\theta$, the result is

$$r = 2\cos 2(-\theta) = 2\cos 2\theta$$

Thus, the graph is symmetric with respect to the polar axis.

The Line $\theta = \pi/2$: If we replace θ by $\pi - \theta$, we obtain

$$r = 2\cos 2(\pi - \theta) = 2\cos(2\pi - 2\theta) = 2\cos(-2\theta) = 2\cos 2\theta$$

Thus, the graph is symmetric with respect to the line $\theta = \pi/2$.

The Pole: Since the graph is symmetric with respect to both the polar axis and the line $\theta = \pi/2$, it must be symmetric with respect to the pole.

Table 4

θ	$r = 2\cos 2\theta$
0	2
$\pi/6$	1
$\pi/4$	0
$\pi/3$	-1
$\pi/2$	-2

Next, we construct Table 4. Due to the symmetry with respect to the polar axis, the line $\theta = \pi/2$, and the pole, we consider only values of θ from 0 to $\pi/2$.

We plot and connect these points in Figure 53(a). Finally, because of symmetry, we reflect this portion of the graph first about the polar axis (x-axis) and then about the line $\theta = \pi/2$ (y-axis) to obtain the complete graph. See Figure 53(b).

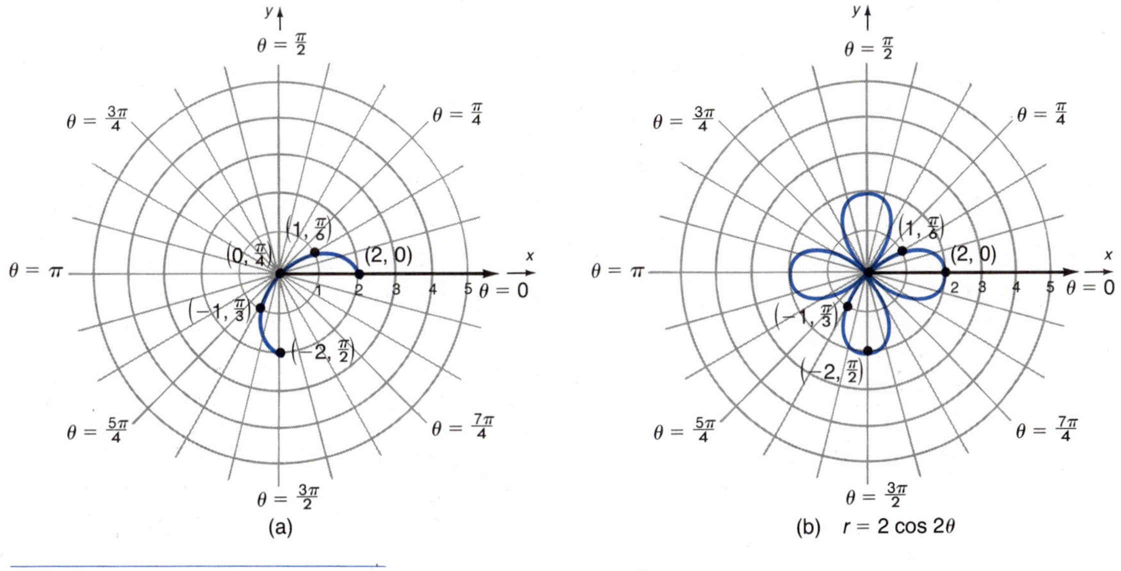

Figure 53

The curve in Figure 53(b) is called a **rose** with four petals. Rose curves are characterized by equations of the form

$$r = a\cos n\theta \qquad r = a\sin n\theta \qquad a > 0$$

and have graphs that are rose-shaped. If n is even, the rose has $2n$ petals; if n is odd, the rose has n petals.

Example 12 Graph the equation: $r^2 = 4\sin 2\theta$

Solution We leave it to you to verify that the graph is symmetric with respect to the pole. Table 5 lists points on the graph for values of $\theta = 0$ through $\theta = \pi/2$. Note that there are no points on the graph for $\pi/2 < \theta < \pi$ (quadrant II), since $\sin 2\theta < 0$ for such values. The points from Table 5 where $r \geq 0$ are plotted in Figure 54(a). The remaining points on the graph may be obtained by using symmetry. Figure 54(b) shows the final graph.

Table 5

θ	$r^2 = 4 \sin 2\theta$	r
0	0	0
$\pi/6$	$2\sqrt{3}$	± 1.9
$\pi/4$	4	± 2
$\pi/3$	$2\sqrt{3}$	± 1.9
$\pi/2$	0	0

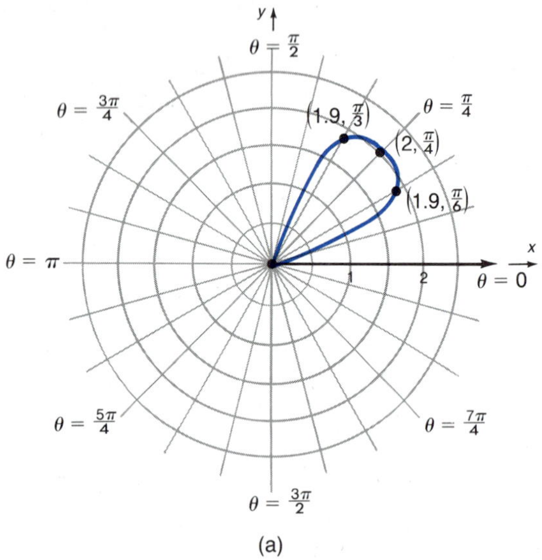

Figure 54

The curve in Figure 54(b) is an example of a **lemniscate**. Lemniscates are characterized by equations of the form

$$r^2 = a^2 \sin 2\theta \qquad r^2 = a^2 \cos 2\theta$$

where $a \neq 0$, and have graphs that are propeller-shaped.

Example 13 Sketch the graph of: $r = e^{\theta/5}$

Solution The tests for symmetry with respect to the pole, the polar axis, and the line $\theta = \pi/2$ fail. Furthermore, there is no number θ for which $r = 0$. Hence, the graph does not pass through the pole. We observe that r is positive for all θ, r increases as θ increases, $r \to 0$ as $\theta \to -\infty$, and $r \to \infty$ as $\theta \to \infty$. With the help of a calculator, we obtain the values in Table 6. See Figure 55 for the graph.

Table 6

θ	$r = e^{\theta/5}$
$-3\pi/2$	0.39
$-\pi$	0.53
$-\pi/2$	0.73
$-\pi/4$	0.85
0	1
$\pi/4$	1.17
$\pi/2$	1.37
π	1.87
$3\pi/2$	2.57
2π	3.51

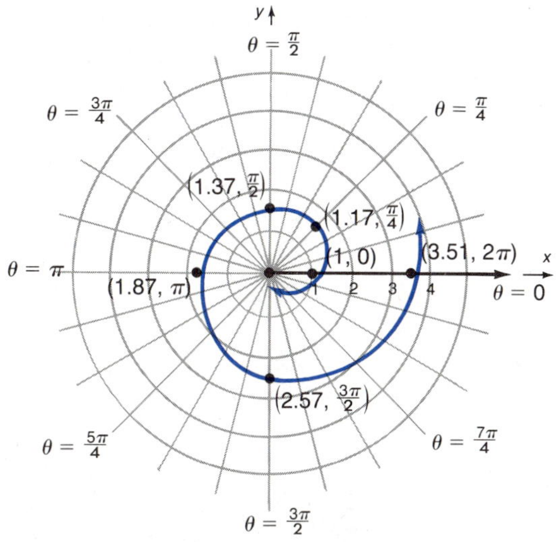

Figure 55
$r = e^{\theta/5}$

The curve in Figure 55 is called a **logarithmic spiral**, since its equation may be written as $\theta = 5 \ln r$ and it spirals infinitely both toward the pole and away from it.

Classification of Polar Equations

The equations of some lines and circles in polar coordinates and their corresponding equations in rectangular coordinates are given in Table 7. Also included are the names and the graphs of a few of the more frequently encountered polar equations.

Table 7

LINES			
Description	Line passing through the pole making an angle α with the polar axis	Vertical line	Horizontal line
Rectangular equation	$y = (\tan \alpha)x$	$x = a$	$y = b$
Polar equation	$\theta = \alpha$	$r \cos \theta = a$	$r \sin \theta = b$
Typical graph	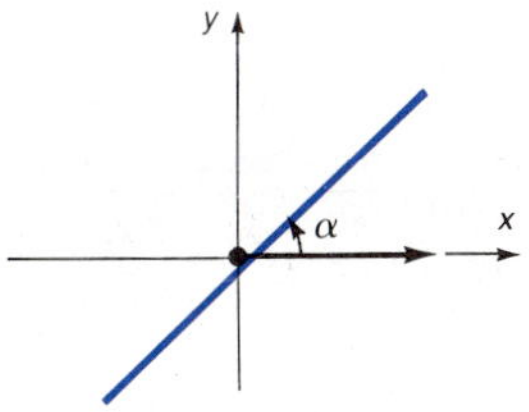	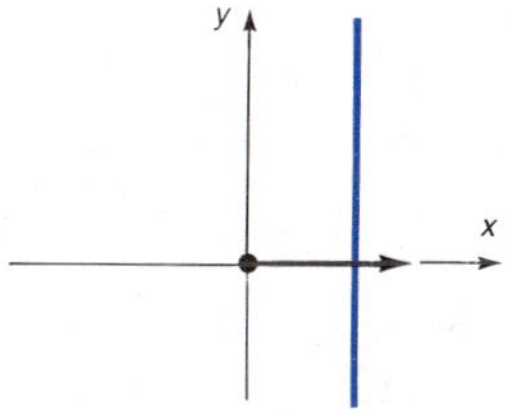	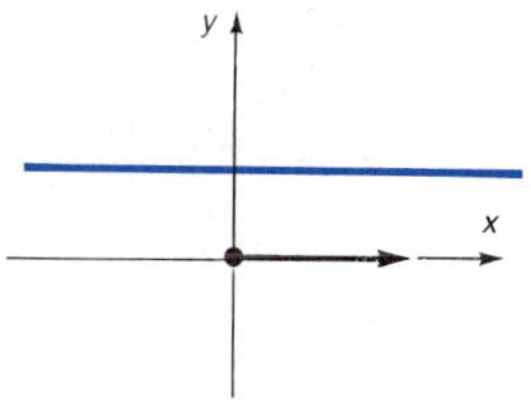

(continued)

Table 7 (continued)

CIRCLES			
Description	Center at the pole, radius a	Passing through the pole, tangent to the line $\theta = \pi/2$, center on the polar axis, radius a	Passing through the pole, tangent to the polar axis, center on the line $\theta = \pi/2$, radius a
Rectangular equation	$x^2 + y^2 = a^2$, $a > 0$	$x^2 + y^2 = \pm 2ax$, $a > 0$	$x^2 + y^2 = \pm 2ay$, $a > 0$
Polar equation	$r = a$, $a > 0$	$r = \pm 2a \cos \theta$, $a > 0$	$r = \pm 2a \sin \theta$, $a > 0$
Typical graph	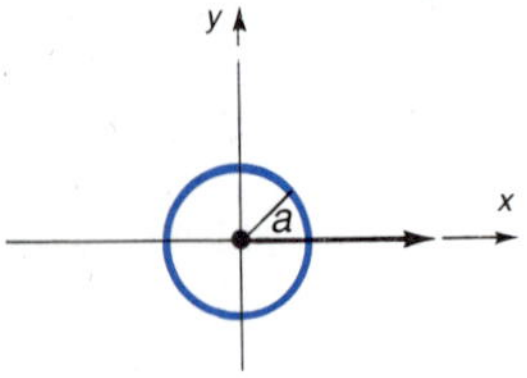	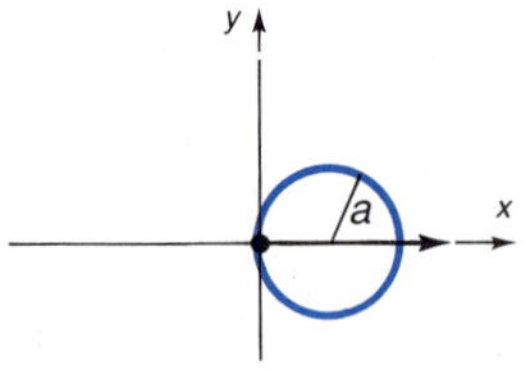	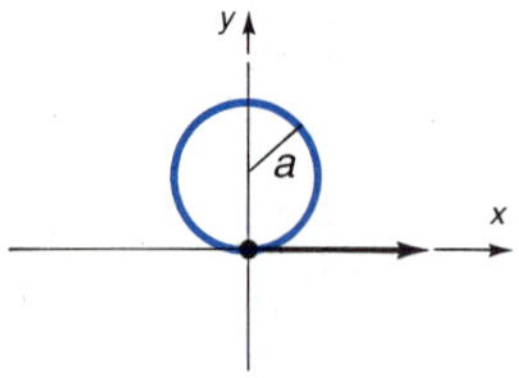
OTHER EQUATIONS			
Name	Cardioid	Limaçon without inner loop	Limaçon with inner loop
Polar equations	$r = a \pm a \cos \theta$, $a > 0$ $r = a \pm a \sin \theta$, $a > 0$	$r = a \pm b \cos \theta$, $0 < b < a$ $r = a \pm b \sin \theta$, $0 < b < a$	$r = a \pm b \cos \theta$, $0 < a < b$ $r = a \pm b \sin \theta$, $0 < a < b$
Typical graph	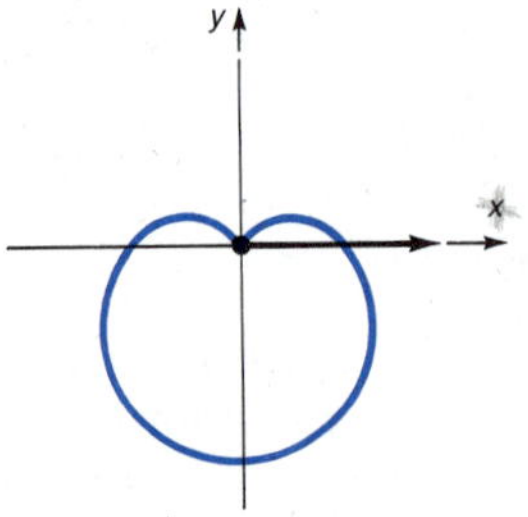	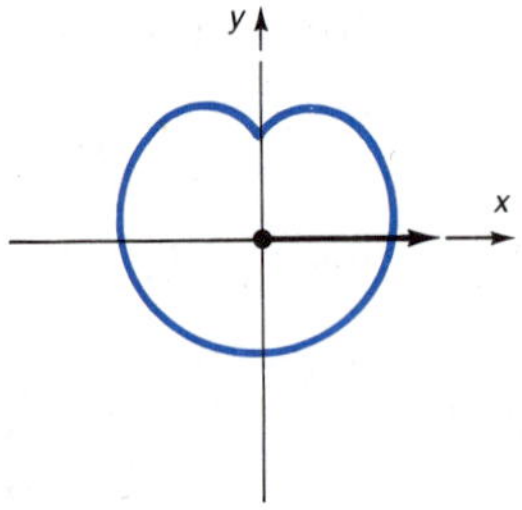	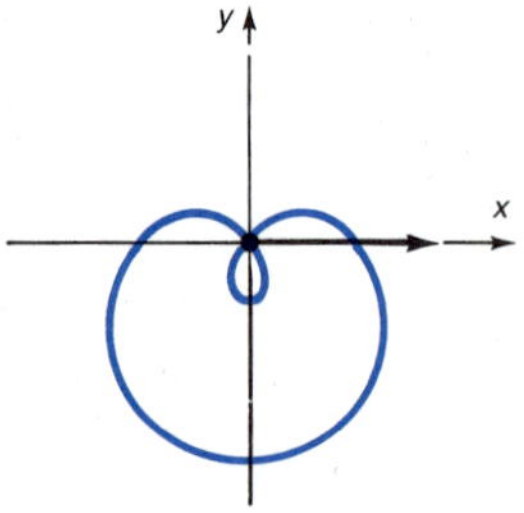
Name	Lemniscate	Rose with three petals	Rose with four petals
Polar equations	$r^2 = a^2 \cos 2\theta$, $a > 0$ $r^2 = a^2 \sin 2\theta$, $a > 0$	$r = a \sin 3\theta$, $a > 0$ $r = a \cos 3\theta$, $a > 0$	$r = a \sin 2\theta$, $a > 0$ $r = a \cos 2\theta$, $a > 0$
Typical graph	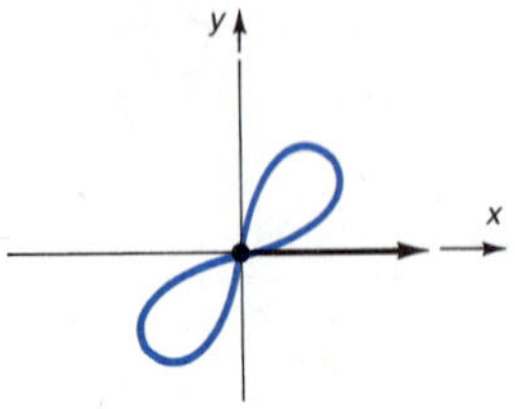	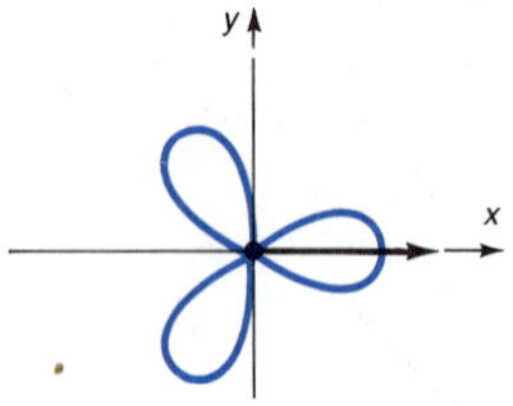	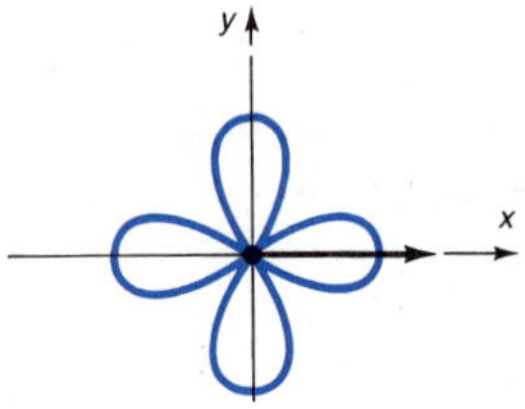

Calculus Comment

For those of you who are planning to study calculus, a comment about one important role of polar equations is in order.

In rectangular coordinates, the equation $x^2 + y^2 = 1$ whose graph is the unit circle, does not define a function. In fact, on the interval $[-1, 1]$ it defines two functions,

$y = \sqrt{1 - x^2}$ Upper semicircle $\qquad y = -\sqrt{1 - x^2}$ Lower semicircle

In polar coordinates, the equation $r = 1$ whose graph is also the unit circle, does define a function. That is, for each choice of θ there is only one corresponding value of r, namely, $r = 1$. Since many uses of calculus require that functions be used, the opportunity to express nonfunctions in rectangular coordinates as functions in polar coordinates becomes extremely useful.

Note also that the vertical line test for functions is valid only for equations in rectangular coordinates.

Historical Comment ■ Polar coordinates seem to have been invented by Jacob Bernoulli (1654–1705) about 1691, although, as with most such ideas, earlier traces of the notion exist. Early users of calculus remained committed to rectangular coordinates, and polar coordinates did not become widely used until the early 1800's. Even then, it was mostly geometers who used them for describing odd curves. Finally, about the mid-1800's, applied mathematicians realized the tremendous simplification polar coordinates make possible in the description of objects with circular or cylindrical symmetry. From then on their use became widespread. ■

EXERCISE 10.5

In Problems 1–16, identify and graph each polar equation.

1. $r = 4$ **2.** $r = 2$ **3.** $\theta = \pi/3$ **4.** $\theta = -\pi/4$

5. $r \sin \theta = 4$ **6.** $r \cos \theta = 4$ **7.** $r \cos \theta = -2$ **8.** $r \sin \theta = -2$

9. $r = 2 \cos \theta$ **10.** $r = 2 \sin \theta$ **11.** $r = -4 \sin \theta$ **12.** $r = -4 \cos \theta$

13. $r \sec \theta = 4$ **14.** $r \csc \theta = 8$ **15.** $r \csc \theta = -2$ **16.** $r \sec \theta = -4$

In Problems 17–36, graph each polar equation. Be sure to test for symmetry.

17. $r = 2 + 2 \cos \theta$ **18.** $r = 1 + \sin \theta$ **19.** $r = 3 - 3 \sin \theta$

20. $r = 2 - 2 \cos \theta$ **21.** $r = 2 + \sin \theta$ **22.** $r = 2 - \cos \theta$

23. $r = 4 - 2 \cos \theta$ **24.** $r = 4 + 2 \sin \theta$ **25.** $r = 1 + 2 \sin \theta$

26. $r = 1 - 2 \sin \theta$ **27.** $r = 2 - 3 \cos \theta$ **28.** $r = 2 + 4 \cos \theta$

29. $r = 3 \cos 2\theta$ **30.** $r = 2 \sin 2\theta$ **31.** $r = 4 \sin 3\theta$

32. $r = 3 \cos 4\theta$ **33.** $r^2 = 9 \cos 2\theta$ **34.** $r^2 = \sin 2\theta$

35. $r = 2^\theta$ **36.** $r = 3^\theta$

In Problems 37–46, graph each polar equation.

37. $r = \dfrac{2}{1 - \cos\theta}$ (parabola)

38. $r = \dfrac{2}{1 - 2\cos\theta}$ (hyperbola)

39. $r = \dfrac{1}{3 - 2\cos\theta}$ (ellipse)

40. $r = \dfrac{1}{1 - \cos\theta}$ (parabola)

41. $r = \theta$, $\theta \geq 0$ (spiral of Archimedes)

42. $r = \dfrac{3}{\theta}$ (reciprocal spiral)

43. $r = \csc\theta - 2$, $0 < \theta < \pi$ (conchoid)

44. $r = \sin\theta\tan\theta$ (cissoid)

45. $r = \tan\theta$ (kappa curve)

46. $r = \cos\dfrac{\theta}{2}$

47. Show that the graph of the equation $r\sin\theta = a$ is a horizontal line a units above the pole if $a > 0$ and $|a|$ units below the pole if $a < 0$.

48. Show that the graph of the equation $r\cos\theta = a$ is a vertical line a units to the right of the pole if $a > 0$ and $|a|$ units to the left of the pole if $a < 0$.

49. Show that the graph of the equation $r = 2a\sin\theta$, $a > 0$, is a circle of radius a with center at $(0, a)$ in rectangular coordinates.

50. Show that the graph of the equation $r = -2a\sin\theta$, $a > 0$, is a circle of radius a with center at $(0, -a)$ in rectangular coordinates.

51. Show that the graph of the equation $r = 2a\cos\theta$, $a > 0$, is a circle of radius a with center at $(a, 0)$ in rectangular coordinates.

52. Show that the graph of the equation $r = -2a\cos\theta$, $a > 0$, is a circle of radius a with center at $(-a, 0)$ in rectangular coordinates.

53. Explain why the following test for symmetry is valid: Replace r by $-r$ and θ by $-\theta$ in a polar equation. If an equivalent equation results, the graph is symmetric with respect to the line $\theta = \pi/2$ (y-axis).
(a) Show that the test on page 583 fails for $r^2 = \cos\theta$, but this new test works.
(b) Show that the test on page 583 works for $r^2 = \sin\theta$, yet this new test fails.

54. Develop a new test for symmetry with respect to the pole.
(a) Find a polar equation for which this new test fails, yet the test on page 583 works.
(b) Find a polar equation for which the test on page 583 fails, yet the new test works.

10.6 The Complex Plane; DeMoivre's Theorem

When we first introduced complex numbers, we were not prepared to give a geometric interpretation of a complex number. Now we are ready. Although there are several such interpretations we could give, the one that follows is the easiest to understand.

A complex number $z = x + yi$ can be interpreted geometrically as the point (x, y) in the xy-plane. Thus, each point in the plane corresponds to a complex number and, conversely, each complex number corresponds to a point in the plane. We shall refer to the collection of such points as the **complex plane**. The x-axis will be referred to as the **real axis**, because any point that lies on the real axis is of the form $z = x + 0i =$

x, a real number. The y-axis is called the **imaginary axis**, because any point that lies on it is of the form $z = 0 + yi = yi$, a pure imaginary number. See Figure 56.

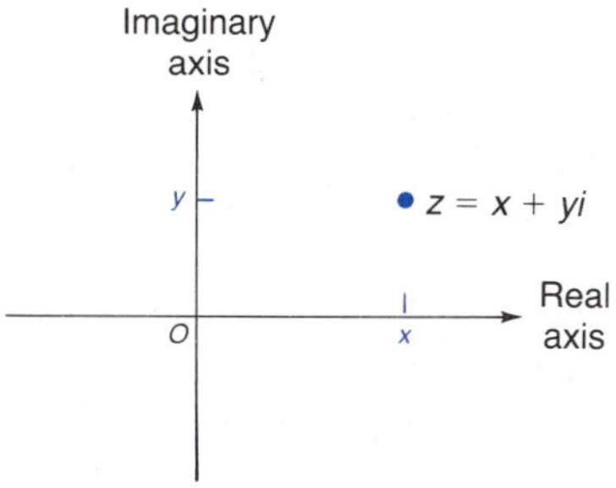

Figure 56
Complex plane

Magnitude

Let $z = x + yi$ be a complex number. The **magnitude** or **modulus** of z, denoted by $|z|$, is defined as the distance from the origin to the point (x, y). Thus,

$$|z| = \sqrt{x^2 + y^2} \qquad (1)$$

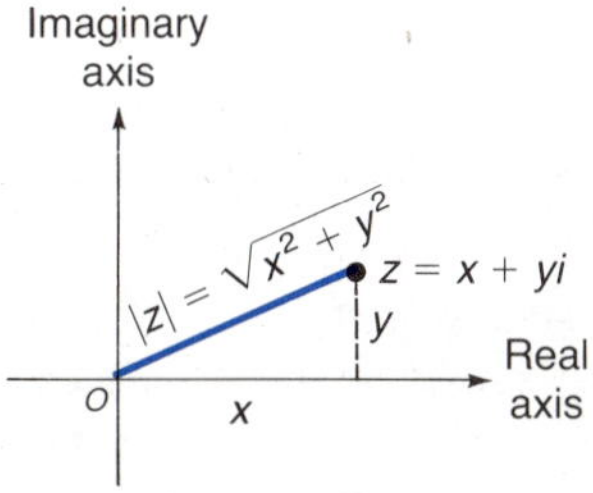

Figure 57

See Figure 57 for an illustration.

This definition for $|z|$ is consistent with the definition for the absolute value of a real number: If $z = x + yi$ is real, then $z = x + 0i$ and

$$|z| = \sqrt{x^2 + 0^2} = \sqrt{x^2} = |x|$$

If $z = x + yi$, its **conjugate**, denoted by $\bar{z}$, is $\bar{z} = x - yi$. Because $z\bar{z} = x^2 + y^2$, it follows from equation (1) that the magnitude of z can be written as

$$|z| = \sqrt{z\bar{z}} \qquad (2)$$

Polar Form of a Complex Number

When a complex number is written in the form $z = x + yi$, we say it is in **rectangular**, or **Cartesian**, **form**, because (x, y) are the rectangular coordinates of the corresponding point in the complex plane. Suppose (r, θ) are the polar coordinates of this point. Then,

$$x = r \cos \theta \qquad y = r \sin \theta \qquad (3)$$

If $r \geq 0$ and $0 \leq \theta < 2\pi$, the complex number $z = x + yi$ may be written in **polar form** as

Polar Form of z

$$z = x + yi = (r \cos \theta) + (r \sin \theta)i = r(\cos \theta + i \sin \theta) \qquad (4)$$

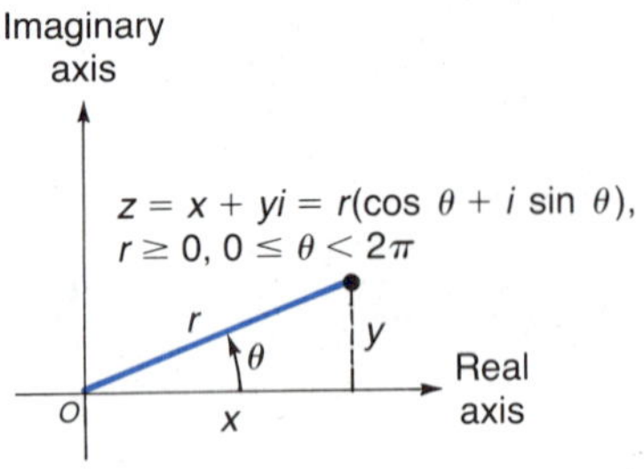

Figure 58

See Figure 58.

If $z = r(\cos \theta + i \sin \theta)$ is the polar form of a complex number, the angle θ, $0 \leq \theta < 2\pi$, is called the **argument of** z. Also, because $r \geq 0$, from equation (3) we have $r = \sqrt{x^2 + y^2}$. Thus, from equation (1) it follows that the magnitude of $z = r(\cos \theta + i \sin \theta)$ is

$$|z| = r$$

Example 1 Plot the point corresponding to $z = \sqrt{3} - i$ in the complex plane, and write an expression for z in polar form.

Solution The point corresponding to $z = \sqrt{3} - i$ has the rectangular coordinates $(\sqrt{3}, -1)$. The point, located in quadrant IV, is plotted in Figure 59. Because $x = \sqrt{3}$ and $y = -1$, it follows that

$$r = \sqrt{x^2 + y^2} = \sqrt{(\sqrt{3})^2 + (-1)^2} = \sqrt{4} = 2$$

and

$$\sin \theta = \frac{y}{r} = \frac{-1}{2} \qquad \cos \theta = \frac{x}{r} = \frac{\sqrt{3}}{2} \qquad 0 \leq \theta < 2\pi$$

Thus, $\theta = 11\pi/6$ and $r = 2$, so the polar form of $z = \sqrt{3} - i$ is

$$z = r(\cos \theta + i \sin \theta) = 2\left(\cos \frac{11\pi}{6} + i \sin \frac{11\pi}{6}\right)$$ ■

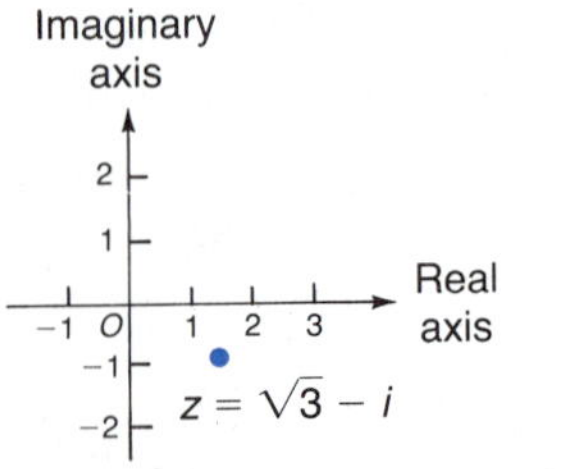

Figure 59

Example 2 Plot the point corresponding to $z = 2(\cos 30° + i \sin 30°)$ in the complex plane, and write an expression for z in rectangular form.

Solution To plot the complex number $z = 2(\cos 30° + i \sin 30°)$, we plot the point whose polar coordinates are $(r, \theta) = (2, 30°)$, as shown in Figure 60. In rectangular form,

$$z = 2(\cos 30° + i \sin 30°) = 2\left(\frac{\sqrt{3}}{2} + \frac{1}{2}i\right) = \sqrt{3} + i$$ ■

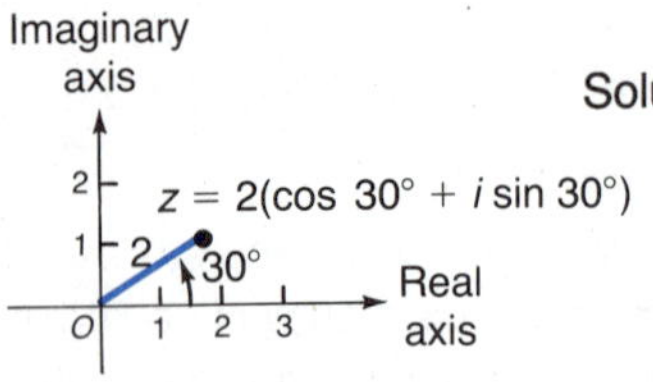

Figure 60

The polar form of a complex number provides an alternative for finding products and quotients of complex numbers.

Theorem Let $z_1 = r_1(\cos\theta_1 + i\sin\theta_1)$ and $z_2 = r_2(\cos\theta_2 + i\sin\theta_2)$ be two complex numbers. Then

$$z_1 z_2 = r_1 r_2[\cos(\theta_1 + \theta_2) + i\sin(\theta_1 + \theta_2)] \tag{5}$$

If $z_2 \neq 0$, then

$$\frac{z_1}{z_2} = \frac{r_1}{r_2}[\cos(\theta_1 - \theta_2) + i\sin(\theta_1 - \theta_2)] \tag{6}$$

Proof We shall prove formula (5). The proof of formula (6) is left as an exercise (see Problem 56 at the end of this section).

$$\begin{aligned} z_1 z_2 &= [r_1(\cos\theta_1 + i\sin\theta_1)][r_2(\cos\theta_2 + i\sin\theta_2)] \\ &= r_1 r_2[(\cos\theta_1 + i\sin\theta_1)(\cos\theta_2 + i\sin\theta_2)] \\ &= r_1 r_2[(\cos\theta_1\cos\theta_2 - \sin\theta_1\sin\theta_2) + i(\cos\theta_1\sin\theta_2 + \sin\theta_1\cos\theta_2)] \\ &= r_1 r_2[(\cos(\theta_1 + \theta_2) + i\sin(\theta_1 + \theta_2)] \end{aligned}$$

■

Because the magnitude of a complex number z is r and its argument is θ, when $z = r(\cos\theta + i\sin\theta)$, we can restate this theorem as follows:

Theorem The magnitude of the product (quotient) of two complex numbers equals the product (quotient) of their magnitudes; the argument of the product (quotient) of two complex numbers equals the sum (difference) of their arguments. ■

Let's look at an example of how this theorem can be used.

Example 3 If $z = 3(\cos 20° + i\sin 20°)$ and $w = 5(\cos 100° + \sin 100°)$, find the following (leave your answers in polar form):

(a) zw (b) z/w

Solution (a)

$$\begin{aligned} zw &= [3(\cos 20° + i\sin 20°)][5(\cos 100° + i\sin 100°)] \\ &= (3 \cdot 5)[\cos(20° + 100°) + i\sin(20° + 100°)] \\ &= 15(\cos 120° + i\sin 120°) \end{aligned}$$

(b)
$$\begin{aligned}\frac{z}{w} &= \frac{3(\cos 20° + i \sin 20°)}{5(\cos 100° + i \sin 100°)}\\ &= \tfrac{3}{5}[\cos(20° - 100°) + i \sin(20° - 100°)]\\ &= \tfrac{3}{5}[\cos(-80°) + i \sin(-80°)]\\ &= \tfrac{3}{5}(\cos 280° + i \sin 280°)\end{aligned}$$
Argument must lie between 0° and 360° ■

DeMoivre's Theorem

DeMoivre's Theorem, stated by Abraham DeMoivre (1667–1754) in 1730 but already known to many people by 1710, is important for the following reason: The fundamental processes of algebra are the four operations of addition, subtraction, multiplication, and division, together with powers and the extraction of roots. DeMoivre's Theorem allows these latter fundamental algebraic operations to be applied to complex numbers.

DeMoivre's Theorem, in its most basic form, is a formula for raising a complex number z to the power n, where $n \geq 1$ is a positive integer. Let's see if we can guess the form of the result.

Let $z = r(\cos \theta + i \sin \theta)$ be a complex number. Then, based on equation (5), we have

$$n = 2: \quad z^2 = r^2(\cos 2\theta + i \sin 2\theta)$$

$$\begin{aligned} n = 3: \quad z^3 &= z^2 \cdot z\\ &= [r^2(\cos 2\theta + i \sin 2\theta)][r(\cos \theta + i \sin \theta)]\\ &= r^3(\cos 3\theta + i \sin 3\theta) \qquad \text{Equation (5)}\end{aligned}$$

$$\begin{aligned} n = 4: \quad z^4 &= z^3 \cdot z\\ &= [r^3(\cos 3\theta + i \sin 3\theta)][r(\cos \theta + i \sin \theta)]\\ &= r^4(\cos 4\theta + i \sin 4\theta) \qquad \text{Equation (5)}\end{aligned}$$

The pattern should now be clear.

Theorem

DeMoivre's Theorem

If $z = r(\cos \theta + i \sin \theta)$ is a complex number, then

$$z^n = r^n(\cos n\theta + i \sin n\theta) \tag{7}$$

where $n \geq 1$ is a positive integer. ■

We shall not prove DeMoivre's Theorem because it requires mathematical induction (which is not discussed until Section 13.1).

Let's look at some examples.

Example 4

$$\begin{aligned}[2(\cos 20° + i \sin 20°)]^3 &= 2^3[\cos(3 \cdot 20°) + i \sin(3 \cdot 20°)] \\ &= 8(\cos 60° + i \sin 60°) \\ &= 8\left(\frac{1}{2} + \frac{\sqrt{3}}{2}i\right) = 4 + 4\sqrt{3}i\end{aligned}$$

■

Example 5 Write $(1 + i)^5$ in the standard form $a + bi$.

Solution To apply DeMoivre's Theorem, we must first write the complex number in polar form. Thus, we begin by writing

$$1 + i = \sqrt{2}\left(\frac{1}{\sqrt{2}} + \frac{1}{\sqrt{2}}i\right) = \sqrt{2}\left(\cos\frac{\pi}{4} + i \sin\frac{\pi}{4}\right)$$

Now,

$$\begin{aligned}(1 + i)^5 &= \left[\sqrt{2}\left(\cos\frac{\pi}{4} + i \sin\frac{\pi}{4}\right)\right]^5 \\ &= (\sqrt{2})^5\left[\cos\left(5 \cdot \frac{\pi}{4}\right) + i \sin\left(5 \cdot \frac{\pi}{4}\right)\right] \\ &= 4\sqrt{2}\left(\cos\frac{5\pi}{4} + i \sin\frac{5\pi}{4}\right) \\ &= 4\sqrt{2}\left[-\frac{1}{\sqrt{2}} + \left(-\frac{1}{\sqrt{2}}\right)i\right] = -4 - 4i\end{aligned}$$

■

C **Example 6** Write $(3 + 4i)^3$ in the standard form $a + bi$.

Solution Again, we start by writing $3 + 4i$ in polar form. This time, we will use degrees for the argument:

$$3 + 4i = 5\left(\frac{3}{5} + \frac{4}{5}i\right) \approx 5(\cos 53.1° + i \sin 53.1°)$$

Now,

$$\begin{aligned}(3 + 4i)^3 &\approx [5(\cos 53.1° + i \sin 53.1°]^3 \\ &= 5^3[\cos(3 \cdot 53.1°) + i \sin(3 \cdot 53.1°)] \\ &= 125(\cos 159.3° + i \sin 159.3°) \\ &\approx 125[-0.935 + i(0.353)] = -116.875 + 44.125i\end{aligned}$$

■

Complex Roots

Let w be a given complex number, and let $n \geq 2$ denote a positive integer. Any complex number z that satisfies the equation

$$z^n = w$$

is called a **complex nth root** of w. In keeping with previous usage, if $n = 2$, the solutions of the equation $z^2 = w$ are called **complex square roots** of w, and if $n = 3$, the solutions of the equation $z^3 = w$ are called **complex cube roots** of w.

Theorem
Finding Complex Roots

Let $w = r(\cos\theta + i\sin\theta)$ be a complex number. If $w \neq 0$, there are n distinct complex nth roots of w, given by the formula

$$z_k = \sqrt[n]{r}\left[\cos\left(\frac{\theta}{n} + \frac{2k\pi}{n}\right) + i\sin\left(\frac{\theta}{n} + \frac{2k\pi}{n}\right)\right] \tag{8}$$

where $k = 0, 1, 2, \ldots, n - 1$.

Proof (Outline)

We shall not prove this result in its entirety. Instead, we shall show only that each z_k in equation (8) obeys the equation $z_k^n = w$ and, hence, each z_k is a complex nth root of w.

$$\begin{aligned} z_k^n &= \left\{\sqrt[n]{r}\left[\cos\left(\frac{\theta}{n} + \frac{2k\pi}{n}\right) + i\sin\left(\frac{\theta}{n} + \frac{2k\pi}{n}\right)\right]\right\}^n \\ &= (\sqrt[n]{r})^n\left[\cos n\left(\frac{\theta}{n} + \frac{2k\pi}{n}\right) + \sin n\left(\frac{\theta}{n} + \frac{2k\pi}{n}\right)\right] \quad \text{DeMoivre's Theorem} \\ &= r[\cos(\theta + 2k\pi) + i\sin(\theta + 2k\pi)] \\ &= r(\cos\theta + i\sin\theta) = w \end{aligned}$$

Thus, each z_k, $k = 0, 1, \ldots, n - 1$, is a complex nth root of w. To complete the proof, we would need to show that each z_k, $k = 0, 1, 2, \ldots, n - 1$, is, in fact, distinct and that there are no complex nth roots of w other than those given by equation (8). ■

Example 7 Find the complex cube roots of $-1 + \sqrt{3}\,i$. Leave your answers in polar form, with θ in degrees.

Solution

First, we express $-1 + \sqrt{3}\,i$ in polar form using degrees:

$$-1 + \sqrt{3}\,i = 2\left(-\frac{1}{2} + \frac{\sqrt{3}}{2}i\right) = 2(\cos 120^\circ + i\sin 120^\circ)$$

The three complex cube roots of $-1 + \sqrt{3}\,i = 2(\cos 120^\circ + i\sin 120^\circ)$ are

$$z_k = \sqrt[3]{2}\left[\cos\left(\frac{120^\circ}{3} + \frac{360^\circ k}{3}\right) + i\sin\left(\frac{120^\circ}{3} + \frac{360^\circ k}{3}\right)\right] \qquad k = 0, 1, 2$$

Thus,

$$z_0 = \sqrt[3]{2}(\cos 40° + i \sin 40°)$$
$$z_1 = \sqrt[3]{2}(\cos 160° + i \sin 160°)$$
$$z_2 = \sqrt[3]{2}(\cos 280° + i \sin 280°)$$

Notice that each of the three complex cube roots of $-1 + \sqrt{3}\,i$ has the same magnitude—namely, $\sqrt[3]{2}$. This means that the points corresponding to each of the cube roots lie the same distance from the origin; and, hence, the three points lie on a circle with center at the origin and radius $\sqrt[3]{2}$. Furthermore, the arguments of these cube roots are 40°, 160°, and 280°, the difference of consecutive pairs being 120°. This means the three points are equally spaced on the circle, as shown in Figure 61. These results are not coincidental. In fact, you are asked to show that these results hold for complex nth roots in Problems 53–55.

Figure 61

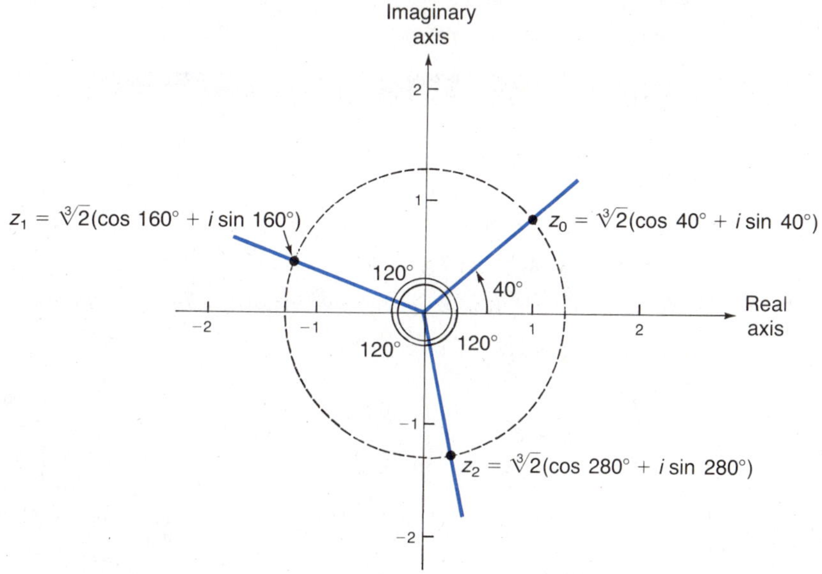

Historical Comment

■ The Babylonians, Greeks, and Arabs considered square roots of negative quantities to be impossible and equations with complex solutions to be unsolvable. The first hint that there was some connection between real solutions of equations and complex numbers came when Girolamo Cardano (1501–1576) and Tartaglia (1499–1557) found *real* roots of cubic equations by taking cube roots of *complex* quantities. For centuries thereafter, mathematicians worked with complex numbers without much belief in their actual existence. In 1673, John Wallis appears to have been the first to suggest the graphical representation of complex numbers, a truly significant idea that was not pursued further until about 1800. Several people, including Karl Friedrich Gauss (1777–1855), then rediscovered the idea, and the graphical representation helped to establish complex numbers as equal members of the number family. In practical applications, complex numbers have found their greatest uses in the area of alternating current, where they are a commonplace tool, and subatomic physics.

Historical Problems

■ **1.** In Problem 81, Exercise 5.5, we found $x = 2$ to be a solution of the cubic equation $x^3 - 6x + 4 = 0$. Recall that we could not apply the Tartaglia–Cardano method, because it led to the cube root of a complex number. Use DeMoivre's Theorem to find the cube root and finish the problem.

2. The quadratic formula will work perfectly well if the coefficients are complex numbers. Solve the following, using DeMoivre's Theorem where necessary. [*Hint:* The answers are "nice."]

(a) $z^2 - (2 + 5i)z - 3 + 5i = 0$ (b) $z^2 - (1 + i)z - 2 - i = 0$ ■

EXERCISE 10.6

In Problems 1–12, write each complex number in polar form. Express the argument in degrees.

1. $1 + i$ **2.** $-1 + i$ **3.** $\sqrt{3} - i$ **4.** $1 - \sqrt{3}i$

5. $-3i$ **6.** -2 **7.** $4 - 4i$ **8.** $9\sqrt{3} + 9i$

C **9.** $3 - 4i$ C **10.** $2 + \sqrt{3}i$ C **11.** $-2 + 3i$ C **12.** $\sqrt{5} - i$

In Problems 13–22, write each complex number in the standard form $a + bi$.

13. $2(\cos 120° + i \sin 120°)$ **14.** $3(\cos 210° + i \sin 210°)$

15. $4\left(\cos \frac{7\pi}{4} + i \sin \frac{7\pi}{4}\right)$ **16.** $2\left(\cos \frac{5\pi}{6} + i \sin \frac{5\pi}{6}\right)$

17. $3\left(\cos \frac{3\pi}{2} + i \sin \frac{3\pi}{2}\right)$ **18.** $4\left(\cos \frac{\pi}{2} + i \sin \frac{\pi}{2}\right)$

C **19.** $0.2(\cos 100° + i \sin 100°)$ C **20.** $0.4(\cos 200° + i \sin 200°)$

C **21.** $2\left(\cos \frac{\pi}{18} + i \sin \frac{\pi}{18}\right)$ C **22.** $3\left(\cos \frac{\pi}{10} + i \sin \frac{\pi}{10}\right)$

In Problems 23–30, find zw and z/w. Leave your answer in polar form.

23. $z = 2(\cos 40° + i \sin 40°)$
$w = 4(\cos 20° + i \sin 20°)$

24. $z = \cos 120° + i \sin 120°$
$w = \cos 100° + i \sin 100°$

25. $z = 3(\cos 130° + i \sin 130°)$
$w = 4(\cos 270° + i \sin 270°)$

26. $z = 2(\cos 80° + i \sin 80°)$
$w = 6(\cos 200° + i \sin 200°)$

27. $z = 2\left(\cos \frac{\pi}{8} + i \sin \frac{\pi}{8}\right)$
$w = 2\left(\cos \frac{\pi}{10} + i \sin \frac{\pi}{10}\right)$

28. $z = 4\left(\cos \frac{3\pi}{8} + i \sin \frac{3\pi}{8}\right)$
$w = 2\left(\cos \frac{9\pi}{16} + i \sin \frac{9\pi}{16}\right)$

29. $z = 2 + 2i$
$w = \sqrt{3} - i$

30. $z = 1 - i$
$w = 1 - \sqrt{3}\,i$

In Problems 31–42, write each expression in the standard form $a + bi$.

31. $[4(\cos 40° + i \sin 40°)]^3$

32. $[3(\cos 80° + i \sin 80°)]^3$

33. $\left[2\left(\cos \frac{\pi}{10} + i \sin \frac{\pi}{10}\right)\right]^5$

34. $\left[\sqrt{2}\left(\cos \frac{5\pi}{16} + i \sin \frac{5\pi}{16}\right)\right]^4$

35. $[\sqrt{3}(\cos 10° + i \sin 10°)]^6$

36. $\left[\frac{1}{2}(\cos 72° + i \sin 72°)\right]^5$

37. $\left[\sqrt{5}\left(\cos \frac{3\pi}{16} + i \sin \frac{3\pi}{16}\right)\right]^4$

38. $\left[\sqrt{3}\left(\cos \frac{5\pi}{18} + i \sin \frac{5\pi}{18}\right)\right]^6$

39. $(1 + i)^5$

40. $(\sqrt{3} - i)^6$

C **41.** $(\sqrt{2} - i)^6$

C **42.** $(1 - \sqrt{5}\,i)^8$

In Problems 43–50, find all the complex roots. Leave your answers in polar form with θ in degrees.

43. The complex cube roots of $1 + i$

44. The complex fourth roots of $\sqrt{3} - i$

45. The complex fourth roots of $4 - 4\sqrt{3}\,i$

46. The complex cube roots of $-8 - 8i$

47. The complex fourth roots of $-16i$

48. The complex cube roots of -8

49. The complex fifth roots of i

50. The complex fifth roots of $-i$

51. Find the four complex roots of unity (1). Plot each one.

52. Find the six complex roots of unity (1). Plot each one.

53. Show that each of the complex nth roots of a nonzero complex number w have the same magnitude.

54. Use the result of Problem 53 to draw the conclusion that each of the complex nth roots lies on a circle with center at the origin. What is the radius of this circle?

55. Refer to Problem 54. Show that the complex nth roots of a nonzero complex number w are equally spaced on the circle.

56. Prove formula (6).

CHAPTER REVIEW

THINGS TO KNOW

Formulas

Law of Sines

$$\frac{\sin \alpha}{a} = \frac{\sin \beta}{b} = \frac{\sin \gamma}{c}$$

Law of Cosines

$$c^2 = a^2 + b^2 - 2ab \cos \gamma$$
$$b^2 = a^2 + c^2 - 2ac \cos \beta$$
$$a^2 = b^2 + c^2 - 2bc \cos \alpha$$

Area of a triangle	$A = \frac{1}{2}bh$ $A = \frac{1}{2}ab \sin \gamma$ $A = \frac{1}{2}bc \sin \alpha$ $A = \frac{1}{2}ac \sin \beta$ $A = \sqrt{s(s-a)(s-b)(s-c)}$, where $s = \frac{1}{2}(a+b+c)$
Relationship between polar coordinates (r, θ) and rectangular coordinates (x, y)	$x = r \cos \theta$, $y = r \sin \theta$
DeMoivre's Theorem	If $z = r(\cos \theta + i \sin \theta)$, then $z^n = r^n(\cos n\theta + i \sin n\theta)$, where $n \geq 1$ is a positive integer
nth root of a complex number	$\sqrt[n]{z} = \sqrt[n]{r}\left[\cos\left(\frac{\theta}{n} + \frac{2k\pi}{n}\right) + i \sin\left(\frac{\theta}{n} + \frac{2k\pi}{n}\right)\right]$, $k = 0, \ldots, n-1$

How To:

Use the Law of Sines to solve a triangle
Use the Law of Cosines to solve a triangle
Find the area of a triangle
Plot polar coordinates
Convert from polar to rectangular coordinates
Convert from rectangular to polar coordinates
Graph polar equations (see Table 7)
Write a complex number in polar form, $z = r(\cos \theta + i \sin \theta)$, $0° \leq \theta < 360°$
Use DeMoivre's Theorem to find complex roots

FILL-IN-THE-BLANK ITEMS

1. If two sides and the angle opposite one of them are known, the Law of ________ is used to determine whether the known information results in no triangle, one triangle, or two triangles.
2. If three sides of a triangle are given, the Law of ________ is used to solve the triangle.
3. If three sides of a triangle are given, ________ Formula is used to find the area of the triangle.
4. In polar coordinates, the origin is called the ________, and the positive x-axis is referred to as the ________ ________.
5. Another representation in polar coordinates for the point $(2, \pi/3)$ is (________, $4\pi/3$).
6. Using polar coordinates (r, θ), the circle $x^2 + y^2 = 2x$ takes the form ________.
7. In a polar equation, replace θ by $-\theta$. If an equivalent equation results, the graph is symmetric with respect to ________ ________.
8. When a complex number z is written in the polar form $z = r(\cos \theta + i \sin \theta)$, the nonnegative number r is the ________ ________ of z, and the angle θ, $0 \leq \theta < 2\pi$, is the ________ of z.

REVIEW EXERCISES

C In Problems 1–20, find the remaining angle(s) and side(s) of each triangle, if it exists. If no triangle exists, say "No triangle."

1. $\alpha = 50°, \beta = 30°, a = 1$
2. $\alpha = 10°, \gamma = 40°, c = 2$
3. $\alpha = 100°, a = 5, c = 2$
4. $a = 2, c = 5, \alpha = 60°$
5. $a = 3, c = 1, \gamma = 110°$
6. $a = 3, c = 1, \gamma = 20°$
7. $a = 3, c = 1, \beta = 100°$
8. $a = 3, b = 5, \beta = 80°$
9. $a = 2, b = 3, c = 1$
10. $a = 10, b = 7, c = 8$
11. $a = 1, b = 3, \gamma = 40°$
12. $a = 4, b = 1, \gamma = 100°$
13. $a = 5, b = 3, \alpha = 80°$
14. $a = 2, b = 3, \alpha = 20°$
15. $a = 1, b = \frac{1}{2}, c = \frac{4}{3}$
16. $a = 3, b = 2, c = 2$
17. $a = 3, \alpha = 10°, b = 4$
18. $a = 4, \alpha = 20°, \beta = 100°$
19. $c = 5, b = 4, \alpha = 70°$
20. $a = 1, b = 2, \gamma = 60°$

C In Problems 21–30, find the area of each triangle.

21. $a = 2, b = 3, \gamma = 40°$
22. $b = 5, c = 4, \alpha = 20°$
23. $b = 4, c = 10, \alpha = 70°$
24. $a = 2, b = 1, \gamma = 100°$
25. $a = 4, b = 3, c = 5$
26. $a = 10, b = 7, c = 8$
27. $a = 4, b = 2, c = 5$
28. $a = 3, b = 2, c = 2$
29. $\alpha = 50°, \beta = 30°, a = 1$
30. $\alpha = 10°, \gamma = 40°, c = 3$

In Problems 31–36, plot each point given in polar coordinates, and find its rectangular coordinates.

31. $(3, \pi/6)$
32. $(4, 2\pi/3)$
33. $(-2, 4\pi/3)$
34. $(-1, 5\pi/4)$
35. $(-3, -\pi/2)$
36. $(-4, -\pi/4)$

In Problems 37–42, the rectangular coordinates of a point are given. Find two pairs of polar coordinates (r, θ) for each point, one with $r > 0$ and the other with $r < 0$. Express θ in radians.

37. $(-3, 3)$
38. $(1, -1)$
39. $(0, -2)$
40. $(2, 0)$
C 41. $(3, 4)$
C 42. $(-5, 12)$

In Problems 43–48, the letters x and y represent rectangular coordinates. Write each equation using polar coordinates (r, θ).

43. $3x^2 + 3y^2 = 6y$
44. $2x^2 - 2y^2 = 5y$
45. $2x^2 - y^2 = \dfrac{y}{x}$
46. $x^2 + 2y^2 = \dfrac{y}{x}$
47. $x(x^2 + y^2) = 4$
48. $y(x^2 - y^2) = 3$

In Problems 49–54, write each polar equation as an equation in rectangular coordinates (x, y).

49. $r = 2 \sin\theta$
50. $3r = \sin\theta$
51. $r = 5$
52. $\theta = \pi/4$
53. $r\cos\theta + 3r\sin\theta = 6$
54. $r^2 \tan\theta = 1$

In Problems 55–60, sketch the graph of each polar equation. Be sure to test for symmetry.

55. $r = 4\cos\theta$
56. $r = 3\sin\theta$
57. $r = 3 - 3\sin\theta$
58. $r = 2 + \cos\theta$
59. $r = 4 - \cos\theta$
60. $r = 1 - 2\sin\theta$

In Problems 61–64, write each complex number in polar form. Express each argument in degrees.

61. $-1 - i$
62. $-\sqrt{3} + i$
C 63. $4 - 3i$
C 64. $3 - 2i$

In Problems 65–70, write each complex number in the standard form $a + bi$.

65. $2(\cos 150° + i \sin 150°)$

66. $3(\cos 60° + i \sin 60°)$

67. $3\left(\cos \frac{2\pi}{3} + i \sin \frac{2\pi}{3}\right)$

68. $4\left(\cos \frac{3\pi}{4} + i \sin \frac{3\pi}{4}\right)$

C **69.** $0.1(\cos 350° + i \sin 350°)$

C **70.** $0.5(\cos 160° + i \sin 160°)$

In Problems 71–76, find zw and z/w. Leave your answers in polar form.

71. $z = \cos 80° + i \sin 80°$
$w = \cos 50° + i \sin 50°$

72. $z = \cos 205° + i \sin 205°$
$w = \cos 85° + i \sin 85°$

73. $z = 3\left(\cos \frac{9\pi}{5} + i \sin \frac{9\pi}{5}\right)$
$w = 2\left(\cos \frac{\pi}{5} + i \sin \frac{\pi}{5}\right)$

74. $z = 2\left(\cos \frac{5\pi}{3} + i \sin \frac{5\pi}{3}\right)$
$w = 3\left(\cos \frac{\pi}{3} + i \sin \frac{\pi}{3}\right)$

75. $z = 5(\cos 10° + i \sin 10°)$
$w = \cos 355° + i \sin 355°$

76. $z = 4(\cos 50° + i \sin 50°)$
$w = \cos 340° + i \sin 340°$

In Problems 77–84, write each expression in the standard form $a + bi$.

77. $[3(\cos 20° + i \sin 20°)]^3$

78. $[2(\cos 50° + i \sin 50°)]^3$

79. $\left[\sqrt{2}\left(\cos \frac{5\pi}{8} + i \sin \frac{5\pi}{8}\right)\right]^4$

80. $\left[2\left(\cos \frac{5\pi}{16} + i \sin \frac{5\pi}{16}\right)\right]^4$

81. $(1 - \sqrt{3}\,i)^6$

82. $(2 - 2i)^8$

C **83.** $(3 + 4i)^4$

C **84.** $(1 - 2i)^4$

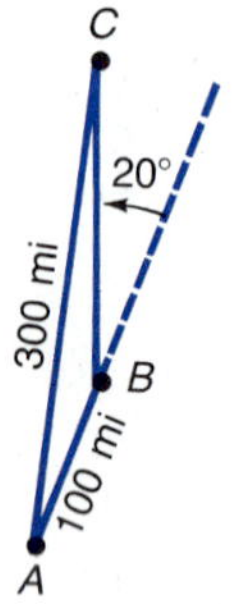

85. Find all the complex cube roots of 27.

86. Find all the complex fourth roots of -16.

C **87.** An airplane flies from city A to city B, a distance of 100 miles, then turns through an angle of 20° and heads toward city C, as indicated in the figure. If the distance from A to C is 300 miles, how far is it from city B to city C?

C **88.** Two cities A and B are 300 miles apart. In flying from city A to city B, a pilot inadvertently took a course that was 5° in error.

(a) If the error was discovered after flying 10 minutes at a constant speed of 420 miles per hour, through what angle should the pilot turn to correct the course? (Consult the figure.)

(b) What new constant speed should be maintained so that no time is lost due to the error? (Assume the speed would have been a constant 420 miles per hour if no error had occurred.)

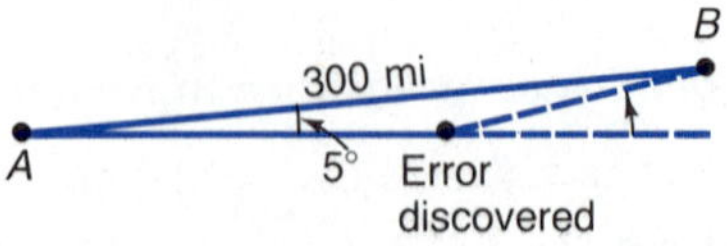

C **89.** The irregular parcel of land shown in the figure below is being sold for \$100 per square foot. What is the cost of this parcel?

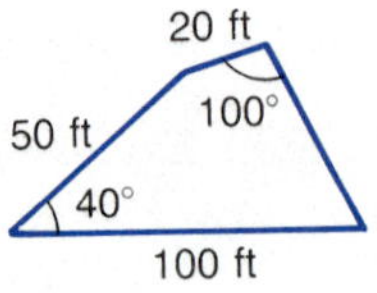

11 Analytic Geometry

Historically, Apollonius (200 BC) was among the first to study *conics* and discover some of their interesting properties. Today, conics are still studied because of their many uses. *Paraboloids of revolution* (parabolas rotated about their axes of symmetry) are used as signal collectors (the satellite dishes used with radar and cable TV, for example), as solar energy collectors, and as reflectors (telescopes, light projection, and so on). The planets circle the Sun in approximately *elliptical* orbits. Elliptical surfaces can be used to reflect signals such as light and sound from one place to another. And *hyperbolas* can be used to determine the positions of ships at sea.

The Greeks used the methods of Euclidean geometry to study conics. We shall use the more powerful methods of analytic geometry, bringing to bear both algebra and geometry, for our study of conics. Thus, we shall give a geometric description of each conic, and then, using rectangular coordinates and the distance formula, we shall find equations that represent conics. We used this same development, you may recall, when we first defined a circle in Section 3.4.

The chapter concludes with a section on equations of conics in polar coordinates, followed by a discussion of plane curves and parametric equations.

11.1 Conics

The word *conic* derives from the word *cone*, which is a geometric figure that can be constructed in the following way: Let a and g be two distinct lines that intersect at a point V. Keep the line a fixed. Now rotate the line g about a while maintaining the same angle between a and g. The collection of points swept out (generated) by the line g is called a (**right circular**) **cone**. See Figure 1. The fixed line a is called the **axis** of the cone; the point V is called its **vertex**; the lines that pass through V and make the same angle with a as g are called **generators** of the cone. Thus, each generator is a line that lies entirely on the cone. The cone consists of two parts, called **nappes**, that intersect at the vertex.

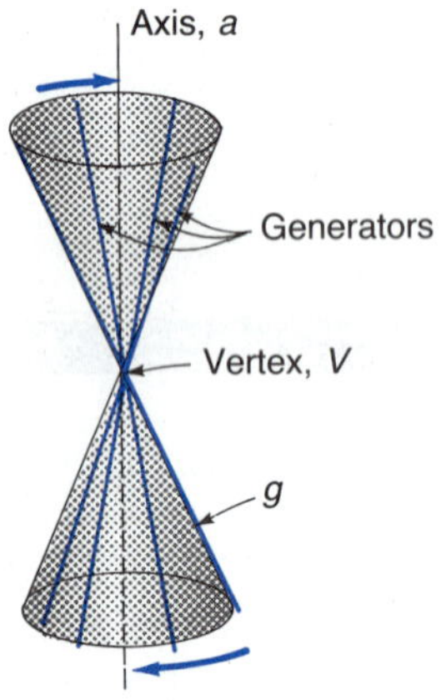

Figure 1

Conics, an abbreviation for **conic sections**, are curves that result from the intersection of a (right circular) cone and a plane. The conics we shall study arise when the plane does not contain the vertex, as shown in Figure 2. These conics are **circles** when the plane is perpendicular to the axis of the cone and intersects each generator; **ellipses** when the plane is tilted slightly so that it intersects each generator, but intersects only one nappe of the cone; **parabolas** when the plane is tilted further so that it is parallel to one (and only one) generator and intersects only one nappe of the cone; and **hyperbolas** when the plane intersects both nappes.

If the plane does contain the vertex, the intersection of the plane and the cone is a point, a line, or a pair of intersecting lines. These are usually called **degenerate conics**.

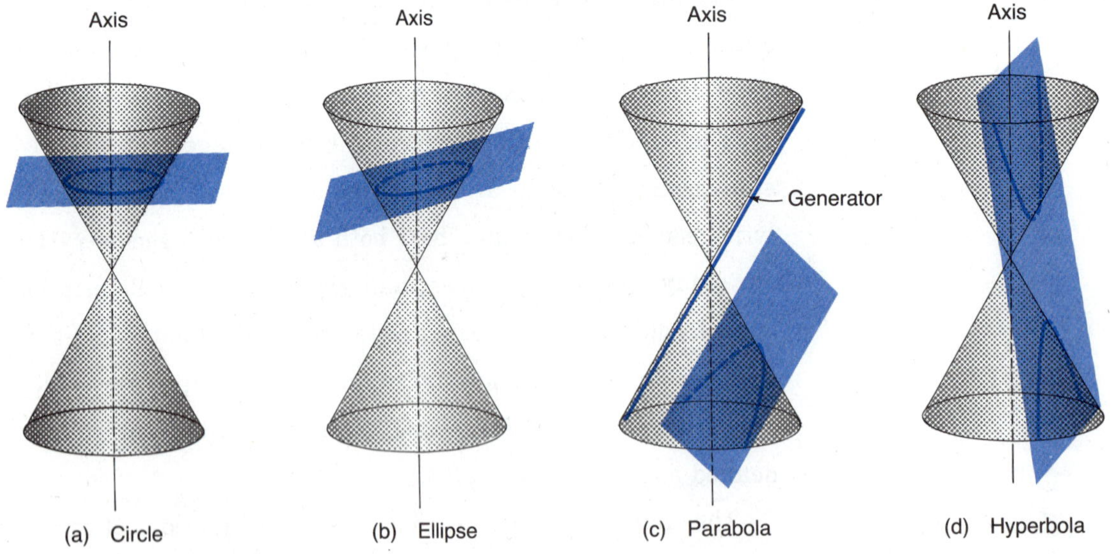

Figure 2

11.2 The Parabola

We stated earlier (Section 5.1) that the graph of a quadratic function is a parabola. In this section, we begin with a geometric definition of parabola.

Parabola

A **parabola** is defined as the collection of all points P in the plane that are the same distance from a fixed point F as they are from a fixed line D. The point F is called the **focus** of the parabola, and the line D is its **directrix**. As a result, a parabola is the set of points P for which

$$d(F, P) = d(P, D) \tag{1}$$

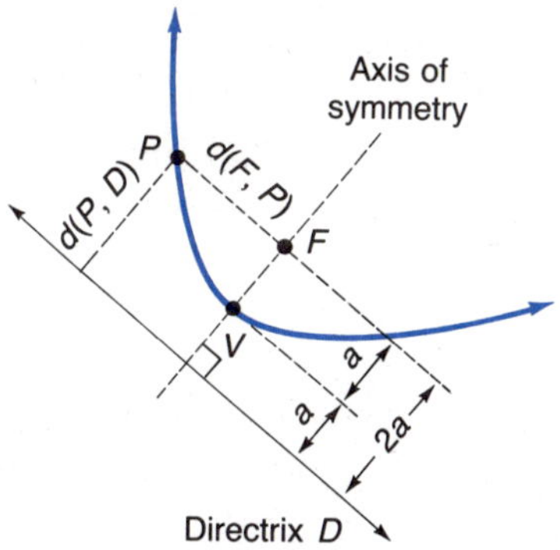

Figure 3

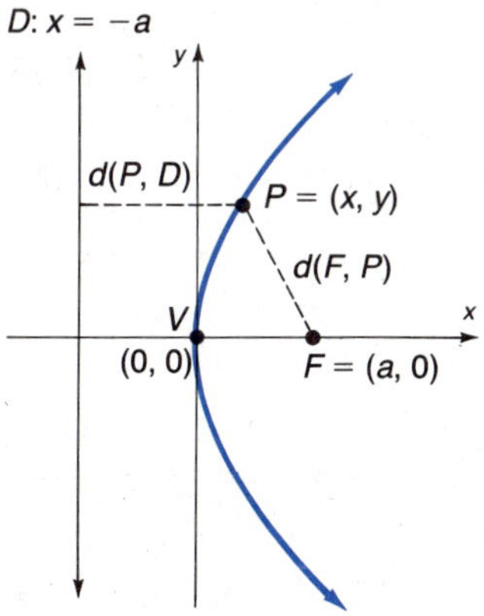

Figure 4
$y^2 = 4ax$

Figure 3 shows a parabola. The line through the focus F and perpendicular to the directrix D is called the **axis of symmetry** of the parabola. The point of intersection of the parabola with its axis of symmetry is called the **vertex** V.

Because the vertex V lies on the parabola, it must satisfy equation (1), namely, $d(F, V) = d(V, D)$. Thus, the vertex is midway between the focus and the directrix. We shall let a equal the distance $d(F, V)$ from F to V. Now we are ready to derive an equation for a parabola. To do this, we use a rectangular system of coordinates, positioned so that the vertex V, focus F, and directrix D of the parabola are conveniently located. If we choose to locate the vertex V at the origin $(0, 0)$, then we can conveniently position the focus F on either the x-axis or the y-axis.

First, we consider the case where the focus F is on the positive x-axis, as shown in Figure 4. Because the distance from F to V is a, the coordinates of F will be $(a, 0)$ with $a > 0$. Similarly, because the distance from V to the directrix D is also a and because D must be perpendicular to the x-axis (since the x-axis is the axis of symmetry), the equation of the directrix D must be $x = -a$. Now, if $P = (x, y)$ is any point on the parabola, then P must obey equation (1), namely,

$$d(F, P) = d(P, D)$$

So, we have

$$\sqrt{(x-a)^2 + y^2} = |x + a| \qquad \text{Use the distance formula.}$$
$$(x-a)^2 + y^2 = (x+a)^2 \qquad \text{Square both sides.}$$
$$x^2 - 2ax + a^2 + y^2 = x^2 + 2ax + a^2$$
$$y^2 = 4ax$$

Theorem The equation of a parabola with vertex at (0, 0), focus at $(a, 0)$, and directrix $x = -a$, $a > 0$, is

Equation of a Parabola; Vertex at (0, 0), Focus at $(a, 0)$, $a > 0$

$$y^2 = 4ax \qquad (2)$$

Example 1 Find an equation of the parabola with vertex at (0, 0) and focus at (3, 0). Graph the equation.

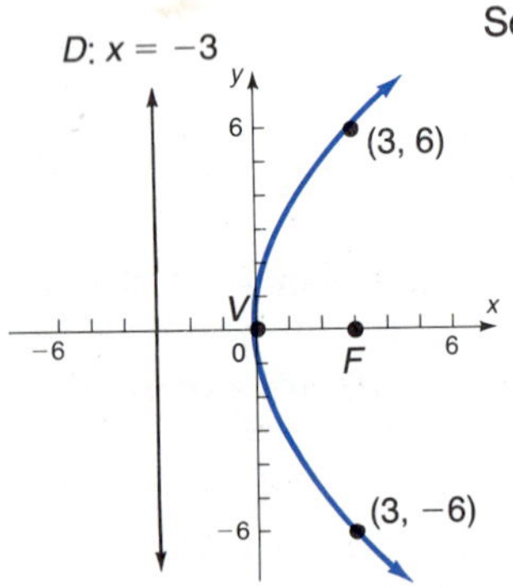

Figure 5
$y^2 = 12x$

Solution The distance from the vertex (0, 0) to the focus (3, 0) is $a = 3$. Based on equation (2), the equation of this parabola is

$$y^2 = 4ax$$
$$y^2 = 12x \quad a = 3$$

To graph this parabola, it is helpful to plot the two points on the graph above and below the focus. To locate them, we let $x = 3$. Then,

$$y^2 = 12x = 36$$
$$y = \pm 6$$

The points on the parabola above and below the focus are (3, −6) and (3, 6). See Figure 5. ■

By reversing the steps, it follows that the graph of an equation of the form of equation (2) is a parabola; its vertex is at (0, 0), its focus is at $(a, 0)$, its directrix is the line $x = -a$, and its axis of symmetry is the x-axis.

In general, the points on a parabola $y^2 = 4ax$ that lie above and below the focus $(a, 0)$ are each at a distance $2a$ from the focus. This follows from the fact that if $x = a$, then $y^2 = 4ax = 4a^2$, or $y = \pm 2a$. The line segment joining these two points is called the **latus rectum**; its length is $4a$.

For the remainder of this section, the direction "Discuss the equation" will mean to find the vertex, focus, and directrix of the parabola and graph it.

Example 2 Discuss the equation: $y^2 = 8x$

Solution The equation $y^2 = 8x$ is of the form $y^2 = 4ax$, where $4a = 8$. Thus, $a = 2$. Consequently, the graph of the equation is a parabola with vertex at (0, 0) and focus on the positive x-axis at (2, 0). The directrix is the vertical line $x = -2$. The two points defining the latus rectum are obtained by letting $x = 2$. Then $y^2 = 16$, or $y = \pm 4$. These points help in graphing

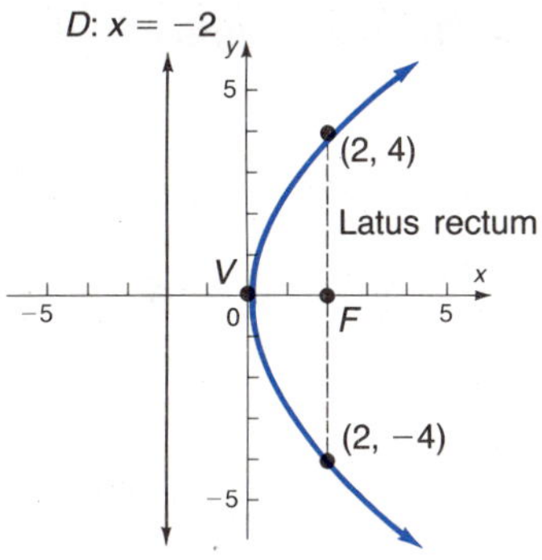

Figure 6
$y^2 = 8x$

the parabola, since they determine the "opening" of the graph. See Figure 6. ■

Recall that we arrived at equation (2) after placing the focus on the positive x-axis. If the focus is placed on the negative x-axis, positive y-axis, or negative y-axis, a different form of the equation for the parabola results. The four forms of the equation of a parabola with vertex at (0, 0) and focus on a coordinate axis a distance a from (0, 0) are given in Table 1 and their graphs are given in Figure 7. Notice that each graph is symmetric with respect to its axis of symmetry.

Table 1 Equations of a Parabola: Vertex at (0, 0); Focus on Axis; $a > 0$

VERTEX	FOCUS	DIRECTRIX	EQUATION	DESCRIPTION
(0, 0)	$(a, 0)$	$x = -a$	$y^2 = 4ax$	Parabola, axis of symmetry is the x-axis, opens to right
(0, 0)	$(-a, 0)$	$x = a$	$y^2 = -4ax$	Parabola, axis of symmetry is the x-axis, opens to left
(0, 0)	$(0, a)$	$y = -a$	$x^2 = 4ay$	Parabola, axis of symmetry is the y-axis, opens up
(0, 0)	$(0, -a)$	$y = a$	$x^2 = -4ay$	Parabola, axis of symmetry is the y-axis, opens down

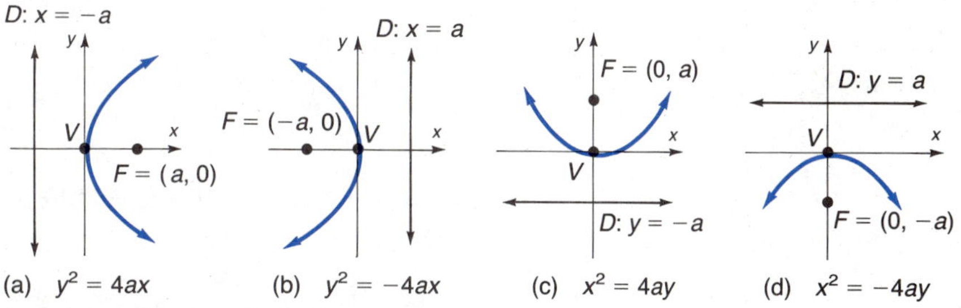

Figure 7

Example 3 Discuss the equation: $x^2 = -12y$

Solution The equation $x^2 = -12y$ is of the form $x^2 = -4ay$, with $a = 3$. Consequently, the graph of the equation is a parabola with vertex at (0, 0), focus at (0, −3), and directrix the line $y = 3$. The parabola opens down, and its axis of symmetry is the y-axis. To obtain the points defining the latus rectum, let $y = -3$. Then $x^2 = 36$, or $x = \pm 6$. See Figure 8. ■

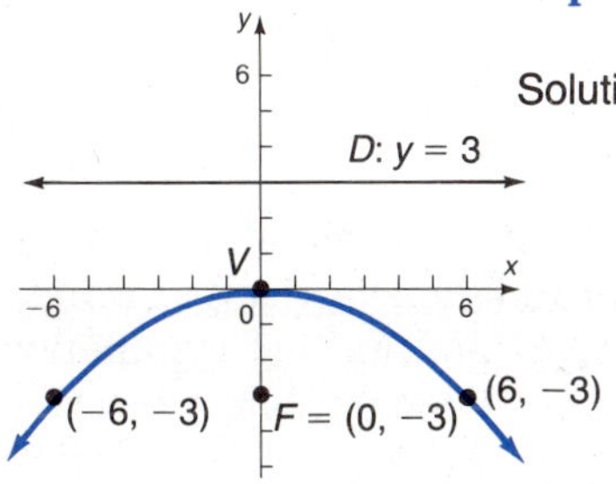

Figure 8
$x^2 = -12y$

Example 4 Find the equation of the parabola with focus at (0, 4) and directrix the line $y = -4$. Graph the equation.

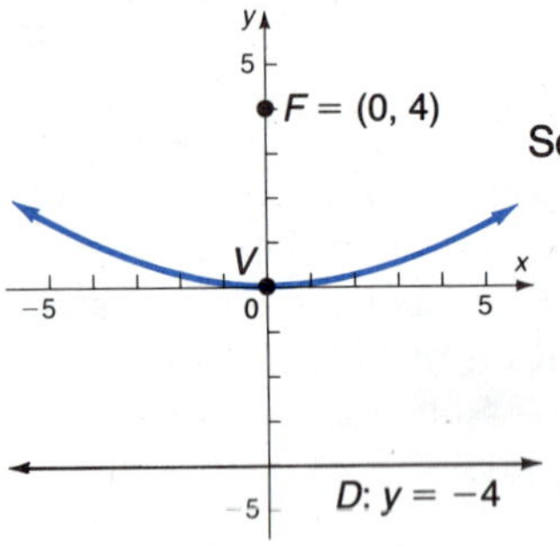

Figure 9
$x^2 = 16y$

Solution A parabola whose focus is at (0, 4) and whose directrix is the horizontal line $y = -4$ will have its vertex at (0, 0). (Do you see why? The vertex is midway between the focus and the directrix.) Thus, the equation is of the form $x^2 = 4ay$, with $a = 4$:

$$x^2 = 16y$$

Figure 9 shows the graph. ■

Example 5 Find the equation of a parabola with vertex at (0, 0) if its axis of symmetry is the x-axis and its graph contains the point $(-\frac{1}{2}, 2)$. Find its focus and directrix, and graph the equation.

Solution Because the vertex is at the origin and the axis of symmetry is the x-axis, we see from Table 1 that the form of the equation is

$$y^2 = kx$$

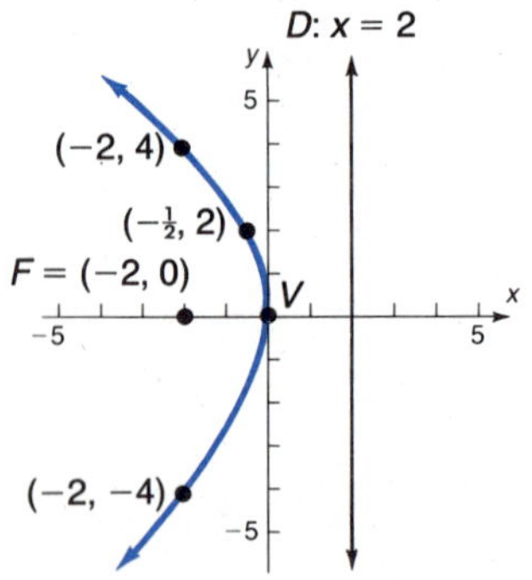

Figure 10
$y^2 = -8x$

Because the point $(-\frac{1}{2}, 2)$ is on the parabola, the coordinates $x = -\frac{1}{2}$, $y = 2$ must satisfy the equation. Putting $x = -\frac{1}{2}$ and $y = 2$ into the equation, we find

$$4 = k\left(-\tfrac{1}{2}\right)$$
$$k = -8$$

Thus, the equation of the parabola is

$$y^2 = -8x$$

Comparing this equation to $y^2 = -4ax$, we find that $a = 2$. The focus is therefore at $(-2, 0)$, and the directrix is the line $x = 2$. Letting $x = -2$, we find $y^2 = 16$ or $y = \pm 4$. The points $(-2, 4)$ and $(-2, -4)$ define the latus rectum. See Figure 10. ■

Vertex at (h, k)

If a parabola with vertex at the origin and axis of symmetry along a coordinate axis is shifted horizontally h units and then vertically k units, the result is a parabola with vertex at (h, k) and axis of symmetry parallel to a coordinate axis. The equations of such parabolas have the same forms as those in Table 1, but with x replaced by $x - h$ and y replaced by $y - k$. Table 2 gives the forms of the equations of such parabolas. Figure 11 illustrates the graphs for $h > 0$, $k > 0$.

Table 2 Parabolas with Vertex at (h, k), Axis of Symmetry Parallel to a Coordinate Axis, $a > 0$

VERTEX	FOCUS	DIRECTRIX	EQUATION	DESCRIPTION
(h, k)	$(h + a, k)$	$x = -a + h$	$(y - k)^2 = 4a(x - h)$	Parabola, axis of symmetry parallel to x-axis, opens to right
(h, k)	$(h - a, k)$	$x = a + h$	$(y - k)^2 = -4a(x - h)$	Parabola, axis of symmetry parallel to x-axis, opens to left
(h, k)	$(h, k + a)$	$y = -a + k$	$(x - h)^2 = 4a(y - k)$	Parabola, axis of symmetry parallel to y-axis, opens up
(h, k)	$(h, k - a)$	$y = a + k$	$(x - h)^2 = -4a(y - k)$	Parabola, axis of symmetry parallel to y-axis, opens down

Figure 11

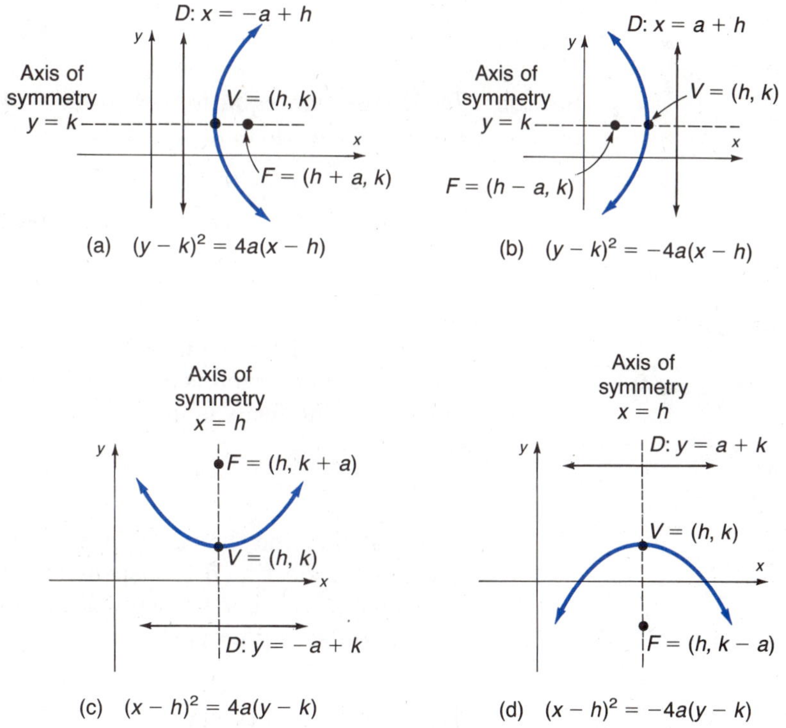

Example 6 Find an equation of the parabola with vertex at $(-2, 3)$ and focus at $(0, 3)$. Graph the equation.

Solution The vertex $(-2, 3)$ and focus $(0, 3)$ both lie on the horizontal line $y = 3$ (the axis of symmetry). The distance a from $(-2, 3)$ to $(0, 3)$ is $a = 2$. Also, because the focus lies to the right of the vertex, we know the parabola opens to the right. Consequently, the form of the equation is

$$(y - k)^2 = 4a(x - h)$$

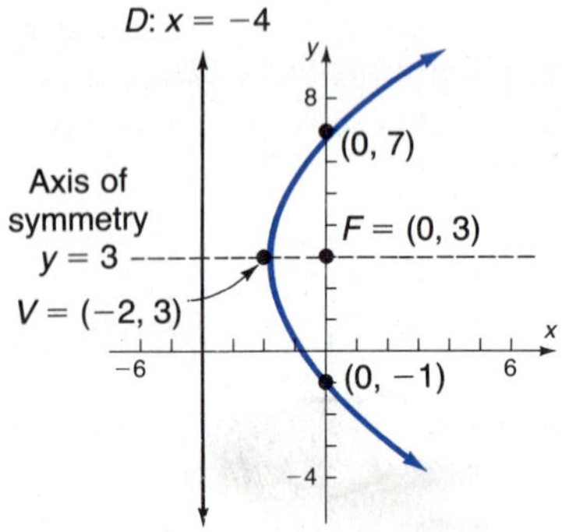

Figure 12
$(y-3)^2 = 8(x+2)$

where $(h, k) = (-2, 3)$ and $a = 2$. Therefore, the equation is

$$(y-3)^2 = 4 \cdot 2[x-(-2)]$$
$$(y-3)^2 = 8(x+2)$$

If $x = 0$, then $(y-3)^2 = 16$. Thus, $y - 3 = \pm 4$ and $y = -1$, $y = 7$. The points $(0, -1)$ and $(0, 7)$ define the latus rectum; the line $x = -4$ is the directrix. See Figure 12. ■

Polynomial equations involving two variables that are quadratic in one variable and linear in the other define parabolas. To discuss this type of equation, we first complete the square of the quadratic variable.

Example 7 Discuss the equation: $x^2 + 4x - 4y = 0$

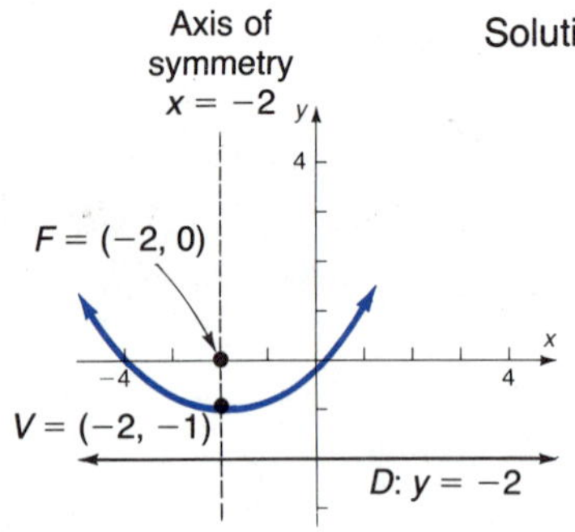

Figure 13
$x^2 + 4x - 4y = 0$

Solution To discuss the equation $x^2 + 4x - 4y = 0$, we complete the square involving the variable x. Thus,

$$x^2 + 4x - 4y = 0$$
$$x^2 + 4x = 4y \quad \text{Isolate the terms involving } x \text{ on the left side.}$$
$$x^2 + 4x + 4 = 4y + 4 \quad \text{Complete the square on the left side.}$$
$$(x+2)^2 = 4(y+1)$$

The equation is of the form $(x-h)^2 = 4a(y-k)$, with $h = -2$, $k = -1$, and $a = 1$. The graph is a parabola with vertex at $(h, k) = (-2, -1)$ that opens up. The focus is at $(-2, 0)$, and the directrix is the line $y = -2$. See Figure 13. ■

Parabolas find their way into many applications. For example, as we discussed in Section 5.1, suspension bridges have cables in the shape of a parabola. Another property of parabolas that is used in applications is their reflecting property.

Reflecting Property

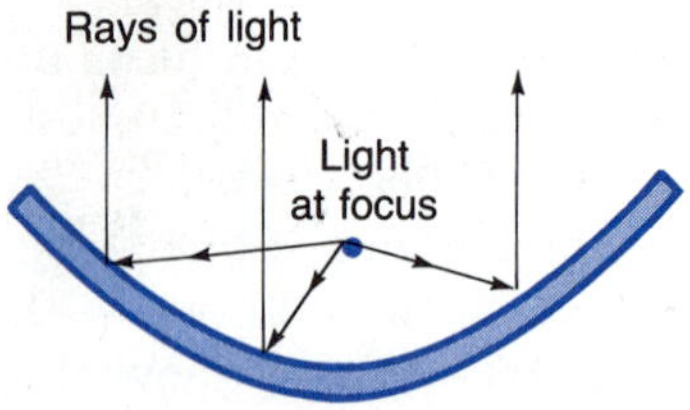

Figure 14
Parabolic mirror

Suppose a mirror is shaped like a **paraboloid of revolution**, a surface formed by rotating a parabola about its axis of symmetry. If a light (or any other emitting source) is placed at the focus of the parabola, all the rays emanating from the light will reflect off the mirror in lines parallel to the axis of symmetry (see Figure 14). This principle is used in the design of flashlights, certain automobile headlights, and other such devices.

Conversely, suppose rays of light (or other signals) emanate from a distant source, so that they are essentially parallel. When these rays strike the surface of a parabolic mirror whose axis of symmetry is parallel to these rays, they are reflected to a single point at the focus. This principle is used in the design of some solar energy devices and cable TV dishes that receive satellite signals.

EXERCISE 11.2

In Problems 1–16, find the equation of the parabola described. Find the two points that define the latus rectum, and graph the equation.

1. Focus at (2, 0); vertex at (0, 0)
2. Focus at (0, 1); vertex at (0, 0)
3. Focus at (0, −3); vertex at (0, 0)
4. Focus at (−4, 0); vertex at (0, 0)
5. Focus at (−2, 0); directrix the line $x = 2$
6. Focus at (0, −1); directrix the line $y = 1$
7. Directrix the line $y = -\frac{1}{2}$; vertex at (0, 0)
8. Directrix the line $x = -\frac{1}{2}$; vertex at (0, 0)
9. Vertex at (2, −3); focus at (2, −5)
10. Vertex at (4, −2); focus at (6, −2)
11. Vertex at (0, 0); axis of symmetry the y-axis; containing the point (2, 3)
12. Vertex at (0, 0); axis of symmetry the x-axis; containing the point (2, 3)
13. Focus at (−3, 4); directrix the line $y = 2$
14. Focus at (2, 4); directrix the line $x = -4$
15. Focus at (−3, −2); directrix the line $x = 1$
16. Focus at (−4, 4); directrix the line $y = -2$

In Problems 17–34, find the vertex, focus, and directrix of each parabola. Graph the equation.

17. $x^2 = 8y$
18. $y^2 = 4x$
19. $y^2 = -16x$
20. $x^2 = -4y$
21. $(y - 2)^2 = 8(x + 1)$
22. $(x + 4)^2 = 16(y + 2)$
23. $(x - 3)^2 = -(y + 1)$
24. $(y + 1)^2 = -4(x - 2)$
25. $(y + 3)^2 = 8(x - 2)$
26. $(x - 2)^2 = 4(y - 3)$
27. $y^2 - 4y + 4x + 4 = 0$
28. $x^2 + 6x - 4y + 1 = 0$
29. $x^2 + 8x = 4y - 8$
30. $y^2 - 2y = 8x - 1$
31. $y^2 + 2y - x = 0$
32. $x^2 - 4x = 2y$
33. $x^2 - 4x = y + 4$
34. $y^2 + 12y = -x + 1$

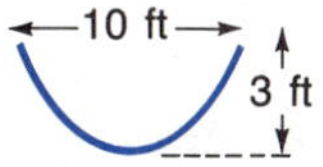

35. A cable TV receiving dish is in the shape of a paraboloid of revolution. Find the location of the receiver, which is placed at the focus, if the dish is 10 feet across at its opening and 3 feet deep. (See the figure.)

36. The reflector of a flashlight is in the shape of a paraboloid of revolution. Its diameter is 4 inches and its depth is 1 inch. How far from the vertex should the light bulb be placed so that the rays will be reflected parallel to the axis?

37. A sealed-beam headlight is in the shape of a paraboloid of revolution. The bulb, which is placed at the focus, is 1 inch from the vertex. If the depth is to be 2 inches, what is the diameter of the headlight at its opening?

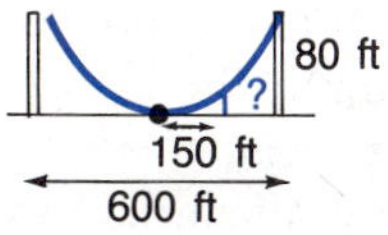

38. The cables of a suspension bridge are in the shape of a parabola, as shown in the figure. The towers supporting the cable are 600 feet apart and 80 feet high. If the cables touch the road surface midway between the towers, what is the height of the cable at a point 150 feet from a tower?

39. Show that an equation of the form

$$Ax^2 + Ey = 0 \qquad A \neq 0, E \neq 0$$

is the equation of a parabola with vertex at (0, 0) and axis of symmetry the y-axis. Find its focus and directrix.

40. Show that an equation of the form

$$Cy^2 + Dx = 0 \qquad C \neq 0, D \neq 0$$

is the equation of a parabola with vertex at (0, 0) and axis of symmetry the x-axis. Find its focus and directrix.

41. Show that the graph of an equation of the form

$$Ax^2 + Dx + Ey + F = 0 \qquad A \neq 0$$

(a) Is a parabola if $E \neq 0$.
(b) Is a vertical line if $E = 0$ and $D^2 - 4AF = 0$.
(c) Is two vertical lines if $E = 0$ and $D^2 - 4AF > 0$.
(d) Contains no points if $E = 0$ and $D^2 - 4AF < 0$.

42. Show that the graph of an equation of the form

$$Cy^2 + Dx + Ey + F = 0 \qquad C \neq 0$$

(a) Is a parabola if $D \neq 0$.
(b) Is a horizontal line if $D = 0$ and $E^2 - 4CF = 0$.
(c) Is two horizontal lines if $D = 0$ and $E^2 - 4CF > 0$.
(d) Contains no points if $D = 0$ and $E^2 - 4CF < 0$.

11.3 The Ellipse

Ellipse An **ellipse** is the collection of all points in the plane, the sum of whose distances from two fixed points, called the **foci**, is a constant.

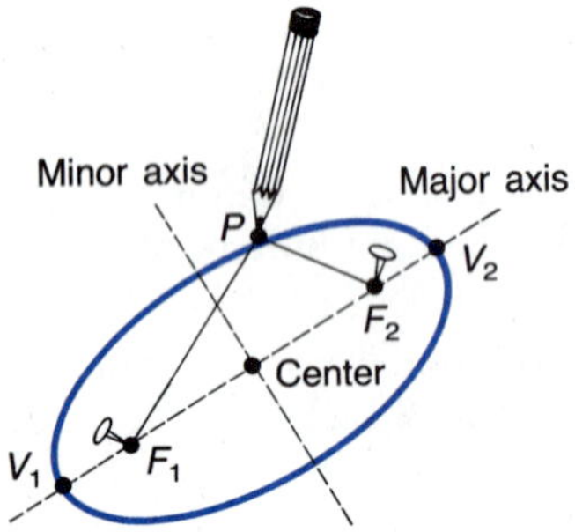

Figure 15

The definition actually contains within it a physical means for drawing an ellipse. Find a piece of string (the length of this string is the constant referred to in the definition). Then take two thumbtacks (the foci) and stick them on a piece of cardboard so that the distance between them is less than the length of the string. Now attach the ends of the string to the thumbtacks and, using the point of a pencil, pull the string taut. Keeping the string taut, rotate the pencil around the two thumbtacks. The pencil traces out an ellipse, as shown in Figure 15.

In Figure 15, the foci are labeled F_1 and F_2. The line containing the foci is called the **major axis**. The midpoint of the line segment joining the foci is called the **center** of the ellipse. The line through the center and perpendicular to the major axis is called the **minor axis**.

The two points of intersection of the ellipse and the major axis are the **vertices**, V_1 and V_2, of the ellipse. The distance from one vertex to the other is called the **length of the major axis**. The ellipse is symmetric with respect to its major axis and with respect to its minor axis.

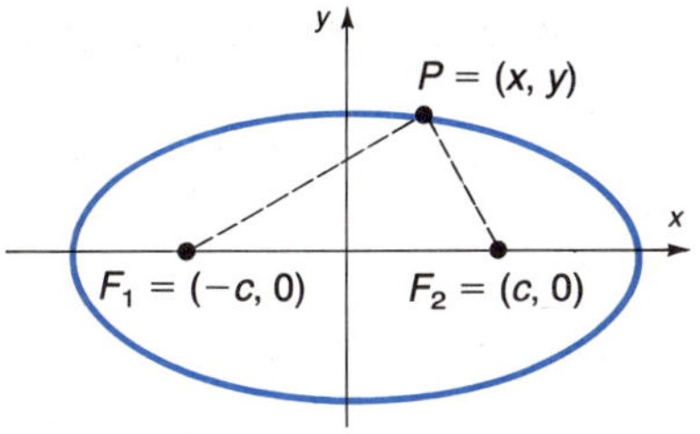

Figure 16
$d(F_1, P) + d(F_2, P) = 2a$

With these ideas in mind, we are now ready to find the equation of an ellipse in a rectangular coordinate system. First, we place the center of the ellipse at the origin. Second, we position the ellipse so that its major axis coincides with a coordinate axis. Suppose the major axis coincides with the x-axis, as shown in Figure 16. If c is the distance from the center to a focus, then one focus will be at $F_1 = (-c, 0)$ and the other at $F_2 = (c, 0)$. As we shall see, it is convenient to let $2a$ denote the constant distance referred to in the definition. Thus, if $P = (x, y)$ is any point on the ellipse, we have

$$d(F_1, P) + d(F_2, P) = 2a$$ Sum of the distances from P to the foci equals a constant

$$\sqrt{(x+c)^2 + y^2} + \sqrt{(x-c)^2 + y^2} = 2a$$ Use the distance formula.

$$\sqrt{(x+c)^2 + y^2} = 2a - \sqrt{(x-c)^2 + y^2}$$ Isolate one radical.

$$(x+c)^2 + y^2 = 4a^2 - 4a\sqrt{(x-c)^2 + y^2} + (x-c)^2 + y^2$$ Square both sides.

$$x^2 + 2cx + c^2 + y^2 = 4a^2 - 4a\sqrt{(x-c)^2 + y^2} + x^2 - 2cx + c^2 + y^2$$

$$4cx - 4a^2 = -4a\sqrt{(x-c)^2 + y^2}$$ Isolate the radical.

$$cx - a^2 = -a\sqrt{(x-c)^2 + y^2}$$ Divide each side by 4.

$$c^2x^2 - 2a^2cx + a^4 = a^2[(x-c)^2 + y^2]$$ Square both sides again.

$$c^2x^2 - 2a^2cx + a^4 = a^2(x^2 - 2cx + c^2 + y^2)$$

$$(c^2 - a^2)x^2 - a^2y^2 = a^2c^2 - a^4$$

$$(a^2 - c^2)x^2 + a^2y^2 = a^2(a^2 - c^2)$$ Multiply each side by -1; factor a^2 on the right side. (1)

To obtain points on the ellipse off the x-axis, it must be that $a > c$. To see why, look again at Figure 16:

$$d(F_1, P) + d(F_2, P) = 2a \quad \text{and} \quad d(F_1, F_2) = 2c$$

But

$$d(F_1, P) + d(F_2, P) > d(F_1, F_2)$$
$$2a > 2c$$
$$a > c$$

Since $a > c$, we also have $a^2 > c^2$, so $a^2 - c^2 > 0$. Let $b^2 = a^2 - c^2$, $b > 0$. Then $a > b$ and equation (1) can be written as

$$b^2x^2 + a^2y^2 = a^2b^2$$

$$\frac{x^2}{a^2} + \frac{y^2}{b^2} = 1$$ Divide each side by a^2b^2.

Theorem An equation of the ellipse with center at (0, 0) and foci at $(-c, 0)$ and $(c, 0)$ is

Equation of an Ellipse; Center at (0, 0); Foci at ($\pm c$, 0); Major Axis along the x-Axis

$$\frac{x^2}{a^2} + \frac{y^2}{b^2} = 1 \qquad \text{where } a > b \text{ and } b^2 = a^2 - c^2 \tag{2}$$

The major axis is the x-axis. ■

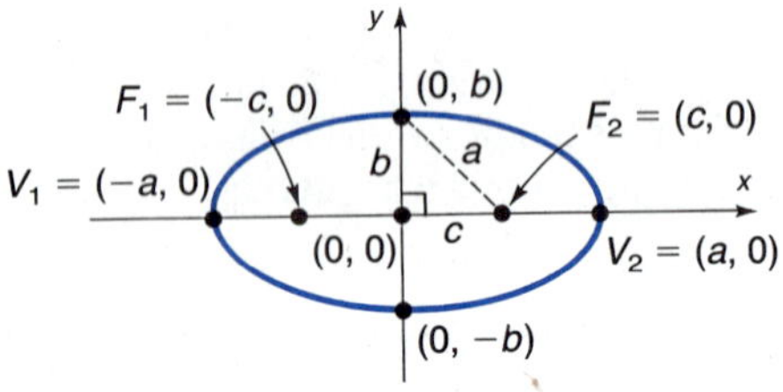

Figure 17

As you can verify, the ellipse defined by equation (2) is symmetric with respect to the x-axis, y-axis, and origin.

To find the vertices of the ellipse defined by equation (2), let $y = 0$. The vertices satisfy the equation $x^2/a^2 = 1$, the solutions of which are $x = \pm a$. Consequently, the vertices of the ellipse given by equation (2) are $V_1 = (-a, 0)$ and $V_2 = (a, 0)$. The y-intercepts of the ellipse, found by letting $x = 0$, have coordinates $(0, -b)$ and $(0, b)$. These four intercepts, $(a, 0)$, $(-a, 0)$, $(0, b)$, and $(0, -b)$, are used to graph the ellipse. See Figure 17.

Notice in Figure 17 the right triangle formed with vertices (0, 0), $(c, 0)$, and $(0, b)$. Because $b^2 = a^2 - c^2$ (or $b^2 + c^2 = a^2$), the distance from the focus at $(c, 0)$ to the point $(0, b)$ is a.

Example 1 Find an equation of the ellipse with center at the origin, one focus at (3, 0), and a vertex at (−4, 0). Graph the equation.

Solution The ellipse has its center at the origin, and the major axis coincides with the x-axis. One focus is at $(c, 0) = (3, 0)$, so $c = 3$. One vertex is at $(-a, 0) = (-4, 0)$, so $a = 4$. From equation (2), it follows that

$$b^2 = a^2 - c^2 = 16 - 9 = 7$$

so an equation of the ellipse is

$$\frac{x^2}{16} + \frac{y^2}{7} = 1$$

Figure 18 shows the graph. ■

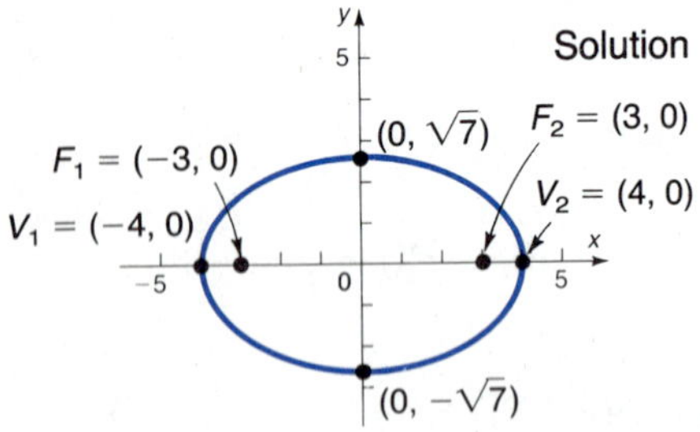

Figure 18 $\frac{x^2}{16} + \frac{y^2}{7} = 1$

An equation of the form of equation (2), with $a > b$, is the equation of an ellipse with center at the origin, foci on the x-axis at $(-c, 0)$ and $(c, 0)$, where $c^2 = a^2 - b^2$, and major axis along the x-axis.

For the remainder of this section, the direction "Discuss the equation" will mean to find the center, major axis, foci, and vertices of the ellipse and graph it.

Example 2 Discuss the equation:

$$\frac{x^2}{25} + \frac{y^2}{9} = 1$$

Solution The given equation is of the form of equation (2), with $a^2 = 25$ and $b^2 = 9$. The equation is that of an ellipse with center at (0, 0) and major axis along the x-axis. The vertices are at $(\pm a, 0) = (\pm 5, 0)$. Because $b^2 = a^2 - c^2$, we find

$$c^2 = a^2 - b^2 = 25 - 9 = 16$$

The foci are at $(\pm c, 0) = (\pm 4, 0)$. Figure 19 shows the graph. ■

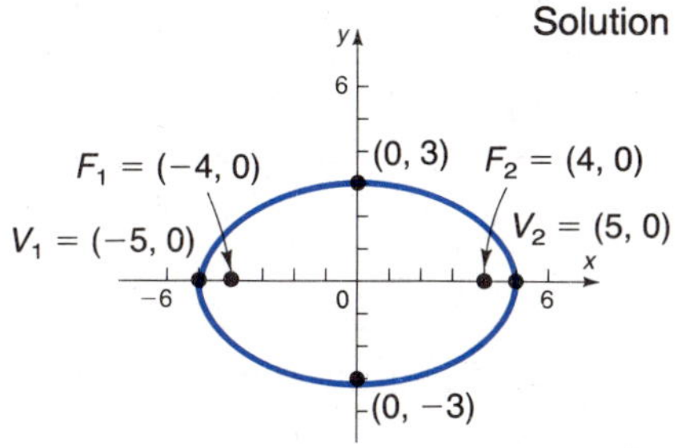

Figure 19 $\frac{x^2}{25} + \frac{y^2}{9} = 1$

Notice in Figures 18 and 19 how we used the intercepts of the equation to graph each ellipse. Following this practice will make it easier for you to obtain an accurate graph of an ellipse.

If the major axis of an ellipse with center at (0, 0) coincides with the y-axis, then the foci are at $(0, -c)$ and $(0, c)$. Using the same steps as before, the definition of an ellipse leads to the following result:

Theorem An equation of the ellipse with center at (0, 0) and foci at $(0, -c)$ and $(0, c)$ is

Equation of an Ellipse; Center at (0, 0); Foci at (0, ±c); Major Axis along the y-Axis

$$\frac{x^2}{b^2} + \frac{y^2}{a^2} = 1 \qquad \text{where } a > b \text{ and } b^2 = a^2 - c^2 \qquad (3)$$

The major axis is the y-axis; the vertices are at $(0, -a)$ and $(0, a)$. ■

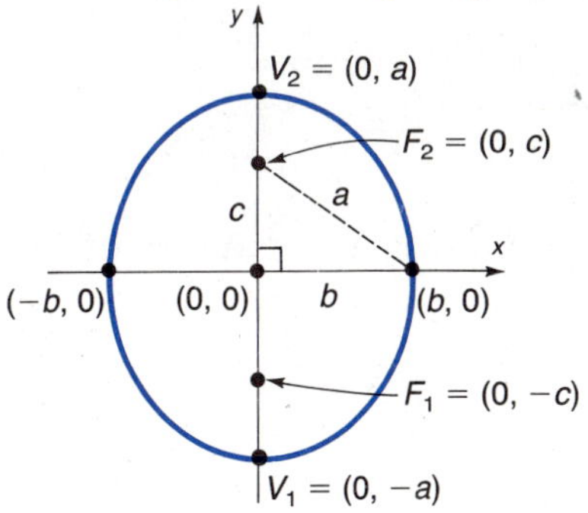

Figure 20

Figure 20 illustrates the graph of such an ellipse. Again, notice the right triangle with vertices at (0, 0), $(b, 0)$, and $(0, c)$.

Look closely at equations (2) and (3). Although they may look alike, there is a difference! In equation (2), the larger number, a^2, is in the denominator of the x^2-term, so that the major axis of the ellipse is along the x-axis. In equation (3), the larger number, a^2, is in the denominator of the y^2-term, so that the major axis is along the y-axis.

Example 3 Discuss the equation:

$$9x^2 + y^2 = 9$$

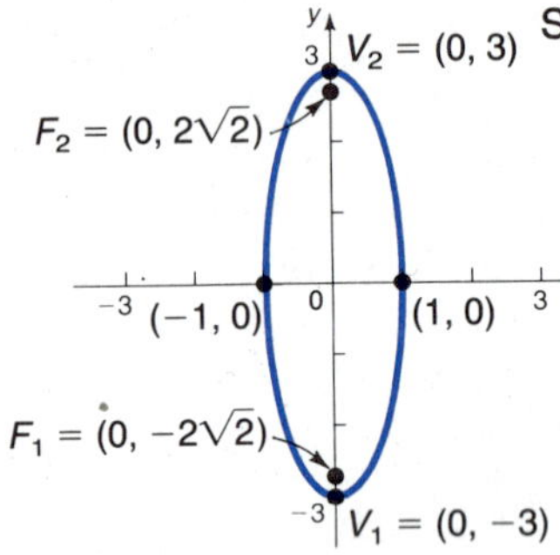

Figure 21
$x^2 + \frac{y^2}{9} = 1$

Solution To put the equation in proper form, we divide each side by 9:

$$x^2 + \frac{y^2}{9} = 1$$

The larger number, 9, is in the denominator of the y^2-term so, based on equation (3), this is the equation of an ellipse with center at the origin and major axis along the y-axis. Also, we conclude that $a^2 = 9$, $b^2 = 1$, and $c^2 = a^2 - b^2 = 9 - 1 = 8$. The vertices are at $(0, \pm a) = (0, \pm 3)$, and the foci are at $(0, \pm c) = (0, \pm 2\sqrt{2})$. The graph is given in Figure 21. ■

Example 4 Find an equation of the ellipse having one focus at (0, 2) and vertices at (0, −3) and (0, 3). Graph the equation.

Solution Because the vertices are at (0, −3) and (0, 3), the center of this ellipse is at the origin. Also, its major axis coincides with the y-axis. The given information also reveals that $c = 2$ and $a = 3$, so $b^2 = a^2 - c^2 = 9 - 4 = 5$. The form of the equation of this ellipse is given by equation (3):

$$\frac{x^2}{b^2} + \frac{y^2}{a^2} = 1$$

$$\frac{x^2}{5} + \frac{y^2}{9} = 1$$

Figure 22 shows the graph. ■

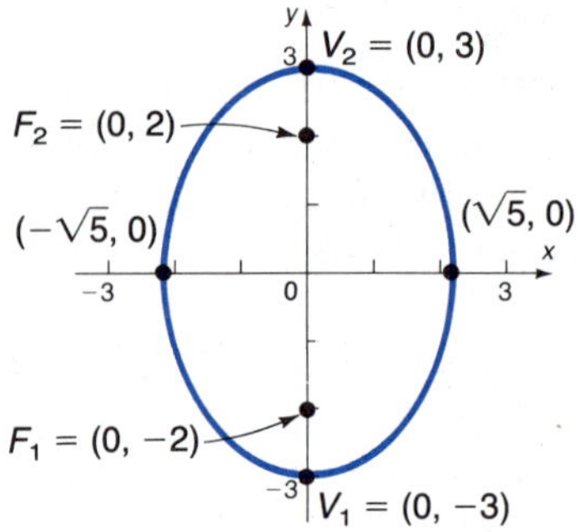

Figure 22
$\frac{x^2}{5} + \frac{y^2}{9} = 1$

The circle may be considered a special kind of ellipse. To see why, let $a = b$ in equation (2) or in equation (3). Then,

$$\frac{x^2}{a^2} + \frac{y^2}{a^2} = 1$$

$$x^2 + y^2 = a^2$$

This is the equation of a circle with center at the origin and radius a. The value of c is

$$c^2 = a^2 - b^2 = 0$$

We conclude that the closer the two foci of an ellipse are, the more the ellipse will look like a circle.

Center at (h, k)

If an ellipse with center at the origin and major axis coinciding with a coordinate axis is shifted horizontally h units and then vertically k units, the result is an ellipse with center at (h, k) and major axis parallel to a coordinate axis. Table 3 gives the forms of the equations of such ellipses, and Figure 23 shows their graphs.

Table 3 Ellipses with Center at (h, k) and Major Axis Parallel to a Coordinate Axis

CENTER	MAJOR AXIS	FOCI	VERTICES	EQUATION
(h, k)	Parallel to x-axis	$(h \pm c, k)$	$(h \pm a, k)$	$\frac{(x-h)^2}{a^2} + \frac{(y-k)^2}{b^2} = 1$, $a > b$ and $b^2 = a^2 - c^2$
(h, k)	Parallel to y-axis	$(h, k \pm c)$	$(h, k \pm a)$	$\frac{(x-h)^2}{b^2} + \frac{(y-k)^2}{a^2} = 1$, $a > b$ and $b^2 = a^2 - c^2$

Figure 23

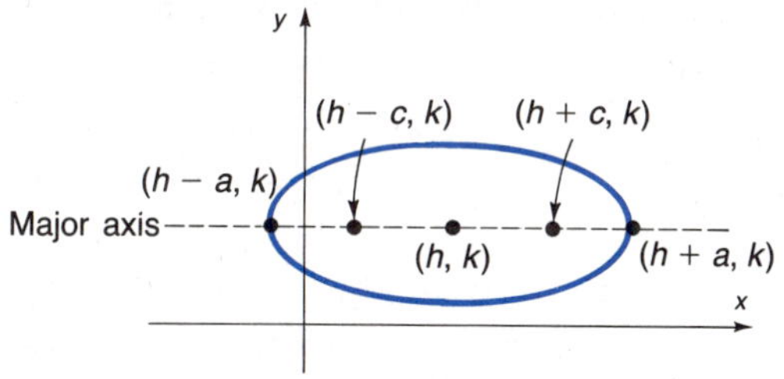

(a) $\frac{(x-h)^2}{a^2} + \frac{(y-k)^2}{b^2} = 1$

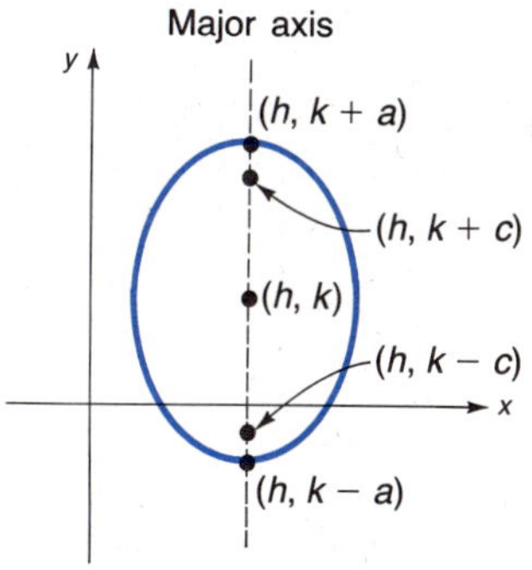

(b) $\frac{(x-h)^2}{b^2} + \frac{(y-k)^2}{a^2} = 1$

Example 5 Find an equation for the ellipse with center at $(2, -3)$, one focus at $(3, -3)$, and one vertex at $(5, -3)$. Graph the equation.

Solution The center is at $(h, k) = (2, -3)$, so $h = 2$ and $k = -3$. The major axis is parallel to the x-axis. The distance from the center $(2, -3)$ to a focus $(3, -3)$ is $c = 1$; the distance from the center $(2, -3)$ to a vertex $(5, -3)$ is $a = 3$. Thus, $b^2 = a^2 - c^2 = 9 - 1 = 8$. The form of the equation is

$$\frac{(x-h)^2}{a^2} + \frac{(y-k)^2}{b^2} = 1 \qquad \text{where } h = 2,\ k = -3,\ a = 3,\ b = 2\sqrt{2}$$

$$\frac{(x-2)^2}{9} + \frac{(y+3)^2}{8} = 1$$

Figure 24 shows the graph.

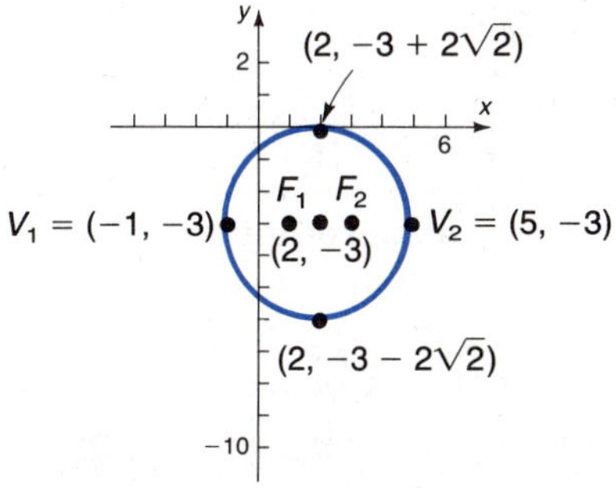

Figure 24 $\frac{(x-2)^2}{9} + \frac{(y+3)^2}{8} = 1$

■

Example 6 Discuss the equation: $4x^2 + y^2 - 8x + 4y + 4 = 0$

Solution We proceed to complete the square in x and in y:

$$4x^2 + y^2 - 8x + 4y + 4 = 0$$
$$4x^2 - 8x + y^2 + 4y = -4$$
$$4(x^2 - 2x) + (y^2 + 4y) = -4$$
$$4(x^2 - 2x + 1) + (y^2 + 4y + 4) = -4 + 4 + 4 \quad \text{Complete each square.}$$
$$4(x - 1)^2 + (y + 2)^2 = 4$$
$$(x - 1)^2 + \frac{(y + 2)^2}{4} = 1 \quad \text{Divide each side by 4.}$$

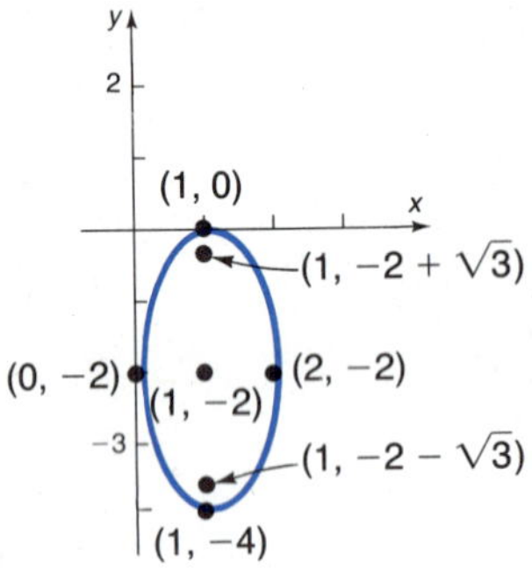

Figure 25
$(x - 1)^2 + \frac{(y + 2)^2}{4} = 1$

This is the equation of an ellipse with center at $(1, -2)$ and major axis parallel to the y-axis. Since $a^2 = 4$ and $b^2 = 1$, we have $c^2 = a^2 - b^2 = 4 - 1 = 3$. The vertices are at $(h, k \pm a) = (1, -2 \pm 2) = (1, 0)$ and $(1, -4)$. The foci are at $(h, k \pm c) = (1, -2 \pm \sqrt{3}) = (1, -2 - \sqrt{3})$ and $(1, -2 + \sqrt{3})$. Figure 25 shows the graph. ■

Applications

Ellipses are found in many applications in science and engineering. For example, the orbits of the planets around the Sun are elliptical, with the Sun's position at a focus. See Figure 26.

Figure 26

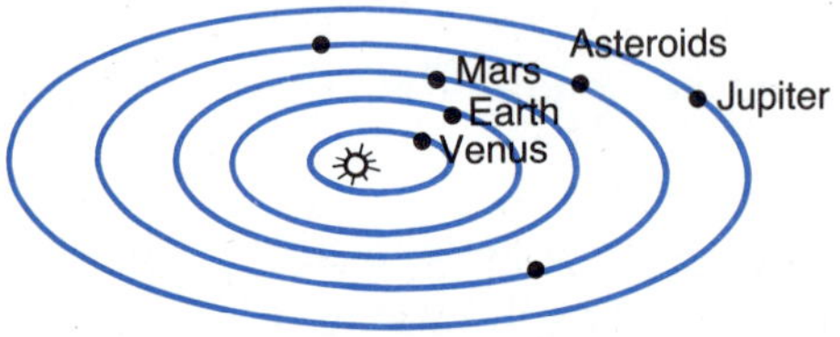

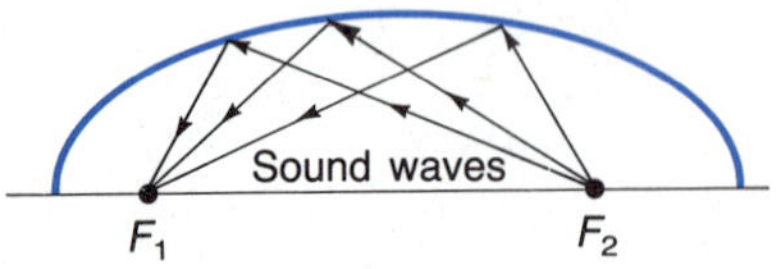

Figure 27
Whispering gallery

Ellipses also have an interesting reflection property. If a source of light (or sound) is placed at one focus, the waves transmitted by the source will reflect off the ellipse and concentrate at the other focus. This is the principle behind "whispering galleries," which are rooms designed with elliptical ceilings. A person standing at one focus of the ellipse can whisper and be heard by a person standing at the other focus, because all the sound waves that reach the ceiling are reflected to the other person. See Figure 27.

Stone and concrete bridges are often shaped as semielliptical arches. Elliptical gears are used in machinery when a variable rate of motion is required.

EXERCISE 11.3

In Problems 1–10, find the vertices and foci of each ellipse. Graph each equation.

1. $\frac{x^2}{9} + \frac{y^2}{4} = 1$ **2.** $\frac{x^2}{16} + \frac{y^2}{4} = 1$ **3.** $\frac{x^2}{9} + \frac{y^2}{25} = 1$

4. $\frac{x^2}{4} + \frac{y^2}{16} = 1$ **5.** $4x^2 + y^2 = 16$ **6.** $x^2 + 9y^2 = 18$

7. $4y^2 + x^2 = 8$ **8.** $4y^2 + 9x^2 = 36$ **9.** $x^2 + y^2 = 16$

10. $x^2 + y^2 = 4$

In Problems 11–20, find an equation for each ellipse. Graph the equation.

11. Center at (0, 0); focus at (3, 0); vertex at (6, 0)
12. Center at (0, 0); focus at (−2, 0); vertex at (3, 0)
13. Center at (0, 0); focus at (0, −4); vertex at (0, 5)
14. Center at (0, 0); focus at (0, 1); vertex at (0, −2)
15. Foci at (±2, 0); length of the major axis is 6
16. Focus at (0, −4); vertices at (0, ±8)
17. Foci at (0, ±3); x-intercepts are ±2
18. Foci at (0, ±2); length of the major axis is 8
19. Center at (0, 0); vertex at (0, 4); $b = 1$
20. Vertices at (±5, 0); $c = 2$

In Problems 21–32, find the center, foci, and vertices of each ellipse. Graph each equation.

21. $\frac{(x-2)^2}{4} + \frac{(y+1)^2}{9} = 1$ **22.** $\frac{(x+4)^2}{9} + \frac{(y+1)^2}{4} = 1$

23. $(x+5)^2 + 4(y-4)^2 = 16$ **24.** $9(x-3)^2 + (y+2)^2 = 18$

25. $x^2 + 4x + 4y^2 - 8y + 4 = 0$ **26.** $x^2 + 3y^2 - 12y + 9 = 0$

27. $2x^2 + 3y^2 - 8x + 6y + 5 = 0$ **28.** $4x^2 + 3y^2 + 8x - 6y = 5$

29. $9x^2 + 4y^2 - 18x + 16y - 11 = 0$ **30.** $x^2 + 9y^2 + 6x - 18y + 9 = 0$

31. $4x^2 + y^2 + 4y = 0$ **32.** $9x^2 + y^2 - 18x = 0$

In Problems 33–42, find an equation for each ellipse. Graph the equation.

33. Center at (2, −2); vertex at (5, −2); focus at (4, −2)
34. Center at (−3, 1); vertex at (−3, 4); focus at (−3, 0)
35. Vertices at (4, 3) and (4, 9); focus at (4, 8)
36. Foci at (1, 2) and (−3, 2); vertex at (−4, 2)
37. Foci at (5, 1) and (−1, 1); length of the major axis is 8
38. Vertices at (2, 5) and (2, −1); $c = 2$
39. Center at (1, 2); focus at (4, 2); contains the point (1, 3)
40. Center at (1, 2); focus at (1, 4); contains the point (2, 2)

41. Center at (1, 2); vertex at (4, 2); contains the point (1, 3)

42. Center at (1, 2); vertex at (1, 4); contains the point (2, 2)

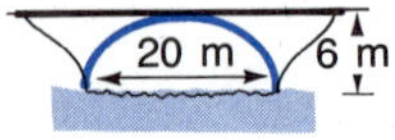

43. An arch in the shape of the upper half of an ellipse is used to support a bridge that is to span a river 20 meters wide. The center of the arch is 6 meters above the center of the river (see the figure). Write an equation for the ellipse in which the x-axis coincides with the water level and the y-axis passes through the center of the arch.

44. The arch of a bridge is a semiellipse with a horizontal major axis. The span is 30 feet, and the top of the arch is 10 feet above the major axis. The roadway is horizontal and is 2 feet above the top of the arch. Find the vertical distance from the roadway to the arch at 5 foot intervals along the roadway.

C **45.** An arch in the form of half an ellipse is 40 feet wide and 15 feet high at the center. Find the height of the arch at intervals of 10 feet along its width.

C **46.** An arch for a bridge over a highway is in the form of half an ellipse. The top of the arch is 20 feet above the ground level (the major axis). The highway has four lanes, each 12 feet wide; a center safety strip 8 feet wide; and two side strips, each 4 feet wide. What should the span of the bridge be (the length of its major axis) if the height 28 feet from the center is to be 13 feet?

47. The **eccentricity** e of an ellipse is defined as the number c/a. Because $a > c$, it follows that $e < 1$. Describe the general shape of an ellipse whose eccentricity is:
(a) Close to 0 (b) Equal to $\frac{1}{2}$ (c) Close to 1

C **48.** The orbit of the Earth is an ellipse with the Sun at one focus. If the length of the **semimajor axis** (half the length of the major axis) is approximately 92 million miles and the eccentricity is $\frac{1}{60}$, find the greatest and least distances of the Earth from the Sun.

49. Show that an equation of the form

$$Ax^2 + Cy^2 + F = 0 \qquad A \neq 0, C \neq 0, F \neq 0$$

where A and C are of the same sign and F is of opposite sign:
(a) Is the equation of an ellipse with center at (0, 0) if $A \neq C$.
(b) Is the equation of a circle with center at (0, 0) if $A = C$.

50. Show that the graph of an equation of the form

$$Ax^2 + Cy^2 + Dx + Ey + F = 0 \qquad A \neq 0, C \neq 0$$

where A and C are of the same sign:
(a) Is an ellipse if $(D^2/4A) + (E^2/4C) - F$ is the same sign as A.
(b) Is a point if $(D^2/4A) + (E^2/4C) - F = 0$.
(c) Contains no points if $(D^2/4A) + (E^2/4C) - F$ is of opposite sign to A.

11.4 The Hyperbola

Hyperbola A **hyperbola** is the collection of all points in the plane, the difference of whose distances from two fixed points, called the **foci**, is a constant.

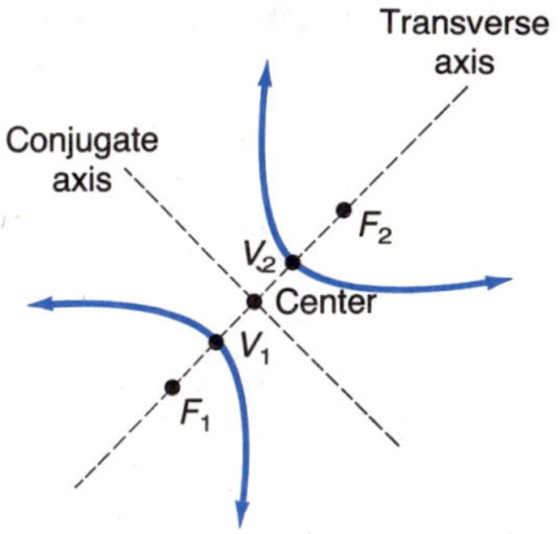

Figure 28

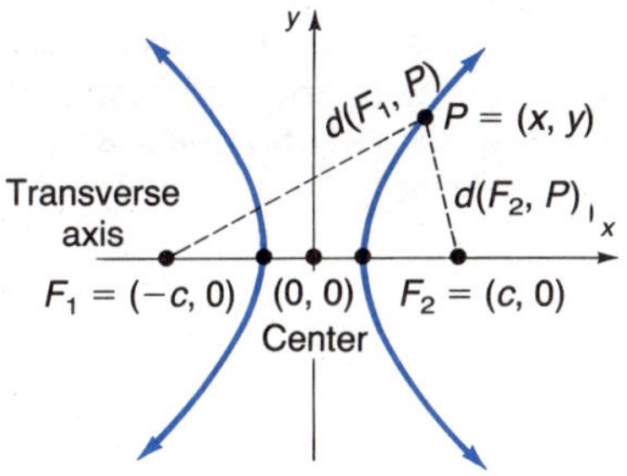

Figure 29

Figure 28 illustrates a hyperbola with foci F_1 and F_2. The line containing the foci is called the **transverse axis**. The midpoint of the line segment joining the foci is called the **center** of the hyperbola. The line through the center and perpendicular to the transverse axis is called the **conjugate axis**. The hyperbola consists of two separate curves, called **branches**, that are symmetric with respect to the transverse axis, conjugate axis, and center. The two points of intersection of the hyperbola and the transverse axis are the **vertices**, V_1 and V_2, of the hyperbola.

With these ideas in mind, we are now ready to find the equation of a hyperbola in a rectangular coordinate system. First, we place the center at the origin. Next, we position the hyperbola so that its transverse axis coincides with a coordinate axis. Suppose the transverse axis coincides with the x-axis, as shown in Figure 29. If c is the distance from the center to a focus, then one focus will be at $F_1 = (-c, 0)$ and the other at $F_2 = (c, 0)$. Now we let the constant difference of the distances from any point $P = (x, y)$ on the hyperbola to the foci F_1 and F_2 be denoted by $\pm 2a$. (If P is on the right branch, the + sign is used; if P is on the left branch, the − sign is used.) The coordinates of P must satisfy the equation

$$d(F_1, P) - d(F_2, P) = \pm 2a \qquad \text{Difference of the distances from } P \text{ to the foci equals } \pm 2a.$$

$$\sqrt{(x + c)^2 + y^2} - \sqrt{(x - c)^2 + y^2} = \pm 2a \qquad \text{Use the distance formula.}$$

$$\sqrt{(x + c)^2 + y^2} = \pm 2a + \sqrt{(x - c)^2 + y^2} \qquad \text{Isolate one radical.}$$

$$(x + c)^2 + y^2 = 4a^2 \pm 4a\sqrt{(x - c)^2 + y^2} + (x - c)^2 + y^2 \qquad \text{Square both sides.}$$

Next, we remove the parentheses:

$$x^2 + 2cx + c^2 + y^2 = 4a^2 \pm 4a\sqrt{(x - c)^2 + y^2} + x^2 - 2cx + c^2 + y^2$$

$$4cx - 4a^2 = \pm 4a\sqrt{(x - c)^2 + y^2} \qquad \text{Isolate the radical.}$$

$$cx - a^2 = \pm a\sqrt{(x - c)^2 + y^2} \qquad \text{Divide each side by 4.}$$

$$(cx - a^2)^2 = a^2[(x - c)^2 + y^2] \qquad \text{Square both sides.}$$

$$c^2x^2 - 2ca^2x + a^4 = a^2(x^2 - 2cx + c^2 + y^2)$$

$$c^2x^2 + a^4 = a^2x^2 + a^2c^2 + a^2y^2$$

$$(c^2 - a^2)x^2 - a^2y^2 = a^2c^2 - a^4$$

$$(c^2 - a^2)x^2 - a^2y^2 = a^2(c^2 - a^2) \qquad (1)$$

To obtain points on the hyperbola off the x-axis, it must be that $a < c$. To see why, look again at Figure 29:

$$d(F_1, P) < d(F_2, P) + d(F_1, F_2) \quad \text{Use triangle } F_1PF_2.$$

$$d(F_1, P) - d(F_2, P) < d(F_1, F_2) \quad P \text{ is on the right branch, so } d(F_1, P) - d(F_2, P) = 2a.$$

$$2a < 2c$$

$$a < c$$

Since $a < c$, we also have $a^2 < c^2$, so $c^2 - a^2 > 0$. Let $b^2 = c^2 - a^2$, $b > 0$. Then equation (1) can be written as

$$b^2x^2 - a^2y^2 = a^2b^2$$

$$\frac{x^2}{a^2} - \frac{y^2}{b^2} = 1$$

To find the vertices of the hyperbola defined by this equation, let $y = 0$. The vertices satisfy the equation $x^2/a^2 = 1$, the solutions of which are $x = \pm a$. Consequently, the vertices of the hyperbola are $V_1 = (-a, 0)$ and $V_2 = (a, 0)$.

Theorem An equation of the hyperbola with center at (0, 0), foci at $(-c, 0)$ and $(c, 0)$, and vertices at $(-a, 0)$ and $(a, 0)$ is

Equation of a Hyperbola; Center at (0, 0); Foci at $(\pm c, 0)$; Vertices at $(\pm a, 0)$; Transverse Axis along x-Axis

$$\frac{x^2}{a^2} - \frac{y^2}{b^2} = 1 \quad \text{where } b^2 = c^2 - a^2 \qquad (2)$$

The transverse axis is the x-axis. ■

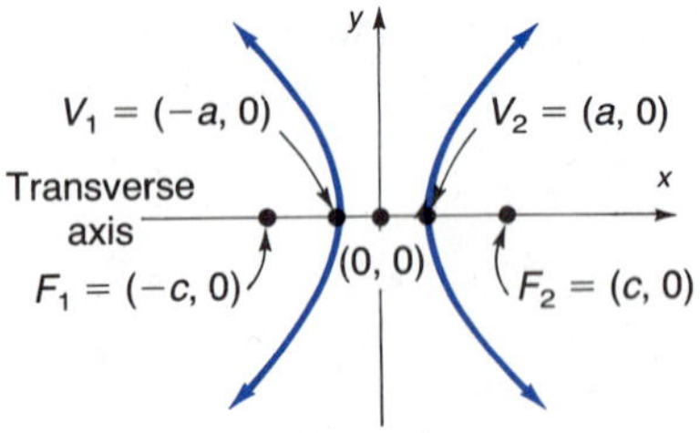

Figure 30
$\frac{x^2}{a^2} - \frac{y^2}{b^2} = 1,\ b^2 = c^2 - a^2$

As you can verify, the hyperbola defined by equation (2) is symmetric with respect to the x-axis, y-axis, and origin. To find the y-intercepts, if any, let $x = 0$ in equation (2). This results in the equation $y^2/b^2 = -1$, which has no solution. We conclude that the hyperbola defined by equation (2) has no y-intercepts. In fact, since $x^2/a^2 - 1 = y^2/b^2$, it follows that $x^2/a^2 \geq 1$. Thus, there are no points on the graph for $-a < x < a$. See Figure 30.

Example 1 Find an equation of the hyperbola with center at the origin, one focus at (3, 0), and one vertex at $(-2, 0)$. Graph the equation.

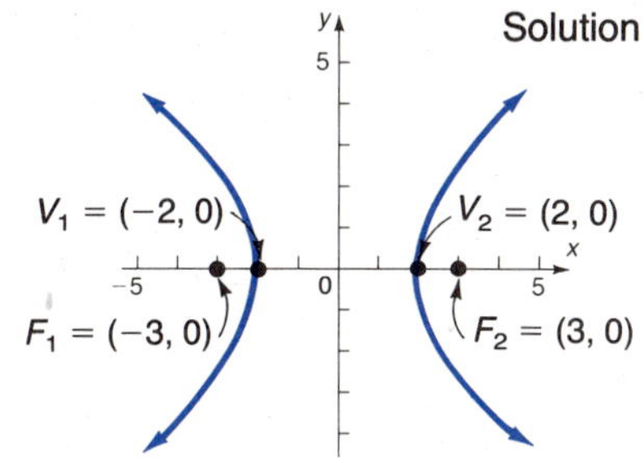

Figure 31
$\frac{x^2}{4} - \frac{y^2}{5} = 1$

Solution The hyperbola has its center at the origin, and the transverse axis coincides with the x-axis. One focus is at $(c, 0) = (3, 0)$, so $c = 3$. One vertex is at $(-a, 0) = (-2, 0)$, so $a = 2$. From equation (2), it follows that $b^2 = c^2 - a^2 = 9 - 4 = 5$, so an equation of the hyperbola is

$$\frac{x^2}{4} - \frac{y^2}{5} = 1$$

See Figure 31. ■

An equation of the form of equation (2) is the equation of a hyperbola with center at the origin, foci on the x-axis at $(-c, 0)$ and $(c, 0)$, where $c^2 = a^2 + b^2$, and transverse axis along the x-axis.

For the remainder of this section, the direction "Discuss the equation" will mean to find the center, transverse axis, vertices, and foci of the hyperbola and graph it.

Example 2 Discuss the equation: $\frac{x^2}{16} - \frac{y^2}{4} = 1$

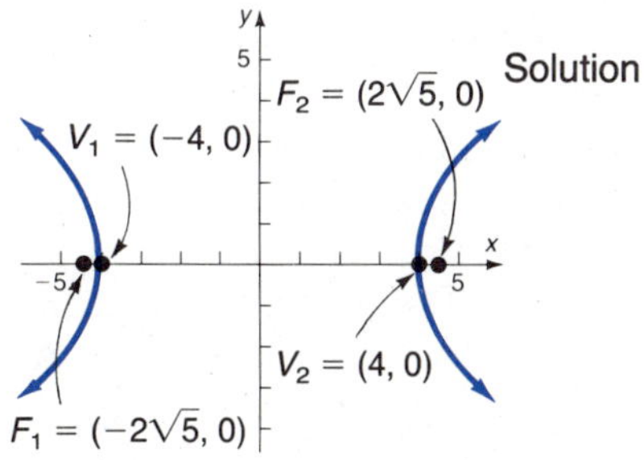

Figure 32
$\frac{x^2}{16} - \frac{y^2}{4} = 1$

Solution The given equation is of the form of equation (2), with $a^2 = 16$ and $b^2 = 4$. Thus, the graph of the equation is a hyperbola with center at $(0, 0)$ and transverse axis along the x-axis. Also, we know that $c^2 = a^2 + b^2 = 16 + 4 = 20$. The vertices are at $(\pm a, 0) = (\pm 4, 0)$, and the foci are at $(\pm c, 0) = (\pm 2\sqrt{5}, 0)$. Figure 32 shows the graph. ■

The next result gives the form of the equation of a hyperbola with center at the origin and transverse axis along the y-axis.

Theorem An equation of the hyperbola with center at $(0, 0)$, foci at $(0, -c)$ and $(0, c)$, and vertices at $(0, -a)$ and $(0, a)$ is

Equation of a Hyperbola; Center at $(0, 0)$; Foci at $(0, \pm c)$; Vertices at $(0, \pm a)$; Transverse Axis along y-Axis

$$\frac{y^2}{a^2} - \frac{x^2}{b^2} = 1 \qquad \text{where } b^2 = c^2 - a^2 \tag{3}$$

The transverse axis is the y-axis. ■

Figure 33 (at the top of the next page) shows the graph of a typical hyperbola defined by equation (3).

Notice the difference in the form of equations (2) and (3). When the y^2-term is subtracted from the x^2-term, the transverse axis is the x-axis.

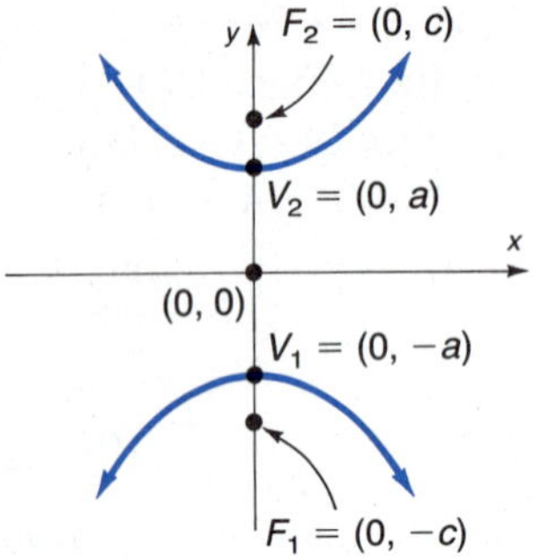

Figure 33
$\dfrac{y^2}{a^2} - \dfrac{x^2}{b^2} = 1$, $b^2 = c^2 - a^2$

When the x^2-term is subtracted from the y^2-term, the transverse axis is the y-axis.

Example 3 Discuss the equation: $y^2 - 4x^2 = 4$

Solution To put the equation in proper form, we divide each side by 4:

$$\frac{y^2}{4} - x^2 = 1$$

Since the x^2-term is subtracted from the y^2-term, the equation is that of a hyperbola with center at the origin and transverse axis along the y-axis. Also, comparing the above equation to equation (3), we find $a^2 = 4$, $b^2 = 1$, and $c^2 = a^2 + b^2 = 5$. The vertices are at $(0, \pm a) = (0, \pm 2)$, and the foci are at $(0, \pm c) = (0, \pm\sqrt{5})$. The graph is given in Figure 34. ■

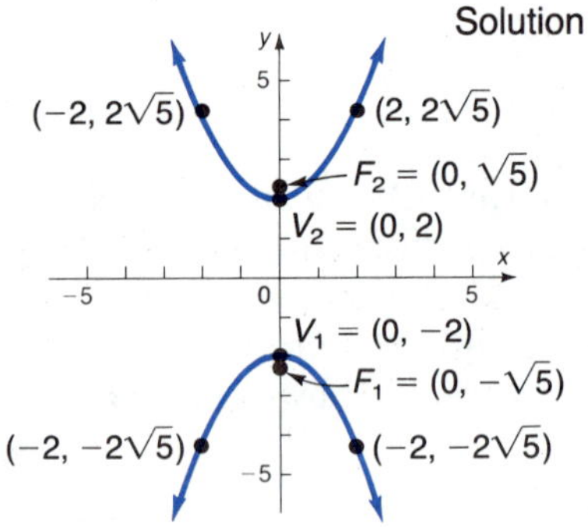

Figure 34
$\dfrac{y^2}{4} - x^2 = 1$

Example 4 Find an equation of the hyperbola having one vertex at (0, 2) and foci at (0, −3) and (0, 3). Graph the equation.

Solution Since the foci are at (0, −3) and (0, 3), the center of the hyperbola is at the origin. Also, the transverse axis is along the y-axis. The given information also reveals that $c = 3$, $a = 2$, and $b^2 = c^2 - a^2 = 9 - 4 = 5$. The form of the equation of the hyperbola is given by equation (3):

$$\frac{y^2}{a^2} - \frac{x^2}{b^2} = 1$$

$$\frac{y^2}{4} - \frac{x^2}{5} = 1$$

See Figure 35. ■

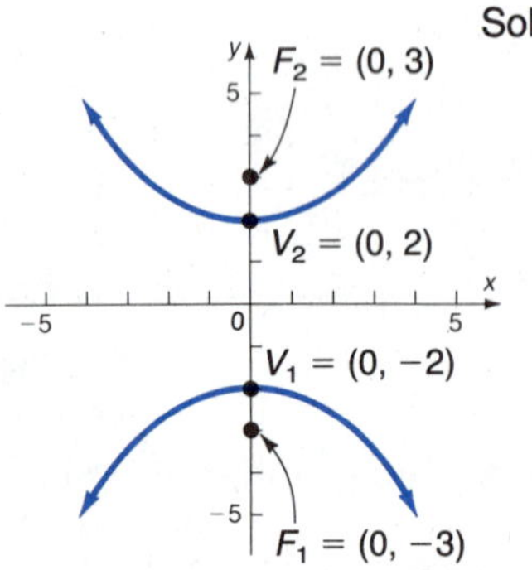

Figure 35
$\dfrac{y^2}{4} - \dfrac{x^2}{5} = 1$

Look at the equations of the hyperbolas in Examples 3 and 4. For the hyperbola in Example 3, $a^2 = 4$ and $b^2 = 1$, so $a > b$; for the hyperbola in Example 4, $a^2 = 4$ and $b^2 = 5$, so $a < b$. We conclude that, for hyperbolas, there are no requirements involving the relative size of a and b. Contrast this situation to the case of an ellipse, in which the relative sizes of a and b dictate which axis is the major axis. Hyperbolas have another feature to distinguish them from ellipses and parabolas: Hyperbolas have asymptotes.

Asymptotes

Recall from Section 5.3 that a horizontal or oblique asymptote of a graph is a line with the property that the distance from the line to points on the graph approaches 0 as $x \to -\infty$ or as $x \to \infty$.

Theorem

Asymptotes of a Hyperbola

The hyperbola $\frac{x^2}{a^2} - \frac{y^2}{b^2} = 1$ has the two oblique asymptotes

$$y = \frac{b}{a}x \quad \text{and} \quad y = -\frac{b}{a}x$$

Proof We begin by solving for y in the equation of the hyperbola:

$$\frac{x^2}{a^2} - \frac{y^2}{b^2} = 1$$

$$\frac{y^2}{b^2} = \frac{x^2}{a^2} - 1$$

$$y^2 = b^2\left(\frac{x^2}{a^2} - 1\right)$$

If $x \neq 0$, we can rearrange the right side in the form

$$y^2 = \frac{b^2x^2}{a^2}\left(1 - \frac{a^2}{x^2}\right)$$

$$y = \pm\frac{bx}{a}\sqrt{1 - \frac{a^2}{x^2}}$$

Now, as $x \to -\infty$ or as $x \to \infty$, the term a^2/x^2 approaches 0, so the expression under the radical approaches 1. Thus, as $x \to -\infty$ or as $x \to \infty$, the value of y approaches $\pm bx/a$; that is, the graph of the hyperbola approaches the lines

$$y = -\frac{b}{a}x \quad \text{and} \quad y = \frac{b}{a}x$$

Thus, these lines are oblique asymptotes of the hyperbola. ■

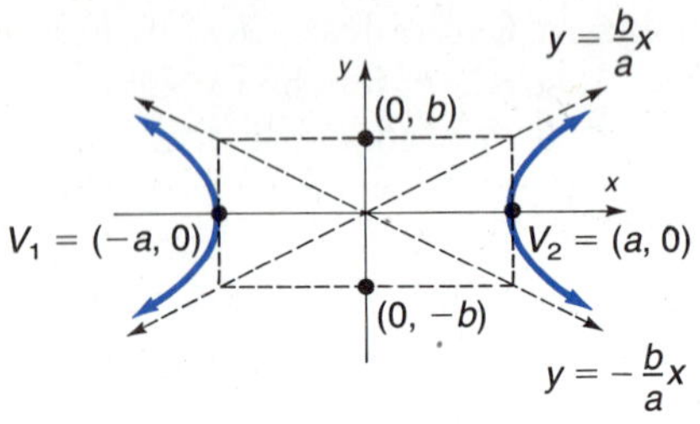

Figure 36
$\frac{x^2}{a^2} - \frac{y^2}{b^2} = 1$

The asymptotes of a hyperbola are not part of the hyperbola, but they do serve as a guide for graphing a hyperbola. For example, suppose we want to graph the equation

$$\frac{x^2}{a^2} - \frac{y^2}{b^2} = 1$$

We begin by plotting the vertices $(-a, 0)$ and $(a, 0)$. Then we plot the points $(0, -b)$ and $(0, b)$, and use these four points to construct a rectangle, as shown in Figure 36. The diagonals of this rectangle have slopes b/a and $-b/a$, and their extensions are the asymptotes $y = (b/a)x$ and $y = -(b/a)x$ of the hyperbola.

Theorem

Asymptotes of a Hypberbola

The hyperbola

$$\frac{y^2}{a^2} - \frac{x^2}{b^2} = 1$$

has the two oblique asymptotes

$$y = \frac{a}{b}x \qquad \text{and} \qquad y = -\frac{a}{b}x$$

■

You are asked to prove this result in Problem 44 at the end of this section.

Example 5 Discuss the equation: $9x^2 - 4y^2 = 36$

Solution First, we divide each side by 36 to put the equation in proper form:

$$\frac{x^2}{4} - \frac{y^2}{9} = 1$$

This is the equation of a hyperbola with center at the origin and transverse axis along the x-axis. Using $a^2 = 4$ and $b^2 = 9$, we find $c^2 = a^2 + b^2 = 13$. The vertices are at $(\pm a, 0) = (\pm 2, 0)$; the foci are at $(\pm c, 0) = (\pm\sqrt{13}, 0)$; and the asymptotes have the equations

$$y = \tfrac{3}{2}x \qquad \text{and} \qquad y = -\tfrac{3}{2}x$$

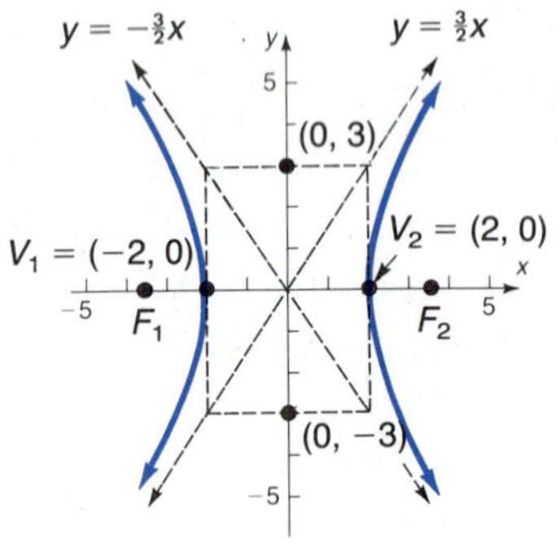

Figure 37
$\frac{x^2}{4} - \frac{y^2}{9} = 1$

Now, form the rectangle containing the points $(\pm a, 0)$ and $(0, \pm b)$, namely $(-2, 0)$, $(2, 0)$, $(0, -3)$, and $(0, 3)$. The extension of the diagonals of this rectangle are the asymptotes. See Figure 37 for the graph. ■

Center at (h, k)

If a hyperbola with center at the origin and transverse axis coinciding with a coordinate axis is shifted horizontally h units and then vertically k units, the result is a hyperbola with center at (h, k) and transverse axis parallel to a coordinate axis. Table 4 gives the forms of the equations of such hyperbolas. See Figure 38 for the graphs.

Table 4 Hyperbolas with Center at (h, k) and Transverse Axis Parallel to a Coordinate Axis

CENTER	TRANSVERSE AXIS	FOCI	VERTICES	EQUATION	ASYMPTOTES
(h, k)	Parallel to x-axis	$(h \pm c, k)$	$(h \pm a, k)$	$\frac{(x-h)^2}{a^2} - \frac{(y-k)^2}{b^2} = 1$, $b^2 = c^2 - a^2$	$y - k = \pm\frac{b}{a}(x - h)$
(h, k)	Parallel to y-axis	$(h, k \pm c)$	$(h, k \pm a)$	$\frac{(y-k)^2}{a^2} - \frac{(x-h)^2}{b^2} = 1$, $b^2 = c^2 - a^2$	$y - k = \pm\frac{a}{b}(x - h)$

Figure 38

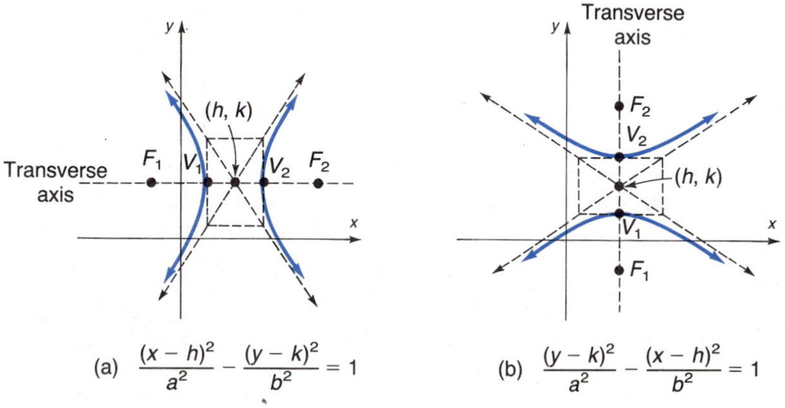

Example 6 Find an equation for the hyperbola with center at $(1, -2)$, one focus at $(4, -2)$, and one vertex at $(3, -2)$. Graph the equation.

Solution The center is at $(h, k) = (1, -2)$, so $h = 1$ and $k = -2$. The transverse axis is parallel to the x-axis. The distance from the center $(1, -2)$ to the focus $(4, -2)$ is $c = 3$; the distance from the center $(1, -2)$ to the vertex $(3, -2)$ is $a = 2$. Thus, $b^2 = c^2 - a^2 = 9 - 4 = 5$. The equation is

$$\frac{(x-h)^2}{a^2} - \frac{(y-k)^2}{b^2} = 1$$

$$\frac{(x-1)^2}{4} - \frac{(y+2)^2}{5} = 1$$

See Figure 39, at the top of the next page.

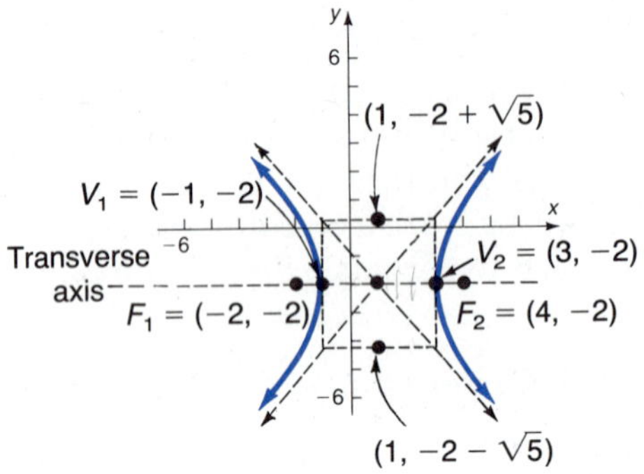

Figure 39
$\frac{(x-1)^2}{4} - \frac{(y+2)^2}{5} = 1$

■

Example 7 Discuss the equation: $-x^2 + 4y^2 - 2x - 16y + 11 = 0$

Solution We complete the squares in x and in y:

$$-x^2 + 4y^2 - 2x - 16y + 11 = 0$$

$$-(x^2 + 2x) + 4(y^2 - 4y) = -11 \qquad \text{Group terms.}$$

$$-(x^2 + 2x + 1) + 4(y^2 - 4y + 4) = -1 + 16 - 11 \qquad \text{Complete each square.}$$

$$-(x+1)^2 + 4(y-2)^2 = 4$$

$$(y-2)^2 - \frac{(x+1)^2}{4} = 1 \qquad \text{Divide by 4.}$$

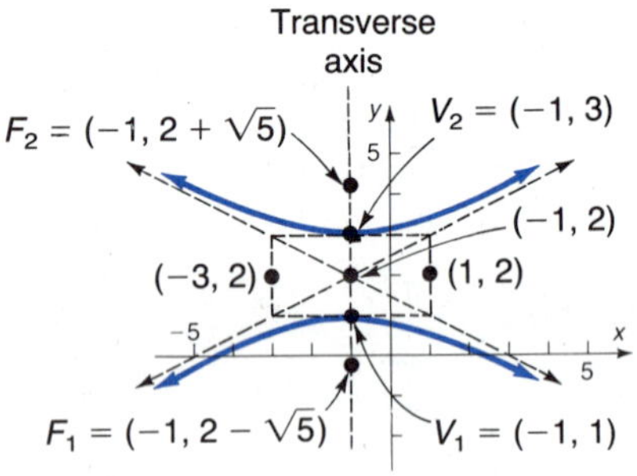

Figure 40
$(y-2)^2 - \frac{(x+1)^2}{4} = 1$

This is the equation of a hyperbola with center at $(-1, 2)$ and transverse axis parallel to the y-axis. Also, $a^2 = 1$ and $b^2 = 4$, so $c^2 = a^2 + b^2 = 5$. The vertices are at $(h, k \pm a) = (-1, 2 \pm 1)$, or $(-1, 1)$ and $(-1, 3)$. The foci are at $(h, k \pm c) = (-1, 2 \pm \sqrt{5})$. The asymptotes are $y - 2 = \frac{1}{2}(x + 1)$ and $y - 2 = -\frac{1}{2}(x + 1)$. Figure 40 shows the graph. ■

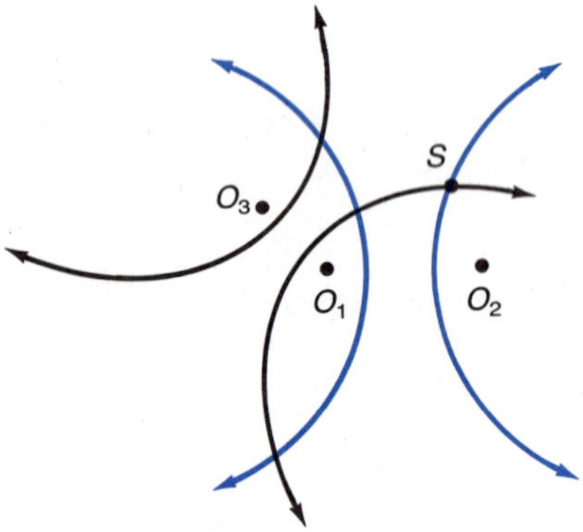

Figure 41

Application

Suppose a gun is fired from an unknown source S. An observer at O_1 hears the report (sound of gun shot) 1 second after another observer at O_2. Because sound travels at about 1100 feet per second, it follows that the point S must be 1100 feet closer to O_2 than to O_1. Thus, S lies on one branch of a hyperbola with foci at O_1 and O_2. (Do you see why? The difference of the distances from S to O_1 and from S to O_2 is the constant 1100.) If a third observer at O_3 hears the same report 2 seconds after O_1 hears it, then S will lie on a branch of a second hyperbola with foci at O_1 and O_3. The intersection of the two hyperbolas will pinpoint the location of S. See Figure 41 for an illustration.

EXERCISE 11.4

In Problems 1–10, find an equation for the hyperbola described. Graph the equation.

1. Center at (0, 0); focus at (4, 0); vertex at (1, 0)
2. Center at (0, 0); focus at (0, 5); vertex at (0, 4)
3. Center at (0, 0); focus at (0, −6); vertex at (0, 4)
4. Center at (0, 0); focus at (−3, 0); vertex at (2, 0)
5. Foci at (−5, 0) and (5, 0); vertex at (3, 0)
6. Focus at (0, 6); vertices at (0, −2) and (0, 2)
7. Vertices at (0, −6) and (0, 6); asymptote the line $y = 2x$
8. Vertices at (−4, 0) and (4, 0); asymptote the line $y = 2x$
9. Foci at (−4, 0) and (4, 0); asymptote the line $y = -x$
10. Foci at (0, −2) and (0, 2); asymptote the line $y = -x$

In Problems 11–18, find the center, transverse axis, vertices, foci, and asymptotes. Graph each equation.

11. $\frac{x^2}{9} - \frac{y^2}{4} = 1$
12. $\frac{y^2}{9} - \frac{x^2}{4} = 1$
13. $4x^2 - y^2 = 16$
14. $y^2 - 4x^2 = 16$
15. $y^2 - 9x^2 = 9$
16. $x^2 - y^2 = 4$
17. $y^2 - x^2 = 25$
18. $2x^2 - y^2 = 4$

In Problems 19–26, find an equation for the hyperbola described. Graph the equation.

19. Center at (4, −1); focus at (7, −1); vertex at (6, −1)
20. Center at (−3, 1); focus at (−3, 6); vertex at (−3, 4)
21. Center at (−3, −4); focus at (−3, −8); vertex at (−3, −2)
22. Center at (1, 4); focus at (−2, 4); vertex at (0, 4)
23. Foci at (3, 7) and (7, 7); vertex at (6, 7)
24. Focus at (−4, 0); vertices at (−4, 4) and (−4, 2)
25. Vertices at (−1, −1) and (3, −1); asymptote the line $(x - 1)/2 = (y + 1)/3$
26. Vertices at (1, −3) and (1, 1); asymptote the line $(x - 1)/2 = (y + 1)/3$

In Problems 27–40, find the center, transverse axis, vertices, foci, and asymptotes. Graph each equation.

27. $\frac{(x-3)^2}{4} - \frac{(y+2)^2}{9} = 1$
28. $\frac{(y+4)^2}{4} - \frac{(x-1)^2}{9} = 1$
29. $(y - 2)^2 - 4(x + 2)^2 = 4$
30. $(x + 4)^2 - 9(y - 3)^2 = 9$
31. $(x + 1)^2 - (y + 2)^2 = 4$
32. $(y - 3)^2 - (x + 2)^2 = 4$
33. $x^2 - y^2 - 2x - 2y - 1 = 0$
34. $y^2 - x^2 - 4y + 4x - 1 = 0$
35. $y^2 - 4x^2 - 4y - 8x - 4 = 0$
36. $2x^2 - y^2 + 4x + 4y - 4 = 0$
37. $4x^2 - y^2 - 24x - 4y + 16 = 0$
38. $2y^2 - x^2 + 2x + 8y + 3 = 0$
39. $y^2 - 4x^2 - 16x - 2y - 19 = 0$
40. $x^2 - 3y^2 + 8x - 6y + 4 = 0$

41. The **eccentricity** e of a hyperbola is defined as the number c/a. Because $c > a$, it follows that $e > 1$. Describe the general shape of a hyperbola whose eccentricity is close to 1. What is the shape if e is very large?

42. A hyperbola for which $a = b$ is called an **equilateral hyperbola**. Find the eccentricity e of an equilateral hyperbola.

43. Two hyperbolas that have the same set of asymptotes are called **conjugate**. Show that the hyperbolas

$$\frac{x^2}{4} - y^2 = 1 \quad \text{and} \quad y^2 - \frac{x^2}{4} = 1$$

are conjugate. Graph each hyperbola.

44. Prove that the hyperbola

$$\frac{y^2}{a^2} - \frac{x^2}{b^2} = 1$$

has the two oblique asymptotes

$$y = \frac{a}{b}x \quad \text{and} \quad y = -\frac{a}{b}x$$

45. Show that the graph of an equation of the form

$$Ax^2 + Cy^2 + F = 0 \qquad A \neq 0, C \neq 0, F \neq 0$$

where A and C are of opposite sign, is a hyperbola with center at $(0, 0)$.

46. Show that the graph of an equation of the form

$$Ax^2 + Cy^2 + Dx + Ey + F = 0 \qquad A \neq 0, C \neq 0$$

where A and C are of opposite sign:

(a) Is a hyperbola if $(D^2/4A) + (E^2/4C) - F \neq 0$.

(b) Is two intersecting lines if $(D^2/4A) + (E^2/4C) - F = 0$.

11.5 Rotation of Axes; General Form of a Conic

In this section, we show that the graph of a general second-degree polynomial containing two variables x and y, namely, an equation of the form

$$Ax^2 + Bxy + Cy^2 + Dx + Ey + F = 0 \tag{1}$$

where A, B, and C are not simultaneously 0, is a conic. We shall not concern ourselves here with the degenerate cases of equation (1), such as $x^2 + y^2 = 0$, whose graph is a single point $(0, 0)$; or $x^2 + 3y^2 + 3 = 0$, whose graph contains no points; or $x^2 - 4y^2 = 0$, whose graph is two lines, $x - 2y = 0$ and $x + 2y = 0$.

We begin with the case where $B = 0$. In this case, the term containing xy is not present, so that equation (1) has the form

$$Ax^2 + Cy^2 + Dx + Ey + F = 0$$

We have already discussed the procedure for identifying the graph of this kind of equation; we complete the squares of the quadratic expressions in x or y, or both. Once this has been done, the conic can be identified by comparing it to one of the forms studied in Sections 11.2–11.4.

In fact, though, we can identify the conic directly from the equation without completing the squares.

Theorem

Identifying Conics without Completing the Squares

Excluding degenerate gases, the equation

$$Ax^2 + Cy^2 + Dx + Ey + F = 0 \qquad (2)$$

(a) Defines a parabola if $AC = 0$.

(b) Defines an ellipse (or a circle) if $AC > 0$.

(c) Defines a hyperbola if $AC < 0$.

Proof (a) If $AC = 0$, then either $A = 0$ or $C = 0$, so that the form of equation (2) is either

$$Ax^2 + Dx + Ey + F = 0 \qquad \text{or} \qquad Cy^2 + Dx + Ey + F = 0$$

Using the results of Problems 41 and 42 in Exercise 11.2, it follows that, except for the degenerate cases, the equation is a parabola.

(b) If $AC > 0$, then A and C are of the same sign. Using the results of Problems 49 and 50 in Exercise 11.3, except for the degenerate cases, the equation is an ellipse if $A \neq C$, or a circle if $A = C$.

(c) If $AC < 0$, then A and C are of opposite sign. Using the results of Problems 45 and 46 in Exercise 11.4, except for the degenerate cases, the equation is a hyperbola. ■

As we mentioned above, we shall not be concerned with the degenerate cases of equation (2). However, in practice, you should be alert to the possibility of degeneracy.

Example 1 Identify each equation without completing the squares.

(a) $3x^2 + 6y^2 + 6x - 12y = 0$ (b) $2x^2 - 3y^2 + 6y + 4 = 0$

(c) $y^2 - 2x + 4 = 0$

Solution (a) We compare the given equation to equation (2) and find that $A = 3$ and $C = 6$. Since $AC = 18 > 0$, the equation is an ellipse.

(b) Here, $A = 2$ and $C = -3$, so $AC = -6 < 0$. The equation is a hyperbola.

(c) Here, $A = 0$ and $C = 1$, so $AC = 0$. The equation is a parabola. ■

Although we can now identify the type of conic represented by any equation of the form of equation (2) without completing the squares, we will still need to complete the squares if we desire additional information about a conic.

Now we turn our attention to equations of the form of equation (1) where $B \neq 0$. To discuss this case, we first need to investigate a new procedure—*rotation of axes.*

Rotation of Axes

In a **rotation of axes**, the origin remains fixed while the x-axis and y-axis are rotated through an angle θ to a new position; the new positions of the x- and y-axes are denoted by x' and y', respectively, as shown in Figure 42(a).

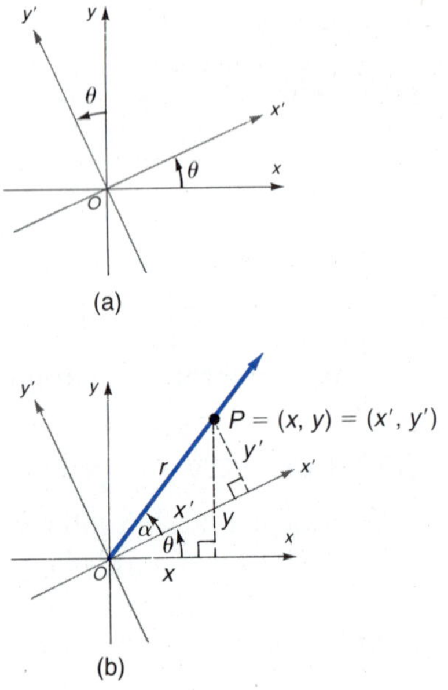

Figure 42

Now look at Figure 42(b). There, the point P has the coordinates (x, y) relative to the xy-plane, while the same point P has coordinates (x', y') relative to the $x'y'$-plane. We seek relationships that will enable us to express x and y in terms of x', y', and θ.

As Figure 42(b) shows, r denotes the distance from the origin O to the point P, and α denotes the angle between the positive x'-axis and the ray from O through P. Then, using the definitions of sine and cosine, we have

$$x' = r\cos\alpha \qquad y' = r\sin\alpha \tag{3}$$

$$x = r\cos(\theta + \alpha) \qquad y = r\sin(\theta + \alpha) \tag{4}$$

Now,

$$\begin{aligned} x &= r\cos(\theta+\alpha) \\ &= r(\cos\theta\cos\alpha - \sin\theta\sin\alpha) && \text{Sum formula} \\ &= (r\cos\alpha)(\cos\theta) - (r\sin\alpha)(\sin\theta) \\ &= x'\cos\theta - y'\sin\theta && \text{By equation (3)} \end{aligned}$$

Similarly,

$$\begin{aligned} y &= r\sin(\theta+\alpha) \\ &= r(\sin\theta\cos\alpha + \cos\theta\sin\alpha) \\ &= x'\sin\theta + y'\cos\theta \end{aligned}$$

Theorem

Rotation Formulas

If the x- and y-axes are rotated through an angle θ, the coordinates (x, y) of a point P relative to the xy-plane and the coordinates (x', y') of the same point relative to the new x'- and y'-axes are related by the formulas

$$x = x'\cos\theta - y'\sin\theta \qquad y = x'\sin\theta + y'\cos\theta \tag{5}$$

■

Example 2 Express the equation $xy = 1$ in terms of new $x'y'$-coordinates by rotating the axes through a 45° angle. Discuss the new equation.

Solution Let $\theta = 45°$ in equation (5). Then

$$x = x'\frac{\sqrt{2}}{2} - y'\frac{\sqrt{2}}{2} = \frac{\sqrt{2}}{2}(x' - y')$$

$$y = x'\frac{\sqrt{2}}{2} + y'\frac{\sqrt{2}}{2} = \frac{\sqrt{2}}{2}(x' + y')$$

Substituting these expressions for x and y in $xy = 1$ gives

$$\left[\frac{\sqrt{2}}{2}(x' - y')\right]\left[\frac{\sqrt{2}}{2}(x' + y')\right] = 1$$

$$\frac{1}{2}(x'^2 - y'^2) = 1$$

$$\frac{x'^2}{2} - \frac{y'^2}{2} = 1$$

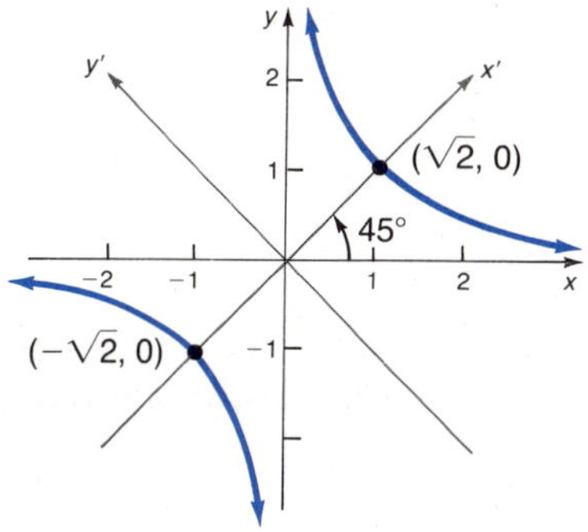

Figure 43

$xy = 1$ or $\frac{x'^2}{2} - \frac{y'^2}{2} = 1$

This is the equation of a hyperbola with center at (0, 0) and transverse axis along the x'-axis. The vertices are at $(\pm\sqrt{2}, 0)$ on the x'-axis; the asymptotes are $y' = x'$ and $y' = -x'$ (which correspond to the original x- and y-axes). See Figure 43 for the graph. ■

As Example 2 illustrates, a rotation of axes through an appropriate angle can transform a second-degree equation in x and y containing an xy-term into one in x' and y' in which no $x'y'$-term appears. In fact, we will show that a rotation of axes through an appropriate angle will transform any equation of the form of equation (1) into an equation in x' and y' without an $x'y'$-term.

To find the formula for choosing an appropriate angle θ through which to rotate the axes, we begin with equation (1),

$$Ax^2 + Bxy + Cy^2 + Dx + Ey + F = 0 \qquad B \neq 0$$

Next, we rotate through an angle θ using rotation formulas (5):

$$A(x' \cos \theta - y' \sin \theta)^2 + B(x' \cos \theta - y' \sin \theta)(x' \sin \theta + y' \cos \theta)$$
$$+ C(x' \sin \theta + y' \cos \theta)^2 + D(x' \cos \theta - y' \sin \theta)$$
$$+ E(x' \sin \theta + y' \cos \theta) + F = 0$$

By expanding and collecting like terms, we obtain

$$\begin{aligned}(A \cos^2 \theta + B \sin \theta \cos \theta + C \sin^2 \theta)x'^2 &+ [B(\cos^2 \theta - \sin^2 \theta) + 2(C - A)(\sin \theta \cos \theta)]x'y' \\ &+ (A \sin^2 \theta - B \sin \theta \cos \theta + C \cos^2 \theta)y'^2 \\ &+ (D \cos \theta + E \sin \theta)x' \\ &+ (-D \sin \theta + E \cos \theta)y' + F = 0\end{aligned} \qquad (6)$$

In equation (6), the coefficient of $x'y'$ is

$$B' = 2(C - A)(\sin\theta\cos\theta) + B(\cos^2\theta - \sin^2\theta)$$

Since we want to eliminate the $x'y'$-term, we select an angle θ so that $B' = 0$. Thus,

$$2(C - A)(\sin\theta\cos\theta) + B(\cos^2\theta - \sin^2\theta) = 0$$

$$(C - A)(\sin 2\theta) + B\cos 2\theta = 0 \quad \text{Double-angle formulas}$$

$$B\cos 2\theta = (A - C)(\sin 2\theta)$$

$$\cot 2\theta = \frac{A - C}{B} \qquad B \neq 0$$

Theorem To transform the equation

$$Ax^2 + Bxy + Cy^2 + Dx + Ey + F = 0 \qquad B \neq 0$$

into an equation in x' and y' without an $x'y'$-term, rotate the axes through an angle θ that satisfies the equation

$$\cot 2\theta = \frac{A - C}{B} \tag{7}$$

■

Equation (7) has an infinite number of solutions for θ. We shall adopt the convention of choosing the acute angle θ that satisfies (7). Then we have the following two possibilities:

If $\cot 2\theta \geq 0$, then $0 < 2\theta \leq \pi/2$ so that $0 < \theta \leq \pi/4$.

If $\cot 2\theta < 0$, then $\pi/2 < 2\theta < \pi$ so that $\pi/4 < \theta < \pi/2$.

Each of these results in a counterclockwise rotation of the axes through an acute angle θ.*

Warning: Be careful if you use a calculator to solve equation (7).

1. If $\cot 2\theta = 0$, then $2\theta = \pi/2$ and $\theta = \pi/4$.
2. If $\cot 2\theta \neq 0$, first find $\cos 2\theta$ (see Section 8.3). Then use the inverse cosine function key(s) to obtain 2θ, $0 < 2\theta < \pi$. Finally, divide by 2 to obtain the correct acute angle θ.

Example 3 Discuss the equation: $x^2 + \sqrt{3}xy + 2y^2 - 10 = 0$

*Any rotation—clockwise or counterclockwise—through an angle θ that satisfies $\cot 2\theta = (A - C)/B$ will eliminate the $x'y'$-term. However, the final form of the transformed equation may be different, depending on the angle chosen.

Solution Since an xy-term is present, we must rotate the axes. Using $A = 1$, $B = \sqrt{3}$, and $C = 2$ in equation (7), the appropriate acute angle θ through which to rotate the axes satisfies the equation

$$\cot 2\theta = \frac{A - C}{B} = \frac{-1}{\sqrt{3}} = \frac{-\sqrt{3}}{3} \qquad 0° < 2\theta < 180°$$

Since $\cot 2\theta = -\sqrt{3}/3$, we find $2\theta = 120°$, so that $\theta = 60°$. Using $\theta = 60°$ in rotation formulas (5), we find

$$x = \frac{1}{2}x' - \frac{\sqrt{3}}{2}y' = \frac{1}{2}(x' - \sqrt{3}y')$$

$$y = \frac{\sqrt{3}}{2}x' + \frac{1}{2}y' = \frac{1}{2}(\sqrt{3}x' + y')$$

Substituting these values into the original equation and simplifying, we have

$$x^2 + \sqrt{3}xy + 2y^2 - 10 = 0$$

$$\frac{1}{4}(x' - \sqrt{3}y')^2 + \sqrt{3}\left[\frac{1}{2}(x' - \sqrt{3}y')\right]\left[\frac{1}{2}(\sqrt{3}x' + y')\right] + 2\left[\frac{1}{4}(\sqrt{3}x' + y')^2\right] = 10$$

Multiply both sides by 4 and expand to obtain

$$x'^2 - 2\sqrt{3}x'y' + 3y'^2 + \sqrt{3}(\sqrt{3}x'^2 - 2x'y' - \sqrt{3}y'^2) + 2(3x'^2 + 2\sqrt{3}x'y' + y'^2) = 40$$

$$10x'^2 + 2y'^2 = 40$$

$$\frac{x'^2}{4} + \frac{y'^2}{20} = 1$$

This is the equation of an ellipse with center at (0, 0) and major axis along the y'-axis. The vertices are at $(0, \pm 2\sqrt{5})$ on the y'-axis. See Figure 44 for the graph.

Figure 44

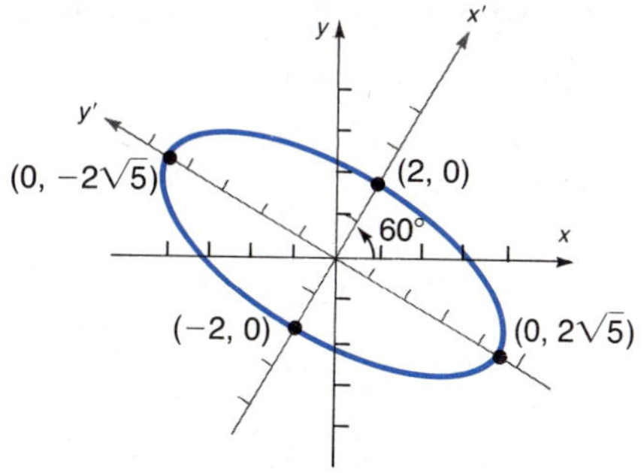

■

In Example 3, the acute angle θ through which to rotate the axes was easy to find because of the numbers we used in the given equation. In general, the equation $\cot 2\theta = (A - C)/B$ will not have such a "nice" solution. As the next example shows, we can still find the appropriate rotation formulas without using a calculator approximation by applying half-angle formulas.

Example 4 Discuss the equation: $4x^2 - 4xy + y^2 + 5\sqrt{5}x + 5 = 0$

Solution Letting $A = 4$, $B = -4$, and $C = 1$ in equation (7), the appropriate angle θ through which to rotate the axes satisfies

$$\cot 2\theta = \frac{A - C}{B} = \frac{3}{-4}$$

In order to use rotation formulas (5), we need to know the values of $\sin\theta$ and $\cos\theta$. Since we seek an acute angle θ, we know that $\sin\theta > 0$ and $\cos\theta > 0$. Thus, we use the half-angle formulas in the form

$$\sin\theta = \sqrt{\frac{1 - \cos 2\theta}{2}} \qquad \cos\theta = \sqrt{\frac{1 + \cos 2\theta}{2}}$$

Now we need to find the value of $\cos 2\theta$. Since $\cot 2\theta = -\frac{3}{4}$ and $\pi/2 < 2\theta < \pi$, it follows that $\cos 2\theta = -\frac{3}{5}$. Thus,

$$\sin\theta = \sqrt{\frac{1 - \cos 2\theta}{2}} = \sqrt{\frac{1 - \left(-\frac{3}{5}\right)}{2}} = \sqrt{\frac{4}{5}} = \frac{2}{\sqrt{5}} = \frac{2\sqrt{5}}{5}$$

$$\cos\theta = \sqrt{\frac{1 + \cos 2\theta}{2}} = \sqrt{\frac{1 + \left(-\frac{3}{5}\right)}{2}} = \sqrt{\frac{1}{5}} = \frac{1}{\sqrt{5}} = \frac{\sqrt{5}}{5}$$

With these values, rotation formulas (5) give us

$$x = \frac{\sqrt{5}}{5}x' - \frac{2\sqrt{5}}{5}y' = \frac{\sqrt{5}}{5}(x' - 2y')$$

$$y = \frac{2\sqrt{5}}{5}x' + \frac{\sqrt{5}}{5}y' = \frac{\sqrt{5}}{5}(2x' + y')$$

Substituting these values in the original equation and simplifying, we obtain

$$4x^2 - 4xy + y^2 + 5\sqrt{5}x + 5 = 0$$

$$4\left[\frac{\sqrt{5}}{5}(x' - 2y')\right]^2 - 4\left[\frac{\sqrt{5}}{5}(x' - 2y')\right]\left[\frac{\sqrt{5}}{5}(2x' + y')\right] + \left[\frac{\sqrt{5}}{5}(2x' + y')\right]^2 + 5\sqrt{5}\left[\frac{\sqrt{5}}{5}(x' - 2y')\right] = -5$$

Multiply both sides by 5 and expand to obtain

$$4(x'^2 - 4x'y' + 4y'^2) - 4(2x'^2 - 3x'y' - 2y'^2) + 4x'^2 + 4x'y' + y'^2 + 25(x' - 2y') = -25$$

$$25y'^2 - 50y' + 25x' = -25$$

$$y'^2 - 2y' + x' = -1$$

$$y'^2 - 2y' + 1 = -x' \quad \text{Complete the square in } y'.$$

$$(y' - 1)^2 = -x'$$

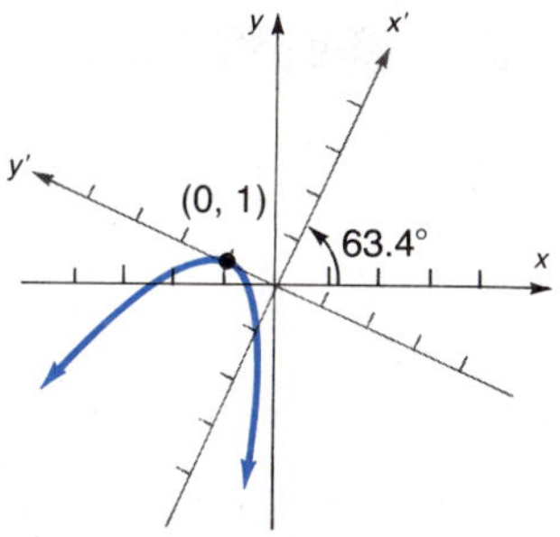

Figure 45

This is the equation of a parabola with vertex at (0, 1) in the $x'y'$-plane. The axis of symmetry is parallel to the x'-axis. Using a calculator to solve $\sin\theta = 2\sqrt{5}/5$, we find that $\theta \approx 63.4°$. See Figure 45 for the graph. ■

Identifying Conics without a Rotation of Axes

Suppose we are required only to identify (rather than discuss) an equation of the form

$$Ax^2 + Bxy + Cy^2 + Dx + Ey + F = 0 \qquad B \neq 0 \tag{8}$$

If we apply rotation formulas (5) to this equation, we obtain an equation of the form

$$A'x'^2 + B'x'y' + C'y'^2 + D'x' + E'y' + F' = 0 \tag{9}$$

where A', B', C', D', E', and F' can be expressed in terms of A, B, C, D, E, F, and the angle θ of rotation (see Problem 43 at the end of this section). It can be shown that the value of $B^2 - 4AC$ in equation (8) and the value of $B'^2 - 4A'C'$ in equation (9) are equal no matter what angle θ of rotation is chosen (see Problem 45). In particular, if the angle θ of rotation satisfies equation (7), then $B' = 0$ in equation (9), and $B^2 - 4AC = -4A'C'$. Since equation (9) then has the form of equation (2),

$$A'x'^2 + C'y'^2 + D'x' + E'y' + F' = 0$$

we can identify it without completing the squares, as we did in the beginning of this section. In fact, now we can identify the conic described by any equation of the form of equation (8) without a rotation of axes.

Theorem

Identifying Conics without a Rotation of Axes

Except for degenerate cases, the equation

$$Ax^2 + Bxy + Cy^2 + Dx + Ey + F = 0$$

(a) Defines a parabola if $B^2 - 4AC = 0$.
(b) Defines an ellipse (or a circle) if $B^2 - 4AC < 0$.
(c) Defines a hyperbola if $B^2 - 4AC > 0$. ■

You are asked to prove this theorem in Problem 46 (at the end of this section).

Example 5 Identify the equation: $8x^2 - 12xy + 17y^2 - 4\sqrt{5}x - 2\sqrt{5}y - 15 = 0$

Solution Here, $A = 8$, $B = -12$, and $C = 17$, so that $B^2 - 4AC = -400$. Since $B^2 - 4AC < 0$, the equation defines an ellipse. ■

EXERCISE 11.5

In Problems 1–10, identify each equation without completing the squares.

1. $x^2 + 4x + y + 3 = 0$
2. $2y^2 - 3y + 3x = 0$
3. $6x^2 + 3y^2 - 12x + 6y = 0$
4. $2x^2 + y^2 - 8x + 4y + 2 = 0$
5. $3x^2 - 2y^2 + 6x + 4 = 0$
6. $4x^2 - 3y^2 - 8x + 6y + 1 = 0$
7. $2y^2 - x^2 - y + x = 0$
8. $y^2 - 8x^2 - 2x - y = 0$
9. $x^2 + y^2 - 8x + 4y = 0$
10. $2x^2 + 2y^2 - 8x + 8y = 0$

In Problems 11–20, determine the appropriate rotation formulas to use so that the new equation contains no xy-term.

11. $x^2 + 4xy + y^2 - 3 = 0$
12. $x^2 - 4xy + y^2 - 3 = 0$
13. $5x^2 + 6xy + 5y^2 - 8 = 0$
14. $3x^2 - 10xy + 3y^2 - 32 = 0$
15. $13x^2 - 6\sqrt{3}xy + 7y^2 - 16 = 0$
16. $11x^2 + 10\sqrt{3}xy + y^2 - 4 = 0$
17. $4x^2 - 4xy + y^2 - 8\sqrt{5}x - 16\sqrt{5}y = 0$
18. $x^2 + 4xy + 4y^2 + 5\sqrt{5}y + 5 = 0$
19. $25x^2 - 36xy + 40y^2 - 12\sqrt{13}x - 8\sqrt{13}y = 0$
20. $34x^2 - 24xy + 41y^2 - 25 = 0$

In Problems 21–32, rotate the axes so that the new equation contains no xy-term. Discuss and graph the new equation. (Refer to Problems 11–20 for Problems 21–30.)

21. $x^2 + 4xy + y^2 - 3 = 0$
22. $x^2 - 4xy + y^2 - 3 = 0$
23. $5x^2 + 6xy + 5y^2 - 8 = 0$
24. $3x^2 - 10xy + 3y^2 - 32 = 0$
25. $13x^2 - 6\sqrt{3}xy + 7y^2 - 16 = 0$
26. $11x^2 + 10\sqrt{3}xy + y^2 - 4 = 0$
27. $4x^2 - 4xy + y^2 - 8\sqrt{5}x - 16\sqrt{5}y = 0$
28. $x^2 + 4xy + 4y^2 + 5\sqrt{5}y + 5 = 0$
29. $25x^2 - 36xy + 40y^2 - 12\sqrt{13}x - 8\sqrt{13}y = 0$
30. $34x^2 - 24xy + 41y^2 - 25 = 0$
31. $16x^2 + 24xy + 9y^2 - 130x + 90y = 0$
32. $16x^2 + 24xy + 9y^2 - 60x + 80y = 0$

In Problems 33–42, identify each equation without applying a rotation of axes.

33. $x^2 + 3xy - 2y^2 + 3x + 2y + 5 = 0$
34. $2x^2 - 3xy + 4y^2 + 2x + 3y - 5 = 0$
35. $x^2 - 7xy + 3y^2 - y - 10 = 0$
36. $2x^2 - 3xy + 2y^2 - 4x - 2 = 0$
37. $9x^2 + 12xy + 4y^2 - x - y = 0$
38. $10x^2 + 12xy + 4y^2 - x - y + 10 = 0$
39. $10x^2 - 12xy + 4y^2 - x - y - 10 = 0$
40. $4x^2 + 12xy + 9y^2 - x - y = 0$
41. $3x^2 - 2xy + y^2 + 4x + 2y - 1 = 0$
42. $3x^2 + 2xy + y^2 + 4x - 2y + 10 = 0$

In Problems 43–46, apply rotation formulas (5) to

$$Ax^2 + Bxy + Cy^2 + Dx + Ey + F = 0$$

to obtain the equation

$$A'x'^2 + B'x'y' + C'y'^2 + D'x' + E'y' + F' = 0$$

43. Express A', B', C', D', E', and F' in terms of A, B, C, D, E, F, and the angle θ of rotation.
44. Show that $A + C = A' + C'$, and thus show that $A + C$ is **invariant**—that is, that its value does not change under a rotation of axes.
45. Refer to Problem 44. Show that $B^2 - 4AC$ is invariant.

46. Prove that, except for degenerate cases, the equation

$$Ax^2 + Bxy + Cy^2 + Dx + Ey + F = 0$$

(a) Defines a parabola if $B^2 - 4AC = 0$.
(b) Defines an ellipse (or a circle) if $B^2 - 4AC < 0$.
(c) Defines a hyperbola if $B^2 - 4AC > 0$.

47. Use rotation formulas (5) to show that distance is invariant under a rotation of axes. That is, show that the distance from $P_1 = (x_1, y_1)$ to $P_2 = (x_2, y_2)$ in the xy-plane equals the distance from $P_1 = (x_1', y_1')$ to $P_2 = (x_2', y_2')$ in the $x'y'$-plane.

48. Show that the graph of the equation $x^{1/2} + y^{1/2} = a^{1/2}$ is part of the graph of a parabola.

11.6 Polar Equations of Conics

In Sections 11.2, 11.3, and 11.4, we gave separate definitions for the parabola, ellipse, and hyperbola, based on geometric properties and the distance formula. In this section, we present an alternative definition that simultaneously defines all these conics. As we shall see, this approach is well-suited to polar coordinate representation. (Refer to Section 10.4.)

Conic

Let D denote a fixed line called the **directrix**; let F denote a fixed point called the **focus**, which is not on D; and let e be a fixed nonnegative number called the **eccentricity**. A **conic** is the set of points P in the plane such that the ratio of the distance from F to P to the distance from P to D equals e. Thus, a conic is the collection of points P for which

$$\frac{d(F, P)}{d(P, D)} = e \qquad (1)$$

If $e = 1$, the conic is a **parabola**.
If $e < 1$, the conic is an **ellipse**.
If $e > 1$, the conic is a **hyperbola**.

Observe that if $e = 1$, the definition of a parabola in equation (1) is exactly the same as the definition used earlier in Section 11.2.

In the case of an ellipse, the **major axis** is a line through the focus perpendicular to the directrix. In the case of a hyperbola, the **transverse axis** is a line through the focus perpendicular to the directrix. For both an ellipse and a hyperbola, the eccentricity e satisfies

$$e = \frac{c}{a} \tag{2}$$

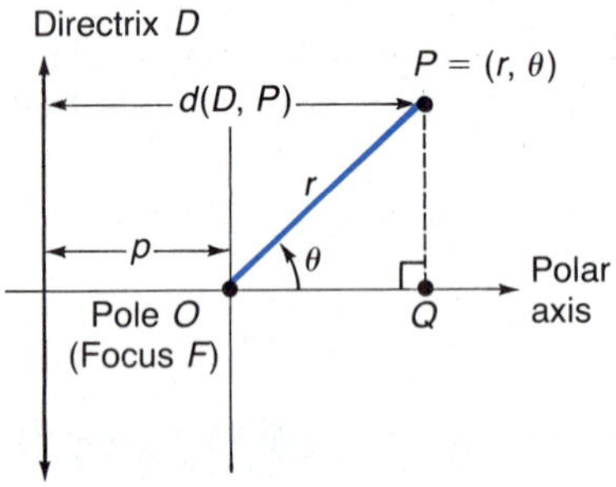

Figure 46

where c is the distance from the center to the focus and a is the distance from the center to a vertex.

Just as we did earlier, we derive equations for the conics in polar coordinates by choosing a convenient position for the focus F and the directrix D. The focus F is positioned at the pole, and the directrix D is either parallel to the polar axis or perpendicular to it.

Suppose we start with the directrix D perpendicular to the polar axis, at a distance p units to the left of the pole (the focus F). See Figure 46.

If $P = (r, \theta)$ is any point on the conic, then by equation (1),

$$\frac{d(F, P)}{d(D, P)} = e \qquad \text{or} \qquad d(F, P) = e \cdot d(D, P) \tag{3}$$

Now we use the point Q obtained by dropping the perpendicular from P to the polar axis to calculate $d(D, P)$:

$$d(D, P) = p + d(O, Q) = p + r \cos \theta$$

Using this expression and the fact that $d(F, P) = d(O, P) = r$ in equation (3), we get

$$\begin{aligned} d(F, P) &= e \cdot d(D, P) \\ r &= e(p + r \cos \theta) \\ r &= ep + er \cos \theta \\ r - er \cos \theta &= ep \\ r(1 - e \cos \theta) &= ep \\ r &= \frac{ep}{1 - e \cos \theta} \end{aligned}$$

Theorem

Polar Equation of a Conic; Focus at Pole; Directrix Perpendicular to Polar Axis a Distance p to the Left of the Pole

The polar equation of a conic with focus at the pole and directrix perpendicular to the polar axis at a distance p to the left of the pole is

$$r = \frac{ep}{1 - e \cos \theta} \tag{4}$$

where e is the eccentricity of the conic. ■

Example 1 Identify and graph the equation: $r = \dfrac{4}{2 - \cos\theta}$

Solution The given equation is not quite in the form of equation (4), since the first term in the denominator is 2 instead of 1. Thus, we divide the numerator and denominator by 2 to obtain

$$r = \frac{2}{1 - \frac{1}{2}\cos\theta}$$

This equation is in the form of equation (4), with

$$e = \tfrac{1}{2} \quad \text{and} \quad ep = \tfrac{1}{2}p = 2$$

Thus, $e = \frac{1}{2}$ and $p = 4$. We conclude that the conic is an ellipse, since $e = \frac{1}{2} < 1$. One focus is at the pole, and the directrix is perpendicular to the polar axis, a distance of 4 units to the left of the pole. It follows that the major axis is along the polar axis. To find the vertices, we let $\theta = 0$ and $\theta = \pi$. Thus, the vertices of the ellipse are $(4, 0)$ and $\left(\frac{4}{3}, \pi\right)$. At the midpoint of the vertices, we locate the center of the ellipse at $\left(\frac{4}{3}, 0\right)$. [Do you see why? The vertices $(4, 0)$ and $\left(\frac{4}{3}, \pi\right)$ in polar coordinates are $(4, 0)$ and $\left(-\frac{4}{3}, 0\right)$ in rectangular coordinates. The midpoint in rectangular coordinates is $\left(\frac{4}{3}, 0\right)$, which is also $\left(\frac{4}{3}, 0\right)$ in polar coordinates.] Thus, a = distance from the center to a vertex $= \frac{8}{3}$. Using $a = \frac{8}{3}$ and $e = \frac{1}{2}$ in equation (2), $e = c/a$, we find $c = \frac{4}{3}$. Finally, using $a = \frac{8}{3}$ and $c = \frac{4}{3}$ in $b^2 = a^2 - c^2$, we have

$$b^2 = a^2 - c^2 = \frac{64}{9} - \frac{16}{9} = \frac{48}{9}$$

$$b = \frac{4\sqrt{3}}{3}$$

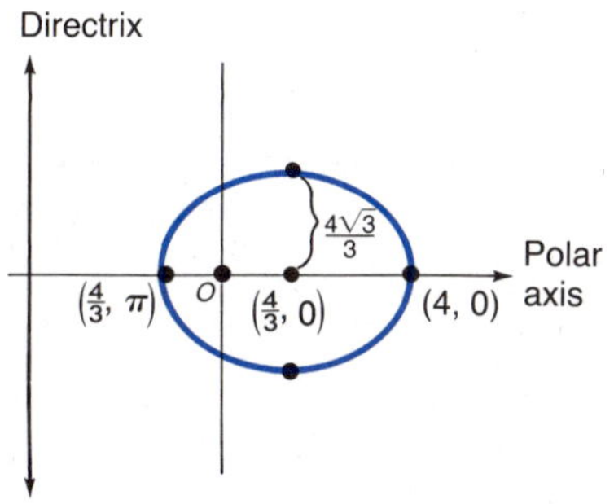

Figure 47

Figure 47 shows the graph. ■

Equation (4) was obtained under the assumption that the directrix was perpendicular to the polar axis at a distance p units to the left of the pole. A similar derivation (see Problem 33 at the end of this section), in which the directrix is perpendicular to the polar axis at a distance p units to the right of the pole, results in the equation

$$r = \frac{ep}{1 + e\cos\theta}$$

In Problems 34 and 35 at the end of this section you are asked to derive the polar equations of conics with focus at the pole and directrix parallel to the polar axis. Table 5 summarizes the polar equations of conics.

Table 5 Polar Equations of Conics (Focus at the Pole, Eccentricity e)

EQUATION	DESCRIPTION
(a) $r = \dfrac{ep}{1 - e\cos\theta}$	Directrix is perpendicular to the polar axis at a distance p units to the left of the pole.
(b) $r = \dfrac{ep}{1 + e\cos\theta}$	Directrix is perpendicular to the polar axis at a distance p units to the right of the pole.
(c) $r = \dfrac{ep}{1 + e\sin\theta}$	Directrix is parallel to the polar axis at a distance p units above the pole.
(d) $r = \dfrac{ep}{1 - e\sin\theta}$	Directrix is parallel to the polar axis at a distance p units below the pole.

ECCENTRICITY

If $e = 1$, the conic is a parabola; the axis of symmetry is perpendicular to the directrix.
If $e < 1$, the conic is an ellipse; the major axis is perpendicular to the directrix.
If $e > 1$, the conic is a hyperbola; the transverse axis is perpendicular to the directrix.

Example 2 Identify and graph the equation: $r = \dfrac{6}{3 + 3\sin\theta}$

Solution To place the equation in proper form, we divide the numerator and denominator by 3 to get

$$r = \frac{2}{1 + \sin\theta}$$

Referring to Table 5, we conclude that this equation is in the form of equation (c) with

$$e = 1 \qquad \text{and} \qquad ep = 2$$

Thus, $e = 1$ and $p = 2$. The conic is a parabola with focus at the pole. The directrix is parallel to the polar axis at a distance 2 units above the pole; the axis of symmetry is perpendicular to the polar axis. The vertex of the parabola is at $(1, \pi/2)$. (Do you see why?) See Figure 48 for the graph. Notice that we plotted two additional points, $(2, 0)$ and $(2, \pi)$, to assist in graphing. ■

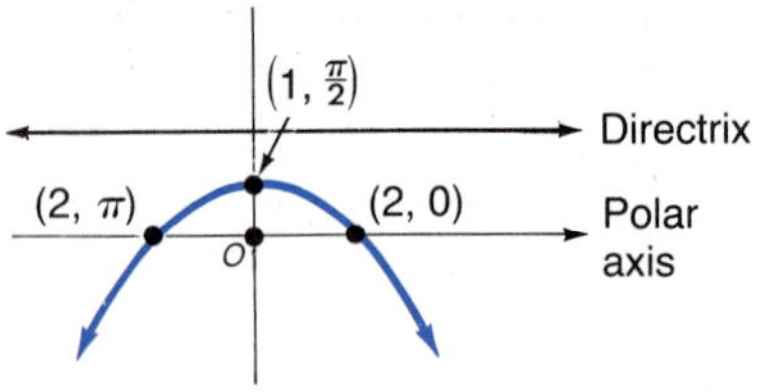

Figure 48

Example 3 Identify and graph the equation: $r = \dfrac{3}{1 + 3\cos\theta}$

Solution This equation is in the form of equation (b) in Table 5. We conclude that

$$e = 3 \quad \text{and} \quad ep = 3p = 3$$

Thus, $e = 3$ and $p = 1$. This is the equation of a hyperbola with a focus at the pole. The directrix is perpendicular to the polar axis, 1 unit to the right of the pole. The transverse axis is along the polar axis. To find the vertices, we let $\theta = 0$ and $\theta = \pi$. Thus, the vertices are $\left(\frac{3}{4}, 0\right)$ and $\left(-\frac{3}{2}, \pi\right)$. The center is at the midpoint of $\left(\frac{3}{4}, 0\right)$ and $\left(-\frac{3}{2}, \pi\right)$, which is $\left(\frac{9}{8}, 0\right)$. Thus, c = distance from the center to a focus $= \frac{9}{8}$. Since $e = 3$, it follows from equation (2), $e = c/a$, that $a = \frac{3}{8}$. Finally, using $a = \frac{3}{8}$ and $c = \frac{9}{8}$ in $b^2 = c^2 - a^2$, we find

$$b^2 = c^2 - a^2 = \frac{81}{64} - \frac{9}{64} = \frac{72}{64} = \frac{9}{8}$$

$$b = \frac{3}{2\sqrt{2}} = \frac{3\sqrt{2}}{4}$$

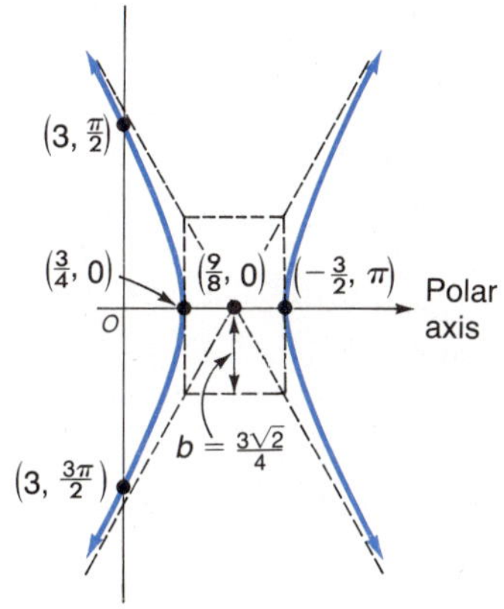

Figure 49

Figure 49 shows the graph. Notice that we plotted two additional points, $(3, \pi/2)$ and $(3, 3\pi/2)$, on the left branch and used symmetry to obtain the right branch. The asymptotes of this hyperbola were found in the usual way by constructing the rectangle shown. ■

Example 4 Convert the polar equation

$$r = \frac{1}{3 - 3\cos\theta}$$

to a rectangular equation.

Solution The strategy here is first to rearrange the equation and square each side, before using the transformation equations:

$$r = \frac{1}{3 - 3\cos\theta}$$

$$3r - 3r\cos\theta = 1$$

$$3r = 1 + 3r\cos\theta \quad \text{Rearrange the equation.}$$

$$9r^2 = (1 + 3r\cos\theta)^2 \quad \text{Square each side.}$$

$$9(x^2 + y^2) = (1 + 3x)^2 \quad \text{Use the transformation equations.}$$

$$9x^2 + 9y^2 = 9x^2 + 6x + 1$$

$$9y^2 = 6x + 1$$

This is the equation of a parabola in rectangular coordinates. ■

EXERCISE 11.6

In Problems 1–6, identify the conic each polar equation represents. Also, give the position of the directrix.

1. $r = \dfrac{1}{1 + \cos\theta}$ **2.** $r = \dfrac{3}{1 - \sin\theta}$ **3.** $r = \dfrac{4}{2 - 3\sin\theta}$

4. $r = \dfrac{2}{1 + 2\cos\theta}$ **5.** $r = \dfrac{3}{4 - 2\cos\theta}$ **6.** $r = \dfrac{6}{8 + 2\sin\theta}$

In Problems 7–16, identify and graph each equation.

7. $r = \dfrac{1}{1 + \cos\theta}$ **8.** $r = \dfrac{3}{1 - \sin\theta}$ **9.** $r = \dfrac{8}{4 + 3\sin\theta}$

10. $r = \dfrac{10}{5 + 4\cos\theta}$ **11.** $r = \dfrac{9}{3 - 6\cos\theta}$ **12.** $r = \dfrac{12}{4 + 8\sin\theta}$

13. $r(3 - 2\sin\theta) = 6$ **14.** $r(2 - \cos\theta) = 2$ **15.** $r = \dfrac{6\sec\theta}{2\sec\theta - 1}$

16. $r = \dfrac{3\csc\theta}{\csc\theta - 1}$

In Problems 17–26, convert each polar equation to a rectangular equation.

17. $r = \dfrac{1}{1 + \cos\theta}$ **18.** $r = \dfrac{3}{1 - \sin\theta}$ **19.** $r = \dfrac{8}{4 + 3\sin\theta}$

20. $r = \dfrac{10}{5 + 4\cos\theta}$ **21.** $r = \dfrac{9}{3 - 6\cos\theta}$ **22.** $r = \dfrac{12}{4 + 8\sin\theta}$

23. $r(3 - 2\sin\theta) = 6$ **24.** $r(2 - \cos\theta) = 2$ **25.** $r = \dfrac{6\sec\theta}{2\sec\theta - 1}$

26. $r = \dfrac{3\csc\theta}{\csc\theta - 1}$

In Problems 27–32, find a polar equation for each conic. For each one, a focus is at the pole.

27. $e = 1$; directrix is parallel to the polar axis 1 unit above the pole

28. $e = 1$; directrix is parallel to the polar axis 2 units below the pole

29. $e = \frac{4}{5}$; directrix is perpendicular to the polar axis 3 units to the left of the pole

30. $e = \frac{2}{3}$; directrix is parallel to the polar axis 3 units above the pole

31. $e = 6$; directrix is parallel to the polar axis 2 units below the pole

32. $e = 5$; directrix is perpendicular to the polar axis 5 units to the right of the pole

33. Derive equation (b) in Table 5: $r = \dfrac{ep}{1 + e\cos\theta}$

34. Derive equation (c) in Table 5: $r = \dfrac{ep}{1 + e\sin\theta}$

35. Derive equation (d) in Table 5: $r = \dfrac{ep}{1 - e\sin\theta}$

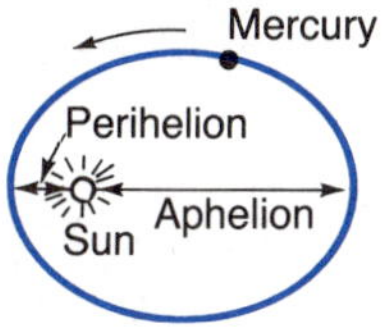

36. The planet Mercury travels around the Sun in an elliptical orbit given approximately by

$$r = \frac{(3.442)10^7}{1 - 0.206 \cos \theta}$$

where r is measured in miles and the Sun is at the pole. Find the distance from Mercury to the Sun at *aphelion* (greatest distance from the Sun) and at *perihelion* (shortest distance from the Sun). See the figure in the margin.

11.7 Plane Curves and Parametric Equations*

Equations of the form $y = f(x)$, where f is a function, have graphs that are intersected no more than once by any vertical line. The graphs of many of the conics and certain other, more complicated graphs do not have this characteristic. Yet each graph, like the graph of a function, is a collection of points (x, y) in the xy-plane; that is, each is a *(plane) curve*. In this section, we discuss another way of representing such graphs.

(Plane) Curve

Let $x = f(t)$ and $y = g(t)$, where f and g are two functions whose common domain is some interval I. The collection of points defined by

$$(x, y) = (f(t), g(t))$$

is called a **(plane) curve**. The equations

$$x = f(t) \qquad y = g(t)$$

where t is in I, are called **parametric equations** of the curve. The variable t is called a **parameter**.

Parametric equations are particularly useful in describing movement along a curve. Suppose a curve is defined by the parametric equations

$$x = f(t) \qquad y = g(t)$$

where f and g are each defined over some interval I. For a given value of t in I, we can find the value of $x = f(t)$ and $y = g(t)$, thus obtaining a point (x, y) on the curve. In fact, as t varies over the interval I in some order, say from left to right, successive values of t give rise to a directed movement along the curve. That is, as t varies over I in some order, the curve is traced out in a certain direction by the corresponding succession of points (x, y). Let's look at an example.

* C Refer to Graphics Calculator Supplement: Activity VII.

Example 1 Discuss the curve defined by the parametric equations

$$x = 3t^2 \qquad y = 2t \qquad -2 \le t \le 2 \tag{1}$$

Solution For each number t, $-2 \le t \le 2$, there corresponds a number x and a number y. For example, when $t = -2$, then $x = 12$ and $y = -4$. When $t = 0$, then $x = 0$ and $y = 0$. Indeed, we can set up a table listing various choices of the parameter t and the corresponding values for x and y, as shown in Table 6. Plotting these points and connecting them with a smooth curve leads to Figure 50.

Table 6

t	x	y	(x, y)
−2	12	−4	(12, −4)
−1	3	−2	(3, −2)
0	0	0	(0, 0)
1	3	2	(3, 2)
2	12	4	(12, 4)

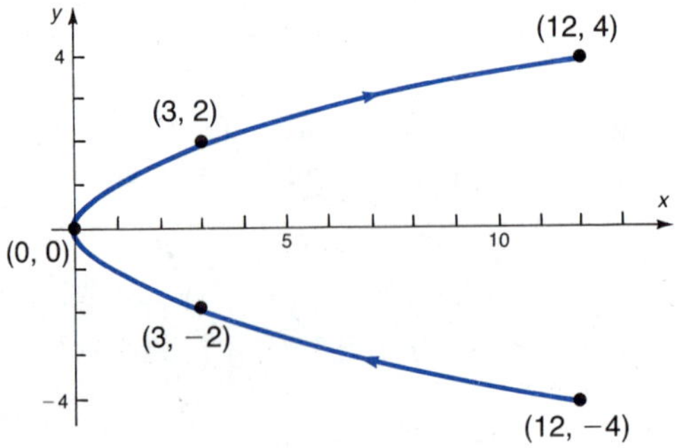

Figure 50

Notice the arrows on the curve in Figure 50. They indicate the direction, or **orientation**, of the curve for increasing values of the parameter t.

Parametric equations defining a curve are not unique. You can verify that the curve in Figure 50 also may be defined by any of the following parametric equations:

$$x = \frac{3t^2}{4} \qquad y = t \qquad -4 \le t \le 4$$

or:

$$x = 3t^2 + 12t + 12 \qquad y = 2t + 4 \qquad -4 \le t \le 0$$

or:

$$x = 3t^{2/3} \qquad y = 2\sqrt[3]{t} \qquad -8 \le t \le 8$$

The curve in Figure 50 should be familiar. To identify it accurately, we find the corresponding rectangular equation by eliminating the parameter t from the parametric equations (1) given in Example 1,

$$x = 3t^2 \qquad y = 2t \qquad -2 \le t \le 2$$

Noting that we can readily solve for t in $y = 2t$, obtaining $t = y/2$, we substitute this expression in the other equation:

$$x = 3t^2 \underset{\uparrow \atop t = \frac{y}{2}}{=} 3\left(\frac{y}{2}\right)^2 = \frac{3y^2}{4}$$

This equation, $x = 3y^2/4$, is the equation of a parabola with vertex at (0, 0) and axis along the x-axis.

Note that the parameterized curve defined by equation (1) and shown in Figure 50 is only a part of the parabola $x = 3y^2/4$. Thus, the graph of the rectangular equation obtained by eliminating the parameter will, in general, contain more points than the original parameterized curve. Care must therefore be taken when a parameterized curve is sketched after eliminating the parameter. Even so, the process of eliminating the parameter t of a parameterized curve in order to identify it accurately is sometimes a better approach than merely plotting points. However, the elimination process sometimes requires a little ingenuity.

Example 2 Find the rectangular equation of the curve whose parametric equations are

$$x = a \cos t \qquad y = a \sin t$$

where $a > 0$ is a constant. Graph this curve, indicating its orientation.

Solution The presence of sines and cosines in the parametric equations suggests that we use a Pythagorean identity. In fact, since

$$\cos t = \frac{x}{a} \qquad \sin t = \frac{y}{a}$$

we find that

$$\cos^2 t + \sin^2 t = 1$$
$$\left(\frac{x}{a}\right)^2 + \left(\frac{y}{a}\right)^2 = 1$$
$$x^2 + y^2 = a^2$$

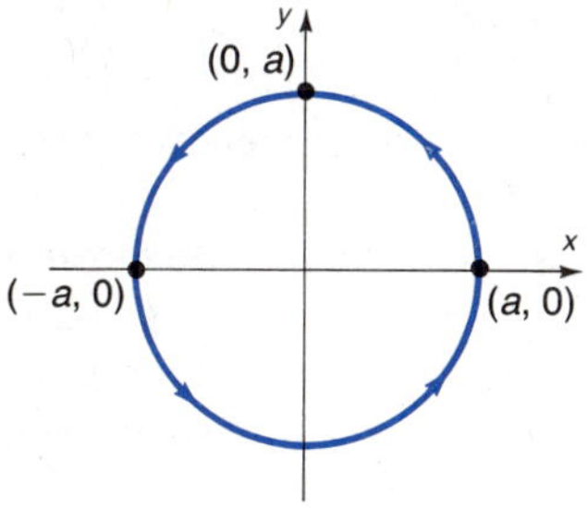

Figure 51

Thus, the curve is a circle with center at (0, 0) and radius a. As the parameter t increases, say, from $t = 0$ [the point $(a, 0)$] to $t = \pi/2$ [the point $(0, a)$] to $t = \pi$ [the point $(-a, 0)$], we see that the corresponding points are traced in a counterclockwise direction around the circle. Hence, the orientation is as indicated in Figure 51. ■

Let's discuss the curve in Example 2 further. The domain of each of the parametric equations is $-\infty < t < \infty$. Thus, the graph in Figure 51 is actually being repeated each time t increases by 2π. If we wanted the curve to consist of exactly 1 revolution in the counterclockwise direction, we could write

$$x = a \cos t \qquad y = a \sin t \qquad 0 \le t \le 2\pi$$

This curve starts at $t = 0$ [the point $(a, 0)$] and, proceeding counterclockwise around the circle, ends at $t = 2\pi$ [also the point $(a, 0)$].

If we wanted the curve to consist of exactly 3 revolutions in the counterclockwise direction, we could write

$$x = a \cos t \qquad y = a \sin t \qquad -2\pi \le t \le 4\pi$$

or

$$x = a \cos t \qquad y = a \sin t \qquad 0 \le t \le 6\pi$$

or

$$x = a \cos t \qquad y = a \sin t \qquad 2\pi \le t \le 8\pi$$

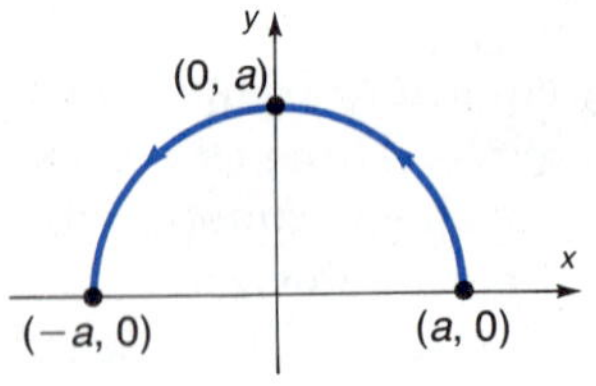

Figure 52

If we wanted the curve to consist of the upper semicircle of radius a, with a counterclockwise orientation, we could write

$$x = a \cos t \qquad y = a \sin t \qquad 0 \le t \le \pi$$

See Figure 52.

If we wanted the curve to consist of the left semicircle of radius a, with a clockwise orientation, we could write

$$x = -a \sin t \qquad y = -a \cos t \qquad 0 \le t \le \pi$$

See Figure 53.

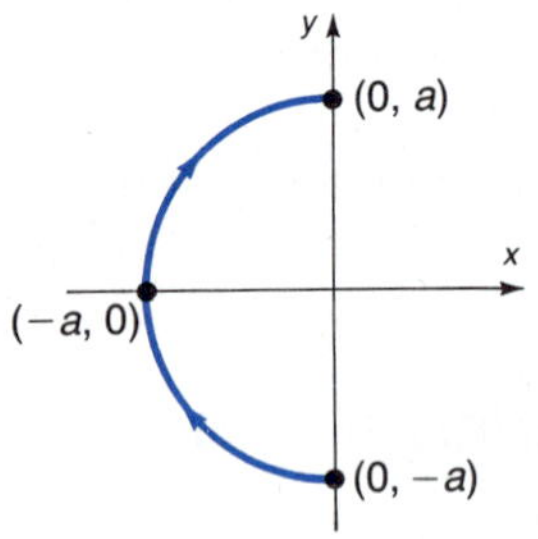

Figure 53

Time as a Parameter

If we think of the parameter t as time, then the parametric equations $x = f(t)$ and $y = g(t)$ of a curve C specify how the x- and y-coordinates of a moving point vary with time.

For example, we can use parametric equations to describe the motion of an object, sometimes referred to as **curvilinear motion**. Using parametric equations, we can specify not only where the object travels—that is, its location (x, y)—but also when it gets there—that is, the time t.

Example 3 Describe the motion of an object that moves along the curve

$$x = 4 \sin t \qquad y = 3 \cos t \qquad 0 \le t \le 2\pi$$

Solution We eliminate the parameter t by using the Pythagorean identity $\sin^2 t + \cos^2 t = 1$, obtaining

$$\left(\frac{x}{4}\right)^2 + \left(\frac{y}{3}\right)^2 = 1$$
$$\frac{x^2}{16} + \frac{y^2}{9} = 1$$

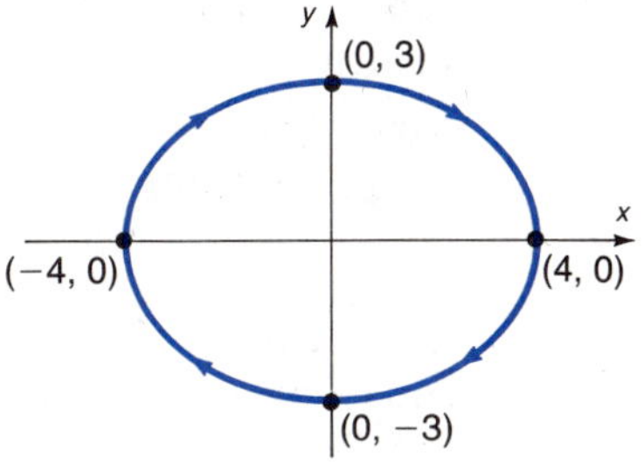

Figure 54

The curve is an ellipse, the center is at (0, 0), the major axis is along the x-axis, and the vertices are at $(\pm 4, 0)$. As t varies from $t = 0$ to $t = 2\pi$, the object moves around the ellipse in a clockwise direction starting at (0, 3), reaching (4, 0) when $t = \pi/2$ and (0, −3) when $t = \pi$, and ending at (0, 3) when $t = 2\pi$. See Figure 54. ■

Finding Parametric Equations

We now take up the question of how to find parametric equations of a given curve.

If the curve is defined by the equation $y = f(x)$, where f is a function, one way of finding parametric equations is simply to let $x = t$. Then $y = f(t)$. Thus,

$$x = t \qquad y = f(t) \qquad t \text{ in the domain of } f$$

are parametric equations of the curve.

Example 4 Find parametric equations for the equation: $y = x^2 - 4$

Solution Let $x = t$. Then the parametric equations are

$$x = t \qquad y = t^2 - 4 \qquad -\infty < t < \infty$$ ■

Another less obvious approach to Example 4 is to let $x = t^3$. Then the parametric equations become

$$x = t^3 \qquad y = t^6 - 4 \qquad -\infty < t < \infty$$

Care must be taken when using this approach, since the substitution for x must be a function that allows x to take on all the values stipulated by the domain of f. Thus, for example, letting $x = t^2$ so that $y = t^4 - 4$ does not result in equivalent parametric equations for $y = x^2 - 4$, since only points for which $x \geq 0$ are obtained.

Example 5 Find parametric equations for the ellipse

$$x^2 + \frac{y^2}{9} = 1$$

where the parameter t is time (in seconds) and:

(a) The motion around the ellipse is clockwise, begins at the point (0, 3), and requires 1 second for a complete revolution

(b) The motion around the ellipse is counterclockwise, begins at the point (1, 0), and requires 2 seconds for a complete revolution

Solution (a) Since the motion begins at the point (0, 3), we want $x = 0$ and $y = 3$ when $t = 0$. Furthermore, since the given equation is an ellipse, we begin by letting

$$x = \sin \omega t \qquad \frac{y}{3} = \cos \omega t$$

for some constant ω. These parametric equations clearly satisfy the equation. Furthermore, with this choice, when $t = 0$, we have $x = 0$ and $y = 3$. For the motion to be clockwise, the motion will have to begin with the value of x increasing and y decreasing as t increases. Thus, $\omega > 0$. Finally, since 1 revolution requires 1 second, then $\omega t = \omega = 2\pi$. Thus, parametric equations that satisfy the conditions stipulated are

$$x = \sin 2\pi t \qquad y = 3 \cos 2\pi t \qquad 0 \le t \le 1 \tag{2}$$

(b) Since the motion begins at the point (1, 0), we want $x = 1$ and $y = 0$ when $t = 0$. Furthermore, since the given equation is an ellipse, we begin by letting

$$x = \cos \omega t \qquad \frac{y}{3} = \sin \omega t$$

for some constant ω. These parametric equations clearly satisfy the equation. Furthermore, with this choice, when $t = 0$, we have $x = 1$ and $y = 0$. For the motion to be counterclockwise, the motion will have to begin with the value of x decreasing and y increasing as t increases. Thus, $\omega > 0$. Finally, since 1 revolution requires 2 seconds, then $\omega t = 2\omega = 2\pi$, so that $\omega = \pi$. Thus, the parametric equations that satisfy the conditions stipulated are

$$x = \cos \pi t \qquad y = 3 \sin \pi t \qquad 0 \le t \le 2 \tag{3}$$

■

Either of equations (2) or (3) can serve as parametric equations for the ellipse $x^2 + (y^2/9) = 1$ given in Example 5. The direction of the motion, the beginning point, and the time for 1 revolution merely serve to help arrive at a particular parametric representation.

The Cycloid

Suppose that a circle rolls along a horizontal line without slipping. As the circle rolls along the line, a point P on the circle will trace out a curve called a **cycloid** (see Figure 55). We now seek parametric equations* for a cycloid.

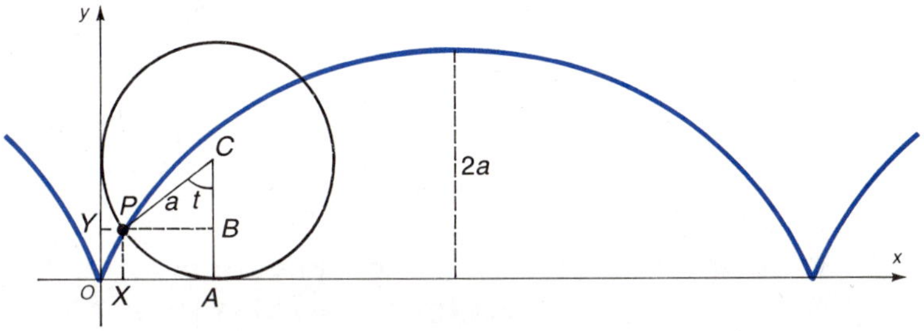

Figure 55
Cycloid

We begin with a circle of radius a and take the fixed line on which the circle rolls as the x-axis. Let the origin be one of the points at which the point P comes in contact with the x-axis. Figure 55 illustrates the position of this point P after the circle has rolled somewhat. The angle t (in radians) measures the angle through which the circle has rolled.

Since we require no slippage, it follows that

$$\text{Arc } AP = d(O, A)$$

Therefore,

$$at = d(O, A)$$

The x-coordinate of the point P is

$$d(O, X) = d(O, A) - d(X, A) = at - a \sin t = a(t - \sin t)$$

The y-coordinate of the point P is equal to

$$d(O, Y) = d(A, C) - d(B, C) = a - a \cos t = a(1 - \cos t)$$

Thus, the parametric equations of the cycloid are

$$x = a(t - \sin t) \qquad y = a(1 - \cos t) \tag{4}$$

*Any attempt to derive the rectangular equation of a cycloid would soon demonstrate how complicated the task is.

Applications to Mechanics

If a is negative in equations (4), we obtain an inverted cycloid, as shown in Figure 56(a). The inverted cycloid occurs as a result of some remarkable applications in the field of mechanics. We shall mention two of them—the *brachistochrone* and the *tautochrone*.*

Figure 56

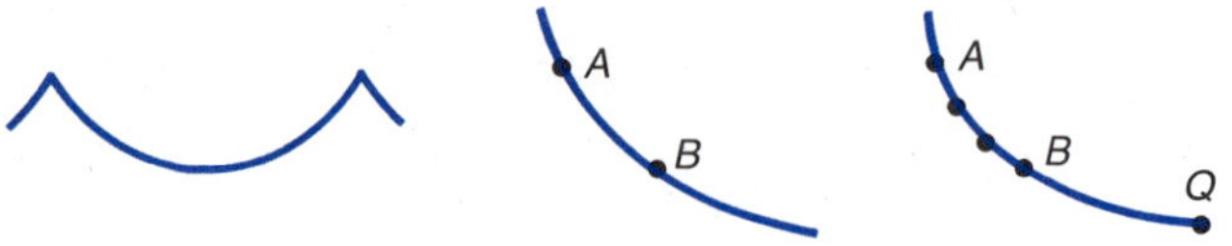

(a) Inverted cycloid (b) Curve of quickest descent (c) All reach Q at the same time

The **brachistochrone** is the curve of quickest descent. If a particle is constrained to follow some path from one point A to a lower point B (not on the same vertical line) and is acted upon only by gravity, the time needed to make the descent is least if the path is an inverted cycloid. See Figure 56(b). This remarkable discovery, which is attributed to many famous mathematicians (including Johann Bernoulli and Blaise Pascal), was a significant step in creating the branch of mathematics known as the *calculus of variations*.

To define the **tautochrone**, let Q be the lowest point on an inverted cycloid. If several particles placed at various positions on an inverted cycloid simultaneously begin to slide down the cycloid, they will reach the point Q at the same time, as indicated in Figure 56(c). The tautochrone property of the cycloid was used by Christian Huygens (1629–1695), the Dutch mathematician, physicist, and astronomer, to construct a pendulum clock with a bob that swings along a cycloid (see Figure 57). In Huygens' clock, the bob was made to swing along a cycloid by suspending the bob on a thin wire constrained by two plates shaped like cycloids. In a clock of this design, the period of the pendulum is independent of its amplitude.

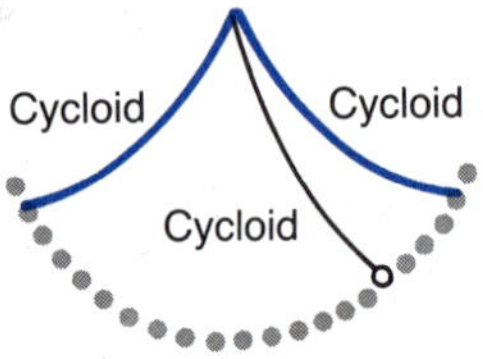

Figure 57
A flexible pendulum constrained by cycloids swings in a cycloid

EXERCISE 11.7

In Problems 1–20, graph the curve whose parametric equations are given, and show its orientation. Find the rectangular equation of each curve.

1. $x = 3t + 2,\quad y = t + 1;\quad 0 \le t \le 4$

2. $x = t - 3,\quad y = 2t + 4;\quad 0 \le t \le 2$

*In Greek, *brachistochrone* means "the shortest time," and *tautochrone* means "equal time."

3. $x = t + 2, \quad y = \sqrt{t}; \qquad t \geq 0$

4. $x = \sqrt{2t}, \quad y = 4t; \quad t \geq 0$

5. $x = t^2 + 4, \quad y = t^2 - 4; \quad -\infty < t < \infty$

6. $x = \sqrt{t} + 4, \quad y = \sqrt{t} - 4; \quad t \geq 0$

7. $x = 3t^2, \quad y = t + 1; \quad -\infty < t < \infty$

8. $x = 2t - 4, \quad y = 4t^2; \quad -\infty < t < \infty$

9. $x = 2e^t, \quad y = 1 + e^t; \quad t \geq 0$

10. $x = e^t, \quad y = e^{-t}; \quad t \geq 0$

11. $x = \sqrt{t}, \quad y = t^{3/2}; \quad t \geq 0$

12. $x = t^{3/2} + 1, \quad y = \sqrt{t}; \quad t \geq 0$

13. $x = 2 \cos t, \quad y = 3 \sin t; \quad 0 \leq t \leq 2\pi$

14. $x = 2 \cos t, \quad y = 3 \sin t; \quad 0 \leq t \leq \pi$

15. $x = 2 \cos t, \quad y = 3 \sin t; \quad -\pi \leq t \leq 0$

16. $x = 2 \cos t, \quad y = \sin t; \quad 0 \leq t \leq \pi/2$

17. $x = \sec t, \quad y = \tan t; \quad 0 \leq t \leq \pi/4$

18. $x = \csc t, \quad y = \cot t; \quad \pi/4 \leq t \leq \pi/2$

19. $x = \sin^3 t, \quad y = \cos^3 t; \quad 0 \leq t \leq 2\pi$

20. $x = t^2, \quad y = \ln t; \quad t > 0$

In Problems 21–24, find two different parametric equations for each rectangular equation.

21. $y = x^3$ **22.** $y = x^4 + 1$ **23.** $x = y^{3/2}$ **24.** $x = \sqrt{y}$

In Problems 25–28, find parametric equations for an object that moves along the ellipse $(x^2/4) + (y^2/9) = 1$ *with the motion described.*

25. The motion begins at (2, 0), is clockwise, and requires 2 seconds for a complete revolution.

26. The motion begins at (0, 3), is clockwise, and requires 1 second for a complete revolution.

27. The motion begins at (0, 3), is counterclockwise, and requires 1 second for a complete revolution.

28. The motion begins at (2, 0), is counterclockwise, and requires 3 seconds for a complete revolution.

In Problems 29 and 30, the parametric equations of four curves are given. Graph each of them, indicating the orientation.

29. C_1: $x = t, \quad y = t^2; \quad -4 \leq t \leq 4$
C_2: $x = \cos t, \quad y = 1 - \sin^2 t; \quad 0 \leq t \leq \pi$
C_3: $x = e^t, \quad y = e^{2t}; \quad 0 \leq t \leq \ln 4$
C_4: $x = \sqrt{t}, \quad y = t; \quad 0 \leq t \leq 16$

30. C_1: $x = t, \quad y = \sqrt{1 - t^2}; \quad -1 \leq t \leq 1$
C_2: $x = \sin t, \quad y = \cos t; \quad 0 \leq t \leq 2\pi$
C_3: $x = \cos t, \quad y = \sin t; \quad 0 \leq t \leq 2\pi$
C_4: $x = \sqrt{1 - t^2}, \quad y = t; \quad -1 \leq t \leq 1$

31. Show that the parametric equations for a line passing through the points (x_1, y_1) and (x_2, y_2) are

$$x = (x_2 - x_1)t + x_1 \qquad y = (y_2 - y_1)t + y_1 \qquad -\infty < t < \infty$$

What is the orientation of this line?

32. The position of a projectile fired with an initial velocity v_0 feet per second and at an angle θ to the horizontal at the end of t seconds is given by the parametric equations

$$x = (v_0 \cos \theta)t \qquad y = (v_0 \sin \theta)t - 16t^2$$

(a) Obtain the rectangular equation of the trajectory and identify the curve.
(b) Show that the projectile hits the ground ($y = 0$) when $t = \frac{1}{16}v_0 \sin \theta$.
(c) How far has the projectile traveled (horizontally) when it strikes the ground?

CHAPTER REVIEW

THINGS TO KNOW

Equations

Parabola	See Tables 1 and 2.	
Ellipse	See Table 3.	
Hyperbola	See Table 4.	
General equation of a conic	$Ax^2 + Bxy + Cy^2 + Dx + Ey + F = 0$	Parabola if $B^2 - 4AC = 0$ Ellipse (or circle) if $B^2 - 4AC < 0$ Hyperbola if $B^2 - 4AC > 0$
Conic in polar coordinates	$\dfrac{d(F, P)}{d(P, D)} = e$	Parabola if $e = 1$ Ellipse if $e < 1$ Hyperbola if $e > 1$
Plane curve	$(x, y) = (f(t), g(t))$, t a parameter	
Polar equations of a conic	See Table 5.	

Definitions

Parabola	Set of points P in the plane for which $d(F, P) = d(P, D)$, where F is the focus and D is the directrix
Ellipse	Set of points P in the plane, the sum of whose distances from two fixed points (the foci) is a constant
Hyperbola	Set of points P in the plane, the difference of whose distances from two fixed points (the foci) is a constant

Formulas

Rotation formulas	$x = x' \cos\theta - y' \sin\theta$ $y = x' \sin\theta + y' \cos\theta$
Angle θ of rotation that eliminates $x'y'$-term	$\cot 2\theta = \dfrac{A - C}{B}$

How To:

Find the vertex, focus, and directrix of a parabola given its equation

Graph a parabola given its equation

Find an equation of a parabola given certain information about the parabola

Find the center, foci, and vertices of an ellipse given its equation
Graph an ellipse given its equation
Find an equation of an ellipse given certain information about the ellipse
Find the center, foci, vertices, and asymptotes of a hyperbola given its equation
Graph a hyperbola given its equation
Find an equation of a hyperbola given certain information about the hyperbola
Identify conics without completing the square
Identify conics without a rotation of axes
Use rotation formulas to transform second-degree equations so that no xy-term is present
Identify and graph conics given by a polar equation
Graph parametric equations
Find the rectangular equation given the parametric equations

FILL-IN-THE-BLANK ITEMS

1. A(n) __________ is the collection of all points in the plane such that the distance from each point to a fixed point equals its distance to a fixed line.
2. A(n) __________ is the collection of all points in the plane, the sum of whose distances from two fixed points is a constant.
3. A(n) __________ is the collection of all points in the plane, the difference of whose distances from two fixed points is a constant.
4. For an ellipse, the foci lie on the __________ axis; for a hyperbola, the foci lie on the __________ axis.
5. For the ellipse $(x^2/9) + (y^2/16) = 1$, the major axis is along the __________.
6. The equations of the asymptotes of the hyperbola $(y^2/9) - (x^2/4) = 1$ are __________ and __________.
7. To transform the equation

$$Ax^2 + Bxy + Cy^2 + Dx + Ey + F = 0 \qquad B \neq 0$$

into one in x' and y' without an $x'y'$-term, rotate the axes through an acute angle θ that satisfies the equation __________.
8. The polar equation

$$r = \frac{8}{4 - 2 \sin \theta}$$

is a conic whose eccentricity is __________. It is a(n) __________ whose directrix is __________ to the polar axis at a distance __________ units __________ the pole.
9. The parametric equations $x = 2 \sin t$ and $y = 3 \cos t$ represent a(n) __________.

REVIEW EXERCISES

In Problems 1–20, identify each equation. If it is a parabola, give its vertex, focus, and directrix; if it is an ellipse, give its center, vertices, and foci; if it is a hyperbola, give its center, vertices, foci, and asymptotes.

1. $y^2 = -16x$ **2.** $16x^2 = y$ **3.** $\frac{x^2}{4} - y^2 = 1$

4. $\frac{x^2}{9} - y^2 = 1$ **5.** $\frac{y^2}{25} + \frac{x^2}{16} = 1$ **6.** $\frac{x^2}{9} + \frac{y^2}{16} = 1$

7. $x^2 + 4y = 4$ **8.** $3y^2 - x^2 = 9$ **9.** $4x^2 - y^2 = 8$

10. $9x^2 + 4y^2 = 36$ **11.** $x^2 - 4x = 2y$ **12.** $2y^2 - 4y = x - 2$

13. $y^2 - 4y - 4x^2 + 8x = 4$ **14.** $4x^2 + y^2 + 8x - 4y + 4 = 0$

15. $4x^2 + 9y^2 - 16x - 18y = 11$ **16.** $4x^2 + 9y^2 - 16x + 18y = 11$

17. $4x^2 - 16x + 16y + 32 = 0$ **18.** $4y^2 + 3x - 16y + 19 = 0$

19. $9x^2 + 4y^2 - 18x + 8y = 23$ **20.** $x^2 - y^2 - 2x - 2y = 1$

In Problems 21–36, obtain an equation of the conic described. Graph the equation.

21. Parabola; focus at $(-2, 0)$; directrix the line $x = 2$

22. Ellipse; center at $(0, 0)$; focus at $(0, 3)$; vertex at $(0, 5)$

23. Hyperbola; center at $(0, 0)$; focus at $(0, 4)$; vertex at $(0, -2)$

24. Parabola; vertex at $(0, 0)$; directrix the line $y = -3$

25. Ellipse; foci at $(-3, 0)$ and $(3, 0)$; vertex at $(4, 0)$

26. Hyperbola; vertices at $(-2, 0)$ and $(2, 0)$; focus at $(4, 0)$

27. Parabola; vertex at $(2, -3)$; focus at $(2, -4)$

28. Ellipse; center at $(-1, 2)$; focus at $(0, 2)$; vertex at $(2, 2)$

29. Hyperbola; center at $(-2, -3)$; focus at $(-4, -3)$; vertex at $(-3, -3)$

30. Parabola; focus at $(3, 6)$; directrix the line $y = 8$

31. Ellipse; foci at $(-4, 2)$ and $(-4, 8)$; vertex at $(-4, 10)$

32. Hyperbola; vertices at $(-3, 3)$ and $(5, 3)$; focus at $(7, 3)$

33. Center at $(-1, 2)$; $a = 3$; $c = 4$; transverse axis parallel to the x-axis

34. Center at $(4, -2)$; $a = 1$; $c = 4$; transverse axis parallel to the y-axis

35. Vertices at $(0, 1)$ and $(6, 1)$; asymptote the line $3y + 2x - 9 = 0$

36. Vertices at $(4, 0)$ and $(4, 4)$; asymptote the line $y + 2x - 10 = 0$

In Problems 37–46, identify each conic without completing the squares and without applying a rotation of axes.

37. $y^2 + 4x + 3y - 8 = 0$ **38.** $2x^2 - y + 8x = 0$

39. $x^2 + 2y^2 + 4x - 8y + 2 = 0$ **40.** $x^2 - 8y^2 - x - 2y = 0$

41. $9x^2 - 12xy + 4y^2 + 8x + 12y = 0$ **42.** $4x^2 + 4xy + y^2 - 8\sqrt{5}x + 16\sqrt{5}y = 0$

43. $4x^2 + 10xy + 4y^2 - 9 = 0$ **44.** $4x^2 - 10xy + 4y^2 - 9 = 0$

45. $x^2 - 2xy + 3y^2 + 2x + 4y - 1 = 0$ **46.** $4x^2 + 12xy - 10y^2 + x + y - 10 = 0$

In Problems 47–52, rotate the axes so that the new equation contains no xy-term. Discuss and graph the new equation.

47. $2x^2 + 5xy + 2y^2 - \frac{9}{2} = 0$

48. $2x^2 - 5xy + 2y^2 - \frac{9}{2} = 0$

49. $6x^2 + 4xy + 9y^2 - 20 = 0$

50. $x^2 + 4xy + 4y^2 + 16\sqrt{5}x - 8\sqrt{5}y = 0$

51. $4x^2 - 12xy + 9y^2 + 12x + 8y = 0$

52. $9x^2 - 24xy + 16y^2 + 80x + 60y = 0$

In Problems 53–58, identify the conic each polar equation represents, and graph it.

53. $r = \dfrac{4}{1 - \cos\theta}$

54. $r = \dfrac{6}{1 + \sin\theta}$

55. $r = \dfrac{6}{2 - \sin\theta}$

56. $r = \dfrac{2}{3 + 2\cos\theta}$

57. $r = \dfrac{8}{4 + 8\cos\theta}$

58. $r = \dfrac{10}{5 + 20\sin\theta}$

In Problems 59–62, convert each polar equation to a rectangular equation.

59. $r = \dfrac{4}{1 - \cos\theta}$

60. $r = \dfrac{6}{2 - \sin\theta}$

61. $r = \dfrac{8}{4 + 8\cos\theta}$

62. $r = \dfrac{2}{3 + 2\cos\theta}$

In Problems 63–68, graph the curve whose parametric equations are given and show its orientation. Find the rectangular equation of each curve.

63. $x = 4t - 2, \quad y = 1 - t; \quad -\infty < t < \infty$

64. $x = 2t^2 + 6, \quad y = 5 - t; \quad -\infty < t < \infty$

65. $x = 3\sin t, \quad y = 4\cos t + 2; \quad 0 \le t \le 2\pi$

66. $x = \ln t, \quad y = t^3; \quad t > 0$

67. $x = \sec^2 t, \quad y = \tan^2 t; \quad 0 \le t \le \pi/4$

68. $x = t^{3/2}, \quad y = 2t + 4; \quad t \ge 0$

69. Find an equation of the hyperbola whose foci are the vertices of the ellipse $4x^2 + 9y^2 = 36$ and whose vertices are the foci of this ellipse.

70. Find an equation of the ellipse whose foci are the vertices of the hyperbola $x^2 - 4y^2 = 16$ and whose vertices are the foci of this hyperbola.

71. Describe the collection of points in a plane so that the distance from each point to the point $(3, 0)$ is three-fourths of its distance from the line $x = \frac{16}{3}$.

72. Describe the collection of points in a plane so that the distance from each point to the point $(5, 0)$ is five-fourths of its distance from the line $x = \frac{16}{5}$.

Systems of Equations and Inequalities

12

In this chapter, we take up the problem of solving equations and inequalities containing two or more variables. As the section titles suggest, there are various ways to solve such problems.

The *method of substitution* for solving equations in several unknowns goes back to ancient times.

The *method of elimination*, though it had existed for centuries, was put into systematic order by Karl Friedrich Gauss (1777–1855) and by Camille Jordan (1838–1922). This method is now used for solving large systems by computer.

The theory of *matrices* was developed in 1857 by Arthur Cayley (1821–1895), though only later were matrices used as we use them in this chapter. Matrices have become a very flexible instrument, useful in almost all areas of mathematics.

The method of *determinants* was invented by Seki Kōwa (1642–1708) in 1683 in Japan and by Gottfried Wilhelm von Leibniz (1646–1716) in 1693 in Germany. Both used them only in relation to linear equations. *Cramer's Rule* is named after Gabriel Cramer (1704–1752) of Switzerland, who popularized the use of determinants for solving linear systems.

12.1 Systems of Linear Equations: Substitution; Elimination

A **system of equations** is a collection of two or more equations, each containing one or more variables. Example 1 gives some samples of systems of equations.

Example 1

(a) $\begin{cases} 2x + y = 5 & (1) \\ -4x + 6y = -2 & (2) \end{cases}$ Two equations containing two variables, x and y

(b) $\begin{cases} x + y^2 = 5 & (1) \\ 2x + y = 4 & (2) \end{cases}$ Two equations containing two variables, x and y

(c) $\begin{cases} x + y + z = 6 & (1) \\ 3x - 2y + 4z = 9 & (2) \\ x - y - z = 0 & (3) \end{cases}$ Three equations containing three variables, x, y, and z

(d) $\begin{cases} x + y + z = 5 & (1) \\ x - y = 2 & (2) \end{cases}$ Two equations containing three variables, x, y, and z

(e) $\begin{cases} x + y + z = 6 & (1) \\ 2x + 2z = 4 & (2) \\ y + z = 2 & (3) \\ x = 4 & (4) \end{cases}$ Four equations containing three variables, x, y, and z ■

We use a brace, as shown above, to remind us that we are dealing with a system of equations. We also will find it convenient to number each equation in the system.

A **solution** of a system of equations consists of values for the variables that reduce each equation of the system to a true statement. To **solve** a system of equations means to find all solutions of the system.

For example, $x = 2$, $y = 1$ is a solution of the system in Example 1(a), because

$$2(2) + 1 = 5 \quad \text{and} \quad -4(2) + 6(1) = -2$$

A solution of the system in Example 1(b) is $x = 1$, $y = 2$, because

$$1 + 2^2 = 5 \quad \text{and} \quad 2(1) + 2 = 4$$

Another solution of the system in Example 1(b) is $x = \frac{11}{4}$, $y = -\frac{3}{2}$, which you can check for yourself. A solution of the system in Example 1(c) is $x = 3$, $y = 2$, $z = 1$, because

$$\begin{cases} 3 + 2 + 1 = 6 & (1) \\ 3(3) - 2(2) + 4(1) = 9 & (2) \\ 3 - 2 - 1 = 0 & (3) \end{cases} \quad x = 3,\ y = 2,\ z = 1$$

Note that $x = 3, y = 3, z = 0$ is not a solution of the system in Example 1(c):

$$\begin{cases} 3 + 3 + 0 = 6 & (1) \quad x = 3, y = 3, z = 0 \\ 3(3) - 2(3) + 4(0) = 3 \neq 9 & (2) \\ 3 - 3 - 0 = 0 & (3) \end{cases}$$

Although these values satisfy equations (1) and (3), they do not satisfy equation (2). Any solution of the system must satisfy *each* equation of the system.

When a system of equations has at least one solution, it is said to be **consistent**; otherwise, it is called **inconsistent**.

An equation in n variables is said to be **linear** if it is equivalent to an equation of the form

$$a_1x_1 + a_2x_2 + \cdots + a_nx_n = b$$

where $x_1, x_2, \ldots, x_n$ are n distinct variables, $a_1, a_2, \ldots, a_n, b$ are constants, and at least one of the a's is not 0.

Some examples of linear equations are

$$2x + 3y = 2 \qquad 5x - 2y + 3z = 10 \qquad 8x + 8y - 2z + 5w = 0$$

If each equation in a system of equations is linear, then we have a **system of linear equations**. Thus, the systems in Examples 1(a), (c), (d), and (e) are linear, whereas the system in Example 1(b) is nonlinear. We concentrate on solving linear systems in Sections 12.1–12.3, and we will take up nonlinear systems in Section 12.4.

Method of Substitution

Let's now discuss one way to solve a system of linear equations.

Example 2 Solve: $\begin{cases} 2x + y = 5 & (1) \\ -4x + 6y = -2 & (2) \end{cases}$

Solution We solve the first equation for y, obtaining

$$y = 5 - 2x \qquad (1)$$

We substitute this result for y in the second equation and solve for the one remaining variable:

$$\begin{aligned} -4x + 6y &= -2 \\ -4x + 6(5 - 2x) &= -2 \\ -4x + 30 - 12x &= -2 \\ -16x &= -32 \\ x &= 2 \end{aligned}$$

Once we know $x = 2$, we can easily find the value of y by **back-substitution**, that is, by substituting 2 for x in one of the original equations. We use the first one:

$$\begin{aligned} 2x + y &= 5 \\ 2(2) + y &= 5 \\ y &= 1 \end{aligned}$$

The solution of the system is $x = 2$, $y = 1$. We checked this solution earlier. [This is the same system as in Example 1(a).] ■

In the solution to Example 2, once we found $x = 2$, we could have found y by using equation (1), the equation we obtained at the beginning when we solved for y:

$$y = 5 - 2x = 5 - 2(2) = 1$$
$$\uparrow$$
$$x = 2$$

The method used to solve the system in Example 2 is called **substitution**. The steps to be used are outlined below.

Steps for Solving by Substitution

STEP 1: Pick one of the equations and solve for one of the variables in terms of the remaining variables.

STEP 2: Substitute the result in the remaining equations.

STEP 3: If one equation in one variable results, solve this equation. Otherwise, repeat Step 1 until a single equation with one variable remains.

STEP 4: Find the values of the remaining variables by back-substitution.

STEP 5: Check the solution found.

Example 3 Solve: $\begin{cases} 3x - 2y = 5 & (1) \\ 5x - y = 6 & (2) \end{cases}$

Solution STEP 1: After looking at the two equations, we conclude that it is easiest to solve for the variable y in equation (2):

$$\begin{aligned} 5x - y &= 6 \\ y &= 5x - 6 \end{aligned}$$

STEP 2: We substitute this result into equation (1) and simplify:

$$\begin{aligned} 3x - 2y &= 5 \\ 3x - 2(5x - 6) &= 5 \\ -7x + 12 &= 5 \\ -7x &= -7 \\ x &= 1 \end{aligned}$$

STEP 3: Because we now have one solution, namely $x = 1$, we proceed to Step 4.

STEP 4: Knowing $x = 1$, we can find y from the equation

$$y = 5x - 6 = 5(1) - 6 = -1$$

STEP 5: *Check:* $\begin{cases} 3(1) - 2(-1) = 3 + 2 = 5 \\ 5(1) - (-1) \quad = 5 + 1 = 6 \end{cases}$

The solution of the system is $x = 1$, $y = -1$. ■

Example 4 Solve: $\begin{cases} 2x - 3y = 7 & (1) \\ 4x + 5y = 3 & (2) \end{cases}$

Solution STEP 1: In looking over the system, we conclude that there is no way to solve for one of the variables without introducing fractions. We solve for the variable x in equation (1):

$$\begin{aligned} 2x - 3y &= 7 \\ 2x &= 3y + 7 \\ x &= \tfrac{3}{2}y + \tfrac{7}{2} \end{aligned}$$

STEP 2: We substitute this result for x in equation (2) and simplify:

$$\begin{aligned} 4x + 5y &= 3 \\ 4\left(\tfrac{3}{2}y + \tfrac{7}{2}\right) + 5y &= 3 \\ 6y + 14 + 5y &= 3 \\ 11y + 14 &= 3 \\ 11y &= -11 \end{aligned}$$

STEP 3: $y = -1$

STEP 4: $x = \frac{3}{2}y + \frac{7}{2} = \frac{3}{2}(-1) + \frac{7}{2} = \frac{4}{2} = 2$

STEP 5: *Check:* $\begin{cases} 2(2) - 3(-1) = 4 + 3 = 7 \\ 4(2) + 5(-1) = 8 - 5 = 3 \end{cases}$

The solution is $x = 2$, $y = -1$. ■

Method of Elimination

A second method for solving a system of linear equations is the *method of elimination.* This method is usually preferred over substitution if substitution leads to fractions or if the system contains more than two variables. Elimination also provides the necessary motivation for solving systems using matrices (the subject of the next section).

The idea behind the method of elimination is to keep replacing the original equations in the system with equivalent equations until a system of equations with an obvious solution is reached. When we proceed in this way we obtain **equivalent systems of equations**. The rules for obtaining equivalent equations are the same as those studied in Chapter 2. However, we may also interchange any two equations of the system and/or replace any equation in the system by the sum or difference of that equation and any other equation in the system.

Rules for Obtaining an Equivalent System of Equations

1. Interchange any two equations of the system.
2. Multiply (or divide) each side of an equation by the same nonzero constant.
3. Replace any equation in the system by the sum (or difference) of that equation and any other equation in the system.

An example will give you the idea. As you work through the example, pay particular attention to the pattern being followed.

Example 5 Solve:
$$\begin{cases} 2x + 3y = 1 & (1) \\ -x + y = -3 & (2) \end{cases}$$

Solution If we multiply each side of equation (2) by 2, we get the equivalent system
$$\begin{cases} 2x + 3y = 1 & (1) \\ -2x + 2y = -6 & (2) \end{cases}$$

If we now replace equation (2) of this system by the sum of the two equations, we get the equivalent system
$$\begin{cases} 2x + 3y = 1 & (1) \\ 5y = -5 & (2) \end{cases}$$

Now we multiply the resulting equation (2) by $\frac{1}{5}$, obtaining the equivalent system
$$\begin{cases} 2x + 3y = 1 & (1) \\ y = -1 & (2) \end{cases}$$

From this system we see the obvious solution $y = -1$. We use this value for y in equation (1) and simplify, to get

$$\begin{aligned} 2x + 3(-1) &= 1 \\ 2x &= 4 \\ x &= 2 \end{aligned}$$

Thus, the solution of the original system is $x = 2$, $y = -1$. We leave it to you to check the solution. ■

The procedure used in Example 5 is called the **method of elimination**. Notice the pattern of the solution. First, we eliminated the variable x from the second equation. Then we back-substituted; that is, we substituted the value found for y back into the first equation to find x.

Let's do another example.

Example 6 Use the method of elimination to solve the system of equations:

$$\begin{cases} \frac{1}{3}x + 5y = -4 & (1) \\ 2x + 3y = 3 & (2) \end{cases}$$

Solution We begin by multiplying each side of equation (1) by 3, in order to remove the fraction $\frac{1}{3}$:

$$\begin{cases} \frac{1}{3}x + 5y = -4 & (1) \\ 2x + 3y = 3 & (2) \end{cases}$$

$$\begin{cases} x + 15y = -12 & (1) \\ 2x + 3y = 3 & (2) \end{cases}$$

Thinking ahead, we decide to multiply each side of equation (1) by -2, because then the sum of the two equations will result in an equation with the variable x eliminated. [Note that we also could multiply each side of equation (1) by 2 and then replace equation (1) by the difference of the two equations. However, because subtracting is more likely to result in a calculation error, we follow the safer practice of adding equations.]

$$\begin{cases} -2x - 30y = 24 & (1) \\ 2x + 3y = 3 & (2) \end{cases}$$

$$\begin{cases} -27y = 27 & (1) \\ 2x + 3y = 3 & (2) \end{cases}$$ Replace equation (1) by the sum of the two equations.

$$\begin{cases} y = -1 & (1) \\ 2x + 3y = 3 & (2) \end{cases}$$ Multiply each side of equation (1) by $-\frac{1}{27}$.

$$\begin{cases} y = -1 & (1) \\ 2x + 3(-1) = 3 & (2) \end{cases} \quad \text{Replace } y \text{ by } -1.$$

$$\begin{cases} y = -1 & (1) \\ 2x = 6 & (2) \end{cases}$$

$$\begin{cases} y = -1 & (1) \\ x = 3 & (2) \end{cases}$$

The solution of the original system is $x = 3$, $y = -1$. ■

Two Linear Equations Containing Two Variables

We can view the problem of solving a system of two linear equations containing two variables as a geometry problem. The graph of each equation in such a system is a straight line. Thus, a system of two equations containing two variables represents a pair of lines. The lines either (*1*) intersect or (*2*) are parallel or (*3*) are **coincident** (that is, identical).

1. If the lines intersect, then the system of equations has one solution, given by the point of intersection. The system is **consistent**.
2. If the lines are parallel, then the system of equations has no solution, because the lines never intersect. The system is **inconsistent**.
3. If the lines are coincident, then the system of equations has infinitely many solutions, represented by the totality of points on the line. The system is **consistent**.

Figure 1 illustrates these conclusions.

Figure 1

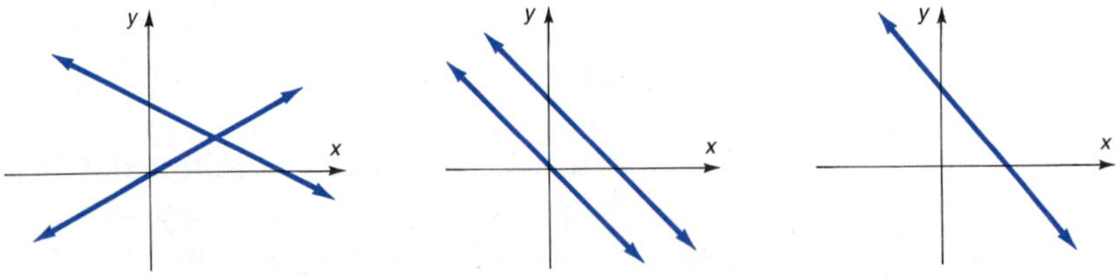

(a) Intersecting lines; system has one solution (b) Parallel lines; system has no solution (c) Coincident lines; system has infinitely many solutions

We have already seen several examples in which the system of equations represents two lines that intersect. The next example illustrates what happens when the method of substitution is used to solve a system of equations that has no solution.

Example 7 Solve: $\begin{cases} 2x + y = 5 & (1) \\ 4x + 2y = 8 & (2) \end{cases}$

Solution We choose to solve equation (1) for y:

$$2x + y = 5$$
$$y = 5 - 2x$$

Substituting in equation (2), we get

$$4x + 2y = 8$$
$$4x + 2(5 - 2x) = 8$$
$$4x + 10 - 4x = 8$$
$$0 \cdot x = -2$$

This equation has no solution. Thus, we conclude that the system itself has no solution and is therefore inconsistent. ■

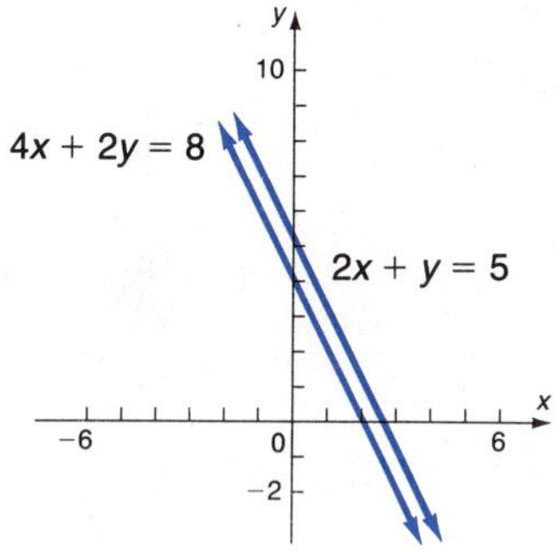

Figure 2

Figure 2 illustrates the pair of lines whose equations form the system in Example 7. Notice that the graphs of the two equations are lines, each with slope -2; one has a y-intercept of 5, the other a y-intercept of 4. Thus, the lines are parallel and have no point of intersection. This geometric statement is equivalent to the algebraic statement that the system has no solution.

The next example is an illustration of a system with infinitely many solutions.

Example 8 Solve: $\begin{cases} 2x + y = 4 & (1) \\ -6x - 3y = -12 & (2) \end{cases}$

Solution We choose to use the method of elimination:

$$\begin{cases} 2x + y = 4 & (1) \\ -6x - 3y = -12 & (2) \end{cases}$$

$$\begin{cases} 6x + 3y = 12 & (1) \\ -6x - 3y = -12 & (2) \end{cases}$$ Multiply each side of equation (1) by 3.

$$\begin{cases} 6x + 3y = 12 & (1) \\ 0 = 0 & (2) \end{cases}$$ Replace equation (2) by the sum of equations (1) and (2).

The original system is thus equivalent to a system containing one equation. This means that any values of x and y for which $6x + 3y = 12$, or equivalently, $2x + y = 4$, are solutions. For example, $x = 2$, $y = 0$; $x = 0$, $y = 4$; $x = -2$, $y = 8$; $x = 4$, $y = -4$; and so on. There are, in fact, infinitely many values of x and y for which $2x + y = 4$, so the original system has infinitely many solutions. We will write the solutions of the original system either as

$$y = 4 - 2x$$

where x can be any real number, or as

$$x = 2 - \frac{1}{2}y$$

where y can be any real number. ■

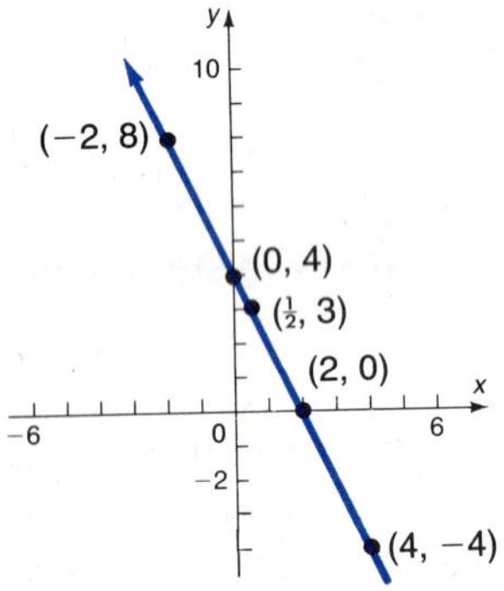

Figure 3
$$\begin{cases} 2x + y = 4 \\ -6x - 3y = -12 \end{cases}$$

Figure 3 illustrates the situation presented in Example 8. Notice that the graphs of the two equations are lines, each with slope -2 and each with y-intercept 4. Thus, the lines are coincident. Notice also that equation (2) in the original system is just -3 times equation (1); again indicating that the two equations are equivalent.

For the system in Example 8, we can write down some of the infinite number of solutions by assigning values to x and then finding $y = 4 - 2x$. Thus:

If $x = 4$, $y = -4$.
If $x = 0$, then $y = 4$.
If $x = \frac{1}{2}$, then $y = 3$.

The pairs (x, y), of course, are points on the line in Figure 3.

Example 9 A-1 Car Hire charges $10 per day plus 12¢ per mile for the rental of a standard car. The same type of car at E-Z Rent costs $12 per day plus 10¢ per mile.

(a) What mileage results in equal rental charges for the two firms?
(b) Use a graph to decide which firm to use if the number of miles to be driven is known and cost is the deciding factor.

Solution (a) Let C denote the cost (in dollars) of each rental, and let x be the number of miles driven. Then we can form the system of equations

$$\begin{cases} C = 10 + 0.12x & \text{For A-1 Car Hire} \\ C = 12 + 0.10x & \text{For E-Z Rent} \end{cases}$$

We solve the system using substitution:

$$\begin{aligned} 10 + 0.12x &= 12 + 0.10x \\ 0.02x &= 2 \\ x &= 100 \text{ miles} \end{aligned}$$

To check, we see whether the costs are equal:

$$C = 10 + 0.12(100) = 10 + 12 = 22$$
$$C = 12 + 0.10(100) = 12 + 10 = 22$$

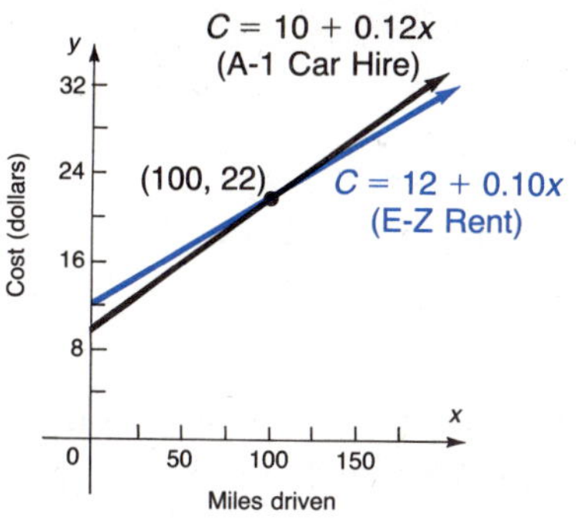

Figure 4

Thus, a daily usage of 100 miles results in equal costs of \$22 for both companies.

(b) We graph each equation in the system, measuring C along the vertical axis, as shown in Figure 4. The point where the lines intersect is the solution we found in part (a). From the graph, it is clear that for $0 < x < 100$, the cost at A-1 Car Hire is less; whereas, for mileage in excess of 100 miles, the cost at E-Z Rent is less. ■

Three Linear Equations Containing Three Variables

Just as with a system of two linear equations containing two variables, a system of three linear equations containing three variables also has either exactly one solution, or no solution, or infinitely many solutions.

Let's see how elimination works on a system of three equations containing three variables.

Example 10 Use the method of elimination to solve the system of equations:

$$\begin{cases} x + y - z = -1 & (1) \\ 4x - 3y + 2z = 16 & (2) \\ 2x - 2y - 3z = 5 & (3) \end{cases}$$

Solution For a system of three equations, we attempt to eliminate one variable at a time, using a pair of equations. We begin by multiplying each side of equation (1) by -2, in anticipation of eliminating the variable x by adding equations (1) and (3):

$$\begin{cases} -2x - 2y + 2z = 2 & (1) \\ 4x - 3y + 2z = 16 & (2) \\ 2x - 2y - 3z = 5 & (3) \end{cases}$$

$$\begin{cases} -2x - 2y + 2z = 2 & (1) \\ 4x - 3y + 2z = 16 & (2) \\ -4y - z = 7 & (3) \end{cases}$$

Replace equation (3) by the sum of equations (1) and (3).

We now eliminate the same variable x from equation (2):

$$\begin{cases} -4x - 4y + 4z = 4 & (1) \\ 4x - 3y + 2z = 16 & (2) \\ -4y - z = 7 & (3) \end{cases}$$

Multiply each side of equation (1) by 2.

$$\begin{cases} -4x - 4y + 4z = 4 & (1) \\ -7y + 6z = 20 & (2) \\ -4y - z = 7 & (3) \end{cases}$$

Replace equation (2) by the sum of equations (1) and (2).

We now eliminate y from equation (3):

$$\begin{cases} -4x - 4y + 4z = 4 & (1) \\ -28y + 24z = 80 & (2) \quad \text{Multiply each side by 4.} \\ 28y + 7z = -49 & (3) \quad \text{Multiply each side by } -7. \end{cases}$$

$$\begin{cases} -4x - 4y + 4z = 4 & (1) \\ -28y + 24z = 80 & (2) \\ 31z = 31 & (3) \quad \text{Replace equation (3) by the sum of equations (2) and (3).} \end{cases}$$

$$\begin{cases} -4x - 4y + 4z = 4 & (1) \\ -28y + 24z = 80 & (2) \\ z = 1 & (3) \quad \text{Multiply each side by } \tfrac{1}{31}. \end{cases}$$

$$\begin{cases} -4x - 4y + 4 = 4 & (1) \quad \text{Back-substitute; replace } z \text{ by 1 in} \\ -28y + 24 = 80 & (2) \quad \text{equations (1) and (2).} \\ z = 1 & (3) \end{cases}$$

$$\begin{cases} -4x - 4y = 0 & (1) \\ y = -2 & (2) \quad \text{Solve equation (2) for } y. \\ z = 1 & (3) \end{cases}$$

$$\begin{cases} -4x + 8 = 0 & (1) \quad \text{Back-substitute; replace } y \text{ by } -2. \\ y = -2 & (2) \\ z = 1 & (3) \end{cases}$$

$$\begin{cases} x = 2 & (1) \\ y = -2 & (2) \\ z = 1 & (3) \end{cases}$$

The solution of the original system is $x = 2$, $y = -2$, $z = 1$. (You should check this.) ■

Look back over the solution given in Example 10. Note the pattern of making equation (3) contain only the variable z, followed by making equation (2) contain only the variable y and equation (1) contain only the variable x. Although which variables to isolate is your choice, the methodology remains the same for all systems.

EXERCISE 12.1

In Problems 1–10, verify that the values of the variables listed are solutions of the system of equations.

1. $\begin{cases} 2x - y = 5 \\ 5x + 2y = 8 \end{cases}$

$x = 2, y = -1$

2. $\begin{cases} 3x + 2y = 2 \\ x - 7y = -30 \end{cases}$

$x = -2, y = 4$

3. $\begin{cases} 3x - 4y = 4 \\ x - 3y = \frac{1}{2} \end{cases}$

$x = 2, y = \frac{1}{2}$

4. $\begin{cases} 2x + y = 0 \\ 5x - 4y = -\frac{13}{2} \end{cases}$ $x = -\frac{1}{2}, y = 1$

5. $\begin{cases} x^2 - y^2 = 3 \\ xy = 2 \end{cases}$ $x = 2, y = 1$

6. $\begin{cases} x^2 - y^2 = 3 \\ xy = 2 \end{cases}$ $x = -2, y = -1$

7. $\begin{cases} \dfrac{x}{1 + x} + 3y = 6 \\ x + 9y^2 = 36 \end{cases}$ $x = 0, y = 2$

8. $\begin{cases} \dfrac{x}{x - 1} + y = 5 \\ 3x - y = 3 \end{cases}$ $x = 2, y = 3$

9. $\begin{cases} 3x + 3y + 2z = 4 \\ x - y - z = 0 \\ 2y - 3z = -8 \end{cases}$ $x = 1, y = -1, z = 2$

10. $\begin{cases} 4x - z = 7 \\ 8x + 5y - z = 0 \\ -x - y + 5z = 6 \end{cases}$ $x = 2, y = -3, z = 1$

In Problems 11–46, solve each system of equations. If the system has no solution, say it is inconsistent. Use either substitution or elimination.

11. $\begin{cases} x + y = 8 \\ x - y = 4 \end{cases}$

12. $\begin{cases} x + 2y = 5 \\ x + y = 3 \end{cases}$

13. $\begin{cases} 5x - y = 13 \\ 2x + 3y = 12 \end{cases}$

14. $\begin{cases} x + 3y = 5 \\ 2x - 3y = -8 \end{cases}$

15. $\begin{cases} 3x = 24 \\ x + 2y = 0 \end{cases}$

16. $\begin{cases} 4x + 5y = -3 \\ -2y = -4 \end{cases}$

17. $\begin{cases} 3x - 6y = 24 \\ 5x + 4y = 12 \end{cases}$

18. $\begin{cases} 2x + 4y = 16 \\ 3x - 5y = -9 \end{cases}$

19. $\begin{cases} 2x + y = 1 \\ 4x + 2y = 6 \end{cases}$

20. $\begin{cases} x - y = 5 \\ -3x + 3y = 2 \end{cases}$

21. $\begin{cases} 2x - 4y = -2 \\ 3x + 2y = 3 \end{cases}$

22. $\begin{cases} 3x + 3y = 3 \\ 4x + 2y = \frac{8}{3} \end{cases}$

23. $\begin{cases} x + 2y = 4 \\ 2x + 4y = 8 \end{cases}$

24. $\begin{cases} 3x - y = 7 \\ 9x - 3y = 21 \end{cases}$

25. $\begin{cases} 2x - 3y = -1 \\ 10x + 10y = 5 \end{cases}$

26. $\begin{cases} 3x - 2y = 0 \\ 5x + 10y = 4 \end{cases}$

27. $\begin{cases} 2x + 3y = 6 \\ x - y = \frac{1}{2} \end{cases}$

28. $\begin{cases} \frac{1}{2}x + y = -2 \\ x - 2y = 8 \end{cases}$

29. $\begin{cases} 2x + 3y = 5 \\ 4x + 6y = 10 \end{cases}$

30. $\begin{cases} 2x + 3y = 5 \\ 4x + 6y = 6 \end{cases}$

31. $\begin{cases} 3x - 5y = 3 \\ 15x + 5y = 21 \end{cases}$

32. $\begin{cases} 2x - y = -1 \\ x + \frac{1}{2}y = \frac{3}{2} \end{cases}$

33. $\begin{cases} x - y = 6 \\ 2x - 3z = 16 \\ 2y + z = 4 \end{cases}$

34. $\begin{cases} 2x + y = -4 \\ -2y + 4z = 0 \\ 3x - 2z = -11 \end{cases}$

35. $\begin{cases} x - 2y + 3z = 7 \\ 2x + y + z = 4 \\ -3x + 2y - 2z = -10 \end{cases}$

36. $\begin{cases} 2x + y - 3z = 0 \\ -2x + 2y + z = -7 \\ 3x - 4y - 3z = 7 \end{cases}$

37. $\begin{cases} x - y - z = 1 \\ 2x + 3y + z = 2 \\ 3x + 2y = 0 \end{cases}$

38. $\begin{cases} 2x - 3y - z = 0 \\ -x + 2y + z = 5 \\ 3x - 4y - z = 1 \end{cases}$

39. $\begin{cases} x - y - z = 1 \\ -x + 2y - 3z = -4 \\ 3x - 2y - 7z = 0 \end{cases}$

40. $\begin{cases} 2x - 3y - z = 0 \\ 3x + 2y + 2z = 2 \\ x + 5y + 3z = 2 \end{cases}$

41. $\begin{cases} 2x - 2y + 3z = 6 \\ 4x - 3y + 2z = 0 \\ -2x + 3y - 7z = 1 \end{cases}$

42. $\begin{cases} 3x - 2y + 2z = 6 \\ 7x - 3y + 2z = -1 \\ 2x - 3y + 4z = 0 \end{cases}$

43. $\begin{cases} x + y - z = 6 \\ 3x - 2y + z = -5 \\ x + 3y - 2z = 14 \end{cases}$

44. $\begin{cases} x - y + z = -4 \\ 2x - 3y + 4z = -15 \\ 5x + y - 2z = 12 \end{cases}$

45. $\begin{cases} x + 2y - z = -3 \\ 2x - 4y + z = -7 \\ -2x + 2y - 3z = 4 \end{cases}$

46. $\begin{cases} x + 4y - 3z = -8 \\ 3x - y + 3z = 12 \\ x + y + 6z = 1 \end{cases}$

47. Solve: $\begin{cases} \dfrac{1}{x} + \dfrac{1}{y} = 8 \\ \dfrac{3}{x} - \dfrac{5}{y} = 0 \end{cases}$

[*Hint:* Let $u = 1/x$ and $v = 1/y$, and solve for u and v. Then $x = 1/u$ and $y = 1/v$.]

48. Solve: $\begin{cases} \dfrac{4}{x} - \dfrac{3}{y} = 0 \\ \dfrac{6}{x} + \dfrac{3}{2y} = 2 \end{cases}$

49. The sum of two numbers is 81. The difference of twice one and three times the other is 62. Find the two numbers.

50. The difference of two numbers is 40. Six times the smaller one less the larger one is 5. Find the two numbers.

51. The perimeter of a rectangular room is 90 feet. Find the dimensions of the room if the length is twice the width.

52. The length of fence required to enclose a rectangular field is 3000 meters. What are the dimensions of the field if it is known that the difference between its length and width is 50 meters?

53. Four large cheeseburgers and two chocolate shakes cost a total of \$7.90. Two shakes cost 15¢ more than one cheeseburger. What is the cost of a cheeseburger? A shake?

54. A movie theater charges \$4.00 for adults and \$1.50 for children under 12. On a day when 325 people paid an admission, the total receipts were \$1025. How many who paid were adults? How many were under 12?

55. With a tail wind, a small Piper aircraft can fly 600 miles in 3 hours. Against this same wind, the Piper can fly the same distance in 4 hours. Find the average wind speed and the average airspeed of the Piper.

56. The average airspeed of a single-engine aircraft is 150 miles per hour. If the aircraft flies the same distance in 2 hours with the wind as it flew in 3 hours against the wind, what was the wind speed?

57. A store sells cashews for \$5.00 per pound and peanuts for \$1.50 per pound. The manager decides to mix 30 pounds of peanuts with some cashews and sell the mixture for \$3.00 per pound. How many pounds of cashews should be mixed with the peanuts so that the mixture will produce the same revenue as would selling the nuts separately?

58. Rework Problem 57 for a mixture using 40 pounds of peanuts.

59. A chemistry laboratory can be used by 38 students at one time. The laboratory has 16 work stations, some set up for 2 students each and the others set up for 3 students each. How many are there of each kind of work station?

60. One group of people purchased 10 hot dogs and 5 soft drinks at a cost of \$12.50. A second group bought 7 hot dogs and 4 soft drinks at a cost of \$9.00. What is the cost of a single hot dog? A single soft drink?

61. The grocery store we use does not mark prices on its goods. My wife went to this store, bought three 1 pound packages of bacon and two cartons of eggs, and paid a total of \$7.45. Not knowing she went to the store, I also went to the same store, purchased two 1 pound packages of bacon and three cartons of eggs, and paid a total of \$6.45. Now we want to return two 1 pound packages of bacon and two cartons of eggs. How much will be refunded?

62. A swimmer requires 3 hours to swim 15 miles downstream. The return trip upstream takes 5 hours. Find the average speed of the swimmer in still water. How fast is the current of the stream? (Assume the speed of the swimmer is the same in each direction.)

63. The sum of three numbers is 48. The sum of the two larger numbers is three times the smallest. The sum of the two smaller numbers is 6 more than the largest. Find the numbers.

64. A coin collection consists of 37 coins—nickels, dimes, and quarters. If the collection has a face value of \$3.25 and there are 5 more dimes than there are nickels, how many of each coin are in the collection?

65. A Broadway theater has 500 seats, divided into orchestra, main, and balcony seating. Orchestra seats sell for \$50, main seats for \$35, and balcony seats for \$25. If all the seats are sold, the gross revenue to the theater is \$17,100. If all the main and balcony seats are sold, but only half the orchestra seats are sold, the gross revenue is \$14,600. How many are there of each kind of seat?

66. Find real numbers b and c such that the parabola $y = x^2 + bx + c$ passes through the points $(1, 3)$ and $(3, 5)$.

67. Find real numbers b and c such that the parabola $y = x^2 + bx + c$ passes through the points $(1, 2)$ and $(-1, 3)$.

68. Find real numbers b and c such that the parabola $y = x^2 + bx + c$ passes through the points (x_1, y_1) and (x_2, y_2).

69. Find real numbers a, b, and c such that the parabola $y = ax^2 + bx + c$ passes through the points $(-1, 4)$, $(2, 3)$, and $(0, 1)$.

70. Find real numbers a, b, and c such that the parabola $y = ax^2 + bx + c$ passes through the points $(-1, -2)$, $(1, -4)$, and $(2, 4)$.

71. Solve: $\begin{cases} y = m_1x + b_1 \\ y = m_2x + b_2 \end{cases}$

where $m_1 \neq m_2$.

72. Solve: $\begin{cases} y = m_1x + b_1 \\ y = m_2x + b_2 \end{cases}$

where $m_1 = m_2 = m$ and $b_1 \neq b_2$.

73. Solve: $\begin{cases} y = m_1x + b_1 \\ y = m_2x + b_2 \end{cases}$

where $m_1 = m_2 = m$ and $b_1 = b_2 = b$.

12.2 Systems of Linear Equations: Matrices

The systematic approach of the method of elimination for solving a system of linear equations provides another method of solution that involves a simplified notation.

Consider the following system of linear equations:

$$\begin{cases} x + 4y = 14 \\ 3x - 2y = 0 \end{cases}$$

If we choose not to write the symbols used for the variables, we can represent this system as

$$\left[\begin{array}{cc|c} 1 & 4 & 14 \\ 3 & -2 & 0 \end{array}\right]$$

where it is understood that the first column represents the coefficients of the variable x, the second column the coefficients of y, and the third column the constants on the right side of the equal signs. The vertical line serves as a reminder of the equal signs. The large square brackets are the traditional symbols used to denote a *matrix* in algebra.

Matrix

A **matrix** is defined as a rectangular array of numbers,

$$\begin{array}{c} \\ \text{Row } 1 \\ \text{Row } 2 \\ \vdots \\ \text{Row } i \\ \vdots \\ \text{Row } m \end{array} \begin{array}{c} \begin{array}{cccccc} \text{Column } 1 & \text{Column } 2 & & \text{Column } j & & \text{Column } n \end{array} \\ \begin{bmatrix} a_{11} & a_{12} & \cdots & a_{1j} & \cdots & a_{1n} \\ a_{21} & a_{22} & \cdots & a_{2j} & \cdots & a_{2n} \\ \vdots & \vdots & & \vdots & & \vdots \\ a_{i1} & a_{i2} & \cdots & a_{ij} & \cdots & a_{in} \\ \vdots & \vdots & & \vdots & & \vdots \\ a_{m1} & a_{m2} & \cdots & a_{mj} & \cdots & a_{mn} \end{bmatrix} \end{array} \qquad (1)$$

Each number a_{ij} of the matrix has two indices: the **row index** i and the **column index** j. The matrix shown in display (1) has m rows and n columns. The numbers a_{ij} are usually referred to as the **entries** of the matrix.

We will discuss matrices in more detail in Chapter 14. For now, we will simply use matrix notation to represent a system of linear equations.

Example 1 Write the matrix representation of each system of equations.

(a) $\begin{cases} 3x - 4y = -6 & (1) \\ 2x - 3y = -5 & (2) \end{cases}$ (b) $\begin{cases} 2x - y + z = 0 & (1) \\ x + z - 1 = 0 & (2) \\ x + 2y - 8 = 0 & (3) \end{cases}$

Solution (a) The matrix representation is

$$\left[\begin{array}{cc|c} 3 & -4 & -6 \\ 2 & -3 & -5 \end{array}\right]$$

(b) Care must be taken that the system is written with the coefficients of all variables present (if any variable is missing, its coefficient is 0) and with all constants to the right of the equal signs. Thus, we need to rearrange the given system as follows:

$$\begin{cases} 2x - y + z = 0 & (1) \\ x + z - 1 = 0 & (2) \\ x + 2y - 8 = 0 & (3) \end{cases}$$

$$\begin{cases} 2x - y + z = 0 & (1) \\ x + 0 \cdot y + z = 1 & (2) \\ x + 2y + 0 \cdot z = 8 & (3) \end{cases}$$

The matrix representation is

$$\left[\begin{array}{ccc|c} 2 & -1 & 1 & 0 \\ 1 & 0 & 1 & 1 \\ 1 & 2 & 0 & 8 \end{array}\right]$$

■

In working with the matrix representation of a system of equations, the matrices found in Example 1,

$$\left[\begin{array}{cc|c} 3 & -4 & -6 \\ 2 & -3 & -5 \end{array}\right] \quad \text{and} \quad \left[\begin{array}{ccc|c} 2 & -1 & 1 & 0 \\ 1 & 0 & 1 & 1 \\ 1 & 2 & 0 & 8 \end{array}\right]$$

usually are referred to as **augmented matrices**. The matrices below, which do not include the constants to the right of the equal signs, are called the **coefficient matrices** of the system:

$$\begin{bmatrix} 3 & -4 \\ 2 & -3 \end{bmatrix} \quad \text{and} \quad \begin{bmatrix} 2 & -1 & 1 \\ 1 & 0 & 1 \\ 1 & 2 & 0 \end{bmatrix}$$

Example 2 Write the system of linear equations corresponding to each augmented matrix.

(a) $\left[\begin{array}{cc|c} 5 & 2 & 13 \\ -3 & 1 & -10 \end{array}\right]$ (b) $\left[\begin{array}{ccc|c} 3 & -1 & -1 & 7 \\ 2 & 0 & 2 & 8 \\ 0 & 1 & 1 & 0 \end{array}\right]$

Solution (a) The matrix has two rows and so represents a system of two equations. The two columns to the left of the vertical bar indicate the system has two variables. If x and y are used to denote these variables, the system of equations is

$$\begin{cases} 5x + 2y = 13 & (1) \\ -3x + y = -10 & (2) \end{cases}$$

(b) This matrix represents a system of three equations containing three variables. If x, y, and z are the three variables, this system is

$$\begin{cases} 3x - y - z = 7 & (1) \\ 2x + 2z = 8 & (2) \\ y + z = 0 & (3) \end{cases}$$

■

Row Operations on a Matrix

Consider the system of equations

$$\begin{cases} 4x - 3y = 11 & (1) \\ 3x + 2y = 4 & (2) \end{cases}$$

We shall use the method of elimination to solve the system. First, multiply each side of equation (2) by −1 and add it to equation (1). Replace equation (1) with the result:

$$\begin{cases} x - 5y = 7 & (1) \\ 3x + 2y = 4 & (2) \end{cases}$$

Multiply each side of equation (1) by −3 and add it to equation (2). Replace equation (2) with the result:

$$\begin{cases} x - 5y = 7 & (1) \\ 0 \cdot x + 17y = -17 & (2) \end{cases}$$

Multiply each side of equation (2) by $\frac{1}{17}$:

$$\begin{cases} x - 5y = 7 & (1) \\ y = -1 & (2) \end{cases}$$

Now we back-substitute $y = -1$ into equation (1) to get

$$x - 5y = 7$$
$$x - 5(-1) = 7$$
$$x = 2$$

The solution of the system is $x = 2$, $y = -1$.

The pattern of solution shown above provides a systematic way to solve any system of equations. The idea is to start with the augmented matrix of the system,

$$\left[\begin{array}{cc|c} 4 & -3 & 11 \\ 3 & 2 & 4 \end{array}\right] \qquad \begin{cases} 4x - 3y = 11 & (1) \\ 3x + 2y = 4 & (2) \end{cases}$$

and eventually arrive at the matrix,

$$\left[\begin{array}{cc|c} 1 & -5 & 7 \\ 0 & 1 & -1 \end{array}\right] \qquad \begin{cases} x - 5y = 7 & (1) \\ y = -1 & (2) \end{cases}$$

Let's go through the procedure again, this time starting with the augmented matrix and keeping the final augmented matrix given above in mind:

$$\left[\begin{array}{cc|c} 4 & -3 & 11 \\ 3 & 2 & 4 \end{array}\right] \qquad \begin{cases} 4x - 3y = 11 & (1) \\ 3x + 2y = 4 & (2) \end{cases}$$

As before, we start by multiplying each side of equation (2) by -1 and adding it to equation (1). This is equivalent to multiplying each entry in the second row of the matrix by -1, adding the result to the corresponding entries in row 1, and replacing row 1 by these entries. The result of this step is that the number 1 appears in row 1, column 1:

$$\left[\begin{array}{cc|c} 1 & -5 & 7 \\ 3 & 2 & 4 \end{array}\right] \qquad \begin{cases} x - 5y = 7 & (1) \\ 3x + 2y = 4 & (2) \end{cases}$$

Multiply each entry in the first row by -3, add the result to the entries in the second row, and replace the second row by these entries. The result of this step is that the number 0 appears in row 2, column 1:

$$\left[\begin{array}{cc|c} 1 & -5 & 7 \\ 0 & 17 & -17 \end{array}\right] \qquad \begin{cases} x - 5y = 7 & (1) \\ 0 \cdot x + 17y = -17 & (2) \end{cases}$$

Multiply each entry in the second row by $\frac{1}{17}$. The result of this step is that the number 1 appears in row 2, column 2:

$$\left[\begin{array}{cc|c} 1 & -5 & 7 \\ 0 & 1 & -1 \end{array}\right] \qquad \begin{cases} x - 5y = 7 & (1) \\ y = -1 & (2) \end{cases}$$

Now that we know $y = -1$, we can back-substitute to find that $x = 2$.

The manipulations just performed on the augmented matrix are called **row operations**. There are three basic row operations:

Row Operations

1. Interchange any two rows.
2. Replace a row by a nonzero multiple of that row.
3. Replace a row by the sum of that row and a constant multiple of some other row.

These three row operations correspond to the three rules given earlier for obtaining an equivalent system of equations. Thus, when a row operation is performed on a matrix, the resulting matrix represents a system of equations equivalent to the system represented by the original matrix.

For example, consider the augmented matrix

$$\left[\begin{array}{cc|c} 1 & 2 & 3 \\ 4 & -1 & 2 \end{array}\right]$$

Suppose we want to apply a row operation to this matrix that results in a matrix whose entry in row 2, column 1 is a 0. The row operation to use is:

Multiply each entry in row 1 by −4 and add the result to the corresponding entries in row 2. (2)

If we use R_2 to represent the new entries in row 2 and we use r_1 and r_2 to represent the original entries in rows 1 and 2, respectively, then we can represent the row operation in statement (2) by

$$R_2 = -4r_1 + r_2$$

Then,

$$\left[\begin{array}{cc|c} 1 & 2 & 3 \\ 4 & -1 & 2 \end{array}\right] \underset{R_2 = -4r_1 + r_2}{\rightarrow} \left[\begin{array}{cc|c} 1 & 2 & 3 \\ -4(1) + 4 & -4(2) + (-1) & -4(3) + 2 \end{array}\right] = \left[\begin{array}{cc|c} 1 & 2 & 3 \\ 0 & -9 & -10 \end{array}\right]$$

As desired, we now have the entry 0 in row 2, column 1.

Example 3 Apply the row operation $R_2 = -3r_1 + r_2$ to the augmented matrix

$$\left[\begin{array}{cc|c} 1 & -2 & 2 \\ 3 & -5 & 9 \end{array}\right]$$

Solution The row operation $R_2 = -3r_1 + r_2$ tells us that the entries in row 2 are to be replaced by the entries obtained after multiplying each entry in row 1 by −3 and adding the result to the corresponding entries in row 2. Thus,

$$\left[\begin{array}{cc|c} 1 & -2 & 2 \\ 3 & -5 & 9 \end{array}\right] \underset{R_2 = -3r_1 + r_2}{\rightarrow} \left[\begin{array}{cc|c} 1 & -2 & 2 \\ -3(1) + 3 & (-3)(-2) + (-5) & -3(2) + 9 \end{array}\right] = \left[\begin{array}{cc|c} 1 & -2 & 2 \\ 0 & 1 & 3 \end{array}\right]$$

■

Example 4 Using the matrix

$$\left[\begin{array}{cc|c} 1 & -2 & 2 \\ 0 & 1 & 3 \end{array}\right]$$

find a row operation that will result in a matrix with a 0 in row 1, column 2.

Solution We want a 0 in row 1, column 2. This result can be accomplished by multiplying row 2 by 2 and adding the result to row 1. That is, we apply the row operation $R_1 = 2r_2 + r_1$:

$$\left[\begin{array}{cc|c} 1 & -2 & 2 \\ 0 & 1 & 3 \end{array}\right] \underset{\substack{\uparrow \\ R_1 = 2r_2 + r_1}}{\rightarrow} \left[\begin{array}{cc|c} 2(0)+1 & 2(1)+(-2) & 2(3)+2 \\ 0 & 1 & 3 \end{array}\right] = \left[\begin{array}{cc|c} 1 & 0 & 8 \\ 0 & 1 & 3 \end{array}\right]$$

■

A word about the notation we have introduced. A row operation such as $R_1 = 2r_2 + r_1$ changes the entries in row 1. Note also that to change the entries in a given row, we multiply the entries in some other row by an appropriate number and add the results to the original entries of the row to be changed.

Now let's see how we use row operations to solve a system of linear equations.

Example 5 Solve: $\begin{cases} 4x + 3y = 11 & (1) \\ x - 3y = -1 & (2) \end{cases}$

Solution First, we write the augmented matrix that represents this system:

$$\left[\begin{array}{cc|c} 4 & 3 & 11 \\ 1 & -3 & -1 \end{array}\right]$$

The first step requires getting the entry 1 in row 1, column 1. An interchange of rows 1 and 2 is the easiest way to do this:

$$\left[\begin{array}{cc|c} 1 & -3 & -1 \\ 4 & 3 & 11 \end{array}\right]$$

Next, we want a 0 under the entry 1 in column 1. We use the row operation $R_2 = -4r_1 + r_2$:

$$\left[\begin{array}{cc|c} 1 & -3 & -1 \\ 4 & 3 & 11 \end{array}\right] \underset{\substack{\uparrow \\ R_2 = -4r_1 + r_2}}{\rightarrow} \left[\begin{array}{cc|c} 1 & -3 & -1 \\ 0 & 15 & 15 \end{array}\right]$$

Now we want the entry 1 in row 2, column 2. We use $R_2 = \frac{1}{15}r_2$:

$$\left[\begin{array}{cc|c} 1 & -3 & -1 \\ 0 & 15 & 15 \end{array}\right] \underset{\substack{\uparrow \\ R_2 = \frac{1}{15}r_2}}{\rightarrow} \left[\begin{array}{cc|c} 1 & -3 & -1 \\ 0 & 1 & 1 \end{array}\right]$$

The second row of the matrix on the right represents the equation $y = 1$. Thus, using $y = 1$, we back-substitute into the equation $x - 3(1) = -1$ (from the first row) to get

$$x - 3(1) = -1$$
$$x = 2$$

The solution of the system is $x = 2$, $y = 1$. ■

The steps we used to solve the system of linear equations in Example 5 can be summarized as follows:

Matrix Method for Solving a System of Linear Equations

STEP 1: Write the augmented matrix that represents the system.

STEP 2: Perform row operations that place the entry 1 in row 1, column 1.

STEP 3: Perform row operations that leave the entry 1 in row 1, column 1 unchanged, while causing 0's to appear below it in column 1.

STEP 4: Perform row operations that place the entry 1 in row 2, column 2 and leave the entries in columns to the left unchanged. If it is impossible to place a 1 in row 2, column 2, then proceed to place a 1 in row 2, column 3. Once a 1 is in place, perform row operations to place 0's under it.

STEP 5: Now repeat Step 4, placing a 1 in the next row, but one column to the right. Continue until the bottom row or the vertical bar is reached.

STEP 6: If any rows are obtained that contain only 0's on the left side of the vertical bar, then place such rows at the bottom of the matrix.

After Steps 1–6 have been completed, the matrix is said to be in **echelon form**. A little thought should convince you that a matrix is in echelon form when:

1. The entry in row 1, column 1 is a 1, and 0's appear below it.
2. The first nonzero entry in each row after the first row is a 1, 0's appear below it, and it appears to the right of the first nonzero entry in any row above.
3. Any rows that contain all 0's to the left of the vertical bar appear at the bottom.

Two advantages of solving a system of equations by writing the augmented matrix in echelon form are:

1. The process is algorithmic; that is, it consists of repetitive steps so that it can be programmed on a computer.
2. The process works on any system of linear equations, no matter how many equations or variables are present.

The next example shows how to write a matrix in echelon form.

Example 6 Solve: $\begin{cases} x - y + z = 8 & (1) \\ 2x + 3y - z = -2 & (2) \\ 3x - 2y - 9z = 9 & (3) \end{cases}$

Solution STEP 1: The augmented matrix of the system is

$$\left[\begin{array}{rrr|r} 1 & -1 & 1 & 8 \\ 2 & 3 & -1 & -2 \\ 3 & -2 & -9 & 9 \end{array}\right]$$

STEP 2: Because the entry 1 is already present in row 1, column 1, we can go to Step 3.

STEP 3: Perform the row operations $R_2 = -2r_1 + r_2$ and $R_3 = -3r_1 + r_3$. Each of these leaves the entry 1 in row 1, column 1 unchanged, while causing 0's to appear under it:

$$\left[\begin{array}{rrr|r} 1 & -1 & 1 & 8 \\ 2 & 3 & -1 & -2 \\ 3 & -2 & -9 & 9 \end{array}\right] \rightarrow \left[\begin{array}{rrr|r} 1 & -1 & 1 & 8 \\ 0 & 5 & -3 & -18 \\ 0 & 1 & -12 & -15 \end{array}\right]$$

$\uparrow$
$R_2 = -2r_1 + r_2$
$R_3 = -3r_1 + r_3$

STEP 4: The easiest way to obtain the entry 1 in row 2, column 2 without altering column 1 is to interchange rows 2 and 3 (another way would be to multiply row 2 by $\frac{1}{5}$, but this introduces fractions):

$$\left[\begin{array}{rrr|r} 1 & -1 & 1 & 8 \\ 0 & 1 & -12 & -15 \\ 0 & 5 & -3 & -18 \end{array}\right]$$

To get 0's under the 1 in row 2, column 2, perform the row operation $R_3 = -5r_2 + r_3$:

$$\left[\begin{array}{rrr|r} 1 & -1 & 1 & 8 \\ 0 & 1 & -12 & -15 \\ 0 & 5 & -3 & -18 \end{array}\right] \rightarrow \left[\begin{array}{rrr|r} 1 & -1 & 1 & 8 \\ 0 & 1 & -12 & -15 \\ 0 & 0 & 57 & 57 \end{array}\right]$$

$\uparrow$
$R_3 = -5r_2 + r_3$

STEP 5: Continuing, we place a 1 in row 3, column 3 by using $R_3 = \frac{1}{57}r_3$:

$$\left[\begin{array}{rrr|r} 1 & -1 & 1 & 8 \\ 0 & 1 & -12 & -15 \\ 0 & 0 & 57 & 57 \end{array}\right] \rightarrow \left[\begin{array}{rrr|r} 1 & -1 & 1 & 8 \\ 0 & 1 & -12 & -15 \\ 0 & 0 & 1 & 1 \end{array}\right]$$

$\uparrow$
$R_3 = \frac{1}{57}r_3$

Because we have reached the bottom row, the matrix is in echelon form and we can stop.

Using $z = 1$ (from the third row of the matrix), we back-substitute to get

$$\begin{cases} x - y + 1 = 8 & \text{From row 1 of the matrix} \\ y - 12(1) = -15 & \text{From row 2 of the matrix} \end{cases}$$

Thus, we get $y = -3$, and back-substituting into $x - y = 7$, we find $x = 4$. The solution of the system is $x = 4$, $y = -3$, $z = 1$. ■

Sometimes, it is advantageous to write a matrix in **reduced echelon form**. In this form, row operations are used to obtain entries that are 0 above (as well as below) the leading 1 in a row. For example, the echelon form obtained in the solution to Example 6 is

$$\left[\begin{array}{rrr|r} 1 & -1 & 1 & 8 \\ 0 & 1 & -12 & -15 \\ 0 & 0 & 1 & 1 \end{array}\right]$$

To write this matrix in reduced echelon form, we proceed as follows:

$$\left[\begin{array}{rrr|r} 1 & -1 & 1 & 8 \\ 0 & 1 & -12 & -15 \\ 0 & 0 & 1 & 1 \end{array}\right] \rightarrow \left[\begin{array}{rrr|r} 1 & 0 & -11 & -7 \\ 0 & 1 & -12 & -15 \\ 0 & 0 & 1 & 1 \end{array}\right] \rightarrow \left[\begin{array}{rrr|r} 1 & 0 & 0 & 4 \\ 0 & 1 & 0 & -3 \\ 0 & 0 & 1 & 1 \end{array}\right]$$

$\uparrow$ $R_1 = r_2 + r_1$ $\qquad$ $\uparrow$ $R_1 = 11r_3 + r_1$, $R_2 = 12r_3 + r_2$

The matrix is now written in reduced echelon form. The advantage of writing the matrix in this form is that the solution to the system, namely, $x = 4, y = -3, z = 1$, is readily found, without the need to back-substitute. Another advantage will be seen in Section 14.1, where the inverse of a matrix is discussed.

The matrix method for solving a system of linear equations also identifies systems that have infinitely many solutions and systems that are inconsistent. Let's see how.

Example 7 Solve:

$$\begin{cases} 6x - y - z = 4 & (1) \\ -12x + 2y + 2z = -8 & (2) \\ 5x + y - z = 3 & (3) \end{cases}$$

Solution We start with the augmented matrix of the system:

$$\left[\begin{array}{rrr|r} 6 & -1 & -1 & 4 \\ -12 & 2 & 2 & -8 \\ 5 & 1 & -1 & 3 \end{array}\right] \rightarrow \left[\begin{array}{rrr|r} 1 & -2 & 0 & 1 \\ -12 & 2 & 2 & -8 \\ 5 & 1 & -1 & 3 \end{array}\right] \rightarrow \left[\begin{array}{rrr|r} 1 & -2 & 0 & 1 \\ 0 & -22 & 2 & 4 \\ 0 & 11 & -1 & -2 \end{array}\right]$$

$\uparrow$ $R_1 = -1r_3 + r_1$ $\qquad$ $\uparrow$ $R_2 = 12r_1 + r_2$, $R_3 = -5r_1 + r_3$

Obtaining a 1 in row 2, column 2 without altering column 1 can be accomplished only by $R_2 = -\frac{1}{22}r_2$ or by $R_3 = \frac{1}{11}r_3$. (Do you see why?) We shall use the first of these:

$$\left[\begin{array}{rrr|r} 1 & -2 & 0 & 1 \\ 0 & -22 & 2 & 4 \\ 0 & 11 & -1 & -2 \end{array}\right] \rightarrow \left[\begin{array}{rrr|r} 1 & -2 & 0 & 1 \\ 0 & 1 & -\frac{1}{11} & -\frac{2}{11} \\ 0 & 11 & -1 & -2 \end{array}\right] \rightarrow \left[\begin{array}{rrr|r} 1 & -2 & 0 & 1 \\ 0 & 1 & -\frac{1}{11} & -\frac{2}{11} \\ 0 & 0 & 0 & 0 \end{array}\right]$$

$\uparrow$ $R_2 = -\frac{1}{22}r_2$ $\qquad$ $\uparrow$ $R_3 = -11r_2 + r_3$

This matrix is in echelon form. Because the bottom row consists entirely of 0's, the system actually consists of only two equations:

$$\begin{cases} x - 2y = 1 & (1) \\ y - \frac{1}{11}z = -\frac{2}{11} & (2) \end{cases}$$

We shall back-substitute the solution for y from the second equation, namely, $y = \frac{1}{11}z - \frac{2}{11}$, to get

$$x = 2y + 1 = 2\left(\frac{1}{11}z - \frac{2}{11}\right) + 1 = \frac{2}{11}z + \frac{7}{11}$$

Thus, the original system is equivalent to the system

$$\begin{cases} x = \frac{2}{11}z + \frac{7}{11} & (1) \\ y = \frac{1}{11}z - \frac{2}{11} & (2) \end{cases}$$

where z can be any real number.

Let's look at the situation. The original system of three equations is equivalent to a system containing two equations. This means any values of x, y, z that satisfy both

$$x = \frac{2}{11}z + \frac{7}{11} \qquad \text{and} \qquad y = \frac{1}{11}z - \frac{2}{11}$$

will be solutions. For example, $z = 0$, $x = \frac{7}{11}$, $y = -\frac{2}{11}$; $z = 1$, $x = \frac{9}{11}$, $y = -\frac{1}{11}$; and $z = -1$, $x = \frac{5}{11}$, $y = -\frac{3}{11}$ are some of the solutions of the original system. There are, in fact, infinitely many values of x, y, and z for which the two equations are satisfied. That is, the original system has infinitely many solutions. We will write the solution of the original system as

$$\begin{cases} x = \frac{2}{11}z + \frac{7}{11} \\ y = \frac{1}{11}z - \frac{2}{11} \end{cases}$$

where z can be any real number.

We also can find the solution by writing the augmented matrix in reduced echelon form. Starting with the echelon form, we have

$$\left[\begin{array}{ccc|c} 1 & -2 & 0 & 1 \\ 0 & 1 & -\frac{1}{11} & -\frac{2}{11} \\ 0 & 0 & 0 & 0 \end{array}\right] \rightarrow \left[\begin{array}{ccc|c} 1 & 0 & -\frac{2}{11} & \frac{7}{11} \\ 0 & 1 & -\frac{1}{11} & -\frac{2}{11} \\ 0 & 0 & 0 & 0 \end{array}\right]$$

↑ $R_1 = 2r_2 + r_1$

The matrix on the right is in reduced echelon form. The corresponding system of equations is

$$\begin{cases} x - \frac{2}{11}z = \frac{7}{11} & (1) \\ y - \frac{1}{11}z = -\frac{2}{11} & (2) \end{cases}$$

or equivalently,

$$\begin{cases} x = \frac{2}{11}z + \frac{7}{11} & (1) \\ y = \frac{1}{11}z - \frac{2}{11} & (2) \end{cases}$$

where z can be any real number. ■

Example 8 Solve: $\begin{cases} x + y + z = 6 \\ 2x - y - z = 3 \\ x + 2y + 2z = 0 \end{cases}$

Solution The augmented matrix is

$$\left[\begin{array}{ccc|c} 1 & 1 & 1 & 6 \\ 2 & -1 & -1 & 3 \\ 1 & 2 & 2 & 0 \end{array}\right] \rightarrow \left[\begin{array}{ccc|c} 1 & 1 & 1 & 6 \\ 0 & -3 & -3 & -9 \\ 0 & 1 & 1 & -6 \end{array}\right] \rightarrow \left[\begin{array}{ccc|c} 1 & 1 & 1 & 6 \\ 0 & 1 & 1 & -6 \\ 0 & -3 & -3 & -9 \end{array}\right] \rightarrow \left[\begin{array}{ccc|c} 1 & 1 & 1 & 6 \\ 0 & 1 & 1 & -6 \\ 0 & 0 & 0 & -27 \end{array}\right]$$

↑ $R_2 = -2r_1 + r_2$, $R_3 = -1r_1 + r_3$ ↑ Interchange rows 2 and 3. ↑ $R_3 = 3r_2 + r_3$

This matrix is in echelon form. The bottom row is equivalent to the equation

$$0x + 0y + 0z = -27$$

which has no solution. Hence, the original system is inconsistent. ■

The matrix method is especially effective for systems of equations for which the number of equations and the number of variables are unequal. Here, too, such a system is either inconsistent or consistent. If it is consistent, it will have either exactly one solution or infinitely many solutions.

Let's look at a system of four equations containing three variables.

Example 9 Solve:
$$\begin{cases} x - 2y + z = 0 & (1)\\ 2x + 2y - 3z = -3 & (2)\\ y - z = -1 & (3)\\ -x + 4y + 2z = 13 & (4)\end{cases}$$

Solution The augmented matrix is

$$\left[\begin{array}{rrr|r} 1 & -2 & 1 & 0\\ 2 & 2 & -3 & -3\\ 0 & 1 & -1 & -1\\ -1 & 4 & 2 & 13\end{array}\right] \to \left[\begin{array}{rrr|r} 1 & -2 & 1 & 0\\ 0 & 6 & -5 & -3\\ 0 & 1 & -1 & -1\\ 0 & 2 & 3 & 13\end{array}\right] \to \left[\begin{array}{rrr|r} 1 & -2 & 1 & 0\\ 0 & 1 & -1 & -1\\ 0 & 6 & -5 & -3\\ 0 & 2 & 3 & 13\end{array}\right]$$

↑ $R_2 = -2r_1 + r_2$, $R_4 = r_1 + r_4$ ↑ Interchange rows 2 and 3.

$$\to \left[\begin{array}{rrr|r} 1 & -2 & 1 & 0\\ 0 & 1 & -1 & -1\\ 0 & 0 & 1 & 3\\ 0 & 0 & 5 & 15\end{array}\right] \to \left[\begin{array}{rrr|r} 1 & -2 & 1 & 0\\ 0 & 1 & -1 & -1\\ 0 & 0 & 1 & 3\\ 0 & 0 & 0 & 0\end{array}\right]$$

↑ $R_3 = -6r_2 + r_3$, $R_4 = -2r_2 + r_4$ ↑ $R_4 = -5r_3 + r_4$

$$\to \left[\begin{array}{rrr|r} 1 & 0 & -1 & -2\\ 0 & 1 & -1 & -1\\ 0 & 0 & 1 & 3\\ 0 & 0 & 0 & 0\end{array}\right] \to \left[\begin{array}{rrr|r} 1 & 0 & 0 & 1\\ 0 & 1 & 0 & 2\\ 0 & 0 & 1 & 3\\ 0 & 0 & 0 & 0\end{array}\right]$$

↑ $R_1 = 2r_2 + r_1$ ↑ $R_1 = r_3 + r_1$, $R_2 = r_3 + r_2$

The matrix is now in reduced echelon form, and we can see that the solution is $x = 1$, $y = 2$, $z = 3$. ■

Example 10 A chemistry laboratory has three containers of nitric acid, HNO_3. One container holds a solution with a concentration of 10% HNO_3, the second holds 20% HNO_3, and the third holds 40% HNO_3. How many liters of each solution should be mixed to obtain 100 liters of a solution whose concentration is 25% HNO_3?

Solution Let x, y, and z represent the number of liters of 10%, 20%, and 40% concentrations of HNO_3, respectively. We want 100 liters in all, and the concentration of HNO_3 from each solution must sum to 25% of 100 liters. Thus, we find that

$$\begin{cases} x + y + z = 100 \\ 0.10x + 0.20y + 0.40z = 0.25(100) \end{cases}$$

Now, the augmented matrix is

$$\left[\begin{array}{ccc|c} 1 & 1 & 1 & 100 \\ 0.10 & 0.20 & 0.40 & 25 \end{array}\right] \underset{R_2 = -0.10r_1 + r_2}{\rightarrow} \left[\begin{array}{ccc|c} 1 & 1 & 1 & 100 \\ 0 & 0.10 & 0.30 & 15 \end{array}\right]$$

$$\underset{R_2 = 10r_2}{\rightarrow} \left[\begin{array}{ccc|c} 1 & 1 & 1 & 100 \\ 0 & 1 & 3 & 150 \end{array}\right] \underset{R_1 = -1r_2 + r_1}{\rightarrow} \left[\begin{array}{ccc|c} 1 & 0 & -2 & -50 \\ 0 & 1 & 3 & 150 \end{array}\right]$$

The matrix is now in reduced echelon form. The final matrix represents the system

$$\begin{cases} x - 2z = -50 & (1) \\ y + 3z = 150 & (2) \end{cases}$$

which has infinitely many solutions given by

$$\begin{cases} x = 2z - 50 & (1) \\ y = -3z + 150 & (2) \end{cases}$$

where z is any real number. However, the practical considerations of this problem require us to restrict the solutions to $x \geq 0$, $y \geq 0$, $z \geq 0$. Furthermore, we require $25 \leq z \leq 50$, because otherwise $x < 0$ or $y < 0$. Some of the possible solutions are given in Table 1. The final determination of what solution the laboratory will pick very likely depends on availability, cost differences, and other considerations.

Table 1

LITERS OF 10% SOLUTION	LITERS OF 20% SOLUTION	LITERS OF 40% SOLUTION	LITERS OF 25% SOLUTION
0	75	25	100
10	60	30	100
12	57	31	100
16	51	33	100
26	36	38	100
38	18	44	100
46	6	48	100
50	0	50	100

■

EXERCISE 12.2

In Problems 1–10, write the matrix representation of the given system of equations.

1. $\begin{cases} x - 3y = 5 \\ 4x + y = 6 \end{cases}$
2. $\begin{cases} 3x + y = 1 \\ x - 2y = 5 \end{cases}$
3. $\begin{cases} 2x + 3y - 6 = 0 \\ 4x - 6y + 2 = 0 \end{cases}$

4. $\begin{cases} 9x - y = 0 \\ 3x - y - 4 = 0 \end{cases}$
5. $\begin{cases} 0.01x - 0.03y = 0.06 \\ 0.13x + 0.10y = 0.20 \end{cases}$
6. $\begin{cases} \frac{4}{3}x - \frac{3}{2}y = \frac{3}{4} \\ -\frac{1}{4}x + \frac{1}{3}y = \frac{2}{3} \end{cases}$

7. $\begin{cases} x - y + z = 10 \\ 3x + 2y = 5 \\ x + y + 2z = 2 \end{cases}$
8. $\begin{cases} 5x - y - z = 0 \\ x + y = 5 \\ 2x - 3z = 2 \end{cases}$
9. $\begin{cases} x + y - z = 2 \\ 3x - 2y = 2 \end{cases}$

10. $\begin{cases} 2x + 3y - 4z = 0 \\ x - 5z + 2 = 0 \end{cases}$

In Problems 11–14, state the row operation used to transform the matrix on the left into the one on the right.

11. $\left[\begin{array}{cc|c} 2 & 3 & 5 \\ 3 & 1 & 4 \end{array}\right] \to \left[\begin{array}{cc|c} -1 & 2 & 1 \\ 3 & 1 & 4 \end{array}\right]$
12. $\left[\begin{array}{cc|c} 4 & 5 & -1 \\ 2 & 2 & 0 \end{array}\right] \to \left[\begin{array}{cc|c} 4 & 5 & -1 \\ 1 & 1 & 0 \end{array}\right]$

13. $\left[\begin{array}{cc|c} 1 & 4 & -3 \\ 0 & 1 & -1 \end{array}\right] \to \left[\begin{array}{cc|c} 1 & 0 & 1 \\ 0 & 1 & -1 \end{array}\right]$
14. $\left[\begin{array}{cc|c} 4 & 8 & -4 \\ 2 & 2 & 0 \end{array}\right] \to \left[\begin{array}{cc|c} 0 & 4 & -4 \\ 2 & 2 & 0 \end{array}\right]$

In Problems 15–20, perform the indicated row operation on the matrix below:

$$\left[\begin{array}{ccc|c} 1 & 2 & 3 & 0 \\ 2 & 4 & 3 & 3 \\ -3 & 2 & 1 & -2 \end{array}\right]$$

15. $R_2 = -2r_1 + r_2$
16. $R_3 = 3r_1 + r_3$
17. $R_2 = -2r_3 + r_2$
18. $R_2 = \frac{1}{2}r_2$
19. $R_3 = r_2 + r_3$
20. $R_1 = r_2 + r_1$

In Problems 21–68, solve each system of equations using matrices (row operations). If the system has no solution, say it is inconsistent.

21. $\begin{cases} x + y = 8 \\ x - y = 4 \end{cases}$
22. $\begin{cases} x + 2y = 5 \\ x + y = 3 \end{cases}$
23. $\begin{cases} 5x - y = 13 \\ 2x + 3y = 12 \end{cases}$

24. $\begin{cases} x + 3y = 5 \\ 2x - 3y = -8 \end{cases}$
25. $\begin{cases} 3x = 24 \\ x + 2y = 0 \end{cases}$
26. $\begin{cases} 4x + 5y = -3 \\ -2y = -4 \end{cases}$

27. $\begin{cases} 3x - 6y = 24 \\ 5x + 4y = 12 \end{cases}$
28. $\begin{cases} 2x + 4y = 16 \\ 3x - 5y = -9 \end{cases}$
29. $\begin{cases} 2x + y = 1 \\ 4x + 2y = 6 \end{cases}$

30. $\begin{cases} x - y = 5 \\ -3x + 3y = 2 \end{cases}$
31. $\begin{cases} 2x - 4y = -2 \\ 3x + 2y = 3 \end{cases}$
32. $\begin{cases} 3x + 3y = 3 \\ 4x + 2y = \frac{8}{3} \end{cases}$

33. $\begin{cases} x + 2y = 4 \\ 2x + 4y = 8 \end{cases}$
34. $\begin{cases} 3x - y = 7 \\ 9x - 3y = 21 \end{cases}$
35. $\begin{cases} 2x - 3y = -1 \\ 10x + 10y = 5 \end{cases}$

36. $\begin{cases} 3x - 2y = 0 \\ 5x + 10y = 4 \end{cases}$
37. $\begin{cases} 2x + 3y = 6 \\ x - y = \frac{1}{2} \end{cases}$
38. $\begin{cases} \frac{1}{2}x + y = -2 \\ x - 2y = 8 \end{cases}$

39. $\begin{cases} 2x + 3y = 5 \\ 4x + 6y = 10 \end{cases}$

40. $\begin{cases} 2x + 3y = 5 \\ 4x + 6y = 6 \end{cases}$

41. $\begin{cases} 3x - 5y = 3 \\ 15x + 5y = 21 \end{cases}$

42. $\begin{cases} 2x - y = -1 \\ x + \frac{1}{2}y = \frac{3}{2} \end{cases}$

43. $\begin{cases} x - y = 6 \\ 2x - 3z = 16 \\ 2y + z = 4 \end{cases}$

44. $\begin{cases} 2x + y = -4 \\ -2y + 4z = 0 \\ 3x - 2z = -11 \end{cases}$

45. $\begin{cases} x - 2y + 3z = 7 \\ 2x + y + z = 4 \\ -3x + 2y - 2z = -10 \end{cases}$

46. $\begin{cases} 2x + y - 3z = 0 \\ -2x + 2y + z = -7 \\ 3x - 4y - 3z = 7 \end{cases}$

47. $\begin{cases} 2x - 2y - 2z = 2 \\ 2x + 3y + z = 2 \\ 3x + 2y = 0 \end{cases}$

48. $\begin{cases} 2x - 3y - z = 0 \\ -x + 2y + z = 5 \\ 3x - 4y - z = 1 \end{cases}$

49. $\begin{cases} -x + y + z = -1 \\ -x + 2y - 3z = -4 \\ 3x - 2y - 7z = 0 \end{cases}$

50. $\begin{cases} 2x - 3y - z = 0 \\ 3x + 2y + 2z = 2 \\ x + 5y + 3z = 2 \end{cases}$

51. $\begin{cases} 2x - 2y + 3z = 6 \\ 4x - 3y + 2z = 0 \\ -2x + 3y - 7z = 1 \end{cases}$

52. $\begin{cases} 3x - 2y + 2z = 6 \\ 7x - 3y + 2z = -1 \\ 2x - 3y + 4z = 0 \end{cases}$

53. $\begin{cases} x + y - z = 6 \\ 3x - 2y + z = -5 \\ x + 3y - 2z = 14 \end{cases}$

54. $\begin{cases} x - y + z = -4 \\ 2x - 3y + 4z = -15 \\ 5x + y - 2z = 12 \end{cases}$

55. $\begin{cases} x + 2y - z = -3 \\ 2x - 4y + z = -7 \\ -2x + 2y - 3z = 4 \end{cases}$

56. $\begin{cases} x + 4y - 3z = -8 \\ 3x - y + 3z = 12 \\ x + y + 6z = 1 \end{cases}$

57. $\begin{cases} 3x + y - z = \frac{2}{3} \\ 2x - y + z = 1 \\ 4x + 2y = \frac{8}{3} \end{cases}$

58. $\begin{cases} x + y = 1 \\ 2x - y + z = 1 \\ x + 2y + z = \frac{8}{3} \end{cases}$

59. $\begin{cases} x + y + z + w = 4 \\ 2x - y + z = 0 \\ 3x + 2y + z - w = 6 \\ x - 2y - 2z + 2w = -1 \end{cases}$

60. $\begin{cases} x + y + z + w = 4 \\ -x + 2y + z = 0 \\ 2x + 3y + z - w = 6 \\ -2x + y - 2z + 2w = -1 \end{cases}$

61. $\begin{cases} x + 2y + z = 1 \\ 2x - y + 2z = 2 \\ 3x + y + 3z = 3 \end{cases}$

62. $\begin{cases} x + 2y - z = 3 \\ 2x - y + 2z = 6 \\ x - 3y + 3z = 4 \end{cases}$

63. $\begin{cases} x - y + z = 5 \\ 3x + 2y - 2z = 0 \end{cases}$

64. $\begin{cases} 2x + y - z = 4 \\ -x + y + 3z = 1 \end{cases}$

65. $\begin{cases} 2x + 3y - z = 3 \\ x - y - z = 0 \\ -x + y + z = 0 \\ x + y + 3z = 5 \end{cases}$

66. $\begin{cases} x - 3y + z = 1 \\ 2x - y - 4z = 0 \\ x - 3y + 2z = 1 \\ x - 2y = 5 \end{cases}$

67. $\begin{cases} 4x + y + z - w = 4 \\ x - y + 2z + 3w = 3 \end{cases}$

68. $\begin{cases} -4x + y = 5 \\ 2x - y + z - w = 5 \\ z + w = 4 \end{cases}$

69. Find the parabola $y = ax^2 + bx + c$ that passes through the points (1, 2), (−2, −7), and (2, −3).

70. Find the parabola $y = ax^2 + bx + c$ that passes through the points (1, −1), (3, −1), and (−2, 14).

71. Find the function $f(x) = ax^3 + bx^2 + cx + d$ for which $f(-3) = -112$, $f(-1) = -2$, $f(1) = 4$, and $f(2) = 13$.

72. Find the function $f(x) = ax^3 + bx^2 + cx + d$ for which $f(-2) = -10$, $f(-1) = 3$, $f(1) = 5$, and $f(3) = 15$.

73. A chemistry laboratory has three containers of sulfuric acid, H_2SO_4. One container holds a solution with a concentration of 15% H_2SO_4, the second holds 25% H_2SO_4, and the third holds 50% H_2SO_4. How many liters of each solution should be mixed to obtain 100 liters of a solution with a concentration of 40% H_2SO_4? Construct a table similar to Table 1 illustrating some of the possible combinations.

74. Three painters, Mike, Dan, and Katy, working together can paint the exterior of a home in 10 hours. Dan and Katy together have painted a similar house in 15 hours. One day, all three worked on this same kind of house for 4 hours, after which Katy left. Mike and Dan required 8 more hours to finish. Assuming no gain or loss in efficiency, how long should it take each person to complete such a job alone?

75. Consider the system of equations

$$\begin{cases} a_1x + b_1y = c_1 \\ a_2x + b_2y = c_2 \end{cases}$$

If $D = a_1b_2 - a_2b_1 \neq 0$, use matrices to show that the solution is

$$x = \frac{1}{D}(c_1b_2 - c_2b_1), \qquad y = \frac{1}{D}(a_1c_2 - a_2c_1)$$

76. For the system in Problem 75, suppose $D = a_1b_2 - a_2b_1 = 0$. Use matrices to show that the system is inconsistent if either $a_1c_2 \neq a_2c_1$ or $b_1c_2 \neq b_2c_1$ and has infinitely many solutions if both $a_1c_2 = a_2c_1$ and $b_1c_2 = b_2c_1$.

12.3 Systems of Linear Equations: Determinants

In the preceding section, we described a method of using matrices to solve any system of linear equations. This section deals with yet another method for solving systems of linear equations; however, it can be used only when the number of equations equals the number of variables. Although the method will work for any system (provided the number of equations equals the number of variables), it is most often used for systems of two equations containing two variables or three equations containing three variables. This method, called *Cramer's Rule*, is based on the concept of a *determinant*.

2 by 2 Determinants

2 by 2 Determinant

If a, b, c, and d are four real numbers, the symbol

$$D = \begin{vmatrix} a & b \\ c & d \end{vmatrix}$$

is called a **2 by 2 determinant**. Its value is the number $ad - bc$; that is,

$$D = \begin{vmatrix} a & b \\ c & d \end{vmatrix} = ad - bc \qquad (1)$$

A device that may be helpful for remembering the value of a 2 by 2 determinant is the following:

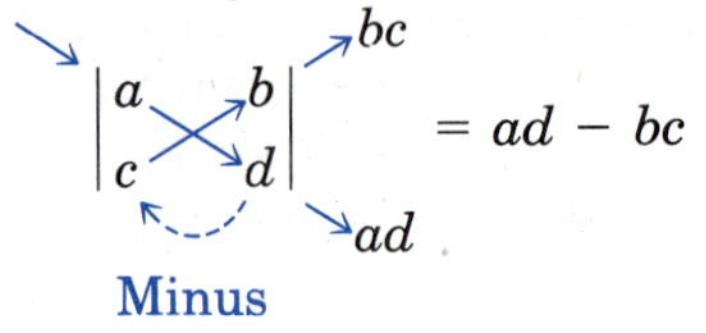

Example 1

$$\begin{vmatrix} 3 & -2 \\ 6 & 1 \end{vmatrix} = (3)(1) - (6)(-2) = 3 - (-12) = 15$$

■

Let's now see the role that a 2 by 2 determinant plays in the solution of a system of two equations containing two variables. Consider the system

$$\begin{cases} ax + by = s & (1) \\ cx + dy = t & (2) \end{cases} \qquad (2)$$

We shall use the method of elimination to solve this system.

Provided $d \neq 0$ and $b \neq 0$, this system is equivalent to the system

$$\begin{cases} adx + bdy = sd & (1) \quad \text{Multiply by } d. \\ bcx + bdy = tb & (2) \quad \text{Multiply by } b. \end{cases}$$

On subtracting the second equation from the first equation, we get

$$\begin{cases} (ad - bc)x + 0 \cdot y = sd - tb & (1) \\ \quad bcx \quad + bdy = tb & (2) \end{cases}$$

Now, the first equation can be rewritten using determinant notation:

$$\begin{vmatrix} a & b \\ c & d \end{vmatrix} x = \begin{vmatrix} s & b \\ t & d \end{vmatrix}$$

If $D = \begin{vmatrix} a & b \\ c & d \end{vmatrix} = ad - bc \neq 0$, we can solve for x to get

$$x = \frac{\begin{vmatrix} s & b \\ t & d \end{vmatrix}}{\begin{vmatrix} a & b \\ c & d \end{vmatrix}} = \frac{\begin{vmatrix} s & b \\ t & d \end{vmatrix}}{D} \qquad (3)$$

Return now to the original system (2). Provided $a \neq 0$ and $c \neq 0$, the system is equivalent to

$$\begin{cases} acx + bcy = cs & (1) \quad \text{Multiply by } c. \\ acx + ady = at & (2) \quad \text{Multiply by } a. \end{cases}$$

On subtracting the first equation from the second equation, we get

$$\begin{cases} acx + bcy = cs & (1) \\ 0 \cdot x + (ad - bc)y = at - cs & (2) \end{cases}$$

The second equation now can be rewritten using determinant notation:

$$\begin{vmatrix} a & b \\ c & d \end{vmatrix} y = \begin{vmatrix} a & s \\ c & t \end{vmatrix}$$

If $D = \begin{vmatrix} a & b \\ c & d \end{vmatrix} = ad - bc \neq 0$, we can solve for y to get

$$y = \frac{\begin{vmatrix} a & s \\ c & t \end{vmatrix}}{\begin{vmatrix} a & b \\ c & d \end{vmatrix}} = \frac{\begin{vmatrix} a & s \\ c & t \end{vmatrix}}{D} \tag{4}$$

Equations (3) and (4) lead us to the following result, called **Cramer's Rule**:

Theorem

Cramer's Rule for Two Equations Containing Two Variables

The solution to the system of equations

$$\begin{cases} ax + by = s & (1) \\ cx + dy = t & (2) \end{cases} \tag{5}$$

is given by

$$x = \frac{\begin{vmatrix} s & b \\ t & d \end{vmatrix}}{\begin{vmatrix} a & b \\ c & d \end{vmatrix}}, \qquad y = \frac{\begin{vmatrix} a & s \\ c & t \end{vmatrix}}{\begin{vmatrix} a & b \\ c & d \end{vmatrix}} \tag{6}$$

provided that

$$D = \begin{vmatrix} a & b \\ c & d \end{vmatrix} = ad - bc \neq 0$$

In the derivation given for Cramer's Rule above, we assumed that none of the numbers a, b, c, and d were 0. In Problem 56 at the end of this section you will be asked to complete the proof under the less stringent conditions that $D = ad - bc \neq 0$.

Now look carefully at the pattern in Cramer's Rule. The denominator in the solution (6) is the determinant of the coefficients of the variables:

$$\begin{cases} ax + by = s \\ cx + dy = t \end{cases} \qquad D = \begin{vmatrix} a & b \\ c & d \end{vmatrix}$$

In the solution for x, the numerator is the determinant, denoted by D_x, formed by replacing the entries in the first column (the coefficients of x) in D by the constants on the right side of the equal sign:

$$D_x = \begin{vmatrix} s & b \\ t & d \end{vmatrix}$$

In the solution for y, the numerator is the determinant, denoted by D_y, formed by replacing the entries in the second column (the coefficients of y) in D by the constants on the right side of the equal sign:

$$D_y = \begin{vmatrix} a & s \\ c & t \end{vmatrix}$$

Cramer's Rule then states that, if $D \neq 0$,

$$x = \frac{D_x}{D}, \qquad y = \frac{D_y}{D} \tag{7}$$

Example 2 Use Cramer's Rule, if applicable, to solve the system

$$\begin{cases} 3x - 2y = 4 & (1) \\ 6x + y = 13 & (2) \end{cases}$$

Solution The determinant D of the coefficients of the variables is

$$D = \begin{vmatrix} 3 & -2 \\ 6 & 1 \end{vmatrix} = (3)(1) - (6)(-2) = 15$$

Because $D \neq 0$, Cramer's Rule (7) can be used:

$$x = \frac{D_x}{D} = \frac{\begin{vmatrix} 4 & -2 \\ 13 & 1 \end{vmatrix}}{15} = \frac{30}{15} = 2, \qquad y = \frac{D_y}{D} = \frac{\begin{vmatrix} 3 & 4 \\ 6 & 13 \end{vmatrix}}{15} = \frac{15}{15} = 1$$

The solution is $x = 2$, $y = 1$. ■

If, in attempting to use Cramer's Rule, the determinant D of the coefficients of the variables is found to equal 0 (so that Cramer's Rule is not applicable), then the system either is inconsistent or has infinitely many solutions. (Refer to Problem 76 in Exercise 12.2.)

3 by 3 Determinants

In order to use Cramer's Rule to solve a system of three equations containing three variables, we need to define a 3 by 3 determinant.

A **3 by 3 determinant** is symbolized by

$$\begin{vmatrix} a_{11} & a_{12} & a_{13} \\ a_{21} & a_{22} & a_{23} \\ a_{31} & a_{32} & a_{33} \end{vmatrix} \tag{8}$$

in which $a_{11}, a_{12}, \ldots$ are real numbers.

As with matrices, we use a double subscript to identify an entry by indicating its row and column numbers. For example, the entry a_{23} is in row 2, column 3.

The value of a 3 by 3 determinant may be defined in terms of 2 by 2 determinants by the following formula:

$$\begin{vmatrix} a_{11} & a_{12} & a_{13} \\ a_{21} & a_{22} & a_{23} \\ a_{31} & a_{32} & a_{33} \end{vmatrix} = a_{11}\begin{vmatrix} a_{22} & a_{23} \\ a_{32} & a_{33} \end{vmatrix} \underset{\text{Minus}}{-} a_{12}\begin{vmatrix} a_{21} & a_{23} \\ a_{31} & a_{33} \end{vmatrix} + a_{13}\begin{vmatrix} a_{21} & a_{22} \\ a_{31} & a_{32} \end{vmatrix} \tag{9}$$

Minus ↓ (the minus sign before a_{12})

↑ 2 by 2 determinant left after removing row and column containing a_{11}

↑ 2 by 2 determinant left after removing row and column containing a_{12}

↑ 2 by 2 determinant left after removing row and column containing a_{13}

Be sure to take note of the minus sign that appears with the second term—it's easy to forget it! Formula (9) is best remembered by noting that each entry in row 1 is multiplied by the 2 by 2 determinant that remains after the row and column containing the entry has been removed, as indicated below:

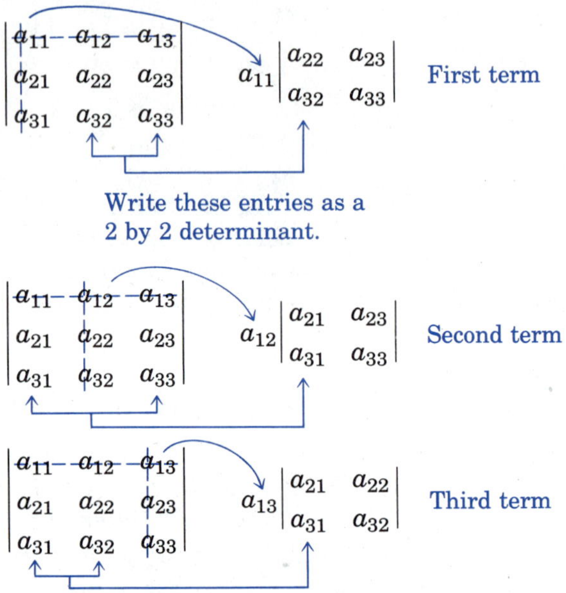

Now insert the minus sign before the middle expression and add:

Minus ↓

$$\begin{vmatrix} a_{11} & a_{12} & a_{13} \\ a_{21} & a_{22} & a_{23} \\ a_{31} & a_{32} & a_{33} \end{vmatrix} = a_{11}\begin{vmatrix} a_{22} & a_{23} \\ a_{32} & a_{33} \end{vmatrix} - a_{12}\begin{vmatrix} a_{21} & a_{23} \\ a_{31} & a_{33} \end{vmatrix} + a_{13}\begin{vmatrix} a_{21} & a_{22} \\ a_{31} & a_{32} \end{vmatrix}$$

Formula (9) exhibits one way to find the value of a 3 by 3 determinant, namely, *by expanding across row 1*. In fact, the expansion can take place across any row or down any column. The terms to be added or subtracted consist of the row (or column) entry times the value of the 2 by 2 determinant that remains after removing the row and column entry. The value of the determinant is found by adding or subtracting the terms according to the following scheme:

$$\begin{matrix} + & - & + \\ - & + & - \\ + & - & + \end{matrix}$$

For example, if we choose to expand down column 2, we obtain

$$\begin{vmatrix} a_{11} & a_{12} & a_{13} \\ a_{21} & a_{22} & a_{23} \\ a_{31} & a_{32} & a_{33} \end{vmatrix} = -a_{12}\begin{vmatrix} a_{21} & a_{23} \\ a_{31} & a_{33} \end{vmatrix} + a_{22}\begin{vmatrix} a_{11} & a_{13} \\ a_{31} & a_{33} \end{vmatrix} - a_{32}\begin{vmatrix} a_{11} & a_{13} \\ a_{21} & a_{23} \end{vmatrix}$$

↑ Expand down column 2 (−, +, −)

If we choose to expand across row 3, we obtain

$$\begin{vmatrix} a_{11} & a_{12} & a_{13} \\ a_{21} & a_{22} & a_{23} \\ a_{31} & a_{32} & a_{33} \end{vmatrix} = a_{31}\begin{vmatrix} a_{12} & a_{13} \\ a_{22} & a_{23} \end{vmatrix} + a_{32}\begin{vmatrix} a_{11} & a_{13} \\ a_{21} & a_{23} \end{vmatrix} - a_{33}\begin{vmatrix} a_{11} & a_{12} \\ a_{21} & a_{22} \end{vmatrix}$$

↑ Expand across row 3 (+, −, +)

It can be shown that the value of a determinant does not depend on the choice of the row or column used in the expansion.

Example 3 Find the value of the 3 by 3 determinant: $\begin{vmatrix} 3 & 4 & -1 \\ 4 & 6 & 2 \\ 8 & -2 & 3 \end{vmatrix}$

Solution We choose to expand across row 1.

Remember the minus sign. ↓

$$\begin{aligned} \begin{vmatrix} 3 & 4 & -1 \\ 4 & 6 & 2 \\ 8 & -2 & 3 \end{vmatrix} &= 3\begin{vmatrix} 6 & 2 \\ -2 & 3 \end{vmatrix} - 4\begin{vmatrix} 4 & 2 \\ 8 & 3 \end{vmatrix} + (-1)\begin{vmatrix} 4 & 6 \\ 8 & -2 \end{vmatrix} \\ &= 3(18 + 4) - 4(12 - 16) + (-1)(-8 - 48) \\ &= 3(22) - 4(-4) + (-1)(-56) \\ &= 66 + 16 + 56 = 138 \end{aligned}$$

■

We could also find the value of the 3 by 3 determinant in Example 3 by expanding down column 3 (the signs are +, −, +):

$$\begin{aligned} \begin{vmatrix} 3 & 4 & -1 \\ 4 & 6 & 2 \\ 8 & -2 & 3 \end{vmatrix} &= (-1)\begin{vmatrix} 4 & 6 \\ 8 & -2 \end{vmatrix} - 2\begin{vmatrix} 3 & 4 \\ 8 & -2 \end{vmatrix} + 3\begin{vmatrix} 3 & 4 \\ 4 & 6 \end{vmatrix} \\ &= -1(-8 - 48) - 2(-6 - 32) + 3(18 - 16) \\ &= 56 + 76 + 6 = 138 \end{aligned}$$

Systems of Three Equations Containing Three Variables

Consider the following system of three equations containing three variables:

$$\begin{cases} a_{11}x + a_{12}y + a_{13}z = c_1 \\ a_{21}x + a_{22}y + a_{23}z = c_2 \\ a_{31}x + a_{32}y + a_{33}z = c_3 \end{cases} \tag{10}$$

It can be shown that if the determinant D of the coefficients of the variables is not 0, that is, if

$$D = \begin{vmatrix} a_{11} & a_{12} & a_{13} \\ a_{21} & a_{22} & a_{23} \\ a_{31} & a_{32} & a_{33} \end{vmatrix} \neq 0$$

then the unique solution of system (10) is given by

Cramer's Rule for Three Equations Containing Three Variables

$$x = \frac{D_x}{D}, \qquad y = \frac{D_y}{D}, \qquad z = \frac{D_z}{D}$$

where

$$D_x = \begin{vmatrix} c_1 & a_{12} & a_{13} \\ c_2 & a_{22} & a_{23} \\ c_3 & a_{32} & a_{33} \end{vmatrix} \qquad D_y = \begin{vmatrix} a_{11} & c_1 & a_{13} \\ a_{21} & c_2 & a_{23} \\ a_{31} & c_3 & a_{33} \end{vmatrix} \qquad D_z = \begin{vmatrix} a_{11} & a_{12} & c_1 \\ a_{21} & a_{22} & c_2 \\ a_{31} & a_{32} & c_3 \end{vmatrix}$$

The similarity of this pattern and the pattern observed earlier for a system of two equations containing two variables should be apparent.

Example 4 Use Cramer's Rule, if applicable, to solve the following system:

$$\begin{cases} 2x + y - z = 3 & (1) \\ -x + 2y + 4z = -3 & (2) \\ x - 2y - 3z = 4 & (3) \end{cases}$$

Solution The value of the determinant D of the coefficients of the variables is

$$D = \begin{vmatrix} 2 & 1 & -1 \\ -1 & 2 & 4 \\ 1 & -2 & -3 \end{vmatrix} = 2\begin{vmatrix} 2 & 4 \\ -2 & -3 \end{vmatrix} - 1\begin{vmatrix} -1 & 4 \\ 1 & -3 \end{vmatrix} + (-1)\begin{vmatrix} -1 & 2 \\ 1 & -2 \end{vmatrix}$$

$$= 2(2) - 1(-1) + (-1)(0)$$

$$= 4 + 1 = 5$$

Because $D \neq 0$, we proceed to find the value of D_x, D_y, and D_z:

$$D_x = \begin{vmatrix} 3 & 1 & -1 \\ -3 & 2 & 4 \\ 4 & -2 & -3 \end{vmatrix} = 3\begin{vmatrix} 2 & 4 \\ -2 & -3 \end{vmatrix} - 1\begin{vmatrix} -3 & 4 \\ 4 & -3 \end{vmatrix} + (-1)\begin{vmatrix} -3 & 2 \\ 4 & -2 \end{vmatrix}$$

$$= 3(2) - 1(-7) + (-1)(-2) = 15$$

$$D_y = \begin{vmatrix} 2 & 3 & -1 \\ -1 & -3 & 4 \\ 1 & 4 & -3 \end{vmatrix} = 2\begin{vmatrix} -3 & 4 \\ 4 & -3 \end{vmatrix} - 3\begin{vmatrix} -1 & 4 \\ 1 & -3 \end{vmatrix} + (-1)\begin{vmatrix} -1 & -3 \\ 1 & 4 \end{vmatrix}$$

$$= 2(-7) - 3(-1) + (-1)(-1)$$

$$= -14 + 3 + 1 = -10$$

$$D_z = \begin{vmatrix} 2 & 1 & 3 \\ -1 & 2 & -3 \\ 1 & -2 & 4 \end{vmatrix} = 2\begin{vmatrix} 2 & -3 \\ -2 & 4 \end{vmatrix} - 1\begin{vmatrix} -1 & -3 \\ 1 & 4 \end{vmatrix} + 3\begin{vmatrix} -1 & 2 \\ 1 & -2 \end{vmatrix}$$

$$= 2(2) - 1(-1) + 3(0) = 5$$

As a result,

$$x = \frac{D_x}{D} = \frac{15}{5} = 3, \qquad y = \frac{D_y}{D} = \frac{-10}{5} = -2, \qquad z = \frac{D_z}{D} = \frac{5}{5} = 1$$

The solution is $x = 3$, $y = -2$, $z = 1$. ■

If the determinant of the coefficients of the variables of a system of three linear equations containing three variables is 0, then Cramer's Rule is not applicable. In such a case, the system either is inconsistent or has infinitely many solutions.

More about Determinants

Determinants have several properties that are sometimes helpful for obtaining their value. We list some of them here.

Theorem The value of a determinant changes sign if any two rows (or any two columns) are interchanged. (11)

Proof for 2 by 2 Determinants

$$\begin{vmatrix} a & b \\ c & d \end{vmatrix} = ad - bc \quad \text{and} \quad \begin{vmatrix} c & d \\ a & b \end{vmatrix} = bc - ad = -(ad - bc)$$ ■

Example 5

$$\begin{vmatrix} 3 & 4 \\ 1 & 2 \end{vmatrix} = 6 - 4 = 2 \qquad \begin{vmatrix} 1 & 2 \\ 3 & 4 \end{vmatrix} = 4 - 6 = -2$$ ■

Theorem If all the entries in any row (or any column) equal 0, the value of the determinant is 0. (12)

Proof Merely expand across the row (or down the column) containing the 0's. ■

Theorem If any two rows (or any two columns) of a determinant have corresponding entries that are equal, the value of the determinant is 0. (13) ■

You are asked to prove this result for a 3 by 3 determinant in which the entries in column 1 equal the entries in column 3 in Problem 59 at the end of this section.

Example 6

$$\begin{vmatrix} 1 & 2 & 3 \\ 1 & 2 & 3 \\ 4 & 5 & 6 \end{vmatrix} = 1\begin{vmatrix} 2 & 3 \\ 5 & 6 \end{vmatrix} - 2\begin{vmatrix} 1 & 3 \\ 4 & 6 \end{vmatrix} + 3\begin{vmatrix} 1 & 2 \\ 4 & 5 \end{vmatrix}$$
$$= 1(-3) - 2(-6) + 3(-3)$$
$$= -3 + 12 - 9 = 0$$ ■

Theorem If any row (or any column) of a determinant is multiplied by a nonzero number k, the value of the determinant is also changed by a factor of k. (14) ■

You are asked to prove this result for a 3 by 3 determinant using row 2 in Problem 58 at the end of this section.

Example 7

$$\begin{vmatrix} 1 & 2 \\ 4 & 6 \end{vmatrix} = 6 - 8 = -2$$
$$\begin{vmatrix} k & 2k \\ 4 & 6 \end{vmatrix} = 6k - 8k = -2k = k(-2) = k\begin{vmatrix} 1 & 2 \\ 4 & 6 \end{vmatrix}$$ ■

Theorem If the entries of any row (or any column) of a determinant are multiplied by a nonzero number k and the result is added to the corresponding entries of another row (or column), the value of the determinant remains unchanged. (15) ■

In Problem 60 at the end of this section, you are asked to prove this result for a 3 by 3 determinant using rows 1 and 2.

Example 8

$$\begin{vmatrix} 3 & 4 \\ 5 & 2 \end{vmatrix} = \begin{vmatrix} -7 & 0 \\ 5 & 2 \end{vmatrix} = -14$$

↑ Multiply row 2 by -2 and add to row 1. ■

EXERCISE 12.3

In Problems 1–10, find the value of each determinant.

1. $\begin{vmatrix} 3 & 1 \\ 4 & 2 \end{vmatrix}$ **2.** $\begin{vmatrix} 6 & 1 \\ 5 & 2 \end{vmatrix}$ **3.** $\begin{vmatrix} 6 & 4 \\ -1 & 3 \end{vmatrix}$ **4.** $\begin{vmatrix} 8 & -3 \\ 4 & 2 \end{vmatrix}$

5. $\begin{vmatrix} -3 & -1 \\ 4 & 2 \end{vmatrix}$ **6.** $\begin{vmatrix} -4 & 2 \\ -5 & 3 \end{vmatrix}$ **7.** $\begin{vmatrix} 3 & 4 & 2 \\ 1 & -1 & 5 \\ 1 & 2 & -2 \end{vmatrix}$ **8.** $\begin{vmatrix} 1 & 3 & -2 \\ 6 & 1 & -5 \\ 8 & 2 & 3 \end{vmatrix}$

9. $\begin{vmatrix} 4 & -1 & 2 \\ 6 & -1 & 0 \\ 1 & -3 & 4 \end{vmatrix}$ **10.** $\begin{vmatrix} 3 & -9 & 4 \\ 1 & 4 & 0 \\ 8 & -3 & 1 \end{vmatrix}$

In Problems 11–38, solve each system of equations using Cramer's Rule, if it is applicable. If it is not, say so.

11. $\begin{cases} x + y = 8 \\ x - y = 4 \end{cases}$ **12.** $\begin{cases} x + 2y = 5 \\ x + y = 3 \end{cases}$ **13.** $\begin{cases} 5x - y = 13 \\ 2x + 3y = 12 \end{cases}$

14. $\begin{cases} x + 3y = 5 \\ 2x - 3y = -8 \end{cases}$ **15.** $\begin{cases} 3x = 24 \\ x + 2y = 0 \end{cases}$ **16.** $\begin{cases} 4x + 5y = -3 \\ -2y = -4 \end{cases}$

17. $\begin{cases} 3x - 6y = 24 \\ 5x + 4y = 12 \end{cases}$ **18.** $\begin{cases} 2x + 4y = 16 \\ 3x - 5y = -9 \end{cases}$ **19.** $\begin{cases} 3x - 2y = 4 \\ 6x - 4y = 0 \end{cases}$

20. $\begin{cases} -x + 2y = 5 \\ 4x - 8y = 6 \end{cases}$ **21.** $\begin{cases} 2x - 4y = -2 \\ 3x + 2y = 3 \end{cases}$ **22.** $\begin{cases} 3x + 3y = 3 \\ 4x + 2y = \frac{8}{3} \end{cases}$

23. $\begin{cases} 2x - 3y = -1 \\ 10x + 10y = 5 \end{cases}$ **24.** $\begin{cases} 3x - 2y = 0 \\ 5x + 10y = 4 \end{cases}$ **25.** $\begin{cases} 2x + 3y = 6 \\ x - y = \frac{1}{2} \end{cases}$

26. $\begin{cases} \frac{1}{2}x + y = -2 \\ x - 2y = 8 \end{cases}$ **27.** $\begin{cases} 3x - 5y = 3 \\ 15x + 5y = 21 \end{cases}$ **28.** $\begin{cases} 2x - y = -1 \\ x + \frac{1}{2}y = \frac{3}{2} \end{cases}$

29. $\begin{cases} x + y - z = 6 \\ 3x - 2y + z = -5 \\ x + 3y - 2z = 14 \end{cases}$ **30.** $\begin{cases} x - y + z = -4 \\ 2x - 3y + 4z = -15 \\ 5x + y - 2z = 12 \end{cases}$ **31.** $\begin{cases} x + 2y - z = -3 \\ 2x - 4y + z = -7 \\ -2x + 2y - 3z = 4 \end{cases}$

32. $\begin{cases} x + 4y - 3z = -8 \\ 3x - y + 3z = 12 \\ x + y + 6z = 1 \end{cases}$ **33.** $\begin{cases} x - 2y + 3z = 1 \\ 3x + y - 2z = 0 \\ 2x - 4y + 6z = 2 \end{cases}$ **34.** $\begin{cases} x - y + 2z = 5 \\ 3x + 2y = 4 \\ -2x + 2y - 4z = -10 \end{cases}$

35. $\begin{cases} x + 2y - z = 0 \\ 2x - 4y + z = 0 \\ -2x + 2y - 3z = 0 \end{cases}$ **36.** $\begin{cases} x + 4y - 3z = 0 \\ 3x - y + 3z = 0 \\ x + y + 6z = 0 \end{cases}$ **37.** $\begin{cases} x - 2y + 3z = 0 \\ 3x + y - 2z = 0 \\ 2x - 4y + 6z = 0 \end{cases}$

38. $\begin{cases} x - y + 2z = 0 \\ 3x + 2y = 0 \\ -2x + 2y - 4z = 0 \end{cases}$

39. Solve: $\begin{cases} \dfrac{1}{x} + \dfrac{1}{y} = 8 \\ \dfrac{3}{x} - \dfrac{5}{y} = 0 \end{cases}$

[*Hint:* Let $u = 1/x$ and $v = 1/y$ and solve for u and v.]

40. Solve: $\begin{cases} \dfrac{4}{x} - \dfrac{3}{y} = 0 \\ \dfrac{6}{x} + \dfrac{3}{2y} = 2 \end{cases}$

In Problems 41–46, solve for x.

41. $\begin{vmatrix} x & x \\ 4 & 3 \end{vmatrix} = 5$

42. $\begin{vmatrix} x & 1 \\ 3 & x \end{vmatrix} = -2$

43. $\begin{vmatrix} x & 1 & 1 \\ 4 & 3 & 2 \\ -1 & 2 & 5 \end{vmatrix} = 2$

44. $\begin{vmatrix} 3 & 2 & 4 \\ 1 & x & 5 \\ 0 & 1 & -2 \end{vmatrix} = 0$

45. $\begin{vmatrix} x & 2 & 3 \\ 1 & x & 0 \\ 6 & 1 & -2 \end{vmatrix} = 7$

46. $\begin{vmatrix} x & 1 & 2 \\ 1 & x & 3 \\ 0 & 1 & 2 \end{vmatrix} = -4x$

In Problems 47–54, use properties of determinants to find the value of each determinant if it is known that

$$\begin{vmatrix} x & y & z \\ u & v & w \\ 1 & 2 & 3 \end{vmatrix} = 4$$

47. $\begin{vmatrix} 1 & 2 & 3 \\ u & v & w \\ x & y & z \end{vmatrix}$

48. $\begin{vmatrix} x & y & z \\ u & v & w \\ 2 & 4 & 6 \end{vmatrix}$

49. $\begin{vmatrix} x & y & z \\ -3 & -6 & -9 \\ u & v & w \end{vmatrix}$

50. $\begin{vmatrix} 1 & 2 & 3 \\ x - u & y - v & z - w \\ u & v & w \end{vmatrix}$

51. $\begin{vmatrix} 1 & 2 & 3 \\ x - 3 & y - 6 & z - 9 \\ 2u & 2v & 2w \end{vmatrix}$

52. $\begin{vmatrix} x & y & z - x \\ u & v & w - u \\ 1 & 2 & 2 \end{vmatrix}$

53. $\begin{vmatrix} 1 & 2 & 3 \\ 2x & 2y & 2z \\ u - 1 & v - 2 & w - 3 \end{vmatrix}$

54. $\begin{vmatrix} x + 3 & y + 6 & z + 9 \\ 3u - 1 & 3v - 2 & 3w - 3 \\ 1 & 2 & 3 \end{vmatrix}$

55. Show that: $\begin{vmatrix} x^2 & x & 1 \\ y^2 & y & 1 \\ z^2 & z & 1 \end{vmatrix} = (y - z)(x - y)(x - z)$

56. Complete the proof of Cramer's Rule for two equations containing two variables. [*Hint:* In system (5), page 693, if $a = 0$, then $b \neq 0$ and $c \neq 0$, so $D = -bc \neq 0$. Now show that equations (6) provide a solution of the system when $a = 0$. There are then three remaining cases: $b = 0$, $c = 0$, and $d = 0$.]

57. Interchange columns 1 and 3 of a 3 by 3 determinant. Show that the value of the new determinant is -1 times the value of the original determinant.

58. Multiply each entry in row 2 of a 3 by 3 determinant by the number k, $k \neq 0$. Show that the value of the new determinant is k times the value of the original determinant.

59. Prove that a 3 by 3 determinant in which the entries in column 1 equal those in column 3 has the value 0.

60. Prove that if row 2 of a 3 by 3 determinant is multiplied by k, $k \neq 0$, and the result is added to the entries in row 1, then there is no change in the value of the determinant.

12.4 Systems of Nonlinear Equations

There is no general methodology for solving a system of nonlinear equations. There are times when substitution is best; other times, elimination is best; and there are times when neither of these methods works. Experience and a certain degree of imagination are your allies here.

Before we begin, two comments are in order:

1. If the system contains two variables and if the equations in the system are easy to graph, then graph them. By graphing each equation in the system, we can get an idea of how many solutions a system has and approximately where they are located.
2. Extraneous solutions can creep in when solving nonlinear systems, so it is imperative that all apparent solutions be checked.

Example 1 Solve the following system of equations:

$$\begin{cases} 3x - y = -2 & (1) \quad \text{A line} \\ 2x^2 - y = 0 & (2) \quad \text{A parabola} \end{cases}$$

Solution First, we notice that the system contains two variables and that we know how to graph each equation. In Figure 5, we see that the system apparently has two solutions.

We shall use substitution to solve the system. Equation (1) is easily solved for y:

$$3x - y = -2$$
$$y = 3x + 2$$

Substituting this value for y in equation (2) gives

$$2x^2 - y = 0$$
$$2x^2 - (3x + 2) = 0$$
$$2x^2 - 3x - 2 = 0$$
$$(2x + 1)(x - 2) = 0$$
$$2x + 1 = 0 \quad \text{or} \quad x - 2 = 0$$
$$x = -\tfrac{1}{2} \qquad\qquad x = 2$$

Using these values for x in $y = 3x + 2$, we find

$$y = 3\left(-\tfrac{1}{2}\right) + 2 = \tfrac{1}{2} \quad \text{or} \quad y = 3(2) + 2 = 8$$

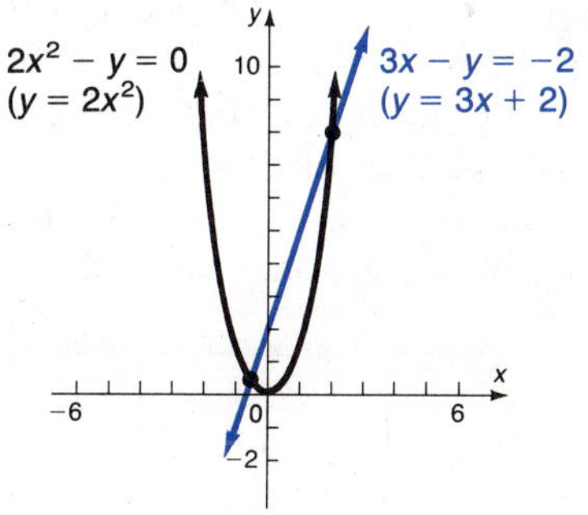

Figure 5

The apparent solutions are $x = -\frac{1}{2}$, $y = \frac{1}{2}$ and $x = 2$, $y = 8$.

Check: For $x = -\frac{1}{2}$, $y = \frac{1}{2}$:

$$3\left(-\tfrac{1}{2}\right) - \tfrac{1}{2} = -\tfrac{3}{2} - \tfrac{1}{2} = -2 \qquad (1)$$
$$2\left(-\tfrac{1}{2}\right)^2 - \tfrac{1}{2} = 2\left(\tfrac{1}{4}\right) - \tfrac{1}{2} = 0 \qquad (2)$$

For $x = 2$, $y = 8$:

$$\begin{cases} 3(2) - 8 = 6 - 8 = -2 & (1) \\ 2(2)^2 - 8 = 2(4) - 8 = 0 & (2) \end{cases}$$

Each solution checks. Now we know the graphs in Figure 5 intersect at $\left(-\frac{1}{2}, \frac{1}{2}\right)$ and at (2, 8). ■

Our next example illustrates how the method of elimination works for nonlinear systems.

Example 2 Solve: $\begin{cases} x^2 + y^2 = 13 & (1) \text{ A circle} \\ x^2 - y = 7 & (2) \text{ A parabola} \end{cases}$

Solution First, we graph each equation, as shown in Figure 6. Based on the graph, we expect four solutions. By subtracting equation (2) from equation (1), the variable x is eliminated, leaving

$$y^2 + y = 6$$

This quadratic equation in y is easily solved by factoring:

$$y^2 + y - 6 = 0$$
$$(y + 3)(y - 2) = 0$$
$$y = -3 \quad \text{or} \quad y = 2$$

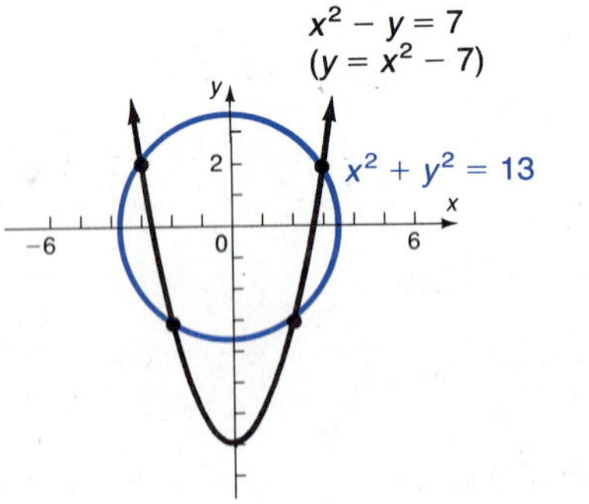

Figure 6

We use these values for y in equation (2) to find x. If $y = 2$, then $x^2 = y + 7 = 9$ and $x = 3$ or -3. If $y = -3$, then $x^2 = y + 7 = 4$ and $x = 2$ or -2. Thus, we have four solutions: $x = 3$, $y = 2$; $x = -3$, $y = 2$; $x = 2$, $y = -3$; and $x = -2$, $y = -3$. You should verify that, in fact, these four solutions also satisfy equation (1), so that all four are solutions of the system. These four solutions, (3, 2), (−3, 2), (2, −3), and (−2, −3), are the four points of intersection of the graphs. Look again at Figure 6. ■

Example 3 Solve: $\begin{cases} x^2 - y^2 = 1 & (1) \\ x^3 - y^2 = x & (2) \end{cases}$

Solution Because the second equation is not so easy to graph, we omit the graphing step. We use elimination, subtracting equation (2) from equation (1), to obtain

$$\begin{aligned} x^2 - x^3 &= 1 - x \\ x^2(1 - x) &= 1 - x \\ x^2(1 - x) - (1 - x) &= 0 \\ (x^2 - 1)(1 - x) &= 0 \end{aligned}$$

$$x^2 - 1 = 0 \quad \text{or} \quad 1 - x = 0$$
$$x = \pm 1 \qquad\qquad x = 1$$

We now use equation (1) to get y. If $x = 1$, then $1 - y^2 = 1$ and $y = 0$. If $x = -1$, then $1 - y^2 = 1$ and $y = 0$. There are two apparent solutions: $x = 1$, $y = 0$ and $x = -1$, $y = 0$. Because each of these solutions also satisfies equation (2), the system has two solutions: $x = 1$, $y = 0$ and $x = -1$, $y = 0$. The graphs of these equations will intersect at (1, 0) and at (−1, 0). ■

Example 4 Solve: $\begin{cases} x^2 + x + y^2 - 3y + 2 = 0 & (1) \\ x + 1 + \dfrac{y^2 - y}{x} = 0 & (2) \end{cases}$

Solution First, we multiply equation (2) by x to eliminate the fraction. The result is an equivalent system because x cannot be 0 [look at equation (2) to see why]:

$$\begin{cases} x^2 + x + y^2 - 3y + 2 = 0 & (1) \\ x^2 + x + y^2 - y = 0 & (2) \end{cases}$$

Now, subtract equation (2) from equation (1) to eliminate x. The result is

$$\begin{aligned} -2y + 2 &= 0 \\ y &= 1 \end{aligned}$$

To find x, we back-substitute $y = 1$ in equation (1):

$$\begin{aligned} x^2 + x + 1 - 3 + 2 &= 0 \\ x^2 + x &= 0 \\ x(x + 1) &= 0 \end{aligned}$$

$$x = 0 \quad \text{or} \quad x = -1$$

Because x cannot be 0, the value $x = 0$ is extraneous, and we discard it. Thus, the solution is $x = -1$, $y = 1$.

Check: We now check $x = -1$, $y = 1$:

$$\begin{cases} (-1)^2 + (-1) + 1^2 - 3(1) + 2 = 1 - 1 + 1 - 3 + 2 = 0 & (1) \\ -1 + 1 + \dfrac{1^2 - 1}{-1} = 0 + \dfrac{0}{-1} = 0 & (2) \end{cases}$$

Thus, the only solution to the system is $x = -1$, $y = 1$. The graphs of these equations will intersect at $(-1, 1)$. ■

Example 5 Solve: $\begin{cases} x^2 - y^2 = 4 & (1) \text{ A hyperbola} \\ y = x^2 & (2) \text{ A parabola} \end{cases}$

Solution Either substitution or elimination can be used here. We use substitution and replace x^2 by y in equation (1). The result is

$$y - y^2 = 4$$
$$y^2 - y + 4 = 0$$

This is a quadratic equation whose discriminant is $1 - 4 \cdot 4 = -15 < 0$. Thus, the equation has no real solutions and, hence, the system is inconsistent. The graphs of these two equations will not intersect. See Figure 7. ■

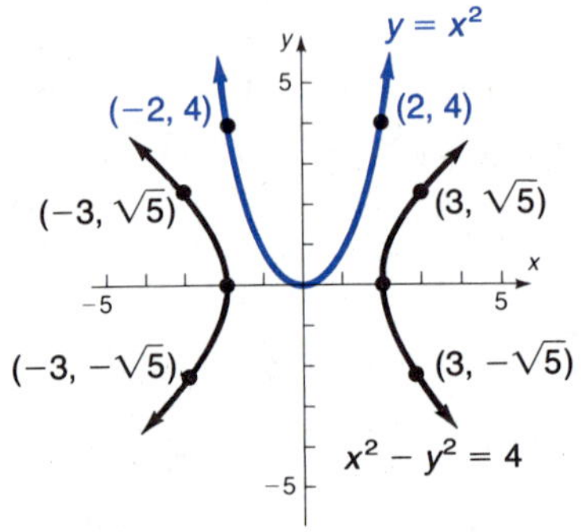

Figure 7

The following examples illustrate two of the more imaginative ways to solve systems of nonlinear equations.

Example 6 Solve: $\begin{cases} 4x^2 - 9xy - 28y^2 = 0 & (1) \\ 16x^2 - 4xy = 16 & (2) \end{cases}$

Solution We take note of the fact that equation (1) can be factored:

$$4x^2 - 9xy - 28y^2 = 0$$
$$(4x + 7y)(x - 4y) = 0$$

This results in the two equations

$$4x + 7y = 0 \quad \text{or} \quad x - 4y = 0$$
$$x = -\tfrac{7}{4}y \qquad\qquad x = 4y$$

We substitute each of these values for x in equation (2):

$$\begin{aligned} 16x^2 - 4xy &= 16 \\ 16\left(-\tfrac{7}{4}y\right)^2 - 4\left(-\tfrac{7}{4}y\right)y &= 16 \\ 49y^2 + 7y^2 &= 16 \\ 56y^2 &= 16 \\ 7y^2 &= 2 \\ y^2 &= \tfrac{2}{7} \end{aligned} \qquad \begin{aligned} 16x^2 - 4xy &= 16 \\ 16(4y)^2 - 4(4y)y &= 16 \\ 16(16y^2) - 16y^2 &= 16 \\ 15y^2 &= 1 \\ y^2 &= \tfrac{1}{15} \end{aligned}$$

Thus, we have

$$y = \pm\sqrt{\frac{2}{7}} = \pm\frac{\sqrt{14}}{7} \qquad y = \pm\frac{\sqrt{15}}{15}$$

$$x = -\frac{7}{4}y = \mp\frac{\sqrt{14}}{4} \qquad x = 4y = \pm\frac{4\sqrt{15}}{15}$$

You should verify for yourself that, in fact, the four solutions $x = -\sqrt{14}/4$, $y = \sqrt{14}/7$; $x = \sqrt{14}/4$, $y = -\sqrt{14}/7$; $x = 4\sqrt{15}/15$, $y = \sqrt{15}/15$; and $x = -4\sqrt{15}/15$, $y = -\sqrt{15}/15$ are actually solutions of the system. ■

Example 7 Solve: $\begin{cases} 3xy - 2y^2 = -2 & (1) \\ 9x^2 + 4y^2 = 10 & (2) \end{cases}$

Solution We multiply equation (1) by 2 and add the result to equation (2) to eliminate the y^2-terms:

$$\begin{cases} 6xy - 4y^2 = -4 & (1) \\ 9x^2 + 4y^2 = 10 & (2) \end{cases}$$

$$9x^2 + 6xy = 6$$

$$3x^2 + 2xy = 2 \quad \text{Divide each side by 3.}$$

Solve for y in this equation, to get

$$y = \frac{2 - 3x^2}{2x} \qquad x \neq 0 \tag{1}$$

Now, substitute for y in equation (2) of the system:

$$9x^2 + 4y^2 = 10$$

$$9x^2 + 4\left(\frac{2 - 3x^2}{2x}\right)^2 = 10$$

$$9x^2 + \frac{(4 - 12x^2 + 9x^4)}{x^2} = 10$$

$$9x^4 + 4 - 12x^2 + 9x^4 = 10x^2$$

$$18x^4 - 22x^2 + 4 = 0$$

$$9x^4 - 11x^2 + 2 = 0$$

This quadratic equation (in x^2) can be factored:

$$(9x^2 - 2)(x^2 - 1) = 0$$

$$9x^2 - 2 = 0 \quad \text{or} \quad x^2 - 1 = 0$$

$$x^2 = \tfrac{2}{9} \qquad x^2 = 1$$

$$x = \pm\frac{\sqrt{2}}{3} \qquad x = \pm 1$$

To find y, we use equation (1):

$$\text{If } x = \frac{\sqrt{2}}{3}: \quad y = \frac{2 - 3x^2}{2x} = \frac{2 - \frac{2}{3}}{2(\sqrt{2}/3)} = \frac{4}{2\sqrt{2}} = \sqrt{2}$$

$$\text{If } x = -\frac{\sqrt{2}}{3}: \quad y = \frac{2 - 3x^2}{2x} = \frac{2 - \frac{2}{3}}{-2(\sqrt{2}/3)} = \frac{4}{-2\sqrt{2}} = -\sqrt{2}$$

$$\text{If } x = 1: \quad y = \frac{2 - 3x^2}{2x} = \frac{2 - 3}{2} = -\frac{1}{2}$$

$$\text{If } x = -1: \quad y = \frac{2 - 3x^2}{2x} = \frac{2 - 3}{-2} = \frac{1}{2}$$

The system has four solutions. Check them for yourself. ■

Historical Comment

■ Recall that, in the beginning of this section, we said imagination and experience are important in solving simultaneous nonlinear equations. Indeed, these kinds of problems lead into some of the deepest and most difficult parts of modern mathematics. Look again at the graphs in Examples 1 and 2 of this section (Figures 5 and 6). We see that Example 1 has two solutions, and Example 2 has four solutions. One might conjecture that the number of solutions is equal to the product of the degrees of the equations involved. This conjecture was indeed made by Etienne Bezout (1739–1783), but working out the details took about 150 years. It turns out that to arrive at the correct number of intersections, we must count not only the complex number intersections, but also those intersections that, in a certain sense, lie at infinity. For example, a parabola and a line lying on the axis of the parabola intersect at the vertex and at infinity. This topic is part of the study of algebraic geometry.

Historial Problem

■ **1.** A papyrus dating back to 1950 BC contains the following problem: A given surface area of 100 units of area shall be represented as the sum of two squares whose sides are to each other as $1 : \frac{3}{4}$. Solve for the sides by solving the system of equations

$$\begin{cases} x^2 + y^2 = 100 \\ x = \frac{3}{4}y \end{cases}$$ ■

EXERCISE 12.4

In Problems 1–12, graph each equation of the system and then solve the system.

1. $\begin{cases} x + 2y + 3 = 0 \\ x^2 + y^2 = 5 \end{cases}$ **2.** $\begin{cases} 3x - y - 3 = 0 \\ x + 2y^2 - 2 = 0 \end{cases}$ **3.** $\begin{cases} x^2 + y^2 = 4 \\ y^2 - x = 4 \end{cases}$

4. $\begin{cases} x^2 + y^2 = 16 \\ x^2 - 2y = 8 \end{cases}$ **5.** $\begin{cases} x^2 + y^2 = 36 \\ x + y = 8 \end{cases}$ **6.** $\begin{cases} x^2 + y^2 = 4 \\ 2x - y + 4 = 0 \end{cases}$

7. $\begin{cases} xy = 4 \\ x^2 + y^2 = 8 \end{cases}$
8. $\begin{cases} x^2 = y \\ xy = 1 \end{cases}$
9. $\begin{cases} x^2 + y^2 = 4 \\ y = x^2 - 9 \end{cases}$

10. $\begin{cases} xy = 1 \\ 2x + 3y = 6 \end{cases}$
11. $\begin{cases} y = x^2 - 4 \\ y = 6x - 13 \end{cases}$
12. $\begin{cases} x^2 + y^2 = 6 \\ xy = 3 \end{cases}$

In Problems 13–42, solve each system. Use any method you wish.

13. $\begin{cases} 2x^2 + y^2 = 18 \\ xy = 4 \end{cases}$
14. $\begin{cases} x^2 - y^2 = 14 \\ x + y = 7 \end{cases}$
15. $\begin{cases} 3x - y = 1 \\ x^2 + 4y^2 = 17 \end{cases}$

16. $\begin{cases} x^2 - 4y^2 = 16 \\ 2y - x = 2 \end{cases}$
17. $\begin{cases} x + y + 1 = 0 \\ x^2 + y^2 + 6y - x = 7 \end{cases}$
18. $\begin{cases} 2x^2 - xy + y^2 = 8 \\ xy = 4 \end{cases}$

19. $\begin{cases} 4x^2 - 3xy + 9y^2 = 15 \\ 2x + 3y = 5 \end{cases}$
20. $\begin{cases} 2y^2 - 3xy + 6y + 2x + 4 = 0 \\ 2x - 3y + 4 = 0 \end{cases}$

21. $\begin{cases} x^2 - 4y^2 + 7 = 0 \\ 3x^2 + y^2 = 31 \end{cases}$
22. $\begin{cases} 3x^2 - 2y^2 + 5 = 0 \\ 2x^2 - y^2 + 2 = 0 \end{cases}$
23. $\begin{cases} 7x^2 - 3y^2 + 5 = 0 \\ 3x^2 + 5y^2 = 12 \end{cases}$

24. $\begin{cases} x^2 - 3y^2 + 1 = 0 \\ 2x^2 - 7y^2 + 5 = 0 \end{cases}$
25. $\begin{cases} x^2 + 2xy = 10 \\ 3x^2 - xy = 2 \end{cases}$
26. $\begin{cases} 5xy + 13y^2 + 36 = 0 \\ xy + 7y^2 = 6 \end{cases}$

27. $\begin{cases} 2x^2 + y^2 = 2 \\ x^2 - 2y^2 + 8 = 0 \end{cases}$
28. $\begin{cases} y^2 - x^2 + 4 = 0 \\ 2x^2 + 3y^2 = 6 \end{cases}$
29. $\begin{cases} x^2 + 2y^2 = 16 \\ 4x^2 - y^2 = 24 \end{cases}$

30. $\begin{cases} 4x^2 + 3y^2 = 4 \\ 2x^2 - 6y^2 = -3 \end{cases}$
31. $\begin{cases} \dfrac{5}{x^2} - \dfrac{2}{y^2} + 3 = 0 \\ \dfrac{3}{x^2} + \dfrac{1}{y^2} = 7 \end{cases}$
32. $\begin{cases} \dfrac{2}{x^2} - \dfrac{3}{y^2} + 1 = 0 \\ \dfrac{6}{x^2} - \dfrac{7}{y^2} + 2 = 0 \end{cases}$

33. $\begin{cases} \dfrac{1}{x^4} + \dfrac{6}{y^4} = 6 \\ \dfrac{2}{x^4} - \dfrac{2}{y^4} = 19 \end{cases}$
34. $\begin{cases} \dfrac{1}{x^4} - \dfrac{1}{y^4} = 1 \\ \dfrac{1}{x^4} + \dfrac{1}{y^4} = 4 \end{cases}$
35. $\begin{cases} x^2 - 3xy + 2y^2 = 0 \\ x^2 + xy = 6 \end{cases}$

36. $\begin{cases} x^2 - xy - 2y^2 = 0 \\ xy + x + 6 = 0 \end{cases}$
37. $\begin{cases} xy - x^2 + 3 = 0 \\ 3xy - 4y^2 = 2 \end{cases}$
38. $\begin{cases} 5x^2 + 4xy + 3y^2 = 36 \\ x^2 + xy + y^2 = 9 \end{cases}$

39. $\begin{cases} x^3 - y^3 = 26 \\ x - y = 2 \end{cases}$
40. $\begin{cases} x^3 + y^3 = 26 \\ x + y = 2 \end{cases}$
41. $\begin{cases} y^2 + y + x^2 - x - 2 = 0 \\ y + 1 + \dfrac{x - 2}{y} = 0 \end{cases}$

42. $\begin{cases} x^3 - 2x^2 + y^2 + 3y - 4 = 0 \\ x - 2 + \dfrac{y^2 - y}{x^2} = 0 \end{cases}$

43. The sum of two numbers is 8 and the sum of their squares is 36. Find the numbers.
44. The sum of two numbers is 24 and the difference of their squares is 48. Find the numbers.
45. The product of two numbers is 7 and the sum of their squares is 50. Find the numbers.
46. The product of two numbers is 12 and the difference of their squares is 7. Find the numbers.
47. The difference of two numbers is the same as their product, and the sum of their reciprocals is 5. Find the numbers.
48. The sum of two numbers is the same as their product, and the difference of their reciprocals is 3. Find the numbers.

49. The perimeter of a rectangle is 16 inches and its area is 15 square inches. What are its dimensions?

50. A wire 60 feet long is cut into two pieces. Is it possible to bend one piece into the shape of a square and the other into the shape of a circle so that the total area enclosed by the two pieces is 100 square feet? If this is possible, find the length of the side of the square and the radius of the circle.

51. A rectangular piece of cardboard, whose area is 216 square centimeters, is made into a box by cutting a 2 centimeter square from each corner and turning up the sides. If the box is to have a volume of 224 cubic centimeters, what size cardboard should you start with?

52. Two circles have perimeters that add up to 12π centimeters and areas that add up to 20π square centimeters. Find the radius of each circle.

53. The altitude of an isosceles triangle drawn to its base is 3 centimeters, and its perimeter is 18 centimeters. Find the length of its base.

54. In a 21 meter race between a tortoise and a hare, the tortoise leaves 9 minutes before the hare. The hare, by running at an average speed of 0.5 meter per hour faster than the tortoise, crosses the finish line 3 minutes before the tortoise. What are the average speeds of the tortoise and the hare?

55. *Descartes' method of equal roots* Descartes' method for finding tangents depends upon the idea that for many graphs, the tangent line at a given point is the *unique* line that intersects the graph at that point only. We will apply his method to find an equation of the tangent line to the parabola $y = x^2$ at the point (2, 4); see the figure. First, we know the equation of the tangent line must be in the form $y = mx + b$. Using the fact that the point (2, 4) is on the line, we can solve for b in terms of m and get the equation $y = mx + (4 - 2m)$. Now we want (2, 4) to be the *unique* solution to the system

$$\begin{cases} y = x^2 \\ y = mx + 4 - 2m \end{cases}$$

From this system, we get $x^2 = mx + 4 - 2m$ or $x^2 - mx + (2m - 4) = 0$. By using the quadratic formula, we get

$$x = \frac{m \pm \sqrt{m^2 - 4(2m - 4)}}{2}$$

In order to obtain a unique solution for x, the two roots must be equal; in other words, the expression $m^2 - 4(2m - 4)$ must be 0. Complete the work to get m, and write an equation of the tangent line.

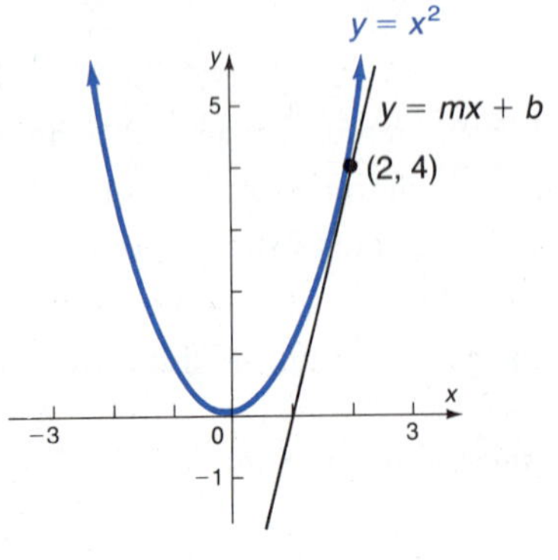

In Problems 56–62, use Descartes' method from Problem 55 to find the equation of the line tangent to each graph at the given point.

56. $x^2 + y^2 = 10$; at $(1, 3)$ **57.** $y = x^2 + 2$; at $(1, 3)$ **58.** $x^2 + y = 5$; at $(-2, 1)$

59. $2x^2 + 3y^2 = 14$; at $(1, 2)$ **60.** $3x^2 + y^2 = 7$; at $(-1, 2)$ **61.** $x^2 - y^2 = 3$; at $(2, 1)$

62. $2y^2 - x^2 = 14$; at $(2, 3)$

63. Find formulas for the length l and width w of a rectangle in terms of its area A and perimeter P.

64. Find formulas for the base b and one of the equal sides l of an isosceles triangle in terms of its altitude h and perimeter P.

In Problems 65–70, graph each equation and find the point(s) of intersection, if any.

65. The line $x + 2y = 0$ and the circle $(x - 1)^2 + (y - 1)^2 = 5$

66. The line $x + 2y + 6 = 0$ and the circle $(x + 1)^2 + (y + 1)^2 = 5$

67. The circle $(x - 1)^2 + (y + 2)^2 = 4$ and the parabola $y^2 + 4y - x + 1 = 0$

68. The circle $(x + 2)^2 + (y - 1)^2 = 4$ and the parabola $y^2 - 2y - x - 5 = 0$

69. The graph of $y = \dfrac{4}{x - 3}$ and the circle $x^2 - 6x + y^2 + 1 = 0$

70. The graph of $y = \dfrac{4}{x + 2}$ and the circle $x^2 + 4x + y^2 - 4 = 0$

71. If r_1 and r_2 are two solutions of a quadratic equation $ax^2 + bx + c = 0$, then it can be shown that

$$r_1 + r_2 = -\frac{b}{a} \quad \text{and} \quad r_1 r_2 = \frac{c}{a}$$

Solve this system of equations for r_1 and r_2.

12.5 Systems of Linear Inequalities

In Chapter 3, we discussed linear equations in two variables—namely, equations of the form

$$Ax + By + C = 0$$

where A and B are not both 0, and A, B, and C are real numbers. The graph of a linear equation is, of course, a line. In this section, we study the graphs of linear inequalities in two variables.

Linear Inequality in Two Variables

A **linear inequality** in two variables is an expression equivalent to one of the following forms:

$$\begin{array}{ll} Ax + By + C < 0 & Ax + By + C > 0 \\ Ax + By + C \le 0 & Ax + By + C \ge 0 \end{array} \qquad (1)$$

where A and B are not both 0, and A, B, and C are real numbers.

The top two inequalities in display (1) are referred to as **strict** inequalities; the remaining two are **nonstrict**. For example, each of the expressions

$$x + 3y - 5 < 0 \qquad \text{and} \qquad 3x - y + 10 \geq 0$$

is a linear inequality. The one on the left is strict; the one on the right is nonstrict, because it includes equality.

The **graph of a linear inequality** in two variables x and y is the set of all points (x, y) whose coordinates satisfy the inequality.

Let's look at an example.

Example 1 Graph the linear inequality: $3x + y - 6 \leq 0$

Solution We begin with the associated problem of the graph of the linear equality

$$3x + y - 6 = 0$$

formed by replacing (for now) the $\leq$ symbol with an $=$ sign. The graph of the linear equation is a line. See Figure 8(a). This line is part of the graph of the inequality we seek because the inequality is nonstrict. (Do you see why? We are seeking points for which $3x + y - 6$ is less than or equal to 0.)

Figure 8

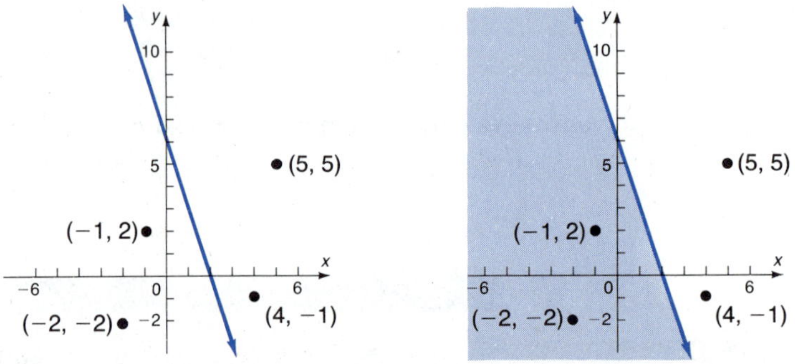

(a) $3x + y - 6 = 0$ (b) Graph of $3x + y - 6 \leq 0$

Now, let's test a few randomly selected points to see whether they belong to the graph of the inequality:

	$3x + y - 6$	CONCLUSION
$(4, -1)$	$3(4) + (-1) - 6 = 5 > 0$	Does not belong to graph
$(5, 5)$	$3(5) + 5 - 6 = 14 > 0$	Does not belong to graph
$(-1, 2)$	$3(-1) + 2 - 6 = -7 < 0$	Belongs to graph
$(-2, -2)$	$3(-2) + (-2) - 6 = -14 < 0$	Belongs to graph

Look again at Figure 8(a). Notice that the two points that belong to the graph both lie on the same side of the line, and the two points that do not belong to the graph lie on the opposite side. As it turns out, this is always the case. Thus, the graph we seek consists of all points that lie on the same side of the line as do $(-1, 2)$ and $(-2, -2)$. The graph we seek is the shaded region in Figure 8(b). ■

The collection of points that comprise the graph of a linear inequality is called a **half-plane**, and the line L is called its **boundary**. See Figure 9. As shown there, if $Ax + By + C = 0$ is the equation of the boundary line, then it divides the plane into two half-planes: one for which $Ax + By + C > 0$, and the other for which $Ax + By + C < 0$.

Figure 9

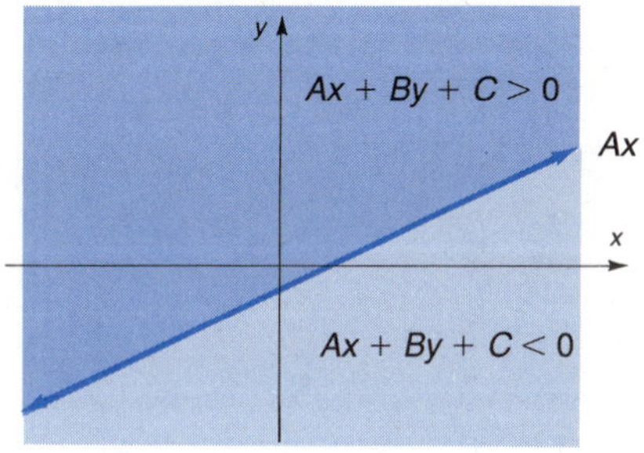

We outline the steps to follow for graphing a linear inequality below:

Steps for Graphing a Linear Inequality

STEP 1: Graph the line L that results from replacing the inequality symbol in the linear inequality with an = sign. This line is the boundary of the graph.

(a) If the linear inequality is strict, the line L will not belong to the graph of the linear inequality. (In this case, we will show the line as a dashed line.)

(b) If the linear inequality is nonstrict, the line L will belong to the graph of the linear inequality. (In this case, we will show the line as a solid line.)

STEP 2: Select a test point P that is not on the line L graphed in Step 1.

STEP 3: (a) If the coordinates of the point P satisfy the linear inequality, then so do all points on the same side of the line L as the point P.

(b) If the coordinates of the point P do not satisfy the linear inequality, then all points on the opposite side of the line L from the point P satisfy the linear inequality.

STEP 4: Shade the half-plane that belongs to the graph.

Example 2 Graph: $3x + y + 3 > 0$

Solution First, we graph the boundary line $3x + y + 3 = 0$; see Figure 10(a). Note that we have used a dashed line because the points on the line do not satisfy the inequality. The point (0, 0) is not on the graph of the boundary line, so we can use it as a test point:

$$(0, 0): \quad 3(0) + 0 + 3 = 3 > 0$$

The origin satisfies the linear inequality. Thus, all points in the half-plane on the same side of the boundary as (0, 0) are solutions. See the shaded region of Figure 10(b).

Figure 10

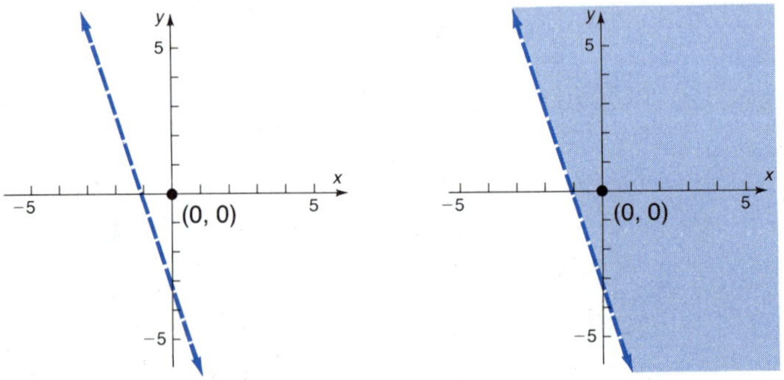

(a) $3x + y + 3 = 0$

(b) Graph of $3x + y + 3 > 0$

Systems of Linear Inequalities in Two Variables

The **graph of a system of linear inequalities** in two variables x and y is the set of all points (x, y) that simultaneously satisfy *each* of the linear inequalities in the system. Thus, the graph of a system of linear inequalities can be obtained by graphing each inequality individually and then determining where, if at all, they intersect.

Example 3 Graph the system: $\begin{cases} x + y \geq 2 \\ 2x - y \leq 4 \end{cases}$

Solution First, we graph the inequality $x + y \geq 2$, as the shaded region in Figure 11(a). Next, we graph the inequality $2x - y \leq 4$ as the shaded region in Figure 11(b). Now, superimpose the two graphs, as shown in Figure 11(c). The points that are in both shaded regions—the overlapping, darker region in Figure 11(c)—are the solutions we seek to the system, because they simultaneously satisfy each linear inequality.

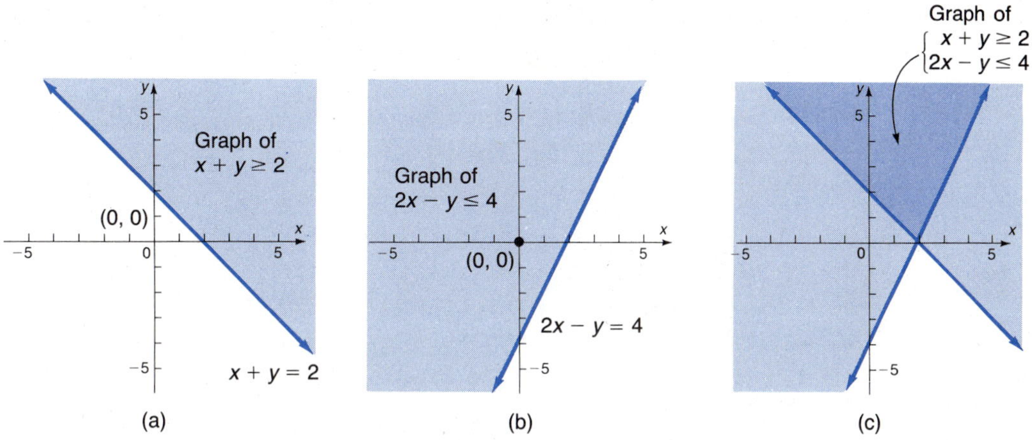

Figure 11

Example 4 Graph the system: $\begin{cases} x + y \leq 2 \\ x + y \geq 0 \end{cases}$

Solution See Figure 12. The overlapping, darker shaded region between the two boundary lines is the graph of the system.

Figure 12

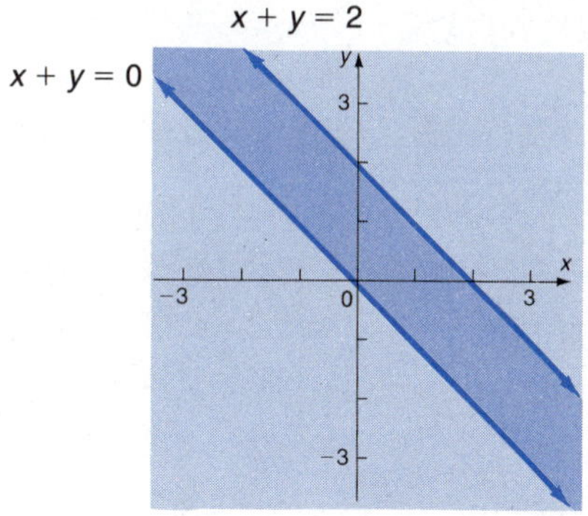

Example 5 Graph the system: $\begin{cases} 2x - y \geq 2 \\ 2x - y \geq 0 \end{cases}$

Solution See Figure 13 (page 716). The overlapping, darker shaded region is the graph of the system. Note that the graph of the system is identical to the graph of the single inequality $2x - y \geq 2$.

Figure 13

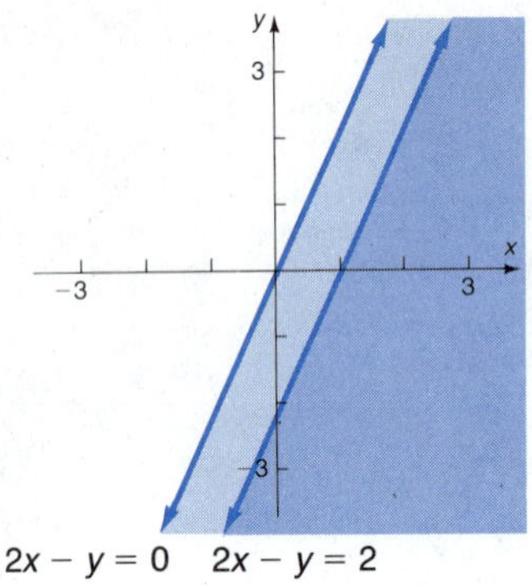

Example 6 Graph the system: $\begin{cases} x + 2y \le 2 \\ x + 2y \ge 6 \end{cases}$

Solution See Figure 14. Because no overlapping region results, there are no points in the xy-plane that simultaneously satisfy each inequality. Hence, the system has no solution.

Figure 14

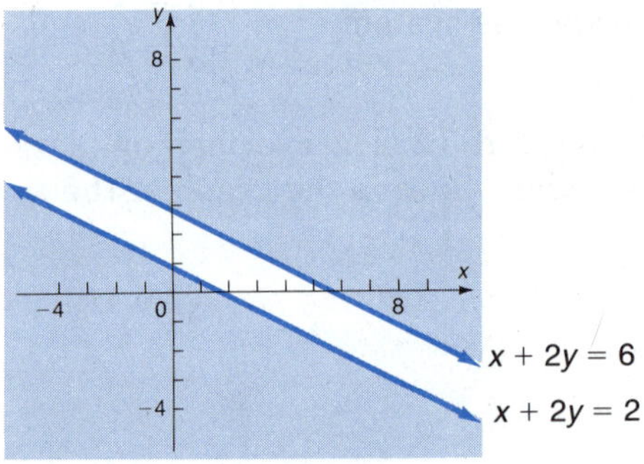

Example 7 Graph the system: $\begin{cases} x + y \ge 3 \\ 2x + y \ge 4 \\ x \ge 0 \\ y \ge 0 \end{cases}$

Solution The two inequalities $x \ge 0$ and $y \ge 0$ require that the graph be in quadrant I. Thus, we concentrate on the other two inequalities. The overlapping, darker shaded region in Figure 15 shows the graph of the system.

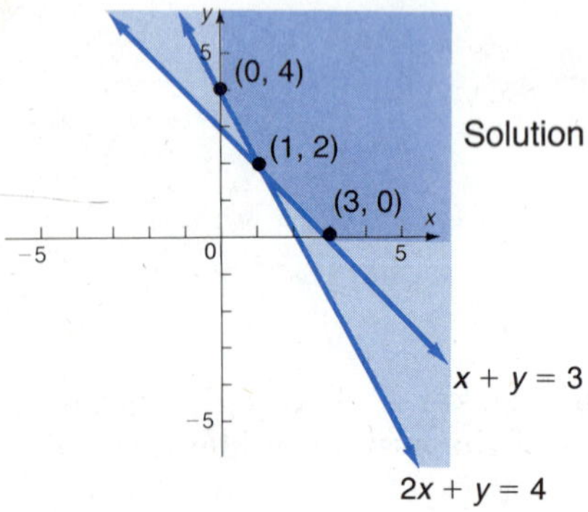

Figure 15

Example 8 Graph the system: $\begin{cases} x + y \le 6 \\ 2x + y \le 8 \\ x \ge 0 \\ y \ge 0 \end{cases}$

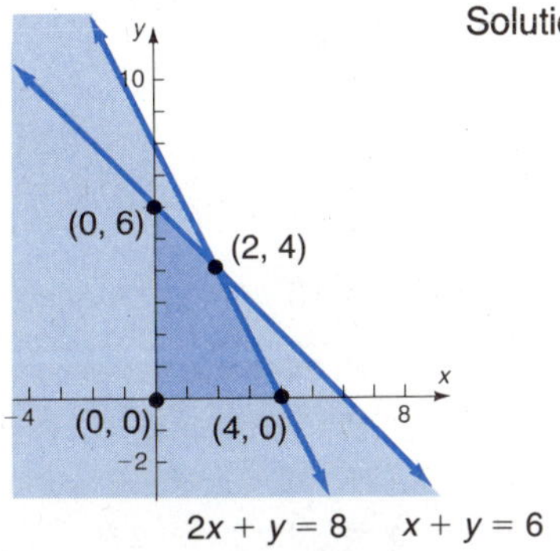

Figure 16

Solution See the overlapping, darker shaded region in Figure 16. Note that the inequalities $x \ge 0$ and $y \ge 0$ again require that the graph of the system be in quadrant I. ■

The graph of the system of linear inequalities in Figure 15 is said to be **unbounded**, because it extends indefinitely in a particular direction. On the other hand, the graph of the system of linear inequalities in Figure 16 is said to be **bounded**, because it can be contained within some circle of sufficiently large radius.

Notice in Figures 15 and 16 that those points belonging to the graph that are also points of intersection of boundary lines have been plotted. Such points are referred to as **vertices** of the graph. Thus, the system graphed in Figure 15 has three vertices: (0, 4), (1, 2), and (3, 0). The system graphed in Figure 16 has four vertices: (0, 0), (0, 6), (2, 4), and (4, 0). To find the vertex (2, 4) in Figure 16, we found the point of intersection of the two lines

$$x + y = 6 \quad \text{and} \quad 2x + y = 8$$

These ideas will be used later in developing a method for solving linear programming problems—one of the important applications of linear inequalities.

EXERCISE 12.5

In Problems 1–12, graph each linear inequality.

1. $x \ge 0$ **2.** $y \ge 0$ **3.** $x + y \ge 4$

4. $x + y \le 2$ **5.** $2x + y \ge 6$ **6.** $3x + 2y \le 6$

7. $4x - y \ge 4$ **8.** $y - 2x \le 2$ **9.** $3x - 2y \ge -6$

10. $4x - y \le 8$ **11.** $3x + 4y \ge 0$ **12.** $2x - 3y \le 0$

In Problems 13–24, graph each system of inequalities.

13. $\begin{cases} x + y \le 2 \\ 2x + y \ge 4 \end{cases}$ **14.** $\begin{cases} 3x - y \ge 6 \\ x + 2y \le 2 \end{cases}$ **15.** $\begin{cases} 2x - y \le 4 \\ 3x + 2y \ge -6 \end{cases}$

16. $\begin{cases} 4x - 5y \le 0 \\ 2x - y \ge 2 \end{cases}$ **17.** $\begin{cases} 2x - 3y \le 0 \\ 3x + 2y \le 6 \end{cases}$ **18.** $\begin{cases} 4x - y \ge 2 \\ x + 2y \ge 2 \end{cases}$

19. $\begin{cases} x - 2y \le 6 \\ 2x - 4y \ge 0 \end{cases}$ **20.** $\begin{cases} x + 4y \le 8 \\ x + 4y \ge 4 \end{cases}$ **21.** $\begin{cases} 2x + y \ge -2 \\ 2x + y \ge 2 \end{cases}$

22. $\begin{cases} x - 4y \le 4 \\ x - 4y \ge 0 \end{cases}$ **23.** $\begin{cases} 2x + 3y \ge 6 \\ 2x + 3y \le 0 \end{cases}$ **24.** $\begin{cases} 2x + y \ge 0 \\ 2x + y \ge 2 \end{cases}$

In Problems 25–34, graph each system of inequalities. Tell whether the graph is bounded or unbounded, and label the vertices.

25. $\begin{cases} x \geq 0 \\ y \geq 0 \\ 2x + y \leq 6 \\ x + 2y \leq 6 \end{cases}$

26. $\begin{cases} x \geq 0 \\ y \geq 0 \\ x + y \geq 4 \\ 2x + 3y \geq 6 \end{cases}$

27. $\begin{cases} x \geq 0 \\ y \geq 0 \\ x + y \geq 2 \\ 2x + y \geq 4 \end{cases}$

28. $\begin{cases} x \geq 0 \\ y \geq 0 \\ 3x + y \leq 6 \\ 2x + y \leq 2 \end{cases}$

29. $\begin{cases} x \geq 0 \\ y \geq 0 \\ x + y \geq 2 \\ 2x + 3y \leq 12 \\ 3x + y \leq 12 \end{cases}$

30. $\begin{cases} x \geq 0 \\ y \geq 0 \\ x + y \geq 2 \\ x + y \leq 10 \\ 2x + y \leq 3 \end{cases}$

31. $\begin{cases} x \geq 0 \\ y \geq 0 \\ x + y \geq 2 \\ x + y \leq 8 \\ 2x + y \leq 10 \end{cases}$

32. $\begin{cases} x \geq 0 \\ y \geq 0 \\ x + y \geq 2 \\ x + y \leq 8 \\ x + 2y \geq 1 \end{cases}$

33. $\begin{cases} x \geq 0 \\ y \geq 0 \\ x + 2y \geq 1 \\ x + 2y \leq 10 \end{cases}$

34. $\begin{cases} x \geq 0 \\ y \geq 0 \\ x + 2y \geq 1 \\ x + 2y \leq 10 \\ x + y \geq 2 \\ x + y \leq 8 \end{cases}$

35. A store that specializes in coffee has available 75 pounds of A grade coffee and 120 pounds of B grade coffee. These will be blended into 1 pound packages as follows: An economy blend that contains 4 ounces of A grade coffee and 12 ounces of B grade coffee, and a superior blend that contains 8 ounces of A grade coffee and 8 ounces of B grade coffee.
(a) Using x to denote the number of packages of the economy blend and y to denote the number of packages of the superior blend, write a system of linear inequalities that describes the possible number of packages of each kind of blend.
(b) Graph the system and show its vertices.

36. Mike's Toy Truck Company manufactures two models of toy trucks, a standard model and a deluxe model. Each standard model requires 2 hours for painting and 3 hours for detail work; each deluxe model requires 3 hours for painting and 4 hours for detail work. There are 2 painters and 3 detail workers employed by the company, and each works 40 hours per week.
(a) Using x to denote the number of standard model trucks and y to denote the number of deluxe model trucks, write a system of linear inequalities that describes the possible number of each model of truck that can be manufactured in a week.
(b) Graph the system and show its vertices.

CHAPTER REVIEW

THINGS TO KNOW

Systems of equations

Systems with no solutions are inconsistent.
Systems with a solution are consistent.
Consistent systems have either a unique solution or an infinite number of solutions.

How To:

Solve a system of linear equations using the method of substitution
Solve a system of linear equations using the method of elimination
Solve a system of linear equations using matrices
Solve a system of linear equations using determinants
Solve a system of nonlinear equations
Graph a system of linear inequalities
Find the vertices of the graph of a system of linear inequalities

FILL-IN-THE-BLANK ITEMS

1. If a system of equations has no solution, it is said to be ________
2. An m by n rectangular array of numbers is called a(n) ________.
3. Cramer's Rule uses ________ to solve a system of linear equations.
4. The matrix used to represent a system of linear equations is called a(n) ________ matrix.
5. The graph of a linear inequality is called a(n) ________.

REVIEW EXERCISES

In Problems 1–20, solve each system of equations using the method of substitution or the method of elimination. If the system has no solution, say it is inconsistent.

1. $\begin{cases} 2x - y = 5 \\ 5x + 2y = 8 \end{cases}$

2. $\begin{cases} 2x + 3y = 2 \\ 7x - y = 3 \end{cases}$

3. $\begin{cases} 3x - 4y = 4 \\ x - 3y = \frac{1}{2} \end{cases}$

4. $\begin{cases} 2x + y = 0 \\ 5x - 4y = -\frac{13}{2} \end{cases}$

5. $\begin{cases} x - 2y - 4 = 0 \\ 3x + 2y - 4 = 0 \end{cases}$

6. $\begin{cases} x - 3y + 5 = 0 \\ 2x + 3y - 5 = 0 \end{cases}$

7. $\begin{cases} y = 2x - 5 \\ x = 3y + 4 \end{cases}$

8. $\begin{cases} x = 5y + 2 \\ y = 5x + 2 \end{cases}$

9. $\begin{cases} x - y + 4 = 0 \\ \frac{1}{2}x + \frac{1}{6}y + \frac{2}{5} = 0 \end{cases}$

10. $\begin{cases} x + \frac{1}{4}y = 2 \\ y + 4x + 2 = 0 \end{cases}$

11. $\begin{cases} x - 2y - 8 = 0 \\ 2x + 2y - 10 = 0 \end{cases}$

12. $\begin{cases} x - 3y + 5 = 0 \\ 2x + 3y - 5 = 0 \end{cases}$

13. $\begin{cases} y - 2x = 11 \\ 2y - 3x = 18 \end{cases}$

14. $\begin{cases} 3x - 4y - 12 = 0 \\ 5x + 2y + 6 = 0 \end{cases}$

15. $\begin{cases} 2x + 3y - 13 = 0 \\ 3x - 2y = 0 \end{cases}$

16. $\begin{cases} 4x + 5y = 21 \\ 5x + 6y = 42 \end{cases}$

17. $\begin{cases} 3x - 2y = 8 \\ x - \frac{2}{3}y = 12 \end{cases}$

18. $\begin{cases} 2x + 5y = 10 \\ 4x + 10y = 15 \end{cases}$

19. $\begin{cases} x + 2y - z = 6 \\ 2x - y + 3z = -13 \\ 3x - 2y + 3z = -16 \end{cases}$

20. $\begin{cases} x + 5y - z = 2 \\ 2x + y + z = 7 \\ x - y + 2z = 11 \end{cases}$

In Problems 21–30, solve each system of equations using matrices. If the system has no solution, say it is inconsistent.

21. $\begin{cases} 3x - 2y = 1 \\ 10x + 10y = 5 \end{cases}$

22. $\begin{cases} 3x + 2y = 6 \\ x - y = -\frac{1}{2} \end{cases}$

23. $\begin{cases} 5x + 6y - 3z = 6 \\ 4x - 7y - 2z = -3 \\ 3x + y - 7z = 1 \end{cases}$

24. $\begin{cases} 2x + y + z = 5 \\ 4x - y - 3z = 1 \\ 8x + y - z = 5 \end{cases}$

25. $\begin{cases} x - 2z = 1 \\ 2x + 3y = -3 \\ 4x - 3y - 4z = 3 \end{cases}$

26. $\begin{cases} x + 2y - z = 2 \\ 2x - 2y + z = -1 \\ 6x + 4y + 3z = 5 \end{cases}$

27. $\begin{cases} x - y + z = 0 \\ x - y - 5z - 6 = 0 \\ 2x - 2y + z - 1 = 0 \end{cases}$

28. $\begin{cases} 4x - 3y + 5z = 0 \\ 2x + 4y - 3z = 0 \\ 6x + 2y + z = 0 \end{cases}$

29. $\begin{cases} x - y - z - t = 1 \\ 2x + y + z + 2t = 3 \\ x - 2y - 2z - 3t = 0 \\ 3x - 4y + z + 5t = -3 \end{cases}$

30. $\begin{cases} x - 3y + 3z - t = 4 \\ x + 2y - z = -3 \\ x + 3z + 2t = 3 \\ x + y + 5z = 6 \end{cases}$

In Problems 31–36, find the value of each determinant.

31. $\begin{vmatrix} 3 & 4 \\ 1 & 3 \end{vmatrix}$

32. $\begin{vmatrix} -4 & 0 \\ 1 & 3 \end{vmatrix}$

33. $\begin{vmatrix} 1 & 4 & 0 \\ -1 & 2 & 6 \\ 4 & 1 & 3 \end{vmatrix}$

34. $\begin{vmatrix} 2 & 3 & 10 \\ 0 & 1 & 5 \\ -1 & 2 & 3 \end{vmatrix}$

35. $\begin{vmatrix} 2 & 1 & -3 \\ 5 & 0 & 1 \\ 2 & 6 & 0 \end{vmatrix}$

36. $\begin{vmatrix} -2 & 1 & 0 \\ 1 & 2 & 3 \\ -1 & 4 & 2 \end{vmatrix}$

In Problems 37–42, use Cramer's Rule, if applicable, to solve each system.

37. $\begin{cases} x - 2y = 4 \\ 3x + 2y = 4 \end{cases}$

38. $\begin{cases} x - 3y = -5 \\ 2x + 3y = 5 \end{cases}$

39. $\begin{cases} 2x + 3y - 13 = 0 \\ 3x - 2y = 0 \end{cases}$

40. $\begin{cases} 3x - 4y - 12 = 0 \\ 5x + 2y + 6 = 0 \end{cases}$

41. $\begin{cases} x + 2y - z = 6 \\ 2x - y + 3z = -13 \\ 3x - 2y + 3z = -16 \end{cases}$

42. $\begin{cases} x - y + z = 8 \\ 2x + 3y - z = -2 \\ 3x - y - 9z = 9 \end{cases}$

In Problems 43–52, solve each system of equations.

43. $\begin{cases} 2x + y + 3 = 0 \\ x^2 + y^2 = 5 \end{cases}$

44. $\begin{cases} x^2 + y^2 = 16 \\ 2x - y^2 = -8 \end{cases}$

45. $\begin{cases} 2xy + y^2 = 10 \\ 3y^2 - xy = 2 \end{cases}$

46. $\begin{cases} 3x^2 - y^2 = 1 \\ 7x^2 - 2y^2 - 5 = 0 \end{cases}$

47. $\begin{cases} x^2 + y^2 = 6y \\ x^2 = 3y \end{cases}$

48. $\begin{cases} 2x^2 + y^2 = 9 \\ x^2 + y^2 = 9 \end{cases}$

49. $\begin{cases} 3x^2 + 4xy + 5y^2 = 8 \\ x^2 + 3xy + 2y^2 = 0 \end{cases}$ **50.** $\begin{cases} 3x^2 + 2xy - 2y^2 = 6 \\ xy - 2y^2 + 4 = 0 \end{cases}$ **51.** $\begin{cases} x^2 - 3x + y^2 + y = -2 \\ \dfrac{x^2 - x}{y} + y + 1 = 0 \end{cases}$

52. $\begin{cases} x^2 + x + y^2 = y + 2 \\ x + 1 = \dfrac{2 - y}{x} \end{cases}$

In Problems 53–58, graph each system of inequalities. Tell whether the graph is bounded or unbounded, and label the vertices.

53. $\begin{cases} -2x + y \le 2 \\ x + y \ge 2 \end{cases}$ **54.** $\begin{cases} x - 2y \le 6 \\ 2x + y \ge 2 \end{cases}$ **55.** $\begin{cases} x \ge 0 \\ y \ge 0 \\ x + y \le 4 \\ 2x + 3y \le 6 \end{cases}$

56. $\begin{cases} x \ge 0 \\ y \ge 0 \\ 3x + y \ge 6 \\ 2x + y \ge 2 \end{cases}$ **57.** $\begin{cases} x \ge 0 \\ y \ge 0 \\ 2x + y \le 8 \\ x + 2y \ge 2 \end{cases}$ **58.** $\begin{cases} x \ge 0 \\ y \ge 0 \\ 3x + y \le 9 \\ 2x + 3y \ge 6 \end{cases}$

59. Find A such that the system of equations below has infinitely many solutions.

$$\begin{cases} 2x + 5y = 5 \\ 4x + 10y = A \end{cases}$$

60. Find A such that the system in Problem 59 is inconsistent.

61. Find the quadratic function $y = ax^2 + bx + c$ that passes through the three points (0, 1), (1, 0), and (−2, 1)

62. Find the general equation of the circle that passes through the three points (0, 1), (1, 0), and (−2, 1). [*Hint:* The general equation of a circle is $x^2 + y^2 + Dx + Ey + F = 0$.]

63. Katy, Mike, Danny, and Colleen agreed to do yard work at home for $45 to be split among them. After they finished, their father determined that Mike deserves twice what Katy gets, Katy and Colleen deserve the same amount, and Danny deserves half of what Katy gets. How much does each one receive?

64. Rework Problem 63 if, after they finished, their father determined that Mike deserves $5 more than Katy, Katy deserves $5 more than Colleen, and Colleen deserves $5 more than Danny.

65. On a flight between Midway Airport in Chicago and Ft. Lauderdale, Florida, a Boeing 737 jet maintains an airspeed of 475 miles per hour. If the trip from Chicago to Ft. Lauderdale takes 2 hours, 30 minutes and the return flight takes 2 hours, 50 minutes, what is the speed of the jet stream? (Assume the speed of the jet stream remains constant at the various altitudes of the plane.)

66. Rework Problem 65 if the airspeed of the jet is 500 miles per hour.

67. If Katy and Mike work together for 1 hour and 20 minutes, they will finish a certain job. If Mike and Danny work together for 1 hour and 36 minutes, the same job can be finished. If Danny and Katy work together, they can complete this job in 2 hours and 40 minutes. How long will it take each of them working alone to finish the job?

68. A chemistry laboratory has three containers of hydrochloric acid, HCl. One container holds a solution with a concentration of 10% HCl, the second holds 25% HCl, and the third holds 40% HCl. How many liters of each should be mixed to obtain 100 liters of a solution with a concentration of 30% HCl? Construct a table showing some of the possible combinations.

69. A small rectangular lot has a perimeter of 68 feet. If its diagonal is 26 feet, what are the dimensions of the lot?

70. The area of a rectangular window is 4 square feet. If the diagonal measures $2\sqrt{2}$ feet, what are the dimensions of the window?

71. A certain right triangle has a perimeter of 14 inches. If the hypotenuse is 6 inches long, what are the lengths of the legs?

72. A certain isosceles triangle has a perimeter of 18 inches. If the altitude is 6 inches, what is the length of the base?

13 Induction; Sequences

Each topic in this chapter is related to the set of natural numbers. *Mathematical induction* is a technique for proving theorems involving the natural numbers. The *Binomial Theorem* is a formula for the expansion of $(x + a)^n$, where n is any natural number. And the last two sections deal with *sequences*, which are functions whose domain is the set of natural numbers. They form the basis for *recursively defined functions* and *recursive procedures* used in computer programming.

This chapter is intended to be introductory. Those of you who pursue mathematics further will use and encounter these topics in greater detail in more advanced courses. If you are interested in computer science, you will see these topics in a discrete mathematics course.

13.1 Mathematical Induction

Mathematical induction is a method for proving that statements involving natural numbers are true for all natural numbers.* For example, the statement, "$2n$ is always an even integer" can be proven true for all natural numbers by using mathematical induction. Also, the statement, "the sum of the first n positive odd integers equals n^2," that is,

$$1 + 3 + 5 + \cdots + (2n - 1) = n^2 \tag{1}$$

can be proved for all natural numbers n by using mathematical induction.

Before stating the method of mathematical induction, let's try to gain a sense of the power of the method. We shall use the statement in equation (1) for this purpose by restating it for various values of $n = 1, 2, 3, \ldots$:

$n = 1$ The sum of the first positive odd integer is 1^2; $1 = 1^2$.

$n = 2$ The sum of the first 2 positive odd integers is 2^2; $1 + 3 = 4 = 2^2$.

$n = 3$ The sum of the first 3 positive odd integers is 3^2; $1 + 3 + 5 = 9 = 3^2$.

$n = 4$ The sum of the first 4 positive odd integers is 4^2; $1 + 3 + 5 + 7 = 16 = 4^2$.

Although from this pattern we might conjecture that statement (1) is true for any choice of n, can we really be sure that it does not fail for some choice of n? The method of proof by mathematical induction will, in fact, prove that the statement is true for all n.

Theorem
The Principle of Mathematical Induction

Suppose the following two conditions are satisfied with regard to a statement about natural numbers:

CONDITION I: The statement is true for the natural number 1.
CONDITION II: If the statement is true for some natural number k, it is also true for the next natural number $k + 1$.

Then the statement is true for all natural numbers. ■

Figure 1

We shall not prove this principle. However, we can provide a physical interpretation that will help us see why the principle works. Think of a collection of natural numbers obeying a statement as a collection of infinitely many dominoes (see Figure 1).

*Recall from Chapter 1 that the natural numbers are the numbers 1, 2, 3, 4, ... In other words, the terms *natural numbers* and *positive integers* are synonymous.

Now, suppose we are told two facts:

1. The first domino is pushed over.
2. If one of the dominoes falls over, say the kth domino, then so will the next one, the $(k + 1)$st domino.

Is it safe to conclude that *all* the dominoes fall over? The answer is yes, because, if the first one falls (Condition I), then the second one does also (by Condition II); and if the second one falls, then so does the third (by Condition II); and so on.

Now let's prove some statements about natural numbers using mathematical induction.

Example 1 Show that the following statement is true for all natural numbers n:

$$1 + 3 + 5 + \cdots + (2n - 1) = n^2 \tag{2}$$

Solution We need to show first that statement (2) holds for $n = 1$. Because $1 = 1^2$, statement (2) is true for $n = 1$. Thus, Condition I holds.

Next, we need to show that Condition II holds. Suppose we know for some k that

$$1 + 3 + \cdots + (2k - 1) = k^2 \tag{3}$$

We wish to show that based on equation (3), statement (2) holds for $k + 1$. Thus, we look at the sum of the first $k + 1$ positive odd integers to determine whether this sum equals $(k + 1)^2$:

$$\begin{aligned} 1 + 3 + \cdots + (2k - 1) + (2k + 1) &= \underbrace{[1 + 3 + \cdots + (2k - 1)]}_{= k^2 \text{ by equation (3)}} + (2k + 1) \\ &= k^2 + (2k + 1) \\ &= k^2 + 2k + 1 = (k + 1)^2 \end{aligned}$$

Conditions I and II are satisfied; thus, by the principle of mathematical induction, statement (2) is true for all natural numbers. ■

Example 2 Show that the following statement is true for all natural numbers n:

$$2^n > n$$

Solution First, we show that the statement $2^n > n$ holds when $n = 1$. Because $2^1 = 2 > 1$, the inequality is true for $n = 1$. Thus, Condition I holds.

Next, we assume, for some natural number k, that $2^k > k$. We wish to show that the formula holds for $k + 1$; that is, we wish to show that $2^{k+1} > k + 1$. Now,

$$2^{k+1} = 2 \cdot 2^k \underset{\substack{\uparrow \\ \text{We know that} \\ 2^k > k.}}{>} 2 \cdot k = k + k \underset{\substack{\uparrow \\ k \geq 1}}{\geq} k + 1$$

Thus, if $2^k > k$, then $2^{k+1} > k + 1$, so Condition II of the principle of mathematical induction is satisfied. Hence, the statement $2^n > n$ is true for all natural numbers n. ■

Example 3 Show that the formula below is true for all natural numbers n:

$$1 + 2 + 3 + \cdots + n = \frac{n(n+1)}{2} \tag{4}$$

Solution First, we show that formula (4) is true when $n = 1$. Because

$$\frac{1(1+1)}{2} = \frac{1(2)}{2} = 1$$

Condition I of the principle of mathematical induction holds.

Next, we assume that formula (4) holds for some k, and we determine whether the formula then holds for $k + 1$. Thus, we assume

$$1 + 2 + 3 + \cdots + k = \frac{k(k+1)}{2} \qquad \text{for some } k \tag{5}$$

Now, we need to show that

$$1 + 2 + 3 + \cdots + k + (k+1) = \frac{(k+1)(k+1+1)}{2} = \frac{(k+1)(k+2)}{2}$$

We do this as follows:

$$\begin{aligned} 1 + 2 + 3 + \cdots + k + (k+1) &= \underbrace{[1 + 2 + 3 + \cdots + k]}_{= \frac{k(k+1)}{2} \text{ by equation (5)}} + (k+1) \\ &= \frac{k(k+1)}{2} + (k+1) \\ &= \frac{k^2 + k + 2k + 2}{2} \\ &= \frac{k^2 + 3k + 2}{2} = \frac{(k+1)(k+2)}{2} \end{aligned}$$

Thus, Condition II also holds. As a result, formula (4) is true for all natural numbers. ■

Example 4 Show that $3^n - 1$ is divisible by 2 for all natural numbers n.

Solution First, we show that the statement is true when $n = 1$. Because $3^1 - 1 = 3 - 1 = 2$ is divisible by 2, the statement is true when $n = 1$. Thus, Condition I is satisfied.

Next, we assume that the statement holds for some k, and we determine whether the statement then holds for $k + 1$. Thus, we assume that $3^k - 1$ is divisible by 2 for some k. We need to show that $3^{k+1} - 1$ is divisible by 2. Now,

$$3^{k+1} - 1 = 3^{k+1} - 3^k + 3^k - 1$$
$$= 3^k(3 - 1) + (3^k - 1) = 3^k \cdot 2 + (3^k - 1)$$

Because $3^k \cdot 2$ is divisible by 2 and $3^k - 1$ is divisible by 2, it follows that $3^k \cdot 2 + (3^k - 1) = 3^{k+1} - 1$ is divisible by 2. Thus, Condition II is also satisfied. As a result, the statement, "$3^n - 1$ is divisible by 2" is true for all natural numbers n. ■

Warning: The conclusion that a statement involving natural numbers is true for all natural numbers is made only after *both* Conditions I and II of the principle of mathematical induction have been satisfied. Problem 27 (below) demonstrates a statement for which only Condition I holds, but the statement is not true for all natural numbers. Problem 28 demonstrates a statement for which only Condition II holds, but the statement is *not* true for any natural number.

EXERCISE 13.1

In Problems 1–26, use the principle of mathematical induction to show that the given statement is true for all natural numbers.

1. $2 + 4 + 6 + \cdots + 2n = n(n + 1)$

2. $1 + 5 + 9 + \cdots + (4n - 3) = n(2n - 1)$

3. $3 + 4 + 5 + \cdots + (n + 2) = \frac{1}{2}n(n + 5)$

4. $3 + 5 + 7 + \cdots + (2n + 1) = n(n + 2)$

5. $2 + 5 + 8 + \cdots + (3n - 1) = \frac{1}{2}n(3n + 1)$

6. $1 + 4 + 7 + \cdots + (3n - 2) = \frac{1}{2}n(3n - 1)$

7. $1 + 2 + 2^2 + \cdots + 2^{n-1} = 2^n - 1$

8. $1 + 3 + 3^2 + \cdots + 3^{n-1} = \frac{1}{2}(3^n - 1)$

9. $1 + 4 + 4^2 + \cdots + 4^{n-1} = \frac{1}{3}(4^n - 1)$

10. $1 + 5 + 5^2 + \cdots + 5^{n-1} = \frac{1}{4}(5^n - 1)$

11. $\dfrac{1}{1 \cdot 2} + \dfrac{1}{2 \cdot 3} + \dfrac{1}{3 \cdot 4} + \cdots + \dfrac{1}{n(n + 1)} = \dfrac{n}{n + 1}$

12. $\dfrac{1}{1 \cdot 3} + \dfrac{1}{3 \cdot 5} + \dfrac{1}{5 \cdot 7} + \cdots + \dfrac{1}{(2n - 1)(2n + 1)} = \dfrac{n}{2n + 1}$

13. $1^2 + 2^2 + 3^2 + \cdots + n^2 = \frac{1}{6}n(n + 1)(2n + 1)$

14. $1^3 + 2^3 + 3^3 + \cdots + n^3 = \frac{1}{4}n^2(n + 1)^2$

15. $4 + 3 + 2 + \cdots + (5 - n) = \frac{1}{2}n(9 - n)$

16. $-2 - 3 - 4 - \cdots - (n + 1) = -\frac{1}{2}n(n + 3)$

17. $1 \cdot 2 + 2 \cdot 3 + 3 \cdot 4 + \cdots + n(n + 1) = \frac{1}{3}n(n + 1)(n + 2)$

18. $1 \cdot 2 + 3 \cdot 4 + 5 \cdot 6 + \cdots + (2n - 1)(2n) = \frac{1}{3}n(n + 1)(4n - 1)$

19. $n^2 + n$ is divisible by 2.

20. $n^3 + 2n$ is divisible by 3.

21. $n^2 - n + 2$ is divisible by 2.

22. $n(n + 1)(n + 2)$ is divisible by 6.

23. If $x > 1$, then $x^n > 1$.

24. If $0 < x < 1$, then $0 < x^n < 1$.

25. $a - b$ is a factor of $a^n - b^n$. [*Hint:* $a^{k+1} - b^{k+1} = a(a^k - b^k) + b^k(a - b)$]

26. $a + b$ is a factor of $a^{2n+1} + b^{2n+1}$.

27. Show that the statement "$n^2 - n + 41$ is a prime number" is true for $n = 1$ but is not true for $n = 41$.

28. Show that the formula

$$2 + 4 + 6 + \cdots + 2n = n^2 + n + 2$$

obeys Condition II of the principle of mathematical induction. That is, show that if the formula is true for some k, it is also true for $k + 1$. Then show that the formula is false for $n = 1$ (or for any other choice of n).

29. Use mathematical induction to prove that if $r \neq 1$, then

$$a + ar + ar^2 + \cdots + ar^{n-1} = a\left(\frac{1 - r^n}{1 - r}\right)$$

30. Use mathematical induction to prove that

$$a + (a + d) + (a + 2d) + \cdots + [a + (n - 1)d] = na + d\left[\frac{n(n - 1)}{2}\right]$$

31. Use mathematical induction to show that the sum of the interior angles of a convex polygon of n sides equals $(n - 2) \cdot 180°$.

32. Use mathematical induction to show that the number of diagonals in a convex polygon of n sides is $\frac{1}{2}n(n - 3)$. [*Hint:* Begin by showing the result is true when $n = 3$ (Condition I).]

13.2 The Binomial Theorem

In Chapter 1, we listed some special products. Among these were formulas for expanding $(x + a)^n$ for $n = 2$ and $n = 3$. The *Binomial Theorem** is a formula for the expansion of $(x + a)^n$ for n any positive integer. If $n = 1$, 2, and 3, the expansion of $(x + a)^n$ is straightforward:

$(x + a)^1 = x + a$	2 terms, beginning with x^1 and ending with a^1
$(x + a)^2 = x^2 + 2ax + a^2$	3 terms, beginning with x^2 and ending with a^2
$(x + a)^3 = x^3 + 3ax^2 + 3a^2x + a^3$	4 terms, beginning with x^3 and ending with a^3

*The name *binomial* derives from the fact that $x + a$ is a binomial—that is, contains two terms.

Notice that each expansion of $(x + a)^n$ begins with x^n and ends with a^n. Also, the number of terms that appear equals $n + 1$. Notice, too, that the degree of each monomial in the expansion equals n. For example, in the expansion of $(x + a)^3$, each monomial $(x^3, 3ax^2, 3a^2x, a^3)$ is of degree 3. As a result, we might conjecture that the expansion of $(x + a)^n$ would look like this:

$$(x + a)^n = x^n + \_ax^{n-1} + \_a^2x^{n-2} + \cdots + \_a^{n-1}x + a^n$$

where the blanks are numbers to be found. This is, in fact, the case, as we shall see shortly.

First, we need to introduce some symbols.

The Factorial Symbol

Factorial Symbol, $n!$

If $n \geq 0$ is an integer, the **factorial symbol $n!$** is defined as follows:

$$0! = 1 \qquad 1! = 1$$
$$n! = n(n-1) \cdot \cdots \cdot 3 \cdot 2 \cdot 1 \qquad \text{if } n \geq 2$$

For example, $2! = 2 \cdot 1 = 2$, $3! = 3 \cdot 2 \cdot 1 = 6$, $4! = 4 \cdot 3 \cdot 2 \cdot 1 = 24$, and so on. Table 1 lists the values of $n!$ for $0 \leq n \leq 6$.

Table 1

n	0	1	2	3	4	5	6
$n!$	1	1	2	6	24	120	720

Because

$$n! = n\underbrace{(n-1)(n-2) \cdot \cdots \cdot 3 \cdot 2 \cdot 1}_{(n-1)!}$$

we can use the formula

$$n! = n(n-1)!$$

to find successive factorials. For example, because $6! = 720$, we have

$$7! = 7 \cdot 6! = 7(720) = 5040$$

and

$$8! = 8 \cdot 7! = 8(5040) = 40{,}320$$

As you can see, factorials increase very rapidly. In fact, 70! is larger than 10^{100} (a *googol*), the largest number most calculators can display.

The Symbol $\binom{n}{j}$

In addition to the factorial symbol, we define the symbol $\binom{n}{j}$.

Symbol $\binom{n}{j}$ If j and n are integers with $0 \le j \le n$, the symbol $\binom{n}{j}$ is defined as

$$\binom{n}{j} = \frac{n!}{j!(n-j)!} \qquad (1)$$

Example 1 Find:

(a) $\binom{3}{1}$ (b) $\binom{4}{2}$ (c) $\binom{8}{7}$

Solution (a) $\binom{3}{1} = \frac{3!}{1!(3-1)!} = \frac{3!}{1!2!} = \frac{3 \cdot 2 \cdot 1}{1(2 \cdot 1)} = \frac{6}{2} = 3$

(b) $\binom{4}{2} = \frac{4!}{2!(4-2)!} = \frac{4!}{2!2!} = \frac{4 \cdot 3 \cdot 2 \cdot 1}{(2 \cdot 1)(2 \cdot 1)} = \frac{24}{4} = 6$

(c) $\binom{8}{7} = \frac{8!}{7!(8-7)!} = \frac{8!}{7!1!} = \frac{8 \cdot \cancel{7!}}{\cancel{7!} \cdot 1!} = \frac{8}{1} = 8$

↑ $8! = 8 \cdot 7!$

■

Two useful formulas involving the symbol $\binom{n}{j}$ are

$$\binom{n}{0} = 1 \quad \text{and} \quad \binom{n}{n} = 1$$

Proof $\binom{n}{0} = \frac{n!}{0!(n-0)!} = \frac{n!}{0!n!} = \frac{1}{1} = 1$

You are asked to show that $\binom{n}{n} = 1$ in Problem 42 at the end of this section. ■

Suppose we arrange the various values of the symbol $\binom{n}{j}$ in a triangular display, as shown below and in Figure 2:

$$\binom{0}{0}$$
$$\binom{1}{0} \quad \binom{1}{1}$$
$$\binom{2}{0} \quad \binom{2}{1} \quad \binom{2}{2}$$
$$\binom{3}{0} \quad \binom{3}{1} \quad \binom{3}{2} \quad \binom{3}{3}$$
$$\binom{4}{0} \quad \binom{4}{1} \quad \binom{4}{2} \quad \binom{4}{3} \quad \binom{4}{4}$$
$$\binom{5}{0} \quad \binom{5}{1} \quad \binom{5}{2} \quad \binom{5}{3} \quad \binom{5}{4} \quad \binom{5}{5}$$

This display is called the **Pascal triangle**, named after Blaise Pascal, a French mathematician.

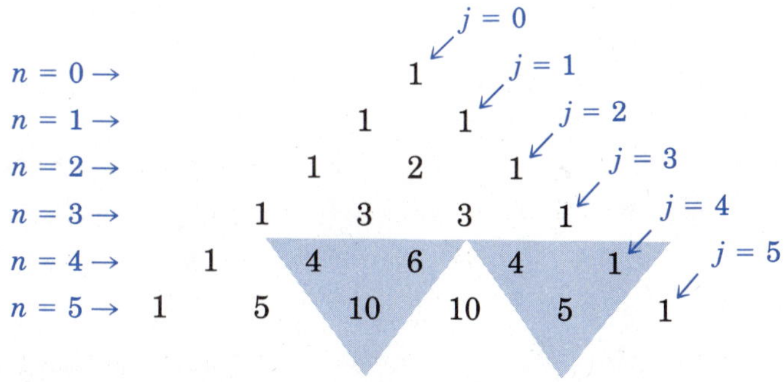

Figure 2
Pascal triangle

The Pascal triangle has 1's down the sides. To get any other entry, merely add the two nearest entries in the row above it. The shaded triangles in Figure 2 serve to illustrate this feature of the Pascal triangle. Based on this feature, the row corresponding to $n = 6$ is found as follows:

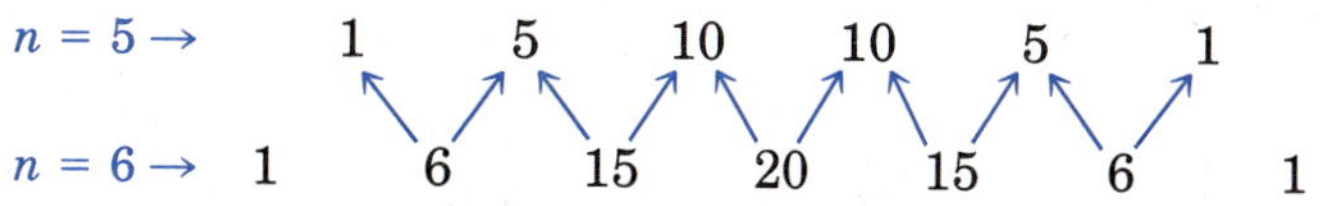

Later, we shall prove that this addition always works (see Example 6).

Although the Pascal triangle provides an interesting and organized display of the symbol $\binom{n}{j}$, in practice it is not all that helpful. For example, if you wanted to know the value of $\binom{12}{5}$, you would need to produce twelve rows of the triangle before seeing the answer. It is much faster instead to use the definition (1).

The Binomial Theorem

Now we are ready to state the **Binomial Theorem**. A proof is given at the end of this section.

Theorem

Binomial Theorem

Let x and a be real numbers. For any positive integer n, we have

$$(x+a)^n = \binom{n}{0}x^n + \binom{n}{1}ax^{n-1} + \cdots + \binom{n}{j}a^jx^{n-j} + \cdots + \binom{n}{n}a^n \qquad (2)$$

■

Now you know why we needed to introduce the symbol $\binom{n}{j}$; these symbols are the numerical coefficients that appear in the expansion of $(x + a)^n$. Because of this, the symbol $\binom{n}{j}$ is called the **binomial coefficient**.

Example 2 Use the Binomial Theorem to expand $(x + 2)^5$.

Solution In the Binomial Theorem, let $a = 2$ and $n = 5$. Then,

$$(x+2)^5 = \binom{5}{0}x^5 + \binom{5}{1}2x^4 + \binom{5}{2}2^2x^3 + \binom{5}{3}2^3x^2 + \binom{5}{4}2^4x + \binom{5}{5}2^5$$

↑ Use equation (2).

$$= 1 \cdot x^5 + 5 \cdot 2x^4 + 10 \cdot 4x^3 + 10 \cdot 8x^2 + 5 \cdot 16x + 1 \cdot 32$$

↑ Use row $n = 5$ of the Pascal triangle or formula (1) for $\binom{n}{j}$.

$$= x^5 + 10x^4 + 40x^3 + 80x^2 + 80x + 32$$

■

Example 3 Expand $(2y - 3)^4$ using the Binomial Theorem.

Solution First, we rewrite the expression $(2y - 3)^4$ as $[2y + (-3)]^4$. Now we use the Binomial Theorem with $n = 4$, $x = 2y$, and $a = -3$:

$$[2y + (-3)]^4 = \binom{4}{0}(2y)^4 + \binom{4}{1}(-3)(2y)^3 + \binom{4}{2}(-3)^2(2y)^2 + \binom{4}{3}(-3)^3 2y + \binom{4}{4}(-3)^4$$

$$= 1 \cdot 16y^4 + 4(-3)8y^3 + 6 \cdot 9 \cdot 4y^2 + 4(-27)2y + 1 \cdot 81$$

↑ Use row $n = 4$ of the Pascal triangle or formula (1) for $\binom{n}{j}$.

$$= 16y^4 - 96y^3 + 216y^2 - 216y + 81$$

In this expansion, note that the signs alternate due to the fact that $a = -3$ is negative. ■

The Binomial Theorem can be used to write a particular term in an expansion without writing the entire expansion. Based on the expansion of $(x + a)^n$, the term containing x^j is

$$\binom{n}{n-j}a^{n-j}x^j \tag{3}$$

Example 4 Find the coefficient of y^8 in the expansion of $(2y + 3)^{10}$.

Solution We use formula (3) with $n = 10$, $a = 3$, $x = 2y$, and $j = 8$. Then the term containing y^8 is

$$\begin{aligned}\binom{10}{10-8}3^{10-8}(2y)^8 &= \binom{10}{2} \cdot 3^2 \cdot 2^8 \cdot y^8\\ &= \frac{10!}{2!8!} \cdot 9 \cdot 2^8y^8\\ &= \frac{10 \cdot 9 \cdot \cancel{8!}}{2!\cancel{8!}} \cdot 9 \cdot 2^8y^8\\ &= 90 \cdot 9 \cdot 2^7y^8 = 103{,}680y^8\end{aligned}$$

■

Example 5 Find the sixth term in the expansion of $(x + 2)^9$.

Solution The sixth term in the expansion of $(x + 2)^9$, which has ten terms total, contains x^4. (Do you see why?) Thus, the sixth term is

$$\binom{9}{9-4}2^{9-4}x^4 = \binom{9}{5}2^5x^4 = \frac{9!}{5!4!} \cdot 32x^4 = 4032x^4$$

■

Next, we show that the "triangular addition" feature of the Pascal triangle illustrated in Figure 2 always works.

Example 6 If n and j are integers with $0 \le j \le n$, show that

$$\binom{n}{j-1} + \binom{n}{j} = \binom{n+1}{j}$$

Solution

$$\begin{aligned}\binom{n}{j-1}+\binom{n}{j} &= \frac{n!}{(j-1)![n-(j-1)]!}+\frac{n!}{j!(n-j)!}\\ &= \frac{n!}{(j-1)!(n-j+1)!}+\frac{n!}{j!(n-j)!}\\ &= \frac{jn!}{j(j-1)!(n-j+1)!}+\frac{(n-j+1)n!}{j!(n-j+1)(n-j)!}\\ &= \frac{jn!}{j!(n-j+1)!}+\frac{(n-j+1)n!}{j!(n-j+1)!}\\ &= \frac{jn!+(n-j+1)n!}{j!(n-j+1)!}\\ &= \frac{n!(j+n-j+1)}{j!(n-j+1)!}\\ &= \frac{n!(n+1)}{j!(n-j+1)!}=\frac{(n+1)!}{j![(n+1)-j]!}=\binom{n+1}{j}\end{aligned}$$

Multiply the first term by j/j and the second term by $(n-j+1)/(n-j+1)$.

Now the denominators are equal.

■

Proof of the Binomial Theorem We use mathematical induction to prove the Binomial Theorem. First, we show that formula (2) is true for $n = 1$:

$$(x+a)^1 = x + a = \binom{1}{0}x^1 + \binom{1}{1}a^1$$

Next, we suppose that formula (2) is true for some k. That is, we assume

$$(x+a)^k = \binom{k}{0}x^k + \binom{k}{1}ax^{k-1} + \cdots + \binom{k}{j-1}a^{j-1}x^{k-j+1} + \binom{k}{j}a^jx^{k-j} + \cdots + \binom{k}{k}a^k \quad (4)$$

Now, we calculate $(x+a)^{k+1}$:

$$\begin{aligned}(x+a)^{k+1} &= (x+a)(x+a)^k = x(x+a)^k + a(x+a)^k\\ &= x\left[\binom{k}{0}x^k + \binom{k}{1}ax^{k-1} + \cdots + \binom{k}{j-1}a^{j-1}x^{k-j+1} + \binom{k}{j}a^jx^{k-j} + \cdots + \binom{k}{k}a^k\right]\\ &\quad + a\left[\binom{k}{0}x^k + \binom{k}{1}ax^{k-1} + \cdots + \binom{k}{j-1}a^{j-1}x^{k-j+1} + \binom{k}{j}a^jx^{k-j} + \cdots + \binom{k}{k-1}a^{k-1}x + \binom{k}{k}a^k\right]\\ &= \binom{k}{0}x^{k+1} + \binom{k}{1}ax^k + \cdots + \binom{k}{j-1}a^{j-1}x^{k-j+2} + \binom{k}{j}a^jx^{k-j+1} + \cdots + \binom{k}{k}a^kx\\ &\quad + \binom{k}{0}ax^k + \binom{k}{1}a^2x^{k-1} + \cdots + \binom{k}{j-1}a^jx^{k-j+1} + \binom{k}{j}a^{j+1}x^{k-j} + \cdots + \binom{k}{k-1}a^kx + \binom{k}{k}a^{k+1}\\ &= \binom{k}{0}x^{k+1} + \left[\binom{k}{1}+\binom{k}{0}\right]ax^k + \cdots + \left[\binom{k}{j}+\binom{k}{j-1}\right]a^jx^{k-j+1} + \cdots + \left[\binom{k}{k}+\binom{k}{k-1}\right]a^kx + \binom{k}{k}a^{k+1}\end{aligned}$$

Use equation (4). (annotation on the second equals sign)

Because

$$\binom{k}{0} = 1 = \binom{k+1}{0}, \quad \binom{k}{1} + \binom{k}{0} \underset{\uparrow}{=} \binom{k+1}{1}, \quad \ldots,$$
Example 6

$$\binom{k}{j} + \binom{k}{j-1} \underset{\uparrow}{=} \binom{k+1}{j}, \quad \ldots, \quad \binom{k}{k} = 1 = \binom{k+1}{k+1}$$
Example 6

we have

$$(x+a)^{k+1} = \binom{k+1}{0}x^{k+1} + \binom{k+1}{1}ax^k + \cdots + \binom{k+1}{j}a^jx^{k-j+1} + \cdots + \binom{k+1}{k+1}a^{k+1}$$

Thus, Conditions I and II of the principle of mathematical induction are satisfied, and formula (2) is therefore true for all n. ■

Historical Commment

■ The case $n = 2$ of the Binomial Theorem, $(a + b)^2$, was known to Euclid in 300 BC, but the general law seems to have been discovered by the Persian mathematician and astronomer Omar Khayyám (1044?–1123?), who is also well known as the author of the *Rubaiyat*, a collection of four-line poems making observations on the human condition. Omar Khayyám did not state the Binomial Theorem explicitly, but he claimed to have a method for extracting third, fourth, fifth roots, and so on. A little study shows that one must know the Binomial Theorem to create such a method.

The heart of the Binomial Theorem is the formula for the numerical coefficients, and, as we saw, they can be written out in a symmetric triangular form. The Pascal triangle appears first in the books of Yang Hui (about 1270) and Chu Shih-chie (1303). Pascal's name is attached to the triangle because of the many applications he made of it, especially to counting and probability. In establishing these results, he was one of the earliest users of mathematical induction.

Many people worked on the proof of the Binomial Theorem, which was finally completed for all n (including complex numbers) by Niels Abel (1802–1829). ■

EXERCISE 13.2

In Problems 1–12, evaluate each expression.

1. $9!$ **2.** $10!$ **3.** $\dfrac{8!5!}{6!6!}$ **4.** $\dfrac{7!6!}{9!8!}$

5. $\dbinom{5}{3}$ **6.** $\dbinom{7}{3}$ **7.** $\dbinom{7}{5}$ **8.** $\dbinom{9}{7}$

9. $\binom{50}{49}$ **10.** $\binom{100}{98}$ **11.** $\binom{1000}{1000}$ **12.** $\binom{1000}{0}$

In Problems 13–24, expand each expression using the Binomial Theorem.

13. $(x + 1)^5$ **14.** $(x - 1)^5$ **15.** $(x - 2)^6$ **16.** $(x + 3)^4$
17. $(3x + 1)^4$ **18.** $(2x + 3)^5$ **19.** $(x^2 + y^2)^5$ **20.** $(x^2 - y^2)^6$
21. $(\sqrt{x} + \sqrt{2})^6$ **22.** $(\sqrt{x} - \sqrt{3})^4$ **23.** $(ax + by)^5$ **24.** $(ax - by)^4$

In Problems 25–38, use the Binomial Theorem to find the indicated coefficient or term.

25. The coefficient of x^6 in the expansion of $(x + 3)^{10}$
26. The coefficient of x^3 in the expansion of $(x - 3)^{10}$
27. The coefficient of x^8 in the expansion of $(2x - 1)^{12}$
28. The coefficient of x^2 in the expansion of $(2x + 1)^{12}$
29. The coefficient of x^7 in the expansion of $(2x + 3)^9$
30. The coefficient of x^2 in the expansion of $(2x - 3)^9$
31. The fifth term in the expansion of $(x + 3)^7$
32. The third term in the expansion of $(x - 3)^7$
C **33.** The third term in the expansion of $(3x - 2)^9$
C **34.** The sixth term in the expansion of $(3x + 2)^8$
35. The coefficient of x^0 in the expansion of $\left(x^2 + \frac{1}{x}\right)^{12}$
36. The coefficient of x^0 in the expansion of $\left(x - \frac{1}{x^2}\right)^9$
37. The coefficient of x^4 in the expansion of $\left(x - \frac{2}{\sqrt{x}}\right)^{10}$
38. The coefficient of x^2 in the expansion of $\left(\sqrt{x} + \frac{3}{\sqrt{x}}\right)^8$

39. Use the Binomial Theorem to find the numerical value of $(1.001)^5$ correct to five decimal places. [*Hint:* $(1.001)^5 = (1 + 10^{-3})^5$]

40. Use the Binomial Theorem to find the numerical value of $(0.998)^6$ correct to five decimal places.

41. Show that $\binom{n}{n} = 1$.

42. Show that, if n and j are integers with $0 \le j \le n$, then

$$\binom{n}{j} = \binom{n}{n-j}$$

Thus, conclude that the Pascal triangle is symmetric with respect to a vertical line drawn from the topmost entry.

43. If n is a positive integer, show that

$$\binom{n}{0} + \binom{n}{1} + \cdots + \binom{n}{n} = 2^n$$

[*Hint:* $2^n = (1 + 1)^n$; now use the Binomial Theorem.]

44. If n is a positive integer, show that

$$\binom{n}{0} - \binom{n}{1} + \binom{n}{2} - \cdots + (-1)^n\binom{n}{n} = 0$$

C **45.** *Stirling's formula* for approximating $n!$ when n is large is given by

$$n! \approx \sqrt{2n\pi}\left(\frac{n}{e}\right)^n\left(1 + \frac{1}{12n - 1}\right)$$

Calculate 12!, 20!, and 25!. Then use Stirling's formula to approximate 12!, 20!, and 25!.

13.3 Sequences

Sequence

A **sequence** is a function whose domain is the set of positive integers.

Because a sequence is a function, it will have a graph. In Figure 3(a), you will recognize the graph of the function $f(x) = 1/x$, $x > 0$. If all the points on this graph were removed except those whose x-coordinates are positive integers—that is, if all points were removed except (1, 1), $\left(2, \frac{1}{2}\right)$, $\left(3, \frac{1}{3}\right)$, and so on—the remaining points would be the graph of the sequence $f(n) = 1/n$, as shown in Figure 3(b).

Figure 3

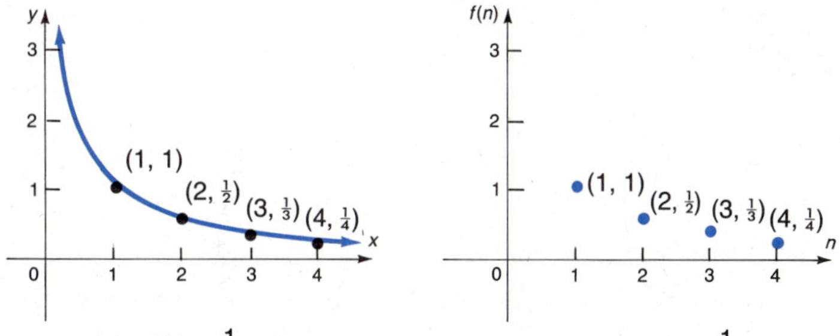

(a) $f(x) = \frac{1}{x}$, $x > 0$ (b) $f(n) = \frac{1}{n}$

A sequence is usually represented by listing its values in order. For example, the sequence whose graph is given in Figure 3(b) might be represented as

$$f(1), f(2), f(3), f(4), \ldots \qquad \text{or} \qquad 1, \tfrac{1}{2}, \tfrac{1}{3}, \tfrac{1}{4}, \ldots$$

The list never ends, as the ellipsis dots indicate. The numbers in this ordered list are called the **terms** of the sequence.

In dealing with sequences, we usually use subscripted letters, for example, a_1 to represent the first term, a_2 for the second term, a_3 for the third term, and so on. Thus, for the sequence $f(n) = 1/n$, we write

$$\begin{aligned} a_1 &= f(1) = 1 \\ a_2 &= f(2) = \tfrac{1}{2} \\ a_3 &= f(3) = \tfrac{1}{3} \\ a_4 &= f(4) = \tfrac{1}{4} \\ &\vdots \\ a_n &= f(n) = \frac{1}{n} \\ &\vdots \end{aligned}$$

In other words, we usually do not use the traditional function notation $f(n)$ for sequences. For this particular sequence, we have a rule for the nth term, namely, $a_n = 1/n$, so it is easy to find any term of the sequence.

When a formula for the nth term of a sequence is known, rather than write out the terms of the sequence, we usually represent the entire sequence by placing braces around the formula for the nth term. For example, the sequence whose nth term is $b_n = \left(\frac{1}{2}\right)^n$ may be represented as

$$\{b_n\} = \{(\tfrac{1}{2})^n\}$$

or by

$$\begin{aligned} b_1 &= \tfrac{1}{2} \\ b_2 &= \tfrac{1}{4} \\ b_3 &= \tfrac{1}{8} \\ &\vdots \\ b_n &= (\tfrac{1}{2})^n \\ &\vdots \end{aligned}$$

Example 1 Write down the first six terms of the sequence below, and graph it.

$$\{a_n\} = \left\{\frac{n-1}{n}\right\}$$

Solution

$$\begin{aligned} a_1 &= 0 \\ a_2 &= \tfrac{1}{2} \\ a_3 &= \tfrac{2}{3} \\ a_4 &= \tfrac{3}{4} \\ a_5 &= \tfrac{4}{5} \\ a_6 &= \tfrac{5}{6} \end{aligned}$$

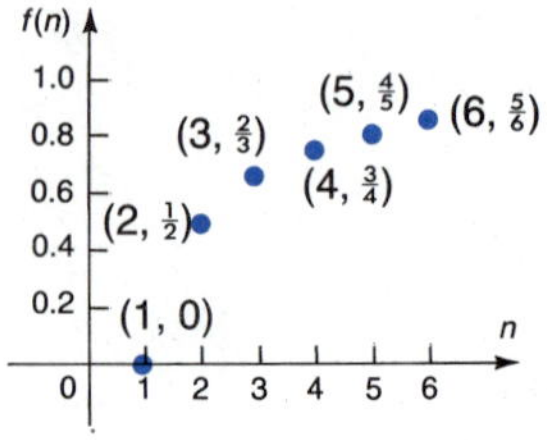

See Figure 4.

Figure 4 ■

Example 2 Write down the first six terms of the sequence below, and graph it.

$$\{b_n\} = \left\{(-1)^{n-1}\left(\frac{2}{n}\right)\right\}$$

Solution

$$b_1 = 2$$
$$b_2 = -1$$
$$b_3 = \tfrac{2}{3}$$
$$b_4 = -\tfrac{1}{2}$$
$$b_5 = \tfrac{2}{5}$$
$$b_6 = -\tfrac{1}{3}$$

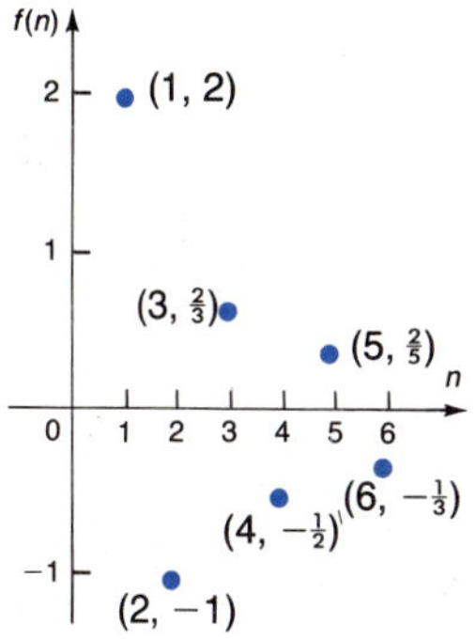

Figure 5

See Figure 5. ■

Example 3 Write down the first six terms of the sequence below, and graph it.

$$\{c_n\} = \begin{cases} n & \text{if } n \text{ is even} \\ 1/n & \text{if } n \text{ is odd} \end{cases}$$

Solution

$$c_1 = 1$$
$$c_2 = 2$$
$$c_3 = \tfrac{1}{3}$$
$$c_4 = 4$$
$$c_5 = \tfrac{1}{5}$$
$$c_6 = 6$$

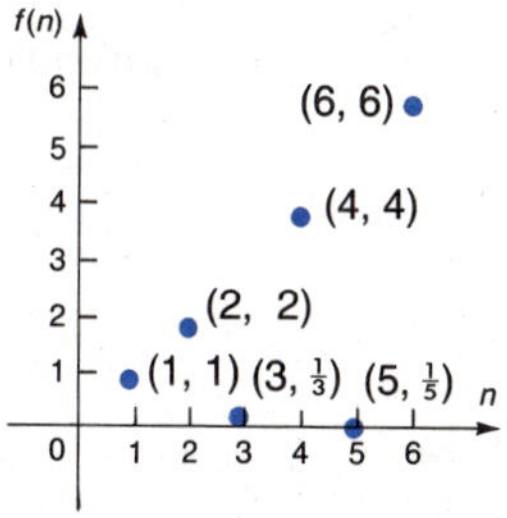

Figure 6

See Figure 6. ■

Sometimes, a sequence is indicated by an observed pattern in the first few terms that makes it possible to infer the makeup of the nth term. In the example that follows, a sufficient number of terms of the sequence is given so that a natural choice for the nth term is suggested.

Example 4

(a) $e, \frac{e^2}{2}, \frac{e^3}{3}, \frac{e^4}{4}, \ldots$ $\quad a_n = \frac{e^n}{n}$

(b) $1, \frac{1}{3}, \frac{1}{9}, \frac{1}{27}, \ldots$ $\quad b_n = \frac{1}{3^{n-1}}$

(c) $1, 3, 5, 7, \ldots$ $\quad c_n = 2n - 1$

(d) $1, 4, 9, 16, 25, \ldots$ $\quad d_n = n^2$

(e) $1, -\frac{1}{2}, \frac{1}{3}, -\frac{1}{4}, \frac{1}{5}, \ldots$ $\quad e_n = (-1)^{n+1}\left(\frac{1}{n}\right)$ ■

Notice that in the sequence $\{e_n\}$ in Example 4(e) the signs of the terms **alternate**. When this occurs, we use factors such as $(-1)^{n+1}$, which equals 1 if n is odd and -1 if n is even, or $(-1)^n$, which equals -1 if n is odd and 1 if n is even.

Recursion Formulas

A second way of defining a sequence is to assign a value to the first (or the first few) terms and specify the nth term by a formula or equation that involves one or more of the terms preceding it. Sequences defined this way are said to be defined **recursively**, and the rule or formula is called a **recursive formula**.

Example 5 Write down the first five terms of the recursively defined sequence given below.

$$s_1 = 1, \qquad s_n = 4s_{n-1}$$

Solution The first term is given as $s_1 = 1$. To get the second term, we use $n = 2$ in the formula to get $s_2 = 4s_1 = 4 \cdot 1 = 4$. To get the third term, we use $n = 3$ in the formula to get $s_3 = 4s_2 = 4 \cdot 4 = 16$. To get a new term requires that we know the value of the preceding term. The first five terms are

$$\begin{aligned} s_1 &= 1 \\ s_2 &= 4 \cdot 1 = 4 \\ s_3 &= 4 \cdot 4 = 16 \\ s_4 &= 4 \cdot 16 = 64 \\ s_5 &= 4 \cdot 64 = 256 \end{aligned}$$

■

Example 6 Write down the first five terms of the recursively defined sequence given below.

$$u_1 = 1, \qquad u_2 = 1, \qquad u_{n+2} = u_n + u_{n+1}$$

Solution We are given the first two terms. To get the third term requires that we know each of the previous two terms. Thus,

$$\begin{aligned} u_1 &= 1 \\ u_2 &= 1 \\ u_3 &= u_1 + u_2 = 2 \\ u_4 &= u_3 + u_2 = 2 + 1 = 3 \\ u_5 &= u_4 + u_3 = 3 + 2 = 5 \end{aligned}$$

■

Example 7 Write down the first five terms of the recursively defined sequence given below.

$$f_1 = 1, \qquad f_{n+1} = (n + 1)f_n$$

Solution Here,

$$\begin{aligned} f_1 &= 1 \\ f_2 &= 2f_1 = 2 \cdot 1 = 2 \\ f_3 &= 3f_2 = 3 \cdot 2 = 6 \\ f_4 &= 4f_3 = 4 \cdot 6 = 24 \\ f_5 &= 5f_4 = 5 \cdot 24 = 120 \end{aligned}$$

■

The sequence defined in Example 6 is called a **Fibonacci sequence**, and the terms of this sequence are called **Fibonacci numbers**. These numbers appear in a wide variety of applications (see Problems 55 and 56 at the end of this section). You should recognize the nth term of the sequence in Example 7 as n factorial.

Adding the First n Terms of a Sequence; Summation Notation

It is often important to be able to find the sum of the first n terms of a sequence $\{a_n\}$, namely,

$$a_1 + a_2 + a_3 + \cdots + a_n \tag{1}$$

Rather than write down all these terms, we introduce a more concise way to express the sum, called **summation notation**. Using summation notation we would write the sum (1) as

$$a_1 + a_2 + a_3 + \cdots + a_n = \sum_{k=1}^{n} a_k$$

The symbol Σ (a stylized version of the Greek letter sigma, which is an S in our alphabet) is simply an instruction to sum, or add up, the terms. The integer k is called the **index** of the sum; it tells you where to start the sum and where to end it. Therefore, the expression

$$\sum_{k=1}^{n} a_k \tag{2}$$

is an instruction to add the terms a_k of the sequence $\{a_n\}$ from $k = 1$ through $k = n$. We read expression (2) as, "the sum of a_k from $k = 1$ to $k = n$."

Example 8 Write out each sum.

(a) $\sum_{k=1}^{n} \frac{1}{k}$ (b) $\sum_{k=1}^{n} k!$

Solution (a) $\sum_{k=1}^{n} \frac{1}{k} = \frac{1}{1} + \frac{1}{2} + \frac{1}{3} + \cdots + \frac{1}{n}$ (b) $\sum_{k=1}^{n} k! = 1! + 2! + \cdots + n!$ ■

Example 9 Express each sum using summation notation.

(a) $1^2 + 2^2 + 3^2 + \cdots + n^2$ (b) $1 + \frac{1}{2} + \frac{1}{4} + \frac{1}{8} + \cdots + \frac{1}{2^{n-1}}$

Solution (a) The sum $1^2 + 2^2 + 3^2 + \cdots + n^2$ has n terms, each of the form k^2, and starts at $k = 1$ and ends at $k = n$. Thus,

$$1^2 + 2^2 + 3^2 + \cdots + n^2 = \sum_{k=1}^{n} k^2$$

(b) The sum

$$1 + \frac{1}{2} + \frac{1}{4} + \frac{1}{8} + \cdots + \frac{1}{2^{n-1}}$$

has n terms, each of the form $1/2^{k-1}$, and starts at $k = 1$ and ends at $k = n$. Thus,

$$1 + \frac{1}{2} + \frac{1}{4} + \frac{1}{8} + \cdots + \frac{1}{2^{n-1}} = \sum_{k=1}^{n} \frac{1}{2^{k-1}}$$ ■

Letters other than k may be used as the index. For example,

$$\sum_{j=1}^{n} j! \quad \text{and} \quad \sum_{i=1}^{n} i!$$

each represent the same sum as the one given in Example 8(b).

Next, we list some properties of summation:

Theorem

Properties of Sequences

If $\{a_n\}$ and $\{b_n\}$ are two sequences and c is a real number, then

1. $\sum_{k=1}^{n} ca_k = c\sum_{k=1}^{n} a_k$
2. $\sum_{k=1}^{n} (a_k + b_k) = \sum_{k=1}^{n} a_k + \sum_{k=1}^{n} b_k$
3. $\sum_{k=1}^{n} (a_k - b_k) = \sum_{k=1}^{n} a_k - \sum_{k=1}^{n} b_k$
4. $\sum_{k=1}^{n} a_k = \sum_{k=1}^{j} a_k + \sum_{k=j+1}^{n} a_k$, when $1 < j < n$ ■

Although we shall not prove these properties, the proofs are based on properties of real numbers.

The index of summation need not always begin at 1 nor end at n; for example,

$$\sum_{k=0}^{n-1} \frac{1}{2^k} = 1 + \frac{1}{2} + \frac{1}{4} + \cdots + \frac{1}{2^{n-1}}$$

EXERCISE 13.3

In Problems 1–12, write down the first five terms of each sequence.

1. $\{n\}$
2. $\{n^2 + 1\}$
3. $\left\{\frac{n}{n+1}\right\}$
4. $\left\{\frac{2n+3}{2n-1}\right\}$
5. $\{(-1)^{n+1}n^2\}$
6. $\left\{(-1)^{n-1}\left(\frac{n}{2n-1}\right)\right\}$
7. $\left\{\frac{2^n}{3^n+1}\right\}$
8. $\left\{\left(\frac{4}{3}\right)^n\right\}$
9. $\left\{\frac{(-1)^n}{(n+1)(n+2)}\right\}$
10. $\left\{\frac{3^n}{n}\right\}$
11. $\left\{\frac{n}{e^n}\right\}$
12. $\left\{\frac{n^2}{2^n}\right\}$

In Problems 13–20, the pattern given continues. Write down the nth term of each sequence suggested by the pattern.

13. $\frac{1}{2}, \frac{2}{3}, \frac{3}{4}, \frac{4}{5}, \ldots$
14. $\frac{1}{1 \cdot 2}, \frac{1}{2 \cdot 3}, \frac{1}{3 \cdot 4}, \frac{1}{4 \cdot 5}, \ldots$
15. $1, \frac{1}{2}, \frac{1}{4}, \frac{1}{8}, \ldots$
16. $\frac{2}{3}, \frac{4}{9}, \frac{8}{27}, \frac{16}{81}, \ldots$
17. $1, -1, 1, -1, 1, -1, \ldots$
18. $1, \frac{1}{2}, 3, \frac{1}{4}, 5, \frac{1}{6}, 7, \frac{1}{8}, \ldots$
19. $1, -2, 3, -4, 5, -6, \ldots$
20. $2, -4, 6, -8, 10, \ldots$

In Problems 21–34, a sequence is defined recursively. Write down the first five terms.

21. $a_1 = 1; \quad a_{n+1} = 2 + a_n$
22. $a_1 = 3; \quad a_{n+1} = 5 - a_n$
23. $a_1 = -2; \quad a_{n+1} = n + a_n$
24. $a_1 = 1; \quad a_{n+1} = n - a_n$
25. $a_1 = 5; \quad a_{n+1} = 2a_n$
26. $a_1 = 2; \quad a_{n+1} = -a_n$
27. $a_1 = 3; \quad a_{n+1} = \frac{a_n}{n}$
28. $a_1 = -2; \quad a_{n+1} = n + 3a_n$
29. $a_1 = 1; \quad a_2 = 2; \quad a_{n+2} = a_n a_{n+1}$
30. $a_1 = -1; \quad a_2 = 1; \quad a_{n+2} = a_{n+1} + na_n$
31. $a_1 = A; \quad a_{n+1} = a_n + d$
32. $a_1 = A; \quad a_{n+1} = ra_n, \quad r \neq 0$
33. $a_1 = \sqrt{2}; \quad a_{n+1} = \sqrt{2 + a_n}$
34. $a_1 = \sqrt{2}; \quad a_{n+1} = \sqrt{a_n/2}$

In Problems 35–44, write out each sum.

35. $\sum_{k=1}^{n} (k + 1)$
36. $\sum_{k=1}^{n} (2k - 1)$
37. $\sum_{k=1}^{n} \frac{k^2}{2}$
38. $\sum_{k=1}^{n} (k + 1)^2$
39. $\sum_{k=0}^{n} \frac{1}{3^k}$
40. $\sum_{k=0}^{n} \left(\frac{3}{2}\right)^k$
41. $\sum_{k=0}^{n-1} \frac{1}{3^{k+1}}$
42. $\sum_{k=0}^{n-1} (2k + 1)$
43. $\sum_{k=2}^{n} (-1)^k \ln k$
44. $\sum_{k=3}^{n} (-1)^{k+1} 2^k$

In Problems 45–54, express each sum using summation notation.

45. $1 + 2 + 3 + \cdots + n$

46. $1^3 + 2^3 + 3^3 + \cdots + n^3$

47. $\frac{1}{2} + \frac{2}{3} + \frac{3}{4} + \cdots + \frac{n}{n+1}$

48. $1 + 3 + 5 + 7 + \cdots + (2n - 1)$

49. $1 - \frac{1}{3} + \frac{1}{9} - \frac{1}{27} + \cdots + (-1)\left(\frac{1}{3^n}\right)$

50. $\frac{2}{3} - \frac{4}{9} + \frac{8}{27} - \cdots + (-1)^{n+1}\left(\frac{2}{3}\right)^n$

51. $3 + \frac{3^2}{2} + \frac{3^3}{3} + \cdots + \frac{3^n}{n}$

52. $\frac{1}{e} + \frac{2}{e^2} + \frac{3}{e^3} + \cdots + \frac{n}{e^n}$

53. $a + (a + d) + (a + 2d) + \cdots + (a + nd)$

54. $a + ar + ar^2 + \cdots + ar^{n-1}$

55. A colony of rabbits begins with one pair of mature rabbits, which will produce a pair of offspring (one male, one female) each month. Assume that all rabbits mature in 1 month and produce a pair of offspring (one male, one female) after 2 months. If no rabbits ever die, how many pairs of mature rabbits are there after 7 months? [*Hint:* A Fibonacci sequence models this colony. Do you see why?]

56. Let

$$u_n = \frac{(1 + \sqrt{5})^n - (1 - \sqrt{5})^n}{2^n\sqrt{5}}$$

define the nth term of a sequence.
(a) Show that $u_1 = 1$ and $u_2 = 1$. (b) Show that $u_{n+2} = u_{n+1} + u_n$.
(c) Draw the conclusion that $\{u_n\}$ is a Fibonacci sequence.

In Problems 57 and 58, we use the fact that in some programming languages it is possible to have a function subroutine include a call to itself.*

57. *Programming exercise* Write a program that accepts a positive integer as input and prints the number and its factorial. Use a recursively defined function; that is, use a function subroutine that calls itself.

58. *Programming exercise* Write a program that accepts a positive integer N as input and outputs the Nth Fibonacci number. Use a recursively defined subroutine.

59. Divide Pascal's triangle using diagonal lines as shown below. Find the sum of the numbers in each of these diagonal rows. Do you recognize this sequence?

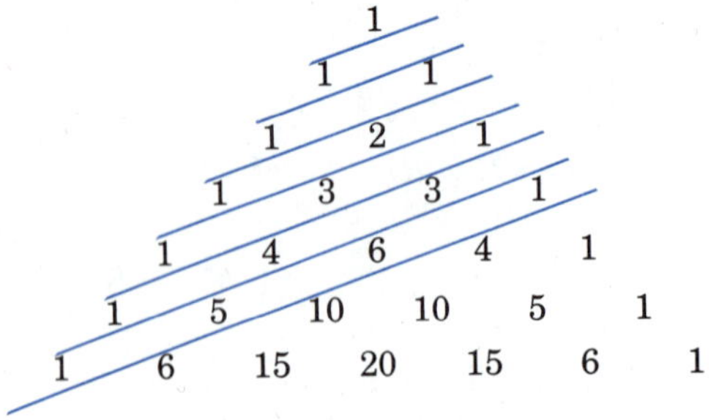

*Pascal, PL/1, ALGOL, and Logo, to name a few.

13.4 Arithmetic and Geometric Sequences; Geometric Series

Arithmetic Sequences

When the difference between successive terms of a sequence is always the same number, the sequence is called **arithmetic**. Thus, an arithmetic sequence,* may defined recursively as $a_1 = a$, $a_{n+1} - a_n = d$, or as

Arithmetic Sequence

$$a_1 = a, \quad a_{n+1} = a_n + d \tag{1}$$

where $a = a_1$ and d are real numbers. The number a is the first term, and the number d is called the **common difference**.

Thus, the terms of an arithmetic sequence with first term a and common difference d follow the pattern

$$a, \quad a + d, \quad a + 2d, \quad a + 3d, \quad \cdots$$

Example 1 Show that the sequence below is arithmetic. Find the first term and the common difference.

$$\{s_n\} = \{3n + 5\}$$

Solution The first term is $s_1 = 3 \cdot 1 + 5 = 8$. The $(n + 1)$st and nth terms of the sequence $\{s_n\}$ are

$$s_{n+1} = 3(n + 1) + 5 = 3n + 8 \qquad \text{and} \qquad s_n = 3n + 5$$

Their difference is

$$s_{n+1} - s_n = (3n + 8) - (3n + 5) = 8 - 5 = 3$$

Thus, the difference of two successive terms does not depend on n; the common difference is 3, and the sequence is arithmetic. ■

Example 2 Show that the sequence below is arithmetic. Find the first term and the common difference.

$$\{t_n\} = \{4 - n\}$$

Solution The first term is $t_1 = 4 - 1 = 3$. The $(n + 1)$st and nth terms are

$$t_{n+1} = 4 - (n + 1) = 3 - n \qquad \text{and} \qquad t_n = 4 - n$$

*Sometimes called an **arithmetic progression**.

Their difference is

$$t_{n+1} - t_n = (3 - n) - (4 - n) = 3 - 4 = -1$$

The difference of two successive terms does not depend on n; it always equals the same number, -1. Hence, $\{t_n\}$ is an arithmetic sequence whose common difference is -1. ■

Suppose a is the first term of an arithmetic sequence whose common difference is d. We seek a formula for the nth term, a_n. To see the pattern, we write down the first few terms:

$$\begin{aligned}
a_1 &= a + 0 \cdot d \\
a_2 &= a + d = a + 1 \cdot d \\
a_3 &= a_2 + d = (a + d) + d = a + 2 \cdot d \\
a_4 &= a_3 + d = (a + 2 \cdot d) + d = a + 3 \cdot d \\
a_5 &= a_4 + d = (a + 3 \cdot d) + d = a + 4 \cdot d \\
&\vdots \\
a_n &= a_{n-1} + d = [a + (n - 2)d] + d = a + (n - 1)d
\end{aligned}$$

We are led to the following result:

Theorem For an arithmetic sequence $\{a_n\}$ whose first term is a and whose common difference is d, the nth term is determined by the formula

nth Term of an Arithmetic Sequence

$$a_n = a + (n - 1)d \qquad (2)$$

Proof A proof of this result requires mathematical induction. If $n = 1$, we have

$$a_1 = a + (1 - 1)d = a + 0 \cdot d = a$$

Thus, Condition I of the principle of mathematical induction is satisfied. Suppose for some k that

$$a_k = a + (k - 1)d \qquad (3)$$

We want to show that

$$a_{k+1} = a + [(k + 1) - 1]d = a + kd$$

We do this as follows:

$$a_{k+1} = a_k + d = [a + (k - 1)d] + d = [a + kd - d] + d = a + kd$$

↑ Equation (3) (under the second equals sign)

↑ Definition of arithmetic sequence, equation (1) (under the first equals sign)

Thus, Condition II of the principle of mathematical induction is also satisfied. The formula $a_n = a + (n - 1)d$ therefore is true for all n. ■

Example 3 Find the 13th term of the arithmetic sequence: 2, 6, 10, 14, 18, . . .

Solution The first term of this arithmetic sequence is $a = 2$, and the common difference is 4. By formula (2), the nth term is

$$a_n = 2 + (n - 1)4$$

Hence, the 13th term is

$$a_{13} = 2 + 12 \cdot 4 = 50$$ ■

Example 4 The 8th term of an arithmetic sequence is 75, and the 20th term is 39. Find the first term and the common difference. Give a recursive formula for the sequence.

Solution By equation (2), we know that

$$\begin{cases} a_8 = a + 7d = 75 \\ a_{20} = a + 19d = 39 \end{cases}$$

This is a system of two linear equations containing two variables, which we can solve by elimination. Thus, subtracting the second equation from the first equation, we get

$$\begin{aligned} -12d &= 36 \\ d &= -3 \end{aligned}$$

With $d = -3$, we find $a = 75 - 7d = 75 - 7(-3) = 96$. A recursive formula for this sequence is

$$a_1 = 96, \qquad a_{n+1} = a_n - 3$$ ■

Based on formula (2), a general formula for the sequence $\{a_n\}$ in Example 4 is

$$a_n = a + (n - 1)d = 96 + (n - 1)(-3) = 99 - 3n$$

Adding the First n Terms of an Arithmetic Sequence

The next result gives a formula for finding the sum of the first n terms of an arithmetic sequence.

Theorem Let $\{a_n\}$ be an arithmetic sequence with first term a and common difference d. The sum S_n of the first n terms of $\{a_n\}$ is

Sum of n Terms of an Arithmetic Sequence

$$S_n = \frac{n}{2}[2a + (n - 1)d] = \frac{n}{2}(a + a_n) \qquad (4)$$

Proof

$$\begin{aligned}
S_n &= a + (a + d) + (a + 2d) + \cdots + [a + (n - 1)d] \\
&= \underbrace{(a + a + \cdots + a)}_{n \text{ terms}} + [d + 2d + \cdots + (n - 1)d] \\
&= na + d[1 + 2 + \cdots + (n - 1)] \\
&= na + d\left[\frac{n(n - 1)}{2}\right] && \text{From Example 3, page 726} \\
&= na + \frac{n}{2}(n - 1)d && \text{Factor out } n/2. \\
&= \frac{n}{2}[2a + (n - 1)d] \\
&= \frac{n}{2}[a + a + (n - 1)d] && \text{Formula (2)} \\
&= \frac{n}{2}(a + a_n)
\end{aligned}$$

■

This proof also can be done using the principle of mathematical induction. (Look back at Problem 30 in Exercise 13.1.)

Formula (4) provides two ways to find the sum of the first n terms of an arithmetic sequence. Notice that one involves the first term and common difference, while the other involves the first term and the nth term. Use whichever form is easier.

Example 5 Find the sum S_n of the first n terms of the sequence $\{3n + 5\}$; that is, find

$$8 + 11 + 14 + \cdots + (3n + 5)$$

Solution The sequence $\{3n + 5\}$ is an arithmetic sequence with first term $a = 8$ and nth term $(3n + 5)$. To find the sum S_n, we use formula (4):

$$S_n = \frac{n}{2}(a + a_n) = \frac{n}{2}[8 + (3n + 5)] = \frac{n}{2}(3n + 13)$$

■

Geometric Sequences

When the ratio of successive terms of a sequence is always the same nonzero number, the sequence is called **geometric**. Thus, a geometric sequence* may be defined recursively as $a_1 = a$, $a_{n+1}/a_n = r$, or as

Geometric Sequence

$$a_1 = a, \quad a_{n+1} = ra_n \tag{5}$$

where $a = a_1$ and $r \neq 0$ are real numbers. The number a is the first term, and the nonzero number r is called the **common ratio**.

Thus, the terms of a geometric sequence with first term a and common ratio r follow the pattern

$$a, \quad ar, \quad ar^2, \quad ar^3, \quad \ldots$$

Example 6 Show that the sequence below is geometric. Find the first term and the common ratio.

$$\{s_n\} = 2^{-n}$$

Solution The first term is $s_1 = 2^{-1} = \frac{1}{2}$. The $(n + 1)$st and nth terms of the sequence $\{s_n\}$ are

$$s_{n+1} = 2^{-(n+1)} \qquad \text{and} \qquad s_n = 2^{-n}$$

Their ratio is

$$\frac{s_{n+1}}{s_n} = \frac{2^{-(n+1)}}{2^{-n}} = 2^{-n-1+n} = 2^{-1} = \frac{1}{2}$$

Because the ratio of successive terms is a nonzero number independent of n, the sequence $\{s_n\}$ is geometric with common ratio $\frac{1}{2}$. ■

Example 7 Show that the sequence below is geometric. Find the first term and the common ratio.

$$\{t_n\} = \{4^n\}$$

Solution The first term is $t_1 = 4^1 = 4$. The $(n + 1)$st and nth terms are

$$t_{n+1} = 4^{n+1} \qquad \text{and} \qquad t_n = 4^n$$

Their ratio is

$$\frac{t_{n+1}}{t_n} = \frac{4^{n+1}}{4^n} = 4$$

Thus, $\{t_n\}$ is a geometric sequence with common ratio 4. ■

*Sometimes called a **geometric progression**.

Suppose a is the first term of a geometric sequence with common ratio $r \neq 0$. We seek a formula for the nth term a_n. To see the pattern, we write down the first few terms:

$$\begin{aligned} a_1 &= 1 \cdot a = ar^0 \\ a_2 &= ra = ar^1 \\ a_3 &= ra_2 = r(ar) = ar^2 \\ a_4 &= ra_3 = r(ar^2) = ar^3 \\ a_5 &= ra_4 = r(ar^3) = ar^4 \\ &\vdots \\ a_n &= ra_{n-1} = r(ar^{n-2}) = ar^{n-1} \end{aligned}$$

We are led to the following result:

Theorem For a geometric sequence $\{a_n\}$ whose first term is a and whose common ratio is r, the nth term is determined by the formula

nth Term of a Geometric Sequence

$$a_n = ar^{n-1} \qquad r \neq 0 \tag{6}$$

■

A proof of this result requires mathematical induction. See Problem 96 at the end of this section.

Example 8 Find the 9th term of the geometric sequence: $2, \frac{2}{3}, \frac{2}{9}, \frac{2}{27}, \ldots$

Solution The first term of this geometric sequence is $a = 2$, and the common ratio is $\frac{1}{3}$. (Use $\frac{2}{3}/2 = \frac{1}{3}$, or $\frac{2}{9}/\frac{2}{3} = \frac{1}{3}$, or any two successive terms.) By formula (6), the nth term is

$$a_n = 2\left(\frac{1}{3}\right)^{n-1}$$

Hence, the 9th term is

$$a_9 = 2\left(\frac{1}{3}\right)^8 = \frac{2}{3^8} = \frac{2}{6561}$$

■

Adding the First n Terms of a Geometric Sequence

The next result gives us a formula for finding the sum of the first n terms of a geometric sequence.

Theorem Let $\{a_n\}$ be a geometric sequence with first term a and common ratio r. The sum S_n of the first n terms of $\{a_n\}$ is

Sum of n Terms of a Geometric Sequence

$$S_n = a\left(\frac{1 - r^n}{1 - r}\right) \qquad r \neq 0, 1 \tag{7}$$

Proof

$$S_n = a + ar + \cdots + ar^{n-1} \tag{8}$$

Multiply each side by r to obtain

$$rS_n = ar + ar^2 + \cdots + ar^n \tag{9}$$

Now, subtract (9) from (8). The result is

$$S_n - rS_n = a - ar^n$$
$$(1 - r)S_n = a(1 - r^n)$$

Since $r \neq 1$, we can solve for S_n:

$$S_n = a\left(\frac{1 - r^n}{1 - r}\right)$$

■

This proof also can be done using the principle of mathematical induction. (Look back at Problem 29 in Exercise 13.1.)

Example 9 Find the sum S_n of the first n terms of the sequence $\{(\frac{1}{2})^n\}$; that is, find

$$\frac{1}{2} + \frac{1}{4} + \frac{1}{8} + \cdots + \left(\frac{1}{2}\right)^n$$

Solution The sequence $\{(\frac{1}{2})^n\}$ is a geometric sequence with $a = \frac{1}{2}$ and $r = \frac{1}{2}$. The sum S_n we seek is the sum of the first n terms of the sequence, so we use formula (7) to get

$$\begin{aligned} S_n = \sum_{k=1}^{n} \left(\frac{1}{2}\right)^k &= \frac{1}{2} + \frac{1}{4} + \frac{1}{8} + \cdots + \left(\frac{1}{2}\right)^n \\ &= \frac{1}{2}\left[\frac{1 - (\frac{1}{2})^n}{1 - \frac{1}{2}}\right] \\ &= \frac{1}{2}\left[\frac{1 - (\frac{1}{2})^n}{\frac{1}{2}}\right] \\ &= 1 - \left(\frac{1}{2}\right)^n \end{aligned}$$

■

Geometric Series

An infinite sum of the form

$$a + ar + ar^2 + \cdots + ar^{n-1} + \cdots$$

with first term a and common ratio r, is called an **infinite geometric series**, and is denoted by

$$\sum_{k=1}^{\infty} ar^{k-1}$$

Based on formula (7), the sum S_n of the first n terms of a geometric series is

$$S_n = a\left(\frac{1 - r^n}{1 - r}\right) = \frac{a}{1 - r} - \frac{ar^n}{1 - r} \qquad (10)$$

If this finite sum S_n approaches a number L as $n \to \infty$, then we call L the **sum of the infinite geometric series**, and we write

$$L = \sum_{k=1}^{\infty} ar^{k-1}$$

Theorem If $|r| < 1$, the sum of the infinite geometric series $\sum_{k=1}^{\infty} ar^{k-1}$ is

Sum of an Infinite Geometric Series

$$\sum_{k=1}^{\infty} ar^{k-1} = \frac{a}{1 - r} \qquad (11)$$

Intuitive Proof Since $|r| < 1$, it follows that $|r^n|$ approaches 0 as $n \to \infty$. Then, based on formula (10), the sum S_n approaches $a/(1 - r)$ as $n \to \infty$. ■

Example 10 Find the sum of the geometric series: $2 + \frac{4}{3} + \frac{8}{9} + \cdots$

Solution The first term is $a = 2$ and the common ratio is

$$r = \frac{\frac{4}{3}}{2} = \frac{4}{6} = \frac{2}{3}$$

Since $|r| < 1$, we use formula (11) to find that

$$2 + \frac{4}{3} + \frac{8}{9} + \cdots = \frac{2}{1 - \frac{2}{3}} = 6$$ ■

Example 11 Find: $\sum_{k=1}^{\infty} 5\left(\frac{1}{2}\right)^{k-1}$

Solution This geometric series has first term $a = 5$ and common ratio $r = \frac{1}{2}$. Since $|r| < 1$, its sum is

$$\sum_{k=1}^{\infty} 5\left(\frac{1}{2}\right)^{k-1} = \frac{5}{1 - \frac{1}{2}} = 10$$

■

Example 12 Show that the repeating decimal 0.999. . . equals 1.

Solution

$$0.999\ldots = \frac{9}{10} + \frac{9}{100} + \frac{9}{1000} + \cdots$$

Thus, 0.999. . . is a geometric series with first term $\frac{9}{10}$ and common ratio $\frac{1}{10}$. Hence,

$$0.999\ldots = \frac{\frac{9}{10}}{1 - \frac{1}{10}} = \frac{\frac{9}{10}}{\frac{9}{10}} = 1$$

■

Historical Comment

■ Sequences are among the oldest objects of mathematical investigation, having been studied for over 3500 years. After the initial steps, however, little progress was made until about 1600.

Arithmetic and geometric sequences appear in the Rhind papyrus, a mathematical text containing 85 problems copied around 1650 BC by the Egyptian scribe Ahmes from an earlier work (see the Historical Problems below). Fibonacci (AD 1220) wrote about problems similar to those found in the Rhind papyrus, leading one to suspect that Fibonacci may have had material available that is now lost. This material would have been in the non-Euclidean Greek tradition of Heron (about AD 75) and Diophantus (about AD 250). One problem, again modified slightly, is still with us in the familiar puzzle rhyme "As I was going to St. Ives . . ." (see Historical Problem 2).

The Rhind papyrus indicates that the Egyptians knew how to add up the terms of an arithmetic or geometric sequence, as did the Babylonians. The rule for summing up a geometric sequence is found in Euclid's *Elements* (book IX, 35, 36), where, like all of Euclid's algebra, it is presented in a geometric form.

Investigations of other kinds of sequences began in the 1500's, when algebra became sufficiently developed to handle the more complicated problems. The development of calculus in the 1600's added a powerful new tool, especially for finding the sum of infinite series, and the subject continues to flourish today.

Historical Problems

■ 1. *Arithmetic sequence problem from the Rhind papyrus* (*statement modified slightly for clarity*) One hundred loaves of bread are to be divided among five people so that the amounts they receive form an arithmetic sequence. The first two together receive one-seventh of what the last three receive. How many does each receive? [*Partial answer:* First person receives $1\frac{2}{3}$ loaves.]

2. The following old English children's rhyme resembles one of the Rhind papyrus problems:

As I was going to St. Ives
I met a man with seven wives
Each wife had seven sacks
Each sack had seven cats
Each cat had seven kits [kittens]
Kits, cats, sacks, wives,
How many were going to St. Ives?

(a) Assuming that the speaker and the cat fanciers met by traveling in opposite directions, what is the answer?
(b) How many kittens are being transported?
(c) Kits, cats, sacks, wives; how many? [*Hint:* It is easier to include the man, find the sum with the formula, and then subtract 1 for the man.] ■

EXERCISE 13.4

In Problems 1–10, an arithmetic sequence is given. Find the common difference and write out the first four terms.

1. $\{n + 5\}$ **2.** $\{n - 3\}$ **3.** $\{2n - 5\}$ **4.** $\{3n + 1\}$ **5.** $\{6 - 2n\}$
6. $\{4 - 2n\}$ **7.** $\left\{\frac{1}{2} - \frac{1}{3}n\right\}$ **8.** $\left\{\frac{2}{3} + \frac{n}{4}\right\}$ **9.** $\{\ln 3^n\}$ **10.** $\{e^{\ln n}\}$

In Problems 11–20, a geometric sequence is given. Find the common ratio and write out the first four terms.

11. $\{2^n\}$ **12.** $\{(-4)^n\}$ **13.** $\left\{-3\left(\frac{1}{2}\right)^n\right\}$ **14.** $\left\{\left(\frac{5}{2}\right)^n\right\}$ **15.** $\left\{\frac{2^{n-1}}{4}\right\}$
16. $\left\{\frac{3^n}{9}\right\}$ **17.** $\{2^{n/3}\}$ **18.** $\{3^{2n}\}$ **19.** $\left\{\frac{3^{n-1}}{2^n}\right\}$ **20.** $\left\{\frac{2^n}{3^{n-1}}\right\}$

In Problems 21–34, determine whether the given sequence is arithmetic, geometric, or neither. If the sequence is arithmetic, find the common difference; if it is geometric, find the common ratio.

21. $\{n + 2\}$ **22.** $\{2n - 5\}$ **23.** $\{2n^2\}$ **24.** $\{3n^2 + 1\}$
25. $\{3 - \frac{2}{3}n\}$ **26.** $\{8 - \frac{3}{4}n\}$ **27.** 1, 3, 6, 10, . . . **28.** 2, 4, 6, 8, . . .
29. $\{(\frac{2}{3})^n\}$ **30.** $\{(\frac{5}{4})^n\}$ **31.** −1, −2, −4, −8, . . .
32. 1, 1, 2, 3, 5, 8, . . . **33.** $\{3^{n/2}\}$ **34.** $\{(-1)^n\}$

In Problems 35–42, find the fifth term and the nth term of the arithmetic sequence whose initial term a and common difference d are given.

35. $a = 1;\ d = 2$ **36.** $a = -1;\ d = 3$ **37.** $a = 5;\ d = -3$
38. $a = 6;\ d = -2$ **39.** $a = 0;\ d = \frac{1}{2}$ **40.** $a = 1;\ d = -\frac{1}{3}$
41. $a = \sqrt{2};\ d = \sqrt{2}$ **42.** $a = 0;\ d = \pi$

In Problems 43–50, find the fifth term and the nth term of the geometric sequence whose initial term a and common ratio r are given.

43. $a = 1; \quad r = 2$
44. $a = -1; \quad r = 3$
45. $a = 5; \quad r = -1$
46. $a = 6; \quad r = -2$
47. $a = 0; \quad r = \frac{1}{2}$
48. $a = 1; \quad r = -\frac{1}{3}$
49. $a = \sqrt{2}; \quad r = \sqrt{2}$
50. $a = 0; \quad r = 1/\pi$

In Problems 51–56, find the indicated term in each arithmetic sequence.

51. 12th term of $2, 4, 6, \ldots$
52. 8th term of $-1, 1, 3, \ldots$
53. 10th term of $1, -2, -5, \ldots$
54. 9th term of $5, 0, -5, \ldots$
55. 8th term of $a, a + b, a + 2b, \ldots$
56. 7th term of $2\sqrt{5}, 4\sqrt{5}, 6\sqrt{5}, \ldots$

In Problems 57–62, find the indicated term in each geometric sequence.

57. 7th term of $1, \frac{1}{2}, \frac{1}{4}, \ldots$
58. 8th term of $1, 3, 9, \ldots$
59. 9th term of $1, -1, 1, \ldots$
60. 10th term of $-1, 2, -4, \ldots$
61. 8th term of $0.4, 0.04, 0.004, \ldots$
62. 7th term of $0.1, 1.0, 10.0, \ldots$

In Problems 63–70, find the first term and the common difference of the arithmetic sequence described.

63. 8th term is 8; 20th term is 44
64. 4th term is 3; 20th term is 35
65. 9th term is -5; 15th term is 31
66. 8th term is 4; 18th term is -96
67. 15th term is 0; 40th term is -50
68. 5th term is -2; 13th term is 30
69. 14th term is -1; 18th term is -9
70. 12th term is 4; 18th term is 28

In Problems 71–80, find the sum.

71. $6 + 11 + 16 + \cdots + (1 + 5n)$
72. $-2 + 1 + 4 + \cdots + (3n - 5)$
73. $\frac{1}{4} + \frac{2}{4} + \frac{2^2}{4} + \frac{2^3}{4} + \cdots + \frac{2^{n-1}}{4}$
74. $\frac{3}{9} + \frac{3^2}{9} + \frac{3^3}{9} + \cdots + \frac{3^n}{9}$
75. $\frac{2}{3} + \left(\frac{2}{3}\right)^2 + \left(\frac{2}{3}\right)^3 + \cdots + \left(\frac{2}{3}\right)^n$
76. $4 + 4 \cdot 3 + 4 \cdot 3^2 + \cdots + 4 \cdot 3^{n-1}$
77. $2 + 4 + 6 + \cdots + 2n$
78. $1 + 3 + 5 + \cdots + (2n - 1)$
79. $-1 - 2 - 4 - 8 - \cdots - (2^{n-1})$
80. $2 + 0 - 2 - 4 - \cdots - (2n - 4)$

In Problems 81–90, find the sum of each infinite geometric series.

81. $1 + \frac{1}{3} + \frac{1}{9} + \cdots$
82. $2 + \frac{4}{3} + \frac{8}{9} + \cdots$
83. $8 + 4 + 2 + \cdots$
84. $6 + 2 + \frac{2}{3} + \cdots$
85. $2 - \frac{1}{2} + \frac{1}{8} - \frac{1}{32} + \cdots$
86. $1 - \frac{3}{4} + \frac{9}{16} - \frac{27}{64} + \cdots$
87. $\sum_{k=1}^{\infty} 3\left(\frac{1}{4}\right)^{k-1}$
88. $\sum_{k=1}^{\infty} 2\left(\frac{1}{3}\right)^{k-1}$
89. $\sum_{k=1}^{\infty} 6\left(-\frac{2}{3}\right)^{k-1}$
90. $\sum_{k=1}^{\infty} 4\left(-\frac{1}{2}\right)^{k-1}$

C **91.** A ball is dropped from a height of 20 feet. Each time it strikes the ground, it bounces up to $\frac{3}{4}$ of the previous height. What height will the ball bounce up to after it strikes the ground for the third time? What is its height after it strikes the ground for the nth time?

C **92.** Your friend has just been hired at an annual salary of \$20,000. If she expects to receive annual increases of 4%, what will her salary be as she begins her fifth year?

C **93.** Compute the distance the ball in Problem 91 has traveled when it strikes the ground for the fifth time.

C **94.** A rich man promises to give you \$1000 on September 1, 1990. Each day thereafter he will give you $\frac{9}{10}$ of what he gave you the previous day. What is the first date on which the amount you receive is less than 1¢? How much have you received when this happens?

C **95.** In an old fable, a commoner who had just saved the king's life was told he could ask the king for any just reward. Being a shrewd man, the commoner said, "A simple wish, sire. Place one grain of wheat on the first square of a chessboard, two grains on the second square, four grains on the third square, continuing until you have filled the board. This is all I seek." Compute the total number of grains needed to do this to see why the request, seemingly simple, could not be granted. (A chessboard consists of $8 \times 8 = 64$ squares.)

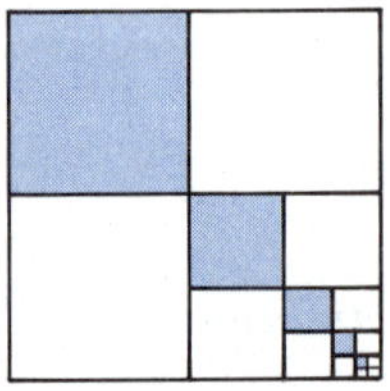

96. Use mathematical induction to prove that for a geometric sequence $\{a_n\}$ with common ratio r, the nth term a_n satisfies the formula

$$a_n = ar^{n-1}$$

97. Look at the figure in the margin. What fraction of the square is eventually shaded if the indicated shading process continues indefinitely?

CHAPTER REVIEW

THINGS TO KNOW

Principle of mathematical induction	Condition I: The statement is true for the natural number 1. Condition II: If the statement is true for some natural number k, it is also true for $k + 1$. Then the statement is true for all natural numbers.
Factorials	$0! = 1,\ 1! = 1,\ n! = n(n-1)\cdot\cdots\cdot 3\cdot 2\cdot 1$ if $n \geq 2$
Binomial coefficient	$\binom{n}{j} = \dfrac{n!}{j!(n-j)!}$
Pascal triangle	See Figure 2.
Binomial Theorem	$(x + a)^n = \binom{n}{0}x^n + \binom{n}{1}ax^{n-1} + \cdots + \binom{n}{j}a^jx^{n-j} + \cdots + \binom{n}{n}a^n$
Sequence	A function whose domain is the set of positive integers.
Arithmetic sequence	$a_1 = a,\ a_{n+1} = a_n + d$; where a = first term, d = common difference, $a_n = a + (n-1)d$
Sum of the first n terms of an arithmetic sequence	$S_n = \frac{n}{2}[2a + (n-1)d] = \frac{n}{2}(a + a_n)$

Geometric sequence	$a_1 = a$, $a_{n+1} = ra_n$; where a = first term, r = common ratio, $a_n = ar^{n-1}$, $r \neq 0$
Sum of the first n terms of a geometric sequence	$S_n = a\left(\dfrac{1 - r^n}{1 - r}\right)$, $r \neq 0, 1$
Infinite geometric series	$a + ar + \cdots + ar^{n-1} + \cdots = \sum_{k=1}^{\infty} ar^{k-1}$
Sum of an infinite geometric series	$\sum_{k=1}^{\infty} ar^{k-1} = \dfrac{a}{1 - r}$, $\lvert r \rvert < 1$

How To:

Prove statements about natural numbers using mathematical induction
Apply the Binomial Theorem
Write down the terms of a sequence
Use summation notation
Identify an arithmetic sequence
Find the sum of the first n terms of an arithmetic sequence
Identify a geometric sequence
Find the sum of the first n terms of a geometric sequence
Find the sum of an infinite geometric series

FILL-IN-THE-BLANK ITEMS

1. The ________ ________ is a triangular display of the binomial coefficients.
2. $\binom{6}{2} =$ ________
3. A(n) ________ is a function whose domain is the set of positive integers.
4. In a(n) ________ sequence, the difference between successive terms is always the same number.
5. In a(n) ________ sequence, the ratio of successive terms is always the same number.

REVIEW EXERCISES

In Problems 1–6, use the principle of mathematical induction to show that the given statement is true for all natural numbers.

1. $3 + 6 + 9 + \cdots + 3n = \dfrac{3n}{2}(n + 1)$
2. $2 + 6 + 10 + \cdots + (4n - 2) = 2n^2$
3. $2 + 6 + 18 + \cdots + 2 \cdot 3^{n-1} = 3^n - 1$
4. $3 + 6 + 12 + \cdots + 3 \cdot 2^{n-1} = 3(2^n - 1)$

5. $1^2 + 4^2 + 7^2 + \cdots + (3n - 2)^2 = \frac{1}{2}n(6n^2 - 3n - 1)$

6. $1 \cdot 3 + 2 \cdot 4 + 3 \cdot 5 + \cdots + n(n + 2) = \frac{n}{6}(n + 1)(2n + 7)$

In Problems 7–10, evaluate each expression.

7. $5!$ 8. $6!$ 9. $\binom{5}{2}$ 10. $\binom{8}{6}$

In Problems 11–14, expand each expression using the Binomial Theorem.

11. $(x + 2)^4$ 12. $(x - 3)^5$ 13. $(2x + 3)^5$ 14. $(3x - 4)^4$

15. Find the coefficient of x^7 in the expansion of $(x + 2)^9$.
16. Find the coefficient of x^3 in the expansion of $(x - 3)^8$.
17. Find the coefficient of x^2 in the expansion of $(2x + 1)^7$.
18. Find the coefficient of x^6 in the expansion of $(2x + 1)^8$.

In Problems 19–26, write down the first five terms of each sequence.

19. $\left\{(-1)^n\left(\frac{n+1}{n+2}\right)\right\}$ 20. $\{(-1)^{n+1}(2n - 3)\}$

21. $\left\{\frac{2^n}{n^2}\right\}$ 22. $\left\{\frac{e^n}{n}\right\}$

23. $a_1 = 3;\quad a_{n+1} = \frac{2}{3}a_n$ 24. $a_1 = 4;\quad a_{n+1} = -\frac{1}{4}a_n$

25. $a_1 = 2;\quad a_{n+1} = 2 - a_n$ 26. $a_1 = -3;\quad a_{n+1} = 4 + a_n$

In Problems 27–38, determine whether the given sequence is arithmetic, geometric, or neither. If the sequence is arithmetic, find the common difference and the sum of the first n terms. If the sequence is geometric, find the common ratio and the sum of the first n terms.

27. $\{n + 3\}$ 28. $\{4n + 1\}$ 29. $\{2n^3\}$

30. $\{2n^2 - 1\}$ 31. $\{2^{3n}\}$ 32. $\{3^{2n}\}$

33. $0, 4, 8, 12, \ldots$ 34. $1, -3, -7, -11, \ldots$ 35. $3, \frac{3}{2}, \frac{3}{4}, \frac{3}{8}, \frac{3}{16}, \ldots$

36. $5, -\frac{5}{3}, \frac{5}{9}, -\frac{5}{27}, \frac{5}{81}, \ldots$ 37. $\frac{2}{3}, \frac{3}{4}, \frac{4}{5}, \frac{5}{6}, \ldots$ 38. $\frac{3}{2}, \frac{5}{4}, \frac{7}{6}, \frac{9}{8}, \frac{11}{10}, \ldots$

In Problems 39–44, find the indicated term in each sequence.

39. 7th term of $3, 7, 11, 15, \ldots$ 40. 7th term of $1, -1, -3, -5, \ldots$

41. 11th term of $1, \frac{1}{10}, \frac{1}{100}, \ldots$ 42. 11th term of $1, 2, 4, 8, \ldots$

43. 9th term of $\sqrt{2}, 2\sqrt{2}, 3\sqrt{2}, \ldots$ 44. 9th term of $\sqrt{2}, 2, 2^{3/2}, \ldots$

In Problems 45–48, find a general formula for each arithmetic sequence.

45. 7th term is 31; 20th term is 96 46. 8th term is -20; 17th term is -47

47. 10th term is 0; 18th term is 8 48. 12th term is 30; 22nd term is 50

In Problems 49–54, find the sum of each infinite geometric series.

49. $3 + 1 + \frac{1}{3} + \frac{1}{9} + \cdots$ 50. $2 + 1 + \frac{1}{2} + \frac{1}{4} + \cdots$

51. $2 - 1 + \frac{1}{2} - \frac{1}{4} + \cdots$ 52. $6 - 4 + \frac{8}{3} - \frac{16}{9} + \cdots$

53. $\sum_{k=1}^{\infty} 4\left(\frac{1}{2}\right)^{k-1}$ 54. $\sum_{k=1}^{\infty} 3\left(-\frac{3}{4}\right)^{k-1}$

14 Miscellaneous Topics

This chapter contains four topics. They are independent of each other and may be covered in any order.

The first section treats a matrix as an algebraic concept, introducing the ideas of addition and multiplication as well as citing some of the properties of matrix algebra.

The second section introduces *linear programming*, a modern application of linear inequalities to certain types of problems. This topic is particularly useful for students interested in operations research.

The third section, *partial fraction decomposition*, provides an application of systems of equations. This particular application is one that is used later in the study of integral calculus.

Finally, the last two sections provide an introduction to the notion of a *vector* and some of its applications.

14.1 Matrix Algebra

In Section 12.2, we defined a matrix as an array of real numbers and used an augmented matrix to represent a system of linear equations. There is, however, a branch of mathematics, called **linear algebra**, that deals with matrices in such a way that an algebra of matrices is permitted. In this section, we provide a survey of how this **matrix algebra** is developed.

Before getting started, we restate the definition of a matrix.

Matrix

A **matrix** is defined as a rectangular array of numbers:

$$\begin{bmatrix} a_{11} & a_{12} & \cdots & a_{1j} & \cdots & a_{1n} \\ a_{21} & a_{22} & \cdots & a_{2j} & \cdots & a_{2n} \\ \vdots & \vdots & & \vdots & & \vdots \\ a_{i1} & a_{i2} & \cdots & a_{ij} & \cdots & a_{in} \\ \vdots & \vdots & & \vdots & & \vdots \\ a_{m1} & a_{m2} & \cdots & a_{mj} & \cdots & a_{mn} \end{bmatrix}$$

(jth column; ith row)

Each number a_{ij} of the matrix has two indices: the **row index** i and the **column index** j. The matrix shown above has m rows and n columns. The $m \cdot n$ numbers a_{ij} are usually referred to as the **entries** of the matrix.

Let's begin with an example that illustrates how matrices can be used to conveniently represent an array of information.

Example 1 In a survey of 1000 people, the following information was obtained:

200 males	Thought federal defense spending was too high
150 males	Thought federal defense spending was too low
45 males	Had no opinion
315 females	Thought federal defense spending was too high
125 females	Thought federal defense spending was too low
165 females	Had no opinion

We can arrange the above data in a rectangular array as follows:

	TOO HIGH	TOO LOW	NO OPINION
MALES	200	150	45
FEMALES	315	125	165

or as

$$\begin{bmatrix} 200 & 150 & 45 \\ 315 & 125 & 165 \end{bmatrix}$$

This matrix has two rows (representing males and females) and three columns (representing "too high," "too low," and "no opinion"). ■

The matrix we developed in Example 1 has 2 rows and 3 columns. In general, a matrix with m rows and n columns is called an ***m* by *n* matrix**. Thus, the matrix we developed in Example 1 is a 2 by 3 matrix. Notice that an m by n matrix will contain $m \cdot n$ entries.

If an m by n matrix has the same number of rows as columns, that is, if $m = n$, then the matrix is referred to as a **square matrix**.

Example 2 (a) $\begin{bmatrix} 5 & 0 \\ -6 & 1 \end{bmatrix}$ A 2 by 2 square matrix (b) $[1 \quad 0 \quad 3]$ A 1 by 3 matrix

(c) $\begin{bmatrix} 6 & -2 & 4 \\ 4 & 3 & 5 \\ 8 & 0 & 1 \end{bmatrix}$ A 3 by 3 square matrix ■

Equality and Addition of Matrices

We begin our discussion of matrix algebra by defining what is meant by two matrices being equal, and then defining the operations of addition and subtraction. It is important to note that these definitions require each matrix to have the same number of rows *and* the same number of columns as a prerequisite for equality and for addition and subtraction.

We usually represent matrices by capital letters, such as A, B, C, and so on.

Equal Matrices Two m by n matrices A and B are said to be **equal**, written as

$$A = B$$

provided each entry a_{ij} in A is equal to the corresponding entry b_{ij} in B.

For example,

$$\begin{bmatrix} 2 & 1 \\ 0.5 & -1 \end{bmatrix} = \begin{bmatrix} \sqrt{4} & 1 \\ \frac{1}{2} & -1 \end{bmatrix} \quad \text{and} \quad \begin{bmatrix} 3 & 2 & 1 \\ 0 & 1 & -2 \end{bmatrix} = \begin{bmatrix} \sqrt{9} & \sqrt{4} & 1 \\ 0 & 1 & \sqrt[3]{-8} \end{bmatrix}$$

$$\begin{bmatrix} 4 & 1 \\ 6 & 1 \end{bmatrix} \neq \begin{bmatrix} 4 & 0 \\ 6 & 1 \end{bmatrix}$$ Because the entries in row 1, column 2 are not equal

$$\begin{bmatrix} 4 & 1 & 2 \\ 6 & 1 & 2 \end{bmatrix} \neq \begin{bmatrix} 4 & 1 & 2 & 3 \\ 6 & 1 & 2 & 4 \end{bmatrix}$$ Because the matrix on the left is 2 by 3 and the matrix on the right is 2 by 4

If each of A and B is an m by n matrix (and each therefore contains $m \cdot n$ entries), the statement $A = B$ actually represents a system of $m \cdot n$ ordinary equations. We will make use of this fact a little later.

Suppose A and B represent two m by n matrices. We define their **sum** $\boldsymbol{A + B}$ to be the m by n matrix formed by adding the corresponding entries a_{ij} of A and b_{ij} of B. The **difference** $\boldsymbol{A - B}$ is defined as the m by n matrix formed by subtracting the entries b_{ij} in B from the corresponding entries a_{ij} in A. Addition and subtraction of matrices is allowed only for matrices having the same number m of rows and the same number n of columns. Thus, for example, a 2 by 3 matrix and a 2 by 4 matrix cannot be added or subtracted.

Example 3 Suppose

$$A = \begin{bmatrix} 2 & 4 & 8 & -3 \\ 0 & 1 & 2 & 3 \end{bmatrix} \quad \text{and} \quad B = \begin{bmatrix} -3 & 4 & 0 & 1 \\ 6 & 8 & 2 & 0 \end{bmatrix}$$

Find:

(a) $A + B$ (b) $A - B$

Solution (a)

$$A + B = \begin{bmatrix} 2 & 4 & 8 & -3 \\ 0 & 1 & 2 & 3 \end{bmatrix} + \begin{bmatrix} -3 & 4 & 0 & 1 \\ 6 & 8 & 2 & 0 \end{bmatrix}$$

$$= \begin{bmatrix} 2+(-3) & 4+4 & 8+0 & -3+1 \\ 0+6 & 1+8 & 2+2 & 3+0 \end{bmatrix}$$ Add corresponding entries.

$$= \begin{bmatrix} -1 & 8 & 8 & -2 \\ 6 & 9 & 4 & 3 \end{bmatrix}$$

(b)

$$A - B = \begin{bmatrix} 2 & 4 & 8 & -3 \\ 0 & 1 & 2 & 3 \end{bmatrix} - \begin{bmatrix} -3 & 4 & 0 & 1 \\ 6 & 8 & 2 & 0 \end{bmatrix}$$

$$= \begin{bmatrix} 2-(-3) & 4-4 & 8-0 & -3-1 \\ 0-6 & 1-8 & 2-2 & 3-0 \end{bmatrix}$$ Subtract corresponding entries.

$$= \begin{bmatrix} 5 & 0 & 8 & -4 \\ -6 & -7 & 0 & 3 \end{bmatrix}$$ ■

Many of the algebraic properties of sums of real numbers are also true for sums of matrices. Suppose A, B, and C are m by n matrices. Then matrix addition is **commutative**. That is,

Commutative Property

$$A + B = B + A$$

Matrix addition is also **associative**. That is,

Associative Property

$$(A + B) + C = A + (B + C)$$

Although we shall not prove these results, the proofs, as the following example illustrates, are based on the commutative and associative properties for real numbers.

Example 4

$$\begin{bmatrix} 2 & 3 & -1 \\ 4 & 0 & 7 \end{bmatrix} + \begin{bmatrix} -1 & 2 & 1 \\ 5 & -3 & 4 \end{bmatrix} = \begin{bmatrix} 2+(-1) & 3+2 & -1+1 \\ 4+5 & 0+(-3) & 7+4 \end{bmatrix}$$
$$= \begin{bmatrix} -1+2 & 2+3 & 1+(-1) \\ 5+4 & -3+0 & 4+7 \end{bmatrix}$$
$$= \begin{bmatrix} -1 & 2 & 1 \\ 5 & -3 & 4 \end{bmatrix} + \begin{bmatrix} 2 & 3 & -1 \\ 4 & 0 & 7 \end{bmatrix}$$ ■

A matrix whose entries are all equal to 0 is called a **zero matrix**. Each of the following matrices is a zero matrix:

$\begin{bmatrix} 0 & 0 \\ 0 & 0 \end{bmatrix}$ 2 by 2 square zero matrix $\begin{bmatrix} 0 & 0 & 0 \\ 0 & 0 & 0 \end{bmatrix}$ 2 by 3 zero matrix $\begin{bmatrix} 0 & 0 & 0 \end{bmatrix}$ 1 by 3 zero matrix

Zero matrices have properties similar to the real number 0. Thus, if A is an m by n matrix and 0 is an m by n zero matrix, then

$$A + 0 = A$$

In other words, the zero matrix is the additive identity in matrix algebra.

We also can multiply a matrix by a real number. If k is a real number and A is an m by n matrix, the matrix kA is the m by n matrix formed by multiplying each entry in A by k. The number k is sometimes referred to as a **scalar**, and the matrix kA is called a **scalar multiple** of A.

Example 5 Suppose

$$A = \begin{bmatrix} 3 & 1 & 5 \\ -2 & 0 & 6 \end{bmatrix} \qquad B = \begin{bmatrix} 4 & 1 & 0 \\ 8 & 1 & -3 \end{bmatrix} \qquad C = \begin{bmatrix} 9 & 0 \\ -3 & 6 \end{bmatrix}$$

Find:

(a) $4A$ (b) $\frac{1}{3}C$ (c) $3A - 2B$

Solution (a) $4A = 4\begin{bmatrix} 3 & 1 & 5 \\ -2 & 0 & 6 \end{bmatrix} = \begin{bmatrix} 4\cdot 3 & 4\cdot 1 & 4\cdot 5 \\ 4(-2) & 4\cdot 0 & 4\cdot 6 \end{bmatrix} = \begin{bmatrix} 12 & 4 & 20 \\ -8 & 0 & 24 \end{bmatrix}$

(b) $\frac{1}{3}C = \frac{1}{3}\begin{bmatrix} 9 & 0 \\ -3 & 6 \end{bmatrix} = \begin{bmatrix} \frac{1}{3}\cdot 9 & \frac{1}{3}\cdot 0 \\ \frac{1}{3}(-3) & \frac{1}{3}\cdot 6 \end{bmatrix} = \begin{bmatrix} 3 & 0 \\ -1 & 2 \end{bmatrix}$

(c)
$$\begin{aligned} 3A - 2B &= 3\begin{bmatrix} 3 & 1 & 5 \\ -2 & 0 & 6 \end{bmatrix} - 2\begin{bmatrix} 4 & 1 & 0 \\ 8 & 1 & -3 \end{bmatrix} \\ &= \begin{bmatrix} 3\cdot 3 & 3\cdot 1 & 3\cdot 5 \\ 3(-2) & 3\cdot 0 & 3\cdot 6 \end{bmatrix} - \begin{bmatrix} 2\cdot 4 & 2\cdot 1 & 2\cdot 0 \\ 2\cdot 8 & 2\cdot 1 & 2(-3) \end{bmatrix} \\ &= \begin{bmatrix} 9 & 3 & 15 \\ -6 & 0 & 18 \end{bmatrix} - \begin{bmatrix} 8 & 2 & 0 \\ 16 & 2 & -6 \end{bmatrix} \\ &= \begin{bmatrix} 9-8 & 3-2 & 15-0 \\ -6-16 & 0-2 & 18-(-6) \end{bmatrix} \\ &= \begin{bmatrix} 1 & 1 & 15 \\ -22 & -2 & 24 \end{bmatrix} \end{aligned}$$

■

We list below some of the algebraic properties of scalar multiplication.

Let h and k be real numbers, and let A and B be m by n matrices. Then:

Properties of Scalar Multiplication

$$k(hA) = (kh)A$$
$$(k + h)A = kA + hA$$
$$k(A + B) = kA + kB$$

The proofs of these properties are based on properties of real numbers. For example, if A and B are 2 by 2 matrices, then

$$\begin{aligned} k(A+B) &= k\left(\begin{bmatrix} a_{11} & a_{12} \\ a_{21} & a_{22} \end{bmatrix} + \begin{bmatrix} b_{11} & b_{12} \\ b_{21} & b_{22} \end{bmatrix}\right) = k\begin{bmatrix} a_{11}+b_{11} & a_{12}+b_{12} \\ a_{21}+b_{21} & a_{22}+b_{22} \end{bmatrix} \\ &= \begin{bmatrix} k(a_{11}+b_{11}) & k(a_{12}+b_{12}) \\ k(a_{21}+b_{21}) & k(a_{22}+b_{22}) \end{bmatrix} = \begin{bmatrix} ka_{11}+kb_{11} & ka_{12}+kb_{12} \\ ka_{21}+kb_{21} & ka_{22}+kb_{22} \end{bmatrix} \\ &= \begin{bmatrix} ka_{11} & ka_{12} \\ ka_{21} & ka_{22} \end{bmatrix} + \begin{bmatrix} kb_{11} & kb_{12} \\ kb_{21} & kb_{22} \end{bmatrix} \\ &= k\begin{bmatrix} a_{11} & a_{12} \\ a_{21} & a_{22} \end{bmatrix} + k\begin{bmatrix} b_{11} & b_{12} \\ b_{21} & b_{22} \end{bmatrix} = kA + kB \end{aligned}$$

Multiplication of Matrices

Unlike the straightforward definition for adding two matrices, the definition for multiplying two matrices is not what one might expect. In preparation for this definition, we need the following definitions:

A **row vector** R is a 1 by n matrix

$$R = [r_1 \quad r_2 \quad \cdots \quad r_n]$$

A **column vector** C is an n by 1 matrix

$$C = \begin{bmatrix} c_1 \\ c_2 \\ \vdots \\ c_n \end{bmatrix}$$

The **product** RC of R times C is defined as the number

$$RC = [r_1 \quad r_2 \quad \cdots \quad r_n] \begin{bmatrix} c_1 \\ c_2 \\ \vdots \\ c_n \end{bmatrix} = r_1c_1 + r_2c_2 + \cdots + r_nc_n$$

Notice that a row vector and a column vector can be multiplied only if they contain the same number of entries.

Example 6 If $R = [3 \quad -5 \quad 2]$ and $C = \begin{bmatrix} 3 \\ 4 \\ -5 \end{bmatrix}$, then

$$RC = [3 \quad -5 \quad 2] \begin{bmatrix} 3 \\ 4 \\ -5 \end{bmatrix} = 3 \cdot 3 + (-5)4 + 2(-5)$$

$$= 9 - 20 - 10 = -21$$ ■

Let's look at an application of the product of a row vector by a column vector.

Example 7 A clothing store sells men's shirts for \$25, silk ties for \$8, and wool suits for \$300. Last month, the store had sales consisting of 100 shirts, 200 ties, and 50 suits. What was the total revenue due to these sales?

Solution We set up a row vector R to represent the price of each item and a column vector C to represent the corresponding number of items sold.

Then,

$$R = \overset{\text{Price}}{\begin{array}{ccc} \text{Shirts} & \text{Ties} & \text{Suits} \\ [\;25 & 8 & 300\;] \end{array}} \qquad C = \overset{\substack{\text{Number}\\\text{sold}}}{\begin{bmatrix} 100 \\ 200 \\ 50 \end{bmatrix}} \begin{array}{l} \text{Shirts} \\ \text{Ties} \\ \text{Suits} \end{array}$$

The total revenue obtained is the product RC. That is,

$$RC = [25 \quad 8 \quad 300]\begin{bmatrix} 100 \\ 200 \\ 50 \end{bmatrix}$$

$$= \underbrace{25 \cdot 100}_{\text{Shirt revenue}} + \underbrace{8 \cdot 200}_{\text{Tie revenue}} + \underbrace{300 \cdot 50}_{\text{Suit revenue}} = \underbrace{\$19{,}100}_{\text{Total revenue}} \quad ■$$

The definition for multiplying two matrices is based on the definition of a row vector times a column vector.

Product Let A denote an m by r matrix and let B denote an r by n matrix. The **product** AB is defined as the m by n matrix whose entry in row i, column j is the product of the ith row of A and the jth column of B.

The definition of the product AB of two matrices A and B, in this order, requires that the number of columns of A equal the number of rows of B; otherwise, no product is defined:

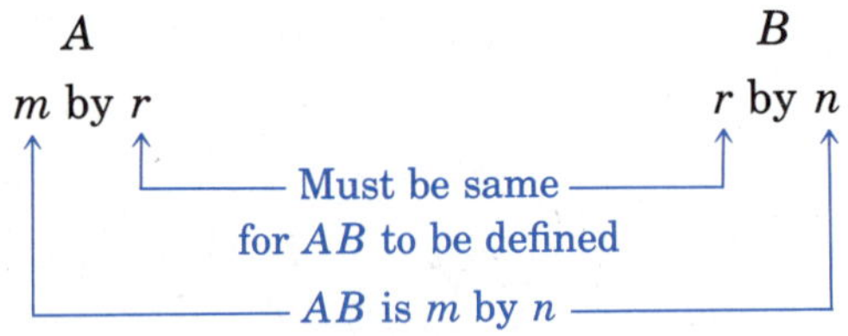

An example will help clarify the definition.

Example 8 Find the product AB if

$$A = \begin{bmatrix} 2 & 4 & -1 \\ 5 & 8 & 0 \end{bmatrix} \quad \text{and} \quad B = \begin{bmatrix} 2 & 5 & 1 & 4 \\ 4 & 8 & 0 & 6 \\ -3 & 1 & -2 & -1 \end{bmatrix}$$

Solution First, we note that A is 2 by 3 and B is 3 by 4 so that the product AB is defined, and will be a 2 by 4 matrix. Suppose we want the entry in row 2, column 3 of AB. To find it, we find the product of the row vector from row 2 of A and the column vector from column 3 of B, namely,

$$\underset{\text{Row 2 of } A}{[5 \quad 8 \quad 0]}\underset{\text{Column 3 of } B}{\begin{bmatrix} 1 \\ 0 \\ -2 \end{bmatrix}} = 5 \cdot 1 + 8 \cdot 0 + 0(-2) = 5$$

So far, we have

$$AB = \begin{bmatrix} - & - & - & - \\ - & - & 5 & - \end{bmatrix} \leftarrow \text{Row 2}$$

(Column 3 ↓ marks the third column.)

Now, to find the entry in row 1, column 4 of AB, we find the product of row 1 of A and column 4 of B:

$$\underset{\text{Row 1 of } A}{[2 \quad 4 \quad -1]}\underset{\text{Column 4 of } B}{\begin{bmatrix} 4 \\ 6 \\ -1 \end{bmatrix}} = 2 \cdot 4 + 4 \cdot 6 + (-1)(-1) = 33$$

Continuing in this fashion, we find AB:

$$AB = \begin{bmatrix} 2 & 4 & -1 \\ 5 & 8 & 0 \end{bmatrix}\begin{bmatrix} 2 & 5 & 1 & 4 \\ 4 & 8 & 0 & 6 \\ -3 & 1 & -2 & -1 \end{bmatrix}$$

$$= \begin{bmatrix} \text{Row 1 of } A \text{ times column 1 of } B & \text{Row 1 of } A \text{ times column 2 of } B & \text{Row 1 of } A \text{ times column 3 of } B & \text{Row 1 of } A \text{ times column 4 of } B \\ \text{Row 2 of } A \text{ times column 1 of } B & \text{Row 2 of } A \text{ times column 2 of } B & \text{Row 2 of } A \text{ times column 3 of } B & \text{Row 2 of } A \text{ times column 4 of } B \end{bmatrix}$$

$$= \begin{bmatrix} 2\cdot2+4\cdot4+(-1)(-3) & 2\cdot5+4\cdot8+(-1)1 & 2\cdot1+4\cdot0+(-1)(-2) & 33 \text{ (from above)} \\ 5\cdot2+8\cdot4+0(-3) & 5\cdot5+8\cdot8+0\cdot1 & 5 \text{ (from above)} & 5\cdot4+8\cdot6+0(-1) \end{bmatrix}$$

$$= \begin{bmatrix} 23 & 41 & 4 & 33 \\ 42 & 89 & 5 & 68 \end{bmatrix}$$

■

Notice that for the matrices given in Example 8, the product BA is not defined, because B is 3 by 4 and A is 2 by 3. Another result that can occur when multiplying two matrices is illustrated in the next example.

Example 9 If

$$A = \begin{bmatrix} 2 & 1 & 3 \\ 1 & -1 & 0 \end{bmatrix} \quad \text{and} \quad B = \begin{bmatrix} 1 & 0 \\ 2 & 1 \\ 3 & 2 \end{bmatrix}$$

find:

(a) AB (b) BA

Solution (a) $AB = \begin{bmatrix} 2 & 1 & 3 \\ 1 & -1 & 0 \end{bmatrix} \begin{bmatrix} 1 & 0 \\ 2 & 1 \\ 3 & 2 \end{bmatrix} = \begin{bmatrix} 13 & 7 \\ -1 & -1 \end{bmatrix}$

(b) $BA = \begin{bmatrix} 1 & 0 \\ 2 & 1 \\ 3 & 2 \end{bmatrix} \begin{bmatrix} 2 & 1 & 3 \\ 1 & -1 & 0 \end{bmatrix} = \begin{bmatrix} 2 & 1 & 3 \\ 5 & 1 & 6 \\ 8 & 1 & 9 \end{bmatrix}$ ■

Notice in Example 9 that AB is 2 by 2 and BA is 3 by 3. Thus, it is possible for both AB and BA to be defined, yet be unequal. In fact, even if A and B are both n by n matrices, so that AB and BA are each defined and n by n, AB and BA will nearly always be unequal.

Example 10 If

$$A = \begin{bmatrix} 2 & 1 \\ 0 & 4 \end{bmatrix} \quad \text{and} \quad B = \begin{bmatrix} -3 & 1 \\ 1 & 2 \end{bmatrix}$$

find:

(a) AB (b) BA

Solution (a) $AB = \begin{bmatrix} 2 & 1 \\ 0 & 4 \end{bmatrix} \begin{bmatrix} -3 & 1 \\ 1 & 2 \end{bmatrix}$

$$= \begin{bmatrix} 2(-3) + 1 \cdot 1 & 2 \cdot 1 + 1 \cdot 2 \\ 0(-3) + 4 \cdot 1 & 0 \cdot 1 + 4 \cdot 2 \end{bmatrix} = \begin{bmatrix} -5 & 4 \\ 4 & 8 \end{bmatrix}$$

(b) $BA = \begin{bmatrix} -3 & 1 \\ 1 & 2 \end{bmatrix} \begin{bmatrix} 2 & 1 \\ 0 & 4 \end{bmatrix}$

$$= \begin{bmatrix} (-3)2 + 1 \cdot 0 & (-3)1 + 1 \cdot 4 \\ 1 \cdot 2 + 2 \cdot 0 & 1 \cdot 1 + 2 \cdot 4 \end{bmatrix} = \begin{bmatrix} -6 & 1 \\ 2 & 9 \end{bmatrix}$$ ■

The preceding examples demonstrate that an important property of real numbers, the commutative property of multiplication, is not shared by matrices. Thus, in general:

Matrix multiplication is not commutative.

Below, we list two of the properties of real numbers that are shared by matrices. Assuming each product and sum is defined, we have:

Associative Property

$$A(BC) = (AB)C$$

Distributive Property

$$A(B + C) = AB + AC$$

The Identity Matrix

For an n by n square matrix, the entries located in row i, column i, $1 \le i \le n$, are called the **diagonal entries**. An n by n square matrix whose diagonal entries are 1's, while all other entries are 0's, is called the **identity matrix I_n**. For example,

$$I_2 = \begin{bmatrix} 1 & 0 \\ 0 & 1 \end{bmatrix} \qquad I_3 = \begin{bmatrix} 1 & 0 & 0 \\ 0 & 1 & 0 \\ 0 & 0 & 1 \end{bmatrix}$$

and so on.

Example 11 Let

$$A = \begin{bmatrix} -1 & 2 & 0 \\ 0 & 1 & 3 \end{bmatrix} \quad \text{and} \quad B = \begin{bmatrix} 3 & 2 \\ 4 & 6 \\ 5 & 2 \end{bmatrix}$$

Find:

(a) AI_3 (b) I_2A (c) BI_2

Solution (a) $AI_3 = \begin{bmatrix} -1 & 2 & 0 \\ 0 & 1 & 3 \end{bmatrix} \begin{bmatrix} 1 & 0 & 0 \\ 0 & 1 & 0 \\ 0 & 0 & 1 \end{bmatrix} = \begin{bmatrix} -1 & 2 & 0 \\ 0 & 1 & 3 \end{bmatrix} = A$

(b) $I_2A = \begin{bmatrix} 1 & 0 \\ 0 & 1 \end{bmatrix} \begin{bmatrix} -1 & 2 & 0 \\ 0 & 1 & 3 \end{bmatrix} = \begin{bmatrix} -1 & 2 & 0 \\ 0 & 1 & 3 \end{bmatrix} = A$

(c) $BI_2 = \begin{bmatrix} 3 & 2 \\ 4 & 6 \\ 5 & 2 \end{bmatrix} \begin{bmatrix} 1 & 0 \\ 0 & 1 \end{bmatrix} = \begin{bmatrix} 3 & 2 \\ 4 & 6 \\ 5 & 2 \end{bmatrix} = B$ ■

Example 11 demonstrates the following property:
If A is an m by n matrix, then

$$I_m A = A \qquad \text{and} \qquad A I_n = A$$

If A is an n by n square matrix, then $AI_n = I_nA = A$.

Thus, an identity matrix has properties analogous to those of the real number 1. In other words, the identity matrix is a multiplicative identity in matrix algebra.

The Inverse of a Matrix

Inverse Matrix Let A be a square n by n matrix. If there exists an n by n matrix A^{-1} for which

$$AA^{-1} = A^{-1}A = I_n$$

then A^{-1} is called the **inverse** of the matrix A.

As we shall soon see, not every square matrix has an inverse. When a matrix A does have an inverse A^{-1}, then A is said to be **nonsingular**.

Example 12 Show that the inverse of

$$A = \begin{bmatrix} 3 & 1 \\ 2 & 1 \end{bmatrix} \quad \text{is} \quad A^{-1} = \begin{bmatrix} 1 & -1 \\ -2 & 3 \end{bmatrix}$$

Solution We need to show that $AA^{-1} = A^{-1}A = I_2$.

$$AA^{-1} = \begin{bmatrix} 3 & 1 \\ 2 & 1 \end{bmatrix}\begin{bmatrix} 1 & -1 \\ -2 & 3 \end{bmatrix} = \begin{bmatrix} 3 \cdot 1 + 1(-2) & 3(-1) + 1 \cdot 3 \\ 2 \cdot 1 + 1(-2) & 2(-1) + 1 \cdot 3 \end{bmatrix}$$

$$= \begin{bmatrix} 1 & 0 \\ 0 & 1 \end{bmatrix} = I_2$$

$$A^{-1}A = \begin{bmatrix} 1 & -1 \\ -2 & 3 \end{bmatrix}\begin{bmatrix} 3 & 1 \\ 2 & 1 \end{bmatrix} = \begin{bmatrix} 3 - 2 & 1 - 1 \\ -6 + 6 & -2 + 3 \end{bmatrix} = \begin{bmatrix} 1 & 0 \\ 0 & 1 \end{bmatrix} = I_2 \quad ■$$

We now show one way to find the inverse of

$$A = \begin{bmatrix} 3 & 1 \\ 2 & 1 \end{bmatrix}$$

Suppose A^{-1} is given by

$$A^{-1} = \begin{bmatrix} x & y \\ z & w \end{bmatrix} \tag{1}$$

where x, y, z, and w are four variables. Based on the definition of an inverse, if, indeed, A has an inverse, then we have

$$AA^{-1} = I_2$$
$$\begin{bmatrix} 3 & 1 \\ 2 & 1 \end{bmatrix}\begin{bmatrix} x & y \\ z & w \end{bmatrix} = \begin{bmatrix} 1 & 0 \\ 0 & 1 \end{bmatrix}$$
$$\begin{bmatrix} 3x + z & 3y + w \\ 2x + z & 2y + w \end{bmatrix} = \begin{bmatrix} 1 & 0 \\ 0 & 1 \end{bmatrix}$$

Because corresponding entries must be equal, it follows that this matrix equation is equivalent to four ordinary equations:

$$\begin{cases} 3x + z = 1 \\ 2x + z = 0 \end{cases} \qquad \begin{cases} 3y + w = 0 \\ 2y + w = 1 \end{cases}$$

The augmented matrix of each system is

$$\left[\begin{array}{cc|c} 3 & 1 & 1 \\ 2 & 1 & 0 \end{array}\right] \qquad \left[\begin{array}{cc|c} 3 & 1 & 0 \\ 2 & 1 & 1 \end{array}\right] \tag{2}$$

The usual procedure would be to transform each augmented matrix into reduced echelon form. Notice, though, that the left sides of the augmented matrices are equal, so the same row operations (see Section 12.2, page 679) can be used to reduce each one. Thus, we find it more efficient to combine the two augmented matrices (2) into a single matrix, as shown below, and then transform it into reduced echelon form:

$$\left[\begin{array}{cc|cc} 3 & 1 & 1 & 0 \\ 2 & 1 & 0 & 1 \end{array}\right]$$

Now we attempt to transform the left side into an identity matrix:

$$\left[\begin{array}{cc|cc} 3 & 1 & 1 & 0 \\ 2 & 1 & 0 & 1 \end{array}\right] \to \left[\begin{array}{cc|cc} 1 & 0 & 1 & -1 \\ 2 & 1 & 0 & 1 \end{array}\right]$$
$$\uparrow \; R_1 = -1r_2 + r_1$$
$$\to \left[\begin{array}{cc|cc} 1 & 0 & 1 & -1 \\ 0 & 1 & -2 & 3 \end{array}\right] \tag{3}$$
$$\uparrow \; R_2 = -2r_1 + r_2$$

Matrix (3) is in reduced echelon form. Now we reverse the earlier step of combining the two augmented matrices in (2), and write the single matrix (3) as two augmented matrices:

$$\left[\begin{array}{cc|c} 1 & 0 & 1 \\ 0 & 1 & -2 \end{array}\right] \quad \text{and} \quad \left[\begin{array}{cc|c} 1 & 0 & -1 \\ 0 & 1 & 3 \end{array}\right]$$

We conclude from these matrices that $x = 1$, $z = -2$, and $y = -1$, $w = 3$. Substituting these values into matrix (1), we find

$$A^{-1} = \begin{bmatrix} 1 & -1 \\ -2 & 3 \end{bmatrix}$$

Notice in matrix (3) that the 2 by 2 matrix to the right of the vertical bar is, in fact, the inverse of A. Also notice that the identity matrix I_2 is the matrix that appears to the left of the vertical bar. These observations and the procedures followed above will work in general.

Procedure for Finding the Inverse of a Nonsingular Matrix

To find the inverse of an n by n nonsingular matrix A, proceed as follows:

STEP 1: Form the matrix $[A \mid I_n]$.
STEP 2: Transform the matrix $[A \mid I_n]$ into reduced echelon form.
STEP 3: The reduced echelon form of $[A \mid I_n]$ will contain the identity matrix I_n on the left of the vertical bar; the n by n matrix on the right of the vertical bar is the inverse of A.

In other words, if A is nonsingular, we begin with the matrix $[A \mid I_n]$ and, after transforming it into reduced echelon form, we end up with the matrix $[I_n \mid A^{-1}]$.

Let's look at another example.

Example 13 The matrix

$$A = \begin{bmatrix} 1 & 1 & 0 \\ -1 & 3 & 4 \\ 0 & 4 & 3 \end{bmatrix}$$

is nonsingular. Find its inverse.

Solution First, we form the matrix

$$[A \mid I_3] = \left[\begin{array}{ccc|ccc} 1 & 1 & 0 & 1 & 0 & 0 \\ -1 & 3 & 4 & 0 & 1 & 0 \\ 0 & 4 & 3 & 0 & 0 & 1 \end{array}\right]$$

Next, we use row operations to transform $[A \mid I_3]$ into reduced echelon form:

$$\left[\begin{array}{rrr|rrr} 1 & 1 & 0 & 1 & 0 & 0 \\ -1 & 3 & 4 & 0 & 1 & 0 \\ 0 & 4 & 3 & 0 & 0 & 1 \end{array}\right] \to \left[\begin{array}{rrr|rrr} 1 & 1 & 0 & 1 & 0 & 0 \\ 0 & 4 & 4 & 1 & 1 & 0 \\ 0 & 4 & 3 & 0 & 0 & 1 \end{array}\right] \to \left[\begin{array}{rrr|rrr} 1 & 1 & 0 & 1 & 0 & 0 \\ 0 & 1 & 1 & \frac{1}{4} & \frac{1}{4} & 0 \\ 0 & 4 & 3 & 0 & 0 & 1 \end{array}\right]$$

$\uparrow$ $R_2 = r_2 + r_1$ $\qquad$ $\uparrow$ $R_2 = \frac{1}{4}r_2$

$$\to \left[\begin{array}{rrr|rrr} 1 & 0 & -1 & \frac{3}{4} & -\frac{1}{4} & 0 \\ 0 & 1 & 1 & \frac{1}{4} & \frac{1}{4} & 0 \\ 0 & 0 & -1 & -1 & -1 & 1 \end{array}\right] \to \left[\begin{array}{rrr|rrr} 1 & 0 & -1 & \frac{3}{4} & -\frac{1}{4} & 0 \\ 0 & 1 & 1 & \frac{1}{4} & \frac{1}{4} & 0 \\ 0 & 0 & 1 & 1 & 1 & -1 \end{array}\right]$$

$\uparrow$ $R_1 = -1r_2 + r_1$, $R_3 = -4r_2 + r_3$ $\qquad$ $\uparrow$ $R_3 = -1r_3$

$$\to \left[\begin{array}{rrr|rrr} 1 & 0 & 0 & \frac{7}{4} & \frac{3}{4} & -1 \\ 0 & 1 & 0 & -\frac{3}{4} & -\frac{3}{4} & 1 \\ 0 & 0 & 1 & 1 & 1 & -1 \end{array}\right]$$

$\uparrow$ $R_1 = r_1 + r_3$, $R_2 = -1r_3 + r_2$

The matrix $[A \mid I_3]$ is now in reduced echelon form, and the identity matrix I_3 is on the left of the vertical bar. Hence, the inverse of A is

$$A^{-1} = \begin{bmatrix} \frac{7}{4} & \frac{3}{4} & -1 \\ -\frac{3}{4} & -\frac{3}{4} & 1 \\ 1 & 1 & -1 \end{bmatrix}$$

You can (and should) verify that this is the correct inverse by showing that $AA^{-1} = A^{-1}A = I_3$. ■

If transforming the matrix $[A \mid I_n]$ into reduced echelon form does not result in the identity matrix I_n to the left of the vertical bar, then A has no inverse. The next example demonstrates such a matrix.

Example 14 Show that the matrix below has no inverse.

$$A = \begin{bmatrix} 4 & 6 \\ 2 & 3 \end{bmatrix}$$

Solution Proceeding as in Example 13, we form the matrix

$$[A \mid I_2] = \left[\begin{array}{rr|rr} 4 & 6 & 1 & 0 \\ 2 & 3 & 0 & 1 \end{array}\right]$$

Now, we use row operations to transform $[A \mid I_2]$ into reduced echelon form:

$$[A \mid I_2] = \left[\begin{array}{cc|cc} 4 & 6 & 1 & 0 \\ 2 & 3 & 0 & 1 \end{array}\right] \underset{R_1 = \frac{1}{4}r_1}{\rightarrow} \left[\begin{array}{cc|cc} 1 & \frac{3}{2} & \frac{1}{4} & 0 \\ 2 & 3 & 0 & 1 \end{array}\right] \underset{R_2 = -2r_1 + r_2}{\rightarrow} \left[\begin{array}{cc|cc} 1 & \frac{3}{2} & \frac{1}{4} & 0 \\ 0 & 0 & -\frac{1}{2} & 1 \end{array}\right]$$

The matrix $[A \mid I_2]$ is sufficiently reduced for us to see that the identity matrix does not appear to the left of the vertical bar. We conclude that A has no inverse. ■

Solving Systems of Equations

Inverse matrices can be used to solve systems of equations in which the number of equations is the same as the number of variables.

Example 15 Solve the system of equations: $\begin{cases} x + y = 3 \\ -x + 3y + 4z = -3 \\ 4y + 3z = 2 \end{cases}$

Solution If we let

$$A = \begin{bmatrix} 1 & 1 & 0 \\ -1 & 3 & 4 \\ 0 & 4 & 3 \end{bmatrix}, \quad X = \begin{bmatrix} x \\ y \\ z \end{bmatrix}, \quad \text{and} \quad B = \begin{bmatrix} 3 \\ -3 \\ 2 \end{bmatrix}$$

then the original system of equations can be written compactly as the matrix equation

$$AX = B \tag{4}$$

We know from Example 13 that the matrix A has the inverse A^{-1}, so we multiply each side of equation (4) by A^{-1}:

$$\begin{aligned} AX &= B \\ A^{-1}(AX) &= A^{-1}B \\ (A^{-1}A)X &= A^{-1}B && \text{Associative property for multiplication} \\ I_3X &= A^{-1}B && \text{Definition of inverse matrix} \\ X &= A^{-1}B && \text{Property of identity matrix} \end{aligned}$$

Now, we use this to find $X = \begin{bmatrix} x \\ y \\ z \end{bmatrix}$:

$$X = \begin{bmatrix} x \\ y \\ z \end{bmatrix} = A^{-1}B = \begin{bmatrix} \frac{7}{4} & \frac{3}{4} & -1 \\ -\frac{3}{4} & -\frac{3}{4} & 1 \\ 1 & 1 & -1 \end{bmatrix} \begin{bmatrix} 3 \\ -3 \\ 2 \end{bmatrix} = \begin{bmatrix} 1 \\ 2 \\ -2 \end{bmatrix}$$

↑ Example 13

Thus, $x = 1$, $y = 2$, $z = -2$. ■

The method used in Example 15 to solve a system of equations is particularly useful when it is necessary to solve several systems of equations in which the constants appearing to the right of the equal signs change, while the coefficients of the variables on the left side remain the same. See Problems 31–50 for some illustrations.

Historical Comment

■ Matrices were invented in 1857 by Arthur Cayley (1821–1895) as a way of efficiently computing the result of substituting one linear system into another (see Historical Problem 2). The resulting system had incredible richness, in the sense that a very wide variety of mathematical systems could be mimicked by the matrices. Cayley and his friend J. J. Sylvester (1814–1897) spent much of the rest of their lives elaborating the theory. The torch was then passed to G. Frobenius (1848–1917), whose deep investigations established a central place for matrices in modern mathematics. In 1924, rather to the surprise of physicists, it was found that matrices (with complex numbers in them) were exactly the right tool for describing the behavior of atomic systems. Today, matrices are used in a wide variety of applications.

Historical Problems

■ **1.** *Matrices and complex numbers* Frobenius emphasized in his research how matrices could be used to mimic other mathematical systems. Here, we mimic the behavior of complex numbers using matrices. Mathematicians call this an *isomorphism.*

Complex number ⟷ Matrix

$$a + bi \longleftrightarrow \begin{bmatrix} a & b \\ -b & a \end{bmatrix}$$

Note that the complex number can be read off the top line of the matrix. Thus,

$$2 + 3i \longleftrightarrow \begin{bmatrix} 2 & 3 \\ -3 & 2 \end{bmatrix} \quad \text{and} \quad \begin{bmatrix} 4 & -2 \\ 2 & 4 \end{bmatrix} \longleftrightarrow 4 - 2i$$

(a) Find the matrices corresponding to $2 - 5i$ and $1 + 3i$.
(b) Multiply the two matrices.
(c) Find the corresponding complex number for the matrix found in part (b).

(d) Multiply $2 - 5i$ by $1 + 3i$. The result should be the same as that found in part (c).

The process also works for addition and subtraction. Try it for yourself.

2. *Cayley's definition of matrix multiplication* Cayley invented matrix multiplication to simplify the following problem:

$$\begin{cases} u = ar + bs \\ v = cr + ds \end{cases} \qquad \begin{cases} x = ku + lv \\ y = mu + nv \end{cases}$$

(a) Find x and y in terms of r and s by substituting u and v from the first system of equations into the second system of equations.

(b) Use the result of part (a) to find the 2 by 2 matrix A in

$$\begin{bmatrix} x \\ y \end{bmatrix} = A\begin{bmatrix} r \\ s \end{bmatrix}$$

(c) Now look at the following way to do it: Write the equations in matrix form,

$$\begin{bmatrix} u \\ v \end{bmatrix} = \begin{bmatrix} a & b \\ c & d \end{bmatrix}\begin{bmatrix} r \\ s \end{bmatrix} \qquad \begin{bmatrix} x \\ y \end{bmatrix} = \begin{bmatrix} k & l \\ m & n \end{bmatrix}\begin{bmatrix} u \\ v \end{bmatrix}$$

so

$$\begin{bmatrix} x \\ y \end{bmatrix} = \begin{bmatrix} k & l \\ m & n \end{bmatrix}\begin{bmatrix} a & b \\ c & d \end{bmatrix}\begin{bmatrix} r \\ s \end{bmatrix}$$

Do you see how Cayley defined matrix multiplication? ■

EXERCISE 14.1

In Problems 1–16, use the matrices below to compute the given expression.

$$A = \begin{bmatrix} 0 & 3 & -5 \\ 1 & 2 & 6 \end{bmatrix} \qquad B = \begin{bmatrix} 4 & 1 & 0 \\ -2 & 3 & -2 \end{bmatrix} \qquad C = \begin{bmatrix} 4 & 1 \\ 6 & 2 \\ -2 & 3 \end{bmatrix}$$

1. $A + B$ **2.** $A - B$ **3.** $4A$ **4.** $-3B$

5. $3A - 2B$ **6.** $2A + 4B$ **7.** AC **8.** BC

9. CA **10.** CB **11.** $C(A + B)$ **12.** $(A + B)C$

13. $AC - 3I_2$ **14.** $CA + 5I_3$ **15.** $CA - CB$ **16.** $AC + BC$

In Problems 17–20, compute each product.

17. $\begin{bmatrix} 2 & -2 \\ 1 & 0 \end{bmatrix}\begin{bmatrix} 2 & 1 & 4 & 6 \\ 3 & -1 & 3 & 2 \end{bmatrix}$

18. $\begin{bmatrix} 4 & 1 \\ 2 & 1 \end{bmatrix}\begin{bmatrix} -6 & 6 & 1 & 0 \\ 2 & 5 & 4 & -1 \end{bmatrix}$

19. $\begin{bmatrix} 1 & 0 & 1 \\ 2 & 4 & 1 \\ 3 & 6 & 1 \end{bmatrix}\begin{bmatrix} 1 & 3 \\ 6 & 2 \\ 8 & -1 \end{bmatrix}$

20. $\begin{bmatrix} 4 & -2 & 3 \\ 0 & 1 & 2 \\ -1 & 0 & 1 \end{bmatrix}\begin{bmatrix} 2 & 6 \\ 1 & -1 \\ 0 & 2 \end{bmatrix}$

In Problems 21–30, each matrix is nonsingular. Find the inverse of each matrix. Be sure to check your answer.

21. $\begin{bmatrix} 2 & 1 \\ 1 & 1 \end{bmatrix}$ **22.** $\begin{bmatrix} 3 & -1 \\ -2 & 1 \end{bmatrix}$ **23.** $\begin{bmatrix} 6 & 5 \\ 2 & 2 \end{bmatrix}$ **24.** $\begin{bmatrix} -4 & 1 \\ 6 & -2 \end{bmatrix}$

25. $\begin{bmatrix} 2 & 1 \\ a & a \end{bmatrix}, \quad a \neq 0$ **26.** $\begin{bmatrix} b & 3 \\ b & 2 \end{bmatrix}, \quad b \neq 0$ **27.** $\begin{bmatrix} 1 & -1 & 1 \\ 0 & -2 & 1 \\ -2 & -3 & 0 \end{bmatrix}$ **28.** $\begin{bmatrix} 1 & 0 & 2 \\ -1 & 2 & 3 \\ 1 & -1 & 0 \end{bmatrix}$

29. $\begin{bmatrix} 1 & 1 & 1 \\ 3 & 2 & -1 \\ 3 & 1 & 2 \end{bmatrix}$ **30.** $\begin{bmatrix} 3 & 3 & 1 \\ 1 & 2 & 1 \\ 2 & -1 & 1 \end{bmatrix}$

In Problems 31–50, use the inverses found in Problems 21–30 to solve each system of equations.

31. $\begin{cases} 2x + y = 8 \\ x + y = 5 \end{cases}$ **32.** $\begin{cases} 3x - y = 8 \\ -2x + y = 4 \end{cases}$ **33.** $\begin{cases} 2x + y = 0 \\ x + y = 5 \end{cases}$

34. $\begin{cases} 3x - y = 4 \\ -2x + y = 5 \end{cases}$ **35.** $\begin{cases} 6x + 5y = 7 \\ 2x + 2y = 2 \end{cases}$ **36.** $\begin{cases} -4x + y = 0 \\ 6x - 2y = 14 \end{cases}$

37. $\begin{cases} 6x + 5y = 13 \\ 2x + 2y = 5 \end{cases}$ **38.** $\begin{cases} -4x + y = 5 \\ 6x - 2y = -9 \end{cases}$ **39.** $\begin{cases} 2x + y = -3 \\ ax + ay = -a \end{cases} \quad a \neq 0$

40. $\begin{cases} bx + 3y = 2b + 3 \\ bx + 2y = 2b + 2 \end{cases} \quad b \neq 0$ **41.** $\begin{cases} 2x + y = 7/a \\ ax + ay = 5 \end{cases} \quad a \neq 0$ **42.** $\begin{cases} bx + 3y = 14 \\ bx + 2y = 10 \end{cases} \quad b \neq 0$

43. $\begin{cases} x - y + z = 0 \\ -2y + z = -1 \\ -2x - 3y = -5 \end{cases}$ **44.** $\begin{cases} x + 2z = 6 \\ -x + 2y + 3z = -5 \\ x - y = 6 \end{cases}$ **45.** $\begin{cases} x - y + z = 2 \\ -2y + z = 2 \\ -2x - 3y = \frac{1}{2} \end{cases}$

46. $\begin{cases} x + 2z = 2 \\ -x + 2y + 3z = -\frac{3}{2} \\ x - y = 2 \end{cases}$ **47.** $\begin{cases} x + y + z = 9 \\ 3x + 2y - z = 8 \\ 3x + y + 2z = 1 \end{cases}$ **48.** $\begin{cases} 3x + 3y + z = 8 \\ x + 2y + z = 5 \\ 2x - y + z = 4 \end{cases}$

49. $\begin{cases} x + y + z = 2 \\ 3x + 2y - z = \frac{7}{3} \\ 3x + y + 2z = \frac{10}{3} \end{cases}$ **50.** $\begin{cases} 3x + 3y + z = 1 \\ x + 2y + z = 0 \\ 2x - y + z = 4 \end{cases}$

In Problems 51–56, show that each matrix has no inverse.

51. $\begin{bmatrix} 4 & 2 \\ 2 & 1 \end{bmatrix}$ **52.** $\begin{bmatrix} -3 & \frac{1}{2} \\ 6 & 1 \end{bmatrix}$ **53.** $\begin{bmatrix} 15 & 3 \\ 10 & 2 \end{bmatrix}$

54. $\begin{bmatrix} -3 & 0 \\ 4 & 0 \end{bmatrix}$ **55.** $\begin{bmatrix} -3 & 1 & -1 \\ 1 & -4 & -7 \\ 1 & 2 & 5 \end{bmatrix}$ **56.** $\begin{bmatrix} 1 & 1 & -3 \\ 2 & -4 & 1 \\ -5 & 7 & 1 \end{bmatrix}$

57. The Acme Steel Company is a producer of stainless steel and aluminum containers. On a certain day, the following stainless steel containers were manufactured: 500 with 10 gallons capacity, 350 with 5 gallons capacity, and 400 with 1 gallon capacity. On the same day, the following aluminum containers were manufactured: 700 with 10 gallons capacity, 500 with 5 gallons capacity, and 850 with 1 gallon capacity.

(a) Find a 2 by 3 matrix representing the above data. Find a 3 by 2 matrix to represent the same data.
(b) If the amount of material used in the 10 gallon containers is 15 pounds, the amount used in the 5 gallon containers is 8 pounds, and the amount used in the 1 gallon containers is 3 pounds, find a 3 by 1 matrix representing the amount of material.
(c) Multiply the 2 by 3 matrix found in part (a) and the 3 by 1 matrix found in part (b) to get a 2 by 1 matrix showing the day's usage of material.
(d) If stainless steel costs Acme \$0.10 per pound and aluminum costs \$0.05 per pound, find a 1 by 2 matrix representing cost.
(e) Multiply the matrices found in parts (c) and (d) to determine what the total cost of the day's production was.

58. A car dealership has two locations, one in a city and the other in the suburbs. In January, the city location sold 400 subcompacts, 250 intermediate-size cars, and 50 station wagons; in February, it sold 350 subcompacts, 100 intermediates, and 30 station wagons. At the suburban location in January, 450 subcompacts, 200 intermediates, and 140 station wagons were sold. In February, the suburban location sold 350 subcompacts, 300 intermediates, and 100 station wagons.
(a) Find 2 by 3 matrices that summarize the sales data for each location for January and February (one matrix for each month).
(b) Use matrix addition to obtain total sales for the 2 month period.
(c) The profit on each kind of car is: \$100 per subcompact, \$150 per intermediate, and \$200 per station wagon. Find a 3 by 1 matrix representing this profit.
(d) Multiply the matrices found in parts (b) and (c) to get a 2 by 1 matrix showing the profit at each location.

59. Consider the 2 by 2 square matrix

$$A = \begin{bmatrix} a & b \\ c & d \end{bmatrix}$$

If $D = ad - bc \neq 0$, show that A is nonsingular and that

$$A^{-1} = \frac{1}{D}\begin{bmatrix} d & -b \\ -c & a \end{bmatrix}$$

14.2 Linear Programming

Certain problems are best viewed as a system of working parts (consisting of capital, raw materials, a labor force, and so on) that are to be allocated according to prescribed limitations and needs, with the ultimate goal in mind of achieving a certain objective, such as maximizing profit or minimizing costs. If such a system can be quantified—that is, represented by mathematical equations and/or inequalities—it may be possible to devise a computational procedure for identifying the best way of achieving the goal. Such procedures are usually referred to as *mathematical programs.*

Indeed, if the system can be represented by a system of linear inequalities and if the goal can be expressed as that of minimizing or maximizing a linear expression, subject to the inequalities, then the procedure for solving the problem is called a **linear program**. Historically, linear programming evolved as a technique for solving problems involving resource allocation of goods and materials for the U.S. Air Force during World War II. Today, linear programming techniques are used to solve a wide variety of problems, such as optimizing airline scheduling, establishing telephone lines, and many others.

Every linear programming problem requires that a certain linear expression, called the **objective function**, be maximized (or minimized). However, such problems further require that this maximization (or minimization) occur under certain conditions, or **constraints**, that can be expressed as linear inequalities. Although many practical linear programming problems involve systems of several hundred linear inequalities containing several hundred variables, we shall limit our discussion to problems containing only two variables, because we can solve such problems using graphing techniques.*

Linear Programming Problem

A **linear programming problem** in two variables x and y consists of maximizing (or minimizing) a linear objective function

$$z = Ax + By \qquad A \text{ and } B \text{ are real numbers, not both } 0$$

subject to certain conditions, or constraints, expressible as linear inequalities in x and y.

In order to maximize (or minimize) the quantity $z = Ax + By$, we need to identify points (x, y) that make the expression for z the largest (or smallest) possible. But not all points (x, y) are eligible; only those that also satisfy each linear inequality (constraint) can be used. We refer to each point (x, y) that satisfies the system of linear inequalities (the constraints) as a **feasible solution**. Thus, in a linear programming problem, we seek the feasible solution(s) that maximizes (or minimizes) the objective function.

Example 1 Consider the linear programming problem

$$\text{Maximize} \quad z = x + 3y$$

subject to the constraints

$$x \geq 0, \quad y \geq 0, \quad x + y \leq 6, \quad x \leq 4$$

*The **simplex method** is a way to solve linear programming problems involving many inequalities and variables. This method was developed by George Dantzig in 1946 and is particularly well-suited for computerization. In 1984, Narendra Karmarkar of Bell Laboratories discovered a way of solving large linear programming problems that improves on the simplex method.

(a) Graph the constraints.

(b) Graph the objective function for $z = 0, 9, 18, 21$.

Solution (a) The constraints are the system of linear inequalities

$$\begin{cases} x \geq 0 \\ y \geq 0 \\ x + y \leq 6 \\ x \leq 4 \end{cases}$$

The graph of this system is the darker shaded region shown in Figure 1.

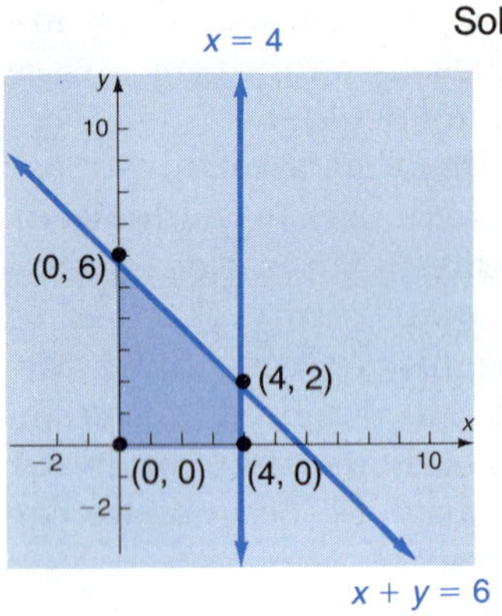

Figure 1

(b) For $z = 0$, the objective function is the line $0 = x + 3y$, or $y = -\frac{1}{3}x$. For $z = 9$, the objective function is the line $9 = x + 3y$, or $y = -\frac{1}{3}x + 3$. For $z = 18$, the objective function is the line $18 = x + 3y$, or $y = -\frac{1}{3}x + 6$. For $z = 21$, the objective function is the line $21 = x + 3y$, or $y = -\frac{1}{3}x + 7$. Figure 2 shows the graphs.

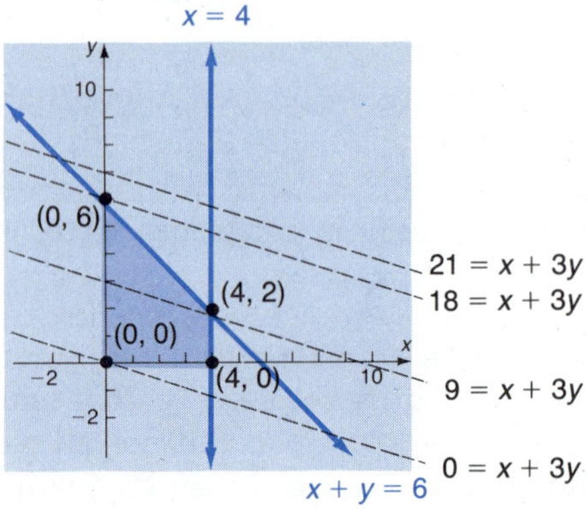

Figure 2

■

A **solution** to a linear programming problem consists of the feasible solutions that maximize (or minimize) the objective function, together with the corresponding values of the objective function. One condition for a linear programming problem in two variables to have a solution is that the graph of the feasible solutions be bounded. (Refer to page 717.)

If none of the feasible solutions maximizes (or minimizes) the objective function or if there are no feasible solutions, then the linear programming problem has no solution.

Consider the linear programming problem stated in Example 1, and look again at Figure 2. The feasible solutions are the points that lie inside the darker shaded region. For example, (2, 2) is a feasible solution, as is (4, 2), (0, 6), (0, 0), etc. To find the solution of the problem requires that we find a feasible solution that makes z, where $z = x + 3y$, as large as possible. Notice that as z increases in value from $z = 0$ to $z = 9$ to

$z = 18$ to $z = 21$, we obtain a collection of parallel lines. Furthermore, notice that the largest value of z that can be obtained while feasible solutions are present is $z = 18$, which corresponds to the line $18 = x + 3y$. Any larger value of z results in a line that does not pass through any feasible solutions. Finally, notice that the feasible solution that yields $z = 18$ is the point $(0, 6)$, a vertex. These observations form the basis of the following result, which we state without proof.

Theorem

Location of Solution of a Linear Programming Problem

If a linear programming problem has a solution, it is located at a vertex of the graph of the feasible solutions.

If a linear programming problem has multiple solutions, at least one of them is located at a vertex of the graph of the feasible solutions.

In either case, the corresponding value of the objective function is unique. ■

We shall not consider here linear programming problems that have no solution. As a result, we can outline the procedure for solving a linear programming problem as follows:

Procedure for Solving a Linear Programming Problem

STEP 1: Write an expression for the quantity to be maximized (or minimized). This expression is the objective function.

STEP 2: Write all the constraints as a system of linear inequalities and graph the system.

STEP 3: List the vertices of the graph of the feasible solutions.

STEP 4: List the corresponding values of the objective function at each vertex. The largest (or smallest) of these is the solution.

Example 2 Minimize the expression

$$z = 2x + 3y$$

subject to the constraints

$$y \le 5, \quad x \le 6, \quad x + y \ge 2, \quad x \ge 0, \quad y \ge 0$$

Solution The objective function is $z = 2x + 3y$. We seek the smallest value of z that can occur if x and y are solutions of the system of linear inequalities

$$\begin{cases} y \le 5 \\ x \le 6 \\ x + y \ge 2 \\ x \ge 0 \\ y \ge 0 \end{cases}$$

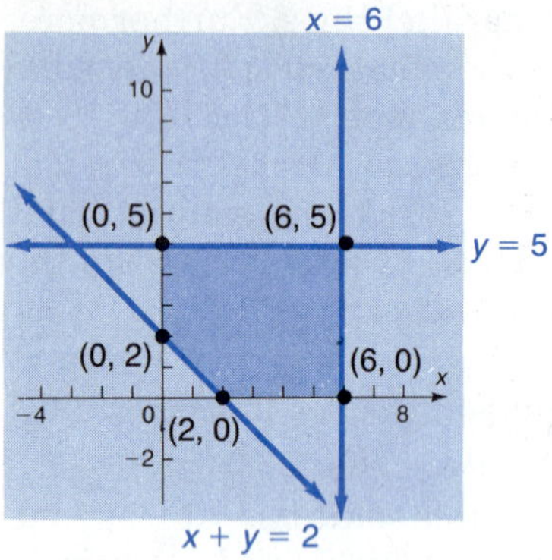

Figure 3

The graph of this system (the feasible solutions) is shown as the darker shaded region in Figure 3. We have also plotted the vertices. Table 1 lists the vertices and the corresponding values of the objective function. From the table, we can see that the minimum value of z is 4, and it occurs at the point (2, 0).

Table 1

VERTEX (x, y)	VALUE OF THE OBJECTIVE FUNCTION $z = 2x + 3y$
(0, 2)	$z = 2(0) + 3(2) = 6$
(0, 5)	$z = 2(0) + 3(5) = 15$
(6, 5)	$z = 2(6) + 3(5) = 27$
(6, 0)	$z = 2(6) + 3(0) = 12$
(2, 0)	$z = 2(2) + 3(0) = 4$

■

Example 3 At the end of every month, after filling orders for its regular customers, a coffee company has some pure Colombian coffee and some special-blend coffee remaining. The practice of the company has been to package a mixture of the two coffees into 1 pound packages as follows: a low-grade mixture containing 4 ounces of Colombian coffee and 12 ounces of special-blend coffee and a high-grade mixture containing 8 ounces of Colombian and 8 ounces of special-blend coffee. A profit of \$0.30 per package is made on the low-grade mixture, whereas a profit of \$0.40 per package is made on the high-grade mixture. This month, 120 pounds of special-blend coffee and 100 pounds of pure Colombian coffee remain. How many packages of each mixture should be prepared to achieve a maximum profit? Assume all packages prepared can be sold.

Solution We begin by assigning symbols for the two variables:

x = Number of packages of the low-grade mixture

y = Number of packages of the high-grade mixture

If P denotes the profit, then

$$P = \$0.30x + \$0.40y$$

This expression is the objective function. We seek to maximize P subject to certain constraints on x and y. Because x and y represent numbers of packages, the only meaningful values for x and y are nonnegative integers. Thus, we have the two constraints

$$x \geq 0, \quad y \geq 0 \quad \text{Nonnegative constraints}$$

We also have only so much of each type of coffee available. For example, the total amount of Colombian coffee used in the two mixtures cannot exceed 100 pounds, or 1600 ounces. Because we use 4 ounces in each low-grade package and 8 ounces in each high-grade package, we are led to the constraint

$$4x + 8y \leq 1600 \quad \text{Colombian coffee constraint}$$

Similarly, the supply of 120 pounds, or 1920 ounces, of special-blend coffee leads to the constraint

$$12x + 8y \leq 1920 \quad \text{Special-blend coffee constraint}$$

The linear programming problem may be stated as:

$$\text{Maximize} \quad P = 0.3x + 0.4y$$

subject to the constraints

$$x \geq 0, \quad y \geq 0, \quad 4x + 8y \leq 1600, \quad 12x + 8y \leq 1920$$

The graph of the constraints (the feasible solutions) is illustrated in Figure 4. We list the vertices and evaluate the objective function at each vertex. In Table 2, we can see that the maximum profit, \$84, is achieved with 40 packages of the low-grade mixture and 180 packages of the high-grade mixture.

Figure 4

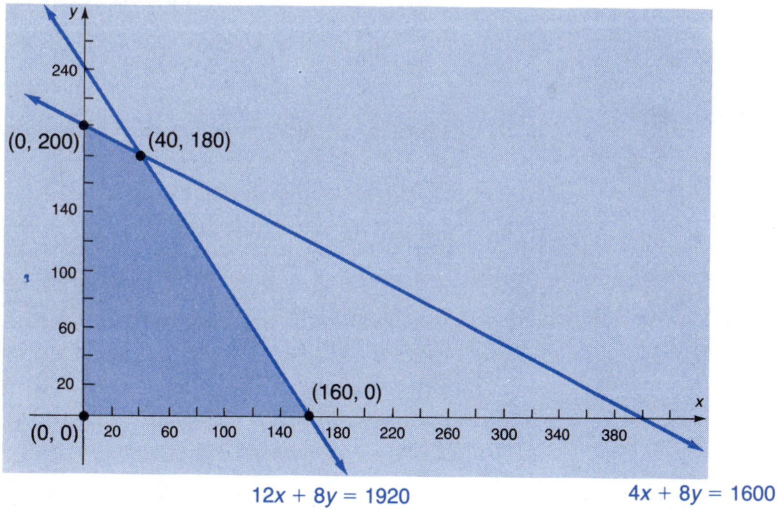

Table 2

VERTEX	VALUE OF PROFIT
(x, y)	$P = 0.3x + 0.4y$
(0, 0)	$P = 0$
(0, 200)	$P = 0.3(0) + 0.4(200) = \$80$
(40, 180)	$P = 0.3(40) + 0.4(180) = \$84$
(160, 0)	$P = 0.3(160) + 0.4(0) = \$48$

■

Example 4 A retired couple has up to \$50,000 to place in fixed-income securities. Their banker suggests two securities to them: one is a AAA bond that yields 12% per annum, the other is a Certificate of Deposit (CD) that yields 9%. After careful consideration of the alternatives, the couple decides to place at most \$20,000 in the AAA bond and at least \$15,000 in the CD. They also instruct the banker to place at least as much in the CD as in the AAA bond. How should the banker proceed to maximize the return on their investment?

Solution The variables are named as

$$x = \text{Amount invested in the AAA bond}$$
$$y = \text{Amount invested in the CD}$$

The return R on investment is given by

$$R = 0.12x + 0.09y$$

This expression is to be maximized (the objective function). The conditions imposed on the variables x and y are:

$x \geq 0, \quad y \geq 0$	Nonnegative constraints
$x + y \leq 50{,}000$	Up to \$50,000 to invest
$x \leq 20{,}000$	Place at most \$20,000 in the AAA bond.
$y \geq 15{,}000$	Place at least \$15,000 in the CD.
$y \geq x$	Place at least as much in the CD as in the bond.

The linear programming problem may be stated as.

$$\text{Maximize} \quad R = 0.12x + 0.09y$$

subject to the contraints

$$x \geq 0, \quad y \geq 0, \quad x + y \leq 50{,}000, \quad x \leq 20{,}000 \quad y \geq 15{,}000, \quad y \geq x$$

The graph of the constraints (the feasible solutions) is illustrated in Figure 5. Table 3 lists the vertices and the corresponding values of the objective function. The maximum return on investment is \$5,100, achieved by placing \$20,000 in the AAA bond and \$30,000 in the CD.

Table 3

VERTEX	RETURN ON INVESTMENT	
(x, y)	$R = 0.12x + 0.09y$	
(0, 15)	$R = 0.12(0) + 0.09(15) = 1.35$ thousand dollars	minimum
(15, 15)	$R = 0.12(15) + 0.09(15) = 3.15$ thousand dollars	
(20, 20)	$R = 0.12(20) + 0.09(20) = 4.2$ thousand dollars	
(20, 30)	$R = 0.12(20) + 0.09(30) = 5.1$ thousand dollars	max
(0, 50)	$R = 0.12(0) + 0.09(50) = 4.5$ thousand dollars	

Figure 5

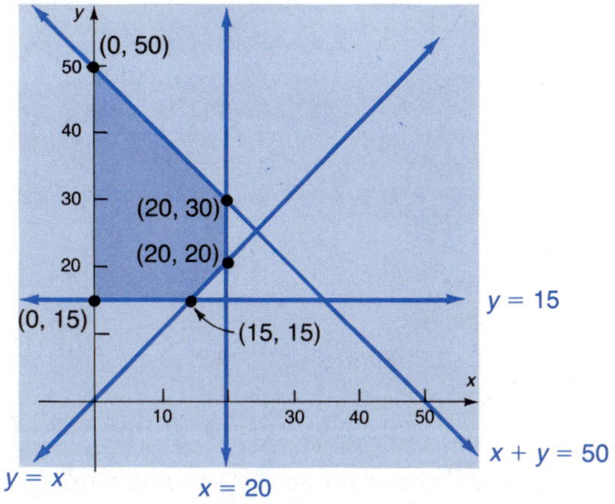

Units in thousands

EXERCISE 14.2

The figure below illustrates the graph of the feasible solutions of a linear programming problem.

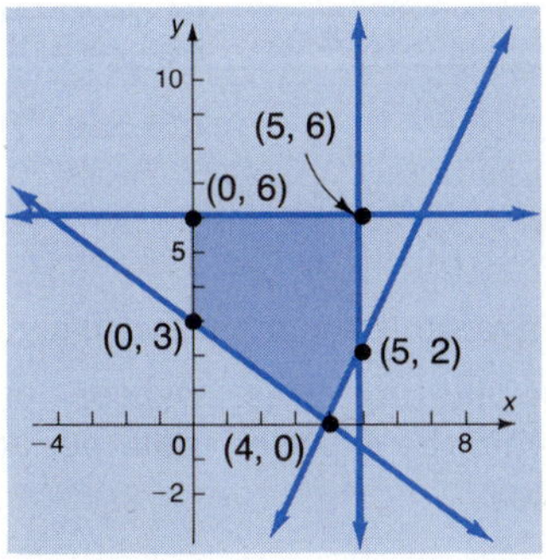

In Problems 1–6, find the maximum and minimum value of the given objective function.

1. $z = x + y$ **2.** $z = 2x + 3y$ **3.** $z = x + 10y$

4. $z = 10x + y$ **5.** $z = 5x + 7y$ **6.** $z = 7x + 5y$

In Problems 7–16, solve each linear programming problem.

7. Maximize $z = 2x + y$
subject to $x \geq 0, \quad y \geq 0, \quad x + y \leq 6, \quad x + y \geq 1$

8. Maximize $z = x + 3y$
subject to $x \geq 0, \quad y \geq 0, \quad x + y \geq 3, \quad x \leq 5, \quad y \leq 7$

9. Minimize $z = 2x + 5y$
subject to $x \geq 0, \quad y \geq 0, \quad x + y \geq 2, \quad x \leq 5, \quad y \leq 3$

10. Minimize $z = 3x + 4y$
subject to $x \geq 0, \quad y \geq 0, \quad 2x + 3y \geq 6, \quad x + y \leq 8$

11. Maximize $z = 3x + 5y$
subject to $x \geq 0, \quad y \geq 0, \quad x + y \geq 2, \quad 2x + 3y \leq 12, \quad 3x + 2y \leq 12$

12. Maximize $z = 5x + 3y$
subject to $x \geq 0, \quad y \geq 0, \quad x + y \geq 2, \quad x + y \leq 8, \quad 2x + y \leq 10$

13. Minimize $z = 5x + 4y$
subject to $x \geq 0, \quad y \geq 0, \quad x + y \geq 2, \quad 2x + 3y \leq 12, \quad 3x + y \leq 12$

14. Minimize $z = 2x + 3y$
subject to $x \geq 0, \quad y \geq 0, \quad x + y \geq 3, \quad x + y \leq 9, \quad 2x + 3y \leq 6$

15. Maximize $z = 5x + 2y$
subject to $x \geq 0, \quad y \geq 0, \quad x + y \leq 10, \quad 2x + y \geq 10, \quad x + 2y \geq 10$

16. Maximize $z = 2x + 4y$
subject to $x \geq 0, \quad y \geq 0, \quad 2x + y \geq 4, \quad x + y \leq 9$

17. Rework Example 3 if the profit on the low-grade mixture is \$0.40 per package and the profit on the high-grade mixture is \$0.30 per package.

18. Rework Example 4 if the retired couple no longer requires that at most \$20,000 be placed in the AAA bond.

19. A manufacturer of skis produces two types: downhill and cross-country. Use the following table to determine how many of each kind of ski should be produced to achieve a maximum profit. What is the maximum profit?

	DOWNHILL	CROSS-COUNTRY	MAXIMUM TIME AVAILABLE
MANUFACTURING TIME PER SKI	2 hours	1 hour	40 hours
FINISHING TIME PER SKI	1 hour	1 hour	32 hours
PROFIT PER SKI	\$70	\$50	

20. Rework Problem 19 if the maximum time available for manufacturing is 48 hours.

21. A farmer has 70 acres of land available for planting either soybeans or wheat. The cost of preparing the soil, the workdays required, and the expected profit per acre planted for each type of crop are given in the following table:

	SOYBEANS	WHEAT
PREPARATION COST PER ACRE	\$60	\$30
WORKDAYS REQUIRED PER ACRE	3	4
PROFIT PER ACRE	\$180	\$100

The farmer cannot spend more than \$1800 in preparation costs nor more than a total of 120 workdays. How many acres of each crop should be planted in order to maximize the profit? What is the maximum profit?

22. Rework Problem 21 if the farmer is willing to spend no more than \$2400 on preparation.

23. An investment broker is instructed by her client to invest up to \$20,000, some in a CD yielding 9% per annum and some in Treasury bills yielding 7% per annum. The client wants to invest at least \$8000 in T bills and no more than \$12,000 in the CD. The client also insists that the amount invested in T bills must equal or exceed the amount placed in the CD. How much should the broker recommend the client place in each type of investment if the objective is to maximize return on investment?

24. Rework Problem 23 if the client insists instead that the amount invested in T bills must not exceed the amount placed in the CD.

25. A factory manufactures two kinds of ice skates: racing skates and figure skates. The racing skate requires 6 work-hours in the fabrication department, whereas the figure skates require 4 work-hours there. The racing skates require 1 work-hour in the finishing department, whereas the figure skates require 2 work-hours there. The fabricating department has available at most 120 work-hours per day, and the finishing department has no more than 40 work-hours per day available. If the profit on each racing skate is \$10 and the profit on each figure skate is \$12, how many of each should be manufactured each day to maximize profit? (Assume all skates made are sold.)

26. A factory manufactures two kinds of ceramic figurines: a dancing girl and a mermaid, each requiring three processes—molding, painting, and glazing. The daily labor available for molding is no more than 90 work-hours; labor available for painting does not exceed 120 work-hours; and labor available for glazing is no more than 60 work-hours. The dancing girl requires 3 work-hours for molding, 6 work-hours for painting, and 2 work-hours for glazing. The mermaid requires 3 work-hours for molding, 4 work-hours for painting, and 3 work-hours for glazing. If the profit on each figurine is \$25 for dancing girls and \$30 for mermaids, how many of each should be produced each day to maximize profit? If management decides to produce the number of each figurine that maximizes profit, determine which of these processes has excess work-hours assigned to it.

27. An airline has two classes of service: first class and coach. Management's experience has been that each aircraft should have at least 8 but not more than 16 first-class seats and at least 80 but not more than 120 coach seats. Management has further decided the ratio of first class to coach should never exceed $\frac{1}{12}$. With how many of each type of seat should an aircraft be configured to maximize revenue? [*Hint:* Assume the airline charges \$$C$ for a coach seat and \$$C$ + \$$F$ for a first-class seat, $C > 0, F > 0$.]

28. Rework Problem 27 if management decides instead that the ratio of first class to coach should never exceed $\frac{1}{8}$.

29. A farm that specializes in raising frying chickens supplements the regular chicken feed with four vitamins. The owner wants the supplemental food to contain at least 50 units of vitamin I, 90 units of vitamin II, 60 units of vitamin III, and 100 units of vitamin IV per 100 ounces of feed. Two supplements are available: supplement A, which contains 5 units of vitamin I, 25 units of vitamin II, 10 units of vitamin III, and 35 units of vitamin IV per ounce; and supplement B, which contains 25 units of vitamin I, 10 units of vitamin II, 10 units of vitamin III, and 20 units of vitamin IV per ounce. If supplement A costs \$0.06 per ounce and supplement B costs \$0.08 per ounce, how much of each supplement should the manager of the farm buy to add to each 100 ounces of feed in order to keep the total cost at a minimum, while still meeting the owner's vitamin specifications?

14.3 Partial Fraction Decomposition

Consider the problem of adding the two fractions

$$\frac{3}{x+4} \quad \text{and} \quad \frac{2}{x-3}$$

The result is

$$\frac{3}{x+4}+\frac{2}{x-3}=\frac{3(x-3)+2(x+4)}{(x+4)(x-3)}=\frac{5x-1}{x^2+x-12}$$

The reverse procedure, of starting with the rational expression $(5x - 1)/(x^2 + x - 12)$ and writing it as the sum (or difference) of the two simpler fractions $3/(x + 4)$ and $2/(x - 3)$, is referred to as **partial fraction decomposition**, and the two simpler fractions are called **partial fractions**. Decomposing a rational expression into a sum of partial fractions is important in solving certain types of calculus problems. This section presents a systematic way to decompose rational expressions.

We begin by recalling that a rational expression is the ratio of two polynomials, say, P and $Q \neq 0$, that have no common factors. Recall also that a rational expression P/Q is called **proper** if the degree of the polynomial in the numerator is less than the degree of the polynomial in the denominator. Otherwise, the rational expression is termed **improper**.

Because any improper rational expression can be reduced by long division to a mixed form consisting of the sum of a polynomial and a proper rational expression, we shall restrict the discussion that follows to proper rational expressions.

The partial fraction decomposition of the rational expression P/Q depends on the factors of the denominator Q. Recall (from Section 5.5) that any polynomial whose coefficients are real numbers can be factored (over the real numbers) into products of linear and/or irreducible quadratic factors. Thus, the denominator Q of the rational expression P/Q will contain only factors of one or both of the following types:

1. *Linear factors* of the form $x - a$, where a is a real number.
2. *Irreducible quadratic factors* of the form $ax^2 + bx + c$, where a, b, and c are real numbers and $ax^2 + bx + c$ cannot be written as the product of two linear factors with real coefficients.

As it turns out, there are four cases to be examined. We begin with the case for which Q has only nonrepeated linear factors.

CASE 1: Q has only nonrepeated linear factors.

Under the assumption that Q has only nonrepeated linear factors, the polynomial Q has the form

$$Q(x) = (x - a_1)(x - a_2) \cdot \cdots \cdot (x - a_n)$$

where none of the numbers $a_1, a_2, \ldots, a_n$ are equal. In this case, the partial fraction decomposition of P/Q is of the form

$$\frac{P(x)}{Q(x)} = \frac{A_1}{x - a_1} + \frac{A_2}{x - a_2} + \cdots + \frac{A_n}{x - a_n} \tag{1}$$

where the numbers $A_1, A_2, \ldots, A_n$ are to be determined.

We show how to find these numbers in the example that follows.

Example 1 Write the partial fraction decomposition of: $\dfrac{x}{x^2 - 5x + 6}$

Solution First, we factor the denominator,

$$x^2 - 5x + 6 = (x - 2)(x - 3)$$

and conclude that the denominator contains only nonrepeated linear factors. Then, we decompose the rational expression according to equation (1):

$$\frac{x}{x^2 - 5x + 6} = \frac{A}{x - 2} + \frac{B}{x - 3} \tag{2}$$

where A and B are to be determined. To find A and B, we clear the fractions by multiplying each side by $(x - 2)(x - 3) = x^2 - 5x + 6$. The result is

$$x = A(x - 3) + B(x - 2)$$

or

$$x = (A + B)x + (-3A - 2B)$$

This equation is an identity in x. Thus, we may equate the coefficients of like powers of x to get

$$\begin{cases} 1 = A + B \\ 0 = -3A - 2B \end{cases}$$

Equate coefficients of x: $1x = (A + B)x$

Equate coefficients of x^0, the constants: $0x^0 = (-3A - 2B)x^0$

This system of two equations containing two variables, A and B, can be solved using whatever method you wish. Solving it, we get

$$A = -2, \qquad B = 3$$

Thus, from equation (2), the partial fraction decomposition is

$$\frac{x}{x^2 - 5x + 6} = \frac{-2}{x - 2} + \frac{3}{x - 3}$$

Check: The decomposition can be checked by adding the fractions:

$$\frac{-2}{x-2} + \frac{3}{x-3} = \frac{-2(x-3) + 3(x-2)}{(x-2)(x-3)} = \frac{x}{(x-2)(x-3)}$$

$$= \frac{x}{x^2 - 5x + 6}$$ ■

CASE 2: Q has repeated linear factors.

If the polynomial Q has a repeated factor, say, $(x-a)^n$, $n \geq 2$ an integer, then, in the partial fraction decomposition of P/Q, we allow for the terms

$$\frac{A_1}{x-a} + \frac{A_2}{(x-a)^2} + \cdots + \frac{A_n}{(x-a)^n}$$

where the numbers $A_1, A_2, \ldots, A_n$ are to be determined.

Example 2 Write the partial fraction decomposition of: $\dfrac{x+2}{x^3 - 2x^2 + x}$

Solution First, we factor the denominator,

$$x^3 - 2x^2 + x = x(x^2 - 2x + 1) = x(x-1)^2$$

and find that the denominator has the nonrepeated linear factor x and the repeated linear factor $(x-1)^2$. By Case 1, we must allow for the term A/x in the decomposition; and, by Case 2, we must allow for the terms $B/(x-1) + C/(x-1)^2$ in the decomposition. Thus, we write

$$\frac{x+2}{x^3 - 2x^2 + x} = \frac{A}{x} + \frac{B}{x-1} + \frac{C}{(x-1)^2} \tag{3}$$

Again, we clear fractions by multiplying each side by $x^3 - 2x^2 + x = x(x-1)^2$. The result is the identity

$$\begin{aligned} x + 2 &= A(x-1)^2 + Bx(x-1) + Cx \\ &= (A+B)x^2 + (-2A - B + C)x + A \end{aligned}$$

Equating coefficients of like powers of x, we obtain the system

$$\begin{cases} A + B = 0 & \text{The coefficient of } x^2 \text{ on the left is 0.} \\ -2A - B + C = 1 \\ A = 2 \end{cases}$$

The solution of this system is $A = 2, B = -2, C = 3$. From equation (3), the partial fraction decomposition is

$$\frac{x + 2}{x^3 - 2x^2 + x} = \frac{2}{x} + \frac{-2}{x - 1} + \frac{3}{(x - 1)^2}$$ ■

The numbers to be found in the partial fraction decomposition can sometimes be found more readily by using suitable choices for x (which may include complex numbers) in the identity obtained after fractions have been cleared. In Example 2, the identity obtained after clearing fractions is

$$x + 2 = A(x - 1)^2 + Bx(x - 1) + Cx \tag{4}$$

If we let $x = 0$ in this expression, the terms containing B and C drop out, leaving $2 = A(-1)^2$, or $A = 2$. Similarly, if we let $x = 1$, the terms containing A and B drop out, leaving $3 = C$. Thus, equation (4) becomes

$$x + 2 = 2(x - 1)^2 + Bx(x - 1) + 3x$$

Now, let $x = 2$ (any choice other than 0 or 1 will work as well). The result is

$$\begin{aligned} 4 &= 2(1)^2 + B(2)(1) + 3(2) \\ 2B &= 4 - 2 - 6 = -4 \\ B &= -2 \end{aligned}$$

As before, we have $A = 2$, $B = -2$, and $C = 3$.

We use this method in the next example.

Example 3 Write the partial fraction decomposition of: $\dfrac{x^3 - 8}{x^2(x - 1)^3}$

Solution The denominator contains the repeated linear factor x^2 and the repeated linear factor $(x - 1)^3$. Thus, the partial fraction decomposition takes the form

$$\frac{x^3 - 8}{x^2(x - 1)^3} = \frac{A}{x} + \frac{B}{x^2} + \frac{C}{x - 1} + \frac{D}{(x - 1)^2} + \frac{E}{(x - 1)^3} \tag{5}$$

As before, we clear fractions and obtain the identity

$$x^3 - 8 = Ax(x - 1)^3 + B(x - 1)^3 + Cx^2(x - 1)^2 + Dx^2(x - 1) + Ex^2 \tag{6}$$

Let $x = 0$. (Do you see why this choice was made?) Then,

$$\begin{aligned} -8 &= B(-1) \\ B &= 8 \end{aligned}$$

Now, let $x = 1$ in equation (6). Then,

$$-7 = E$$

Use $B = 8$ and $E = -7$ in equation (6), and collect like terms:

$$x^3 - 8 = Ax(x-1)^3 + 8(x-1)^3 + Cx^2(x-1)^2 + Dx^2(x-1) - 7x^2$$

$$x^3 - 8 - 8(x^3 - 3x^2 + 3x - 1) + 7x^2 = Ax(x-1)^3 + Cx^2(x-1)^2 + Dx^2(x-1)$$

$$-7x^3 + 31x^2 - 24x = x(x-1)[A(x-1)^2 + Cx(x-1) + Dx]$$

$$x(x-1)(-7x+24) = x(x-1)[A(x-1)^2 + Cx(x-1) + Dx]$$

$$-7x + 24 = A(x-1)^2 + Cx(x-1) + Dx \qquad (7)$$

We now work with equation (7). Let $x = 0$. Then,

$$24 = A$$

Now, let $x = 1$ in equation (7). Then,

$$17 = D$$

Use $A = 24$ and $D = 17$ in equation (7), and collect like terms:

$$-7x + 24 = 24(x-1)^2 + Cx(x-1) + 17x$$

$$-24x^2 + 48x - 24 - 17x - 7x + 24 = Cx(x-1)$$

$$-24x^2 + 24x = Cx(x-1)$$

$$-24x(x-1) = Cx(x-1)$$

$$-24 = C$$

We now know all the numbers A, B, C, D, and E, so, from equation (5), we have the decomposition

$$\frac{x^3-8}{x^2(x-1)^3} = \frac{24}{x} + \frac{8}{x^2} + \frac{-24}{x-1} + \frac{17}{(x-1)^2} + \frac{-7}{(x-1)^3}$$ ■

The method employed in Example 3, although somewhat tedious, is still preferable to solving the system of five equations containing five variables that the expansion of equation (5) leads to.

The final two cases involve irreducible quadratic factors. As mentioned in Section 5.5, a quadratic factor is irreducible if it cannot be factored into linear factors with real coefficients. A quadratic expression $ax^2 + bx + c$ is irreducible whenever $b^2 - 4ac < 0$. For example, $x^2 + x + 1$ and $x^2 + 4$ are irreducible.

CASE 3: Q contains a nonrepeated irreducible quadratic factor.

If Q contains a nonrepeated irreducible quadratic factor of the form $ax^2 + bx + c$, then, in the partial fraction decomposition of P/Q, allow for the term

$$\frac{Ax + B}{ax^2 + bx + c}$$

where the numbers A and B are to be determined.

Example 4 Write the partial fraction decomposition of: $\dfrac{3x - 5}{x^3 - 1}$

Solution We factor the denominator,

$$x^3 - 1 = (x - 1)(x^2 + x + 1)$$

and find that it has a nonrepeated linear factor $x - 1$ and a nonrepeated irreducible quadratic factor $x^2 + x + 1$. Thus, by Case 1, we allow for the term $A/(x - 1)$ and, by Case 3, we allow for the term $(Bx + C)/(x^2 + x + 1)$. Hence, we write

$$\frac{3x - 5}{x^3 - 1} = \frac{A}{x - 1} + \frac{Bx + C}{x^2 + x + 1} \tag{8}$$

We clear fractions by multiplying each side of equation (8) by $x^3 - 1 = (x - 1)(x^2 + x + 1)$ to obtain

$$3x - 5 = A(x^2 + x + 1) + (Bx + C)(x - 1) \tag{9}$$

Collecting like terms, we arrive at

$$3x - 5 = (A + B)x^2 + (A - B + C)x + (A - C)$$

Equating coefficients, we obtain the system

$$\begin{cases} A + B = 0 \\ A - B + C = 3 \\ A - C = -5 \end{cases}$$

The solution is $A = -\frac{2}{3}$, $B = \frac{2}{3}$, $C = \frac{13}{3}$. Thus, from equation (8), we see that

$$\frac{3x - 5}{x^3 - 1} = \frac{-\frac{2}{3}}{x - 1} + \frac{\frac{2}{3}x + \frac{13}{3}}{x^2 + x + 1}$$ ■

An alternative to solving the system of three equations containing three variables in Example 4 would be to let $x = 1$ in equation (9) to get $-\frac{2}{3} = A$. Using this value in equation (9), collecting terms, and then equating coefficients is a little faster. Try it for yourself.

CASE 4: Q contains repeated irreducible quadratic factors.

If the polynomial Q contains a repeated irreducible quadratic factor $(ax^2 + bx + c)^n$, $n \geq 2$, n an integer, then, in the partial fraction decomposition of P/Q, allow for the terms

$$\frac{A_1x + B_1}{ax^2 + bx + c} + \frac{A_2x + B_2}{(ax^2 + bx + c)^2} + \cdots + \frac{A_nx + B_n}{(ax^2 + bx + c)^n}$$

where the numbers $A_1, B_1, A_2, B_2, \ldots, A_n, B_n$ are to be determined.

Example 5 Write the partial fraction decomposition of: $\dfrac{x^3 + x^2}{(x^2 + 4)^2}$

Solution The denominator contains the repeated irreducible quadratic factor $(x^2 + 4)^2$, so we write

$$\frac{x^3 + x^2}{(x^2 + 4)^2} = \frac{Ax + B}{x^2 + 4} + \frac{Cx + D}{(x^2 + 4)^2} \tag{10}$$

We clear fractions to obtain

$$x^3 + x^2 = (Ax + B)(x^2 + 4) + Cx + D$$

Collecting like terms yields

$$x^3 + x^2 = Ax^3 + Bx^2 + (4A + C)x + D + 4B$$

Equating coefficients, we arrive at the system

$$\begin{cases} A = 1 \\ B = 1 \\ 4A + C = 0 \\ D + 4B = 0 \end{cases}$$

The solution is $A = 1$, $B = 1$, $C = -4$, $D = -4$. Hence, from equation (10),

$$\frac{x^3 + x^2}{(x^2 + 4)^2} = \frac{x + 1}{x^2 + 4} + \frac{-4x - 4}{(x^2 + 4)^2}$$ ■

EXERCISE 14.3

In Problems 1–8, tell whether the given rational expression is proper or improper. If improper, rewrite it as the sum of a polynomial and a proper rational expression.

1. $\dfrac{x}{x^2 - 1}$ **2.** $\dfrac{5x + 2}{x^3 - 1}$ **3.** $\dfrac{x^2 + 5}{x^2 - 4}$ **4.** $\dfrac{3x^2 - 2}{x^2 - 1}$

5. $\dfrac{5x^3 + 2x - 1}{x^2 - 4}$ **6.** $\dfrac{3x^4 + x^2 - 2}{x^3 + 8}$ **7.** $\dfrac{x(x - 1)}{(x + 4)(x - 3)}$ **8.** $\dfrac{2x(x^2 + 4)}{x^2 + 1}$

In Problems 9–42, write the partial fraction decomposition of each rational expression.

9. $\dfrac{4}{x(x - 1)}$ **10.** $\dfrac{3x}{(x + 2)(x - 1)}$ **11.** $\dfrac{1}{x(x^2 + 1)}$

12. $\dfrac{1}{(x + 1)(x^2 + 4)}$ **13.** $\dfrac{x}{(x - 1)(x - 2)}$ **14.** $\dfrac{3x}{(x + 2)(x - 4)}$

15. $\dfrac{x^2}{(x - 1)^2(x + 1)}$ **16.** $\dfrac{x + 1}{x^2(x - 2)}$ **17.** $\dfrac{1}{x^3 - 8}$

18. $\dfrac{2x + 4}{x^3 - 1}$ **19.** $\dfrac{x^2}{(x - 1)^2(x + 1)^2}$ **20.** $\dfrac{x + 1}{x^2(x - 2)^2}$

21. $\dfrac{x - 3}{(x + 2)(x + 1)^2}$ **22.** $\dfrac{x^2 + x}{(x + 2)(x - 1)^2}$ **23.** $\dfrac{x + 4}{x^2(x^2 + 4)}$

24. $\dfrac{10x^2 + 2x}{(x - 1)^2(x^2 + 2)}$ **25.** $\dfrac{x^2 + 2x + 3}{(x + 1)(x^2 + 2x + 4)}$ **26.** $\dfrac{x^2 - 11x - 18}{x(x^2 + 3x + 3)}$

27. $\dfrac{x}{(3x - 2)(2x + 1)}$ **28.** $\dfrac{1}{(2x + 3)(4x - 1)}$ **29.** $\dfrac{x}{x^2 + 2x - 3}$

30. $\dfrac{x^2 - x - 8}{(x + 1)(x^2 + 5x + 6)}$ **31.** $\dfrac{x^2 + 2x + 3}{(x^2 + 4)^2}$ **32.** $\dfrac{2x + 1}{(x^2 + 16)^2}$

33. $\dfrac{7x + 3}{x^3 - 2x^2 - 3x}$ **34.** $\dfrac{x^5 + 1}{x^6 - x^4}$ **35.** $\dfrac{x^2}{x^3 - 4x^2 + 5x - 2}$

36. $\dfrac{x^2 + 1}{x^3 + x^2 - 5x + 3}$ **37.** $\dfrac{x^3}{(x^2 + 16)^3}$ **38.** $\dfrac{x^2}{(x^2 + 4)^3}$

39. $\dfrac{4}{2x^2 - 5x - 3}$ **40.** $\dfrac{4x}{2x^2 + 3x - 2}$ **41.** $\dfrac{2x + 3}{x^4 - 9x^2}$

42. $\dfrac{x^2 + 9}{x^4 - 2x^2 - 8}$

14.4

Vectors

In simple terms, a **vector** (derived from the Latin *vehere*, meaning "to carry") is a quantity that has both magnitude and direction. For a vector in the plane, which is the only type we shall discuss, it is convenient to represent a vector by using an arrow. The length of the arrow represents the **magnitude** of the vector, and the arrowhead indicates the **direction** of the vector.

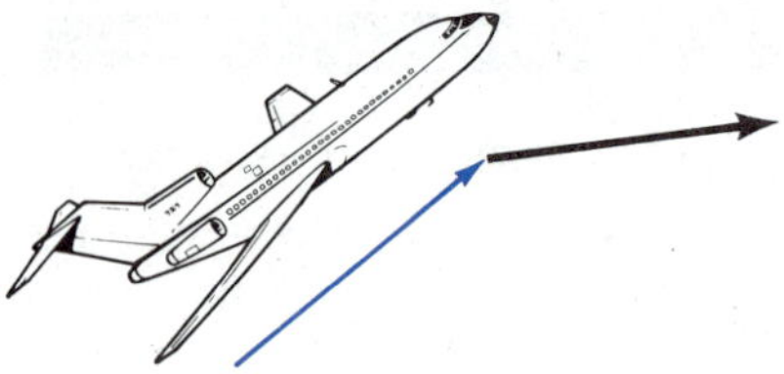

Figure 6

Many quantities in physics can be represented by vectors. For example, the velocity of an aircraft can be represented by an arrow that points in the direction of movement; the length of the arrow represents speed. Thus, if the aircraft speeds up, we lengthen the arrow; if the aircraft changes direction, we introduce an arrow in the new direction. See Figure 6. Based on this representation, it is not surprising that vectors and directed line segments are somehow related.

Directed Line Segments

If P and Q are two distinct points in the xy-plane, there is exactly one line containing both P and Q. The points on that part of the line that joins P to Q, including P and Q, form what is called the **line segment** $\overline{PQ}$. If we order the points so that they proceed from P to Q, we have a **directed line segment** from P to Q, which we denote by $\overrightarrow{PQ}$. In a directed line segment $\overrightarrow{PQ}$, we call P the **initial point** and Q the **terminal point**, as indicated in Figure 7.

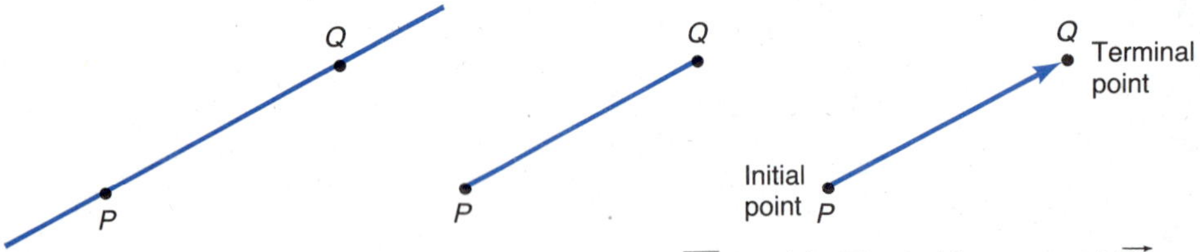

(a) Line containing P and Q (b) Line segment $\overline{PQ}$ (c) Directed line segment $\overrightarrow{PQ}$

Figure 7

The magnitude of the directed line segment $\overrightarrow{PQ}$ is the distance from the point P to the point Q, that is, it is the length of the line segment. The direction of $\overrightarrow{PQ}$ is from P to Q. If a vector **v*** has the same magnitude and the same direction as the directed line segment $\overrightarrow{PQ}$, then we write

$$\mathbf{v} = \overrightarrow{PQ}$$

The vector **v** whose magnitude is 0 is called the **zero vector, 0**. The zero vector is assigned no direction.

Two vectors **v** and **w** are **equal**, written

$$\mathbf{v} = \mathbf{w}$$

if they have the same magnitude and the same direction.

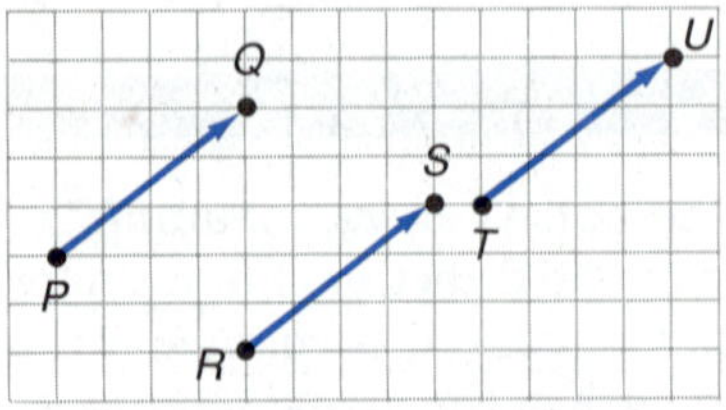

Figure 8

For example, the vectors shown in Figure 8 have the same magnitude and the same direction, so they are equal, even though they have different initial points and different terminal points. As a result, we find it useful to think of a vector simply as an arrow, keeping in mind that two arrows (vectors) are equal if they have the same direction and the same magnitude (length).

*Boldface letters will be used to denote vectors, in order to distinguish them from numbers. For handwritten work, an arrow is placed over the letter to signify a vector.

Adding Vectors

The **sum v + w** of two vectors is defined as follows: We position the vectors **v** and **w** so that the terminal point of **v** coincides with the initial point of **w**, as shown in Figure 9. The vector **v + w** is then the unique vector whose initial point coincides with the initial point of **v** and whose terminal point coincides with the terminal point of **w**.

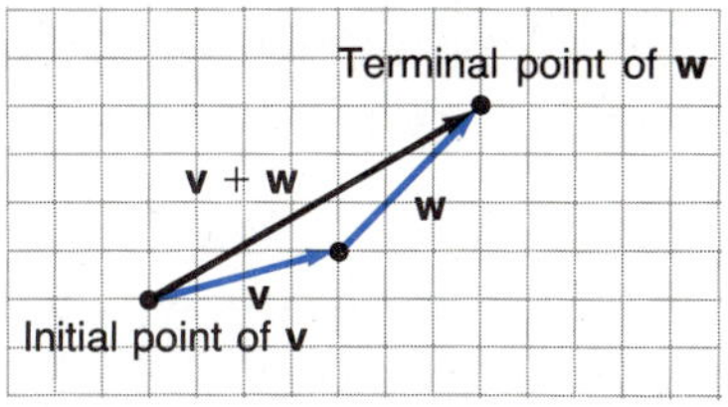

Figure 9

Vector addition is **commutative**. That is, if **v** and **w** are any two vectors, then

Commutative Property

$$\mathbf{v} + \mathbf{w} = \mathbf{w} + \mathbf{v}$$

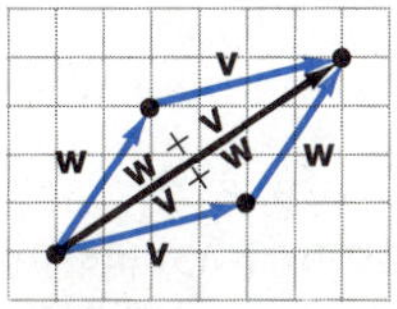

Figure 10

Figure 10 illustrates this fact. (Observe that the commutative property is another way of saying that opposite sides of a parallelogram are equal and parallel.)

Vector addition is also **associative**. That is, if **u**, **v**, and **w** are vectors, then

Associative Property

$$\mathbf{u} + (\mathbf{v} + \mathbf{w}) = (\mathbf{u} + \mathbf{v}) + \mathbf{w}$$

Figure 11 illustrates the associative property for vectors.

Figure 11
(u + v) + w = u + (v + w)

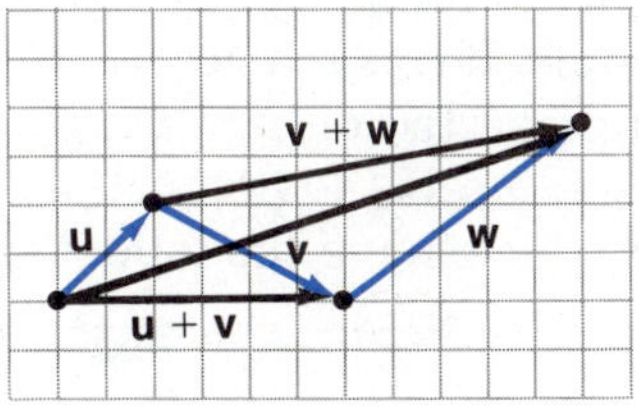

The zero vector has the property that

Identity Property

$$\mathbf{v} + \mathbf{0} = \mathbf{0} + \mathbf{v} = \mathbf{v}$$

for any vector **v**.

If **v** is a vector, then −**v** is the vector having the same magnitude as **v**, but whose direction is opposite to **v**, as shown in Figure 12.

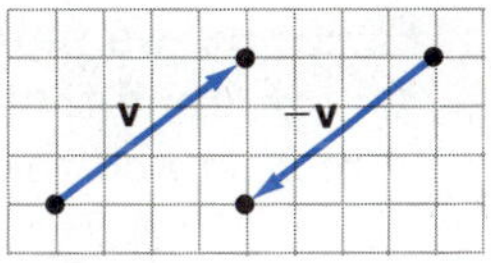

Figure 12

Furthermore,

Additive Inverse Property

$$\mathbf{v} + (-\mathbf{v}) = \mathbf{0}$$

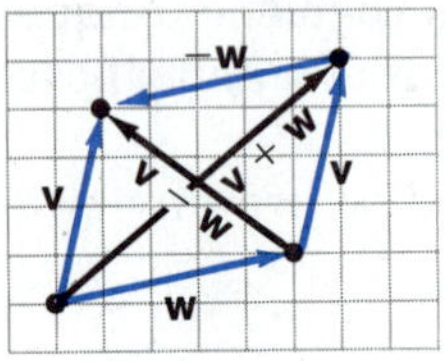

Figure 13

If $\mathbf{v}$ and $\mathbf{w}$ are two vectors, we define the **difference** $\mathbf{v} - \mathbf{w}$ as

$$\mathbf{v} - \mathbf{w} = \mathbf{v} + (-\mathbf{w})$$

Figure 13 illustrates the relationships among $\mathbf{v}$, $\mathbf{w}$, $\mathbf{v} + \mathbf{w}$, and $\mathbf{v} - \mathbf{w}$.

Multiplying Vectors by Numbers

When dealing with vectors, we refer to real numbers as **scalars**. Scalars are quantities that have only magnitude. Examples from physics of scalar quantities are temperature, speed, and time. We now define how to multiply a vector by a scalar.

Scalar Product

If α is a scalar and $\mathbf{v}$ is a vector, the **scalar product** $\alpha\mathbf{v}$ is defined as:

1. If $\alpha > 0$, the product $\alpha\mathbf{v}$ is the vector whose magnitude is α times the magnitude of $\mathbf{v}$ and whose direction is the same as $\mathbf{v}$.
2. If $\alpha < 0$, the product $\alpha\mathbf{v}$ is the vector whose magnitude is $|\alpha|$ times the magnitude of $\mathbf{v}$ and whose direction is opposite that of $\mathbf{v}$.
3. If $\alpha = 0$ or if $\mathbf{v} = \mathbf{0}$, then $\alpha\mathbf{v} = \mathbf{0}$.

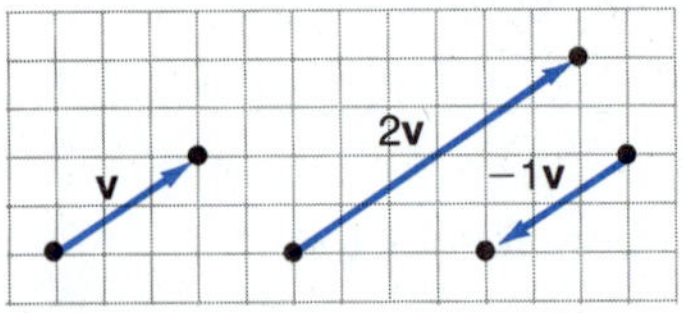

Figure 14

See Figure 14 for some illustrations.

For example, if $\mathbf{a}$ is the acceleration of an object of mass m due to a force $\mathbf{F}$ being exerted on it, then by Newton's Second Law of Motion, $\mathbf{F} = m\mathbf{a}$. Here, $m\mathbf{a}$ is the product of the scalar m and the vector $\mathbf{a}$.

Scalar products have the following properties:

Properties of Scalar Products

$$0\mathbf{v} = \mathbf{0} \qquad 1\mathbf{v} = \mathbf{v} \qquad -1\mathbf{v} = -\mathbf{v}$$
$$(\alpha + \beta)\mathbf{v} = \alpha\mathbf{v} + \beta\mathbf{v} \qquad \alpha(\mathbf{v} + \mathbf{w}) = \alpha\mathbf{v} + \alpha\mathbf{w}$$
$$\alpha(\beta\mathbf{v}) = (\alpha\beta)\mathbf{v}$$

Example 1 Use the vectors illustrated in Figure 15 to graph each expression.

Figure 15

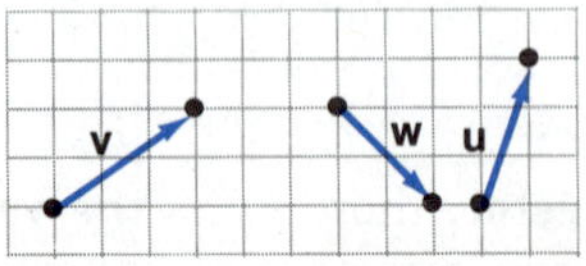

(a) $\mathbf{v} - \mathbf{w}$ (b) $2\mathbf{v} + 3\mathbf{w}$ (c) $2\mathbf{v} - \mathbf{w} + 3\mathbf{u}$

Solution

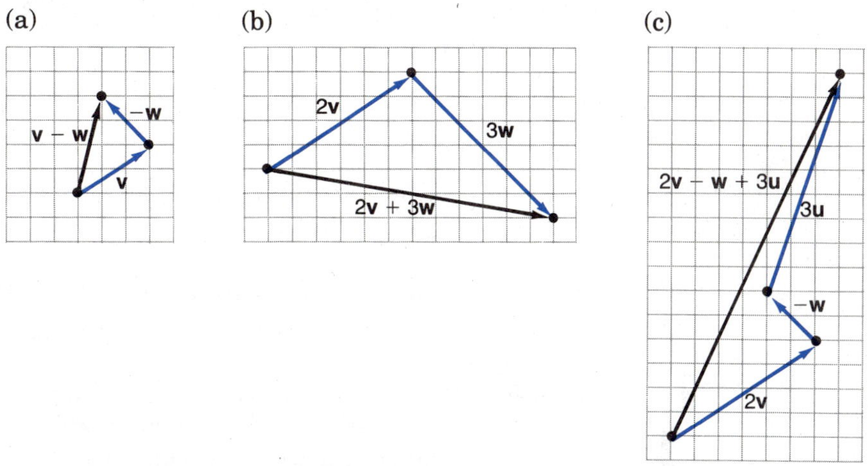

■

Magnitudes of Vectors

If **v** is a vector, we use the symbol $\|\mathbf{v}\|$ to represent the **magnitude** of **v**. Since $\|\mathbf{v}\|$ equals the length of a directed line segment, it follows that $\|\mathbf{v}\|$ has the following properties:

Theorem

Properties of $\|\mathbf{v}\|$

If **v** is a vector and if α is a scalar, then

(a) $\|\mathbf{v}\| \geq 0$ (b) $\|\mathbf{v}\| = 0$ if and only if $\mathbf{v} = \mathbf{0}$

(c) $\|-\mathbf{v}\| = \|\mathbf{v}\|$ (d) $\|\alpha\mathbf{v}\| = |\alpha|\|\mathbf{v}\|$ ■

Property (a) is a consequence of the fact that distance is a nonnegative number. Property (b) follows, because the length of the directed line segment $\overrightarrow{PQ}$ is positive unless P and Q are the same point, in which case the length is 0. Property (c) follows, because the length of the line segment $\overline{PQ}$ equals the length of the line segment $\overline{QP}$. Property (d) is a direct consequence of the definition of a scalar product.

Unit Vector

A vector **v** for which $\|\mathbf{v}\| = 1$ is called a **unit vector.**

To compute the magnitude and direction of a vector, we need an algebraic way of representing vectors.

Representing Vectors in the Plane

We use a rectangular coordinate system to represent vectors in the plane. Let **i** denote a unit vector whose direction is along the positive x-axis; let **j** denote a unit vector whose direction is along the positive y-axis. If **v** is a vector with initial point at the origin O and terminal point at $P = (a, b)$, then we can represent **v** in terms of the vectors **i** and **j** as

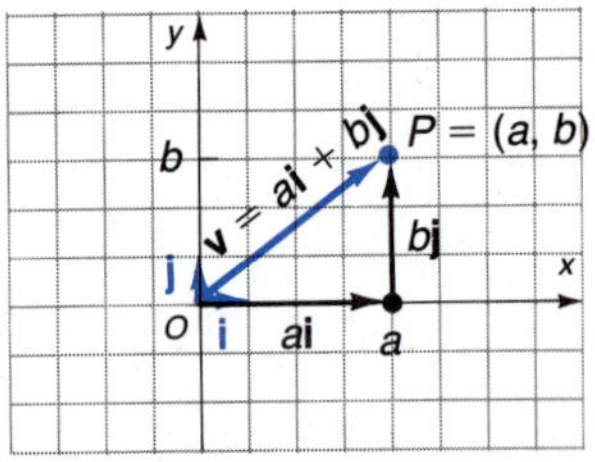

Figure 16

$$\mathbf{v} = a\mathbf{i} + b\mathbf{j}$$

See Figure 16. The scalars a and b are called the **components** of the vector $\mathbf{v} = a\mathbf{i} + b\mathbf{j}$, a being the component in the direction $\mathbf{i}$ and b being the component in the direction $\mathbf{j}$.

A vector whose initial point is at the origin is called a **position vector**. The next result states that any vector whose initial point is not at the origin is equal to a unique position vector.

Theorem Suppose $\mathbf{v}$ is a vector with initial point $P_1 = (x_1, y_1)$, not necessarily the origin, and terminal point $P_2 = (x_2, y_2)$, so that $\mathbf{v} = \overrightarrow{P_1P_2}$. Then $\mathbf{v}$ is equal to the position vector whose components are

$$\mathbf{v} = (x_2 - x_1)\mathbf{i} + (y_2 - y_1)\mathbf{j}$$

■

To see why this is true, look at Figure 17. Triangle OPA and triangle P_1P_2Q are congruent. (Do you see why?) The line segments have the same magnitude, so $d(O, P) = d(P_1, P_2)$; and they have the same direction, so $\angle POA = \angle P_2P_1Q$. Since the triangles are right triangles, we have Angle–Side–Angle. Thus, it follows that corresponding sides are equal. As a result, $x_2 - x_1 = a$ and $y_2 - y_1 = b$, so that $\mathbf{v}$ may be written as

$$\mathbf{v} = a\mathbf{i} + b\mathbf{j} = (x_2 - x_1)\mathbf{i} + (y_2 - y_1)\mathbf{j} \tag{1}$$

Figure 17
$\mathbf{v} = a\mathbf{i} + b\mathbf{j} = (x_2 - x_1)\mathbf{i} + (y_2 - y_1)\mathbf{j}$

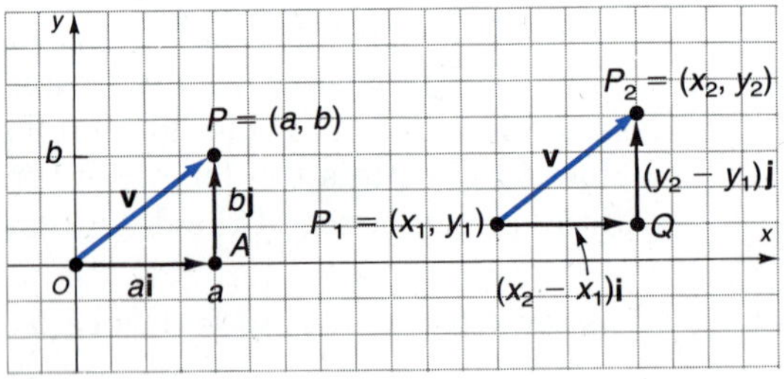

Example 2 Find the position vector of the vector $\mathbf{v} = \overrightarrow{P_1P_2}$ if $P_1 = (-1, 2)$ and $P_2 = (4, 6)$.

Solution By equation (1), the position vector equal to $\mathbf{v}$ is

$$\mathbf{v} = [4 - (-1)]\mathbf{i} + (6 - 2)\mathbf{j} = 5\mathbf{i} + 4\mathbf{j}$$

See Figure 18.

Figure 18

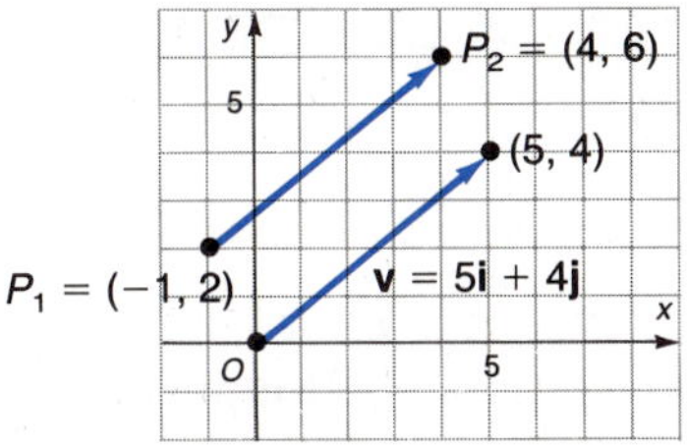

■

Two position vectors **v** and **w** are equal if and only if the terminal point of **v** is the same as the terminal point of **w**. This leads to the following result:

Theorem

Equality of Vectors

Two vectors **v** and **w** are equal if and only if their corresponding components are equal. That is:

If $\mathbf{v} = a_1\mathbf{i} + b_1\mathbf{j}$ and $\mathbf{w} = a_2\mathbf{i} + b_2\mathbf{j}$,
then $\mathbf{v} = \mathbf{w}$ if and only if $a_1 = a_2$ and $b_1 = b_2$.

■

Because of the above result, we can replace any vector (directed line segment) by a unique position vector, and vice versa. This flexibility is one of the main reasons for the wide use of vectors. Unless otherwise specified, from now on, the term *vector* will mean the unique position vector equal to it.

Next, we define addition, subtraction, scalar product, and magnitude in terms of the components of a vector.

Let $\mathbf{v} = a_1\mathbf{i} + b_1\mathbf{j}$ and $\mathbf{w} = a_2\mathbf{i} + b_2\mathbf{j}$ be two vectors, and let α be a scalar. Then:

$$\mathbf{v} + \mathbf{w} = (a_1 + a_2)\mathbf{i} + (b_1 + b_2)\mathbf{j} \tag{2}$$

$$\mathbf{v} - \mathbf{w} = (a_1 - a_2)\mathbf{i} + (b_1 - b_2)\mathbf{j} \tag{3}$$

$$\alpha\mathbf{v} = (\alpha a_1)\mathbf{i} + (\alpha b_1)\mathbf{j} \tag{4}$$

$$\|\mathbf{v}\| = \sqrt{a_1^2 + b_1^2} \qquad (5)$$

These definitions are compatible with the geometric ones given earlier in this section. See Figure 19.

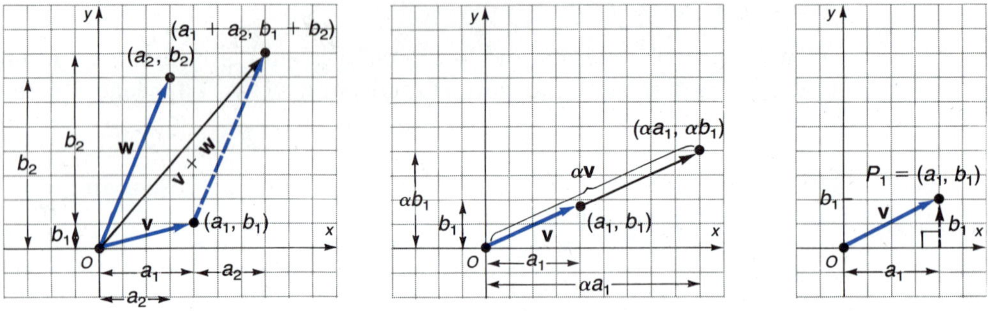

(a) Illustration of property (2)

(b) Illustration of property (4)

(c) Illustration of property (5): $\|\mathbf{v}\|$ = Distance from O to P_1 $= \sqrt{a_1^2 + b_1^2}$

Figure 19

Thus, to add two vectors, simply add corresponding components. To subtract two vectors, subtract corresponding components.

Example 3 If $\mathbf{v} = 2\mathbf{i} + 3\mathbf{j}$ and $\mathbf{w} = 3\mathbf{i} - 4\mathbf{j}$, find:

(a) $\mathbf{v} + \mathbf{w}$ (b) $\mathbf{v} - \mathbf{w}$

Solution (a) $\mathbf{v} + \mathbf{w} = (2\mathbf{i} + 3\mathbf{j}) + (3\mathbf{i} - 4\mathbf{j}) = (2 + 3)\mathbf{i} + (3 - 4)\mathbf{j}$
$= 5\mathbf{i} - \mathbf{j}$

(b) $\mathbf{v} - \mathbf{w} = (2\mathbf{i} + 3\mathbf{j}) - (3\mathbf{i} - 4\mathbf{j}) = (2 - 3)\mathbf{i} + [3 - (-4)]\mathbf{j}$
$= -\mathbf{i} + 7\mathbf{j}$ ■

Example 4 If $\mathbf{v} = 2\mathbf{i} + 3\mathbf{j}$ and $\mathbf{w} = 3\mathbf{i} - 4\mathbf{j}$, find:

(a) $3\mathbf{v}$ (b) $2\mathbf{v} - 3\mathbf{w}$ (c) $\|\mathbf{v}\|$

Solution (a) $3\mathbf{v} = 3(2\mathbf{i} + 3\mathbf{j}) = 6\mathbf{i} + 9\mathbf{j}$

(b) $2\mathbf{v} - 3\mathbf{w} = 2(2\mathbf{i} + 3\mathbf{j}) - 3(3\mathbf{i} - 4\mathbf{j}) = 4\mathbf{i} + 6\mathbf{j} - 9\mathbf{i} + 12\mathbf{j}$
$= -5\mathbf{i} + 18\mathbf{j}$

(c) $\|\mathbf{v}\| = \|2\mathbf{i} + 3\mathbf{j}\| = \sqrt{2^2 + 3^2} = \sqrt{13}$ ■

Recall that a unit vector $\mathbf{v}$ is one for which $\|\mathbf{v}\| = 1$. In many applications, it is useful to be able to find a unit vector that has the same direction as a given vector $\mathbf{v}$.

Theorem

Unit Vector in Direction of v

For any nonzero vector $\mathbf{v}$, the vector

$$\mathbf{u} = \frac{\mathbf{v}}{\|\mathbf{v}\|}$$

is a unit vector that has the same direction as $\mathbf{v}$.

Proof Let $\mathbf{v} = a\mathbf{i} + b\mathbf{j}$. Then $\|\mathbf{v}\| = \sqrt{a^2 + b^2}$ and

$$\mathbf{u} = \frac{\mathbf{v}}{\|\mathbf{v}\|} = \frac{a\mathbf{i} + b\mathbf{j}}{\sqrt{a^2 + b^2}} = \frac{a}{\sqrt{a^2 + b^2}}\mathbf{i} + \frac{b}{\sqrt{a^2 + b^2}}\mathbf{j}$$

The vector $\mathbf{u}$ is in the same direction as $\mathbf{v}$, since $\|\mathbf{v}\| > 0$, and

$$\|\mathbf{u}\| = \sqrt{\frac{a^2}{a^2 + b^2} + \frac{b^2}{a^2 + b^2}} = \sqrt{\frac{a^2 + b^2}{a^2 + b^2}} = 1$$

Thus, $\mathbf{u}$ is a unit vector in the direction of $\mathbf{v}$. ■

Example 5 Find a unit vector in the same direction as $\mathbf{v} = 4\mathbf{i} - 3\mathbf{j}$.

Solution We find $\|\mathbf{v}\|$ first:

$$\|\mathbf{v}\| = \|4\mathbf{i} - 3\mathbf{j}\| = \sqrt{16 + 9} = 5$$

Now, we multiply $\mathbf{v}$ by the scalar $1/\|\mathbf{v}\| = \frac{1}{5}$. The result is

$$\frac{\mathbf{v}}{\|\mathbf{v}\|} = \frac{4\mathbf{i} - 3\mathbf{j}}{5} = \frac{4}{5}\mathbf{i} - \frac{3}{5}\mathbf{j}$$

Check: This vector is, in fact, a unit vector because $\left(\frac{4}{5}\right)^2 + \left(-\frac{3}{5}\right)^2 = \frac{16}{25} + \frac{9}{25} = \frac{25}{25} = 1$. ■

Applications

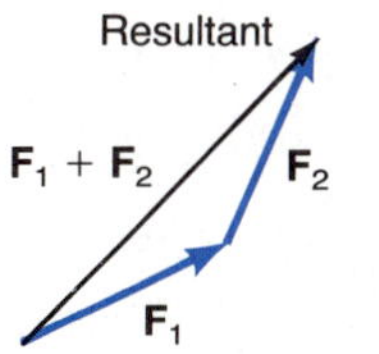

Figure 20

Forces provide an example of physical quantities that may be conveniently represented by vectors; two forces "combine" the way vectors "add." How do we know this? Laboratory experiments bear it out. Thus, if $\mathbf{F}_1$ and $\mathbf{F}_2$ are two forces simultaneously acting on an object, the vector sum $\mathbf{F}_1 + \mathbf{F}_2$ is the force that produces the same effect on the object as that obtained when the forces $\mathbf{F}_1$ and $\mathbf{F}_2$ act on the object. The force $\mathbf{F}_1 + \mathbf{F}_2$ is sometimes called the **resultant** of $\mathbf{F}_1$ and $\mathbf{F}_2$. See Figure 20.

Two important applications of the resultant of two vectors occur with aircraft flying in the presence of a wind and with boats cruising across a river with a current. For example, consider the velocity of wind acting on the velocity of an airplane (see Figure 21, at the top of the next page). Suppose $\mathbf{w}$ is a vector describing the velocity of the wind; that is, $\mathbf{w}$ represents the direction and speed of the wind. If $\mathbf{v}$ is the velocity of the airplane in the absence of wind (called its **velocity relative to the air**), then $\mathbf{v} + \mathbf{w}$ is the vector equal to the actual velocity of the airplane (called its **velocity relative to the ground**).

Figure 21

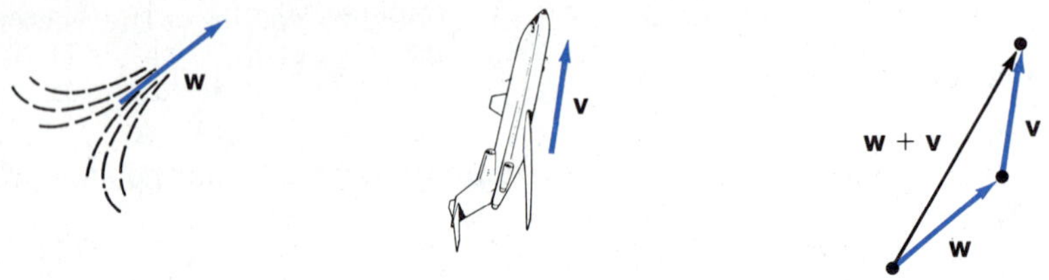

(a) Velocity **w** of wind relative to the ground (b) Velocity **v** of airplane relative to air (c) Resultant **w** + **v** equals velocity of airplane relative to the ground

Our next example illustrates this use of vectors in navigation.

Example 6 A Boeing 737 aircraft maintains a constant airspeed of 500 miles per hour in the direction due south. The velocity of the jet stream is 80 miles per hour in a northeasterly direction.

(a) Find a unit vector having northeast as direction.

(b) Find a vector 80 units in magnitude having the same direction as the unit vector found in part (a).

C (c) Find the actual speed of the aircraft relative to the ground.

Solution We set up a coordinate system in which north (N) is along the positive y-axis. See Figure 22. Let

$$\mathbf{v}_a = \text{Velocity of aircraft in the air} = -500\mathbf{j}$$
$$\mathbf{v}_g = \text{Velocity of aircraft relative to ground}$$
$$\mathbf{v}_w = \text{Velocity of jet stream}$$

Figure 22

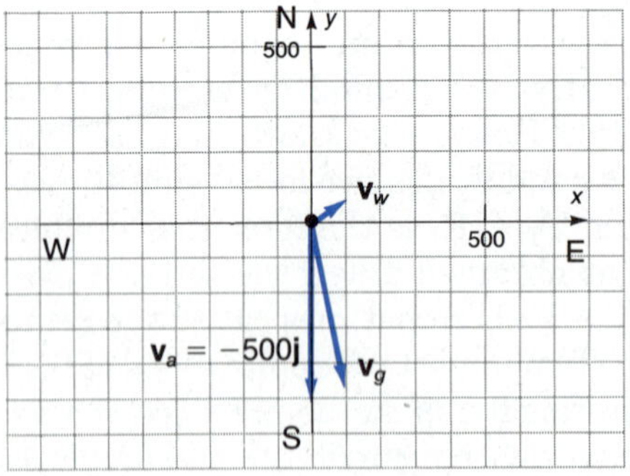

(a) A vector having northeast as direction is $\mathbf{i} + \mathbf{j}$. The unit vector in this direction is

$$\frac{\mathbf{i}+\mathbf{j}}{\|\mathbf{i}+\mathbf{j}\|} = \frac{\mathbf{i}+\mathbf{j}}{\sqrt{1+1}} = \frac{1}{\sqrt{2}}(\mathbf{i}+\mathbf{j})$$

(b) The velocity $\mathbf{v}_w$ of the jet stream is a vector with magnitude 80 in the direction of $(1/\sqrt{2})(\mathbf{i} + \mathbf{j})$. Thus,

$$\mathbf{v}_w = \frac{80}{\sqrt{2}}(\mathbf{i} + \mathbf{j}) = 40\sqrt{2}(\mathbf{i} + \mathbf{j})$$

(c) The velocity $\mathbf{v}_g$ of the aircraft relative to the ground is the resultant of the vectors $\mathbf{v}_a$ and $\mathbf{v}_w$. Thus,

$$\begin{aligned}\mathbf{v}_g = \mathbf{v}_a + \mathbf{v}_w &= -500\mathbf{j} + 40\sqrt{2}(\mathbf{i} + \mathbf{j}) \\ &= 40\sqrt{2}\,\mathbf{i} + (40\sqrt{2} - 500)\mathbf{j}\end{aligned}$$

The actual speed of the aircraft is

$$\|\mathbf{v}_g\| = \sqrt{(40\sqrt{2})^2 + (40\sqrt{2} - 500)^2} \approx 447 \text{ miles per hour} \quad ■$$

Historical Comment

■ The history of vectors is surprisingly complicated for such a natural concept. In the *xy*-plane, complex numbers do a good job of imitating vectors. About 1840, mathematicians became interested in finding a system that would do for three dimensions what the complex numbers do for two dimensions. Hermann Grassmann (1809–1877), in Germany, and William Rowan Hamilton (1805–1865), in Ireland, both attempted to find solutions.

Hamilton's system was the *quaternions*, which are best thought of as a real number plus a vector, and do for four dimensions what complex numbers do for two dimensions. In this system the order of multiplication matters; that is, $\mathbf{ab} \neq \mathbf{ba}$. Hamilton spent the rest of his life working out quaternion theory and trying to get it accepted in applied mathematics, but he encountered fierce resistance due to the complicated nature of quaternion multiplication. In the work with quaternions, two products of vectors emerged, the scalar (or dot) and the vector (or cross) products.

Grassmann fared even worse than Hamilton; if people did not like Hamilton's work, at least they understood it. Grassmann's abstract style, though easily read today, was almost impenetrable during the previous century, and only a few of his ideas were appreciated. Among those few were the same scalar and vector products that Hamilton had found.

About 1880, the American physicist Josiah Willard Gibbs (1839–1903) worked out an algebra involving only the simplest concepts—the vectors and the two products. He then added some calculus, and the resulting system was simple, flexible, and well-adapted to expressing a large number of physical laws. This system remains in use essentially unchanged. Hamilton's and Grassmann's more extensive systems each gave birth to much interesting mathematics, but little of this mathematics is seen at elementary levels. ■

EXERCISE 14.4

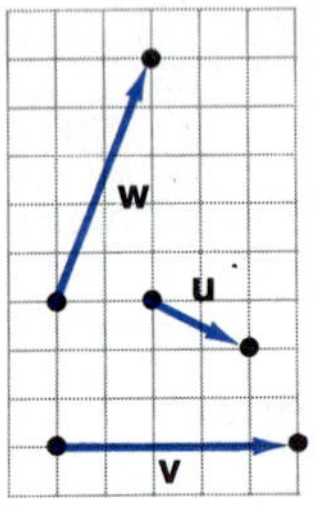

In Problems 1–8, use the vectors in the figure in the margin to graph each expression.

1. $\mathbf{v} + \mathbf{w}$
2. $\mathbf{u} + \mathbf{v}$
3. $3\mathbf{v}$
4. $4\mathbf{w}$
5. $\mathbf{v} - \mathbf{w}$
6. $\mathbf{u} - \mathbf{v}$
7. $3\mathbf{v} + \mathbf{u} - 2\mathbf{w}$
8. $2\mathbf{u} - 3\mathbf{v} + \mathbf{w}$

In Problems 9–16, use the figure below to find each vector.

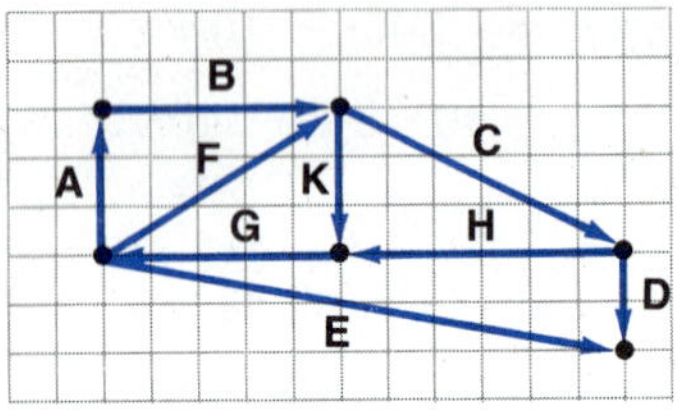

9. $\mathbf{x}$, if $\mathbf{x} + \mathbf{B} = \mathbf{F}$
10. $\mathbf{x}$, if $\mathbf{x} + \mathbf{D} = \mathbf{E}$
11. $\mathbf{C}$ in terms of $\mathbf{E}$, $\mathbf{D}$, and $\mathbf{F}$
12. $\mathbf{G}$ in terms of $\mathbf{C}$, $\mathbf{D}$, $\mathbf{E}$, and $\mathbf{K}$
13. $\mathbf{E}$ in terms of $\mathbf{G}$, $\mathbf{H}$, and $\mathbf{D}$
14. $\mathbf{E}$ in terms of $\mathbf{A}$, $\mathbf{B}$, $\mathbf{C}$, and $\mathbf{D}$
15. $\mathbf{x}$, if $\mathbf{x} = \mathbf{A} + \mathbf{B} + \mathbf{K} + \mathbf{G}$
16. $\mathbf{x}$, if $\mathbf{x} = \mathbf{A} + \mathbf{B} + \mathbf{C} + \mathbf{H} + \mathbf{G}$
17. If $\|\mathbf{v}\| = 4$, what is $\|3\mathbf{v}\|$?
18. If $\|\mathbf{v}\| = 2$, what is $\|-4\mathbf{v}\|$?

In Problems 19–26, the vector $\mathbf{v}$ *has initial point P and terminal point Q. Write* $\mathbf{v}$ *in the form* $a\mathbf{i} + b\mathbf{j}$*; that is, find the position vector* $\mathbf{v} = \overrightarrow{PQ}$.

19. $P = (0, 0)$; $Q = (3, 4)$
20. $P = (0, 0)$; $Q = (-3, -5)$
21. $P = (3, 2)$; $Q = (5, 6)$
22. $P = (-3, 2)$; $Q = (6, 5)$
23. $P = (-2, -1)$; $Q = (6, -2)$
24. $P = (-1, 4)$; $Q = (6, 2)$
25. $P = (1, 0)$; $Q = (0, 1)$
26. $P = (1, 1)$; $Q = (2, 2)$

In Problems 27–32, find $\|\mathbf{v}\|$.

27. $\mathbf{v} = 3\mathbf{i} - 4\mathbf{j}$
28. $\mathbf{v} = -5\mathbf{i} + 12\mathbf{j}$
29. $\mathbf{v} = \mathbf{i} - \mathbf{j}$
30. $\mathbf{v} = -\mathbf{i} - \mathbf{j}$
31. $\mathbf{v} = -2\mathbf{i} + 3\mathbf{j}$
32. $\mathbf{v} = 6\mathbf{i} + 2\mathbf{j}$

In Problems 33–38, find each quantity if $\mathbf{v} = 3\mathbf{i} - 5\mathbf{j}$ *and* $\mathbf{w} = -2\mathbf{i} + 3\mathbf{j}$.

33. $2\mathbf{v} + 3\mathbf{w}$
34. $3\mathbf{v} - 2\mathbf{w}$
35. $\|\mathbf{v} - \mathbf{w}\|$
36. $\|\mathbf{v} + \mathbf{w}\|$
37. $\|\mathbf{v}\| - \|\mathbf{w}\|$
38. $\|\mathbf{v}\| + \|\mathbf{w}\|$

In Problems 39–44, find the unit vector having the same direction as $\mathbf{v}$.

39. $\mathbf{v} = 5\mathbf{i}$
40. $\mathbf{v} = -3\mathbf{j}$
41. $\mathbf{v} = 3\mathbf{i} - 4\mathbf{j}$
42. $\mathbf{v} = -5\mathbf{i} + 12\mathbf{j}$
43. $\mathbf{v} = \mathbf{i} - \mathbf{j}$
44. $\mathbf{v} = 2\mathbf{i} - \mathbf{j}$
45. Find a vector $\mathbf{v}$ whose magnitude is 4 and whose component in the $\mathbf{i}$ direction is twice the component in the $\mathbf{j}$ direction.
46. Find a vector $\mathbf{v}$ whose magnitude is 3 and whose component in the $\mathbf{i}$ direction is equal to its component in the $\mathbf{j}$ direction.
47. If $\mathbf{v} = 2\mathbf{i} - \mathbf{j}$ and $\mathbf{w} = x\mathbf{i} + 3\mathbf{j}$, find all numbers x for which $\|\mathbf{v} + \mathbf{w}\| = 5$.
48. If $P = (-3, 1)$ and $Q = (x, 4)$, find all numbers x such that the vector represented by $\overrightarrow{PQ}$ has length 5.

C **49.** An airplane has an airspeed of 500 kilometers per hour in an easterly direction. If the wind velocity is 60 kilometers per hour in a northwesterly direction, find the speed of the airplane relative to the ground.

C **50.** After 1 hour in the air, an airplane arrives at a point 200 miles due south of its departure point. If there was a steady wind of 30 miles per hour from the northwest during the entire flight, what was the average airspeed of the airplane?

C **51.** An airplane travels in a northwesterly direction at a constant ground speed of 250 miles per hour, due to an easterly wind of 50 miles per hour. How fast would the plane have gone if there had been no wind?

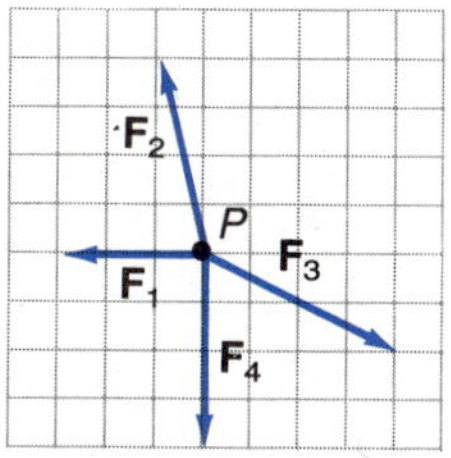

52. A small motorboat in still water maintains a speed of 10 miles per hour. In heading directly across a river (that is, perpendicular to the current) whose current is 4 miles per hour, what will be the true speed of the motorboat?

53. Show on the graph in the margin the force needed to prevent an object at P from moving.

14.5 The Dot Product

The definition for a product of two vectors is somewhat unexpected. However, such a product has meaning in many geometric and physical applications.

Dot Product v · w

If $\mathbf{v} = a_1\mathbf{i} + b_1\mathbf{j}$ and $\mathbf{w} = a_2\mathbf{i} + b_2\mathbf{j}$ are two vectors, the **dot product** $\mathbf{v} \cdot \mathbf{w}$ is defined as

$$\mathbf{v} \cdot \mathbf{w} = a_1a_2 + b_1b_2 \qquad (1)$$

Example 1 If $\mathbf{v} = 2\mathbf{i} - 3\mathbf{j}$ and $\mathbf{w} = 5\mathbf{i} + 3\mathbf{j}$, find:

(a) $\mathbf{v} \cdot \mathbf{w}$ (b) $\mathbf{w} \cdot \mathbf{v}$ (c) $\mathbf{v} \cdot \mathbf{v}$

(d) $\mathbf{w} \cdot \mathbf{w}$ (e) $\|\mathbf{v}\|$ (f) $\|\mathbf{w}\|$

Solution

(a) $\mathbf{v} \cdot \mathbf{w} = 2(5) + (-3)3 = 1$ (b) $\mathbf{w} \cdot \mathbf{v} = 5(2) + 3(-3) = 1$

(c) $\mathbf{v} \cdot \mathbf{v} = 2(2) + (-3)(-3) = 13$ (d) $\mathbf{w} \cdot \mathbf{w} = 5(5) + 3(3) = 34$

(e) $\|\mathbf{v}\| = \sqrt{2^2 + (-3)^2} = \sqrt{13}$ (f) $\|\mathbf{w}\| = \sqrt{5^2 + 3^2} = \sqrt{34}$ ■

Since the dot product $\mathbf{v} \cdot \mathbf{w}$ of two vectors $\mathbf{v}$ and $\mathbf{w}$ is a real number (scalar), we sometimes refer to it as the **scalar product**.

Properties

The results obtained in Example 1 suggest some general properties.

Theorem If **u**, **v**, and **w** are vectors, then

Commutative Property	$\mathbf{u} \cdot \mathbf{v} = \mathbf{v} \cdot \mathbf{u}$	(2)
Distributive Property	$\mathbf{u} \cdot (\mathbf{v} + \mathbf{w}) = \mathbf{u} \cdot \mathbf{v} + \mathbf{u} \cdot \mathbf{w}$	(3)
	$\mathbf{v} \cdot \mathbf{v} = \|\mathbf{v}\|^2$	(4)
	$\mathbf{0} \cdot \mathbf{v} = 0$	(5)

Proof We shall prove properties (2) and (4) here, and leave properties (3) and (5) as exercises (see Problems 29 and 30 at the end of this section).

To prove property (2), we let $\mathbf{u} = a_1\mathbf{i} + b_1\mathbf{j}$ and $\mathbf{v} = a_2\mathbf{i} + b_2\mathbf{j}$. Then

$$\mathbf{u} \cdot \mathbf{v} = a_1a_2 + b_1b_2 = a_2a_1 + b_2b_1 = \mathbf{v} \cdot \mathbf{u}$$

To prove property (4), we let $\mathbf{v} = a\mathbf{i} + b\mathbf{j}$. Then

$$\mathbf{v} \cdot \mathbf{v} = a^2 + b^2 = \|\mathbf{v}\|^2$$ ■

One use of the dot product is to calculate the angle between two vectors.

Angle between Vectors

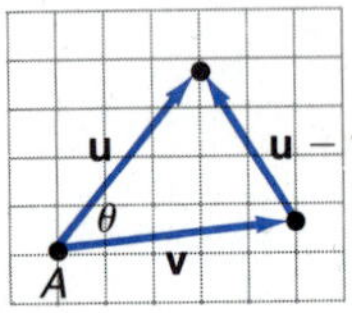

Figure 23

Let **u** and **v** be two vectors with the same initial point A. Then the vectors **u**, **v**, and $\mathbf{u} - \mathbf{v}$ form a triangle. The angle θ at vertex A of the triangle is the **angle between the vectors u and v**. See Figure 23. We wish to find a formula for calculating the angle θ.

The sides of the triangle are of lengths $\|\mathbf{v}\|$, $\|\mathbf{u}\|$, and $\|\mathbf{u} - \mathbf{v}\|$; and θ is the included angle between the sides of length $\|\mathbf{v}\|$ and $\|\mathbf{u}\|$. The Law of Cosines (Section 10.2) can be used to find the cosine of the included angle:

$$\|\mathbf{u} - \mathbf{v}\|^2 = \|\mathbf{u}\|^2 + \|\mathbf{v}\|^2 - 2\|\mathbf{u}\|\,\|\mathbf{v}\| \cos\theta$$

Now we use property (4) to rewrite this equation in terms of dot products:

$$(\mathbf{u} - \mathbf{v}) \cdot (\mathbf{u} - \mathbf{v}) = \mathbf{u} \cdot \mathbf{u} + \mathbf{v} \cdot \mathbf{v} - 2\|\mathbf{u}\|\,\|\mathbf{v}\| \cos\theta \qquad (6)$$

and then apply the distributive property (3) twice on the left side to obtain

$$\begin{aligned}(\mathbf{u}-\mathbf{v})\cdot(\mathbf{u}-\mathbf{v}) &= \mathbf{u}\cdot(\mathbf{u}-\mathbf{v}) - \mathbf{v}\cdot(\mathbf{u}-\mathbf{v}) \\ &= \mathbf{u}\cdot\mathbf{u} - \mathbf{u}\cdot\mathbf{v} - \mathbf{v}\cdot\mathbf{u} + \mathbf{v}\cdot\mathbf{v} \qquad (7)\\ &\underset{\uparrow}{=} \mathbf{u}\cdot\mathbf{u} + \mathbf{v}\cdot\mathbf{v} - 2\mathbf{u}\cdot\mathbf{v}\end{aligned}$$

Property (2)

Combining equations (6) and (7), we have

$$\mathbf{u}\cdot\mathbf{u} + \mathbf{v}\cdot\mathbf{v} - 2\mathbf{u}\cdot\mathbf{v} = \mathbf{u}\cdot\mathbf{u} + \mathbf{v}\cdot\mathbf{v} - 2\|\mathbf{u}\|\,\|\mathbf{v}\|\cos\theta$$
$$\mathbf{u}\cdot\mathbf{v} = \|\mathbf{u}\|\,\|\mathbf{v}\|\cos\theta$$

Thus, we have proved the following result:

Theorem

Angle between Vectors

If **u** and **v** are two nonzero vectors, the angle θ, $0 \le \theta \le \pi$, between **u** and **v** is determined by the formula

$$\cos\theta = \frac{\mathbf{u}\cdot\mathbf{v}}{\|\mathbf{u}\|\,\|\mathbf{v}\|} \qquad (8)$$

■

Example 2 Find the cosine of the angle θ between the vectors

$$\mathbf{u} = 4\mathbf{i} - 3\mathbf{j} \qquad \text{and} \qquad \mathbf{v} = 5\mathbf{i} + 12\mathbf{j}$$

Solution We compute the quantities $\mathbf{u}\cdot\mathbf{v}$, $\|\mathbf{u}\|$, and $\|\mathbf{v}\|$:

$$\mathbf{u}\cdot\mathbf{v} = 4(5) + (-3)(12) = -16$$
$$\|\mathbf{u}\| = \sqrt{4^2 + (-3)^2} = 5$$
$$\|\mathbf{v}\| = \sqrt{5^2 + 12^2} = 13$$

By formula (8), if θ is the angle between **u** and **v**, then

$$\cos\theta = \frac{\mathbf{u}\cdot\mathbf{v}}{\|\mathbf{u}\|\,\|\mathbf{v}\|} = \frac{-16}{5(13)} = -\frac{16}{65}$$

Using a calculator, we find that $\theta \approx 104°$. See Figure 24. ■

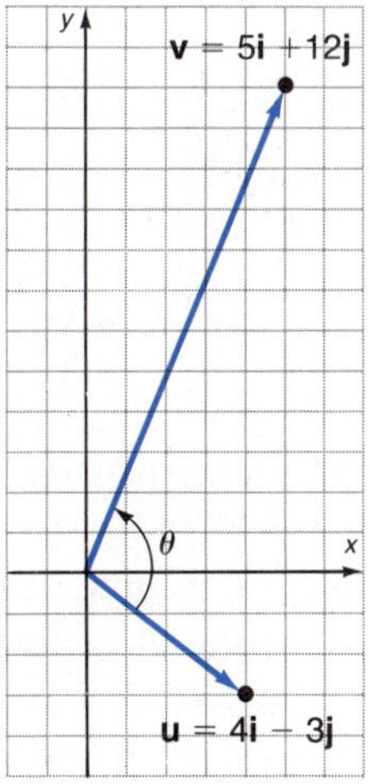

Figure 24

C **Example 3** A Boeing 737 aircraft maintains a constant airspeed of 500 miles per hour in the direction due south. The velocity of the jet stream is 80 miles per hour in a northeasterly direction. Find the actual direction of the aircraft relative to the ground.

Solution This is the same information given in Example 6 of Section 14.4. We repeat the figure from that example as Figure 25.

Figure 25

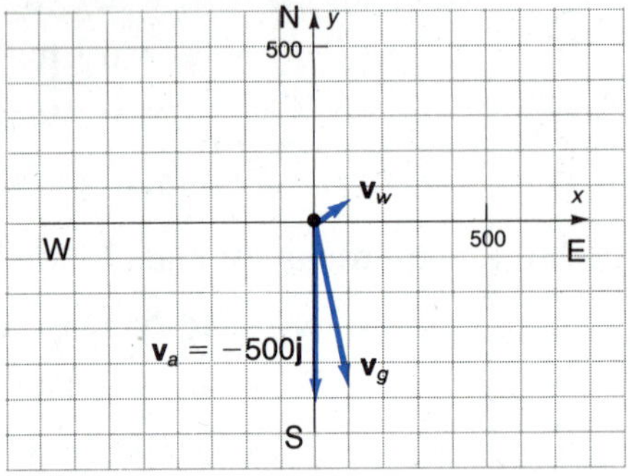

We have already calculated the velocity of the aircraft relative to the ground to be

$$\mathbf{v}_g = 40\sqrt{2}\,\mathbf{i} + (40\sqrt{2} - 500)\mathbf{j}$$

The angle θ between $\mathbf{v}_g$ and the vector $\mathbf{v}_a = -500\mathbf{j}$ (the velocity of the aircraft in the air) is determined by the equation

$$\cos\theta = \frac{\mathbf{v}_g \cdot \mathbf{v}_a}{\|\mathbf{v}_g\|\,\|\mathbf{v}_a\|} = \frac{(40\sqrt{2} - 500)(-500)}{(447)(500)} \approx 0.9920$$

$$\theta \approx 7.2°$$

The direction of the aircraft relative to the ground is about 7.2° east of south. ■

Parallel and Orthogonal Vectors

Two vectors $\mathbf{v}$ and $\mathbf{w}$ are said to be **parallel** if there is a nonzero scalar α so that $\mathbf{v} = \alpha\mathbf{w}$. In this case, the angle θ between $\mathbf{v}$ and $\mathbf{w}$ is 0 or π.

Example 4 The vectors $\mathbf{v} = 3\mathbf{i} - \mathbf{j}$ and $\mathbf{w} = 6\mathbf{i} - 2\mathbf{j}$ are parallel, since $\mathbf{v} = \frac{1}{2}\mathbf{w}$. Furthermore, since

$$\cos\theta = \frac{\mathbf{v} \cdot \mathbf{w}}{\|\mathbf{v}\|\,\|\mathbf{w}\|} = \frac{18 + 2}{\sqrt{10}\sqrt{40}} = \frac{20}{\sqrt{400}} = 1$$

the angle θ between $\mathbf{v}$ and $\mathbf{w}$ is 0. ■

If the angle θ between two nonzero vectors $\mathbf{v}$ and $\mathbf{w}$ is $\pi/2$, the vectors $\mathbf{v}$ and $\mathbf{w}$ are called **orthogonal**.*

*_Orthogonal_, _perpendicular_, and _normal_ are all terms that mean "meet at a right angle." It is customary to refer to two vectors as being _orthogonal_, two lines as being _perpendicular_, and a line and a plane or a vector and a plane as being _normal_.

It follows from formula (8) that if $\mathbf{v}$ and $\mathbf{w}$ are orthogonal, then $\mathbf{v} \cdot \mathbf{w} = 0$, since $\cos(\pi/2) = 0$.

On the other hand, if $\mathbf{v} \cdot \mathbf{w} = 0$, then either $\mathbf{v} = \mathbf{0}$ or $\mathbf{w} = \mathbf{0}$ or $\cos\theta = 0$. In the latter case, $\theta = \pi/2$ and $\mathbf{v}$ and $\mathbf{w}$ are orthogonal. If $\mathbf{v}$ or $\mathbf{w}$ is the zero vector, then since the zero vector has no specific direction, we adopt the convention that the zero vector is orthogonal to every vector.

Theorem Two vectors $\mathbf{v}$ and $\mathbf{w}$ are orthogonal if and only if

$$\mathbf{v} \cdot \mathbf{w} = 0$$

■

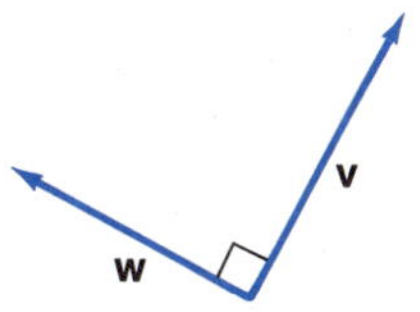

Figure 26

See Figure 26.

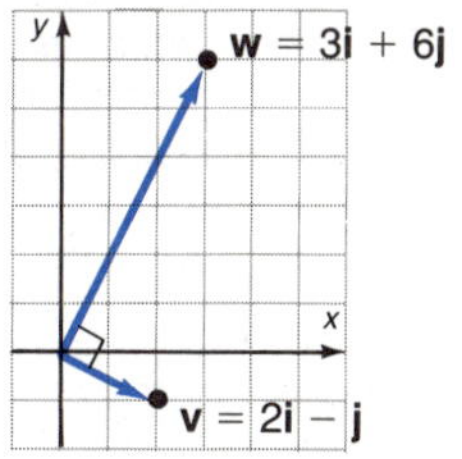

Figure 27

Example 5 The vectors

$$\mathbf{v} = 2\mathbf{i} - \mathbf{j} \qquad \text{and} \qquad \mathbf{w} = 3\mathbf{i} + 6\mathbf{j}$$

are orthogonal, since

$$\mathbf{v} \cdot \mathbf{w} = 6 - 6 = 0$$

See Figure 27. ■

Projection of a Vector onto Another Vector

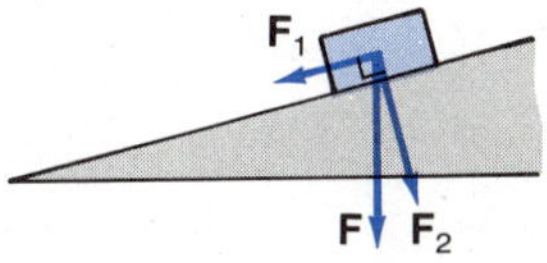

Figure 28

In many physical applications, it is necessary to find "how much" of a vector is applied in a given direction. Look at Figure 28. The force $\mathbf{F}$ due to gravity is pulling straight down (toward the center of the Earth) on the block. To study the effect of gravity on the block, it is necessary to determine how much of $\mathbf{F}$ is actually pushing the block down the incline ($\mathbf{F}_1$) and how much is pressing the block against the incline ($\mathbf{F}_2$), at a right angle to the incline. Knowing the **decomposition** of $\mathbf{F}$ often will allow us to determine when friction is overcome and the block will slide down the incline.

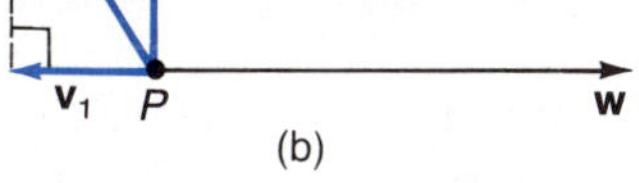

Figure 29

Suppose $\mathbf{v}$ and $\mathbf{w}$ are two nonzero vectors with the same initial point P. We seek to decompose $\mathbf{v}$ into two vectors: $\mathbf{v}_1$, which is parallel to $\mathbf{w}$; and $\mathbf{v}_2$, which is orthogonal to $\mathbf{w}$. See Figure 29. The vector $\mathbf{v}_1$ is called the **vector projection of v onto w** and is denoted by $\text{proj}_{\mathbf{w}}\,\mathbf{v}$.

The vector $\mathbf{v}_1$ is obtained as follows: From the terminal point of $\mathbf{v}$ drop a perpendicular to the line containing $\mathbf{w}$. The vector $\mathbf{v}_1$ is the vector from P to the foot of this perpendicular. The vector $\mathbf{v}_2$ is given by $\mathbf{v}_2 = \mathbf{v} - \mathbf{v}_1$. Note that $\mathbf{v} = \mathbf{v}_1 + \mathbf{v}_2$, $\mathbf{v}_1$ is parallel to $\mathbf{w}$, and $\mathbf{v}_2$ is orthogonal to $\mathbf{w}$. This is the decomposition of $\mathbf{v}$ we wanted.

Now we seek a formula for $\mathbf{v}_1$ that is based on a knowledge of the vectors $\mathbf{v}$ and $\mathbf{w}$. Since $\mathbf{v} = \mathbf{v}_1 + \mathbf{v}_2$, we have

$$\mathbf{v} \cdot \mathbf{w} = (\mathbf{v}_1 + \mathbf{v}_2) \cdot \mathbf{w} = \mathbf{v}_1 \cdot \mathbf{w} + \mathbf{v}_2 \cdot \mathbf{w} \qquad (9)$$

Since $\mathbf{v}_2$ is orthogonal to $\mathbf{w}$, we have $\mathbf{v}_2 \cdot \mathbf{w} = 0$. Since $\mathbf{v}_1$ is parallel to $\mathbf{w}$, we have $\mathbf{v}_1 = \alpha\mathbf{w}$ for some scalar α. Thus, equation (9) can be written as

$$\mathbf{v} \cdot \mathbf{w} = \alpha\mathbf{w} \cdot \mathbf{w} = \alpha\|\mathbf{w}\|^2$$

$$\alpha = \frac{\mathbf{v} \cdot \mathbf{w}}{\|\mathbf{w}\|^2}$$

Thus,

$$\mathbf{v}_1 = \alpha\mathbf{w} = \frac{\mathbf{v} \cdot \mathbf{w}}{\|\mathbf{w}\|^2}\mathbf{w}$$

Theorem If $\mathbf{v}$ and $\mathbf{w}$ are two nonzero vectors, the vector projection of $\mathbf{v}$ onto $\mathbf{w}$ is

$$\text{proj}_{\mathbf{w}}\, \mathbf{v} = \frac{\mathbf{v} \cdot \mathbf{w}}{\|\mathbf{w}\|^2}\mathbf{w}$$

The decomposition of $\mathbf{v}$ into $\mathbf{v}_1$ and $\mathbf{v}_2$, where $\mathbf{v}_1$ is parallel to $\mathbf{w}$ and $\mathbf{v}_2$ is perpendicular to $\mathbf{w}$, is

$$\mathbf{v}_1 = \text{proj}_{\mathbf{w}}\, \mathbf{v} = \frac{\mathbf{v} \cdot \mathbf{w}}{\|\mathbf{w}\|^2}\mathbf{w} \qquad \mathbf{v}_2 = \mathbf{v} - \mathbf{v}_1 \qquad (10)$$

■

Example 6 Find the vector projection of $\mathbf{v} = \mathbf{i} + 3\mathbf{j}$ onto $\mathbf{w} = \mathbf{i} + \mathbf{j}$. Decompose $\mathbf{v}$ into two vectors $\mathbf{v}_1$ and $\mathbf{v}_2$, where $\mathbf{v}_1$ is parallel to $\mathbf{w}$ and $\mathbf{v}_2$ is orthogonal to $\mathbf{w}$.

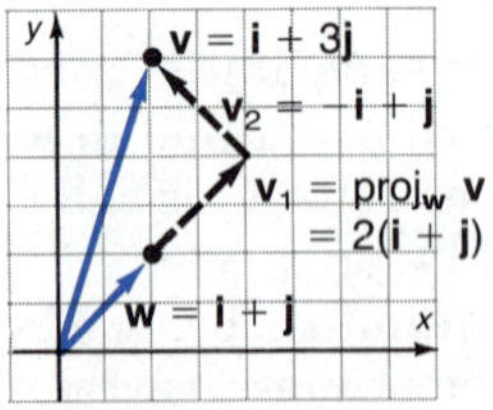

Figure 30

Solution We use formulas (10):

$$\mathbf{v}_1 = \text{proj}_{\mathbf{w}}\mathbf{v} = \frac{\mathbf{v} \cdot \mathbf{w}}{\|\mathbf{w}\|^2}\mathbf{w} = \frac{1 + 3}{(\sqrt{2})^2}\mathbf{w} = 2\mathbf{w} = 2(\mathbf{i} + \mathbf{j})$$

$$\mathbf{v}_2 = \mathbf{v} - \mathbf{v}_1 = (\mathbf{i} + 3\mathbf{j}) - 2(\mathbf{i} + \mathbf{j}) = -\mathbf{i} + \mathbf{j}$$

See Figure 30. ■

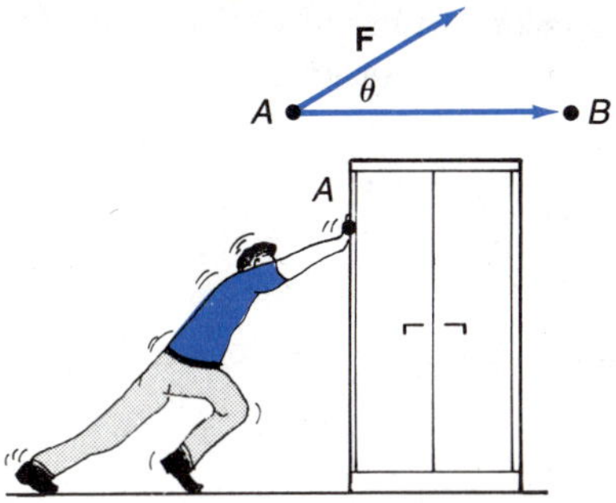

Figure 31

Work Done by a Constant Force

In elementary physics, the **work** W done by a constant force $\mathbf{F}$ in moving an object from a point A to a point B is defined as

$$W = (\text{Magnitude of force})(\text{Distance}) = \|\mathbf{F}\| \, \|\overrightarrow{AB}\|$$

In this definition, it is assumed that the force $\mathbf{F}$ is applied along the line of motion. If the constant force $\mathbf{F}$ is not along the line of motion, but, instead, is at an angle θ to the direction of motion, as illustrated in Figure 31, then the **work W done by F** in moving an object from A to B is defined as

$$W = \mathbf{F} \cdot \overrightarrow{AB} \tag{11}$$

This definition is compatible with the force times distance definition given above, since

$$\begin{aligned} W &= (\text{Amount of force in direction of } \overrightarrow{AB})(\text{Distance}) \\ &= \|\text{proj}_{\overrightarrow{AB}}\, \mathbf{F}\| \, \|\overrightarrow{AB}\| = \frac{\mathbf{F} \cdot \overrightarrow{AB}}{\|\overrightarrow{AB}\|^2} \|\overrightarrow{AB}\| \, \|\overrightarrow{AB}\| = \mathbf{F} \cdot \overrightarrow{AB} \end{aligned}$$

Example 7 Find the work done by a force of 5 pounds acting in the direction $\mathbf{i} + \mathbf{j}$ in moving an object 1 foot from (0, 0) to (1, 0).

Solution First, we must express the force $\mathbf{F}$ as a vector. Since the force has magnitude 5 and direction $\mathbf{i} + \mathbf{j}$, it can be written as

$$\mathbf{F} = \frac{5}{\sqrt{2}}(\mathbf{i} + \mathbf{j})$$

The line of motion of the object is along $\overrightarrow{AB} = \mathbf{i}$. The work W is therefore

$$W = \mathbf{F} \cdot \overrightarrow{AB} = \frac{5}{\sqrt{2}}(\mathbf{i} + \mathbf{j}) \cdot \mathbf{i} = \frac{5}{\sqrt{2}} \text{ foot-pounds}$$ ■

Example 8 Figure 32(a) shows a girl pulling a wagon with a force of 50 pounds. How much work is done in moving the wagon 100 feet if the handle makes an angle of 30° with the ground?

Figure 32

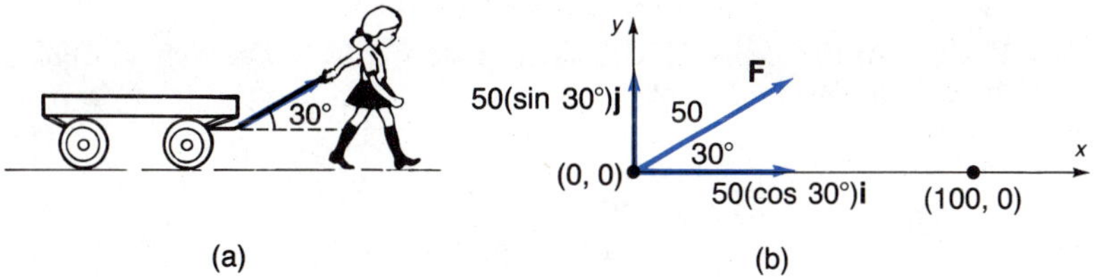

Solution We position the vectors in a coordinate system in such a way that the wagon is moved from (0, 0) to (100, 0). Thus, the motion is along $\overrightarrow{AB} = 100\mathbf{i}$. The force vector **F**, as shown in Figure 32(b), is

$$\mathbf{F} = 50(\cos 30°)\mathbf{i} + 50(\sin 30°)\mathbf{j} = 25\sqrt{3}\,\mathbf{i} + 25\mathbf{j}$$

By formula (11), the work W done is

$$W = \mathbf{F} \cdot \overrightarrow{AB} = (25\sqrt{3}\,\mathbf{i} + 25\mathbf{j}) \cdot 100\mathbf{i} = 2500\sqrt{3} \text{ foot-pounds}$$ ■

Historical Problem

■ 1. We stated in an earlier Historical Comment that complex numbers were used as vectors in the plane before the general notion of vector was clarified. Suppose we make the correspondence

$$\text{Vector} \longleftrightarrow \text{Complex number}$$
$$a\mathbf{i} + b\mathbf{j} \longleftrightarrow a + bi$$
$$c\mathbf{i} + d\mathbf{j} \longleftrightarrow c + di$$

Show that

$$(a\mathbf{i} + b\mathbf{j}) \cdot (c\mathbf{i} + d\mathbf{j}) = \text{Real part}[(\overline{a + bi})(c + di)]$$

This is how the dot product was found originally. The imaginary part is also interesting. It is a determinant (see Section 12.3) and represents the area of the parallelogram whose edges are the vectors. This is close to some of Hermann Grassmann's ideas and is also connected with the scalar triple product of three-dimensional vectors. ■

EXERCISE 14.5

In Problems 1–10, find the dot product $\mathbf{v} \cdot \mathbf{w}$ *and the cosine of the angle between* **v** *and* **w**.

1. $\mathbf{v} = \mathbf{i} - \mathbf{j}, \quad \mathbf{w} = \mathbf{i} + \mathbf{j}$
2. $\mathbf{v} = \mathbf{i} + \mathbf{j}, \quad \mathbf{w} = -\mathbf{i} + \mathbf{j}$
3. $\mathbf{v} = 2\mathbf{i} + \mathbf{j}, \quad \mathbf{w} = \mathbf{i} + 2\mathbf{j}$
4. $\mathbf{v} = 2\mathbf{i} + 2\mathbf{j}, \quad \mathbf{w} = \mathbf{i} + 2\mathbf{j}$
5. $\mathbf{v} = \sqrt{3}\mathbf{i} - \mathbf{j}, \quad \mathbf{w} = \mathbf{i} + \mathbf{j}$
6. $\mathbf{v} = \mathbf{i} + \sqrt{3}\mathbf{j}, \quad \mathbf{w} = \mathbf{i} - \mathbf{j}$
7. $\mathbf{v} = 3\mathbf{i} + 4\mathbf{j}, \quad \mathbf{w} = 4\mathbf{i} + 3\mathbf{j}$
8. $\mathbf{v} = 3\mathbf{i} - 4\mathbf{j}, \quad \mathbf{w} = 4\mathbf{i} - 3\mathbf{j}$
9. $\mathbf{v} = 4\mathbf{i}, \quad \mathbf{w} = \mathbf{j}$
10. $\mathbf{v} = \mathbf{i}, \quad \mathbf{w} = -3\mathbf{j}$
11. Find a such that the angle between $\mathbf{v} = a\mathbf{i} - \mathbf{j}$ and $\mathbf{w} = 2\mathbf{i} + 3\mathbf{j}$ is $\pi/2$.
12. Find b such that the angle between $\mathbf{v} = \mathbf{i} + \mathbf{j}$ and $\mathbf{w} = \mathbf{i} + b\mathbf{j}$ is $\pi/2$.

In Problems 13–18, decompose **v** *into two vectors* $\mathbf{v}_1$ *and* $\mathbf{v}_2$, *where* $\mathbf{v}_1$ *is parallel to* **w** *and* $\mathbf{v}_2$ *is orthogonal to* **w**.

13. $\mathbf{v} = 2\mathbf{i} - 3\mathbf{j}, \quad \mathbf{w} = \mathbf{i} - \mathbf{j}$
14. $\mathbf{v} = -3\mathbf{i} + 2\mathbf{j}, \quad \mathbf{w} = 2\mathbf{i} + \mathbf{j}$
15. $\mathbf{v} = \mathbf{i} - \mathbf{j}, \quad \mathbf{w} = \mathbf{i} + 2\mathbf{j}$
16. $\mathbf{v} = 2\mathbf{i} - \mathbf{j}, \quad \mathbf{w} = \mathbf{i} - 2\mathbf{j}$
17. $\mathbf{v} = 3\mathbf{i} + \mathbf{j}, \quad \mathbf{w} = -2\mathbf{i} - \mathbf{j}$
18. $\mathbf{v} = \mathbf{i} - 3\mathbf{j}, \quad \mathbf{w} = 4\mathbf{i} - \mathbf{j}$

C **19.** A DC-10 jumbo jet maintains an airspeed of 550 miles per hour in a southwesterly direction. The velocity of the jet stream is a constant 80 miles per hour from the west. Find the actual speed and direction of the aircraft.

C **20.** The pilot of an aircraft wishes to head directly east, but is faced with a wind speed of 40 miles per hour from the northwest. If the pilot maintains an airspeed of 250 miles per hour, what compass heading should be maintained? What is the actual speed of the aircraft?

C **21.** A small stream has a constant current of 3 kilometers per hour. At what angle to a boat dock should a motorboat—capable of maintaining a constant speed of 20 kilometers per hour—be headed in order to reach a point directly opposite the dock? If the stream is $\frac{1}{2}$ kilometer wide, how long will it take to cross?

C **22.** Repeat Problem 21 if the current is 5 kilometers per hour.

C **23.** A river is 500 meters wide and has a current of 1 kilometer per hour. If Marsha can swim at a rate of 2 kilometers per hour, at what angle to the shore should she swim if she wishes to cross the river to a point directly opposite? How long will it take to swim across the river?

C **24.** An airplane travels 200 miles due west and then 150 miles 60° north of west. Determine the resultant displacement.

C **25.** Find the work done by a force of 3 pounds acting in the direction $2\mathbf{i} + \mathbf{j}$ in moving an object 2 feet from (0, 0) to (0, 2).

26. Find the work done by a force of 1 pound acting in the direction $2\mathbf{i} + 2\mathbf{j}$ in moving an object 3 feet from (0, 0) to (1, 2).

27. A wagon is pulled horizontally by exerting a force of 20 pounds on the handle at an angle of 30° with the horizontal. How much work is done in moving the wagon 100 feet?

28. Find the acute angle that a constant unit force vector makes with the positive x-axis if the work done by the force in moving a particle from (0, 0) to (4, 0) equals 2.

29. Prove the distributive property, $\mathbf{u} \cdot (\mathbf{v} + \mathbf{w}) = \mathbf{u} \cdot \mathbf{v} + \mathbf{u} \cdot \mathbf{w}$.

30. Prove property (5), $\mathbf{0} \cdot \mathbf{v} = 0$.

31. If $\mathbf{v}$ is a unit vector and the angle between $\mathbf{v}$ and $\mathbf{i}$ is α, show that $\mathbf{v} = \cos \alpha \mathbf{i} + \sin \alpha \mathbf{j}$.

32. Suppose that $\mathbf{v}$ and $\mathbf{w}$ are unit vectors. If the angle between $\mathbf{v}$ and $\mathbf{i}$ is α and if the angle between $\mathbf{w}$ and $\mathbf{i}$ is β, use the idea of the dot product $\mathbf{v} \cdot \mathbf{w}$ to prove that

$$\cos(\alpha - \beta) = \cos \alpha \cos \beta + \sin \alpha \sin \beta$$

33. Show that the projection of $\mathbf{v}$ onto $\mathbf{i}$ is $(\mathbf{v} \cdot \mathbf{i})\mathbf{i}$. In fact, show that we can always write a vector $\mathbf{v}$ as

$$\mathbf{v} = (\mathbf{v} \cdot \mathbf{i})\mathbf{i} + (\mathbf{v} \cdot \mathbf{j})\mathbf{j}$$

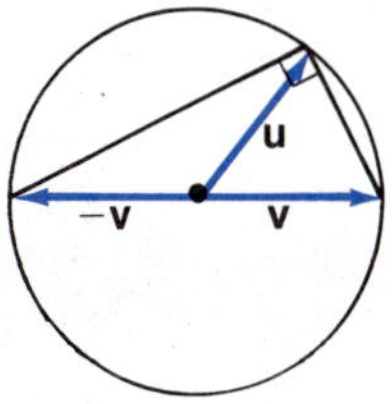

34. (a) If $\mathbf{u}$ and $\mathbf{v}$ have the same magnitude, then show that $\mathbf{u} + \mathbf{v}$ and $\mathbf{u} - \mathbf{v}$ are orthogonal. (b) Use this to prove that an angle inscribed in a semicircle is a right angle (see the figure).

35. Let $\mathbf{v}$ and $\mathbf{w}$ denote two nonzero vectors. Show that the vector $\mathbf{v} - \alpha\mathbf{w}$ is orthogonal to $\mathbf{w}$ if $\alpha = (\mathbf{v} \cdot \mathbf{w})/\|\mathbf{w}\|^2$.

36. Let $\mathbf{v}$ and $\mathbf{w}$ denote two nonzero vectors. Show that the vectors $\|\mathbf{w}\|\mathbf{v} + \|\mathbf{v}\|\mathbf{w}$ and $\|\mathbf{w}\|\mathbf{v} - \|\mathbf{v}\|\mathbf{w}$ are orthogonal.

37. In the definition of work given in this section, what is the work done if $\mathbf{F}$ is orthogonal to $\overrightarrow{AB}$?

38. Prove the **polarization identity:** $\|\mathbf{u} + \mathbf{v}\|^2 - \|\mathbf{u} - \mathbf{v}\|^2 = 4(\mathbf{u} \cdot \mathbf{v})$

CHAPTER REVIEW

THINGS TO KNOW

Matrix	Rectangular array of numbers, called entries
m by n matrix	Matrix with m rows and n columns
Identity matrix I	Square matrix whose diagonal entries are 1's, while all other entries are 0's
Inverse of a matrix	A^{-1} is the inverse of A if $AA^{-1} = A^{-1}A = I$
Nonsingular matrix	A matrix that has an inverse
Linear programming problem	Maximize (or minimize) a linear objective function, $z = Ax + By$, subject to certain conditions, or constraints, expressible as linear inequalities in x and y.
Feasible solution	A point (x, y) that satisfies the constraints of a linear programming problem
Location of solution	If a linear programming problem has a solution, it is located at a vertex of the graph of the feasible solutions. If a linear programming problem has multiple solutions, at least one of them is located at a vertex of the graph of the feasible solutions. In either case, the corresponding value of the objective function is unique.
Vector	Quantity having magnitude and direction; equivalent to a directed line segment $\overrightarrow{PQ}$
Position vector	Vector whose initial point is at the origin
Unit vector	Vector whose magnitude is 1
Dot product	If $\mathbf{v} = a_1\mathbf{i} + b_1\mathbf{j}$ and $\mathbf{w} = a_2\mathbf{i} + b_2\mathbf{j}$, then $\mathbf{v} \cdot \mathbf{w} = a_1a_2 + b_1b_2$.
Angle θ between two nonzero vectors $\mathbf{u}$ and $\mathbf{v}$	$\cos\theta = \dfrac{\mathbf{u} \cdot \mathbf{v}}{\|\mathbf{u}\|\,\|\mathbf{v}\|}$

How To:

Recognize equal matrices

Add and subtract matrices

Multiply matrices

Find the inverse of a nonsingular matrix

Solve a system of equations using the inverse of a matrix

Solve a linear programming problem (see Steps 1–4, page 781)

Write the partial fraction decomposition of a rational expression

Add and subtract vectors

Form scalar multiples of vectors
Find the magnitude of a vector
Solve problems involving vectors
Find the dot product of two vectors
Find the angle between two vectors
Determine whether two vectors are parallel
Determine whether two vectors are orthogonal
Find the vector projection of **v** onto **w**

FILL-IN-THE-BLANK ITEMS

1. A matrix B for which $AB = I_n$, the identity matrix, is called the ________ of A.
2. *True or false* The statement that matrix multiplication is commutative is ________.
3. In the algebra of matrices, the matrix that has properties similar to the number 1 is called the ________ matrix.
4. A linear programming problem requires that a linear expression, called the ________ ________, be maximized or minimized.
5. Each point that satisfies the constraints of a linear programming problem is called a(n) ________ ________.
6. A rational function is called ________ if the degree of its numerator is less than the degree of its denominator.
7. A vector whose magnitude is 1 is called a(n) ________ vector.
8. If the angle between two vectors **v** and **w** is $\pi/2$, then the dot product $\mathbf{v} \cdot \mathbf{w}$ equals ________.

REVIEW EXERCISES

In Problems 1–8, use the matrices below to compute each expression.

$$A = \begin{bmatrix} 1 & 0 \\ 2 & 4 \\ -1 & 2 \end{bmatrix} \quad B = \begin{bmatrix} 4 & -3 & 0 \\ 1 & 1 & -2 \end{bmatrix} \quad C = \begin{bmatrix} 3 & -4 \\ 1 & 5 \\ 5 & -2 \end{bmatrix}$$

1. $A + C$
2. $A - C$
3. $6A$
4. $-4B$
5. AB
6. BA
7. CB
8. BC

In Problems 9–14, find the inverse of each matrix, if there is one. If there is not an inverse, say the matrix is singular.

9. $\begin{bmatrix} 4 & 6 \\ 1 & 3 \end{bmatrix}$
10. $\begin{bmatrix} -3 & 2 \\ 1 & -2 \end{bmatrix}$
11. $\begin{bmatrix} 1 & 3 & 3 \\ 1 & 2 & 1 \\ 1 & -1 & 2 \end{bmatrix}$
12. $\begin{bmatrix} 3 & 1 & 2 \\ 3 & 2 & -1 \\ 1 & 1 & 1 \end{bmatrix}$
13. $\begin{bmatrix} 4 & -8 \\ -1 & 2 \end{bmatrix}$
14. $\begin{bmatrix} -3 & 1 \\ -6 & 2 \end{bmatrix}$

In Problems 15–20, solve each linear programming problem.

15. Maximize $z = 3x + 4y$
subject to $x \geq 0,\quad y \geq 0,\quad 3x + 2y \geq 6,\quad x + y \leq 8$

16. Maximize $z = 2x + 4y$
subject to $x \geq 0,\quad y \geq 0,\quad x + y \leq 6,\quad x \geq 2$

17. Minimize $z = 3x + 5y$
subject to $x \geq 0,\quad y \geq 0,\quad x + y \geq 1,\quad 3x + 2y \leq 12,\quad x + 3y \leq 12$

18. Minimize $z = 3x + y$
subject to $x \geq 0,\quad y \geq 0,\quad x \leq 8,\quad y \leq 6,\quad 2x + y \geq 4$

19. Maximize $z = 5x + 4y$
subject to $x \geq 0,\quad y \geq 0,\quad x + 2y \geq 2,\quad 3x + 4y \leq 12,\quad y \geq x$

20. Maximize $z = 4x + 5y$
subject to $x \geq 0,\quad y \geq 0,\quad 2x + 3y \geq 6,\quad x \geq y,\quad 2x + y \leq 12$

In Problems 21–30, write the partial fraction decomposition of each rational expression.

21. $\dfrac{6}{x(x-4)}$ **22.** $\dfrac{x}{(x+2)(x-3)}$ **23.** $\dfrac{x-4}{x^2(x-1)}$ **24.** $\dfrac{2x-6}{(x-2)^2(x-1)}$

25. $\dfrac{x}{(x^2+9)(x+1)}$ **26.** $\dfrac{3x}{(x-2)(x^2+1)}$ **27.** $\dfrac{x^3}{(x^2+4)^2}$ **28.** $\dfrac{x^3+1}{(x^2+16)^2}$

29. $\dfrac{x^2}{(x^2+1)(x^2-1)}$ **30.** $\dfrac{4}{(x^2+4)(x^2-1)}$

In Problems 31–34, the vector $\mathbf{v}$ *is represented by the directed line segment* $\overrightarrow{PQ}$. *Write* $\mathbf{v}$ *in the form* $a\mathbf{i} + b\mathbf{j}$ *and find* $\|\mathbf{v}\|$.

31. $P = (1, -2);\quad Q = (3, -6)$ **32.** $P = (-3, 1);\quad Q = (4, -2)$

33. $P = (0, -2);\quad Q = (-1, 1)$ **34.** $P = (3, -4);\quad Q = (-2, 0)$

In Problems 35–42, use the vectors $\mathbf{v} = -2\mathbf{i} + \mathbf{j}$ *and* $\mathbf{w} = 4\mathbf{i} - 3\mathbf{j}$.

35. Find $4\mathbf{v} - 3\mathbf{w}$. **36.** Find $-\mathbf{v} + 2\mathbf{w}$. **37.** Find $\|\mathbf{v}\|$. **38.** Find $\|\mathbf{v} + \mathbf{w}\|$.

39. Find $\|\mathbf{v}\| + \|\mathbf{w}\|$. **40.** Find $\|2\mathbf{v}\| - 3\|\mathbf{w}\|$.

41. Find a unit vector having the same direction as $\mathbf{v}$.

42. Find a unit vector having the opposite direction of $\mathbf{w}$.

In Problems 43–46, find the dot product $\mathbf{v} \cdot \mathbf{w}$ *and the cosine of the angle between* $\mathbf{v}$ *and* $\mathbf{w}$.

43. $\mathbf{v} = -2\mathbf{i} + \mathbf{j},\quad \mathbf{w} = 4\mathbf{i} - 3\mathbf{j}$ **44.** $\mathbf{v} = 3\mathbf{i} - \mathbf{j},\quad \mathbf{w} = \mathbf{i} + \mathbf{j}$

45. $\mathbf{v} = \mathbf{i} - 3\mathbf{j},\quad \mathbf{w} = -\mathbf{i} + \mathbf{j}$ **46.** $\mathbf{v} = \mathbf{i} + 4\mathbf{j},\quad \mathbf{w} = 3\mathbf{i} - 2\mathbf{j}$

47. Find the vector projection of $\mathbf{v} = 2\mathbf{i} + 3\mathbf{j}$ onto $\mathbf{w} = 3\mathbf{i} + \mathbf{j}$.

48. Find the vector projection of $\mathbf{v} = -\mathbf{i} + 2\mathbf{j}$ onto $\mathbf{w} = 3\mathbf{i} - \mathbf{j}$.

C **49.** Find the angle between the vectors $\mathbf{v} = 3\mathbf{i} - 4\mathbf{j}$ and $\mathbf{w} = 12\mathbf{i} - 5\mathbf{j}$.

C **50.** Find the angle between the vectors $\mathbf{v} = \mathbf{i} - \mathbf{j}$ and $\mathbf{w} = 2\mathbf{i} + \mathbf{j}$.

C **51.** A swimmer can maintain a constant speed of 5 miles per hour. If the swimmer heads directly across a river that has a current moving at the rate of 2 miles per hour, what is the actual speed of the swimmer? (See the figure.) If the river is 1 mile wide, how far downstream will the swimmer end up from the point directly across the river?

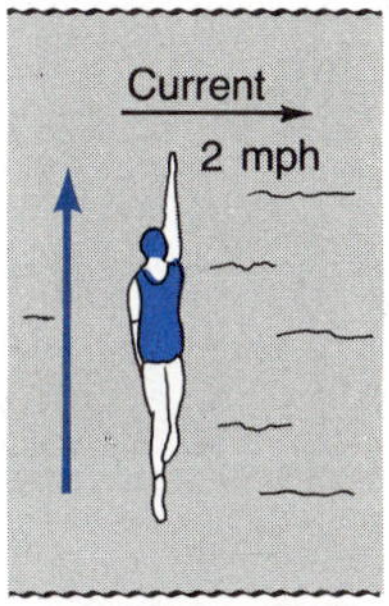

C **52.** A small motorboat is moving at a true speed of 11 miles per hour in a southerly direction. The current is known to be from the northeast at 3 miles per hour. What is the speed of the motorboat relative to the water? In what direction does the compass indicate the boat is headed?

53. A factory produces gasoline engines and diesel engines. Each week the factory is obligated to deliver at least 20 gasoline engines and at least 15 diesel engines. Due to physical limitations, however, the factory cannot make more than 60 gasoline engines nor more than 40 diesel engines. Finally, to prevent layoffs, a total of at least 50 engines must be produced. If gasoline engines cost \$450 each to produce and diesel engines cost \$550 each to produce, how many of each should be produced per week to minimize the cost? What is the excess capacity of the factory; that is, how many of each kind of engine are being produced in excess of the number the factory is obligated to deliver?

54. Rework Problem 53 if the factory is obligated to deliver at least 25 gasoline engines and at least 20 diesel engines.

C **55.** A stream 1 kilometer wide has a constant current of 5 kilometers per hour. At what angle to the shore should a person head a boat that is capable of maintaining a constant speed of 15 kilometers per hour, in order to reach a point directly opposite?

C **56.** An airplane has an airspeed of 500 kilometers per hour in a northerly direction. The wind velocity is 60 kilometers per hour in a southeasterly direction. Find the actual speed and direction of the plane relative to the ground.

Sets; Counting; Probability

15

The first three sections of this chapter deal with techniques and formulas for counting the number of objects in a set, a part of the branch of mathematics called *combinatorics*. These formulas are used in computer science to analyze algorithms and recursive functions and to study stacks and queues. They are also used to determine *probabilities*, the likelihood that a certain outcome of a random experiment will occur, which is the subject of the fourth section in this chapter.

Applications of the topics in this chapter can be found in fields including engineering, business and economics, the social sciences, and the biological sciences. You will see these topics again in more detail if you take a course in finite mathematics or discrete mathematics.

15.1 Sets and Counting

Sets

A **set** is a well-defined collection of distinct objects. The objects of a set are called its **elements**. By **well-defined**, we mean that there is a rule that enables us to determine whether a given object is an element of the set. If a set has no elements, it is called the **empty set**, or **null set**, and is denoted by the symbol $\varnothing$.

Because the elements of a set are distinct, we never repeat elements. Thus, we would never write $\{1, 2, 3, 2\}$; the correct listing is $\{1, 2, 3\}$. Furthermore, because a set is a collection, the order in which the elements are listed is immaterial. Thus, $\{1, 2, 3\}$, $\{1, 3, 2\}$, $\{2, 1, 3\}$, and so on, all represent the same set.

Example 1 Write the set consisting of the possible outcomes from tossing a coin twice. Use H for "heads" and T for "tails."

Solution In tossing a coin twice, we can get heads each time, HH; or heads the first time and tails the second, HT; or tails the first time and heads the second, TH; or tails each time, TT. Because no other possibilities exist, the set of outcomes is

$$\{\text{HH, HT, TH, TT}\}$$

If two sets A and B have precisely the same elements, then we say that A and B are **equal** and write $A = B$.

If each element of a set A is also an element of a set B, then we say that A is a **subset** of B and write $A \subseteq B$.

If $A \subseteq B$ and $A \neq B$, then we say that A is a **proper subset** of B and write $A \subset B$.

Thus, if $A \subseteq B$, every element in set A is also in set B, but B may or may not have additional elements. If $A \subset B$, every element in A is also in B, and B has at least one element not found in A.

Finally, we agree that the empty set is a subset of every set; that is,

$$\varnothing \subseteq A \qquad \text{for any set } A$$

Example 2 Consider the sets

$$A = \{1, 2, 3, 4, 5, 6, 7, 8, 9, 10\}$$
$$B = \{1, 3, 5, 7, 9\}$$
$$C = \{3, 5, 7\}$$

Some possible relationships involving these sets are:

1. $B \subseteq A$, because every element in B is also in A.
2. $\varnothing \subseteq C$, because the empty set is a subset of any set.
3. $C \subseteq B$, because every element in C is also in B.
4. $C \subset B$, because every element in C is also in B, and B has elements not found in C. ■

Example 3 Write down all the subsets of the set $\{a, b, c\}$.

Solution To organize our work, we write down all the subsets with no elements, then those with one element, then those with two elements, and finally those with three elements. These will give us all the subsets—do you see why?

0 ELEMENTS	1 ELEMENT	2 ELEMENTS	3 ELEMENTS
$\varnothing$	$\{a\}, \{b\}, \{c\}$	$\{a, b\}, \{b, c\}, \{a, c\}$	$\{a, b, c\}$

■

Intersection, Union, and Complement

Intersection; Union

If A and B are sets, the **intersection** of A with B, denoted $A \cap B$, is the set consisting of elements that belong to *both A and B*. The **union** of A with B, denoted $A \cup B$, is the set consisting of elements that belong to *either A or B, or both.*

Example 4 Let $A = \{1, 3, 5, 8\}$, $B = \{3, 5, 7\}$, and $C = \{2, 4, 6, 8\}$. Find:

(a) $A \cap B$ (b) $A \cup B$ (c) $B \cap (A \cup C)$

Solution (a) $A \cap B = \{1, 3, 5, 8\} \cap \{3, 5, 7\} = \{3, 5\}$

(b) $A \cup B = \{1, 3, 5, 8\} \cup \{3, 5, 7\} = \{1, 3, 5, 7, 8\}$

(c)
$$\begin{aligned} B \cap (A \cup C) &= \{3, 5, 7\} \cap [\{1, 3, 5, 8\} \cup \{2, 4, 6, 8\}] \\ &= \{3, 5, 7\} \cap \{1, 2, 3, 4, 5, 6, 8\} \\ &= \{3, 5\} \end{aligned}$$
■

Usually, in working with sets, we designate a **universal set**—namely, the set consisting of all the elements we wish to consider. Once a universal set has been designated, we can consider elements of the universal set not found in a given set.

Complement

If A is a set, the **complement** of A, denoted A', is the set consisting of all the elements not in A.

Example 5 If the universal set is $U = \{1, 2, 3, 4, 5, 6, 7, 8, 9\}$, and if $A = \{1, 3, 5, 7, 9\}$, then $A' = \{2, 4, 6, 8\}$. ■

Notice that: $A \cup A' = U$ and $A \cap A' = \varnothing$

Venn Diagrams

Figure 1

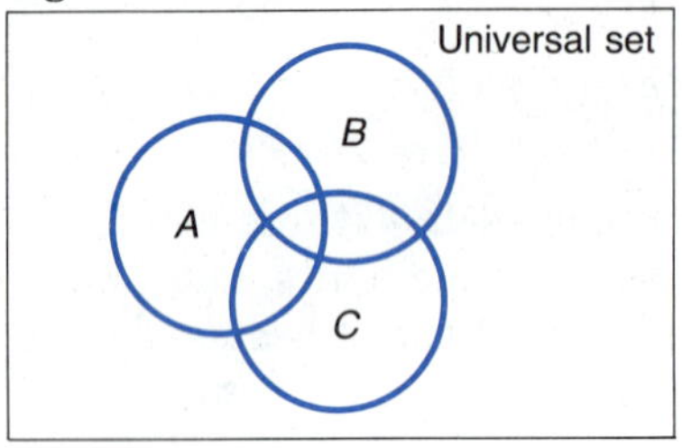

It is often helpful to draw pictures of sets. Such pictures, called **Venn diagrams**, represent sets as circles enclosed in a rectangle, which represents the universal set. Such diagrams are often helpful to visualize various relationships among sets. See Figure 1.

If we know that $A \subseteq B$, we might use the Venn diagram in Figure 2(a). If we know that A and B have no elements in common, that is, if $A \cap B = \varnothing$, we might use the Venn diagram in Figure 2(b).

Figure 2

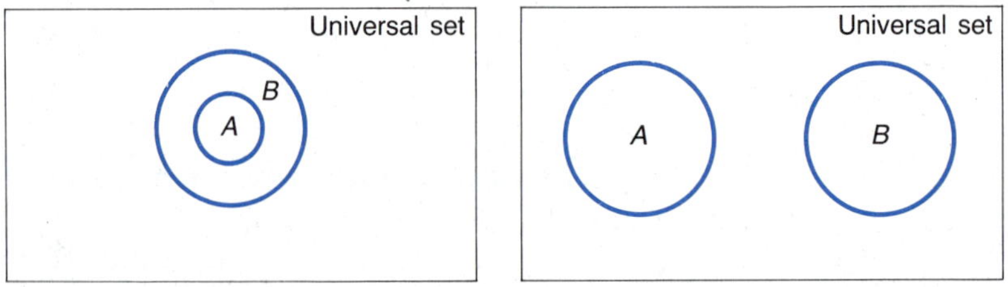

(a) $A \subseteq B$ (b) $A \cap B = \varnothing$

Figures 3(a), 3(b), and 3(c) use Venn diagrams to illustrate the definitions of intersection, union, and complement, respectively.

Figure 3

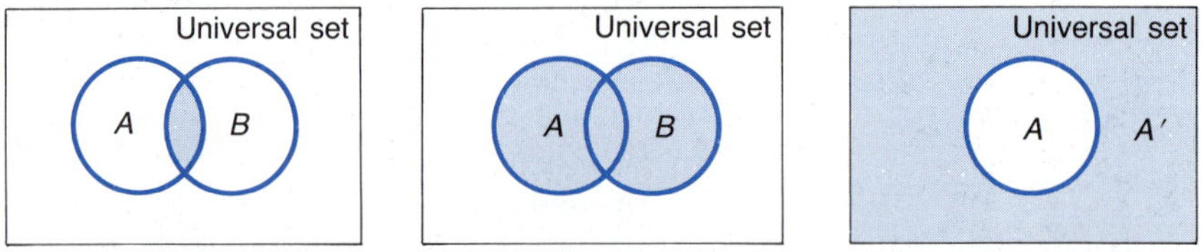

(a) $A \cap B$ (b) $A \cup B$ (c) A'

Counting

As you count the number of students in a classroom or the number of pennies in your pocket, what you are really doing is matching, on a one-to-one basis, each object to be counted with the counting numbers 1, 2, 3, . . . , n, for some number n. If a set A matched up in this fashion with the set $\{1, 2, \ldots, 25\}$, you would conclude that there are 25 elements in the set A. We use the notation $n(A) = 25$ to indicate that there are 25 elements in the set A.

Because the empty set has no elements, we write

$$n(\varnothing) = 0$$

If the number of elements in a set is a nonnegative integer, we say the set is **finite**. Otherwise, it is **infinite**. We shall concern ourselves only with finite sets.

From Example 3, we can see that a set with 3 elements has $2^3 = 8$ subsets. In fact, it can be shown that a set with n elements has exactly 2^n subsets. This fact has an important application to computers, which we take up at the end of this section.

Example 6 In a survey of 100 college students, 35 were registered in College Algebra, 52 were registered in Introduction to Computer Science, and 18 were in both courses. How many were registered in neither course?

Solution First, we let

A = Set of students in College Algebra
B = Set of students in Introduction to Computer Science

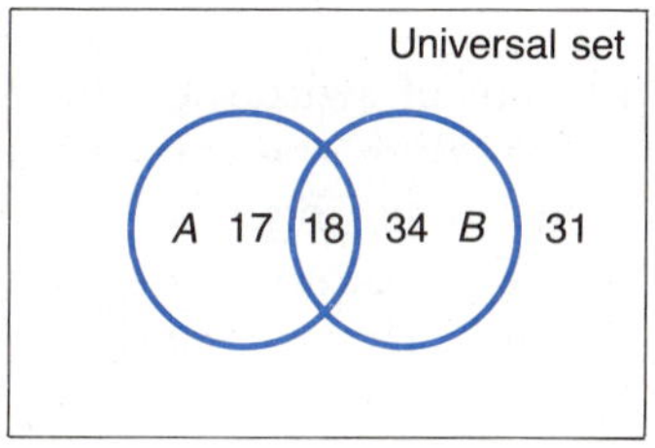

Figure 4

Then the information tells us that

$$n(A) = 35 \qquad n(B) = 52 \qquad n(A \cap B) = 18$$

Refer to Figure 4. Do you see how the numerical entries were determined? Based on the diagram, we conclude that $17 + 18 + 34 = 69$ students were registered in at least one of the two courses. Since 100 students were surveyed, it follows that $100 - 69 = 31$ were registered in neither course. ■

The conclusions drawn in Example 6 lead us to formulate a general counting formula. If we count the elements in each of two sets A and B, we necessarily count twice any elements that are in both A and B, that is, those elements in $A \cap B$. Thus, to count correctly the elements that are in A or B, that is, to find $n(A \cup B)$, we need to subtract those in $A \cap B$ from $n(A) + n(B)$.

Theorem If A and B are finite sets, then

Counting Formula

$$n(A \cup B) = n(A) + n(B) - n(A \cap B) \tag{1}$$

■

A special case of the counting formula (1) occurs if A and B have no elements in common. In this case, $A \cap B = \varnothing$ so that $n(A \cap B) = 0$.

Theorem If two sets A and B have no elements in common, then

Addition Principle of Counting

$$n(A \cup B) = n(A) + n(B) \tag{2}$$

■

Example 7 A certain code is to consist of either a letter of the alphabet or a digit, but not both. How many codes are possible?

Solution Let the sets A and B be defined as

$$A = \text{Set of letters in the alphabet}$$
$$B = \text{Set of digits } \{0, 1, 2, \ldots, 9\}$$

Then,

$$n(A) = 26 \qquad n(B) = 10$$

Because letters and digits are different, $A \cap B = \varnothing$. The number of ways either a letter or a digit can be chosen is, therefore,

$$n(A \cup B) = n(A) + n(B) = 26 + 10 = 36$$

■

Application to Computers

Information is stored in a computer using a series of switches, which may be on or off. These switches are denoted by either the number 0 (off) or the number 1 (on); these numbers are the binary digits, or **bits**. A **register** holds a certain fixed number of bits. For example, the Z-80 microprocessor has 8 bit registers; the PDP-11 minicomputer has 16 bit registers; and the IBM-370 computer has 32 bit registers. Thus, a Z-80 register may hold an entry that looks like this: 01111001 (8 bits). We wish to find out how many different representations are possible in a given register.

We proceed in steps, first looking at a hypothetical 3 bit register. Look again at the solution to Example 3, and arrange all the subsets of $\{a, b, c\}$ as shown in Table 1. As the table illustrates, the number of subsets of a set with 3 elements equals the number of different representations in a 3 bit register. A set with n elements has 2^n subsets; thus, an n bit register has 2^n representations. So an 8 bit register can hold $2^8 = 256$ different symbols, a 16 bit register can hold $2^{16} = 65{,}536$ different symbols, and a 32 bit register can hold $2^{32} \approx 4.3 \times 10^9$ different symbols.

Table 1

a	b	c	SUBSET
0	0	0	$\varnothing$
1	0	0	$\{a\}$
0	1	0	$\{b\}$
0	0	1	$\{c\}$
1	1	0	$\{a, b\}$
0	1	1	$\{b, c\}$
1	0	1	$\{a, c\}$
1	1	1	$\{a, b, c\}$

EXERCISE 15.1

In Problems 1–10, use $A = \{1, 3, 5, 7, 9\}$, $B = \{1, 5, 6, 7\}$, *and* $C = \{1, 2, 4, 6, 8, 9\}$ *to find each set.*

1. $A \cup B$ **2.** $A \cup C$ **3.** $A \cap B$ **4.** $A \cap C$
5. $(A \cup B) \cap C$ **6.** $(A \cap C) \cup (B \cap C)$ **7.** $(A \cap B) \cup C$ **8.** $(A \cup B) \cup C$
9. $(A \cup C) \cap (B \cup C)$ **10.** $(A \cap B) \cap C$

In Problems 11–20, use U = *Universal set* $= \{0, 1, 2, 3, 4, 5, 6, 7, 8, 9\}$, $A = \{1, 3, 4, 5, 9\}$, $B = \{2, 4, 6, 7, 8\}$, *and* $C = \{1, 3, 4, 6\}$ *to find each set.*

11. A' **12.** C' **13.** $(A \cap B)'$ **14.** $(B \cup C)'$

15. $A' \cup B'$ **16.** $B' \cap C'$ **17.** $(A \cap C')'$ **18.** $(B' \cup C)'$

19. $(A \cup B \cup C)'$ **20.** $(A \cap B \cap C)'$

21. Write down all the subsets of $\{a, b, c, d\}$.

22. Write down all the subsets of $\{a, b, c, d, e\}$.

23. If $n(A) = 15$, $n(B) = 20$, and $n(A \cap B) = 10$, find $n(A \cup B)$.

24. If $n(A) = 20$, $n(B) = 40$, and $n(A \cup B) = 35$, find $n(A \cap B)$.

25. If $n(A \cup B) = 50$, $n(A \cap B) = 10$, and $n(B) = 20$, find $n(A)$.

26. If $n(A \cup B) = 60$, $n(A \cap B) = 40$, and $n(A) = n(B)$, find $n(A)$.

In Problems 27–34, use the information given in the figure below.

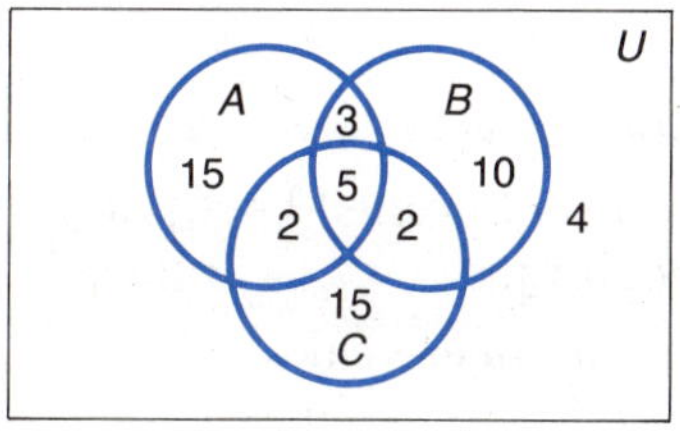

27. How many are in set A?

28. How many are in set B?

29. How many are in A or B?

30. How many are in A and B?

31. How many are in A but not C?

32. How many are not in A?

33. How many are in A and B and C?

34. How many are in A or B or C?

35. In a consumer survey of 500 people, 200 indicated they would be buying a major appliance within the next month; 150 indicated they would buy a car, and 25 said they would purchase both a major appliance and a car. How many will purchase neither? How many will purchase only a car?

36. In a student survey, 200 indicated they would attend Summer Session I and 150 indicated Summer Session II. If 75 students plan to attend both summer sessions and 275 indicated they would attend neither session, how many students participated in the survey?

37. In a survey of 100 investors in the stock market,

50 owned shares in IBM
40 owned shares in AT&T
45 owned shares in GE
20 owned shares in both IBM and AT&T
20 owned shares in both IBM and GE
15 owned shares in both AT&T and GE
5 owned shares in all three

(a) How many of the investors surveyed did not have shares in any of the three companies?
(b) How many owned just IBM shares?
(c) How many owned just GE shares?
(d) How many owned neither IBM nor GE?
(e) How many owned either IBM or AT&T but no GE?

38. Human blood is classified as either Rh+ or Rh−. Blood is also classified by type: A, if it contains an A antigen; B, if it contains a B antigen; AB, if it contains both A and B antigens; and O, if it contains neither antigen. Draw a Venn diagram illustrating the various blood types. Based on this classification, how many different kinds of blood are there?

15.2 Permutations

Counting plays a major role in many diverse areas, such as probability, statistics, and computer science. In this section and the next, we shall look at special types of counting problems and develop general formulas for solving them.

We begin with an example that will demonstrate a general counting principle.

Example 1 The fixed-price dinner at a restaurant provides the following choices:

Appetizer:	soup, salad, or cole slaw
Entree:	baked chicken, broiled beef patty, baby beef liver, or roast beef au jus
Dessert:	ice cream or cheese cake

How many different meals can be ordered?

Solution Ordering such a meal requires three separate decisions:

CHOOSE AN APPETIZER	CHOOSE AN ENTREE	CHOOSE A DESSERT
3 choices	4 choices	2 choices

Look at the **tree diagram** in Figure 5. We see that, for each choice of appetizer, there are 4 choices of entrees. And for each of these $3 \cdot 4 = 12$ choices, there are 2 choices for dessert. Thus, there are a total of

$$3 \cdot 4 \cdot 2 = 24$$

different meals that can be ordered. ■

Example 1 illustrates a general counting principle.

Theorem

Multiplication Principle of Counting

If a task consists of a sequence of choices in which there are p selections for the first choice, q selections for the second choice, r selections for the third choice, and so on, then the task of making these selections can be done in

$$p \cdot q \cdot r \cdot \cdots$$

different ways. ■

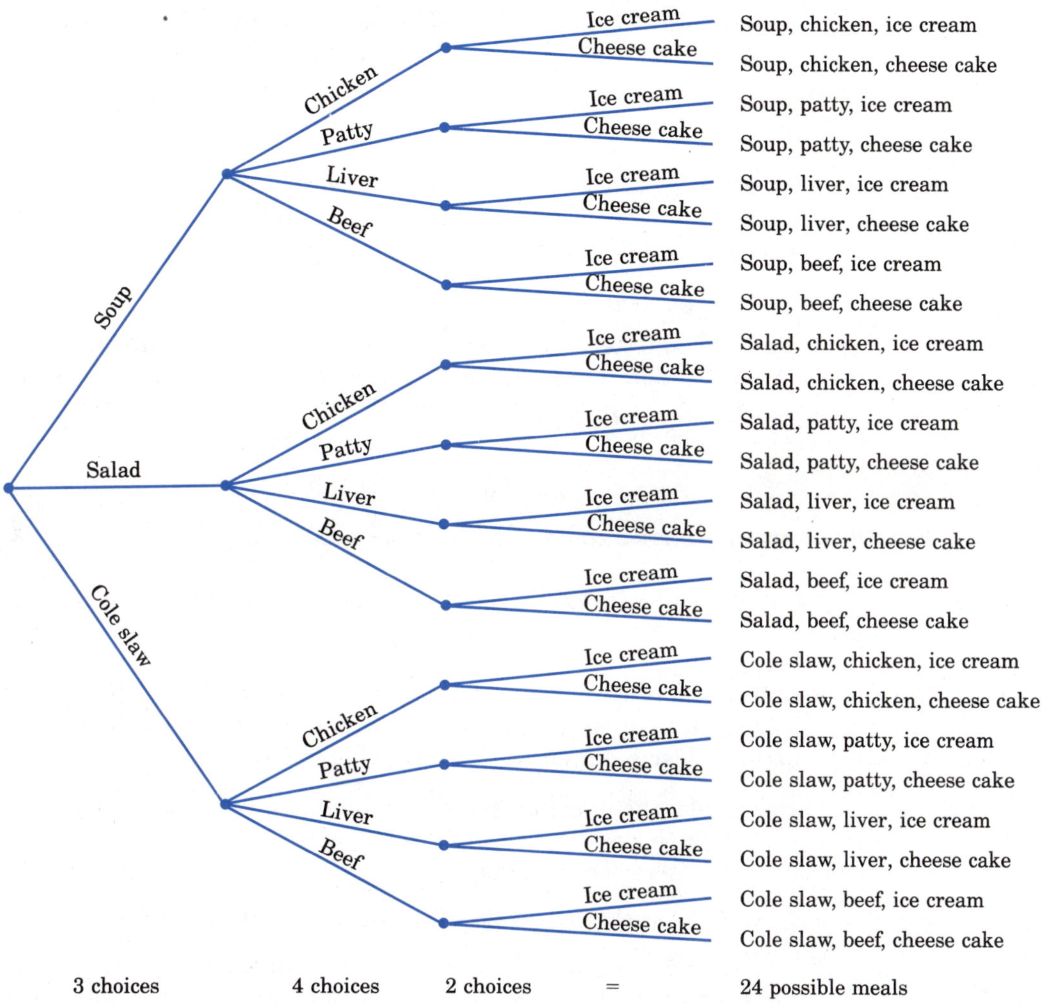

Figure 5

Example 2 The International Airline Transportation Association (IATA) assigns three-letter codes to represent airport locations. For example, JFK represents Kennedy International in New York. How many different airport codes are possible?

Solution The task consists of making three selections. Each selection requires choosing a letter of the alphabet (26 choices). Thus, by the multiplication principle, there are

$$26 \cdot 26 \cdot 26 = 17{,}576$$

different airport codes. ■

In Example 2, we were allowed to repeat a letter. For example, a valid airport code is FLL (Ft. Lauderdale International Airport), in which the letter L appears twice. In the next example, such repetition is not allowed.

Example 3 Suppose we wish to establish a three-letter code using any of the 26 letters of the alphabet, but we require that no letter be used more than once. How many different three-letter codes are there?

Solution The task consists of making three selections. The first selection requires choosing from 26 letters. Because no letter can be used more than once, the second selection requires choosing from 25 letters. The third selection requires choosing from 24 letters. (Do you see why?) By the multiplication principle, there are

$$26 \cdot 25 \cdot 24 = 15{,}600$$

different three-letter codes with no letter repeated. ■

Example 3 illustrates a type of counting problem referred to as a *permutation.*

Permutation

A **permutation** is an ordered arrangement of n distinct objects without repetitions. The symbol $P(n, r)$ represents the number of permutations of n distinct objects, taken r at a time, where $r \leq n$.

For example, the question posed in Example 3 asks for the number of ways the 26 letters of the alphabet can be arranged using three nonrepeated letters. The answer is

$$P(26, 3) = 26 \cdot 25 \cdot 24 = 15{,}600$$

To arrive at a formula for $P(n, r)$, we note that the task of obtaining an ordered arrangement of n objects in which only $r \leq n$ of them are used, without repeating any of them, requires making r selections. For the first selection, there are n choices; for the second selection, there are $n - 1$ choices; for the third selection, there are $n - 2$ choices; . . . ; for the rth selection, there are $n - (r - 1)$ choices. By the multiplication principle, we have

$$\begin{aligned} P(n, r) &= \overset{\text{1st}}{n} \cdot \overset{\text{2nd}}{(n-1)} \cdot \overset{\text{3rd}}{(n-2)} \cdot \cdots \cdot \overset{r\text{th}}{[n-(r-1)]} \\ &= n \cdot (n-1) \cdot (n-2) \cdot \cdots \cdot (n-r+1) \end{aligned}$$

This formula for $P(n, r)$ can be compactly written using factorial notation (see Section 13.2):

$$P(n, r) = n \cdot (n-1) \cdot (n-2) \cdot \cdots \cdot (n-r+1)$$
$$= n \cdot (n-1) \cdot (n-2) \cdot \cdots \cdot (n-r+1) \cdot \frac{(n-r) \cdot \cdots \cdot 3 \cdot 2 \cdot 1}{(n-r) \cdot \cdots \cdot 3 \cdot 2 \cdot 1} = \frac{n!}{(n-r)!}$$

Theorem

Number of Permutations of n Distinct Objects Taken r at a Time

The number of different arrangements of n objects using $r \le n$ of them, in which

1. the n objects are distinct,
2. once an object is used it cannot be repeated, and
3. order is important,

is given by the formula

$$P(n, r) = \frac{n!}{(n-r)!} \qquad (1)$$

■

Example 4 Evaluate:

(a) $P(7, 3)$ (b) $P(6, 1)$ (c) $P(4, 4)$

Solution We shall work each problem in two ways.

(a) $P(7, 3) = \underbrace{7 \cdot 6 \cdot 5}_{\text{3 factors}} = 210$

$$P(7, 3) = \frac{7!}{(7-3)!} = \frac{7!}{4!} = \frac{7 \cdot 6 \cdot 5 \cdot \cancel{4!}}{\cancel{4!}} = 210$$

(b) $P(6, 1) = \underbrace{6}_{\text{1 factor}} = 6$

$$P(6, 1) = \frac{6!}{(6-1)!} = \frac{6!}{5!} = \frac{6 \cdot \cancel{5!}}{\cancel{5!}} = 6$$

(c) $P(4, 4) = \underbrace{4 \cdot 3 \cdot 2 \cdot 1}_{\text{4 factors}} = 24$

$$P(4, 4) = \frac{4!}{(4-4)!} = \frac{4!}{0!} = \frac{4 \cdot 3 \cdot 2 \cdot 1}{1} = 24$$

■

Example 5 In how many ways can 5 people be lined up?

Solution The 5 people are obviously distinct. Once a person is in line, that person will not be repeated elsewhere in the line; and, in lining up people, order is important. Thus, we have a permutation of 5 objects taken 5 at a time. We can line up the 5 people in

$$P(5, 5) = 5! = 120 \text{ ways}$$

■

EXERCISE 15.2

1. A man has 5 shirts and 3 ties. How many different shirt and tie combinations can he wear?
2. A woman has 3 blouses and 5 skirts. How many different outfits can she wear?
3. How many two-letter codes can be formed using the letters A, B, C, D? Repeated letters are allowed.
4. How many two-letter codes can be formed using the letters A, B, C, D, E? Repeated letters are allowed.
5. How many three-digit numbers can be formed using the digits 0 and 1? Repeated digits are allowed.
6. How many three-digit numbers can be formed using the digits 0, 1, 2, 3, 4, 5, 6, 7, 8, 9? Repeated digits are allowed.
7. C How many arrangements of answers are possible for a true/false test with 10 questions?
8. C How many arrangements of answers are possible in a multiple-choice test with 5 questions, each of which has 4 possible answers?
9. How many four-digit numbers can be formed using the digits 0, 1, 2, 3, 4, 5, 6, 7, 8, 9, if the first digit cannot be 0? Repeated digits are allowed.
10. How many five-digit numbers can be formed using the digits 0, 1, 2, 3, 4, 5, 6, 7, 8, 9, if the first digit cannot be 0 or 1? Repeated digits are allowed.

In Problems 11–18, find the value of each permutation.

11. $P(6, 2)$
12. $P(7, 2)$
13. $P(5, 5)$
14. $P(4, 4)$
15. $P(8, 0)$
16. $P(9, 0)$
17. $P(8, 3)$
18. $P(8, 5)$
19. In how many ways can 4 people be lined up?
20. In how many ways can 5 books be arranged on a shelf?
21. How many different three-letter codes are there if only the letters A, B, C, D, E can be used and no letter can be used more than once?
22. How many different four-letter codes are there if only the letters A, B, C, D, E, F can be used and no letter can be used more than once?
23. How many arrangements are there of the letters in the word MONEY?
24. How many arrangements are there of the digits in the number 51,342?
25. C In the American Baseball League, a designated hitter may be used. How many batting orders are possible for a manager to use? (There are 9 regular players on a team.)
26. C In the National Baseball League, the pitcher usually bats ninth. If this is the case, how many batting orders are possible for a manager to use?
27. C How many different license plates can be issued if every plate consists of two letters, without repetition, followed by any four-digit number that does not start with a 0?
28. C Answer Problem 27 if repetition of the letters is allowed.

15.3 Combinations

In a permutation, order is important; for example, the arrangements ABC, CAB, BAC, . . . are considered different arrangements of the letters A, B, and C. In many situations, though, order is unimportant. For example, in the card game of poker, the order in which the cards are received does not matter—it is the *combination* of the cards that matters.

Combination

A **combination** is an arrangement, without regard to order, of n distinct objects without repetitions. The symbol $C(n, r)$ represents the number of combinations of n distinct objects taken r at a time, where $r \leq n$.

Example 1 List all the combinations of the 4 objects a, b, c, d taken 2 at a time. What is $C(4, 2)$?

Solution One combination of a, b, c, d taken 2 at a time is

$$ab$$

The object ba is excluded, because order is not important in a combination. The list of all such combinations (convince yourself of this) is

$$ab, \quad ac, \quad ad, \quad bc, \quad bd, \quad cd$$

Thus,

$$C(4, 2) = 6$$

■

We can find a formula for $C(n, r)$ by noting that the only difference between a permutation and a combination is that we disregard order in combinations. Thus, to determine $C(n, r)$, we need only eliminate from the formula for $P(n, r)$ the number of permutations that were simply rearrangments of a given set of r objects. But that is easily determined from the formula for $P(n, r)$ by calculating $P(r, r) = r!$ So, if we divide $P(n, r)$ by $r!$, we will have the desired formula for $C(n, r)$:

$$C(n, r) = \frac{P(n, r)}{r!} \underset{\uparrow}{=} \frac{n!/(n-r)!}{r!} = \frac{n!}{(n-r)!r!}$$

Use formula (1), Section 15.2.

We have proved the following result.

Theorem

Number of Combinations of n Distinct Objects Taken r at a Time

The number of different arrangements of n objects using $r \leq n$ of them, in which

1. the n objects are distinct,
2. once an object is used, it cannot be repeated, and
3. order is not important

is given by the formula

$$C(n, r) = \frac{n!}{(n-r)!r!} \qquad (1)$$

■

Based on formula (1) above, we discover that the symbol $C(n, r)$ and the symbol $\binom{n}{r}$ for the binomial coefficients are, in fact, the same. Thus, the Pascal triangle (see Section 13.2) can be used to find the value of $C(n, r)$. However, because it is more practical and convenient, we shall use formula (1) instead.

Example 2 Use formula (1) to find the value of each expression:

(a) $C(3, 1)$ (b) $C(6, 3)$ (c) $C(n, n)$ (d) $C(n, 0)$

Solution

(a) $C(3, 1) = \dfrac{3!}{(3-1)!1!} = \dfrac{3!}{2!1!} = \dfrac{3 \cdot \cancel{2} \cdot 1}{\cancel{2} \cdot 1 \cdot 1} = 3$

(b) $C(6, 3) = \dfrac{6!}{(6-3)!3!} = \dfrac{6 \cdot 5 \cdot 4 \cdot \cancel{3!}}{3! \cdot \cancel{3!}} = \dfrac{\cancel{6} \cdot 5 \cdot 4}{\cancel{6}} = 20$

(c) $C(n, n) = \dfrac{n!}{(n-n)!n!} = \dfrac{\cancel{n!}}{0!\cancel{n!}} = \dfrac{1}{1} = 1$

(d) $C(n, 0) = \dfrac{n!}{(n-0)!0!} = \dfrac{\cancel{n!}}{\cancel{n!}0!} = \dfrac{1}{1} = 1$ ■

Example 3 How many different committees of 3 people can be formed from a pool of 7 people?

Solution The 7 people are, of course, distinct. More important, though, is the observation that the order of being selected on a committee is not significant. Thus, the problem asks for the number of combinations of 7 objects taken 3 at a time, namely,

$$C(7, 3) = \frac{7!}{4!3!} = \frac{7 \cdot 6 \cdot 5 \cdot \cancel{4!}}{\cancel{4!}3!} = \frac{7 \cdot \cancel{6} \cdot 5}{\cancel{6}} = 35$$ ■

C **Example 4** Using a regular deck of 52 cards, how many different 5-card hands can be dealt?

Solution As mentioned earlier, it is the combination, not the order, of the cards that is important. Thus, we seek the number

$$C(52, 5) = \frac{52!}{47!5!} = \frac{52 \cdot 51 \cdot 50 \cdot 49 \cdot 48 \cdot \cancel{47!}}{\cancel{47!}5!}$$

$$= \frac{52 \cdot 51 \cdot 50 \cdot 49 \cdot 48}{5 \cdot 4 \cdot 3 \cdot 2 \cdot 1} = 2{,}598{,}960$$ ■

Example 5 In how many ways can a committee consisting of 2 faculty members and 3 students be formed if there are 6 faculty members and 10 students eligible to serve on the committee?

Solution The problem can be separated into two parts: the number of ways the faculty members can be chosen—namely, $C(6, 2)$—and the number of ways the student members can be chosen—namely, $C(10, 3)$. By the multiplication principle, the committee can be formed in

$$C(6, 2) \cdot C(10, 3) = \frac{6!}{4!2!} \cdot \frac{10!}{7!3!} = \frac{6 \cdot 5 \cdot \cancel{4!}}{\cancel{4!}2!} \cdot \frac{10 \cdot 9 \cdot 8 \cdot \cancel{7!}}{\cancel{7!}3!}$$

$$= \frac{30}{2} \cdot \frac{720}{6} = 1800 \text{ ways}$$ ■

EXERCISE 15.3

1. List all the combinations of the 5 objects a, b, c, d, e taken 3 at a time. What is $C(5, 3)$?
2. List all the combinations of the 5 objects a, b, c, d, e taken 2 at a time. What is $C(5, 2)$?
3. List all the combinations of the 4 objects 1, 2, 3, 4 taken 3 at a time. What is $C(4, 3)$?
4. List all the combinations of the 6 objects 1, 2, 3, 4, 5, 6 taken 3 at a time. What is $C(6, 3)$?

In Problems 5–12, use formula (1) *to find the indicated number of combinations.*

5. $C(8, 2)$
6. $C(8, 6)$
7. $C(6, 4)$
8. $C(6, 2)$
9. $C(15, 15)$
10. $C(18, 1)$
11. C $C(26, 13)$
12. C $C(18, 9)$
13. In how many ways can a committee of 4 students be formed from a pool of 7 students?
14. In how many ways can a committee of 3 professors be formed from a department having 8 professors?
15. C How many different 13-card bridge hands are possible? (A bridge hand consists of 13 cards dealt from a regular deck of 52 cards.)
16. C How many different 7-card stud hands are possible? (A 7-card stud hand consists of 7 cards dealt from a regular deck of 52 cards.)
17. A student dance committee is to be formed consisting of 2 boys and 3 girls. If the membership is to be chosen from 4 boys and 8 girls, how many different committees are possible?
18. A baseball team has 15 members. Four of the players are pitchers, and the remaining 11 members can play any position. How many different teams of 9 players can be formed?

C **19.** The student relations committee of a college consists of 2 administrators, 3 faculty members, and 5 students. There are 4 administrators, 8 faculty members, and 20 students eligible to serve. How many different committees are possible?

C **20.** A defensive football squad consists of 25 players. Of these, 10 are linemen, 10 are linebackers, and 5 are safeties. How many different teams of 5 linemen, 3 linebackers, and 3 safeties can be formed?

21. A jar contains 6 red balls and 4 green balls. In how many ways can 4 balls be chosen such that exactly 2 are green?

22. Answer Problem 21 if the 4 balls are to be chosen such that exactly 1 is green.

23. *Programming exercise* When both n and r are large, finding $C(n, r)$ on a computer may lead to integers too large to compute. To avoid this, we can approximate the values of $C(n, r)$. One way to do this is the following:

$$C(40, 20) = \frac{40!}{20!20!} = \frac{40 \cdot 39 \cdot 38 \cdot \cdots \cdot 21}{20 \cdot 19 \cdot 18 \cdot \cdots \cdot 1} = \frac{40}{20} \cdot \frac{39}{19} \cdot \frac{38}{18} \cdot \cdots \cdot \frac{21}{1}$$
$$\approx 2.000 \cdot 2.053 \cdot 2.111 \cdot \cdots \cdot 21.000 = 1.3784652 \times 10^{11}$$

(a) Write a program that inputs two integers N and R and computes $C(N, R)$ using formula (1).
(b) Use the program to determine where overflow occurs on your computer.
(c) Write a program that inputs two integers N and R and computes $C(N, R)$ by the approximation technique shown above.
(d) Compare the answers found in parts (a) and (c).

15.4 Probability

Probability is an area of mathematics that deals with experiments that yield random results yet admit a certain regularity. Such experiments do not always produce the same result or outcome, so the result of any one observation is not predictable. However, the results of the experiment over a long period do produce regular patterns that enable us to predict with remarkable accuracy.

Example 1 In tossing a fair coin, we know that the outcome is either a head or a tail. On any particular throw, we cannot predict what will happen, but, if we toss the coin many times, we observe that the number of times a head comes up is approximately equal to the number of times we get a tail. It seems reasonable, therefore, to assign a probability of $\frac{1}{2}$ that a head comes up and a probability of $\frac{1}{2}$ that a tail comes up. ■

Probabilistic Models

The discussion in Example 1 constitutes the construction of a **probabilistic model** for the experiment of tossing a fair coin once. A probabilistic model has two components: a sample space and an assignment of probabilities. A **sample space** S is a set whose elements represent all the logical possibilities that can occur as a result of the experiment. Each element of S is called an **outcome**. To each outcome, we assign a number, called the **probability** of that outcome, which has two properties:

1. Each probability is nonnegative.
2. The sum of all the probabilities equals 1.

Thus, if a probabilistic model has the sample space

$$S = \{e_1, e_2, \ldots, e_n\}$$

where $e_1, e_2, \ldots, e_n$ are the possible outcomes, and if $P(e_1), P(e_2), \ldots, P(e_n)$ denote the respective probabilities of these outcomes, then

$$P(e_1) \geq 0, \quad P(e_2) \geq 0, \quad \ldots, \quad P(e_n) \geq 0 \tag{1}$$

$$P(e_1) + P(e_2) + \cdots + P(e_n) = 1 \tag{2}$$

Let's look at an example.

Example 2 An experiment consists of rolling a fair die once.* Construct a probabilistic model for this experiment.

Solution A sample space S consists of all the logical possibilities that can occur. Because rolling the die will result in one of six faces showing, the sample space S consists of

$$S = \{1, 2, 3, 4, 5, 6\}$$

Because the die is fair, one face is no more likely to occur than another. As a result, our assignment of probabilities is

$$P(1) = \tfrac{1}{6} \qquad P(2) = \tfrac{1}{6}$$
$$P(3) = \tfrac{1}{6} \qquad P(4) = \tfrac{1}{6}$$
$$P(5) = \tfrac{1}{6} \qquad P(6) = \tfrac{1}{6}$$

■

Example 3 An experiment consists of tossing a fair die and then a fair coin. Construct a probabilistic model for this experiment.

*A die is a cube with each face having either 1, 2, 3, 4, 5, or 6 dots on it.

Solution A tree diagram is helpful in listing all the possible outcomes. See Figure 6. The sample space consists of the outcomes

$$S = \{1H, 1T, 2H, 2T, 3H, 3T, 4H, 4T, 5H, 5T, 6H, 6T\}$$

The die and the coin are fair; thus, no one outcome is more likely to occur than another. As a result, we assign the probability $\frac{1}{12}$ to each of the 12 outcomes. ■

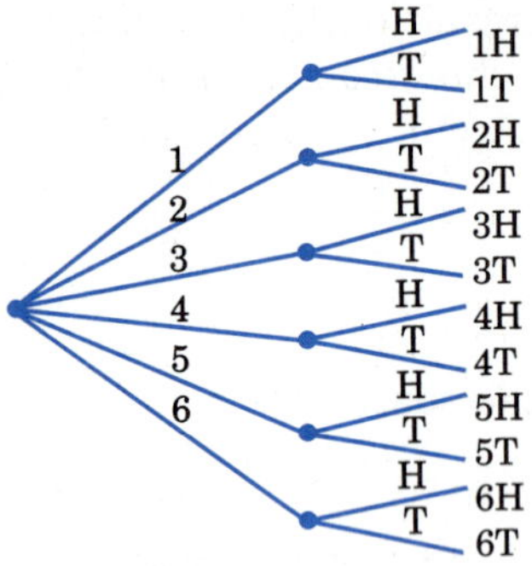

Figure 6

In working with probabilistic models, the term **event** is used to describe a set of possible outcomes of the experiment. Thus, an event E is some subset of the sample space S. The **probability of an event** E, denoted by $P(E)$, is defined as the sum of the probabilities of the outcomes in E.

Example 4 For the experiment described in Example 3, what is the probability that an even number followed by a head occurs?

Solution The event E, an even number followed by a head, consists of

$$E = \{2H, 4H, 6H\}$$

The probability of E is

$$P(E) = P(2H) + P(4H) + P(6H) = \tfrac{1}{12} + \tfrac{1}{12} + \tfrac{1}{12} = \tfrac{1}{4}$$ ■

Equally Likely Outcomes

When the same probability is assigned to each outcome of the sample space, the experiment is said to have **equally likely outcomes**.

Theorem

Probability for Equally Likely Outcomes

If an experiment has n equally likely outcomes, and if the number of ways an event E can occur is m, then the probability of E is

$$P(E) = \frac{\text{Number of ways that } E \text{ can occur}}{\text{Number of all logical possibilities}} = \frac{m}{n}$$

Thus, if S is the sample space of this experiment, then

$$P(E) = \frac{n(E)}{n(S)}$$

■

Based on this result, an alternative method of solution of Example 4 is

$$P(E) = \frac{n(E)}{n(S)} = \frac{3}{12} = \frac{1}{4}$$

Example 5 What is the probability of selecting a ten, jack, queen, or king from a well-shuffled regular deck of 52 cards?*

Solution We assume no one card is more likely to be selected than another. Then, the outcomes are equally likely. The sample space S consists of the 52 cards, including 4 tens, 4 jacks, 4 queens, and 4 kings. The event E consists of a ten (4 possibilities), a jack (4 possibilities), a queen (4 possibilities), or a king (4 possibilities). Thus, $n(S) = 52$ and $n(E) = 16$. As a result, the probability of E is

$$P(E) = \frac{n(E)}{n(S)} = \frac{16}{52} = \frac{4}{13} \approx 0.3077$$ ■

Example 6 In the game of "craps," two fair dice are rolled. If the total of the faces equals 7 or 11, you win. If the totals are 2, 3, or 12, you have thrown craps and you lose. In all other cases, you throw again.

(a) What is the probability you will win?
(b) What is the probability you will lose?
(c) What is the probability you will need to throw again?

Solution We begin by constructing a probabilistic model for the experiment. The tree diagram in Figure 7 (page 840) will help us see all the logical probabilities. Because the dice are fair, no one of the 36 possible outcomes in the sample space S is more likely to occur than any other. Thus, we have equally likely outcomes with $n(S) = 36$.

(a) The event E, "the dice total 7 or 11," consists of the outcomes

$$E = \{(1, 6), (2, 5), (3, 4), (4, 3), (5, 2), (6, 1), (5, 6), (6, 5)\}$$

Because $n(E) = 8$, we have

$$P(E) = \frac{n(E)}{n(S)} = \frac{8}{36} = \frac{2}{9} \approx 0.222$$

(b) The event F, "the dice total 2, 3, or 12," consists of the outcomes

$$F = \{(1, 1), (1, 2), (2, 1), (6, 6)\}$$

*A regular deck of 52 cards consists of four suits (clubs, diamonds, hearts, spades), each containing 13 cards (ace, 2, 3, 4, 5, 6, 7, 8, 9, 10, jack, queen, king).

Figure 7

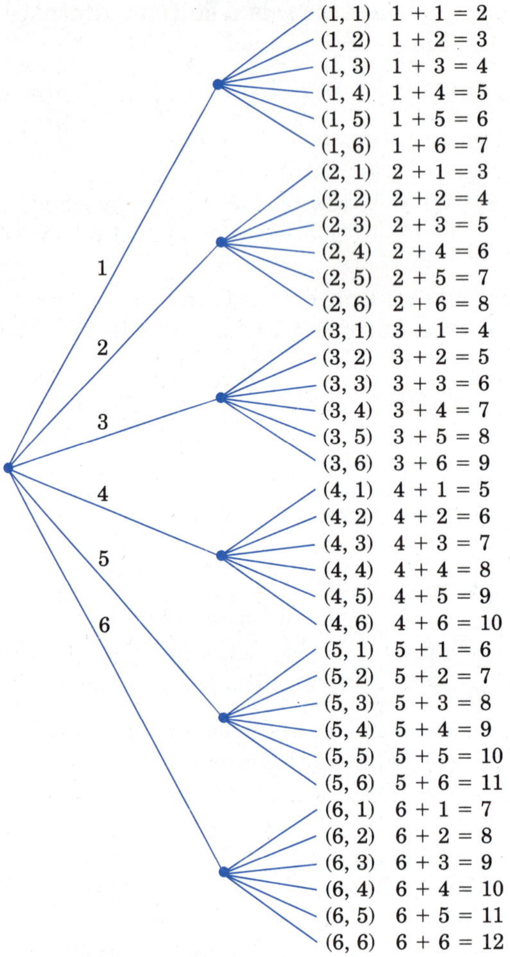

Because $n(F) = 4$,

$$P(F) = \frac{n(F)}{n(S)} = \frac{4}{36} = \frac{1}{9} \approx 0.111$$

(c) The number of possibilities that require that you throw again is $36 - n(E) - n(F) = 36 - 8 - 4 = 24$. Thus, the probability that another throw is required is

$$\frac{24}{36} = \frac{2}{3} \approx 0.667$$

■

Applications Involving Permutations and Combinations

C **Example 7** After shuffling a regular deck of 52 cards, what is the probability that, when they are all dealt, the first 4 cards dealt are aces?

Solution The sample space S consists of all the arrangements of 52 cards taken 52 at a time. Thus,

$$n(S) = P(52, 52) = 52!$$

The event E requires that the first 4 cards dealt be aces. Thus, in part, E consists of all arrangements of 4 cards taken 4 at a time, namely, $P(4, 4) = 4!$. We now need to count the number of possible arrangements of the remaining 48 cards, namely, $P(48, 48) = 48!$. By the multiplication principle, the number of elements in E is $n(E) = 4! \cdot 48!$. Because all the outcomes are equally likely, we have

$$P(E) = \frac{n(E)}{n(S)} = \frac{4! \cdot 48!}{52!} = \frac{24}{52 \cdot 51 \cdot 50 \cdot 49} \approx 0.000003693$$ ■

C **Example 8** In the poker game of 5-card stud, what is the probability of being dealt a:

(a) Royal flush (ace, king, queen, jack, ten of the same suit)?
(b) Flush (5 cards of the same suit)?

(In 5-card stud, 5 cards are dealt—no replacements allowed—from a regular deck of 52 cards.)

Solution (a) Because a regular deck contains 4 suits (clubs, diamonds, hearts, spades), there are 4 ways to be dealt a royal flush. All outcomes are equally likely. Thus we can use the result of Example 4 to find

$$\text{Probability(Royal flush)} = \frac{4}{C(52, 5)} = \frac{4}{2{,}598{,}960} \approx 0.0000015$$

(b) The number of ways a flush can be obtained in a given suit equals the number of combinations of 13 cards taken 5 at a time, namely, $C(13, 5)$. Because there are 4 possible suits, we use the multiplication principle and compute the total number of flushes as

$$4 \cdot C(13, 5) = 4 \cdot \frac{13!}{8!5!} = \frac{\cancel{4} \cdot 13 \cdot 12 \cdot 11 \cdot 10 \cdot 9 \cdot \cancel{8!}}{\cancel{8!} \cdot 5 \cdot \cancel{4} \cdot 3 \cdot 2 \cdot 1} = 5148$$

From this number, though, we need to subtract the royal flushes (4) and the straight flushes (36).* Thus, the number of flushes is $5148 - 4 - 36 = 5108$. Because all outcomes are equally likely, we find

$$\text{Probability(Flush)} = \frac{5108}{C(52, 5)} = \frac{5108}{2{,}598{,}960} \approx 0.0020$$ ■

*A straight flush consists of 5 consecutive cards of the same suit. Besides royal flushes, there are 9 ways {A, 1, 2, 3, 4}, . . . , {9, 10, J, Q, K}, in each of 4 suits, for a total of 36 ways, to get a straight flush. Royal flushes and straight flushes are better poker hands than regular flushes since they are more difficult to obtain.

Historical Comments

■ Set theory, counting, and probability first took form as a systematic theory in the exchange of letters (1654) between Pierre de Fermat (1601–1665) and Blaise Pascal (1623–1662). They discussed the problem of how to divide the stakes in a game that is interrupted before completion, knowing how many points each player needs to win. Fermat solved the problem by listing all possibilities and counting the favorable ones, whereas Pascal made use of the triangle that now bears his name. As mentioned in the text, the entries in Pascal's triangle are equivalent to $C(n, r)$. This recognition of the role of $C(n, r)$ in counting is the foundation of all further developments.

The first book on probability, the work of Christian Huygens (1629–1695), appeared in 1657. In it, the notion of mathematical expectations is explored. This allows the calculation of the profit or loss a gambler may expect, knowing the probabilities involved in the game (see the Historical Problems that follow).

It is interesting to note that Girolamo Cardano (1501–1576) wrote a treatise on probability, but it was not published until 1663 in Cardano's collected works, and this was too late to have any effect on the development of the theory.

In 1713, the posthumously published *Ars Conjectandi* of Jacob Bernoulli gave the theory the form it would have until 1900. In the current century, both combinatorics (counting) and probability have undergone rapid development due to the use of computers.

A final comment about notation. The notations $C(n, r)$ and $P(n, r)$ are variants of a form of notation developed in England after 1830. The notation $\binom{n}{r}$ for $C(n, r)$ goes back to Leonhard Euler (1707–1783), but is now losing ground because it has no clearly related symbolism of the same type for permutations. The set symbols $\cup$ and $\cap$ were introduced by Giuseppe Peano (1858–1932) in 1888 in a slightly different context. The inclusion symbol $\subset$ was introduced by E. Schroeder (1841–1902) about 1890. The treatment of set theory in the text is due to George Boole (1815–1864), who wrote $A + B$ for $A \cup B$ and AB for $A \cap B$ (statisticians still use AB for $A \cap B$).

Historical Problems

■ **1.** *The problem discussed by Fermat and Pascal* A game between two equally skilled players, A and B, is interrupted when A needs 2 points to win and B needs 3 points. In what proportion should the stakes be divided? [*Note:* If each play results in 1 point for either player, at most four more plays will decide the game.]

(a) *Fermat's solution* List all possible outcomes that will end the game to form the sample space (for example, $ABAA$, $ABBB$, etc.). The probabilities for A to win and B to win then determine how the stakes should be divided.

(b) *Pascal's solution* Use combinations to determine the number of ways the 2 points needed for A to win could occur in four plays. Then use combinations to determine the number of ways the 3 points needed for B to win could occur. This is trickier than it looks, since A can win with 2 points in either two plays, three plays, or four plays. Compute the probabilities and compare with the results in part (a).

2. *Huygens' mathematical expectation* In a game with n possible outcomes with probabilities $p_1, p_2, \ldots, p_n$, suppose that the *net* winnings are $w_1, w_2, \ldots, w_n$, respectively. Then the mathematical expectation is

$$E = p_1w_1 + p_2w_2 + \cdots + p_nw_n$$

The number E represents the profit or loss per game in the long run. The following problems are a modification of those of Huygens:

(a) A fair die is tossed. A gambler wins \$3 if he throws a 6 and \$6 if he throws a 5. What is his expectation? [*Note:* $w_1 = w_2 = w_3 = w_4 = 0$]

(b) A gambler plays the same game as in part (a), but now the gambler must pay \$1 to play. This means $w_5 = \$5$, $w_6 = \$2$, and $w_1 = w_2 = w_3 = w_4 = -\$1$. What is the expectation? ■

EXERCISE 15.4

In Problems 1–6, construct a probabilistic model for each experiment.

1. Tossing a fair coin twice

2. Tossing two fair coins once

3. Tossing two fair coins, then a fair die

4. Tossing a fair coin, a fair die, and then a fair coin

5. Tossing three fair coins once

6. Tossing one fair coin three times

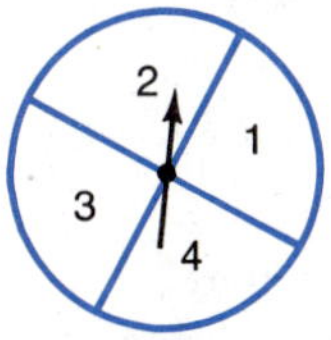

Spinner I

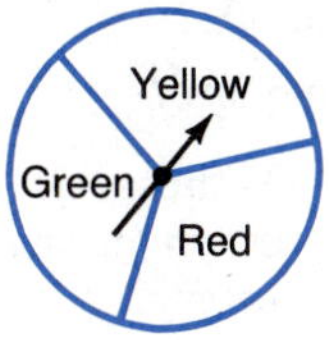

Spinner II

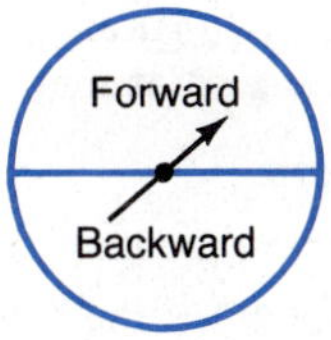

Spinner III

In Problems 7–12, use the spinners shown in the margin, and construct a probabilistic model for each experiment.

7. Spin spinner I, then spinner II. What is the probability of getting a 2 or a 4, followed by Red?

8. Spin spinner III, then spinner II. What is the probability of getting Forward, followed by Yellow or Green?

9. Spin spinner I, then II, then III. What is the probability of getting a 1, followed by Red or Green, followed by Backward?

10. Spin spinner II, then I, then III. What is the probability of getting Yellow, followed by a 2 or a 4, followed by Forward?

11. Spin spinner I twice, then spinner II. What is the probability of getting a 2, followed by a 2 or a 4, followed by Red or Green?

12. Spin spinner III, then spinner I twice. What is the probability of getting Forward, followed by a 1 or a 3, followed by a 2 or a 4?

13. What is the probability of selecting a face card (jack, queen, or king) from a well-shuffled deck of 52 cards?
14. What is the probability of selecting a red 3 or a black face card from a well-shuffled deck of 52 cards?
15. What is the probability of throwing a 6 or an 8 in a game of craps?
16. What is the probability of throwing a 5 or a 9 in a game of craps?

In Problems 17–20, a golf ball is selected at random from a container. If the container has 9 white balls, 8 green ones, and 3 orange ones, find the probability of each event.

17. The golf ball is white.
18. The golf ball is green.
19. The golf ball is white or green.
20. The golf ball is not white.

Problems 21–24 are based on a consumer survey of annual incomes in 100 households. The following table gives the data:

Income	\$0–9999	\$10,000–\$19,999	\$20,000–\$29,999	\$30,000–\$39,999	\$40,000 or more
Number of households	5	35	30	20	10

21. What is the probability that a household has an annual income of \$30,000 or more?
22. What is the probability that a household has an annual income between \$10,000 and \$29,999, inclusive?
23. What is the probability that a household has an annual income less than \$20,000?
24. What is the probability that a household has an annual income of \$20,000 or more?

C 25. In a certain lottery, there are ten balls, numbered 0, 1, 2, 3, 4, 5, 6, 7, 8, 9. Of these, five are drawn in the correct order. If you pick five numbers that match those drawn in the correct order, you win \$1,000,000. What is the probability of winning such a lottery?

C 26. Rework Problem 25 if only four balls are drawn.

C 27. A committee of 6 people is to be chosen at random from a group of 14 people consisting of 2 supervisors, 5 skilled laborers, and 7 unskilled laborers. What is the probability that the committee chosen consists of 2 skilled and 4 unskilled laborers?

C 28. Rework Problem 27 if the committee chosen consists only of unskilled laborers.

C 29. What is the probability of being dealt a straight in a game of 5-card stud? (A straight consists of 5 consecutive cards, such as 3, 4, 5, 6, 7, of any suit. The ace may be used as a 1 or as a card following the king.)

C 30. What is the probability of being dealt four of a kind in a game of 5-card stud?

CHAPTER REVIEW

THINGS TO KNOW

Set		Well-defined collection of distinct objects, called elements
Null set	$\varnothing$	Set that has no elements
Equality	$A = B$	A and B have the same elements.
Subset	$A \subseteq B$	Each element of A is also an element of B.
Intersection	$A \cap B$	Set consisting of elements that belong to both A and B
Union	$A \cup B$	Set consisting of elements that belong to either A or B, or both
Universal set	U	Set consisting of all the elements we wish to consider
Complement	A'	Set consisting of elements of the universal set that are not in A
Finite set		The number of elements in the set is a nonnegative integer.
Infinite set		A set that is not finite
Counting formula	$n(A \cup B) = n(A) + n(B) - n(A \cap B)$	
Addition principle		If $A \cap B = \varnothing$, then $n(A \cup B) = n(A) + n(B)$.
Multiplication principle		If a task consists of a sequence of choices in which there are p selections for the first choice, q selections for the second choice, and so on, then the task of making these selections can be done in $p \cdot q \cdot \cdots$ ways.
Permutation	$P(n, r) = n(n-1) \cdot \cdots \cdot n-(r-1) = \dfrac{n!}{(n-r)!}$	An ordered arrangement of n distinct objects without repetition
Combination	$C(n, r) = \dfrac{P(n, r)}{r!} = \dfrac{n!}{(n-r)!r!}$	An arrangement, without regard to order, of n distinct objects without repetition
Sample space		Set whose elements represent all the logical possibilities that can occur as a result of an experiment
Probability		A number assigned to each outcome of a sample space; the sum of all the probabilities of the outcomes equals 1.
Equally likely outcomes	$P(E) = \dfrac{n(E)}{n(S)}$	The same probability is assigned to each outcome.

How To:

Find unions, intersections, and complements of sets
Use Venn diagrams to illustrate sets
Recognize a permutation problem
Recognize a combination problem
Solve certain probability problems
Count the elements in a sample space
Draw a tree diagram

FILL-IN-THE-BLANK ITEMS

1. The __________ of A with B consists of all elements either in A or in B; the __________ of A with B consists of all elements both in A and in B.
2. $P(5, 2) =$ __________; $C(5, 2) =$ __________
3. A(n) __________ is an ordered arrangement of n distinct objects.
4. A(n) __________ is an arrangement of n distinct objects without regard to order.
5. When the same probability is assigned to each outcome of a sample space, the experiment is said to have __________ __________ outcomes.

REVIEW EXERCISES

In Problems 1–8, use U = *Universal set* = $\{1, 2, 3, 4, 5, 6, 7, 8, 9\}$, $A = \{1, 3, 5, 7\}$, $B = \{3, 5, 6, 7, 8\}$, *and* $C = \{2, 3, 7, 8, 9\}$ *to find each set.*

1. $A \cup B$
2. $B \cup C$
3. $A \cap C$
4. $A \cap B$
5. $A' \cup B'$
6. $B' \cap C'$
7. $(B \cap C)'$
8. $(A \cup B)'$
9. If $n(A) = 8$, $n(B) = 12$, and $n(A \cap B) = 3$, find $n(A \cup B)$.
10. If $n(A) = 12$, $n(A \cup B) = 30$, and $n(A \cap B) = 6$, find $n(B)$.

In Problems 11–16, use the information supplied in the following figure:

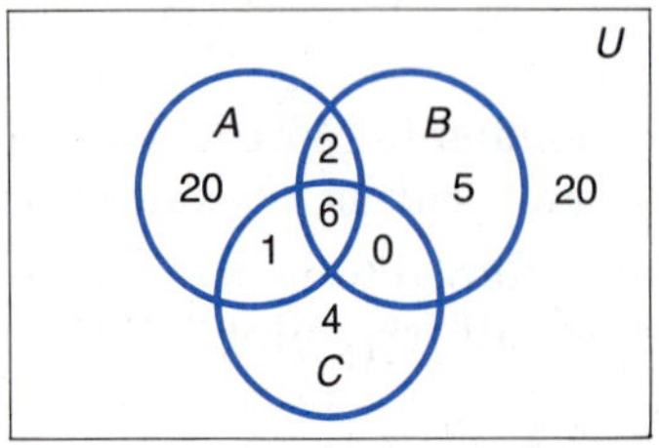

11. How many are in A?
12. How many are in A or B?
13. How many are in A and C?
14. How many are not in B?
15. How many are neither in A nor in C?
16. How many are in B but not in C?

In Problems 17–20, find the value of each expression.

17. $P(8, 3)$ **18.** $P(7, 3)$ **19.** $C(8, 3)$ **20.** $C(7, 3)$

21. A clothing store sells pure wool and polyester/wool suits. Each suit comes in 3 colors and 10 sizes. How many suits are required for a complete assortment?

22. In connecting a certain electrical device, 5 wires are to be connected to 5 different terminals. How many different wirings are possible if 1 wire is connected to each terminal?

23. On a given day, the American Baseball League schedules 7 games. How many different outcomes are possible, assuming each game is played to completion?

24. On a given day, the National Baseball League schedules 6 games. How many different outcomes are possible, assuming each game is played to completion?

25. If 4 people enter a bus having 9 vacant seats, in how many ways can they be seated?

26. How many different arrangements are there of the letters in the word ROSE?

27. In how many ways can a squad of 4 relay runners be chosen from a track team of 8 runners?

28. A professor has 10 similar problems to put on a test with 3 problems. How many different tests can she design?

29. In how many different ways can the 14 baseball teams in the American League be paired without regard to which team is at home?

30. Rework Problem 29 for the National League, which has 12 teams.

31. Rework Problem 29 if the pairings are with regard to which team is at home.

32. Rework Problem 30 if the pairings are with regard to which team is at home.

33. From a box containing three 40 watt bulbs, six 60 watt bulbs, and eleven 75 watt bulbs, a bulb is drawn at random. What is the probability that the bulb is 40 watts? What is the probability that it is not a 75 watt bulb?

34. You have four \$1 bills, three \$5 bills, and two \$10 bills in your wallet. If you pick a bill at random, what is the probability it will be a \$1 bill?

35. Each of the letters in the word ROSE is written on an index card and the cards are then shuffled. What is the probability that, when the cards are dealt out, they spell the word ROSE?

36. Each of the numbers 1, 2, . . . , 100 is written on an index card and the cards are then shuffled. If a card is selected at random, what is the probability that the number on the card is divisible by 5? What is the probability that the card selected either is a 1 or names a prime number?

Tables

Table I Exponential Functions

x	e^x	e^{-x}	x	e^x	e^{-x}	x	e^x	e^{-x}
0.00	1.0000	1.0000	0.50	1.6487	0.6065	3.0	20.086	0.0498
0.01	1.0101	0.9901	0.55	1.7333	0.5769	3.1	22.198	0.0450
0.02	1.0202	0.9802	0.60	1.8221	0.5488	3.2	24.533	0.0408
0.03	1.0305	0.9705	0.65	1.9155	0.5220	3.3	27.113	0.0369
0.04	1.0408	0.9608	0.70	2.0138	0.4966	3.4	29.964	0.0334
0.05	1.0513	0.9512	0.75	2.1170	0.4724	3.5	33.115	0.0302
0.06	1.0618	0.9418	0.80	2.2255	0.4493	3.6	36.598	0.0273
0.07	1.0725	0.9324	0.85	2.3396	0.4274	3.7	40.447	0.0247
0.08	1.0833	0.9331	0.90	2.4596	0.4066	3.8	44.701	0.0224
0.09	1.0942	0.9139	0.95	2.5857	0.3867	3.9	49.402	0.0202
0.10	1.1052	0.9048	1.0	2.7183	0.3679	4.0	54.598	0.0183
0.11	1.1163	0.8958	1.1	3.0042	0.3329	4.1	60.340	0.0166
0.12	1.1275	0.8869	1.2	3.3201	0.3012	4.2	66.686	0.0150
0.13	1.1388	0.8781	1.3	3.6693	0.2725	4.3	73.700	0.0136
0.14	1.1503	0.8694	1.4	4.0552	0.2466	4.4	81.451	0.0123
0.15	1.1618	0.8607	1.5	4.4817	0.2231	4.5	90.017	0.0111
0.16	1.1735	0.8521	1.6	4.9530	0.2019	4.6	99.484	0.0101
0.17	1.1853	0.8437	1.7	5.4739	0.1827	4.7	109.95	0.0091
0.18	1.1972	0.8353	1.8	6.0496	0.1653	4.8	121.51	0.0082
0.19	1.2092	0.8270	1.9	6.6859	0.1496	4.9	134.29	0.0074
0.20	1.2214	0.8187	2.0	7.3891	0.1353	5.0	148.41	0.0067
0.21	1.2337	0.8106	2.1	8.1662	0.1225	5.5	244.69	0.0041
0.22	1.2461	0.8025	2.2	9.0250	0.1108	6.0	403.43	0.0025
0.23	1.2586	0.7945	2.3	9.9742	0.1003	6.5	665.14	0.0015
0.24	1.2712	0.7866	2.4	11.023	0.0907	7.0	1096.6	0.0009
0.25	1.2840	0.7788	2.5	12.182	0.0821	7.5	1808.0	0.0006
0.30	1.3499	0.7408	2.6	13.464	0.0743	8.0	2981.0	0.0003
0.35	1.4191	0.7047	2.7	14.880	0.0672	8.5	4914.8	0.0002
0.40	1.4918	0.6703	2.8	16.445	0.0608	9.0	8103.1	0.0001
0.45	1.5683	0.6376	2.9	18.174	0.0550	10.0	22026	0.00005

Table II Common Logarithms

x	0	1	2	3	4	5	6	7	8	9
1.0	.0000	.0043	.0086	.0128	.0170	.0212	.0253	.0294	.0334	.0374
1.1	.0414	.0453	.0492	.0531	.0569	.0607	.0645	.0682	.0719	.0755
1.2	.0792	.0828	.0864	.0899	.0934	.0969	.1004	.1038	.1072	.1106
1.3	.1139	.1173	.1206	.1239	.1271	.1303	.1335	.1367	.1399	.1430
1.4	.1461	.1492	.1523	.1553	.1584	.1614	.1644	.1673	.1703	.1732
1.5	.1761	.1790	.1818	.1847	.1875	.1903	.1931	.1959	.1987	.2014
1.6	.2041	.2068	.2095	.2122	.2148	.2175	.2201	.2227	.2253	.2279
1.7	.2304	.2330	.2355	.2380	.2405	.2430	.2455	.2480	.2504	.2529
1.8	.2553	.2577	.2601	.2625	.2648	.2672	.2695	.2718	.2742	.2765
1.9	.2788	.2810	.2833	.2856	.2878	.2900	.2923	.2945	.2967	.2989
2.0	.3010	.3032	.3054	.3075	.3096	.3118	.3139	.3160	.3181	.3201
2.1	.3222	.3243	.3263	.3284	.3304	.3324	.3345	.3365	.3385	.3404
2.2	.3424	.3444	.3464	.3483	.3502	.3522	.3541	.3560	.3579	.3598
2.3	.3617	.3636	.3655	.3674	.3692	.3711	.3729	.3747	.3766	.3784
2.4	.3802	.3820	.3838	.3856	.3874	.3892	.3909	.3927	.3945	.3962
2.5	.3979	.3997	.4014	.4031	.4048	.4065	.4082	.4099	.4116	.4133
2.6	.4150	.4166	.4183	.4200	.4216	.4232	.4249	.4265	.4281	.4298
2.7	.4314	.4330	.4346	.4362	.4378	.4393	.4409	.4425	.4440	.4456
2.8	.4472	.4487	.4502	.4518	.4533	.4548	.4564	.4579	.4594	.4609
2.9	.4624	.4639	.4654	.4669	.4683	.4698	.4713	.4728	.4742	.4757
3.0	.4771	.4786	.4800	.4814	.4829	.4843	.4857	.4871	.4886	.4900
3.1	.4914	.4928	.4942	.4955	.4969	.4983	.4997	.5011	.5024	.5038
3.2	.5051	.5065	.5079	.5092	.5105	.5119	.5132	.5145	.5159	.5172
3.3	.5185	.5198	.5211	.5224	.5237	.5250	.5263	.5276	.5289	.5302
3.4	.5315	.5328	.5340	.5353	.5366	.5378	.5391	.5403	.5416	.5428
3.5	.5441	.5453	.5465	.5478	.5490	.5502	.5514	.5527	.5539	.5551
3.6	.5563	.5575	.5587	.5599	.5611	.5623	.5635	.5647	.5658	.5670
3.7	.5682	.5694	.5705	.5717	.5729	.5740	.5752	.5763	.5775	.5786
3.8	.5798	.5809	.5821	.5832	.5843	.5855	.5866	.5877	.5888	.5899
3.9	.5911	.5922	.5933	.5944	.5955	.5966	.5977	.5988	.5999	.6010
4.0	.6021	.6031	.6042	.6053	.6064	.6075	.6085	.6096	.6107	.6117
4.1	.6128	.6138	.6149	.6160	.6170	.6180	.6191	.6201	.6212	.6222
4.2	.6232	.6243	.6253	.6263	.6274	.6284	.6294	.6304	.6314	.6325
4.3	.6335	.6345	.6355	.6365	.6375	.6385	.6395	.6405	.6415	.6425
4.4	.6435	.6444	.6454	.6464	.6474	.6484	.6493	.6503	.6513	.6522
4.5	.6532	.6542	.6551	.6561	.6571	.6580	.6590	.6599	.6609	.6618
4.6	.6628	.6637	.6646	.6656	.6665	.6675	.6684	.6693	.6702	.6712
4.7	.6721	.6730	.6739	.6749	.6758	.6767	.6776	.6785	.6794	.6803
4.8	.6812	.6821	.6830	.6839	.6848	.6857	.6866	.6875	.6884	.6893
4.9	.6902	.6911	.6920	.6928	.6937	.6946	.6955	.6964	.6972	.6981
5.0	.6990	.6998	.7007	.7016	.7024	.7033	.7042	.7050	.7059	.7067
5.1	.7076	.7084	.7093	.7101	.7110	.7118	.7126	.7135	.7143	.7152
5.2	.7160	.7168	.7177	.7185	.7193	.7202	.7210	.7218	.7226	.7235
5.3	.7243	.7251	.7259	.7267	.7275	.7284	.7292	.7300	.7308	.7316
5.4	.7324	.7332	.7340	.7348	.7356	.7364	.7372	.7380	.7388	.7396

x	0	1	2	3	4	5	6	7	8	9
5.5	.7404	.7412	.7419	.7427	.7435	.7443	.7451	.7459	.7466	.7474
5.6	.7482	.7490	.7497	.7505	.7513	.7520	.7528	.7536	.7543	.7551
5.7	.7559	.7566	.7574	.7582	.7589	.7597	.7604	.7612	.7619	.7627
5.8	.7634	.7642	.7649	.7657	.7664	.7672	.7679	.7686	.7694	.7701
5.9	.7709	.7716	.7723	.7731	.7738	.7745	.7752	.7760	.7767	.7774
6.0	.7782	.7789	.7796	.7803	.7810	.7818	.7825	.7832	.7839	.7846
6.1	.7853	.7860	.7868	.7875	.7882	.7889	.7896	.7903	.7910	.7917
6.2	.7924	.7931	.7938	.7945	.7952	.7959	.7966	.7973	.7980	.7987
6.3	.7993	.8000	.8007	.8014	.8021	.8028	.8035	.8041	.8048	.8055
6.4	.8062	.8069	.8075	.8082	.8089	.8096	.8102	.8109	.8116	.8122
6.5	.8129	.8136	.8142	.8149	.8156	.8162	.8169	.8176	.8182	.8189
6.6	.8195	.8202	.8209	.8215	.8222	.8228	.8235	.8241	.8248	.8254
6.7	.8261	.8267	.8274	.8280	.8287	.8293	.8299	.8306	.8312	.8319
6.8	.8325	.8331	.8338	.8344	.8351	.8357	.8363	.8370	.8376	.8382
6.9	.8388	.8395	.8401	.8407	.8414	.8420	.8426	.8432	.8439	.8445
7.0	.8451	.8457	.8463	.8470	.8476	.8482	.8488	.8494	.8500	.8506
7.1	.8513	.8519	.8525	.8531	.8537	.8543	.8549	.8555	.8561	.8567
7.2	.8573	.8579	.8585	.8591	.8597	.8603	.8609	.8615	.8621	.8627
7.3	.8633	.8639	.8645	.8651	.8657	.8663	.8669	.8675	.8681	.8686
7.4	.8692	.8698	.8704	.8710	.8716	.8722	.8727	.8733	.8739	.8745
7.5	.8751	.8756	.8762	.8768	.8774	.8779	.8785	.8791	.8797	.8802
7.6	.8808	.8814	.8820	.8825	.8831	.8837	.8842	.8848	.8854	.8859
7.7	.8865	.8871	.8876	.8882	.8887	.8893	.8899	.8904	.8910	.8915
7.8	.8921	.8927	.8932	.8938	.8943	.8949	.8954	.8960	.8965	.8971
7.9	.8976	.8982	.8987	.8993	.8998	.9004	.9009	.9015	.9020	.9025
8.0	.9031	.9036	.9042	.9047	.9053	.9058	.9063	.9069	.9074	.9079
8.1	.9085	.9090	.9096	.9101	.9106	.9112	.9117	.9122	.9128	.9133
8.2	.9138	.9143	.9149	.9154	.9159	.9165	.9170	.9175	.9180	.9186
8.3	.9191	.9196	.9201	.9206	.9212	.9217	.9222	.9227	.9232	.9238
8.4	.9243	.9248	.9253	.9258	.9263	.9269	.9274	.9279	.9284	.9289
8.5	.9294	.9299	.9304	.9309	.9315	.9320	.9325	.9330	.9335	.9340
8.6	.9345	.9350	.9355	.9360	.9365	.9370	.9375	.9380	.9385	.9390
8.7	.9395	.9400	.9405	.9410	.9415	.9420	.9425	.9430	.9435	.9440
8.8	.9445	.9450	.9455	.9460	.9465	.9469	.9474	.9479	.9484	.9489
8.9	.9494	.9499	.9504	.9509	.9513	.9518	.9523	.9528	.9533	.9538
9.0	.9542	.9547	.9552	.9557	.9562	.9566	.9571	.9576	.9581	.9586
9.1	.9590	.9595	.9600	.9605	.9609	.9614	.9619	.9624	.9628	.9633
9.2	.9638	.9643	.9647	.9652	.9657	.9661	.9666	.9671	.9675	.9680
9.3	.9685	.9689	.9694	.9699	.9703	.9708	.9713	.9717	.9722	.9727
9.4	.9731	.9736	.9741	.9745	.9750	.9754	.9759	.9763	.9768	.9773
9.5	.9777	.9782	.9786	.9791	.9795	.9800	.9805	.9809	.9814	.9818
9.6	.9823	.9827	.9832	.9836	.9841	.9845	.9850	.9854	.9859	.9863
9.7	.9868	.9872	.9877	.9881	.9886	.9890	.9894	.9899	.9903	.9908
9.8	.9912	.9917	.9921	.9926	.9930	.9934	.9939	.9943	.9948	.9952
9.9	.9956	.9961	.9965	.9969	.9974	.9978	.9983	.9987	.9991	.9996

Table III Natural Logarithms

x	$\ln x$	x	$\ln x$	x	$\ln x$
		4.5	1.5041	9.0	2.1972
0.1	−2.3026	4.6	1.5261	9.1	2.2083
0.2	−1.6094	4.7	1.5476	9.2	2.2192
0.3	−1.2040	4.8	1.5686	9.3	2.2300
0.4	−0.9163	4.9	1.5892	9.4	2.2407
0.5	−0.6931	5.0	1.6094	9.5	2.2513
0.6	−0.5108	5.1	1.6292	9.6	2.2618
0.7	−0.3567	5.2	1.6487	9.7	2.2721
0.8	−0.2231	5.3	1.6677	9.8	2.2824
0.9	−0.1054	5.4	1.6864	9.9	2.2925
1.0	0.0000	5.5	1.7047	10	2.3026
1.1	0.0953	5.6	1.7228	11	2.3979
1.2	0.1823	5.7	1.7405	12	2.4849
1.3	0.2624	5.8	1.7579	13	2.5649
1.4	0.3365	5.9	1.7750	14	2.6391
1.5	0.4055	6.0	1.7918	15	2.7081
1.6	0.4700	6.1	1.8083	16	2.7726
1.7	0.5306	6.2	1.8245	17	2.8332
1.8	0.5878	6.3	1.8405	18	2.8904
1.9	0.6419	6.4	1.8563	19	2.9444
2.0	0.6931	6.5	1.8718	20	2.9957
2.1	0.7419	6.6	1.8871	25	3.2189
2.2	0.7885	6.7	1.9021	30	3.4012
2.3	0.8329	6.8	1.9169	35	3.5553
2.4	0.8755	6.9	1.9315	40	3.6889
2.5	0.9163	7.0	1.9459	45	3.8067
2.6	0.9555	7.1	1.9601	50	3.9120
2.7	0.9933	7.2	1.9741	55	4.0073
2.8	1.0296	7.3	1.9879	60	4.0943
2.9	1.0647	7.4	2.0015	65	4.1744
3.0	1.0986	7.5	2.0149	70	4.2485
3.1	1.1314	7.6	2.0281	75	4.3175
3.2	1.1632	7.7	2.0412	80	4.3820
3.3	1.1939	7.8	2.0541	85	4.4427
3.4	1.2238	7.9	2.0669	90	4.4998
3.5	1.2528	8.0	2.0794	100	4.6052
3.6	1.2809	8.1	2.0919	110	4.7005
3.7	1.3083	8.2	2.1041	120	4.7875
3.8	1.3350	8.3	2.1163	130	4.8676
3.9	1.3610	8.4	2.1282	140	4.9416
4.0	1.3863	8.5	2.1401	150	5.0106
4.1	1.4110	8.6	2.1518	160	5.0752
4.2	1.4351	8.7	2.1633	170	5.1358
4.3	1.4586	8.8	2.1748	180	5.1930
4.4	1.4816	8.9	2.1861	190	5.2470

Table IV Trigonometric Functions

DEGREES	RADIANS	sin	cos	tan	cot		
0	0.0000	0.0000	1.0000	0.0000		1.5708	90
1	0.0175	0.0175	0.9998	0.0175	57.290	1.5533	89
2	0.0349	0.0349	0.9994	0.0349	28.636	1.5359	88
3	0.0524	0.0523	0.9986	0.0524	19.081	1.5184	87
4	0.0698	0.0698	0.9976	0.0699	14.301	1.5010	86
5	0.0873	0.0872	0.9962	0.0875	11.430	1.4835	85
6	0.1047	0.1045	0.9945	0.1051	9.5144	1.4661	84
7	0.1222	0.1219	0.9925	0.1228	8.1443	1.4486	83
8	0.1396	0.1392	0.9903	0.1405	7.1154	1.4312	82
9	0.1571	0.1564	0.9877	0.1584	6.3138	1.4137	81
10	0.1745	0.1736	0.9848	0.1763	5.6713	1.3963	80
11	0.1920	0.1908	0.9816	0.1944	5.1446	1.3788	79
12	0.2094	0.2079	0.9781	0.2126	4.7046	1.3614	78
13	0.2269	0.2250	0.9744	0.2309	4.3315	1.3439	77
14	0.2443	0.2419	0.9703	0.2493	4.0108	1.3265	76
15	0.2618	0.2588	0.9659	0.2679	3.7321	1.3090	75
16	0.2793	0.2756	0.9613	0.2867	3.4874	1.2915	74
17	0.2967	0.2924	0.9563	0.3057	3.2709	1.2741	73
18	0.3142	0.3090	0.9511	0.3249	3.0777	1.2566	72
19	0.3316	0.3256	0.9455	0.3443	2.9042	1.2392	71
20	0.3491	0.3420	0.9397	0.3640	2.7475	1.2217	70
21	0.3665	0.3584	0.9336	0.3839	2.6051	1.2043	69
22	0.3840	0.3746	0.9272	0.4040	2.4751	1.1868	68
23	0.4014	0.3907	0.9205	0.4245	2.3559	1.1694	67
24	0.4189	0.4067	0.9135	0.4452	2.2460	1.1519	66
25	0.4363	0.4226	0.9063	0.4663	2.1445	1.1345	65
26	0.4538	0.4384	0.8988	0.4877	2.0503	1.1170	64
27	0.4712	0.4540	0.8910	0.5095	1.9626	1.0996	63
28	0.4887	0.4695	0.8829	0.5317	1.8807	1.0821	62
29	0.5061	0.4848	0.8746	0.5543	1.8040	1.0647	61
30	0.5236	0.5000	0.8660	0.5774	1.7321	1.0472	60
31	0.5411	0.5150	0.8572	0.6009	1.6643	1.0297	59
32	0.5585	0.5299	0.8480	0.6249	1.6003	1.0123	58
33	0.5760	0.5446	0.8387	0.6494	1.5399	0.9948	57
34	0.5934	0.5592	0.8290	0.6745	1.4826	0.9774	56
35	0.6109	0.5736	0.8192	0.7002	1.4280	0.9599	55
36	0.6283	0.5878	0.8090	0.7265	1.3764	0.9425	54
37	0.6458	0.6018	0.7986	0.7536	1.3270	0.9250	53
38	0.6632	0.6157	0.7880	0.7813	1.2799	0.9076	52
39	0.6807	0.6293	0.7771	0.8098	1.2349	0.8901	51
40	0.6981	0.6428	0.7660	0.8391	1.1918	0.8727	50
41	0.7156	0.6561	0.7547	0.8693	1.1504	0.8552	49
42	0.7330	0.6691	0.7431	0.9004	1.1106	0.8378	48
43	0.7505	0.6820	0.7314	0.9325	1.0724	0.8203	47
44	0.7679	0.6947	0.7193	0.9657	1.0355	0.8029	46
45	0.7854	0.7071	0.7071	1.0000	1.0000	0.7854	45
		cos	sin	cot	tan	RADIANS	DEGREES

Answers

CHAPTER 1 EXERCISE 1.1

1. 0.666. . . **3.** 0.125 **5.** −1.6 **7.** 0.111. . . **9.** 0.16 **11.** −0.428571428571. . . **13.** 2 **15.** −1 **17.** $\frac{13}{3}$ **19.** −11 **21.** 11 **23.** −4 **25.** $\frac{7}{2}$ **27.** $\frac{52}{15}$ **29.** $\frac{81}{10}$ **31.** $3x + 12$ **33.** $x^2 - 3x$ **35.** $x^2 + 6x + 8$ **37.** $x^2 - x - 2$ **39.** $x^2 - 10x + 16$ **41.** $x^2 - 4$ **43.** No; $2 - 3 \neq 3 - 2$ **45.** No; $\frac{2}{3} \neq \frac{3}{2}$ **47.** Symmetric property **49.**

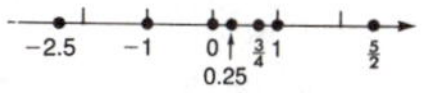

51. $x > 0$ **53.** $x < 2$ **55.** $|x| \leq 1$ **57.** Positive **59.** Positive **61.** Positive **63.** Positive **65.** 0 **67.** 1 **69.** 5 **71.** 1 **73.** 23 **75.** 7 **77.** $\{x|x \neq 3\}$ **79.** $\{x|x \neq -4\}$ **81.** $\{x|x \neq 0, x \neq 1\}$ **83.** $\{x|x \neq 0, x \neq -2\}$ **85.** All real numbers **87.** $A = lw$ **89.** $C = \pi d$ **91.** $A = \frac{\sqrt{3}}{4}x^2$ **93.** $V = \frac{4}{3}\pi r^3$ **95.** $V = x^3$ **97.** $C = 0°$ **99.** $C = 20°$ **101.** (a) $C = \$6000$ (b) $C = \$8000$ **103.** \$0.35 **105.** $a = 4, b = 5, c = 2, d = 5$ **107.** $a = 2, b = 3, c = 1$ **109.** $a = 5$ **111.** 4950 **113.** −2450 **115.** 200

EXERCISE 1.2

1. (a) 18.953 (b) 18.952 **3.** (a) 28.653 (b) 28.653 **5.** (a) 0.063 (b) 0.062 **7.** (a) 9.999 (b) 9.998 **9.** (a) 0.429 (b) 0.428 **11.** (a) 34.733 (b) 34.733 **13.** 72.42 **15.** 30.66 **17.** 111.17 **19.** 11.03 **21.** 3.32 **23.** 0.59 **25.** 21.58 **27.** 43.90 **29.** 0.30 **31.** 22.68 **33.** $\frac{290}{1{,}000{,}000}$; 0.00029 **35.** 800

EXERCISE 1.3

1. 1 **3.** $\frac{1}{16}$ **5.** $\frac{4}{9}$ **7.** $\frac{1}{8}$ **9.** $\frac{1}{4}$ **11.** $\frac{1}{9}$ **13.** 8 **15.** $\frac{9}{4}$ **17.** 324 **19.** $\frac{4}{81}$ **21.** y^2 **23.** $\frac{y}{x^2}$ **25.** $\frac{1}{64x^6}$ **27.** $\frac{-4}{x}$ **29.** 5 **31.** $\frac{1}{x^3y}$ **33.** $\frac{1}{xy}$ **35.** $\frac{y}{x}$ **37.** $\frac{25y^2}{16x^2}$ **39.** $\frac{1}{x} + \frac{1}{y^2}$ **41.** $\frac{1}{x^3y^3z^2}$ **43.** $\frac{-8x^3}{9yz^2}$ **45.** $\frac{y^2}{x^2 + y^2}$ **47.** $\frac{y^3}{x^8}$ **49.** $\frac{16x^2}{9y^2}$ **51.** $\frac{y}{x^3}$ **53.** x^4y **55.** $\frac{y^2}{x^2}$ **57.** 1, −11 **59.** 37,073.984 **61.** 0.004 **63.** 481.890 **65.** 0.000 **67.** 4.542×10^2 **69.** 1.3×10^{-2} **71.** 3.2155×10^4 **73.** 4.23×10^{-4} **75.** 21,500 **77.** 0.001214 **79.** 110,000,000 **81.** 0.081 **83.** 5.8657×10^{12} mi

EXERCISE 1.4

1. Yes; 2 **3.** Yes; 0 **5.** No **7.** Yes; 3 **9.** No **11.** $x^2 + 4x + 3$ **13.** $x^3 - 5x^2 + 6x + 9$ **15.** $6x^5 + 5x^4 + 3x^2 + x$ **17.** $9x^2 - 7x - 5$ **19.** $-2x^3 + 18x^2 - 18$ **21.** $2x^2 - 4x + 6$ **23.** $15y^2 - 18y + 3$ **25.** $2ax + a^2$ **27.** $3ax^2 + 3a^2x + a^3$ **29.** $x^2 + x - 12$ **31.** $2x^2 + 17x + 8$ **33.** $-3x^2 - 11x + 4$ **35.** $12x^2 - 11x + 2$ **37.** $x^3 - 1$ **39.** $2x^3 + 3x^2 - 4x - 6$ **41.** $6x^2 - x - 15$ **43.** $2x^3 + 7x^2 + 7x + 5$ **45.** $6x^3 - 11x^2 + 9x - 2$ **47.** $x^4 - x^2 + 2x - 1$ **49.** $6x^4 - 11x^3 + 23x^2 - 16x + 16$ **51.** $2x^5 - 6x^4 - x^3 + 13x^2 - 5$ **53.** $x^3 + 3x^2 - 4$ **55.** $2x^3 - x^2 - 4x + 3$ **57.** $x^3 - 6x^2 + 11x - 6$ **59.** $6x^2 + 2$

61. $x^4 - 2x^2 + 1$ **63.** $x^6 - 4x^5 + 6x^4 - 6x^3 + 4x^2 - 2x + 1$ **65.** $8x^3 + 36x^2 + 54x + 27$ **67.** $8x^3 - 36ax^2 + 54a^2x - 27a^3$ **69.** $x^2 - xy - 2y^2$ **71.** $x^2 - 2y^2 + 3x + y$ **73.** $-4xy$ **75.** $2x^2 - 2xy + 13y^2$ **77.** $2x^2 + 2y^2 + 2z^2 + 2xy + 2yz$ **79.** $x^3 - y^3$ **81.** $6x^4 + x^3y + 6x^2y^2 + xy^3 + 2y^4$ **83.** $x^2 - y^2 - z^2 - 2yz$

85. The degree of the product equals the degree of the product of the leading terms:

$$(a_nx^n + a_{n-1}x^{n-1} + \cdots + a_1x + a_0)(b_mx^m + b_{m-1}x^{m-1} + \cdots + b_1x + b_0) = a_nb_mx^{n+m} + (a_nb_{m-1} + a_{n-1}b_m)x^{n+m-1} + \cdots$$

87. (2) $(x - a)(x + a) = x(x + a) - a(x + a) = x^2 + ax - ax - a^2 = x^2 - a^2$
(3a) $(x + a)^2 = (x + a)(x + a) = x(x + a) + a(x + a) = x^2 + 2ax + a^2$
(3b) $(x - a)^2 = (x - a)(x - a) = x(x - a) - a(x - a) = x^2 - 2ax + a^2$
(6) $(x - a)(x^2 + ax + a^2) = x(x^2 + ax + a^2) - a(x^2 + ax + a^2) = x^3 + ax^2 + a^2x - ax^2 - a^2x - a^3 = x^3 - a^3$

89. $(x + a)^4 = (x + a)^2(x + a)^2 = (x^2 + 2ax + a^2)(x^2 + 2ax + a^2) = x^4 + 4ax^3 + 6a^2x^2 + 4a^3x + a^4$

EXERCISE 1.5

1. $3(x + 2)$ **3.** $a(x^2 + 1)$ **5.** $x(x^2 + x + 1)$ **7.** $2(x^2 + x + 1)$ **9.** $3xy(x - 2y + 4)$ **11.** $(x - 1)(x + 1)$ **13.** $(2x - 1)(2x + 1)$ **15.** $(x + 2)(x + 5)$ **17.** $(x - 7)(x - 3)$ **19.** $(x - 8)(x + 1)$ **21.** $(x + 1)^2$ **23.** $(x + 2)^2$ **25.** $(x - 5)(x + 3)$ **27.** $3(x + 2)(x - 6)$ **29.** $y^2(y + 5)(y + 6)$ **31.** $(4x + 1)^2$ **33.** $(2x + 3)^2$ **35.** $a(x - 9a)(x + 5a)$ **37.** $(x - 3)(x^2 + 3x + 9)$ **39.** $(2x + 3)(4x^2 - 6x + 9)$ **41.** $2x(x + 2)(x^2 - 2x + 4)$ **43.** $(3x + 1)(x + 1)$ **45.** $(x^2 + 9)(x + 3)(x - 3)$ **47.** $(x - 1)^2(x^2 + x + 1)^2$ **49.** $x^5(x - 1)(x + 1)$ **51.** $(2z + 3)(z + 1)$ **53.** $(4x - 3)^2$ **55.** $(4x - 5)(4x + 1)$ **57.** $(2y + 5)(2y + 3)$ **59.** $9(2x - 3)(x + 1)$ **61.** $2(4x^2 + x + 3)$ **63.** Prime **65.** $2x(2x + 1)(x - 3)$ **67.** $(x^2 - x + 1)(x^2 + x + 1)$ **69.** $(3x - 1)(3x + 1)(x^2 + 1)$ **71.** $(x + 3)(x - 6)$ **73.** $(x + 2)(x - 3)$ **75.** $(x + 5)(3x + 13)$ **77.** $(x - 1)(x + 1)(x + 2)$ **79.** $(x - 1)(x + 1)(x^2 - x + 1)$ **81.** $(x^2 + 1)(x + 2)(x^2 - 2x + 4)$

83. The possibilities are $(x \pm 1)(x \pm 4) = x^2 \pm 5x + 4$ or $(x \pm 2)(x \pm 2) = x^2 \pm 4x + 4$, none of which equal $x^2 + 4$.

EXERCISE 1.6

1. $\frac{5}{x - 2}$ **3.** $\frac{x}{3}$ **5.** $\frac{4}{2x - 1}$ **7.** $\frac{y + 5}{2}$ **9.** $x + 5$ **11.** $\frac{x - 2}{x + 3}$ **13.** $-(x + 7)$ **15.** $\frac{2x - 5}{x - 4}$ **17.** $\frac{x - 7}{x + 7}$ **19.** $\frac{6(x + 1)}{x^2 + 2x + 4}$ **21.** $\frac{3(x - 3)}{5x}$ **23.** $\frac{x(2x - 1)}{x + 4}$ **25.** $\frac{8}{3x}$ **27.** $\frac{x - 3}{x + 7}$ **29.** $\frac{(x - 4)(x + 3)}{(x - 1)(2x + 1)}$ **31.** $\frac{3x - 2}{3x + 2}$ **33.** $\frac{x^2(5x - 1)^2}{(x - 1)(3x + 2)^2}$ **35.** $\frac{2x - 3}{x - 3}$ **37.** $\frac{x + 9}{2x - 1}$ **39.** $\frac{4 - x}{x - 2}$ **41.** $\frac{3(x + 3)}{(x - 1)(x + 2)}$ **43.** $\frac{3x^2 - 2x - 3}{(x + 1)(x - 1)}$ **45.** $\frac{-8x}{(x + 2)(x - 2)}$ **47.** $\frac{2(x^2 - 2)}{x(x - 2)(x + 2)}$ **49.** $\frac{2x^3 - 2x^2 + 2x - 1}{x(x - 1)^2}$ **51.** $\frac{x^3 - 7x^2 + 3x + 5}{(x + 1)(x - 1)(x - 2)}$ **53.** $\frac{x^2 - 2x - 1}{x(x - 1)(x + 1)}$ **55.** $(x - 2)(x + 2)(x + 1)$ **57.** $x(x - 1)(x + 1)$ **59.** $x^3(2x - 1)^2$ **61.** $x(x - 1)^2(x + 1)(x^2 + x + 1)$ **63.** $\frac{5x}{(x - 6)(x - 1)(x + 4)}$ **65.** $\frac{2(x + 4)}{(x - 2)(x + 2)(x + 3)}$ **67.** $\frac{5x + 1}{(x - 1)^2(x + 1)^2}$ **69.** $\frac{-x^2 + 3x + 13}{(x - 2)(x + 1)(x + 4)}$ **71.** $\frac{x^3 - 2x^2 + 4x + 3}{x^2(x + 1)(x - 1)}$ **73.** $\frac{-1}{x(x + h)}$ **75.** $\frac{(x - 1)(x + 1)}{x^2 + 1}$ **77.** $\frac{(x - 1)(-x^2 + 3x + 3)}{x^2 + x - 1}$ **79.** $\frac{2(5x - 1)}{(x - 2)(x + 1)^2}$ **81.** $\frac{-2x(x^2 - 2)}{(x + 2)(x^2 - x - 3)}$ **83.** $\frac{-1}{x - 1}$ **85.** $\frac{4}{(x + h + 2)(x + 2)}$

87. a: 1, 2, 3, 5, 8, 13, . . . ; b: 1, 1, 2, 3, 5, 8, . . . ; c: 0, 1, 1, 2, 3, 5, . . .
(These are the numbers of a Fibonacci sequence, which we discuss in Chapter 9.)

EXERCISE 1.7

1. 8 **3.** 3 **5.** -1 **7.** $\frac{1}{2}$ **9.** $3|x|$ **11.** $2(1 + x)$ **13.** $2\sqrt{2}$ **15.** $2x\sqrt[3]{2x}$ **17.** x **19.** $\frac{4}{3}x\sqrt{2}$ **21.** x^3y^2 **23.** x^2y **25.** $6\sqrt{x}$ **27.** $6x\sqrt{x}$ **29.** 1 **31.** $\frac{2}{3xy^2}$ **33.** $15\sqrt[3]{3}$ **35.** $\frac{1}{x(2x + 3)}$ **37.** $\sqrt[24]{x}$ **39.** 1 **41.** $4\sqrt{2}$ **43.** $4\sqrt{2}$ **45.** $\sqrt[3]{2}$ **47.** $(x + 5)(x - 3)\sqrt{2x}$ **49.** $(-x - 5y)\sqrt[3]{2xy}$ **51.** $12\sqrt{3}$ **53.** $2\sqrt{3}$ **55.** $12(4 + \sqrt{7})$ **57.** $x - 2\sqrt{x} + 1$ **59.** $x - 3\sqrt[3]{x^2} + 3\sqrt[3]{x} - 1$ **61.** $4x + 4\sqrt{xy} - 15y$ **63.** $\frac{-x^2}{\sqrt{1 - x^2}}$ **65.** $\frac{\sqrt{2}}{2}$ **67.** $\frac{-\sqrt{15}}{5}$ **69.** $\frac{\sqrt{x}}{x}$ **71.** $\frac{\sqrt{3}(5 + \sqrt{2})}{23}$ **73.** $\frac{-19 + 8\sqrt{5}}{41}$ **75.** $5 - 2\sqrt{6}$ **77.** $\frac{\sqrt{x} - 2}{x - 4}$ **79.** $\frac{2\sqrt{x} - 1}{4x - 1}$ **81.** $\frac{2x + h - 2\sqrt{x(x + h)}}{h}$ **83.** $\frac{\sqrt[3]{x^2} - \sqrt[3]{x} + 1}{x + 1}$ **85.** $\frac{-\sqrt[3]{3xy}}{3xy^2}$ **87.** $\frac{1}{\sqrt{x + h} + \sqrt{x}}$

EXERCISE 1.8

1. 4 **3.** -3 **5.** 64 **7.** $\frac{1}{27}$ **9.** $\frac{27\sqrt{2}}{32}$ **11.** $\frac{27\sqrt{2}}{32}$ **13.** 8 **15.** 8 **17.** 27 **19.** $\frac{1}{2}$ **21.** $x^{7/12}$ **23.** xy^2 **25.** $x^{4/3}y^{5/3}$ **27.** $\frac{8x^{3/2}}{y^{1/4}}$ **29.** $\frac{x^{11}}{y^{13}}$
31. $\frac{3x+2}{(1+x)^{1/2}}$ **33.** $\frac{x(3x^2+2)}{(x^2+1)^{1/2}}$ **35.** $\frac{22x+5}{10\sqrt{x-5}\sqrt{4x+3}}$ **37.** $\frac{2+x}{2(1+x)^{3/2}}$ **39.** $\frac{4-x}{(x+4)^{3/2}}$ **41.** $\frac{1}{x^2(x^2-1)^{1/2}}$ **43.** $\frac{1-3x^2}{2\sqrt{x}(1+x^2)^2}$
45. $\frac{1}{2}(5x+2)(x+1)^{1/2}$ **47.** $2x^{1/2}(3x-4)(x+1)$ **49.** $(x^2+4)^{1/3}(11x^2+12)$ **51.** $(3x+5)^{1/3}(2x+3)^{1/2}(17x+27)$

53.
$$\begin{aligned}\sqrt{1+\left[\frac{1}{2}\left(x^3-\frac{1}{x^3}\right)\right]^2} &= \sqrt{1+\frac{1}{4}\left(x^6-2+\frac{1}{x^6}\right)}\\ &= \sqrt{1+\frac{1}{4}\left(\frac{x^{12}-2x^6+1}{x^6}\right)}\\ &= \sqrt{\frac{4x^6+x^{12}-2x^6+1}{4x^6}}\\ &= \frac{1}{2}\sqrt{\frac{x^{12}+2x^6+1}{x^6}}\\ &= \frac{1}{2}\sqrt{x^6+2+\frac{1}{x^6}}\\ &= \frac{1}{2}\sqrt{\left(x^3+\frac{1}{x^3}\right)^2}\\ &= \frac{1}{2}\left(x^3+\frac{1}{x^3}\right)\end{aligned}$$

55.
$$\begin{aligned}1+\left[\frac{1}{2}\left(x^2-\frac{1}{x^2}\right)\right]^2 &= 1+\frac{1}{4}\left(x^4-2+\frac{1}{x^4}\right)\\ &= 1+\frac{x^8-2x^4+1}{4x^4}\\ &= \frac{4x^4+x^8-2x^4+1}{4x^4}\\ &= \frac{1}{4}\left(\frac{x^8+2x^4+1}{x^4}\right)\\ &= \frac{1}{4}\left(x^4+2+\frac{1}{x^4}\right)\\ &= \frac{1}{4}\left(x^2+\frac{1}{x^2}\right)^2\end{aligned}$$

EXERCISE 1.9

1. $8+5i$ **3.** $-7+6i$ **5.** $-6-11i$ **7.** $6-18i$ **9.** $6+4i$ **11.** $10-5i$ **13.** 37 **15.** $\frac{6}{5}+\frac{8}{5}i$ **17.** $1-2i$ **19.** $\frac{5}{2}-\frac{7}{2}i$
21. $-\frac{1}{2}+(\sqrt{3}/2)i$ **23.** $2i$ **25.** $\frac{2}{5}-\frac{11}{5}i$ **27.** $\frac{7}{50}-\frac{1}{50}i$ **29.** $-i$ **31.** i **33.** -6 **35.** $-10i$ **37.** $-2+2i$ **39.** 0 **41.** $\frac{1}{2}-\frac{1}{2}i$ **43.** 0
45. 6 **47.** 25 **49.** $36+23i$ **51.** $11-7i$ **53.** $z+\bar{z}=(a+bi)+(a-bi)=2a;\ z-\bar{z}=(a+bi)-(a-bi)=2bi$
55. $\overline{z+w}=\overline{(a+bi)+(c+di)}=\overline{(a+c)+(b+d)i}=(a+c)-(b+d)i=(a-bi)+(c-di)=\bar{z}+\bar{w}$

EXERCISE 1.10

1. 13 **3.** 26 **5.** 25 **7.** Yes; 5 **9.** Not a right triangle **11.** Yes; 25 **13.** Not a right triangle **15.** 46.7 mi **17.** 3 mi
19. $a^2+b^2=(m^2-n^2)^2+(2mn)^2=m^4-2m^2n^2+n^4+4m^2n^2=m^4+2m^2n^2+n^4=(m^2+n^2)^2=c^2$
The conclusion follows from the converse of the Pythagorean Theorem.

FILL-IN-THE-BLANK ITEMS

1. symmetric **2.** commutative **3.** distributive **4.** base; exponent **5.** 3; leading **6.** index; radicand
7. real; imaginary; imaginary unit

REVIEW EXERCISES

1. -16 **3.** $\frac{1}{6}$ **5.** $\frac{87}{8}$ **7.** -2 **9.** $\frac{1}{128}$ **11.** 16 **13.** $\frac{3\sqrt[3]{3}}{4}$ **15.** $\frac{1}{2}$ **17.** 3 **19.** 4 **21.** $\frac{y^2}{x^2}$ **23.** $\frac{1}{x^5y}$ **25.** $\frac{125}{x^2y}$ **27.** $\frac{x^2}{16y^3}$
29. $2x^3-13x^2+19x-6$ **31.** $9x^3-20x^2+6x+13$ **33.** $6x^2-11x-10$ **35.** $x^3+6x^2+11x+6$ **37.** $x^2+2xy-8y^2$
39. $(x+7)(x-2)$ **41.** $(3x+2)(2x-3)$ **43.** $3(x+2)(x-7)$ **45.** $(2x+1)(4x^2-2x+1)$ **47.** $(2x+3)(x-1)(x+1)$
49. $(5x-2)(5x+2)$ **51.** Prime **53.** $\frac{2x+7}{x-2}$ **55.** $\frac{3(3x-1)}{(x+3)(3x+1)}$ **57.** $\frac{4x}{(x+1)(x-1)}$ **59.** $\frac{x^2+17x+2}{(x-2)(x+2)^2}$ **61.** $\frac{2x+1}{x+1}$
63. $\frac{2\sqrt{3}}{3}$ **65.** $-2(1+\sqrt{2})$ **67.** $\frac{-(3+\sqrt{5})}{2}$ **69.** $\frac{2(1+x^2)}{(2+x^2)^{1/2}}$ **71.** $\frac{x(3x+16)}{2(x+4)^{3/2}}$ **73.** $4-7i$ **75.** $-3+2i$ **77.** $\frac{9}{10}-\frac{3}{10}i$ **79.** 1
81. $-46+9i$

CHAPTER 2 EXERCISE 2.1

1. 6 **3.** -2 **5.** 4 **7.** $\frac{5}{4}$ **9.** -4 **11.** 3 **13.** -1 **15.** $-\frac{4}{3}$ **17.** -18 **19.** -3 **21.** $-\frac{3}{4}$ **23.** -16 **25.** 2 **27.** 0.5 **29.** $\frac{46}{5}$ **31.** 2 **33.** 2 **35.** -1 **37.** 3 **39.** 2 **41.** 21 **43.** 1 **45.** $-\frac{20}{39}$ **47.** 6 **49.** $-\frac{11}{6}$ **51.** -6 **53.** $x = \dfrac{b+c}{a}$ **55.** $x = \dfrac{abc}{a+b}$ **57.** $x = a^2$ **59.** $a = 3$ **61.** $R = \dfrac{R_1R_2}{R_1 + R_2}$ **63.** $R = \dfrac{mv^2}{F}$ **65.** $r = \dfrac{S-a}{S}$
67. In step (7) we divided by $x - 2$. Since $x = 2$ from step (1), we actually divided by 0.

EXERCISE 2.2

1. $A = \pi r^2$; r = Radius, A = Area **3.** $A = s^2$; A = Area, s = Length of a side **5.** $F = ma$; F = Force, m = Mass, a = Acceleration **7.** $W = Fd$; W = Work, F = Force, d = Distance **9.** 41, 42 **11.** 13, 14 **13.** 65, 25 **15.** \$8.50 per hr **17.** 5 touchdowns **19.** Length = 14 ft; Width = 6 ft **21.** Length = 10 ft; Width = 5 ft **23.** \$31,250 in bonds; \$18,750 in a certificate **25.** \$8000 **27.** 40 lb **29.** 30 cc of 15% HCl; 70 cc of 5% HCl **31.** 3260 adults **33.** \$110,000; \$16,500 **35.** \$28 **37.** 12 min **39.** 85 **41.** At tight end's 45 yd line **43.** $\frac{2}{3}$ gal **45.** 2.14 mph **47.** 9:45 AM **49.** 1 hr **51.** $\frac{1}{3}$ mi; 2 min **53.** Slower car's average speed: 60 mph; faster car's average speed: 70 mph; distance traveled: 210 mi by each car

EXERCISE 2.3

1. $\{0, 4\}$ **3.** $\{-3, 3\}$ **5.** $\{-4, 3\}$ **7.** $\{-\frac{1}{2}, 3\}$ **9.** $\{-\frac{1}{2}, \frac{1}{3}\}$ **11.** $\{3, 4\}$ **13.** $\frac{3}{2}$ **15.** $\{-\frac{2}{3}, \frac{3}{2}\}$ **17.** $\{-\frac{2}{3}, \frac{3}{2}\}$ **19.** $\{-\frac{3}{4}, 2\}$ **21.** $\{2 - \sqrt{2}, 2 + \sqrt{2}\}$ **23.** $\left\{\dfrac{5-\sqrt{29}}{2}, \dfrac{5+\sqrt{29}}{2}\right\}$ **25.** $\{1, \frac{3}{2}\}$ **27.** No real solution **29.** $\left\{\dfrac{-1-\sqrt{5}}{4}, \dfrac{-1+\sqrt{5}}{4}\right\}$ **31.** $\{0, \frac{9}{4}\}$ **33.** $\frac{1}{3}$ **35.** $\{1 - \sqrt{2}, 1 + \sqrt{2}\}$ **37.** $\left\{\dfrac{1-\sqrt{33}}{8}, \dfrac{1+\sqrt{33}}{8}\right\}$ **39.** No real solution **41.** $\{0.59, 3.41\}$ **43.** $\{-2.80, 1.07\}$ **45.** $\{-0.85, 1.17\}$ **47.** No real solution **49.** $\{-\sqrt{6}, \sqrt{6}\}$ **51.** $\frac{1}{4}$ **53.** $\{-\frac{3}{5}, \frac{5}{2}\}$ **55.** $\{-\frac{1}{2}, \frac{2}{3}\}$ **57.** $\left\{\dfrac{-\sqrt{2}+2}{2}, \dfrac{-\sqrt{2}-2}{2}\right\}$ **59.** $\left\{\dfrac{-1-\sqrt{17}}{2}, \dfrac{-1+\sqrt{17}}{2}\right\}$ **61.** No real solution **63.** Repeated real solution **65.** Two unequal real solutions **67.** 4 **69.** $\frac{1}{16}$ **71.** $\frac{1}{9}$ **73.** $\{-7, 3\}$ **75.** $\{-\frac{1}{4}, \frac{3}{4}\}$ **77.** $\left\{\dfrac{-1-\sqrt{7}}{6}, \dfrac{-1+\sqrt{7}}{6}\right\}$ **79.** 11, 13 **81.** 7 m by 8 m **83.** 4 ft by 4 ft **85.** (a) 6 sec (b) 5 sec **87.** 10 days **89.** 175 boxes **91.** 2.56 ft **93.** Length = 11.55 cm; Width = 6.55 cm; Thickness = 3 cm (as given) **95.** 2.71 ft **97.** 5 mph **99.** 6 in. by 8 in. **101.** $\dfrac{-b+\sqrt{b^2-4ac}}{2a} + \dfrac{-b-\sqrt{b^2-4ac}}{2a} = \dfrac{-2b}{2a} = \dfrac{-b}{a}$ **103.** $k = \frac{1}{2}$ or $k = -\frac{1}{2}$ **105.** $ax^2 + bx + c = 0$, $x = \dfrac{-b \pm \sqrt{b^2-4ac}}{2a}$; $ax^2 - bx + c = 0$, $x = \dfrac{b \pm \sqrt{(-b)^2-4ac}}{2a}$ **107.** 36 **109.** 49.5 mph

EXERCISE 2.4

1. $2i$ **3.** $5i$ **5.** $5i$ **7.** $-15 + 8i$ **9.** $12 + 5i$ **11.** $\{-2i, 2i\}$ **13.** $\{-4, 4\}$ **15.** $\{3 - 2i, 3 + 2i\}$ **17.** $\{3 - i, 3 + i\}$ **19.** $\{\frac{1}{4} - \frac{1}{4}i, \frac{1}{4} + \frac{1}{4}i\}$ **21.** $\{-\frac{1}{5} - \frac{2}{5}i, -\frac{1}{5} + \frac{2}{5}i\}$ **23.** $\left\{-\dfrac{1}{2} - \dfrac{\sqrt{3}}{2}i, -\dfrac{1}{2} + \dfrac{\sqrt{3}}{2}i\right\}$ **25.** $\{2, -1 - \sqrt{3}i, -1 + \sqrt{3}i\}$ **27.** $\{-2, 2, -2i, 2i\}$ **29.** $\{-2i, 2i, -3i, 3i\}$ **31.** Two complex solutions that are not real **33.** Two unequal real solutions **35.** A repeated real solution **37.** $2 - 3i$

EXERCISE 2.5

1. 1 **3.** No real solution **5.** -13 **7.** 3 **9.** 2 **11.** $-\frac{8}{5}$ **13.** 2 **15.** $\{-1, 3\}$ **17.** $\{1, 5\}$ **19.** 1 **21.** 5 **23.** 9 **25.** $\{-3, 3\}$ **27.** $\{0, 16\}$ **29.** 9 **31.** $\{-\frac{3}{2}, 2\}$ **33.** $\{-2, -1, 1, 2\}$ **35.** 1 **37.** $\left\{\left(\dfrac{9-\sqrt{17}}{8}\right)^4, \left(\dfrac{9+\sqrt{17}}{8}\right)^4\right\}$ **39.** $\{\sqrt{2}, \sqrt{3}\}$ **41.** $\{-4, 1\}$ **43.** $\{-2, -\frac{1}{2}\}$ **45.** $\{-\frac{3}{2}, \frac{1}{3}\}$ **47.** $\{-\frac{1}{8}, 27\}$ **49.** $\{-2, -\frac{4}{5}\}$ **51.** $\{0, 64\}$ **53.** $\{0, 2\}$ **55.** $\{-4, 0, 3\}$ **57.** $\{-1\}$ **59.** $\{-2, 2, 3\}$ **61.** $\{-1, 1\}$ **63.** $\{0.34, 11.66\}$ **65.** $\{-1.03, 1.03\}$ **67.** $\{-1.85, 0.17\}$ **69.** 230 ft

EXERCISE 2.6

1. $x \geq -1$

3. $x \geq 4$ and $x < 6$

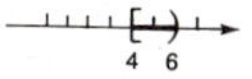

5. $x \leq 0$ or $x < 6$

7. There are no numbers for which $x \leq -2$ and $x > 1$.

9. $x \leq -2$ or $x > 1$

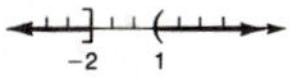

11. Negative **13.** Negative **15.** Positive **17.** Positive **19.** Positive

21. $[0, 4]$

23. $[4, 6)$

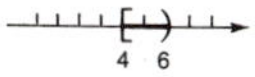

25. $[2, \infty)$

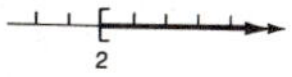

27. $(-\infty, -4)$

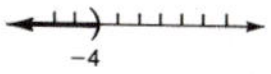

29. $2 \leq x \leq 5$

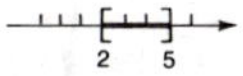

31. $-3 < x < -2$

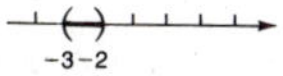

33. $x \geq 4$

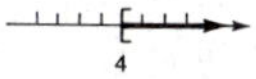

35. $x < -1$

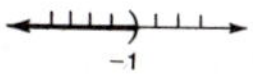

37. $a \leq b,\ c > 0$;
$$a - b \leq 0$$
$$(a - b)c \leq 0(c)$$
$$ac - bc \leq 0$$
$$ac \leq bc$$

39. $\dfrac{a+b}{2} - a = \dfrac{a+b-2a}{2} = \dfrac{b-a}{2} > 0$; therefore, $a < \dfrac{a+b}{2}$

$b - \dfrac{a+b}{2} = \dfrac{2b-a-b}{2} = \dfrac{b-a}{2} > 0$; therefore, $b > \dfrac{a+b}{2}$

41. $(\sqrt{ab})^2 - a^2 = ab - a^2 = a(b-a) > 0$; thus, $(\sqrt{ab})^2 > a^2$ and $\sqrt{ab} > a$

$b^2 - (\sqrt{ab})^2 = b^2 - ab = b(b-a) > 0$; thus, $b^2 > (\sqrt{ab})^2$ and $b > \sqrt{ab}$

43. $h - a = \dfrac{2ab}{a+b} - a = \dfrac{ab - a^2}{a+b} = \dfrac{a(b-a)}{a+b} > 0$; thus, $h > a$

$b - h = b - \dfrac{2ab}{a+b} = \dfrac{b^2 - ab}{a+b} = \dfrac{b(b-a)}{a+b} > 0$; thus, $h < b$

45. $21 < \text{Age} < 30$ **47.** (a) Male ≥ 73.4 (b) Female ≥ 79.7 (c) Female; by 6.3 yr

EXERCISE 2.7

1. (a) $0 < 2$ (b) $-2 < 0$ (c) $9 < 15$ (d) $-6 > -10$

3. (a) $2x - 2 < -1$ (b) $2x - 4 < -3$ (c) $6x + 3 < 6$ (d) $-4x - 2 > -4$

5. $\{x \mid x < 4\}$ or $(-\infty, 4)$

7. $\{x \mid x \geq -1\}$ or $[-1, \infty)$

9. $\{x \mid x > 3\}$ or $(3, \infty)$

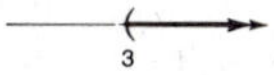

11. $\{x \mid x \geq 2\}$ or $[2, \infty)$

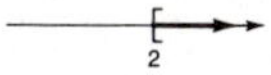

13. $\{x \mid x > -6\}$ or $(-6, \infty)$

15. $\{x \mid x \leq \frac{2}{3}\}$ or $\left(-\infty, \frac{2}{3}\right]$

17. $\{x \mid x < -20\}$ or $(-\infty, -20)$

19. $\{x \mid x \geq \frac{4}{3}\}$ or $\left[\frac{4}{3}, \infty\right)$

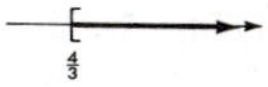

21. $\{x \mid 3 \leq x \leq 5\}$ or $[3, 5]$

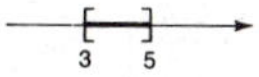

23. $\{x \mid -\frac{1}{3} \leq x \leq \frac{7}{3}\}$ or $\left[-\frac{1}{3}, \frac{7}{3}\right]$

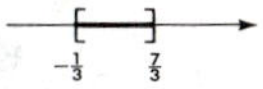

25. $\{x \mid -\frac{11}{2} < x < \frac{1}{2}\}$ or $\left(-\frac{11}{2}, \frac{1}{2}\right)$

27. $\{x \mid -6 < x < 0\}$ or $(-6, 0)$

29. $\{x \mid x < -5\}$ or $(-\infty, -5)$

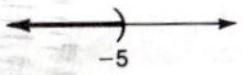

31. $\{x \mid x \geq -1\}$ or $[-1, \infty)$

33. $\{x \mid \frac{1}{2} \leq x < \frac{5}{4}\}$ or $\left[\frac{1}{2}, \frac{5}{4}\right)$

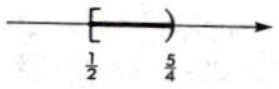

35. $\{x \mid x < -\frac{1}{2}\}$ or $\left(-\infty, -\frac{1}{2}\right)$

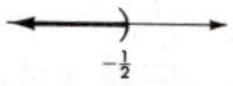

37. $\{x \mid x > 3\}$ or $(3, \infty)$

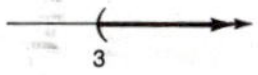

39. $\{x \mid x > 3\}$ or $(3, \infty)$

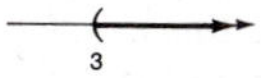

41. $\{x \mid x \geq 2\}$ or $[2, \infty)$

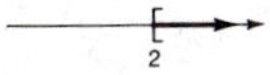

43. From 1600 to 2400 cc, inclusive **45.** From \$45,000 to \$95,000, inclusive; from 5% to 8.6%, inclusive

47. From \$87.71 to \$101.71, inclusive **49.** From 675.48 kWhr to 2500.86 kWhr, inclusive

51. From \$7457.63 to \$7857.14, inclusive **53.** Fifth test score ≥ 74 **55.** From 12 to 20 gal, inclusive

EXERCISE 2.8

1. $\{x \mid -1 < x < 3\}$ **3.** $\{x \mid x < 0 \text{ or } x > 4\}$ **5.** $\{x \mid -3 < x < 3\}$ **7.** $\{x \mid x < -4 \text{ or } x > 3\}$ **9.** $\{x \mid -\frac{1}{2} < x < 3\}$ **11.** $\{x \mid x < 3 \text{ or } x > 4\}$ **13.** No real solution **15.** $\{x \mid x < -\frac{2}{3} \text{ or } x > \frac{3}{2}\}$ **17.** $\{x \mid x > 1\}$ **19.** $\{x \mid x < 1 \text{ or } 2 < x < 3\}$ **21.** $\{x \mid -2 < x < 0 \text{ or } x > 4\}$ **23.** $\{x \mid -1 < x < 0 \text{ or } x > 1\}$ **25.** $\{x \mid x > 1\}$ **27.** $\{x \mid x < -1 \text{ or } x > 1\}$ **29.** $\{x \mid x < -1 \text{ or } x > 1\}$ **31.** $\{x \mid x < -1 \text{ or } 0 < x < 1\}$ **33.** $\{x \mid -1 < x < 1 \text{ or } x \geq 2\}$ **35.** $\{x \mid x < -\frac{2}{3} \text{ or } 0 < x < \frac{3}{2}\}$ **37.** $\{x \mid x < 2\}$ **39.** $\{x \mid -2 < x \leq 9\}$ **41.** $\{x \mid x < -3 \text{ or } 3 < x < 15\}$ **43.** $\{x \mid x < -3 \text{ or } -1 < x < 1 \text{ or } x > 2\}$ **45.** $\{x \mid x < -5 \text{ or } -4 < x < -3 \text{ or } x > 1\}$ **47.** $\{x \mid x > 4\}$ **49.** $\{x \mid x \leq -3 \text{ or } x \geq 3\}$ **51.** $2 < t < 3$ **53.** $10 \leq x \leq 50$ **55.** $b - a = (\sqrt{b} - \sqrt{a})(\sqrt{b} + \sqrt{a})$; since $a \geq 0$ and $b \geq 0$, then $\sqrt{a} \geq 0$ and $\sqrt{b} \geq 0$ so that $\sqrt{b} + \sqrt{a} \geq 0$; thus, $b - a \geq 0$ is equivalent to $\sqrt{b} - \sqrt{a} \geq 0$ and $a \leq b$ is equivalent to $\sqrt{a} \leq \sqrt{b}$ **57.** $k < 1$

EXERCISE 2.9

1. $\{-3, 3\}$ **3.** $\{-4, 1\}$ **5.** $\{-1, \frac{3}{2}\}$ **7.** $\{-4, 4\}$ **9.** 2 **11.** $\{-12, 12\}$ **13.** $\{-\frac{36}{5}, \frac{24}{5}\}$ **15.** No real solution **17.** $\{-2, 2\}$ **19.** $\{-1, 3\}$ **21.** $\{-2, -1, 0, 1\}$ **23.** $(-3, 3)$ **25.** $(-\infty, -4) \cup (4, \infty)$ **27.** $(1, 3)$ **29.** $\left[-\frac{2}{3}, 2\right]$ **31.** $(-\infty, -1] \cup [3, \infty)$ **33.** $\left(-1, \frac{3}{2}\right)$

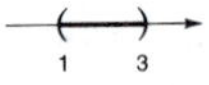

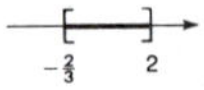

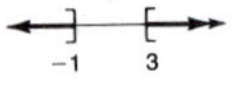

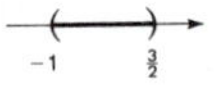

35. $(-\infty, -1) \cup (2, \infty)$ **37.** $(-\infty, -2) \cup (-2, \infty)$ **39.** $(-\infty, \infty)$ **41.** $(0.49, 0.51)$ **43.** $\left|\frac{a}{b}\right| = \sqrt{\left(\frac{a}{b}\right)^2} = \sqrt{\frac{a^2}{b^2}} = \frac{\sqrt{a^2}}{\sqrt{b^2}} = \frac{|a|}{|b|}$

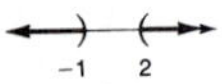

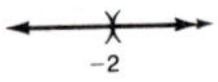

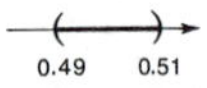

45. $(a + b)^2 = a^2 + 2ab + b^2 \leq |a|^2 + 2|a||b| + |b|^2 = [|a| + |b|]^2$; therefore, $\sqrt{(a + b)^2} \leq \sqrt{(|a| + |b|)^2}$ or $|a + b| \leq |a| + |b|$ **47.** $|x - 2| < \frac{1}{2}$; $\frac{3}{2} < x < \frac{5}{2}$ **49.** $|x + 3| > 2$; $x < -5$ or $x > -1$ **51.** $|x - 98.6| \geq 1.5$; $x \leq 97.1$°F or $x \geq 100.1$°F **53.** $x^2 - a < 0$; $(x - \sqrt{a})(x + \sqrt{a}) < 0$; therefore, $-\sqrt{a} < x < \sqrt{a}$ **55.** $-1 < x < 1$ **57.** $x \geq 3$ or $x \leq -3$ **59.** $-4 \leq x \leq 4$ **61.** $x > 2$ or $x < -2$

FILL-IN-THE-BLANK ITEMS

1. equivalent **2.** identity **3.** add; $\frac{25}{4}$ **4.** discriminant; negative **5.** $-a$ **6.** double; multiplicity 2 **7.** extraneous **8.** negative

REVIEW EXERCISES

1. -9 **3.** 6 **5.** $\frac{1}{5}$ **7.** 5 **9.** No real solution **11.** $\frac{11}{8}$ **13.** $\{-2, \frac{3}{2}\}$ **15.** $\left\{\frac{1 - \sqrt{13}}{4}, \frac{1 + \sqrt{13}}{4}\right\}$ **17.** $\{-3, 3\}$ **19.** No real solution **21.** $\{-2, -1, 1, 2\}$ **23.** 2 **25.** 0 **27.** $\frac{\sqrt{5}}{2}$ **29.** $\{-\frac{1}{8}, 1\}$ **31.** $\{-1, \frac{1}{2}\}$ **33.** $\left\{\frac{m}{1 - n}, \frac{m}{1 + n}\right\}$ **35.** $\left\{\frac{-9b}{5a}, \frac{2b}{a}\right\}$ **37.** $-\frac{9}{5}$ **39.** $\{x \mid x \geq 4\}$ **41.** $\{x \mid -\frac{31}{2} \leq x \leq \frac{33}{2}\}$ **43.** $\{x \mid -23 < x < -7\}$ **45.** $\{x \mid -4 < x < \frac{3}{2}\}$ **47.** $\{x \mid -2 < x \leq 4\}$ **49.** $\{x \mid x < 1 \text{ or } x > 2\}$ **51.** $\{x \mid 1 < x < 2 \text{ or } x > 3\}$ **53.** $\{x \mid x < -4 \text{ or } 2 < x < 4 \text{ or } x > 6\}$ **55.** $\{x \mid -\frac{3}{2} < x < -\frac{7}{6}\}$ **57.** $\{x \mid x \leq -1 \text{ or } x \geq 6\}$ **59.** $\left\{\frac{-1 - \sqrt{3}i}{2}, \frac{-1 + \sqrt{3}i}{2}\right\}$ **61.** $\left\{\frac{-1 - \sqrt{17}}{4}, \frac{-1 + \sqrt{17}}{4}\right\}$ **63.** $\left\{\frac{1 - \sqrt{11}i}{2}, \frac{1 + \sqrt{11}i}{2}\right\}$ **65.** $\left\{\frac{1 - \sqrt{23}i}{2}, \frac{1 + \sqrt{23}i}{2}\right\}$ **67.** $\{-\sqrt{2}, \sqrt{2}, -2i, 2i\}$ **69.** 3300 ft **71.** 616 mi **73.** A little less than 1 hr, 35 min **75.** 190.67 ft **77.** 36; \$13.40

CHAPTER 3 EXERCISE 3.1

1. (a) Quadrant II
(b) Positive x-axis
(c) Quadrant III
(d) Quadrant I
(e) Negative y-axis
(f) Quadrant IV

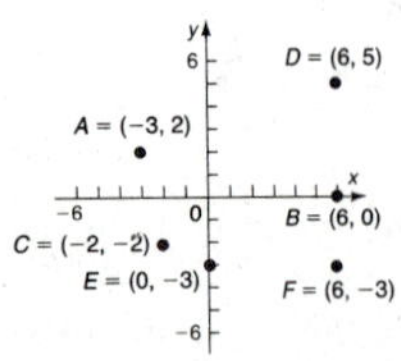

3. The points will be on a vertical line that is 2 units to the right of the y-axis.

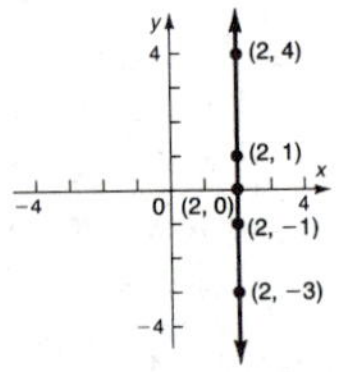

5. 5 **7.** $\sqrt{85}$ **9.** $2\sqrt{5}$ **11.** 2.625 **13.** $\sqrt{a^2 + b^2}$

15. $d(A, B) = \sqrt{13}$
$d(B, C) = \sqrt{13}$
$d(A, C) = \sqrt{26}$
$(\sqrt{13})^2 + (\sqrt{13})^2 = (\sqrt{26})^2$
Area $= \frac{13}{2}$ square units

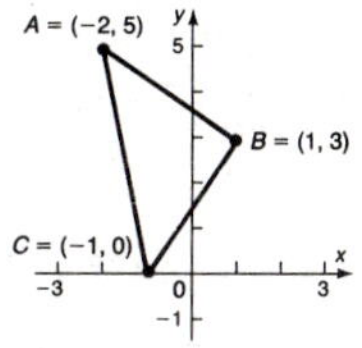

17. $d(A, B) = \sqrt{130}$
$d(B, C) = \sqrt{26}$
$d(A, C) = \sqrt{104}$
$(\sqrt{26})^2 + (\sqrt{104})^2 = (\sqrt{130})^2$
Area $= 26$ square units

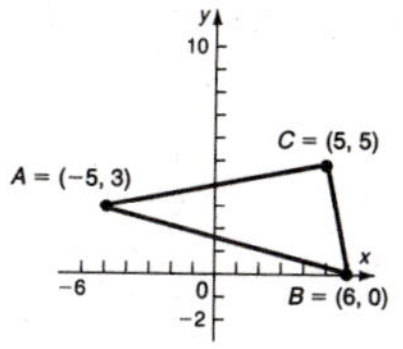

19. $d(A, B) = 4$
$d(A, C) = 5$
$d(B, C) = \sqrt{41}$
$4^2 + 5^2 = 16 + 25 = (\sqrt{41})^2$
Area $= 10$ square units

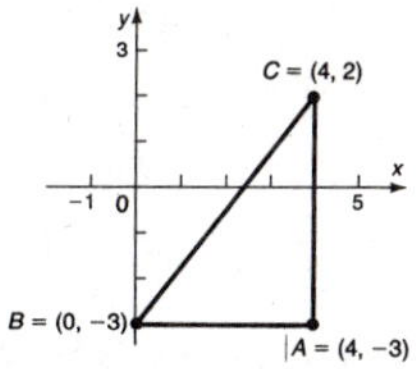

21. (2, 2); (2, −4) **23.** (−2, 0); (6, 0) **25.** $\left(3, -\frac{3}{2}\right)$ **27.** $\left(\frac{3}{2}, 1\right)$ **29.** (5, −1) **31.** (1.05, 0.7) **33.** $(a/2, b/2)$ **35.** $\sqrt{73}$; $2\sqrt{13}$; 5
37. $d(P_1, P_2) = 6$; $d(P_2, P_3) = 4$; $d(P_1, P_3) = 2\sqrt{13}$; right triangle
39. $d(P_1, P_2) = \sqrt{68}$; $d(P_2, P_3) = \sqrt{34}$; $d(P_1, P_3) = \sqrt{34}$; isosceles right triangle
41. $\frac{x - x_1}{x_2 - x_1} = r$ and $\frac{y - y_1}{y_2 - y_1} = r$; thus, $x = x_1 + r(x_2 - x_1)$ and $y = y_1 + r(y_2 - y_1)$ **43.** P_2 **45.** (9, 8) **47.** $90\sqrt{2} \approx 127.28$ ft
49. $M_1 = (s/2, s/2)$; $M_2 = (s/2, s/2)$ **51.** $d = 50t$

EXERCISE 3.2

1.

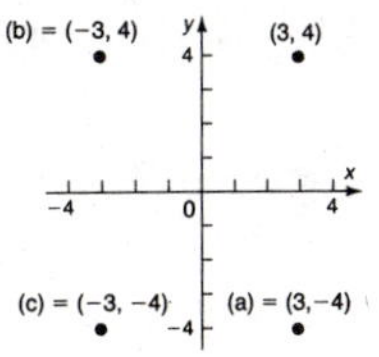

3.

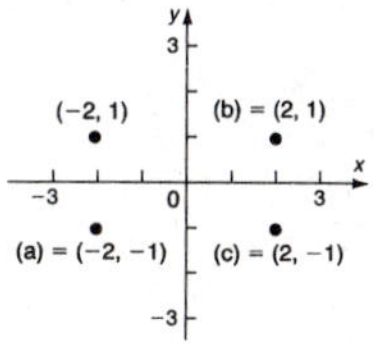

5.

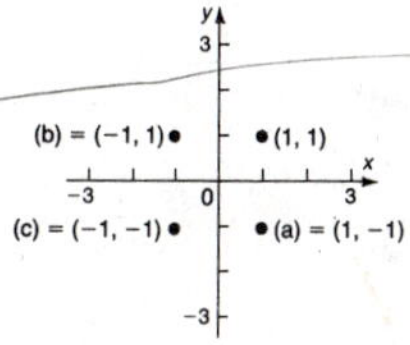

7.

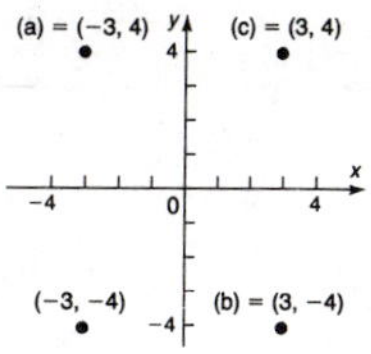

9.

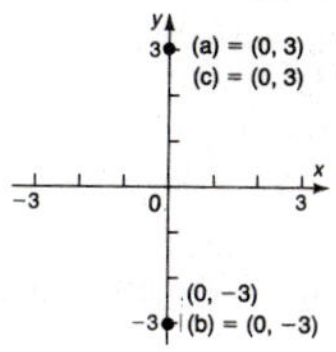

11. $y = 3x + 2$

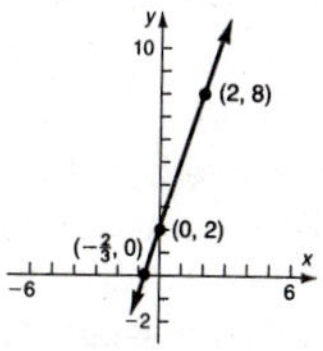

13. $y = -2x + 1$

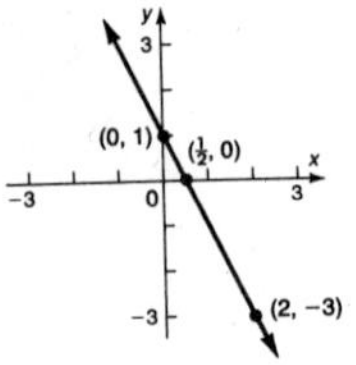

15. $3x - 2y + 6 = 0$

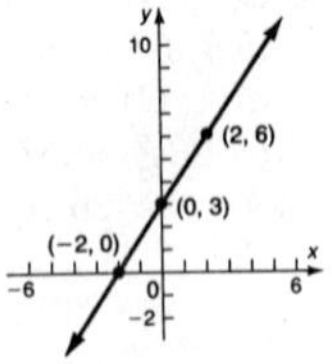

17. $y = 2x^2$

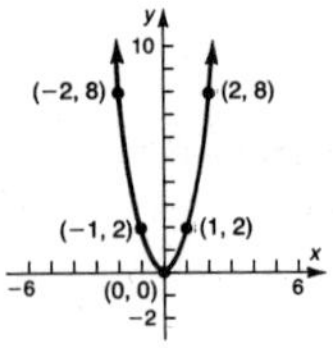

19. $y = -x^2$

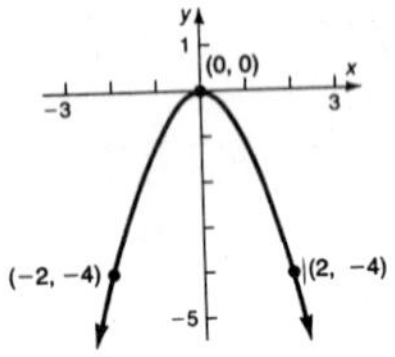

21. $y = x^2 + 3$

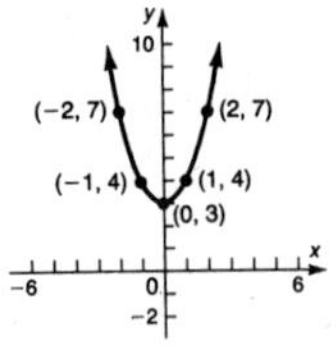

23. $y = x^3 - 1$

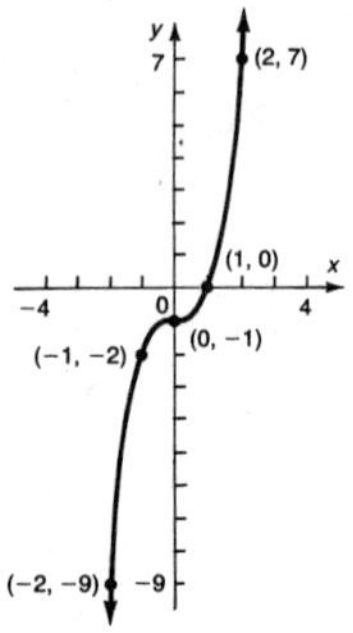

25. $x^2 = 4y$

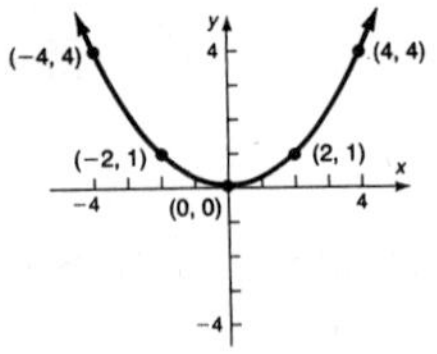

27. $x^2 = y + 1$

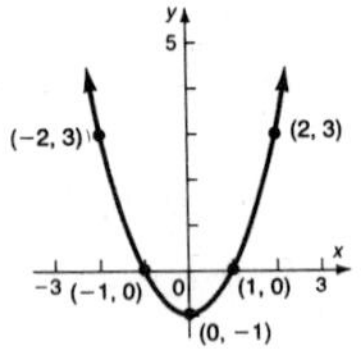

29. $y = \sqrt{x}$

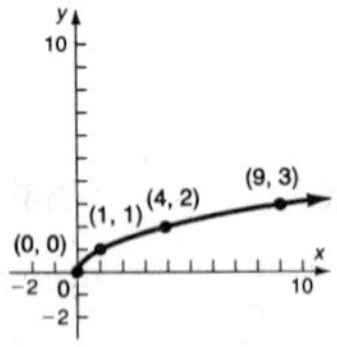

31. $y = \sqrt{x - 1}$

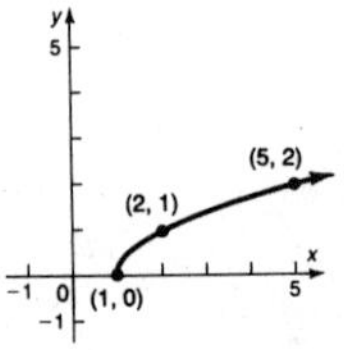

33. $y = \dfrac{1}{x - 2}$

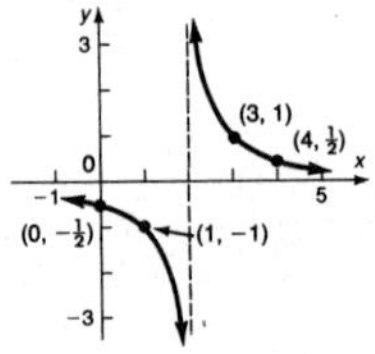

35.

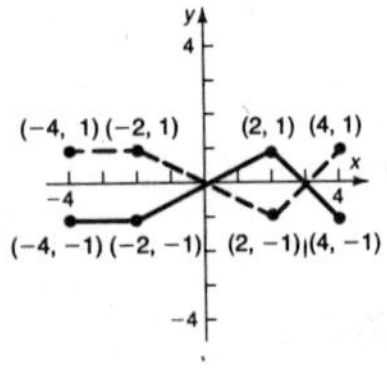

37.

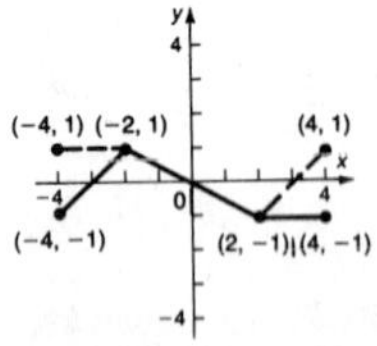

39.

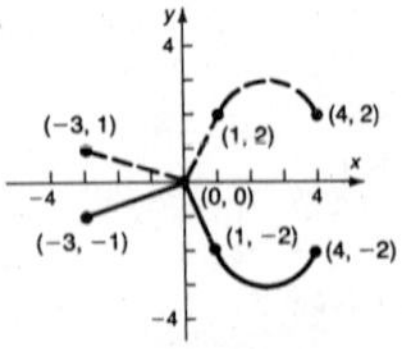

41.

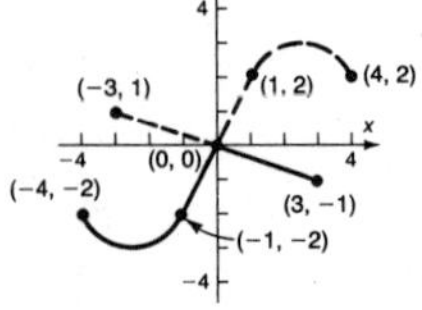

43. (0, 0); symmetric with respect to the y-axis **45.** (0, 0); symmetric with respect to the origin
47. (0, 9), (3, 0), (−3, 0); symmetric with respect to the y-axis
49. (0, 2), (0, −2), (3, 0), (−3, 0); symmetric with respect to the x-axis, y-axis, and origin **51.** (0, −27), (3, 0); no symmetry
53. (0, −4), (4, 0), (−1, 0); no symmetry **55.** (0, 0); symmetric with respect to the origin

EXERCISE 3.3

1. Slope = 3

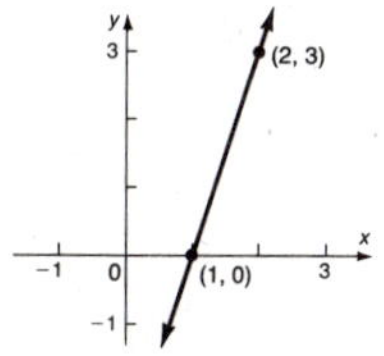

3. Slope = $-\frac{1}{2}$

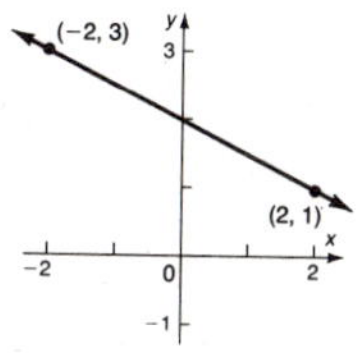

5. Slope = 0

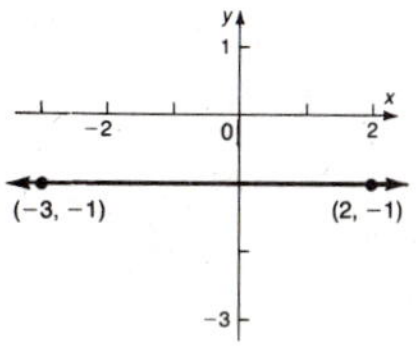

7. Slope undefined

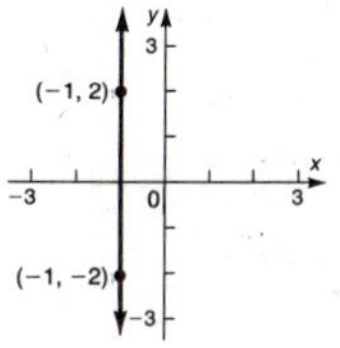

9. Slope = $\dfrac{\sqrt{3}-3}{1-\sqrt{2}} \approx 3.06$

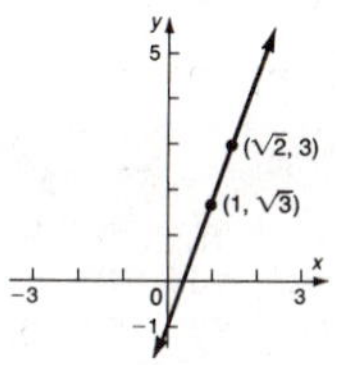

11.

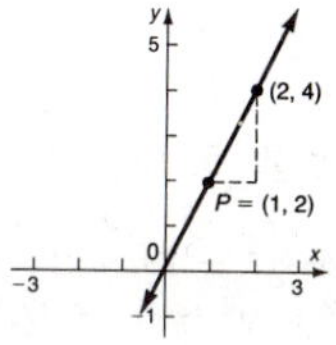

13.

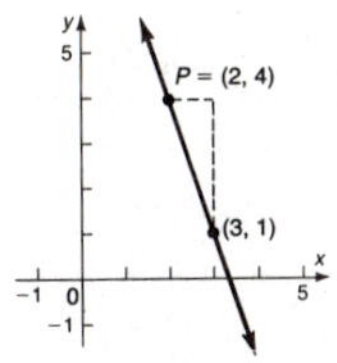

15.

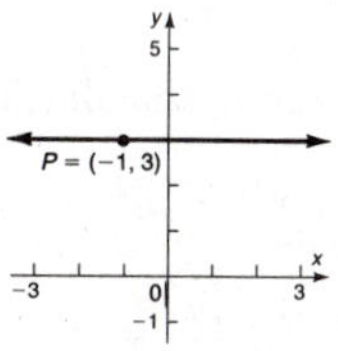

17.

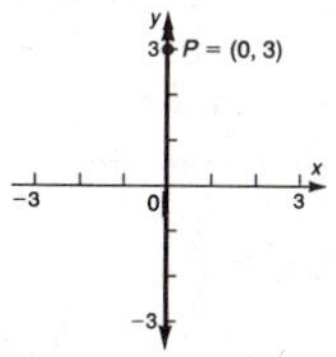

19. $2x - y + 7 = 0$ **21.** $2x + 3y + 1 = 0$ **23.** $x - 2y + 5 = 0$ **25.** $3x + y - 3 = 0$ **27.** $x - 2y - 2 = 0$ **29.** $x - 1 = 0$

31. Slope = 2; y-intercept = 3

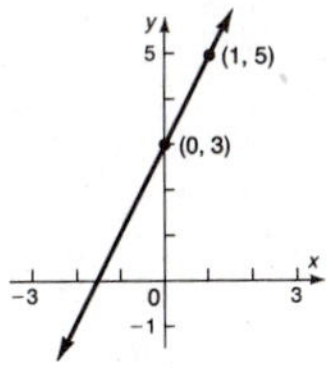

33. Slope = 2; y-intercept = -2

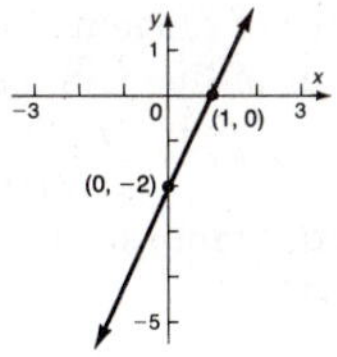

35. Slope = $\frac{2}{3}$; y-intercept = -2

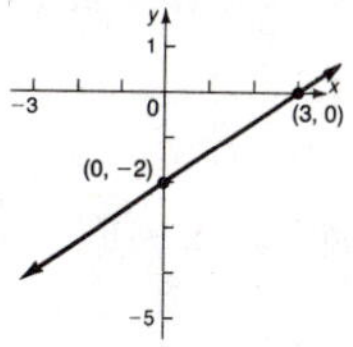

37. Slope = -1; y-intercept = 1

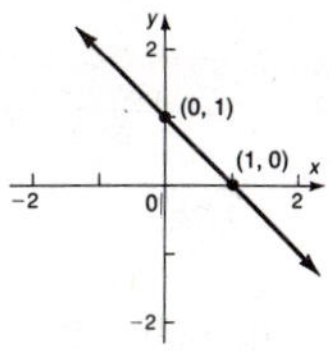

39. Slope undefined; no y-intercept

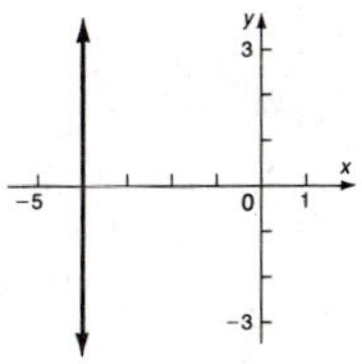

41. Slope = $\frac{3}{2}$; y-intercept = 0

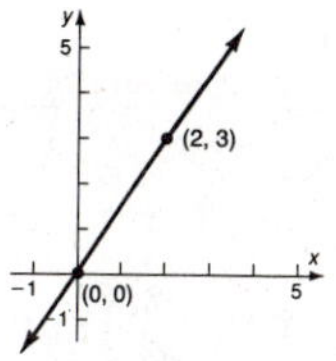

43. $x - 2y = 0$ **45.** $x + y - 2 = 0$ **47.** $y = 0$ **49.** °C = $\frac{5}{9}$(°F − 32); approx. 21° C

51. $P = 0.5x - 100$

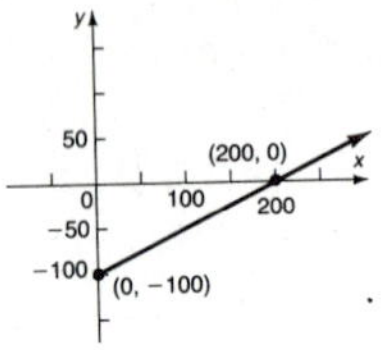

53. $(a, 0)$ and $(0, b)$ are points on the line;

$$\text{Slope} = \frac{b}{-a}$$

$$y - 0 = -\frac{b}{a}(x - a)$$

$$y = -\frac{bx}{a} + b$$

$$\frac{bx}{a} + y = b$$

$$\frac{x}{a} + \frac{y}{b} = 1$$

EXERCISE 3.4

1. (a) 2 (b) $-\frac{1}{2}$ **3.** (a) $-\frac{1}{2}$ (b) 2 **5.** (a) $\frac{1}{2}$ (b) -2 **7.** (a) $-\frac{3}{5}$ (b) $\frac{5}{3}$ **9.** (a) Undefined (b) 0 **11.** $3x - y + 5 = 0$ **13.** $2x - y = 0$ **15.** $x - 4 = 0$ **17.** $2x + y = 0$ **19.** $x - 2y + 3 = 0$ **21.** $y - 4 = 0$ **23.** $(x - 1)^2 + (y + 1)^2 = 1$; $x^2 + y^2 - 2x + 2y + 1 = 0$ **25.** $x^2 + (y - 2)^2 = 4$; $x^2 + y^2 - 4y = 0$ **27.** $(x - 4)^2 + (y + 3)^2 = 25$; $x^2 + y^2 - 8x + 6y = 0$ **29.** $x^2 + y^2 = 4$; $x^2 + y^2 - 4 = 0$ **31.** $r = 2$; $(h, k) = (0, 0)$ **33.** $r = 2$; $(h, k) = (3, 0)$ **35.** $r = 3$; $(h, k) = (-2, 2)$ **37.** $r = \frac{1}{2}$; $(h, k) = \left(\frac{1}{2}, -1\right)$ **39.** $r = 5$; $(h, k) = (3, -2)$ **41.** $x^2 + y^2 - 13 = 0$ **43.** $x^2 + y^2 - 4x - 6y + 4 = 0$ **45.** $x^2 + y^2 + 2x - 6y + 5 = 0$ **47.** $P_1 = (-2, 5)$, $P_2 = (1, 3)$, $m_1 = -\frac{2}{3}$; $P_2 = (1, 3)$, $P_3 = (-1, 0)$, $m_2 = \frac{3}{2}$; because $m_1m_2 = -1$, the lines are perpendicular; the points P_1, P_2, and P_3 thus form a right triangle. **49.** $P_1 = (-1, 0)$, $P_2 = (2, 3)$, $m_1 = 1$; $P_3 = (1, -2)$, $P_4 = (4, 1)$, $m_2 = 1$; $P_1 = (-1, 0)$, $P_3 = (1, -2)$, $m_3 = -1$; $P_2 = (2, 3)$, $P_4 = (4, 1)$, $m_4 = -1$; opposite sides are parallel, adjacent sides are perpendicular; the points form a rectangle. **51.** 12 square units **53.** 4 square units **55.** 5 square units **57.** Refer to Figure 40: $m_1m_2 = -1$; $d(A, B) = \sqrt{(m_2 - m_1)^2}$; $d(O, A) = \sqrt{1 + m_2^2}$; $d(O, B) = \sqrt{1 + m_1^2}$. Now show that $[d(O, B)]^2 + [d(O, A)]^2 = [d(A, B)]^2$.

59. (a)

$$x^2 + (mx + b)^2 = r^2$$
$$(1 + m^2)x^2 + 2mbx + b^2 - r^2 = 0$$
$$\text{One solution if and only if discriminant} = 0$$
$$(2mb)^2 - 4(1 + m^2)(b^2 - r^2) = 0$$
$$-4b^2 + 4r^2 + 4m^2r^2 = 0$$
$$r^2(1 + m^2) = b^2$$

(b)

$$x = \frac{-2mb}{2(1 + m^2)} = \frac{-2mb}{2b^2/r^2} = \frac{-r^2m}{b}$$
$$y = m\left(\frac{-r^2m}{b}\right) + b = \frac{-r^2m^2}{b} + b$$
$$= \frac{-r^2m^2 + b^2}{b} = \frac{r^2}{b}$$

(c) Slope of tangent line $= m$
Slope of line joining center to point of tangency $= \dfrac{r^2/b}{-r^2m/b} = -\dfrac{1}{m}$

61. $\sqrt{2}x + 4y - 11\sqrt{2} + 12 = 0$ **63.** $x + 5y + 13 = 0$ **65.** All have the same slope, 2; the lines are parallel.

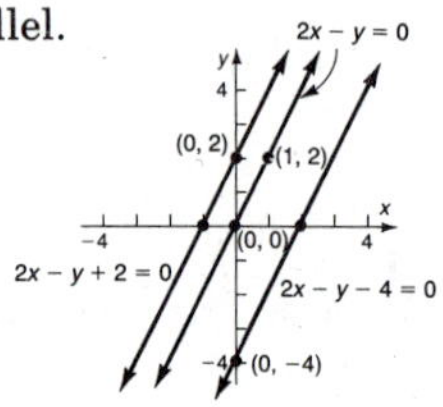

EXERCISE 3.5

1. $y = \frac{1}{2}x$ **3.** $A = \pi x^2$ **5.** $F = \dfrac{250}{d^2}$ **7.** $z = \frac{1}{5}(x^2 + y^2)$ **9.** $M = \dfrac{9d^2}{2\sqrt{x}}$ **11.** $T^2 = 64\dfrac{a^3}{d^2}$ **13.** $V = \dfrac{4\pi}{3}r^3$ **15.** $A = \frac{1}{2}bh$ **17.** $V = \pi r^2h$ **19.** $F = 6.67 \times 10^{-11}\left(\dfrac{mM}{d^2}\right)$ **21.** 144 ft; 2 sec **23.** 2.25 **25.** 45.45 lb **27.** $3\sqrt[3]{\frac{3}{4}} \approx 2.73$ in. **29.** 900 ft-lb **31.** 384 psi **33.** $\frac{720}{49} \approx 14.69$ ohms **35.** $v = \sqrt{gr}$ **37.** 18,001 mph **39.** Approx. 505 mi **41.** $F = \dfrac{mv^2}{r}$ **43.** By 21% **45.** 9 times

FILL-IN-THE-BLANK ITEMS

1. x-coordinate; y-coordinate **2.** midpoint **3.** y-axis **4.** circle; radius; center **5.** undefined; 0 **6.** $m_1 = m_2$; $m_1 \cdot m_2 = -1$ **7.** $kx^2y^3/\sqrt{t}$

REVIEW EXERCISES

1. $2x + y - 3 = 0$ **3.** $x + 3 = 0$ **5.** $x + 5y + 10 = 0$ **7.** $2x - 3y + 19 = 0$ **9.** $-x + y + 4 = 0$

11.

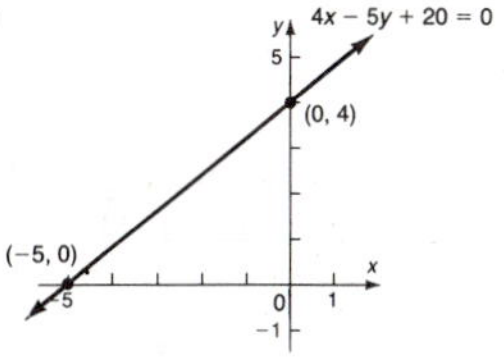

13.

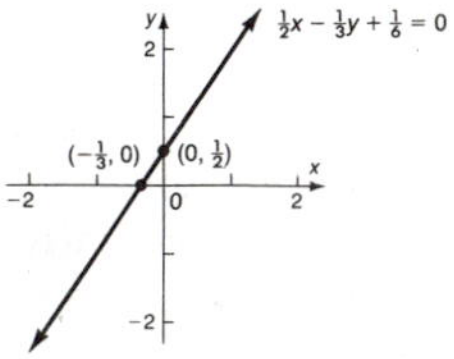

15.

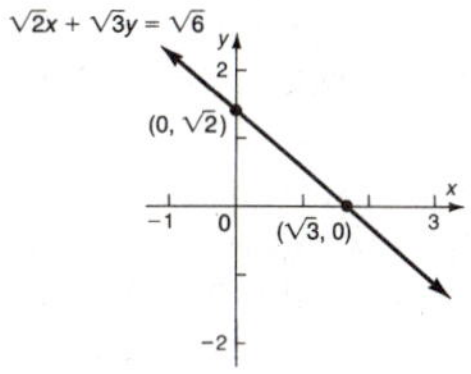

17. Center $(1, -2)$; radius $= 3$ **19.** Center $(1, -2)$; radius $= \sqrt{5}$ **21.** Slope $= \frac{1}{2}$; distance $= 2\sqrt{5}$; midpoint $= (-1, 3)$

23.

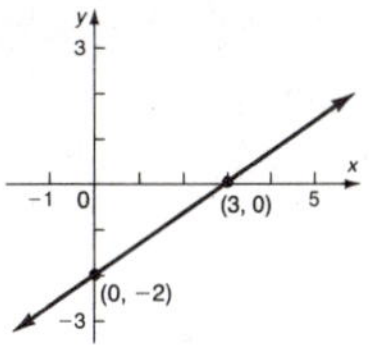

25.

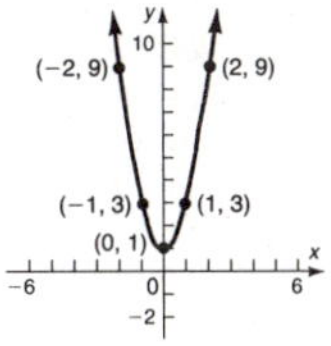

27.

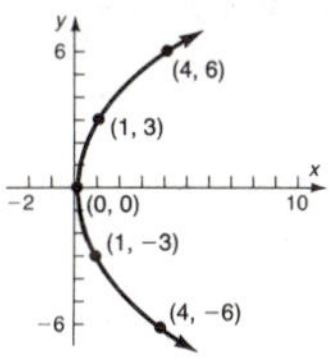

29.

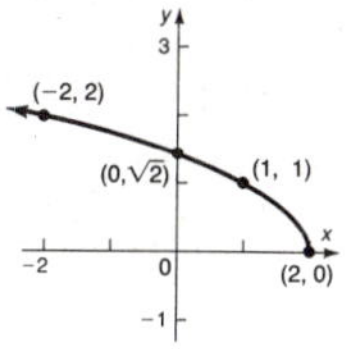

31. Intercept: $(0, 0)$; symmetric with respect to the x-axis
33. Intercepts: $(0, -1)$, $(0, 1)$, $\left(\frac{1}{2}, 0\right)$, $\left(-\frac{1}{2}, 0\right)$; symmetric with respect to the x-axis, y-axis, and origin
35. Intercept: $(0, 1)$; symmetric with respect to the y-axis **37.** Intercepts: $(0, 0)$, $(0, -2)$, $(-1, 0)$; no symmetry
39. $A = ks^2 = \dfrac{\sqrt{3}}{4}s^2$; $s = \dfrac{8}{\sqrt[4]{3}} \approx 6.08$ cm **41.** $a \approx 36$ million miles
43. $M = \left(\dfrac{a}{2}, \dfrac{b}{2}\right)$; $d(O, M) = \frac{1}{2}\sqrt{a^2 + b^2}$; $d(A, M) = \frac{1}{2}\sqrt{a^2 + b^2}$; $d(B, M) = \frac{1}{2}\sqrt{a^2 + b^2}$ **45.** $d(A, B) = \sqrt{13}$; $d(B, C) = \sqrt{13}$
47. Slope using A and B is -1; slope using B and C is -1 **49.** Center $(1, -2)$; radius $= 4\sqrt{2}$; $x^2 + y^2 - 2x + 4y - 27 = 0$

CHAPTER 4 EXERCISE 4.1

1. (a) -4 (b) -5 (c) -9 (d) -12 **3.** (a) 0 (b) $\frac{1}{2}$ (c) $-\frac{1}{2}$ (d) $\frac{2}{5}$ **5.** (a) 4 (b) 5 (c) 5 (d) 6
7. (a) $-\frac{1}{5}$ (b) $-\frac{3}{2}$ (c) $\frac{1}{8}$ (d) 5 **9.** (a) $-2x + 3$ (b) $-2x - 3$ (c) $4x + 3$ (d) $2x - 3$ (e) $\dfrac{3x + 2}{x}$ (f) $\dfrac{1}{2x + 3}$
11. (a) $2x^2 - 4$ (b) $-2x^2 + 4$ (c) $8x^2 - 4$ (d) $2x^2 - 12x + 14$ (e) $\dfrac{2 - 4x^2}{x^2}$ (f) $\dfrac{1}{2x^2 - 4}$
13. (a) $-x^3 + 3x$ (b) $-x^3 + 3x$ (c) $8x^3 - 6x$ (d) $x^3 - 9x^2 + 24x - 18$ (e) $\dfrac{1}{x^3} - \dfrac{3}{x}$ (f) $\dfrac{1}{x^3 - 3x}$
15. (a) $-\dfrac{x}{x^2 + 1}$ (b) $-\dfrac{x}{x^2 + 1}$ (c) $\dfrac{2x}{4x^2 + 1}$ (d) $\dfrac{x - 3}{x^2 - 6x + 10}$ (e) $\dfrac{x}{x^2 + 1}$ (f) $\dfrac{x^2 + 1}{x}$
17. (a) $|x|$ (b) $-|x|$ (c) $2|x|$ (d) $|x - 3|$ (e) $\dfrac{1}{|x|}$ (f) $\dfrac{1}{|x|}$ **19.** All real numbers **21.** All real numbers **23.** $\{x \mid x \neq -1, x \neq 1\}$
25. $\{x \mid x \neq 0\}$ **27.** $\{x \mid x \geq 4\}$ **29.** $(-\infty, -3] \cup [3, \infty)$ **31.** $(-\infty, 1) \cup [2, \infty)$ **33.** 0 **35.** -3 **37.** $6x + 3h - 2$
39. $3x^2 + 3xh + h^2 - 1$ **41.** $-\dfrac{1}{x(x + h)}$ **43.** $\dfrac{\sqrt{x + h} - \sqrt{x}}{h} \cdot \dfrac{\sqrt{x + h} + \sqrt{x}}{\sqrt{x + h} + \sqrt{x}} = \dfrac{h}{h(\sqrt{x + h} + \sqrt{x})} = \dfrac{1}{\sqrt{x + h} + \sqrt{x}}$ **45.** $A = -4$
47. $A = -4$ **49.** $A = 8$; undefined at 3 **51.** (a) 15.1 m, 14.07 m, 12.94 m, 11.72 m (b) 2.02 sec **53.** $A(x) = \frac{1}{2}x^2$ **55.** $G(x) = 5x$
57. $A(x) = (7 - 2x)(11 - 2x)$, $0 \leq x \leq \frac{7}{2}$, $0 \leq A \leq 77$ **59.** 1.11 sec; 0.46 sec **61.** h

EXERCISE 4.2

1. $f(0) = 3; f(2) = 4$ **3.** Positive **5.** $-3, 6,$ and 10 **7.** $\{x \mid -6 \le x \le 11\}$ **9.** $-6 \le x \le 2, 8 \le x \le 11$ **11.** 3 times
13. (a) No (b) -3 (c) 14 (d) $\{x \mid x \ne 6\}$ **15.** (a) Yes (b) $\frac{8}{17}$ (c) $-1, 1$ (d) All real numbers **17.** Not a function
19. Function (a) Domain: $\{x \mid -\pi \le x \le \pi\}$; Range: $\{y \mid -1 \le y \le 1\}$ (b) Increasing on the interval $[-\pi, 0]$; decreasing on the interval $[0, \pi]$ (c) Even (d) Intercepts: $(-\pi/2, 0), (\pi/2, 0), (0, 1)$
21. Not a function
23. Function (a) Domain: $\{x \mid x > 0\}$; Range: all real numbers (b) Increasing on its domain (c) Neither (d) Intercept: $(1, 0)$
25. Function (a) Domain: all real numbers; Range: $\{y \mid y \le 2\}$ (b) Increasing on the interval $(-\infty, -1]$; decreasing on the interval $[1, \infty)$; constant on the interval $[-1, 1]$ (c) Even (d) Intercepts: $(-3, 0), (3, 0), (0, 2)$
27. Function (a) Domain: $\{x \mid -4 \le x < 4\}$; Range: $\{-2, 0, 2, 3\}$ (b) Constant on $[-4, -2), [-2, 0), [0, 2), [2, 4)$ (c) Neither (d) Intercepts: $(x, 0)$ for $0 \le x < 2$
29. Function (a) Domain: $\{x \mid -4 \le x < 4\}$; Range: $\{y \mid -3 \le y \le -2, -1 \le y \le 0, 1 \le y \le 2\}$ (b) Increasing on the interval $[-4, -1)$; decreasing on the interval $[1, 4)$; constant on the interval $[-1, 1)$ (c) Neither (d) Intercepts: $(2, 0), (0, 1)$
31. Odd **33.** Even **35.** Odd **37.** Neither **39.** Even **41.** Odd **43.** At most one

45. (a) $(-\infty, \infty)$
(b) $(0, -3), (1, 0)$
(d) $(-\infty, \infty)$
(c)
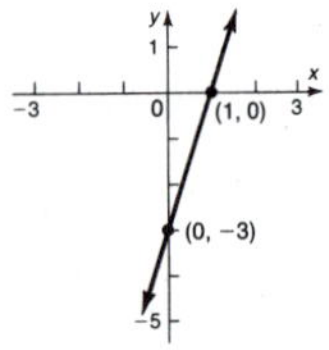

47. (a) $(-\infty, \infty)$
(b) $(-2, 0), (2, 0), (0, -4)$
(d) $[-4, \infty)$
(c)
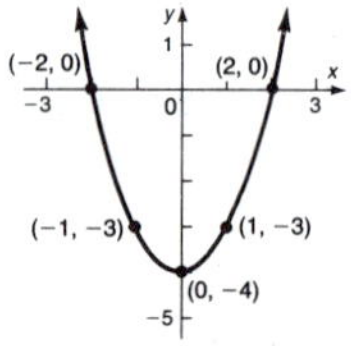

49. (a) $(-\infty, \infty)$
(b) $(0, 0)$
(d) $(-\infty, 0]$
(c)
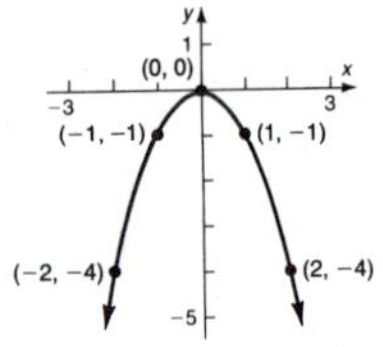

51. (a) $[2, \infty)$
(b) $(2, 0)$
(d) $[0, \infty)$
(c)
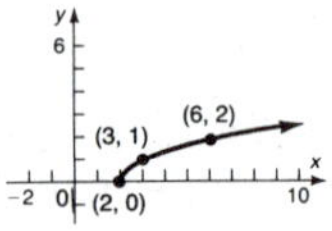

53. (a) $(-\infty, 2]$
(b) $(2, 0), (0, \sqrt{2})$
(d) $[0, \infty)$
(c)
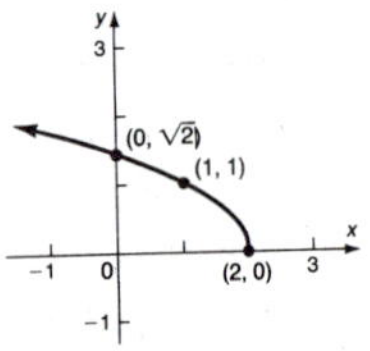

55. (a) $(-\infty, \infty)$
(b) $(0, 3)$
(d) $[3, \infty)$
(c)
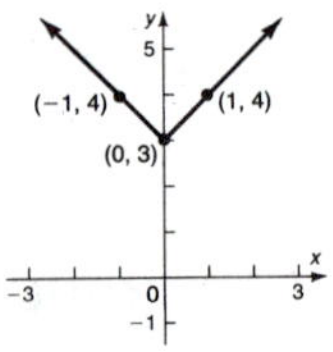

57. (a) $(-\infty, \infty)$
(b) $(0, 0)$
(d) $(-\infty, 0]$
(c)
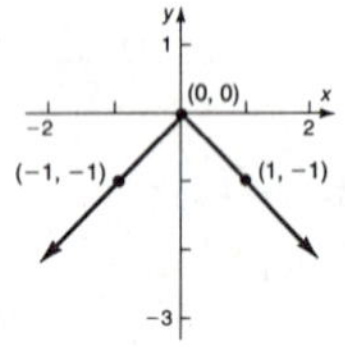

59. (a) $(-\infty, \infty)$
(b) $(0, 0)$
(d) $(-\infty, \infty)$
(c)
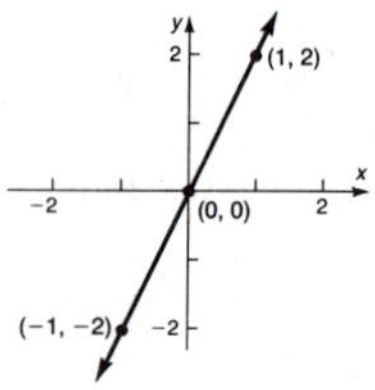

61. (a) $(-\infty, \infty)$
(b) $(0, 0), (-1, 0)$
(d) $(-\infty, \infty)$
(c)
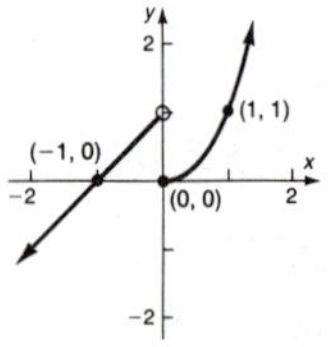

63. (a) $[-2, \infty)$
(b) $(0, 1)$
(d) $(0, \infty)$
(c)
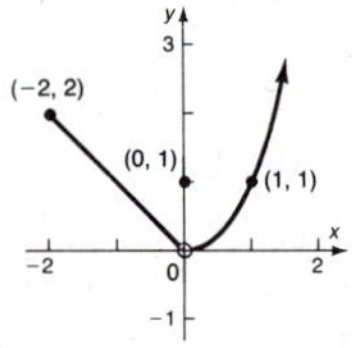

65. (a) $(-\infty, \infty)$ (b) $(0, 1)$ (c)

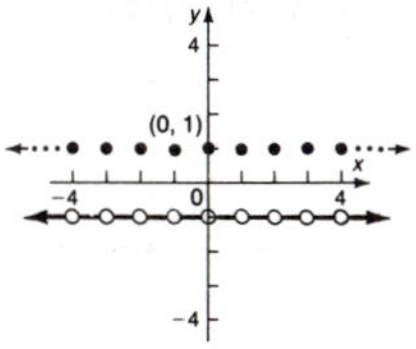

(d) $\{-1, 1\}$

67. (a) $(-\infty, \infty)$ (b) $(x, 0)$ for $0 \le x < 1$ (c)

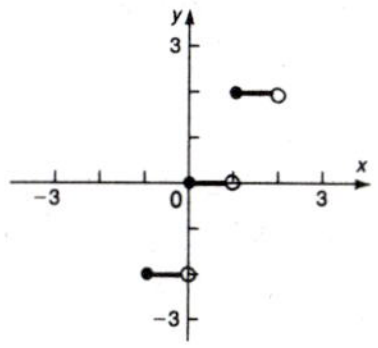

(d) Set of even integers

69. Function **71.** Function **73.** Not a function **75.** Function

77. $f(x) = \begin{cases} -x & \text{if } -1 \le x \le 0 \\ \frac{1}{2}x & \text{if } 0 < x \le 2 \end{cases}$ (Other answers are possible.) **79.** $f(x) = \begin{cases} -x & \text{if } x \le 0 \\ -x + 2 & \text{if } 0 < x \le 2 \end{cases}$ (Other answers are possible.)

81. No; $f(-2) = 8$ and $f(2) = 6$

83. (a) $E(-x) = \frac{1}{2}[f(-x) + f(x)] = E(x)$ (b) $O(-x) = \frac{1}{2}[f(-x) - f(x)] = -\frac{1}{2}[f(x) - f(-x)] = -O(x)$
(c) $E(x) + O(x) = \frac{1}{2}[f(x) + f(-x)] + \frac{1}{2}[f(x) - f(-x)] = f(x)$ (d) Combine the results of parts (a), (b), and (c).

EXERCISE 4.3

1.

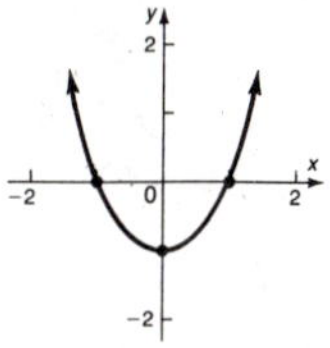

3.

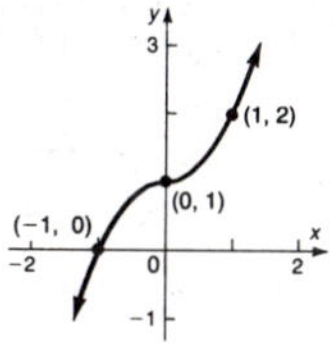

5.

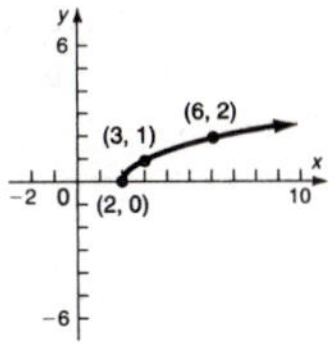

7.

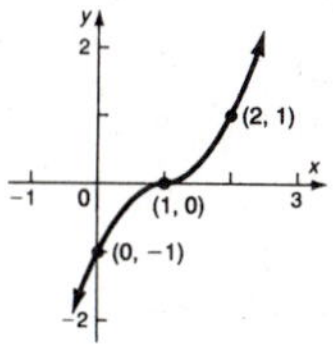

9.

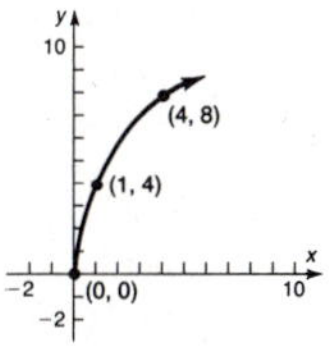

11.

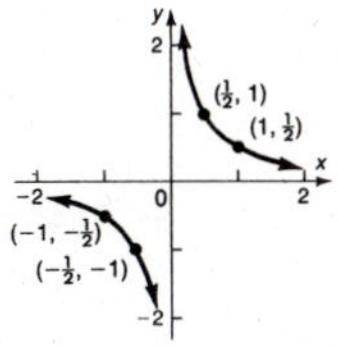

13.

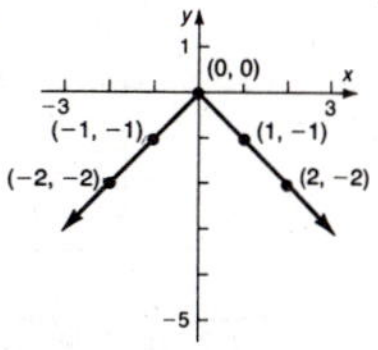

15.

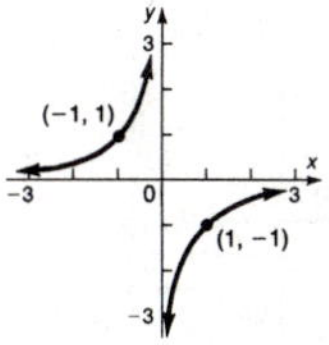

17.

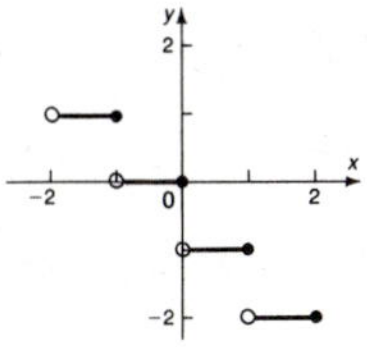

19.

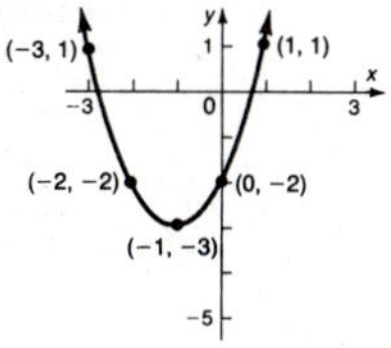

21.

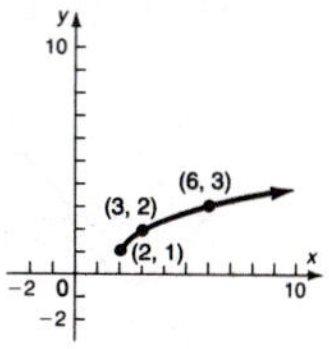

23.

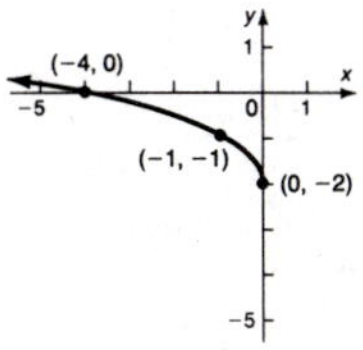

25.

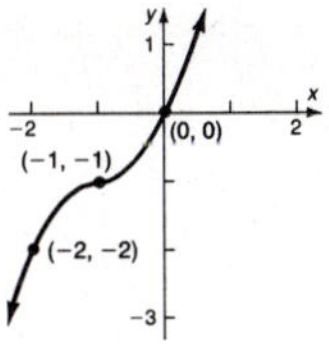

27.

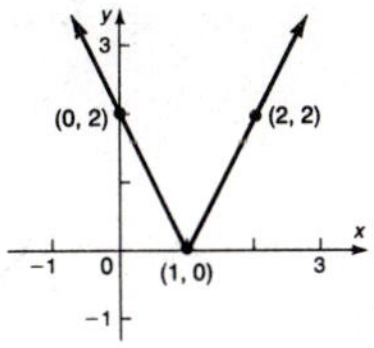

29.

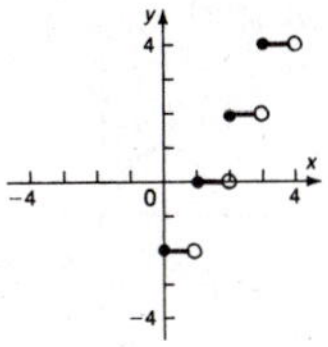

31. (a) $F(x) = f(x) + 3$

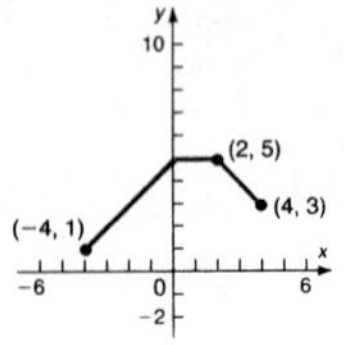

(b) $G(x) = f(x + 2)$

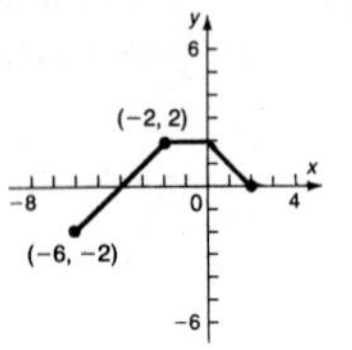

(c) $P(x) = -f(x)$

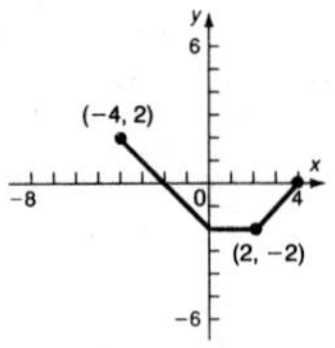

(d) $Q(x) = \frac{1}{2}f(x)$

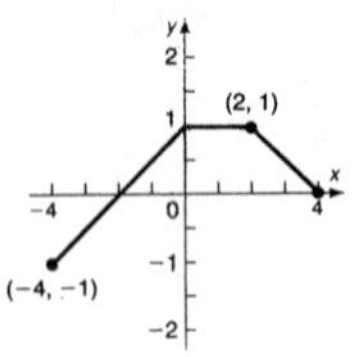

(e) $g(x) = f(-x)$

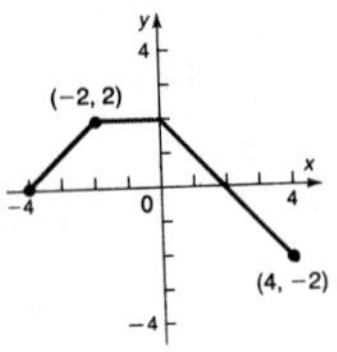

(f) $h(x) = 3f(x)$

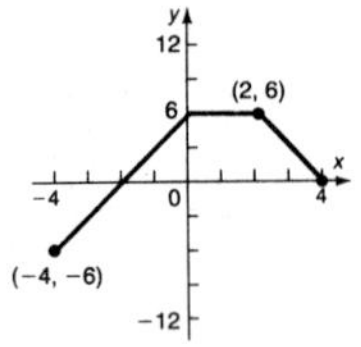

33. (a) $F(x) = f(x) + 3$

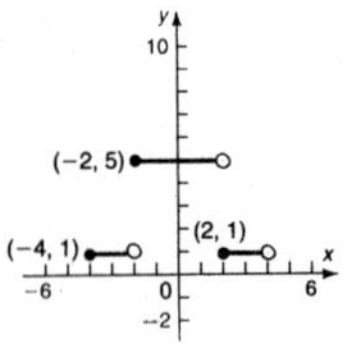

(b) $G(x) = f(x + 2)$

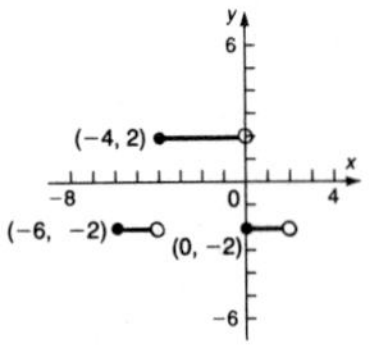

(c) $P(x) = -f(x)$

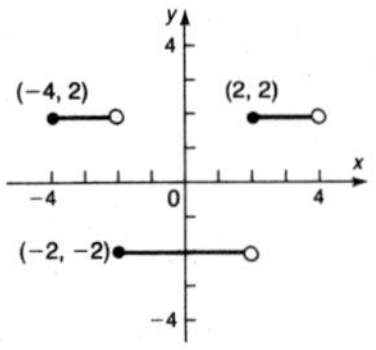

(d) $Q(x) = \frac{1}{2}f(x)$

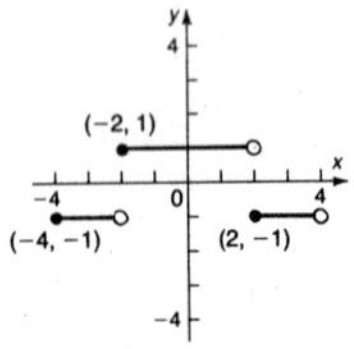

(e) $g(x) = f(-x)$

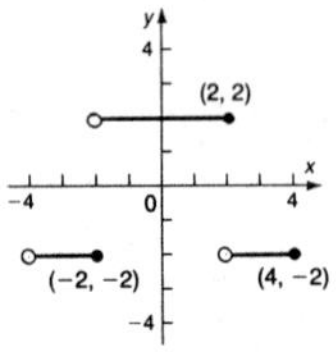

(f) $h(x) = 3f(x)$

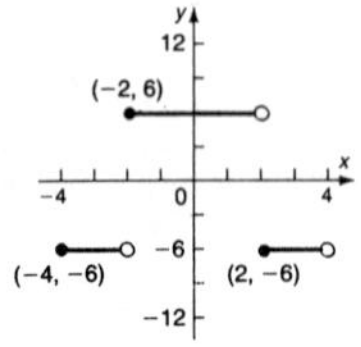

35. (a) $F(x) = f(x) + 3$

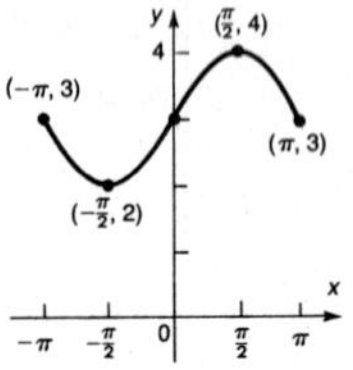

(b) $G(x) = f(x + 2)$

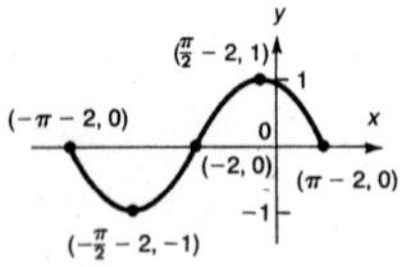

(c) $P(x) = -f(x)$

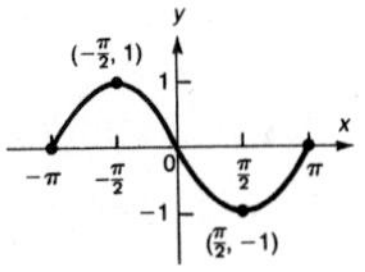

(d) $Q(x) = \frac{1}{2}f(x)$

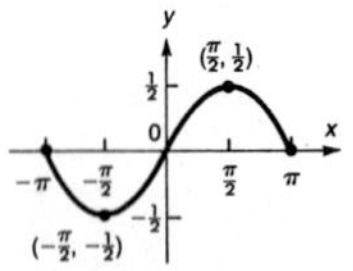

(e) $g(x) = f(-x)$

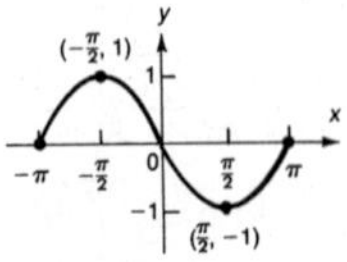

(f) $h(x) = 3f(x)$

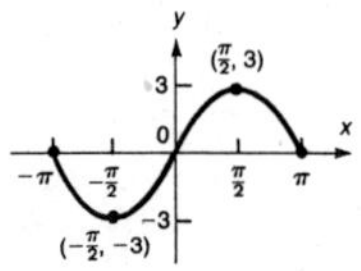

37. $f(x) = (x + 1)^2 - 1$

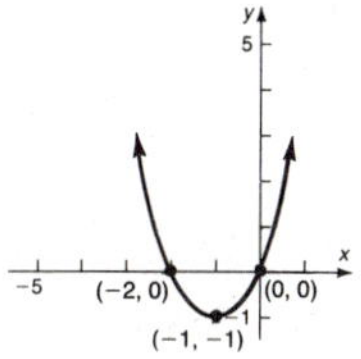

39. $f(x) = (x - 4)^2 - 15$

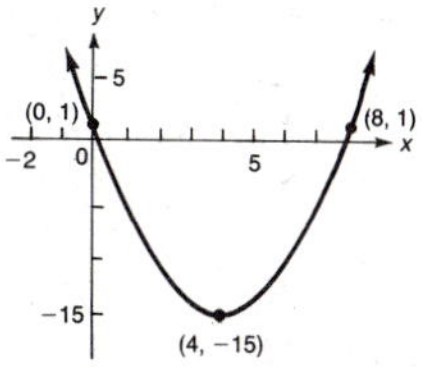

41. $f(x) = (x + \frac{1}{2})^2 + \frac{3}{4}$

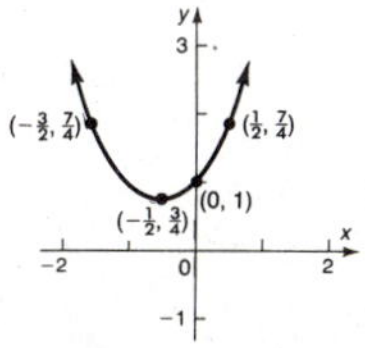

43.

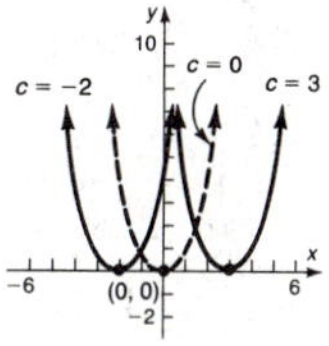

45.

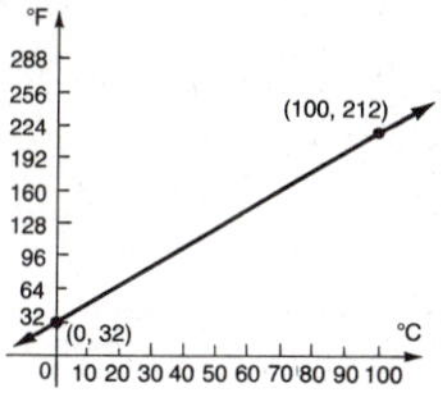

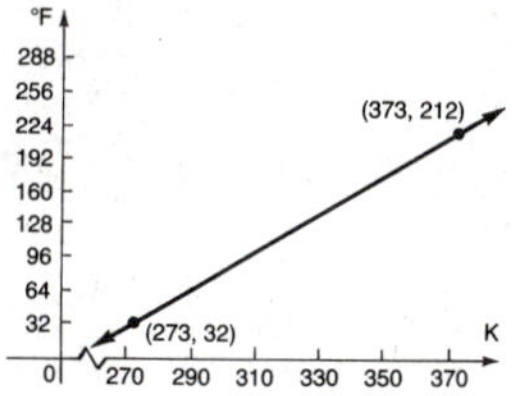

47. (a)

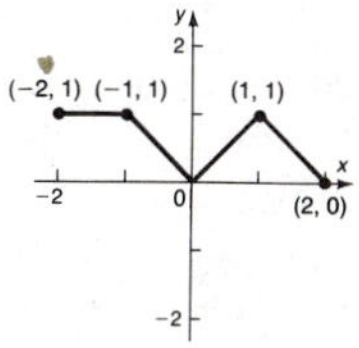

(b)

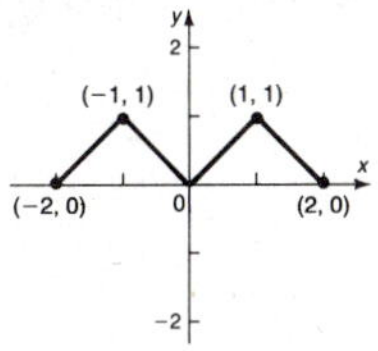

EXERCISE 4.4

1. (a) $(f + g)(x) = 5x - 1$; all real numbers (b) $(f - g)(x) = x - 7$; all real numbers
(c) $(f \cdot g)(x) = 6x^2 + x - 12$; all real numbers (d) $\left(\frac{f}{g}\right)(x) = \frac{3x - 4}{2x + 3}$; all real numbers x except $x = -\frac{3}{2}$

3. (a) $(f + g)(x) = 2x^2 + x - 1$; all real numbers (b) $(f - g)(x) = -2x^2 + x - 1$; all real numbers
(c) $(f \cdot g)(x) = 2x^3 - 2x^2$; all real numbers (d) $\left(\frac{f}{g}\right)(x) = \frac{x - 1}{2x^2}$; all real numbers x except $x = 0$

5. (a) $(f + g)(x) = \sqrt{x} + 3x - 5$; $x \geq 0$ (b) $(f - g)(x) = \sqrt{x} - 3x + 5$; $x \geq 0$ (c) $(f \cdot g)(x) = 3x\sqrt{x} - 5\sqrt{x}$; $x \geq 0$
(d) $\left(\frac{f}{g}\right)(x) = \frac{\sqrt{x}}{3x - 5}$; $x \geq 0$, $x \neq \frac{5}{3}$

7. (a) $(f + g)(x) = 1 + \frac{2}{x}$; $x \neq 0$ (b) $(f - g)(x) = 1$; $x \neq 0$ (c) $(f \cdot g)(x) = \frac{1}{x} + \frac{1}{x^2}$; $x \neq 0$ (d) $\left(\frac{f}{g}\right)(x) = x + 1$; $x \neq 0$

9. (a) $(f + g)(x) = \frac{3x + 3}{3x - 2}$; $x \neq \frac{2}{3}$ (b) $(f - g)(x) = \frac{x + 3}{3x - 2}$; $x \neq \frac{2}{3}$ (c) $(f \cdot g)(x) = \frac{2x^2 + 3x}{(3x - 2)^2}$; $x \neq \frac{2}{3}$ (d) $\left(\frac{f}{g}\right)(x) = \frac{2x + 3}{x}$; $x \neq 0, \frac{2}{3}$

11. $g(x) = 5 - \frac{7}{2}x$ **13.**

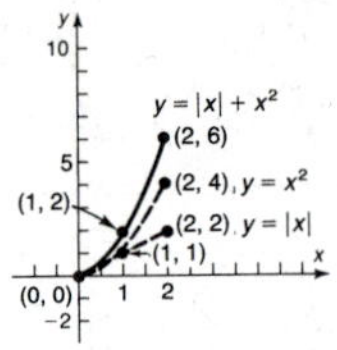

15.

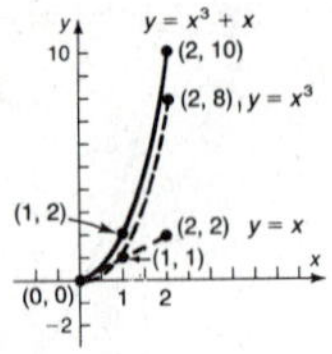

17. (a) 98 (b) 49 (c) 4 (d) 4 **19.** (a) 97 (b) $-\frac{163}{2}$ (c) 1 (d) $-\frac{3}{2}$ **21.** (a) $2\sqrt{2}$ (b) $2\sqrt{2}$ (c) 1 (d) 0
23. (a) $\frac{1}{17}$ (b) $\frac{1}{5}$ (c) 1 (d) $\frac{1}{2}$ **25.** (a) $\frac{3}{5}$ (b) $\sqrt{15}/5$ (c) $\frac{12}{13}$ (d) 0
27. (a) $(f \circ g)(x) = 6x + 1$ (b) $(g \circ f)(x) = 6x + 3$ (c) $(f \circ f)(x) = 4x + 3$ (d) $(g \circ g)(x) = 9x$
29. (a) $(f \circ g)(x) = 3x^2 + 1$ (b) $(g \circ f)(x) = 9x^2 + 6x + 1$ (c) $(f \circ f)(x) = 9x + 4$ (d) $(g \circ g)(x) = x^4$
31. (a) $(f \circ g)(x) = \sqrt{x^2 - 1}$ (b) $(g \circ f)(x) = x - 1$ (c) $(f \circ f)(x) = \sqrt[4]{x}$ (d) $(g \circ g)(x) = x^4 - 2x^2$
33. (a) $(f \circ g)(x) = \dfrac{1 - x}{1 + x}$ (b) $(g \circ f)(x) = \dfrac{x + 1}{x - 1}$ (c) $(f \circ f)(x) = -\dfrac{1}{x}$ (d) $(g \circ g)(x) = x$
35. (a) $(f \circ g)(x) = x$ (b) $(g \circ f)(x) = |x|$ (c) $(f \circ f)(x) = x^4$ (d) $(g \circ g)(x) = \sqrt[4]{x}$
37. (a) $(f \circ g)(x) = \dfrac{1}{4x + 9}$ (b) $(g \circ f)(x) = \dfrac{2}{2x + 3} + 3 = \dfrac{6x + 11}{2x + 3}$ (c) $(f \circ f)(x) = \dfrac{2x + 3}{6x + 11}$ (d) $(g \circ g)(x) = 4x + 9$
39. (a) $(f \circ g)(x) = acx + ad + b$ (b) $(g \circ f)(x) = acx + bc + d$ (c) $(f \circ f)(x) = a^2x + ab + b$
(d) $(g \circ g)(x) = c^2x + cd + d$
41. $(f \circ g)(x) = f(g(x)) = f(\frac{1}{3}x) = 3(\frac{1}{3}x) = x$; $(g \circ f)(x) = g(f(x)) = g(3x) = \frac{1}{3}(3x) = x$
43. $(f \circ g)(x) = f(\sqrt[3]{x}) = (\sqrt[3]{x})^3 = x$; $(g \circ f)(x) = g(x^3) = \sqrt[3]{x^3} = x$
45. $(f \circ g)(x) = f(\frac{1}{2}(x + 6)) = 2[\frac{1}{2}(x + 6)] - 6 = x + 6 - 6 = x$; $(g \circ f)(x) = g(2x - 6) = \frac{1}{2}(2x - 6 + 6) = x$
47. $(f \circ g)(x) = f\left(\dfrac{1}{a}(x - b)\right) = a\left[\dfrac{1}{a}(x - b)\right] + b = x$; $(g \circ f)(x) = g(ax + b) = \dfrac{1}{a}\left(ax + b - b\right) = x$
49. $(f \circ g)(x) = 11$; $(g \circ f)(x) = 2$ **51.** $(f \circ (g \circ h))(x) = 5 - 3x + 4\sqrt{1 - 3x}$
53. $((f + g) \circ h)(x) = 9x^2 - 6x + 3 + \sqrt{1 - 3x}$
55. $F = f \circ g$ **57.** $H = h \circ f$ **59.** $q = f \circ h$ **61.** $P = f \circ f$ **63.** $f(x) = x^3$; $g(x) = 2x + 5$ **65.** $f(x) = \sqrt{x}$; $g(x) = x^2 + x + 1$
67. $f(x) = x^2$; $g(x) = 1 - \dfrac{1}{x^2}$ **69.** $f(x) = [\![x]\!]$; $g(x) = x^2 + 1$ **71.** $S(r(t)) = \frac{16}{9}\pi t^6$ **73.** $C(N(t)) = 5000 + 600{,}000t - 30{,}000t^2$

EXERCISE 4.5

1. One-to-one
3. Not one-to-one
5. One-to-one

7.

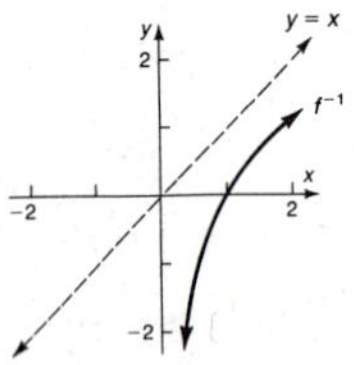

9.

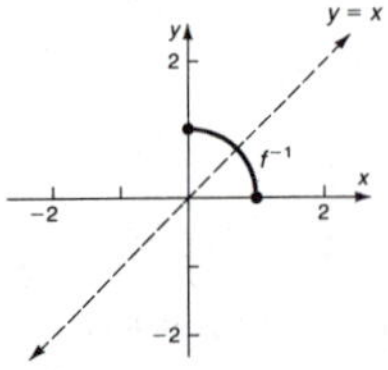

11.

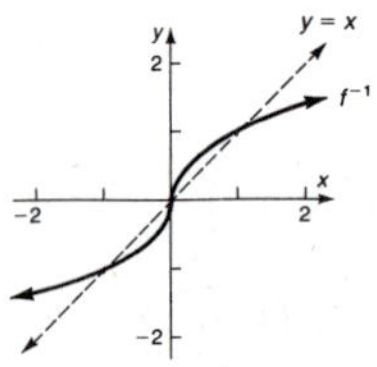

13. $f(g(x)) = f\left(\frac{1}{3}(x + 4)\right) = 3\left[\frac{1}{3}(x + 4)\right] - 4 = x$; $g(f(x)) = g(3x - 4) = \frac{1}{3}[(3x - 4) + 4] = x$

15. $f(g(x)) = 4\left[\dfrac{x}{4} + 2\right] - 8 = x$; $g(f(x)) = \dfrac{4x - 8}{4} + 2 = x$

17. $f(g(x)) = (\sqrt[3]{x + 8})^3 - 8 = x$; $g(f(x)) = \sqrt[3]{(x^3 - 8) + 8} = x$ **19.** $f(g(x)) = \dfrac{1}{1/x} = x$; $g(f(x)) = \dfrac{1}{1/x} = x$

21. $f(g(x)) = \dfrac{2\left(\dfrac{4x - 3}{2 - x}\right) + 3}{\dfrac{4x - 3}{2 - x} + 4} = \dfrac{8x - 3x}{-3 + 8} = x$; $g(f(x)) = \dfrac{4\left(\dfrac{2x + 3}{x + 4}\right) - 3}{2 - \dfrac{2x + 3}{x + 4}} = x$

23. $f^{-1}(x) = \frac{1}{2}x$

$f(f^{-1}(x)) = 2\left(\frac{1}{2}x\right) = x$

$f^{-1}(f(x)) = \frac{1}{2}(2x) = x$

Domain f = Range $f^{-1} = (-\infty, \infty)$

Range f = Domain $f^{-1} = (-\infty, \infty)$

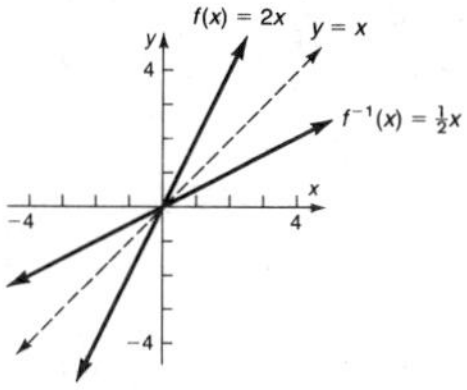

25. $f^{-1}(x) = \dfrac{x}{4} - \dfrac{1}{2}$

$f(f^{-1}(x)) = 4\left(\dfrac{x}{4} - \dfrac{1}{2}\right) + 2 = x$

$f^{-1}(f(x)) = \dfrac{4x + 2}{4} - \dfrac{1}{2} = x$

Domain f = Range $f^{-1} = (-\infty, \infty)$

Range f = Domain $f^{-1} = (-\infty, \infty)$

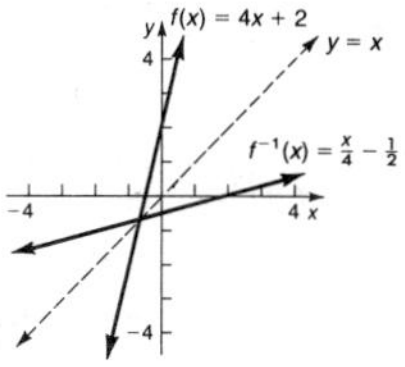

27. $f^{-1}(x) = \sqrt[3]{x + 1}$

$f(f^{-1}(x)) = (\sqrt[3]{x + 1})^3 - 1 = x$

$f^{-1}(f(x)) = \sqrt[3]{x^3 - 1 + 1} = x$

Domain f = Range $f^{-1} = (-\infty, \infty)$

Range f = Domain $f^{-1} = (-\infty, \infty)$

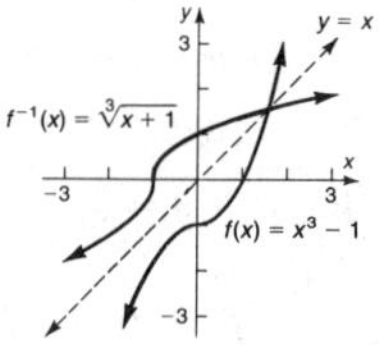

29. $f^{-1}(x) = \sqrt{x - 4}$

$f(f^{-1}(x)) = (\sqrt{x - 4})^2 + 4 = x$

$f^{-1}(f(x)) = \sqrt{(x^2 + 4) - 4} = \sqrt{x^2} = |x| = x$

Domain f = Range $f^{-1} = [0, \infty)$

Range f = Domain $f^{-1} = [4, \infty)$

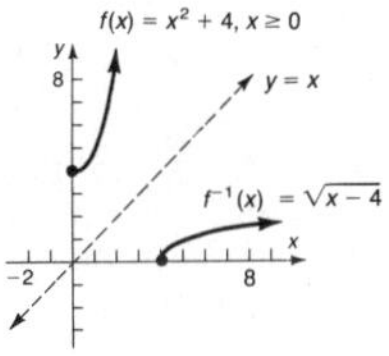

31. $f^{-1}(x) = \dfrac{4}{x}$

$f(f^{-1}(x)) = \dfrac{4}{4/x} = x$

$f^{-1}(f(x)) = \dfrac{4}{4/x} = x$

y = 4/X

Domain f = Range f^{-1} = All real numbers except 0

Range f = Domain f^{-1} = All real numbers except 0

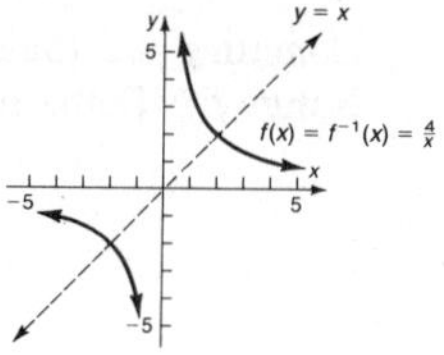

33. $f^{-1}(x) = \dfrac{2x + 1}{x}$

$f(f^{-1}(x)) = \dfrac{1}{\dfrac{2x + 1}{x} - 2} = x$

$f^{-1}(f(x)) = \dfrac{2\left(\dfrac{1}{x - 2}\right) + 1}{\dfrac{1}{x - 2}} = x$

Domain f = Range f^{-1} = All real numbers except 2

Range f = Domain f^{-1} = All real numbers except 0

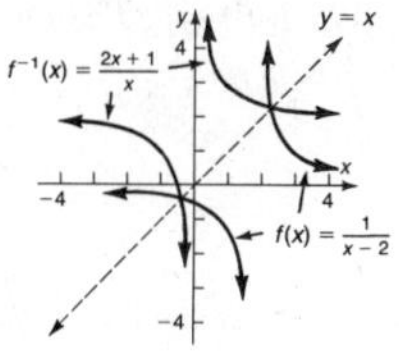

35. $f^{-1}(x) = \dfrac{1 - 3x}{x}$

$f(f^{-1}(x)) = \dfrac{1}{3 + \dfrac{1 - 3x}{x}} = x$

$f^{-1}(f(x)) = \dfrac{1 - 3\left(\dfrac{1}{3 + x}\right)}{\dfrac{1}{3 + x}} = x$

Domain f = Range f^{-1} = All real numbers except -3

Range f = Domain f^{-1} = All real numbers except 0

37. $f^{-1}(x) = \sqrt{x} - 2$

$f(f^{-1}(x)) = (\sqrt{x} - 2 + 2)^2 = x$

$f^{-1}(f(x)) = \sqrt{(x + 2)^2} - 2 = |x + 2| - 2 = x$

Domain f = Range $f^{-1} = [-2, \infty)$

Range f = Domain $f^{-1} = [0, \infty)$

39. $f^{-1}(x) = \dfrac{x}{x-2}$

$$f(f^{-1}(x)) = \frac{2\dfrac{x}{x-2}}{\dfrac{x}{x-2} - 1} = x$$

$$f^{-1}(f(x)) = \frac{\dfrac{2x}{x-1}}{\dfrac{2x}{x-1} - 2} = x$$

Domain f = Range f^{-1} = All real numbers except 1
Range f = Domain f^{-1} = All real numbers except 2

41. $f^{-1}(x) = \dfrac{3x+4}{2x-3}$

$$f(f^{-1}(x)) = \frac{3\left(\dfrac{3x+4}{2x-3}\right) + 4}{2\left(\dfrac{3x+4}{2x-3}\right) - 3} = x$$

$$f^{-1}(f(x)) = \frac{3\left(\dfrac{3x+4}{2x-3}\right) + 4}{2\left(\dfrac{3x+4}{2x-3}\right) - 3} = x$$

Domain f = Range f^{-1} = All real numbers except $\frac{3}{2}$
Range f = Domain f^{-1} = All real numbers except $\frac{3}{2}$

43. $f^{-1}(x) = \dfrac{-2x+3}{x-2}$

$$f(f^{-1}(x)) = \frac{2\left(\dfrac{-2x+3}{x-2}\right) + 3}{\dfrac{-2x+3}{x-2} + 2} = x$$

$$f^{-1}(f(x)) = \frac{-2\left(\dfrac{2x+3}{x+2}\right) + 3}{\dfrac{2x+3}{x+2} - 2} = x$$

Domain f = Range f^{-1} = All real numbers except -2
Range f = Domain f^{-1} = All real numbers except 2

45. $f^{-1}(x) = \dfrac{x^3}{8}$

$$f(f^{-1}(x)) = 2\sqrt[3]{\frac{x^3}{8}} = x$$

$$f^{-1}(f(x)) = \frac{(2\sqrt[3]{x})^3}{8} = x$$

Domain f = Range f^{-1} = $(-\infty, \infty)$
Range f = Domain f^{-1} = $(-\infty, \infty)$

47. $f^{-1}(x) = \dfrac{1}{m}(x - b),\ m \neq 0$

49. No; whenever x and $-x$ are in the domain of f, two equal y values, $f(x)$ and $f(-x)$, are present.

51. Quadrant I **53.** $f(x) = |x|,\ x \geq 0$ is one-to-one; this may be written as $f(x) = x$; $f^{-1}(x) = x$

55. $f(g(x)) = \frac{9}{5}\left[\frac{5}{9}(x - 32)\right] + 32 = x$; $g(f(x)) = \frac{5}{9}\left[\left(\frac{9}{5}x + 32\right) - 32\right] = x$ **57.** $l(T) = gT^2/4\pi^2$

59. $f^{-1}(x) = \dfrac{-dx + b}{cx - a}$; $f = f^{-1}$ if $a = -d$

EXERCISE 4.6

1. $V(r) = 2\pi r^3$ **3.** $R(x) = -\frac{1}{5}x^2 + 100x$ **5.** $R(x) = -\frac{1}{20}x^2 + 5x$ **7.** $A(x) = -x^2 + 50x;\ 0 < x < 50$

9. (a) $C(x) = x$ (b) $A(x) = \dfrac{x^2}{4\pi}$ **11.** $A(x) = \frac{1}{2}x^4$ **13.** (a) $d(x) = \sqrt{x^4 - 7x^2 + 16}$ (b) $d(0) = 4$ (c) $d(1) = \sqrt{10}$

15. $d(x) = \sqrt{x^2 - x + 1}$ **17.** $d(t) = 5\sqrt{89}t$ **19.** $V(x) = x(24 - 2x)^2$ **21.** $A(x) = 2x^2 + \dfrac{40}{x}$

23. $A(x) = x(16 - x^2)$; Domain $= \{x \mid 0 \leq x \leq 4\}$ **25.** (a) $A(x) = 4x(4 - x^2)^{1/2}$ (b) $p(x) = 4x + 4(4 - x^2)^{1/2}$

27. $C(r) = 12\pi r^2 + \dfrac{4000}{r}$ **29.** $A(x) = x^2 + \dfrac{25 - 20x + 4x^2}{\pi}$; Domain $= \{x \mid 0 < x < 2.5\}$

31. (a) $C(x) = 10x + 14\sqrt{x^2 - 10x + 29}$
(b) Domain $= \{x \mid 0 \leq x \leq 5\}$ (c) $C(1) \approx \$72.61$; $C(2) \approx \$70.48$; $C(3) \approx \$69.60$; $C(4) \approx \$71.30$

33. (a) $A(r) = 2r^2$ (b) $p(r) = 6r$ **35.** $A(x) = \left(\dfrac{\pi}{3} - \dfrac{\sqrt{3}}{4}\right)x^2$

37. $$C = \begin{cases} 95 & \text{if } x = 7 \\ 119 & \text{if } 7 < x \leq 8 \\ 143 & \text{if } 8 < x \leq 9 \\ 167 & \text{if } 9 < x \leq 10 \\ 190 & \text{if } 10 < x \leq 14 \end{cases}$$

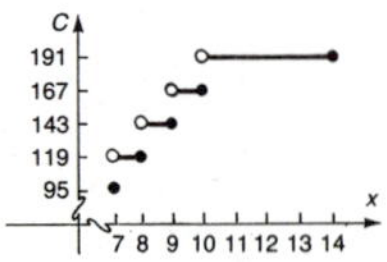

39. $V(h) = \dfrac{\pi}{48}h^3$

41. Schedule X: $y = f(x) = \begin{cases} 0.15x & \text{if } 0 < x \le 17{,}850 \\ 2677.50 + 0.28(x - 17{,}850) & \text{if } 17{,}850 < x \le 43{,}150 \\ 9761.50 + 0.33(x - 43{,}150) & \text{if } 43{,}150 < x \le 89{,}560 \end{cases}$

Schedule Y − 1; $y = f(x) = \begin{cases} 0.15x & \text{if } 0 < x \le 29{,}750 \\ 4462.50 + 0.28(x - 29{,}750) & \text{if } 29{,}750 < x \le 71{,}900 \\ 16{,}264.50 + 0.33(x - 71{,}900) & \text{if } 71{,}900 < x \le 149{,}250 \end{cases}$

FILL-IN-THE-BLANK ITEMS

1. independent; dependent **2.** vertical **3.** even; odd **4.** horizontal; right **5.** $g(f(x)) = (g \circ f)(x)$ **6.** one-to-one **7.** $y = x$

REVIEW EXERCISES

1. $f(x) = -2x + 6$ **3.** $A = 11$ **5.** (a) B, C, D (b) D

7. (a) $f(-x) = \dfrac{-x}{x^2 - 4}$ (b) $-f(x) = \dfrac{-x}{x^2 - 4}$ (c) $f(x + 2) = \dfrac{x + 2}{x^2 + 4x}$ (d) $f(x - 2) = \dfrac{x - 2}{x^2 - 4x}$

9. (a) $f(-x) = \sqrt{x^2 - 4}$ (b) $-f(x) = -\sqrt{x^2 - 4}$ (c) $f(x + 2) = \sqrt{x^2 + 4x}$ (d) $f(x - 2) = \sqrt{x^2 - 4x}$

11. (a) $f(-x) = -\dfrac{x^2 - 4}{x^2}$ (b) $-f(x) = -\dfrac{x^2 - 4}{x^2}$ (c) $f(x + 2) = \dfrac{x^2 + 4x}{x^2 + 4x + 4}$ (d) $f(x - 2) = \dfrac{x^2 - 4x}{x^2 - 4x + 4}$ **13.** Odd

15. Even **17.** Neither **19.** $\{x \mid x \neq -2, x \neq 2\}$ **21.** $(-\infty, 2]$ **23.** $(0, \infty)$ **25.** $\{x \mid x \neq -3, x \neq 1\}$ **27.** $[-1, \infty)$ **29.** $[0, \infty)$

31. (a) $(-\infty, \infty)$
(b) $(0, -4), (-4, 0), (4, 0)$
(d) $[-4, \infty)$
(c)
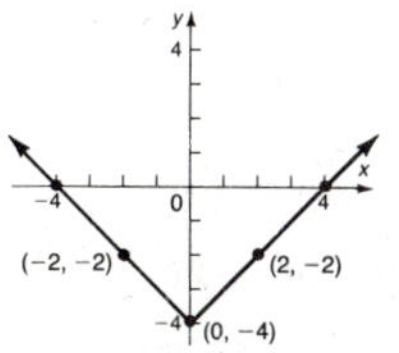

33. (a) $(-\infty, \infty)$
(b) $(0, 0)$
(d) $(-\infty, 0]$
(c)
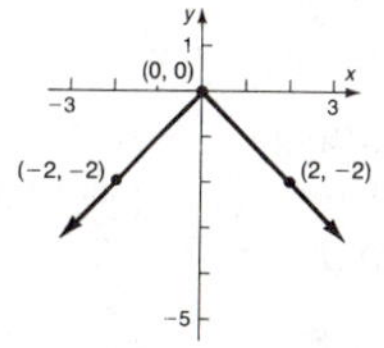

35. (a) $[1, \infty)$
(b) $(1, 0)$
(d) $[0, \infty)$
(c)
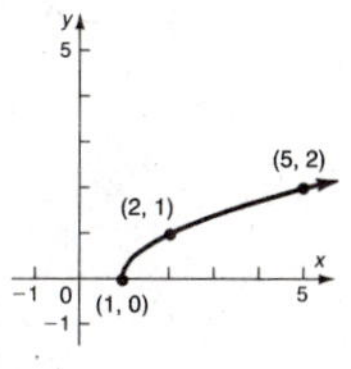

37. (a) $(-\infty, 1]$
(b) $(1, 0), (0, 1)$
(d) $[0, \infty)$
(c)
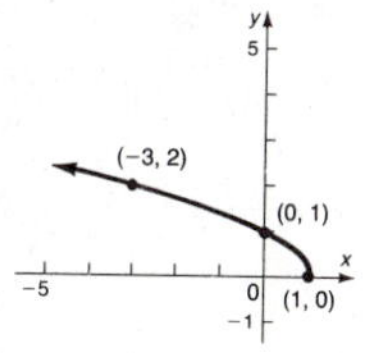

39. (a) $(-\infty, \infty)$
(b) $(0, 4), (2, 0)$
(d) $(-\infty, \infty)$
(c)
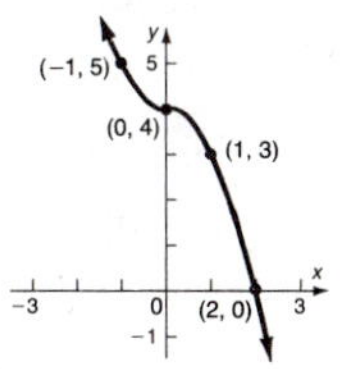

41. (a) $(-\infty, \infty)$
(b) $(0, 3)$
(d) $[2, \infty)$
(c)
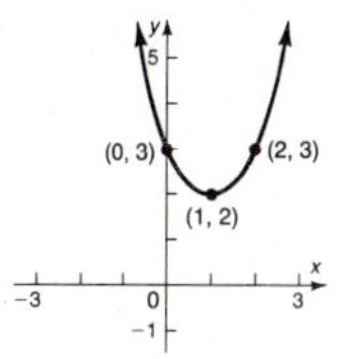

43. (a) $(-\infty, \infty)$
(b) $(0, 0)$
(d) $(-\infty, \infty)$
(c)
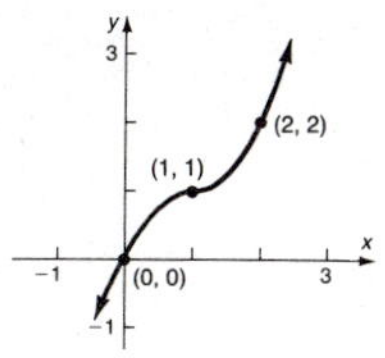

45. (a) $(0, \infty)$
(b) None
(d) $(0, \infty)$
(c)
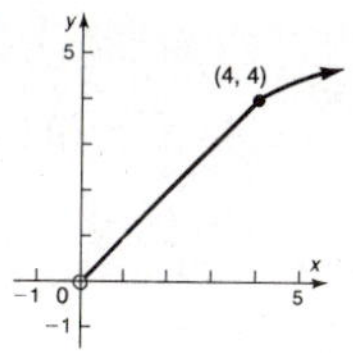

47. (a) $\{x \mid x \neq 1\}$ (b) $(0, 0)$ (d) $\{y \mid y \neq 1\}$ (c)

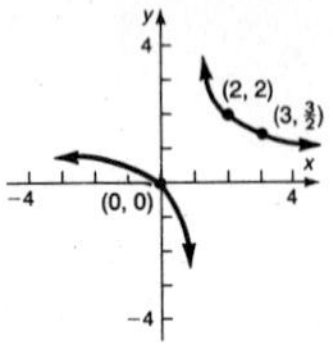

49. (a) $(-\infty, \infty)$ (b) $-1 < x \leq 0$ are x-intercepts (d) Set of integers (c)

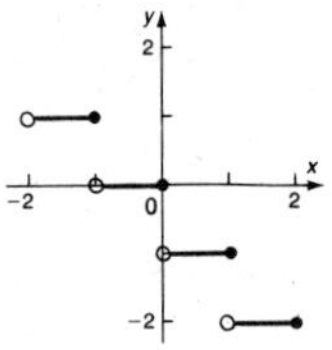

51. $f^{-1}(x) = \dfrac{2x + 3}{x - 2}$

$$f(f^{-1}(x)) = \frac{2\left(\frac{2x + 3}{x - 2}\right) + 3}{\frac{2x + 3}{x - 2} - 2} = x$$

$$f^{-1}(f(x)) = \frac{2\left(\frac{2x + 3}{x - 2}\right) + 3}{\frac{2x + 3}{x - 2} - 2} = x$$

Domain f = Range f^{-1} = All real numbers except 2
Range f = Domain f^{-1} = All real numbers except 2

53. $f^{-1}(x) = \dfrac{x + 1}{x}$

$$f(f^{-1}(x)) = \frac{1}{\frac{x + 1}{x} - 1} = x$$

$$f^{-1}(f(x)) = \frac{\frac{1}{x - 1} + 1}{\frac{1}{x - 1}} = x$$

Domain f = Range f^{-1} = All real numbers except 1
Range f = Domain f^{-1} = All real numbers except 0

55. $f^{-1}(x) = \dfrac{27}{x^3}$

$$f(f^{-1}(x)) = \frac{3}{(27/x^3)^{1/3}} = x$$

$$f^{-1}(f(x)) = \frac{27}{(3/x^{1/3})^3} = x$$

Domain f = Range f^{-1} = All real numbers except 0
Range f = Domain f^{-1} = All real numbers except 0

57. (a) -26 (b) -241 (c) 16 (d) -1

59. (a) $\sqrt{11}$ (b) 1 (c) $\sqrt{\sqrt{6} + 2}$ (d) 19 **61.** (a) $\frac{1}{20}$ (b) $-\frac{13}{8}$ (c) $\frac{400}{1601}$ (d) -17

63. $(f \circ g)(x) = \dfrac{-3x}{3x + 2}$; $(g \circ f)(x) = \dfrac{6 - x}{x}$; $(f \circ f)(x) = \dfrac{3x - 2}{2 - x}$; $(g \circ g)(x) = 9x + 8$

65. $(f \circ g)(x) = 27x^2 + 3|x| + 1$; $(g \circ f)(x) = 3|3x^2 + x + 1|$; $(f \circ f)(x) = 3(3x^2 + x + 1)^2 + 3x^2 + x + 2$; $(g \circ g)(x) = 9|x|$

67. $(f \circ g)(x) = \dfrac{1 + x}{1 - x}$; $(g \circ f)(x) = \dfrac{x - 1}{x + 1}$; $(f \circ f)(x) = x$; $(g \circ g)(x) = x$

69. (a)

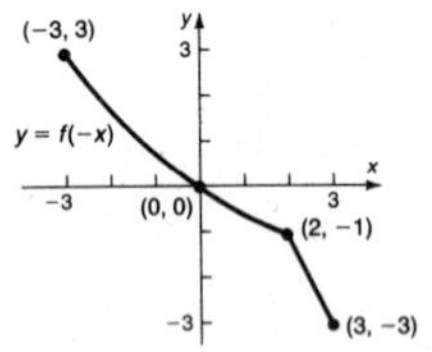

(b)

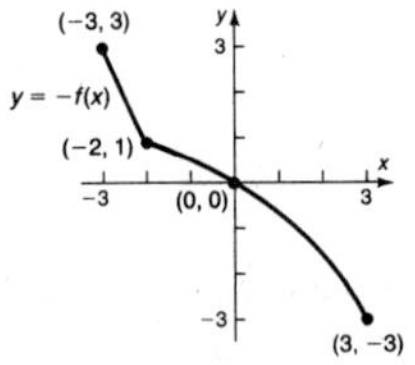

(c)

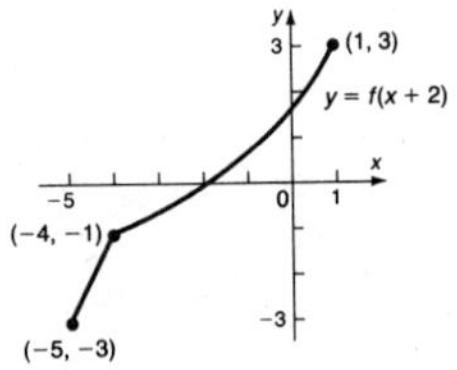

(d)

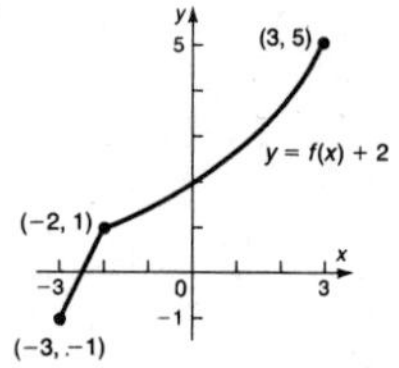

(e)

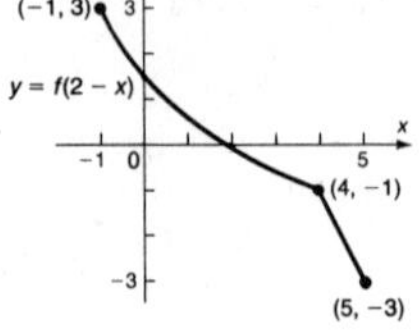

(f)

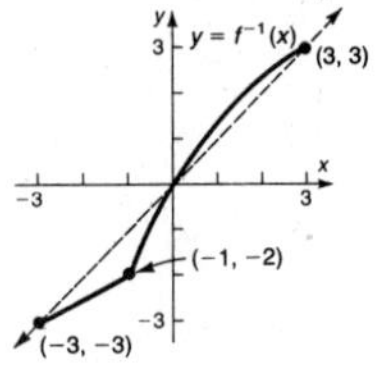

71. $T(h) = -0.0025h + 30$

CHAPTER 5 EXERCISE 5.1

1.

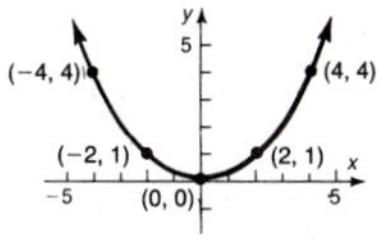

3.

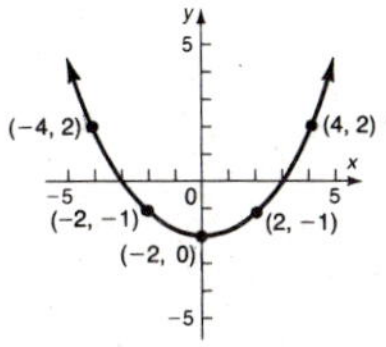

5.

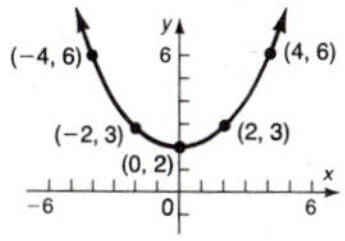

7.

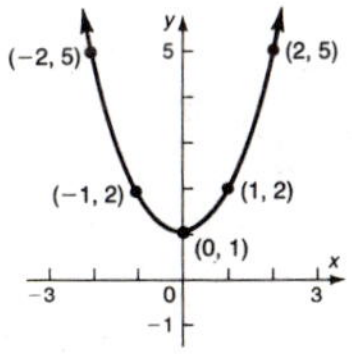

9.

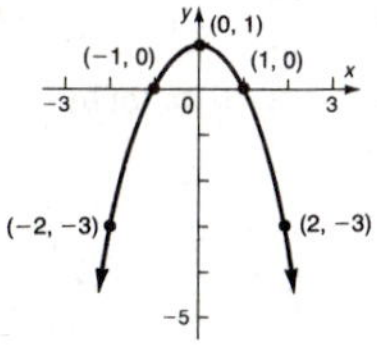

11.

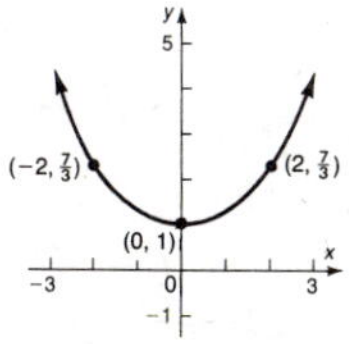

13. $f(x) = (x + 2)^2 - 2$

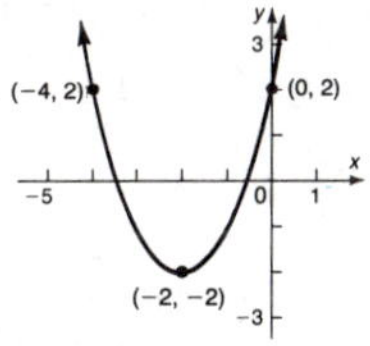

15. $f(x) = 2(x - 1)^2 - 1$

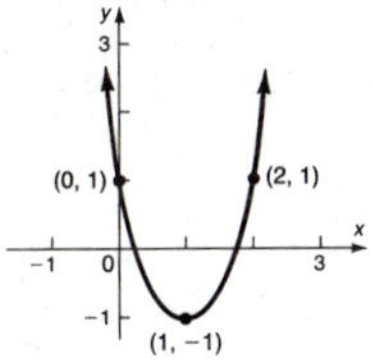

17. $f(x) = -(x + 1)^2 + 1$

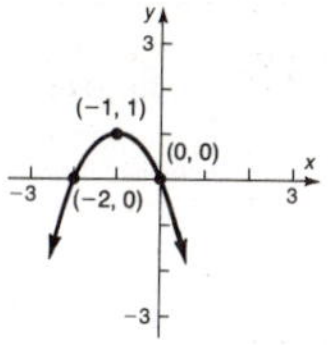

19. $f(x) = \frac{1}{2}(x + 1)^2 - \frac{3}{2}$

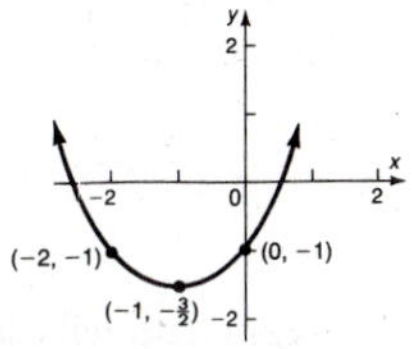

21.

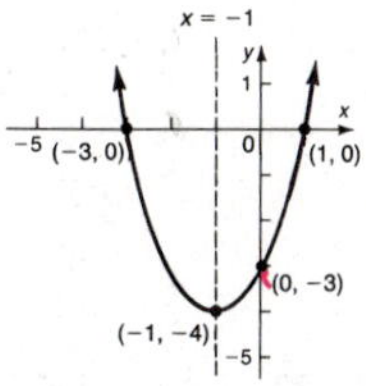

23.

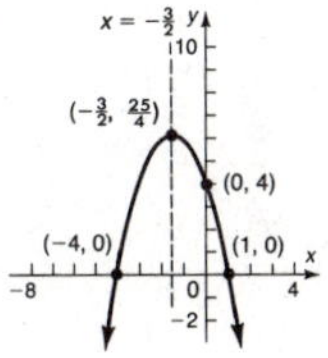

25.

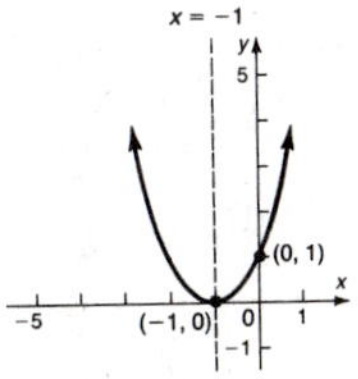

27.

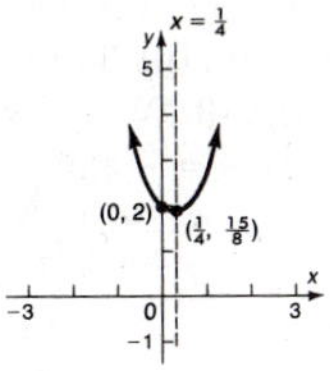

29.

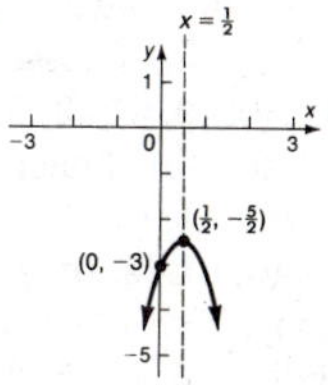

31.

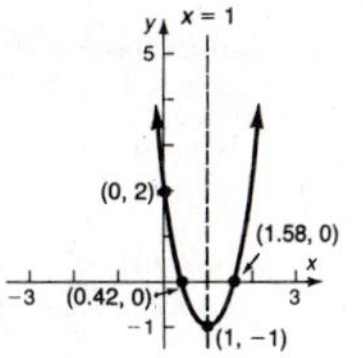

33.

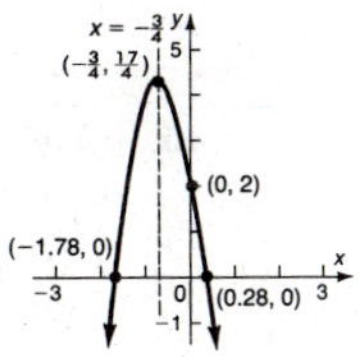

35. Minimum value; −9
37. Maximum value; 21
39. Maximum value; 13

41. Opens up; vertex at $(-1, f(-1))$; axis of symmetry $x = -1$

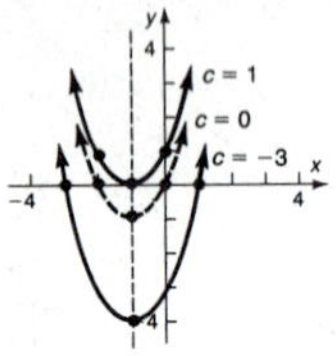

43. 15, 15 **45.** Price: \$500; maximum revenue: \$1,000,000 **47.** 625 ft^2; 25 ft by 25 ft **49.** 2,000,000 m^2 **51.** 37,500,000 m^2 **53.** 80 members **55.** 8 PM **57.** $\frac{64}{3}$ m **59.** 3 in.

61. Width $= \dfrac{40}{\pi + 4} \approx 5.6$ ft; Length ≈ 2.8 ft **63.** $x = \dfrac{16}{6 - \sqrt{3}} \approx 3.75$ ft; other side ≈ 2.38 ft **65.** $a = 6, b = 0, c = 2$; $f(x) = 6x^2 + 2$

67. $\dfrac{a}{2}$ **69.** $\left.\begin{aligned} ah^2 - bh + c &= y_0 \\ c &= y_1 \\ ah^2 + bh + c &= y^2 \end{aligned}\right\}$ $\left.\begin{aligned} y_0 + y_2 &= 2ah^2 + 2c \\ 4y_1 &= 4c \end{aligned}\right\}$ Area $= \dfrac{h}{3}(2ah^2 + 6c) = \dfrac{h}{3}(y_0 + 4y_1 + y_2)$

EXERCISE 5.2

1. Yes; degree 3 **3.** Yes; degree 2 **5.** No; x is raised to the -1 power. **7.** No; x is raised to the $\frac{3}{2}$ power. **9.** Yes; degree 4

11.

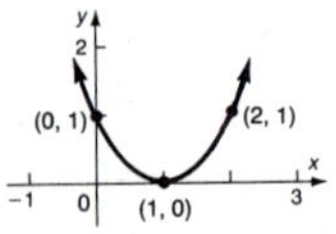

13.

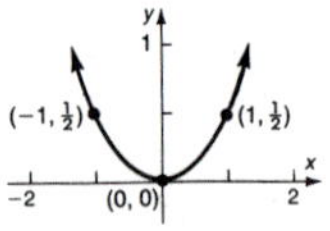

15.

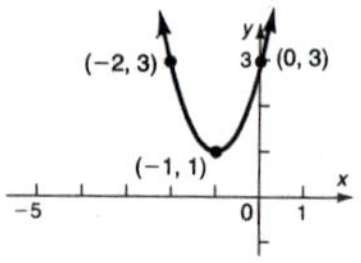

17.

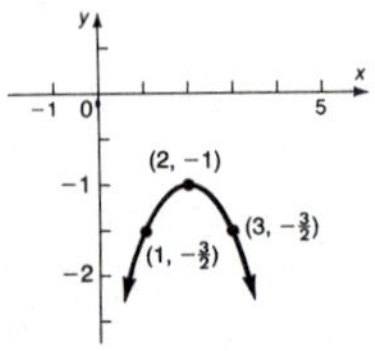

19. 4, multiplicity 1; -5, multiplicity 2 **21.** 2, multiplicity 3 **23.** $-\frac{1}{2}$, multiplicity 2 **25.** 5, multiplicity 3; -4, multiplicity 2

27. No real zeros **29.** (a) $(0, 1)$; $(1, 0)$ (b) Touches at $(1, 0)$ (c) $y = x^2$ (d) One

31. (a) $(0, 0)$; $(1, 0)$ (b) Touches at $(0, 0)$; crosses at $(1, 0)$ (c) $y = x^3$ (d) Two

33. (a) $(0, 0)$; $(-4, 0)$ (b) Crosses at $(0, 0)$ and $(-4, 0)$ (c) $y = 6x^4$ (d) Three

35. (a) $(0, -6)$; $(1, 0)$; $(2, 0)$; $(3, 0)$ (b) Crosses at $(1, 0)$, $(2, 0)$, and $(3, 0)$ (c) $y = x^3$ (d) Two

37. (a) $(0, 0)$; $(-2, 0)$ (b) Touches at $(0, 0)$; crosses at $(-2, 0)$ (c) $y = -4x^3$ (d) Two

39. (a) $(0, 0)$; $(2, 0)$; $(-4, 0)$
(b) Crosses at $(0, 0)$, $(2, 0)$, and $(-4, 0)$
(c) Below x-axis: $x < -4$, $0 < x < 2$
Above x-axis: $-4 < x < 0$, $x > 2$
(d)

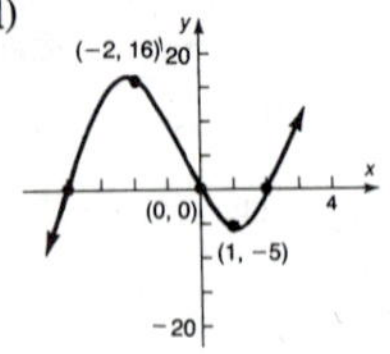

41. (a) $(0, 0)$; $(3, 0)$
(b) Touches at $(0, 0)$, crosses at $(3, 0)$
(c) Below x-axis: $x < 3$
Above x-axis: $x > 3$
(d)

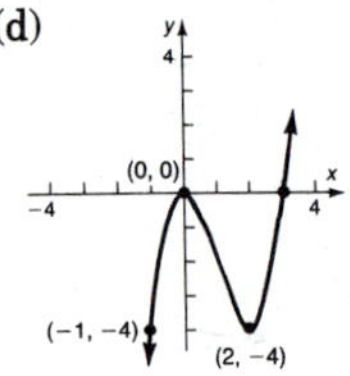

43. (a) $(0, 0)$; $(-2, 0)$; $(2, 0)$
(b) Crosses at $(-2, 0)$, $(0, 0)$, and $(2, 0)$
(c) Above x-axis: $x < -2$, $0 < x < 2$
Below x-axis: $-2 < x < 0$, $x > 2$
(d)

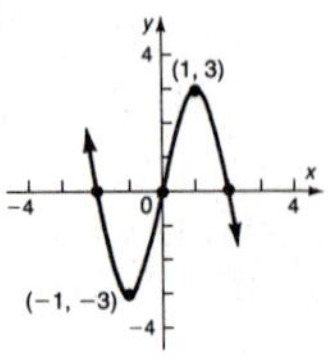

45. (a) $(-1, 0)$; $(0, 0)$; $(3, 0)$
(b) Crosses at $(-1, 0)$, $(0, 0)$, and $(3, 0)$
(c) Below x-axis: $x < -1$, $0 < x < 3$
Above x-axis: $-1 < x < 0$, $x > 3$
(d)

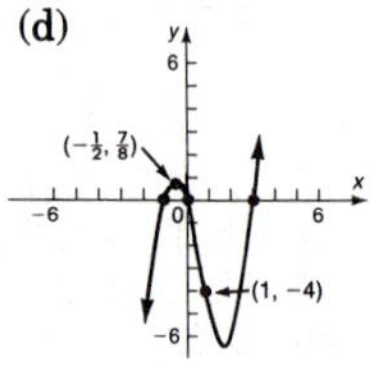

47. (a) $(-2, 0)$; $(0, 0)$; $(2, 0)$
(b) Crosses at $(-2, 0)$ and $(2, 0)$; touches at $(0, 0)$
(c) Below x-axis: $-2 < x < 2$
Above x-axis: $x < -2$, $x > 2$
(d)

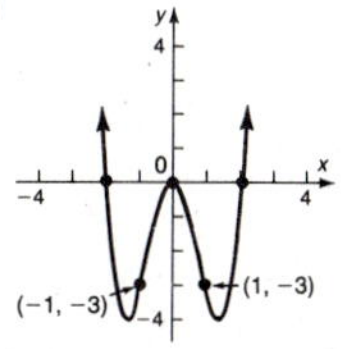

49. (a) $(0, 0)$; $(2, 0)$
(b) Touches at $(0, 0)$ and $(2, 0)$
(c) Above x-axis for all x except 0 and 2
(d)

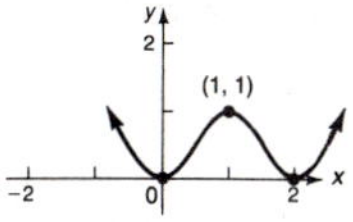

51. (a) $(-1, 0)$; $(0, 0)$; $(3, 0)$
(b) Crosses at $(-1, 0)$ and $(3, 0)$; touches at $(0, 0)$
(c) Below x-axis: $-1 < x < 3$
Above x-axis: $x < -1$, $x > 3$
(d)

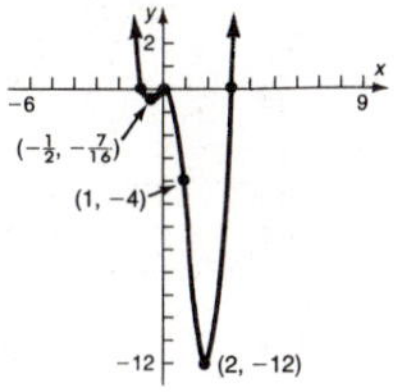

53. (a) $(-2, 0)$; $(0, 0)$; $(4, 0)$; $(6, 0)$
(b) Crosses at $(-2, 0)$, $(0, 0)$, $(4, 0)$, and $(6, 0)$
(c) Below x-axis: $-2 < x < 0$, $4 < x < 6$
Above x-axis: $x < -2$, $0 < x < 4$, $x > 6$
(d)

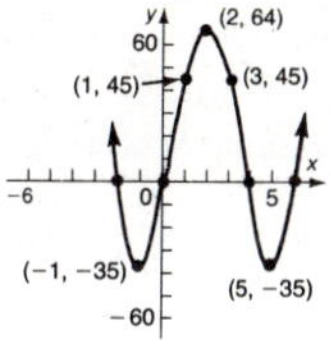

55. (a) $(0, 0)$; $(2, 0)$
(b) Crosses at $(2, 0)$; touches at $(0, 0)$
(c) Below x-axis: $x < 2$
Above x-axis: $x > 2$

(d)

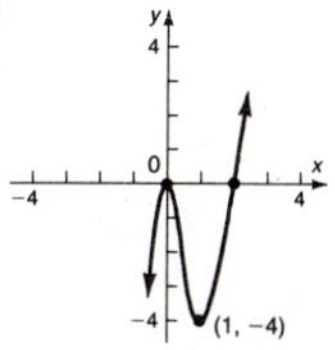

EXERCISE 5.3

1. All real numbers except 2 **3.** All real numbers except 2 and -1 **5.** All real numbers except $-\frac{1}{2}$ and 3
7. All real numbers except 2 **9.** All real numbers

11.

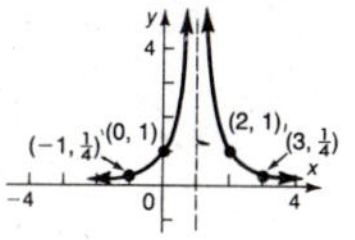

13.

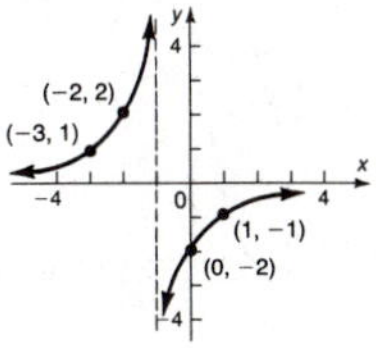

15.

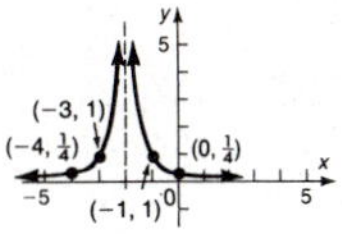

17.

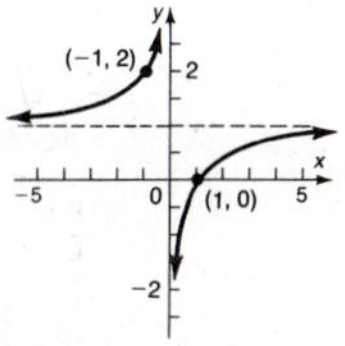

19.

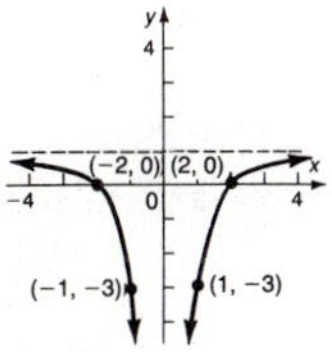

21. Horizontal asymptote: $y = 1$; vertical asymptote: $x = -1$ **23.** No asymptotes
25. Horizontal asymptote: $y = 0$; vertical asymptotes: $x = 1$, $x = -1$ **27.** Horizontal asymptote: $y = 0$; vertical asymptote: $x = 0$
29. Oblique asymptote: $y = 3x$; vertical asymptote: $x = 0$

31. 1. x-intercept: -1; no y-intercept
2. No symmetry
3. Vertical asymptotes: $x = 0$, $x = -4$
4. Horizontal asymptote: $y = 0$, intersected at $(-1, 0)$
5. $x < -4$: below x-axis
$-4 < x < -1$: above x-axis
$-1 < x < 0$: below x-axis
$x > 0$: above x-axis

6.
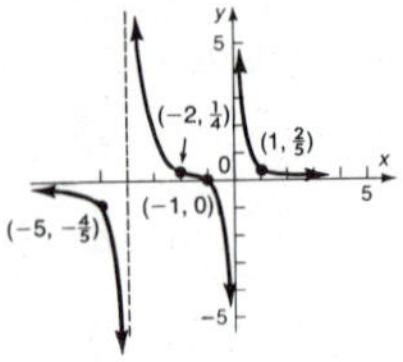

33. 1. x-intercept: -1; y-intercept: $\frac{3}{4}$
2. No symmetry
3. Vertical asymptote: $x = -2$
4. Horizontal asymptote: $y = \frac{3}{2}$, not intersected
5. $x < -2$: above x-axis
$-2 < x < -1$: below x-axis
$x > -1$: above x-axis

6.
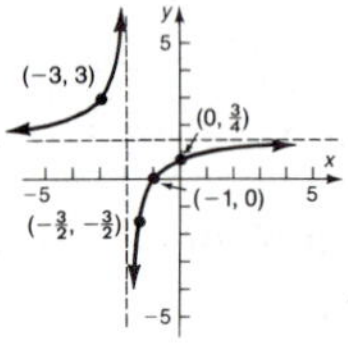

35. 1. No x-intercept; y-intercept: $-\frac{3}{4}$
2. Symmetric with respect to y-axis
3. Vertical asymptotes: $x = 2$, $x = -2$
4. Horizontal asymptote: $y = 0$, not intersected
5. $x < -2$: above x-axis
$-2 < x < 2$: below x-axis
$x > 2$: above x-axis

6.
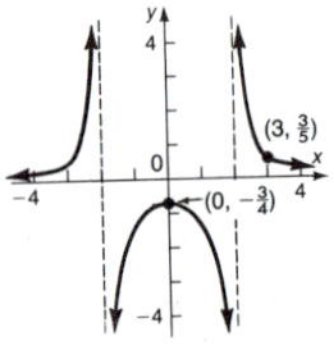

37. 1. No x-intercept; y-intercept: -1
2. Symmetric with respect to y-axis
3. Vertical asymptotes: $x = -1$, $x = 1$
4. No horizontal or oblique asymptotes
5. $x < -1$: above x-axis
$-1 < x < 1$: below x-axis
$x > 1$: above x-axis

6.
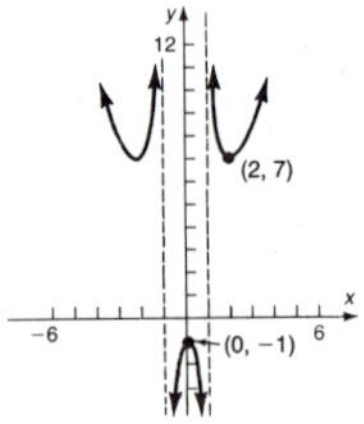

39. 1. x-intercept: 1; y-intercept: $\frac{1}{9}$
2. No symmetry
3. Vertical asymptotes: $x = 3$, $x = -3$
4. Oblique asymptote: $y = x$, intersected at $\left(\frac{1}{9}, \frac{1}{9}\right)$
5. $x < -3$: below x-axis
$-3 < x < 1$: above x-axis
$1 < x < 3$: below x-axis
$x > 3$: above x-axis

6.
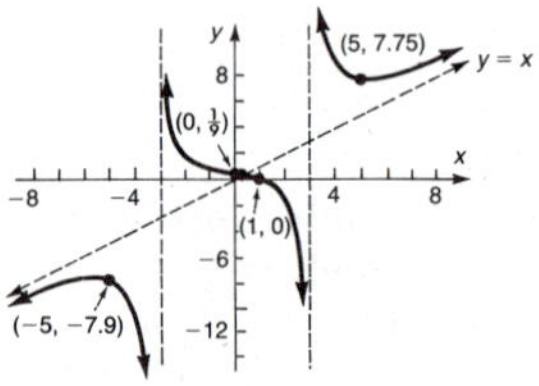

41. 1. Intercept $(0, 0)$
2. No symmetry
3. Vertical asymptotes: $x = 2$, $x = -3$
4. Horizontal asymptote: $y = 1$, intersected at $(6, 1)$
5. $x < -3$: above x-axis
$-3 < x < 0$: below x-axis
$0 < x < 2$: below x-axis
$x > 2$: above x-axis

6.
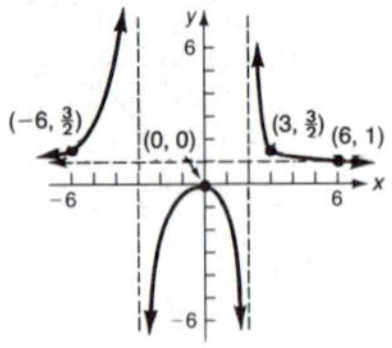

43. 1. Intercept $(0, 0)$
2. Symmetry with respect to origin
3. Vertical asymptotes: $x = -2$, $x = 2$
4. Horizontal asymptote: $y = 0$, intersected at $(0, 0)$
5. $x < -2$: below x-axis
$-2 < x < 0$: above x-axis
$0 < x < 2$: below x-axis
$x > 2$: above x-axis

6.
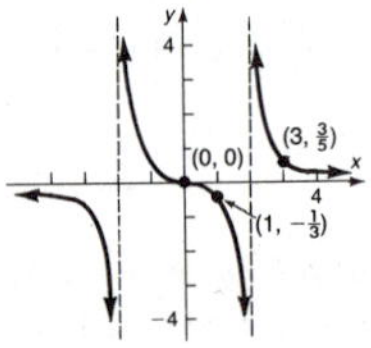

45. 1. No x-intercept; y-intercept: $\frac{3}{4}$
2. No symmetry
3. Vertical asymptotes: $x = -2$, $x = 1$, $x = 2$
4. Horizontal asymptote: $y = 0$, not intersected
5. $x < -2$: below x-axis
$-2 < x < 1$: above x-axis
$1 < x < 2$: below x-axis
$x > 2$: above x-axis
6.
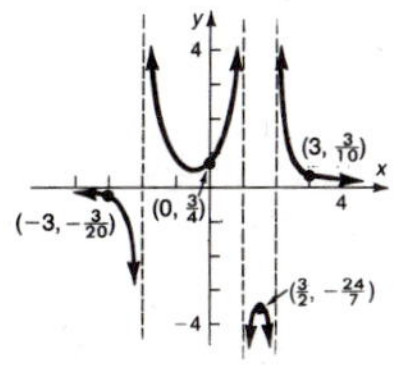

47. 1. x-intercepts: -1, 1; y-intercept: $\frac{1}{4}$
2. Symmetric with respect to y-axis
3. Vertical asymptotes: $x = -2$, $x = 2$
4. Horizontal asymptote: $y = 0$, intersected at $(-1, 0)$ and $(1, 0)$
5. $x < -2$: above x-axis
$-2 < x < -1$: below x-axis
$-1 < x < 1$: above x-axis
$1 < x < 2$: below x-axis
$x > 2$: above x-axis
6.
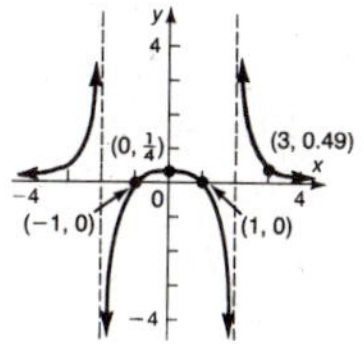

49. 1. x-intercepts: -1, 4; y-intercept: -2
2. No symmetry
3. Vertical asymptote: $x = -2$
4. Oblique asymptote: $y = x - 5$, not intersected
5. $x < -2$: below x-axis
$-2 < x < -1$: above x-axis
$-1 < x < 4$: below x-axis
$x > 4$: above x-axis
6.
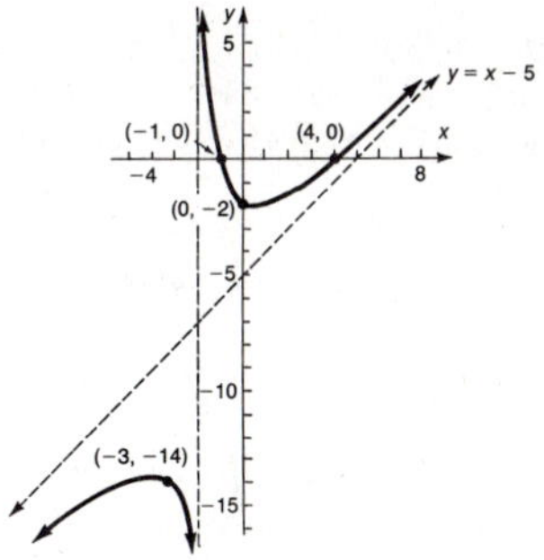

51. 1. x-intercepts: -4, 3; y-intercept: 3
2. No symmetry
3. Vertical asymptote: $x = 4$
4. Oblique asymptote: $y = x + 5$, not intersected
5. $x < -4$: below x-axis
$-4 < x < 3$: above x-axis
$3 < x < 4$: below x-axis
$x > 4$: above x-axis
6.
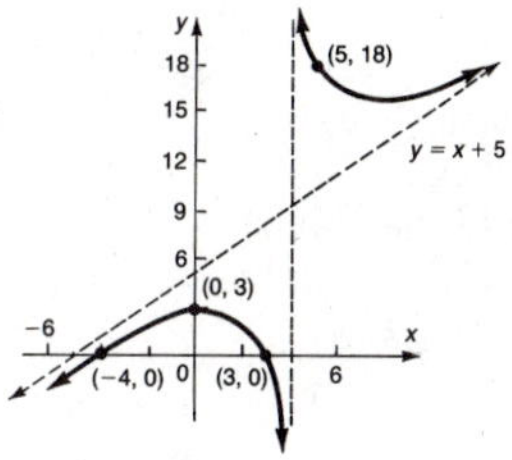

53. 1. x-intercepts: -4, 3; y-intercept: -6
2. No symmetry
3. Vertical asymptote: $x = -2$
4. Oblique asymptote: $y = x - 1$, not intersected
5. $x < -4$: below x-axis
$-4 < x < -2$: above x-axis
$-2 < x < 3$: below x-axis
$x > 3$: above x-axis
6.
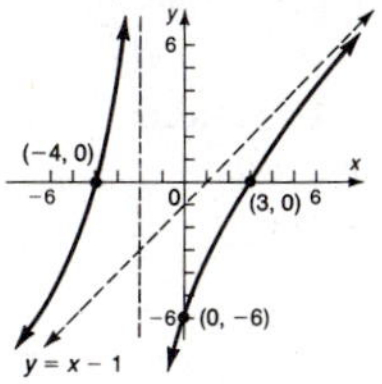

55. 1. x-intercepts: 0, 1; y-intercept: 0
2. No symmetry
3. Vertical asymptote: $x = -3$
4. Horizontal asymptote: $y = 1$, not intersected
5. $x < -3$: above x-axis
$-3 < x < 0$: below x-axis
$0 < x < 1$: above x-axis
$x > 1$: above x-axis
6.
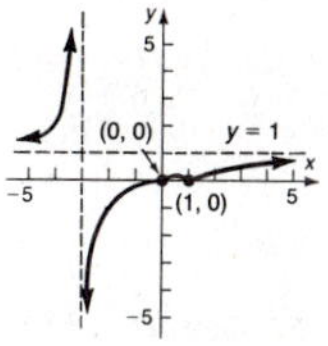

57. 4 must be a zero of the denominator; hence, $x - 4$ must be a factor.

EXERCISE 5.4

1. $q(x) = 4x - 6$; $r(x) = x + 7$ **3.** $q(x) = -2x^3 + 7x^2 - 24x + 72$; $r(x) = -214$ **5.** $q(x) = 3x^2 - 5x$; $r(x) = -6x^2 + 12x - 4$
7. $q(x) = -x^3 - x$; $r(x) = -x + 1$ **9.** $q(x) = x^3 + cx^2 + c^2x + c^3$; $r(x) = 0$ **11.** $q(x) = \frac{5}{2}x$; $r(x) = \frac{1}{2}x + 3$
13. $q(x) = 3x^2 + 11x + 32$; $R = 99$ **15.** $q(x) = x^4 - 3x^3 + 5x^2 - 15x + 46$; $R = -138$
17. $q(x) = 4x^5 + 4x^4 + x^3 + x^2 + 2x + 2$; $R = 7$ **19.** $q(x) = 0.1x^2 - 0.11x + 0.321$; $R = -0.3531$
21. $q(x) = x^4 + x^3 + x^2 + x + 1$; $R = 0$ **23.** No; $f(2) = 29$ **25.** Yes; $f(2) = 0$ **27.** Yes; $f(-3) = 0$ **29.** No; $f(-4) = -19$
31. Yes; $f(\frac{1}{2}) = 0$ **33.** 41 **35.** -4 **37.** 15 **39.** $[(3x + 2)x - 3]x + 3$ **41.** $[(3x - 6)x \cdot x - 5]x + 10$
43. $(3x \cdot x \cdot x - 82)x \cdot x \cdot x + 27$ **45.** $[(4x \cdot x - 64)x \cdot x + 1]x \cdot x - 15$ **47.** $[(2x - 1)x \cdot x + 2]x - 1$ **49.** 7.464 **51.** -0.1472
53. -105.738 **55.** -134.326 **57.** 3.8192 **59.** $k = 5$ **61.** -7 **63.** If $f(x) = x^n - c^n$, then $f(c) = c^n - c^n = 0$.
65. (a) 201,498 nanoseconds (b) 101,499 nanoseconds

EXERCISE 5.5

1. 2 or 0 positive; 1 negative **3.** 2 or 0 positive; 2 or 0 negative **5.** 2 or 0 positive; 1 negative **7.** 2 or 0 positive; 2 or 0 negative
9. 0 positive; 3 or 1 negative **11.** 1 positive; 1 negative **13.** ± 1 **15.** $\pm 1, \pm 3$ **17.** $\pm\frac{1}{2}, \pm 1$ **19.** $\pm\frac{1}{3}, \pm\frac{2}{3}, \pm 1, \pm 2$
21. $\pm\frac{1}{2}, \pm 1, \pm 2, \pm 4$ **23.** $\pm\frac{1}{6}, \pm\frac{1}{3}, \pm\frac{1}{2}, \pm\frac{2}{3}, \pm 1, \pm 2$ **25.** $-3, -1, 2$; $f(x) = (x + 3)(x + 1)(x - 2)$ **27.** $\frac{1}{2}$; $f(x) = 2\left(x - \frac{1}{2}\right)(x^2 + 1)$
29. $-1, 1$; $f(x) = (x + 1)(x - 1)(x^2 + 2)$ **31.** $-\frac{1}{2}, \frac{1}{2}$; $f(x) = 4\left(x + \frac{1}{2}\right)\left(x - \frac{1}{2}\right)(x^2 + 2)$ **33.** $-2, -1, 1, 1$; $f(x) = (x + 2)(x + 1)(x - 1)^2$
35. $-\sqrt{2}/2, \sqrt{2}/2, 2$; $f(x) = 4(x + \sqrt{2}/2)(x - \sqrt{2}/2)(x - 2)\left(x^2 + \frac{1}{2}\right)$ **37.** $\{-1, 2\}$ **39.** $\{\frac{2}{3}, -1 + \sqrt{2}, -1 - \sqrt{2}\}$ **41.** $\{\frac{1}{3}, \sqrt{5}, -\sqrt{5}\}$
43. $\{-3, -2\}$ **45.** $-\frac{1}{3}$

47. $\left(\frac{1}{2}, 0\right)$; $(0, -2)$
$x < \frac{1}{2}$, $f(0) = -1$, below x-axis
$x > \frac{1}{2}$, $f(1) = 2$, above x-axis

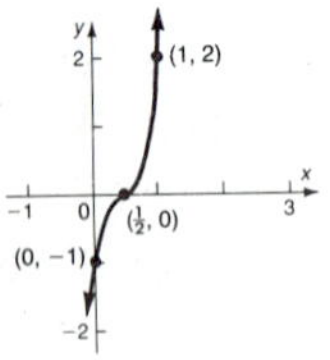

49. $(-1, 0)$; $(1, 0)$; $(0, -2)$
$x < -1$, $f(-2) = 18$, above x-axis
$-1 < x < 1$, $f(0) = -2$, below x-axis
$x > 1$, $f(2) = 18$, above x-axis
(symmetric with respect to the y-axis)

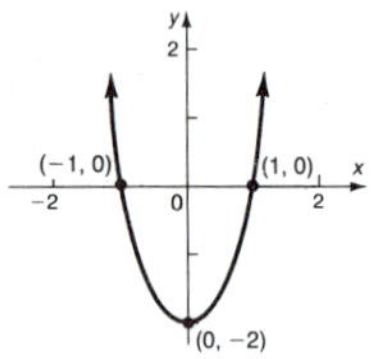

51. $\left(-\frac{1}{2}, 0\right)$; $\left(\frac{1}{2}, 0\right)$; $(0, -2)$
$x < -\frac{1}{2}$, $f(-1) = 9$, above x-axis
$-\frac{1}{2} < x < \frac{1}{2}$, $f(0) = -2$, below x-axis
$x > \frac{1}{2}$, $f(1) = 9$, above x-axis
(symmetric with respect to the y-axis)

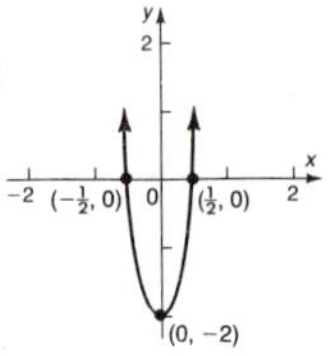

53. $(-1, 0)$; $(-2, 0)$; $(1, 0)$; $(0, 2)$
$x < -2$, $f(-3) = 32$, above x-axis
$-2 < x < -1$, $f\left(-\frac{3}{2}\right) = -\frac{25}{16}$, below x-axis
$-1 < x < 1$, $f(0) = 2$, above x-axis
$x > 1$, $f(2) = 12$, above x-axis

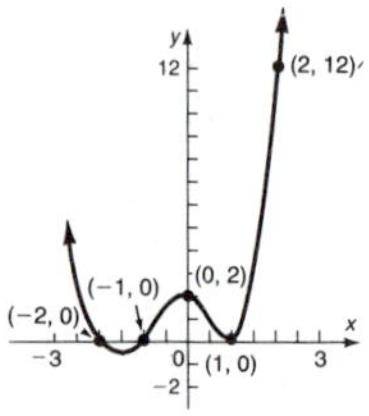

55. $(2, 0)$; $(-\sqrt{2}/2, 0)$; $(\sqrt{2}/2, 0)$; $(0, 2)$
$x < -\sqrt{2}/2$, $f(-1) = -9$, below x-axis
$-\sqrt{2}/2 < x < \sqrt{2}/2$, $f(0) = 2$, above x-axis
$\sqrt{2}/2 < x < 2$, $f(1) = -3$, below x-axis
$x > 2$, $f(3) = 323$, above x-axis

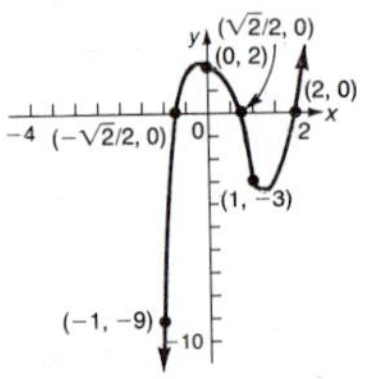

57. $\{-1, 1, -2i, 2i\}$

59. $\left\{-1, 1, \dfrac{-1 - \sqrt{3}i}{2}, \dfrac{-1 + \sqrt{3}i}{2}\right\}$

61. $\left\{-2, 2, \dfrac{-3 - \sqrt{3}i}{2}, \dfrac{-3 + \sqrt{3}i}{2}\right\}$

63. $\{1, -i, i\}$
65. No (use the Rational Zeros Theorem)
67. No (use the Rational Zeros Theorem)
69. 7 in.

71. All the potential rational zeros are integers. Hence, r either is an integer or is not a rational root (and is therefore irrational).

73.

$$y^3 + by^2 + cy + d = 0$$
$$\left(x - \frac{b}{3}\right)^3 + b\left(x - \frac{b}{3}\right)^2 + c\left(x - \frac{b}{3}\right) + d = 0$$
$$x^3 - \frac{3b}{3}x^2 + 3\left(\frac{b^2}{9}\right)x - \frac{b^3}{27} + b\left(x^2 - \frac{2bx}{3} + \frac{b^2}{9}\right) + cx - \frac{bc}{3} + d = 0$$
$$x^3 - \frac{b}{3}x + cx - \frac{b^3}{27} + \frac{b^3}{9} - \frac{bc}{3} + d = 0$$
$$x^3 + \left(c - \frac{b}{3}\right)x + \left(\frac{2b^3}{27} - \frac{bc}{3} + d\right) = 0$$

75. $K = \frac{-p}{3H}$

$$H^3 + \left(\frac{-p}{3H}\right)^3 = -q$$
$$H^6 + qH^3 - \frac{p^3}{27} = 0$$
$$H^3 = \frac{-q \pm \sqrt{q^2 + \frac{4p^3}{27}}}{2} \quad \text{Choose + sign}$$
$$H = \sqrt[3]{\frac{-q}{2} + \sqrt{\frac{q^2}{4} + \frac{p^3}{27}}}$$

77. $x = H + K$; now use results from Problems 75 and 76 **79.** $p = 3$, $q = -14$; $x = \sqrt[3]{7 + 5\sqrt{2}} + \sqrt[3]{7 - 5\sqrt{2}}$
81. $p = -6$, $q = 4$; $x = \sqrt[3]{-2 + \sqrt{4 - 8}} + \sqrt[3]{-2 - \sqrt{4 - 8}} = \sqrt[3]{-2 + 2i} + \sqrt[3]{-2 - 2i}$; or $x^3 - 6x + 4 = (x - 2)(x^2 + 2x - 2) = 0$; $x = 2$; $x = \frac{-2 \pm \sqrt{4 + 8}}{2} = -1 \pm \sqrt{3}$

EXERCISE 5.6

1. $f(0) = -1$; $f(1) = 10$ **3.** $f(-5) = -58$; $f(-4) = 2$ **5.** $f(1.4) = -0.17536$; $f(1.5) = 1.40625$ **7.** -1 and 1 **9.** -5 and -2
11. -5 and 2 **13.** Between 1.15 and 1.16 **15.** Between 2.53 and 2.54 **17.** Between 0.21 and 0.22 **19.** Between -4.05 and -4.04

EXERCISE 5.7

1. $1 + i$ **3.** $-i, 1 - i$ **5.** $-i, -2i$ **7.** $-i$ **9.** $2 - i, -3 + i$
11. Zeros that are complex numbers must occur in conjugate pairs; or a polynomial with real coefficients of odd degree must have at least one real zero.
13. If the remaining zero were a complex number, then its conjugate would also be a zero.
15. $1, -\frac{1}{2} + \frac{\sqrt{3}}{2}i, -\frac{1}{2} - \frac{\sqrt{3}}{2}i$ **17.** $-3 + i$ **19.** $-1 + 5i$ **21.** $-4 + 4i$ **23.** $-18 - 16i$ **25.** $16 - 18i$ **27.** $38 + 31i$
29. $z^3 + (-7 - 2i)z^2 + (15 + 12i)z - 9 - 18i$ **31.** $z^3 - 3z^2 + (3 - i)z - 2 + 2i$
33. $z^4 + (2i - 6)z^3 + (8 - 12i)z^2 + (6 + 18i)z - 9$

FILL-IN-THE-BLANK ITEMS

1. parabola; vertex **2.** Remainder; Dividend **3.** $f(c)$ **4.** $f(c) = 0$ **5.** zero **6.** three; one; two; no **7.** $\pm 1, \pm\frac{1}{2}$ **8.** $y = 1$
9. $x = -1$ **10.** $3 - 4i$

REVIEW EXERCISES

1.

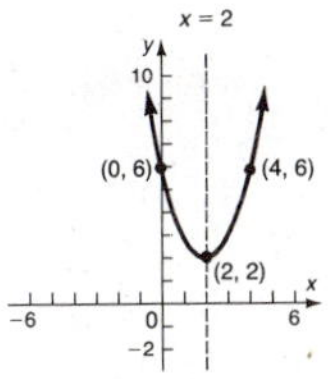

3.

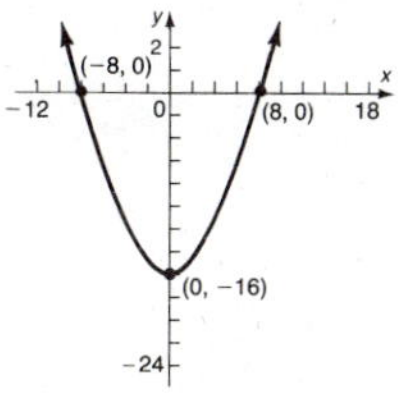

5.

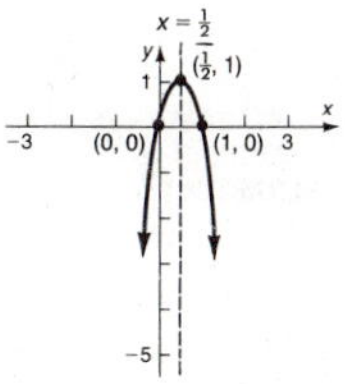

7. 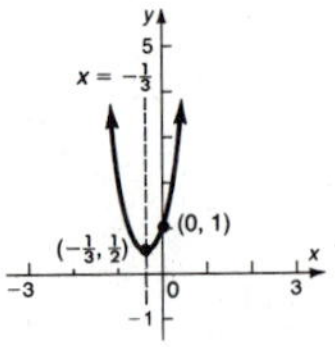

9. 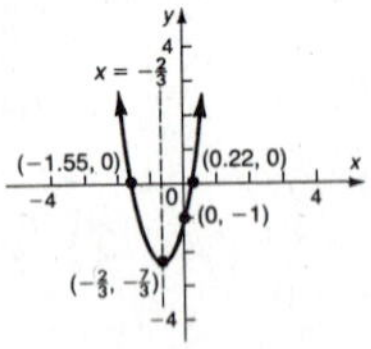

11. 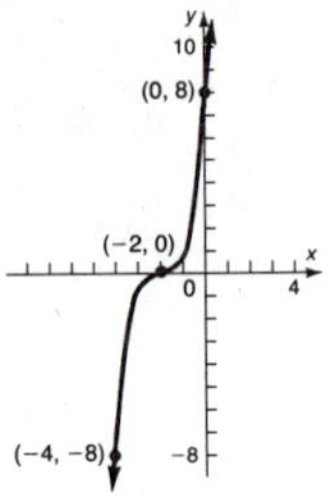

13. 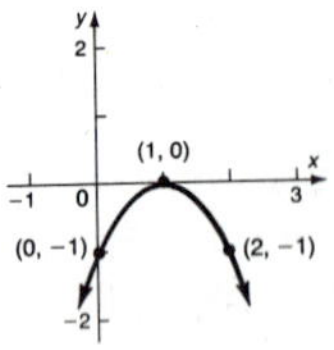

15. 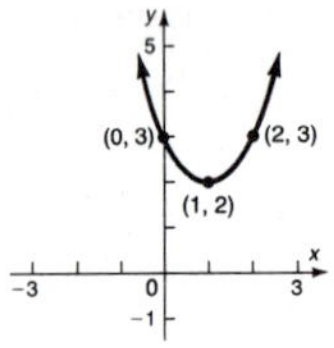

17. Minimum value; 1 **19.** Maximum value; 12 **21.** Maximum value; 20

23. (a) $(-4, 0)$; $(-2, 0)$; $(0, 0)$
(b) Crosses at $(-4, 0)$, $(-2, 0)$, and $(0, 0)$
(c) Below x-axis: $x < -4$, $-2 < x < 0$
Above x-axis: $-4 < x < -2$, $x > 0$
(d) 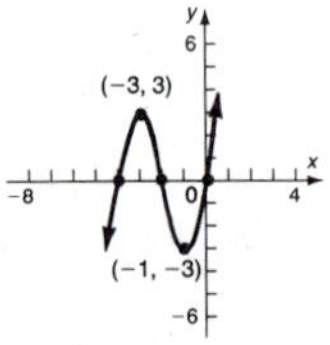

25. (a) $(-4, 0)$; $(2, 0)$; $(0, 16)$
(b) Crosses at $(-4, 0)$; touches at $(2, 0)$
(c) Below x-axis: $x < -4$
Above x-axis: $x > -4$
(d) 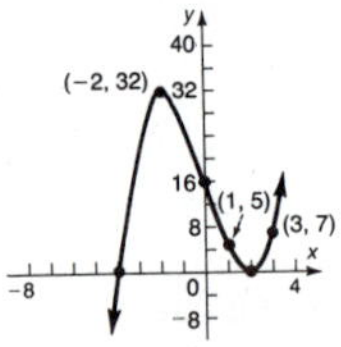

27. (a) $(0, 0)$; $(4, 0)$
(b) Crosses at $(4, 0)$; touches at $(0, 0)$
(c) Below x-axis: $x < 4$
Above x-axis: $x > 4$
(d) 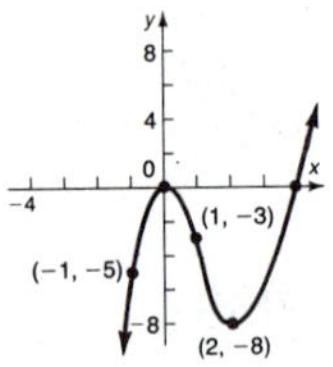

29. (a) $(-3, 0)$; $(-1, 0)$; $(1, 0)$; $(0, 3)$
(b) Crosses at $(-3, 0)$ and $(-1, 0)$; touches at $(1, 0)$
(c) Below x-axis: $-3 < x < -1$
Above x-axis: $x < -3$; $x > -1$
(d) 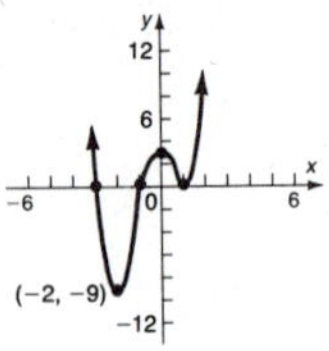

31. 1. x-intercept: 3; no y-intercept
2. No symmetry
3. Vertical asymptote: $x = 0$
4. Horizontal asymptote: $y = 2$, not intersected
5. $x < 0$: above x-axis
$0 < x < 3$: below x-axis
$x > 3$: above x-axis

6. 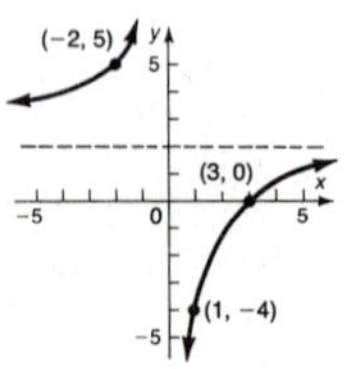

33. 1. x-intercept: -2; no y-intercept
2. No symmetry
3. Vertical asymptotes: $x = 0$, $x = 2$
4. Horizontal asymptote: $y = 0$, intersected at $(-2, 0)$
5. $x < -2$: below x-axis
$-2 < x < 0$: above x-axis
$0 < x < 2$: below x-axis
$x > 2$: above x-axis

6.
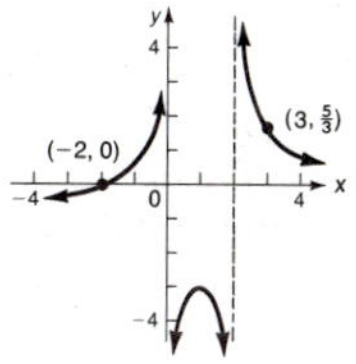

35. 1. Intercept $(0, 0)$
2. No symmetry
3. Vertical asymptote: $x = 1$
4. Horizontal asymptote: $y = 1$, intersected at $\left(\frac{1}{2}, 1\right)$
5. $x < 0$: above x-axis
$0 < x < 1$: above x-axis
$x > 1$: above x-axis

6.
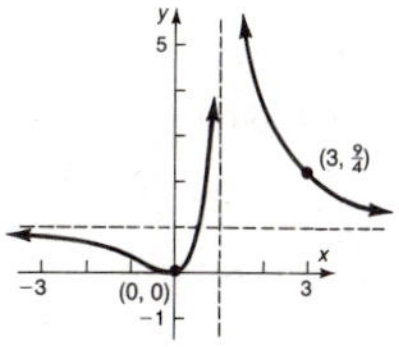

37. 1. Intercept $(0, 0)$
2. Symmetric with respect to the origin
3. Vertical asymptotes: $x = -2$, $x = 2$
4. Oblique asymptote: $y = x$, intersected at $(0, 0)$
5. $x < -2$: below x-axis
$-2 < x < 0$: above x-axis
$0 < x < 2$: below x-axis
$x > 2$: above x-axis

6.
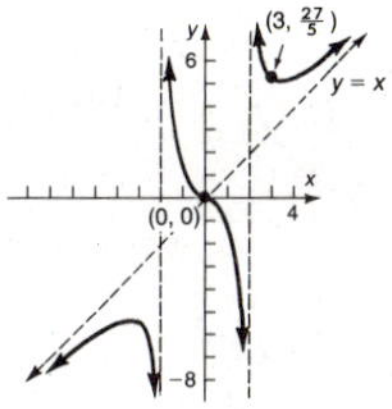

39. 1. Intercept $(0, 0)$
2. No symmetry
3. Vertical asymptote: $x = 1$
4. No oblique or horizontal asymptote
5. $x < 0$: above x-axis
$0 < x < 1$: above x-axis
$x > 1$: above x-axis

6.
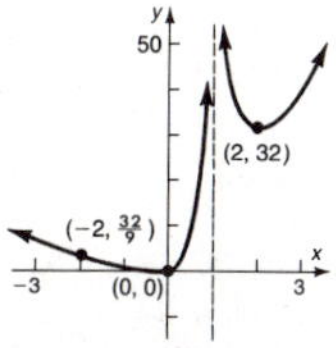

41. $q(x) = 8x^2 + 6x + 7$; $R = 3$ **43.** $q(x) = x^3 - 4x^2 + 8x - 15$; $R = 29$ **45.** $f(4) = 47{,}105$ **47.** 5, 3, or 1 positive; 1 negative
49. $\pm\frac{1}{12}, \pm\frac{1}{6}, \pm\frac{1}{4}, \pm\frac{1}{3}, \pm\frac{1}{2}, \pm\frac{3}{4}, \pm 1, \pm\frac{3}{2}, \pm 3$ **51.** $-2, 1, 4$; $f(x) = (x + 2)(x - 1)(x - 4)$
53. $\frac{1}{2}$, multiplicity 2; -2; $f(x) = 4\left(x - \frac{1}{2}\right)^2(x + 2)$ **55.** 2, multiplicity 2; $f(x) = (x - 2)^2(x^2 + 5)$ **57.** $\{-3, 2\}$ **59.** $\{-3, -1, -\frac{1}{2}, 1\}$

61. x-intercepts: $-2, 1, 4$
y-intercept: 8
Above x-axis: $-2 < x < 1$, $x > 4$
Below x-axis: $x < -2$, $1 < x < 4$

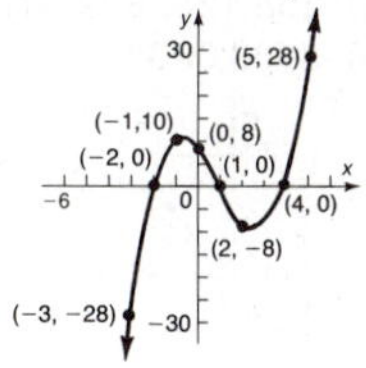

63. x-intercepts: $-2, \frac{1}{2}$
y-intercept: 2
Above x-axis: $x > -2$
Below x-axis: $x < -2$

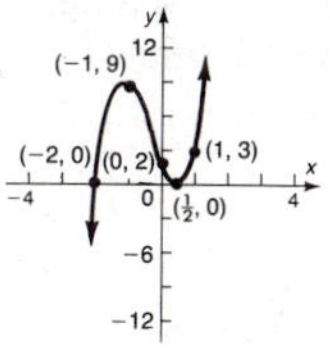

65. x-intercept: 2
y-intercept: 20
Above x-axis: all x

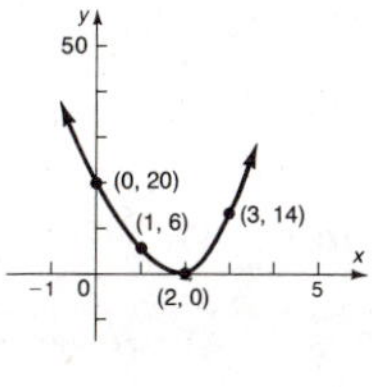

67. x-intercepts: $-3, 2$
y-intercept: -6
Above x-axis: $x < -3$, $x > 2$
Below x-axis: $-3 < x < 2$

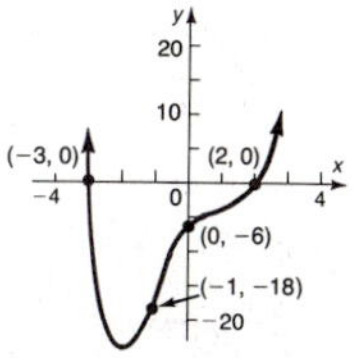

69. x-intercepts: $-3, -1, -\frac{1}{2}, 1$
y-intercept: -3
Above x-axis: $x < -3$, $-1 < x < -\frac{1}{2}$, $x > 1$
Below x-axis: $-3 < x < -1$, $-\frac{1}{2} < x < 1$

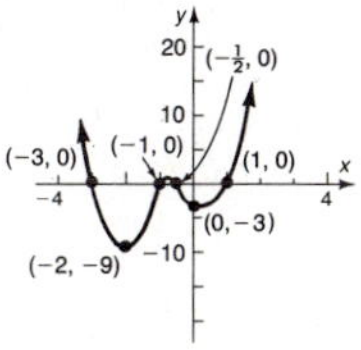

71. $f(0) = -1; f(1) = 1$ **73.** $f(0) = -1; f(1) = 1$ **75.** -2 and 2 **77.** -2 and 5 **79.** Between 1.52 and 1.53
81. Between 0.93 and 0.94 **83.** $1 - i$ **85.** $-i, 1 - i$ **87.** $f(z) = z^4 - (4 + i)z^3 + (5 + 4i)z^2 - (2 + 5i)z + 2i$
89. $f(z) = z^3 - (6 + i)z^2 + (11 + 5i)z - 6 - 6i$ **91.** $q(x) = x^2 + 5x + 6; R = 0$ **93.** $\{-3, 2\}$ **95.** $\{\frac{1}{3}, 1, -i, i\}$
97. $f(x) = [(8x - 2)x + 1]x - 4; f(1.5) = 20$ **99.** $f(x) = [(x - 2)x \cdot x + 1]x - 1; f(1.5) = -1.1875$
101. 1 is an upper bound; -2 is a lower bound **103.** (2, 2) **105.** 3.6 ft

CHAPTER 6 EXERCISE 6.1

1. (a) 11.211578 (b) 11.587251 (c) 11.663882 (d) 11.664753 **3.** (a) 8.8152409 (b) 8.8213533 (c) 8.8244111 (d) 8.8249778
5. (a) 21.216638 (b) 22.216690 (c) 22.440403 (d) 22.459158

7.

x	-4	-2	-1	0	1	2	4
$f(x) = (\sqrt{2})^x$	$\frac{1}{4}$	$\frac{1}{2}$	$\sqrt{2}/2$	1	$\sqrt{2}$	2	4

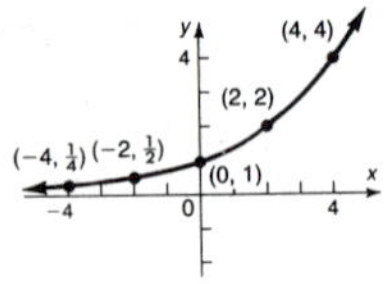

9.
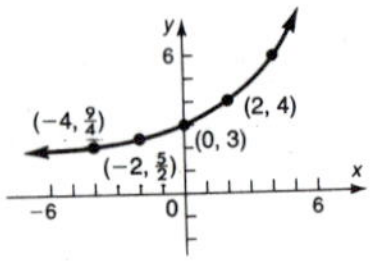

11.
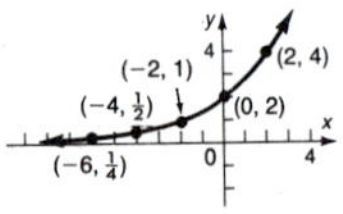

13.
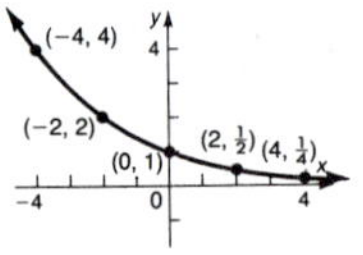

15.
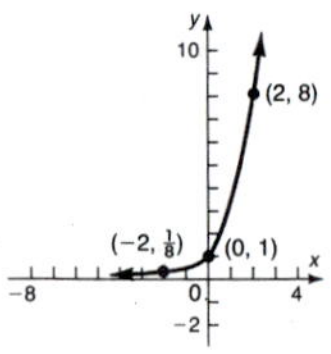

17.
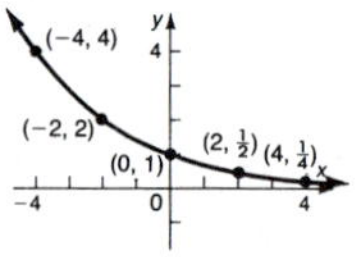

19.
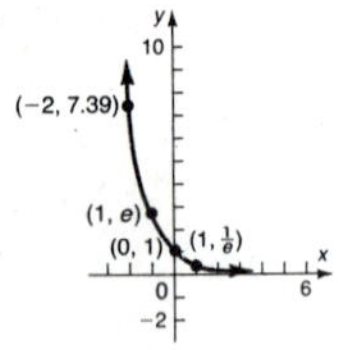

21.
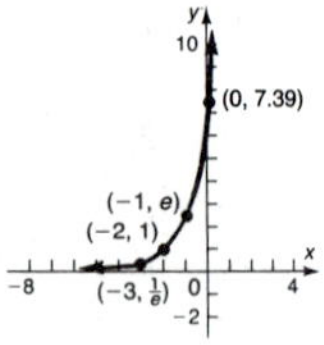

23. $\frac{1}{2}$ **25.** $\{0, -\sqrt{2}, \sqrt{2}\}$ **27.** $\left\{1 + \frac{\sqrt{6}}{3}, 1 - \frac{\sqrt{6}}{3}\right\}$ **29.** 0 **31.** 2 **33.** 0
35. (a) 56.47%; 68.17% (b) 70% **37.** (a) 76.47%; 88.17% (b) 90%

39. (a) 0.2376 amp; 7.5854 amp
(b) 12 amp
(c)
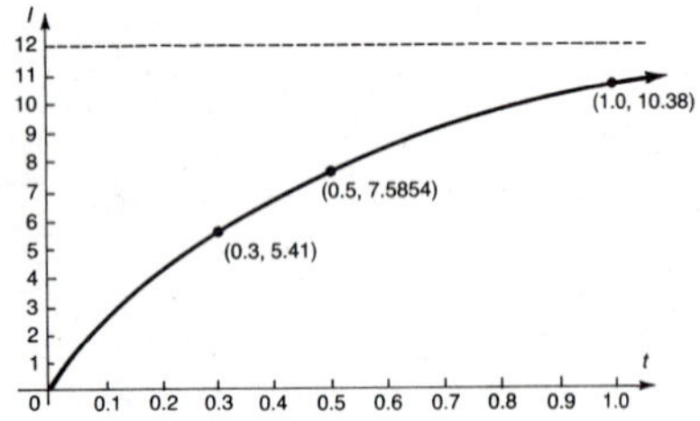

41. (a) \$98,125; \$99,941.41
(b) \$100,000
(c)
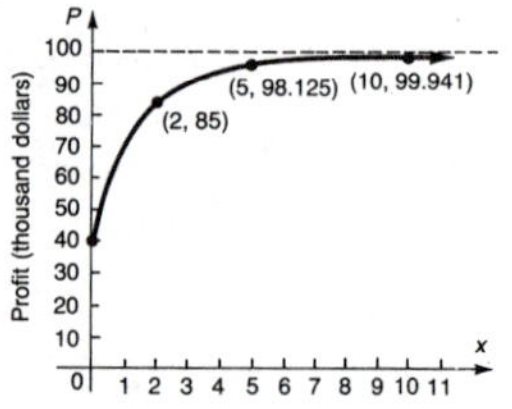

43. 12.696481 **45.** (a) $\sinh(-x) = \frac{1}{2}(e^{-x} - e^{x}) = -\frac{1}{2}(e^{x} - e^{-x}) = -\sinh x$
(b)
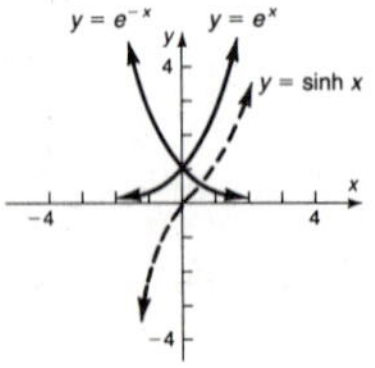

47. $\frac{f(x + h) - f(x)}{h} = \frac{a^{x+h} - a^x}{h} = \frac{a^x a^h - a^x}{h} = \frac{a^x(a^h - 1)}{h}$

49. $f(-x) = a^{-x} = \frac{1}{a^x} = \frac{1}{f(x)}$

51. $f(1) = 5, f(2) = 17, f(3) = 257, f(4) = 65{,}537;$
$f(5) = 4{,}294{,}967{,}297 = 641 \times 6{,}700{,}417$

EXERCISE 6.2

1. \$117.29 **3.** \$640.04 **5.** \$697.09 **7.** \$12.46 **9.** \$125.23 **11.** \$85.26 **13.** \$860.72 **15.** \$473.65 **17.** \$59.71 **19.** \$361.93
21. (a) \$1364.62 (b) \$907.18 **23.** You have \$11,632.73; your friend has \$10,947.89.
25. The \$1000 invested at 10% becomes \$1349.86. It is better to receive \$1000 now.
27. (a) Interest is \$30,000. (b) Interest is \$38,613.59. (c) Interest is \$37,752.73. Simple interest at 12% per annum is best.
29. Quarterly compounding **31.** $A = Pe^{rt}$

$$P = \frac{A}{e^{rt}}$$

$$P = Ae^{-rt}$$

EXERCISE 6.3

1. $\{x|x < 3\}$ **3.** All real numbers except 0 **5.** $\{x|x < -2 \text{ or } x > 3\}$ **7.** $\{x|x > 0, x \neq 1\}$ **9.** $\{x|x < -1 \text{ or } x > 0\}$ **11.** $2^3 = 8$
13. $a^6 = 3$ **15.** $3^x = 2$ **17.** $2^{1.3} = M$ **19.** $(\sqrt{2})^x = \pi$ **21.** $e^x = 4$ **23.** $2 = \log_3 9$ **25.** $2 = \log_a 1.6$ **27.** $2 = \log_{1.1} M$
29. $x = \log_2 7.2$ **31.** $\sqrt{2} = \log_x \pi$ **33.** $x = \ln 8$ **35.** 0 **37.** 2 **39.** -4 **41.** $\frac{1}{2}$ **43.** 4 **45.** $\frac{1}{2}$ **47.** $\sqrt{2}$

49.

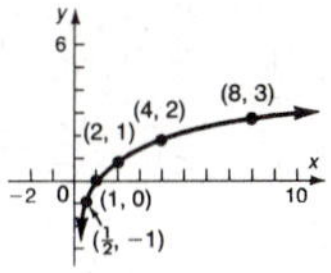

x	$f(x)$
0.5	−1
1	0
2	1
3	1.59
4	2
5	2.32
6	2.59
7	2.81
8	3

51.

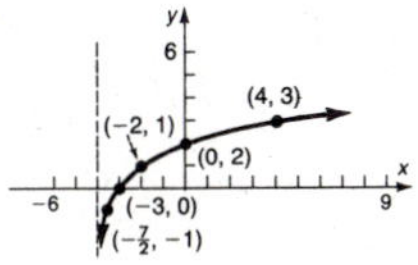

53.

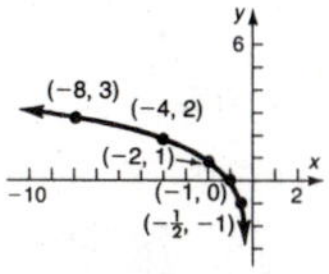

55.

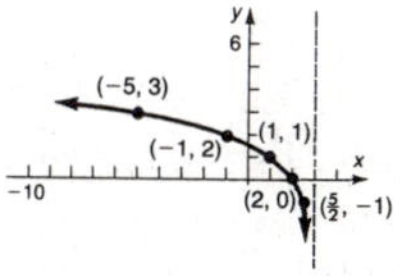

57.

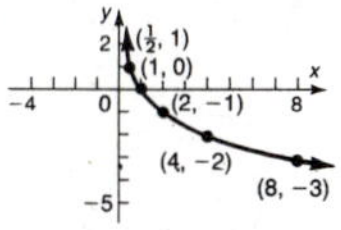

59.

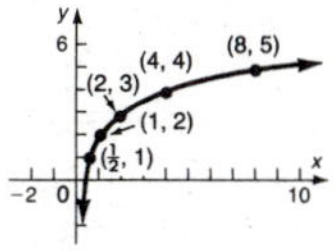

61.

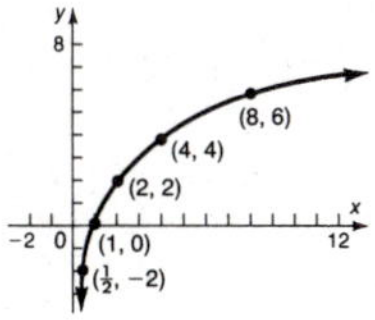

63.

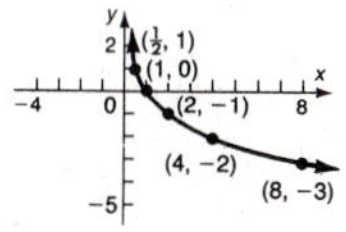

EXERCISE 6.4

1. $a + b$ **3.** $b - a$ **5.** $a + 1$ **7.** $2a + b$ **9.** $\frac{1}{5}(a + 2b)$ **11.** $\frac{b}{a}$ **13.** $2 \ln x + \frac{1}{2}\ln(1 - x)$ **15.** $3 \log_2 x - \log_2(x - 3)$
17. $\log x + \log(x + 2) - 2 \log(x + 3)$ **19.** $\frac{1}{3}\ln(x - 2) + \frac{1}{3}\ln(x + 1) - \frac{2}{3}\ln(x + 4)$ **21.** $\ln 5 + \ln x + \frac{1}{2}\ln(1 - 3x) - 3 \ln(x - 4)$
23. $\log_5 u^3v^4$ **25.** $-\frac{5}{2}\log_{1/2} x$ **27.** $-2 \ln(x - 1)$ **29.** $\log_2[x(3x - 2)^4]$ **31.** $\log_a\left[\frac{25x^6}{(2x + 3)^{1/2}}\right]$ **33.** 81 **35.** 3 **37.** 5 **39.** 2
41. $\{-2, 4\}$ **43.** 21 **45.** $\frac{9}{2}$ **47.** 2 **49.** -1 **51.** 1

53. $\log_a(x+\sqrt{x^2-1})+\log_a(x-\sqrt{x^2-1})=\log_a[(x+\sqrt{x^2-1})(x-\sqrt{x^2-1})]=\log_a[x^2-(x^2-1)]=\log_a 1=0$
55. 2.7712437 **57.** −3.880058 **59.** 5.6147098 **61.** 0.8735685 **63.** 3.3219281 **65.** −0.0876781 **67.** 0.3065736
69. 1.3559551 **71.** 0 **73.** 0.5337408
75. Domain of f: all real numbers except 0; domain of g: all positive real numbers; the equality holds only for $x>0$.
77. $y=f(x)=\log_a x$; $a^y=x$; $\left(\frac{1}{a}\right)^{-y}=x$; $-y=\log_{1/a} x$; $-f(x)=\log_{1/a} x$ **79.** $f(AB)=\log_a AB=\log_a A+\log_a B=f(A)+f(B)$
81. $y=Cx$ **83.** $y=Cx(x+1)$ **85.** $y=Ce^{3x}$ **87.** $y=Ce^{-4x}+3$ **89.** $y=\dfrac{\sqrt[3]{C}(2x+1)^{1/6}}{(x+4)^{1/9}}$ **91.** 3 **93.** 1
95. 104.32 mo; 103.97 mo **97.** 61.02 mo; 60.82 mo
99. If $A=\log_a M$ and $B=\log_a N$, then $a^A=M$ and $a^B=N$. Then $\log_a(M/N)=\log_a(a^A/a^B)=\log_a a^{A-B}=A-B=\log_a M-\log_a N$.

EXERCISE 6.5

1. 70 decibels **3.** 111.76 decibels **5.** 10 W/m^2 **7.** 4.0 on the Richter scale
9. 70,794.58 mm; the San Francisco earthquake was 11.22 times as intense as the one in Mexico City.
11. 3229.54 m above sea level

EXERCISE 6.6

1. 34.7 days; 69.3 days **3.** 28.4 yr **5.** 94.4 yr **7.** 5832; 3.9 days **9.** 25,198 **11.** 9.797 g **13.** 9727 years ago **15.** 5:18 PM
17. 18.63°C; 25.1°C **19.** 7.34 kg; 76.6 hr **21.** 28.15%; 0.24; 0.15 **23.** 0.2695 sec; 0.8959 sec
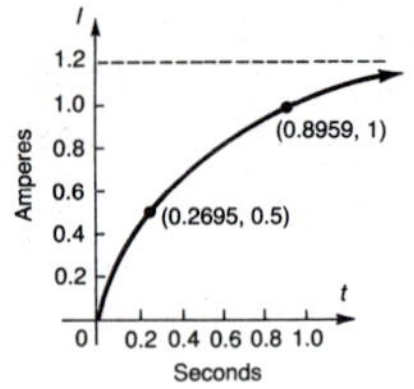

25. $k\approx 0.02107$; 38 words; 54 words; 109 min **27.** 26.5 days

FILL-IN-THE-BLANK ITEMS

1. (0, 1) **2.** 1 **3.** 4 **4.** sum **5.** 1 **6.** 7 **7.** $x>0$ **8.** (1, 0) **9.** 1 **10.** 7

REVIEW EXERCISES

1. −3 **3.** $\sqrt{2}$ **5.** 0.4 **7.** $\frac{25}{4}\log_4 x$ **9.** $\ln\left[\frac{1}{(x+1)^2}\right]=-2\ln(x+1)$ **11.** $\log\left(\frac{4x^3}{[(x+3)(x-2)]^{1/2}}\right)$ **13.** $y=Ce^{2x^2}$ **15.** $y=(Ce^{3x^2})^2$
17. $y=\sqrt{e^{x+C}+9}$ **19.** $y=\ln(x^2+4)-C$

21.
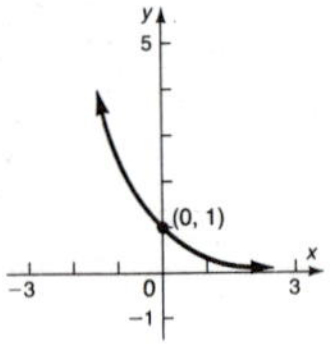

23.
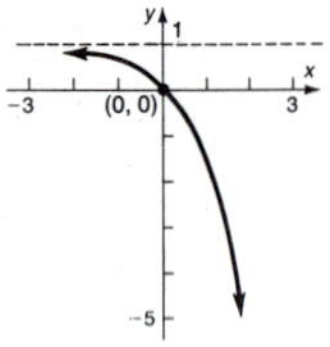

25.
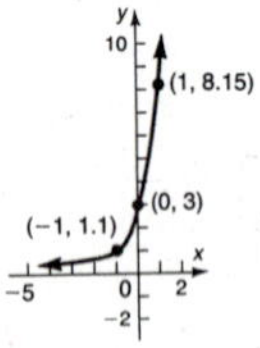

27.
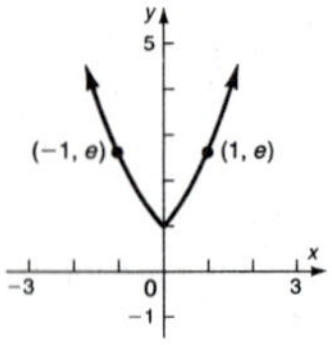

29.
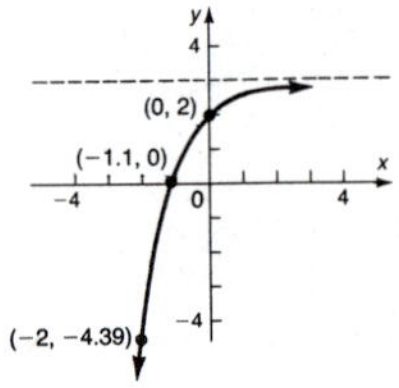

31. $\frac{1}{4}$ **33.** $\left\{\frac{-1-\sqrt{3}}{2}, \frac{-1+\sqrt{3}}{2}\right\}$ **35.** $\frac{1}{4}$ **37.** 5 **39.** 4.301 **41.** $\frac{12}{5}$ **43.** $\{-3, \frac{1}{2}\}$ **45.** −1 **47.** −0.6094 **49.** −9.3274
51. 10.436% **53.** 80 decibels **55.** 24,203 years ago

CHAPTER 7 EXERCISE 7.1

1.
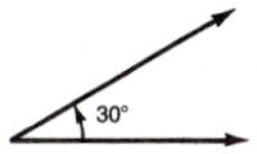

3.
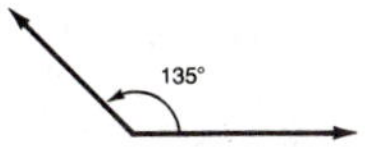

5.
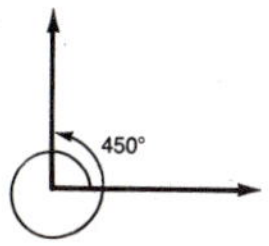

7.
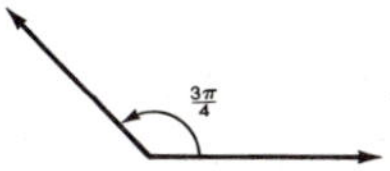

9.
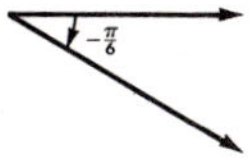

11.
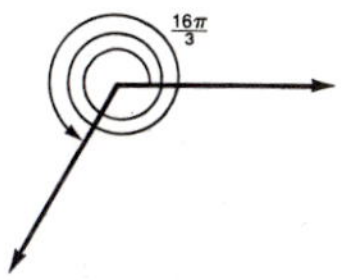

13. $\pi/6$ **15.** $4\pi/3$ **17.** $-\pi/3$ **19.** $5\pi/18$ **21.** $5\pi/4$ **23.** 60° **25.** −225° **27.** 630° **29.** 15° **31.** 120° **33.** 5 m **35.** 6 ft **37.** 0.6 radian **39.** $\pi/3 \approx 1.047$ in. **41.** 0.2967 **43.** −0.6981 **45.** 2.1817 **47.** 5.9341 **49.** 179.9° **51.** 587.3° **53.** 114.6° **55.** 362.1° **57.** 40.1736° **59.** 1.0342° **61.** 9.1525° **63.** 40°19′12″ **65.** 18°15′18″ **67.** 19°59′24″ **69.** $3\pi \approx 9.4248$ in.; $5\pi \approx 15.7080$ in. **71.** $\omega = \frac{1}{60}$ radian/sec; $v = \frac{1}{12}$ cm/sec **73.** 452.5 rpm **75.** 37.7 in. **77.** 2292 mph **79.** $\frac{3}{4}$ rpm **81.** 2.86 mph **83.** 1.152 mi

EXERCISE 7.2

1. $\sin\theta = \frac{5}{13}$; $\cos\theta = \frac{12}{13}$; $\tan\theta = \frac{5}{12}$; $\csc\theta = \frac{13}{5}$; $\sec\theta = \frac{13}{12}$; $\cot\theta = \frac{12}{5}$
3. $\sin\theta = 2\sqrt{13}/13$; $\cos\theta = 3\sqrt{13}/13$; $\tan\theta = \frac{2}{3}$; $\csc\theta = \sqrt{13}/2$; $\sec\theta = \sqrt{13}/3$; $\cot\theta = \frac{3}{2}$
5. $\sin\theta = \sqrt{3}/2$; $\cos\theta = \frac{1}{2}$; $\tan\theta = \sqrt{3}$; $\csc\theta = 2\sqrt{3}/3$; $\sec\theta = 2$; $\cot\theta = \sqrt{3}/3$
7. $\sin\theta = \sqrt{6}/3$; $\cos\theta = \sqrt{3}/3$; $\tan\theta = \sqrt{2}$; $\csc\theta = \sqrt{6}/2$; $\sec\theta = \sqrt{3}$; $\cot\theta = \sqrt{2}/2$
9. $\sin\theta = \sqrt{5}/5$; $\cos\theta = 2\sqrt{5}/5$; $\tan\theta = \frac{1}{2}$; $\csc\theta = \sqrt{5}$; $\sec\theta = \sqrt{5}/2$; $\cot\theta = 2$
11. $\tan\theta = \sqrt{3}/3$; $\csc\theta = 2$; $\sec\theta = 2\sqrt{3}/3$; $\cot\theta = \sqrt{3}$ **13.** $\tan\theta = \sqrt{2}/4$; $\csc\theta = 3$; $\sec\theta = 3\sqrt{2}/4$; $\cot\theta = 2\sqrt{2}$
15. $\cos\theta = \sqrt{2}/2$; $\tan\theta = 1$; $\csc\theta = \sqrt{2}$; $\sec\theta = \sqrt{2}$; $\cot\theta = 1$
17. $\sin\theta = 2\sqrt{2}/3$; $\tan\theta = 2\sqrt{2}$; $\csc\theta = 3\sqrt{2}/4$; $\sec\theta = 3$; $\cot\theta = \sqrt{2}/4$
19. $\sin\theta = \sqrt{5}/5$; $\cos\theta = 2\sqrt{5}/5$; $\csc\theta = \sqrt{5}$; $\sec\theta = \sqrt{5}/2$; $\cot\theta = 2$
21. $\sin\theta = 2\sqrt{2}/3$; $\cos\theta = \frac{1}{3}$; $\tan\theta = 2\sqrt{2}$; $\csc\theta = 3\sqrt{2}/4$; $\cot\theta = \sqrt{2}/4$
23. $\sin\theta = \sqrt{6}/3$; $\cos\theta = \sqrt{3}/3$; $\csc\theta = \sqrt{6}/2$; $\sec\theta = \sqrt{3}$; $\cot\theta = \sqrt{2}/2$ **25.** 1 **27.** 1 **29.** 0 **31.** 0 **33.** 1 **35.** 0 **37.** 0 **39.** 1 **41.** (a) $\frac{1}{2}$ (b) $\frac{3}{4}$ (c) 2 (d) 2 **43.** (a) 17 (b) $\frac{1}{4}$ (c) 4 (d) $\frac{17}{16}$ **45.** (a) $\frac{1}{4}$ (b) 15 (c) 4 (d) $\frac{16}{15}$
47. (a) 0.7880 (b) 0.7813 (c) 1.2799 (d) 1.2690 (e) 1.6243 (f) 0.7880 (g) 0.6157 (h) 1.2799 **49.** 0.6 **51.** 20°

53. (a) $|OA| = |OC| = 1$; angle OAC + angle OAC + 180° − θ = 180°; angle $OAC = \dfrac{\theta}{2}$

(b) $\sin\theta = \dfrac{|CD|}{|OC|} = |CD|$; $\cos\theta = \dfrac{|OD|}{|OC|} = |OD|$ (c) $\tan\dfrac{\theta}{2} = \dfrac{|CD|}{|AD|} = \dfrac{\sin\theta}{1 + |OD|} = \dfrac{\sin\theta}{1 + \cos\theta}$

55. $h = x\tan\theta$ and $h = (1 - x)\tan n\theta$; thus, $x\tan\theta = (1 - x)\tan n\theta$

$$x = \frac{\tan n\theta}{\tan\theta + \tan n\theta}$$

57. (a) Area $\triangle OAC = \frac{1}{2}|OC||AC| = \frac{1}{2}\cdot\dfrac{|OC|}{1}\cdot\dfrac{|AC|}{1} = \frac{1}{2}\sin\alpha\cos\alpha$

(b) Area $\triangle OCB = \frac{1}{2}|BC||OC| = \frac{1}{2}|OB|^2\dfrac{|BC|}{|OB|}\cdot\dfrac{|OC|}{|OB|} = \frac{1}{2}|OB|^2\sin\beta\cos\beta$

(c) Area $\triangle OAB = \frac{1}{2}|BD||OA| = \frac{1}{2}|OB|\dfrac{|BD|}{|OB|} = \frac{1}{2}|OB|\sin(\alpha + \beta)$

(d) $\dfrac{\cos\alpha}{\cos\beta} = \dfrac{|OC|/1}{|OC|/|OB|} = |OB|$

(e) Use the hint and above results.

59. $\sin\alpha = \tan\alpha\cos\alpha = \cos\beta\cos\alpha = \cos\beta\tan\alpha = \sin\beta;$

$$\sin^2\alpha + \cos^2\alpha = 1$$
$$\sin^2\alpha + \tan^2\beta = 1$$
$$\sin^2\alpha + \frac{\sin^2\beta}{\cos^2\beta} = 1$$
$$\sin^2\alpha + \frac{\sin^2\alpha}{1-\sin^2\alpha} = 1$$
$$\sin^2\alpha - \sin^4\alpha + \sin^2\alpha = 1 - \sin^2\alpha$$
$$\sin^4\alpha - 3\sin^2\alpha + 1 = 0$$
$$\sin^2\alpha = \frac{3 \pm \sqrt{5}}{2}$$
$$\sin^2\alpha = \frac{3 - \sqrt{5}}{2}$$
$$\sin\alpha = \sqrt{\frac{3-\sqrt{5}}{2}}$$

61. Since $a^2 + b^2 = c^2$, $a > 0$, $b > 0$, then $0 < a^2 < c^2$ or $0 < a < c$. Thus, $0 < a/c < 1$ and $0 < \sin\theta < 1$.

EXERCISE 7.3

1. $\sin 45° = \sqrt{2}/2$; $\cos 45° = \sqrt{2}/2$; $\tan 45° = 1$; $\cot 45° = 1$; $\sec 45° = \sqrt{2}$; $\csc 45° = \sqrt{2}$
3. $\sin 60° = \sqrt{3}/2$; $\cos 60° = \frac{1}{2}$; $\tan 60° = \sqrt{3}$; $\cot 60° = \sqrt{3}/3$; $\sec 60° = 2$; $\csc 60° = 2\sqrt{3}/3$
5. 10 **7.** $\sqrt{2} - \frac{4\sqrt{3}}{3}$ **9.** $-\frac{8}{3}$ **11.** 1 **13.** 0 **15.** $\sqrt{3}/2$ **17.** $\frac{1}{2}$ **19.** $\frac{3}{4}$ **21.** $\sqrt{3}$ **23.** $\sqrt{3}/4$ **25.** 0.4695 **27.** 0.3839
29. 1.3250 **31.** 0.3640 **33.** 0.3090 **35.** 3.7321 **37.** 1.0353 **39.** 5.6713 **41.** 0.8415 **43.** 0.0175 **45.** 0.9304 **47.** 0.3093
49. 0.4695 **51.** 0.3839 **53.** 0.7547 **55.** 0.3640 **57.** 0.6428 **59.** 0.1763 **61.** 0.5000 **63.** 1.4280
65. $R \approx 310.56$ ft; $H \approx 77.64$ ft **67.** $R \approx 19{,}542$ m; $H \approx 2278$ m **69.** (a) 1.2 sec (b) 1.12 sec (c) 1.2 sec

71.

θ	0.5	0.4	0.2	0.1	0.01	0.001	0.0001	0.00001
$\sin\theta$	0.4794	0.3894	0.1987	0.0998	0.0100	0.0010	0.0001	0.00001
$\frac{\sin\theta}{\theta}$	0.9589	0.9735	0.9933	0.9983	1.0000	1.0000	1.0000	1.0000

$(\sin\theta)/\theta$ approaches 1 as $\theta \to 0$

73. 1 **75.** $\sqrt{2}/2$

EXERCISE 7.4

1. $a \approx 28.36$, $c \approx 28.79$ **3.** $b \approx 5.03$, $c \approx 7.83$ **5.** $a \approx 0.705$, $c \approx 4.06$ **7.** $b \approx 10.72$, $c \approx 11.83$ **9.** $b \approx 3.08$, $a \approx 8.46$
11. 1.72 in., 2.46 in. **13.** 5.52 in. or 11.83 in. **15.** 70 ft **17.** 985.9 ft **19.** 20 m **21.** 76.34 mph **23.** 137 m **25.** 20.67 ft
27. 449.36 ft **29.** 30 ft **31.** 530 ft **33.** 555 ft

EXERCISE 7.5

1. $\sin\theta = \frac{4}{5}$; $\cos\theta = -\frac{3}{5}$; $\tan\theta = -\frac{4}{3}$; $\csc\theta = \frac{5}{4}$; $\sec\theta = -\frac{5}{3}$; $\cot\theta = -\frac{3}{4}$
3. $\sin\theta = -3\sqrt{13}/13$; $\cos\theta = 2\sqrt{13}/13$; $\tan\theta = -\frac{3}{2}$; $\csc\theta = -\sqrt{13}/3$; $\sec\theta = \sqrt{13}/2$; $\cot\theta = -\frac{2}{3}$
5. $\sin\theta = -\sqrt{2}/2$; $\cos\theta = -\sqrt{2}/2$; $\tan\theta = 1$; $\csc\theta = -\sqrt{2}$; $\sec\theta = -\sqrt{2}$; $\cot\theta = 1$
7. $\sin\theta = -2\sqrt{13}/13$; $\cos\theta = -3\sqrt{13}/13$; $\tan\theta = \frac{2}{3}$; $\csc\theta = -\sqrt{13}/2$; $\sec\theta = -\sqrt{13}/3$; $\cot\theta = \frac{3}{2}$
9. II **11.** IV **13.** IV **15.** II **17.** 30° **19.** 60° **21.** 30° **23.** $\pi/4$ **25.** $\pi/3$ **27.** 45° **29.** $\pi/3$ **31.** 60° **33.** $\sqrt{2}/2$ **35.** 1
37. 1 **39.** $\sqrt{3}$ **41.** $\sqrt{2}/2$ **43.** 0 **45.** $\sqrt{2}$ **47.** $\sqrt{3}/3$ **49.** $\frac{1}{2}$ **51.** $\sqrt{2}/2$ **53.** -2 **55.** $-\sqrt{3}$ **57.** $\sqrt{2}/2$ **59.** $\sqrt{3}/3$ **61.** $\frac{1}{2}$
63. $-\sqrt{3}/2$ **65.** $-\sqrt{3}$ **67.** $\sqrt{2}$

69. $\cos\theta = -\frac{5}{13}$; $\tan\theta = -\frac{12}{5}$; $\csc\theta = \frac{13}{12}$; $\sec\theta = -\frac{13}{5}$; $\cot\theta = -\frac{5}{12}$
71. $\sin\theta = -\frac{3}{5}$; $\tan\theta = \frac{3}{4}$; $\csc\theta = -\frac{5}{3}$; $\sec\theta = -\frac{5}{4}$; $\cot\theta = \frac{4}{3}$
73. $\cos\theta = -\frac{12}{13}$; $\tan\theta = -\frac{5}{12}$; $\csc\theta = \frac{13}{5}$; $\sec\theta = -\frac{13}{12}$; $\cot\theta = -\frac{12}{5}$
75. $\sin\theta = 2\sqrt{2}/3$; $\tan\theta = -2\sqrt{2}$; $\csc\theta = 3\sqrt{2}/4$; $\sec\theta = -3$; $\cot\theta = -\sqrt{2}/4$
77. $\cos\theta = -\sqrt{5}/3$; $\tan\theta = -2\sqrt{5}/5$; $\csc\theta = \frac{3}{2}$; $\sec\theta = -3\sqrt{5}/5$; $\cot\theta = -\sqrt{5}/2$
79. $\sin\theta = -\sqrt{3}/2$; $\cos\theta = \frac{1}{2}$; $\tan\theta = -\sqrt{3}$; $\csc\theta = -2\sqrt{3}/3$; $\cot\theta = -\sqrt{3}/3$
81. $\sin\theta = -\frac{3}{5}$; $\cos\theta = -\frac{4}{5}$; $\csc\theta = -\frac{5}{3}$; $\sec\theta = -\frac{5}{4}$; $\cot\theta = \frac{4}{3}$
83. $\sin\theta = \sqrt{10}/10$; $\cos\theta = -3\sqrt{10}/10$; $\csc\theta = \sqrt{10}$; $\sec\theta = -\sqrt{10}/3$; $\cot\theta = -3$
85. 0 **87.** −0.1 **89.** 3 **91.** 5 **93.** 0

EXERCISE 7.6

1. $\sin t = \frac{1}{2}$; $\cos t = \sqrt{3}/2$; $\tan t = \sqrt{3}/3$; $\csc t = 2$; $\sec t = 2\sqrt{3}/3$; $\cot t = \sqrt{3}$
3. $\sin t = -\sqrt{3}/2$; $\cos t = -\frac{1}{2}$; $\tan t = \sqrt{3}$; $\csc t = -2\sqrt{3}/3$; $\sec t = -2$; $\cot t = \sqrt{3}/3$
5. $\sin t = \frac{4}{5}$; $\cos t = -\frac{3}{5}$; $\tan t = -\frac{4}{3}$; $\csc t = \frac{5}{4}$; $\sec t = -\frac{5}{3}$; $\cot t = -\frac{3}{4}$
7. $\sin t = -2\sqrt{5}/5$; $\cos t = \sqrt{5}/5$; $\tan t = -2$; $\csc t = -\sqrt{5}/2$; $\sec t = \sqrt{5}$; $\cot t = -\frac{1}{2}$
9. $\cos\theta = -\sqrt{5}/3$; $\tan\theta = -2\sqrt{5}/5$; $\csc\theta = \frac{3}{2}$; $\sec\theta = -3\sqrt{5}/5$; $\cot\theta = -\sqrt{5}/2$
11. $\sin\theta = -\sqrt{5}/5$; $\cos\theta = -2\sqrt{5}/5$; $\csc\theta = -\sqrt{5}$; $\sec\theta = -\sqrt{5}/2$; $\cot\theta = 2$
13. $\cos\theta = \sqrt{15}/4$; $\tan\theta = -\sqrt{15}/15$; $\csc\theta = -4$; $\sec\theta = 4\sqrt{15}/15$; $\cot\theta = -\sqrt{15}$
15. $\sin\theta = -2\sqrt{2}/3$; $\cos\theta = \frac{1}{3}$; $\tan\theta = -2\sqrt{2}$; $\csc\theta = -3\sqrt{2}/4$; $\cot\theta = -\sqrt{2}/4$
17. All real numbers **19.** At odd multiples of $\pi/2$ **21.** At odd multiples of $\pi/2$ **23.** $[-1, 1]$ **25.** $(-\infty, \infty)$
27. $(-\infty, -1] \cup [1, \infty)$ **29.** Odd; yes; origin **31.** Odd; yes; origin **33.** Even; yes; y-axis
35. Let $P = (x, y)$ be the point on the unit circle that corresponds to t. Consider the equation $\tan t = y/x = a$. Then $y = ax$. But $x^2 + y^2 = 1$ so that $x^2 + a^2x^2 = 1$. Thus, $x = \pm 1/\sqrt{1 + a^2}$ and $y = \pm a/\sqrt{1 + a^2}$; that is, for any real number a, there is a point $P = (x, y)$ on the unit circle for which $\tan t = a$. In other words, $-\infty < \tan t < \infty$, and the range of the tangent function is the set of all real numbers.
37. Suppose there is a number p, $0 < p < 2\pi$, for which $\sin(\theta + p) = \sin\theta$ for all θ. If $\theta = 0$, then $\sin(0 + p) = \sin p = \sin 0 = 0$; so that $p = \pi$. If $\theta = \pi/2$, then $\sin(\pi/2 + p) = \sin(\pi/2)$. But $p = \pi$. Thus, $\sin(3\pi/2) = -1 = \sin(\pi/2) = 1$. This is impossible. The smallest positive number p for which $\sin(\theta + p) = \sin\theta$ for all θ is therefore $p = 2\pi$.
39. $\sec\theta = 1/(\cos\theta)$; since $\cos\theta$ has period 2π, so does $\sec\theta$
41. If $P = (a, b)$ is the point on the unit circle corresponding to θ, then $Q = (-a, -b)$ is the point on the unit circle corresponding to $\theta + \pi$. Thus, $\tan(\theta + \pi) = (-b)/(-a) = b/a = \tan\theta$; that is, the period of the tangent function is π.
43. $m = \dfrac{\sin\theta - 0}{\cos\theta - 0} = \dfrac{\sin\theta}{\cos\theta} = \tan\theta$

FILL-IN-THE-BLANK ITEMS

1. angle; initial side; terminal side **2.** radians **3.** π **4.** complementary **5.** cosine **6.** standard position **7.** 45° **8.** 2π; π

REVIEW EXERCISES

1. $3\pi/4$ **3.** $\pi/10$ **5.** 135° **7.** −450° **9.** $\frac{1}{2}$ **11.** $3\sqrt{2}/2 - 4\sqrt{3}/3$ **13.** $-3\sqrt{2} - 2\sqrt{3}$ **15.** 3 **17.** 0 **19.** 0 **21.** 1 **23.** 1
25. 1 **27.** −1 **29.** 1 **31.** $\cos\theta = \frac{3}{5}$, $\tan\theta = -\frac{4}{3}$, $\csc\theta = -\frac{5}{4}$, $\sec\theta = \frac{5}{3}$, $\cot\theta = -\frac{3}{4}$
33. $\sin\theta = -\frac{12}{13}$, $\cos\theta = -\frac{5}{13}$, $\csc\theta = -\frac{13}{12}$, $\sec\theta = -\frac{13}{5}$, $\cot\theta = \frac{5}{12}$
35. $\sin\theta = \frac{3}{5}$, $\cos\theta = -\frac{4}{5}$, $\tan\theta = -\frac{3}{4}$, $\csc\theta = \frac{5}{3}$, $\cot\theta = -\frac{4}{3}$
37. $\cos\theta = -\frac{5}{13}$, $\tan\theta = -\frac{12}{5}$, $\csc\theta = \frac{13}{12}$, $\sec\theta = -\frac{13}{5}$, $\cot\theta = -\frac{5}{12}$
39. $\cos\theta = \frac{12}{13}$, $\tan\theta = -\frac{5}{12}$, $\csc\theta = -\frac{13}{5}$, $\sec\theta = \frac{13}{12}$, $\cot\theta = -\frac{12}{5}$
41. $\sin\theta = -\sqrt{10}/10$, $\cos\theta = -3\sqrt{10}/10$, $\csc\theta = -\sqrt{10}$, $\sec\theta = -\sqrt{10}/3$, $\cot\theta = 3$
43. $\sin\theta = -2\sqrt{2}/3$, $\cos\theta = \frac{1}{3}$, $\tan\theta = -2\sqrt{2}$, $\csc\theta = -3\sqrt{2}/4$, $\cot\theta = -\sqrt{2}/4$
45. $\sin\theta = \sqrt{5}/5$, $\cos\theta = -2\sqrt{5}/5$, $\tan\theta = -\frac{1}{2}$, $\csc\theta = \sqrt{5}$, $\sec\theta = -\sqrt{5}/2$
47. $\pi/3$ ft **49.** 114.59 revolutions per hour **51.** 839 ft **53.** 23.32 ft **55.** 2.15 mi

CHAPTER 8 EXERCISE 8.1

1. 0 **3.** 0 **5.** $-\pi/2 \le x \le \pi/2$ **7.** $y = \cos x$; $y = \sec x$ **9.** 1 **11.** 0, π, 2π **13.** $-3\pi/2$, $-\pi/2$, $\pi/2$, $3\pi/2$
15. $-3\pi/2$, $-\pi/2$, $\pi/2$, $3\pi/2$ **17.** $\sec x = 1$ for $x = -2\pi, 0, 2\pi$; $\sec x = -1$ for $x = -\pi, \pi$
19. $\sin x = 1$ for $x = -3\pi/2, \pi/2$; $\sin x = -1$ for $x = -\pi/2, 3\pi/2$

21.
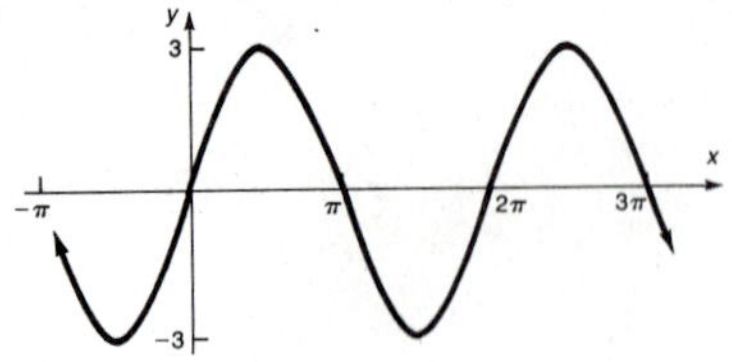

23.
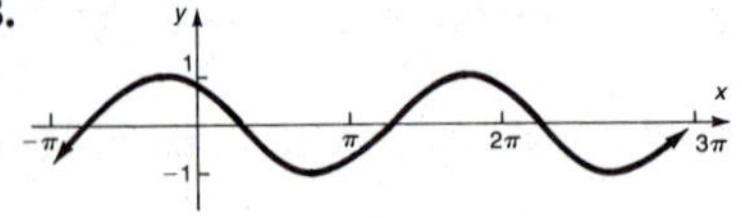

25.
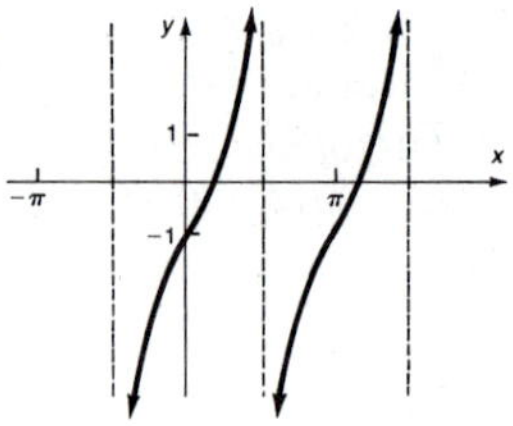

27.
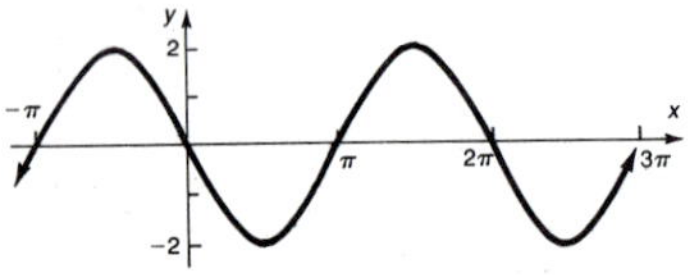

29.
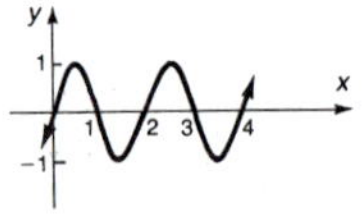

31.
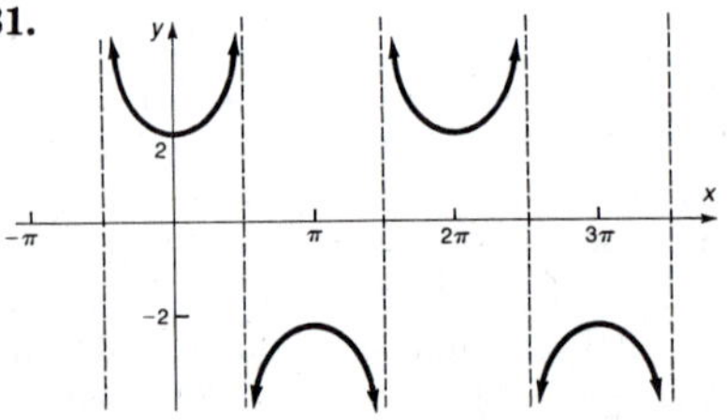

33.
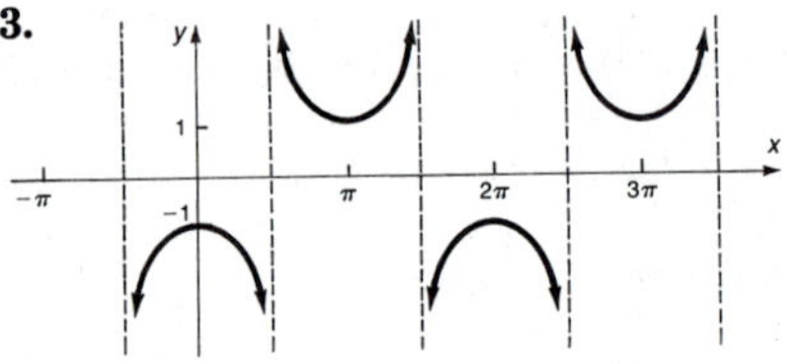

35.
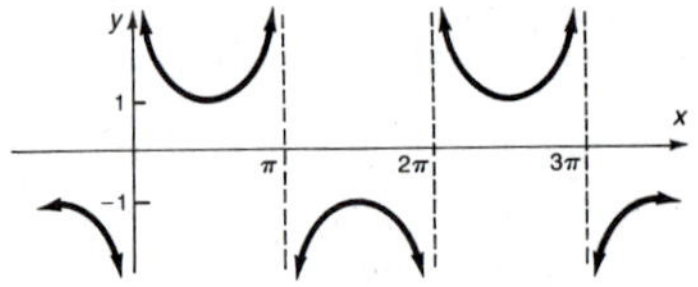

37.
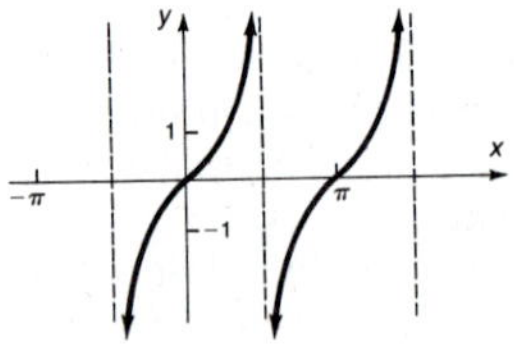

39.
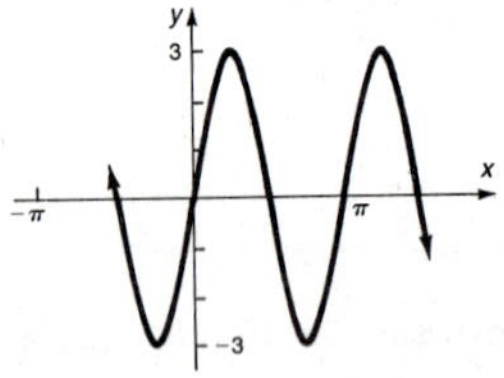

41.
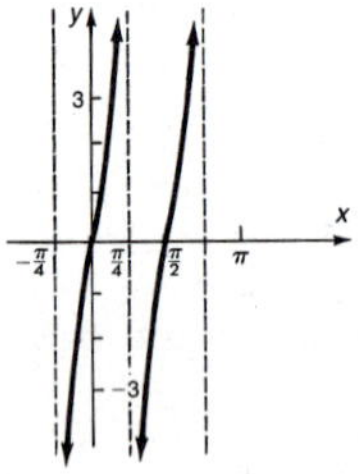

43.
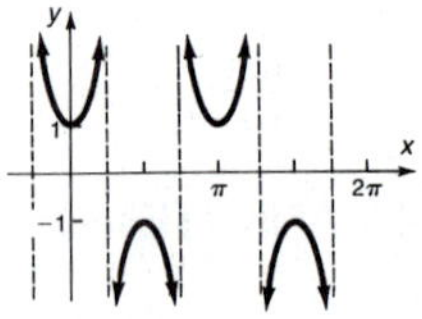

45.
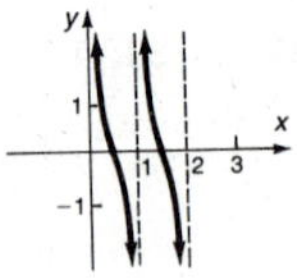

47.
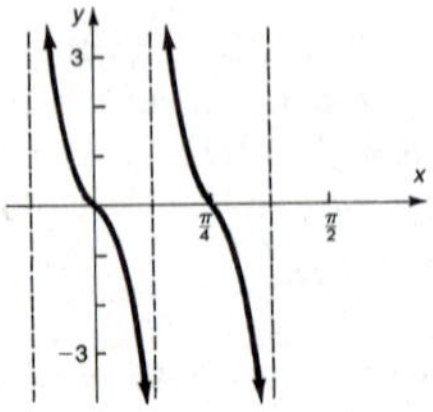

49.
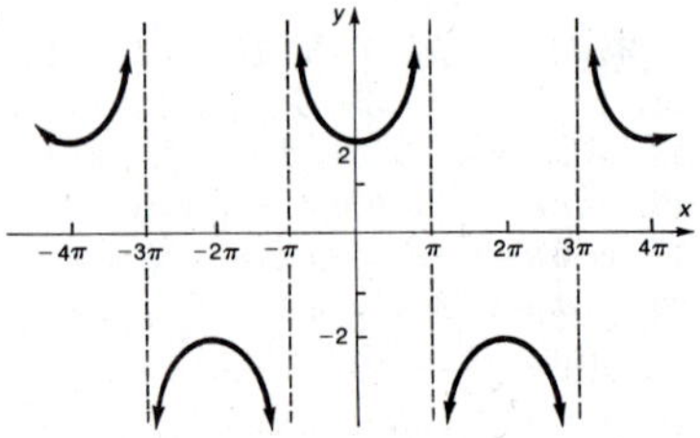

51. The graphs are identical; yes.

EXERCISE 8.2

1. Amplitude = 2; Period = 2π **3.** Amplitude = 4; Period = π **5.** Amplitude = 6; Period = 2 **7.** Amplitude = $\frac{1}{2}$; Period = $4\pi/3$ **9.** Amplitude = $\frac{5}{3}$; Period = 3 **11.** F **13.** A **15.** H **17.** C **19.** J

21.

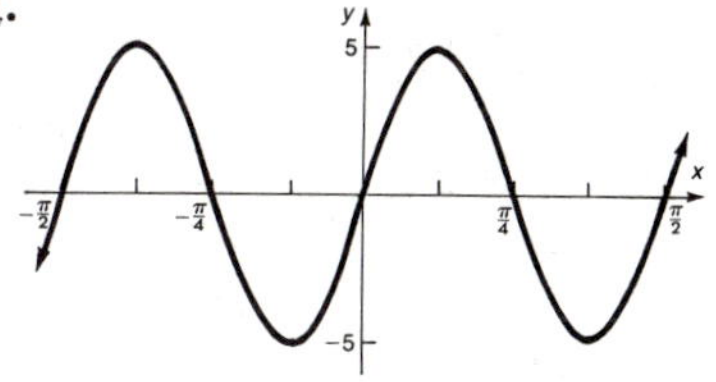

23.

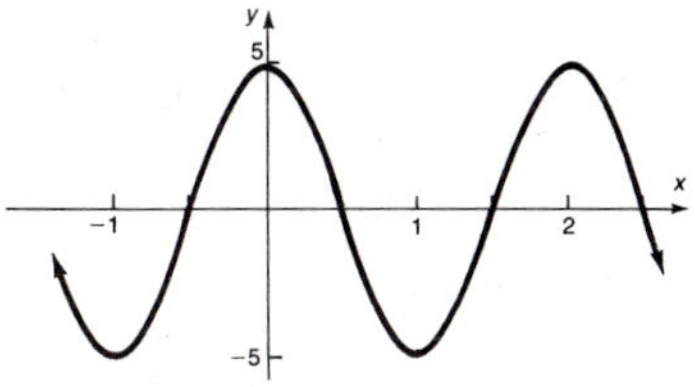

25.

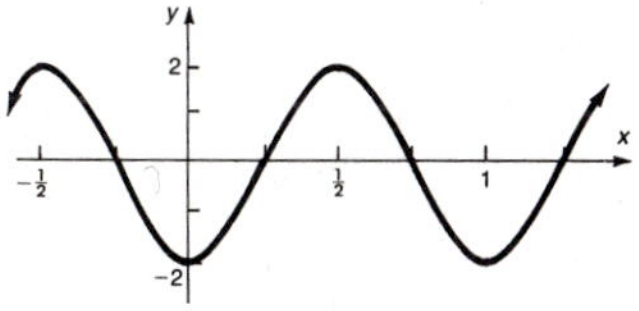

27.

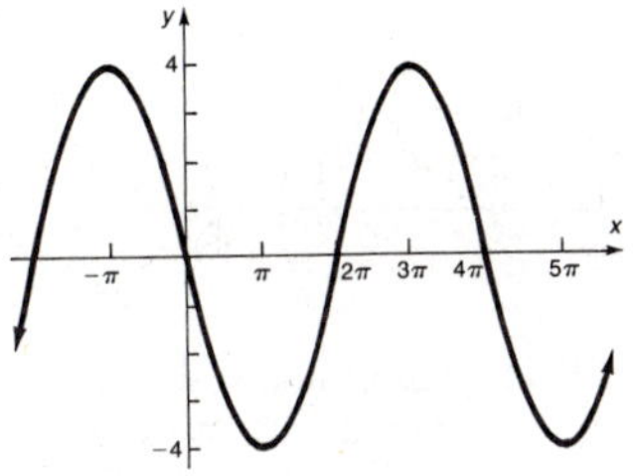

29.

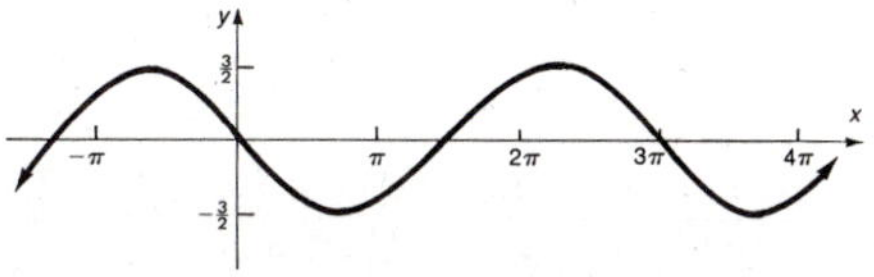

31. $y = 5 \cos \frac{\pi}{4}x$
33. $y = -3 \cos \frac{1}{2}x$
35. $y = \frac{3}{4} \sin 2\pi x$
37. $y = -\sin \frac{3}{2}x$
39. $y = -2 \cos \frac{3\pi}{2}x$

41. Amplitude = 4
Period = π
Phase shift = $\pi/2$

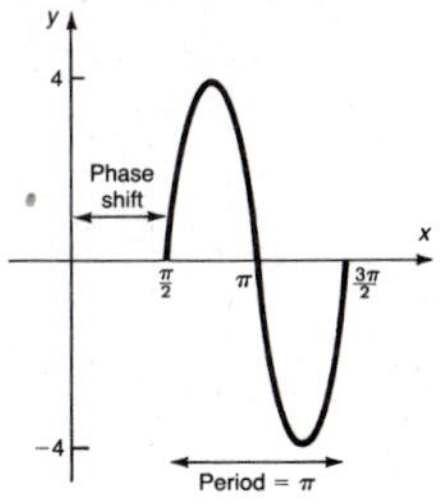

43. Amplitude = 2
Period = $2\pi/3$
Phase shift = $-\pi/6$

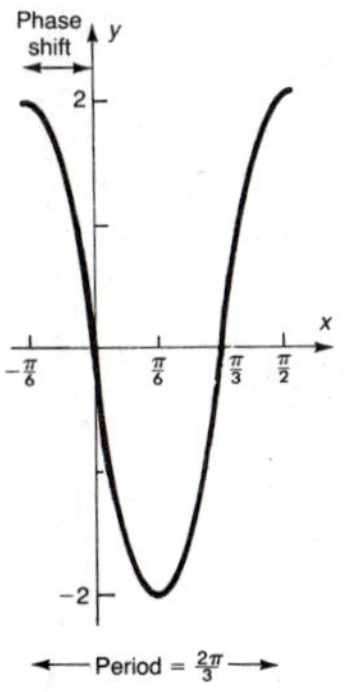

45. Amplitude = 3
Period = π
Phase shift = $-\pi/4$

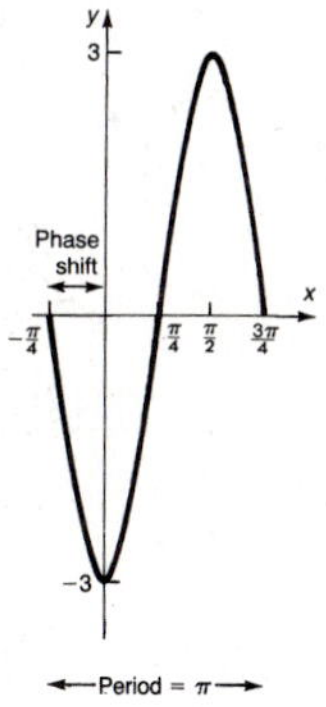

EXERCISE 8.3

1. 0 **3.** $-\pi/2$ **5.** 0 **7.** $\pi/4$ **9.** $\pi/3$ **11.** $5\pi/6$ **13.** 0.1002 **15.** 1.3734 **17.** 0.5054 **19.** -0.3805 **21.** -0.1203 **23.** 1.0799 **25.** $\sqrt{2}/2$ **27.** $-\sqrt{3}/3$ **29.** 2 **31.** $\sqrt{2}$ **33.** $-\sqrt{2}/2$ **35.** $2\sqrt{3}/3$ **37.** $\sqrt{2}/4$ **39.** $\sqrt{5}/2$ **41.** $-\sqrt{14}/2$ **43.** $-3\sqrt{10}/10$ **45.** $\sqrt{5}$ **47.** 0.5779 **49.** 0.0995 **51.** 0.5708 **53.** 0.4271 **55.** 0.3655

57. Let $\theta = \tan^{-1} v$. Then $\tan \theta = v$, $-\pi/2 < \theta < \pi/2$. Now, $\sec \theta > 0$ and $\tan^2 \theta + 1 = \sec^2 \theta$. Thus, $\sec \theta = \sec(\tan^{-1} v) = \sqrt{1 + v^2}$.

59. Let $\theta = \cos^{-1} v$. Then $\cos \theta = v$, $0 \le \theta \le \pi$, and $\tan(\cos^{-1} v) = \tan \theta = \dfrac{\sin \theta}{\cos \theta} = \dfrac{\sqrt{1 - \cos^2 \theta}}{\cos \theta} = \dfrac{\sqrt{1 - v^2}}{v}$.

61. Let $\theta = \sin^{-1} v$. Then $\sin \theta = v$, $-\pi/2 \le \theta \le \pi/2$, and $\cos(\sin^{-1} v) = \cos \theta = \sqrt{1 - \sin^2 \theta} = \sqrt{1 - v^2}$.
63. Let $\alpha = \sin^{-1} v$ and $\beta = \cos^{-1} v$. Then $\sin \alpha = v = \cos \beta$, so α and β are complementary angles. Thus, $\alpha + \beta = \pi/2$.
65. 1.32 **67.** 0.46 **69.** −0.34 **71.** 2.72 **73.** 77 in. **75.** $-1 \le x \le 1$

77.

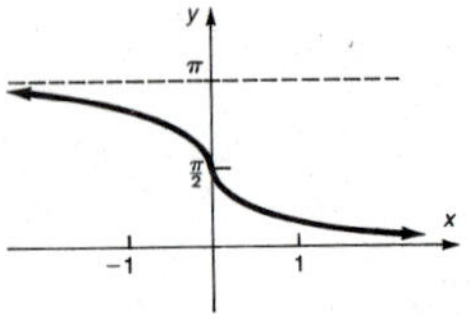

79.

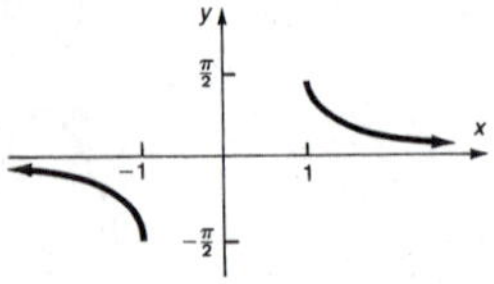

FILL-IN-THE-BLANK ITEMS

1. $y = 3 \sin \pi x$ **2.** 3; $\pi/3$ **3.** $-1 \le x \le 1$; $-\pi/2 \le y \le \pi/2$ **4.** 0

REVIEW EXERCISES

1. Amplitude = 4; Period = 2π **3.** Amplitude = 8; Period = 4

5. Amplitude = 4
Period = $2\pi/3$
Phase shift = 0

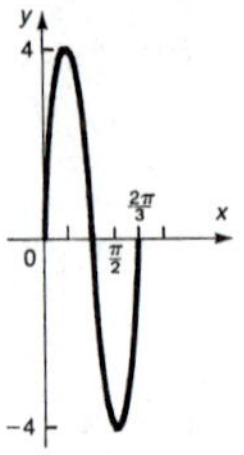

7. Amplitude = 2
Period = 4
Phase shift = $-1/\pi$

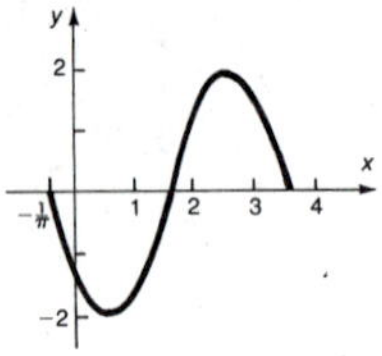

9. Amplitude = $\frac{1}{2}$
Period = $4\pi/3$
Phase shift = $2\pi/3$

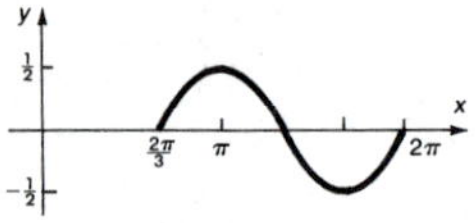

11. Amplitude = $\frac{2}{3}$
Period = 2
Phase shift = $6/\pi$

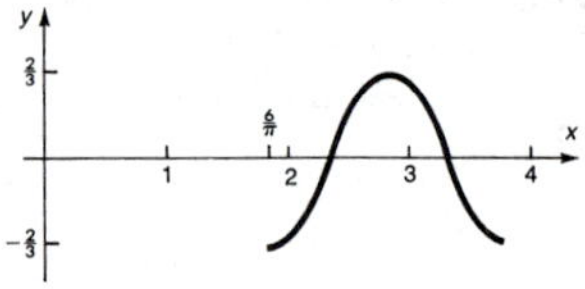

13. $y = 5 \cos \dfrac{x}{4}$

15. $y = -6 \cos \dfrac{\pi}{4} x$

17.

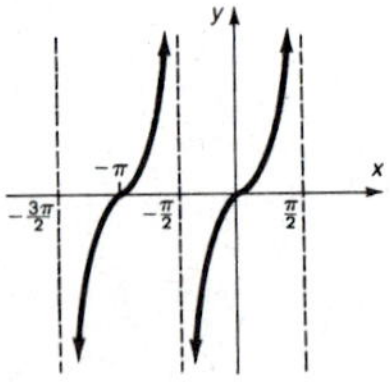

19.

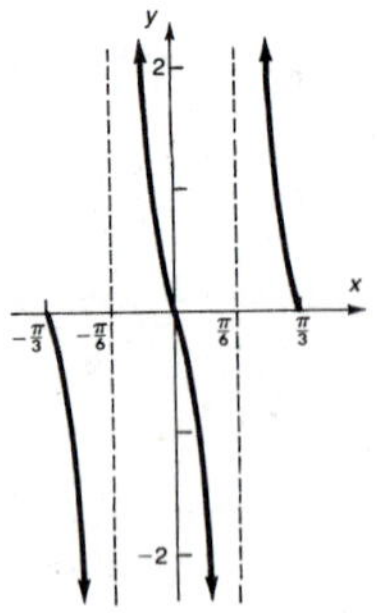

21. $\pi/2$
23. $\pi/4$
25. $5\pi/6$
27. $\sqrt{2}/2$
29. $-\sqrt{3}$
31. $2\sqrt{3}/3$
33. $\frac{3}{5}$
35. $-\frac{4}{3}$

CHAPTER 9 EXERCISE 9.1

1. $\csc \theta \cdot \cos \theta = \dfrac{1}{\sin \theta} \cdot \cos \theta = \dfrac{\cos \theta}{\sin \theta} = \cot \theta$ **3.** $1 + \tan^2(-\theta) = 1 + (-\tan \theta)^2 = 1 + \tan^2 \theta = \sec^2 \theta$

5. $\cos \theta(\tan \theta + \cot \theta) = \cos \theta\left(\dfrac{\sin \theta}{\cos \theta} + \dfrac{\cos \theta}{\sin \theta}\right) = \cos \theta\left(\dfrac{\sin^2 \theta + \cos^2 \theta}{\cos \theta \sin \theta}\right) = \dfrac{1}{\sin \theta} = \csc \theta$

7. $\tan\theta\cot\theta - \cos^2\theta = \dfrac{\sin\theta}{\cos\theta}\cdot\dfrac{\cos\theta}{\sin\theta} - \cos^2\theta = 1 - \cos^2\theta = \sin^2\theta$ **9.** $(\sec\theta - 1)(\sec\theta + 1) = \sec^2\theta - 1 = \tan^2\theta$

11. $(\sec\theta + \tan\theta)(\sec\theta - \tan\theta) = \sec^2\theta - \tan^2\theta = 1$

13. $\sin^2\theta(1 + \cot^2\theta) = \sin^2\theta\csc^2\theta = \sin^2\theta\left(\dfrac{1}{\sin^2\theta}\right) = 1$

15. $(\sin\theta + \cos\theta)^2 + (\sin\theta - \cos\theta)^2 = \sin^2\theta + 2\sin\theta\cos\theta + \cos^2\theta + \sin^2\theta - 2\sin\theta\cos\theta + \cos^2\theta = \sin^2\theta + \cos^2\theta + \sin^2\theta + \cos^2\theta = 1 + 1 = 2$

17. $\sec^4\theta - \sec^2\theta = \sec^2\theta(\sec^2\theta - 1) = (1 + \tan^2\theta)\tan^2\theta = \tan^4\theta + \tan^2\theta$

19. $\sec\theta - \tan\theta = \dfrac{1}{\cos\theta} - \dfrac{\sin\theta}{\cos\theta} = \dfrac{1 - \sin\theta}{\cos\theta}\cdot\dfrac{1 + \sin\theta}{1 + \sin\theta} = \dfrac{1 - \sin^2\theta}{\cos\theta(1 + \sin\theta)} = \dfrac{\cos^2\theta}{\cos\theta(1 + \sin\theta)} = \dfrac{\cos\theta}{1 + \sin\theta}$

21. $3\sin^2\theta + 4\cos^2\theta = 3\sin^2\theta + 3\cos^2\theta + \cos^2\theta = 3(\sin^2\theta + \cos^2\theta) + \cos^2\theta = 3 + \cos^2\theta$

23. $1 - \dfrac{\cos^2\theta}{1 + \sin\theta} = 1 - \dfrac{1 - \sin^2\theta}{1 + \sin\theta} = 1 - (1 - \sin\theta) = \sin\theta$ **25.** $\dfrac{1 + \tan\theta}{1 - \tan\theta} = \dfrac{1 + \dfrac{1}{\cot\theta}}{1 - \dfrac{1}{\cot\theta}} = \dfrac{\dfrac{\cot\theta + 1}{\cot\theta}}{\dfrac{\cot\theta - 1}{\cot\theta}} = \dfrac{\cot\theta + 1}{\cot\theta - 1}$

27. $\dfrac{\sec\theta}{\csc\theta} + \dfrac{\sin\theta}{\cos\theta} = \dfrac{1/\cos\theta}{1/\sin\theta} + \tan\theta = \dfrac{\sin\theta}{\cos\theta} + \tan\theta = \tan\theta + \tan\theta = 2\tan\theta$

29. $\dfrac{1 + \sin\theta}{1 - \sin\theta} = \dfrac{1 + \dfrac{1}{\csc\theta}}{1 - \dfrac{1}{\csc\theta}} = \dfrac{\dfrac{\csc\theta + 1}{\csc\theta}}{\dfrac{\csc\theta - 1}{\csc\theta}} = \dfrac{\csc\theta + 1}{\csc\theta - 1}$

31. $\dfrac{1 - \sin\theta}{\cos\theta} + \dfrac{\cos\theta}{1 - \sin\theta} = \dfrac{(1 - \sin\theta)^2 + \cos^2\theta}{\cos\theta(1 - \sin\theta)} = \dfrac{1 - 2\sin\theta + \sin^2\theta + \cos^2\theta}{\cos\theta(1 - \sin\theta)} = \dfrac{2 - 2\sin\theta}{\cos\theta(1 - \sin\theta)} = \dfrac{2(1 - \sin\theta)}{\cos\theta(1 - \sin\theta)} = \dfrac{2}{\cos\theta} = 2\sec\theta$

33. $\dfrac{\sin\theta}{\sin\theta - \cos\theta} = \dfrac{1}{\dfrac{\sin\theta - \cos\theta}{\sin\theta}} = \dfrac{1}{1 - \dfrac{\cos\theta}{\sin\theta}} = \dfrac{1}{1 - \cot\theta}$

35. $(\sec\theta - \tan\theta)^2 = \sec^2\theta - 2\sec\theta\tan\theta + \tan^2\theta = \dfrac{1}{\cos^2\theta} - \dfrac{2\sin\theta}{\cos^2\theta} + \dfrac{\sin^2\theta}{\cos^2\theta} = \dfrac{1 - 2\sin\theta + \sin^2\theta}{\cos^2\theta} = \dfrac{(1 - \sin\theta)^2}{1 - \sin^2\theta} = \dfrac{(1 - \sin\theta)^2}{(1 - \sin\theta)(1 + \sin\theta)} = \dfrac{1 - \sin\theta}{1 + \sin\theta}$

37. $\dfrac{\cos\theta}{1 - \tan\theta} + \dfrac{\sin\theta}{1 - \cot\theta} = \dfrac{\cos\theta}{1 - \dfrac{\sin\theta}{\cos\theta}} + \dfrac{\sin\theta}{1 - \dfrac{\cos\theta}{\sin\theta}} = \dfrac{\cos\theta}{\dfrac{\cos\theta - \sin\theta}{\cos\theta}} + \dfrac{\sin\theta}{\dfrac{\sin\theta - \cos\theta}{\sin\theta}} = \dfrac{\cos^2\theta}{\cos\theta - \sin\theta} + \dfrac{\sin^2\theta}{\sin\theta - \cos\theta} = \dfrac{\cos^2\theta - \sin^2\theta}{\cos\theta - \sin\theta} = \dfrac{(\cos\theta - \sin\theta)(\cos\theta + \sin\theta)}{\cos\theta - \sin\theta} = \sin\theta + \cos\theta$

39. $\tan\theta + \dfrac{\cos\theta}{1 + \sin\theta} = \dfrac{\sin\theta}{\cos\theta} + \dfrac{\cos\theta}{1 + \sin\theta} = \dfrac{\sin\theta(1 + \sin\theta) + \cos^2\theta}{\cos\theta(1 + \sin\theta)} = \dfrac{\sin\theta + \sin^2\theta + \cos^2\theta}{\cos\theta(1 + \sin\theta)} = \dfrac{\sin\theta + 1}{\cos\theta(1 + \sin\theta)} = \dfrac{1}{\cos\theta} = \sec\theta$

41. $\dfrac{\tan\theta + \sec\theta - 1}{\tan\theta - \sec\theta + 1} = \dfrac{\tan\theta + (\sec\theta - 1)}{\tan\theta - (\sec\theta - 1)}\cdot\dfrac{\tan\theta + (\sec\theta - 1)}{\tan\theta + (\sec\theta - 1)} = \dfrac{\tan^2\theta + 2\tan\theta(\sec\theta - 1) + \sec^2\theta - 2\sec\theta + 1}{\tan^2\theta - (\sec^2\theta - 2\sec\theta + 1)} = \dfrac{\sec^2\theta - 1 + 2\tan\theta(\sec\theta - 1) + \sec^2\theta - 2\sec\theta + 1}{\sec^2\theta - 1 - \sec^2\theta + 2\sec\theta - 1} = \dfrac{2\sec^2\theta - 2\sec\theta + 2\tan\theta(\sec\theta - 1)}{-2 + 2\sec\theta} = \dfrac{2\sec\theta(\sec\theta - 1) + 2\tan\theta(\sec\theta - 1)}{2(\sec\theta - 1)} = \dfrac{2(\sec\theta - 1)(\sec\theta + \tan\theta)}{2(\sec\theta - 1)} = \sec\theta + \tan\theta$

43. $\dfrac{\tan\theta - \cot\theta}{\tan\theta + \cot\theta} = \dfrac{\dfrac{\sin\theta}{\cos\theta} - \dfrac{\cos\theta}{\sin\theta}}{\dfrac{\sin\theta}{\cos\theta} + \dfrac{\cos\theta}{\sin\theta}} = \dfrac{\dfrac{\sin^2\theta - \cos^2\theta}{\cos\theta\sin\theta}}{\dfrac{\sin^2\theta + \cos^2\theta}{\cos\theta\sin\theta}} = \dfrac{\sin^2\theta - \cos^2\theta}{1} = \sin^2\theta - \cos^2\theta$

45. $\dfrac{\tan\theta-\cot\theta}{\tan\theta+\cot\theta}=\dfrac{\dfrac{\sin\theta}{\cos\theta}-\dfrac{\cos\theta}{\sin\theta}}{\dfrac{\sin\theta}{\cos\theta}+\dfrac{\cos\theta}{\sin\theta}}=\dfrac{\dfrac{\sin^2\theta-\cos^2\theta}{\cos\theta\sin\theta}}{\dfrac{\sin^2\theta+\cos^2\theta}{\cos\theta\sin\theta}}=\sin^2\theta-\cos^2\theta=\sin^2\theta-(1-\sin^2\theta)=2\sin^2\theta-1$

47. $\dfrac{\sec\theta+\tan\theta}{\cot\theta+\cos\theta}=\dfrac{\dfrac{1}{\cos\theta}+\dfrac{\sin\theta}{\cos\theta}}{\dfrac{\cos\theta}{\sin\theta}+\dfrac{\cos\theta\sin\theta}{\sin\theta}}=\dfrac{\dfrac{1+\sin\theta}{\cos\theta}}{\dfrac{\cos\theta+\cos\theta\sin\theta}{\sin\theta}}=\dfrac{1+\sin\theta}{\cos\theta}\cdot\dfrac{\sin\theta}{\cos\theta(1+\sin\theta)}=\dfrac{\sin\theta}{\cos\theta}\cdot\dfrac{1}{\cos\theta}=\tan\theta\sec\theta$

49. $\dfrac{1-\tan^2\theta}{1+\tan^2\theta}=\dfrac{1-\tan^2\theta}{\sec^2\theta}=\dfrac{1}{\sec^2\theta}-\dfrac{\tan^2\theta}{\sec^2\theta}=\cos^2\theta-\dfrac{\sin^2\theta/\cos^2\theta}{1/\cos^2\theta}=\cos^2\theta-\sin^2\theta=\cos^2\theta-(1-\cos^2\theta)=2\cos^2\theta-1$

51. $\dfrac{\sec\theta-\csc\theta}{\sec\theta\csc\theta}=\dfrac{\dfrac{1}{\cos\theta}-\dfrac{1}{\sin\theta}}{\dfrac{1}{\cos\theta}\cdot\dfrac{1}{\sin\theta}}=\dfrac{\dfrac{\sin\theta-\cos\theta}{\cos\theta\sin\theta}}{\dfrac{1}{\cos\theta\sin\theta}}=\sin\theta-\cos\theta$

53. $\sec\theta-\cos\theta=\dfrac{1}{\cos\theta}-\dfrac{\cos^2\theta}{\cos\theta}=\dfrac{1-\cos^2\theta}{\cos\theta}=\dfrac{\sin^2\theta}{\cos\theta}=\sin\theta\cdot\dfrac{\sin\theta}{\cos\theta}=\sin\theta\tan\theta$

55. $\dfrac{1}{1-\sin\theta}+\dfrac{1}{1+\sin\theta}=\dfrac{1+\sin\theta+1-\sin\theta}{(1+\sin\theta)(1-\sin\theta)}=\dfrac{2}{1-\sin^2\theta}=\dfrac{2}{\cos^2\theta}=2\sec^2\theta$

57. $\dfrac{\sec\theta}{1-\sin\theta}=\dfrac{\sec\theta}{1-\sin\theta}\cdot\dfrac{1+\sin\theta}{1+\sin\theta}=\dfrac{\sec\theta(1+\sin\theta)}{1-\sin^2\theta}=\dfrac{\sec\theta(1+\sin\theta)}{\cos^2\theta}=\dfrac{1+\sin\theta}{\cos^3\theta}$

59. $\dfrac{(\sec\theta-\tan\theta)^2+1}{\csc\theta(\sec\theta-\tan\theta)}=\dfrac{\sec^2\theta-2\sec\theta\tan\theta+\tan^2\theta+1}{\dfrac{1}{\sin\theta}\left(\dfrac{1}{\cos\theta}-\dfrac{\sin\theta}{\cos\theta}\right)}=\dfrac{2\sec^2\theta-2\sec\theta\tan\theta}{\dfrac{1}{\sin\theta}\left(\dfrac{1-\sin\theta}{\cos\theta}\right)}=\dfrac{\dfrac{2}{\cos^2\theta}-\dfrac{2\sin\theta}{\cos^2\theta}}{\dfrac{1-\sin\theta}{\sin\theta\cos\theta}}=$

$\dfrac{2-2\sin\theta}{\cos^2\theta}\cdot\dfrac{\sin\theta\cos\theta}{1-\sin\theta}=\dfrac{2(1-\sin\theta)}{\cos\theta}\cdot\dfrac{\sin\theta}{1-\sin\theta}=\dfrac{2\sin\theta}{\cos\theta}=2\tan\theta$

61. $\dfrac{\sin\theta+\cos\theta}{\cos\theta}-\dfrac{\sin\theta-\cos\theta}{\sin\theta}=\dfrac{\sin\theta(\sin\theta+\cos\theta)-\cos\theta(\sin\theta-\cos\theta)}{\cos\theta\sin\theta}=\dfrac{\sin^2\theta+\sin\theta\cos\theta-\sin\theta\cos\theta+\cos^2\theta}{\cos\theta\sin\theta}=$

$\dfrac{1}{\cos\theta\sin\theta}=\sec\theta\csc\theta$

63. $\dfrac{\sin^3\theta+\cos^3\theta}{\sin\theta+\cos\theta}=\dfrac{(\sin\theta+\cos\theta)(\sin^2\theta-\sin\theta\cos\theta+\cos^2\theta)}{\sin\theta+\cos\theta}=\sin^2\theta+\cos^2\theta-\sin\theta\cos\theta=1-\sin\theta\cos\theta$

65. $\dfrac{\cos^2\theta-\sin^2\theta}{1-\tan^2\theta}=\dfrac{\cos^2\theta-\sin^2\theta}{1-\dfrac{\sin^2\theta}{\cos^2\theta}}=\dfrac{\cos^2\theta-\sin^2\theta}{\dfrac{\cos^2\theta-\sin^2\theta}{\cos^2\theta}}=\cos^2\theta$

67. $\dfrac{(2\cos^2\theta-1)^2}{\cos^4\theta-\sin^4\theta}=\dfrac{[2\cos^2\theta-(\sin^2\theta+\cos^2\theta)]^2}{(\cos^2\theta-\sin^2\theta)(\cos^2\theta+\sin^2\theta)}=\cos^2\theta-\sin^2\theta=(1-\sin^2\theta)-\sin^2\theta=1-2\sin^2\theta$

69. $\dfrac{1+\sin\theta+\cos\theta}{1+\sin\theta-\cos\theta}=\dfrac{(1+\sin\theta)+\cos\theta}{(1+\sin\theta)-\cos\theta}\cdot\dfrac{(1+\sin\theta)+\cos\theta}{(1+\sin\theta)+\cos\theta}=\dfrac{1+2\sin\theta+\sin^2\theta+2(1+\sin\theta)(\cos\theta)+\cos^2\theta}{1+2\sin\theta+\sin^2\theta-\cos^2\theta}=$

$\dfrac{1+2\sin\theta+\sin^2\theta+2(1+\sin\theta)(\cos\theta)+(1-\sin^2\theta)}{1+2\sin\theta+\sin^2\theta-(1-\sin^2\theta)}=\dfrac{2+2\sin\theta+2(1+\sin\theta)(\cos\theta)}{2\sin\theta+2\sin^2\theta}=$

$\dfrac{2(1+\sin\theta)+2(1+\sin\theta)(\cos\theta)}{2\sin\theta(1+\sin\theta)}=\dfrac{2(1+\sin\theta)(1+\cos\theta)}{2\sin\theta(1+\sin\theta)}=\dfrac{1+\cos\theta}{\sin\theta}$

71. $(a\sin\theta+b\cos\theta)^2+(a\cos\theta-b\sin\theta)^2=a^2\sin^2\theta+2ab\sin\theta\cos\theta+b^2\cos^2\theta+a^2\cos^2\theta-2ab\sin\theta\cos\theta+b^2\sin^2\theta=$
$a^2(\sin^2\theta+\cos^2\theta)+b^2(\cos^2\theta+\sin^2\theta)=a^2+b^2$

73. $\dfrac{\tan\alpha+\tan\beta}{\cot\alpha+\cot\beta}=\dfrac{\tan\alpha+\tan\beta}{\dfrac{1}{\tan\alpha}+\dfrac{1}{\tan\beta}}=\dfrac{\tan\alpha+\tan\beta}{\dfrac{\tan\beta+\tan\alpha}{\tan\alpha\tan\beta}}=(\tan\alpha+\tan\beta)\cdot\dfrac{\tan\alpha\tan\beta}{\tan\alpha+\tan\beta}=\tan\alpha\tan\beta$

75. $(\sin\alpha+\cos\beta)^2+(\cos\beta+\sin\alpha)(\cos\beta-\sin\alpha)=(\sin^2\alpha+2\sin\alpha\cos\beta+\cos^2\beta)+(\cos^2\beta-\sin^2\alpha)=$
$2\cos^2\beta+2\sin\alpha\cos\beta=2\cos\beta(\cos\beta+\sin\alpha)$

77. $\ln|\sec\theta|=\ln|\cos\theta|^{-1}=-\ln|\cos\theta|$

79. $\ln|1+\cos\theta|+\ln|1-\cos\theta|=\ln(|1+\cos\theta||1-\cos\theta|)=\ln|1-\cos^2\theta|=\ln|\sin^2\theta|=2\ln|\sin\theta|$

EXERCISE 9.2

1. $\frac{1}{4}(\sqrt{6}+\sqrt{2})$ **3.** $\frac{1}{4}(\sqrt{2}-\sqrt{6})$ **5.** $-\frac{1}{4}(\sqrt{2}+\sqrt{6})$ **7.** $\dfrac{\sqrt{3}-1}{1+\sqrt{3}}=2-\sqrt{3}$ **9.** $-\frac{1}{4}(\sqrt{6}+\sqrt{2})$

11. $\dfrac{4}{\sqrt{6}+\sqrt{2}}=\sqrt{6}-\sqrt{2}$ **13.** $\frac{1}{2}$ **15.** 0 **17.** 1 **19.** −1 **21.** $-\sqrt{3}/2$ **23.** (a) $\dfrac{2\sqrt{5}}{25}$ (b) $\dfrac{11\sqrt{5}}{25}$ (c) $\dfrac{2\sqrt{5}}{5}$ (d) 2

25. (a) $\dfrac{4-3\sqrt{3}}{10}$ (b) $\dfrac{-3-4\sqrt{3}}{10}$ (c) $\dfrac{4+3\sqrt{3}}{10}$ (d) $\dfrac{4+3\sqrt{3}}{4\sqrt{3}-3}=\dfrac{25\sqrt{3}+48}{39}$

27. (a) $-\dfrac{1}{26}(5+12\sqrt{3})$ (b) $\dfrac{1}{26}(12-5\sqrt{3})$ (c) $-\dfrac{1}{26}(5-12\sqrt{3})$ (d) $\dfrac{-5+12\sqrt{3}}{12+5\sqrt{3}}=\dfrac{-240+169\sqrt{3}}{69}$

29. $\sin\left(\dfrac{\pi}{2}+\theta\right)=\sin\dfrac{\pi}{2}\cos\theta+\cos\dfrac{\pi}{2}\sin\theta=1\cdot\cos\theta+0\cdot\sin\theta=\cos\theta$

31. $\sin(\pi-\theta)=\sin\pi\cos\theta-\cos\pi\sin\theta=0\cdot\cos\theta-(-1)\sin\theta=\sin\theta$

33. $\sin(\pi+\theta)=\sin\pi\cos\theta+\cos\pi\sin\theta=0\cdot\cos\theta+(-1)\sin\theta=-\sin\theta$

35. $\tan(\pi-\theta)=\dfrac{\tan\pi-\tan\theta}{1+\tan\pi\tan\theta}=\dfrac{0-\tan\theta}{1+0}=-\tan\theta$

37. $\sin\left(\dfrac{3\pi}{2}+\theta\right)=\sin\dfrac{3\pi}{2}\cos\theta+\cos\dfrac{3\pi}{2}\sin\theta=(-1)\cos\theta+0\cdot\sin\theta=-\cos\theta$

39. $\sin(\alpha+\beta)+\sin(\alpha-\beta)=\sin\alpha\cos\beta+\cos\alpha\sin\beta+\sin\alpha\cos\beta-\cos\alpha\sin\beta=2\sin\alpha\cos\beta$

41. $\dfrac{\sin(\alpha+\beta)}{\sin\alpha\cos\beta}=\dfrac{\sin\alpha\cos\beta+\cos\alpha\sin\beta}{\sin\alpha\cos\beta}=\dfrac{\sin\alpha\cos\beta}{\sin\alpha\cos\beta}+\dfrac{\cos\alpha\sin\beta}{\sin\alpha\cos\beta}=1+\cot\alpha\tan\beta$

43. $\dfrac{\cos(\alpha+\beta)}{\cos\alpha\cos\beta}=\dfrac{\cos\alpha\cos\beta-\sin\alpha\sin\beta}{\cos\alpha\cos\beta}=\dfrac{\cos\alpha\cos\beta}{\cos\alpha\cos\beta}-\dfrac{\sin\alpha\sin\beta}{\cos\alpha\cos\beta}=1-\tan\alpha\tan\beta$

45. $\dfrac{\sin(\alpha+\beta)}{\sin(\alpha-\beta)}=\dfrac{\sin\alpha\cos\beta+\cos\alpha\sin\beta}{\sin\alpha\cos\beta-\cos\alpha\sin\beta}=\dfrac{\dfrac{\sin\alpha\cos\beta+\cos\alpha\sin\beta}{\cos\alpha\cos\beta}}{\dfrac{\sin\alpha\cos\beta-\cos\alpha\sin\beta}{\cos\alpha\cos\beta}}=\dfrac{\dfrac{\sin\alpha\cos\beta}{\cos\alpha\cos\beta}+\dfrac{\cos\alpha\sin\beta}{\cos\alpha\cos\beta}}{\dfrac{\sin\alpha\cos\beta}{\cos\alpha\cos\beta}-\dfrac{\cos\alpha\sin\beta}{\cos\alpha\cos\beta}}=\dfrac{\tan\alpha+\tan\beta}{\tan\alpha-\tan\beta}$

47. $\cot(\alpha+\beta)=\dfrac{\cos(\alpha+\beta)}{\sin(\alpha+\beta)}=\dfrac{\cos\alpha\cos\beta-\sin\alpha\sin\beta}{\sin\alpha\cos\beta+\cos\alpha\sin\beta}=\dfrac{\dfrac{\cos\alpha\cos\beta-\sin\alpha\sin\beta}{\sin\alpha\sin\beta}}{\dfrac{\sin\alpha\cos\beta+\cos\alpha\sin\beta}{\sin\alpha\sin\beta}}=\dfrac{\dfrac{\cos\alpha\cos\beta}{\sin\alpha\sin\beta}-\dfrac{\sin\alpha\sin\beta}{\sin\alpha\sin\beta}}{\dfrac{\sin\alpha\cos\beta}{\sin\alpha\sin\beta}+\dfrac{\cos\alpha\sin\beta}{\sin\alpha\sin\beta}}=$
$\dfrac{\cot\alpha\cot\beta-1}{\cot\beta+\cot\alpha}$

49. $\sec(\alpha+\beta)=\dfrac{1}{\cos(\alpha+\beta)}=\dfrac{1}{\cos\alpha\cos\beta-\sin\alpha\sin\beta}=\dfrac{\dfrac{1}{\sin\alpha\sin\beta}}{\dfrac{\cos\alpha\cos\beta-\sin\alpha\sin\beta}{\sin\alpha\sin\beta}}=\dfrac{\dfrac{1}{\sin\alpha}\cdot\dfrac{1}{\sin\beta}}{\dfrac{\cos\alpha\cos\beta}{\sin\alpha\sin\beta}-\dfrac{\sin\alpha\sin\beta}{\sin\alpha\sin\beta}}=$
$\dfrac{\csc\alpha\csc\beta}{\cot\alpha\cot\beta-1}$

51. $\sin(\alpha-\beta)\sin(\alpha+\beta)=(\sin\alpha\cos\beta-\cos\alpha\sin\beta)(\sin\alpha\cos\beta+\cos\alpha\sin\beta)=\sin^2\alpha\cos^2\beta-\cos^2\alpha\sin^2\beta=$
$(\sin^2\alpha)(1-\sin^2\beta)-(1-\sin^2\alpha)(\sin^2\beta)=\sin^2\alpha-\sin^2\beta$

53. $\sin(\theta+k\pi)=\sin\theta\cos k\pi+\cos\theta\sin k\pi=(\sin\theta)(-1)^k+(\cos\theta)(0)=(-1)^k\cdot\sin\theta$, k any integer

55. $\sqrt{3}/2$ **57.** $\sqrt{3}/2$ **59.** −1 **61.** $-\frac{24}{25}$ **63.** $-\frac{33}{65}$ **65.** $\frac{65}{63}$ **67.** $\frac{1}{39}(48-25\sqrt{3})$

69. $\dfrac{\sin(x+h)-\sin x}{h}=\dfrac{\sin x\cos h+\cos x\sin h-\sin x}{h}=\dfrac{\cos x\sin h-(\sin x)(1-\cos h)}{h}=\cos x\cdot\dfrac{\sin h}{h}-\sin x\cdot\dfrac{1-\cos h}{h}$

71. $\sin(\sin^{-1}u+\cos^{-1}u)=\sin(\sin^{-1}u)\cos(\cos^{-1}u)+\cos(\sin^{-1}u)\sin(\cos^{-1}u)=(u)(u)+\sqrt{1-u^2}\sqrt{1-u^2}=u^2+1-u^2=1$

73. $\tan\dfrac{\pi}{2}$ is not defined; $\tan\left(\dfrac{\pi}{2}-\theta\right)=\dfrac{\sin\left(\dfrac{\pi}{2}-\theta\right)}{\cos\left(\dfrac{\pi}{2}-\theta\right)}=\dfrac{\cos\theta}{\sin\theta}=\cot\theta$

75. $\tan\theta=\tan(\theta_2-\theta_1)=\dfrac{\tan\theta_2-\tan\theta_1}{1+\tan\theta_1\tan\theta_2}=\dfrac{m_2-m_1}{1+m_1m_2}$

EXERCISE 9.3

1. (a) $\frac{24}{25}$ (b) $\frac{7}{25}$ (c) $\sqrt{10}/10$ (d) $3\sqrt{10}/10$ **3.** (a) $\frac{24}{25}$ (b) $-\frac{7}{25}$ (c) $2\sqrt{5}/5$ (d) $-\sqrt{5}/5$

5. (a) $-2\sqrt{2}/3$ (b) $\frac{1}{3}$ (c) $\sqrt{\dfrac{3+\sqrt{6}}{6}}$ (d) $\sqrt{\dfrac{3-\sqrt{6}}{6}}$ **7.** (a) $4\sqrt{2}/9$ (b) $-\frac{7}{9}$ (c) $\sqrt{3}/3$ (d) $\sqrt{6}/3$

9. (a) $-\frac{4}{5}$ (b) $\frac{3}{5}$ (c) $\sqrt{\dfrac{5+2\sqrt{5}}{10}}$ (d) $\sqrt{\dfrac{5-2\sqrt{5}}{10}}$ **11.** $\dfrac{\sqrt{2-\sqrt{2}}}{2}$ **13.** $1-\sqrt{2}$ **15.** $-\dfrac{\sqrt{2+\sqrt{3}}}{2}$

17. $\dfrac{2}{\sqrt{2+\sqrt{2}}} = (2-\sqrt{2})\sqrt{2+\sqrt{2}}$ **19.** $-\dfrac{\sqrt{2-\sqrt{2}}}{2}$

21. $\sin^4\theta = (\sin^2\theta)^2 = \left(\dfrac{1-\cos 2\theta}{2}\right)^2 = \dfrac{1}{4}(1-2\cos 2\theta+\cos^2 2\theta) = \dfrac{1}{4}-\dfrac{1}{2}\cos 2\theta+\dfrac{1}{4}\cos^2 2\theta = \dfrac{1}{4}-\dfrac{1}{2}\cos 2\theta+\dfrac{1}{4}\left(\dfrac{1+\cos 4\theta}{2}\right) =$
$\dfrac{1}{4}-\dfrac{1}{2}\cos 2\theta+\dfrac{1}{8}+\dfrac{1}{8}\cos 4\theta = \dfrac{3}{8}-\dfrac{1}{2}\cos 2\theta+\dfrac{1}{8}\cos 4\theta$

23. $\sin 4\theta = \sin 2(2\theta) = 2\sin 2\theta\cos 2\theta = (4\sin\theta\cos\theta)(1-2\sin^2\theta) = 4\sin\theta\cos\theta - 8\sin^3\theta\cos\theta =$
$(\cos\theta)(4\sin\theta - 8\sin^3\theta)$

25. $16\sin^5\theta - 20\sin^3\theta + 5\sin\theta$ **27.** $\dfrac{1-\cos\theta}{\sin\theta} = \dfrac{1-\cos 2(\theta/2)}{\sin 2(\theta/2)} = \dfrac{1-[1-2\sin^2(\theta/2)]}{2\sin(\theta/2)\cos(\theta/2)} = \dfrac{2\sin^2(\theta/2)}{2\sin(\theta/2)\cos(\theta/2)} = \tan\dfrac{\theta}{2}$

29. $\cos^4\theta - \sin^4\theta = (\cos^2\theta+\sin^2\theta)(\cos^2\theta-\sin^2\theta) = \cos 2\theta$

31. $\cot 2\theta = \dfrac{1}{\tan 2\theta} = \dfrac{1}{\dfrac{2\tan\theta}{1-\tan^2\theta}} = \dfrac{1-\tan^2\theta}{2\tan\theta} = \dfrac{1-\dfrac{1}{\cot^2\theta}}{2\left(\dfrac{1}{\cot\theta}\right)} = \dfrac{\dfrac{\cot^2\theta-1}{\cot^2\theta}}{\dfrac{2}{\cot\theta}} = \dfrac{\cot^2\theta-1}{\cot^2\theta}\cdot\dfrac{\cot\theta}{2} = \dfrac{\cot^2\theta-1}{2\cot\theta}$

33. $\sec 2\theta = \dfrac{1}{\cos 2\theta} = \dfrac{1}{2\cos^2\theta-1} = \dfrac{1}{\dfrac{2}{\sec^2\theta}-1} = \dfrac{1}{\dfrac{2-\sec^2\theta}{\sec^2\theta}} = \dfrac{\sec^2\theta}{2-\sec^2\theta}$ **35.** $\cos^2 2\theta - \sin^2 2\theta = \cos 2(2\theta) = \cos 4\theta$

37. $\dfrac{\cos 2\theta}{1+\sin 2\theta} = \dfrac{\cos^2\theta-\sin^2\theta}{1+2\sin\theta\cos\theta} = \dfrac{(\cos\theta-\sin\theta)(\cos\theta+\sin\theta)}{\sin^2\theta+\cos^2\theta+2\sin\theta\cos\theta} = \dfrac{(\cos\theta-\sin\theta)(\cos\theta+\sin\theta)}{(\sin\theta+\cos\theta)(\sin\theta+\cos\theta)} = \dfrac{\cos\theta-\sin\theta}{\cos\theta+\sin\theta} =$
$\dfrac{\dfrac{\cos\theta-\sin\theta}{\sin\theta}}{\dfrac{\cos\theta+\sin\theta}{\sin\theta}} = \dfrac{\dfrac{\cos\theta}{\sin\theta}-\dfrac{\sin\theta}{\sin\theta}}{\dfrac{\cos\theta}{\sin\theta}+\dfrac{\sin\theta}{\sin\theta}} = \dfrac{\cot\theta-1}{\cot\theta+1}$

39. $\sec^2\dfrac{\theta}{2} = \dfrac{1}{\cos^2(\theta/2)} = \dfrac{1}{\dfrac{1+\cos\theta}{2}} = \dfrac{2}{1+\cos\theta}$

41. $\cot^2\dfrac{\theta}{2} = \dfrac{1}{\tan^2(\theta/2)} = \dfrac{1}{\dfrac{1-\cos\theta}{1+\cos\theta}} = \dfrac{1+\cos\theta}{1-\cos\theta} = \dfrac{1+\dfrac{1}{\sec\theta}}{1-\dfrac{1}{\sec\theta}} = \dfrac{\dfrac{\sec\theta+1}{\sec\theta}}{\dfrac{\sec\theta-1}{\sec\theta}} = \dfrac{\sec\theta+1}{\sec\theta}\cdot\dfrac{\sec\theta}{\sec\theta-1} = \dfrac{\sec\theta+1}{\sec\theta-1}$

43. $\dfrac{1-\tan^2(\theta/2)}{1+\tan^2(\theta/2)} = \dfrac{1-\dfrac{1-\cos\theta}{1+\cos\theta}}{1+\dfrac{1-\cos\theta}{1+\cos\theta}} = \dfrac{\dfrac{1+\cos\theta-(1-\cos\theta)}{1+\cos\theta}}{\dfrac{1+\cos\theta+1-\cos\theta}{1+\cos\theta}} = \dfrac{2\cos\theta}{1+\cos\theta}\cdot\dfrac{1+\cos\theta}{2} = \cos\theta$

45. $\dfrac{\sin 3\theta}{\sin\theta} - \dfrac{\cos 3\theta}{\cos\theta} = \dfrac{\sin 3\theta\cos\theta - \cos 3\theta\sin\theta}{\sin\theta\cos\theta} = \dfrac{\sin(3\theta-\theta)}{\frac{1}{2}(2\sin\theta\cos\theta)} = \dfrac{2\sin 2\theta}{\sin 2\theta} = 2$

47. $\tan 3\theta = \tan(\theta+2\theta) = \dfrac{\tan\theta+\tan 2\theta}{1-\tan\theta\tan 2\theta} = \dfrac{\tan\theta+\dfrac{2\tan\theta}{1-\tan^2\theta}}{1-\dfrac{\tan\theta(2\tan\theta)}{1-\tan^2\theta}} = \dfrac{\tan\theta-\tan^3\theta+2\tan\theta}{1-\tan^2\theta-2\tan^2\theta} = \dfrac{3\tan\theta-\tan^3\theta}{1-3\tan^2\theta}$

49.
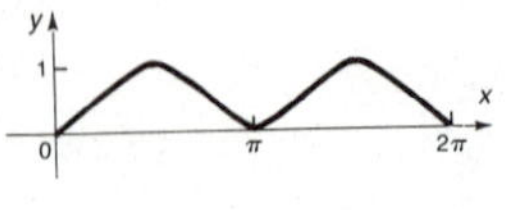

51. $\sin\dfrac{\pi}{24} = \dfrac{\sqrt{2}}{4}\sqrt{4-\sqrt{6}-\sqrt{2}}$; $\cos\dfrac{\pi}{24} = \dfrac{\sqrt{2}}{4}\sqrt{4+\sqrt{6}+\sqrt{2}}$

53. $\sqrt{3}/2$ **55.** $\frac{7}{25}$ **57.** $\frac{24}{7}$ **59.** $\frac{24}{25}$

61. $\frac{1}{5}$ **63.** $\frac{25}{7}$ **65.** 4

67. $\sin^3\theta + \sin^3(\theta + 120°) + \sin^3(\theta + 240°) = \sin^3\theta + (\sin\theta\cos 120° + \cos\theta\sin 120°)^3 + (\sin\theta\cos 240° + \cos\theta\sin 240°)^3 =$
$\sin^3\theta + \left(-\frac{1}{2}\sin\theta + \frac{\sqrt{3}}{2}\cos\theta\right)^3 + \left(-\frac{1}{2}\sin\theta - \frac{\sqrt{3}}{2}\cos\theta\right)^3 =$
$\sin^3\theta + \frac{1}{8}(3\sqrt{3}\cos^3\theta - 9\cos^2\theta\sin\theta + 3\sqrt{3}\cos\theta\sin^2\theta - \sin^3\theta) -$
$\frac{1}{8}(\sin^3\theta + 3\sqrt{3}\sin^2\theta\cos\theta + 9\sin\theta\cos^2\theta + 3\sqrt{3}\cos^3\theta) = \frac{3}{4}\sin^3\theta - \frac{9}{4}\cos^2\theta\sin\theta = \frac{3}{4}[\sin^3\theta - 3\sin\theta(1 - \sin^2\theta)] =$
$\frac{3}{4}(4\sin^3\theta - 3\sin\theta) = -\frac{3}{4}\sin 3\theta$

69. $\frac{1}{2}(\ln|1 - \cos 2\theta| - \ln 2) = \ln\left(\frac{|1 - \cos 2\theta|}{2}\right)^{1/2} = \ln|\sin^2\theta|^{1/2} = \ln|\sin\theta|$

EXERCISE 9.4

1. $\frac{1}{2}(\cos 2\theta - \cos 6\theta)$ **3.** $\frac{1}{2}(\sin 6\theta + \sin 2\theta)$ **5.** $\frac{1}{2}(\cos 8\theta + \cos 2\theta)$ **7.** $\frac{1}{2}(\cos\theta - \cos 3\theta)$ **9.** $\frac{1}{2}(\sin 2\theta + \sin\theta)$ **11.** $2\sin\theta\cos 3\theta$
13. $2\cos 3\theta\cos\theta$ **15.** $2\sin 2\theta\cos\theta$ **17.** $2\sin\theta\sin(\theta/2)$ **19.** $\frac{\sin\theta + \sin 3\theta}{2\sin 2\theta} = \frac{2\sin 2\theta\cos(-\theta)}{2\sin 2\theta} = \cos(-\theta) = \cos\theta$

21. $\frac{\sin 4\theta + \sin 2\theta}{\cos 4\theta + \cos 2\theta} = \frac{2\sin 3\theta\cos\theta}{2\cos 3\theta\cos\theta} = \frac{\sin 3\theta}{\cos 3\theta} = \tan 3\theta$ **23.** $\frac{\cos\theta - \cos 3\theta}{\sin\theta + \sin 3\theta} = \frac{-2\sin 2\theta\sin(-\theta)}{2\sin 2\theta\cos(-\theta)} = \frac{\sin\theta}{\cos\theta} = \tan\theta$

25. $\sin\theta(\sin\theta + \sin 3\theta) = \sin\theta[2\sin 2\theta\cos(-\theta)] = 2\sin 2\theta\sin\theta\cos\theta = \cos\theta(2\sin 2\theta\sin\theta) = \cos\theta[2\cdot\frac{1}{2}(\cos\theta - \cos 3\theta)] =$
$\cos\theta(\cos\theta - \cos 3\theta)$

27. $\frac{\sin 4\theta + \sin 8\theta}{\cos 4\theta + \cos 8\theta} = \frac{2\sin 6\theta\cos(-2\theta)}{2\cos 6\theta\cos(-2\theta)} = \frac{\sin 6\theta}{\cos 6\theta} = \tan 6\theta$

29. $\frac{\sin 4\theta + \sin 8\theta}{\sin 4\theta - \sin 8\theta} = \frac{2\sin 6\theta\cos(-2\theta)}{2\sin(-2\theta)\cos 6\theta} = \frac{\sin 6\theta}{\cos 6\theta}\cdot\frac{\cos 2\theta}{-\sin 2\theta} = \tan 6\theta(-\cot 2\theta) = -\frac{\tan 6\theta}{\tan 2\theta}$

31. $$\frac{\sin\alpha + \sin\beta}{\sin\alpha - \sin\beta} = \frac{2\sin\frac{\alpha+\beta}{2}\cos\frac{\alpha-\beta}{2}}{2\sin\frac{\alpha-\beta}{2}\cos\frac{\alpha+\beta}{2}} = \frac{\sin\frac{\alpha+\beta}{2}}{\cos\frac{\alpha+\beta}{2}}\cdot\frac{\cos\frac{\alpha-\beta}{2}}{\sin\frac{\alpha-\beta}{2}} = \tan\frac{\alpha+\beta}{2}\cot\frac{\alpha-\beta}{2}$$

33. $$\frac{\sin\alpha + \sin\beta}{\cos\alpha + \cos\beta} = \frac{2\sin\frac{\alpha+\beta}{2}\cos\frac{\alpha-\beta}{2}}{2\cos\frac{\alpha+\beta}{2}\cos\frac{\alpha-\beta}{2}} = \frac{\sin\frac{\alpha+\beta}{2}}{\cos\frac{\alpha+\beta}{2}} = \tan\frac{\alpha+\beta}{2}$$

35. $1 + \cos 2\theta + \cos 4\theta + \cos 6\theta = (1 + \cos 6\theta) + (\cos 2\theta + \cos 4\theta) = 2\cos^2 3\theta + 2\cos 3\theta\cos(-\theta) = 2\cos 3\theta(\cos 3\theta + \cos\theta) =$
$2\cos 3\theta(2\cos 2\theta\cos\theta) = 4\cos\theta\cos 2\theta\cos 3\theta$

37. $\sin 2\alpha + \sin 2\beta + \sin 2\gamma = 2\sin(\alpha+\beta)\cos(\alpha-\beta) + \sin 2\gamma = 2\sin(\alpha+\beta)\cos(\alpha-\beta) + 2\sin\gamma\cos\gamma =$
$2\sin(\pi-\gamma)\cos(\alpha-\beta) + 2\sin\gamma\cos\gamma = 2\sin\gamma\cos(\alpha-\beta) + 2\sin\gamma\cos\gamma = 2\sin\gamma[\cos(\alpha-\beta) + \cos\gamma] =$
$2\sin\gamma\left(2\cos\frac{\alpha-\beta+\gamma}{2}\cos\frac{\alpha-\beta-\gamma}{2}\right) = 4\sin\gamma\cos\frac{\pi-2\beta}{2}\cos\frac{2\alpha-\pi}{2} = 4\sin\gamma\cos\left(\frac{\pi}{2}-\beta\right)\cos\left(\alpha-\frac{\pi}{2}\right) =$
$4\sin\gamma\sin\beta\sin\alpha$

39.
$$\sin(\alpha-\beta) = \sin\alpha\cos\beta - \cos\alpha\sin\beta$$
$$\sin(\alpha+\beta) = \sin\alpha\cos\beta + \cos\alpha\sin\beta$$
$$\sin(\alpha-\beta) + \sin(\alpha+\beta) = 2\sin\alpha\cos\beta$$
$$\sin\alpha\cos\beta = \frac{1}{2}[\sin(\alpha+\beta) + \sin(\alpha-\beta)]$$

41. $2\cos\frac{\alpha+\beta}{2}\cos\frac{\alpha-\beta}{2} = 2\cdot\frac{1}{2}\left[\cos\left(\frac{\alpha+\beta}{2} + \frac{\alpha-\beta}{2}\right) + \cos\left(\frac{\alpha+\beta}{2} - \frac{\alpha-\beta}{2}\right)\right] = \cos\frac{2\alpha}{2} + \cos\frac{2\beta}{2} = \cos\alpha + \cos\beta$

EXERCISE 9.5

1. $\theta = \pi/6, 5\pi/6$ **3.** $\theta = 5\pi/6, 11\pi/6$ **5.** $\theta = \pi/2, 3\pi/2$ **7.** $\theta = \pi/2, 7\pi/6, 11\pi/6$ **9.** $\theta = 3\pi/4, 7\pi/4$
11. $\theta = 4\pi/9, 8\pi/9, 16\pi/9$ **13.** $\theta = 0.4115168, \pi - 0.4115168$ **15.** $\theta = 1.3734008, \pi + 1.3734008$
17. $\theta = 2.6905658, 2\pi - 2.6905658$ **19.** $\theta = 1.8234766, 2\pi - 1.8234766$ **21.** $\theta = \pi/4, 5\pi/4$ **23.** $\theta = 0, \pi/3, \pi, 5\pi/3$
25. $\theta = \pi/2, 3\pi/2$ **27.** $\theta = 0, 2\pi/3, 4\pi/3$ **29.** $\theta = 0, \pi/3, 2\pi/3, \pi/2, 4\pi/3, \pi, 5\pi/3, 3\pi/2$
31. $\theta = 0, \pi/5, 2\pi/5, 3\pi/5, 4\pi/5, \pi, 6\pi/5, 7\pi/5, 8\pi/5, 9\pi/5$ **33.** $\theta = \pi/6, 5\pi/6, 3\pi/2$ **35.** $\theta = \pi/6, 5\pi/6, 3\pi/2$
37. $\theta = \pi/3, 5\pi/3$ **39.** No real solutions **41.** No real solutions **43.** $\theta = \pi/2, 7\pi/6$ **45.** $\theta = 0, \pi/3, \pi, 5\pi/3$ **47.** 28.9°

49. Yes; it varies from 1.27 to 1.34 **51.** 1.47

53. If θ is the original angle of incidence and ϕ is the angle of refraction, then $(\sin\theta)/(\sin\phi) = n_2$. The angle of incidence of the emerging beam is also ϕ, and the index of refraction is $1/n_2$. Thus, θ is the angle of refraction of the emerging beam.

FILL-IN-THE-BLANK ITEMS

1. identity; conditional **2.** $-$ **3.** $+$ **4.** $\sin^2\theta$; $2\cos^2\theta$; $2\sin^2\theta$ **5.** $1 - \cos\alpha$

REVIEW EXERCISES

1. $\tan\theta\cot\theta - \sin^2\theta = 1 - \sin^2\theta = \cos^2\theta$ **3.** $\cos^2\theta(1 + \tan^2\theta) = \cos^2\theta\sec^2\theta = 1$

5. $4\cos^2\theta + 3\sin^2\theta = \cos^2\theta + 3(\cos^2\theta + \sin^2\theta) = 3 + \cos^2\theta$

7. $\dfrac{1-\cos\theta}{\sin\theta} + \dfrac{\sin\theta}{1-\cos\theta} = \dfrac{(1-\cos\theta)^2 + \sin^2\theta}{\sin\theta(1-\cos\theta)} = \dfrac{1 - 2\cos\theta + \cos^2\theta + \sin^2\theta}{\sin\theta(1-\cos\theta)} = \dfrac{2(1-\cos\theta)}{\sin\theta(1-\cos\theta)} = 2\csc\theta$

9. $\dfrac{\cos\theta}{\cos\theta - \sin\theta} = \dfrac{\dfrac{\cos\theta}{\cos\theta}}{\dfrac{\cos\theta - \sin\theta}{\cos\theta}} = \dfrac{1}{1 - \dfrac{\sin\theta}{\cos\theta}} = \dfrac{1}{1 - \tan\theta}$

11. $\dfrac{\csc\theta}{1+\csc\theta} = \dfrac{\dfrac{1}{\sin\theta}}{1 + \dfrac{1}{\sin\theta}} = \dfrac{1}{1+\sin\theta} = \dfrac{1}{1+\sin\theta}\cdot\dfrac{1-\sin\theta}{1-\sin\theta} = \dfrac{1-\sin\theta}{1-\sin^2\theta} = \dfrac{1-\sin\theta}{\cos^2\theta}$

13. $\csc\theta - \sin\theta = \dfrac{1}{\sin\theta} - \sin\theta = \dfrac{1-\sin^2\theta}{\sin\theta} = \dfrac{\cos^2\theta}{\sin\theta} = \cos\theta\cdot\dfrac{\cos\theta}{\sin\theta} = \cos\theta\cot\theta$

15. $\dfrac{1-\sin\theta}{\sec\theta} = \cos\theta(1-\sin\theta)\cdot\dfrac{1+\sin\theta}{1+\sin\theta} = \dfrac{\cos\theta(1-\sin^2\theta)}{1+\sin\theta} = \dfrac{\cos^3\theta}{1+\sin\theta}$

17. $\cot\theta - \tan\theta = \dfrac{\cos\theta}{\sin\theta} - \dfrac{\sin\theta}{\cos\theta} = \dfrac{\cos^2\theta - \sin^2\theta}{\sin\theta\cos\theta} = \dfrac{1 - 2\sin^2\theta}{\sin\theta\cos\theta}$

19. $\dfrac{\cos(\alpha+\beta)}{\cos\alpha\sin\beta} = \dfrac{\cos\alpha\cos\beta - \sin\alpha\sin\beta}{\cos\alpha\sin\beta} = \dfrac{\cos\alpha\cos\beta}{\cos\alpha\sin\beta} - \dfrac{\sin\alpha\sin\beta}{\cos\alpha\sin\beta} = \cot\beta - \tan\alpha$

21. $\dfrac{\cos(\alpha-\beta)}{\cos\alpha\cos\beta} = \dfrac{\cos\alpha\cos\beta + \sin\alpha\sin\beta}{\cos\alpha\cos\beta} = \dfrac{\cos\alpha\cos\beta}{\cos\alpha\cos\beta} + \dfrac{\sin\alpha\sin\beta}{\cos\alpha\cos\beta} = 1 + \tan\alpha\tan\beta$

23. $(1+\cos\theta)\left(\tan\dfrac{\theta}{2}\right) = \left(2\cos^2\dfrac{\theta}{2}\right)\dfrac{\sin(\theta/2)}{\cos(\theta/2)} = 2\sin\dfrac{\theta}{2}\cos\dfrac{\theta}{2} = \sin\theta$

25. $2\cot\theta\cot 2\theta = 2\left(\dfrac{\cos\theta}{\sin\theta}\right)\left(\dfrac{\cos 2\theta}{\sin 2\theta}\right) = \dfrac{2\cos\theta(\cos^2\theta - \sin^2\theta)}{2\sin^2\theta\cos\theta} = \dfrac{\cos^2\theta - \sin^2\theta}{\sin^2\theta} = \cot^2\theta - 1$

27. $1 - 8\sin^2\theta\cos^2\theta = 1 - 2(2\sin\theta\cos\theta)^2 = 1 - 2\sin^2 2\theta = \cos 4\theta$

29. $\dfrac{\sin 2\theta + \sin 4\theta}{\cos 2\theta + \cos 4\theta} = \dfrac{2\sin 3\theta\cos(-\theta)}{2\cos 3\theta\cos(-\theta)} = \tan 3\theta$

31. $\dfrac{\cos 2\theta - \cos 4\theta}{\cos 2\theta + \cos 4\theta} - \tan\theta\tan 3\theta = \dfrac{-2\sin 3\theta\sin(-\theta)}{2\cos 3\theta\cos(-\theta)} - \tan\theta\tan 3\theta = \tan 3\theta\tan\theta - \tan\theta\tan 3\theta = 0$

33. $\frac{1}{4}(\sqrt{6} - \sqrt{2})$ **35.** $\frac{1}{4}(\sqrt{6} - \sqrt{2})$ **37.** $\frac{1}{2}$ **39.** $\sqrt{\dfrac{2-\sqrt{2}}{2+\sqrt{2}}} = \sqrt{2} - 1$

41. (a) $-\frac{33}{65}$ (b) $-\frac{56}{65}$ (c) $-\frac{63}{65}$ (d) $\frac{33}{56}$ (e) $\frac{24}{25}$ (f) $\frac{119}{169}$ (g) $5\sqrt{26}/26$ (h) $2\sqrt{5}/5$

43. (a) $-\frac{16}{65}$ (b) $-\frac{63}{65}$ (c) $-\frac{56}{65}$ (d) $\frac{16}{63}$ (e) $\frac{24}{25}$ (f) $\frac{119}{169}$ (g) $\sqrt{26}/26$ (h) $-\sqrt{10}/10$

45. (a) $-\frac{63}{65}$ (b) $\frac{16}{65}$ (c) $\frac{33}{65}$ (d) $-\frac{63}{16}$ (e) $\frac{24}{25}$ (f) $-\frac{119}{169}$ (g) $2\sqrt{13}/13$ (h) $-\sqrt{10}/10$

47. (a) $(-\sqrt{3} - 2\sqrt{2})/6$ (b) $(1 - 2\sqrt{6})/6$ (c) $(-\sqrt{3} + 2\sqrt{2})/6$ (d) $(-\sqrt{3} - 2\sqrt{2})/(1 - 2\sqrt{6}) = (8\sqrt{2} + 9\sqrt{3})/23$ (e) $-\sqrt{3}/2$ (f) $-\frac{7}{9}$ (g) $\sqrt{3}/3$ (h) $\sqrt{3}/2$

49. (a) 1 (b) 0 (c) $-\frac{1}{9}$ (d) Not defined (e) $4\sqrt{5}/9$ (f) $-\frac{1}{9}$ (g) $\sqrt{30}/6$ (h) $-\sqrt{6}\sqrt{3 - \sqrt{5}}/6$

51. $\theta = \pi/3, 5\pi/3$ **53.** $\theta = 3\pi/4, 5\pi/4$ **55.** $\theta = 3\pi/4, 7\pi/4$ **57.** $\theta = 0, \pi/2, \pi, 3\pi/2$ **59.** $\theta = 1.1197695, \pi - 1.1197695$

61. $\theta = 0, \pi$ **63.** $\theta = 0, 2\pi/3, \pi, 4\pi/3$ **65.** $\theta = 0, \pi/6, 5\pi/6$ **67.** $\theta = \pi/6, \pi/2, 5\pi/6$ **69.** $\theta = \pi/2, \pi$

CHAPTER 10 EXERCISE 10.1

1. $a = 3.2262$, $b = 3.5490$, $\alpha = 40°$ **3.** $a = 3.2501$, $c = 4.2265$, $\beta = 45°$ **5.** $\gamma = 95°$, $c = 9.8618$, $a = 6.3633$
7. $\alpha = 40°$, $a = 2$, $c = 3.0642$ **9.** $\gamma = 120°$, $b = 1.0642$, $c = 2.6946$ **11.** $\alpha = 100°$, $a = 5.2401$, $c = 0.9240$
13. $\beta = 40°$, $a = 5.6382$, $b = 3.8567$ **15.** $\gamma = 100°$, $a = 1.3054$, $b = 1.3054$ **17.** One triangle; $\beta = 30.7°$, $\gamma = 99.3°$, $c = 3.8647$
19. One triangle; $\gamma = 36.2°$, $\alpha = 43.8°$, $a = 3.5141$ **21.** No triangle
23. Two triangles; $\gamma_1 = 30.9°$, $\alpha_1 = 129.1°$, $a_1 = 9.0760$ or $\gamma_2 = 149.1°$, $\alpha_2 = 10.9°$, $a_2 = 2.2115$ **25.** No triangle
27. Two triangles; $\alpha_1 = 57.7°$, $\beta_1 = 97.3°$, $b_1 = 2.3470$ or $\alpha_2 = 122.3°$, $\beta_2 = 32.7°$, $b_2 = 1.2783$
29. (a) Station Able is 94.9 mi from the ship; Station Baker is 82.8 mi from the ship. (b) Approx. 25 min **31.** 149 ft
33. 381.7 ft **35.** (a) 169 mi (b) 161.3° **37.** 84.7°; 183.7 ft

39. $$\frac{a+b}{c} = \frac{a}{c} + \frac{b}{c} = \frac{\sin\alpha}{\sin\gamma} + \frac{\sin\beta}{\sin\gamma} = \frac{\sin\alpha + \sin\beta}{\sin\gamma} = \frac{2\sin\frac{\alpha+\beta}{2}\cos\frac{\alpha-\beta}{2}}{2\sin\frac{\gamma}{2}\cos\frac{\gamma}{2}} = \frac{\sin\left(\frac{\pi}{2} - \frac{\gamma}{2}\right)\cos\frac{\alpha-\beta}{2}}{\sin\frac{\gamma}{2}\cos\frac{\gamma}{2}} = \frac{\cos\frac{1}{2}(\alpha-\beta)}{\sin\frac{1}{2}\gamma}$$

41. $$a = \frac{b\sin\alpha}{\sin\beta} = \frac{b\sin[180° - (\beta+\gamma)]}{\sin\beta} = \frac{b}{\sin\beta}(\sin\beta\cos\gamma + \cos\beta\sin\gamma) = b\cos\gamma + \frac{b\sin\gamma}{\sin\beta}\cos\beta = b\cos\gamma + c\cos\beta$$

43. $\sin\beta = \sin(\text{Angle } AB'C) = b/2r$; $\dfrac{\sin\beta}{b} = \dfrac{1}{2r}$; the result follows using the Law of Sines

EXERCISE 10.2

1. $b = 2.9473$, $\alpha = 28.7°$, $\gamma = 106.3°$ **3.** $c = 3.7478$, $\alpha = 32.1°$, $\beta = 52.9°$ **5.** $\alpha = 48.5°$, $\beta = 38.6°$, $\gamma = 92.9°$
7. $\alpha = 127.2°$, $\beta = 32.1°$, $\gamma = 20.7°$ **9.** $c = 2.5720$, $\alpha = 48.6°$, $\beta = 91.4°$ **11.** $a = 2.9930$, $\beta = 19.2°$, $\gamma = 80.8°$
13. $b = 4.1357$, $\alpha = 43.0°$, $\gamma = 27.0°$ **15.** $c = 1.6905$, $\alpha = 65.0°$, $\beta = 65.0°$ **17.** $\alpha = 67.4°$, $\beta = 90°$, $\gamma = 22.6°$
19. $\alpha = 60°$, $\beta = 60°$, $\gamma = 60°$ **21.** $\alpha = 33.6°$, $\beta = 62.2°$, $\gamma = 84.3°$ **23.** $\alpha = 97.9°$, $\beta = 52.4°$, $\gamma = 29.7°$ **25.** 70.7521 ft
27. (a) 12° (b) 220.8 mph **29.** (a) 63.7 ft (b) 66.8 ft (c) 92.5° **31.** (a) 492.6 ft (b) 269.3 ft **33.** 342.3 ft

35. $$\cos\frac{\gamma}{2} = \sqrt{\frac{1+\cos\gamma}{2}} = \sqrt{\frac{1 + \frac{a^2+b^2-c^2}{2ab}}{2}} = \sqrt{\frac{2ab + a^2 + b^2 - c^2}{4ab}} = \sqrt{\frac{(a+b)^2 - c^2}{4ab}} =$$
$$\sqrt{\frac{(a+b+c)(a+b-c)}{4ab}} = \sqrt{\frac{2s(2s-2c)}{4ab}} = \sqrt{\frac{s(s-c)}{ab}}$$

37. $$\frac{\cos\alpha}{a} + \frac{\cos\beta}{b} + \frac{\cos\gamma}{c} = \frac{b^2+c^2-a^2}{2abc} + \frac{a^2+c^2-b^2}{2abc} + \frac{a^2+b^2-c^2}{2abc} = \frac{b^2+c^2-a^2+a^2+c^2-b^2+a^2+b^2-c^2}{2abc} = \frac{a^2+b^2+c^2}{2abc}$$

EXERCISE 10.3

1. 2.8284 **3.** 2.9886 **5.** 14.9812 **7.** 9.5623 **9.** 3.8567 **11.** 1.4772 **13.** 2.8191 **15.** 1.5321 **17.** 30 **19.** 1.7321
21. 19.8997 **23.** 19.8100 **25.** $5446.38 **27.** $A = \frac{1}{2}ab\sin\gamma = \frac{1}{2}a\sin\gamma\left(\frac{a\sin\beta}{\sin\alpha}\right) = \frac{a^2\sin\beta\sin\gamma}{2\sin\alpha}$ **29.** 0.9216 **31.** 2.2748
33. 5.4362 **35.** 0.8391

EXERCISE 10.4

1.
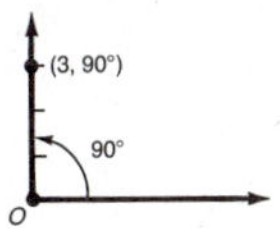

3.

5.
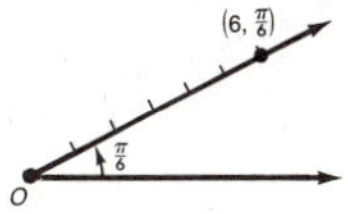

7. 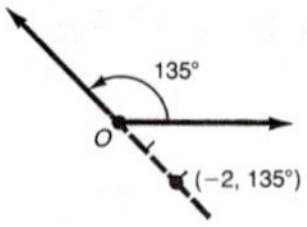

9. 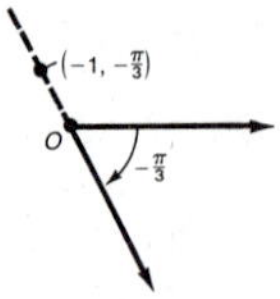

11. 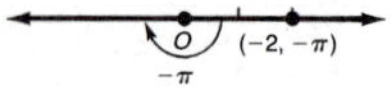

13. 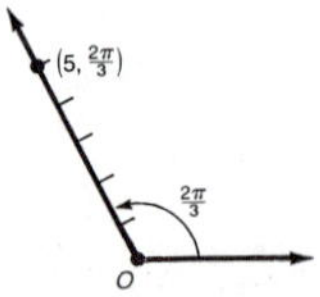

(a) $(5, -4\pi/3)$
(b) $(-5, 5\pi/3)$
(c) $(5, 8\pi/3)$

15. 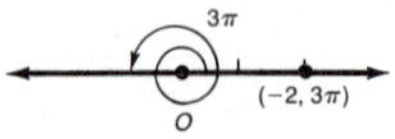

(a) $(2, -2\pi)$
(b) $(-2, \pi)$
(c) $(2, 2\pi)$

17. 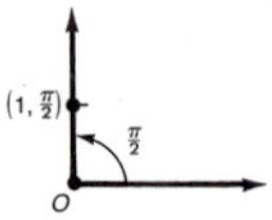

(a) $(1, -3\pi/2)$
(b) $(-1, 3\pi/2)$
(c) $(1, 5\pi/2)$

19. 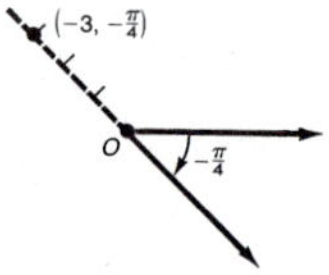

(a) $(3, -5\pi/4)$
(b) $(-3, 7\pi/4)$
(c) $(3, 11\pi/4)$

21. $(0, 3)$ **23.** $(-2, 0)$ **25.** $(-3\sqrt{3}, 3)$ **27.** $(\sqrt{2}, -\sqrt{2})$ **29.** $(-\frac{1}{2}, \sqrt{3}/2)$ **31.** $(2, 0)$ **33.** $(3, 0)$ **35.** $(-1, 0)$ **37.** $(\sqrt{2}, -\pi/4)$
39. $(2, \pi/6)$ **41.** $r^2 = \frac{3}{2}$ **43.** $r^2 \cos^2\theta - 4r \sin\theta = 0$ **45.** $r^2 \sin 2\theta = 1$ **47.** $r \cos\theta = 4$
49. $x^2 + y^2 - x = 0$ or $\left(x - \frac{1}{2}\right)^2 + y^2 = \frac{1}{4}$ **51.** $(x^2 + y^2)^{3/2} - x = 0$ **53.** $x^2 + y^2 = 4$ **55.** $y^2 = 8(x + 2)$
57. $d = \sqrt{(r_2 \cos\theta_2 - r_1 \cos\theta_1)^2 + (r_2 \sin\theta_2 - r_1 \sin\theta_1)^2} = \sqrt{r_1^2 + r_2^2 - 2r_1r_2 \cos(\theta_2 - \theta_1)}$

EXERCISE 10.5

1. Circle, radius 4, center at pole

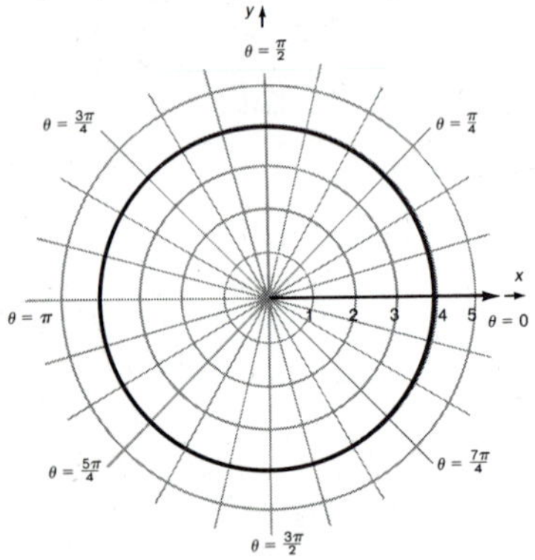

3. Line through pole, making an angle of $\pi/3$ with polar axis

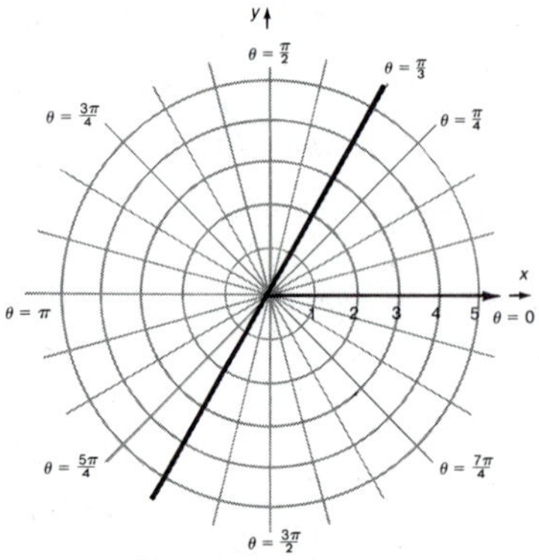

5. Horizontal line 4 units above the pole

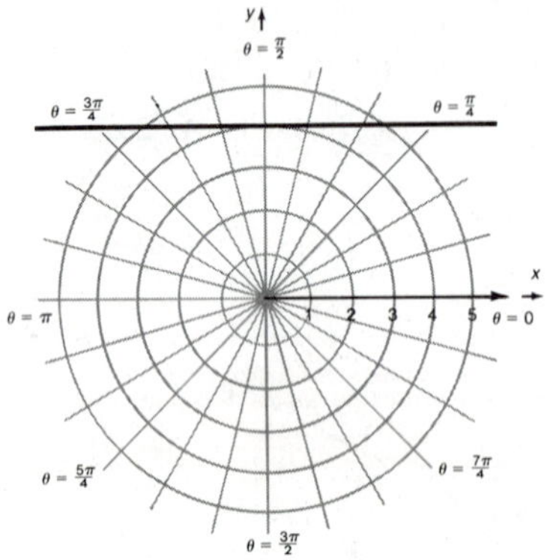

7. Vertical line 2 units to the left of the pole

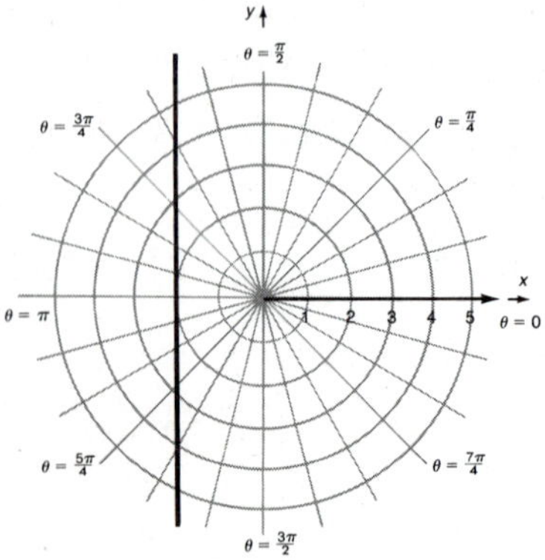

9. Circle, radius 1, center at (1, 0) in rectangular coordinates

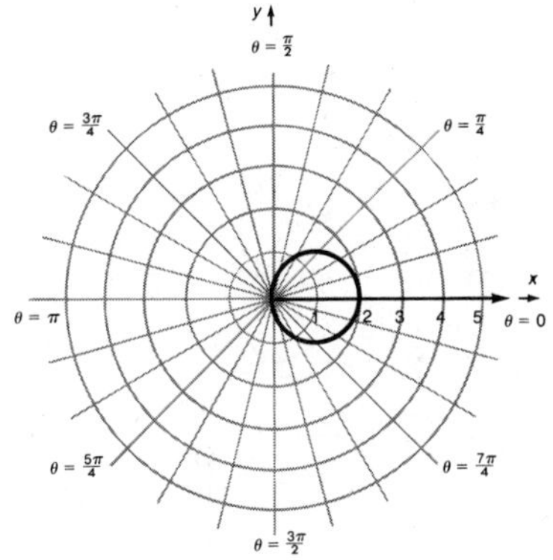

11. Circle, radius 2, center at (0, −2) in rectangular coordinates

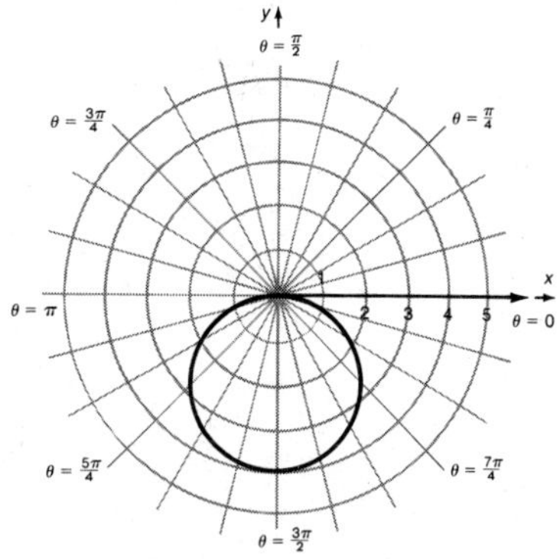

13. Circle, radius 2, center at (2, 0) in rectangular coordinates

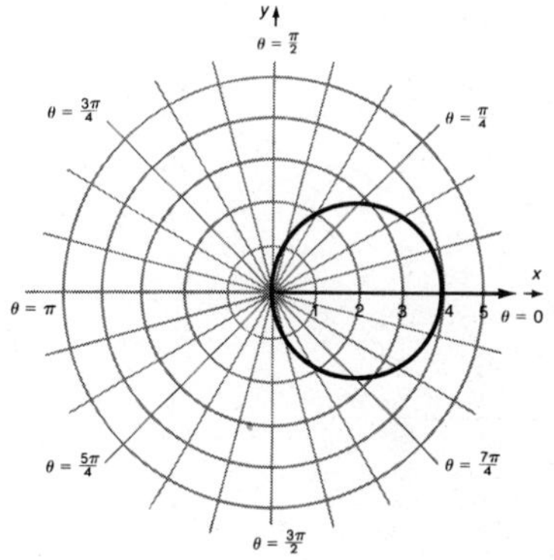

15. Circle, radius 1, center at (0, −1) in rectangular coordinates

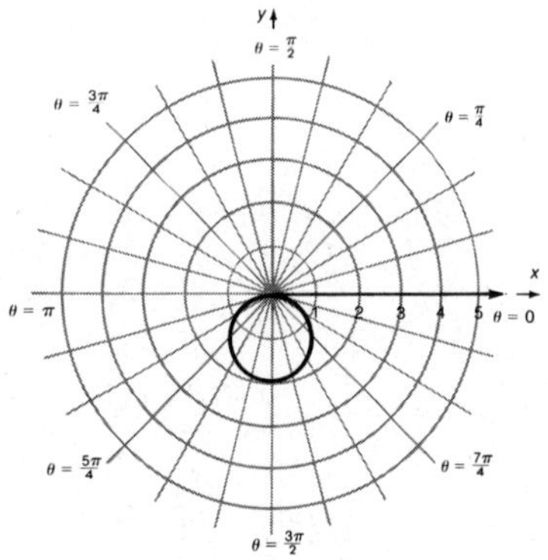

17.

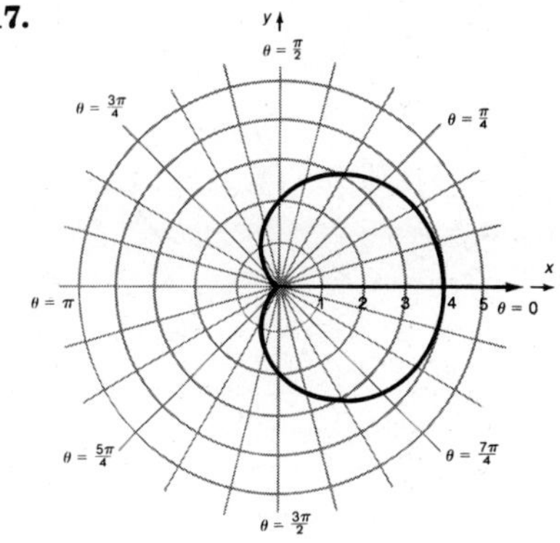

19.

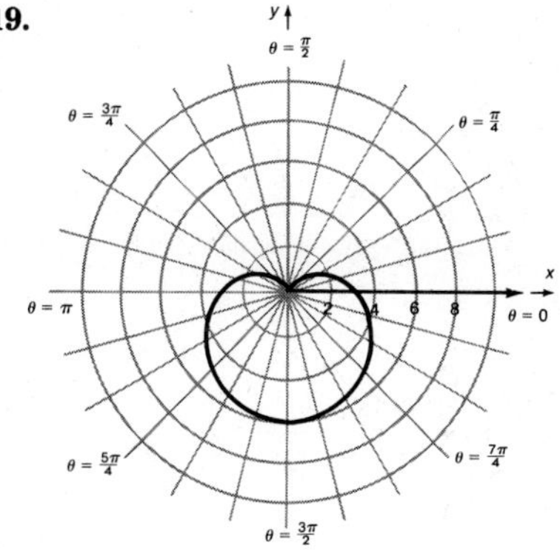

21.

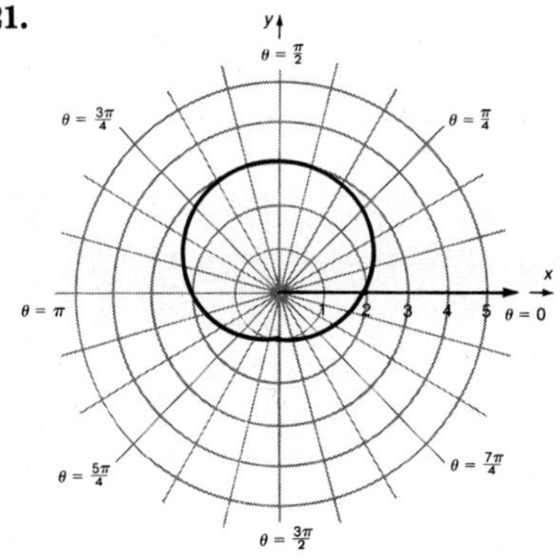

23.

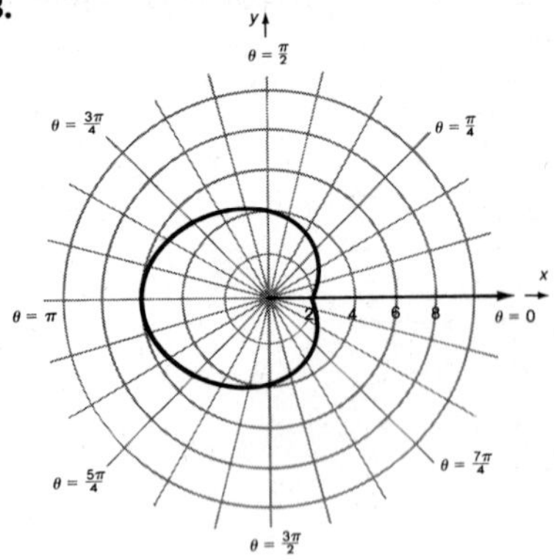

25.

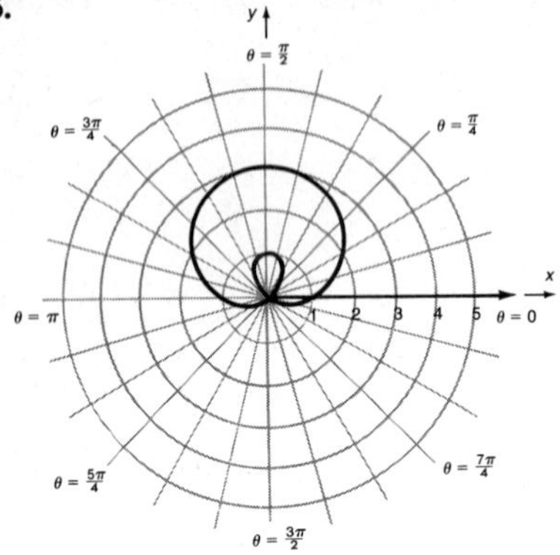

27.

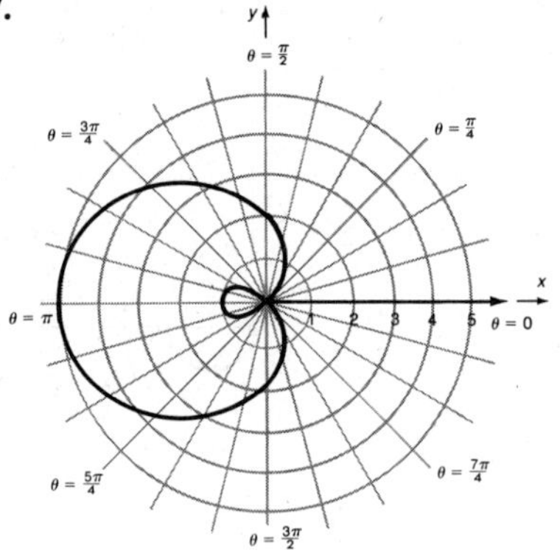

29.

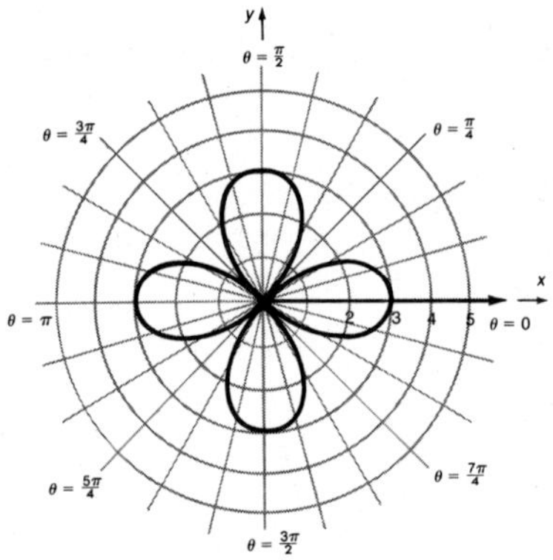

31.

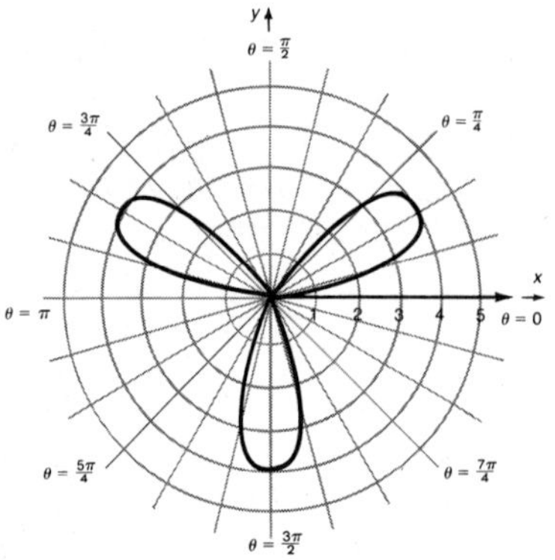

33.

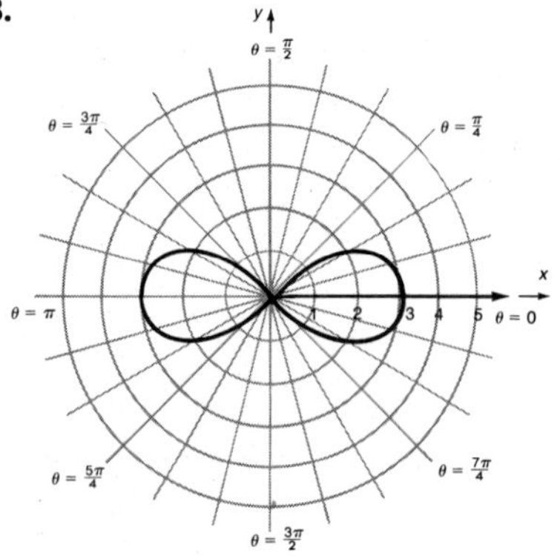

35.

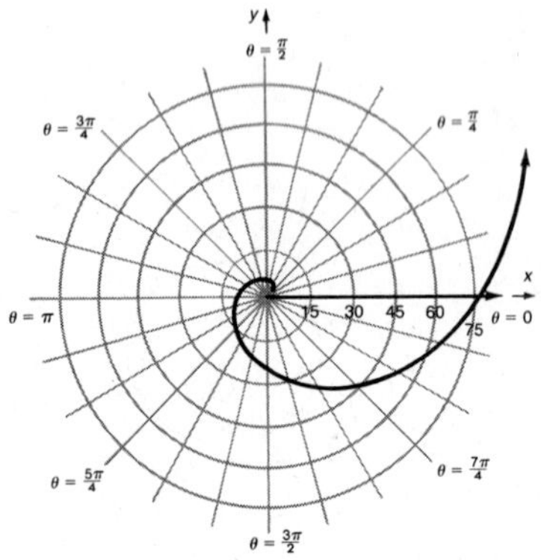

37.

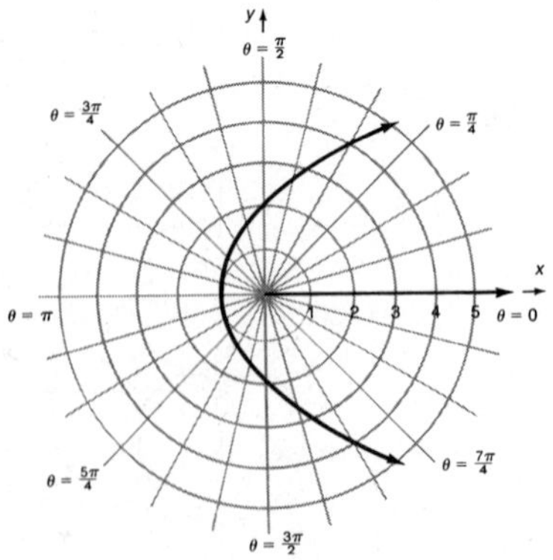

39.

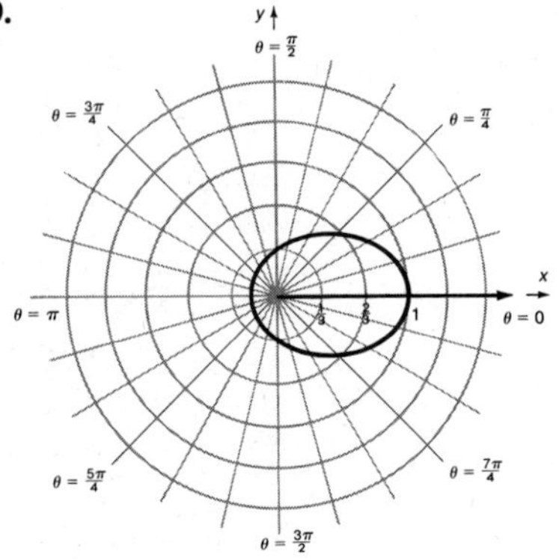

41.

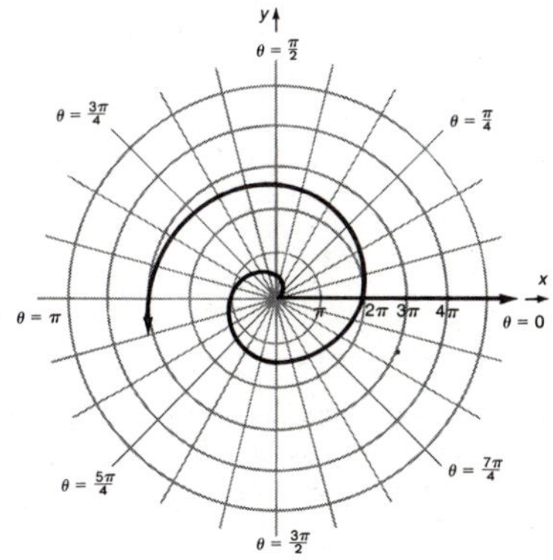

43.

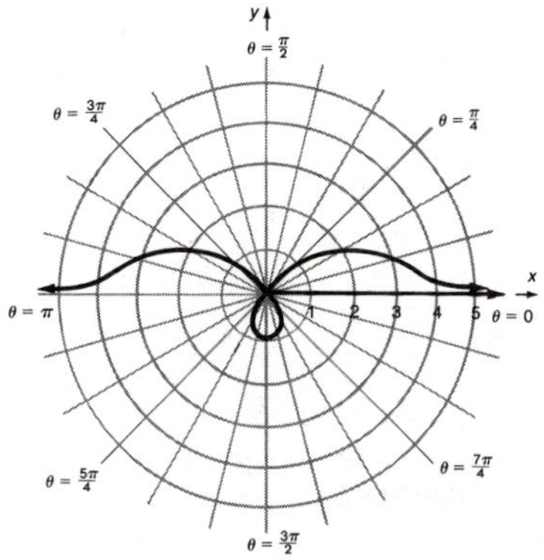

45.

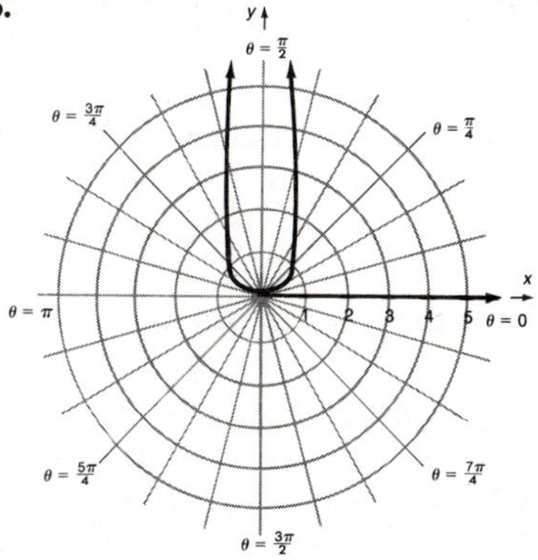

47. $r \sin \theta = a$
$y = a$

49.
$$r = 2a \sin \theta$$
$$r^2 = 2ar \sin \theta$$
$$x^2 + y^2 = 2ay$$
$$x^2 + y^2 - 2ay = 0$$
$$x^2 + (y - a)^2 = a^2$$
Circle, radius a, center at $(0, a)$ in rectangular coordinates

51.
$$r = 2a \cos \theta$$
$$r^2 = 2ar \cos \theta$$
$$x^2 + y^2 = 2ax$$
$$x^2 - 2ax + y^2 = 0$$
$$(x - a)^2 + y^2 = a^2$$
Circle, radius a, center at $(a, 0)$ in rectangular coordinates

53. (a) $r^2 = \cos \theta$: $r^2 = \cos(\pi - \theta)$, $r^2 = -\cos \theta$. Not equivalent; test fails. $(-r)^2 = \cos(-\theta)$, $r^2 = \cos \theta$. New test works

(b) $r^2 = \sin \theta$: $r^2 = \sin(\pi - \theta)$, $r^2 = \sin \theta$. Test works. $(-r)^2 = \sin(-\theta)$, $r^2 = -\sin \theta$. New test fails

EXERCISE 10.6

1. $\sqrt{2}(\cos 45° + i \sin 45°)$ **3.** $2(\cos 330° + i \sin 330°)$ **5.** $3(\cos 270° + i \sin 270°)$ **7.** $4\sqrt{2}(\cos 315° + i \sin 315°)$
9. $5(\cos 306.9° + i \sin 306.9°)$ **11.** $\sqrt{13}(\cos 123.7° + i \sin 123.7°)$ **13.** $-1 + \sqrt{3}\,i$ **15.** $2\sqrt{2} - 2\sqrt{2}\,i$ **17.** $-3i$
19. $-0.035 + 0.197i$ **21.** $1.97 + 0.347i$ **23.** $zw = 8(\cos 60° + i \sin 60°)$; $z/w = \frac{1}{2}(\cos 20° + i \sin 20°)$
25. $zw = 12(\cos 40° + i \sin 40°)$; $z/w = \frac{3}{4}(\cos 220° + i \sin 220°)$ **27.** $zw = 4\left(\cos \frac{9\pi}{40} + i \sin \frac{9\pi}{40}\right)$; $\frac{z}{w} = \cos \frac{\pi}{40} + i \sin \frac{\pi}{40}$
29. $zw = 4\sqrt{2}(\cos 15° + i \sin 15°)$; $z/w = \sqrt{2}(\cos 75° + i \sin 75°)$ **31.** $32(-1 + \sqrt{3}\,i)$ **33.** $32i$ **35.** $\frac{27}{2}(1 + \sqrt{3}\,i)$
37. $\frac{25\sqrt{2}}{2}(-1 + i)$ **39.** $4(-1 - i)$ **41.** $-23 + 14.15i$
43. $\sqrt[6]{2}(\cos 15° + i \sin 15°)$, $\sqrt[6]{2}(\cos 135° + i \sin 135°)$, $\sqrt[6]{2}(\cos 255° + i \sin 255°)$
45. $\sqrt[4]{8}(\cos 75° + i \sin 75°)$, $\sqrt[4]{8}(\cos 165° + i \sin 165°)$, $\sqrt[4]{8}(\cos 255° + i \sin 255°)$, $\sqrt[4]{8}(\cos 345° + i \sin 345°)$
47. $2(\cos 67.5° + i \sin 67.5°)$, $2(\cos 157.5° + i \sin 157.5°)$, $2(\cos 247.5° + i \sin 247.5°)$, $2(\cos 337.5° + i \sin 337.5°)$
49. $\cos 18° + i \sin 18°$, $\cos 90° + i \sin 90°$, $\cos 162° + i \sin 162°$, $\cos 234° + i \sin 234°$, $\cos 306° + i \sin 306°$
51. $1, i, -1, -i$

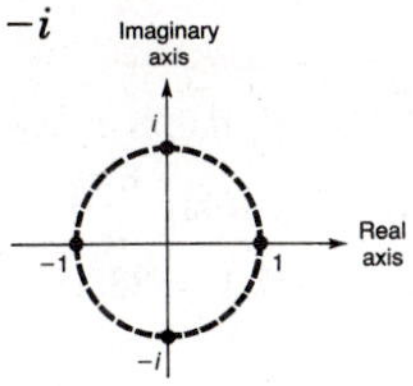

53. Look at formula (8); $|z_k| = \sqrt[n]{r}$ for all k.
55. Look at formula (8). The z_k are spaced apart by an angle of $2\pi/n$.

FILL-IN-THE-BLANK ITEMS

1. Sines **2.** Cosines **3.** Heron's **4.** pole; polar axis **5.** -2 **6.** $r = 2\cos\theta$ **7.** the polar axis (x-axis)
8. magnitude or modulus; argument

REVIEW EXERCISES

1. $\gamma = 100°$, $b = 0.6527$, $c = 1.2856$ **3.** $\beta = 56.8°$, $\gamma = 23.2°$, $b = 4.2484$ **5.** No triangle **7.** $b = 3.3229$, $\alpha = 62.8°$, $\gamma = 17.2°$
9. No triangle **11.** $c = 2.3246$, $\alpha = 16.1°$, $\beta = 123.9°$ **13.** $\beta = 36.2°$, $\gamma = 63.8°$, $c = 4.5555$ **15.** $\alpha = 39.6°$, $\beta = 18.5°$, $\gamma = 121.9°$
17. Two triangles: $\beta_1 = 13.4°$, $\gamma_1 = 156.6°$, $c_1 = 6.8576$; $\beta_2 = 166.6°$, $\gamma_2 = 3.4°$, $c_2 = 1.0209$ **19.** $a = 5.2268$, $\beta = 46°$, $\gamma = 64°$
21. 1.93 **23.** 18.79 **25.** 6 **27.** 3.80 **29.** 0.32

31. $(3\sqrt{3}/2, \frac{3}{2})$

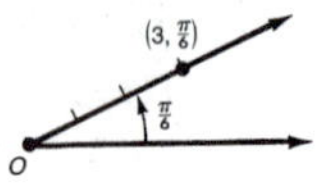

33. $(1, \sqrt{3})$

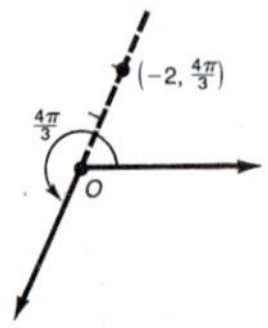

35. $(0, 3)$

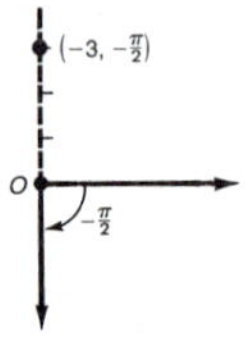

37. $(3\sqrt{2}, 3\pi/4)$, $(-3\sqrt{2}, -\pi/4)$ **39.** $(2, -\pi/2)$, $(-2, \pi/2)$ **41.** $(5, 0.93)$, $(-5, 4.07)$ **43.** $3r^2 - 6r\sin\theta = 0$
45. $r^2(2 - 3\sin^2\theta) - \tan\theta = 0$ **47.** $r^3\cos\theta = 4$ **49.** $x^2 + y^2 - 2y = 0$ **51.** $x^2 + y^2 = 25$ **53.** $x + 3y = 6$

55. Circle; radius 2, center at (2, 0) in rectangular coordinates

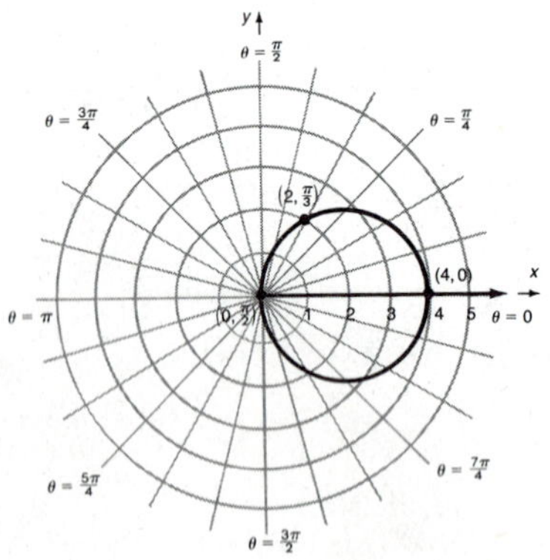

57.

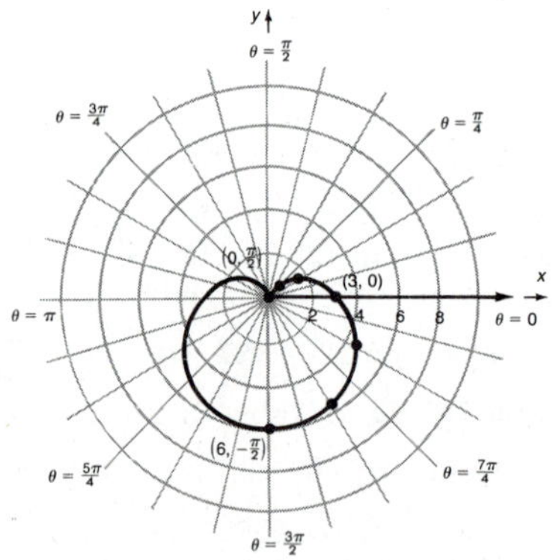

59.

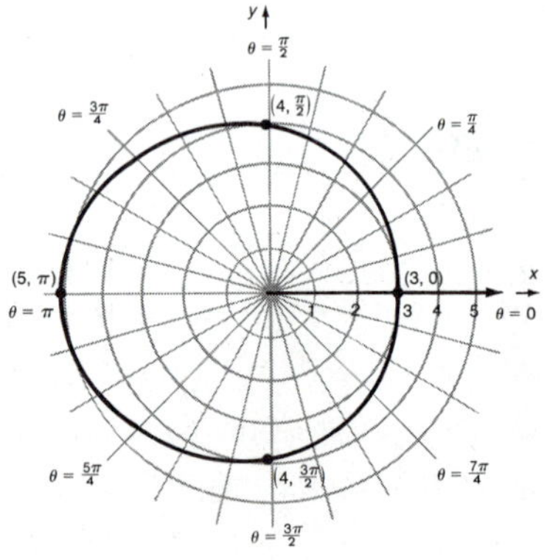

61. $\sqrt{2}(\cos 225° + i\sin 225°)$ **63.** $5(\cos 323.1° + i\sin 323.1°)$ **65.** $-\sqrt{3} + i$ **67.** $-\frac{3}{2} + (3\sqrt{3}/2)i$ **69.** $0.098 - 0.017i$
71. $zw = \cos 130° + i\sin 130°$; $z/w = \cos 30° + i\sin 30°$ **73.** $zw = 6(\cos 0 + i\sin 0)$; $\frac{z}{w} = \frac{3}{2}\left(\cos\frac{8\pi}{5} + i\sin\frac{8\pi}{5}\right)$
75. $zw = 5(\cos 5° + i\sin 5°)$; $z/w = 5(\cos 15° + i\sin 15°)$ **77.** $\frac{27}{2}(1 + \sqrt{3}i)$ **79.** $4i$ **81.** 64 **83.** $-527.1 - 335.8i$
85. 3, $3(\cos 120° + i\sin 120°)$, $3(\cos 240° + i\sin 240°)$ **87.** 204.1 mi **89.** $222,980

CHAPTER 11 EXERCISE 11.2

1. $y^2 = 8x$

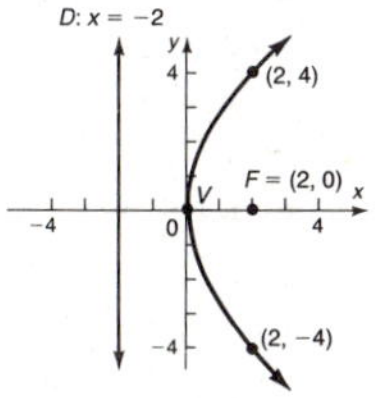

3. $x^2 = -12y$

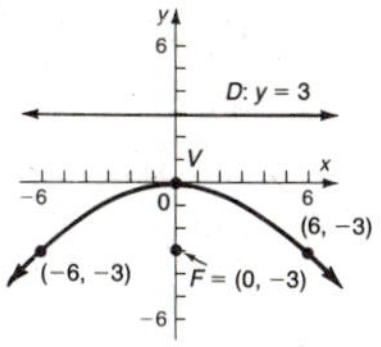

5. $y^2 = -8x$

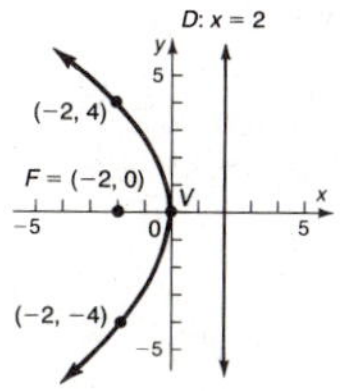

7. $x^2 = 2y$

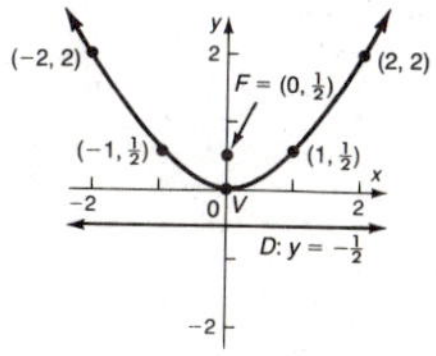

9. $(x - 2)^2 = -8(y + 3)$

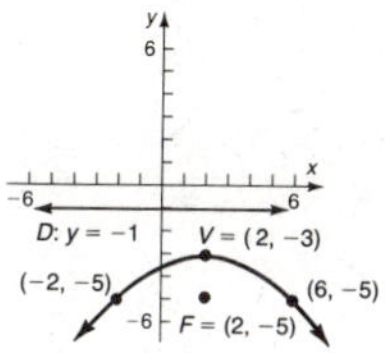

11. $x^2 = \frac{4}{3}y$

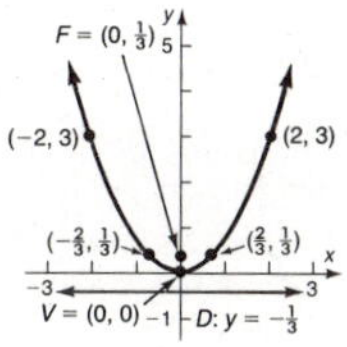

13. $(x + 3)^2 = 4(y - 3)$

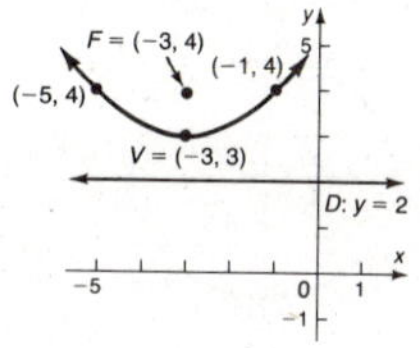

15. $(y + 2)^2 = -8(x + 1)$

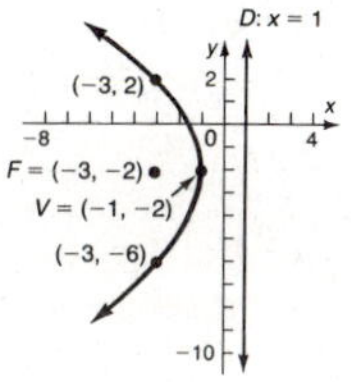

17.

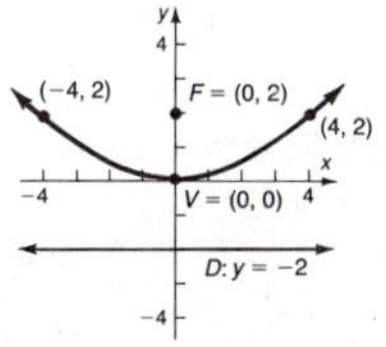

19.

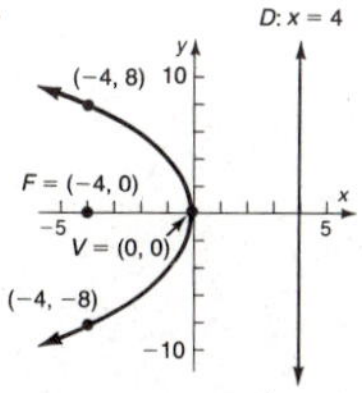

21.

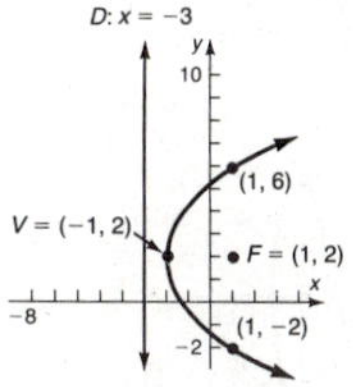

23.

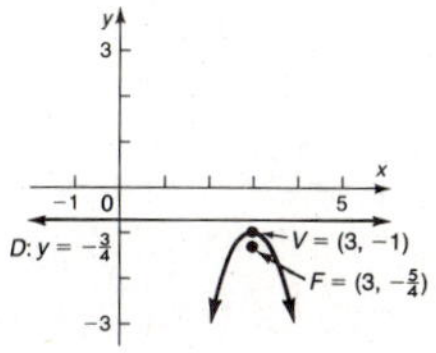

25.

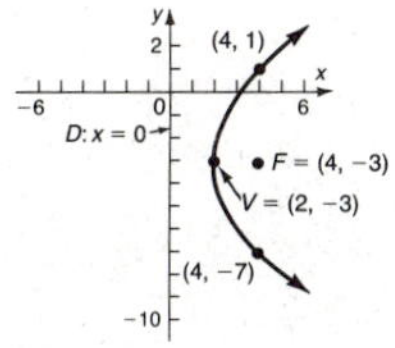

27.

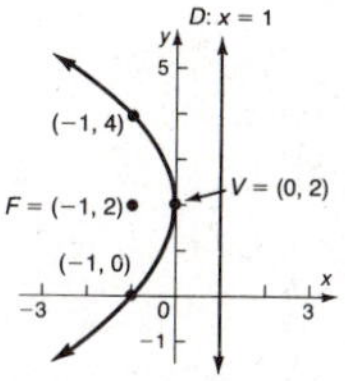

29.

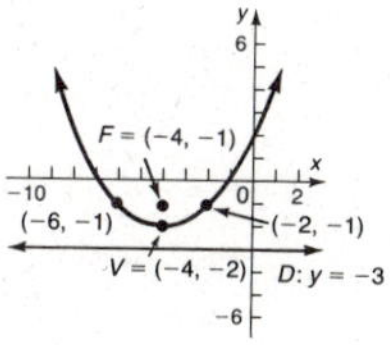

31.

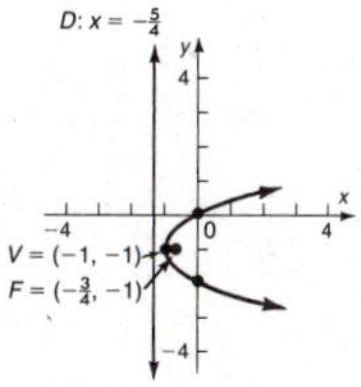

33.

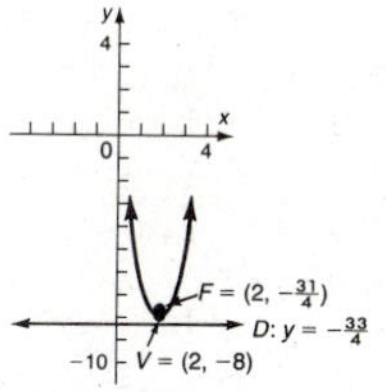

35. 2.083 ft from the vertex **37.** $4\sqrt{2}$ in.

39. $Ax^2 + Ey = 0$

$$x^2 = -\frac{E}{A}y$$

This is the equation of a parabola with vertex at (0, 0) and axis of symmetry the y-axis. The focus is at $(0, -E/4A)$; the directrix is the line $y = E/4A$. The parabola opens up if $-E/A > 0$ and down if $-E/A < 0$.

41. $Ax^2 + Dx + Ey + F = 0$

$$Ax^2 + Dx = -Ey - F$$

$$x^2 + \frac{D}{A}x = -\frac{E}{A}y - \frac{F}{A}$$

$$\left(x + \frac{D}{2A}\right)^2 = -\frac{E}{A}y - \frac{F}{A} + \frac{D^2}{4A^2}$$

$$\left(x + \frac{D}{2A}\right)^2 = -\frac{E}{A}y + \frac{D^2 - 4AF}{4A^2}$$

(a) If $E \neq 0$, then the equation may be written as

$$\left(x + \frac{D}{2A}\right)^2 = -\frac{E}{A}\left(y - \frac{D^2 - 4AF}{4AE}\right)$$

This is the equation of a parabola with vertex at $(-D/2A, (D^2 - 4AF)/4AE)$ and axis of symmetry parallel to the y-axis.

(b)–(d) If $E = 0$, the graph of the equation contains no points if $D^2 - 4AF < 0$, is a single vertical line if $D^2 - 4AF = 0$, and is two vertical lines if $D^2 - 4AF > 0$.

EXERCISE 11.3

1.

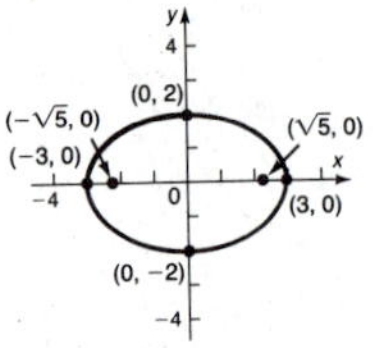

3.

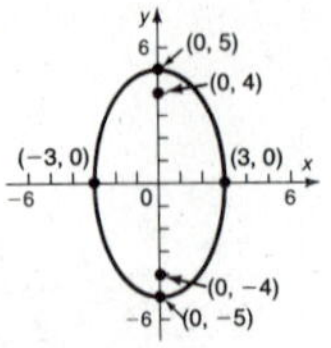

5. $\dfrac{x^2}{4} + \dfrac{y^2}{16} = 1$

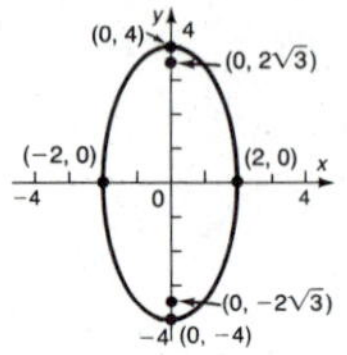

7. $\dfrac{x^2}{8} + \dfrac{y^2}{2} = 1$

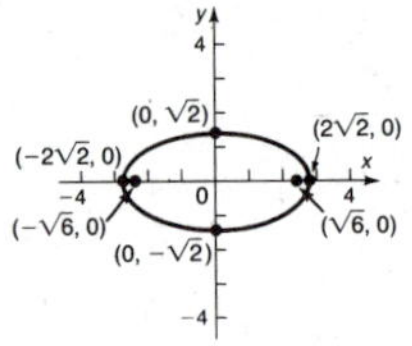

9.

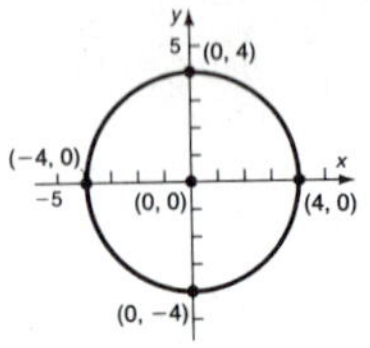

11. $\dfrac{x^2}{36} + \dfrac{y^2}{27} = 1$

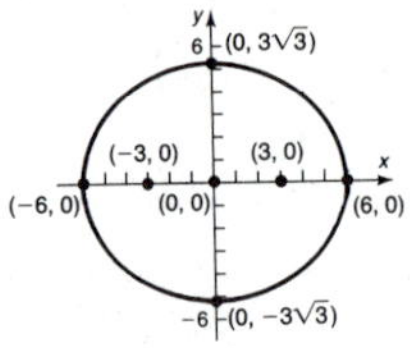

13. $\dfrac{x^2}{9} + \dfrac{y^2}{25} = 1$

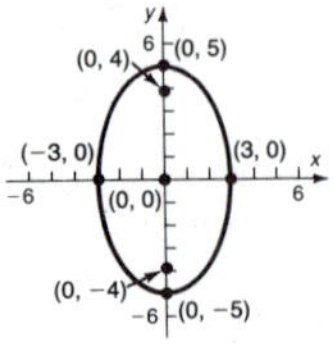

15. $\dfrac{x^2}{9} + \dfrac{y^2}{5} = 1$

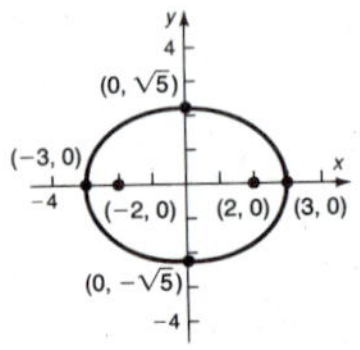

17. $\dfrac{x^2}{4} + \dfrac{y^2}{13} = 1$

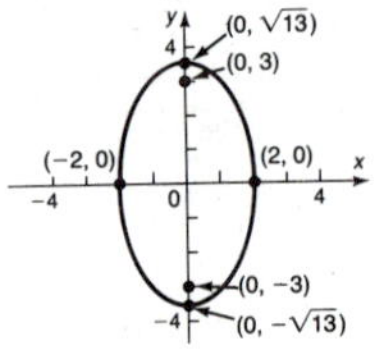

19. $x^2 + \dfrac{y^2}{16} = 1$

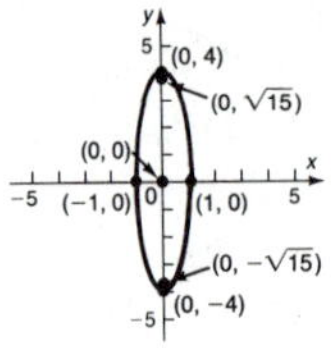

21.

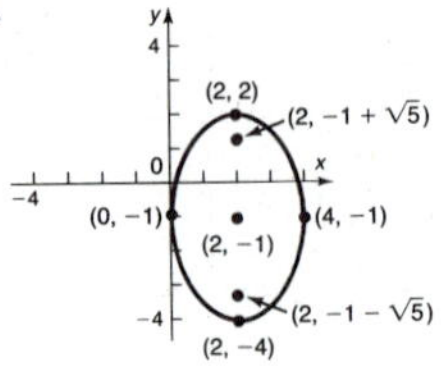

23. $\dfrac{(x+5)^2}{16} + \dfrac{(y-4)^2}{4} = 1$

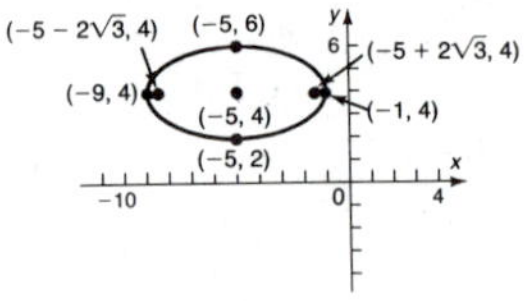

25. $\dfrac{(x+2)^2}{4} + (y-1)^2 = 1$

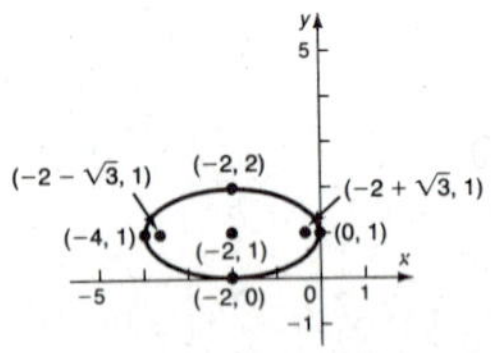

27. $\dfrac{(x-2)^2}{3} + \dfrac{(y+1)^2}{2} = 1$

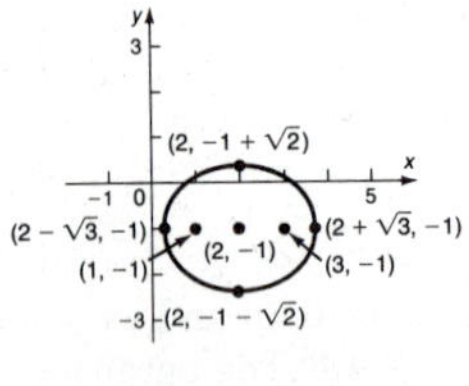

29. $\dfrac{(x-1)^2}{4} + \dfrac{(y+2)^2}{9} = 1$

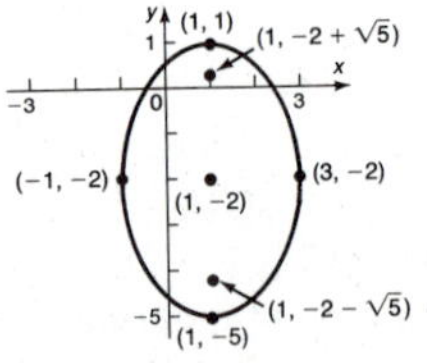

31. $x^2 + \dfrac{(y+2)^2}{4} = 1$

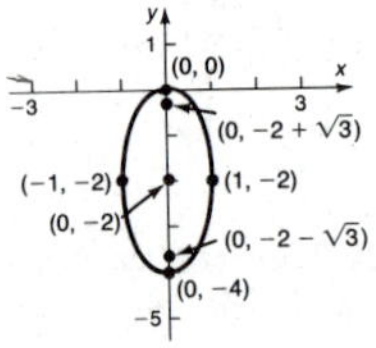

33. $\dfrac{(x-2)^2}{9} + \dfrac{(y+2)^2}{5} = 1$

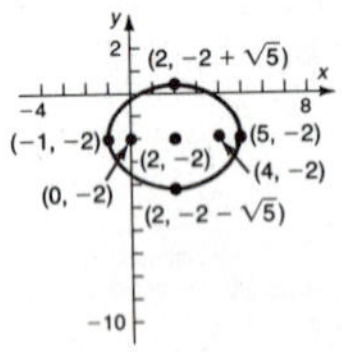

35. $\dfrac{(x-4)^2}{5} + \dfrac{(y-6)^2}{9} = 1$

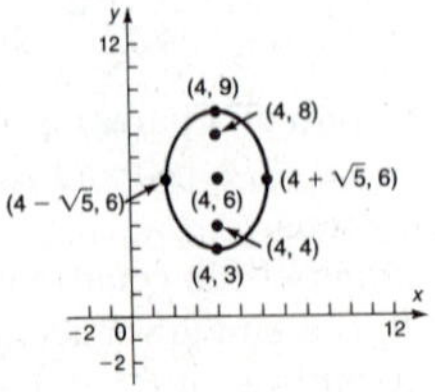

37. $\dfrac{(x-2)^2}{16} + \dfrac{(y-1)^2}{7} = 1$

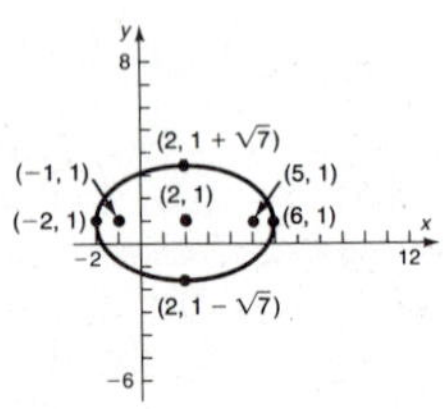

39. $\frac{(x-1)^2}{10} + (y-2)^2 = 1$

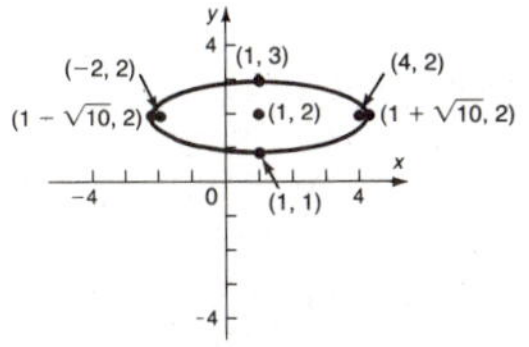

41. $\frac{(x-1)^2}{9} + (y-2)^2 = 1$

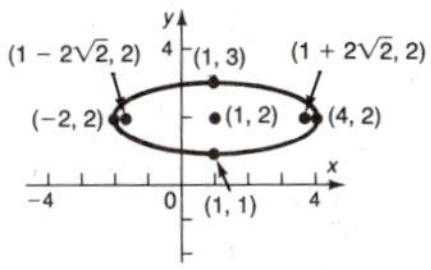

43. $\frac{x^2}{100} + \frac{y^2}{36} = 1$ **45.** 0, ≈12.99, 15, ≈12.99, 0

47. (a) The ellipse is close to a circle. (b) The ellipse is oval.
(c) The ellipse is elongated, with the length of the minor axis small in comparison to the length of the major axis.

49. (a) $Ax^2 + Cy^2 + F = 0$
$Ax^2 + Cy^2 = -F$
If A and C are of the same sign and F is of opposite sign, then the equation takes the form $x^2/(-F/A) + y^2/(-F/C) = 1$, where $-F/A$ and $-F/C$ are positive. This is the equation of an ellipse with center at (0, 0).
(b) If $A = C$, the equation may be written as $x^2 + y^2 = -F/A$. This is the equation of a circle with center at (0, 0) and radius equal to $\sqrt{-F/A}$.

EXERCISE 11.4

1. $x^2 - \frac{y^2}{15} = 1$

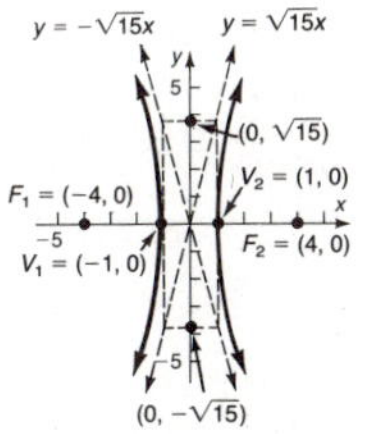

3. $\frac{y^2}{16} - \frac{x^2}{20} = 1$

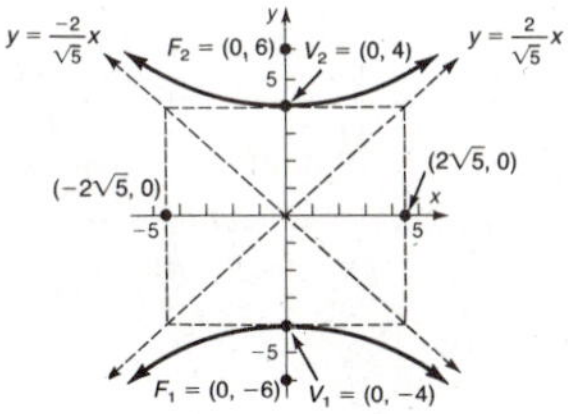

5. $\frac{x^2}{9} - \frac{y^2}{16} = 1$

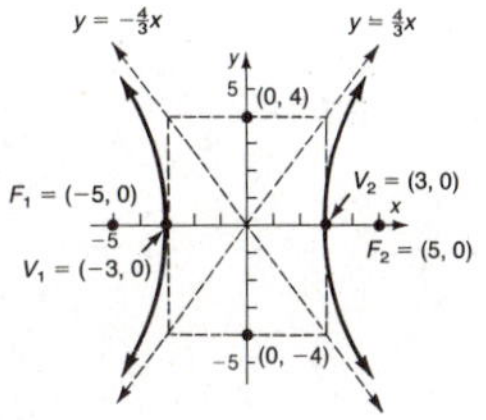

7. $\frac{y^2}{36} - \frac{x^2}{9} = 1$

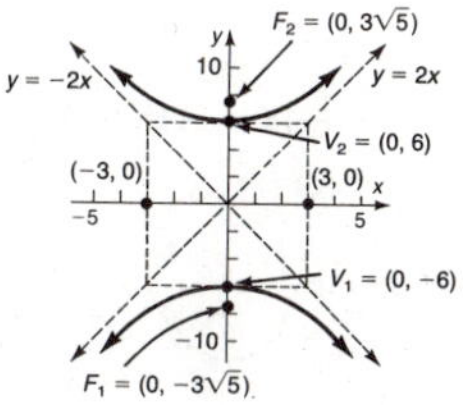

9. $\frac{x^2}{8} - \frac{y^2}{8} = 1$

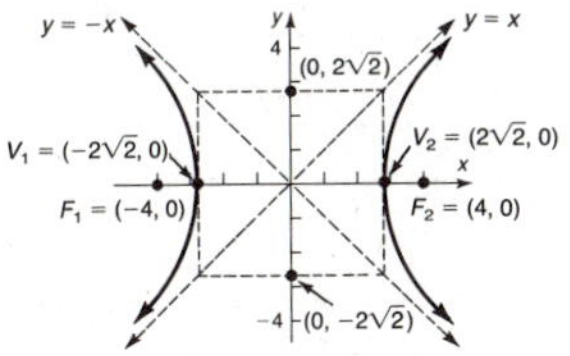

11.

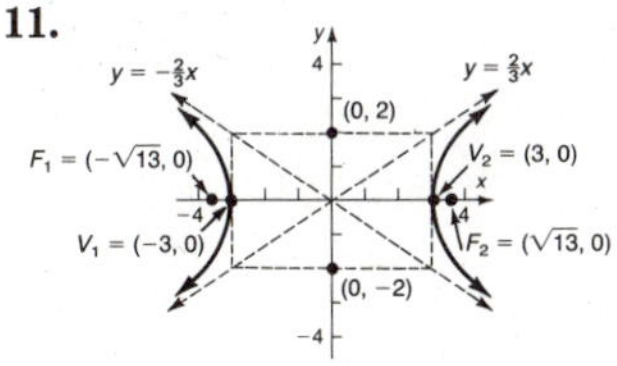

13. $\frac{x^2}{4} - \frac{y^2}{16} = 1$

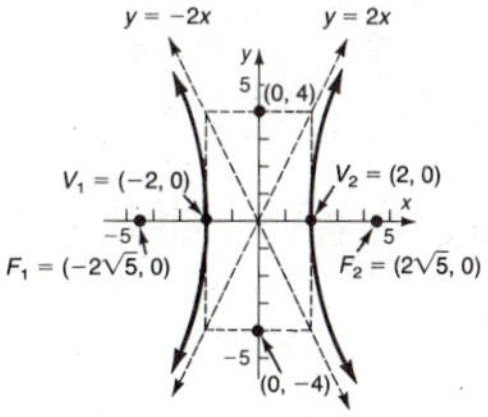

15. $\frac{y^2}{9} - x^2 = 1$

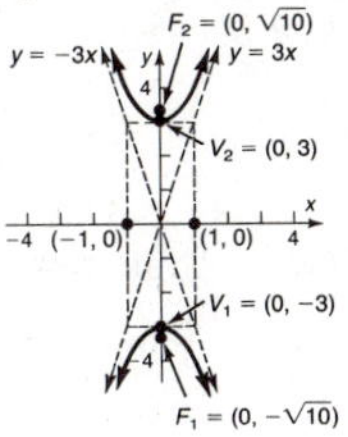

17. $\frac{y^2}{25} - \frac{x^2}{25} = 1$

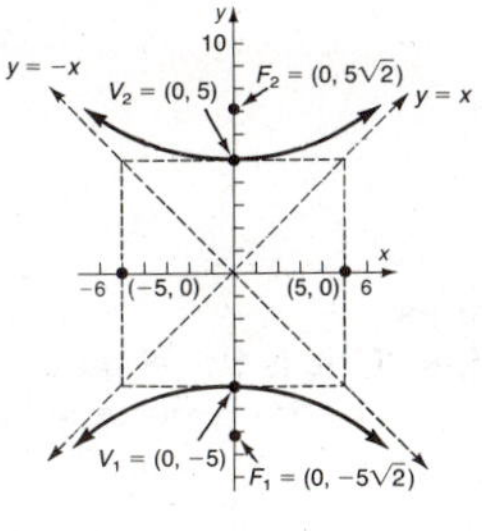

19. $\frac{(x-4)^2}{4} - \frac{(y+1)^2}{5} = 1$

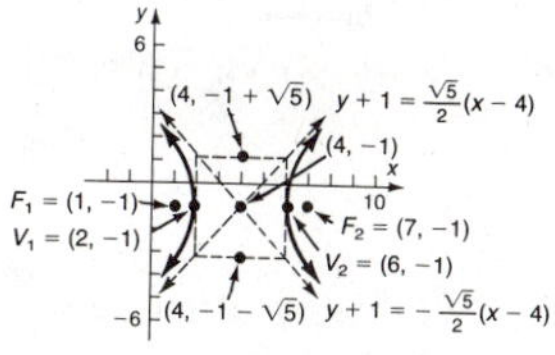

21. $\frac{(y+4)^2}{4} - \frac{(x+3)^2}{12} = 1$

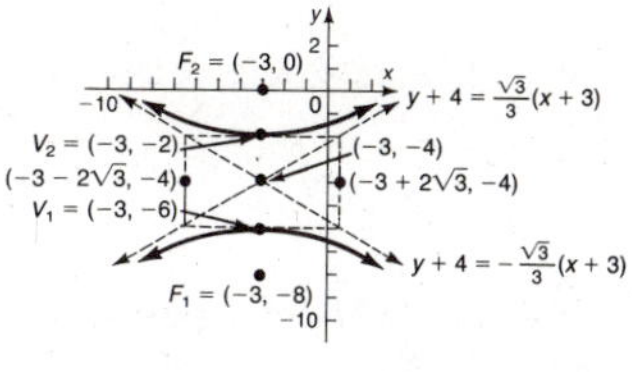

23. $(x-5)^2 - \dfrac{(y-7)^2}{3} = 1$

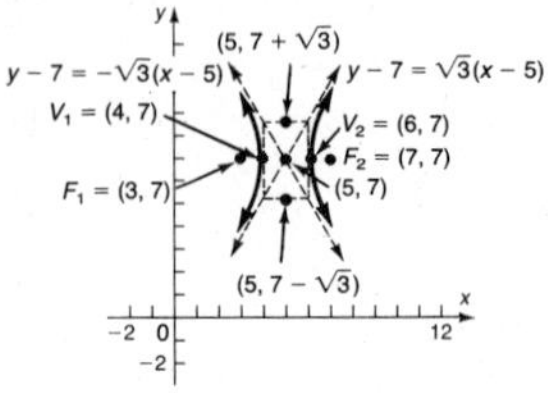

25. $\dfrac{(x-1)^2}{4} - \dfrac{(y+1)^2}{9} = 1$

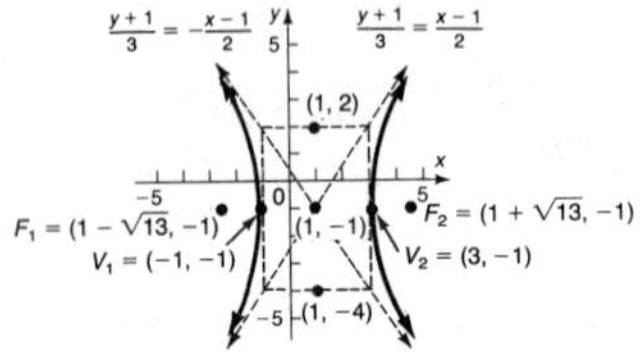

27.

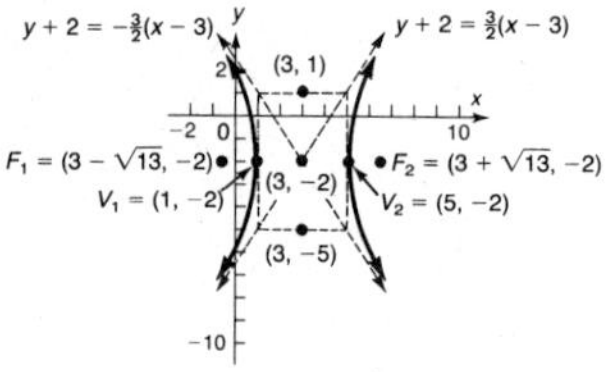

29. $\dfrac{(y-2)^2}{4} - (x+2)^2 = 1$

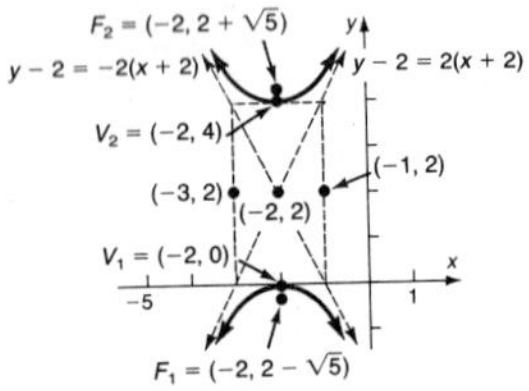

31. $\dfrac{(x+1)^2}{4} - \dfrac{(y+2)^2}{4} = 1$

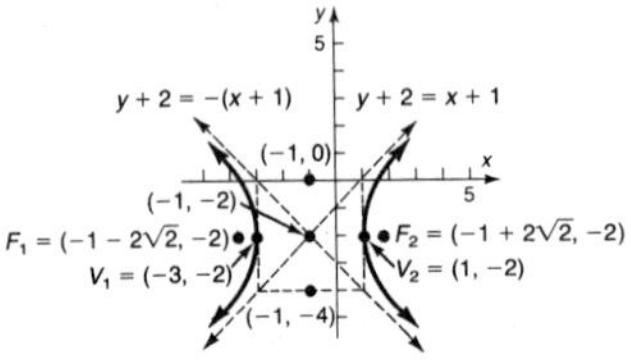

33. $(x-1)^2 - (y+1)^2 = 1$

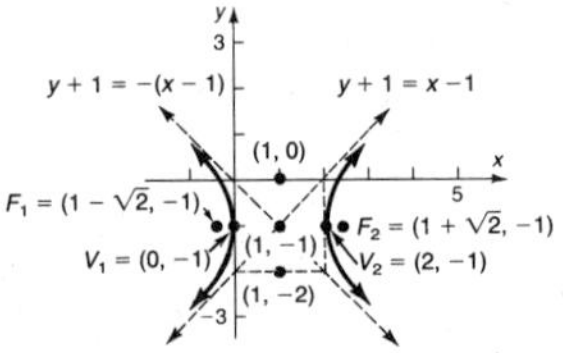

35. $\dfrac{(y-2)^2}{4} - (x+1)^2 = 1$

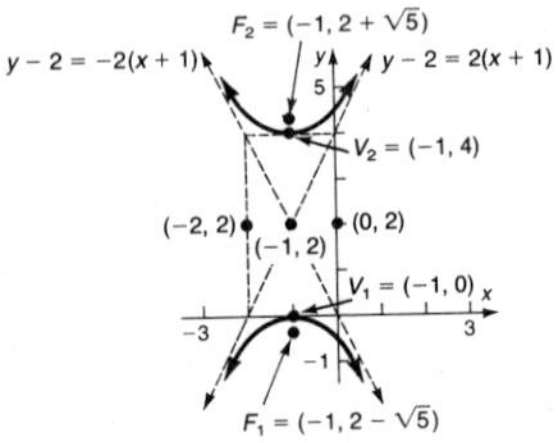

37. $\dfrac{(x-3)^2}{4} - \dfrac{(y+2)^2}{16} = 1$

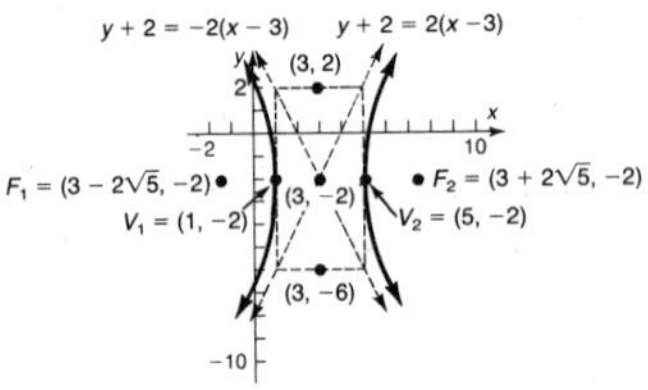

39. $\dfrac{(y-1)^2}{4} - (x+2)^2 = 1$

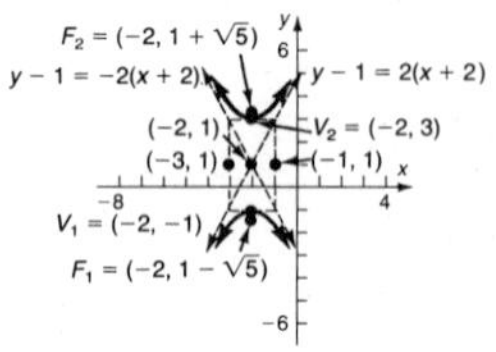

41. If e is close to 1, narrow hyperbola; if e is very large, wide hyperbola

43. $\dfrac{x^2}{4} - y^2 = 1$; asymptotes $y = \pm\frac{1}{2}x$

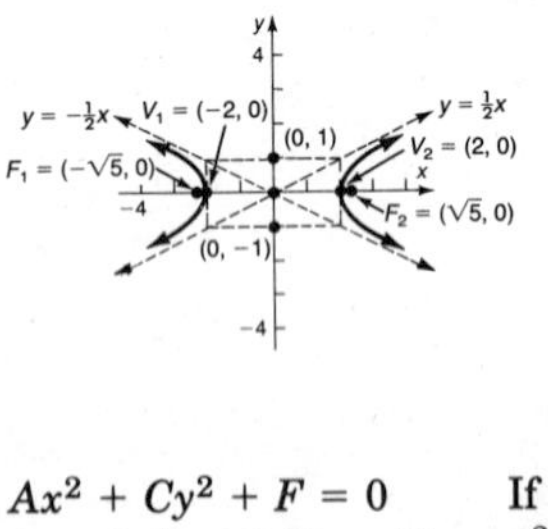

$y^2 - \dfrac{x^2}{4} = 1$; asymptotes $y = \pm\frac{1}{2}x$

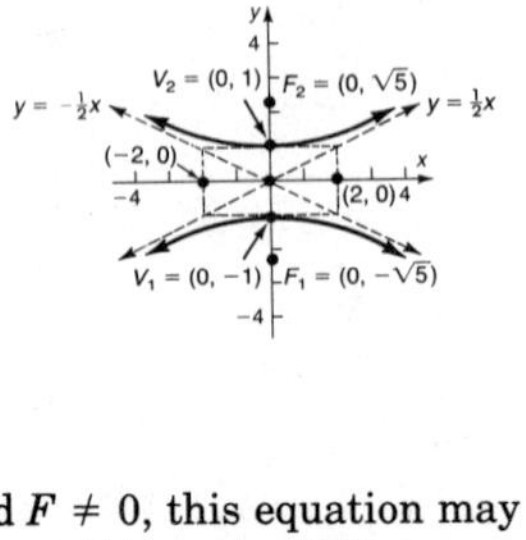

45. $Ax^2 + Cy^2 + F = 0$
$Ax^2 + Cy^2 = -F$

If A and C are of opposite sign and $F \neq 0$, this equation may be written as $x^2/(-F/A) + y^2/(-F/C) = 1$, where $-F/A$ and $-F/C$ are opposite in sign. This is the equation of a hyperbola with center at (0, 0). The transverse axis is the x-axis if $-F/A > 0$; the transverse axis is the y-axis if $-F/A < 0$.

EXERCISE 11.5

1. Parabola **3.** Ellipse **5.** Hyperbola **7.** Hyperbola **9.** Circle **11.** $x = \sqrt{2}/2(x' - y'),\ y = \sqrt{2}/2(x' + y')$
13. $x = \sqrt{2}/2(x' - y'),\ y = \sqrt{2}/2(x' + y')$ **15.** $x = \frac{1}{2}(x' - \sqrt{3}y'),\ y = \frac{1}{2}(\sqrt{3}x' + y')$
17. $x = \sqrt{5}/5(x' - 2y'),\ y = \sqrt{5}/5(2x' + y')$ **19.** $x = \sqrt{13}/13(3x' - 2y'),\ y = \sqrt{13}/13(2x' + 3y')$

21. $\theta = 45°$ (see Problem 11)
$x'^2 - \dfrac{y'^2}{3} = 1$
Hyperbola
Center at origin
Transverse axis is the x'-axis
Vertices at $(\pm 1, 0)$

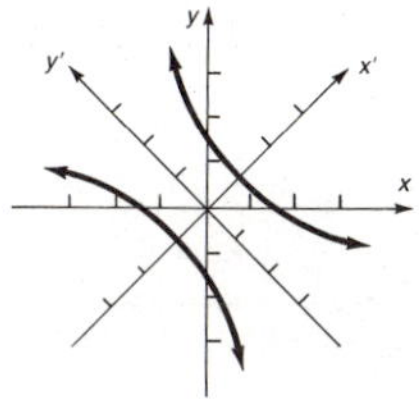

23. $\theta = 45°$ (see Problem 13)
$x'^2 + \dfrac{y'^2}{4} = 1$
Ellipse
Center at (0, 0)
Major axis is the y'-axis
Vertices at $(0, \pm 2)$

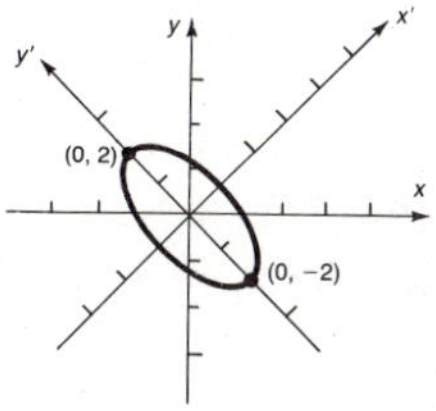

25. $\theta = 60°$ (see Problem 15)
$\dfrac{x'^2}{4} + y'^2 = 1$
Ellipse
Center at (0, 0)
Major axis is the x'-axis
Vertices at $(\pm 2, 0)$

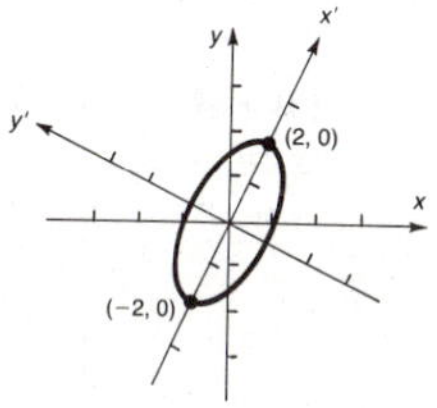

27. $\theta \approx 63°$ (see Problem 17)
$y'^2 = 8x'$
Parabola
Vertex at (0, 0)
Focus at (2, 0)

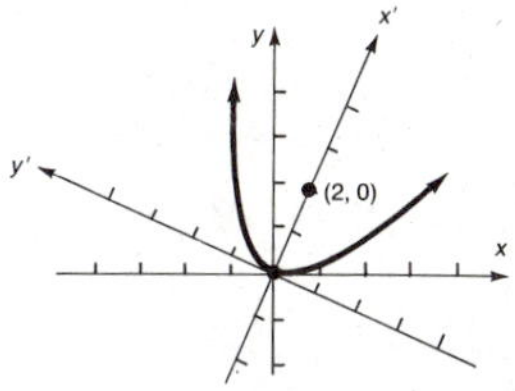

29. $\theta \approx 34°$ (see Problem 19)
$\dfrac{(x' - 2)^2}{4} + y'^2 = 1$
Ellipse
Center at (2, 0)
Major axis is the x'-axis
Vertices at (4, 0) and (0, 0)

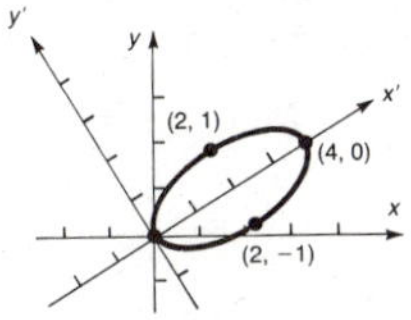

31. $\cot 2\theta = \frac{7}{24};\ \theta \approx 37°$
$(x' - 1)^2 = -6\left(y' - \frac{1}{6}\right)$
Parabola
Vertex at $\left(1, \frac{1}{6}\right)$
Focus at $\left(1, -\frac{4}{3}\right)$

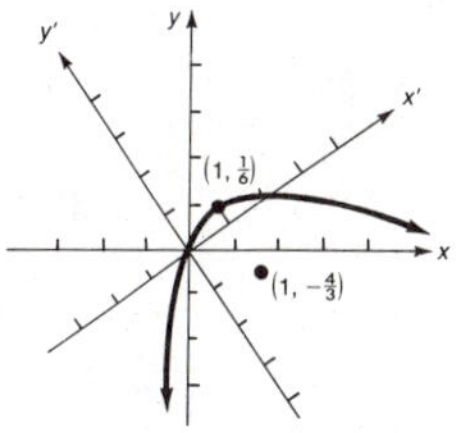

33. Hyperbola **35.** Hyperbola **37.** Parabola **39.** Ellipse **41.** Ellipse
43. Refer to equation (6):

$$\begin{aligned} A' &= A\cos^2\theta + B\sin\theta\cos\theta + C\sin^2\theta \\ B' &= B(\cos^2\theta - \sin^2\theta) + 2(C - A)(\sin\theta\cos\theta) \\ C' &= A\sin^2\theta - B\sin\theta\cos\theta + C\cos^2\theta \\ D' &= D\cos\theta + E\sin\theta \\ E' &= -D\sin\theta + E\cos\theta \\ F' &= F \end{aligned}$$

45. Use Problem 43 to find $B'^2 - 4A'C'$. After much cancellation, $B'^2 - 4A'C' = B^2 - 4AC$.
47. Use formulas (5) to find $d^2 = (x_2 - x_1)^2 + (y_2 - y_1)^2$. After simplifying, $(x_2 - x_1)^2 + (y_2 - y_1)^2 = (x_2' - x_1')^2 + (y_2' - y_1')^2$.

EXERCISE 11.6

1. Parabola; directrix is perpendicular to the polar axis 1 unit to the right of the pole.
3. Hyperbola; directrix is parallel to the polar axis $\frac{4}{3}$ units below the pole.
5. Ellipse; directrix is perpendicular to the polar axis $\frac{3}{2}$ units to the left of the pole.

7. Parabola; directrix is perpendicular to the polar axis 1 unit to the right of the pole.

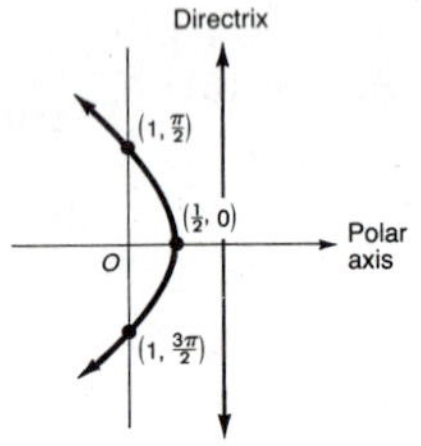

9. Ellipse; directrix is parallel to the polar axis $\frac{8}{3}$ units above the pole.

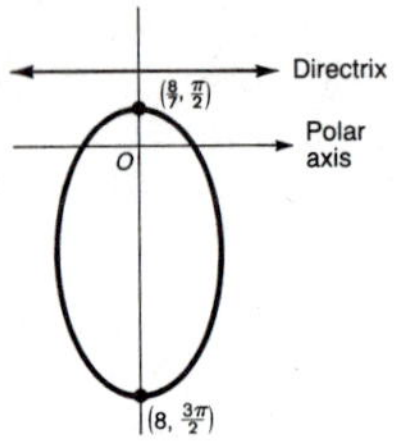

11. Hyperbola; directrix is perpendicular to the polar axis $\frac{3}{2}$ units to the left of the pole.

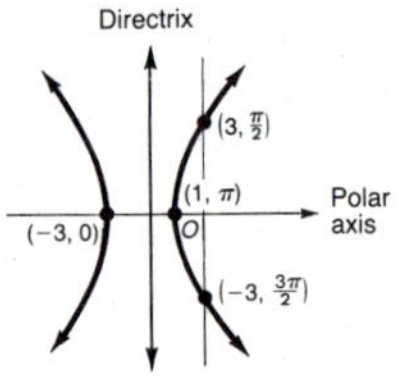

13. Ellipse; directrix is parallel to the polar axis 3 units below the pole; vertices are at $(6, \pi/2)$ and $\left(\frac{6}{5}, 3\pi/2\right)$.

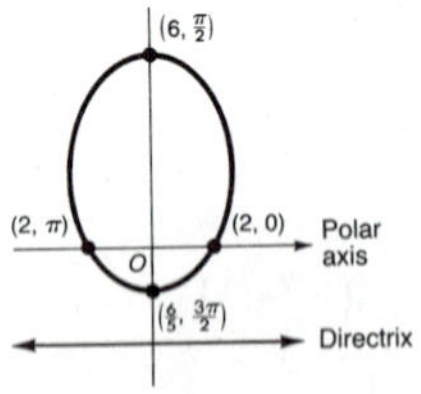

15. Ellipse; directrix is perpendicular to the polar axis 6 units to the left of the pole; vertices are at $(6, 0)$ and $(2, \pi)$.

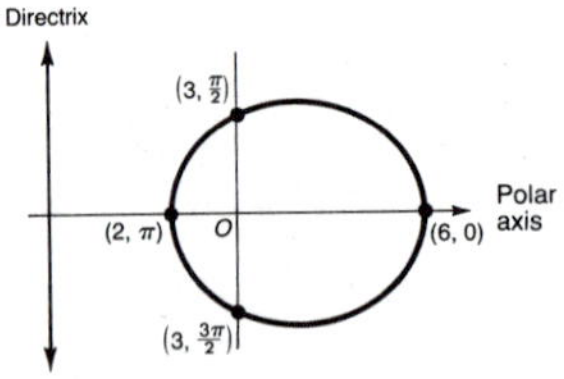

17. $y^2 + 2x - 1 = 0$
19. $16x^2 + 7y^2 + 48y - 64 = 0$
21. $3x^2 - y^2 + 12x + 9 = 0$
23. $9x^2 + 5y^2 - 24y - 36 = 0$
25. $3x^2 + 4y^2 - 12x - 36 = 0$
27. $r = 1/(1 + \sin\theta)$
29. $r = 12/(5 - 4\cos\theta)$
31. $r = 12/(1 - 6\sin\theta)$
33. Use $d(D, P) = p - r\cos\theta$ in the derivation of equation (a) in Table 5.
35. Use $d(D, P) = p + r\sin\theta$ in the derivation of equation (a) in Table 5.

EXERCISE 11.7

1. $x - 3y + 1 = 0$

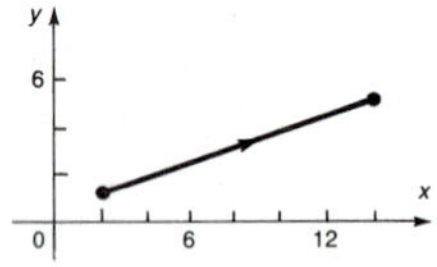

3. $y = \sqrt{x - 2}$

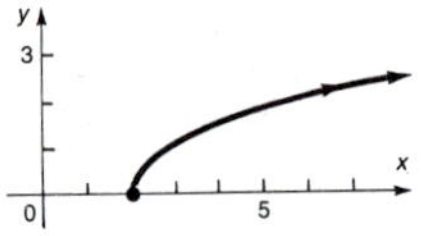

5. $x = y + 8$

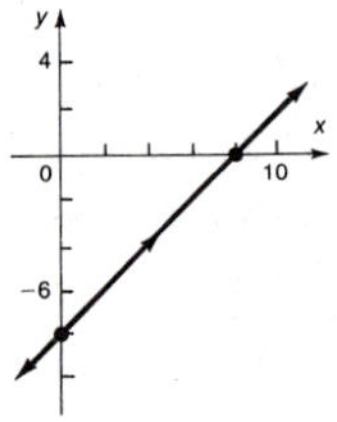

7. $x = 3(y - 1)^2$

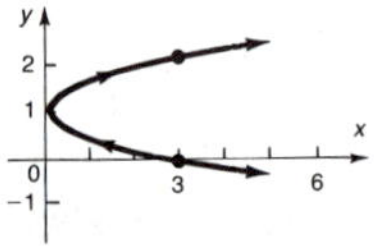

9. $2y = 2 + x$

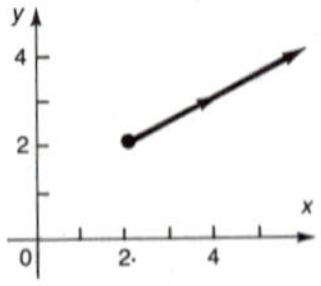

11. $y = x^3$

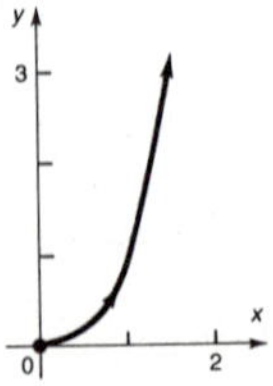

13. $\frac{x^2}{4} + \frac{y^2}{9} = 1$

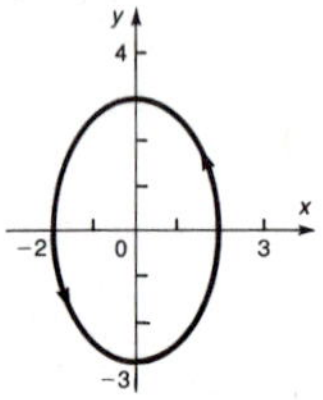

15. $\frac{x^2}{4} + \frac{y^2}{9} = 1$

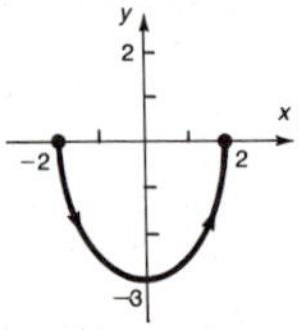

17. $x^2 - y^2 = 1$

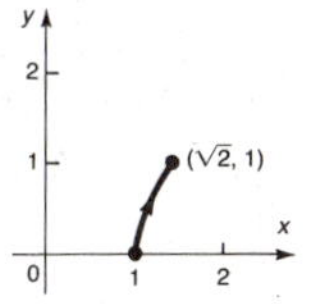

19. $x^{2/3} + y^{2/3} = 1$

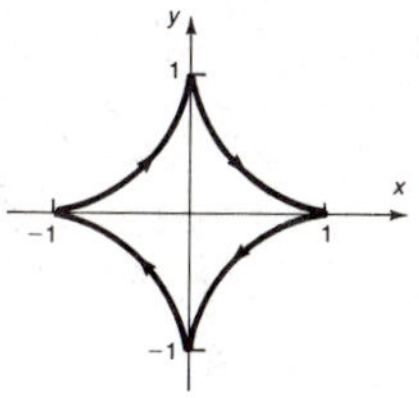

21. $x = t, y = t^3; x = \sqrt[3]{t}, y = t$
23. $x = t, y = t^{2/3}; x = t^{3/2}, y = t$
25. $x = 2\cos \pi t, y = -3\sin \pi t, 0 \le t \le 2$
27. $x = -2\sin 2\pi t, y = 3\cos 2\pi t, 0 \le t \le 1$

29.

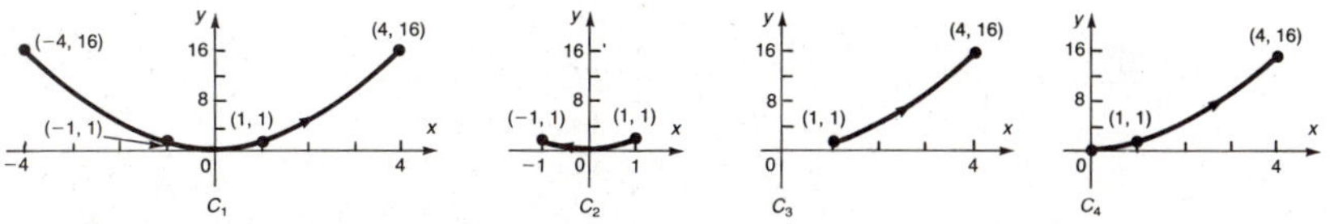

31. The orientation is from (x_1, y_1) to (x_2, y_2).

FILL-IN-THE-BLANK ITEMS

1. parabola **2.** ellipse **3.** hyperbola **4.** major; transverse **5.** y-axis **6.** $y/3 = x/2$; $y/3 = -x/2$ **7.** $\cot 2\theta = (A - C)/B$ **8.** $\frac{1}{2}$; ellipse; parallel; 4; below **9.** ellipse

REVIEW EXERCISES

1. Parabola; vertex (0, 0), focus (−4, 0), directrix $x = 4$
3. Hyperbola; center (0, 0), vertices (2, 0) and (−2, 0), foci $(\sqrt{5}, 0)$ and $(-\sqrt{5}, 0)$, asymptotes $y = \frac{1}{2}x$ and $y = -\frac{1}{2}x$
5. Ellipse; center (0, 0), vertices (0, 5) and (0, −5), foci (0, 3) and (0, −3)
7. $x^2 = -4(y - 1)$: Parabola; vertex (0, 1), focus (0, 0), directrix $y = 2$
9. $\frac{x^2}{2} - \frac{y^2}{8} = 1$: Hyperbola; center (0, 0), vertices $(\sqrt{2}, 0)$ and $(-\sqrt{2}, 0)$, foci $(\sqrt{10}, 0)$ and $(-\sqrt{10}, 0)$, asymptotes $y = 2x$ and $y = -2x$
11. $(x - 2)^2 = 2(y + 2)$: Parabola; vertex (2, −2), focus $(2, -\frac{3}{2})$, directrix $y = -\frac{5}{2}$
13. $\frac{(y - 2)^2}{4} - (x - 1)^2 = 1$: Hyperbola; center (1, 2), vertices (1, 4) and (1, 0), foci $(1, 2 + \sqrt{5})$ and $(1, 2 - \sqrt{5})$, asymptotes $y - 2 = \pm 2(x - 1)$
15. $\frac{(x - 2)^2}{9} + \frac{(y - 1)^2}{4} = 1$: Ellipse; center (2, 1), vertices (5, 1) and (−1, 1), foci $(2 + \sqrt{5}, 1)$ and $(2 - \sqrt{5}, 1)$
17. $(x - 2)^2 = -4(y + 1)$: Parabola; vertex (2, −1), focus (2, −2), directrix $y = 0$
19. $\frac{(x - 1)^2}{4} + \frac{(y + 1)^2}{9} = 1$: Ellipse; center (1, −1), vertices (1, 2) and (1, −4), foci $(1, -1 + \sqrt{5})$ and $(1, -1 - \sqrt{5})$

21. $y^2 = -8x$

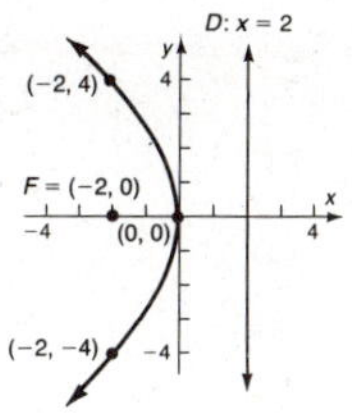

23. $\frac{y^2}{4} - \frac{x^2}{12} = 1$

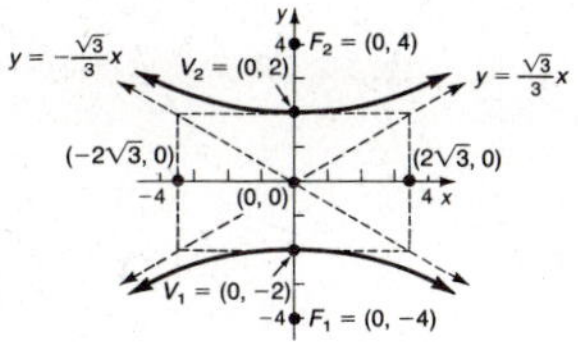

25. $\frac{x^2}{16} + \frac{y^2}{7} = 1$

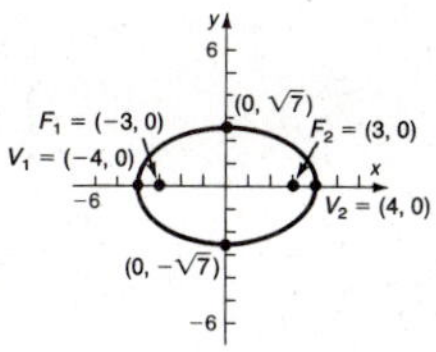

27. $(x - 2)^2 = -4(y + 3)$

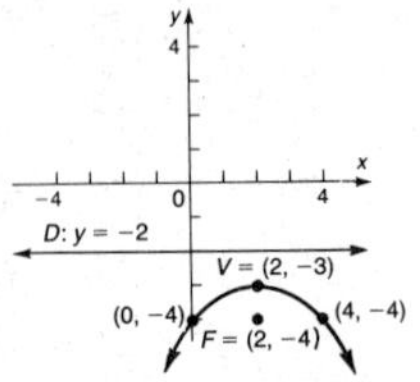

29. $(x + 2)^2 - \dfrac{(y + 3)^2}{3} = 1$

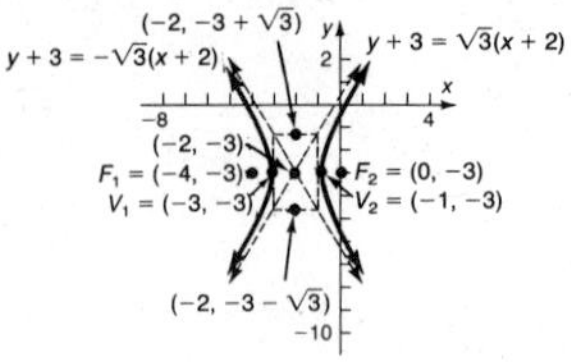

31. $\dfrac{(x + 4)^2}{16} + \dfrac{(y - 5)^2}{25} = 1$

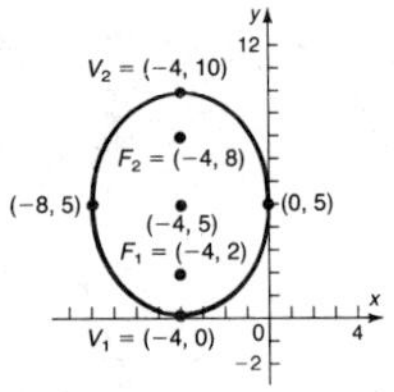

33. $\dfrac{(x + 1)^2}{9} - \dfrac{(y - 2)^2}{7} = 1$

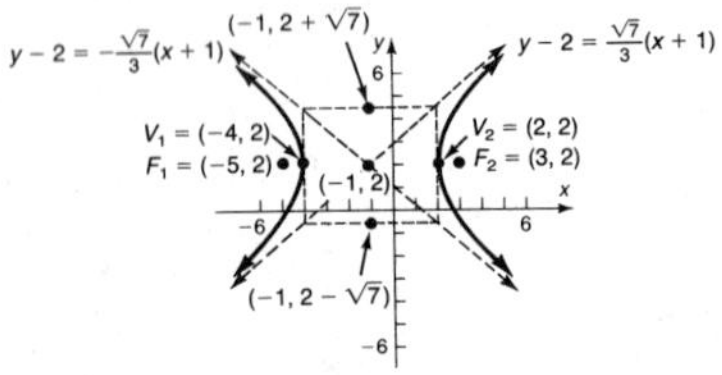

35. $\dfrac{(x - 3)^2}{9} - \dfrac{(y - 1)^2}{4} = 1$

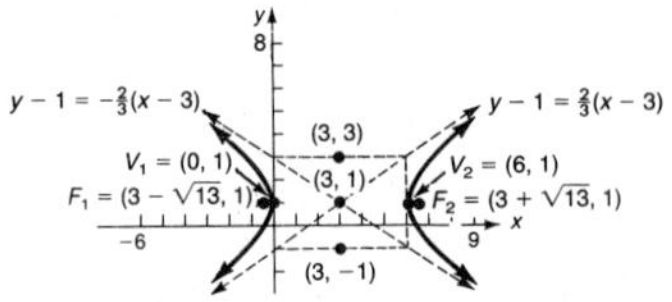

37. Parabola
39. Ellipse
41. Parabola
43. Hyperbola
45. Ellipse

47. $x'^2 - \dfrac{y'^2}{9} = 1$

Hyperbola
Center at the origin
Transverse axis the x'-axis
Vertices at $(\pm 1, 0)$

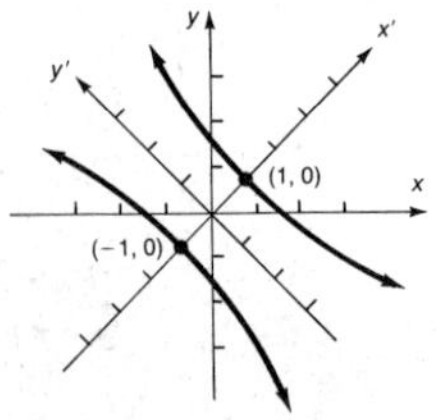

49. $\dfrac{x'^2}{2} + \dfrac{y'^2}{4} = 1$

Ellipse
Center at origin
Major axis the y'-axis
Vertices at $(0, \pm 2)$

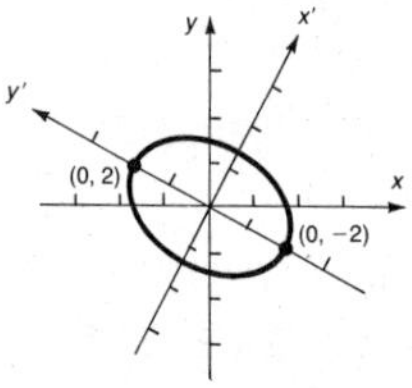

51. $y'^2 = -\dfrac{4\sqrt{13}}{13}x'$

Parabola
Vertex at the origin
Focus on the x'-axis at $(-\sqrt{13}/13, 0)$

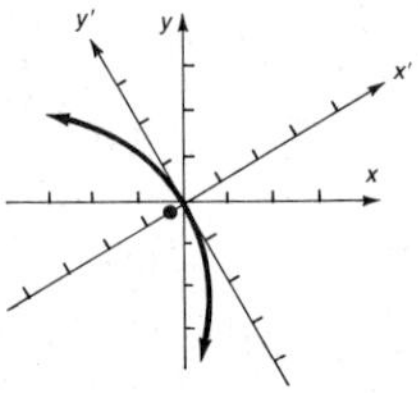

53. Parabola; directrix is perpendicular to the polar axis 4 units to the left of the pole.

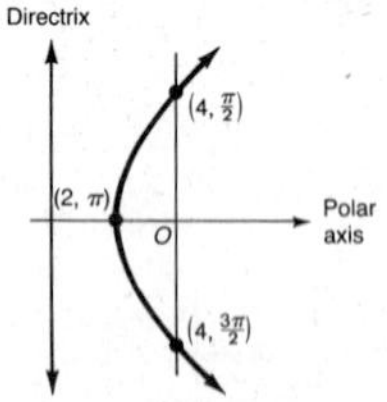

55. Ellipse; directrix is parallel to the polar axis 6 units below the pole; vertices are $(6, \pi/2)$ and $(2, 3\pi/2)$.

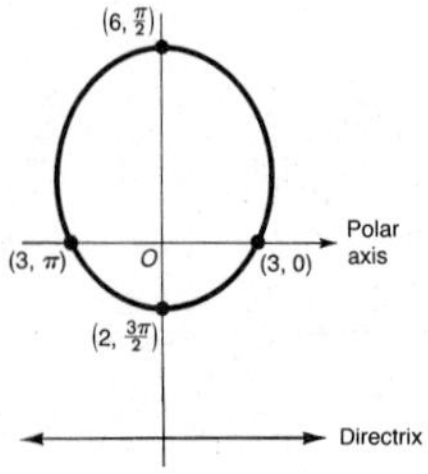

57. Hyperbola; directrix is perpendicular to the polar axis 1 unit to the right of the pole; vertices are at $\left(\frac{2}{3}, 0\right)$ and $(-2, \pi)$.

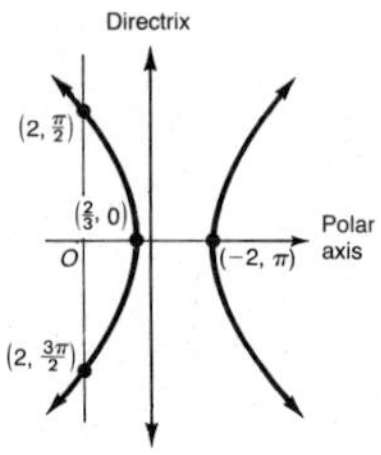

59. $y^2 - 8x - 16 = 0$ **61.** $3x^2 - y^2 - 8x + 4 = 0$

63. $x + 4y = 2$

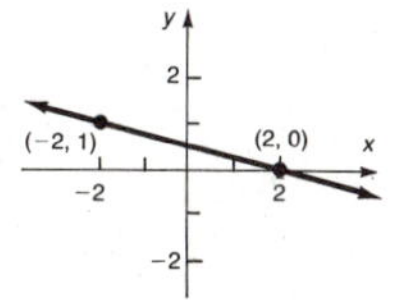

65. $\dfrac{x^2}{9} + \dfrac{(y-2)^2}{16} = 1$

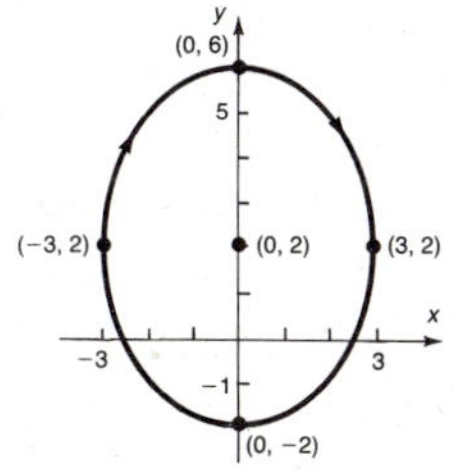

67. $1 + y = x$

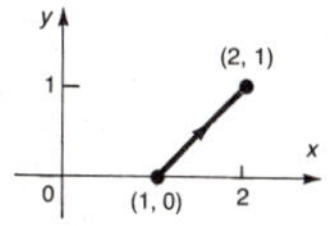

69. $\dfrac{x^2}{5} - \dfrac{y^2}{4} = 1$ **71.** The ellipse $\dfrac{x^2}{16} + \dfrac{y^2}{7} = 1$

CHAPTER 12 EXERCISE 12.1

1. $2(2) - (-1) = 5$ and $5(2) + 2(-1) = 8$ **3.** $3(2) - 4(\frac{1}{2}) = 4$ and $2 - 3(\frac{1}{2}) = \frac{1}{2}$ **5.** $2^2 - 1^2 = 3$ and $(2)(1) = 2$
7. $\dfrac{0}{1+0} + 3(2) = 6$ and $0 + 9(2)^2 = 36$ **9.** $3(1) + 3(-1) + 2(2) = 4$, $1 - (-1) - 2 = 0$, and $2(-1) - 3(2) = -8$ **11.** $x = 6, y = 2$
13. $x = 3, y = 2$ **15.** $x = 8, y = -4$ **17.** $x = 4, y = -2$ **19.** Inconsistent **21.** $x = \frac{1}{2}, y = \frac{3}{4}$ **23.** $x = 4 - 2y$, y is any real number
25. $x = \frac{1}{10}, y = \frac{2}{5}$ **27.** $x = \frac{3}{2}, y = 1$ **29.** $y = \frac{5}{3} - \frac{2}{3}x$, x is any real number **31.** $x = \frac{4}{3}, y = \frac{1}{5}$ **33.** $x = 8, y = 2, z = 0$
35. $x = 2, y = -1, z = 1$ **37.** Inconsistent
39. $x = 5z - 2$, $y = 4z - 3$ where z is any real number, or $x = \frac{5}{4}y + \frac{7}{4}$, $z = \frac{1}{4}y + \frac{3}{4}$ where y is any real number, or $y = \frac{4}{5}x - \frac{7}{5}$, $z = \frac{1}{5}x + \frac{2}{5}$ where x is any real number
41. Inconsistent **43.** $x = 1, y = 3, z = -2$ **45.** $x = -3, y = \frac{1}{2}, z = 1$ **47.** $x = \frac{1}{5}, y = \frac{1}{3}$ **49.** 20 and 61 **51.** 15 ft by 30 ft
53. Cheeseburger \$1.55; shake \$0.85 **55.** Average wind speed 25 mph; average airspeed 175 mph **57.** 22.5 lb
59. 10 for 2 students and 6 for 3 students **61.** \$5.56 **63.** 12, 15, 21 **65.** 100 orchestra, 210 main, and 190 balcony seats
67. $b = -\frac{1}{2}, c = \frac{3}{2}$ **69.** $a = \frac{4}{3}, b = -\frac{5}{3}, c = 1$ **71.** $x = \dfrac{b_1 - b_2}{m_2 - m_1}, y = \dfrac{m_2 b_1 - m_1 b_2}{m_2 - m_1}$ **73.** $y = mx + b$, x is any real number

EXERCISE 12.2

1. $\left[\begin{array}{cc|c} 1 & -3 & 5 \\ 4 & 1 & 6 \end{array}\right]$ **3.** $\left[\begin{array}{cc|c} 2 & 3 & 6 \\ 4 & -6 & -2 \end{array}\right]$ **5.** $\left[\begin{array}{cc|c} 0.01 & -0.03 & 0.06 \\ 0.13 & 0.10 & 0.20 \end{array}\right]$ **7.** $\left[\begin{array}{ccc|c} 1 & -1 & 1 & 10 \\ 3 & 2 & 0 & 5 \\ 1 & 1 & 2 & 2 \end{array}\right]$ **9.** $\left[\begin{array}{ccc|c} 1 & 1 & -1 & 2 \\ 3 & -2 & 0 & 2 \end{array}\right]$

11. $R_1 = -1r_2 + r_1$ **13.** $R_1 = -4r_2 + r_1$ **15.** $\left[\begin{array}{ccc|c} 1 & 2 & 3 & 0 \\ 0 & 0 & -3 & 3 \\ -3 & 2 & 1 & -2 \end{array}\right]$ **17.** $\left[\begin{array}{ccc|c} 1 & 2 & 3 & 0 \\ 8 & 0 & 1 & 7 \\ -3 & 2 & 1 & -2 \end{array}\right]$ **19.** $\left[\begin{array}{ccc|c} 1 & 2 & 3 & 0 \\ 2 & 4 & 3 & 3 \\ -1 & 6 & 4 & 1 \end{array}\right]$

21. $x = 6, y = 2$ **23.** $x = 3, y = 2$ **25.** $x = 8, y = -4$ **27.** $x = 4, y = -2$ **29.** Inconsistent **31.** $x = \frac{1}{2}, y = \frac{3}{4}$
33. $x = 4 - 2y$, y is any real number **35.** $x = \frac{1}{10}, y = \frac{2}{5}$ **37.** $x = \frac{3}{2}, y = 1$ **39.** $x = \frac{5}{2} - \frac{3}{2}y$, y is any real number
41. $x = \frac{4}{3}, y = \frac{1}{5}$ **43.** $x = 8, y = 2, z = 0$ **45.** $x = 2, y = -1, z = 1$ **47.** Inconsistent
49. $x = 5z - 2, y = 4z - 3$ or $x = \frac{5}{4}y + \frac{7}{4}, z = \frac{1}{4}y + \frac{3}{4}$ or $y = \frac{4}{5}x - \frac{7}{5}, z = \frac{1}{5}x + \frac{2}{5}$ **51.** Inconsistent **53.** $x = 1, y = 3, z = -2$
55. $x = -3, y = \frac{1}{2}, z = 1$ **57.** $x = \frac{1}{3}, y = \frac{2}{3}, z = 1$ **59.** $x = 1, y = 2, z = 0, w = 1$ **61.** $y = 0, z = 1 - x$, x is any real number
63. $x = 2, y = z - 3$, z is any real number **65.** $x = \frac{13}{9}, y = \frac{7}{18}, z = \frac{19}{18}$
67. $x = \frac{7}{5} - \frac{3}{5}z - \frac{2}{5}w, y = -\frac{8}{5} + \frac{7}{5}z + \frac{13}{5}w$, where z and w are any real numbers **69.** $y = -2x^2 + x + 3$ **71.** $f(x) = 3x^3 - 4x^2 + 5$

73. x = liters of 15% H_2SO_4, y = liters of 25% H_2SO_4, z = liters of 50% H_2SO_4: $\begin{cases} x = \frac{5}{2}z - 150 \\ y = 250 - \frac{7}{2}z \end{cases}$

15%	25%	50%	40%
0	40	60	100
10	26	64	100
20	12	68	100

75. If $a_1 \neq 0$,

$$\left[\begin{array}{cc|c} a_1 & b_1 & c_1 \\ a_2 & b_2 & c_2 \end{array}\right] \to \left[\begin{array}{cc|c} 1 & \dfrac{b_1}{a_1} & \dfrac{c_1}{a_1} \\ a_2 & b_2 & c_2 \end{array}\right] \to \left[\begin{array}{cc|c} 1 & \dfrac{b_1}{a_1} & \dfrac{c_1}{a_1} \\ 0 & \dfrac{-a_2b_1}{a_1} + b_2 & \dfrac{-a_2c_1}{a_1} + c_2 \end{array}\right] \to \left[\begin{array}{cc|c} 1 & \dfrac{b_1}{a_1} & \dfrac{c_1}{a_1} \\ 0 & \dfrac{-a_2b_1 + b_2a_1}{a_1} & \dfrac{-a_2c_1 + c_2a_1}{a_1} \end{array}\right]$$

$$\to \left[\begin{array}{cc|c} 1 & \dfrac{b_1}{a_1} & \dfrac{c_1}{a_1} \\ 0 & 1 & \dfrac{-a_2c_1 + c_2a_1}{a_1} \cdot \dfrac{a_1}{-a_2b_1 + b_2a_1} \end{array}\right] \to \left[\begin{array}{cc|c} 1 & \dfrac{b_1}{a_1} & \dfrac{c_1}{a_1} \\ 0 & 1 & \dfrac{-a_2c_1 + c_2a_1}{-a_2b_1 + b_2a_1} \end{array}\right] \to \left[\begin{array}{cc|c} 1 & 0 & \dfrac{-b_1c_2 + b_2c_1}{-a_2b_1 + b_2a_1} \\ 0 & 1 & \dfrac{-a_2c_1 + c_2a_1}{-a_2b_1 + b_2a_1} \end{array}\right]$$

$$x = \frac{1}{a_1b_2 - a_2b_1}(c_1b_2 - c_2b_1) = \frac{1}{D}(c_1b_2 - c_2b_1), \quad y = \frac{1}{a_1b_2 - a_2b_1}(a_1c_2 - a_2c_1) = \frac{1}{D}(a_1c_2 - a_2c_1)$$

If $a_1 = 0$, then $a_2 \neq 0$, $b_1 \neq 0$, and

$$\left[\begin{array}{cc|c} 0 & b_1 & c_1 \\ a_2 & b_2 & c_2 \end{array}\right] \to \left[\begin{array}{cc|c} a_2 & b_2 & c_2 \\ 0 & b_1 & c_1 \end{array}\right] \to \left[\begin{array}{cc|c} 1 & \dfrac{b_2}{a_2} & \dfrac{c_2}{a_2} \\ 0 & b_1 & c_1 \end{array}\right] \to \left[\begin{array}{cc|c} 1 & \dfrac{b_2}{a_2} & \dfrac{c_2}{a_2} \\ 0 & 1 & \dfrac{c_1}{b_1} \end{array}\right] \to \left[\begin{array}{cc|c} 1 & 0 & \dfrac{c_2}{a_2} - \dfrac{b_2c_1}{a_2b_1} = \dfrac{c_1b_2 - c_2b_1}{-a_2b_1} \\ 0 & 1 & \dfrac{c_1}{b_1} = \dfrac{-a_2c_1}{-a_2b_1} \end{array}\right]$$

EXERCISE 12.3

1. 2 **3.** 22 **5.** -2 **7.** 10 **9.** -26 **11.** $x = 6, y = 2$ **13.** $x = 3, y = 2$ **15.** $x = 8, y = -4$ **17.** $x = 4, y = -2$
19. Not applicable **21.** $x = \frac{1}{2}, y = \frac{3}{4}$ **23.** $x = \frac{1}{10}, y = \frac{2}{5}$ **25.** $x = \frac{3}{2}, y = 1$ **27.** $x = \frac{4}{3}, y = \frac{1}{5}$ **29.** $x = 1, y = 3, z = -2$
31. $x = -3, y = \frac{1}{2}, z = 1$ **33.** Not applicable **35.** $x = 0, y = 0, z = 0$ **37.** Not applicable **39.** $x = \frac{1}{5}, y = \frac{1}{3}$ **41.** -5 **43.** $\frac{13}{11}$
45. 0 or -9 **47.** -4 **49.** 12 **51.** 8 **53.** 8

55. $$\begin{vmatrix} x^2 & x & 1 \\ y^2 & y & 1 \\ z^2 & z & 1 \end{vmatrix} = x^2\begin{vmatrix} y & 1 \\ z & 1 \end{vmatrix} - x\begin{vmatrix} y^2 & 1 \\ z^2 & 1 \end{vmatrix} + \begin{vmatrix} y^2 & y \\ z^2 & z \end{vmatrix} = x^2(y - z) - x(y^2 - z^2) + yz(y - z)$$
$$= (y - z)[x^2 - x(y + z) + yz] = (y - z)[(x^2 - xy) - (xz - yz)] = (y - z)[x(x - y) - z(x - y)] = (y - z)(x - y)(x - z)$$

57. $$\begin{vmatrix} a_{13} & a_{12} & a_{11} \\ a_{23} & a_{22} & a_{21} \\ a_{33} & a_{32} & a_{31} \end{vmatrix} = a_{13}(a_{22}a_{31} - a_{32}a_{21}) - a_{12}(a_{23}a_{31} - a_{33}a_{21}) + a_{11}(a_{23}a_{32} - a_{33}a_{22})$$
$$= -[a_{11}(a_{22}a_{33} - a_{32}a_{23}) - a_{12}(a_{21}a_{33} - a_{31}a_{23}) + a_{13}(a_{21}a_{32} - a_{31}a_{22})] = -\begin{vmatrix} a_{11} & a_{12} & a_{13} \\ a_{21} & a_{22} & a_{23} \\ a_{31} & a_{32} & a_{33} \end{vmatrix}$$

59. $$\begin{vmatrix} a_{11} & a_{12} & a_{11} \\ a_{21} & a_{22} & a_{21} \\ a_{31} & a_{32} & a_{31} \end{vmatrix} = a_{11}(a_{22}a_{31} - a_{32}a_{21}) - a_{12}(a_{21}a_{31} - a_{31}a_{21}) + a_{11}(a_{21}a_{32} - a_{31}a_{22})$$
$$= a_{11}a_{22}a_{31} - a_{11}a_{32}a_{21} - a_{12}(0) + a_{11}a_{21}a_{32} - a_{11}a_{31}a_{22} = 0$$

EXERCISE 12.4

1. $x = 1, y = -2; x = -\frac{11}{5}, y = -\frac{2}{5}$

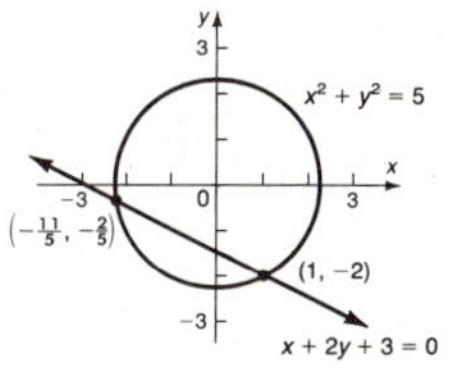

3. $x = 0, y = 2; x = 0, y = -2;$ $x = -1, y = \sqrt{3}; x = -1, y = -\sqrt{3}$

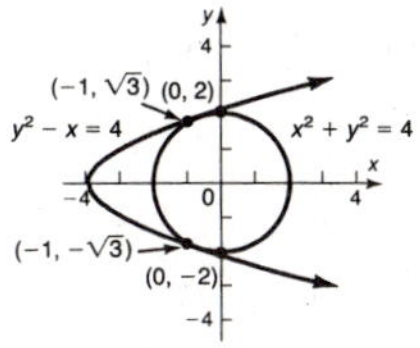

5. $x = 4 - \sqrt{2}, y = 4 + \sqrt{2};$ $x = 4 + \sqrt{2}, y = 4 - \sqrt{2}$

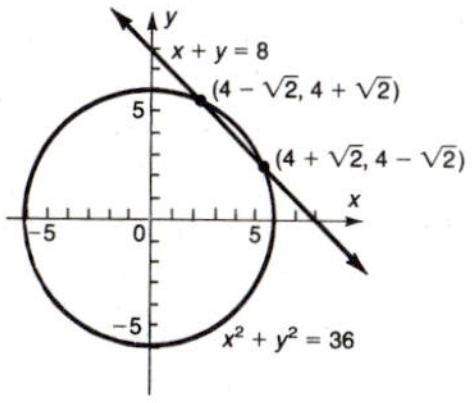

7. $x = 2, y = 2; x = -2, y = -2$

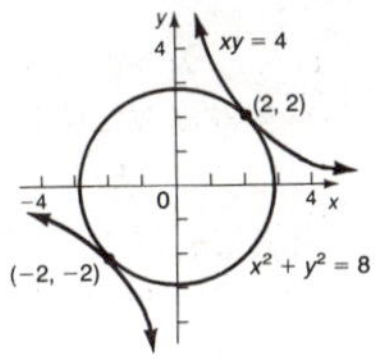

9. Inconsistent

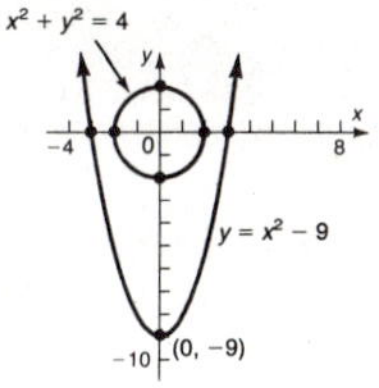

11. $x = 3, y = 5$

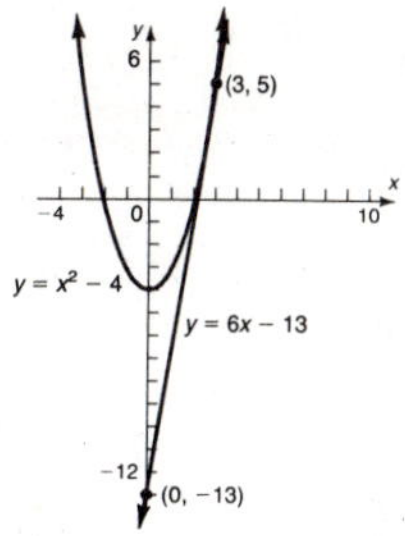

13. $x = 1, y = 4; x = -1, y = -4; x = 2\sqrt{2}, y = \sqrt{2}; x = -2\sqrt{2}, y = -\sqrt{2}$ **15.** $x = 1, y = 2; x = -\frac{13}{37}, y = -\frac{76}{37}$
17. $x = 4, y = -5; x = -\frac{3}{2}, y = \frac{1}{2}$ **19.** $x = 2, y = \frac{1}{3}; x = \frac{1}{2}, y = \frac{4}{3}$ **21.** $x = 3, y = 2; x = 3, y = -2; x = -3, y = 2; x = -3, y = -2$
23. $x = \frac{1}{2}, y = \frac{3}{2}; x = \frac{1}{2}, y = -\frac{3}{2}; x = -\frac{1}{2}, y = \frac{3}{2}; x = -\frac{1}{2}, y = -\frac{3}{2}$ **25.** $x = \sqrt{2}, y = 2\sqrt{2}; x = -\sqrt{2}, y = -2\sqrt{2}$
27. No solution; system is inconsistent **29.** $x = \frac{8}{3}, y = 2\sqrt{10}/3; x = -\frac{8}{3}, y = 2\sqrt{10}/3; x = \frac{8}{3}, y = -2\sqrt{10}/3; x = -\frac{8}{3}, y = -2\sqrt{10}/3$
31. $x = 1, y = \frac{1}{2}; x = -1, y = \frac{1}{2}; x = 1, y = -\frac{1}{2}; x = -1, y = -\frac{1}{2}$ **33.** No solution; system is inconsistent
35. $x = 2, y = 1; x = -2, y = -1; x = \sqrt{3}, y = \sqrt{3}; x = -\sqrt{3}, y = -\sqrt{3}$
37. $x = 3, y = 2; x = -3, y = -2; x = 2, y = \frac{1}{2}; x = -2, y = -\frac{1}{2}$ **39.** $x = 3, y = 1; x = -1, y = -3$
41. $x = 0, y = -2; x = 0, y = 1; x = 2, y = -1$ **43.** $4 + \sqrt{2}$ and $4 - \sqrt{2}$ **45.** 1 and 7; −1 and −7 **47.** $\frac{1}{2}$ and $\frac{1}{3}$
49. 5 in. by 3 in. **51.** 12 cm by 18 cm **53.** 8 cm **55.** $y = 4x - 4$ **57.** $y = 2x + 1$ **59.** $y = -\frac{1}{3}x + \frac{7}{3}$ **61.** $y = 2x - 3$
63. $l = \dfrac{P + \sqrt{P^2 - 16A}}{4}; w = \dfrac{P - \sqrt{P^2 - 16A}}{4}$

65.

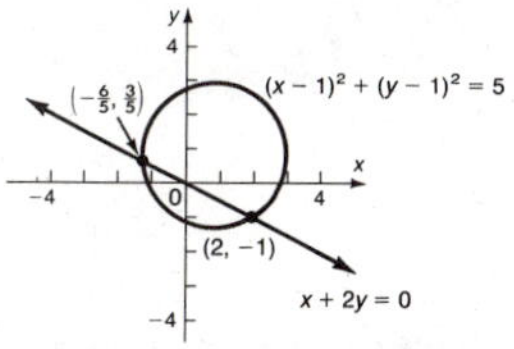

67. Solutions: $(0, -\sqrt{3} - 2)$, $(0, \sqrt{3} - 2)$, (1, 0), (1, −4)

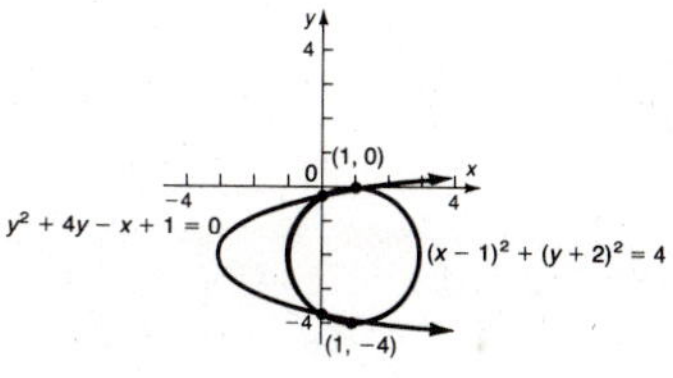

69.

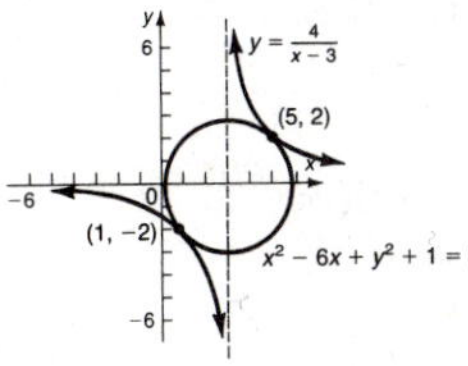

71. $r_1 = \dfrac{-b + \sqrt{b^2 - 4ac}}{2a}$

$r_2 = \dfrac{-b - \sqrt{b^2 - 4ac}}{2a}$

EXERCISE 12.5

1.

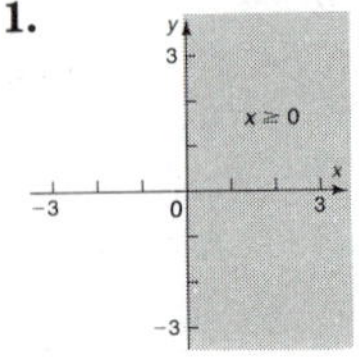

3.

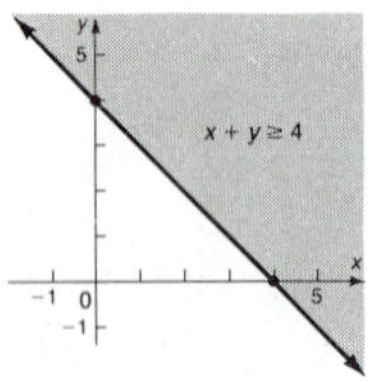

5.

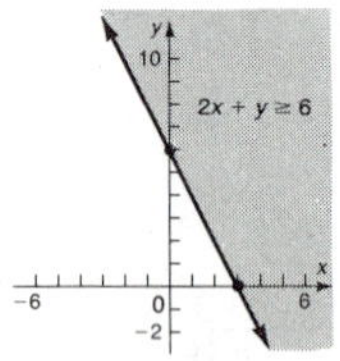

7.

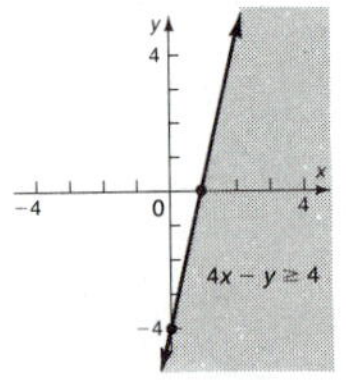

9.

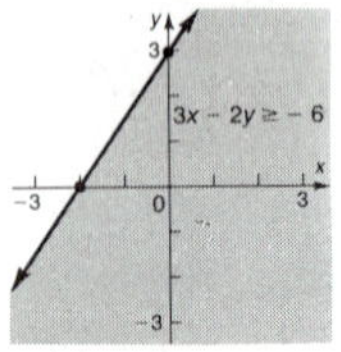

11.

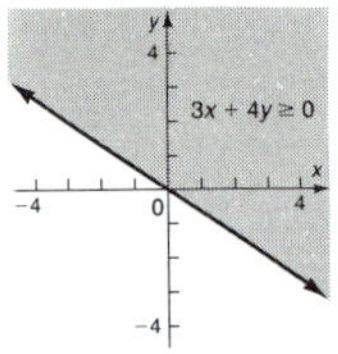

13.

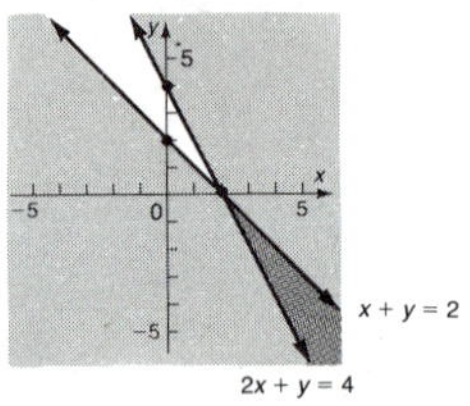

15.

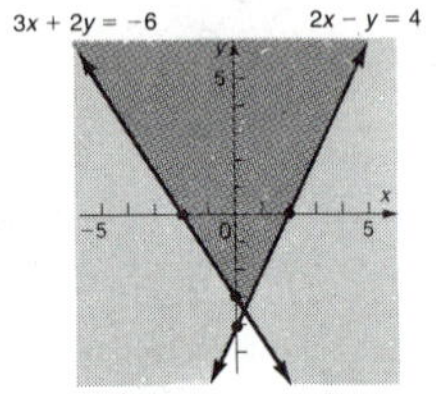

17.

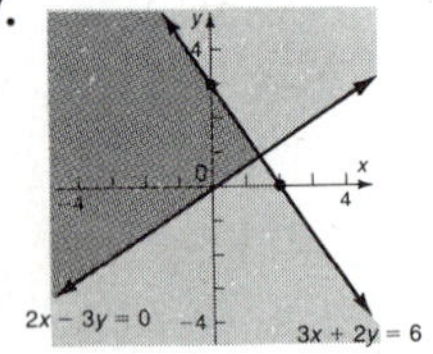

19.

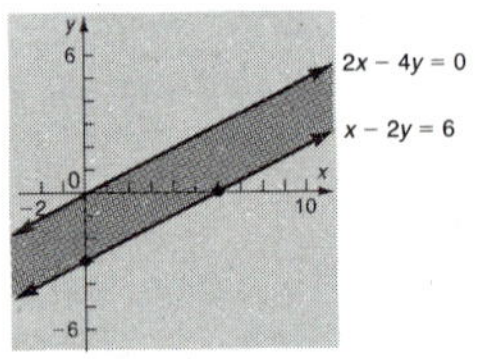

21.

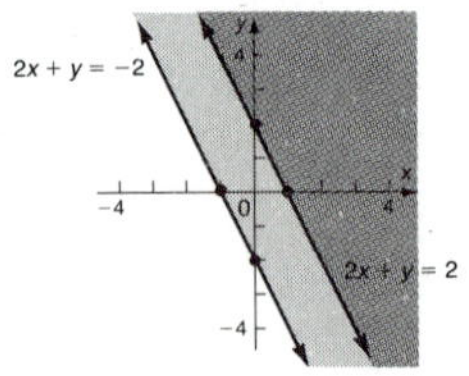

23. No solution

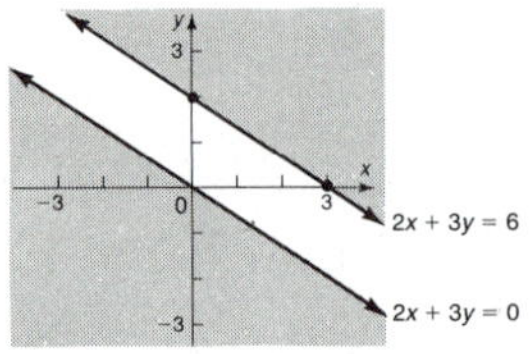

25. Bounded; vertices (0, 0), (3, 0), (2, 2), (0, 3)

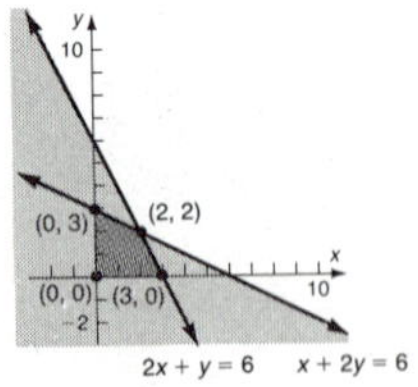

27. Unbounded; vertices (2, 0), (0, 4)

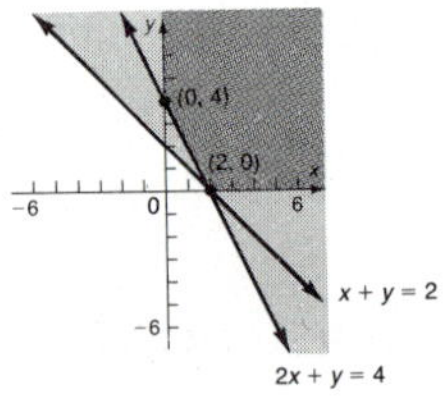

29. Bounded; vertices (2, 0), (4, 0), $\left(\frac{24}{7}, \frac{12}{7}\right)$, (0, 4), (0, 2)

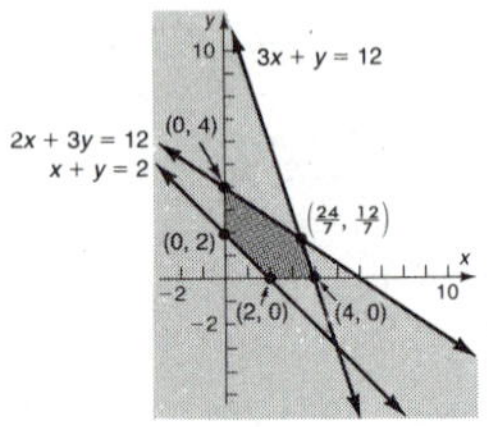

31. Bounded; vertices (2, 0), (5, 0), (2, 6), (0, 8), (0, 2)

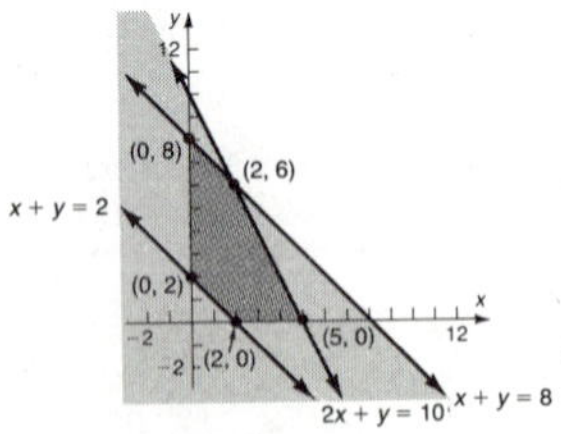

33. Bounded; vertices (1, 0), (10, 0), (0, 5), $\left(0, \frac{1}{2}\right)$

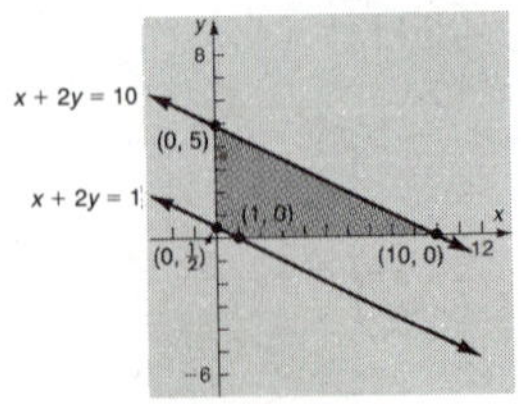

35.

$$\begin{cases} x \geq 0 \\ y \geq 0 \\ x + 2y \leq 300 \\ 3x + 2y \leq 480 \end{cases}$$

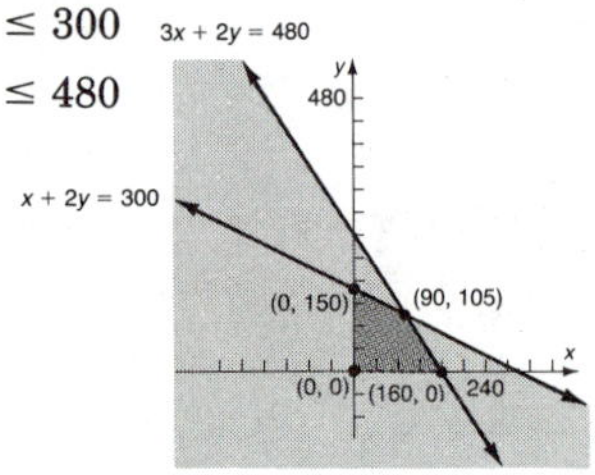

FILL-IN-THE-BLANK ITEMS

1. inconsistent **2.** matrix **3.** determinants **4.** augmented **5.** half-plane

REVIEW EXERCISES

1. $x = 2, y = -1$ **3.** $x = 2, y = \frac{1}{2}$ **5.** $x = 2, y = -1$ **7.** $x = \frac{11}{5}, y = -\frac{3}{5}$ **9.** $x = -\frac{8}{5}, y = \frac{12}{5}$ **11.** $x = 6, y = -1$ **13.** $x = -4, y = 3$
15. $x = 2, y = 3$ **17.** Inconsistent **19.** $x = -1, y = 2, z = -3$ **21.** $x = \frac{2}{5}, y = \frac{1}{10}$ **23.** $x = \frac{1}{2}, y = \frac{2}{3}, z = \frac{1}{6}$
25. $x = -\frac{1}{2}, y = -\frac{2}{3}, z = -\frac{3}{4}$ **27.** $z = -1, x = y + 1$, y any real number **29.** $x = 1, y = 2, z = -3, t = 1$ **31.** 5 **33.** 108
35. -100 **37.** $x = 2, y = -1$ **39.** $x = 2, y = 3$ **41.** $x = -1, y = 2, z = -3$ **43.** $x = -\frac{2}{5}, y = -\frac{11}{5}$; $x = -2, y = 1$
45. $x = 2\sqrt{2}, y = \sqrt{2}$; $x = -2\sqrt{2}, y = -\sqrt{2}$ **47.** $x = 0, y = 0$; $x = 3, y = 3$; $x = -3, y = 3$
49. $x = \sqrt{2}, y = -\sqrt{2}$; $x = -\sqrt{2}, y = \sqrt{2}$; $x = \frac{4}{3}\sqrt{2}, y = -\frac{2}{3}\sqrt{2}$; $x = -\frac{4}{3}\sqrt{2}, y = \frac{2}{3}\sqrt{2}$ **51.** $x = 1, y = -1$

53. Unbounded; vertex (0, 2)

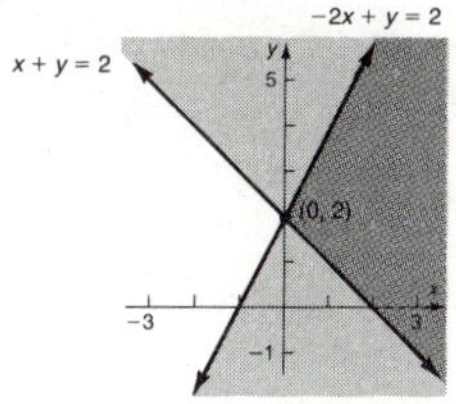

55. Bounded; vertices (0, 0), (0, 2), (3, 0)

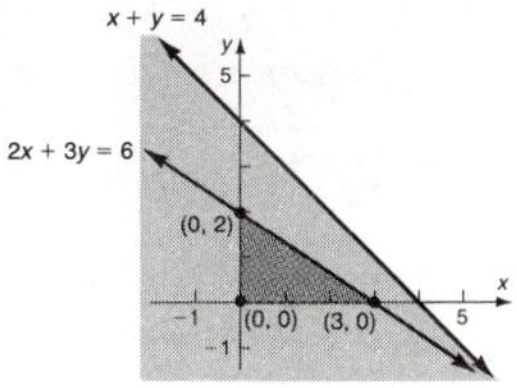

57. Bounded; vertices (0, 1), (0, 8), (4, 0), (2, 0)

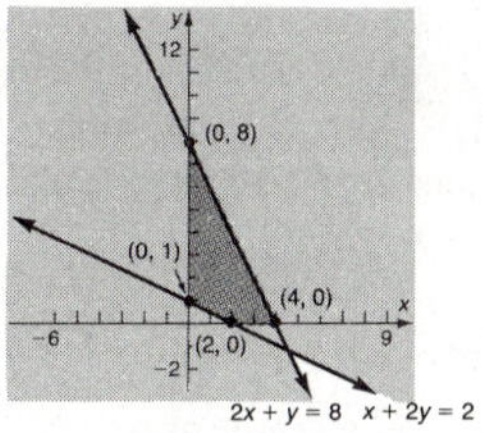

59. 10 **61.** $y = -\frac{1}{3}x^2 - \frac{2}{3}x + 1$ **63.** Katy gets \$10, Mike gets \$20, Danny gets \$5, Colleen gets \$10. **65.** 29.69 mph
67. Katy: 4 hr; Mike: 2 hr; Danny: 8 hr **69.** 24 ft by 10 ft **71.** $4 + \sqrt{2}$ in. and $4 - \sqrt{2}$ in.

CHAPTER 13 EXERCISE 13.1

1. (I) $n = 1$: $2 \cdot 1 = 2$ and $1(1 + 1) = 2$
(II) If $2 + 4 + 6 + \cdots + 2k = k(k + 1)$, then $2 + 4 + 6 + \cdots + 2k + 2(k + 1) = (2 + 4 + 6 + \cdots + 2k) + 2(k + 1) = k(k + 1) + 2(k + 1) = k^2 + 3k + 2 = (k + 1)(k + 2)$.

3. (I) $n = 1$: $1 + 2 = 3$ and $\frac{1}{2}(1)(1 + 5) = \frac{1}{2}(6) = 3$
(II) If $3 + 4 + 5 + \cdots + (k + 2) = \frac{1}{2}k(k + 5)$, then $3 + 4 + 5 + \cdots + (k + 2) + [(k + 1) + 2] = [3 + 4 + 5 + \cdots + (k + 2)] + (k + 3) = \frac{1}{2}k(k + 5) + k + 3 = \frac{1}{2}(k^2 + 7k + 6) = \frac{1}{2}(k + 1)(k + 6)$.

5. (I) $n = 1$: $3 \cdot 1 - 1 = 2$ and $\frac{1}{2}(1)[3(1) + 1] = \frac{1}{2}(4) = 2$
(II) If $2 + 5 + 8 + \cdots + (3k - 1) = \frac{1}{2}k(3k + 1)$, then $2 + 5 + 8 + \cdots + (3k - 1) + [3(k + 1) - 1] = [2 + 5 + 8 + \cdots + (3k - 1)] + 3k + 2 = \frac{1}{2}k(3k + 1) + (3k + 2) = \frac{1}{2}(3k^2 + 7k + 4) = \frac{1}{2}(k + 1)(3k + 4)$.

7. (I) $n = 1$: $2^{1-1} = 1$ and $2^1 - 1 = 1$
(II) If $1 + 2 + 2^2 + \cdots + 2^{k-1} = 2^k - 1$, then $1 + 2 + 2^2 + \cdots + 2^{k-1} + 2^{(k+1)-1} = (1 + 2 + 2^2 + \cdots + 2^{k-1}) + 2^k = 2^k - 1 + 2^k = 2(2^k) - 1 = 2^{k+1} - 1$.

9. (I) $n = 1$: $4^{1-1} = 1$ and $\frac{1}{3}(4^1 - 1) = \frac{1}{3}(3) = 1$
(II) If $1 + 4 + 4^2 + \cdots + 4^{k-1} = \frac{1}{3}(4^k - 1)$, then $1 + 4 + 4^2 + \cdots + 4^{k-1} + 4^{(k+1)-1} = (1 + 4 + 4^2 + \cdots + 4^{k-1}) + 4^k = \frac{1}{3}(4^k - 1) + 4^k = \frac{1}{3}[4^k - 1 + 3(4^k)] = \frac{1}{3}[4(4^k) - 1] = \frac{1}{3}(4^{k+1} - 1)$.

11. (I) $n = 1$: $\frac{1}{1 \cdot 2} = \frac{1}{2}$ and $\frac{1}{1 + 1} = \frac{1}{2}$
(II) If $\frac{1}{1 \cdot 2} + \frac{1}{2 \cdot 3} + \frac{1}{3 \cdot 4} + \cdots + \frac{1}{k(k + 1)} = \frac{k}{k + 1}$, then $\frac{1}{1 \cdot 2} + \frac{1}{2 \cdot 3} + \frac{1}{3 \cdot 4} + \cdots + \frac{1}{k(k + 1)} + \frac{1}{(k + 1)[(k + 1) + 1]} = \left[\frac{1}{1 \cdot 2} + \frac{1}{2 \cdot 3} + \frac{1}{3 \cdot 4} + \cdots + \frac{1}{k(k + 1)}\right] + \frac{1}{(k + 1)(k + 2)} = \frac{k}{k + 1} + \frac{1}{(k + 1)(k + 2)} = \frac{k + 1}{k + 2}$.

13. (I) $n = 1$: $1^2 = 1$ and $\frac{1}{6} \cdot 1 \cdot 2 \cdot 3 = 1$
(II) If $1^2 + 2^2 + 3^2 + \cdots + k^2 = \frac{1}{6}k(k + 1)(2k + 1)$, then $1^2 + 2^2 + 3^2 + \cdots + k^2 + (k + 1)^2 =$ $(1^2 + 2^2 + 3^2 + \cdots + k^2) + (k + 1)^2 = \frac{1}{6}k(k + 1)(2k + 1) + (k + 1)^2 = \frac{1}{6}(2k^3 + 9k^2 + 13k + 6) =$ $\frac{1}{6}(k + 1)(k + 2)(2k + 3)$.

15. (I) $n = 1$: $5 - 1 = 4$ and $\frac{1}{2}(9 - 1) = \frac{1}{2} \cdot 8 = 4$
(II) If $4 + 3 + 2 + \cdots + (5 - k) = \frac{1}{2}k(9 - k)$, then $4 + 3 + 2 + \cdots + (5 - k) + 5 - (k + 1) =$ $[4 + 3 + 2 + \cdots + (5 - k)] + 5 - (k + 1) = \frac{1}{2}k(9 - k) + 4 - k = \frac{1}{2}(-k^2 + 7k + 8) = \frac{1}{2}(8 - k)(k + 1) =$ $\frac{1}{2}(k + 1)[9 - (k + 1)]$.

17. (I) $n = 1$: $1 \cdot (1 + 1) = 2$ and $\frac{1}{3} \cdot 1 \cdot 2 \cdot 3 = 2$
(II) If $1 \cdot 2 + 2 \cdot 3 + 3 \cdot 4 + \cdots + k(k + 1) = \frac{1}{3}k(k + 1)(k + 2)$, then $1 \cdot 2 + 2 \cdot 3 + 3 \cdot 4 + \cdots + k(k + 1) + (k + 1)(k + 2) = [1 \cdot 2 + 2 \cdot 3 + 3 \cdot 4 + \cdots + k(k + 1)] + (k + 1)(k + 2) =$ $\frac{1}{3}k(k + 1)(k + 2) + (k + 1)(k + 2) = \frac{1}{3}(k + 1)(k + 2)(k + 3)$.

19. (I) $n = 1$: $1^2 + 1 = 2$ is divisible by 2.
(II) If $k^2 + k$ is divisible by 2, then $(k + 1)^2 + (k + 1) = k^2 + 2k + 1 + k + 1 = (k^2 + k) + 2k + 2$. Since $k^2 + k$ is divisible by 2 and $2k + 2$ is divisible by 2, therefore, $(k + 1)^2 + k + 1$ is divisible by 2.

21. (I) $n = 1$: $1^2 - 1 + 2 = 2$ is divisible by 2.
(II) If $k^2 - k + 2$ is divisible by 2, then $(k + 1)^2 - (k + 1) + 2 = k^2 + 2k + 1 - k - 1 + 2 = (k^2 - k + 2) + 2k$. Since $k^2 - k + 2$ is divisible by 2 and $2k$ is divisible by 2, therefore, $(k + 1)^2 - (k + 1) + 2$ is divisible by 2.

23. (I) $n = 1$: If $x > 1$, then $x^1 = x > 1$.
(II) Assume, for any natural number k, that if $x > 1$, then $x^k > 1$. Show that if $x > 1$, then $x^{k+1} > 1$:
$$x^{k+1} = x^k \cdot x^1 \underset{\substack{\uparrow \\ x^k > 1}}{>} 1 \cdot x = x > 1$$

25. (I) $n = 1$: $a - b$ is a factor of $a^1 - b^1 = a - b$.
(II) If $a - b$ is a factor of $a^k - b^k$, show that $a - b$ is a factor of $a^{k+1} - b^{k+1}$: $a^{k+1} - b^{k+1} = a(a^k - b^k) + b^k(a - b)$. Since $a - b$ is a factor of $a^k - b^k$ and $a - b$ is a factor of $a - b$, therefore, $a - b$ is a factor of $a^{k+1} - b^{k+1}$.

27. $n = 1$: $1^2 - 1 + 41 = 41$ is a prime number.
$n = 41$: $41^2 - 41 + 41 = 1681 = 41^2$ is not prime.

29. (I) $n = 1$: $ar^{1-1} = a \cdot 1 = a$ and $a \cdot \dfrac{1 - r^1}{1 - r} = a$, because $r \neq 1$.
(II) If $a + ar + ar^2 + \cdots + ar^{k-1} = a\left(\dfrac{1 - r^k}{1 - r}\right)$, then $a + ar + ar^2 + \cdots + ar^{k-1} + ar^{(k+1)-1} =$
$$(a + ar + ar^2 + \cdots + ar^{k-1}) + ar^k = a\left(\frac{1 - r^k}{1 - r}\right) + ar^k = \frac{a(1 - r^k) + ar^k(1 - r)}{1 - r} = \frac{a - ar^k + ar^k - ar^{k+1}}{1 - r} = a\left(\frac{1 - r^{k+1}}{1 - r}\right).$$

31. (I) $n = 3$: The sum of the angles of a triangle is $(3 - 2) \cdot 180° = 180°$.
(II) Assume for any k that the sum of the angles of a convex polygon of k sides is $(k - 2) \cdot 180°$. A convex polygon of $k + 1$ sides consists of a convex polygon of k sides plus a triangle (see the illustration). The sum of the angles is $(k - 2) \cdot 180° + 180° = (k - 1) \cdot 180°$. Since Conditions I and II have been met, the result follows.

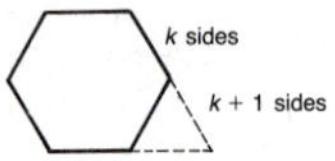

EXERCISE 13.2

1. 362,880 **3.** $\frac{28}{3}$ **5.** 10 **7.** 21 **9.** 50 **11.** 1 **13.** $x^5 + 5x^4 + 10x^3 + 10x^2 + 5x + 1$
15. $x^6 - 12x^5 + 60x^4 - 160x^3 + 240x^2 - 192x + 64$ **17.** $81x^4 + 108x^3 + 54x^2 + 12x + 1$
19. $x^{10} + 5y^2x^8 + 10y^4x^6 + 10y^6x^4 + 5y^8x^2 + y^{10}$ **21.** $x^3 + 6\sqrt{2}x^{5/2} + 30x^2 + 40\sqrt{2}x^{3/2} + 60x + 24\sqrt{2}x^{1/2} + 8$
23. $(ax)^5 + 5by(ax)^4 + 10(by)^2(ax)^3 + 10(by)^3(ax)^2 + 5(by)^4(ax) + (by)^5$ **25.** 17,010 **27.** 126,720 **29.** 41,472 **31.** $2835x^3$
33. $314{,}928x^7$ **35.** 495 **37.** 3360 **39.** 1.00501 **41.** $\binom{n}{n} = \dfrac{n!}{n!(n - n)!} = \dfrac{n!}{n!0!} = \dfrac{n!}{n!} = 1$
43. $2^n = (1 + 1)^n = \binom{n}{0}1^n + \binom{n}{1}(1)(1)^{n-1} + \cdots + \binom{n}{n}1^n = \binom{n}{0} + \binom{n}{1} + \cdots + \binom{n}{n}$
45. $12! = 4.790016 \times 10^8$, $20! = 2.432302 \times 10^{18}$, $25! = 1.551121 \times 10^{25}$; $12! \approx 4.790139724 \times 10^8$, $20! \approx 2.432924 \times 10^{18}$, $25! \approx 1.5511299 \times 10^{25}$

EXERCISE 13.3

1. 1, 2, 3, 4, 5 **3.** $\frac{1}{2}, \frac{2}{3}, \frac{3}{4}, \frac{4}{5}, \frac{5}{6}$ **5.** 1, −4, 9, −16, 25 **7.** $\frac{1}{2}, \frac{2}{5}, \frac{2}{7}, \frac{8}{41}, \frac{8}{61}$ **9.** $-\frac{1}{6}, \frac{1}{12}, -\frac{1}{20}, \frac{1}{30}, -\frac{1}{42}$ **11.** $1/e, 2/e^2, 3/e^3, 4/e^4, 5/e^5$
13. $n/(n+1)$ **15.** $1/2^{n-1}$ **17.** $(-1)^{n+1}$ **19.** $(-1)^{n+1}n$ **21.** $a_1 = 1, a_2 = 3, a_3 = 5, a_4 = 7, a_5 = 9$
23. $a_1 = -2, a_2 = -1, a_3 = 1, a_4 = 4, a_5 = 8$ **25.** $a_1 = 5, a_2 = 10, a_3 = 20, a_4 = 40, a_5 = 80$
27. $a_1 = 3, a_2 = 3, a_3 = \frac{3}{2}, a_4 = \frac{1}{2}, a_5 = \frac{1}{8}$ **29.** $a_1 = 1, a_2 = 2, a_3 = 2, a_4 = 4, a_5 = 8$
31. $a_1 = A, a_2 = A + d, a_3 = A + 2d, a_4 = A + 3d, a_5 = A + 4d$
33. $a_1 = \sqrt{2}, a_2 = \sqrt{2+\sqrt{2}}, a_3 = \sqrt{2+\sqrt{2+\sqrt{2}}}, a_4 = \sqrt{2+\sqrt{2+\sqrt{2+\sqrt{2}}}}, a_5 = \sqrt{2+\sqrt{2+\sqrt{2+\sqrt{2+\sqrt{2}}}}}$
35. $2 + 3 + 4 + \cdots + (n+1)$ **37.** $\frac{1}{2} + 2 + \frac{9}{2} + \cdots + \frac{n^2}{2}$ **39.** $1 + \frac{1}{3} + \frac{1}{9} + \cdots + \frac{1}{3^n}$ **41.** $\frac{1}{3} + \frac{1}{9} + \cdots + \frac{1}{3^n}$
43. $\ln 2 - \ln 3 + \ln 4 - \cdots + (-1)^n \ln n$ **45.** $\sum_{k=1}^{n} k$ **47.** $\sum_{k=1}^{n} \frac{k}{k+1}$ **49.** $\sum_{k=0}^{n} (-1)^k\left(\frac{1}{3^k}\right)$ **51.** $\sum_{k=1}^{n} \frac{3^k}{k}$ **53.** $\sum_{k=0}^{n} (a + kd)$ **55.** 21
59. A Fibonacci sequence

EXERCISE 13.4

1. $d = 1$; 6, 7, 8, 9 **3.** $d = 2$; −3, −1, 1, 3 **5.** $d = -2$; 4, 2, 0, −2 **7.** $d = -\frac{1}{3}; \frac{1}{6}, -\frac{1}{6}, -\frac{1}{2}, -\frac{5}{6}$
9. $d = \ln 3$; $\ln 3, 2\ln 3, 3\ln 3, 4\ln 3$ **11.** $r = 2$; 2, 4, 8, 16 **13.** $r = \frac{1}{2}; -\frac{3}{2}, -\frac{3}{4}, -\frac{3}{8}, -\frac{3}{16}$ **15.** $r = 2; \frac{1}{4}, \frac{1}{2}, 1, 2$
17. $r = 2^{1/3}; 2^{1/3}, 2^{2/3}, 2, 2^{4/3}$ **19.** $r = \frac{3}{2}; \frac{1}{2}, \frac{3}{4}, \frac{9}{8}, \frac{27}{16}$ **21.** Arithmetic; $d = 1$ **23.** Neither **25.** Arithmetic; $d = -\frac{2}{3}$ **27.** Neither
29. Geometric; $r = \frac{2}{3}$ **31.** Geometric; $r = 2$ **33.** Geometric; $r = 3^{1/2}$ **35.** $a_5 = 9; a_n = 2n - 1$ **37.** $a_5 = -7; a_n = 8 - 3n$
39. $a_5 = 2; a_n = \frac{1}{2}(n-1)$ **41.** $a_5 = 5\sqrt{2}; a_n = \sqrt{2}n$ **43.** $a_5 = 16; a_n = 2^{n-1}$ **45.** $a_5 = 5; a_n = (-1)^{n-1}(5)$ **47.** $a_5 = 0; a_n = 0$
49. $a_5 = 4\sqrt{2}; a_n = (\sqrt{2})^n$ **51.** $a_{12} = 24$ **53.** $a_{10} = -26$ **55.** $a_8 = a + 7b$ **57.** $a_7 = \frac{1}{64}$ **59.** $a_9 = 1$ **61.** $a_8 = 0.00000004$
63. $a_1 = -13; d = 3$ **65.** $a_1 = -53; d = 6$ **67.** $a_1 = 28; d = -2$ **69.** $a_1 = 25; d = -2$ **71.** $\frac{1}{2}n(5n+7)$ **73.** $-\frac{1}{4}(1 - 2^n)$
75. $2\left[1 - \left(\frac{2}{3}\right)^n\right]$ **77.** $n(1+n)$ **79.** $1 - 2^n$ **81.** $\frac{3}{2}$ **83.** 16 **85.** $\frac{8}{5}$ **87.** 4 **89.** $\frac{18}{5}$ **91.** $\left(\frac{3}{4}\right)^3 \cdot 20 = \frac{135}{16}$ ft; $\left(\frac{3}{4}\right)^n \cdot 20$ ft
93. About 102 ft **95.** 1.845×10^{19} **97.** $\frac{1}{3}$

FILL-IN-THE-BLANK ITEMS

1. Pascal triangle **2.** 15 **3.** sequence **4.** arithmetic **5.** geometric

REVIEW EXERCISES

1. (I) $n = 1$: $3 \cdot 1 = 3$ and $\frac{3 \cdot 1}{2}(2) = 3$

(II) If $3 + 6 + 9 + \cdots + 3k = \frac{3k}{2}(k+1)$, then $3 + 6 + 9 + \cdots + 3k + 3(k+1) = (3 + 6 + 9 + \cdots + 3k) + (3k+3) = \frac{3k}{2}(k+1) + (3k+3) = \frac{3k^2}{2} + \frac{9k}{2} + \frac{6}{2} = \frac{3}{2}(k^2 + 3k + 2) = \frac{3}{2}(k+1)(k+2)$.

3. (I) $n = 1$: $2 \cdot 3^{1-1} = 2$ and $3^1 - 1 = 2$

(II) If $2 + 6 + 18 + \cdots + 2 \cdot 3^{k-1} = 3^k - 1$, then $2 + 6 + 18 + \cdots + 2 \cdot 3^{k-1} + 2 \cdot 3^{(k+1)-1} = (2 + 6 + 18 + \cdots + 2 \cdot 3^{k-1}) + 2 \cdot 3^k = 3^k - 1 + 2 \cdot 3^k = 3 \cdot 3^k - 1 = 3^{k+1} - 1$.

5. (I) $n = 1$: $1^2 = 1$ and $\frac{1}{2}(6 - 3 - 1) = \frac{1}{2}(2) = 1$

(II) If $1^2 + 4^2 + 7^2 + \cdots + (3k-2)^2 = \frac{1}{2}k(6k^2 - 3k - 1)$, then $1^2 + 4^2 + 7^2 + \cdots + (3k-2)^2 + [3(k+1) - 2]^2 = [1^2 + 4^2 + 7^2 + \cdots + (3k-2)^2] + (3k+1)^2 = \frac{1}{2}k(6k^2 - 3k - 1) + (3k+1)^2 = \frac{1}{2}(6k^3 + 15k^2 + 11k + 2) = \frac{1}{2}(k+1)(6k^2 + 9k + 2) = \frac{1}{2}(k+1)[6(k+1)^2 - 3(k+1) - 1]$.

7. 120 **9.** 10 **11.** $x^4 + 8x^3 + 24x^2 + 32x + 16$ **13.** $32x^5 + 240x^4 + 720x^3 + 1080x^2 + 810x + 243$ **15.** 144 **17.** 84
19. $-\frac{2}{3}, \frac{3}{4}, -\frac{4}{5}, \frac{5}{6}, -\frac{6}{7}$ **21.** $2, 1, \frac{8}{9}, 1, \frac{32}{25}$ **23.** $3, 2, \frac{4}{3}, \frac{8}{9}, \frac{16}{27}$ **25.** 2, 0, 2, 0, 2 **27.** Arithmetic; $d = 1$; $\frac{n}{2}(7+n)$ **29.** Neither
31. Geometric; $r = 8$; $\frac{8}{7}(8^n - 1)$ **33.** Arithmetic; $d = 4$; $2n(n-1)$ **35.** Geometric; $r = \frac{1}{2}$; $6\left[1 - \left(\frac{1}{2}\right)^n\right]$ **37.** Neither **39.** 27
41. $\left(\frac{1}{10}\right)^{10}$ **43.** $9\sqrt{2}$ **45.** $5n - 4$ **47.** $n - 10$ **49.** $\frac{9}{2}$ **51.** $\frac{4}{3}$ **53.** 8

CHAPTER 14 EXERCISE 14.1

1. $\begin{bmatrix} 4 & 4 & -5 \\ -1 & 5 & 4 \end{bmatrix}$ **3.** $\begin{bmatrix} 0 & 12 & -20 \\ 4 & 8 & 24 \end{bmatrix}$ **5.** $\begin{bmatrix} -8 & 7 & -15 \\ 7 & 0 & 22 \end{bmatrix}$ **7.** $\begin{bmatrix} 28 & -9 \\ 4 & 23 \end{bmatrix}$ **9.** $\begin{bmatrix} 1 & 14 & -14 \\ 2 & 22 & -18 \\ 3 & 0 & 28 \end{bmatrix}$ **11.** $\begin{bmatrix} 15 & 21 & -16 \\ 22 & 34 & -22 \\ -11 & 7 & 22 \end{bmatrix}$

13. $\begin{bmatrix} 25 & -9 \\ 4 & 20 \end{bmatrix}$ **15.** $\begin{bmatrix} -13 & 7 & -12 \\ -18 & 10 & -14 \\ 17 & -7 & 34 \end{bmatrix}$ **17.** $\begin{bmatrix} -2 & 4 & 2 & 8 \\ 2 & 1 & 4 & 6 \end{bmatrix}$ **19.** $\begin{bmatrix} 9 & 2 \\ 34 & 13 \\ 47 & 20 \end{bmatrix}$ **21.** $\begin{bmatrix} 1 & -1 \\ -1 & 2 \end{bmatrix}$ **23.** $\begin{bmatrix} 1 & -\frac{5}{2} \\ -1 & 3 \end{bmatrix}$ **25.** $\begin{bmatrix} 1 & -1/a \\ -1 & 2/a \end{bmatrix}$

27. $\begin{bmatrix} 3 & -3 & 1 \\ -2 & 2 & -1 \\ -4 & 5 & -2 \end{bmatrix}$ **29.** $\begin{bmatrix} -\frac{5}{7} & \frac{1}{7} & \frac{3}{7} \\ \frac{9}{7} & \frac{1}{7} & -\frac{4}{7} \\ \frac{3}{7} & -\frac{2}{7} & \frac{1}{7} \end{bmatrix}$ **31.** $x = 3, y = 2$ **33.** $x = -5, y = 10$ **35.** $x = 2, y = -1$ **37.** $x = \frac{1}{2}, y = 2$

39. $x = -2, y = 1$ **41.** $x = 2/a, y = 3/a$ **43.** $x = -2, y = 3, z = 5$ **45.** $x = \frac{1}{2}, y = -\frac{1}{2}, z = 1$ **47.** $x = -\frac{34}{7}, y = \frac{85}{7}, z = \frac{12}{7}$

49. $x = \frac{1}{3}, y = 1, z = \frac{2}{3}$ **51.** $\left[\begin{array}{cc|cc} 4 & 2 & 1 & 0 \\ 2 & 1 & 0 & 1 \end{array}\right] \to \left[\begin{array}{cc|cc} 1 & \frac{1}{2} & \frac{1}{4} & 0 \\ 2 & 1 & 0 & 1 \end{array}\right] \to \left[\begin{array}{cc|cc} 1 & \frac{1}{2} & \frac{1}{4} & 0 \\ 0 & 0 & -\frac{1}{2} & 1 \end{array}\right]$

53. $\left[\begin{array}{cc|cc} 15 & 3 & 1 & 0 \\ 10 & 2 & 0 & 1 \end{array}\right] \to \left[\begin{array}{cc|cc} 1 & \frac{1}{5} & \frac{1}{15} & 0 \\ 10 & 2 & 0 & 1 \end{array}\right] \to \left[\begin{array}{cc|cc} 1 & \frac{1}{5} & \frac{1}{15} & 0 \\ 0 & 0 & -\frac{2}{3} & 1 \end{array}\right]$

55. $\left[\begin{array}{ccc|ccc} -3 & 1 & -1 & 1 & 0 & 0 \\ 1 & -4 & -7 & 0 & 1 & 0 \\ 1 & 2 & 5 & 0 & 0 & 1 \end{array}\right] \to \left[\begin{array}{ccc|ccc} 1 & 2 & 5 & 0 & 0 & 1 \\ 1 & -4 & -7 & 0 & 1 & 0 \\ -3 & 1 & -1 & 1 & 0 & 0 \end{array}\right] \to \left[\begin{array}{ccc|ccc} 1 & 2 & 5 & 0 & 0 & 1 \\ 0 & -6 & -12 & 0 & 1 & -1 \\ 0 & 7 & 14 & 1 & 0 & 3 \end{array}\right]$

$\to \left[\begin{array}{ccc|ccc} 1 & 2 & 5 & 0 & 0 & 1 \\ 0 & 1 & 2 & 0 & -\frac{1}{6} & \frac{1}{6} \\ 0 & 1 & 2 & \frac{1}{7} & 0 & \frac{3}{7} \end{array}\right] \to \left[\begin{array}{ccc|ccc} 1 & 2 & 5 & 0 & 0 & 1 \\ 0 & 1 & 2 & 0 & -\frac{1}{6} & \frac{1}{6} \\ 0 & 0 & 0 & \frac{1}{7} & \frac{1}{6} & \frac{11}{42} \end{array}\right]$

57. (a) $\begin{bmatrix} 500 & 350 & 400 \\ 700 & 500 & 850 \end{bmatrix}$; $\begin{bmatrix} 500 & 700 \\ 350 & 500 \\ 400 & 850 \end{bmatrix}$ (b) $\begin{bmatrix} 15 \\ 8 \\ 3 \end{bmatrix}$ (c) $\begin{bmatrix} 11{,}500 \\ 17{,}050 \end{bmatrix}$ (d) $[0.10 \quad 0.05]$ (e) \$2002.50

59. If $a \neq 0$, $\left[\begin{array}{cc|cc} a & b & 1 & 0 \\ c & d & 0 & 1 \end{array}\right] \to \left[\begin{array}{cc|cc} 1 & \frac{b}{a} & \frac{1}{a} & 0 \\ c & d & 0 & 1 \end{array}\right] \to \left[\begin{array}{cc|cc} 1 & \frac{b}{a} & \frac{1}{a} & 0 \\ 0 & \frac{-cb + da}{a} & -\frac{c}{a} & 1 \end{array}\right] \to \left[\begin{array}{cc|cc} 1 & \frac{b}{a} & \frac{1}{a} & 0 \\ 0 & 1 & \frac{-c}{-cb + da} & \frac{a}{-cb + da} \end{array}\right]$

$\to \left[\begin{array}{cc|cc} 1 & 0 & \frac{d}{ad - bc} & \frac{-b}{ad - bc} \\ 0 & 1 & \frac{-c}{ad - bc} & \frac{a}{ad - bc} \end{array}\right]$. Therefore, $A^{-1} = \frac{1}{D}\begin{bmatrix} d & -b \\ -c & a \end{bmatrix}$.

If $a = 0$, $\left[\begin{array}{cc|cc} 0 & b & 1 & 0 \\ c & d & 0 & 1 \end{array}\right] \to \left[\begin{array}{cc|cc} c & d & 0 & 1 \\ 0 & b & 1 & 0 \end{array}\right] \to \left[\begin{array}{cc|cc} 1 & \frac{d}{c} & 0 & \frac{1}{c} \\ 0 & b & 1 & 0 \end{array}\right] \to \left[\begin{array}{cc|cc} 1 & 0 & \frac{-d}{cb} & \frac{1}{c} \\ 0 & 1 & \frac{1}{b} & 0 \end{array}\right] \to \left[\begin{array}{cc|cc} 1 & 0 & \frac{d}{-bc} & \frac{-b}{-bc} \\ 0 & 1 & \frac{-c}{-bc} & 0 \end{array}\right]$.

Since $a = 0$, $D = ad - bc = -bc$, so $A^{-1} = \frac{1}{D}\begin{bmatrix} d & -b \\ -c & a \end{bmatrix}$.

EXERCISE 14.2

1. Maximum value is 11; minimum value is 3. **3.** Maximum value is 65; minimum value is 4.
5. Maximum value is 67; minimum value is 20. **7.** The maximum value of z is 12, and it occurs at the point (6, 0).
9. The minimum value of z is 4, and it occurs at the point (2, 0).

11. The maximum value of z is 20, and it occurs at the point (0, 4).
13. The minimum value of z is 8, and it occurs at the point (0, 2).
15. The maximum value of z is 50, and it occurs at the point (10, 0).
17. The maximum profit is \$70. It is achieved with 40 packages of the low-grade mixture and 180 packages of the high-grade mixture.
19. 8 downhill, 24 cross-country; \$1760 **21.** 24 acres of soybeans, 12 acres of wheat; \$5520
23. \$10,000 in the CD; \$10,000 in T bills **25.** 10 racing skates, 15 figure skates **27.** 10 first-class, 120 coach
29. 5 oz supplement A, 1 oz supplement B

EXERCISE 14.3

1. Proper **3.** Improper; $1 + \frac{9}{x^2 - 4}$ **5.** Improper; $5x + \frac{22x - 1}{x^2 - 4}$ **7.** Improper; $1 + \frac{-2(x - 6)}{(x + 4)(x - 3)}$ **9.** $\frac{-4}{x} + \frac{4}{x - 1}$
11. $\frac{1}{x} + \frac{-x}{x^2 + 1}$ **13.** $\frac{-1}{x - 1} + \frac{2}{x - 2}$ **15.** $\frac{\frac{1}{4}}{x + 1} + \frac{\frac{3}{4}}{x - 1} + \frac{\frac{1}{2}}{(x - 1)^2}$ **17.** $\frac{\frac{1}{12}}{x - 2} + \frac{-\frac{1}{12}(x + 4)}{x^2 + 2x + 4}$
19. $\frac{\frac{1}{4}}{(x - 1)} + \frac{\frac{1}{4}}{(x - 1)^2} - \frac{\frac{1}{4}}{x + 1} + \frac{\frac{1}{4}}{(x + 1)^2}$ **21.** $\frac{-5}{x + 2} + \frac{5}{x + 1} + \frac{-4}{(x + 1)^2}$ **23.** $\frac{\frac{1}{4}}{x} + \frac{1}{x^2} - \frac{\frac{1}{4}(x + 4)}{x^2 + 4}$ **25.** $\frac{\frac{2}{3}}{x + 1} + \frac{\frac{1}{3}(x + 1)}{x^2 + 2x + 4}$
27. $\frac{\frac{2}{7}}{3x - 2} + \frac{\frac{1}{7}}{2x + 1}$ **29.** $\frac{\frac{3}{4}}{x + 3} + \frac{\frac{1}{4}}{x - 1}$ **31.** $\frac{1}{x^2 + 4} + \frac{2x - 1}{(x^2 + 4)^2}$ **33.** $\frac{-1}{x} + \frac{2}{x - 3} + \frac{-1}{x + 1}$ **35.** $\frac{4}{x - 2} + \frac{-3}{x - 1} + \frac{-1}{(x - 1)^2}$
37. $\frac{x}{(x^2 + 16)^2} + \frac{-16x}{(x^2 + 16)^3}$ **39.** $\frac{-\frac{8}{7}}{2x - 3} + \frac{\frac{4}{7}}{x + 1}$ **41.** $\frac{-\frac{2}{9}}{x} - \frac{\frac{1}{3}}{x^2} + \frac{\frac{1}{6}}{x - 3} + \frac{\frac{1}{18}}{x + 3}$

EXERCISE 14.4

1.

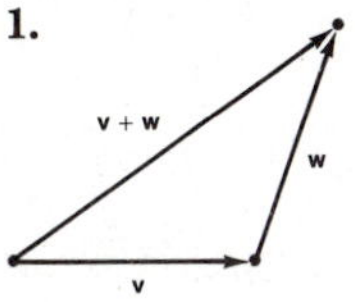

3.

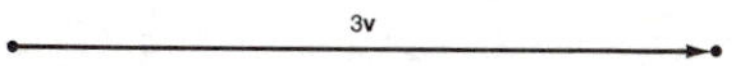

5.

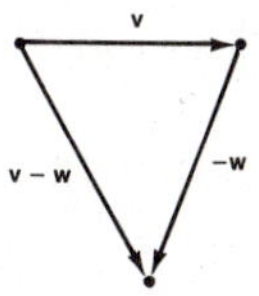

7.

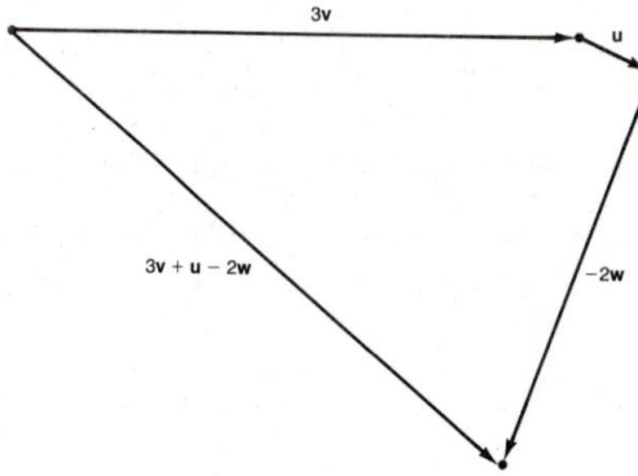

9. $\mathbf{x} = \mathbf{A}$ **11.** $\mathbf{C} = -\mathbf{F} + \mathbf{E} - \mathbf{D}$ **13.** $\mathbf{E} = -\mathbf{G} - \mathbf{H} + \mathbf{D}$ **15.** $\mathbf{x} = \mathbf{0}$ **17.** 12 **19.** $\mathbf{v} = 3\mathbf{i} + 4\mathbf{j}$ **21.** $\mathbf{v} = 2\mathbf{i} + 4\mathbf{j}$ **23.** $\mathbf{v} = 8\mathbf{i} - \mathbf{j}$
25. $\mathbf{v} = -\mathbf{i} + \mathbf{j}$ **27.** 5 **29.** $\sqrt{2}$ **31.** $\sqrt{13}$ **33.** $-\mathbf{j}$ **35.** $\sqrt{89}$ **37.** $\sqrt{34} - \sqrt{13}$ **39.** $\mathbf{i}$ **41.** $\frac{3}{5}\mathbf{i} - \frac{4}{5}\mathbf{j}$ **43.** $\frac{\sqrt{2}}{2}\mathbf{i} - \frac{\sqrt{2}}{2}\mathbf{j}$
45. $\mathbf{v} = \frac{8\sqrt{5}}{5}\mathbf{i} + \frac{4\sqrt{5}}{5}\mathbf{j}$ or $\mathbf{v} = -\frac{8\sqrt{5}}{5}\mathbf{i} - \frac{4\sqrt{5}}{5}\mathbf{j}$ **47.** $\{-2 + \sqrt{21}, -2 - \sqrt{21}\}$ **49.** 460 kph **51.** 218 mph **53.**

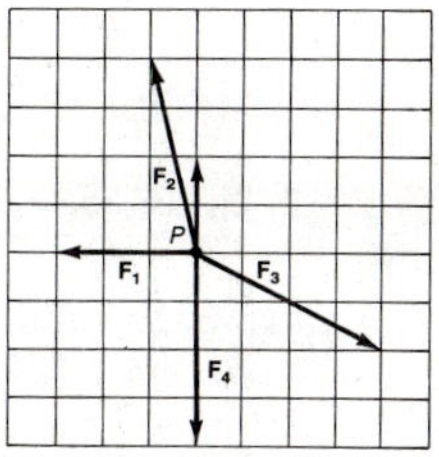

EXERCISE 14.5

1. 0; 0 **3.** 4; $\frac{4}{5}$ **5.** $\sqrt{3} - 1$; $(\sqrt{6} - \sqrt{2})/4$ **7.** 24; $\frac{24}{25}$ **9.** 0; 0 **11.** $\frac{3}{2}$ **13.** $\mathbf{v}_1 = \text{proj}_{\mathbf{w}}\,\mathbf{v} = \frac{5}{2}(\mathbf{i} - \mathbf{j})$, $\mathbf{v}_2 = -\frac{1}{2}\mathbf{i} - \frac{1}{2}\mathbf{j}$
15. $\mathbf{v}_1 = \text{proj}_{\mathbf{w}}\,\mathbf{v} = -\frac{1}{5}(\mathbf{i} + 2\mathbf{j})$, $\mathbf{v}_2 = \frac{6}{5}\mathbf{i} - \frac{3}{5}\mathbf{j}$ **17.** $\mathbf{v}_1 = \text{proj}_{\mathbf{w}}\,\mathbf{v} = \frac{7}{5}(2\mathbf{i} + \mathbf{j})$, $\mathbf{v}_2 = \frac{1}{5}\mathbf{i} - \frac{2}{5}\mathbf{j}$ **19.** 496.7 mph; 51.5° south of west
21. 8.7° off direct heading into the current; 1.5 min **23.** 60°; 17.32 min **25.** 2.68 ft-lb **27.** 1732 ft-lb
29. Let $\mathbf{u} = a_1\mathbf{i} + b_1\mathbf{j}$, $\mathbf{v} = a_2\mathbf{i} + b_2\mathbf{j}$, $\mathbf{w} = a_3\mathbf{i} + b_3\mathbf{j}$. Compute $\mathbf{u} \cdot (\mathbf{v} + \mathbf{w})$ and $\mathbf{u} \cdot \mathbf{v} + \mathbf{u} \cdot \mathbf{w}$.

31. $\cos\alpha = \dfrac{\mathbf{v} \cdot \mathbf{i}}{\|\mathbf{v}\|\,\|\mathbf{i}\|} = \mathbf{v} \cdot \mathbf{i}$; if $\mathbf{v} = x\mathbf{i} + y\mathbf{j}$, then $\mathbf{v} \cdot \mathbf{i} = x = \cos\alpha$ and $\mathbf{v} \cdot \mathbf{j} = y = \cos\left(\dfrac{\pi}{2} - \alpha\right) = \sin\alpha$.

33. $\mathbf{v} = a\mathbf{i} + b\mathbf{j}$; $\text{proj}_{\mathbf{i}}\,\mathbf{v} = \dfrac{\mathbf{v} \cdot \mathbf{i}}{\|\mathbf{i}\|^2}\mathbf{i} = (\mathbf{v} \cdot \mathbf{i})\mathbf{i}$; $\mathbf{v} \cdot \mathbf{i} = a$, $\mathbf{v} \cdot \mathbf{j} = b$, so $\mathbf{v} = (\mathbf{v} \cdot \mathbf{i})\mathbf{i} + (\mathbf{v} \cdot \mathbf{j})\mathbf{j}$

35. $(\mathbf{v} - \alpha\mathbf{w}) \cdot \mathbf{w} = \mathbf{v} \cdot \mathbf{w} - \alpha\mathbf{w} \cdot \mathbf{w} = \mathbf{v} \cdot \mathbf{w} - \alpha\|\mathbf{w}\|^2 = \mathbf{v} \cdot \mathbf{w} - \dfrac{\mathbf{v} \cdot \mathbf{w}}{\|\mathbf{w}\|^2}\|\mathbf{w}\|^2 = 0$ **37.** 0

FILL-IN-THE-BLANK ITEMS

1. inverse **2.** False **3.** identity **4.** objective function **5.** feasible solution **6.** proper **7.** unit **8.** 0

REVIEW EXERCISES

1. $\begin{bmatrix} 4 & -4 \\ 3 & 9 \\ 4 & 0 \end{bmatrix}$ **3.** $\begin{bmatrix} 6 & 0 \\ 12 & 24 \\ -6 & 12 \end{bmatrix}$ **5.** $\begin{bmatrix} 4 & -3 & 0 \\ 12 & -2 & -8 \\ -2 & 5 & -4 \end{bmatrix}$ **7.** $\begin{bmatrix} 8 & -13 & 8 \\ 9 & 2 & -10 \\ 18 & -17 & 4 \end{bmatrix}$ **9.** $\begin{bmatrix} \frac{1}{2} & -1 \\ -\frac{1}{6} & \frac{2}{3} \end{bmatrix}$ **11.** $\begin{bmatrix} -\frac{5}{7} & \frac{9}{7} & \frac{3}{7} \\ \frac{1}{7} & \frac{1}{7} & -\frac{2}{7} \\ \frac{3}{7} & -\frac{4}{7} & \frac{1}{7} \end{bmatrix}$ **13.** Singular

15. The maximum value is 32 when $x = 0$ and $y = 8$. **17.** The minimum value is 3 when $x = 1$ and $y = 0$.
19. The maximum value is $\frac{108}{7}$ when $x = \frac{12}{7}$ and $y = \frac{12}{7}$. **21.** $\dfrac{-\frac{3}{2}}{x} + \dfrac{\frac{3}{2}}{x - 4}$ **23.** $\dfrac{-3}{x - 1} + \dfrac{3}{x} + \dfrac{4}{x^2}$ **25.** $\dfrac{-\frac{1}{10}}{x + 1} + \dfrac{\frac{1}{10}x + \frac{9}{10}}{x^2 + 9}$
27. $\dfrac{x}{x^2 + 4} - \dfrac{4x}{(x^2 + 4)^2}$ **29.** $\dfrac{\frac{1}{2}}{x^2 + 1} + \dfrac{\frac{1}{4}}{x - 1} - \dfrac{\frac{1}{4}}{x + 1}$ **31.** $\mathbf{v} = 2\mathbf{i} - 4\mathbf{j}$; $\|\mathbf{v}\| = 2\sqrt{5}$ **33.** $\mathbf{v} = -\mathbf{i} + 3\mathbf{j}$; $\|\mathbf{v}\| = \sqrt{10}$
35. $-20\mathbf{i} + 13\mathbf{j}$ **37.** $\sqrt{5}$ **39.** $\sqrt{5} + 5 \approx 7.24$ **41.** $\dfrac{-2\sqrt{5}}{5}\mathbf{i} + \dfrac{\sqrt{5}}{5}\mathbf{j}$ **43.** $\mathbf{v} \cdot \mathbf{w} = -11$; $\cos\theta = -11\sqrt{5}/25$
45. $\mathbf{v} \cdot \mathbf{w} = -4$; $\cos\theta = -2\sqrt{5}/5$ **47.** $\text{proj}_{\mathbf{w}}\,\mathbf{v} = \frac{9}{10}(3\mathbf{i} + \mathbf{j})$ **49.** 30.5° **51.** $\sqrt{29} \approx 5.39$ mph; 0.4 mi
53. 35 gasoline engines, 15 diesel engines; 15 gasoline engines, 0 diesel engines **55.** At an angle of 70.5° to the shore

CHAPTER 15 EXERCISE 15.1

1. {1, 3, 5, 6, 7, 9} **3.** {1, 5, 7} **5.** {1, 6, 9} **7.** {1, 2, 4, 5, 6, 7, 8, 9} **9.** {1, 2, 4, 5, 6, 7, 8, 9} **11.** {0, 2, 6, 7, 8}
13. {0, 1, 2, 3, 5, 6, 7, 8, 9} **15.** {0, 1, 2, 3, 5, 6, 7, 8, 9} **17.** {0, 1, 2, 3, 4, 6, 7, 8} **19.** {0}
21. ∅, {*a*}, {*b*}, {*c*}, {*d*}, {*a*, *b*}, {*a*, *c*}, {*a*, *d*}, {*b*, *c*}, {*b*, *d*}, {*c*, *d*}, {*a*, *b*, *c*}, {*b*, *c*, *d*}, {*a*, *c*, *d*}, {*a*, *b*, *d*}, {*a*, *b*, *c*, *d*} **23.** 25 **25.** 40
27. 25 **29.** 37 **31.** 18 **33.** 5 **35.** 175; 125 **37.** (a) 15 (b) 15 (c) 15 (d) 25 (e) 40

EXERCISE 15.2

1. 15 **3.** 16 **5.** 8 **7.** 1024 **9.** 9000 **11.** 30 **13.** 120 **15.** 1 **17.** 336 **19.** 24 **21.** 60 **23.** 120 **25.** 362,880 **27.** 5,850,000

EXERCISE 15.3

1. (*a*, *b*, *c*), (*a*, *b*, *d*), (*a*, *b*, *e*), (*a*, *c*, *d*), (*a*, *c*, *e*), (*a*, *d*, *e*), (*b*, *c*, *d*), (*b*, *c*, *e*), (*b*, *d*, *e*), (*c*, *d*, *e*); 10
3. (1, 2, 3), (2, 3, 4), (1, 2, 4), (1, 3, 4); 4 **5.** 28 **7.** 15 **9.** 1 **11.** 10,400,600 **13.** 35 **15.** 6.3501355×10^{11} **17.** 336
19. 5,209,344 **21.** 90

EXERCISE 15.4

1. $S = \{HH, HT, TH, TT\}$; $P(HH) = \frac{1}{4}$, $P(HT) = \frac{1}{4}$, $P(TH) = \frac{1}{4}$, $P(TT) = \frac{1}{4}$
3. S = {HH1, HH2, HH3, HH4, HH5, HH6, HT1, HT2, HT3, HT4, HT5, HT6, TH1, TH2, TH3, TH4, TH5, TH6, TT1, TT2, TT3, TT4, TT5, TT6}; each outcome has the probability of $\frac{1}{24}$.
5. S = {HHH, HHT, HTH, HTT, THH, THT, TTH, TTT}; each outcome has the probability of $\frac{1}{8}$.
7. S = {1 Yellow, 1 Red, 1 Green, 2 Yellow, 2 Red, 2 Green, 3 Yellow, 3 Red, 3 Green, 4 Yellow, 4 Red, 4 Green}; each outcome has the probability of $\frac{1}{12}$; thus, $P(2 \text{ Red}) + P(4 \text{ Red}) = \frac{1}{12} + \frac{1}{12} = \frac{1}{6}$.
9. S = {1 Yellow Forward, 1 Yellow Backward, 1 Red Forward, 1 Red Backward, 1 Green Forward, 1 Green Backward, 2 Yellow Forward, 2 Yellow Backward, 2 Red Forward, 2 Red Backward, 2 Green Forward, 2 Green Backward, 3 Yellow Forward, 3 Yellow Backward, 3 Red Forward, 3 Red Backward, 3 Green Forward, 3 Green Backward, 4 Yellow Forward, 4 Yellow Backward, 4 Red Forward, 4 Red Backward, 4 Green Forward, 4 Green Backward}; each outcome has the probability of $\frac{1}{24}$; thus, $P(1 \text{ Red Backward}) + P(1 \text{ Green Backward}) = \frac{1}{24} + \frac{1}{24} = \frac{1}{12}$.
11. S = {11 Red, 11 Yellow, 11 Green, 12 Red, 12 Yellow, 12 Green, 13 Red, 13 Yellow, 13 Green, 14 Red, 14 Yellow, 14 Green, 21 Red, 21 Yellow, 21 Green, 22 Red, 22 Yellow, 22 Green, 23 Red, 23 Yellow, 23 Green, 24 Red, 24 Yellow, 24 Green, 31 Red, 31 Yellow, 31 Green, 32 Red, 32 Yellow, 32 Green, 33 Red, 33 Yellow, 33 Green, 34 Red, 34 Yellow, 34 Green, 41 Red, 41 Yellow, 41 Green, 42 Red, 42 Yellow, 42 Green, 43 Red, 43 Yellow, 43 Green, 44 Red, 44 Yellow, 44 Green}; each outcome has the probability of $\frac{1}{48}$; thus, E = {22 Red, 22 Green, 24 Red, 24 Green}; $P(E) = n(E)/n(S) = \frac{4}{48} = \frac{1}{12}$.
13. $\frac{3}{13}$ **15.** $\frac{5}{18}$ **17.** $\frac{9}{20}$ **19.** $\frac{17}{20}$ **21.** $\frac{3}{10}$ **23.** $\frac{2}{5}$ **25.** 0.000033068 **27.** Approx. 0.11655 **29.** Approx. 0.0039

FILL-IN-THE-BLANK ITEMS

1. union; intersection **2.** 20; 10 **3.** permutation **4.** combination **5.** equally likely

REVIEW EXERCISES

1. {1, 3, 5, 6, 7, 8} **3.** {3, 7} **5.** {1, 2, 4, 6, 8, 9} **7.** {1, 2, 4, 5, 6, 9} **9.** 17 **11.** 29 **13.** 7 **15.** 25 **17.** 336 **19.** 56 **21.** 60 **23.** 128 **25.** 3024 **27.** 70 **29.** 91 **31.** 182 **33.** $\frac{3}{20}$, $\frac{9}{20}$ **35.** $\frac{1}{24}$

Index